GEOGRAPHY IN AMERICA

GARY L. GAILE
University of Colorado, Boulder

CORT J. WILLMOTT
University of Delaware

MERRILL PUBLISHING COMPANY
A Bell & Howell Information Company
Columbus   Toronto   London   Melbourne

Published by Merrill Publishing Company
A Bell & Howell Information Company
Columbus, Ohio 43216

This book was set in Garamond.

Administrative Editor: Wendy W. Jones
Production Coordinator: Linda Kauffman Peterson
Art Coordinator: Vincent A. Smith
Cover Designer: Brian Deep
Text Designer: Anne Daly

Copyright © 1989, by Merrill Publishing Company. All rights reserved. No part of this book may be reproduced in any form, electronic or mechanical, including photocopy, recording, or any information storage and retrieval system, without permission in writing from the publisher. "Merrill Publishing Company" and "Merrill" are registered trademarks of Merrill Publishing Company.

Library of Congress Catalog Card Number: 89–60382
International Standard Book Number: 0–675–20648–0
Printed in the United States of America
1 2 3 4 5 6 7 8 9—92 91 90 89

To
William A. V. Clark
Werner H. Terjung
William L. Thomas, Jr.
and
The UCLA Department
of Geography

Contributors

Stuart C. Aitken
San Diego State University

Shakib Al-Khameri
University of Kentucky

Roger G. Barry
University of Colorado, Boulder

Thomas J. Bassett
University of Illinois, Urbana

Duane D. Baumann
Southern Illinois University, Carbondale

John E. Benhart
Shippensburg University

William H. Berentsen
University of Connecticut

William B. Beyers
University of Washington

William R. Black
Indiana University

Howard Botts
University of Wisconsin, Whitewater

Waltraud A. R. Brinkmann
University of Wisconsin, Madison

Ray Bromley
State University of New York, Albany

Lawrence A. Brown
The Ohio State University

Marilyn A. Brown
Oak Ridge National Laboratory

Karl W. Butzer
University of Texas, Austin

Martin Cadwallader
University of Wisconsin, Madison

James Campbell
Virginia Polytechnic Institute

Nicholas R. Chrisman
University of Washington

Susan Christopherson
Cornell University

William A. V. Clark
University of California, Los Angeles

Craig E. Colton
Illinois State Museum

Ronald U. Cooke
University College, London

David J. Cowen
University of South Carolina

Susan L. Cutter
Rutgers University

John E. Damron
Mary Washington University

George J. Demko
U.S. Department of State

David G. Dickason
Western Michigan University

Lary Dilsaver
University of South Alabama

Contributors

Mona Domosh
Loughborough University

Jeff Dozier
University of California, Santa Barbara

Ashok Dutt
University of Akron

Robert J. Earickson
University of Maryland/Baltimore County

Carville Earle
Louisiana State University

Jacque L. Emel
Clark University

Rodney A. Erickson
Pennsylvania State University

Jack Estes
University of California, Santa Barbara

Peter F. Fisher
Kent State University

Kenneth E. Foote
University of Texas, Austin

Luisa M. Freeman
Freeman and Associates, New York

Gary L. Gaile
University of Colorado, Boulder

J. H. Galloway
University of Toronto

Lay J. Gibson
University of Arizona

Patricia Gober
Arizona State University

Peter Goheen
Queens University

Stephen M. Golant
University of Florida

Reginald G. Golledge
University of California, Santa Barbara

Michael F. Goodchild
University of California, Santa Barbara

James M. Goodman
University of Oklahoma

Michael R. Greenberg
Rutgers University

Ronald Grim
U.S. Library of Congress

Eve C. Gruntfest
University of Colorado, Colorado Springs

James A. Hafner
University of Massachusetts

Peter Hall
University of California, Berkeley

R. Cole Harris
University of British Columbia

Sean Hartnett
University of Wisconsin, Eau Claire

Martha L. Henderson
University of Minnesota, Duluth

Geoffrey G. D. Hewings
University of Illinois, Urbana

A. David Hill
University of Colorado, Boulder

Katherine K. Hirschboeck
Louisiana State University

Michael Hodgson
University of Colorado, Boulder

Peter J. Hugill
Texas A&M University

John R. Jensen
University of South Carolina

Stephen C. Jett
University of California, Davis

Richard C. Jones
University of Texas, San Antonio

P. P. Karan
University of Kentucky

Jeanne Kay
University of Utah

A. Jon Kimerling
Oregon State University

Andrew Kirby
University of Arizona

David B. Knight
Carleton University

Lisa A. LaPrairie
University of Colorado, Boulder

James Lemon
University of Toronto

Nancy Davis Lewis
University of Hawaii, Manoa

C. P. Lo
University of Georgia

Kamlesh Lulla
N.A.S.A. Johnson Space Center

Contributors

Edward J. Malecki
University of Florida

David M. Mark
State University of New York, Buffalo

Richard A. Marston
University of Wyoming

Sallie A. Marston
University of Arizona

Olen Paul Matthews
University of Idaho

Jonathan D. Mayer
University of Washington

James E. McConnell
State University of New York, Buffalo

Kevin McHugh
Arizona State University

Melinda S. Meade
University of North Carolina, Chapel Hill

James Merchant
University of Kansas

Judith W. Meyer
University of Connecticut

James K. Mitchell
Rutgers University

Lisle S. Mitchell
University of South Carolina

Robert D. Mitchell
University of Maryland

Burrell E. Montz
State University of New York, Binghamton

Debnath Mookherjee
Western Washington University

Richard L. Morrill
University of Washington

Darrell Napton
Southwest Texas State University

John Odland
Indiana University

John E. Oliver
Indiana State University

Kenji K. Oshiro
Wright State University

Robert C. Ostergren
University of Wisconsin, Madison

Philip R. Pryde
San Diego State University

John N. Rayner
The Ohio State University

John Rees
University of North Carolina, Greensboro

David R. Reynolds
University of Iowa

David Robinson
Syracuse University

Peter A. Rogerson
State University of New York, Buffalo

William D. Romey
St. Lawrence University

Graham D. Rowles
University of Kentucky

Lester B. Rowntree
San Jose State University

James L. Sell
University of Arizona

Nanda Shrestha
University of Wisconsin, Whitewater

Ray Smith
University of California, Santa Barbara

Richard V. Smith
Miami University

Barry D. Solomon
U. S. Environmental Protection Agency

Ihor Stebelsky
University of Windsor

Doug Stow
San Diego State University

Alan Strahler
Boston University

S. Martin Taylor
McMaster University

Graham A. Tobin
University of Minnesota, Duluth

George Towers
University of Arizona

Stanley W. Trimble
University of California, Los Angeles

Thomas T. Veblen
University of Colorado, Boulder

Richard Walker
University of California, Berkeley

David Ward
University of Wisconsin, Madison

Contributors

Roy Welch
University of Georgia

Niels West
University of Rhode Island

Stephen E. White
Kansas State University

Cort J. Willmott
University of Delaware

Dick G. Winchell
Eastern Washington University

William Wyckoff
Montana State University

Foreword

This volume, *Geography in America,* edited by Gary L. Gaile and Cort J. Willmott, is one of the most interesting and insightful books about American Geography ever published. Although it is only a snapshot at a given time, it is an extraordinarily detailed view of what American Geography is all about and what American geographers are doing. The contributing authors are the leaders of the specialty groups of the Association of American Geographers. These groups are indeed the *cores* of geographic research in the United States; they represent the pulse of American geographical research and related activity. Unlike the slick inventories and overviews of intellectual disciplines, in which a small group synthesizes and condenses trends in a field, this work is a grass-roots, factory-floor version of a discipline in America. This "from-the-bottom up" approach provides the reader with a view of what's going on in an incredible range of geographic subfields. Topics cover large, mainstream specialties, such as urban and industrial geography, as well as more focused aspects of the discipline, such as "Geography from the Left" and "Canadian Geography." The material includes recent directions in remote sensing and historical geography, as well as exciting developments in cartography and geographic information systems. In short, just about everything you ever wanted to know about American Geography is here for the reading.

One is struck, in reading *Geography in America,* by the robustness of the discipline and particularly by its multifaceted nature. Despite all rumors to the contrary, American Geography is alive, well, and active. Although there have been some difficult times for the discipline in the past, when a number of departments in American universities were closed, we have been witnessing a resurgence of commitment—from the University of Southern California to Swarthmore College. The contents of the pages that follow help to explain this renewed interest in the venerable old art and science of Geography. The editors are to be commended for carrying out the Herculean task of organizing, administering, and editing the material for this collection. I am already looking forward to its sequel.

George J. Demko, Geographer
U.S. Department of State, Washington, D.C.

Preface

Geography in America was conceived near the shores of Lake Mendota on the campus of the University of Wisconsin–Madison, on the patio of the University's Rathskeller. Following the 1985 Annual Meeting of the Association of American Geographers (AAG) in Detroit, the coeditors met with other geographers at this famous venue to consider events just past.

Among the topics discussed was the expanding role of the specialty groups within the AAG. For several years, these groups had organized a significant number of the paper sessions at the Annual Meeting, and there was every indication that their influence was increasing.

It was also clear that the specialty groups had not attained their potential. While they enhanced within-group viability and communications, there was a larger, though undeveloped, cohesive role that they could play within the Association. They should serve as conduits for better communication among specialty groups, among geographers. Early examples of this new role include jointly sponsored paper sessions (by two or more specialty groups) at the Annual Meetings, and the specialty-group chairs' luncheon which takes place at the Annual Meeting and serves as an informal senate of the groups.

Another way to improve communications was also explored at the informal meeting in Madison. This involved using the specialty-group framework to survey and report on the state and future of Geography. It was felt that specialty-group outlooks were representative of grass-roots or collective views of the discipline. In that specialty groups are identified and developed by geographers themselves, they approach the imperfect but wholesome structure characteristic of successful democracies. Specialty-group authors for this publication could be chosen by the membership by democratic means. Participation would be encouraged through the publicity of elections and the project, itself. There was a strong likelihood that a specialty-group-based format would net a relatively comprehensive and unbiased synthesis. Balanced coverage would be maintained by assigning page-length restrictions based on specialty-group size and levels of activity at the Annual Meetings. The very diversity of specialty groups would ensure that the broad spectrum of approaches to the study of Geography would be represented. Furthermore, communications among AAG members would be promoted through the presentation and discussion of draft specialty-group syntheses at an Annual Meeting.

With these considerations in mind, the *Geography in America* project was formally launched and publicized at the 1987 Annual Meeting in Portland, with many groups electing authors at that time. Paper sessions to report preliminary thoughts to the groups and to the Association at large were organized and became reality at the 1988 Annual Meeting in Phoenix. Specialty-group presentations and ensuing debates were often lively, and many syntheses were revised on the basis of these discussions.

Quality was a major goal of this project. Written versions of the specialty-group syntheses were subjected to external reviews. Based on these external reviews, as well as on editorial and within-group peer reviews, the authors were asked to make appropriate revisions. While a concerted effort was made to attain comprehensiveness, quality first and foremost dictated the inclusion of a specialty-group contribution within *Geography in America*.

Geography in America is the written formalization of the process by which the AAG membership explored the state and future of Geography. We, the coeditors, believe the final product is one of which its contributors can be proud.

ACKNOWLEDGMENTS

Geography in America required the cooperation of numerous people, and we are most pleased to be able to formally acknowledge our debt to them. The AAG specialty groups themselves deserve the lion's share of our thanks for their overwhelming support and participation. Many long hours were spent by specialty-group authors and other members to produce high-quality and comprehensive essays. The AAG, too, played an important role in the project by alternately encouraging and questioning our efforts. On occasion, their pointed queries spurred us to clarify our goals and procedures. *Geography in America* is better for their concerns.

George Demko, past president of the AAG and a friend, saw the project as a sign of new scholarship and institutional structure. It was George's belief in our vision and his supportive words before the AAG Council that paved the way for official AAG "encouragement" and for *Geography in America,* itself. George, please accept our heartfelt thanks, as well as some small part of the blame.

Merrill Publishing Company agreed to our many pleas concerning timetables. Administrative Editor Wendy Jones and Product Manager David Garza suffered through our anxieties and helped promote the work. The production and manufacturing staff, including Tanya Tiberi, JoEllen Gohr, Linda Peterson, Diane Jordan, and others, kept the massive ball rolling. Fred Schroyer copyedited with a lightning and frightening blue pencil. To all of you, we are most grateful.

We would also like to acknowledge those publishers who allowed the use of copyrighted graphic material. These are *Cartographica, The Cartographic Journal,* and *Social Science and Medicine.*

Specialty-group presentations at the Annual Meeting in Phoenix were greatly assisted by the able guidance of the following session chairs, who have our gratitude:

Lyn Brown	Ross MacKinnon
Bill Clark	Mike McNulty
Bill Denevan	Risa Palm
Nick Helburn	Alan Strahler
Diana Liverman	Waldo Tobler

External reviewers also deserve special commendation. They worked diligently and quickly and provided the indispensable insights that were required to ensure the quality of the chapters, and thereby the volume itself. They are the following:

Deuce Aguado	Dean Hanink	Phil Muehrcke
Larry Bourne	Susan Hanson	Sam Natoli
Ronald Boyce	Rudi Hartmann	Duane Nellis
Harold Brodsky	Charles Heatwole	Allen Noble
Ray Bromley	Dave Hill	Karl Nordstrom
Bill Clark	Jim Huff	John O'Laughlin
Gordon Clark	John Hunter	Martin Pasqualetti
Bob Cromley	Jeanne Kay	Rutherford Platt
Michael Dear	Andy Kirby	Tom Poiker
Ken Dewey	C. Gregory Knight	Douglas Richardson
Gary Dunbar	Victor Konrad	Clint Rowe
Nancy Ettlinger	Tom Kontuly	Tom Saarinen
Pete Fielding	Mickey Lauria	Eric Sheppard
Thomas Frank	Mike Libbee	Colin Thorn
Owen Furuseth	Laurence Ma	Tom Vale
Gerald Galloway, Jr.	Alan MacEachren	Burke Vanderhill
Pat Gober	Mike McNulty	Ingolf Vogeler
Robert Gohstand	Russ Mather	Phil Wagner
Reg Golledge	Vernon Meentemeyer	Marv Waterstone
Frank Gossette	Tom Meierding	Tom Wilbanks
Sam Goward	Janice Monk	Craig ZumBrunnen
Will Graf	Mark Monmonier	

The Departments of Geography at the University of Colorado at Boulder and the University of Delaware lent now-heralded support. Joanne, Krista, Annette, Kathy, and Scott—thanks for those long hours of no fun at all. Helen Ruth, Bonnie, Carol, and Karen—thanks as well. Respective department chairs David Greenland and Russ Mather bent (but did not break) under the demands placed on the resources under their control. We are grateful.

Whoever they are, we would like to thank the people responsible for BITNET. They made the mass of communications between the coeditors, and among the coeditors and the authors, timely and manageable. The "computer age" made this project's many drafts, revisions, and communications immensely more efficient and expedient.

As geographers, we believe it is important to acknowledge *places* that influenced this work, with regard to either the production thereof or the respite therefrom. In addition to the above-mentioned Lake Mendota venue and our respective departments, we would like to acknowledge the intangible contributions of Boulder, Colorado; Lewes and Newark, Delaware; Nairobi, Kenya; Phoenix, Arizona; the road through Taos and the Peak-To-Peak Highway; and the Boulderado, Quinns, the Deer Park, and the "Baboon." The Continental Divide also played an editorial role—manuscripts edited on the deck overlooking the mountains were required to maintain one editor's unDivided attention.

We most happily acknowledge our friends who helped us, tolerated us, or simply distracted us: the Fergusons, Deuce Aguado, Spud and Mavis Murphy, Lefty Westside, Andy Weyl-Mark Diamond and the Jazz Showcase, Tim Harris, Stu Lilly, Mike McNulty, Steve Peterson, Jeff, Wes and Jan, Tim and Jim, Cousin Brucie, Jay and Marie, Buzz and Elayne, the Wards, Mike Rich, Smokie, Perfect, Tom Waits, the Four Directions, Orchestra King Mama, and the people of Kenya.

Finally, our families want us back. They earned an elephant's share of gratitude for their support and their tolerance of the unremitting unsocial behavior caused by the hours that *Geography in America* consumed. Susan, Pat, Abby, and Julia—we love you dearly and we'll be home more—soon.

We must say, however, we really enjoyed our work.

GLG, Boulder
CJW, Newark

Contents

INTRODUCTION   xxiii

Foundations of Modern American Geography   xxiv

    Reflections of Ourselves   xxv
    Unity Amidst Diversity   xxvii
    Logic of Specialization   xxxii
    Policies and Applications   xxxiii
    Vital Signs   xxxiii
    Overview of *Geography in America*   xxxiv
    Building on Geography's Foundations   xlii

Geography in American Education   1

    School Geography, Academic Geography,
       and the Social Studies   2
    Criticisms in Context   14
    Signs of a Renaissance?   18
    Concluding Scenario   19

ENVIRONMENTAL PROCESSES AND RESOURCES   27

Biogeography   28

    Development of Biogeography Within U.S. Geography   30

Climatology   47

    Modeling in Climatology   47
    Synoptic and Dynamic Climatology   51

Climate Change 54
Applied Climatology 57

Geomorphology 70

Publications 72
Important Contributions by American
    Geographers in Geomorphology 75
Professional Affiliations for American Geomorphologists 84
Geomorphology in Education 88

Energy Geography 95

Research Since TMI and the 1979–81 Oil Price Shock 96
Fossil Fuels, Nuclear Power, and Electric Utilities 98
Renewable-Energy Sources 102
Energy Conservation 103
Integrated Energy Analysis 105
Critical Challenges and Future Prospects 106

Water Resources 112

Hydrologic Research 113
Water-Quality Research 116
Water-Management Research:
    Urban Supply and Demand 119
Flood-Hazard Research 123
Groundwater-Resources Research 126
Law and Water Resources 130
Future Directions in Water-Resources Research 132

Coastal and Marine Geography 141

Coastal and Marine Geography 142
Coastal Geomorphology 143
Marine Recreation 146
Marine Boundaries 149
Marine Transportation 151
The Future 151

HISTORICAL AND CULTURAL CONTRIBUTIONS TO GEOGRAPHIC UNDERSTANDING 155

Historical Geography 156

Theory and Paradigms 157
North American Perspectives 162
Perspectives Abroad 171
Tools of the Trade 174
Needs and Opportunities 177

Cultural Ecology — 192

- Unifying Threads and Themes   192
- Cultural Ecology as a New Paradigm   193
- Complex Interrelationships   195
- Interdisciplinary Connections   197
- Culture Ecological Research   199

Cultural Geography — 209

- An Overview of Issues in Cultural Geography   210

Environmental Perception and Behavioral Geography — 218

- Human Behavior and Spatial Cognition   220
- Person-Environment and Ecological Considerations   223
- Landscape Perception and Experience   225
- Individual, Social, and Cultural Comparative Research   228
- Conclusions   231

Geographical Research on Native Americans — 239

- Native American Geographic Research   239
- Native American Geographic Literature   240
- The Future Focus of Native American Research   246
- Conclusion   249

ANALYSIS AND MANAGEMENT OF SOCIETAL GROWTH AND CHANGE — 257

Population Geography — 258

- Residential Mobility   260
- Urban Housing and Households   263
- Counterurbanization   266
- Internal Migration   269
- International Migration   272
- Population and Development   275
- Future Directions   281

Industrial Geography — 290

- Locational Shifts in Manufacturing   291
- Research on the Producer Services   294
- High Technology and Industrial Innovation   297
- International Trade and Foreign Investment   301
- Labor and Labor Markets   304
- Models for Regional Industrial Analysis   306
- Industrial Geography and Policy Analysis   310

Transportation Geography 316

Paradigms Since 1954   317
Content of the Field   318
General Findings   321
Perspectives on Research in the Field   327
The Future   327

Contemporary Agriculture and Rural Land Use 333

Context   333
Current Themes in Contemporary Rural and Agricultural Research   334
Theory, Methodology, and Technology   342
Conclusion   345

Regional Development and Planning 351

Geography and Planning: An Evolving Interface   353
Five Key Interdisciplinary Careers   356
Planning Theory: Contributions by Geographers   361
Research Focusing upon the United States   366
International and Comparative Studies   370
Future Agenda   374

The Geography of Recreation, Tourism, and Sport 387

Historical Background and Fundamental Concepts   387
Research Themes and Contributions   388
The State of the Specialty   400
Challenges   402
Conclusion   404

ASSESSMENT AND MANAGEMENT OF HAZARDS AND INFIRMITY 409

Hazards Research 410

Why Do Geographers Study Hazards?   411
The Status of Hazards Research   412
Changes in the Interpretation of Hazards   415
New Departures in Hazards Research   417
Summary and Conclusions   420

Medical Geography 425

Disease Ecology and Malnutrition   425
Medical and Health Services Location and Utilization   433

Aging and the Aged 451

The Gerontological Literature of Geographers—An Overview   452

The Residential Locations and Migration/Mobility Behavior of
    the Elderly—Patterns, Antecedents, Implications   452
The Residential Environment of Older People   455
Future Directions   458

INTERNATIONAL UNDERSTANDING THROUGH REGIONAL SYNTHESIS 467

Perspectives on Africa in the 1980s 468

Population and Resources   470
Planning African Development   475
Sociospatial Patterns and Processes   478
Climatology   481
Environmental-Resource Management   482
Conclusion   482

Latin America 488

Historical Antecedents   490
Contemporary Complexities   493
A Glimpse Ahead   497

Asia 506

Asian Studies Development in North America   506
East Asia   509
Southeast Asia   517
South Asia   520
Southwest Asia   531
Conclusion   533

Soviet and East Europe 546

Major Research Themes and Published Works
    on the Soviet Union   546
Research Themes on Eastern Europe   551
Research Exchanges with Eastern Europe   553
History of Contacts and Exchanges with the USSR   553
The Journal *Soviet Geography*   556
Summary   556

Study of Canadian Geography 563

Social and Urban Geography   564
Physical Geography and Biogeography   566
Industrial and Agricultural Geography   568
Cultural and Human Geography   569
Perception of Place, Landscape Studies,
    and Environmental Perception   570

Environmental Geography 571
Historical Geography 571
Border Questions 572
Regional Geography 572
Other Areas of Research on Canadian Geography 574
Education 574
Reflections and Conclusions 575

EMERGING PERSPECTIVES ON GEOGRAPHIC INQUIRY   581

Political Geography   582

Changing Definitions 582
Political Geography and Social Theory: Theoretical Debate in the "New" Political Geography 584
Political Geography of Urban Regional Development 589
Political Parties and Elections 594
Local Autonomy and the Local State 599
Nation-Building and National Disintegration 602
International Behaviors 609
Continuing Agendas 610

Geography from the Left   619

Urban Geography 621
Industrial Geography 622
Quantitative Methods 626
Development Studies 627
Environmental and Resource Geography 629
Socialist-Feminist Geography 631
Regional Geography and Locality Studies 632
Political Geography and the State 635
Conclusion 637

The Urban Problematic   651

The Historical Legacy of Urban Geography 651
Analytical Urban Geography 653
Systems of Cities 653
Internal Structures of Cities 655
Residential Structure and Migration Behavior 656
Explanation and Understanding in Urban Geography 657
Urban Historical Geography 659
The Urban Question 666
Summary and Conclusions 669

Geographic Perspectives on Women   673

Gender in a Global Perspective 674
Gender and Work 676

Gender and Landscape 677
Methodological Applications 678
Gender-Balancing the Curriculum 679
New Directions 680

ANALYSIS AND DISPLAY OF GEOGRAPHIC PHENOMENA 685

Cartography 686

Communication and Design Facet 687
Analytical Facet 697
Map Production Facet 706
Historical Facet 709
Research Prospects 712

Mathematical and Statistical Analysis in Human Geography 719

Statistical Models for Spatial Structure
 and Spatial Variation 720
Analysis of Spatial Interaction 725
Models in Economic Location Theory 729
Quantitative Analysis of Spatial Choices
 and Spatial Behavior 730
Computational-Process Models 735
General Concepts of Space 736
Integration with Geographic Systems 737
Directions for the Future 738

Remote Sensing 746

A Definition of Remote Sensing 746
The Remote-Sensing Process 747
Remote-Sensing Applications 753
Remote Sensing's Relations with Cartography and GIS 761
Artificial Intelligence in Remote Sensing 763
Modeling in Remote Sensing 765
The Future of Remote Sensing in Geography 768
Remote Sensing and the Discipline of Geography 768

Geographic Information Systems 776

Origins of GIS 777
DIME Files 779
Defining GIS 783
Driving Forces 785
Current Status and Selected Case Studies 787
Future Trends and Issues 790

Name Index 797

Subject Index 817

INTRODUCTION

Foundations of Modern American Geography

Gary L. Gaile | Cort J. Willmott

Geography is an old interest and field of study. Classical scholars from Eratosthenes to Kant attempted to set down Geography's distinctive place among the scholarly endeavors of their time. They thought it necessary to establish an intellectual framework for interpreting the expanding knowledge of the world and its inhabitants. Much knowledge has been acquired since then, and yet little has changed. Geography continues to be a growing contemporary discipline that is actively acquiring, debating, and applying new outlooks and methods.

Modern Geography's foundations are built on syntheses of traditional approaches to the organization, study, and understanding of geographic processes and phenomena, and new approaches that have been made possible through the coevolution of philosophical, theoretical, technical, and informational advances. It would be difficult, for instance, to overstate the fundamental changes that have occurred and will continue to occur within Geography as a consequence of the ongoing electronic revolution. The purpose of this volume is to chronicle the state of the discipline as it is being practiced in the 1980s, and to explore avenues over which geographic inquiry will likely advance.

The discipline of Geography is difficult to define in a few phrases. Unlike many other scholarly fields, it is not characterized by a discrete subject matter or method or even philosophy. For comparison, consider that Biology, Chemistry, Physics, Computer Science, and Sociology are perceived as well-defined by their subject matter, while Mathematics and Statistics seem relatively unambiguously characterized by method. Recognition by geographers and other scholars that Geography is not so neatly categorized led to several earlier treatises on "the nature of Geography" (cf. Hartshorne 1939, Harvey 1969, Gould 1985), as well as to numerous philosophical debates (e.g., between Harvey and Berry in 1974) and methodological debates (cf. Gould 1984). All have attempted to interpret or prescribe the essence of Geography. Geographers have debated, for instance, whether the core of the discipline is nature-society relationships, regional synthesis, or spatial analysis. While much of the debate has been instrumental in advancing the field, it also has projected a fragmented image to the larger academic community (Nuhfer 1988), and fostered defensiveness among some geographers. The search for distinctiveness on the basis of simple content, method, or philosophy has been unsuccessful because it presupposes that a boundary around the "core" of the discipline exists, and therefore can be articulated.

A simple, easily articulated definition of Geography, consistent with the traditional notions about how the pursuit of knowledge should be compartmentalized, simply does not exist—nor should it. Geography, like History and a few other fields, is not bounded. Such disciplines are set apart by integrative perspectives, themes, or approaches to organization of many interacting processes and phenomena. In the case of History, time underlies historical descriptions and explanations of virtually all types of events. In the case of Geography, place and its dimensions serve as the bases for geographic descriptions and explanations of events. It is the roles that place and its locational attributes play in natural and human processes occurring on the Earth's surface that are at the heart of geographic inquiry and knowledge.

Geographers traditionally have concerned themselves with describing the state of the Earth's surface, using maps as the standard means for storage and communication of such information. But it is an understanding of the dynamic processes occurring in the landscape (which give rise to spatial flows of people, commodities, money, energy, etc.) that underlies any fundamental knowledge of Geography. Only this understanding of spatially dynamic processes can adequately explain past and present patterns on the land, as well as provide the conceptual bases necessary to forecast or plan future geographies.

REFLECTIONS OF OURSELVES

Geographers' reflections on the discipline have been heightened recently with lively debates and analyses of the role of specialization (Goodchild and Janelle 1988a; Buttimer 1988; Marcus 1988; Gatrell 1988; Wheeler 1988). Concern also has been expressed over the very process of self-examination—does it reflect paranoia, smugness, or simple lack of direction? Self-evaluation is natural, and group stock-taking is a to-be-expected extension of this tendency. It may well be desirable and an ". . . essential overhead of academic and professional activity" (Goodchild and Janelle 1988b, 4). Further, better data and the efficiency of computers make this process easier and thus more likely to occur.

External evaluation also has struck human chords. The 1980s, for instance, will be marked in American Geography as the decade when widespread geographic illiteracy was "discovered." Geographers' reactions varied from "it is not our fault" to "what an opportunity!" and they are chronicled by Hill and LaPrairie in the next chapter on Geography in American education. The 1980s also will be remembered as the decade in which several major departments were closed by their university administrations. These events frightened, saddened, and angered many geographers, but most importantly, they instilled a resolve to make Geography a viable and competitive intellectual discipline.

Introspection, while certainly intensified of late, is not new to Geography. There is a rich tradition upon which this book builds, and a synopsis follows.

James and Jones and Fundamental Changes Since 1954

American Geography: Inventory and Prospect is a benchmark volume edited by Preston James and Clarence Jones and published for The Association of American Geographers (AAG) in 1954. It features 26 chapters on geographic specializations authored by a who's who of American geographers. It is interesting to note how this work differs from the present volume. James's and Jones's book contains greater coverage of economic geography, including chapters on the geography of resources and the geography of mineral production. It also covers some areas that are not covered separately in *Geography in America,* notably military geography and field techniques.

Conversely, *Geography in America* includes several specializations that are not reviewed by James and Jones. Interestingly, "culture," a traditional focus within Geography, does not appear in their table of contents. Field techniques, while still a central methodology of many geographers, have been overshadowed by the technological wizardry associated with quantitative methods, remote sensing, and geographic information systems (GIS), all of which are surveyed in *Geography in America*. Applications of geography have extended far beyond the purely economic arena to include considerations of energy, hazards, aging and the aged, regional development and planning, and medical and recreational geography. Geographers also are pursuing ideas arising from the "left" and from gender equality that cast the discipline as a new whole. These changes reflect a dynamic discipline that has avoided entrenchment while maintaining its traditions.

Continuing Self-Examination

More than 10 years after *American Geography: Inventory and Prospect,* Saul Cohen edited *Problems and Trends in American Geography* (1967) from a series of lectures and interviews prepared for the Voice of America. This book's 19 chapters survey the field of Geography and document some of the dynamic changes in its research agenda. Cohen's book also reflects the heightened social consciousness of the 1960s, as it contains chapters on poverty, the inner city, planning, and peace in Vietnam. The diversity of research that is so apparent in *Geography in America* was beginning to develop.

Over a decade later, a retrospective assessment, *The Association of American Geographers: The First Seventy-Five Years 1904–1979,* was written by Preston James and Geoffrey Martin (1978). Their chronicle concluded with the conjecture that:

> Unity of the discipline has given way to ever-increasing diversity. From simplicity came dichotomy, from unity emerged diversity, and from diversity came pluralism. (James and Martin 1978, 200)

James and Martin expressed fears that specialty groups might break away from the core of the discipline:

> Eventually these groups may break away from the parent society, hoping to form a new professional field, and in many cases these separations have been successful. . . .The greatest difficulty occurs perhaps when scholars who use quantitative procedures find no common grounds to justify their continued association with those who use the literary methods (198)

On the one hand, their trepidation was warranted—consider, for instance, the independent climatological and geomorphological groups founded in the United Kingdom, largely by geographers. On the other hand, since the AAG now supports specialty groups to such an extent that "they are at this time remarkably free from many of the practical constraints faced by other groups and divisions" (Goodchild and Janelle 1988b, 9), it is unlikely that sufficient impetus exists for the formation of new geographical societies.

A more personal view of Geography was written in the 1980s by Peter Gould (1985). In it, Gould accuses geographers of work. While it is difficult to ascribe "work" to the obvious joy with which *The Geographer at Work* was written, the "explosion" of the old field of Geography which Gould chronicles over the last 30 years must have been caused by some force—perhaps it *was* work. Gould's Geography has "a number of threads of continuity" with the old discipline, but also recognizes many new fields and approaches which "bear little resemblance to those of the past." Gould's impressions of what geographers do are convincing arguments that not only are geographers working on diverse research problems, but this diversity has kept us healthy, curious, and relevant.

Our colleagues abroad also are diversifying at a brisk rate, as evidenced by their recent state-of-the-discipline volumes (Cori, Fondi, and Zunica 1988; De Moor 1988; Dietvorst and Kwaad 1988; Hobbs and Walmsley 1988; *Revue Belge de Geographie* 1988; Slaymaker and Troughton 1988). Diversification as well as self-examination, it seems, are characteristics of all nationalities of geographers. Through *Progress in Physical Geography* and *Progress in Human Geography,* for example, our British associates make a regular habit of self-examination.

UNITY AMIDST DIVERSITY

Despite the aggressive search for new directions, Geography remains firmly grounded in its traditional concerns. Location and distance remain important. Sense of place remains important. Interaction between nature and society remains important. Regions, too, are still major organizing frameworks. Maps, though often produced by different methods, are still a basic means by which geographic information can be conveyed. Another aspect of Geography, which has changed only slightly in the 35 years since *American Geography: Inventory and Prospect,* is that Geography "is new in the significance of the role it plays; but it is old in terms of its traditional point of view" (James and Jones 1954, 16). To these traditions, however, has been added the plurality that comes with attention to new epistemological approaches, related disciplines, policy concerns, applications, and just plain creative impulses.

Resurgence of the Traditional

Traditional points of view in Geography have evolved through research, but their essence remains. This can best be seen in the resurgence of several traditional fields of Geography which had waned in previous decades.

Renewed dynamism in geographical education can be attributed, in part, to the much-publicized geographic illiteracy that pervades American classrooms. The result has been a marked rise in the quest for information on geographic education and a concomitant increase in "geographic alliances" (Salter 1988). How geographers have responded to this problem and opportunity is documented herein by Hill and LaPrairie. Historical geography is another example of a recast and revitalized subfield. While keeping with tradition, the field has restructured itself through a diversity of epistemological approaches. Political geography, too, has transgressed boundaries and frontiers, and embarked on innovative investigations of the state, local autonomy, elections, and conflicts. It has embraced an array of philosophical directions, and its debates are animated. Chapters by Earle et al. (historical) and Reynolds and Knight (political) chart the courses of these transformations.

Interaction between nature and society was a dominant theme in Geography during the first half of this century, and William Thomas's *Man's Role in Changing the Face of the Earth* (1956) is a landmark of that era. Many geographers who embraced this theme were heavily engaged in the debate with the purveyors of the quantitative revolution (Billinge, Gregory, and Martin 1983). Now this debate has subsided, and the study of nature-society interactions remains central to the discipline. Butzer's chapter on cultural ecology is but one example of our ongoing interest in the nature-society continuum.

While it is clear that geographers wish to maintain broad intellectual ties with their geographical colleagues, there also is concern that insisting upon traditional approaches unnecessarily constrains the discipline. Recall that quantitative geographers waged a "revolution" to overcome the constriction of tradition, and radical geographers are only now gaining territory in their effort to achieve academic equality in Geography.

There seems to be a pervasive fear among "traditionalists" that something sacred will be lost in a changing Geography. *Geography in America* should assist in assuaging that fear. Geography's sacred traditions are shown to be prospering amidst a burgeoning and dynamic discipline! It is our belief that readers of *Geography in America* will be impressed by the excitement of this old and new field.

Epistemological Frontiers

During the 1960s and 1970s, the quantitative revolution in Geography separated geographers into two distinct groups on the basis of their approaches to geographic inquiry. Simplicity characterized the issue (one was either quantitative in method or not), and this perceived simplicity was reinforced by the staunchness with which opposing views were held. But to today's geographer, the choice of approach to knowledge is far from simple. Fortunately, inflexibility among those holding opposing views has diminished and intellectual experimentation is at least tolerated and often encouraged.

Logical positivism and exceptionalism now reside together firmly within traditional geographical epistemology. Radical approaches to geography appeared in the early 1970s (Peet 1977), and they provided a salient diversion. They also resulted in the production of the journal *Antipode*. Since that time, epistemological approaches have multiplied through functionalism, structuralism, humanism, poststructuralism, and structuration, to postmodernism (which is to epistemology what anarchism is to government). Geographers are trying them all!

Innovation in Method

American geographers are making significant contributions to the way in which geographic processes are investigated as well as to the use of procedures for better managing societal growth and change. As mentioned above, fundamental changes have occurred and will continue to take place within Geography as a consequence of the ongoing electronic revolution. Seldom in Geography can innovation be comprehensively characterized as methodological; nevertheless, innovative methods have provided a basis for many geographic "discoveries." Improved methods not only provide the means for innovative problem-solving (applications), but in some cases, for advances in theory or philosophy.

Research into the development and application of new geographic methods is growing at an explosive rate. Advances arising from the electronic revolution (e.g., in computer speed and graphics capability, as well as in remote sensing and other monitoring technologies) are allowing geographers to tackle complex and large-scale problems that were virtually intractable only 10 years ago. Using remote-sensing technology, for instance, geographers are documenting and evaluating large-scale landscape changes such as desertification and deforestation (see the chapter on remote sensing). Four new polar orbiting platforms are planned for the 1990s, and they will even further enhance our ability to monitor, map, investigate, and hopefully manage "global change" (Kotlyakov et al. 1988). Geographers who contribute to and use remote-sensing technologies and data are making, and will continue to make, key contributions to our understanding of large-scale geographic transformations.

Remotely-sensed and other inherently geographic data (e.g., census enumerations) would be of little use without adequate spatial data-handling capabilities. Advances in our theories about spatial processes also are essential to the analysis and interpretation of spatial information. Ongoing research by American cartographers, for instance, is providing necessary algorithms and theory. Cartographers have begun to think of maps as much fuller representations of the Earth's surface than a static image on a piece of paper. Maps have become virtual in that their component parts can reside on a computer and can be rapidly manipulated prior to display. This has created new means of map compilation and design, and the future looks even brighter. Using new technologies, topologic concepts, and new data structures, cartographers also envision the development of digital animation as well as holographic and other three-dimensional realizations of virtually stored maps.

New concepts of space are being investigated by the more mathematically oriented among us, and they are providing new perspectives on spatial data and processes. Fuzzy sets (Leung 1987) and fractals (Goodchild and Mark 1987), for example, are being formalized for many geographic problems. Fractal representations already allow us to reliably describe and evaluate geographic features (e.g., coastal lines) and processes that exhibit variation at multiple geographic scales.

Mathematical geographers are looking increasingly into computational process models as analogs to key spatial processes (Couclelis 1986). Growing interest among mathematically oriented geographers in computationally oriented problems and solutions is exciting inasmuch as their contributions are more likely to be adopted by cartographers, remote-sensing researchers, and geographers interested in GIS—all of whom work primarily in digital media. Digital computing, in general, is inadvertently serving to bring back together geographers of different stripes in the search for comprehensive solutions to multifaceted geographic problems.

A most exciting aspect of the convergence of geographic expertise around computational resources is the recent formalization of a subfield which has as its purpose the development and improvement of geographic information systems (GIS). As described in the chapter on that topic, a GIS is an integrated system for the input, storage, analysis, and output of spatial information, and as such it combines methods and outlooks from many geographic specialty areas. GIS that monitor and forecast agricultural resources through a set of indirect remotely sensed observations and model translations, for instance, are already feasible. Future systems will additionally employ theoretical improvements growing out of artificial-intelligence research (Robinson and Frank 1987) and expert-systems research (Nickerson and Freeman 1986), as well as from the cartographic, remote-sensing, and mathematical work mentioned above.

Perhaps even more intriguing is that concepts and principles arising from topical or regional geographic research can be incorporated into a GIS as a series of rules. While still in its infancy, GIS represents the first viable means to verifiably produce regional syntheses and analyses in accordance with explicit sets of rules. Geographers must continue to work, however, to integrate spatial statistical and other methodologies into GIS in order to enhance intellectual content.

Challenges Facing American Geography

Despite the obvious advances in and enthusiasm for geographic inquiry that run throughout *Geography in America* and the discipline at large, there are several disturbing trends. While aspects of these problems occasionally surface in this volume, the underlying causes are more frequently debated in *The Professional Geographer* and *The AAG Newsletter,* as well as in a variety of forums at the AAG Annual Meeting. One of this book's purposes is to outline those problems that may have far-reaching consequences in order that they may be considered by the geographic community at large.

Perhaps Geography's foremost problem is an undercurrent of intolerance for approaches to geographic inquiry that differ from tradition, and a resistance to change in general, as mentioned above. Geography has a rich tradition in nature-society analyses, regional syntheses, spatial analysis, and physical geography, and this ought to be preserved. Preservation, however, should not be equated with codification. Traditions should grow and change as knowledge and the world around us change. New geographic traditions should be founded when new and pressing geographic problems require innovative solutions. Critical analysis should remain a cornerstone of careful geographic inquiry, but at the same time, geographers should be open to promising new methods, theories, and philosophies. It is innovation built on top of tradition that will keep Geography both viable and vital.

A second issue that concerns many geographers is the prevalence of geographic illiteracy. Geographic illiteracy takes many forms, although Americans' inability to correctly recite or locate geographic features is most often mentioned. There is, however, another aspect of geographic illiteracy that is even more disturbing. This has to do with deficiencies among professional geographers that tend to support perceptions that Geography is a "soft" discipline with little hope of making ground-breaking intellectual contributions or providing competitive professionals to "the real world." Too few geographers have sufficient training in quantitative methods, for instance, to make meaningful contributions to modern cartography, GIS, or remote sensing. Training also is weak in the physical and natural sciences, which translates into occasionally naive analyses of environment and subsequently of nature-society relations. Further, few geographers are fully literate in languages other than English, especially in non-European languages which would enable them to conduct regional syntheses in areas beyond the shrinking influence of European culture. While it is impractical to expect each geographer to be conversant with all these skills, graduate-level study should demand training in an allied quantitative field (e.g., mathematics), in a related hard science, or in a foreign language. Such training, when integrated into the geographic curriculum, will make not only better geographers but better scholars.

A third problem arises from changes in emphases that have taken place over the last three decades. Human geographers have increasingly looked to the allied social sciences and humanities for inspiration, rather than to the physical and natural sciences. (Consider that human-geographic topics receive three times the discourse of physical-geographic topics in this book, *Geography in America*.) Physical geographers, on the other hand, have become even more like their colleagues in the physical and natural sciences. While this divergence has produced geographers who are stronger social or physical scientists, it also has produced a gulf between human and physical geography. Physical geographers, who are in the minority, have expressed concern that the AAG has been overly social-science oriented and has not met the needs of physical and environmental geographers (Minutes of the AAG Council Meetings for Fall 1986 and April 1987). More importantly, this schism undermines the ability of geographers to meaningfully contribute to our understanding of nature-society interactions. Activities currently underway to bring physical geography back into the fold have increased recently, especially under AAG Presidents Abler and Demko, and they should continue to be encouraged.

A fourth problem has been the decline of international and regional interests and expertise among American geographers. This is related to the rise of systematic interests, to the decline in our collective linguistic abilities, to reduced government support, and to the perception that regional geography is not as challenging as some other specialties. Within this volume, for instance, only five chapters are explicitly devoted to regions. However, several chapters—e.g., those on GIS, geography from the left, and regional development and planning—suggest that contemporary geographic methods and outlooks, when combined with, for instance, the growing importance of the Third World and international trade, should revitalize American geographers' interests in regional synthesis. Regional interests and expertise are central to geographic analysis and synthesis, and should be encouraged.

A final problem concerns the diminishing spatial scales at which most American geographers conduct their research. Somewhat akin to their increasing topical specialization has been a trend to focus upon increasingly smaller regions. For example, geographers typically work at community, urban, or ecological scales. One consequence of this small-scale bias is that few American geographers work at national, continental, or global scales. More geographers investigating large-scale patterns and processes are needed if American Geography is to make important contributions to the largest international scientific initiative ever proposed—the International Geosphere Biosphere Program (Kotlyakov et al. 1988). To conduct investigations under this program, geographers also will have to become better acquainted with the special numerical and computational requirements demanded by such large-scale research.

LOGIC OF SPECIALIZATION

Part of the editors' original vision of *Geography in America* included a complete description and evaluation of specialization within Geography and the formation of specialty groups within the AAG. Given the many debates that have occurred over the wisdom of specialization in the discipline (Goodchild and Janelle 1988b), the editors had originally planned to consider both specialization and specialty groups at length within this introduction. However, the recent and comprehensive treatment of these topics by Goodchild and Janelle (1988b) considerably lessens the need for another complete evaluation. The purpose here, then, is to introduce specialization and the AAG specialty groups, and to conjecture about their future impact on American Geography. Readers desiring a more complete presentation are referred to Goodchild and Jannelle (1988a, 1988b), as well as to published comments by Buttimer (1988), Gatrell (1988), Wheeler (1988), and Marcus (1988).

Specialty-group frameworks exist not only within the AAG but within the national geographical organizations of Canada and the United Kingdom, as well as within the International Geographical Union (IGU). They also are common in allied scientific organizations, such as the American Meteorological Society (AMS) and the American Geophysical Union (AGU). It could be argued, in fact, that the specialty-group format came late to the AAG, although the need for such specialization has long been recognized (James and Jones 1954, 16).

Over a decade ago (April 1978), the AAG ad hoc Long-Range Planning Committee (LRPC), which had been formed by the AAG Council nearly two years previously, recommended to Council that specialty groups be formed within the AAG. The LRPC recommendation was adopted by Council, and the path toward specialty-group formation lay open. AAG members rapidly began forming, joining, and actively participating in one or more specialty groups, and by 1988 the AAG specialty-group list contained 40 entries (35 are represented in this volume). The groups are quite active; that is, several hold their own meetings independently of the AAG, many issue a periodic newsletter, and collectively these groups typically organize more than half of the paper and poster sessions at the AAG Annual Meeting. This specialty-group infrastructure has radically transformed the AAG by adding a kind of lower house to the governing infrastructure, an infrastructure that already included elected representation (the AAG President, Vice President, Council, et cetera) as well as an appointed AAG Executive Director and his staff.

Heralded by some and criticized by others, the decision to allow the formation of and support for specialty groups grew out of the necessity for ". . . a flexible framework for accommodating the growth and changing patterns of specialization among geographers" (Goodchild and Janelle 1988b, 3). Formation of AAG specialty groups allowed AAG members to pursue their research interests within the Association—adding their vitality to the AAG—rather than forming external societies, as had occurred in the 1960s and 1970s. Simply put, specialty groups represented a viable compromise between geographers' desire to maintain a collective enterprise and the need to conduct research on meaningful and tractable topical scales.

The prognosis for American Geography and AAG specialty groups is quite encouraging. Specialty groups may gradually increase in number over the next few years, albeit at a decreasing rate. According to statistics presented at the 1987 meeting of specialty-group chairs in Portland, Oregon, nearly all groups (34 of 37 surveyed) have increased their memberships during the mid-1980s. As suggested by the contributions to *Geography in America,* they will facilitate diversity and the innovation it brings, but under the banner of Geography. As their abilities to conduct business and govern themselves mature, they increasingly will seek collaboration with allied specialty groups (this is already beginning to take place, as mentioned in the Preface). Increasing communication should strengthen ties among the various special interests and traditions of Geography and add to the coherence of the discipline. While some geographers' interests may lie at larger or smaller scales than are represented by the specialty groups, or be outside of or incongruent with the current specialty-group structure, the contents of this volume suggest that these groups well serve the great majority of American geographers.

POLICIES AND APPLICATIONS

It is clear that geographers occasionally leave the confines of "the ivory tower." *Geography in America* clearly reflects the welcome fact that geographers are working on a wide variety of applied and policy issues that have "real world" impact and import. For example, Geography's contribution to solving pervasive health problems is carefully documented in the chapter on medical geography. Growing concern for problems affecting the aged also is evident in the chapter on aging and the aged. Hazards geographers have contributed a chapter that details the continuing effort to predict and avoid hazards, and to inform the public about living on a hazardous planet made more hazardous in some cases by human activity. Climatologists are making headlines with their varied predictions of climate change and potential human impact. Planners quietly attempt to develop parts of the world into better places. These and other examples of applied geography represent some of the best—if often unpublicized and unpublished—work Geography has to offer.

VITAL SIGNS

While it is important that geographers are optimistic about the status and future of our discipline, ultimately Geography will be judged on the basis of what it contributes to academia, knowledge, and society. Assessing the view from without, in other words, is likely one of the best single ways to diagnose Geography's health and to make prognoses about its fitness in the years to come. Reliable indicators of how Geography is perceived from without may be difficult to obtain; however, growth and the allocation of resources necessary for that growth (by university administrations, for instance) are unmistakable indications of others' faith in Geography and geographers.

When viewed through such a filter, recent patterns are most encouraging. According to an AAG circulated notice entitled "Changes in Departments of Geography Since

1970" (September 1988), 24 new degree programs have been established since 1970. Moreover, 48 existing programs have expanded, most through the addition of a graduate program. A number of journals also have been founded in recent years, largely by American and European geographers. Some examples are *Urban Geography, Physical Geography, Journal of Biogeography, Journal of Cultural Geography, Earth Surface Processes and Landforms, Journal of Historical Geography,* and the *Journal of Climatology.* Despite this overall growth, the unfortunate news is that, between 1982 and 1986, six university-level geography programs were eliminated.

The 1980s also witnessed the infusion of new capital into graduate-level geography programs by the National Geographic Society (NGS) for the purpose of improving geographic education at the grade- and high-school levels. NGS funds will facilitate the development of contemporary geography lessons, the continuing education of geography teachers, the improved education of aspiring geography teachers, and ties among university-level geography and school teachers and administrators (Salter 1988).

At the National Science Foundation (NSF), funds for research administered under the Geography/Regional Science Program have more than doubled over the last several years, and in fiscal 1988 the budget called for approximately $2.9 million (Abler 1988). Many of these new monies ($1.1 million/year for five years) are committed to the newly established National Center for Geographic Information and Analysis (see the chapter on geographic information systems). The majority of NSF funds, however, support varied geographic research that is conducted in the nation's colleges and universities and research assistantships that lead to the development of future professional geographers.

Hill and LaPrairie's chapter on Geography in American education points to a number of other positive signs and they conclude that, with respect to higher education, Geography "enjoys today a higher status than ever before."

OVERVIEW OF *GEOGRAPHY IN AMERICA*

Geography in America evolved from paper sessions held at the 1988 AAG Annual Meetings and the subsequent sharing of chapter outlines among the authors. It soon became clear that there were several common threads that partially defined clusters of specialties. Subjective groupings of the chapters suggested to the editors seven relatively coherent but overlapping themes around which *Geography in America* could be and is organized. These themes were cast as headings for the volume's major sections.

Using several taxonometric approaches to the classification of specialty groups, Goodchild and Janelle (1988b) also found clusters of specialty groups, albeit only three major ones. They suggest, consistent with Pattison (1964), that:

> The man-land theme is the most central, represented . . . by historical geography and environmental studies, . . . [and there is] the clear existence of spatial and earth-science traditions. (p. 24)

Rationale for a seven-group classification of this volume's chapters arises in large part from an examination of the chapters as well as points made by Goodchild and

Janelle (1988b) and others. The three most homogeneous clusters of chapters are the physically oriented, the methods, and the regional chapters. Common among the physically oriented or "Environmental Processes and Resources" chapters are concerns for the scientific investigation of physical and biological processes that occur in the landscape or the management of environmental resources. Goodchild and Janelle (1988b) also found significant membership affinity (mentioned above) among these groups.

Methods for the "Analysis and Display of Geographic Phenomena" also are related in that they are based largely on spatial statistics and mathematics, and increasingly on computational methods and resources. Goodchild and Janelle (1988b) reported very strong membership links between the Cartography and Remote Sensing specialty groups. It is quite likely that there also are significant membership ties among the Cartography and Remote Sensing groups and the GIS group, the latter of which did not exist in 1984. While Goodchild and Janelle (1988b) found less affinity between the "mathematical models" group and the other methods groups, it is clear that all of these groups now share a compatible quantitative outlook. The chapter on mathematical and statistical analysis in human geography differs from the other methods chapters primarily in that it is less generic. It grew out of several human-geographic subfields (e.g., transportation), and its focus upon human geographic problems has remained.

Homogeneity also exists among the regional chapters. Regional geographers, in virtually all cases, have continued to work toward regional understanding and problem-solving through syntheses of important environmental and social factors. Emphasis, however, is usually placed on the human-geographic aspects of a region.

Somewhat less homogeneous, each of the remaining four sections has common traits that set it apart from the others. Chapters appearing under the heading of "Historical and Cultural Contributions to Geographic Understanding" are largely based in the humanities. Time and culture provide structure and foci, although social-science outlooks also are common, especially in the chapters on environmental perception and behavioral geography and geographical research on Native America.

Contributions contained in the "Analysis and Management of Societal Growth and Change" section are predominantly written from the social-science standpoint. Hypotheses are often precisely stated and tested (frequently statistically), and in several areas (e.g., population and industrial geography), more detailed, process-based models are being developed. Among these chapters, there is a strong interest in the management of human-geographic change in addition to the scientific evaluation of it.

While the three chapters comprising the "Assessment and Management of Hazards and Infirmity" also are firmly grounded in social-science traditions and methods, they emphasize policy responses to environmental influences on the quality of life, health, and mortality. Major themes include the assessment of and planning for natural- or human-induced environmental pressures that may range from low levels of infectious disease (e.g., influenza) to catastrophic technological disasters (e.g., nuclear "incidents").

Chapters presented within the "Emerging Perspectives on Geographic Inquiry" section are dissimilar in subject matter, but all make convincing arguments for new perspectives. Radical and feminist outlooks, as well as their development within Geography, are reviewed. More detailed synopses of each of the book's main sections follow here.

Environmental Processes and Resources

Growing populations, diminishing resources, and environmental degradation all have spawned a renewed interest in the environment. Geographers and the public alike have shared concerns about pollution, soil erosion, deforestation, and climate change. Within Geography, these concerns have translated into a marked resurgence of biogeography, climatology, and geomorphology. Concern for our rapidly changing environment also pervades the larger scientific community, which has responded with a massive international scientific program entitled "The International Geosphere Biosphere Program" in which geographers can play an important role (Kotlyakov et al. 1988). Increasing demands for energy and water resources, as well as for land, also have motivated renewed interest in resource assessment and management.

The three traditional major subfields of physical geography—biogeography, climatology, and geomorphology—are concerned with investigating biophysical processes that take place on the land. Of special interest are human interventions in those processes. Biogeographers concentrate on the varying ranges of plants and animals at a variety of temporal and spatial scales (although few biogeographers are currently working on continental and global scales). Ecosystem and community dynamics are important, but the core of biogeographic contributions lies in vegetation dynamics, vegetation-environment patterns, and human impacts on the biological environment.

Recent prognostications of climate change, especially of global warming and its possible consequences, have heightened both the interest of geographers in climatology as well as public awareness of its importance. Geographical climatology tends to be more quantitative than the other physical geographic specialties. Researchers are developing a wide variety of statistical and numerical models of climatic processes and impacts. Because data sets and numerical models are large, climatologists also require large-scale computing resources. Dynamic climatologists and modelers, by and large, are trying to understand the fundamental biophysical and dynamic processes that control climate. Others are attempting to explain the climate change that appears in the historical and geological records. Synoptic and applied climatologists are working to uncover and predict climate impacts on a wide variety of human and natural systems. The importance of our climatic resource is increasingly being recognized by geographers, other scientists, and the public, and this should continue to strengthen geographical climatology.

Geomorphologists continue to study landforms and soils. Approaches vary from historical analyses of landform and soil change to functional analyses of the salient physical, chemical, and biological processes. They emphasize the relations among biophysical systems, human activities, and changes in the morphology of the Earth's surface. Geomorphology remains largely an empirically based subdiscipline, and therefore more theoreticians are needed to link theory with measurements and applications, and to develop process-response models of geomorphic systems. Advances such as these assure that American geomorphologists will contribute significantly to broad, emerging initiatives in "Earth System Science" and "Global Change" (Kotlyakov et al. 1988).

Assessment and management of energy and water, as well as land, require physical-geographic knowledge about the nature of these resources. Questions such as "how much exists, where is it located, how expensive is it to extract, is it renewable, and can it be preserved?" all require an accurate environmental understanding. Resource geographers, however, also must be conversant with human-geographic aspects of resource management. Management of the resource and policy development demand it. Geographers are especially interested in energy and water resources, as well as the "resource" that the coastal environment offers. Resource-oriented specialties are good examples of the utility of Geography's nature-society tradition.

Energy geographers are evaluating both energy supply and demand issues, although, owing to the current perceptions of surplus—especially in petroleum—their emphasis has shifted to demand. Behavioral aspects of energy consumption, therefore, have become at least as important as the purely economic ones. The global nature of the energy industry also has caused a change in emphasis from local, regional, and national issues to global energy issues. Energy geographers are continuing to evaluate a wide variety of resources and issues (e.g., nuclear and hydroelectric) and will continue to provide a comprehensive nature-society perspective on energy.

Geographers who are studying and managing water resources have relied upon and maintained Geography's nature-society tradition perhaps as much as any other group. The geography of water resources truly bridges the gap between human and physical geography. An understanding of water resources requires grounding in climatology and geomorphology, as well as facility with socioeconomic and behavioral aspects. Legal and policy issues also affect resource use and availability, and decision-making processes must be considered. Not only are water-resource geographers making contributions to the understanding, management, and preservation of this resource, but, like energy geographers, they are demonstrating the value of Geography's nature-society theme to the public and geographers alike.

Coastal and marine geographers are, on the one hand, resource geographers, and on the other hand, regional geographers, because the areas of interest are rather well defined. Interest in coastal and marine geography developed out of competing desires to exploit nearshore or onshore resources, and to preserve the coastal environs. Much of the early geographic research in these regions was geomorphologic. It sought to improve our understanding of coastal change in response to natural and human-induced modifications of the coastal and nearshore environments. Now, however, coastal and marine geographers also are examining marine recreation, marine-boundary issues, marine transportation, and other factors. Coastal and marine geography largely has been an empirically based subfield, but calls for a more-theoretical base suggest that a transition will occur in the years to come.

Historical and Cultural Contributions to Geographic Understanding

A growing tradition of Geography is most evident in the "Historical and Cultural Contributions to Geographic Understanding" section of this volume. The Historical Geography Specialty Group summoned the talents of eighteen of their members to relate the multifaceted work in this dynamic field. While retaining their traditional core of analyzing landscape and culture change, historical geographers are aggressively expanding their perspectives and their spheres of analyses. Historical geography has extended its reach to embrace additional subfields, e.g., political, economic, urban, population, and physical geography.

A diametrically opposite approach was taken by the Cultural Ecology Specialty Group in that one representative, Karl Butzer, was entrusted with the writing. This group represents another traditional interest within Geography—the integration of nature with culture—yet its research simultaneously "draws from. . .[and] represents a break with that tradition," through a paradigm shift that has incorporated scientific methodology and analytical anthropology.

Cultural geography, while also being a traditional subdiscipline, is one of the newest specialty groups in the AAG. The Cultural Geography chapter describes a potpourri of research, and then interestingly critiques its own pluralism, focusing upon the new epistemological diversity of the field.

Environmental perception and behavior studies have "come of age" and have supplied Geography "with many of its most innovative concepts," argues Stuart Aitken and his colleagues. Their chapter chronicles the debates over analytical and epistemological approaches and developments that substantiate above assertions. This research area is vitally important to the recent resurgence of cultural and historical geography.

The last chapter in this section reports on research on a specific cultural group—Native Americans. This small specialty group has produced research "as diverse as the discipline itself." Emphasizing topics such as land use, the environment, poverty, and education, this chapter demonstrates how historical and cultural geographical perspectives can be applied to one of the more intractable and embarrassing problems on this continent.

Analysis and Management of Societal Growth and Change

Human geographers have applied a significant share of their resources toward the analysis and management of societal growth and change. These studies must begin with people, and an understanding of population dynamics is a critical cornerstone of this body of research. Population geography's contributions to the study of demography have been profound. The authors of the Population Geography chapter describe the development of a field which has moved from an initial emphasis on areal differentiation to a focus upon spatial interaction that now includes significant behavioral perspectives. While geographers in this specialty continue to study the spatial variation in fertility and mortality, their work has primarily focused upon analyses of migration. The simultaneous growth of quantitative sophistication and policy relevance in their research distinguishes population geography as a trend-setter in the discipline.

These methodological and policy trends are also being followed by industrial geographers. The scope of this specialty has enlarged considerably upon its traditional base of industrial location. Service and the commercial economy are receiving greater attention, commensurate with their increased importance in the American economy. Research on high technology and international trade also is expanding as America's position in the world economy dramatically changes. The role of labor is being transformed, and it is critiqued and analyzed from more radical perspectives. Regional industrial analysis studies open the door to an understanding of the rapidly changing geography of the American industrial economy. Finally, industrial geographers are becoming increasingly responsible for policy input, as America's spatial economic dynamism becomes more apparent. The Industrial Geography chapter shows why "economics is too important to be left to economists."

Transportation geography has not enjoyed the same dynamism as some of its related specialty areas. While model development has proceeded apace, this research area has been negatively impacted by funding cuts and a data-shortage problem. The electronic revolution may well presage a new era of optimism and productivity for this basic area of geographic inquiry.

"Land" is the ground upon which much of the nature-society work in Geography is based. The Contemporary Agriculture and Rural Land Use Specialty Group emphasizes rural and agricultural use of the land in developed nations. For example, major transformations that rural America is undergoing are of primary interest. Population changes, environmental impacts, and technological innovations are raising provocative research questions about rural America.

A more global approach to societal growth and change is evident in the Regional Development and Planning chapter. Geographers are involved in "designing the future" in almost every country in the world. Attempting to move from a current reality to a desired reality, given the pragmatic constraints of the region and a political economy, provides considerable challenges to this group of geographers. Broad research areas of these geographers span from theory, to nation-building, to locating market centers for development.

A funless society that stays at home is a dreadful prospect. The Geography of Recreation, Tourism, and Sport chapter convinces us that a group of geographers is acting as watchdog against such an outcome. The specialty area could be dismissed by the "ignorati" as "golfers at work" or "professional beach bums." A reading of this chapter, however, will indicate that the work is indeed fascinating and intellectually challenging (if not without fringe benefits). Theoretical insights have begun to emerge to explain why societies recreate the way they do, tour where they tour, and engage in differing sports. A behavioral understanding leads to an ability to better plan for the good times this group studies.

Assessment and Management of Hazards and Infirmity

Everyday life is complicated by the fact that there are a multitude of things that can "get you." Policy-oriented geographers are actively involved in preventing, mitigating, and planning reaction to a wide variety of these perils. The Hazards Research chapter reports on research involving generally acknowledged hazards such as earthquakes, tornadoes, and floods, as well as a growing variety of hazards from drought and famine to AIDS and nuclear war. This relatively small group of geographers has made contributions to disaster research significantly out of proportion to their numbers. Using a policy-oriented approach to nature-society interactions, these geographers have succeeded in calling attention to the neglected human dimensions of hazards, and are leading efforts to diffuse this knowledge globally.

Disease can also "get you." The Medical Geography chapter illustrates how the combination of a spatial perspective and a knowledge of nature-society interactions significantly increases an understanding of disease ecology. Medical geographers not only are active agents in the fight against the spread of many diseases, including AIDS, but also apply a spatial statistical methodology to medical/health systems provision and planning. Location-allocation models are being applied by these geographers to rationally make locational decisions.

If disasters or disease fail to "get you," you can be assured that old age will. Throughout the world, and especially in America, the elderly percentage of the population is increasing. The Aging and the Aged chapter illustrates research analyzing the residential locations and environments of the elderly. This research will continue to grow and assist decision-makers by informing them about the increasingly predictable problems facing this growing segment of our society.

International Understanding Through Regional Synthesis

Geographers continue to ply their trade abroad and to attempt regional syntheses, albeit in diminished numbers. While regional geography remains a tradition within the discipline, not all regions are represented by specialty groups. In particular, research pertaining to the United States, Western Europe, the Pacific, and the Polar regions is considered by American geographers only under the guise of systematic specialty groups.

Africa is in crisis and geographers are responding to the challenge using a variety of approaches. The Perspectives on Africa chapter highlights the use of both the more mainstream "reformist development geography" approach and the more radical "geography of underdevelopment" approach to analyzing these problems. Crises involving food, debt, the environment, population, and energy have all received attention from geographers bent on improving the situation.

Progress in the study of Latin America by geographers has been "relatively desultory," according to David Robinson, despite the increasing impacts—"via drugs, debts, droughts, democracy, or civil-political disorder"—on public perception. Yet a rally of new interest is predicted as geographers are "rediscovering the region." This chapter surveys the current research and suggests 10 tasks worthy of study to revitalize the field.

The Asian Geography chapter incorporates the work of both the Asian and China Specialty Groups. This is an exciting region due to the dynamic changes brought about by the rise of economic power in the Pacific Rim, the advent of the newly industrialized countries (NICs), the rapid developmental rise and innovative policy initiatives of China, the changing political realities of Southeast Asia, and the continuing dilemma of India. Serious gaps identified by the authors in the study of this region should provide future geographers with a wealth of research possibilities.

Contrasting with the primarily Third World nature of the research in the first three chapters of this section, the Soviet and East Europe chapter reflects work on a region that captures the headlines through its new willingness to interact with the U.S. in structuring a new world political order. There is little doubt that the impacts of *glasnost* and *perestroika* are ushering in a new era of interaction and cooperation. Geographers are responding to the challenge by not only expanding their research on the Soviet Union, but also by increasing their interactions and cooperative ventures with Soviet geographers (Kotlyakov et al. 1988). Unfortunately, this extended level of activity has not yet diffused to American geographers' research on Eastern Europe. It is hoped that easing access will lead to a significant increase in research on this region.

Finally, access is quite easy to our northern neighbor. Nevertheless, as the Canadian Geography chapter shows, Canada is not a suburb of the U.S. The Canadian Geography Specialty Group is trying to remedy the fact that it is "an invisible, often unnamed partner in the AAG." With increased trade agreements, a rise in Canadian

Studies programs in U.S. universities, and the efforts of this small group of committed scholars, Canada may one day command a greater interest among American geographers.

Emerging Perspectives on Geographic Inquiry

Inadequate or incomplete answers to geographic questions have provided the impetus for development and application of nontraditional perspectives. Relations among politics, history, society, gender, economics, and geography, in particular, are poorly understood. Pressing human needs, inequities, and egalitarian spirits also have encouraged geographers to look beyond traditional means to explain, for instance, interstate and intrastate variability and the urban landscape.

Geographic investigations and applications of alternative ideologies are especially evident in the Political Geography and Geography from the Left chapters. There is ongoing reexamination of such key concepts as state, society, nationalism, space, and place among both political and socialist geographers. Concern for advancing social theory is evident, and there are strong interests in the production and reproduction of places in the world capitalist economy. Historical materialism remains an important concept among left geographers, although recent initiatives have enhanced the geographical insights offered by classical Marxist theory. The Geography from the Left contribution discusses the varied applications of left views to urban problems, industrial development, and feminist issues.

Greater detail concerning social analyses and theory, on the one hand, and urban and feminist issues on the other, is contained in the chapters titled The Urban Problematic and Geographic Perspectives on Women. Urban geographers are increasingly engaged in theory evaluation and construction after what is described as a "crippling historical legacy" of mere description. Not only are urban geographers developing pertinent social theory; they also have gained a new historical conscience, improved analytical and numerical skills, and an appreciation for legal issues. Evolving theory and new methods have supported the broader view that urbanization and social change develop and are restructured together, and therefore should be studied together.

The feminist perspective, too, improves existing theory by making explicit the influences that gender and gender roles have on geographic processes, patterns, and inferences. Gender-based views of global issues, work, and landscape, for example, yield interpretations that differ considerably from traditional, male-based perspectives. Improved understanding of women's roles portends an improved understanding of Geography.

Analysis and Display of Geographic Phenomena

Rapid and ongoing technological developments, especially in electronics, have provided the resources for revolutionary advancements in geographic methods. Modern computers make possible both statistical and graphical analyses of extremely large databases, as well as numerical simulations of complex geographic processes. Computational speed on many computers now exceeds 100 million instructions per second. The electronic revolution already has irreversibly transformed many cartographers into "geographic information scientists" and is making feasible inferences from remotely sensed data and GIS. Geographers involved in developing analytical expressions of human-geographic processes also are making use of available computational resources, and the number and variety of numerical or "computational" models is increasing.

Cartographers are now constructing and managing digital databases and designing maps on interactive work stations. They also use computer-driven printing and plotting devices to obtain computationally produced maps. While the media are changing, traditional cartographic interests in historical cartography, map design, symbolization, map perception, and analytical cartography have not waned. The new media simply are making the research more efficient, exacting, and far-reaching.

Among those American geographers conducting "Mathematical and Statistical Analysis in Human Geography," research focuses upon developing formal (mathematical) rules for expressing geographical concepts and information. Such formalizations contain explicit reference to location and place. In fact, contributions from quantitative geographers have made possible the extension of conventional statistics and probability theory to locationally dependent data. Mathematical models have the potential to help "bridge . . . the chasms that separate conceptual thinking in different disciplines . . ." and many of the subdisciplines of Geography as well. A most-exciting prospect is the incorporation of these formalizations of spatial theory and processes into GIS.

Remote sensing in Geography is, in essence, an indirect means of obtaining spatial data. Many of these data depict the biophysical character and state of the landscape, as well as transformations caused by natural processes or human activities. Most data are obtained by satellite-borne sensors and then transmitted back to Earth where the information is converted to digital form. Billions of bytes of data are collected every day, and therefore data storage and computational requirements are considerable. Geographers who make use of remotely sensed data then also have key interests in computer resources and methods. They are developing and applying spatial statistics as well as models of pertinent biophysical processes in order to discern rates of landscape change. These geographers will play a central role in monitoring "global change" (Kotlyakov et al. 1988).

Perhaps the most rapidly growing geographic specialty is Geographical Information Systems (GIS). Once again, GIS are integrated computer algorithms "for the input, storage, analysis, and output of spatial information." These systems often contain instruction sets developed by geographers from many different specialties, and therefore GIS may soon make possible automated regional synthesis. Such systems also represent a focus for a vast array of geographic research, and may inadvertently serve to strengthen ties among many of the geographic specialties.

BUILDING ON GEOGRAPHY'S FOUNDATIONS

It is the editors' hope that a reading of *Geography in America* will infuse nongeographers and geographers alike with the sense of accomplishment, optimism, and joy that geographers share in their research. Growing diversity has led to a grass-roots development of specialty groups, and they hold significant promise. Despite some problems, Geography in America is alive and well, and faces an exciting future.

REFERENCES

Abler, R. F. 1988. Awards, rewards, and excellence: Keeping geography alive and well. *The Professional Geographer* 40:135–40.

Berry, B. J. L. 1974. Review of David Harvey's *Social Justice in the City*. *Antipode* 6:142–45.

Billinge, M.; Gregory, D.; and Martin, R. 1983. *Recollections of a revolution: Geography as spatial science.* New York: St. Martin's Press.

Buttimer, A. 1988. Specialization in the structure and organization of geography: Comment on Goodchild/Janelle. *Annals of the Association of American Geographers* 78:534–37.

Cohen, S., ed. 1967. *Problems and trends in American geography*. New York: Basic Books.

Cori, B.; Fondi, M; and Zunica, M., eds. 1988. *Italian geography in the eighties: Selected contributions.* Pisa: Giardini Editori E Stampatori.

Couclelis, H. 1986. A theoretical framework for alternative models of spatial decision and behavior. *Annals of the Association of American Geographers* 76:95–113.

De Moor, G., ed. 1988. *Belgian physical geographers at home and in the world between 1950 and 1985.* Bevas/Sobeg 1.

Dietvorst, A. G. J., and Kwaad, F. J. P. M., eds. 1988. *Geographical research in the Nederlands 1978–1987.* Nederlandse Geografische Studies 64. Amsterdam: Koninklijk Nederlands aardrijkskundig genootschap/International Geographical Union sectie Nederland.

Gatrell, A. 1988. Commentary on 'Specialization in the structure and organization of geography.' *Annals of the Association of American Geographers* 78:538–39.

Goodchild, M. F., and Janelle, D. G. 1988a. Questions regarding 'Specialization in the structure and organization of geography'–A reply. *Annals of the Association of American Geographers* 78:547–49.

———, and ———. 1988b. Specialization in the structure and organization of geography. *Annals of the Association of American Geographers* 78:1–28.

———, and Mark, D. M. 1987. The fractal nature of geographic phenomena. *Annals of the Association of American Geographers* 77:265–78.

Gould, P. 1984. Statistics and human geography. In *Spatial Statistics and Models,* eds. G. L. Gaile and C. J. Willmott, pp. 17–32. Dordrecht: D. Reidel.

———. 1985. *The geographer at work*. Boston: Routledge & Kegan Paul.

Hartshorne, R. 1939. The nature of geography. *Annals of the Association of American Geographers* 29:173–658. (Reissued by the AAG in 1946.)

Harvey, D. 1969. *Explanation in geography*. New York: St. Martin's Press.

———, and Berry, B. J. L. 1974. Discussion. *Antipode* 6:145–49.

Hobbs, J. E., and Walmsley, D. J., eds. 1988. *Australian Geographical Studies* 26(1).

James, P. E., and Jones, C. F., eds. 1954. *American geography: Inventory and prospect*. Syracuse: Syracuse University Press.

———, and Martin, G. J. 1978. *The Association of American Geographers: The first seventy-five years 1904–1979.* Washington: The Association of American Geographers.

Kotlyakov, V. M.; Mather, J. R.; Sdasyuk, G. V.; and White, G. F. 1988. Global change: Geographical approaches (A review). *Proceedings of the National Academy of Sciences* 85:5986–91.

Leung, Y. 1987. On the imprecision of boundaries. *Geographical Analysis* 19:125–51.

Marcus, M. 1988. New twists on the horns of an old dilemma. *Annals of the Association of American Geographers* 78:540–42.

Nickerson, B. G., and Freeman, H. 1986. Development of a rule-based system for automatic map generalization. Proceedings, Second International Symposium on Spatial Data Handling, Seattle, 50–64.

Nuhfer, E. B. 1988. Academic geographers are partly to blame for Americans' ignorance of geography. *The Chronicle of Higher Education* 34:B2.

Pattison, W. D. 1964. The four traditions of geography. *The Journal of Geography* 63:211–16.

Peet, R., ed. 1977. *Radical geography*. Chicago: Maaroufra Press.

Revue Belge de Geographie. 1988. Human geography in Belgium: An epistemological survey with an extensive bibliography. Special issue.

Robinson, V. B., and Frank, A. U. 1987. Expert systems for geographic information systems. *Photogrammetric Engineering and Remote Sensing* 53:1435–41.

Salter, C. L. 1988. "Utilizing non-college geographers to stimulate geographic change: A cost-benefit analysis." Paper presented to the 26th Congress of the International Geographical Union, Sydney, Australia.

Slaymaker, H. O., and Troughton, M. J. 1988. Canadian geography 1984–1988: A review. Ottawa: Canadian Committee, International Geographical Union.

Thomas, William L., Jr. ed. 1956. *Man's role in changing the face of the earth*. Chicago: University of Chicago Press.

Wheeler, J. O. 1988. Diversity, quality, and the core of geography: A comment. *Annals of the Association of American Geographers* 78:543–46.

GEOGRAPHY IN
AMERICAN EDUCATION

Geography in American Education

A. David Hill | Lisa A. LaPrairie

Judging from the amount of published commentary on the subject in geographic journals, the status of geography is intensely interesting to American geographers. Scores of geographers have expressed themselves in print on the topic; indeed, along with "the nature of geography" and the relative status of individual geographers, it is one of our favorite subjects. This is reason enough to include a chapter on the topic in a book on *Geography in America,* but not without trepidation.

Do geographers devote too much attention to the status of their subject? We have no way of knowing whether it concerns us more or less than it does, say, sociologists or economists. In any event, attending to status is natural, for esteem motivates. Still, manifest concern for status embarrasses, because idealistically satisfaction should come from substantive work. Worse, excessive image-building may stimulate "false values," i.e., the subordination of substance to image (Hudson 1984). Accordingly, this chapter's authors—who are devoted geographers—seek to tread a fine line, examining objectively the status of geography in American education without extravagantly gilding the lily of their chosen field.

This chapter deals with the status of geography, both in the schools and in the colleges and universities. Although the Association of American Geographers (AAG) Specialty Group on Geography in Higher Education does not encompass precollegiate geography, the authors have chosen to include it to provide a more comprehensive picture of the status of the discipline in American education. Thus, the chapter begins by relating geography at the elementary and secondary level to academic geography. Unlike academic geography, which has strong subfields of both physical and human geography, school geography has become submerged in the social studies, and very little physical geography has survived. Geography's shifting role in the curriculum over time is noted, but space does not permit a full historical treatment of the subject. Geography's status in higher education is measured primarily in terms of enrollments and degrees conferred in comparison with cognate fields. The chapter then turns to a review of both the numerous criticisms of school and college geography, as well as to the multitude of suggestions for enhancing their status.

GEOGRAPHY IN AMERICAN EDUCATION

Finally, signs of a recent renaissance in geographic education at the elementary and secondary levels are noted. If this renaissance persists, impressive changes are sure to be registered at the collegiate level.

SCHOOL GEOGRAPHY, ACADEMIC GEOGRAPHY, AND THE SOCIAL STUDIES

In 1902, William Morris Davis, "the father of American geography," worried about a lack of "mature geography" in higher education to support school geography (Davis 1902). Geography was widely taught in the schools, and Davis was concerned about the "deficiency of higher learning in geography." (The first graduate department of geography in America, at the University of Chicago, had not yet been established.) Davis saw the progress of geography in the schools impeded by "the want of a well-developed body of higher geographical learning, with respect to which the geography in the schools shall stand only as a beginning" (Davis 1902, 28).

Today, the opposite is true. Geography has made great strides as a university subject since the days of Davis, but school geography has almost disappeared as a separate subject. It was commonly taught in America's schools in the eighteenth and nineteenth centuries. As the classical curriculum of Latin and Greek began to make concessions to modern subjects in the early part of the nineteenth century, geography was one of the first new fields to be required for college entrance (Brubacher and Willis 1968). But today, less than 10% of high school students study geography (Gardner 1986). How did this turnabout happen?

Schooling in the United States is very much a local affair; the Constitution says nothing about education. Even state control of education is uncommon; indeed in some states it is unconstitutional. Locally elected school boards control budgets, personnel, and curriculum. Although no national curriculum is mandated, considerable similarity obtains from state to state, because of the professionalization of education. Colleges of education conduct research, train school teachers and administrators, and grant degrees; state departments of education, staffed by professional educators, license teachers and administrators; professional educators also staff regional accreditation agencies, which conduct periodic assessments of schools. Educators have numerous associations (e.g., for school principals, social studies teachers, etc.) and a very large general organization, the National Education Association (NEA), which has some 1.9 million members. America's schools are largely the product of professional educators. The professionals representing the academic disciplines, such as history and geography, certainly are less influential.

Geography's Shifting Role

Largely owing to the texts of Jedidiah Morse, geography was the most firmly established social study in the elementary curriculum until the 1820s, when American history texts began to appear. Americanism became the main objective of the social studies (Thomas and Brubaker 1971), and history courses became the means by which to achieve this aim. History has dominated the social-studies curriculum ever since, although geography has always had a foothold at the elementary level, where early on it was taught with memorization exercises that were considered ideally suited to the abilities of elementary students (Stoltman 1987). Geography's position was reinforced by the enactment of state laws around 1830 requiring that it be taught in elementary schools (Rumble 1946). Geography is integrated into today's elementary social-studies curriculum, mainly in the form of map skills, but the social studies on the whole are not an important part of that curriculum: two-thirds of primary teachers report that they have insufficient time to teach social studies (Lengel and Superka 1982; Stake and Easley 1978).

Throughout the history of high-school geography, the subject emphasis has continually shifted, and so has the subject's status in the curriculum. The publication of Guyot's *Physical Geography* in 1873 helped to shift the focus of high-school geography from locational to physical geography (Rumble 1946; Rosen 1957). This trend, which had begun in the 1850s, was reinforced in 1892 by the NEA's Committee of Ten, which provided for the first time a definition of the nation's secondary-school curriculum (Robinson and Kirman 1986). This influential committee's report (National Education Association 1894) adopted the radical recommendation of its Conference on Geography, led by William Morris Davis, which served to establish physical geography as the basis for general science teaching and for the teaching of all geography (James and Martin 1981).

After 1908, general-science courses began to replace the highly similar physical-geography courses as both the NEA and the AAG reported growing dissatisfaction with physical geography in the high schools (Rosen 1957). Both organizations recommended that physical geography be replaced by a more-vocational economic and commercial geography designed to meet the needs of terminal students. This was the period of interest in vocational high schools. America's industry was growing rapidly and needed to find new markets. Industries were demanding trained workers, and the school curriculum responded. A curriculum that had largely taught physical geography now began to shift to economic, commercial, and regional studies (Mayo 1965). The percentage of high-school students electing physical geography fell from 21.5% in 1905 to 4.3% in 1922, and to 1.6% by 1934. All geography courses were disappearing from the high schools. Commercial geography was the only survivor, with only 4% of the students in 1934. Geography lost its position as an admission subject in most liberal-arts colleges during this period, no doubt because of its new vocational reputation. In addition, the College Board stopped examinations in geography in 1934 (Rosen 1957).

While geography experienced this precipitous decline in the curriculum, some of its content was being subsumed under a major new curriculum category, the social studies. The social studies did not exist when the Committee of Ten met in 1892, but a 1911 NEA secondary-school curriculum review played a major role in establishing social studies as a field without disciplinary boundaries (U.S. Bureau of Education 1916; Stoltman 1987). The reaction against isolated subjects and the emergence of the social studies was in part a revolt against the rote learning that marked the teaching of history, geography, and civics.

The major impact of the social-studies movement came in the years following World War I (Vuicich and Stoltman 1974). Geography's place in the social studies might have been strengthened at that time, but geographers held that geography was not a social study and refused to participate in the NEA social-studies curriculum-development process of 1916 (James 1969; Stoltman 1987). This left nongeographers to create the geography that was to be included.

The 1916 NEA Committee on the Social Studies set the pattern for high-school social-studies curricula (Robinson and Kirman 1986; Morrissett 1982). Schools readily adopted the NEA model. Most of the educators who developed this model curriculum were trained in history, and thus this was the discipline most emphasized. The pattern set in 1916 persists today, as the comparison in Table 1 shows.

The National Council for the Social Studies (NCSS) recommended a "scope and sequence" for K–12 social studies in 1984, shown in Table 2.

Table 1

Comparison of 1916 NEA Model curriculum and today's typical curriculum

1916 NEA Model	Grade	1982 Pattern
Geography/European History	7	World History/Cultures/Geography
American History	8	U.S. History
Civics	9	Civics/Government/World Cultures
European History	10	World Cultures/History
American History	11	U.S. History
Problems of Democracy	12	American Government/Sociology/Psychology

Source: Morrissett 1982, 33–34.

Table 2

NCSS-recommended scope and sequence for K–12 social studies

Grade	Focus
K	Awareness of self in a social setting
1	The individual in primary social groups: understanding school and family life
2	Meeting basic needs in nearby social groups: the neighborhood
3	Sharing Earth space with others: the community
4	Human life in varied environments: the region
5	People of the Americas: the United States and its close neighbors
6	People and cultures: the Eastern Hemisphere
7	A changing world of many nations: a global view
8	Building a strong free nation: the United States
9	Systems that make a democratic society work: law, justice, and economics
10	Origins of major cultures: a world history
11	The maturing of America: United States history
12	One-year course from electives: ☐ Issues and problems of modern society ☐ Introduction to the social sciences ☐ The arts in human societies ☐ International area studies ☐ Social-science courses: discipline-specific ☐ Supervised experience in community affairs ☐ Local options

Source: National Council for the Social Studies 1984.

The NCSS scope and sequence recommendations clearly draw heavily upon geographic ideas from K–7, but geography is not mentioned in grades 8–12, nor is it listed among the discipline-specific social sciences mentioned for grade 12. Recently, geographic educators have paid specific attention to the "scope and sequence problem" (Stoltman and Libbee 1988).

Enrollment percentages for high-school social-studies courses for selected years from 1928 to 1982 show clearly that history, having 42% or more of total enrollment over the period, dominates the social studies (Figure 1). Political science (govern-

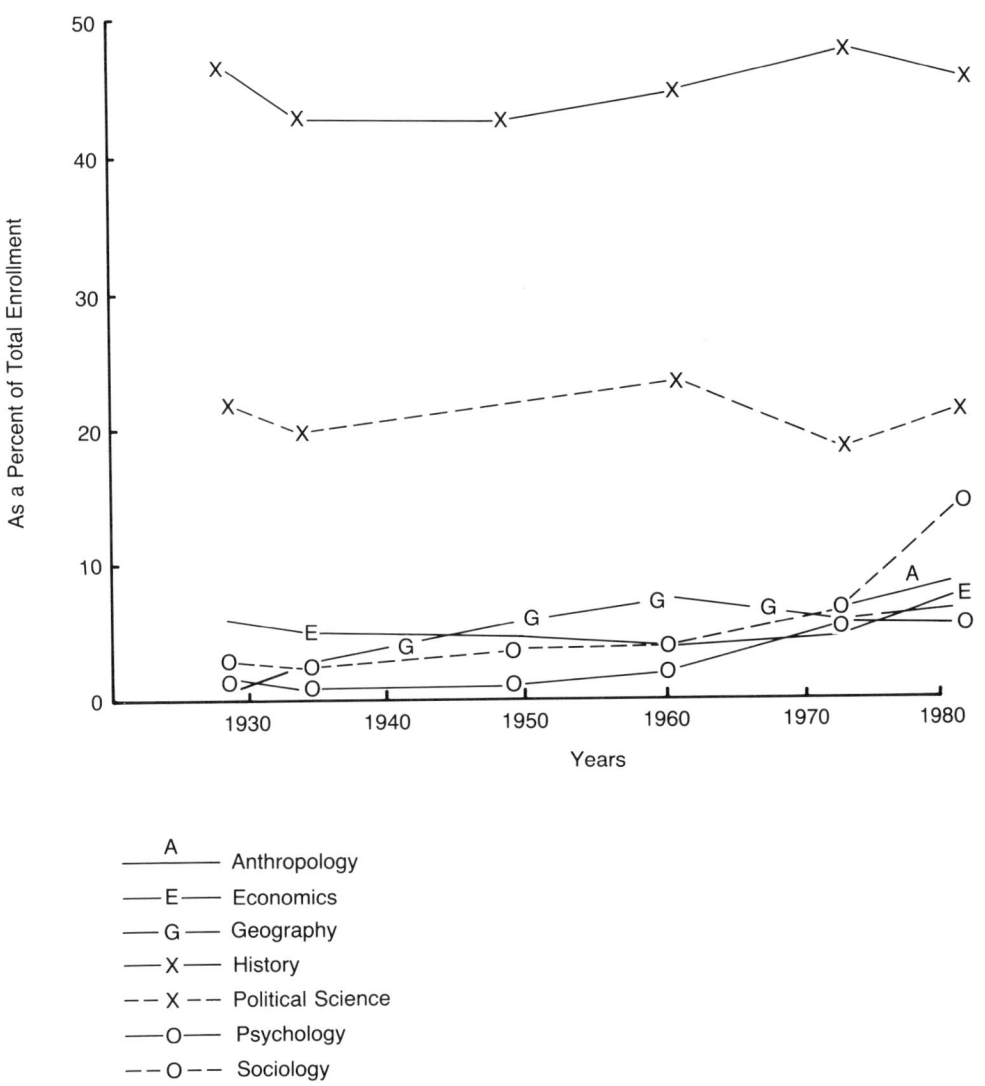

Figure 1
Enrollments in social studies courses, grades 9–12, selected years from 1928 to 1982 (National Center for Education Statistics 1984; Wright 1965)

ment) holds the second position, with about 20% of student enrollment. Less than 10% of students have been enrolled in any of the other social studies during any given year, with the exception of sociology, which jumped to 14.3% in 1982. Although geography accounted for only 6.3% of total enrollments in 1982, its position had improved considerably, having begun the period in 1928 with only 0.3%.

World War II reawakened interest in geography, especially world and regional geography; studies of European countries, cartography, meteorology, and conservation also gained in popularity. This caused the NCSS to reassess geography's role in the high school, and to recommend that it have a stronger position in the social-studies curriculum (Vuicich and Stoltman 1974). New courses listed included Global Geography, World Geography, and Air Age Geography. A 1950s nationwide survey of high schools (National Council of Geography Teachers 1956) reported that 66% of faculty favored geography as a separate subject, and stated that one-third of social studies was allotted to geography.

The late 1950s and 1960s (the "post-Sputnik era") was a period of intense educational reform, characterized by heavy federal funding and the involvement of college and university scientists and social scientists in elementary and secondary school curriculum development. Professional geography's answer for improving geographic education in the schools was the AAG-sponsored High School Geography Project (HSGP), which brought together school teachers and university professors "to prepare an improved course in high school geography" (White 1970). Unfortunately, like all social-science projects of the time, its acceptance and use was minimal (Weiss 1978; Switzer, Mitchell, and Walker 1981). Although the HSGP course, "Geography in an Urban Age," was not widely adopted, the influence of the project can still be seen today in textbooks and materials.

Lack of teacher preparedness was noted as one reason for the limited success of HSGP (Winston 1986), but a fuller understanding may come from an examination of the various roles played by the "two cultures"—professional educators and academics from the disciplines. Geographers rather than professional educators were supported to produce a new high-school geography, but without the strong support of "school people," curriculum reform was likely to fail. The "new social studies" projects were strongly criticized by social-studies specialists (Haas 1977), and some close observers found that social-studies educators had strong predispositions against the curriculum projects (Tucker 1972; Thompson 1973; Wiley 1977; Helburn and Helburn 1979). Wiley (1977) found that for the period 1955–75, geography was not mentioned as a separate high-school requirement by any of the states.

The relationship between the traditional academic disciplines and professional education has not been especially friendly, with the disciplines often viewing education as unrigorous and devoid of content (Morrissett 1968), and education accusing the disciplines of self-promotion at the expense of student interest and the needs of citizenship education. In 1963, the *Bulletin of the Council for Basic Education* opined that:

> At the heart of the desire for reform of the social studies is the conviction that an understanding of citizenship and the values of democracy, and an understanding of the peoples of other nations, are not directly teachable matters but are by-products of knowledge—ordered geographical knowledge, knowledge of the history of one's own country and of the world, knowledge of how government functions. (Smith 1966, 44)

On the other hand, Donald Bragaw, a social-studies advocate, argues that social studies as a school subject "draws upon all the social science disciplines for its main strength"; does not force, as do the universities, the "unnatural separation of the disciplines"; and that the "study of society would not permit a narrower [disciplinary] focus" (Wronski and Bragaw 1986, xii–xiii). However, this same author admits that the two perspectives have not been reconciled:

> It is the balance of knowledge of the specific discipline and the usefulness to overall notions of citizen responsibility that has yet to be achieved in social studies education—and the search continues. (Wronski and Bragaw 1986, xii–xiii)

The geographer's concern over the lack of geography in the high-school curriculum is not unique. Other social sciences have similar concerns (Wronski and Bragaw 1986). Professional education has created "new curriculum areas" that utilize social-science concepts and interests, e.g., multicultural education, world studies, world cultures, and international or development education (Becker 1979). Such areas contend for curriculum time with freestanding courses in the disciplines. One of these that competes with geography is "global education," which has won wide acceptance in the 1980s; indeed, more than 40 states have passed course mandates or resolutions of support for global or international education (King 1987).

Textbooks

Textbooks are an important indicator of educational practice and of the status of geography (Manson 1981; Winston 1984). Since Colonial times, textbook-oriented teaching has dominated classroom practice, and it continues to do so today (Thomas and Brubaker 1971; Lengel and Superka 1982). Curriculum materials dominate 90% of classroom time, with two-thirds of this time devoted primarily to textbooks (Educational Products Information Exchange Institute 1977; Patrick and Hawke 1982).

Owing to the competitiveness of the market, publishers base their books on the dominant pattern, thus reinforcing the curriculum set in 1916 (Lengel and Superka 1982). It is difficult, if not impossible, for a single school district to make drastic changes in the traditional curriculum pattern, since a textbook publisher cannot realistically publish for such a small market. Even if it were feasible for a local school district or even a state to produce a curriculum guide that differed from the traditional pattern, teachers would tend to disregard it, for textbooks are the major influence on teacher course and lesson organization (Stake and Easley 1978).

States that have statewide textbook adoption play an inordinately important role (English 1980). These states, 23 in number, account for about 46% of the national textbook market, and only three of them—California, Texas, and Florida—take 17% of the market (Patrick and Hawke 1982). Thus, what these "adoption states" want in their textbooks plays a major role in determining the content of textbooks throughout the U.S.

This may not be true for geography because of its small market (Sturm and Weiss 1988). A survey (Switzer 1986) showed that 46.7% of textbook publishers produced texts in geography; the percentages for other fields were U.S. history—86.7%, world history—66.7%, economics—40.0%, and sociology—26.7%. Geography content is well represented in the textbooks for lower grades, but is largely replaced by history in the high-school texts (Manson 1981).

Training and Certification

Mayo (1965, 149) reported that certification requirements for secondary teachers were the "gloomiest aspect of geography's status in the United States," and certification requirements largely determine training requirements. Certification is the prerogative of individual states: teachers become certified after they complete a state-approved program in a college or university. There exists no single certification system, and subject-specialization requirements for preservice teachers vary considerably from state to state (Winston 1984).

With the exception of history, most states do not require college credits in specific social sciences to teach social studies. Wiley (1977) noted that secondary social-studies teachers in the 1950s and 1960s were least well-prepared academically in anthropology, economics, geography, and sociology. According to survey results, about 40% of all prospective elementary teachers are not required to take geography courses (Manson 1981), and it is possible to teach high-school geography in some states with as little as 5 college credits in the subject. A 1963 survey (Swain) of geography departments reported that 32% of the departments involved in teacher training required no geography course for social-studies teachers, 29% required one course, and 39% required two courses. When reasons were given for having no geography-course requirements, the most frequent response was that there was no state requirement.

Certification requirements result from compromises among many groups including parents, legislatures, school boards, school administrators, teachers, and academic scholars (Woodring 1984). Professional geography has only very recently developed its own *Draft Suggestions for the Preparation of Elementary and Secondary School Geography Teachers* (Geographic Education National Implementation Project 1987a; Spetz 1988). The document recommends that a high-school teacher have 18–24 semester hours in geography, and that all elementary and high-school social-studies teachers have two or three courses covering physical, cultural, and world regional geography. The importance of these guidelines lies in the fact that formal guideline documents issued by professional associations and learned societies often influence the standards established by the National Council for Accreditation of Teacher Education, or NCATE (National Council for Accreditation of Teacher Education 1982).

Several studies have shown that teachers play the major role in determining what is actually taught in the social-studies classroom, and that the major factor in determining what they teach is their subject-matter interest and preparation (Lengel and Superka 1982; Morrissett 1982; Switzer 1986; Weiss 1978; Patrick and Hawke 1982; Manson 1981). Most social-studies teachers major in history (Pabst 1986).

A national survey of high-school social-studies teachers reported the average number of courses taken in the subject area in which they most-frequently taught: geography had the lowest mean of 10 areas reported (Rutter 1986). A 1984 survey (Farmer) of geography teachers showed that only 6% of them felt most qualified to teach geography, whereas over 60% felt most qualified to teach American or world history. Finally, a recent survey (Cirrincione and Farrell 1988) concluded that teachers strongly support expanding geography in the curriculum, but that, at present, they are insufficiently prepared in the subject.

Advanced Placement

The last time geography was tested by the College Board was in 1934 (Rosen 1957). There is no opportunity at the present time for students to take Advanced Placement (AP) geography. Administered by the College Board, AP enables students matriculating in colleges to receive placement, credit, or both for their high-school courses. Since AP is available in most mainstream subjects (e.g., biology, chemistry, English, history, mathematics, physics, and most-recently political science), some geographers believe it is important for geography to have AP status, if the subject is to gain greater credibility.

Thus, geographers are currently working to achieve this recognition. For this to happen, the College Board must be convinced that geography can generate the minimum number of students to take the examination each year (4000), and that a common first-year college course is offered at the approximately 200 campuses that account for most of the entering students with AP credits. Currently, geography is taught on about 60% of those campuses. A recent GENIP survey indicated that three courses, rather than one course, constitute the most common introductory core courses in geography—world regional, physical, and human/cultural (J. F. Marran pers. com. 1988).

Current Trends

In recent years, more attention has been paid to geography in the schools, and new geography requirements have been set either for school districts or entire states in Arizona; California (Stutz 1985; Salter 1986a, 1986b); Colorado (Morrow-Jones 1986; University of Colorado's. . .1983); Kentucky (Spetz 1986); Michigan (Stoltman and Sweet 1986); Minnesota (Lanegran 1986); Oklahoma; South Dakota (Gritzner 1986); Tennessee (Jumper 1986); Texas (Boehm and Kracht 1986); and Utah (Wahlquist 1986). For example, the Oklahoma legislature enacted a law in 1987 requiring a geography course for all students in either the seventh or eighth grade, beginning with the fall of 1988 (J. M. Goodman, pers. com. 1988). The Illinois State Board of Education has identified world geography as one of five basic proficiencies essential to a sound education in the social studies (Marran 1987). In Oregon, geography is having a resurgence because of a state mandate that requires all students to complete a two-semester global-studies course at the tenth grade (Pabst 1986).

Facing entrenched curricular patterns in the schools, a few college and university geographers have been change agents in bringing attention back to the subject. Some of their colleagues are concerned that professional geography does not have the resources to attend to both school and college geography. This chapter now turns its focus to the latter, and to its articulation with precollegiate geography.

GEOGRAPHY IN HIGHER EDUCATION

Geography has benefitted along with many other subjects from America's unprecedented growth in enrollment in higher education. For example, Brubacher and Willis (1968) noted that the college population increased almost 1000% between 1900 and

1948, while the total population rose only 100%. The American development of geography is often compared unfavorably with its status in Britain (e.g., Abler 1987), yet Britain clearly has had vastly different educational priorities. For example, Brubacher and Willis pointed out that, if in 1948 the same percentage of youth had gone to college in Britain as in America, then enrollment in British higher education should have been approximately 800,000. Actually, it was less than 80,000 (1968).

At the turn of the century, William Morris Davis spoke of the "break between school geography and professional geography," saying that "our professional geographers are all self-made men" (Davis 1895, 146), because there existed little opportunity to study geography in colleges and universities. Today over a half-million students are enrolled in college geography classes; there are about 12,000 undergraduate majors, over 3400 graduate students, and 2500 full-time faculty in American geography (Schwendeman 1987). (Schwendeman's *Directory of College Geography* gives geography a longer-running record of such data than any other discipline. However, it does present difficulties in interpretation and year-to-date comparison.)

Comparative Enrollments and Degrees Conferred

In the last 25 years, along with other subjects, geography experienced the college enrollment build-up of the 1960s, to the peak of the early 1970s, and the decline of the 1980s. Total geography enrollment in 1960 was 336,787; in 1970 it was 762,954. By 1980 it had dropped to 538,880, and appears to have leveled off; enrollment stood at 512,855 in 1986 (Schwendeman 1961, 1970, 1981, 1987).

Graduate enrollments in geography declined 9.2% during the period 1977–84. Other fields experiencing decreases were anthropology (-17%), political science (-1.2%), sociology (-23.7%), and biology (-27.1%). Psychology gained 5.3% and economics 9.8%, but the social sciences as a whole dropped 1.1%. Total graduate enrollment in all sciences increased 10.7% during the period (U.S. Department of Education 1987).

Figures 2, 3, and 4 show data on bachelor's, master's, and doctor's degrees conferred in geography and seven other selected disciplines over the same 25-year period. Data for each discipline is given as a percentage of all eight disciplines. At the bachelor's level (Figure 2), history's dominant position in the 1960s eroded considerably, and appears to be continuing on a downward course. Psychology showed the greatest gain, surpassing history in the 1970s. Political science and economics have also exceeded history in the past decade. Of the eight disciplines, geography has been the steadiest performer throughout the period, albeit ranking in the bottom three along with geology and anthropology, both of which have had more relative change than geography. At the master's and doctor's levels (Figures 3 and 4), the patterns are similar to those for the bachelor's degree. Geography shows the least volatility, and psychology and history have changed the most over the period.

The strength of a field at the secondary-school level (Figure 1) is not always related to its collegiate strength, as measured by bachelor's degrees conferred (Figure 2). For example, psychology is the least-enrolled high-school social-studies subject, but it confers more bachelor's degrees (as well as graduate) than any other social science. History has continued to dominate the high-school curriculum, but at the bachelor's level has plummeted below psychology, political science, and economics.

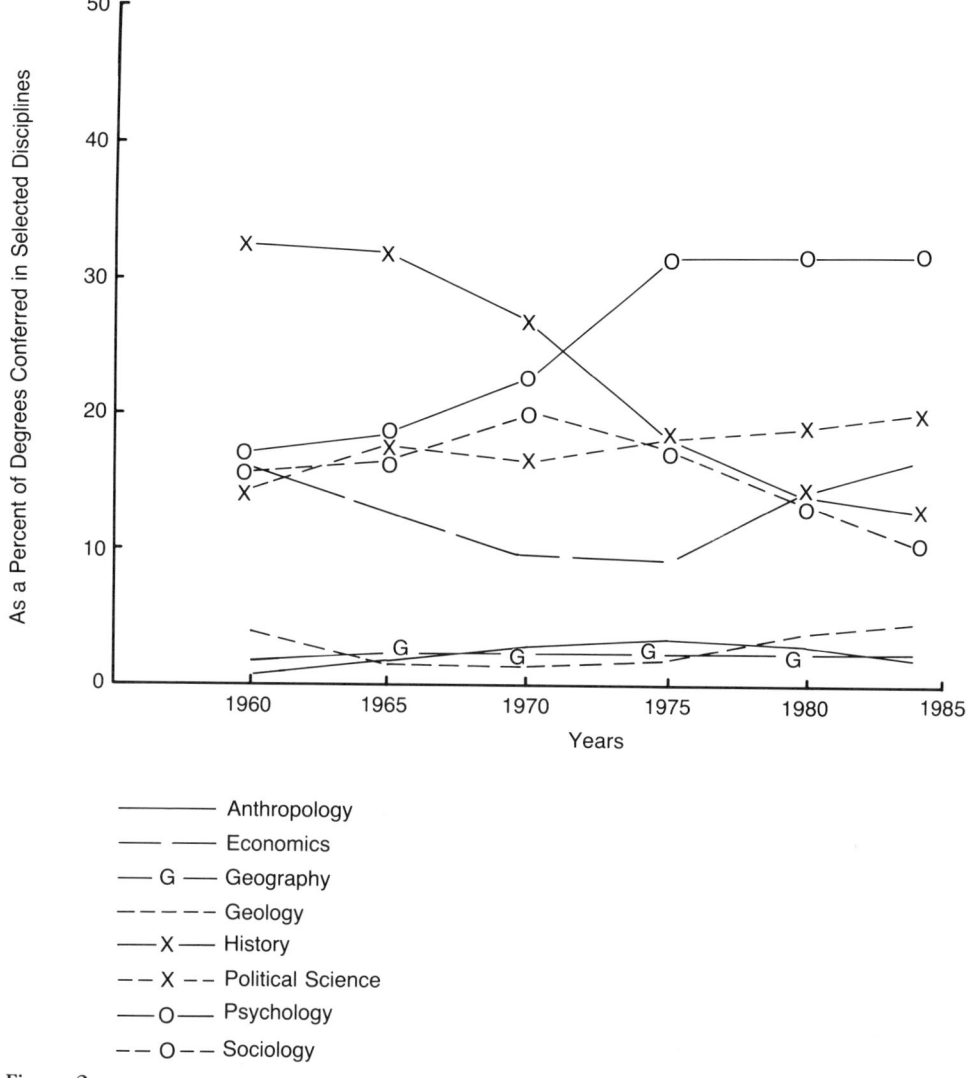

Figure 2
Bachelor's degrees conferred in selected disciplines, 1960–84 (U.S. Department of Health, Education, and Welfare 1960–61, 1965–66, 1970–71, 1975–76; U.S. Department of Education (Center for Statistics) 1983–84, 1987)

Since 1970, sociology has shown a sharp increase in high-school enrollments, but in college it has dropped significantly from ranking third (behind history and psychology) to ranking fifth (behind psychology, political science, economics, and history).

Geography's relative position among high-school social studies improved during the 1950s and 1960s (ranking third behind history and political science in 1960), but declined again (ranking sixth behind history, political science, sociology, anthropology, and economics in 1982). Geography's increase in relative high-school strength in the 1950s and 1960s was not reflected at the college level.

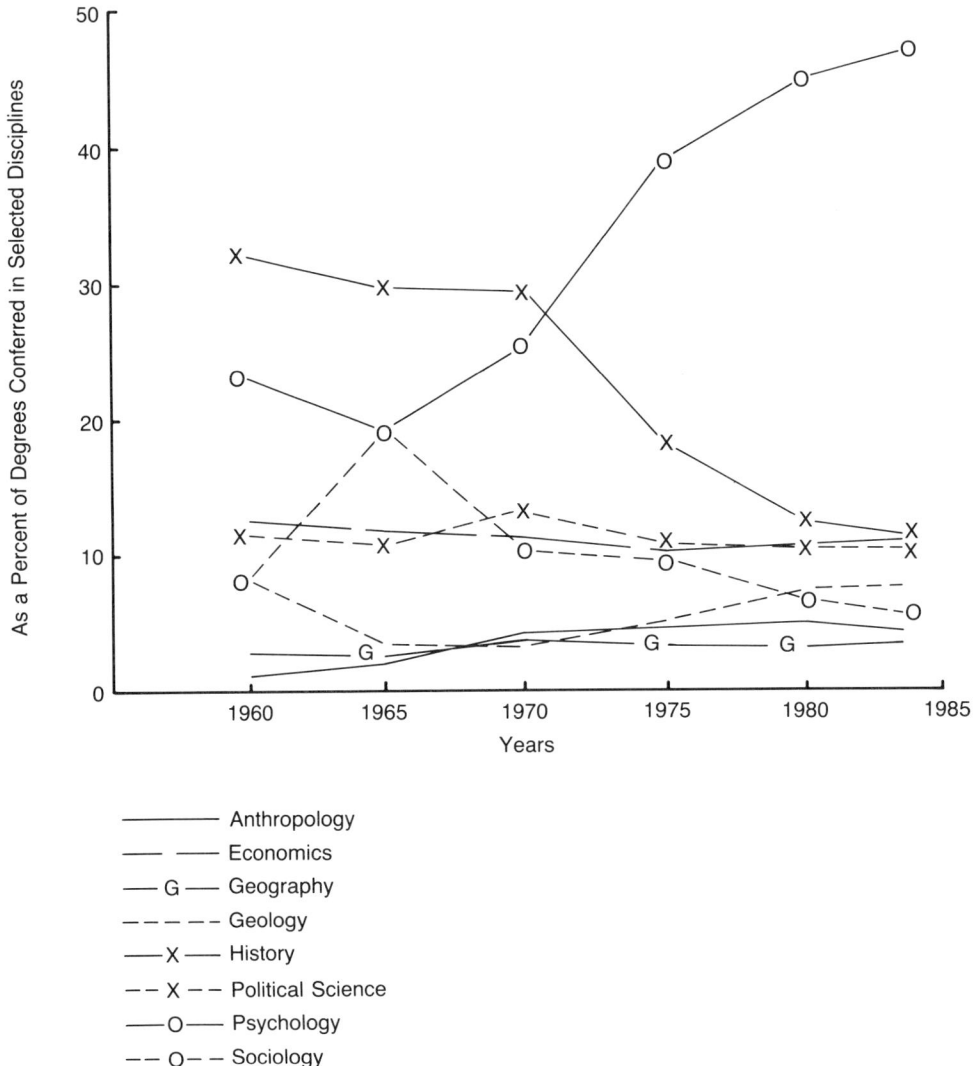

Figure 3
Master's degrees conferred in selected disciplines, 1960–84 (U.S. Department of Health, Education, and Welfare 1960–61, 1965–66, 1970–71, 1975–76; U.S. Department of Education (Center for Statistics) 1983–84, 1987)

Honors

Some would argue that recognition earned by its members is a better indicator of a field's status than relative numbers of students. In American geography, membership in the National Academy of Sciences (NAS) perhaps represents the pinnacle of status. Eight geographers are members. Their names and dates of election (National Academy of Sciences 1987) are:

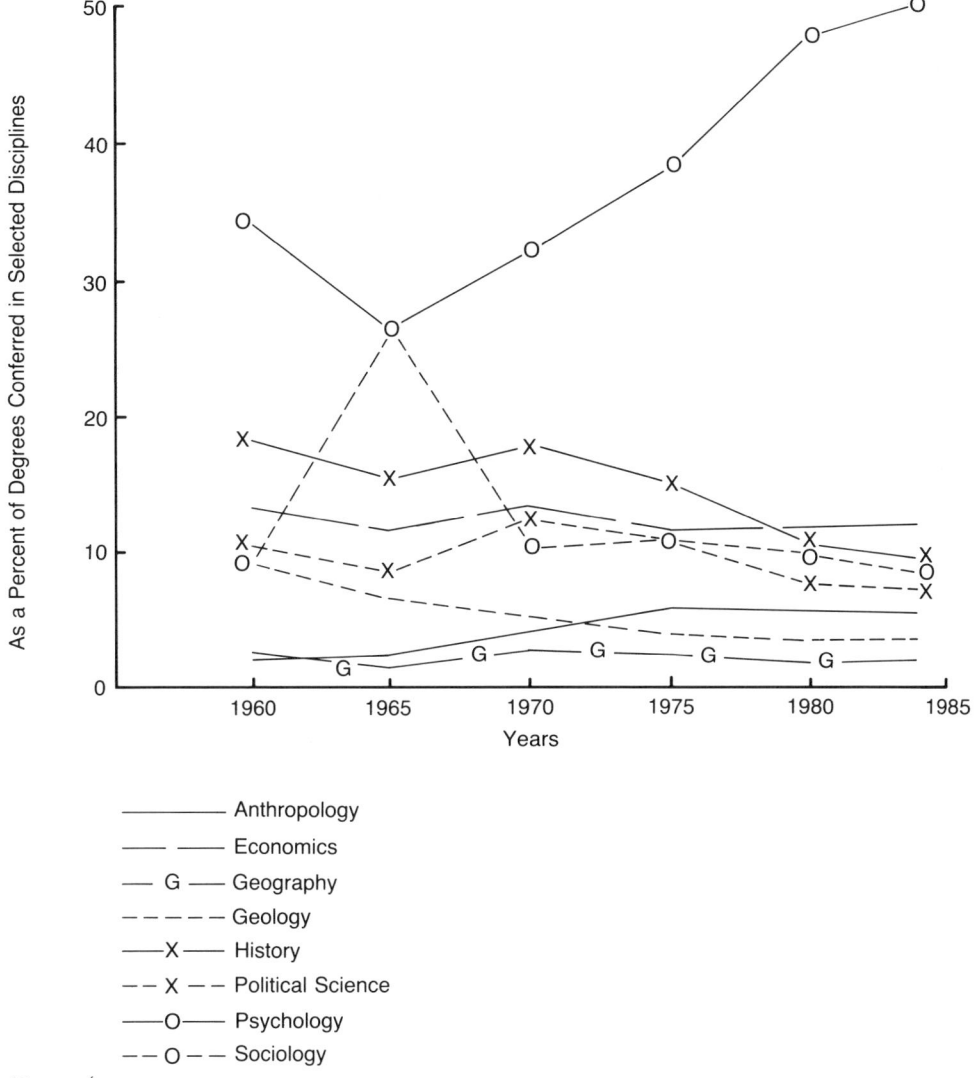

Figure 4
Doctor's degrees conferred in selected disciplines, 1960–84 (U.S. Department of Health, Education, and Welfare 1960–61, 1965–66, 1970–71, 1975–76; U.S. Department of Education (Center for Statistics) 1983–84, 1987)

G. F. White, 1973   J. Wolpert, 1977
B. J. L. Berry, 1975   W. R. Tobler, 1982
R. W. Kates, 1975   W. Isard, 1985
J. R. Borchert, 1976   M. G. Wolman, 1988

NAS membership in June of 1987 totalled 1520, so geographers accounted for 0.5% of the members. Graduate enrollment in geography (3000) accounts for 1% of all graduate enrollment in science (300,000) (U.S. Department of Education 1987). Geography would need to have 15 members in the NAS to make its membership proportional to its relative standing in graduate enrollments.

Of the 2315 members of the American Academy of Arts and Sciences in 1984, five were geographers (C. G. Harris, G. F. White, B. J. L. Berry, M. G. Wolman, and K. W. Butzer) (American Academy of Arts and Sciences 1984).

As of 1974, the fiftieth anniversary of the Guggenheim Memorial Foundation, 38 out of 8395 Fellows were geographers, or about 0.45% (Guggenheim 1975). This same percentage held for the most recent period, 1975–86, during which 17 out of 3748 Fellows were geographers (Judge 1975–86).

CRITICISMS IN CONTEXT

There is no single measure of the status of an academic field. Each observer will have his or her own perspective, yet it is important to remember that, as Abler (1987) noted, geography is a "small discipline." AAG membership in 1985 (5700) was smaller than that of comparable organizations of cognate disciplines: anthropology (8500), economics (20,000), geology (15,500), history (16,000), planning (21,000), political science (12,000), psychology (60,000), and sociology (15,000).

The geographic literature is replete with commentary about how geographic education has "failed" (e.g., Gritzner 1981; Libbee 1984). Yet, similar commentary is made about other fields (Morrissett 1968; Wronski and Bragaw 1986). Some writers note the unsatisfactory relations between elementary/secondary-school and collegiate conditions (e.g., Swain 1963; Ring 1979; Winston 1984). Obviously, university geography must relate initially to students who have very poor backgrounds in the subject; indeed, students are surprised to find geography in the university curriculum, because their elementary and secondary-school images of it greatly diverge from the university approach to the subject (Hubbard and Stoddard 1979; McTeer 1979).

Whatever the school images of geography, there is virtually no predisposition on the part of high-school graduates to select that major in college (Astin et al. 1985). Yet, it can be seen (Figures 1 and 2) that the field of psychology, which is even less represented than geography in the high-school curriculum, is a major university subject. Perhaps inherent human interest in the self and in human behavior explains why 4.4% of incoming freshmen intend to major in psychology, a field with even less high-school exposure than geography, which attracts 0.0% (rounded) of incoming freshmen. Other cognate fields and the percentage of incoming freshmen intending to major in them are anthropology—0.1%, biology—2.7%, economics—0.8%, geology—0.1%, history—1.3%, political science—4.0%, and sociology/social work—1.7% (Astin et al. 1985).

Many observers have complained about the quality of geographic instruction in the schools (Gritzner 1982; Ligocki 1982; Kincheloe 1984), but the same complaint is made about all social-studies instruction (Morrissett 1968; Wronski and Bragaw 1986). The incorporation of geography into the social studies, rather than being taught as a freestanding subject, has been seen by many geographers as an important problem (Pattison 1970; Gritzner 1982; James 1983), but other social scientists have similar complaints about the treatment of their subjects in the schools (e.g., Armento 1986;

Bare 1986; Dynneson 1986; Switzer 1986). Although geography faces competition in the school curriculum from the "new social studies subjects" of environmental, future, global, and international education (Gritzner 1982), so too do other subjects. Nevertheless, it is argued that geography may have a unique problem in that school administrators and others have been confused by the discipline's frequent shifts in emphasis, and by the lack of a widely accepted approach to the subject (Kohn 1982). Also, school teachers and administrators, reflecting on their own experiences as students, think geography is dull (Gritzner 1981).

Complaints about the quality of geographic instruction in the schools are minor when compared with recent findings that the general state of education, particularly at the high-school level, is unacceptably poor (National Commission on Excellence in Education 1983; Holton 1984). Teaching is not attracting large numbers of the best students. For example, a "Survey of Global Understanding" among college students reported that education majors ranked lowest among all majors surveyed in general knowledge about the world (Barrows, Klein, and Clark 1981). Other studies have shown that average SAT scores of teacher-education majors is lower than nearly all other fields.

Geographic education in colleges and universities has not escaped the critics. University geography teachers are accused of failing to give the public what it expects: the study of places (Abler 1987; Gritzner 1981; Libbee 1984; Kincheloe 1984). Also, the content of our courses is seen to be insufficiently relevant to the world (Harper 1985). Increasing specialization and fragmentation away from a cohesive core of the discipline is frequently mentioned as a problem with geography (James 1983; Kincheloe 1984; Abler 1987), which may lead to internal strife in departments and the discipline (Gritzner 1982).

Karan and Mather (1986) fault university geographers for their teaching abilities, and indeed for their lack of commitment to education. Ineffective geography teaching is, they argue, common at the college level. Partly, this is because of the heavy use of graduate students as teachers of introductory courses, and the placing of the heaviest teaching responsibilities upon those having the least experience and achievements. But they also argue that it is because there are simply too many professors who have no commitment to teaching.

Furthermore, Karan and Mather (1986) find a professional "snobbishness" toward teaching within the AAG, a point that has been made elsewhere (Boehm 1984; Gritzner 1982; Olmstead 1987). Partly because of these attitudes, there exists a large gap in understanding of goals, objectives, and working contexts between university geographers and high-school teachers, and there have been few coordinated efforts among them (Kohn 1982; Ligocki 1982). Kohn (1982) finds it unfortunate that Ph.D.-granting departments in geography have few faculty conducting research in geographic education. Some geographers are not sanguine about finding the resolve among their peers to strengthen geographic education (de Souza 1984).

The status of geography in American higher education may indeed suffer from the above-mentioned problems. At the same time, it would be difficult to argue that these problems are unique to geography. The separation of the schools from the universities is common; as we have mentioned earlier, professional education, which has a generally low status in the university, provides that linkage. The attitudes of university geographers toward geographic education are not likely to be much different from the attitudes of other disciplinarians toward education in their own fields (Morrissett 1968).

Although geographers have found much to criticize about themselves, disciplinary self-flagellation may be excessive, if not unhealthy (Gritzner 1981; Abler 1987). In the experience of a recent past president of the AAG, geographers do not particularly suffer from a lack of esteem by people outside the discipline (Abler 1987). Another recent past president of the AAG believed that geography had gained considerably in respect since he entered the field in the 1950s (Morrill 1983). Despite the relative absence of geography departments in the Ivy League, and the recent termination of a few traditionally strong departments (a fact of great concern—e.g., Smith 1987; Tenner 1988), geography is "doing quite well in many AAU [Association of American Universities] institutions" having large undergraduate and graduate programs (Dunbar 1986) as well as in smaller, often "overlooked departments" in private liberal arts colleges and state colleges and universities (de Souza 1981). A recent compilation of changes in geography departments since 1970 listed 24 new programs, 43 expanded programs, and only 6 programs reduced or eliminated (Association of American Geographers 1988). Indeed, one outside observer is impressed with geography's structure in American higher education, especially given its weak position in the schools (Haigh 1982).

Enrollment shifts from the recent round of college general-education curricular changes have more often than not favored geography (Garrett and Hecock 1984). Perhaps this is because the subject's multifaceted nature (Gritzner 1982) and the innate curiosity about environments and other peoples and places (Hudson 1984) make it attractive for liberal education. Although geography's position in both the social and natural sciences sometimes presents curricular planning and organizational problems, it is by no means unique in those terms—anthropology and psychology also have human and physical dimensions. In some institutions, geography lies fully on one side or the other, but in others, it lies in both. Many applied problems have both human and physical dimensions, a condition that *potentially* favors those scholars and scientists equipped to work across this spectrum (Boulding 1966; Branscomb 1986).

OPPORTUNITIES, TASKS, AND CHALLENGES

Geographers talk in terms of opportunities, tasks, and challenges facing the profession, i.e., of ways to enhance the status of the field. Many commentaries take a positive approach to "the problem." Indeed, it is asserted that we know what the problem is, and we know how to solve it (Gritzner 1982; de Souza 1984), but as a matter of fact each individual defines a different "problem" because each focuses only upon some aspect of the status of geography. Thus, challenges, tasks, and opportunities are noted in geographic education, in basic and applied research, and in convincing others that geography is a worthy enterprise.

Few geographers resist the temptation to urge their colleagues to produce more and better products that "command attention": journal articles, books, textbooks, and the like (e.g., Jordan 1988; Abler 1987; Harper 1985; de Souza 1984; Hudson 1984; Morrill 1983). Geographic research should focus upon "real-world" problems, rather than upon approaches to problems (philosophy) (Abler 1987; White 1972; Kates 1987). Others are sure that solid theoretical work is the sine qua non for enhancing the discipline's reputation (Morrill 1987; Palm 1986).

To be successful in higher education, geography should capitalize on its "unifying role" in knowledge (Gritzner 1981). The discipline has an important and distinctive role to play in general education (Geography in Liberal Education Project 1965; Mikesell 1980; Hill 1982; Harper 1982; Rees and Natoli 1982; Morrill 1983, 1984; Garrett and Hecock 1984).

Another task is to develop service courses for other professions, such as business and education (Libbee 1984; Harper 1985). Also, geography has an important role in internationalizing the curriculum (Committee on Geography and International Studies 1982); more world-geography courses should tap the current interest in "global education" (Hill 1981; Libbee 1984, 1988; Harper 1985). Libbee (1984) sees geography thriving at the undergraduate level if it can compete effectively for the "non-traditional student." He would put greater emphasis on high quality, professionally oriented undergraduate and master's programs, but would eliminate small Ph.D. programs and reduce larger ones.

Some argue that geographic education should be given greater priority by the profession; indeed, Harper (1985) urges both the AAG and the National Council for Geographic Education (NCGE) to mobilize the best teaching talent in the profession to work on these challenges. Abler (1987) agrees that the AAG should at least devote "some time" to geographic education, a view apparently not shared by Jordan (1988). Several writers (e.g., Kohn 1982; Manson 1977) have encouraged geographers to conduct research in geographic education, and Abler (1987) urges advancement for those who excel in this work. Good teaching must be appropriately rewarded if we are to improve our role in higher education (Karan and Mather 1986). There is no shortage of suggestions for effective geography teaching (e.g., Karan and Mather 1986; Demko 1986); indeed, one can even find extended profiles of the "good" teacher (e.g., Hill 1972).

In the last quarter-century, the AAG has focused substantial energy on geographic education (Monk 1986). Major efforts were conducted to improve high-school geography (Helburn 1968; White 1961; Patton 1970). The improvement of undergraduate geography curriculum and content was the target for numerous activities and nearly 50 publications over a decade of the Commission on College Geography (e.g., 1967a; 1967b). The philosophy and practice of teaching, especially at the introductory level, received the attention of a series of conferences (Ball et al. 1972), and papers inspired by those conferences (Helburn 1972). Another project was devoted to the development of both a concern and a facility for teaching among geography graduate students (Fink 1978); longitudinal research with the project's participants was reported in a series of papers (Fink 1983, 1984, 1985). These activities and the accumulating experience of the geographers engaged in them marked a new paradigm in geographic education (Hill 1972) and a period of remarkable change (Pattison and Natoli 1977).

National economic stress in the late 1970s and early 1980s led to cutbacks in spending for education; two geography departments (University of Michigan and University of Pittsburgh) were eliminated in the period. This alerted other departments to the potential threat and incited a "survival kit" approach. For example, department chairs held day-long workshops at AAG meetings to discuss strategies. The literature carried many suggestions for "avoiding the demise of academic geography" (e.g., Wilbanks and Libbee 1979; Kish and Ward 1981; Rees and Natoli 1982; Kahn and Vuicich 1984; Jumper 1984; Natoli 1986). One AAG special session on the problems of survival brought an unexpected and—for geography—unfortunate piece of reporting in the

Chronicle of Higher Education (Scully 1982), from which experience some geographers formulated an additional survival strategy: avoid publicizing one's difficulties.

SIGNS OF A RENAISSANCE?

At the same time that college geographers were worrying about the survival of their departments, events were moving in another direction for school geography. Evidence appeared to foreshadow a renaissance in geographic education, particularly at the elementary and secondary levels (Hill 1988). Whether this will have a salutary effect on the status of geography as a whole is arguable, but the potential for such an effect is not. This renaissance began with reports of national commissions and other prestigious surveys concerned with inadequacies in America's educational system, some of which focused upon apparent lack of preparation in geography, foreign languages, and other fields leading to international understanding (National Assessment of Educational Progress 1979; President's Commission on Foreign Language and International Studies 1979; Barrows, Klein, and Clark 1981; National Commission on Excellence in Education 1983; *Dallas Times Herald* 1983; National Governors' Association 1986). The AAG published a timely report on this situation (Committee on Geography and International Studies 1982), followed by a resource book of geographic material that focused upon global knowledge (Natoli and Bond 1985).

Several other surveys (e.g., Kopec 1984; CBS Affiliates Survey 1987) highlighted pervasive high-school and college-student ignorance of world place-location knowledge, a condition termed "geographic illiteracy." Such stories received national-media attention; indeed, media interest in this issue has continued unabated to the present. Although acknowledging that such exposure offers a rare opportunity for achieving public support for the field, geographers expressed concern that this attention unfortunately may perpetuate the elementary, one-dimensional view of the nature of the subject (Schmudde 1987; Cohen 1988). When geographers have had the opportunity to promote the field to nongeographer audiences, they often make this point (Hill 1984, 1988). Another tactic is to call upon the testimony of nongeographers who stress the importance of geography for various pleasures, endeavors, and careers from writing novels to working in the foreign service (e.g., Dr. Studebaker Calls for Geography 1981; Michener 1970; Woodring 1983; Dunford 1988).

The *Guidelines for Geographic Education: Elementary and Secondary Schools* (Joint Committee on Geographic Education 1984) proved to be one of the most-influential documents in geography's recent resurgence. Its significance lay in answering a perennial complaint that the discipline suffered in the schools because it had no generally agreed-upon statement clarifying the essentials of geography for school people (Manson 1981; Kohn 1982). Its five "fundamental themes in geography" and its clear learning outcomes for each theme gave needed direction, which is increasingly being used for curriculum and materials development and for teacher training (Natoli 1988). Some 65,000 copies of this document have been printed, and its wide acceptance has occasioned the development of separate elaborations for the elementary level (Geographic Education National Implementation Project 1987b) and secondary level.

The success of the *Guidelines* contributed to the formation of the Geographic Education National Implementation Project (GENIP) in 1985. This was the first joint venture of all four national geographic organizations–NCGE, AAG, National Geographic Society (NGS), and American Geographic Society (AGS). Its purpose is to improve the status and quality of geographic education in grades K–12 in the U.S., through projects designed to improve teacher preparation, teaching materials, interactions between teachers and university professors, evaluation of materials and programs, and public relations.

Not content with its tremendous success as a purveyor of "popular geography," the National Geographic Society began in 1984 to exert leadership in promoting school geography. Among the four GENIP organizations, NGS has emerged as the most potent change agent because of its ability to make a large financial commitment, to quickly develop and conduct new programs, and because its president and chairman, Gilbert M. Grosvenor, is championing the cause. Speaking to the AAG, he called upon the geographic discipline to join with the NGS to advance the status of the field (1984).

The Society's new Geographic Education Program was spending $4 to $5 million a year in 1987 and 1988. Its centerpiece is a new network of state-based geographic alliances, which began with seven states in January 1986 and had grown to 22 by 1988 (Geographic Education Program 1988). Coordinated by academic geographers, the alliances work at the local level with classroom teachers, academic geographers, administrators, and other citizens to promote geographic education in the schools of their respective regions. NGS conducts a summer institute for teachers in Washington, DC and supports regional institutes, workshops, and conferences organized by the alliances.

A large number of other projects and activities have been started. In January 1988, the National Geographic Society Education Foundation was established with $20 million, and an additional $20 million challenge grant, to support these and other efforts that improve geographic education and increase geographic knowledge (Elliott 1988). It seems fair to say that, at no other time in its history, has geography been the focus of so much attention and the recipient of so many tangible resources.

CONCLUDING SCENARIO

Americans, consummate pragmatists, will judge geography by what it proves it can do to help them improve their lives and their worlds, as they define them. Significant research will be the major criterion of status in academe. Teaching quality will count with students at all levels. Individual perceptions, based on different criteria, are not easily averaged out. Given this caveat, we dare to suggest that geography in higher education, although by no means free of troubles, enjoys today a higher status than ever before. School geography is a different story, yet recent widespread recognition of its inadequacy has generated more public attention about the subject than it has ever received.

The potential of the current campaign to improve school geography augers well for the status of the discipline, yet the difficulties of major school reform should not be underrated. America's massive and fragmented schools are embattled institutions facing diverse pressures from a host of constituencies, of which geography is only a minor one. We have seen how the star of individual disciplines has risen and fallen. Moreover, reform movements come and go, since the public's attention easily flags and activists tire or search for new causes. Finally, the ability of reformers to make changes pales in comparison with the effects of major swings in socioeconomic conditions and national agendas.

In 1989, education is again reasonably high on the national agenda, giving some reason to believe that we may already be in another period of change. If so, and if that period can be sustained, we may reach a time when *each* student will learn *some* geography, and the subject will become prominent and respected in the schools. Then, college geography will face its greatest challenge: many more students with solid geographic educations will be looking to expand their geographic studies in college. Instead of less than 1%, perhaps 2% or even 3% of incoming freshmen might opt to major in geography. Such changes would have marked effects on the entire system.

In this future scenario, the current level of sophistication of college geography courses would become inadequate, because material taught in college introductory courses would be taught at the precollege level. The need for geography-teacher training would place new demands on college faculty. The translation of research into educational materials would constitute a large task. Certainly the need would prevail to clarify the nature and purpose of geography in American education.

Effective demonstration of the power that geographic education can provide to the individual and the society would be required. Surely this would address issues of the quality of life, aesthetics, efficiency, and equity (Helburn 1982), and the need for minority and feminist perspectives in the discipline (Monk 1987), as well as provide the type of preparation required for solving pressing environmental problems of the future (Natoli 1988). As America's experiment with mass education continues, geographers would need to develop a responsiveness to a wider variety of students. The questions "What is worth teaching?" and "What is worth studying?" would demand answers again and again.

These are some of the potential challenges, but there is an awful distance between hopes and realities. If geographers can unite to marshall their scarce resources, if they are able to invent and support devices that systematically encourage cumulative improvement, and if they can convince non-geographers that geography is essential, the status of their field will improve.

REFERENCES

Abler, R. F. 1987. What shall we say? To whom shall we speak? *Annals of the Association of American Geographers* 77:511–24.

American Academy of Arts and Sciences. 1984. *Records of the Academy: 1983–1984.* Cambridge, MA.

Armento, B. J. 1986. Promoting economic literacy. In *Social studies and social sciences: A fifty-year perspective,* eds. S. P. Wronski and D. H. Bragaw. Bulletin No. 78. Washington: National Council for the Social Studies.

Association of American Geographers. 1988. Changes in departments of geography since 1970. Unpublished document.

Astin, A. W., et al. 1985. *The American freshman: National norms for fall 1985.* Los Angeles: University of California at Los Angeles Graduate School of Education.

Ball, J. M., Kurfman, D. G., Lansky, L. M., and Natoli, S. J. 1972. Experiments in teaching College Geography: A report to the profession. *Professional Geographer* 24:350–59.

Bare, J. K. 1986. Teaching psychology in high schools. In *Social studies and social sciences: A fifty-year perspective,* eds. S. P. Wronski and D. H. Bragaw. Bulletin No. 78. Washington: National Council for the Social Studies.

Barrows, T. S., Klein, S. F., and Clark, J. L. D. 1981. *College students' knowledge and beliefs: A survey of global understanding.* New Rochelle, NY: Change Magazine Press.

Becker, J. M. 1979. *Schooling for a global age.* New York: McGraw-Hill.

Boehm, R. G. 1984. On prejudice in geography. *Journal of Geography* 83:52–53.

———, and Kracht, J. B. 1986. Enhancing high school geography in Texas. *Professional Geographer* 38:255–56.

Boulding, K. E. 1966. *The impact of the social sciences.* New Brunswick, NJ: Rutgers University Press.

Branscomb, L. M. 1986. Science in 2006. *American Scientist* 74:650–58.

Brubacher, J. S., and Willis, R. 1968. *Higher education in transition: A history of American colleges and universities, 1936–1968.* New York: Harper & Row.

CBS Affiliates Survey. January, 1987.

Cirrincione, J. M., and Farrell, R. T. 1988. The status of geography in middle/junior and senior high schools. In *Strengthening geography in the social studies,* ed. S. J. Natoli, pp. 11–21. Bulletin No. 81. Washington: National Council for the Social Studies.

Cohen, S. B. 1988. Geography—public awareness and the public arena. *Social Education* 52:248–50.

Commission on College Geography. 1967a. *New approaches in introductory college geography courses.* Publication No. 4. Washington: Association of American Geographers.

———. 1967b. *Introductory geography: Viewpoints and themes.* Publication No. 5. Washington: Association of American Geographers.

Committee on Geography and International Studies. 1982. *Geography and international knowledge.* Washington: Association of American Geographers.

Dallas Times Herald. 1983. American education: The ABCs of failure. December 11, 1983.

Davis, W. M. [1909] 1954. The need of geography in the university. In *Geographical essays* by William Morris Davis, ed. D. W. Johnson, pp. 146–64. New York: Dover.

———. [1909] 1954. The progress of geography in the schools. In *Geographical essays* by William Morris Davis, ed. D. W. Johnson, pp. 23–69. New York: Dover.

Demko, G. J. 1986. Ideas on how to enliven the teaching of geography. *Journal of Geography* 85:246–48.

de Souza, A. 1981. The overlooked departments of geography. *Journal of Geography* 80:170–75.

———. 1984. A crisis in geographical education? *Journal of Geography* 83:3.

Dr. Studebaker calls for geography. June, 1981. NCGE Perspective, 9, no. 5. Address delivered at a national conference of college and university presidents. Baltimore, Maryland. March 3–4, 1942.

Dunbar, G. S. 1986. Geography in the bellwether universities of the United States. *Area* 18:25–33.

Dunford, D. J. 1988. Geographic illiteracy: Secondary school education and foreign affairs. The Senior Seminar, Foreign Service Institute. U. S. Department of State.

Dynneson, T. L. 1986. Trends in precollegiate Anthropology. In *Social studies and social sciences: A fifty-year perspective,* eds. S. P. Wronski and D. H. Bragaw. Bulletin No. 78. Washington: National Council for the Social Studies.

Educational Products Information Exchange Institute. 1977. *Report on a national study of the nature and the quality of instructional materials most used by teachers and learners.* EPIE Report No. 76. New York: EPIE.

Elliott, L. H. 1988. President Grosvenor announces the National Geographic Society Education Foundation. *National Geographic* 173:329A–D.

English, R. 1980. The politics of textbook adoption. *Phi Delta Kappan* December:277.

Farmer, R. 1984. The social studies teacher in the 80's: Report from a national survey. *Social Studies* 75:166–71.

Fink, L. D. 1978. A discipline's experiment in higher education: A report on the TLGG project. *Journal of Geography in Higher Education* 2:77–85.

———. 1983. First year on the faculty: Getting there. *Journal of Geography in Higher Education* 7:45–56.

———. 1984. First year on the faculty: Being there. *Journal of Geography in Higher Education* 8:11–25.

———. 1985. First year on the faculty: The quality of their teaching. *Journal of Geography in Higher Education* 9:129–48.

Gardner, D. P. 1986. Geography in the school curriculum. *Annals of the Association of American Geographers* 76:1–4.

Garrett, M. J., and Hecock, R. D. 1984. General education and geography: A profile of institution types. *Journal of Geography* 83:273–76.

Geographic Education National Implementation Project (GENIP). 1987a. *Draft suggestions for the preparation of elementary and secondary school geography teachers*. Washington: GENIP.

———. 1987b. *K–6 geography: Themes, key ideas, and learning opportunities*. Washington: GENIP.

Geographic Education Program. 1988. Update. Washington: National Geographic Society. Spring: 8.

Geography in Liberal Education Project. 1965. *Geography in undergraduate liberal education*. Washington: Association of American Geographers.

Gritzner, C. F. 1981. Geographic education—where have we failed? *Journal of Geography* 80:264–66.

———. 1982. What is right with geography? *Journal of Geography* 8:237–39.

———. 1986. The South Dakota experience. *Professional Geographer* 38:252–53.

Grosvenor, G. M. 1984. The society and the discipline. *Professional Geographer* 36:413–18.

Guggenheim, John Simon: Memorial Foundation. 1975. *Directory of Fellows 1925–1974*. New York: John Simon Guggenheim Memorial Foundation.

Haas, J. D. 1977. The era of the new social studies. ERIC Clearing House. Boulder, CO: Social Science Education Consortium.

Haigh, M. J. 1982. The crisis in American geography. *Area* 14:85–89.

Harper, R. A. 1982. Geography in general education: the need to focus on the geography of the field. *Journal of Geography* 81:122–39.

———. 1985. New teaching opportunities and challenges for U.S. geography. *Journal of Geography* 84:3–4.

Helburn, N. 1968. The educational objectives of high school Geography. *Journal of Geography* 67:274–81.

———, ed. 1972. *Challenge and change in college geography*. Boulder, CO: Commission on Geographic Education, Association of American Geographers, and the ERIC Clearinghouse for Social Studies/Social Science Education.

———. 1982. Geography and the quality of life. *Annals of the Association of American Geographers* 72:445–56.

———, and Helburn, S. W. 1979. Stability and reform in recent American curriculum. In *Post-war curriculum development, an historical appraisal*, ed. W. E. Marsden, pp. 29-48. Leicester, England: History of Education Society of Great Britain.

Hill, A. D. 1972. Geography and geographic education: Paradigms and prospects. In *Challenge and change in college geography*, ed. N. Helburn. Boulder, CO: Commission on Geographic Education, Association of American Geographers, and ERIC Clearinghouse for Social Studies/Social Sciences Education.

———. 1981. A survey of the global understanding of American college students: A report to geographers. *Professional Geographer* 33:237–45.

———. 1982. Another view of the sixteen-million-hour question. *Professional Geographer* 34:69–72.

———. 1984. *The essentiality of geography in the high schools*. ERIC No. ED 244–873.

———. 1988. The Western PLACE Conference: Teaching models for the renaissance of geography. In *Placing geography in the curriculum: Ideas from the Western PLACE Conference, 1–4*, ed. A. D. Hill. Boulder, CO: Colorado Geographic Alliance and Center for Geographic Education.

Holton, G. 1984. A nation at risk revisited. *Daedalus* 113:1–27.

Hubbard, R., and Stoddard, R. H. 1979. High school students' images of geography: An exploratory analysis. *Journal of Geography* 78:188–94.

Hudson, J. C. 1984. Geography's image crisis. *Journal of Geography* 83:100–101.

James, P. E. 1969. The significance of geography in American education. *Journal of Geography* 68:473–86.

———. 1983. Problems and opportunities in geography. *Journal of Geography* 82:92–93.

———, and Martin, G. J. 1981. *All possible worlds: A history of geographical ideas*. 2d ed. New York: Wiley.

Joint Committee on Geographic Education. 1984. *Guidelines for geographic education: Elementary and secondary schools*. Washington: Association and National Council for Geographic Education.

Jordan, T. G. 1988. President's column. *Association of American Geographers Newsletter* 23:1.

Judge, M. A., ed., 1975–1986. *Reports of the President and the Treasurer: John Simon Guggenheim Memorial Foundation.* New York: John Simon Guggenheim Memorial Foundation.

Jumper, S. R. 1984. Departmental relationships and images within the university. *Journal of Geography in Higher Education* 8:41–47.

———. 1986. The Tennessee experience. *Professional Geographer* 38:254–55.

Kahn, S., and Vuicich, G. 1984. Effective leadership in geography: The role of the department chairperson. *Professional Geographer* 36:158–64.

Karan, P. P., and Mather, C. 1986. The trouble with college geography. *Journal of Geography* 85:95–97.

Kates, R. W. 1987. The human environment: The road not taken, the road still beckoning. *Annals of the Association of American Geographers* 77:525–34.

Kincheloe, J. L. 1984. The trouble with geography. *Social Studies* 75:141–44.

King, D. C. 1987. Delay persists in social studies reform, but signs point to headway just ahead. *Curriculum Update*. Alexandria, VA: Association for Supervision and Curriculum Development.

Kish, G., and Ward, R. M. 1981. A survival package for geography and other endangered disciplines. *Association of American Geographers Newsletter* 16:8.

Kohn, C. F. 1982. Looking back; working ahead. *Journal of Geography* 81:44–46.

Kopec, R. J. 1984. *Geography: No 'where' in North Carolina 1984*. ERIC No. ED 256630.

Lanegran, D. A. 1986. Strengthening geographic education. *Professional Geographer* 38:71.

Lengel, J. G., and Superka, D. P. 1982. Curriculum patterns. In *Social studies in the 1980s: A report of Project Span,* ed. I. Morrissett. Alexandria, Va: Association for Supervision and Curriculum Development.

Libbee, M. 1984. Geography in higher education: Why we're in trouble and how to get out. *Journal of Geography* 83:5–6.

———. 1988. World geography and international understanding. *Journal of Geography* 87:5–12.

Ligocki, C. 1982. High school geography and the need for communication. *Journal of Geography* 81:188–90.

McTeer, J. H. 1979. High school students' attitudes toward geography. *Journal of Geography* 78:55–56.

Manson, G. 1977. The introductory course in geography: A curricular perspective. In *Geographical Horizons,* eds. J. Odland and R.N. Taaffe. Dubuque, IA: Kendall/Hunt.

———. 1981. Notes on the status of geography in American schools. *Journal of Geography* 80:244–48.

Marran, J. F. 1987. Presentation to the AAG Council. Portland, Oregon.

Mayo, W. L. 1965. *The development and status of secondary school geography in the United States and Canada*. Ann Arbor, MI: University Publishers.

Michener, J. A. 1970. The mature social studies teacher. *Social Education* November:764–65.

Mikesell, M. W. 1980. The sixteen-million-hour question. *Professional Geographer* 32:263–68.

Monk, J. 1986. The Association of American Geographers' role in educational leadership: An interview with Sam Natoli. *Journal of Geography in Higher Education* 10:113–31.

———. 1987. Geography meeting its mission. *Journal of Geography* 86:143–47.

Morrill, R. L. 1983. The nature, unity and value of geography. *Professional Geographer* 35:1–8.

———. 1984. The responsibility of geography. *Annals of the Association of American Geographers* 74:1–8.

———. 1987. A theoretical imperative. *Annals of the Association of American Geographers* 77:535–41.

Morrissett, I. 1968. The role of academicians in the education of teachers (draft manuscript). *Report of the U.S. Commissioner of Education on the State of the Education Professions*. Washington: U.S. Office of Education.

———. 1982. Status of social studies: The mid-1980s. In *Council of State Social Studies Specialists (CS4). National Survey: Social Studies Education Kindergarten-grade 12*. Virginia: CS4.

Morrow-Jones, H. A. 1986. The Colorado experience. *Professional Geographer* 38:74–75.

National Academy of Sciences, National Academy of Engineering, Institute of Medicine, National Research Council. 1987. *Organization and Members 1987*. Washington, D. C.

National Assessment of Educational Progress (NAEP). 1979. *Summaries and technical documentation* for performance changes in citizenship and social studies assessment, 1969–1976. Denver, CO: NAEP, 1979.

National Center for Education Statistics. 1984. *Contractor report: A trend study of high school offerings and enrollments: 1972–73 and 1981–82*. Washington: GPO.

National Commission on Excellence in Education. 1983. *A nation at risk: The imperative for educational reform*. Washington: GPO.

National Council for Accreditation of Teacher Education. 1982. *Standards for the accreditation of teacher education*. Washington: NCATE.

National Council for the Social Studies. 1984. In search of a scope and sequence for social studies. *Social Education* 48:249–62.

National Council of Geography Teachers. 1956. The status of geography in the secondary schools of the United States: A report of a survey by a committee of the National Council of Geography Teachers. *Special Publication No. 4*. Chicago: National Council of Geography Teachers.

National Education Association. 1894. *Report of the committee of ten on secondary school social studies*. New York: American Book Company.

National Governors' Association. 1986. *Time for results: The governors' 1991 report on education*. Washington: National Governors' Association.

Natoli, S. J. 1986. The importance of redundancy and eternal vigilance. *Professional Geographer* 38:75–76.

———, ed. 1988. Strengthening geography in the social studies. Bulletin No. 81. Washington: National Council for the Social Studies.

———, and Bond, A. R. eds. 1985. *Geography in internationalizing the undergraduate curriculum*. Washington: Association of American Geographers.

Olmstead, C. W. 1987. Knowing and being who we are. *Journal of Geography* 86:3–4.

Pabst, D. L. 1986. Geography: The forgotten subject. *Principal* 66:22–24.

Palm, R. 1986. Coming home. *Annals of the Association of American Geographers* 76:469–79.

Patrick, J. J., and Hawke, S. D. 1982. Curriculum materials. In *Social studies in the 1980s: A report of Project Span*, ed. I. Morrissett. Alexandria, VA: Association for Supervision and Curriculum Development.

Pattison, W. D. 1970. From the National Council for Geographic Education. In *Focus on geography: 40th Yearbook of the National Council of the Social Studies*, ed. P. Bacon, pp. viii–ix. Washington: NCSS.

———, and Natoli, S. J. 1977. Change in geography: The doing dimension. *Change* 7:21.

Patton, D. J., ed. 1970. *From geographic discipline to inquiring student: Final report on the high school geography project*. Washington: Association of American Geographers.

President's Commission on Foreign Language and International Studies. 1979. *Strength through wisdom: A critique of U.S. capability*. Washington: GPO.

Rees, P. W., and Natoli, S. J. 1982. *Nurturing healthy geography programs in colleges and universities*. Washington: Association of American Geographers.

Ring, N. 1979. Teacher education in geography: A cure for the "grumps"? *Journal of Geography* 78:194–95.

Robinson, P., and Kirman, J. M. 1986. From monopoly to dominance. In *Social studies and social sciences: A fifty-year perspective*, eds. S. P. Wronski and D. H. Bragaw. Bulletin No. 78. Washington: National Council for the Social Studies.

Rosen, S. 1957. A short history of high school geography (to 1936). *Journal of Geography* 56:405–13.

Rumble, H. E. 1946. Early geography instruction in America. *Social Studies* 37:266–68.

Rutter, R. A. 1986. Profile of the profession. *Social Education* 50:252–55.

Salter, C. L. 1986a. Geography and California's educational reform: One approach to a common cause. *Annals of the Association of American Geographers* 76: 5–17.

———. 1986b. Response to Stutz's "enhancing high school geography." *Professional Geographer* 38:72.

Schmudde, T. H. 1987. The image of geography equals the structure of its curriculum and courses. *Journal of Geography* 86:46–47.

Schwendeman, J. R. 1961, 1970, 1981, and 1987. *Directory of college geography of the United States*. Richmond, KY: Eastern Kentucky University.

Scully, M. 1982. Academic geography: Few students, closed departments, fuzzy images. *Chronicle of Higher Education* 24:1.

Smith, M. 1966. *A decade of comment on education 1956–1966: Selections from the bulletin of the Council for Basic Education.* Washington: Council for Basic Education.

Smith, N. 1987. "Academic war over the field of geography:" The elimination of geography at Harvard, 1947–1951. *Annals of the Association of American Geographers* 77:155–72.

Spetz, D. L. 1986. Strengthening geography in Kentucky. *Professional Geographer* 38:73–74.

———. 1988. The preparation of geography teachers. In *Strengthening geography in the social studies,* ed. S. J. Natoli, pp. 51–58. Bulletin No. 81. Washington: National Council for the Social Studies.

Stake, R. E., and Easley, J. A., Jr. 1978. *Case studies in science education.* Washington: National Science Foundation.

Stoltman, J. P. 1987. The "Where" and "Why There" of geography in the K-12 curriculum. A paper presented at the Western PLACE Conference of the Colorado Geographic Alliance, Boulder, CO.

———, and Sweet, J. K. 1986. The Michigan experience in geographical education. *Professional Geographer* 38:73–74.

———, and Libbee, M. 1988. Geography in the social studies scope and sequence. In *Strengthening geography in the social studies,* ed. S. J. Natoli, pp. 42–50. Bulletin No. 81. Washington: National Council for the Social Studies.

Sturm, R., and Weiss, E. T., Jr. 1988. Adoption of high school geography textbooks: The Texas Maverick rejoins the herd. *Social Education* 42:254–57.

Stutz, F. P. 1985. Enhancing high school geography at the local level. *Professional Geographer* 37:391–95.

Swain, G. W., Jr. 1963. The role of college and university geography departments in improving the teaching and course content of high school geography. *Journal of Geography* 62:339–52.

Switzer, T. J. 1986. Teaching Sociology in K-12 Classrooms. In *Social studies and social sciences: A fifty-year perspective,* eds. S. P. Wronski and D. H. Bragaw. Bulletin No. 78. Washington: National Council for the Social Studies.

———, Mitchell, G., and Walker, E., 1981. Undergraduate social studies methods instructors knowledge and use of new curricular methods. *Journal of Social Studies Research* 5:9–18.

Tenner, E. 1988. Harvard, bring back geography! *Harvard Magazine* May–June:27–30.

Thomas, R. M., and Brubaker, D. L. 1971. *Curriculum patterns in elementary social studies.* Belmont, CA: Wadsworth Publishing Company.

Thompson, C. A., Jr. 1973. Secondary social studies methods instruction in the United States and its relationship to the new social studies. Unpublished doctoral dissertation, University of Colorado, Boulder, CO.

Tucker, J. L. 1972. Teacher educators and the "new" social Studies. *Social Education* 36:548–60.

University of Colorado's new admission requirements to include geography. 1983. *Association of American Geographers Newsletter* 18:1.

U.S. Bureau of Education. 1916. *The social studies in secondary education: Report of the committee on the social studies.* Bulletin No. 28. Washington: U.S. Bureau of Education.

U.S. Department of Education (Center for Statistics). 1983–84. *Digest of education statistics.* Washington: GPO.

———. 1987. *Digest of education statistics.* Washington: GPO.

U.S. Department of Health, Education, and Welfare. Various years. Earned degrees conferred: Bachelor's and higher degrees 1960–1961; 1965–66; 1970–71; and 1975–76. Washington: GPO.

Vuicich, G. and Stoltman, J. 1974. *Geography in elementary and secondary education.* Boulder, CO: ERIC Clearinghouse for the Social Studies and Social Science Education Consortium.

Wahlquist, W. L. 1986. The Utah success story. *Professional Geographer* 38:253–54.

Weiss, I. R. 1978. *National survey of science, mathematics, and social studies education.* Research Triangle Park, NC: Center for Educational Research and Evaluation.

White, G. F. 1961. A joint effort to improve high school geography. *Journal of Geography* 60:357–60.

———. 1970. Assessment in midstream. In *From geographic discipline to inquiring student,* ed. D. J. Patton. Washington: Association of American Geographers.

———. 1972. Geography and public policy. *Professional Geographer* 24:101–4.

Wilbanks, T. J., and Libbee, M. 1979. Avoiding the demise of geography in the United States. *Professional Geographer* 31:1–7.

Wiley, K. B. 1977. *The status of pre-collegiate science, mathematics, and social science education: 1955–1975.* Boulder, CO: Social Science Education Consortium.

Winston, B. J. 1984. Teacher education in geography in the United States. In *Teacher education models in geography: An international comparison,* ed. W. Marsden, pp. 133–49. Paris: Papers for the 25th Congress, International Geographical Union.

———. 1986. Teaching and learning in geography. In *Social studies and social sciences: A fifty-year perspective,* eds. S. P. Wronski and D. H. Bragaw. Bulletin No. 78. Washington: National Council for the Social Studies.

Woodring, P. 1983. *The persistent problems of education.* Phi Delta Kappa Education Foundation. Bloomington, IN.

———. 1984. Geography's place in basic education. *Journal of Geography* 83:143–44.

Wright, G. S. 1965. *Subject offerings and enrollments in public secondary schools, 1961–62.* Washington: GPO.

Wronski, S. P., and Bragaw, D. H., eds. 1986. *Social studies and social sciences: A fifty-year perspective.* Bulletin No. 78. Washington: National Council for the Social Studies.

ENVIRONMENTAL PROCESSES AND RESOURCES

Biogeography

Thomas T. Veblen

As a popular and rapidly growing subfield within American geography (Rogers 1983), biogeography encompasses a wide array of research aims. Biogeographers study the constantly changing ranges of plants and animals, over a range of temporal and spatial scales. They also study the structure and dynamics of communities and ecosystems in relation to both natural and anthropogenic processes. As practiced by both biologists and geographers, it is clearly an interdisciplinary subject with foundations in both the biological and Earth sciences.

Although this chapter is concerned with biogeography as practiced by geographers, it is useful first to consider how biogeography is perceived in the field of biology, where it is practiced by many more individuals than it is in geography. To most biologists, *biogeography* connotes one of two types of studies:

- □ biotic distributions at broad scales, and interpretations of the evolutionary and dispersal history of a single taxon or a few taxa; or
- □ biotic distributions at local-to-regional scales, and interpretations of these distributions in relation to contemporary environments and rates of immigration and extinction.

The first type is most frequently connoted by the term biogeography, and is conventionally termed *classical biogeography,* reflecting its continuity of research aims with nineteenth-century biogeographers. The second type is termed *geographical ecology,* reflecting the theoretical direction bestowed by ecologists and population biologists (e.g., MacArthur and Wilson 1967); it often merges with ecology.

For critically commenting on this chapter and/or providing useful information, I thank C. Bahre, W. Baker, S. Beatty, G. Brush, L. Conkey, J. Crowley, J. Franklin, R. Frenkel, G. Gaile, L. Graumlich, K. Hadley, K. Hansen-Bristow, J. Harman, S. Herwitz, E. Hobbs, S. Horn, J. Kay, P. Klein, A. Lara, K. Lulla, G. Malanson, E. McIntire, V. Meentemeyer, M. Merlin, M. Mielke, A. Parker, K. Parker, J. Parsons, G. Rogers, J. Sauer, M. Savage, C. Smith, A. Strahler, A. Taylor, T. Vale, H. Walter, W. Westman, C. Willmott, and K. Young.

Classical biogeography is practiced mainly by systematists, those biologists concerned with the classification, taxonomic affinities, and evolutionary histories of organisms. The techniques employed are phylogenetic and paleontological, and the interpretations of distributions are historical in the sense of evolutionary history and in the sense of the influence of geologic history on biotic ranges.

In contrast to the taxonomic training of classical biogeographers, geographical ecologists are largely trained as ecologists and population biologists. Geographical ecologists are often, but not exclusively, concerned with the numbers of species and species turnover in restricted areas (i.e., islands or island-like habitats). Geographical ecologists seek explanations for patterns of species diversity in terms of nonhistorical variables such as island area, distance from mainland sources of new species, and habitat diversity.

Biogeography and ecology are surprisingly difficult to clearly distinguish. Although ecology is usually defined as the study of interrelationships of organisms with their environment (Barbour, Burk, and Pitts 1987), it also has been defined as "the science that seeks to understand the distribution and abundance of life on earth" (Colinvaux 1986). A concern with spatial pattern is fundamental to both biogeography and ecology. Most ecological studies tend to be conducted over smaller areas, in contrast to the broad scale adopted in many biogeographical studies. However, distinction between the two fields solely on the basis of spatial scale is arbitrary. With the application of remote-sensing techniques to ecological questions of community structure and function (Lulla 1981; Westman 1987), the distinction between biogeography and ecology on the basis of spatial scale will become less valid. In fact, many biologists do not see a need to clearly distinguish between biogeography and ecology (e.g., MacArthur and Wilson 1967).

Only a small proportion of the biogeographic research conducted by geographers can be clearly described as either classical biogeography, with its concern for broadscale biotic distributions and systematic relationships (e.g., Sauer 1972; De Laubenfels 1978; McDonald 1981), or as geographical ecology (e.g., Malanson 1982; Hadley 1987). Despite the nonconformity of most geographical biogeography to biologists' perception of biogeography, geographical biogeography is also fundamentally concerned with spatial pattern. However, in most cases, geographical biogeographers study spatial patterns of communities rather than an individual taxon, and work at local-to-regional scales rather than the continental-to-global scale of classical biogeographers. Similarly, many geographical biogeographers working in paleoecology, bioclimatology, and ecosystem ecology stress abiotic elements rather than the biota. Thus, in terms of research aims and methods, geographical biogeography is more similar to ecology than to classical biogeography.

Some of the differences between geographical biogeography and biological biogeography are reflected in the content of biogeography courses taught in biology versus geography departments. Geographers include much material on community and ecosystem organization and dynamics, which is usually taught in ecology courses. They define biogeography as a component of physical geography, and stress the interactions between the inorganic realm and the biosphere. Thus, in addition to the unifying theme of evolutionary theory, which is typical of a biology course, geographers stress the Earth-system perspective of physical geography. Linkages among atmosphere, lithosphere, and biosphere are emphasized in terms of the flow of energy and materials, not just at the local scale of the ecologist but also at global scales.

Also, in contrast to biological biogeography courses, geographers devote substantial attention to the interactions of humans with plants and animals. This includes topics such as the role of humans as dispersal agents, plant and animal domestication, and human agency in modifying the structure and composition of communities. Although systems perspectives and a concern with human impacts are also common in biology curricula, they are usually included in ecology rather than biogeography courses.

DEVELOPMENT OF BIOGEOGRAPHY WITHIN U.S. GEOGRAPHY

Most articles dealing with the role of biogeography in American geography have focused on particular approaches offered as fruitful lines of research for geographers. For example, Bennett (1960) and Simoons (1974) identified research opportunities in the context of cultural zoogeography dealing with the domestication and dispersal of animals. Kellman (1969) encouraged geographers to abandon descriptive and deductive approaches to the study of vegetation in favor of more analytical and inductive approaches. In contrast, Fosberg (1976) argued that geographical biogeographers should stress regional description of biota. Ecosystem structure and function and the analysis of biogeochemical cycles have been suggested as profitable approaches by Gersmehl (1976) and Meentemeyer and Elton (1977). Vale and Parker (1980) identified research opportunities for biogeographers mainly in the context of the ecology of contemporary communities and the study of human impacts.

Attempts at comprehensive reviews of biogeographic research are relatively few, and all are out-of-date. Raup (1942), in a review of plant geography, in the *Annals of the Association of American Geographers* could find no significant body of pertinent literature produced by geographers. Similarly, reviews of plant and animal geography of the 1950s lamented the lack of biogeographic publications by American geographers (Stuart 1954; Küchler 1954; Bennett 1960). Some of the pre-1960 work of cultural-historical geographers on the role of human agency in modifying landscapes was biogeographical in nature (e.g., Sauer 1947; Clark 1949), but it was overlooked in these early reviews. The pre-1960 invisibility of biogeography in American geography, with the exception of Küchler's (1956) vegetation mapping, contrasts dramatically with the recent high productivity in this subfield.

How did the specialty evolve from a state of such little activity in the 1950s to one of high productivity in less than 30 years? Consideration of the historical antecedents of the current group of biogeographers will at least partially answer this question. There have been four key influences since the mid-1950s:

- stimulus provided by a few influential faculty in leading geography departments;
- attraction of ecologically oriented students by the environmental movement of the late 1960s and 1970s;

- incorporation of ecological perspectives (e.g., Odum 1959) into geographical curricula in the 1960s; and
- general resurgence of physical geography beginning in the 1970s.

The latter three influences are well appreciated and do not require further elaboration. The first, however, requires some historical explanation.

Early in the development of American academic geography there occurred substantial cross-fertilization and professional interaction with the discipline of ecology. For example, E. Huntington was a charter member, and later the second president, of the Ecological Society of America (McIntosh 1985). W. M. Davis, I. Bowman, and H. L. Shantz were well known among ecologists during the World War I era (James and Martin 1978). Influential ecologists who were charter members of the Association of American Geographers included C. C. Adams, F. E. Clements, H. C. Cowles, and C. H. Merriam (James and Martin 1978).

Probably the most significant influence of an ecologist on geography occurred through the interaction of C. O. Sauer with H. C. Cowles, the founder of physiographic plant ecology. Cowles, before completing his doctorate in botany at the University of Chicago in 1898, studied physiography under T. C. Chamberlin in the Geology Department and was greatly influenced by Davisian geomorphology (Rogers and Robertson 1986). Cowles remained at Chicago, where he taught plant ecology to the young Sauer. In his later years, Sauer in conversation frequently acknowledged the botanist's influence on his thinking about vegetation as a dynamic phenomenon.

Sauer's (1925, 1956) concern with landscape change, particularly under the influence of human activities, was a precursor to many of the research interests pursued by contemporary biogeographers. Sauer, with his student and long-time colleague James J. Parsons, stimulated in many students an interest in the theme of human/biota interaction. Between 1947 and 1983 at least 25 Ph.D. dissertations at Berkeley focused on biogeographical topics (Parsons and Vonnegut 1983). Although the subsequent careers of many of these students developed within cultural geography, others have pursued research aims and used methodologies more typical of ecology and physical geography.

The concentration of several biogeographers and ecologists in the Department of Geography of the University of California at Los Angeles since the early 1970s also had a major impact. C. Bennett and J. Sauer are long-time members of that Department; they were trained in zoology and botany, respectively. In the 1970s they were joined by W. Westman and H. Walter, trained in ecology and ornithology, respectively, and in 1981 by S. Beatty, also trained in ecology. During the 1970s and early 1980s approximately 20 Ph.D. dissertations at UCLA focused on biogeographical topics (J. Sauer and H. Walter, pers. com.).

BIOGEOGRAPHIC RESEARCH BY CONTEMPORARY GEOGRAPHERS

In the following review, typical examples of recent publications by biogeographers were selected to reflect the nature and extent of biogeographic research conducted by contemporary American geographers. A few of the authors are based at Canadian institutions, but have had significant involvement with U.S. biogeographers.

Human/Biota Interactions

Many geographers focus their research on the theme of human/biota interactions. These studies range from examination of plant and animal domestication to the role of humans in modifying plant and animal communities. Such studies are conducted both by geographers who would describe their interests as lying principally in cultural geography, as well as by physically oriented biogeographers.

Numerous geographers conduct research on plants and animals in the context of ethnoecology or ethnobiology. The focus in these studies may be either on the cultural significance of plant and animal use, or on the modification of the taxon's genotype as the consequence of human selection. (The former are included in the chapter on cultural ecology.) There is a long tradition of interest in the processes of domestication of plants and animals and the origins of agriculture (e.g., Sauer 1952), and this continues to be an active area of research (e.g., Kimber 1970, 1978; Johnson 1973; Johannessen 1966, 1981; Merlin 1982; Sauer and Kaplan 1969; Ruddle et al. 1978). The importance of many nonagricultural plants to native societies for nutritional, medicinal, and religious purposes has been revealed in numerous ethnobotanical studies by geographers (e.g., Fredrich 1978; Bahre and Bradbury 1980; Kay 1982; Rees 1976; Gade 1979; Gordon 1982; Pennington 1973).

Considerable biogeographic research deals with human impacts on plant and animal distributions and the structure and composition of communities. The introduction of alien plants and their impacts on native vegetation have been investigated by numerous geographers. Parsons (1972) documented the spread of African pasture grasses throughout the neotropics, and their ecological and economic importance to the region. The spread of weeds associated with European settlement of the Americas has been studied in California by Frenkel (1970) and Wester (1980), in the Pacific Northwest by Frenkel (1974) and by Frenkel and Boss (1988), and in Hawaii by Wester and Juvik (1983). A major conclusion of these studies is that successful invasion by alien plants occurs mainly in native communities that are chronically disturbed by humans (but, see Sauer 1988).

Landscape modification by feral livestock and introduced game animals has been studied in Texas by Doughty (1983), in California by Hobbs (1986), in Australia by McKnight (1976), and in New Zealand by Veblen and Stewart (1982b). Landscape changes related to the combined effects of livestock, logging, burning, and fuelwood-gathering have been historically documented in the western United States by Bahre and Bradbury (1978) and Crowley (1975), in north-central Chile by Bahre (1979), and in the Canary Islands by Parsons (1981).

Biogeographers frequently conduct their research in the context of conservation and resource-management issues (e.g., Beiswenger 1986; Cole 1981; Doughty 1975; Smith 1981; Kay 1985; Frenkel 1980). For example, geographers have made important contributions to the literature on the causes and consequences of tropical deforestation (Parsons 1976; Denevan 1981; Hecht 1981; Smith 1982; Sternberg 1987). Similarly, urban biogeography, which examines the relation of plant communities in urban areas to both intentional human management and inadvertent human impacts, is becoming increasingly important (Hobbs in press, 1988; Schmid 1975; Rowntree 1984).

Vegetation Dynamics

Many, if not most, geographical biogeographers study some aspect of vegetation dynamics, which is the study of the processes and patterns of successional change and regeneration in plant communities. A major focus in vegetation dynamics is the role of disturbance in shaping community composition and structure (Pickett and White 1985). Much of the work of geographers on vegetation dynamics examines vegetation responses to disturbance and the spatial and temporal characteristics of *disturbance regimes*.

Geography has a long tradition of research on characterization of disturbance regimes. This includes the physical geographer's concern with magnitudes and frequencies of physical-disturbance phenomena such as floods, mass movements, and snow avalanches, all of which are common disturbance agents (e.g., Wolman and Miller 1960). The tradition of work on magnitude and frequencies is reflected in the interest of contemporary biogeographers in disturbance regimes. Similarly, contemporary geographers' research on human modification of disturbance regimes stems from an earlier concern with human agency (e.g., Sauer 1956).

Geographers have actively investigated the impact of a wide range of disturbances on plant communities. Changes in fire regimes and the effects of fires on vegetation have been particularly attractive research themes for geographers (Vale 1982). The possible role of human-set fires in maintaining tropical savannas has been a major theme in biogeography (Sauer 1956; Hills 1969). Kellman and his associates have revealed some of the complexities of fire, nutrient, and vegetation interactions in tropical savannas (Kellman 1985; Kellman, Miyanishi, and Hiebert 1985).

In Mediterranean-type climate regions, geographers have studied fire behavior and vegetation response in California chaparral (Minnich 1983, 1987b), in coastal sage communities in California (Westman 1979, 1981; Malanson 1984; Malanson and Westman 1985; Westman and O'Leary 1986), and in garrigue in southern France (Malanson and Trabaud 1988). In temperate forests the effects of fire on stand structure and succession have been investigated in the western U.S. by Parker and Parker (1983), Vale (1979), Veblen (1986a), and Veblen and Lorenz (1986), and in the southern Andes by Veblen and Lorenz (1987). Fire in the Sonoran Desert has been studied by Brown and Minnich (1986) and Rogers (1986). Together, these studies have demonstrated the pervasive influence of fire on plant community composition and structure.

Butler (1985) and Malanson and Butler (1986) investigated avalanche frequencies and vegetation responses in the subalpine zone of the northern Rockies. Veblen et al. (1981) studied stand recovery following avalanches in subalpine *Nothofagus* forests in the southern Andes. Martin and Johnson (1987) and Hupp and Osterkamp (1985) investigated vegetation responses to stream-course changes and flooding. Veblen et al. (1981) demonstrated the controlling influences of earthquake-triggered mass movements on the structure of temperate rain forests in the southern Andes of Chile. In the eastern U.S., Conkey (1987) is investigating the possible influences of forest disturbance (including air pollution) and climatic variability on forest structure and productivity.

In addition to studying the effects of exogenous disturbance, geographers have also investigated the role of small-scale endogenous disturbance in the form of treefalls in temperate forests. Beatty (1984) analyzed the responses of understory plants to the mound-and-pit microtopography created by treefalls in the eastern deciduous forests of the U.S. Veblen (1985) in the southern Andes and Taylor and Zisheng (1988, n.d.) in central China investigated responses of understory bamboos to treefalls and their inhibitory influences on tree regeneration. Succession and regeneration dynamics in coniferous forests of the western U.S. have been studied by A. J. Parker (1986), Frenkel and Heinitz (1987), Baker (1988), Vale (1977), and Veblen (1986b). The role of seed budgets and seed dispersal in vegetation dynamics has been investigated by Kellman (1978), K. C. Parker (1987b), Lee (1981), and Young, Ewel, and Brown (1987).

Recent tree invasion of grasslands and shrublands, often resulting in dramatic landscape changes, has become a topic of substantial research interest among geographers. Vale (1981, 1987) documented a widespread pattern of invasion of subalpine meadows by conifers in the western cordilleras of the U.S. over the past 100 years. Similar patterns of tree and/or shrub invasion have been demonstrated for montane meadows in the central and northern Rockies (Veblen and Lorenz 1986; Butler 1986), for shrublands and grasslands of the Great Basin (Vale 1975; Rogers 1982), for coastal grasslands in North America (Rogers et al. 1985), and for grasslands and shrublands in northern Patagonia (Veblen and Lorenz 1988). Possible explanations for these invasions have involved the combined influences of fire suppression, recent climatic variability, or changes in livestock pressure.

Several geographers have investigated tree regeneration and the stability of treelines in the ecotonal areas between forests and either alpine or arctic tundra. Minnich (1984) investigated the complex interaction of snow drifting and persistence with stand structure of pine krummholz in southern California. Elliott (1979) and Elliott-Fisk (1983) demonstrated that trees are successfully regenerating under modern climatic conditions at the arctic treeline of Labrador.

In the Colorado Front Range, a paucity of tree seedlings in the forest tundra ecotone and slow rates of tree regeneration at disturbed sites led Ives and Hansen-Bristow (1983) to hypothesize that the upper limit of tree growth has been lowered due to climatic cooling over the past millennium. Shankman (1984), however, concluded that tree regeneration following fire at the upper forest limit, although slow, is gradually permitting the reestablishment of a forest cover.

Hansen-Bristow (1986) and Hansen-Bristow, Ives, and Wilson (1988) studied changes in tree phenology and growth in relation to climate variability along an elevational gradient from forest through the forest/tundra ecotone in the Front Range. In subalpine forests in New Zealand, Veblen and Stewart (1982a) investigated a similar hypothesis of tree-regeneration failure, but found no evidence of climatically induced regeneration failure.

Vegetation-Environment Relations

In contrast to most of the above studies, which focus on changes in community composition and structure that occur over time, numerous geographers have investigated vegetational changes that occur over space. In most of these studies, the objective is to relate changes in community parameters such as composition, structure, and physiognomy to variation in the physical environment. The spatial scale employed in these studies ranges from stand-scale to continental or global scales.

Beatty (1987a, 1987b) demonstrated a fine-scale heterogeneity of soil properties in California chaparral that may be the consequence of plant influences and which may contribute to species coexistence. In mixed stands of deciduous and evergreen *Nothofagus* in the southern Andes, Veblen, Veblen, and Schlegel (1979) demonstrated that frequencies and sizes of understory plants differ on microsites beneath the deciduous versus the evergreen trees. Other studies at an intrastand scale include Herwitz and Olsvig-Whittaker's (n.d.) work on growth and rooting patterns of shrubs in relation to substrate variation, and Perez's (1987) work on giant rosette distribution in relation to soil moisture in Andean paramos.

At a landscape scale, geographers have used ordination and gradient analysis to investigate the variation in vegetation patterns in relation to environmental gradients in numerous locales: eastern deciduous forests (Strahler 1977, 1978b; Brush, Lenk, and Smith 1980; Dodge and Harman 1985), western coniferous forests (Westman 1975; A. J. Parker 1982a, 1982b), wetlands of the Pacific Northwest (Eilers 1979), California shrublands (Westman 1981, 1983) and Rocky Mountain alpine vegetation (Baker 1983). Studies of vegetation-environment patterns at landscape scale and regional scales also include those by Harman (1970) in Michigan, A. J. Parker (1987) in the eastern U.S., Sauer (1982) and Kimber (1988) in tropical coastal areas, Holland (1980) in eastern Canada, Brown and Gersmehl (1985) in the Great Plains, Holland, Steyn, and Fuggle (1977) in South Africa, and Minnich (1987a) in Baja California.

Physiographic plant geography is the study of plant distributions at the scale of landforms, involving spatial analysis of vegetation in relation to physiographic processes (Zimmerman and Thom 1982). Methodologically, physiographic plant geography relies heavily on mapping of vegetation and landforms, and the demonstration of covariation in both phenomena. Examples include the work of Hack and Goodlett (1960) on the forests of the central Appalachians, Thom, Wright, and Coleman (1975) on vegetation patterns and tropical shoreline development, Brush, Lenk, and Smith (1980) on the forests of Maryland, and Zimmerman (1969) on vegetation patterns in the semiarid southwestern U.S. The interactions of plant ecological and geomorphic processes have also been investigated by Hupp (1986), Price (1971b), Baker (n.d.), Hansen-Bristow and Price (1985), Herwitz (n.d.), Malanson and Kay (1980), and Westman (1975).

At a global scale, Box (1981) has done definitive work, quantitatively relating vegetation physiognomy to macroclimate. Mapping vegetation at regional and continental scales continues to be an important objective for a few geographers (Küchler 1967; Bailey and Hogg 1986); such mapping is now being revolutionized by the application of remote-sensing techniques (Lulla 1981; Strahler 1981; Franklin 1986; and see chapter on remote sensing). Geographers have a long tradition of working on the larger-scale relationships of vegetation and climate patterns (e.g., Thornthwaite 1948; Mather and Yoshioka 1968); this work continues, based on remotely sensed data (Matthews 1983; Willmott and Klink 1986).

Ecosystem Structure and Function

Nutrient fluxes in tropical savannas, particularly in relation to fire, have been extensively investigated by Kellman, Miyanishi, and Hiebert (1985). Westman has investigated nutrient cycling in coniferous forests of California (Westman 1978b) and in

Australian eucalypt forests (Westman 1978a). Herwitz (1986, 1987) documented the influence of tropical rain forest vegetation on rainfall interception processes and intrasystem nutrient cycles. Geographers have studied primary productivity in California coniferous forests (Westman 1975), in coastal wetlands (Eilers 1979), in bamboo communities (Veblen, Schlegel, and Escobar 1980; Taylor and Zisheng 1987), and in Australian eucalypt forests and scrub (Holland 1969).

Modeling of primary productivity and other ecosystem processes has attracted substantial attention among geographers. At global and continental scales, Meentemeyer, Box and their associates have modeled several functional aspects of ecosystems and their rates in relation to climate, including primary productivity (Box 1978), litter production (Meentemeyer, Box, and Thompson 1982), litter decomposition (Meentemeyer 1984), and soil carbon (Meentemeyer, Gardner, and Box 1985). These studies also usually involve the creation of maps of predicted ecosystem properties and rates derived from climatic driving variables. Smith and Klinger (1985) proposed a global model of aboveground-belowground phytomass ratios in relation to environmental gradients.

Geographers' research on remote-sensing techniques to estimate ecosystem parameters at the scale of biomes is described in the chapter on remote sensing.

Zoogeography and Animal Ecology

Compared to the number of geographers who conduct research on vegetation, the number who conduct research on animals is rather small. Walter (1979a, 1979b) has investigated the microspatial distributions of falcons in relation to their autecology and to palearctic-African bird-migration systems. Vale, Parker, and Parker (1982) used published bird censuses from numerous localities to analyze the relationships between bird communities and vegetation structure in the U.S. In the southwestern U.S., K. C. Parker (1986, 1987a) has studied partitioning of foraging space and nesting sites and bird-community diversity.

Mielke (1977) documented the complex impacts of gophers on vegetation and soils in North America. Price (1971a) demonstrated the microtopographic effects of ground squirrels in the alpine tundra. Smith (1983a) showed that there are spatial variations in snowshoe hare population cycles. Beiswenger (1978) related the distribution of toads in Wyoming to their responses to laboratory gradients of temperature. McDonald (1981) has comprehensively analyzed the classification and evolution of the North American bison.

Paleoecology

Geographers who conduct research to determine the nature of paleoenvironments use various techniques, ranging from palynological methods to dendroclimatic methods (e.g., Bartlein, Prentice, and Webb 1986; Byrne, Adam, and Luther 1981; Horn 1985; Graumlich 1987; Conkey 1986). Such studies provide vital linkages of biogeography to other aspects of physical geography. Although many biogeographers are actively working in Quaternary studies, most would describe their interests as being fundamentally in paleoclimatology. To avoid duplication with the chapter on climatology, in the present review studies have been selected in which the main research interest is on past changes in biota, rather than in the physical environment.

Using paleontological techniques, Johnson (1978, 1980) documented human impacts upon the Pleistocene zoogeography of large mammals (mainly mammoths). McDonald (1981) also used paleontological techniques in his examination of the Quaternary zoogeography of the North American bison.

Geographers have used palynology to document the incidence of prehistoric fire in California (Byrne 1978), human impacts on estuarine communities (Brush and Davis 1984; Davis 1985), and postglacial vegetation changes at numerous sites in North America (e.g., Liu and Lam 1985; Kearny and Luckman 1983; MacDonald 1987). Palynological work in the western Amazon by Liu and Colinvaux (1985; n.d.) reveals substantial vegetation change in relation to both Holocene climatic fluctuation and disturbance by nonclimatic factors.

Methodological Research

Getis and Franklin (1987) and Harvey, Davis, and Gale (1988) have proposed new techniques of spatial pattern analysis for application to biogeographical problems. Geographers have developed new or modified existing techniques for sampling and analyzing vegetation (Strahler 1977, 1978a; Westman 1980; Westman and O'Leary 1986; Parker 1982b). In zoogeography, Smith (1983b) developed new techniques for classifying global mammalian faunal regions. The application of remote-sensing techniques to biogeographical and ecological problems is a major and promising development in which geographers are particularly active (Lulla 1981; Westman 1987; and see chapter on remote sensing).

Present Status and Future Directions

Rogers (1983) compiled several measures that show the academic status of biogeography within American geography to be healthy. Between 1965 and 1981, the biogeography-course percentage of total geography enrollment approximately doubled. Similarly, the percentage of universities and colleges offering at least one biogeography course more than tripled, and total enrollment more than doubled.

The high research productivity of the subfield is another measure of its health. Biogeographers are publishing with increasing frequency in prestigious journals such as the *Annals of the Association of American Geographers, Journal of Ecology, Vegetatio, Ecology,* and *Quaternary Research.* For example, in 1987 in the *Annals,* seven of the 35 articles were on biogeographic topics (Beatty 1987b; Graumlich 1987; Martin and Johnson 1987; Minnich 1987b; A. J. Parker 1987; Omernik 1987; and Vale 1987). It is important that biogeographers continue to publish their work, both in traditional geographical outlets and in the outlets of cognate fields.

The diversity of research aims pursued by geographical biogeographers has scattered research efforts across a great range of topics. Although individual geographers are making important contributions to global modeling, ethnobiogeography, Quaternary studies, and vegetation dynamics, geographers account for only small percentages of the practitioners in these interdisciplinary fields. Identification of common goals is needed, to link the work of biogeographers to that of other geographers, as well as to that of scientists in cognate fields.

Within geography, biogeography often provides opportunities for the bridging of human and physical geography. This linkage occurs mainly in studies where historical-cultural approaches are combined with methods from ecology and physical geography to analyze the effects of human activities on vegetation. Similarly, ethnobiogeographical research is strongly linked with cultural geography and cultural ecology. Important linkages exist among biogeography, climatology, and geomorphology in the form of research on climate and biota interactions and the interactions of geomorphic and ecological processes.

The increasing concern among scientists from a wide range of disciplines for the future habitability of the Earth provides important opportunities for biogeographers, as well as other geographers, to enhance the visibility of geography. The overall aim of the International Geosphere-Biosphere Program (IGBP) is to provide a comprehensive understanding of the processes linking the atmosphere, oceans, and land as they are affected by human activities (Mooney 1988). Such an objective has long been at the heart of geography, and geographers should play a vital role in global-change research.

For biogeographers, the current interest in landscape ecology among biologists, planners, and resource managers poses a major opportunity. The term *landscape ecology* was coined by the German geographer Carl Troll in 1950. Although it has been a popular field among European geographers, planners, and ecologists for over 20 years, some American biologists regard it as a new subfield of ecology (e.g., Urban, O'Neill, and Shugart 1987). Much of the research by contemporary biogeographers on vegetation-environment patterns, vegetation dynamics, and human impacts on biota is directly relevant to landscape ecology.

In its focus on landscape patterns and processes over a range of spatial and temporal scales, often influenced by human activities, landscape ecology is not original. For example, in community ecology there has long been an interest in vegetation analysis at a landscape scale (e.g. Whittaker 1956; Curtis 1959). Concern with the role of natural and anthropogenic disturbance in creating landscape patterns has long been fundamental to geography. Similarly, a hierarchical perspective, or the importance of spatial scale and different levels of organization, has long been emphasized in both ecology (Odum 1959) and geomorphology (Schumm and Lichty 1965). Landscape ecology is, in fact, an amalgamation of concepts and principles which have long been fundamental to ecology and geography.

The challenge posed to geographers by landscape ecology is one of educating ecologists to the work of geographers. Reciprocal interaction of biologists and geographers through publication in each other's journals and participation in professional meetings already exists, of course, but should be intensified. Given the core of biogeographic research on vegetation dynamics, vegetation-environment patterns, and human impacts, landscape ecology can provide for much mutually beneficial interdisciplinary interaction.

REFERENCES

Bahre, C. J. 1979. *Destruction of the natural vegetation of north-central Chile*. University of California Publication in Geography, 23. Berkeley: University of California Press.

———, and Bradbury, D. E. 1980. Manufacture of mescal in Sonora, Mexico. *Economic Botany* 34:391–400.

Bailey, R. G., and Hogg, H. C. 1986. A world ecoregions map for resource reporting. *Environmental Conservation* 13:195–202.

Baker, W. L. 1983. Alpine vegetation of Wheeler Peak, New Mexico: Gradient analysis, classification, and biogeography. *Arctic and Alpine Research* 15:223–40.

———. 1988. Size-class structure of contiguous riparian woodlands along a Rocky Mountain river. *Physical Geography* 9:1–14.

———. N.d. Effects of macro-environment on riparian vegetation in western Colorado. *Annals of the Association of American Geographers*. In press.

Barbour, M. G.; Burk, J. H.; and Pitts, W. D. 1987. *Terrestrial Plant Ecology*. Menlo Park, CA: Benjamin/Cummings Publishing Co.

Bartlein, P. J.; Prentice, I. C.; and Webb, III, T. 1986. Climatic response surfaces from pollen data for some eastern North America taxa. *Journal of Biogeography* 13:35–57.

Beatty, S. W. 1984. Influence of microtopography and canopy species on spatial patterns of forest understory plants. *Ecology* 65:1406–19.

———. 1987a. Origin and role of soil variability in southern California chaparral. *Physical Geography* 8:1–17.

———. 1987b. Spatial distributions of *Adenostoma* species in southern California chaparral: An analysis of niche separation. *Annals of the Association of American Geographers* 77:255–64.

Beiswenger, R. E. 1978. Responses of *Bufo* tadpoles to laboratory gradients of temperature. *Journal of Herpetology* 12:499–504.

———. 1986. An endangered species, the Wyoming toad *Bufo hemiophrys baxteri:* The importance of an early warning system. *Biological Conservation* 37:59–71.

Bennett, C. F., Jr. 1960. Cultural animal geography: An inviting field of research. *Professional Geographer* 12:12–14.

Box, E. O. 1978. Geographical dimensions of terrestrial net and gross primary production. *Radiation and Environmental Biophysics* 15:305–22.

———. 1981. *Macroclimate and plant forms: An introduction to predictive modeling in phytogeography*. Tasks for Vegetation Science, 1. The Hague: Dr. W. Junk.

Brown, D. A., and Gersmehl, P. J. 1985. Migration models for grasses in the American midcontinent. *Annals of the Association of American Geographers* 75:383–94.

Brown, D. E., and Minnich, R. A. 1986. Fire and changes in creosote bush scrub of the western Sonoran desert, California. *American Midland Naturalist* 116:411–22.

Brush, G. S., and Davis, F. W. 1984. Stratigraphic evidence of human disturbance in an estuary. *Quaternary Research* 22:91–108.

———; Lenk, C.; and Smith, J. 1980. The natural forests of Maryland: An explanation of the vegetation map of Maryland. *Ecological Monographs* 50:77–92.

Butler, D. R. 1985. Vegetational and geomorphic change on snow avalanche paths, Glacier National Park, Montana, U.S.A. *Great Basin Naturalist* 45:313–17.

———. 1986. Conifer invasion of subalpine meadows, central Lemhi Mountains, Idaho. *Northwest Science* 60:160–73.

Byrne, R. 1978. Fossil record discloses wildfire history. *California Agriculture* 32:13–14.

———; Adam, D.; and Luther, E. 1981. A late Pleistocene/Holocene pollen record from Laguna de la Trancas, Santa Cruz County, California. *Madrono* 28:255–72.

Clark, A. H. 1949. *The invasion of New Zealand by people, plants and animals: The South Island*. New Brunswick, NJ: Rutgers University Press.

Cole, D.N. 1981. Vegetational changes associated with recreational use and fire suppression in the Eagle Cap Wilderness, Oregon: Some management implications. *Biological Conservation* 20:247–70.

Colinvaux, P. 1986. *Ecology*. New York: Wiley.

Conkey, L. E. 1986. Red spruce tree-ring widths and densities in eastern North America as indicators of past climate. *Quaternary Research* 26:232–43.

———. 1987. Red spruce tree-ring density and growth decline. In *Proceedings of the international symposium on ecological aspects of tree-ring analysis*, eds. G. C. Jacoby, Jr. and J. W. Hornbeck, pp. 382–91. Washington: U.S. Department of Energy.

Crowley, J. M. 1975. Ranching in the mountain parks of Colorado. *Geographical Review* 65:445–60.

Curtis, J. T. 1959. *The vegetation of Wisconsin: An ordination of plant communities*. Madison, WI: University of Wisconsin Press.

Davis, F. W. 1985. Historical changes in submerged macrophyte communities of upper Cheapeake Bay. *Ecology* 66:981–93.

De Laubenfels, D. J. 1978. The genus *Prumnopitys* (Podocarpaceae) in Malesia. *Blumea* 24:189–90.

Denevan, W. M. 1981. Swiddens and cattle versus forest: The imminent demise of the Amazon rainforest reexamined. *Studies in Third World Societies* 13:25–44.

Dodge, S. L., and Harman, J. R. 1985. Soil, subsoil, and forest composition in south-central Michigan, U.S.A. *Physical Geography* 6:85–100.

Doughty, R. W. 1975. *Feather fashions and bird preservation: A study in nature protection.* Berkeley: University of California Press.

———. 1983. *Man and wildlife in Texas: Environmental change and conservation.* College Station, TX: Texas A & M University Press.

Eilers, E. H. 1979. Production ecology in an Oregon coastal salt marsh. *Estuarine and Coastal Marine Science* 8:399–410.

Elliott, D. L. 1979. The current regenerative capacity of the northern Canadian trees, Keewatin, N.W.T., Canada: Some preliminary observations. *Arctic and Alpine Research* 11:243–51.

Elliott-Fisk, D. L. 1983. The stability of the northern Canadian tree limit. *Annals of the Association of American Geographers* 73:560–76.

Fosberg, F. R. 1976. Geography, ecology, and biogeography. *Annals of the Association of American Geographers* 66:117–28.

Franklin, J. 1986. Thematic mapper analysis of coniferous forest structure and composition. *International Journal of Remote Sensing* 7:1287–1301.

Fredrich, B. 1978. Dooryard medicinal plants of St. Lucia. *Association of Pacific Coast Geographers Yearbook* 40:65–78.

Frenkel, R. E. 1970. *Ruderal vegetation along some California roadsides.* University of California Publication in Geography, 29. Berkeley: University of California Press.

———. 1974. Floristic changes along Everitt Memorial Highway, Mount Shasta, California. *Wasmann Journal of Biology* 32:105–36.

———. 1980. Natural area inventory and assessment: Blacklock Point, Oregon. *Association of Pacific Coast Geographers Yearbook* 42:119–29.

———, and Heinitz, E. F. 1987. Composition and structure of Oregon ash (*Fraxinus latifolia*) forest in William L. Finley National Wildlife Refuge, Oregon. *Northwest Science* 61:203–12.

———, and Boss, T. R. 1988. Introduction, establishment and spread of *Spartina repens* on Cox Island, Siusla Estuary, Oregon. *Wetlands* 8:1–17.

Gade, D. W. 1979. Petitgrain from *Citrus aurantium*: Essential oil of Paraguay. *Economic Botany* 33:63–71.

Gersmehl, P. J. 1976. An alternative biogeography. *Annals of the Association of American Geographers* 66:223–41.

Getis, A., and Franklin, J. 1987. Second-order neighborhood analysis of mapped point patterns. *Ecology* 68:473–77.

Gordon, B. L. 1982. *A Panama forest and shore: Natural history and Amerindian culture in Bocas del Toro.* Pacific Grove, CA: Boxwood Press.

Graumlich, L. J. 1987. Precipitation variation in the Pacific northwest (1675–1975) as reconstructed from tree rings. *Annals of the Association of American Geographers* 77:19–29.

Hack, J. T., and Goodlett, J. C. 1960. Geomorphology and forest ecology of a mountain region in the central Appalachians. *USGS Professional Paper* 347:1–67.

Hadley, K. S. 1987. Vascular alpine plant distributions within the central and southern Rocky Mountains, U.S.A. *Arctic and Alpine Research* 19:242–51.

Hansen-Bristow, K. J. 1986. Influence of increasing elevation on growth characteristics at timberline. *Canadian Journal of Botany* 64:2517–23.

———, and Price, L. W. 1985. Turf-banked terraces in the Olympic Mountains, Washington. *Arctic and Alpine Research* 17:261–70.

———; Ives, J. D.; and Wilson, J. P. 1988. Climatic variability and tree response within a mountain ecotone. *Annals of the Association of American Geographers* 78:505–19.

Harman, J. R. 1970. Forest and climatic gradients along the southeast shoreline of Lake Michigan. *Annals of the Association of American Geographers* 60:456–65.

Harvey, L. E.; Davis, F. W.; and Gale, N. 1988. The analysis of class dispersion patterns using matrix comparisons. *Ecology* 69:537–42.

Hecht, S. B. 1981. Deforestation in the Amazon basin: Magnitude and dynamics. *Studies in Third World Societies* 13:61–101.

Herwitz, S. 1986. Episodic stemflow inputs of magnesium and potassium to a tropical rainforest floor during heavy rainfall events. *Oecologia* 70:423–25.

———. 1987. Raindrop impact and water flow on the vegetative surfaces of trees and the effects on stemflow and throughfall generation. *Earth Surface Processes* 12:425–32.

———. N.d. Buttresses of tropical rainforest trees influence hillslope processes. *Earth Surface Processes*. In press.

———, and Olsvig-Whittaker, L. N.d. Preferential upslope growth of *Zygophyllum dumosum* Boiss. (Zygophyllaceae) roots into bedrock fissures in the northern Negev desert. *Journal of Biogeography*. In press.

Hills, T. L. 1969. Savanna landscapes of the Amazon Basin. *McGill University Savanna Research Series* 14:1–41.

Hobbs, E. R. 1986. Characterizing the boundary between California annual grassland and coastal sage scrub with differential profiles. *Vegetatio* 65:115–26.

———. 1988. Using ordination to analyze the composition and structure of urban forest islands. *Forest Ecology and Management* 23:139–58.

———. 1988. Species richness of urban forest patches and implications for urban landscape diversity. *Landscape Ecology*. In press.

Holland, P. G. 1969. Weight dynamics of *Eucalyptus* in the mallee vegetation of southeast Australia. *Ecology* 50:212–19.

———. 1980. Trout lily in Nova Scotia: An assessment of the status of its geographic range. *Journal of Biogeography* 7:261–67.

———; Steyn, D. G.; and Fuggle, R. F. 1977. Habitat occupation by *Aloe ferox* Mill. (Liliaceae) in relation to topographic variations in direct beam solar radiation income. *Journal of Biogeography* 7:61–72.

Horn, S. 1985. Estudio palinologico preliminar de dos nucleos cortos del Golfo Dulce, Costa Rica. *Anales Escuela Nacional de Ciencias Biologicas* 29:57–70.

Hupp, C. R. 1986. The headward extent of fluvial landforms and associated vegetation on Massanutten Mountain, Virginia. *Earth Surface Processes* 11:545–55.

———, and Osterkamp, W. R. 1985. Bottomland vegetation distribution along Passage Creek, Virginia in relation to fluvial landforms. *Ecology* 66:670–81.

Ives, J. D., and Hansen-Bristow, K. J. 1983. Stability and instability of natural and modified upper timberline landscapes in the Colorado Rocky Mountains, U.S.A. *Mountain Research and Development* 3:149–55.

James, P. E., and Martin, G. J. 1978. *The Association of American Geographers: The first seventy-five years 1904–1979*. Washington: Association of American Geographers.

Johannessen, C. W. 1966. The domestication process in trees reproduced by seeds: The pejibaye palm in Costa Rica. *Geographical Review* 56:363–87.

———. 1981. Domestication process of maize continues in Guatemala. *Economic Botany* 36:84–99.

Johnson, D. L. 1978. The origin of island mammoths and the Quaternary land bridge history of the northern Channel Islands, California. *Quaternary Research* 10:204–25.

———. 1980. Problems in the land vertebrate zoogeography of certain islands and the swimming powers of elephants. *Journal of Biogeography* 7:383–98.

Johnson, D. V. 1973. Geography and ecology of native cashew in northeastern Brazil. *Revista Brasileira de Biologia* 33:485–94.

Kay, J. 1982. The ecological basis of Menominee ethnobotany. *Journal of Cultural Geography* 2:1–12.

———. 1985. Native Americans in the fur trade and wildlife depletion. *Environmental Review* 9:118–30.

Kearny, M. S., and Luckman, B. H. 1983. Holocene vegetational and climatic history of Tonquin Pass, British Columbia. *Canadian Journal of Earth Science* 20:776–86.

Kellman, M. C. 1969. A critique of some geographical approaches to the study of vegetation. *Professional Geographer* 21:11–14.

———. 1978. Microdistribution of viable weed seed in two tropical soils. *Journal of Biogeography* 5:291–300.

———. 1985. Nutrient retention by savanna ecosystems, III. Response to artificial loading. *Journal of Ecology* 73:963–72.

———; Miyanishi, K.; and Hiebert, P. 1985. Nutrient retention by savanna ecosystems, II. Retention after fire. *Journal of Ecology* 73:953–62.

Kimber, C. T. 1970. Blue mahoe, a case of incipient plant domestication. *Economic Botany* 24:233–40.

———. 1978. Folk context for plant domestication: Or the dooryard garden revisited. *Anthropological Journal of Canada* 16:2–11.

———. 1988. *Martinique revisited: The changing plant geographies of a West Indian Island*. College Station, TX: Texas A & M University Press.

Küchler, A. W. 1954. Plant geography. In *American geography: Inventory and Prospect,* eds. P. E. James and C. F. Jones, pp. 428–41. Syracuse, NY: Syracuse University Press.

———. 1956. Classification and purpose in vegetation maps. *Geographical Review* 46:155–67.

———. 1967. *Vegetation mapping*. New York: Ronald Press.

Lee, M. A. B. 1981. Seed dispersal in hybrid salsola. *Great Basin Naturalist* 41:370–76.

Liu, K. B., and Colinvaux, P. A. 1985. Forest changes in the Amazon Basin in the last glacial maxium. *Nature* 318:556–57.

———, and ———. N.d. A 5200-year history of Amazon rainforest. *Journal of Biogeography*. In press.

———, and Lam, N. S. N. 1985. Paleovegetational reconstruction based on modern and fossil pollen data: An application of discriminant analysis. *Annals of the Association of American Geographers* 75:115–30.

Lulla, K. 1981. Remote sensing in ecological studies. *Canadian Journal of Remote Sensing* 7:97–107.

MacArthur, R. H., and Wilson, E. O. 1967. *The theory of island biogeography*. Princeton: Princeton University Press.

MacDonald, G. M. 1987. Postglacial vegetation history of the Mackenzie River basin. *Quaternary Research* 28:245–62.

Malanson, G. P. 1982. The assembly of hanging gardens: Effects of age, area, and location. *American Naturalist* 119:145–50.

———. 1984. Fire history and patterns of Venturan subassociations of Californian coastal sage scrub. *Vegetatio* 57: 121–28.

———, and Kay, J. 1980. Flood frequency and the assemblage of dispersal types in hanging gardens of the Narrows, Zion National Park, Utah. *Great Basin Naturalist* 40:365–71.

———, and Butler, D. R. 1986. Floristic patterns on avalanche paths in the northern Rocky Mountains, U.S.A. *Physical Geography* 7:231–38.

———, and Westman, W. E. 1985. Post-fire succession in Californian coastal sage scrub: The role of continual basal sprouting. *American Midland Naturalist* 113:309–18.

———, and Trabaud, L. 1988. Vigour of resprouting by *Quercus coccifera* L. *Journal of Ecology* 76:1–15.

Martin, C. W., and Johnson, W. C. 1987. Historical channel narrowing and riparian vegetation expansion in the Medicine Lodge River Basin, Kansas, 1871–1983. *Annals of the Association of American Geographers* 77:436–49.

Mather, J. R., and Yoshioka, G. A. 1968. The role of climate in the distribution of vegetation. *Annals of the Association of American Geographers* 58:29–41.

Matthews, E. 1983. Global vegetation and land use: New high-resolution data bases for climate studies. *Journal of Climate and Applied Meteorology* 22:474–87.

McDonald, J. N. 1981. *North American Bison*. Berkeley: University of California Press.

McIntosh, R. P. 1985. *The background of ecology: Concept and theory*. Cambridge: Cambridge University Press.

McKnight, T. L. 1976. *Friendly vermin: A survey of feral livestock in Australia*. University of California Publication in Geography, 21. Berkeley: University of California Press.

Meentemeyer, V. 1984. The geography of organic decomposition rates. *Annals of the Association of American Geographers* 74:551–60.

———, and Elton, W. 1977. The potential implementation of biogeochemical cycles in biogeography. *Professional Geographer* 29:266–71.

———; Box, E. O.; and Thompson, R. 1982. World patterns and amounts of terrestrial plant litter production. *Bioscience* 32:125–28.

———; Gardner, J.; and Box, E. O. 1985. World patterns and amounts of detrital soil carbon. *Earth Surface Processes* 10:557–67.

Merlin, M. D. 1982. The origins and dispersal of true taro, *Colocasia esculenta* L. *Native Planters* 1:6–17.

Mielke, M. W. 1977. Mound building by pocket gophers (Geomyidae): Their impact on soils and vegetation in North America. *Journal of Biogeography* 4:171–80.

Minnich, R. A. 1983. Fire mosaics in southern California and northern Baja California. *Science* 219:1287–94.

———. 1984. Snow drifting and timberline dynamics on Mount San Gorgonio, California, U.S.A. *Arctic and Alpine Research* 16:395–412.

———. 1987a. The distribution of forest trees in northern Baja California, Mexico. *Madrono* 34:98–127.

———. 1987b. Fire behavior in southern California chaparral before fire control: The Mount Wilson burns at the turn of the century. *Annals of the Association of American Geographers* 77:599–618.

Mooney, H. A. 1988. Ecologists and the global change program. *Trends in Ecology and Evolution* 3:4–6.

Odum, E. P. 1959. *Fundamentals of ecology*. Philadelphia: W. B. Saunders Co.

Omernik, J. M. 1987. Ecoregions of the conterminous United States. *Annals of the Association of American Geographers* 77:118–25.

Parker, A. J. 1982a. Environmental and compositional ordinations of conifer forests in Yosemite National Park, California. *Madrono* 29:109–18.

———. 1982b. The topographic relative moisture index: An approach to soil-moisture assessment in mountain terrain. *Physical Geography* 3:160–68.

———. 1986. Persistence of lodgepole pine forests in central Sierra Nevada. *Ecology* 67:1560–67.

———. 1987. Structural and functional features of mesophytic forests in the east-central United States. *Annals of the Association of American Geographers* 77:423–35.

———, and Parker, K. C. 1983. Comparative successional roles of trembling aspen and lodgepole pine in the southern Rocky Mountains. *Great Basin Naturalist* 43:447–55.

Parker, K. C. 1986. Partitioning of foraging space and next sites in a desert shrubland bird community. *American Midland Naturalist* 115:225–67.

———. 1987a. Avian nesting habits and vegetation structure. *Professional Geographer* 39:47–58.

———. 1987b. Seedcrop characteristics and minimum reproductive size of organ pipe cactus in southern Arizona. *Madrono* 34:294–303.

Parsons, J. J. 1972. Spread of African pasture grasses to the American tropics. *Journal of Range Management* 25:12–17.

———. 1976. Forest to pasture: Development or destruction. *Revista de Biologia Tropical* 24:121–38.

———. 1981. Human influences on the pine and laurel forests of the Canary Islands. *Geographical Review* 71:253–71.

———, and Vonnegut, N., eds. 1983. *Sixty years of Berkeley Geography: 1923–1983*. Berkeley: Department of Geography, University of California.

Pennington, C. 1973. Plantas medicinales utilizadas por el Pima montanes de Chihuahua. *America Indigena* 33:213–32.

Perez, F. L. 1987. Soil moisture and the upper altitudinal limit of giant paramo rosettes. *Journal of Biogeography* 14:173–86.

Pickett, S. T. A., and White, P. S., eds. 1985. *The ecology of natural disturbance and patch dynamics*. Orlando: Academic Press.

Price, L. W. 1971a. Geomorphic effect of the arctic ground squirrel in an alpine environment. *Geografiska Annaler* A53:100–106.

———. 1971b. Vegetation, microtopography, and depth of active layer on different exposures in subarctic alpine tundra. *Ecology* 52:638–47.

Raup, H. M. 1942. Trends in geographic botany. *Annals of the Association of American Geographers* 32:319–54.

Rees, J. D. 1976. The Oaxaca Christmas plant market. *Journal of the Bromeliad Society* 26:223–32.

Rogers, G. 1982. *Then and now: A photographic history of vegetation change in the central Great Basin desert*. Salt Lake City: University of Utah Press.

———. 1983. Growth of biogeography in Canadian and U.S. geography departments. *Professional Geographer* 35:219–26.

———. 1986. Comparison of fire occurrence in desert and nondesert vegetation in Tonto National Forests, Arizona. *Madrono* 33:278–83.

———; Robertson, J.; Solecki, W.; and Vint, M. 1985. Rate of grassland replacement by *Myrica pensylvanica,* Floyd Bennett Field, Gateway National Recreation Areas. *Bulletin of the Torrey Botanical Club* 112:74–78.

———, and Robertson, J. 1986. Henry Chandler Cowles: 1869–1939. *Geographers: Biobibliographical Studies* 10:29–33.

Rowntree, R. A. 1984. Forest canopy cover and land use in four eastern United States cities. *Urban Ecology* 8:55–67.

Ruddle, K.; Johnson, D.; Townsend, P. K.; and Rees, J. D. 1978. *Palm Sago: A tropical starch for marginal lands*. Honolulu: University of Hawaii Press.

Sauer, C. O. 1925. *The morphology of landscape*. University of California Publication in Geography, 2. Berkeley: University of California Press.

———. 1947. Early relations of man to plants. *Geographical Review* 37:1–25.

———. 1952. *Agricultural origins and dispersals*. Cambridge: MIT Press.

———. 1956. The agency of man on the earth. In *Man's role in changing the face of the earth*, ed. W. L. Thomas, pp. 49–69. Chicago: University of Chicago.

Sauer, J. D. 1972. Revision of *Stenotaphrum* with attention to its historical geography. *Brittonia* 24:202–22.

———. 1982. *Cayman Islands seashore vegetation: A study in comparative biogeography*. University of California Publication in Geography, 25. Berkeley: University of California Press.

———. 1988. *Plant migration: The dynamics of geographic patterning in seed plants*. Berkeley: University of California Press.

———, and Kaplan, L. 1969. Canavalia beans in American prehistory. *American Antiquity* 34:417–24.

Schumm, S. A., and Lichty, R. W. 1965. Time, space and causality in geomorphology. *American Journal of Science* 263:110–19.

Schmid, J. A. 1975. *Urban vegetation: A review and Chicago case study*. University of Chicago Department of Geography Research Paper No. 161. Chicago: University of Chicago.

Shankman, D. 1984. Tree regeneration following fire as evidence of timberline stability in the Colorado Front Range, U.S.A. *Arctic and Alpine Research* 16:413–17.

Simoons, F. J. 1974. Contemporary research themes in the cultural geography of domesticated animals. *Geographical Review* 64:557–76.

Smith, C. 1983a. Spatial trends in Canadian snowshoe hare (*Lepus americanus*) population cycles. *Canadian Field Naturalist* 97:151–60.

———. 1983b. A system of world mammal faunal regions. I. Logical and statistical derivation of the regions. *Journal of Biogeography* 10:455–66.

Smith, J. M. B., and Klinger, L. F. 1985. Aboveground-belowground phytomass ratios in Venezuelan paramo vegetation and their significance. *Arctic and Alpine Research* 17:189–98.

Smith, N. 1981. Caimans, capybaras, otters, manatees, and man in Amazonia. *Biological Conservation* 19:177–87.

———. 1982. Colonization lessons from a tropical forest. *Science* 214:755–61.

Sternberg, H. O'R. 1987. Aggravation of floods in the Amazon River as a consequence of deforestation? *Geografiska Annaler* A69:201–19.

Strahler, A. H. 1977. Response of woody species to site factors in Maryland, U.S.A.: Evaluation of sampling plans and continuous and binary measurement techniques. *Vegetatio* 35: 1–19.

———. 1978a. Binary discriminant analysis: A new method for investigating species-environment relationships. *Ecology* 59:108–16.

———. 1978b. Response of woody species to site factors of slope angle, rocky type, and topographic position in Maryland as evaluated by binary discriminant analysis. *Journal of Biogeography* 5:403–23.

———. 1981. Stratification of natural vegetation for forest and rangeland inventory using Landsat digital imagery and collateral data. *International Journal of Remote Sensing* 2:15–41.

Stuart, L. C. 1954. Animal geography. In *American geography: Inventory and prospect,* eds. P. E. James and C. F. Jones, pp. 442–51.

Taylor, A. H., and Zisheng, Q. 1987. Culm dynamics and dry-matter production of bamboos in the Wolong and Tangjiahe giant panda reserves, Sichuan, China. *Journal of Applied Ecology* 24:419–33.

———, and ———. 1988. Regeneration of *Sinarundinaria fangiana,* a bamboo, from seed in the Wolong giant panda reserve, Sichuan, China. *American Journal of Botany* 75: 1065–73.

———, and ———. N.d. Regeneration patterns in old-growth *Abies-Betula* forests in the Wolong Natural Reserve, Sichuan, China. *Journal of Ecology*. In press.

Thom, B. G.; Wright, L. D.; and Coleman, J. M. 1975. Mangrove ecology and deltaic-estuarine geomorphology: Cambridge Gulf-Ord River, Western Australia. *Journal of Ecology* 63:203–32.

Thornthwaite, C. W. 1948. An approach toward a rational classification of climate. *Geographical Review* 38:55–94.

Urban, D. L.; O'Neill, R. V. O.; and Shugart, H. H., Jr. 1987. A hierarchical perspective can help scientists understand spatial patterns. *Bioscience* 37:119–27.

Vale, T. R. 1975. Presettlement vegetation in the sagebrush-grass area of the Intermountain West. *Journal of Range Management* 28:32–36.

———. 1977. Forest changes in the Warner Mountains, California. *Annals of the Association of American Geographers* 67:28–45.

———. 1979. Coulter pine and wild fire on Mount Diablo, California. *Madrono* 26:135–40.

———. 1981. Tree invasion of montane meadows in Oregon. *American Midland Naturalist* 105:61–69.

———. 1982. *Plants and people: Vegetation change in North America*. Washington: Association of American Geographers.

———. 1987. Vegetation change and park purposes in the high elevations of Yosemite National Park, California. *Annals of the Association of American Geographers* 77:1–18.

———, and Parker, A. J. 1980. Biogeography: Research opportunities for geographers. *Professional Geographer* 32:149–57.

———; Parker, A. J.; Parker, K. C. 1982. Bird communities and vegetation structure in the United States. *Annals of the Association of American Geographers* 72:120–30.

Veblen, T. T. 1985. Forest development in treefall gaps in the temperate rain forests of Chile. *National Geographic Society Research Reports* 1:162–83.

———. 1986a. Age and size structure of subalpine forests in the Colorado Front Range. *Bulletin of the Torrey Botanical Club* 113:225–40.

———. 1986b. Treefalls and the coexistence of conifers in subalpine forests of the central Rockies. *Ecology* 67:644–49.

———; Veblen, A. T.; and Schlegel, F. M. 1979. Understory patterns in mixed evergreen-deciduous *Nothofagus* forests in Chile. *Journal of Ecology* 60:937–45.

———; Schlegel, F. M.; and Escobar, B. 1980. Dry matter production of two species of bamboos (*Chusquea culeou* and *C. tenuiflora*) in south-central Chile. *Journal of Ecology* 68:397–404.

———; Donoso, C.; Schlegel, F. M.; and Escobar, B. 1981. Forest dynamics in south-central Chile. *Journal of Biogeography* 8:211–47.

———, and Stewart, G. H. 1982a. On the conifer regeneration gap in New Zealand: The dynamics of *Libocedrus bidwillii* stands on South Island. *Journal of Ecology* 70:413–36.

———. 1982b. The effects of introduced wild animals on New Zealand forests. *Annals of the Association of American Geographers* 72:372–97.

———, and Lorenz, D. C. 1986. Anthropogenic disturbance and recovery patterns in montane forests, Colorado Front Range. *Physical Geography* 7:1–24.

———, and ———. 1987. Post-fire stand development of *Austrocedrus-Nothofagus* forests in Patagonia. *Vegetatio* 73:113–26.

———, and ———. 1988. Recent vegetation changes along the forest/steppe ecotone in northern Patagonia. *Annals of the Association of American Geographers* 78:93–111.

Walter, H. 1979a. *Eleonora's falcon: Adaptations to prey and habitat in a social raptor*. Chicago: University of Chicago Press.

———. 1979b. The sooty falcon (*Falco concolor*) in Oman: Results of a breeding survey, 1978. *Journal of Oman Studies* 5:9–59.

Wester, L. 1980. Composition of native grasslands in the San Joaquin Valley, California. *Madrono* 28:231–41.

———, and Juvik, J. O. 1983. Roadside plant communities on Mauna Loa, Hawaii. *Journal of Biogeography* 10:307–16.

Westman, W. E. 1975. Edaphic climax pattern of the pygmy forest region of California. *Ecological Monographs* 45:109–35.

———. 1978a. Inputs and cycling of mineral nutrients in a coastal subtropical eucalypt forest. *Journal of Ecology* 66:513–31.

———. 1978b. Patterns of nutrient flow in the pygmy forest region of northern California. *Vegetatio* 36:1–17.

———. 1979. A potential role of coastal sage scrub understories in the recovery of chaparral after fire. *Madrono* 26:64–68.

———. 1980. Gaussian analysis: Identifying environmental factors influencing bell-shaped species distributions. *Ecology* 61:733–39.

———. 1981. Factors influencing the distribution of Californian coastal sage scrub. *Ecology* 62:439–55.

———. 1983. Xeric Mediterranean-type shrubland associations of Alta and Baja California and the community/continuum debate. *Vegetatio* 52:3–19.

———. 1987. Monitoring the environment by remote sensing. *Trends in Ecology and Evolution* 2:333–37.

———, and O'Leary, J. F. 1986. Measures of resilience: The response of coastal sage scrub to fire. *Vegetatio* 65:179–89.

Whittaker, R. H. 1956. Vegetation of the Great Smoky Mountains. *Ecological Monographs* 26:1–80.

Willmott, C. J., and Klink, K. 1986. A representation of the terrestrial biosphere for use in global climate studies. *Proceedings of the ISLSCP conference, Rome, Italy, 2–6 Dec 1985*. ESA SP-248.

Wolman, M. G., and Miller, J. P. 1960. Magnitude and frequency of forces in geomorphic processes. *Journal of Geology* 68:54–74.

Young, K. R.; Ewel, J. J.; and Brown, B. J. 1987. Seed dynamics during forest succession in Costa Rica. *Vegetatio* 71:157–73.

Zimmerman, R. C. 1969. Plant ecology of an arid basin Tres Alamos-Redington area southeastern Arizona. *U.S. Geological Survey Professional Paper* 485–D.

———, and Thom, B. G. 1982. Physiographic plant geography. *Progress in Physical Geography* 6:45–59.

Climatology

John E. Oliver | Roger G. Barry | Waltraud A. R. Brinkmann | John N. Rayner

Over the last two decades (1970s–1980s) the frequency of extreme events, together with speculations concerning climatic variability and possible climatic change, has created both a popular awareness of, and renewed broad scientific interest in climatic research. American geographers have made significant contributions to the research literature.

While the research interests of geographical climatologists range across the entire spectrum of climatology, only selected components of their work are presented here. Geographical climatology is divided into four topics for the purposes of this discussion. Each of these encompasses broad areas of climatology, and they are themselves interrelated. *Modeling* involves all aspects of climate and leads naturally to a discussion of the dynamic atmosphere, often in terms of *Synoptic* and *Dynamic Climatology*. These two topics, in turn, provide basic methodologies for understanding and interpreting present climate, its variability, and the third topic, *Climate Change*. They also yield the necessary baseline information that enables analysis of practical problems in *Applied Climatology*.

MODELING IN CLIMATOLOGY

For the climatologist, the 1990s promise to be one of the most exciting periods in history. The tools are at hand for explaining the climate of a place or region, in the sense that the controlling and influencing factors may be thoroughly analyzed and their relative importance elucidated. Of the many necessary components for this major advance, modeling will play a central role.

The topic of modeling is broad and heterogeneous (Willmott 1987a). However, as argued by Oke (1982), the ultimate goal of scientific enquiry is the construction of largely deterministic, process-response models to predict the behavior of phenomena. These are of necessity numerical models which are based upon the underlying physics. Other models are therefore excluded from this discussion.

The aim of modeling or simulation in the current context is understanding, prediction, and control. *Understanding* comes first, with creation of the model at the research frontier. Later, the model may be relegated to a role as a teaching tool. (A word of caution: as with any tool, using it without a thorough understanding of its properties may lead to serious errors in conclusions.) *Prediction* involves time periods ranging from seconds to geological eras. Frequently, for climatologists, a model's value is in its diagnostic possibilities. Thus, for example, Howarth (1986) was able to analyze the water-vapor flux for the whole southern hemisphere, based upon weather-prediction model output data on a regular grid. The study could not have been conducted at the same scale with observational data, which are extremely sparse in that hemisphere.

Intentional *control* of climate is still very much in its infancy. Weather modification has experienced only limited success so far, and it has raised many unresolved legal, social, political, and ethical questions. The importance of having realistic climatic models is that the consequences of altering controlling factors may be investigated theoretically. The large volume of current research into inadvertent modification (CO_2, nuclear winter, ozone, etc.) may be classified under this rubric (discussed below).

Elements of Modeling

Critical elements involved in modeling and simulation, such as differencing, error propagation, stability, convergence, grids, initial conditions, boundary conditions, parameterization, and validation, have been outlined by various authors (Rayner 1984; Pielke 1984; Washington and Parkinson 1986). Clearly, mathematics is necessary prerequisite training for a climatologist. However, no longer is calculus sufficient. Experience with many other branches of applied mathematics is important, such as finite-difference and finite-element methods and various transformations, including spectral techniques. Physics, chemistry, and computer science are required. The list does not stop here, and after years of pleas (Hare 1960; Thornthwaite 1961) more and more climatologists in geography are receiving such training. No longer can meteorologists cheerfully make blanket statements that "...one year of differential and integral calculus will eliminate the geographical curriculum from consideration for professional accreditation" (Baum 1953).

Parameterization is still a major underdeveloped area of modeling. It is applied where the basic physics are too complex, or their inclusion is too costly, or the systems are smaller than the grid-cell size can resolve. Willmott (1984), for example, looked specifically at parameterization of the seasonal snow-cover cycle. Good summaries are given for general-circulation models by Smagorinsky (1982), and for energy-balance climate models by Shine and Henderson-Sellers (1983). Saltzman (1985) points out that, for paleoclimatic modeling, "we must parameterize not only sub-synoptic phenomena but all phenomena up to and including inter-annual variation."

Because in complex models the parameterization factors often involve "guesstimate constants," they are frequently manipulated in conjunction with one another to produce realistic results. This raises the most perplexing issue of validating models. Not only is there a problem of measuring the accuracy of results (Willmott 1981; Willmott et al. 1985), but there is also a concern over whether errors in parameterizations might be offsetting one another.

Specification of *boundary conditions* is one area where geographers contribute significantly to modeling. For example, Wilson and Henderson-Sellers (1985) have created a database of land cover and soils, and Scharfer et al. (1987) have done the same for Arctic albedo.

Models

Rayner (1984) chose to discuss models in relation to atmospheric *element* and to *scale*. Although the present review focuses on scale, the element grouping still occupies a significant number of geographers: in particular, models of radiation flux have been created or used by Arnfield (1982b, 1983), Davies, Schertzer, and Nunez (1975), Dozier (1980), Munro and Young (1982), Nunez (1980), Raphael and Hay (1984), Rowe and Willmott (1984), and Suckling and Hay (1976). Temperature remains central to many surface heat-balance studies, including those of Arnfield (1982a), Outcalt (1971) and Terjung (many papers, e.g., Terjung and O'Rourke 1984). Evapotranspiration has also been simulated by Terjung et al. (1984) and Sellers and Lockwood (1981).

However, classification of models according to *scale* appears to be a more-satisfactory division, because the majority of models encompass many elements and processes. A partial discussion of the scale of atmospheric phenomena is given by Steyn, Oke, and Hay (1981), who follow in part earlier work by Smagorinsky, details of which are given in a more recent publication (Smagorinsky 1982).

Emphasis on modeling in the atmospheric-science community has been placed on the global scale. The original impetus, of course, came from the need for numerical weather prediction. The current status of this subject is covered by Brown (1987). In recent years these models have been adapted for investigation of climate. Building such comprehensive models has been a long task involving teams of researchers. Since few institutions employ more than one or two climatologists, not many names of geographers are associated with the development of global models. An exception is Parkinson (Parkinson and Washington 1979; Parkinson and Herman 1980; Washington and Parkinson 1986). A review of general circulation models is given by Manabe (1985).

Less comprehensive are the one- and two-dimensional global models discussed by Shine and Henderson-Sellers (1983). Cerveny and Balling (1984a, 1984b, 1985), based on the work of W. D. Sellers, developed a simple one-dimensional energy-balance model from which they were able to simulate average latitudinal temperature and meridional and zonal wind speeds.

Anthes (1983) has reviewed mid-latitude regional-scale models, and Pielke (1984) has published an excellent book on the mesoscale, including a list of research groups. Atkinson (1983) and Maddox (1987) have also reviewed current models at this scale. In geography, only Hobgood (1986a, 1986b; Hobgood and Cerveny 1988) seems to have developed a three-dimensional system in this category.

Geographers have been most active at the one-dimensional terrain and urban levels. Outcalt (1971, 1972) was the first to develop a surface-climate simulator. He and his students, especially Dozier, have continued to study terrain, either in high latitudes or high elevations. One of Outcalt's most recent papers investigates ice growth and decay in the St. Lawrence River (Greene and Outcalt 1985). Models of urban settings have been created by Arnfield (1976, 1982a), Henderson-Sellers (1980), and Terjung and his students (e.g. Terjung and O'Rourke 1981).

Another approach to classifying models is to consider general *topical* groupings. Under this category we may consider models that are used to investigate atmospheric effects on vegetation and animals. Terjung and his students have applied their models to studying vegetation and crop productivity, and human heat budgets and stress (e.g., Burt et al. 1981; Burt, O'Rourke, and Terjung 1982; Hayes et al. 1982; O'Rourke and Terjung 1981a, 1981b, 1981c). Arnfield (1986) has performed a detailed analysis of estimating diffuse irradiance on organisms.

By far the largest use of General Circulation Models (GCMs) today is in sensitivity analysis; that is, a model is used experimentally to investigate how the atmosphere and climate might respond to a change in some factor. A significant problem here is the validation of such experiments. As indicated, parameterization may contain unknown errors; of particular concern are feedback mechanisms and coupling between the atmosphere and the Earth (water, ice, land, and biosphere). For example, Barry, Henderson-Sellers, and Shine (1984) and Shine, Henderson-Sellers, and Barry (1984) have shown that feedback between the cryosphere and high-latitude cloudiness has a potentially significant impact on determining albedo at high latitudes. As Saltzman (1985) points out, the average amount of out-of-equilibrium energy required to melt the Wisconsin ice between 18,000 BP and 8000 BP is $10^{-1} Wm^{-2}$. This "slow physics" can easily be accounted for from ocean storage, and is well below the accuracy of observational and model data. Hence it may not be possible to explain some forms of climatic change with the current climate models.

The earliest geographers to use a GCM were Williams, Barry, and Washington (1974) and Williams (1975), to simulate atmospheric conditions at the height of the Wisconsin glaciation. Williams (1978a, 1978b) also compared the results with other GCMs. Other geographers who have used a GCM are Parkinson (Parkinson and Kellogg 1979) and Bach (1984), who simulated the results of CO_2 increases on climate. Carbon dioxide changes are still of great concern, and research has been reviewed recently by Schlesinger and Mitchell (1987) for simulations on a variety of models including GCMs, energy-balance models, and radiative-convective models. They found that, whereas these models generally showed an increase in temperature with increases in CO_2, the variation was so great, and the unanalyzed feedback processes so important, that no clear conclusions could be drawn.

Another topic of current research interest is the influence of sea-surface temperatures. GCM studies show some correlation in tropical areas, but again, the relationship for mid-latitudes is uncertain.

The Future

As indicated, climate-modeling research is on the verge of a potentially exciting decade:

1. Our current understanding of atmospheric processes is quite sophisticated. Many problems exist, but the questions to be investigated are defined.
2. More climatologists are being better-trained in the necessary mathematics, physics, etc.
3. Enormous computing power is becoming universally available (Raveché, Lawrie, and Despain 1987), ranging from desk-top processors to supercomputers. Breakthroughs in hardware and software can be expected.

4. Computer graphics and associated software analysis techniques are at hand:
   a. Geographers should be well aware of developments in geographic information systems (see chapter on this topic). These provide a whole new array of techniques which climatologists can use for data analysis and assimilation.
   b. Significant advances are in the developmental stage for visualizing fluid flow (McCormick, DeFanti, and Brown 1987). These should provide new insights into the systems which climatologists study.

SYNOPTIC AND DYNAMIC CLIMATOLOGY

Synoptic climatology is the study of local and regional climates, in terms of the properties and motion of the overlying atmosphere (Barry 1987). Its primary information source is the synoptic weather map and corresponding upper-air charts. The first text on synoptic climate (Barry and Perry 1973) helped such research achieve recognition as a subdisciplinary field. Since its publication, there has been no slackening of activity in this field, with geographers playing a major role.

Dynamic climatology is concerned with the large-scale mechanisms that determine global atmospheric circulation and its principal modes of behavior. More recently, the broader term *climate dynamics,* which treats the total climate system, has come into wide use. It incorporates intercomparison of the results of large-scale modeling and empirical analyses.

The subtopics to be examined here are data sources, synoptic climatological methods, applications, and dynamic climatology.

Data Sources

Prior to the early 1970s, data for synoptic climatological studies were extracted manually from time series of pressure and height charts. Subsequently, the increasing availability of gridded data sets has greatly aided the researcher—for example, those produced by the National Meteorological Center, now available for 1946–85 on a single CD/ROM (Mass et al. 1987). Using such data, catalogs of the daily frequency of synoptic weather patterns have been developed for many regional applications (Moritz 1979; Bradley and England 1979; Barry, Kiladis, and Bradley 1981; Yarnal 1985a, 1985b). Products from operational meteorological analyses of such variables as vertical velocity and moisture remain to be explored (Finger et al. 1985).

An alternative source of information is provided by satellite imagery. However, the availability of images having adequate spatial and temporal resolution dates only from the early 1970s. Visible and infrared imagery have been used for climatological studies of southern-hemisphere cyclones, synoptic-scale atmosphere/ice interactions, the summer monsoon of the U.S. Southwest, polar lows (by Carleton 1981, 1985a, 1985b, 1985c) and for synoptic analysis of arctic clouds (by Barry et al. 1987).

Analogous to the transition to gridded pressure fields, we can expect to see a shift toward the use of gridded satellite products in the near future. Such data are currently available for sea-ice products from the Electrically Scanning Microwave Radiometer (1973–76) and Scanning Multichannel Microwave Radiometer (1978–87), and will be prepared from the Special Sensor Microwave Imager launched in June

1988 (Weaver, Morris, and Barry 1987). Gridded maps of hemispheric snow cover, derived from satellite analysis, are also available (Dewey and Heim 1982). Other similar products can be expected in the future.

Synoptic Climatological Methods

The last decade has seen significant advances in synoptic climatological methods over the earlier, largely subjective classification procedures (Barry and Perry 1973). This has been facilitated by the availability of computers and packaged programs.

First, there has been important progress in evaluating "objective" synoptic typing procedures. For example, Key and Crane (1986), Yarnal (1984a), and Yarnal and White (1987) provide sample results illustrating the effects of scale of analysis, number of classes defined, limits on sample size of classes, and so on. Willmott (1987b) demonstrates the virtual identity of the procedure employed in the map-correlation technique—originally proposed by Lund (1963)—and the more recent "sums of squares" method developed by Kirchhofer (1974) and used by Moritz (1979), Bradley and England (1979), and Barry, Kiladis, and Bradley (1981).

Second, new techniques of clustering (Kalkstein, Tan, and Skindlov 1987) and rotated principal-component analysis (Cohen 1983) have been used in synoptic climatological typing applications.

Third, diagnostic analysis techniques have begun to be applied to synoptic climatological case studies. For example, LeDrew (1984, 1985) analyzes contributions of local forcings and large-scale external processes to the spatial field of vertical velocity in arctic synoptic systems, and Keen (1980) examines cyclonic activity in terms of vorticity flux divergence at the 500 mb level in a study of regional circulation anomalies for the eastern Canadian Arctic.

Finally, the applicability of synoptic classifications for intercomparison of observed pressure fields and simulated pressure fields (derived from the Goddard Institute of Space Sciences climate model) is illustrated by Crane and Barry (1988). This suggests a potentially valuable methodology for assessing the capability of general circulation models to reproduce realistic synoptic-scale variability.

Applications

The primary objective of most synoptic climatological analyses is to aid in the interpretation of atmospheric circulation controls on some aspect of regional climate and its spatial and temporal variability. This analysis may be directed at specific climatic elements; examples include precipitation conditions (Alijani and Harman 1985; Harrington and Brown 1985; Carleton 1987), dust storms (Brazel and Nickling 1986), and air quality (Muller and Jackson 1985; Kalkstein and Corrigan 1986).

Such applications have also been extended to studies of the climatic controls of other physical variables. Alt (1987), for example, related extreme glacier mass-balance conditions on the Queen Elizabeth Islands to three generalized synoptic patterns, while Yarnal (1984b) used a more detailed synoptic classification to examine temporal variations in mass-balance conditions on a British Columbian glacier. Analogous studies of sea-ice variability in relation to synoptic regimes have been made by Crane (1978) for Davis Strait, and Rogers (1978) for the Beaufort Sea. Other studies on

Antarctic sea-ice/cyclone interactions have been made by Carleton (1983), Howarth (1983), and Rogers (1983). Airflow trajectories related to synoptic types have also been used to examine the probable frequency of exotic pollen deposition from the boreal forest in the eastern Canadian Arctic (Barry, Elliott, and Crane 1981).

Apart from synoptic analyses of atmospheric circulation patterns, similar classification procedures can be applied to the spatial occurrence of climatic regimes as shown by Nicholson (1981) for West Africa, using composite maps of wet-year and dry-year rainfall to identify the dominant patterns.

Studies focusing on the time domain have sought to use synoptic analyses to assess the seasonal structure of the climate of a region, and/or its interannual variability and longer-term trends. The idea that certain types of weather are more common in particular months, and even around certain dates, has a long history. Such weather singularities and natural seasons for the U.S. have been reexamined by Kalnicky (1987) using eigenvectors of the hemispheric circulation types originally identified by B. Dzerdzeevski in the Soviet Union. In addition to developing a new calendar of singularities, Kalnicky showed the secular variability of both the length and character of the seasons.

In a synoptic approach to climate change studies, Barry, Kiladis, and Bradley (1981) determined the relative contributions of temporal changes in synoptic-type frequency and of the characteristics of individual types to observed changes in temperature and precipitation in the western U.S. since 1899. They concluded that changes in both type frequency and type characteristics contributed to the temperature and precipitation changes. The role of circulation changes in modifying climatic regimes has also been examined by Diaz (1986), using seasonal empirical orthogonal functions of 700 mb heights, temperature, and precipitation fields over North America.

Dynamic Climatology

Research on large-scale global circulation patterns has become an area of renewed activity in recent years, and several geographers have contributed to this effort. Examples are studies of circulation variability and its significance for climatic anomalies (Rogers 1981b, 1983), and of the large-scale transport of atmospheric water vapor (Howarth 1986). Teleconnection patterns in pressure fields, such as the North Atlantic and North Pacific oscillations (Rogers 1981a), have been examined, principally through analysis of composite fields of pressure and height anomalies (van Loon and Rogers 1978; Meehl and van Loon 1979; Rogers 1984; Moses et al. 1987). The El Niño/ southern oscillation phenomenon (Yarnal 1985b) and Asian monsoon (Meehl 1987) have also received attention in view of their significance for climatic anomalies on a global scale (Caviedes 1984; Rogers 1984; Yarnal 1985a; Yarnal and Kiladis 1985; Yarnal and Diaz 1986).

These studies of the major modes of atmospheric circulation and their variability have an important bearing on theoretical studies of the atmosphere. A key issue is whether there exists a unique climatic state for a given set of external boundary conditions (a transitive system), or whether there may be more than one climatic state (an intransitive system), as discussed by Lorenz (1976). This unresolved question is at the root of studies of atmospheric blocking (Knox and Hay 1985; Tarleton 1987), and of modeling investigations of climate stability (Barry 1979).

Outlook

The ready availability of global data sets of meteorological variables from surface and satellite observations, on space and time scales appropriate for synoptic research, provides the essential basis for continued activity in synoptic and dynamic climatology. Beyond the study of surface-atmosphere interactions and applications of synoptic analogs in climate-change research, there is likely to be growing interest in examining synoptic and global aspects of atmospheric chemistry and biogeochemistry, in connection with the International Geosphere-Biosphere Program and studies of global change (Yarnal et al. 1987). There is also likely to be a closer interaction among empirical research, theoretical meteorology, and climate-modeling studies.

CLIMATE CHANGE

Climate-change research has two goals: description of temporal and spatial patterns of change, and ultimately, identification of causal mechanisms. An exciting surge of interest in this area of climatology is presently occurring as a result of the development of new techniques of data collection and analysis, and of advances in the understanding of causal mechanisms. Geographers have contributed in many ways to this rapid growth, so this review is restricted to examples of recent work of North Americans having a *geography* affiliation.

Evidence of Climate Change

Geographers have long contributed to the important task of evaluating the instrumental record to document short-term variability. Studies of regional anomaly patterns include those for Arizona (Johnson 1980), the Great Lakes region (Brinkmann 1983), Illinois (Changnon 1983), northern North America (Diaz 1986), and western Canada (Singh and Powell 1986). Results of such studies show that secular trends in surface temperature and precipitation have not been geographically uniform in *magnitude* or in *direction,* and that this is due to changes in atmospheric circulation features. Changes in location and amplitude of the standing waves and associated changes in cyclone and anticyclone activities have given rise to regional differences in surface trends in middle and high latitudes (van Loon and Williams 1977; Reitan 1979; Rogers 1983; Harman 1987). In low latitudes, shifts in the ITC and changes in the Hadley regime have occurred (Nicholson 1981; Byrne, Granger, and Monteverdi 1982).

Most regional studies do, however, agree on the *timing* of major changes—the 1920s and 1950s. Major shifts in the hemispheric circulation regime occurred at those times (rather than around 1940, the time of a major change in the hemispheric *temperature* trend—Blasing and Lofgren 1980; Brinkmann 1981; Kalnicky 1987). This reflects perhaps an "almost intransitive" nature of the climate system. Such a system, in response to some gradual forcing, will persist in one pattern or state until this is no longer dynamically possible, and a change to a new state occurs.

The circulation shift of the 1950s was, however, not a return to the pre-1920s pattern, but a shift to a new pattern. The pre-1920s and post-1950s segments of local climate records are therefore not necessarily identical; nor does a division based on the timing of the change in the hemispheric-temperature trend—into pre-1940s and post-1940s—necessarily result in segments that are similar but opposite in sign.

Our growing vulnerability to climate fluctuations has helped focus attention on such critical topics as climate variability (Skaggs 1978; Granger 1979; Mearns, Katz, and Schneider 1984) and growing-season length (Nkemdirim and Venkatesan 1985; Skaggs and Baker 1985; Suckling 1986). For instance, increased variability is believed to be associated with hemispherically cool periods, which have tended to be periods of unusually low temperatures in high latitudes. (Temperature changes in low latitudes have tended to be small.) This leads to an increase in the exchange of energy across latitudes, with more warm-air advection to higher latitudes than usual, and more cold-air advection to lower latitudes than usual, and consequently to an increase in weather extremes.

There is, however, little evidence in support of this. For a large portion of the U.S., for instance, variability decreased between the hemispherically warmer period of the 1930s and 1940s and the cooler period of the 1950s and 1960s. And many records have already been broken during this present warm decade. The reason for this discrepancy between expectation and observation is that the mean hemispheric temperature masks the complexity of the real climate system (Brinkmann 1985). The hemispheric surface temperature is, for instance, not a perfect indicator of upper-level temperatures, and thus of the true heat content of the atmosphere.

The use of hemispheric temperature as an indicator of the length of the growing season has been shown to be another fallacy: the length of the local growing season is determined by the local temperature trend, specifically that for spring and fall, and not by the hemispheric trend (Brinkmann 1985). And, the two are not necessarily the same, as discussed above.

Reconstructing Past Climates. A second area of climate change to which geographers have contributed significantly is the reconstruction of past climates. The use of historical records (e.g., Lawson and Stockton 1981; Rannie 1983; Rowntree 1985; Catchpole and Halpin 1987) is limited by their relative shortness and sparseness. But paleoclimatic interpretations of physical and biological indicators of long-term climate change have been numerous. Recent examples include speleothem isotope records of temperature (Harmon et al. 1978), African lake levels (Butzer 1980), arctic tree line (Elliott-Fisk 1983), ice core records of precipitation (Thompson et al. 1985), and combinations of indicators to describe the climate history of a region (Davis 1977; Kay and Johnson 1981; Jones, Spayne, and Schenk 1984).

Discontinuities in the land record, caused by erosion, have been a problem in the reconstruction of very long-term climate changes. The relatively continuous ocean record is now providing a framework for evaluating the land record (e.g. Madsen and Currey 1979; Oviatt, McCoy, and Reider 1987; Dorn et al. 1987).

A significant recent advance in climate reconstruction has been the development of "transfer functions"—statistical (regression-type) techniques that translate primary proxy data into quantitative climate estimates. Geographers have been very quick in adopting these to reconstruct past climates from various sources: tree-ring width (Kuivinen and Lawson 1982; Graumlich 1987; Stahle and Cleaveland 1988), tree-ring density (Conkey 1986), and isotope content (Luckman and Kearney 1986), as well as from pollen data (Kay and Andrews 1983; Bartlein, Webb, and Fleri 1984), soil characteristics (Sorenson 1977), and isotopes in ice cores (Lawson, Kuivinen, and Balling 1982).

Work has begun on higher-level climate estimates synthesized from site-specific data to provide "time-slice" climate maps for large areas, mainly under the leadership of two international, multi-institutional projects—CLIMAP and COHMAP (Webb, Bartlein, and Kutzback 1987).

Limitations of Proxy Indicators. With the growing use of proxy indicators has come a growing awareness of their limitations, such as over- and under-representation, lagged responses, and "no-analog" situations. Holliday (1987), for instance, has questioned the late-Pleistocene boreal forest reconstruction for the Southern High Plains, which is based on types of pollen that tend to be more-easily preserved and able to travel long distances, but which is not supported by other types of proxy indicators.

Another limitation is the response of many indicators to more than one environmental variable. For example, the oxygen-isotope record in deep-sea cores responds to the removal of water and thus to the volume of ice on land, as well as to local water temperatures. This ice-volume/sea-level record is now being reappraised by glacial and coastal geomorphologists (e.g., Occhietti 1983).

Another example is tree-ring width, which in most areas represents a response to evapotranspiration. The reconstruction of summer precipitation from trees in the Great Lakes region, for instance, is poor at times of cool droughts when the trees are able to compensate for the lack of precipitation (Brinkmann 1988).

There is also the question of the seasonality of proxy indicators, versus the seasonality of climate change. It has recently been shown that the orbital-parameter values during the early Holocene—a time considered to have been the warmest of the past 10,000 years—were such that insolation was indeed above the present level in summer, but below the present level in winter. Many proxy indicators, particularly biological ones, respond to *summer* conditions. Paleoclimatologists have barely begun to expand upon the concept of seasonality (Knox 1988).

Causes of Climate Change

The immediate causes of temperature and precipitation variations are variations in atmospheric-circulation patterns. Geographers have played a major role in elucidating these dynamic relationships—a primary object of synoptic climatology. Circulation patterns prior to the period of instrumentation have been developed using analog methods (e.g., Nicholson and Flohn 1980; Barry 1982) and numerical modeling (discussed above).

Several exciting developments in the understanding of causal mechanisms of climate change have occurred during recent decades. These include the discovery of the importance of orbital-frequency rhythms, of a sulfuric acid link between volcanic eruptions and climate, of a solar-climate oscillation on the order of perhaps 1000 years, and of a solar cycle in stratospheric winds. The involvement of geographers in this aspect of climate change has, however, been relatively small. The few exceptions include the work on volcanic eruptions/climate links by Bryson and Goodman (1980) and Hirschboeck (1980), and some climate-modeling work discussed earlier.

The risk of an approaching carbon dioxide-induced climatic change has made this the most-widely discussed factor. But predictions of the future "greenhouse" climate are plagued with numerous uncertainties. What, for instance, caused the mean hemispheric temperature to *decrease* after 1940, while the atmospheric carbon dioxide concentration continued to *increase*? Because of a lack of an unequivocal signal in the temperature record, and because of problems with the representativeness of the hemispheric mean, easily monitored climate parameters that are expected to give early signals of the greenhouse effect are receiving attention—for example, the freeze-up and break-up of lakes (Barry 1985). Efforts are also being made to improve the database that is used to compute the hemispheric and global means (Jones et al. 1986; Bradley et al. 1987).

Future Research

Geographers have contributed significantly to many areas in the study of climate change. A few important topics were touched upon only briefly or not at all in this review, because the involvement of geographers in these areas has so far been relatively small. These include a better understanding of the climate message in proxy indicators (including seasonality), whether and why abrupt climate changes have occurred, and examinations of causal mechanisms and their interactions. Geographers have much to contribute to these and other important topics. The decade ahead will be an exciting one for climate-change research.

APPLIED CLIMATOLOGY

Definitions of applied climatology range from those in which it is a narrowly circumscribed provision of the operational needs of agricultural and technological pursuits (Landsberg and Jacobs 1951), to a broader inclusion of the operational needs of the physical environment and the influence of climate upon it (Oliver 1973). In terms of current trends, the former is considered too restrictive, while components of the latter are addressed in specialty subjects such as climatic geomorphology and urban climatology.

Smith (1987) provides a current definition: applied climatology concerns "the use of both archived and real-time climatic information to solve a variety of social, economic, and environmental problems." This definition corresponds closely to one of the basic categories encompassed by the National Climate Program Act, the goal of which is to "assist the Nation and the world to understand and respond to natural and man induced climate processes and their implications" (U.S. Congress 1978). One category of the program, Climate Impacts Assessment (CIA), calls for identification of procedures to evaluate climate's effects on society, the economy, and the environment.

The multidisciplinary aspects of applied climatology are illustrated in many ways. Entries in *The Encyclopedia of Climatology* (Oliver and Fairbridge 1987) relate climate to a host of subjects, including architecture, art, crime, therapy, geomorphology, and soils. The interrelationships between meteorology and the human environment are dealt with at length in Houghton (1985). The Climate Applications Branch of the Canadian Climate Center has an active research program which produces many items of high utility (e.g., Thomas 1981). An annual review of international research in applied climatology is provided in *Progress in Physical Geography* (Musk 1987).

Realms of Study

A mere listing of the realms of study of applied climatology does not indicate the thrust of recent research in the field. Together with traditional studies, the use of climatic data to produce scenarios, models, and climate-impact studies has taken on appreciable importance. A text published by the Scientific Committee on Problems of the Environment (Kates, Ausubel, and Berberian 1985) deals specifically with such research. The book's organization, appropriately supplemented, provides an orderly framework in which to view the realms of applied climatology (Table 1). The divisions shown are largely self-explanatory and are discussed below; the "technical" use of climatic data is grouped under the title "Technoclimatology," a term used by Landsberg (1958) and reintroduced here. Table 1 cannot show the dimensions of time and scale that are basic to applied climatic studies, but these factors are well illustrated in the citations presented.

The Biophysical Environment

Physical geographers have long studied the biophysical environment, and this realm is of major importance in applied climatic studies. Climatic studies related to agriculture exemplify the various scales of investigation used in applied climatology. The spatial investigations vary from global assessments (Hekstra and Liverman 1986), to subcontinental (Liverman et al. 1986) and local (Skaggs and Baker 1985). Similarly, the time scale deals with the past, present, and future, as illustrated by the research of Warrick (1980), Parry and Carter (1985), and Hare (1980). Topics assessed within these frameworks are themselves varied, ranging from the climatic inputs into agribusiness (Lamb, Sonka, and Changnon 1985) to the role of climatic variations in agricultural productivity (Sonka et al. 1982).

Climatic aspects of water resources have also been afforded considerable attention (see the chapter on water resources). Studies have considered almost all atmospheric components of the hydrologic cycle, including snowmelt (Butler 1986; Gardner 1986), changing precipitation patterns (Cohen 1986), extreme events (Posey 1980), and hydroclimatic modeling (Shelton 1985). Water-balance studies continue to provide significant information, as shown by Mather (1978) and Muller (1982). Comprehensive surveys of the nature of climate/hydrologic research have been provided by Miller (1977) and others.

In recent years, however, the widely publicized human suffering associated with world droughts has led to an emphasis upon applied climatic studies associated with

Table 1
Applied climatology

Realm	Examples of Systems Related to Climate
Biophysical Environment	Agriculture, animal husbandry, fisheries, water resources, energy resources.
Socioeconomic Environment	Health, nutrition, population, social conditions, economic conditions, perception.
Technoclimatology	Design specifications, including housing, clothing, air-pollution control, construction, aviation, transportation.
Integrated Assessments	Models and scenarios (e.g. socioeconomic/biophysical interaction), impact studies.

Source: After Kates, Ausubel, and Berberian 1985.

drought. Wilhite, Easterling, and Ward (1987) have produced a comprehensive account of drought that contains both meaningful reviews of methodology and specialized case studies of selected world areas. Perspective on drought in the U.S. is exemplified by McGregor (1985) and Shelton (1984). As part of the thrust toward a better understanding of drought, studies on desertification and land degradation also appear frequently in the literature (see *Climatic Change* 9 (1): 2).

Geographers have effectively researched the applied climatology of a number of controversial issues associated with biophysical problems of the contemporary scene. Acid rain (Davis 1984; Hare 1984) is a significant example. There is little doubt that other topics—with the impact of ozone depletion foremost—will be a focus of applied climatic research.

Energy resources also provide a topic for applied climatic research. Solar-energy availability (Burt 1985), residential energy demands (Suckling and Stackhouse 1983), and energy conservation (Sanderson 1983) provide apt examples. Atomic energy has attracted appreciable attention, especially since the wide publicity of nuclear power station accidents. Interest in this topic has certainly been enhanced since the introduction of the concept of "nuclear winter" (Schumacher, Balling, and Cerveny 1984; Bach 1986).

The Socioeconomic Environment

In recent years, researchers have been increasingly concerned with the impacts of climatic extremes upon the socioeconomic environment. For example, Changnon (1979) studied the impacts of severe winters, McFarlane and Waller (1976) examined hot summers, and Riebsame et al. (1986) investigated a variety of hazards. Other avenues of research include assessment of climate in relation to tourism, crime (Harries, Stadler, and Zdorkowski 1985), consideration of various bioclimatological indices (Kalkstein and Valimont 1986), and human responses to future climates (Butzer 1983). The great surge of perception studies in geography of the 1970s did not impinge greatly upon work in applied climatology. While there are notable exceptions (Kates et al. 1984; Taylor, Stewart, and Downton 1987), they are surprisingly few in relation to the potential.

Recent studies of climatic impacts upon social conditions have tended to deal with effects of climatic variations in underdeveloped areas. The impacts of recent extensive droughts have, as already noted, been the topic of many publications. Climatic aspects of the economic environment are often subsumed under studies of individual industries or activities, rather than specifically as studies of economics. For example, articles deal with the potential costs of a climatic event, rather than a study of the impact upon an entire region. The dynamics of development have been related to climate (Biswas 1984), as have the potential impacts of climatic change. More-detailed studies that include the direct costs are often found in technoclimatic research.

Technoclimatology

Technoclimatology is the use of climatic data as input into a given technical problem. Not surprisingly, the academic literature on the topic is quite limited. Much of the work completed is used on a local basis for solving a particular problem. Often, the applied data are used in legal situations, in suits ranging from flooding problems to roof collapse resulting from snowfall. The need in such areas is considerable, and has given rise to specialists who label themselves *forensic climatologists*.

Despite the often rather specialized nature of technoclimatology, there are published works that provide overviews of the field (e.g., Thomas 1981; Oliver 1981). Studies are sometimes reported in publications resulting from *Applied Geography* symposia.

Depending upon their nature, air-pollution studies may fall under the rubric of technoclimatology. For example, the concept of the airshed (Greenland and Carleton 1982) provides an areal basis for applied studies, while methodologies for dealing with specific air pollution scenarios are to be found in a number of sources (Kalkstein and Corrigan 1986; Muller and Jackson 1985).

Integrated Assessments

One of the most marked features of published research in recent years is the increasing interest in what may be termed *integrated studies*. In these, interconnections are sought between identified events in terms of their potential causation and eventual impact. Such research often requires the construction of models or scenarios. Examples include potential impacts resulting from El Niño/southern oscillation events (Glantz, Katz, and Krenz 1987) and reconstruction and assessment of historic eras.

Climatic change and variation provide the basis for a number of integrated assessments. Clearly, the scenario "What would happen if climate were to change. . ." presents many potential avenues for research, as seen in the wide variety of papers. For example, Warrick and Riebsame (1983) deal with the possible impacts of climatic change caused by addition of carbon dioxide to the atmosphere; Bowden et al. (1981) examine the influence of climatic fluctuations on human populations; while Liverman (1987) relates a global food model to possible climatic changes.

The charge of the National Climate Program Act has resulted in many symposia, meetings, and taskforces. Most of these are held under titles other than geography, but the contributions of geographers are significant. In many instances, the integrated approach to climate-impact studies requires an umbrella organization to bring together researchers from many disciplines. This trend certainly will continue, and it is anticipated that American geographers will be active participants in the research.

Future Trends

The future holds many exciting possibilities for the applied climatologist. With energy and water shortages looming, research in the biophysical sector will undoubtedly continue. The role of climate in economic decision making and government policymaking presents limitless opportunities for research, not only in impact studies, but also in climatic-change information, to evaluate how future climates will modify or influence environments and activities. There is little doubt that integrated studies will emerge as a main research thrust. The key to such advancement is interaction between climatologists and specialists in other disciplines. This is already evident in the current literature; it will become more so in the years ahead.

CONCLUSIONS

As clearly indicated in this chapter, climatic research over the past decade is notable for both quality and quantity. The surge of interest in climate and the immediacy of problems that need to be solved causes both the present and the future to offer exciting prospects for geographers engaged in climatic research. Suggestions relating to future research and trends within each area of specialization have been presented, and need not be reiterated.

Of significance, however, is the common theme that identifies and calls for close interaction between climatology and related subjects. Ranging from awareness of developments in GIS to interaction with scientists of the International Geosphere-Biosphere Program, climatologists have the opportunity to play a significant role, not only in geography but in a wide range of topical research. It is a challenging future.

REFERENCES

Alijani, B., and Harman, J. R. 1985. Synoptic climatology of precipitation in Iran. *Annals of the Association of American Geographers* 75:404–16.

Alt, B. T. 1987. Developing synoptic analogs for extreme mass balance conditions on Queen Elizabeth Island ice caps. *Journal of Climate and Applied Meteorology* 26, 1605–23.

Anthes, R. A. 1983. Review of Regional models of the atmosphere in middle latitudes. *Monthly Weather Review* 111:1306–35.

Arnfield, A. J. 1976. Numerical modelling of urban surface radiative parameters. In *Papers in Climatology: the Cam Allen Memorial Volume,* ed. J. A. Davis, *Discussion Paper No. 7.* Hamilton, Ontario: Department of Geography, McMaster University.

———. 1982a. An approach to the estimation of the surface radiative properties and radiation budgets of cities. *Physical Geography* 3:97–112.

———. 1982b. Estimation of diffuse irradiance on sloping, obstructed surfaces: an error analysis. *Archiv für Meteorologie, Geophysik, und Bioklimatologie* B30, 303–320.

———. 1983. Modeling the diffuse components of radiation budgets: an examination of the validity of the isotropic assumption. *Modeling and Simulation* 14:677–82.

———. 1986. Estimating diffuse irradiance on organisms: a comparison of results from isotropic and anisotropic sky radiance models. *International Journal of Biometeorology* 30:201–22.

Atkinson, B. W. 1983. Numerical modelling of thermally-driven, mesoscale air flows involving the planetary boundary layer. *Progress in Physical Geography* 7:177–209.

Bach, W. 1984. CO_2-sensitivity experiments using general circulation models. *Progress in Physical Geography* 8:583–609.

———. 1986. The acid rain/carbon dioxide threat-control strategy. *Geojournal* 10:293–352.

Barry, R. G. 1979. Recent advances in climate theory based on simple climate models. *Progress in Physical Geography* 3:119–31.

———. 1982. Approaches to reconstructing the climate of the steppe-tundra biome. In *Paleoecology of Beringia,* eds. D. M. Hopkins, J. V. Matthews Jr., C. E. Schwegger, and S. B. Young. New York: Academic Press. 195–204.

———. 1985. The cryosphere and climate change. In *Detecting the Climatic Effects of Increasing Carbon Dioxide,* eds. M. C. MacCracken and F. M. Luther, pp. 111–48. DOE/ER-0235. Washington: U.S. Department of Energy.

———. 1987. Synoptic climatology. In *The Encyclopedia of Climatology,* eds. J. E. Oliver and R. W. Fairbridge, pp. 823–28. New York: Van Nostrand Reinhold.

———, and Perry, A. H. 1973. *Synoptic Climatology: Methods and Applications.* London: Methuen.

———; Elliott, D. L.; and Crane, R. G. 1981. The palaeoclimatic interpretation of exotic pollen peaks in Holocene records from the Eastern Canadian Arctic. *Review of Palaeobotany and Palynology* 33:153–67.

———; Kiladis, G. N.; and Bradley, R. S. 1981. Synoptic climatology of the western United States in relation to climatic fluctuations during the twentieth century. *Journal of Climatology* 1:97–113.

———; Henderson-Sellers, A.; and Shine, K. P. 1984. Climate sensitivity and the marginal cryosphere. In *Climate Processes and Climate Sensitivity,* eds. J. E. Hansen and T. Takahashi, pp. 221–37. Washington: American Geophysical Union.

———; Crane, R. G.; Schweiger, A.; and Newell, J. 1987. Arctic cloudiness in spring from satellite imagery. *Journal of Climatology* 7:423–51.

Bartlein, P. J.; Webb, III, T.; and Fleri, E. 1984. Holocene climatic change in the northern Midwest: Pollen-derived estimates. *Quaternary Research* 22:361–74.

Baum, W. A. 1953. Accrediting problems in meteorology. *Bulletin of the American Meteorological Society* 34:319–20.

Biswas, A. K., ed. 1984. *Climate and development*. Dublin: Tycooly International Publishers.

Blasing, T. J., and Lofgren, G. R. 1980. Seasonal climatic anomaly types for the North Pacific sector and western North America. *Monthly Weather Review* 108:700–19.

Bowden, M. J.; Kates, R. W.; Kay, P. A.; Riebsame, W. E.; Warwick, R. A.; Johnson, D. L.; Gould, H. A.; and Weiner, D. 1981. The effect of climate fluctuations on human populations: Two hypotheses. In *Climate and history,* ed. T. M. Wigley, pp. 497–513. Cambridge: Cambridge Univ. Press.

Bradley, R. S., and England, J. 1979. Synoptic climatology of the Canadian High Arctic. *Geografiska Annaler* 61:187–201.

Bradley, R. S.; Diaz, H. F.; Eischeid, J. K.; Jones, P. D.; Kelly, P. M.; and Goodess, C. M. 1987. Precipitation fluctuations over Northern Hemisphere land areas since the mid-19th century. *Science* 237:171–75.

Brazel, A. J., and Nickling, W. G. 1986. The relationship of weather types to dust storm generation in Arizona (1965–1980). *Journal of Climatology* 6:255–75.

Brinkmann, W. A. R. 1981. Sea level pressure patterns over eastern North America, 1899–1976. *Monthly Weather Review* 109:1305–18.

———. 1983. Secular variations of surface temperature and precipitation patterns over the Great Lakes region. *Journal of Climatology* 3:167–77.

———. 1985. The Northern Hemisphere temperature curve: Representativeness and interpretive fallacies. *Physical Geography* 5:165–85.

———. 1988. Comparison of two indicators of climate change. In *Proceedings of the annual meeting of the Association of American Geographers,* p. 20. Washington: Association of American Geographers.

Brown, J. A. 1987. Operational numerical prediction. *Reviews of Geophysics* 25:312–22.

Bryson, R. A., and Goodman, B. M. 1980. Volcanic activity and climatic changes. *Science* 207:1041–44.

Burt, J. E. 1985. Interpolation errors and spatial resolution of the United States radiation network. *Physical Geography* 6:230–46.

———; Hayes, J. T.; O'Rourke, P. A.; Terjung, W. H.; and Todhunter, P. E. 1981. A parametric crop water use model. *Water Resources Research* 17:1095–1108.

———; O'Rourke, P. A.; and Terjung, W. H. 1982. Viewfactors leading to the simulation of human heat stress and radiant exchanges: an algorithm. *Archiv füer Meteorologie, Geophysik, und Bioklimatologie* B30:321–31.

Butler, D. R. 1986. Snow-avalanche hazards in Glacier National Park, Montana: Meteorological and climatological aspects. *Physical Geography* 7:72–87.

Butzer, K. W. 1980. The Holocene lake plain of north Rudolph, East Africa. *Physical Geography* 1:42–58.

———. 1983. Human response to environmental change in the perspective of future, global climate. *Quaternary Research* 19:279–92.

Byrne, R.; Granger, O.; and Monteverdi, J. 1982. Recent rainfall trends on the margins of the subtropical deserts: A comparison of selected Northern Hemisphere regions. *Quaternary Research* 17:14–25.

Carleton, A. M. 1981. Monthly variability of satellite-derived cyclonic activity for the Southern Hemisphere winter. *Journal of Climatology* 1:21–38.

———. 1983. Variations in Antarctic sea ice conditions and relations with Southern Hemisphere cyclone activity winters 1973–77. *Archiv füer Meteorologie, Geophysik, und Bioklimatologie* B32:1–22.

———. 1985a. Satellite climatological aspects of the "polar low" and "instant occlusion." *Tellus*. Dynamic Meteorology and Oceanography A37:433–50.

———. 1985b. Synoptic and satellite aspects of the Southwestern U.S. summer "monsoon." *Journal of Climatology* 5:389–402.

———. 1985c. Synoptic cryosphere-atmosphere interactions in the Northern Hemisphere from DMSP Image Analysis. *International Journal of Remote Sensing* 6:239–61.

———. 1987. Summer circulation climate of the American Southwest, 1945–1984. *Annals of the Association of American Geographers* 77:619–34.

Catchpole, A. J. W., and Halpin, J. 1987. Measuring summer sea ice severity in eastern Hudson Bay 1751–1870. *Canadian Geographer* 31:233–44.

Caviedes, C. N. 1984. El Niño 1982–83. *Geographical Review* 74:267–90.

Cerveny, R. S., and Balling, R. C. 1984a. CONSTABLE: A simple one-dimensional climate model for climatologists in geography. *Professional Geographer* 36:188–96.

———, and ———. 1984b. Numerical experiments with the global energy balance model CONSTABLE. *Modeling and Simulation* 15:321–26.

———, and ———. 1985. Energy balance climate modeling: An alternate dynamic parameterization. *Journal of Climatology* 5:423–31.

Changnon, S. A. 1979. How a severe winter impacts on individuals. *Bulletin of the American Meteorological Society* 60:110–14.

———. 1983. Trends in floods and related climate conditions in Illinois. *Climatic Change* 5:341–63.

Cohen, S. J. 1983. Classification of 500 mb height anomalies using obliquely rotated principal components. *Journal of Climate and Applied Meteorology* 22:1975–88.

———. 1986. Climatic change, population growth and their effects on Lakes water supplies. *Professional Geographer* 38:317–23.

Conkey, L. E. 1986. Red spruce tree-ring widths and densities in eastern North America as indicators of past climate. *Quaternary Research* 26:232–43.

Crane, R. G. 1978. Seasonal variations of sea ice extent in the Davis Strait-Labrador Sea area and relationships with synoptic-scale atmospheric circulation. *Arctic* 31:434–47.

———, and Barry, R. G. 1988. Comparison of the MSL synoptic pressure patterns of the Arctic as observed and simulated by the GISS General Circulation Model. *Meteorology and Atmospheric Physics*. In press.

Davies, J. A.; Schertzer, W.; and Nunez, M. 1975. Estimating global radiation. *Boundary-Layer Meteorology* 9:33–52.

Davis, A. M. 1977. The prairie-deciduous forest ecotone in the upper Middle West. *Annals of the Association of American Geographers* 67:204–13.

Davis, J. K. 1984. Environmental law and the export of pollution. *Journal of Geography* 83:154–58.

Dewey, K. F., and Heim, R., Jr. 1982. A digital archive of Northern Hemisphere snow cover November 1966 through December 1980. *Bulletin of the American Meteorological Society* 63:1132–41.

Diaz, H. F. 1986. An analysis of twentieth century climate fluctuations in northern North America. *Journal of Climate and Applied Meteorology* 25:1625–57.

Dorn, R. I.; Turrin, B. D.; Jull, A. J. T.; Linick, T. W.; and Donahue, D. J. 1987. Radiocarbon and cation-ratio ages for rock varnish on Tioga and Tahoe morainal boulders of Pine Creek, eastern Sierra Nevada, California, and their paleoclimatic implications. *Quaternary Research* 28:38–49.

Dozier, J. 1980. A clear-sky spectral solar radiation model for snow-covered mountainous terrain. *Water Resources Research* 16:709–18.

Elliott-Fisk, D. L. 1983. The stability of the northern Canadian tree limit. *Annals of the Association of American Geographers* 73:560–76.

Finger, F. G.; Laver, J. D.; Bergman, K. H.; and Patterson, V. L. 1985. The Climate Analysis Center's user information service. *Bulletin of the American Meteorological Society* 66:413–20.

Gardner, J. S. 1986. Snow as a resource and hazard in early twentieth-century mining. *Canadian Geographer* 30:217–28.

Glantz, M.; Katz, R. W.; and Krenz, M. 1987. *The societal impacts associated with the 1982–83 worldwide climate anomalies.* Environmental and Societal Impacts Group. Boulder, CO: National Center for Atmospheric Research.

Granger, O. 1979. Increasing variability in California precipitation. *Annals of the Association of American Geographers* 69:533–43.

Graumlich, L. J. 1987. Precipitation variation in the Pacific Northwest (1675–1975) as reconstructed from tree rings. *Annals of the Association of American Geographers* 77:19–29.

Greene, G. M., and Outcalt, S. I. 1985. A simulation model of river ice cover thermodynamics. *Cold Regions Science and Technology* 10:251–62.

Greenland, D., and Carleton, A. M. 1982. The "Airshed" concept and its application in complex terrain. *Physical Geography* 3:169–79.

Hare, F. K. 1960. Climatology and the geography student. *Canadian Geographer* 16.

———. 1980. Climate and agriculture: The uncertain future. *Journal of Soil and Water Conservation* 35:112–15.

———. 1984. Changing climate and human response: The impacts of recent events upon climatology. *Geoforum* 15:383–94.

Harman, J. R. 1987. Mean monthly North American anticyclone frequencies, 1950–79. *Monthly Weather Review* 115:2840–48.

Harmon, R. S.; Thompson, P.; Schwarcz, H. P.; and Ford, D. C. 1978. Late Pleistocene paleoclimates of North America as inferred from stable isotope studies of speleothems. *Quaternary Research* 9:54–70.

Harries, K. D.; Stadler, S.; and Zdorkowski, T. 1985. Seasonality and assault: Explorations in interneighborhood variations, Dallas, 1980. *Annals of the Association of American Geographers* 74:590–604.

Harrington, J. A., and Brown, B. J. 1985. A synoptic climatology of depressions in warm season precipitation profiles from the upper Middle West. *Physical Geography* 5:186–97.

Hayes, J. T.; O'Rourke, P. A.; Terjung, W. H.; and Todhunter, P. E. 1982. A feasible crop yield model for worldwide international food production. *International Journal of Biometeorology* 26:239–57.

Hekstra, G. P., and Liverman, D. M. 1986. Global food futures and desertification. *Canadian Geographer* 9:59–66.

Henderson-Sellers, A. 1980. A simple numerical simulation of urban mixing depths. *Journal of Applied Meteorology* 19:215–18.

Hirschboeck, K. K. 1980. A new worldwide chronology of volcanic eruptions. *Palaeogeography, Palaeoclimatology, Palaeoecology* 29:223–41.

Hobgood, J. S. 1986a. The influence of relative humidity on the development of tropical cyclones. *Physical Geography* 7:283–91.

———. 1986b. A possible mechanism for the diurnal oscillations of tropical cyclones. *Journal of the Atmospheric Sciences* 43:2901–22.

———, and Cerveny, R. S. 1988. Ice-age hurricanes and tropical storms. *Nature* 333:243–45.

Holliday, V. T. 1987. A reexamination of late-Pleistocene boreal forest reconstructions for the southern High Plains. *Quaternary Research* 28:238–44.

Houghton, D. D., ed. 1985. *Handbook of applied meteorology*. New York: Wiley.

Howarth, D. A. 1983. An analysis of the variability of cyclones around Antarctica and their relationship to sea-ice extent. *Annals of the Association of American Geographers* 73:519–37.

———. 1986. An analysis of the water vapor flux divergence field over the Southern Hemisphere. *Annals of the Association of American Geographers* 76:190–207.

Johnson, D. M. 1980. An index of Arizona summer rainfall developed through eigenvector analysis. *Journal of Applied Meteorology* 19:849–56.

Jones, J. R.; Spayne, R. W.; and Schenk, R. 1984. Early Holocene conditions at the western margin of Cape Cod Bay. *Physical Geography* 4:56–65.

Jones, P. D.; Raper, S. C. B.; Bradley, R. S.; Diaz, H. F.; Kelly, P. M.; and Wigley, T. M. L. 1986. Northern Hemisphere surface air temperature variations: 1851–1984. *Journal of Climate and Applied Meteorology* 25:161–79.

Kalkstein, L. S., and Corrigan, P. 1986. A synoptic climatological approach for geographical analysis: assessment of sulfur dioxide concentrations. *Annals of the Association of American Geographers* 76:381–95.

———, L. S., and Valimont, K. M. 1986. An evaluation of summer discomfort in the United States using a relative climatological index. *Bulletin of the American Meteorological Society* 67:842–48.

———; Tan, G.; and Skindlov, J. A. 1987. An evaluation of three clustering procedures for use in synoptic climatological classification. *Journal of Climate and Applied Meteorology* 26:717–30.

Kalnicky, R. A. 1987. Seasons, singularities, and climatic changes over the mid latitudes of the Northern Hemisphere during 1899–1969. *Journal of Climate and Applied Meteorology* 26:1496–1510.

Kates, R. W.; Changnon, S. A.; Karl, T. R.; Riebsame, W. E.; and Easterling, W. E. 1984. *The climate impact, perception and adjustment experiment (CLIMPAX)*. Worcester, MA: Climate and Society Research Group, Center for Technology, Environment and Development, Clark University.

———; Ausubel, J. H.; and Berberian, J. H., eds. 1985. *Climate impact assessment: Studies of the interaction of climate and society*. New York: Wiley, on behalf of the Scientific Committee on Problems of the Environment (SCOPE), Vol. 27.

Kay, P. A., and Andrews, J. T. 1983. Re-evaluation of pollen-climate transfer functions in Keewatin, Northern Canada. *Annals of the Association of American Geographers* 73:550–59.

———, and Johnson, D. L. 1981. Estimation of Tigris-Euphrates streamflow from regional paleoenvironmental proxy data. *Climatic Change* 3:251–63.

Keen, R. A. 1980. Temperature and circulation anomalies in the eastern Canadian Arctic, summer 1946–76. Occasional Paper No. 34, Boulder, CO: Institute for Arctic and Alpine Research, University of Colorado.

Key, J., and Crane, R. G. 1986. A comparison of synoptic classification schemes based on 'objective' procedures. *Journal of Climatology* 6:375–88.

Kirchhofer, W. 1974. Classification of European 500 mb patterns. *Schweizer Meteorologische Zentralanstalt*, Arbeits Nr. 43, Zurich: 16 pp. (Report from occasional series.)

Knox, J. C. 1988. Climatic influences on upper Mississippi valley floods. In *Flood Geomorphology,* eds. V. R. Baker, R. C. Kochel, and P. C. Patton, pp. 279–300. New York: Wiley.

Knox, J. L., and Hay, J. E. 1985. Blocking signatures in the Northern Hemisphere: frequency distribution and interpretation. *Journal of Climatology* 5:1–6.

Kuivinen, K. C., and Lawson, M. P. 1982. Dendroclimatic analysis of birch in south Greenland. *Arctic and Alpine Research* 14:243–50.

Lamb, P. J.; Sonka, S. T.; and Changnon, S. A. 1985. Use of climate information by the U. S. agribusiness. *NOAA Technical Report*. Washington: NOAA.

Landsberg, H. 1958. *Physical Climatology*. Dubois, PA: Gray Printing Co.

———, and Jacobs, W. C. 1951. Applied climatology. In *Compendium of Meteorology,* ed. T. F. Malone, pp. 976–92. Boston: American Meteorological Society.

Lawson, M. P.; Kuivinen, K. C.; and Balling, R. C., Jr. 1982. Analysis of the climatic signal in the South Dome, Greenland ice core. *Climatic Change* 4:375–84.

———, and Stockton, C. W. 1981. Desert myth and climatic reality. *Annals of the Association of American Geographers* 71:527–35.

LeDrew, E. F. 1984. The role of local heat sources in synoptic activity within the Polar Basin. *Atmosphere-Ocean* 22:309–27.

———. 1985. The dynamic climatology of the Beaufort to Laptev Sea sector of the Polar Basin for the winters of 1975 and 1976. *Journal of Climatology* 5:253–72.

Liverman, D. M. 1987. Forecasting the impact of climate on food systems: model testing and model linkage. *Climatic Change* 11:267–85.

———; Terjung, W. H.; Hayes, J. T.; and Mearns, L. O. 1986. Climatic change and grain corn yields in the North American Great Plains. *Canadian Geographer* 9:193–220.

Lorenz, E. W. 1976. Non-deterministic theories of climatic change. *Quaternary Research* 6:495–507.

Luckman, B. H., and Kearney, M. S. 1986. Reconstruction of Holocene changes in alpine vegetation and climate in the Maligne Range, Jasper National Park, Alberta. *Quaternary Research* 26:244–61.

Lund, I. A. 1963. Map-pattern classification by statistical methods. *Journal of Applied Meteorology* 2:5665.

Maddox, R. A. 1987. Mesoscale and severe storm meteorology. *Reviews of Geophysics and Space Physics* 25:329–56.

Madsen, D. B., and Currey, D. R. 1979. Late Quaternary glacial and vegetation changes, Little Cottonwood Canyon Area, Wasatch Mountains, Utah. *Quaternary Research* 12:254–70.

Manabe, S., ed. 1985. Issues in atmospheric and oceanic modeling. Part A: Climate dynamics, p. 591. Part B: Weather dynamics, p. 432. *Advances in Geophysics* 28.

Mass, C. F.; Edmon, H. J.; Friedman, H. L.; Cheney, M. E.; and Recker, E. E. 1987. The use of compact discs for the storage of large meteorological and oceanographic data sets. *Bulletin of the American Meteorological Society* 68:1556–58.

Mather, J. R. 1978. *The climatic water balance in environmental analysis*. Lexington, MA: D. C. Heath.

McFarlane, A., and Waller, R. E. 1976. Short term increases in mortality during heatwaves. *Nature* 264:434–36.

McCormick, B. H.; DeFanti, T. A.; and Brown, M. D., eds. 1987. Visualization in scientific computing. Special Interest Group on Computer Graphics of the Association for Computing Machinery. *Computers and Graphics* 21:6.

McGregor, K. M. 1985. The tourism climatic index: a method for evaluating world climates for tourism. *Canadian Geographer* 29:220–33.

Mearns, L. O.; Katz, R. W.; and Schneider, S. H. 1984. Extreme high-temperature events: Changes in their probabilities with changes in mean temperature. *Journal of Climate and Applied Meteorology* 23:1601–13.

Meehl, G. A. 1987. Interactions between the Asian monsoons, the tropical Pacific, and the Southern Hemisphere extratropics. Cooperative Thesis No. 106, University of Colorado and NCAR, Boulder, CO.

———, and van Loon, H. 1979. The seesaw in winter temperatures between Greenland and northern Europe. Part III. Teleconnections with lower latitudes. *Monthly Weather Review* 107:1095–1106.

Miller, D. H. 1977. *Water at the surface of the earth: An introduction to ecosystem hydrodynamics.* New York: Academic Press.

Moritz, R. E. 1979. Synoptic climatology of the Beaufort Sea coast, Alaska. Occasional Paper No. 30, Boulder, CO: Institute for Arctic and Alpine Research, University of Colorado.

Moses, T.; Kiladis, G. N.; Diaz, H. F.; and Barry, R. G. 1987. Characteristics and frequency of reversals in mean sea level pressure in the North Atlantic sector and their relationship to long-term temperature trends. *Journal of Climatology* 7:13–30.

Muller, R. A. 1982. The water budget as a tool for inventory and analysis of factors effecting variability and change of river regimen. In *The environment: Chinese and American views,* eds. L. C. J. Ma and A. Noble, pp. 171–86. New York: Methuen.

———, and Jackson, A. L. 1985. Estimates of climatic air quality potential at Shreveport, Louisiana. *Journal of Climate and Applied Meteorology* 24:293–301.

Munro, D. S., and Young, G. J. 1982. An operational net shortwave radiation model for glacier basins. *Water Resources Research* 18:220–30.

Musk, L. F. 1987. Applied climatology. *Progress in Physical Geography* 11:370–83.

Nicholson, S. E. 1981. Rainfall and atmospheric circulation during drought periods and wetter years in West Africa. *Monthly Weather Review* 109:2191–208.

———, and Flohn, H. 1980. African environmental and climatic changes and the general atmospheric circulation in late Pleistocene and Holocene. *Climatic Change* 2:313–48.

Nkemdirim, L. C., and Venkatesan, D. 1985. The nature of the distribution in space and time of the length of the annual frost free season in Canada, east of the Rockies. *Geografiska Annaler* 67A:1–12.

Nunez, M. 1980. The calculation of solar and net radiation in mountainous terrain. *Journal of Biogeography* 7:173–86.

O'Rourke, P. A., and Terjung, W. H. 1981a. Canopy leaf temperatures and energy budget components affected by cloud types and amounts. *Archiv füer Meteorologie, Geophysik, und Bioklimatologie* B29:265–80.

———, and ———. 1981b. Modelling of influence of cloud amounts and types on leaf net photosynthetic rates inside a mature canopy. *Photosynthetica* 15:317–29.

———, and ———. 1981c. Relative influence of city structure on canopy photosynthesis. *International Journal of Biometeorology* 25:1–19.

Occhietti, S. 1983. Laurentide ice sheet: Oceanic and climatic implications. *Palaeogeography, Palaeoclimatology, Palaeoecology* 44:1–22.

Oke, T. R. 1982. The energetic bases of the urban heat island. *Quarterly Journal of the Royal Meteorological Society* 108:1–24.

Oliver, J. E. 1973. *Climate and man's environment: An introduction to applied Climatology.* New York: Wiley.

———. 1981. *Climatology: Selected applications.* New York: Wiley.

———, and Fairbridge, R. W., eds. 1987. *Encyclopedia of Climatology.* New York: Van Nostrand Reinhold.

Outcalt, S. I. 1971. A numerical surface climate simulator. *Geographical Analysis* 3:379–93.

———. 1972. The development and application of a simple digital surface-climate simulator. *Journal of Applied Meteorology* 11:629–36.

Oviatt, C. G.; McCoy, W. D.; and Reider, R. G. 1987. Evidence for a shallow early or middle Wisconsin-age lake in the Bonneville Basin, Utah. *Quaternary Research* 27:248–62.

Parkinson, C. L., and Washington, W. M. 1979. A large scale numerical model of sea ice. *Journal of Geophysical Research* 84:311–37.

———, and Kellogg, W. W. 1979. Arctic sea ice decay simulated for a CO_2-induced temperature rise. *Climatic Change* 2:149–62.

———, and Herman, G. P. 1980. Sea-ice simulations based on fields generated by the GLAS GCM. *Monthly Weather Review* 108:2080–91.

Parry, M. L., and Carter, T. R. 1985. The effect of climatic variations on agricultural risk. *Climatic Change* 7:95–110.

Pielke, R. A. 1984. *Mesoscale Meteorological Model.* Orlando: Academic.

Posey, C. 1980. Heatwave. *Weatherwise* 33:112–16.

Rannie, W. F. 1983. Breakup and freezeup of the Red River at Winnipeg, Manitoba Canada in the 19th century and some climatic implications. *Climatic Change* 5:283–96.

Raphael, C., and Hay, J. E. 1984. An assessment of models which use satellite data to estimate solar irradiance of the earth's surface. *Journal of Climate and Applied Meteorology* 23:832–44.

Raveché, H. J.; Lawrie, D. H.; and Despain, A. M. 1987. *A National Computing Initiative: The Agenda for Leadership.* Philadelphia: Society for Industrial and Applied Mathematics.

Rayner, J. N. 1984. Simulation models in climatology. In *Spatial Statistics and Models,* ed. G. L. Gaile and C. J. Willmott, pp. 417–42. Boston: D. Reidel.

Reitan, C. H. 1979. Trends in the frequencies of cyclone activity over North America. *Monthly Weather Review* 107:1684–88.

Riebsame, W. E.; Diaz, H. F.; Moses, T.; and Price, M. 1986. The social burden of weather and climatic hazards. *Bulletin of the American Meteorological Society* 67:1378–85.

Rogers, J. C. 1978. Meteorological factors affecting interannual variability of summertime ice extent in the Beaufort Sea. *Monthly Weather Review* 106:890–97.

———. 1981a. The North Pacific Oscillation. *Journal of Climatology* 1:39–57.

———. 1981b. Spatial variability of seasonal sea level pressure and 500 mb height anomalies. *Monthly Weather Review* 109:2093–2106.

———. 1983. Spatial variability of Antarctic temperature anomalies and their association with the Southern Hemisphere atmospheric circulation. *Annals of the Association of American Geographers* 73:502–18.

———. 1984. The association between the North Atlantic Oscillation and the Southern Oscillation in the Northern Hemisphere. *Monthly Weather Review* 112:1999–2015.

Rowe, C. M., and Willmott, C. J. 1984. Solar irradiance on flat-plate collectors in urban environments. *Solar Energy* 33:343–51.

Rowntree, L. B. 1985. A crop-based rainfall chronology for pre-instrumental record southern California. *Climatic Change* 7:327–41.

Saltzman, B. 1985. Paleoclimatic modelling. In *Paleoclimate Analysis and Modeling,* ed. S. D. Hecht, pp. 341–96. New York: Wiley.

Sanderson, M. 1983. Heating degree day research in Alberta: Residents conserve natural gas. *Professional Geographer* 35:437–40.

Scharfer, G.; Barry, R. G.; Robinson, D. A.; Kukla, G.; and Servese, M. C. 1987. Large-scale patterns of snow melt on arctic sea ice mapped from meteorological satellite imagery. *Annals of Glaciology* 9:1–6.

Schlesinger, M. E., and Mitchell, J. F. B. 1987. Climate model simulations of the equilibrium climate response to increased carbon dioxide. *Reviews of Geophysics and Space Physics* 25:760–98.

Schumacher, J. A.; Balling, R. C.; and Cerveny, R. S. 1984. Impacts of nuclear wars: results from an energy balance global model. *Physical Geography* 5:199–205.

Sellers, P. J., and Lockwood, J. G. 1981. A computer simulation of the effects of differing crop types on the water balance of small catchments over long time periods. *Quarterly Journal of the Royal Meteorological Society* 107:395–414.

Shelton, M. L. 1984. Hydroclimatic analysis of severe drought in the Sacramento River basin, California. *Physical Geography* 5:262–86.

———. 1985. Modeling hydroclimatic processes in large watersheds. *Annals of the Association of American Geographers* 75:185–202.

Shine, K. P., and Henderson-Sellers, A. 1983. Modelling climate and the nature of climate models: a review. *Journal of Climatology* 3:81–94.

Shine, K. P.; Henderson-Sellers, A.; and Barry, R. G. 1984. Albedo-climate feedback: The importance of cloud and cryosphere rainability. In *New Perspectives in Climate Models,* ed. A. L. Berger, pp. 135–55, New York: Elsevier.

Singh, T., and Powell, J. M. 1986. Climatic variation and trends in the boreal forest region of western Canada. *Climatic Change* 8:267–78.

Skaggs, R. H. 1978. Climatic change and persistence in western Kansas. *Annals of the Association of American Geographers* 68:73–80.

———, and Baker, D. G. 1985. Fluctuations in the length of the growing season in Minnesota. *Climatic Change* 7:327–41.

Smagorinsky, J. 1982. Large-scale climate modelling and small-scale physical processes. In *Land Surface Processes in Atmospheric General Circulation Models,* ed. P. S. Eagleson, pp. 3–18. Cambridge: Cambridge University Press.

Smith, K. 1987. Applied Climatology. In *The Encyclopedia of Climatology,* eds. J. E. Oliver and R. W. Fairbridge, pp. 64–68. New York: Van Nostrand Reinhold.

Sonka, S. T.; Lamb, P. J.; Changnon, S. A.; and Wiboonpongse, A. 1982. Can climate forecasts for the growing season be valuable for crop producers: Some general considerations and an Illinois pilot study. *Journal of Applied Meteorology* 21:471–6.

Sorenson, C. J. 1977. Reconstructed Holocene bioclimates. *Annals of the Association of American Geographers* 67:214–22.

Stahle, D. W., and Cleaveland, M. K. 1988. Texas drought history reconstructed and analyzed from 1698 to 1980. *Journal of Climate* 1:59–74.

Steyn, D. G.; Oke, T. R.; and Hay, J. E. 1981. On scales in meteorology and climatology. *Climatological Bulletin* (McGill University): 30.

Suckling, P. W. 1986. Fluctuations of last spring freeze dates in the southeastern United States. *Physical Geography* 7:239–45.

———, and Hay, J. E. 1976. Modelling direct, diffuse, and total solar radiation for cloudless days. *Atmosphere* 14:298–308.

———, and Stackhouse, L. L. 1983. Impact of climatic variability on residential electrical energy consumption in the Eastern United States. *Archiv für Meteorologie, Geophysik, und Bioklimatologie* B33:219–27.

Tarleton, L. F. 1987. Persistence characteristics of 500 mb blocking and zonal flows in the mid-latitudes of the Northern and Southern Hemispheres. Ph.D. diss., University of Colorado, Boulder.

Taylor, J. G.; Stewart, T. R.; and Downton, M. W. 1987. Perceptions of drought in the Ogallala Aquifer region of the Western U. S. Great Plains. In *Planning for Drought,* eds. D. A. Wilhite, W. E. Easterling, and D. A. Wood. Boulder, CO: Westview Press.

Terjung, W. H.; Ji, H-Y; Hayes, J. T.; O'Rourke, P. A.; and Todhunter, P. E. 1984. Actual and potential yield for rainfed and irrigated maize in China. *International Journal of Biometeorology* 28:115–35.

———, and O'Rourke, P. A. 1981. Energy input and resultant surface temperatures for individual urban interfaces, selected latitudes and seasons. *Archiv füer Meteorologie, Geophysik, und Bioklimatologie* B29:1–22.

———, and ———. 1984. Summer energy budget simulation for a latitudinal transect along the east coast of the Americas. *Journal of Climatology* 4:111–22.

Thomas, M. K. 1981. *The nature and scope of climatic applications.* Downsview, Ontario: Atmospheric Environment Service, Canadian Climate Center Report 81–5.

Thompson, L. G.; Mosley-Thompson, E.; Bolzan, J. F.; and Koci, B. R. 1985. A 1500-year record of tropical precipitation in ice cores from the Quelccaya ice cap, Peru. *Science* 229:971–73.

Thornthwaite, C. W. 1961. The task ahead. *Annals of the Association of American Geographers* 51:345–56.

U.S. Congress. 1978. *The National Climate Program Act.* Conference report to accompany HR6669. Washington: U.S. Government Printing Office.

van Loon, H., and Rogers, J. C. 1978. The seesaw in winter temperatures between Greenland and Northern Europe. Part I: General Description. *Monthly Weather Review* 106:296–310.

———, and Williams, J. 1977. The connection between trends of mean temperature and circulation at the surface: Part IV. Comparison of the surface changes in the Northern Hemisphere with the upper air and with the Antarctic winter. *Monthly Weather Review* 105:636–47.

Warrick, R. A. 1980. Drought in the Great Plains: A case study on climate and society in the U.S.A. In *Climatic constraints and human activities,* ed. J. Ausebel and A. K. Biswas. Oxford: Pergamon Press.

———, and Riebsame, W. E. 1983. Societal response to CO_2 induced climatic change: Opportunities for research. In *Social science research and climatic change: An interdisciplinary appraisal,* eds. R. S. Chan, E. M. Boulding, and S. H. Schneider. Dordrecht, Holland: Reidel.

Washington, W. M., and Parkinson, C. L. 1986. *An Introduction to Three-Dimensional Climate Modeling.* Mill Valley, CA: University Science Books.

Weaver, R.; Morris, C.; and Barry, R. G. 1987. Passive microwave data for snow and ice research: planned products from the DMSP SSM/I system. *EOS* 68(39):769, 776–7.

Webb, III, T.; Bartlein, P. J.; and Kutzbach, J. E. 1987. Climatic change in eastern North America during the past 18,000 years: Comparisons of pollen data with model results. In *The Geology of North America. Vol. K-3, North America and adjacent oceans during the last deglaciation,* eds. W. F. Ruddiman and H. E. Wright, Jr., pp. 447–62. Boulder, CO: Geological Society of America.

Wilhite, D. A.; Easterling, W. E.; and Ward, D. A. 1987. *Planning for drought: Toward a reduction of societal vulnerability* Boulder, CO: Westview Press.

Williams, J. 1975. The influence of snowcover on the atmospheric circulation and its role in climatic change: An analysis based on results from the NCAR global circulation model. *Journal of Applied Meteorology* 14:137–52.

———. 1978a. A brief comparison of model simulations of glacial period maximum atmospheric circulation. *Palaeogeography, Palaeoclimatology, Palaeoecology* 25:191–8.

———. 1978b. The use of numerical models in studying climatic changes. In *Climatic Change,* ed. J. Gribbin, pp. 178–90. Cambridge: Cambridge University Press.

———; Barry, R. G.; and Washington, W. M. 1974. Simulation of the atmospheric circulation using the NCAR global circulation model with Ice Age boundary conditions. *Journal of Applied Meteorology* 13:305–317.

Willmott, C. J. 1981. On the validation of models. *Physical Geography* 2:184–94.

———. 1984. A primer on the representation of the terrestrial seasonal snow cyle within GCM's. *Modeling and Simulation* 15:327–32.

———. 1987a. Climatic Models. In *Encyclopedia of Climatology,* eds. J. E. Oliver and R. W. Fairbridge, pp. 584–90. New York: Van Nostrand Reinhold.

———. 1987b. Synoptic weather-map classification: Correlation versus sums-of-squares. *Professional Geographer* 39:205–7.

———; Ackleson, S. G.; Davis, R. E.; Feddema, J. J.; Klink, K. M.; Legates, D. R.; O'Donnell, J.; and Rowe, C. M. 1985. Statistics for the evaluation and comparison of models. *Journal of Geophysical Research* 90:8995–9005.

Wilson, M. F., and Henderson-Sellers, A. 1985. A global archive of land cover and soils data for use in general circulation climate models. *Journal of Climatology* 5:119–43.

Yarnal, B. 1984a. The effect of weather map scale on the results of a synoptic climatology. *Journal of Climatology* 4:482–93.

———. 1984b. Synoptic-scale atmospheric circulation over British Columbia in relation to the mass balance of Sentinel Glacier. *Annals of the Association of American Geographers* 74:375–92.

———. 1985a. Extratropical teleconnections with El Niño/Southern Oscillation (ENSO) events. *Progress in Physical Geography* 9:315–52.

———. 1985b. A 500 mb synoptic climatology of Pacific Northwest Coast winters in relation to climatic variability, 1948–1949. *Journal of Climatology*: 237–52.

———; Crane, R. G.; Carleton, A. M.; and Kalkstein, L. S. 1987. A new challenge for climate studies in geography. *Professional Geographer* 39(4):465–73.

———, and Diaz, H. F. 1986. Relationships between extremes of the Southern Oscillation and the winter climate of the Anglo-American Pacific Coast. *Journal of Climatology* 6:197–219.

———, and Kiladis, G. N. 1985. Tropical teleconnections associated with El Niño/Southern Oscillation (ENSO) events. *Progress in Physical Geography* 9:524–58.

———, and White, D. A. 1987. Subjectivity in a computer-assisted synoptic climatology. Part 1: Classification Results. *Journal of Climatology* 7:119–28.

Geomorphology

Richard A. Marston

Geomorphology is the science that studies landforms and soils in a historical and functional context, including morphological elements, physical and chemical processes, and materials of composition. If training in geography provides a special opportunity for making contributions to geomorphology, it is in studies that examine interactions between geomorphic and related biophysical systems, human interference in geomorphic systems, and predictive spatial (dynamic and statistical) modeling of morphological, cascading, and process-response systems in geomorphology. These opportunities closely parallel the "integration," "synthesis," and "prediction," respectively, described by Orme (1985) as the preferred goals of the geographer's science.

This chapter describes the impact of American geographers on the discipline of geomorphology. This is accomplished through:

- inventory of research in geomorphology published by American geographers;
- citation of selected works by American geographers in areas of geomorphology where a geographer's perspective has proved useful in the development of fundamental concepts; and
- discussion of the participation by geographers in professional geomorphological organizations.

This is not intended as a historical review of geographical geomorphology, but rather a report on the current aims, methods, and central ideas, as well as possible future directions that geographers may pursue. Trends in the studies of soils and physical

---

[1] Of the 1435 members in the Quaternary Geology and Geomorphology Division of the GSA (October 1987 data), approximately 8% are foreign-based, and 4% are geographers, according to the current chair of the Division. Of the 268 members in the Geomorphology Specialty Group of the AAG (May 1987 data), 3.4% are foreign-based, and 4.9% are geologists. Students comprise approximately 20% of the membership of both groups. Another 63 American geographers claim geomorphology as a specialty in the *Guide to Departments of Geography in the United States and Canada, 1987–88*. Therefore, geologists number 976 (79.4%) and geographers number 253 (20.6%) of the total number of 1229 geomorphologists practicing in the U.S. When foreign-based members of each group are included, geologists and geographers comprise 80.6% and 19.4%, respectively, of the total number of geomorphologists belonging to either the GSA or AAG.

environments of the Quaternary are included, reflecting the natural ties of these fields to geomorphology. American geographers in geomorphology are defined as those who practice their profession in the United States, a definition that includes approximately 5% of the total AAG membership and 21% of all practicing geomorphologists in the United States.[1]

Separating the contributions of American geographers from the contributions of geomorphologists having backgrounds other than geography, and from geographers who practice geomorphology outside the U.S., is a task that has not been pursued in the literature. This is a separation that is somewhat artificial in light of the increasing interaction between geography and other disciplines, and the increasing opportunities for exchange of information between colleagues in countries worldwide. Indeed, Vitek (1988) argued that institutions (professional societies, universities, and informal groups) and the transfer of enthusiasm and ideas by influential practitioners have become more important factors in the advance of the discipline than one's particular field of training.

A significant portion of the research by American geographers in geomorphology is concerned with topics that are not geographical in the sense of integration, synthesis, and predictive spatial modeling. Nevertheless, it is critical to solidifying the position of geomorphology within geography to identify a sample of those works that pursue the integration-synthesis-prediction themes, among other important contributions.

A number of manuscripts have been published recently that address various aspects of the field: the status of geomorphology (Pitty 1982, 1985; Hart 1986; Bashenina et al. 1987; Cooke 1987; Tricart 1987); the function of geomorphology within physical geography (Gregory 1985); and the position of geomorphology (and other subfields) in geography (Johnston 1985). Baker (1986) and Ritter (1988) provide the perspective of American geologists on the nature of geomorphology. Walker and Orme (1986) offer some clues as to the nature of geomorphology at an international level in the 1980s, based on themes evident in the First International Conference on Geomorphology. These reviews all paint an optimistic future for the discipline of geomorphology, but the influence of American geographers is not apparent.

The themes, developments, potentials, and needs of geomorphology in geography were stated at the beginning of the decade by Graf and his colleagues, although the contributions by American geographers were not singled out. Graf et al. (1980) applauded the new pluralism in geomorphology and pointed out the following major developments, all of which remain valid late in the decade:

> . . . an increased dependence on field research (an old tradition revisited), more realistic expectations from research tools, a resurgence of interest in man-land relationships with a concomitant dependence on the historical approach, an expanded appreciation of the hydrologic cycle, a reinvestigation of morphogenetic regions, new interest in planetary surfaces other than the earth's, more detailed investigations of event magnitude and frequency, and an involvement with applied problems.

In another key review of the discipline, Graf (1984) reported on the uneven distribution of field localities for major research efforts in American geomorphology since 1817. He warned of the dangers of applying geomorphic theories developed from research in one region to investigations in another region without recognition of and adjustment for this spatial bias, particularly with respect to mountain regions and fluvial systems.

Costa and Graf (1984) examined the numbers, professional affiliations, publications and locations of geomorphologists and Quaternary geologists in the United States, a review that provided the first statement of the relative contribution of geographers to geomorphology.

PUBLICATIONS

An update of the work by Costa and Graf (1984) reveals the number of geomorphology papers published in the period 1976–86, in refereed journals, printed in English, with international circulation, and containing at least one geomorphology paper per year (Table 1). The journals are ranked according to the "impact factor" calculated by the *Science Citation Index* and *Social Science Citation Index* for 1986. Note that the data do not distinguish between American and foreign-based geomorphologists.

The journals most heavily used by geographical geomorphologists are not among those with the highest impact factors, which limits the visibility of research. Unchanged from the earlier survey by Costa and Graf (1984), the *Professional Geogra-*

Table 1
Published geomorphology papers, 1976–86

Journal	Impact Factor[a]	By Geographer	By Geologist	Others	Sub-total	All Papers	% Geomorphology
Geology	2.182	17	73	18	108	1499	7.2
GSA Bulletin	2.163	51	150	26	227	1591	14.3
American J. Science	1.859	9	5	5	19	485	3.9
Journal of Geology	1.827	16	29	9	54	545	9.9
Quaternary Research	1.750	8	28	16	52	457	11.4
Annals AAG	1.397	29	6	1	36	341	10.6
Water Resour. Research	1.389	32	19	75	126	1826	6.9
Geogr. Annaler (A)	1.000	165	19	33	217	236	91.9
Arctic & Alpine Resear.	0.987	55	25	33	113	397	28.5
Earth Surf. Form & Pro.	0.817	215	53	96	364	387	94.1
Zeit. Geomorphologie	0.730	165	66	75	306	341	89.7
Prof. Geographer	0.678	13	0	0	13	399	3.2
Prog. in Phys. Geogr.[b]	0.661	130	7	21	158	275	57.5
Journal of Hydrology	0.586	44	6	35	85	1461	5.8
Env. Geol. & Water Sci.	0.432	4	23	26	53	209	25.4
Catena	0.193	79	7	22	108	222	48.6
Physical Geography[c]	0.097	38	12	0	50	120	41.6
TOTALS	(Mean = 1.103)	1070	528	491	2089	10,791	19.4

Source: Data for 1976–80 period from Costa and Graf (1984).
[a]Impact factor = number of citations from journal in 1984–85 divided by total number of citable items in that journal in 1984–85, from the *Science Citation Index* and *Social Science Citation Index* for 1986.
[b]Started publication in 1977.
[c]Started publication in 1980.

pher continued to publish a disproportionately low number of geomorphology papers, while the *Annals* published a disproportionately high number when compared to the percentage of AAG members who specialize in geomorphology. Geographers (American and foreign-based) accounted for 51.2% of the geomorphology publications sampled, a figure that is magnified when compared to their percentage of the practicing geomorphologists. American geographers contribute relatively few articles to the international journals, relying more heavily on those published in the U.S.

The topical areas of research by American geographers in geomorphology were investigated by examining *Geographical Abstracts* (formerly *Geo Abstracts*), *Part A: Landforms and the Quaternary, Part B: Climatology and Hydrology* (Hydrology sections only), and *Part E: Sedimentology*. This bibliographic service provides a compilation of abstracts from formal publications having international coverage. Abstracts are classified by major subtopics in geomorphology.

The names of 262 American geographers in geomorphology were checked for entries in *Parts A, B,* and *E* of *Geographical Abstracts* for the years 1980–87. Of the 262 names, only 147 were referenced at least once in the eight years. Many of the others are recent Ph.D.s in geography, or among those that list geomorphology as a specialty but are not members of the AAG Geomorphology Specialty Group. A total of 22,139 abstracts were compiled in *Geographical Abstracts, Part A* over this time period, and American geographers account for 322 entries (1.45%). An additional 347 abstracts by American geographers in geomorphology were cited in the hydrology and sedimentology parts. Twenty-two researchers account for 50% of the papers published by American geographers.

The classification of entries provides an indication of where American geographers are making contributions to the field, in the sense of quantity of published research (Figure 1). Quaternary studies that emphasize physical environments, slope studies, and fluvial studies account for the greatest concentration of effort. Receiving slightly less emphasis have been publications on periglacial form and process; runoff and hydraulics; beaches, barrier islands, and other coasts; glacial landforms and sediments; weathering and related pedogenesis; soils; and regional physiography.

Topics most closely associated with sedimentology have received some attention by American geographers. The published research in karst has been dominated by a few researchers, notably M. J. Day (e.g., 1983, 1985). A relatively small effort has been forthcoming in research involving neotectonics and structural control, deltas, estuaries, tidal flats, glaciology, and geomorphological mapping. The low number of entries for applied geomorphology is an artifact of the classification scheme used by the editors of *Geographical Abstracts*.

Research by American geographers on volcanic form and process has been rather sparse, in spite of the opportunities afforded by the 1980 eruption of Mount St. Helens. Rosenfeld (1980) and Rosenfeld and Cooke (1982) outlined the preeruption sequence of events, the events of 18 May (directional blast, landslides and debris flows, pyroclastic activity, formation of the lava dome), and posteruption landscape development. Yamaguichi (1984) used tree coring to date previous eruptions of Mount St. Helens at A.D. 1800 and A.D. 1480 that produced distinct tephra layers. Mount St. Helens was the site of the 1986 Conference of the American Geomorphological Field Group, generating a renewed interest in the recovery of landscapes disturbed by catastrophism.

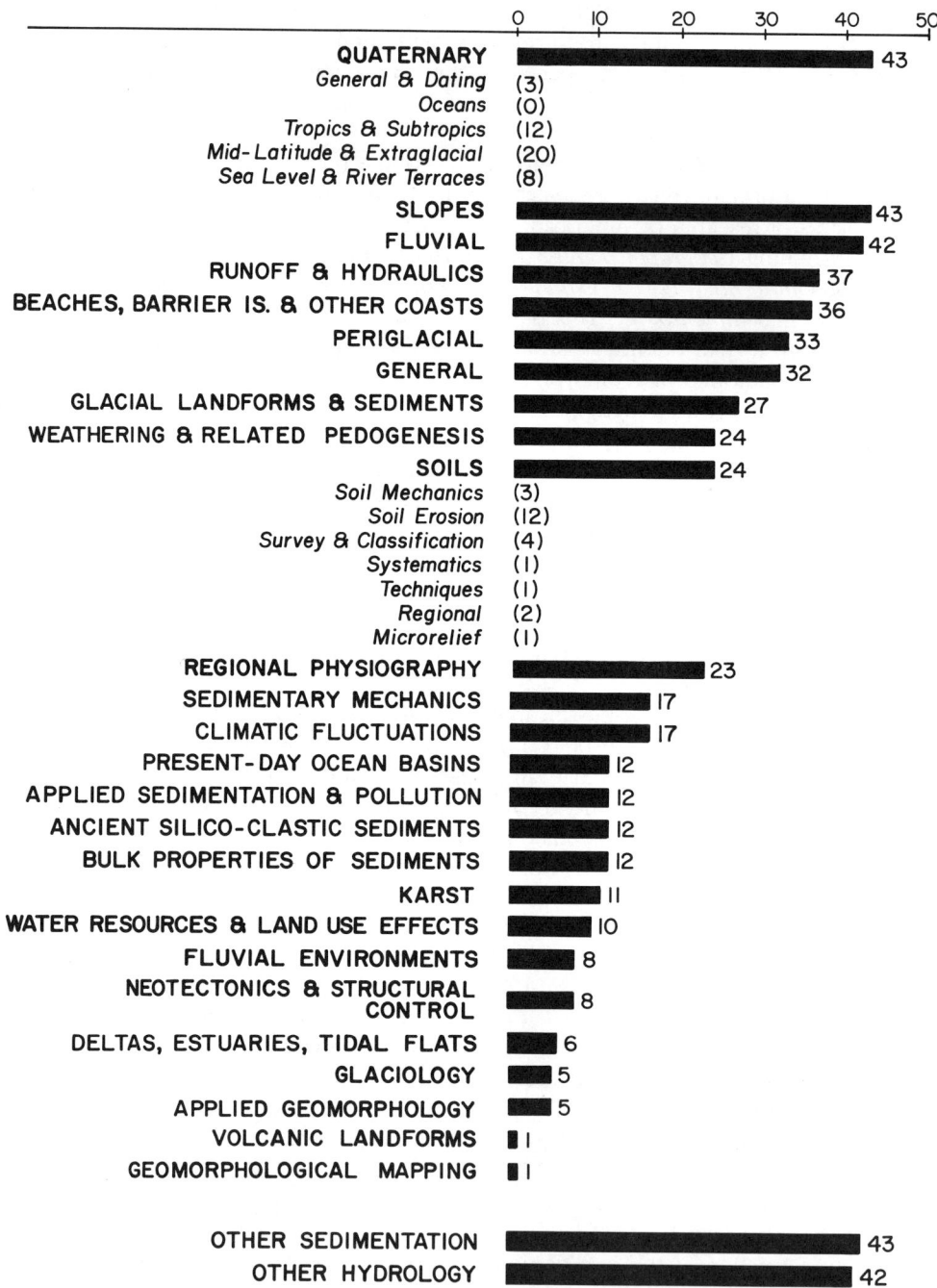

Figure 1
Number of abstracts authored by American geographers from 1980–87 in *Geographical Abstracts, Part A: Landforms and the Quaternary, Part B: Climatology and Hydrology,* and *Part E: Sedimentology*

A most-welcome addition to the literature on periglacial environments is the recent book edited by Giardino, Shroder, and Vitek (1987), *Rock Glaciers*. It is the most comprehensive work to date on the topic, with chapters by American geographers on a review of the knowledge base (Vitek and Giardino 1987b), rock glaciers as part of the alpine sediment cascade (Olyphant 1987), stratigraphy (Morris 1987), site characteristics and rock-glacier morphometry (Parson 1987), techniques of analysis (Shroder and Giardino 1987), movement dynamics (Shroder 1987), and geological engineering aspects of rock glaciers (Giardino and Vick 1987). The book also contains an extensive bibliography (Vitek and Giardino 1987a) that adds to its value as a guide to future research on what may be the largest landform of unconsolidated debris in alpine environment.

IMPORTANT CONTRIBUTIONS BY AMERICAN GEOGRAPHERS IN GEOMORPHOLOGY

It is possible to identify several areas where American geographers have had a significant impact on current thinking in geomorphology. These areas include the refinement and verification of geomorphic paradigms; the link among measurement, theory, and application in geomorphology (borrowing the theme of an IGU commission); separating the natural variation in geomorphic systems from the fluctuations triggered by human activities; and landscape ecology. It is also important to give credit to American geographers who have made important contributions to the development of techniques in geomorphology, regional geomorphology, and advances in Quaternary studies. These areas are addressed below.

Refinement and Verification of Paradigms in Geomorphology

Four paradigms affect the way in which geomorphologists in North America view landscapes (after Schumm 1977):

1. *Uniformity*—the laws of physics and chemistry control the operation of geomorphic processes today as they always have in the past.
2. *Landform Evolution*—landforms result from the interplay of the resisting framework, driving forces, and time.
3. *Complexity*—the interpretation of present-day landforms is complicated by changes in the resisting framework and driving forces over time, causing similar landforms to develop from different initial conditions, and landforms to respond to a change in the resisting framework and/or driving forces in opposite ways at different times.
4. *Thresholds*—abrupt changes may occur during landscape development, as geomorphic thresholds are exceeded; thresholds are values involving processes and/or forms that, when exceeded, initiate an episode of accelerated landscape change.

The paradigm of uniformity was developed in other branches of physical geology. It was extended to geomorphology by Strahler (1952, 1980), who is widely acknowledged as the father of modern quantitative geomorphology for his works on the dynamic basis of geomorphology and the logical extension of systems theory to the field. Dury (1980) underlined the need for catastrophism as part of uniformity.

With regard to landform evolution, great advances in understanding the sediment cascade for various geomorphic systems have followed from particularly innovative or rigorous research strategies. Notable work has been done by Trimble (1981) and Trimble and Lund (1982) in humid-region fluvial systems; by Graf (1987b) in a dryland river system; by Caine (1984) in alpine sectors in the Colorado Rocky Mountains; by Weirich (1985, 1986b) in high-energy glacial lakes; and by Nordstrom, McCluskey, and Rosen (1986), Allen (1988), and Sherman (1988) in sandy beach environments. These studies serve to illustrate several points. First, improved instrumentation must remain a priority among geomorphologists as part of the greater need to improve empirical coefficients and to test theories of sediment transport. Second, systematic data collection over a longer time period will yield more-meaningful results in geomorphic systems, where feedbacks and response times play an important role. Third, repeat photography and historical records of reservoir sedimentation offer attractive tools for drainage basin scale studies of the sediment cascade.

The complex response in geomorphology was expertly illustrated by Cooke and Reeves (1976), who proposed a model for arroyo development in the American Southwest. They demonstrated that arroyos in coastal California and southern Arizona are similar in form, but were generated by contrasting scenarios of environmental change. Complexity becomes the paramount consideration in Quaternary studies, as described below.

The thresholds concept was developed in the 1970s, and was quickly used to support episodic models of landscape development (e.g., dynamic metastable equilibrium). But recent work by American geographers has complicated this link. Coates and Vitek (1980a) presented an excellent summary of the development of the thresholds concept, with examples to illustrate the various types, and their social relevance. Salisbury (1980) found that thresholds must be used to explain why the merging of two streams does not always produce a predictable change in valley form.

However, reliance on thresholds to explain episodic landscape change has been tempered by Howard (1982), Graf (1983), and Rhoads (1988). They pointed out that significant landscape change can occur without transgressing thresholds, as disequilibrium is translated spatially throughout the fluvial system. Moreover, Costa and Cleaves (1984) showed that equilibrium and episodic landforms can be found in the same modern landscape.

The Link Among Measurement, Theory, and Application

Fluvial geomorphology continues to be the focus of much effort by American geographers. It is a subfield where American geographers have attained a high degree of success in achieving the synthesis, integration, and prediction aspects of the science. For example, the AAG Resource Publication by Graf (1985a) linked measurement and theory regarding biogeochemical processes in the Colorado River Basin with application to river-basin management issues.

The book by Mueller (1975) explored problems in political geography along the U.S.–Mexico border caused by geomorphic instability of the Rio Grande. He demonstrated that misunderstanding of the processes of channel change in an arid-region river led to long-standing problems of border demarcation. Toy and Hadley (1987) demonstrated the linkages among geomorphic principles, environmental impacts caused by human activities, and the effectiveness of reclamation practices.

General reviews of opportunities in applied physical geography and applied geomorphology were presented by Marcus (1979b) and Costa and Fleisher (1984), respectively.

Coastal geomorphologists have added to the success in linking measurement, theory, and application. Terich and colleagues (Komar, Lizarrage-Arciniega, and Terich 1976) used measurements of beach accretion and erosion in a computer simulation model to demonstrate that jetties can cause considerable shoreline change, even in areas of zero net littoral transport. Their findings have been used to guide development along sandy beaches of the Oregon coast.

Separating Natural and Human-Triggered Variation in Geomorphic Systems

The most difficult problem facing geomorphologists may be separating the natural variation in geomorphic systems from fluctuations triggered by human activities. The techniques that are used to make the distinction must be sufficiently refined to detect change over multiple spatial and temporal scales. Graf (1979) combined photogrammetric measurements from historical photographs with field measurements to judge the impact of gold and silver mining on mountain stream systems in Colorado. Threshold values of erosive force were surpassed in response to changes in general basin vegetation cover, valley-floor vegetation, channel slope, width, and roughness, all subject to human impact.

A study by Marston and Lloyd (1985) utilized the water-budget approach to explain changes in water supply, and to isolate the effects of channel modifications along the Rio Grande below El Paso, Texas.

The effect of silvicultural activities on channel equilibrium and sediment storage in forest streams can be assessed using the methods described by Marston (1982). Sternberg (1987) attempted to separate the influence of deforestation on flooding and channel changes in the Amazon River from the influence of rainfall variations and neotectonics.

The differential effect of military maneuvers on aeolian transport was the subject of a paper by Marston (1986b). Walker and Mossa (1986) used 10 case studies along the shoreline of Japan to contrast the influence of human modifications with that of tectonic processes, tsunamis, and storm surges.

Overall, however, the human-environment (synthesis) theme is not as common in the published work of American geographers in geomorphology as one might expect, given the training of geographers. A survey of consulting activity would probably reveal the true extent of this type of work.

Landscape Ecology and Earth-System Science

The application of concepts and techniques from geomorphology to ecosystem studies and terrain analysis represents a great opportunity for the discipline, given the current need for an interdisciplinary approach to complex environmental problems.

Several examples are available of such research, conducted by American geographers in geomorphology.

Caine and his coworkers (Swanson et al. 1988) claimed that ecosystem behavior can be predicted by a better understanding of how landforms affect those processes. They noted that landform-ecosystem interactions may take multiple forms, and that patterns imposed by one set of interactions may be overridden by another set. The link between landform stability and ecosystem development remains to be quantified. Marston (1989, in press) has traced research on the link between sediment transport and nutrient export from agricultural and forested ecosystems. No consistent proportion of nitrogen or phosphorus is released by either sediment loss or dissolution, undermining the usefulness of any nonpoint-source models that rely on this assumption.

Stream ecologists have suddenly realized that fluvial geomorphologists can provide insights regarding physical habitat features. Working in an interdisciplinary group, Dixon and his coworkers (Brussock, Brown, and Dixon 1985) proposed a regional classification of stream habitat. It is based on channel form and regular changes along the longitudinal profile in physical features that comprise rearing habitat (e.g., pool-riffle ratios) and spawning habitat (e.g., gravel bars).

The extent to which geomorphologists take advantage of this opportunity will depend on their initiative and skill in working with other physical geographers and researchers in engineering and the other natural sciences.

The National Aeronautics and Space Administration (NASA) convened an Earth System Sciences Committee (1986). It has recommended a coordinated sequence of specialized space-research missions for studies of Earth-system processes, and an interdisciplinary program of basic Earth-system research in conjunction with other federal agencies. According to the report, among the many areas to which geographical geomorphologists can make a contribution are studies of sedimentary processes and biogeochemical cycles. Other opportunities exist for the pursuit of Earth-system science through the International Geosphere-Biosphere Program (IGBP) of the International Council of Scientific Unions and the Annual Workshop in Earth System Science held at Pennsylvania State University.

A volume is being prepared jointly by the Institute of Geography at the USSR Academy of Sciences and U.S. geographers, summarizing what is known about the geography of change in global-resource systems and major scientific questions within the goals of IGBP to which geographers may contribute. Some interest has also been expressed for the formation of an "Earth Systems" Specialty Group within the AAG (Borchert 1987). The potential exists for major contributions to these efforts by geomorphologists, by extending their interest in "mega-scale" geomorphology.

Geomorphic Techniques

Geomorphologists have developed a wide range of techniques for data collection and analysis, including field, laboratory, and numerical techniques (Table 2). Many of these have been adopted from cognate fields, but all have been used by American geographers in geomorphology. A sampling follows.

Several advances in field instrumentation by American geographers in geomorphology have been notable. Leatherman (1978) has devised a low-cost and effective

Table 2
Geomorphic techniques used by geographers

Field Techniques		
Air reconnaissance	Chemical tracers	Sand/pebble tracers
Augering	Isotope tracers	Water tracing
Automatic weather stations	Particulate tracers	Mass-balance studies
Dune sand movement	Particle/stone shape	Petrology
Field mapping	Resistivity measurements	Rock jointing
Soil/rock creep measurement	Seabed drifters	Runoff plots
Frost heave	Suspended sediment sampling	Soil moisture/tension/pressure
Leveling/surveying	Bedload measurement	Throughflow measurement
Dendrochronology	Solute sampling; conductivity	Infiltration measurement
Lichenometry	Gravel/pebble size	Aqualung diving
Discharge measurement	Seismic survey	Photoelastic-stress analysis
Macrofossils	Sounding (lake/sea)	Peat boring
Macrofabric of tills	Stone counts	Timelapse photography
Dye tracers	Stone orientation	
Laboratory Techniques		
Air-photo interpretation	Heavy-mineral analysis	Soil thin sections
Landsat interpretation	Isotope techniques	Microfossils
Radar-imagery analysis	Hardware simulation	Wave-tank models
Carbon-14 dating	Morphometry	X-ray powder photography
Chemical analysis: sediments	Pollen analysis	X-ray diffraction
Chemical analysis: water	Particle size: sieve	X-ray fluorescence
Cold laboratory experiments	Particle size: silt/clay	Isotope analysis of water
Digitizer	Coulter counter	Spectrophotometry
DTA	Index properties (Atterberg)	Till microfabric
Electron microscopy	Direct shear box	Quantimet analysis
Experimental rock weathering	Triaxial tests	
Flume studies	Solution experiments	
Numerical Techniques		
Experimental design	Spectral analysis	Directional statistics
Analysis of variance	Stochastic processes	Information-content statistics
Correlation structures	Trend surface analysis	Differential equations
Simple regression	Computer mapping	Mass/energy-balance equations
Multiple regression	Mathematical modeling	Thermodynamic models
Canonical correlation	Digital simulation	Mainframe computers
PCA/Factor analysis	Analogue modeling	Microcomputers
Serial correlation	Spatial analysis	

Source: British Geomorphological Research Group International Research Register.

sand trap for coastal studies (Nordstrom, McCluskey, and Rosen 1986) and desert dune studies (Marston 1986b). Toy (1983) developed an instrument for measuring small-scale changes in elevation, as might be required in studies of aeolian erosion-deposition or frost heave. Graf (1985b) outlined methods for making geomorphic measurements from ground-based photographs, a technique he utilized effectively in other published research (Graf 1978, 1979). Mandel, Sorenson, and Jackson (1985) described the use of erosion pins to estimate erosion on drastically disturbed lands. Day (1984) compared rates of erosion for various carbonate rocks from a wide range of locations through the use of "erosion weight-loss tablets." Weirich (1984, 1986a) designed a network of optical and thermal sensors to document the internal characteristics of turbidity and density currents in lakes.

Field techniques need greater utilization in two respects. First, a strong need exists for initiating and maintaining long-term field monitoring of geomorphic systems, in the tradition of experimental watershed studies that flourished in the 1960s. The data set compiled by Caine (1984) over two decades has been invaluable in this regard. The annual cycle common to many funding programs has been a deterrent to long-term research. Reoccupying old study sites is an alternate strategy in some cases, if a mechanism exists to recover former field stations. The ongoing work by Marston (1986b) in desert dunes and by Weirich (1987) with fire-related debris flows are examples.

Second, greater use should be made of controlled field-scale experiments. Laboratory studies traditionally yield results that alone have limited applicability at the landscape scale. Field studies, without a control on process rates, often frustrate researchers who wait for the "characteristic" event to occur. Results with rainfall simulators (e.g., Luk, Abrahams, and Parsons 1986; Abrahams, Parsons and Luk 1988; Dolan and Marston 1989) and with portable wind tunnels are particularly encouraging.

The most-recognized contribution by American geographers to the advancement of dating techniques is certainly the body of published literature by Dorn and Oberlander; their 1982 paper was cited for the GSA Kirk Bryan Award. They have been responsible for some remarkable developments in the use of rock varnish as an absolute dating tool and as a record of environmental change (climatic and tectonic). Dorn and Oberlander were the first to identify rock-varnish bacteria in the laboratory. They have developed two new rock-varnish dating techniques, radiocarbon dating of organics extracted from rock varnish, and cation-ratio dating. They also have utilized stable carbon isotopes, micromorphology of dust layers in rock varnish, and microchemical laminations in rock-varnish layers as indicators of fluctuations in climate.

Most recently, Dorn (1988) has used rock varnish on Quaternary alluvial fans in Death Valley to reveal three cycles of fan development controlled by climatic change and tectonic activity. Dorn received the 1988 G. K. Gilbert Award for this latest effort. Geomorphologists worldwide are looking to Dorn and his colleagues for further developments in environmental reconstruction using rock varnish.

Several other studies can be cited to illustrate the range of applications in geomorphology using dating techniques. Dendrogeomorphology has been used as a tool in the analysis of flooding, mass movement, and rock glaciers (Butler 1979; Shroder and Giardino 1987). Marcus and Marcus (1980) demonstrated the usefulness of stratigraphic and radiocarbon data from a drained tarn in the reconstruction of Holocene climates. Butler, Sorenson, and Dort (1983) combined geomorphic, stratigraphic, palynologic, and pedologic evidence to discriminate among various types and ages of morainic deposits.

Three-dimensional fabric analysis has been applied to a wide range of sedimentary environments by American geographers in geomorphology. For example, it has been applied to sorted stone stripes (Nelson 1982), solifluction lobes (Nelson 1985), hillslope colluvium (Mills 1983), debris flows from Mount St. Helens (Mills 1984), glacial till (Mark 1974), and rock glaciers (Giardino and Vitek 1985). Eigenvector analysis of fabric data holds promise as a means of differentiating various types of cobble-boulder-size deposits when the origin is uncertain.

A very thorough study by Dixon, Thorn, and Darmody (1984) of chemical weathering on a nunatak of the Juneau Icefield pointed out the importance of dissolution and clay-mineral transformation in periglacial environments. A convenient chronosequence was provided by collecting soil data on successively lower berm levels, which were carved into the nunatak during progressive downwasting of the adjacent Taku Glacier. Dixon and his coworkers utilized a variety of laboratory techniques to analyze chemical alteration of nunatak gruss and soils, including atomic-absorption spectrophotometry (AA), X-ray diffractometry, and scanning electron microscopy (SEM). Changes in the molar ratios of mobile-to-resistant oxides were used to detect contrasting degrees of weathering. The work also highlighted the possible significance of aeolian inputs of fines to account for textural and mineralogical anomalies observed in the near-surface soil profile.

Laity (1983) also used SEM interpretations to judge the importance of diagenetic controls on groundwater sapping and valley formation in the Colorado Plateau region.

Morphometric analyses of drainage basins received considerable attention in the early 1980s. Factors controlling the direction, density and pattern of channel networks have been the subject of benchmark papers by Abrahams (e.g., 1980a, 1980b, 1983, 1984b) and Abrahams and Ponczynski (1984). This coherent body of work was recognized by the AAG Geomorphology Specialty Group when they presented the G. K. Gilbert Award to Abrahams in 1985. However, early enthusiasm for the use of drainage-basin morphometric variables to estimate water and sediment production has been tempered by the realization that the present-day hillslope and channel network morphology may include relict components which do not contribute to modern water and sediment cascades.

Remote-sensing techniques for geomorphologists have been outlined by Rosenfeld (1984) and are common tools employed by geographers. Expectations remain high for new applications in geomorphology as new sensors with improved technology are launched. In particular, the remote-sensing platforms now being designed by NASA for the Earth Orbiting Space Station of the 1990s will afford opportunities for study of mega-scale geomorphology, including catastrophic geomorphic events and tectonic geomorphology. However, the traditional reliance on expensive and time-consuming field research will not be replaced (Graf et al. 1980).

Geographers have been somewhat reluctant to utilize numerical techniques, given their strong tradition of field work. But some researchers have recognized the benefits of both. Computer-aided mapping has been utilized more and more for simulation work (e.g., Band 1985) and erosion mapping (e.g., Marston 1986a; Dolan and

Marston 1989). A recent book by Kirkby et al. (1987) reviewed computer simulation in physical geography, including the important contributions by American geographers. Cluster analysis has been applied to pollen counts to establish the spatial extent of distinct pollen assemblages (Elliott-Fisk et al. 1983).

Trend-surface analysis has been applied to a wide range of geomorphic environments. For instance, trend-surface mapping of buried organic horizons shows promise as a tool for recognizing paleosurfaces and separating drift sheets in areas with limited subsurface data (Rhoads, Rieck, and Winters 1984). Meierding (1982) used the technique to reconstruct the equilibrium-line altitudes of Pleistocene glaciers, and Jones and Cameron (1977) analyzed shifts in particle-size characteristics of barrier-island sands. The ready availability of microcomputers to researchers with the software discussed above will contribute to greater use of numerical modeling in the future.

Development of Theory in Geomorphology

Geomorphology has been described as a "derivative science, borrowing techniques and generalizations from other sciences" (Graf et al. 1980). The belief exists among some geomorphologists that theoretical work is tangential to the mainstream of geomorphology. However, a book by Thorn (1988) should help dispel these notions. This introductory text makes the convincing argument that field work is made more valuable when guided and based upon an established theoretical foundation. Elsewhere, Thorn has outlined advances in the development of theory with regard to ergodic reasoning, time and space in geomorphology (see Thorn 1982), and landscape evolution.

Geomorphic models have been described as theories or hypotheses about system form, and/or process, and/or behavior (Woldenberg 1985). By this definition, much of the effort in geomorphology could be considered as modeling. However, a relatively small effort to date has been put forward by geographers seeking a theoretical basis for form and process in geomorphic systems. Studies of channel networks by faculty at SUNY–Buffalo and their colleagues are significant exceptions to this trend. For instance, optimality in network branching phenomena in nature has been addressed by drawing analogies between fluvial systems and biological systems (Roy and Woldenberg 1982; Woldenberg and Horsfield 1983, 1986). Woldenberg (1969, 1979) has also added to our understanding of spatial hierarchies in geomorphology, again with ties to biological systems.

However, it appears that greater explanation of channel-network development at the landscape scale is being achieved by presuming the operation of stochastic processes than by seeking the deterministic explanation from exact physical-chemical laws (see, e.g., Abrahams 1984a; Abrahams and Mark 1986). Deterministic models will continue to work best for small-scale studies, but empiricism without prediction is still favored by too many geomorphologists when their deterministic-modeling efforts meet with frustration.

The application of fractal concepts to geomorphic systems has been reviewed by Goodchild and Mark (1987). Fractal surfaces are comprised of values that are dependent on neighboring values at all scales. Fractal geometry deserves more attention from geomorphologists as a tool for analyzing various topics—thresholds of erosion by overland flow, scales of roughness in hillslope morphology, and numerous others.

Fractals have already been used with some success in describing coastlines and river courses. Software is becoming more readily available for research and teaching purposes (e.g., Kirkby et al. 1987).

The Spatial Theme in Geographical Geomorphology

Baker (1986) has noted that regional studies of geomorphology were a major focus of geographers until the middle of this century. Comprehensive works on regional geomorphology have been sparse in the geomorphic literature in North America during recent years, until the appearance of the GSA Centennial Special Volume edited by Graf (1987a). Nine of the 13 chapters were authored or coauthored by American geographers. This excellent compendium avoided descriptive geomorphology in favor of an emphasis on modern process and form, with some reference to earlier time frames, including the Quaternary. The volume represents an assessment of geomorphologic theory and a review of selected geomorphic research problems, organized by physiographic province. It represents a modern approach to regional synthesis in geomorphology, departing from the more descriptive "physiography" of the first half of this century.

Mountain regions have received considerable attention from American geographers in geomorphology, and this attention is deserved. Mountain regions are the setting for polygenetic landforms and extremes in rates of geomorphic processes. Applied geomorphologists are noting that mountain regions are being subjected to population pressures in many regions of the world: for recreational development and silviculture in the developed world, and for subsistence farming in the developing world. The text by Price (1981), *Mountains and Man,* provides a particularly useful synthesis of environmental perception, geomorphology, soils, biogeography, land use, and human-environment relationships.

Barsch and Caine (1984) presented a journal-length synopsis of mountain geomorphology, focusing on the high frequency of catastrophic events and the high potential for accelerated erosion where mountain terrain is impacted by human activity. Caine (1983) also authored a book that epitomizes the "new" process-geomorphology approach to regional studies, summarizing 20 years of work in the mountains of northeastern Tasmania.

A special issue of the journal *Mountain Research and Development* (Ives and Ives 1987) was devoted to an analysis of the "Theory of Himalayan Environmental Degradation" which has caused so much alarm and interregional conflict. This excellent work explored whether current mountain land-use practices produce the downstream destruction accredited to them, and if so, what mitigation measures can be pursued. The papers in this volume succeed in exposing the problems that result when wide-ranging hypotheses are presented as fact, with little geomorphic field data to support them.

Quaternary Studies

Reconstructing Quaternary environments has been a major focus of interest by American geographers in geomorphology (Figure 1). To construct a reliable hypothesis for past environmental change, the most successful researchers rely on a wide range of correlative evidence. Work by Holliday (1985a, 1985b) at the Lubbock Lake site in Texas has significance because the sequence at that site is one of the most tightly integrated records of late Quaternary human occupance, sedimentation, and soil formation to be documented in North America. Elsewhere in the southern Great Plains, Hall and Lintz (1984) used a sequence of radiocarbon-dated buried trees, buried soils, a carbonate zone, and molluscan fauna to reconstruct climatic variations.

Butzer's (1981) work in Spain deserves attention because of the varied evidence he utilizes to reconstruct a complex sequence of environmental phases: external-sediment flux in karst cave columns, cave slope and roof rubble, chemical precipitates, cryoturbations, fine soil derivatives, and cultural components.

Several studies illustrate the complex nature of the environmental change that Quaternary scientists are trying to detect and measure. In a study of toposequences of soils under a subalpine forest, Reider (1983) found that the soils are in a state of disequilibrium with the current environment because of oscillating ecotones during the Pleistocene and Holocene. Knox, McDowell, and Johnson (1981) reconstructed the magnitude and causes of floods over the last 10,000 years for the Driftless Area of Wisconsin, an area where controversy remains regarding the effects of continental glaciation. Working in the same area, McDowell (1983) demonstrated the episodic nature of stream response to changes in climate and vegetation during the Holocene.

Miller (1985) has utilized 40 years of data from the Juneau Icefield to reconstruct paleoenvironments in southeast Alaska during late Neoglacial times. The study was exceptional in the way it used multidisciplinary data sets to eventually explain glacier response to the complex interactions of astronomical, geophysical, atmospheric, oceanic, and anthropogenic factors.

PROFESSIONAL AFFILIATIONS FOR AMERICAN GEOMORPHOLOGISTS

The principal professional organizations for American geographers in geomorphology are the Geomorphology Specialty Group of the AAG, the Quaternary Geology and Geomorphology Division of the GSA, the American Quaternary Association, and the Soil Science Society of America. The American Geomorphological Field Group and Friends of the Pleistocene hold field-oriented meetings.

Ample opportunities for interaction with foreign colleagues across the subfields of geomorphology is provided by the Annual Binghamton Geomorphology Symposia, the British Geomorphological Field Group, the Guelph Symposia on Geomorphology, the International Conferences on Geomorphology, the International Quaternary Association, the International Geographical Union, the International Symposium on Erosion and Sedimentation in the Pacific Rim, and the International Union of Geological Sciences. In addition, a great number—too numerous to list—of one-time topical symposia that are of interest to geomorphologists are offered each year.

The AAG Geomorphology Specialty Group

The Geomorphology Specialty Group of the AAG provides a forum for 268 members (May 1987 data), ranking ninth in size among the 38 specialty groups in the association. Geomorphologists present a large number of papers at the annual meetings of the AAG (Table 3), approximately 90% of which are by American geographers. The specialty group also sponsors six or more special sessions each year. Each year since 1983, the AAG Geomorphology Specialty Group has presented the G. K. Gilbert Award for Excellence in Geomorphic Research to the author(s) of a significant recent contribution to the published literature in the field of geomorphology (Table 4).

The Geomorphology Specialty Group was formed through the efforts of J. D. Vitek and C. E. Thorn in the late 1970s, when geomorphologists perceived that they were not being included in AAG activities in proportion to their number within the association (Costa and Graf 1984). This problem is beginning to be addressed by the AAG through an increase in the number of physical geographers nominated for AAG offices and awards. For instance, the Warren J. Nystrom Award for a paper based on dissertation research in geography by a recent Ph.D. recipient has been awarded by the AAG to five American geographers in geomorphology: R. S. Hayden in 1980, R. A. Marston in 1981, F. H. Weirich in 1984, K. K. Hirschboeck in 1987, and D. W. May in 1988.

The GSA Quaternary Geology and Geomorphology Division

The Quaternary Geology and Geomorphology Division of the GSA provides a forum for 1,435 members (October 1987), ranking third in size among the 10 divisions of the GSA. A large number of geomorphology papers are presented at the annual meetings (Table 3), approximately 10% by American geographers.

Table 3

Geomorphology papers presented at AAG and GSA meetings

Year	AAG	GSA	Year	AAG	GSA	Year	AAG	GSA
1976	16	44	1980	61	77	1984	86	81
1977	56	30	1981	71	72	1985	60	95
1978	76	49	1982	74	66	1986	72	89
1979	62	62	1983	70	47	1987	86	124
						Mean	65.8	69.7

Source: Data for the period 1976–80 from Costa and Graf (1984).

Table 4

Recipients of the G. K. Gilbert Award, presented by the AAG Geomorphology Specialty Group

Year	Recipient	Contribution to Geomorphology
1983	J. Ross Mackay	Permafrost studies in the Mackenzie River delta area
1984	Will Graf[a]	Fluvial processes in the southwest United States
1985	Athol Abrahams[a]	Channel networks
1986	Karl Butzer[a]	Archaeology as human ecology
1987	Derek Ford	Karst geomorphology and Quaternary chronologies in the Castleguard Cave area of the Canadian Rockies
1988	Ron Dorn[a]	The use of rock varnish to distinguish three episodes of alluvial fan development in Death Valley

[a]American geographers

The division presents several prestigious awards, with an occasional American geographer as a recipient. The Distinguished Career Award was first presented in 1986. The Kirk Bryan Award is presented for an outstanding paper in Quaternary geology and geomorphology; the 1986 award was presented to R. I. Dorn and T. M. Oberlander for their 1982 paper, "Rock Varnish."

The Gladys W. Cole Memorial Research Award is used to support the investigation of the geomorphology of semiarid and arid terrains in the U.S. and Mexico. It was awarded to W. L. Graf in 1984 to support his study of radionuclides in dryland rivers, and to A. D. Abrahams in 1985 for use in his studies of rock-slope form in the Mojave Desert (see Abrahams and Parsons 1987).

The Robert K. Fahnestock Award is presented to the applicant having the best proposal in sediment transport or related aspects of fluvial geomorphology. It was awarded to R. Andrle in 1987 for his study of pools and riffles in low-sinuosity alluvial streams.

The division also awards two Mackin Grants each year for support of graduate-student research in Quaternary geology and geomorphology, one award to a master's student and one to a doctoral student. Mark Gonzalez (University of Wisconsin–Madison) and Dorothy Sack (University of Utah) are two geography students in American universities who have been selected for Mackin Grants in recent years.

The Binghamton Symposia in Geomorphology

The First Geomorphology Symposia in 1970 was organized and hosted by Donald R. Coates and Marie Morisawa at SUNY–Binghamton. After a long tenure on that campus, the annual event has been moved to other locales, but the name "Binghamton Symposia in Geomorphology" has become associated with this distinguished international, interdisciplinary meeting. American geographers have contributed 9.5% of the papers delivered and later published by the symposia (Table 5). Four of the 18 volumes have utilized American geographers as editors: *Thresholds in Geomorphology* by Coates and Vitek (1980b), *Space and Time in Geomorphology* by Thorn (1982), *Models in Geomorphology* by Woldenberg (1985), and *Hillslope Processes* by Abrahams (1986).

International Geomorphology Activities

The International Geographical Union holds a congress every four years. Immediately before the main congress, and in years between the main congresses, IGU commissions, working groups, and study groups meet at separate venues to examine past and ongoing research in the field and to hold more formal paper sessions. The commissions, working groups, and study groups operating as of 1988 that are of interest to geomorphologists are listed below, with the date they were established (Walker 1987):

Measurement, Theory, and Application in Geomorphology (1968)
Mountain Geoecology (1976)
Coastal Environment (1952)
The Significance of Periglacial Phenomena (1949)
International Hydrological Programme (1964)
Geomorphology of River and Coastal Plains (1980)

Table 5
The Binghamton Symposia in geomorphology

No.	Year Held	Topic	Editor(s)	Contributors: Amer; Can Geographers	Total
1	1970	Environmental Geomorphology	D. R. Coates	2	18
2	1971	Quantitative Geomorphology	M. Morisawa	2	12
3	1972	Coastal Geomorphology	D. R. Coates	2	16
4	1973	Fluvial Geomorphology	M. Morisawa	1	20
5	1974	Glacial Geomorphology	D. R. Coates	0	16
6	1975	Theories of Landform Development	W. N. Melhorn and R. C. Flemal	2	16
7	1976	Geomorphology & Engineering	D. R. Coates	0	20
8	1977	Geomorphology in Arid Regions	D. O. Doehring	1	19
9	1978	Thresholds in Geomorphology	D. R. Coates and J. D. Vitek[a]	2	29
10	1979	Adjustments of the Fluvial System	D. D. Rhodes and E. J. Williams	2	25
11	1980	Applied Geomorphology	R. J. Craig and J. L. Craft	1	33
12	1981	Space & Time in Geomorphology	C. E. Thorn[a]	4	20
13	1982	Groundwater as a Geomorphic Agent	R. G. LaFleur	1	19
14	1983	Models in Geomorphology	M. J. Woldenberg[a]	1	31
15	1984	Tectonic Geomorphology	M. Morisawa and J. T. Hack	3	22
16	1985	Hillslope Processes	A. D. Abrahams[a]	2	28
17	1986	Aeolian Geomorphology	W. G. Nickling	8	27
18	1987	Catastrophic Flooding	L. Mayer and D. Nash	4	27

[a]American geographers

> Morphotectonics (1980)
> Geomorphological Survey and Mapping (1968)
> Man's Impact on Karst Areas (1984)

At the 1988 IGC in Sydney, 12 of the 112 geomorphology papers/posters presented were authored by American geographers.

Much excitement has been generated over the prospect of an international organization for geomorphologists. In September 1985, the British Geomorphological Research Group hosted the First International Conference on Geomorphology in Manchester, England. It was attended by 675 geomorphologists from 51 countries. Most in attendance agreed that an international organization was desirable (Sugden 1987). A committee was formed to explore the ramifications of establishing an organization, and its findings will be reported at the Second Conference, to be held in West Germany in 1989.

Meanwhile, a Newsletter and World Directory of Geomorphologists are being developed. The Proceedings of the First Conference have been published by Wiley and Sons (Gardiner 1987) in two volumes, totaling 2590 pages. Abrahams (1987), Psuty and Allen (1987), Sherman (Sherman and Greenwood 1987), and Walker (1987) each published a paper on behalf of American geographers from a total of 78 presentations by American geomorphologists.

Walker and Orme (1986) have summarized the prominent themes of the meeting and arrived at two main conclusions: first, geomorphology remains a field-oriented discipline, although laboratory experiments and theory development are increasing in importance. Second, geomorphology is a global science, in the sense of interest worldwide, and participation by American geographers must be increased.

GEOMORPHOLOGY IN EDUCATION

The decline of both field geology and the number of geomorphologists associated with academic geology units has been noted by Costa and Graf (1984). At the same time, geomorphology has gained in strength within geography departments at the university level in the U.S. Vitek (1988) ascribes this shift to the dynamic growth in the petroleum industry, hydrogeology, sedimentology, and engineering geology, which has attracted geology students away from potential careers in geomorphology.

To explain the lack of support for geomorphology by other geology faculty, Ritter (1988) cited a "lack of unity" in the discipline caused by the dichotomy of purpose (historical versus process-oriented studies) and vagueness of paradigms. Actually, the dichotomy of purpose to which Ritter refers is part of the pluralism claimed as a strength of the discipline by others. Moreover, paradigms in geomorphology have been stated most succinctly by a geologist (Schumm 1977). His article is addressed to other geoscientists who may "cling to ideas which lend themselves to easy pedagogy but are possibly wrong and certainly do not reflect geomorphic science as it is today or what it will be like in the future."

Meanwhile, several generations of students in geography have emerged as research professors in geomorphology, thanks to a tradition of inspiration that began with their mentors during the "quantitative revolution" in geomorphology in the 1950s. Marcus (1979a) noted that "employment opportunities have been good [and] basic and applied research has thrived" for physical geographers. In terms of employment, 74% of American geographers in geomorphology are employed by universities, 11% are employed in the private sector (including self-employed), and 9% are employed by federal and state governmental agencies. Given this situation, it is somewhat surprising that, of the 191 U.S. universities that list department specialties in the 1987–88 *Guide to Departments of Geography in the United States and Canada,* only 72 list geomorphology and 33 list soils as department specialties.

A glaring omission by American geographers has been their failure to author new introductory texts in geomorphology. The most recent effort was by Butzer (1976), distinguished from other texts by weaving dynamic geomorphology into a regional framework. However, the themes in geomorphology toward which American geomorphologists are making significant contributions are not well represented in the texts that are receiving widest adoption—those authored by geologists. Similarly, American geographers contribute few articles on the teaching of geomorphology to the *Journal of Geography* and the *Journal of Geological Education.*

EVALUATION AND PROSPECTS

American geographers working in geomorphology today benefit from paradigms that provide a broad scientific foundation for the discipline. American geographers in geomorphology have never before had so many opportunities for presentation and publication of research. Geographers are publishing at a high rate compared to the overall number of geomorphologists, and they are making important contributions beyond the scope of geographic inquiry.

While gaining new perspectives from greater international collaboration and interaction with geomorphologists from other backgrounds, geographical geomorphologists need to leave an imprint on the study of landforms and soils that reflects their training as geographers. Geomorphology is undergoing a reorganization that may mask the identity of American geographers within geomorphology, but the need will remain for certain contributions that are best supplied by those possessing a background in geography. The synthesis, integration, and predictive spatial modeling that mark geomorphic research by geographers are imprints that should be emphasized at a time when pluralism exists in the discipline. Unfortunately, geographers are missing an opportunity to leave a distinct imprint on the study of landforms and soils by often pursuing research that duplicates the objectives of geologic inquiry.

The strongest contributions by American geographers to the discipline will be in studies that refine existing paradigms; link measurement, theory, and application; separate natural and human-triggered variation in geomorphic systems; and pursue landscape ecology and Earth-system science. Geographers will continue to contribute their fair share toward the development of research techniques in geomorphology and toward the differentiation of geomorphic systems in time and space (i.e., Quaternary studies and regional geomorphology).

More theoreticians are needed among the ranks of American geographers in geomorphology. Long-term field monitoring of geomorphic systems is desired to better understand adjustment of form and process over the scale of decades. And greater reliance on controlled field-scale experiments is urged, to achieve more meaningful results at a scale that can link laboratory and landscape scales of understanding in geomorphology.

New introductory texts in geomorphology, authored by American geographers, are desired to advance the themes of geographical geomorphology in education. Finally, American geographers in geomorphology must seek more interaction with their colleagues on an international level and publish more in the high-impact-factor journals to ensure their efforts receive the widest recognition.

REFERENCES

Abrahams, A. D. 1980a. Channel link density and ground slope. *Annals of the Association of American Geographers* 70:80–93.

———. 1980b. Divide angles and their relation to interior link lengths in natural channel networks. *Geographical Analysis* 12:157–71.

———. 1983. Geological controls on the topologic properties of some trellis channel networks. *Bulletin of the Geological Society of America* 94:80–91.

———. 1984a. Channel networks: A geomorphological perspective. *Water Resources Research* 20:161–88.

———. 1984b. Tributary development along winding streams and valleys. *American Journal of Science* 284:863–92.

———, ed. 1986. *Hillslope processes.* Boston: Allen & Unwin.

———. 1987. Channel network topology: Regular or random? In *International Geomorphology 1986, Part II, Proceedings of the First International Conference on Geomorphology,* ed. V. Gardiner, pp. 145–58. Chichester: Wiley.

———, and Ponczynski, J. J. 1984. Drainage density in relation to precipitation intensity in the U.S.A. *Journal of Hydrology* 75:383–88.

———, and Mark, D. M. 1986. The random topology model of channel networks: Bias in statistical tests. *Professional Geographer* 38:77–81.

———, and Parsons, A. J. 1987. Identification of strength equilibrium rock slopes: further statistical considerations. *Earth Surface Processes* 12:631–35.

———, Parsons, A. J., and Luk, S-H. 1988. Hydrologic and sediment responses to simulated rainfall on desert hillslopes in southern Arizona. *Catena* 15:103–17.

Allen, J. R. 1988. Nearshore sediment transport. *Geographical Review* 78:148–57.

Baker, V. R. 1986. Introduction: Regional landform analysis. In *Geomorphology from space,* eds. N. M. Short and R. W. Blair, pp. 1–11. Washington: National Aeronautics and Space Administration.

Band, L. E. 1985. Simulation of slope development and the magnitude and frequency of overland flow erosion in an abandoned hydraulic gold mine. In *Models in geomorphology,* ed. M. J. Woldenberg, pp. 191–211. Boston: Allen & Unwin.

Barsch, D., and Caine, N. 1984. The nature of mountain geomorphology. *Mountain Research and Development* 4:287–98.

Bashenina, N. V.; Leontjev, O. K.; Piotrovsky, M. V.; and Simonov, Y. G. 1987. A step forward in the new geomorphology. In *International Geomorphology 1986, Part I, Proceedings of the First International Conference on Geomorphology,* ed. V. Gardiner, pp. 25–31. Chichester: Wiley.

Borchert, J. 1987. Session on the International Geosphere-Biosphere Program at the AAG Portland meetings. *International Geographical Union Bulletin* 37:42–43.

Brussock, P. P.; Brown, A. V.; and Dixon, J. C. 1985. Channel form and stream ecosystem models. *Water Resources Bulletin* 21:859–66.

Butler, D. R. 1979. Dendrogeomorphological analyses of flooding and mass movement, Ram Plateau, Mackenzie Mountains, Northwest Territories. *Canadian Geographer* 23:62–65.

———; Sorenson, C. J.; and Dort, W. 1983. Differentiation of morainic deposits based on geomorphic, stratigraphic, palynologic, and pedologic evidence, Lemhi Mountains, Idaho. In *Tills and related deposits,* eds. E. B. Evenson, C. Schlkuchter, and J. Rabassa, pp. 373–80. Rotterdam: Balkema.

Butzer, K. W. 1976. *Geomorphology from the earth.* New York: Harper & Row.

———. 1981. Cave sediments, Upper Pleistocene stratigraphy and Mousterian facies in Cantabrian Spain. *Journal of Archaeological Science* 8:133–83.

Caine, N. 1983. *The mountains of northeastern Tasmania: a study of alpine geomorphology.* Rotterdam: Balkema.

———. 1984. Elevational contrasts in contemporary geomorphic activity in the Colorado Front Range. *Studia Geomorphologica Carpatho-Balcanica* 18:5–31.

Coates, D. R., and Vitek, J. D. 1980a. Perspectives on geomorphic thresholds. In *Thresholds in Geomorphology,* eds. D. R. Coates and J. D. Vitek, pp. 3–24. London: Allen & Unwin.

———, and ———, eds. 1980b. *Thresholds in geomorphology.* London: Allen & Unwin.

Cooke, R. U. 1987. The use of geomorphology. In *International Geomorphology 1986, Part II, Proceedings of the First International Conference on Geomorphology,* ed. V. Gardiner, pp. 63–82. Chichester: Wiley.

———, and Reeves, R. W. 1976. *Arroyos and environmental change in the American Southwest.* Oxford: Clarendon Press.

Costa, J. E., and Cleaves, E. T. 1984. The Piedmont landscape of Maryland: a new look at an old problem. *Earth Surface Processes* 9:59–74.

———, and Fleisher, P. J., eds. 1984. *Developments and applications of geomorphology.* Berlin: Springer-Verlag.

———, and Graf, W. L. 1984. The geography of geomorphologists in the United States. *Professional Geographer* 36:82–89.

Day, M. J. 1983. Slope form and process in cockpit karst in Belize. In *New directions in karst,* eds. K. Paterson and M. M. Sweeting, pp. 363–82. Norwich: Geo Books.

———. 1984. Carbonate erosion rates in southwestern Wisconsin. *Physical Geography* 5:142–49.

———. 1985. Limestone valley systems in north central Jamaica. *Caribbean Geography* 2:16–32.

Dixon, J. C.; Thorn, C. E.; and Darmody, R. G. 1984. Chemical weathering processes on the Vantage Peak Nunatak, Juneau Icefield, southern Alaska. *Physical Geography* 5:111–31.

Dolan, L. S., and Marston, R. A. 1989. Computer assisted mapping of sediment production from an arid watershed. *Journal of Soil And Water Conservation.* In press.

Dorn, R. I. 1988. A rock varnish interpretation of alluvial-fan development in Death Valley, California. *National Geographic Research* 4:56–73.

———, and Oberlander, T. M. 1982. Rock varnish. *Progress in Physical Geography* 6:317–67.

Dury, G. H. 1980. Neocatastrophism: a further look. *Progress in Physical Geography* 4:391–413.

Earth System Sciences Committee. 1986. *Earth system science overview: a program for change.* Washington: National Aeronautics and Space Administration.

Elliott-Fisk, D. E.; Andrews, J. T.; Short, S. K.; and Mode, W. N. 1983. Isopoll maps and an analysis of the distribution of the modern pollen rain, eastern and central northern Canada. *Geographie Physique et Quaternaire* 36:91–108.

Gardiner, V., ed. 1987. *International Geomorphology 1986, Part II, Proceedings of the First International Conference on Geomorphology.* Chichester: Wiley.

Giardino, J. R., and Vick, S. G. 1987. Geologic engineering aspects of rock glaciers. In *Rock glaciers,* eds. J. R. Giardino, J. F. Shroder, and J. D. Vitek, pp. 265–88. Boston: Allen & Unwin.

———, and Vitek, J. D. 1985. A statistical interpretation of the fabric of a rock glacier. *Arctic and Alpine Research* 17:165–77.

———; Shroder, J. F.; and Vitek, J. D., eds. 1987. *Rock glaciers.* Boston: Allen & Unwin.

Goodchild, M. F., and Mark, D. M. 1987. The fractal nature of geographic phenomena. *Annals of the Association of American Geographers* 77:265–78.

Graf, W. L. 1978. Fluvial adjustments to the spread of tamarisk in the Colorado Plateau Region. *Bulletin of the Geological Society of America* 89:1491–1501.

———. 1979. Mining and channel response. *Annals of the Association of American Geographers* 69:262–75.

———. 1983. Downstream changes in stream power in the Henry Mountains, Utah. *Annals of the Association of American Geographers* 73:373–87.

———. 1984. The geography of American field geography. *Professional Geographer* 36:78–82.

———. 1985a. The Colorado River: instability and basin management. *Resource Publications in Geography* 1984/2. Washington: Association of American Geographers.

———. 1985b. Geomorphic measurements from ground-based photographs. In *Geomorphology: themes and trends,* ed. A. F. Pitty, pp. 211–25. Totowa, NJ: Barnes & Noble.

———, ed. 1987a. *Geomorphic systems of North America.* Centennial Special Volume 2. Boulder, CO: Geological Society of America.

———. 1987b. Late Holocene sediment storage in canyons of the Colorado Plateau. *Bulletin of the Geological Society of America* 99:261–71.

———; Trimble, S. W.; Toy, T. J.; and Costa, J. E. 1980. Geographic geomorphology in the eighties. *Professional Geographer* 32:279–84.

Gregory, K. J. 1985. *The nature of physical geography.* London: Edward Arnold.

Hall, S. A., and Lintz, C. 1984. Buried trees, water table fluctuations, and 3000 years of changing climate in west-central Oklahoma. *Quaternary Research* 22:129–33.

Hart, M. G. 1986. *Geomorphology: pure and applied.* London: Allen & Unwin.

Holliday, V. T. 1985a. Archaeological geology of the Lubbock Lake site, southern High Plains of Texas. *Bulletin of the Geological Society of America* 96:1483–92.

———. 1985b. Morphology of late Holocene soils at the Lubbock Lake archaeological site, Texas. *Soil Science Society of America* 49:938–46.

Howard, A. D. 1982. Equilibrium and time scales in geomorphology: Application to sand-bed alluvial streams. *Earth Surface Processes* 7:303–25.

Ives, J. D., and Ives, P., ed. 1987. The Himalaya-Ganges problem, proceedings of the Mohonk Mountain

Conference. *Mountain Research and Development* 7:181–344.

Johnston, R. J., ed. 1985. *The future of geography*. London: Methuen.

Jones, J. R., and Cameron, B. 1977. Landward migration of barrier island sands under stable sea level conditions: Plum Island, Massachusetts. *Journal of Sedimentary Petrology* 47:1475–83.

Kirkby, M. J.; Naden, P. S.; Burt, T. P.; and Butcher, D. P. 1987. *Computer simulation in physical geography*. Chichester: Wiley.

Knox, J. C.; McDowell, P. F.; and Johnson, W. C. 1981. Holocene fluvial stratigraphy and climatic change in the Driftless Area, Wisconsin. In *Quaternary paleoclimate,* ed. W. C. Mahaney, pp. 107–27. Norwich: Geo Books.

Komar, P. D.; Lizarraga-Arciniega, J. R.; and Terich, T. A. 1976. Oregon coast shoreline changes due to jetties. *Journal of the Waterways* 102:13–30. (Harbors and Coastal Engineering Division, American Society of Civil Engineers).

Laity, J. E. 1983. Diagenetic controls on groundwater sapping and valley formation, Colorado Plateau, revealed by optical and electron microscopy. *Physical Geography* 4:103–25.

Leatherman, S. P. 1978. A new eolian sand trap design. *Sedimentology* 25:303–6.

Luk, S-H.; Abrahams, A. D.; and Parsons, A. J. 1986. A simple rainfall simulator and trickle system for hydro-geomorphological experiments. *Physical Geography* 7:344–56.

McDowell, P. F. 1983. Evidence of stream response to Holocene climatic change in a small Wisconsin watershed. *Quaternary Research* 19:100–16.

Mandel, R. D., Sorenson, C. J., and Jackson, D. W. 1985. Estimating erosion rates and sediment yields on drastically disturbed lands. In *Papers and proceedings of the applied geography conference,* ed. J. W. Frazer, pp. 65–74. Binghamton, NY: Department of Geography, State University of New York.

Marcus, M. G. 1979a. Coming full circle: Physical geography in the twentieth century. *Annals of the Association of American Geographers* 69:521–32.

———. 1979b. The range of opportunity in applied physical geography. *Geographical Survey* 8:3–4.

———, and Marcus, W. A. 1980. Deglaciation and early Holocene history of the Lake Emma cirque basin, San Juan Mountains, Colorado. In *Late and post-glacial oscillations of glaciers: glacial and periglacial forms,* ed. H. Schroeder-Lanz, pp. 357–70. Trier: Balkema.

Mark, D. M. 1974. On the interpretation of till fabrics. *Geology* 2:101–4.

Marston, R. A. 1982. The geomorphic significance of log steps in forest streams. *Annals of the Association of American Geographers* 72:99–108.

———. 1986a. *Dust storm potential in the Chihuahuan Desert*. Contributed Papers of the Second Symposium on Resources of the Chihuahuan Desert Region: United States and Mexico No. 1. Alpine: Chihuahuan Desert Research Institute.

———. 1986b. Maneuver-caused wind erosion impacts, south-central New Mexico. In *Aeolian Geomorphology,* ed. W. G. Nickling, pp. 273–90. London: Allen & Unwin.

———. 1989. Particulate and dissolved losses of nitrogen and phosphorus from forest and agricultural soils. *Progress in Physical Geography* 13. In press.

———, and Lloyd, W. J. 1985. River budget for the Rio Grande, El Paso-Juarez Valley. *Journal of Arid Environments* 8:109–19.

Meierding, T. C. 1982. Late Pleistocene glacial equilibrium-line altitudes in the Colorado Front Range: a comparison of methods. *Quaternary Research* 18:289–310.

Miller, M. M. 1985. Recent climatic variations, their causes and Neogene perspectives. In *Late Cenozoic history of the Pacific Northwest,* pp. 357–414. San Francisco: Pacific Division of the American Association for the Advancement of Science.

Mills, H. H. 1983. Clast-fabric strength in hillslope colluvium as a function of slope angle. *Geografiska Annaler* A65:255–62.

———. 1984. Clast orientation in Mount St. Helens debris-flow deposits, North Fork Toutle River, Washington. *Journal of Sedimentary Petrology* 54:626–34.

Morris, S. E. 1987. Regional and topoclimatic implications of rock glacier stratigraphy. In *Rock glaciers,* eds. J. R. Giardino, J. F. Shroder, and J. D. Vitek, pp. 107–26. Boston: Allen & Unwin.

Mueller, J. E. 1975. *Restless river: international law and the behavior of the Rio Grande*. El Paso: Texas Western Press.

Nelson, F. E. 1982. Sorted stripe macrofabrics. *Geografiska Annaler,* A64:25–33.

———. 1985. A preliminary investigation of solifluction macrofabrics. *Catena* 12:23–33.

Nordstrom, K. F., McCluskey, J. M., and Rosen, P. S. 1986. Aeolian processes and dune characteristics of a developed shoreline: Westhampton Beach, New York. In *Aeolian Geomorphology,* ed. W. G. Nickling, pp. 131–47. London: Allen & Unwin.

Olyphant, G. A. 1987. Rock glacier response to abrupt changes in talus production. In *Rock glaciers,* eds. J. R. Giardino, J. F. Shroder, and J. D. Vitek, pp. 55–64. Boston: Allen & Unwin.

Orme, A. 1985. Understanding and predicting the physical world. In *The future of geography,* ed. R. J. Johnston, pp. 258–75. London: Methuen.

Parson, C. G. 1987. Rock glaciers and site characteristics on the Blanca Massif, Colorado. In *Rock glaciers,* eds. J. R. Giardino, J. F. Shroder, and J. D. Vitek, pp. 127–44. Boston: Allen & Unwin.

Pitty, A. F. 1982. *The nature of geomorphology.* London: Methuen.

———. 1985. *Geomorphology: themes and trends.* Totowa, NJ: Barnes & Noble.

Price, L. W. 1981. *Mountains and man: A study of process and environment.* Berkeley: University of California Press.

Psuty, N. P., and Allen, J. R. 1987. Analysis of dune crestline changes, Fire Island, New York, U.S.A. In *International Geomorphology 1986, Part I, Proceedings of the First International Conference on Geomorphology,* ed. V. Gardiner, pp. 1169–83. Chichester: Wiley.

Reider, R. G. 1983. A soil catena in the Medicine Bow Mountains, Wyoming, USA, with reference to paleoenvironment influences. *Arctic and Alpine Research* 15:181–92.

Rhoads, B. L. 1988. Mutual adjustments between process and form in a desert mountain fluvial system. *Annals of the Association of American Geographers* 78:271–87.

———, Rieck, R. L., and Winters, H. A. 1984. Trend surface analysis of glacially buried Pleistocene organic deposits in central Michigan. *Professional Geographer* 36:64–73.

Ritter, D. F. 1988. Landscape analysis and the search for geomorphic unity. *Bulletin of the Geological Society of America* 100:160–71.

Rosenfeld, C. L. 1980. Observations on the Mount St. Helens eruption. *American Scientist* 68:494–509.

———. 1984. Remote sensing techniques for geomorphologists. In *Developments and applications of geomorphology,* eds. J. E. Costa and P. J. Fleisher, pp. 1–37. Berlin: Springer-Verlag.

———, and Cooke, R. 1982. *Earthfire: The eruption of Mount St. Helens.* Cambridge: MIT Press.

Roy, A. G., and Woldenberg, M. J. 1982. A generalization of the optimal models of arterial branching. *Bulletin of Mathematical Biology* 44:349–60.

Salisbury, N. E. 1980. Thresholds and valley widths in the South River Basin, Iowa. In *Thresholds in Geomorphology,* eds. D. R. Coates and J. D. Vitek, pp. 103–30. London: Allen & Unwin.

Schumm, S. A. 1977. *The Fluvial System.* New York: Wiley.

Sherman, D. J. 1988. Empirical evaluation of longshore-current models. *Geographical Review* 78:158–68.

———, and Greenwood, B. 1987. Bedform controls on longshore current velocity. In *International Geomorphology 1986, Part I, Proceedings of the First International Conference on Geomorphology,* ed. V. Gardiner, pp. 1145–68. Chichester: Wiley.

Shroder, J. F. 1987. Rock glaciers and slope failures: High Plateaus and La Sal Mountains, Colorado Plateau, Utah, USA. In *Rock glaciers,* eds. J. R. Giardino, J. F. Shroder, and J. D. Vitek, pp. 193–238. Boston: Allen & Unwin.

———, and Giardino, J. R. 1987. Analysis of rock glaciers in Utah and Colorado, USA, using dendrogeomorphological techniques. In *Rock glaciers,* eds. J. R. Giardino, J. F. Shroder, and J. D. Vitek, pp. 151–60. Boston: Allen & Unwin.

Sternberg, H. O'R. 1987. Aggravation of floods in the Amazon River as a consequence of deforestation? *Geografiska Annaler* A69:201–19.

Strahler, A. N. 1952. Dynamic basis of geomorphology. *Bulletin of the Geological Society of America* 63:923–28.

———. 1980. Systems theory in physical geography. *Physical Geography* 1:1–27.

Sugden, D. E. 1987. A possible international geomorphological organization: survey results. In *International Geomorphology 1986, Part II, Proceedings of the First International Conference on Geomorphology,* ed. V. Gardiner, pp. 3–10. Chichester: Wiley.

Swanson, F. J.; Kratz, T. K.; Caine, N.; and Woodmansee, R. G. 1988. Landform effects on ecosystem patterns and processes. *Bioscience* 38:92–98.

Thorn, C. E., ed. 1982. *Space and time in geomorphology*. London: Allen & Unwin.

———. 1988. *Introduction to theoretical geomorphology*. Boston: Allen & Unwin.

Toy, T. J. 1983. A linear erosion/elevation measuring instrument (LEMI). *Earth Surface Processes* 8:313–22.

———, and Hadley, R. F. 1987. *Geomorphology and reclamation of disturbed lands*. Orlando: Academic Press.

Tricart, J. L. F. 1987. Geomorphology for the future: geomorphology for development and development for geomorphology. In *International Geomorphology 1986, Part II, Proceedings of the First International Conference on Geomorphology,* ed. V. Gardiner, pp. 63–82. Chichester: Wiley.

Trimble, S. W. 1981. Changes in sediment storage in the Coon Creek Basin, Driftless Area, Wisconsin, 1853–1975. *Science* 214:181–83.

———, and Lund, S. W. 1982. *Soil conservation and the reduction of erosion and sedimentation in the Coon Creek Basin, Wisconsin*. U.S. Geological Survey Professional Paper 1234. Washington: U.S. Government Printing Office.

Vitek, J. D. 1988. A perspective on geomorphology in the twentieth century: Links to the past and present. In *History of geomorphology,* ed. K. J. Tinkler. Boston: Allen & Unwin. In press.

———, and Giardino, J. R. 1987a. Rock glacier bibliography. In *Rock glaciers,* eds. J. R. Giardino, J. F. Shroder, and J. D. Vitek, pp. 305–44. Boston: Allen & Unwin.

———, and ———. 1987b. Rock glaciers: A review of the knowledge base. In *Rock glaciers,* eds. J. R. Giardino, J. F. Shroder, and J. D. Vitek, pp. 1–26. Boston: Allen & Unwin.

Walker, H. J. 1987. Potentials for international collaboration in geomorphological research. In *International Geomorphology 1986, Part I, Proceedings of the First International Conference on Geomorphology,* ed. V. Gardiner, pp. 11–24. Chichester: Wiley.

———, and Mossa, J. 1986. Human modification of the shoreline of Japan. *Physical Geography* 7:116–39.

———, and Orme, A. 1986. International geomorphology in the 1980s. *Zeitschrift füer Geomorphologie* 30:503–11.

Weirich, F. H. 1984. Turbidity currents: monitoring their occurrence and movement with a three-dimensional sensor network. *Science* 224:384–87.

———. 1985. Sediment budget for a high energy glacial lake. *Geografiska Annaler* A67:83–99.

———. 1986a. The record of density-induced underflows in a glacial lake. *Sedimentology* 33:261–77.

———. 1986b. A study of the nature and incidence of density currents in a shallow glacial lake. *Annals of the Association of American Geographers* 76:396–413.

———. 1987. Sediment transport and deposition by fire-related debris flows in southern California. In *Proceedings of the International Symposium on Erosion and Sedimentation in the Pacific Rim,* International Association of Hydrological Sciences Publication 165:283–84. Wallingford, Oxfordshire, U.K.

Woldenberg, M. J. 1969. Spatial order in fluvial systems: Horton's laws derived from mixed hexagonal hierarchies of drainage basin areas. *Bulletin of the Geological Society of America* 80:97–112.

———. 1979. A periodic table of spatial hierarchies. In *Philosophy in geography,* eds. S. Gale and G. Olsson, pp. 429–56. Dordrecht: Reidel.

———, ed. 1985. *Models in Geomorphology*. Boston: Allen & Unwin.

———, and Horsfield, K. 1983. Finding the optimal lengths for three branches at a junction. *Journal of Theoretical Biology* 104:301–18.

———. 1986. Relation of branching angles to optimality for four cost principles. *Journal of Theoretical Biology* 122:187–204.

Yamaguichi, D. K. 1984. New tree-ring dates for recent eruptions of Mount St. Helens. *Quaternary Research* 20:246–50.

Energy Geography

Barry D. Solomon | Marilyn A. Brown | Luisa M. Freeman

Energy studies in American geography have evolved in close association with national and international energy conditions. The turbulence of the oil market over the past two decades, combined with two major nuclear power-plant accidents and the growing evidence for energy-related global environmental impacts, have pulled energy geographers in many directions. There has been a shift of emphasis from energy-supply to energy-demand issues, from purely economic to more-behavioral approaches, and from a focus upon local, regional, and national impacts to a growing interest in the global consequences of energy use. The breadth of training that characterizes the discipline has allowed geographers to make significant contributions to these evolving research and public-policy debates.

Petroleum is the most-important commodity in international trade. In 1987 it accounted for well over 40% of total U.S. energy use, approximately one-fourth of our trade deficit, and 10% of total U.S. imports. As a result, fluctuations in the availability and price of crude oil have been primary determinants of the public's (and our profession's) concern for energy issues. The 1973–74 Arab oil embargo and the rise of crude oil prices again in 1979–81 brought national attention to the energy "crisis" and attracted geographers to this field. By the end of the second oil shock, the U.S. was consuming no more energy than it did in 1973 and significantly less than in 1979, due primarily to improvements in the efficiency of energy use. There was a simultaneous shift to geographic research on renewable energy development and energy conservation.

By 1985, a buyers' market for oil and gas had emerged, and in 1986 crude oil prices plummeted to below their 1979 levels. The public image of an energy crisis

---

Helpful comments on a previous version of this chapter were provided by Cutler Cleveland, Gary Gaile, Ed Hillsman, Don Jones, Steve Lonergan, Mike Pasqualetti, John Sorensen, Frank Southworth, Tom Wilbanks, Cort Willmott, and two anonymous reviewers. Oak Ridge National Laboratory is operated by Martin Marietta Energy Systems, Inc. for the U.S. Department of Energy under Contract No. DE-AC05-84OR21400. The views and opinions expressed in this chapter are those of the authors and not necessarily those of the U.S. Government.

was dispelled, funding for energy research declined, and many geographers turned their attention away from energy issues.

The public visibility of nuclear power has provided a second major backdrop against which energy geographers have defined their interests. The March 1979 accident at the Three Mile Island (TMI) nuclear power plant led to numerous studies of evacuation behavior in areas surrounding nuclear reactors, of radioactive-waste disposal, and of nuclear-plant decommissioning. The April 1986 disaster at the Chernobyl nuclear plant rekindled the profession's interest in nuclear-power issues. The number of geographers actively engaged in energy research, however, is still well below the 1979 level.

Energy issues cut across virtually all areas of geography, addressing themes and questions that are broadly geographic in nature:

- *Location and spatial distribution*—where are our energy resources, and where are their markets? Where should we locate power plants, and where should we store nuclear wastes?
- *Spatial allocation and movement*—how can energy supplies best reach their markets? What are the energy implications of alternative transportation systems?
- *Diffusion*—how do people react to different energy options, and how do new energy technologies gain acceptance? What is the potential for energy-efficiency improvements as a major supply option?
- *Regions*—what are the regional consequences of changes in energy supplies, demand, technologies, and policy?
- *Regional economic development*—how is the demand for energy affected by development, and how are development paths affected by energy supplies? What are the regional economic and environmental impacts of alternative energy futures?
- *Environment*—what are the environmental consequences of energy development and use?

These themes characterize the energy research agenda of American geographers as well as the agenda of our European counterparts. They have also been addressed by researchers in allied fields. While this chapter focuses upon the contributions of American geographers over the last decade—reflecting the scope of *Geography in America*—our attention occasionally turns to this broader context of related research. The main body of this chapter provides a review of the major research contributions of American geographers, primarily as reflected in the literature. Our approach is to consider major "fuel" types in a generally chronological fashion, since fuels (including conservation) still drive the energy-policy debate in America. Integrated into this review is a discussion of related public-policy contributions. We end by addressing the critical challenges and future prospects for energy geographers.

RESEARCH SINCE TMI AND THE 1979–81 OIL PRICE SHOCK

Geographers have made important contributions to problem solving regarding all major energy resources, with no one dominant topic area. Our contributions may not be widely recognized, however, for no single major U.S. geography journal has published a large number of energy-related articles during this past decade (although *Economic Geography* and *The Professional Geographer* have served as important outlets). Energy research by British geographers has apparently enjoyed greater popularity, as witnessed by the relative abundance of articles found in *Geography* and *The Geographical Magazine* (Table 1).

Before the TMI accident, landmark studies in the geography of energy appeared in the form of books by Gerald Manners, Peter Odell, and Earl Cook, along with three articles in the September 1971 *Scientific American* by William Kemp, Earl Cook, and Daniel Luten (only Cook and Luten are from the United States). These early works placed the central research foci for energy studies within economic geography, but since then almost every conceivable geographic perspective on energy has been taken, with a discernible shift in the 1980s toward behavioral approaches. These perspectives have ranged from traditional cartography (Cuff and Young 1986) and regional studies (as exhibited in major energy books on the Soviet Union by Dienes and Shabad 1979, China by Smil 1987, Britain by Fernie 1980, and all of Europe by Hoffman 1985), to sophisticated economic-modeling exercises described by Lakshmanan (1983) and Lakshmanan, Anderson, and Jourabchi (1984). Many other approaches and applications have been discussed by Hoare (1979), Wilbanks (1982), and in Calzonetti and Solomon (1985).

Table 1
Energy articles in selected geography journals, 1980–87

Journal	1980	1981	1982	1983	1984	1985	1986	1987	Total
Annals of the Association of American Geographers	1	0	0	0	1	3	0	2	7
Economic Geography	0	2	0	5	1	0	2	1	11
The Professional Geographer	0	1	0	3	3	1	2	1	11
The Geographical Review	2	2	2	1	0	1	0	1	9
Geographical Analysis	1	0	0	0	0	1	0	0	2
The Canadian Geographer (Canada)	1	0	1	2	1	3	1	0	9
The Geographical Magazine[a] (Britain)	3	4	1	0	4	3	4	5	24
Geography[a] (Britain)	3	7	2	1	6	2	5	3	29
Applied Geography (Britain)	—[b]	1	0	0	2	1	3	1	8
Total	11	17	6	12	18	15	17	14	110

Note: This compilation omits commentary and book reviews.
[a]The number of articles in these foreign outlets is inflated because of the much-larger number of papers (many of them short) appearing in these journals each year.
[b]Applied Geography began publication in 1981.

FOSSIL FUELS, NUCLEAR POWER, AND ELECTRIC UTILITIES

Petroleum

Geographers have made significant contributions to problem-solving on all major conventional energy resources. They have also been well represented in public agencies (e.g., U.S. Department of Energy, U.S. Department of the Interior, Synthetic Fuels Corporation, and public utility commissions), energy committees of the National Academy of Sciences/National Research Council, and in the private sector. Yet petroleum resources (oil and gas) figure prominently, both in energy policy debates and in the research of energy geographers. Attention has focused primarily upon issues surrounding production or use of crude oil and natural gas in specific regions (although Odell's 1986 volume on oil and world political economy remains a classic). The article by Lins (1979) on the Kenai Peninsula of Alaska is a prime example. Sheskin and Osleeb (1982) analyzed the possible penetration of Mexican natural gas into U.S. markets using a modified out-of-kilter algorithm, and found serious pipeline-capacity shortages. While these studies are important, U.S. geographers have generally neglected such topics as the oil price-induced economic depression currently plaguing the Southwest, increased oil production and trade with Latin America and Canada, and the shift of refining and petroleum-feedstock processing for bulk chemicals from the U.S. to other oil-producing countries.

Several studies in petroleum geography have focused upon Europe's North Sea region. G. Manners (1984) has analyzed many of the benefits and costs of North Sea oil development, especially on employment. Most of these impacts are concentrated in Britain and Norway. Environmental-planning issues accompanying North Sea oil development have been considered by I. Manners (1982), who conducted similar studies for the Texas and Louisiana Gulf coast. Sommers (1986) has also looked beyond the North Sea and examined physical and other issues surrounding expanded oil and gas development in the northern high latitudes, where ice and rough weather are major impediments to energy development.

American geographers have conducted few regional studies of oil and gas in the Middle East, but they have analyzed in detail the energy resources of the Soviet Union, the world's leading petroleum producer. Natural gas has emerged as the Soviet's fuel of choice, and domestic production grew 48% between 1980 and 1985. Changes in gas distribution and flows in the USSR have been analyzed by Green and Sagers (1985); yet lack of flexibility in Soviet oil refining and lack of extensive gas storage capacity have limited the substitutability of gas for lower-priority uses of Soviet oil (Sagers and Tretyakova 1986). Sagers and Green (1986a, 1986b) have modeled the transport of Soviet fossil fuels and electricity in great detail, while emphasizing the inefficiencies and bottlenecks. Dienes has also taken a comprehensive look at the development of Soviet fossil fuels (1983), but has emphasized the oil industry (1987).

Many of the studies by geographers on refined petroleum products in the U.S. have been led by Greene, and these have generally relied upon econometric analysis. After related work demonstrated the importance of the spatial aspects of gasoline demand and consumer gasoline use (e.g., Greene 1980; Zelinsky and Sly 1984), a closer examination was made of the effect of the 1979 summer oil shortage on U.S. state gasoline demand (Greene and Chen 1983). Indeed, Greene's statistical analyses were used by the Carter White House in the late 1970s. Similar work was done on the regional demand for heavy-truck diesel fuel in the United States (Greene 1984).

Cleveland is one of the few geographers to use the concept of "embodied energy" (i.e., the direct and indirect energy required to produce various goods and services) and the technique of net energy analysis (Cleveland et al. 1984). His physically based perspective uses the concepts of energy and natural-resource quality to characterize and compare economic processes, and to expose the physical underpinnings of human economic existence. A net energy analysis of U.S. oil and gas production found that drilling for these fossil fuels would become a net energy loser very early in the next century (Hall and Cleveland 1981). Subsequent work (Cleveland and Costanza 1984) applied the same general method to geopressurized gas along the Gulf Coast; it was shown to be undesirable for near-term exploitation.

Coal

Geographers' interest in coal resources has concentrated on specific coal basins or projects in the U.S. and Britain. Within the U.S., such studies have focused upon resource-development issues in the West (e.g., Science and Public Policy Program 1981), where the abundant low-sulfur coals are easily surface-mined in large quantities at low unit cost. These and other desirable properties of western coals have accounted for the West's growing share of national coal production. An ultimate constraint on massive western coal output is the region's arid climate and competition for water use with irrigated agriculture. These trade-offs have been examined by Georgianna and Haynes (1981) and Hickcox (1980). Numerous regulatory obstacles have also slowed the siting of coal-fired power plants in the West (Calzonetti 1981). Yet, the federal government continues to actively encourage western coal leasing; Jones et al. (1986), in an analysis of tract rents, concluded that the government is charging an economically defensible royalty rate for these coals. Isserman and Merrifield (1987) showed how "quasi-experimental" control-group methods can be used to assess the socioeconomic impacts of coal development on boom towns and larger areas. Solomon and Pyrdol (1986) used a joint cluster-regression analysis method to delineate high- and low-sulfur coal market regions.

Coal development in the eastern U.S. has a longer history than in the West, and a more pronounced cultural geography. The greater Appalachian region is well known for its over-dependence on the coal-resource delivery system—from mining, production, transport, and conversion to consumption. Marsh (1987) provides a telling description of the complex loyalty of residents to their environment in the anthracite region of northeastern Pennsylvania, a loyalty that has remained well after the material benefits have been virtually exhausted. Economic geographers have investigated shipment patterns of eastern coals. Osleeb and Ratick (1983) have studied transshipment problems involving railroads and ports, and have extended their analysis to the capacity of U.S. ports and railways in the East to handle increased coal exports.

American geographers have also conducted research on the coal industry in Britain, which has a history similar to that of the eastern U.S. A good overview of recent research in both countries is provided by Spooner and Calzonetti (1984). The British coal industry and its contemporary regional geography have been aptly characterized by G. Manners (1981) and Spooner (1981), who illustrate that Britain's National Coal

Board has been investing in productive central coal fields, and disinvesting in declining peripheral areas.

Synthetic Fuels

A brief spurt of interest in synthetic fuels, based mostly on coal, occurred in the U.S. and Canada in the late 1970s and early 1980s. Interest in these new fuel sources was encouraged by massive federal subsidies. Geographic research on synthetic fuels has focused upon the environmental issues and impacts of coal-liquefaction projects in the East (Fowler, Randolph, and Gordon 1983; Calzonetti 1981), and oil shale development in the West. Lonergan (1985a) evaluated the regional impacts of syncrude development in Alberta, Canada, which are similar to those foreseen for synthetic-fuels projects in the U.S. (i.e., boom-town effects and high resource requirements). Wilbanks (1987b) sees poor prospects for synthetic fuels in the short term, but has outlined options to facilitate their development; he has twice testified before Congress specifically on oil shale.

Nuclear Power

Despite the small contribution of nuclear power to the nation's energy budget, it has been the focus of a great deal of research by geographers over the past decade. This keen interest can be attributed to our special skills in the analysis of energy-facility siting and of technological hazards. Geographers have addressed public acceptance of nuclear power, reactor siting, impacts, emergency planning, radioactive-waste disposal, and reactor decommissioning (Pasqualetti, 1986a). Two recently edited books, with most of their contributions by geographers, reflect the richness of approaches that have been taken to the subject (Pasqualetti and Pijawka 1984; Blowers and Pepper 1987).

U.S. opposition to nuclear power has been strong and has emerged from the nation's weapons legacy and lingering problems of waste management (Cook 1982). Opposition is greatest among the most highly educated and wealthiest constituency—those who are most likely to actively question the safety of nuclear-reactor sitings (Kasperson et al. 1980). In response to negative public opinion in England, the British government held an extensive public inquiry on the proposed pressurized water reactor at Sizewell (O'Riordan 1984). While a few foreign countries are proceeding with vigorous nuclear-power development, no new nuclear plants have been sited in the U.S. in over a decade, and more than 100 have been cancelled.

Before the accident at TMI, geographers were involved in the development of power-plant siting methods that integrated socioeconomic and environmental parameters, relying on multicriteria techniques (Dobson 1979; Cohon et al. 1980). Since then, attention has turned away from specific reactor-siting studies (as so few new reactors are being built), but siting issues are still very much alive. Openshaw, a leading British geographer, argued for the adoption of more sophisticated nuclear-reactor siting standards and methods in Britain (Openshaw 1986). Analogous quantitative studies have been made of the socioeconomic impacts of siting. For example, Pijawka and Chalmers (1983) have developed an input-output analysis of nuclear-reactor sitings. Sorensen et al. (1987) combined several research techniques to examine the impacts of restarting the undamaged reactor unit at TMI. The latter study delved into complex psychological and emotional issues, but found that reactor startup would not have major effects on most residents.

The impetus for the involvement of geographers in emergency planning for nuclear power-plant accidents was the 1979 accident at TMI. Zeigler, Johnson, and Brunn (1981) documented that many more people evacuated the area than were ordered to do so by the Governor of Pennsylvania; they termed this an "evacuation shadow" phenomenon. The response occurred despite the clear link between residents' proximity to TMI and their propensity to evacuate (Cutter and Barnes 1982). This suggests that emergency-response planners should anticipate hypervigilant behavior by many people at future accident sites (Johnson and Zeigler 1983).

While emergency planning and warning systems have improved in the U.S. since TMI, the applicable regulations of the Nuclear Regulatory Commission are *not* based on spatial and behavioral considerations (Sorensen 1984). Moreover, Cutter (1984) has argued that it would be impossible to quickly evacuate communities at several of the nation's existing reactor sites in the event of a major accident, because of local population densities and distributions and adverse weather. Finally, Johnson (1985) reviews most of this research in geography and related fields, and suggests improvements to the federal regulation of radiological emergency preparedness.

Radioactive-waste disposal is an area of increased research interest among political geographers. Kasperson (1983) provides an assessment of social and institutional uncertainties in recent radioactive-waste programs. Kasperson, Derr, and Kates (1983) also consider a variety of social-equity principles that should be addressed by the U.S. government, and suggest how these principles might be applied. Their work was influential with the National Academy of Sciences, where Kasperson and two other geographers served on a nuclear-waste panel of the National Research Council. An application of equity principles was made by Solomon and Cameron (1985), who proposed a constitutional-choice decision model to site radioactive-waste repositories. The U.S., of course, is not the only nation with this problem; Solomon and Shelley (1988) reviewed the status of nuclear-waste management around the world.

Another recent topic of interest is nuclear power-plant decommissioning. Relevant geographic issues include the environmental and socioeconomic impacts of plant retirement, transport of radioactive wastes (mostly low-level) to a repository, and post-decommissioning land use and siting (Solomon 1984). Since only the smallest nuclear reactors have thus far been decommissioned, electric utilities have been very slow to consider these issues. This is expected to change quickly, however, as large nuclear plants are to be decommissioned in the near future in both Britain and the U.S. (Pasqualetti 1988).

Other research on nuclear power has addressed the use of this energy source in Europe, particularly Britain (Fernie 1980, 77–104; Openshaw 1986), France (Boyle and Robinson 1987), and the Soviet Union (Pryde 1979). These nations have ambitious nuclear-power development plans, even after the disaster at Chernobyl.

Power-Plant Siting

Geographers have considered electric-utility issues that are applicable to several energy sources, especially coal and nuclear power. These issues include the power-plant siting methods discussed earlier, though regional siting-analysis models are less appropriate today now that U.S. electric utilities have generally abandoned or deferred new construction (Hillsman, Alvic, and Church 1988). Calzonetti, Sayre, and Spooner (1987) also examined the implementation failure of site-suitability analysis in Maryland. Their work highlighted the real-world difficulties of finding sites in an era of lowered energy demand. Glut in electricity supply is of course not universal. Populous regions such as the Northeast can benefit from electricity surpluses elsewhere. Calzonetti, Mann, and Witt (1986) considered the prospects for growth in interregional electricity trade, and concluded that careful planning can increase the markets. Lonergan (1985b) proposed "robustness" as a criterion in energy models and policies. His application to Ontario Hydro showed that existing nuclear expansion plans are not robust, while construction of small-scale hydroelectric plants would provide a better cost hedge against uncertain future demand.

Pollution Issues

Public attention has recently turned to electric power-related pollution issues. Acid rain and deposition, the "greenhouse effect," and destruction of the ozone layer all result from fossil-fuel combustion. A few geographers have made major contributions to the debate on acid-rain control. Smil (1985) has drawn attention to the poor quality of the baseline data, and the difficulty of quantifying acid rain-damage relationships. Topics such as these on global energy-environmental issues may well dominate the research agendas of energy geographers in the 1990s.

RENEWABLE-ENERGY SOURCES

Geographers have conducted research on all the major sources of renewable energy: solar, biomass, geothermal, and hydroelectric. Despite its minor contribution to the U.S. energy budget, renewable energy has had a special appeal to the profession. This attraction probably derives from renewable energy's use of local and regional resources, the resulting local control and independence, reduced environmental impacts and pollution, and use of seasonal and daily climatic and biological patterns long-studied by geographers. The public, too, favors renewable-energy development, though by no means is there unanimity or a strong demand for renewable technologies (Gruntfest and Eichen 1984; Jackson 1988). Several themes emerge from geographic research on renewables, including the analysis of regional-resource availability, land-use planning and management, and public-policy analysis.

Of the four major sources of renewable energy, geographers have focused the most attention upon solar energy (Pryde 1983; Sawyer 1986). Climatologists have modeled and simulated insolation rates, among other types of solar-energy analyses. Several of these contributions are discussed in the climatology chapter of this book. Another theme has been the process of adoption of the technology. Sawyer, Sorrentino, and Wirtshafter (1984) described the pattern of adoption of solar water-heating and space-heating in the U.S. while Sawyer and Wirtshafter (1984) and Sawyer and Friedlander (1983) assessed the ability of federal and state renewable-energy tax incentives to stimulate commercialization. Finally, Pryde (1985) has examined the land requirements of direct and indirect solar electricity facilities, as have Pasqualetti and Miller (1984) and Smil (1984).

Geographic research on biomass energy has been diverse. Smith (1981) and Smil (1983) have written comprehensive monographs on the subject, drawing on international experiences with fuelwood and other biomass fuels. The relationship between fuelwood use and deforestation in developing countries has been intensively studied, as have possible policy options (Allen and Barnes 1985; Whitney, Dufournaud, and Murck 1987; Hosier 1988). For instance, Jones (1988) estimates that 50% of technical fuel savings from stove-efficiency improvements are lost through increased fuel use that is induced by income and price effects. Economic, environmental, and locational issues concerning biomass energy development have also been examined using multi-objective goal-programming (Cocklin, Lonergan, and Smit 1986) and Thünen land-use models (Jones and Krummel 1985). Alcohol fuels in particular are often touted as an attractive alternative to gasoline for fueling the transportation sector. Solomon (1980) has responded with a net energy analysis of grain-ethanol production in the U.S. (for usage in gasohol) which showed a net energy gain for the process.

Geographic research on geothermal energy (sometimes considered to be nonrenewable) has been primarily limited, this past decade, to the work of Pasqualetti. Air, water, and noise pollution, subsidence, induced seismicity, blowouts, and other ecological problems encountered at each of the world's geothermal generating plants are reviewed by Pasqualetti (1980); siting issues and a land-use planning procedure for geothermal development are discussed by Pasqualetti (1986b).

Hydroelectric energy has received little attention from geographers, despite the many climatic, land-use, and economic-development issues that surround it. Sternberg (1983) has examined the impacts of hydro projects on the development of Amazonia, while Hamley (1983) has assessed the La Grande hydroelectric complex in Quebec. There appears to have been no published geographic research on hydroelectricity in the U.S. since TMI, other than the brief consideration by Pryde (1983, 1985) and Sawyer (1986).

ENERGY CONSERVATION

Spatial and other aspects of improved energy efficiency have been major themes of the profession's recent energy research (since the Arab oil embargo). Contributions by geographers to the understanding of energy conservation have spanned each of the three end-use sectors: buildings, transportation, and industry.

Research on energy conservation in buildings has capitalized on two traditional areas of inquiry: decision processes and the operation of the housing market. Within these two traditions, behavioral geographers have made important contributions.

Borrowing behavioral concepts and methods previously used in the analysis of environmental hazards and innovation diffusion, geographers have helped explain the choices people make that affect their consumption of residential energy. Attitudes, beliefs, values, personality, habit, and information flows all have been shown to be significant predictors of home energy-conservation behavior (Macey and Brown 1983). Perhaps more significant is the fact that many of the same concepts are able to explain why different population subgroups consume different amounts of energy and are more or less prone to purchase energy-efficient technologies. Insufficient knowledge, limited planning horizons, and a strong desire for convenience and comfort have been found to be important barriers to improved energy efficiency on the part of elderly and low-income households (Berry and Brown 1988; Brown and Rollison 1985).

Other behavioral-energy research is described in the June 1988 special issue of *The Canadian Geographer* organized by Jackson and the 1988 special issue of *Energy and Buildings* edited by Brown and Keating. Interestingly, ethnic and cultural differences in energy use have not been examined by geographers, despite the discipline's firm grounding in cultural studies and the significant cross-cultural differences in usage patterns.

At a larger scale, geographers have studied many land-use planning aspects of energy conservation in buildings and communities. For instance, the influence of energy on patterns of land use has been examined (Keyes 1980). White (1985) has shown that there is a substantial increment in prices for homes with a basic set of energy-efficiency improvements, yet homebuyers describe the energy efficiency of housing as a low priority in the purchase of a new residence. In general, the transportation time and energy costs of alternative locations have had greater influence over consumer location decisions.

The impact of settlement patterns on energy requirements and opportunities has also been studied (Owens 1986). Here the approach involves simulations and detailed energy-cost calculations. Curtis et al. (1984) described five energy-conservation land-use planning strategies for municipalities. Brown and White (1988) illustrate some of the advantages and disadvantages of implementing energy-conservation programs at the community scale in densely settled communities. High energy density (energy consumption per unit area) also decreases the cost of district heating and cooling and waste-heat utilization, but it may adversely affect the ability to use solar power.

There are many fruitful areas for further research. With the exception of Cullen and Johnson's (1984) study of the distribution of energy assistance needs across U.S. cities, few geographers have addressed spatial patterns of energy conservation in buildings at a national or regional scale. For instance, describing and explaining broad patterns of the adoption of conservation measures might identify barriers and incentives to further penetration. Energy conservation in apartment and commercial buildings also warrants attention, perhaps using the behavioral concepts and methods that have been applied to the study of home energy use. Finally, climatologists could undoubtedly improve upon the current understanding of how weather affects energy consumption in buildings.

A few geographers have conducted extensive research on energy conservation in transportation. Their research has addressed three types of issues: (1) transportation planning at the metropolitan scale; (2) spatial dimensions of fuel use at the national and state level; and (3) transportation energy use in developing countries.

At the urban and metropolitan scale, geographers have examined the fuel efficiency of alternative highway modes. For example, Southworth and Janson (1982) and Southworth and Westbrook (1985) have researched the energy implications of high-occupancy vehicle (HOV) use in metropolitan areas in the U.S., culminating in the development of an HOV-lane simulation model (Janson et al. 1987). Southworth, Papathanassopoulos, and Zavattero (1981) have conducted an empirical study of the effect of truck-route circuity on the direct and indirect energy consumed by Chicago's urban trucking industry. The study highlights the wide range of pick-up-and-drop activity in urban-goods movement, creating significantly different energy consumption per vehicle miles of travel for different commodity types. Finally, Beaumont and Keys (1982) have looked at the possible impacts of energy on future transport patterns.

In a review of vehicle miles of travel and fuel-use forecasting in the U.S., Southworth and Peterson (1989) note the paucity of spatial and socioeconomic disaggregation of national statistics on transportation fuel use. To compensate for these data deficiencies, they have developed a method that determines geographical variations in fuel use and vehicle miles of travel (VMT). Their method involves a log-linear model for forecasting VMT, cross-classified by state, type of highway, and type of vehicle.

Several geographers have recently been involved in transportation energy planning in developing countries. These studies have generally not emphasized geographic questions. For instance, Greene, Meddeb, and Liu (1986) developed a model of vehicle stock evolution and used the model to analyze the effects of hypothetical conservation and fuel-switching initiatives in Tunisia. A related study was conducted for Costa Rica, where recommendations for energy conservation focused primarily upon highway transport and the identification of efficient practices for vehicle maintenance and operation (Greene et al. 1987). A common focus of these studies has been the development of microcomputer and spreadsheet software for use by local government officials.

Given that highway consumption dominates petroleum fuel use, it is not surprising that transportation geographers have limited their research to the highway-vehicle modes. Nevertheless, there is a need for research on the energy implications of nonhighway vehicle modes (i.e., air, rail, and water). Geographers could also fruitfully turn their attention to the relationship between population cohorts, their travel behavior, and the resulting fuel use.

Very few geographic studies have been conducted on energy conservation in industry. Two studies that are particularly noteworthy are parallel-site screening analyses and socioeconomic-impact assessments for industrial cogeneration in Chicago (Fowler, Baougher, and Jansen 1984) and Leeds, Britain (Beaumont and Keys 1981). The latter work emphasized the need to go beyond conventional financial project evaluations, because of critical socioeconomic and environmental benefits resulting from cogeneration, as with most conservation technologies. A popular conservation option in Europe and Japan, and increasingly in the U.S., cogeneration of process steam and electricity could eliminate the need for many new electric power plants. Geographers have also looked at the energy implications of recycling materials (Hannon and Brodrick 1982).

INTEGRATED ENERGY ANALYSIS

The types of fuels that regions and nations produce and consume have broad implications for their economic strength and environmental quality (Lakshmanan 1981). The unequal distribution of the burden of environmental deterioration, and the imbalance between energy supplies and other resources, provide tremendous potential for future political conflict. Integrated energy analysis offers methods and models to resolve these conflicts by clarifying the regional interactions and the relations among energy, economy, and the environment (Lakshmanan 1983; Lakshmanan and Bolton 1986).

Lakshmanan spearheaded the development of one of the most widely used integrated energy models—the Strategic Environmental Assessment System (SEAS) (Ratick and Lakshmanan 1983; Lakshmanan and Nijkamp 1983). SEAS is an integrated series of computer-implemented models and databases organized in a framework for conditional analysis of how economic technology and demographic activities affect and are dependent upon the physical environment. Originally developed in the early 1970s for the U.S. Environmental Protection Agency, SEAS has since been expanded and used extensively for assessing environmental and energy policies in the U.S.

Principles of integrated energy-policy analysis have also been developed and operationalized by a team of social scientists at Oak Ridge National Laboratory, under the leadership of Wilbanks. In 1979 he led an impact-assessment effort that contributed significantly to the development of the U.S. National Energy Plan II. More recently, Wilbanks and his colleagues have examined the energy needs of developing countries with support from the Agency for International Development. This national energy-planning experience has shown that major institutional limitations must be overcome in order to implement plans (Wilbanks 1987a).

CRITICAL CHALLENGES AND FUTURE PROSPECTS

Energy geography has reached a crossroads in its development. After acquiring considerable expertise and recognition over the last decade and a half, it now faces a public apathy that inhibits the recruitment of new members and the retention of its experienced core. It is only a matter of time before energy issues regain prominence in the public-policy arena, and if energy geography does not remain vigorous, it will be unprepared to play a central role when that time comes. It will lose time in retraining, retooling, and recruiting.

On a positive note, the emergence of several centers of excellence offers energy geography a strength in numbers that was not present in the 1970s. While many colleagues are widely dispersed, two centers exist where roles for energy geographers are abundant and thriving. The U.S. Department of Energy's Oak Ridge National Laboratory in Tennessee employs more than a dozen geographers, and has one of the largest concentrations of applied geographers outside of the Washington, DC area. Another cluster of energy geographers is located at Boston University, in its Department of Geography and Center for Energy and Environmental Studies.

Despite the public apathy toward energy issues, there are many critical research, teaching, and policy challenges that face energy geographers today. Energy contributes to many problems of national and international significance. For instance:

- ☐ In the short run, much of the world is vulnerable to Middle Eastern petroleum politics; in the long run, the world will experience a potentially tumultuous transition to nonfossil sources of energy services.
- ☐ U.S. international competitiveness is handicapped by weak energy productivity, and regional energy economies are at the mercy of international trade decisions.

- Nuclear power and radioactive-waste disposal face public opposition that threatens the viability of a strong contribution by nuclear power to the U.S. energy budget.
- The potential link between fossil-fuel combustion (especially coal and oil) and global climatic damage may threaten the fossil-fuel option.

The diversity and breadth of methodological and theoretical expertise offered by energy geographers indicates the potential magnitude of the contributions we could make to solving these problems.

In 1981 Wilbanks bleakly noted that:

> In the more than seven years since the oil embargo, we have been unable to determine what to do with synfuels, how to proceed with nuclear energy, how hard to push conservation, how much to help solar energy, how to balance energy objectives with economic and environmental objectives, which research and development (R & D) to support, when to tax, and where to regulate.

After seven more years, these issues remain unresolved. All of these decisions could benefit from advances in geographic theory, methodology, application, and practice. The critical challenges have been exposed; clearly, much is at stake and the future prospects are great.

REFERENCES

Allen, J. C., and Barnes, D. F. 1985. The causes of deforestation in developing countries. *Annals of the Association of American Geographers* 75:163–84.

Beaumont, J. R., and Keys, P. 1981. Combined heat and power generation schemes. *Environment and Planning A* 13:623–34.

———, and ———. 1982. *Future cities: Spatial analysis of energy issues*. Chichester: Research Studies Press.

Berry, L. G., and Brown, M. A. 1988. Participation of the elderly in residential conservation programs. *Energy Policy* 16:152–63.

Blowers, A., and Pepper, D., eds. 1987. *Nuclear power in crisis: Politics and planning for the nuclear state*. London: Croom Helm.

Boyle, M. J., and Robinson, M. E. 1987. Nuclear energy in France. In *Nuclear power in crisis: Politics and planning for the nuclear state*, eds. A. Blowers and D. Pepper, pp. 55–84. London: Croom Helm.

Brown, M. A., and Rollison, P. A. 1985. The residential energy consumption of low-income and elderly households. *Energy Systems and Policy* 9:271–302.

———, and White, D. L. 1988. Stimulating energy conservation by sharing the savings. *Environment and Planning A* 20:517–34.

Calzonetti, F. J., with Eckert, M. S. 1981. *Finding a place for energy*. Washington: Association of American Geographers.

———, and Solomon, B. D., eds. 1985. *Geographical dimensions of energy*. Dordrecht, Holland: D. Reidel.

———; Mann, P. C.; and Witt, T. S. 1986. U.S. power plant location and fuel mix. *Energy Policy* 14:528–41.

———; Sayre, G. S.; and Spooner, D. 1987. A reassessment of site suitability analysis for power plant siting. *Applied Geography* 7:223–41.

Cleveland, C. J., and Costanza, R. 1984. Net energy analysis of geopressurized gas resources in the U.S. Gulf Coast region. *Energy* 9:35–51.

———; Hall, C. A. S.; and Kaufman, R. 1984. Energy and the U.S. economy. *Science* 225:890–97.

Cocklin, C.; Lonergan, S. C.; and Smit, B. 1986. Assessing options in resource use for renewable energy through multiobjective goal programming. *Environment and Planning A* 18:1323–38.

Cohon, J. L. et al. 1980. Application of a multiobjective facility location model to power plant siting in a six state region of the U.S. *Computers and Operations Research* 7:107–23.

Cook, E. 1982. The role of history in the acceptance of nuclear power. *Social Science Quarterly* 63:3–15.

Cuff, D. J. and Young, W. J. 1986. *The United States energy atlas*. 2d ed. New York: Macmillan.

Cullen, B. T., and Johnson, J. H. 1984. Energy assistance for the poor. *Energy* 9:571–81.

Curtis, F. A. et al. 1984. Energy conservation and land use planning. *International Journal of Energy Research* 8:369–74.

Cutter, S. L. 1984. Emergency preparedness and planning for nuclear power plant accidents. *Applied Geography* 4:235–45.

———, and Barnes, K. 1982. Evacuation behavior and Three Mile Island. *Disasters* 6:116–24.

Dienes, L. 1983. Soviet energy policy and the fossil fuels. In *Soviet natural resources in the world economy*, eds. R. G. Jensen, T. Shabad, and A. W. Wright, pp. 275–95. Chicago: University of Chicago Press.

———. 1987. The Soviet oil industry in the twelfth five-year plan. *Soviet Geography. Review and Translations*. 28:617–55.

———, and Shabad, T. 1979. *The Soviet energy system*. Silver Spring, MD: Winston & Sons.

Dobson, J. E. 1979. A regional screening procedure for land use suitability analysis. *Geographical Review* 69:224–34.

Fernie, J. 1980. *A geography of energy in the United Kingdom*. London: Longman.

Fowler, G. L.; Baougher, A. H.; and Jansen, S. D. 1984. Evaluating sites for industrial cogeneration in Chicago. *Energy* 9:97–101.

———; Randolph, J. C.; and Gordon, S. I. 1983. Environmental impacts of synthetic liquid fuels development in the Ohio River Basin. In *Beyond the urban fringe*, eds. R. H. Platt and G. Macinko, pp. 335–54. Minneapolis: University of Minnesota Press.

Georgianna, T. D., and Haynes, K. E. 1981. Competition for water resources. *Economic Geography* 57:225–37.

Green, M. B., and Sagers, M. J. 1985. Changes in Soviet natural gas flows. *Professional Geographer* 37:310–19.

Greene, D. L. 1980. The spatial dimension of gasoline demand. *Geographical Survey* 9(2):19–28.

———. 1984. A derived demand model of regional highway diesel fuel use. *Transportation Research, Part B* 18:43–61.

———, and Chen, C. K. 1983. A time series analysis of state gasoline demand, 1975-1980. *Professional Geographer* 35:40–51.

———; Meddeb, N.; and Liu, J. T. 1986. Vehicle stock modeling of highway energy use. *Energy Policy* 14:437–46.

——— et al. 1987. Road transport energy conservation in Costa Rica. *Energy* 12:1299–1308.

Gruntfest, E. C., and Eichen, M. A. 1984. Public opposition to large scale soft technology in Vermont. *Journal of Environmental Systems* 14:137–46.

Hall, C. A. S., and Cleveland, C. J. 1981. Petroleum drilling and production in the United States. *Science* 211:576–79.

Hamley, W. 1983. Hydroelectric development in the James Bay region, Quebec. *Geographical Review* 73:110–12.

Hannon, B. and Brodrick, J. R. 1982. Steel recycling and energy conservation. *Science* 216:485–91.

Hickcox, D. H. 1980. Water rights, allocation, and conflicts in the Tongue River Basin, southeastern Montana. *Water Resources Bulletin* 16:797–803.

Hillsman, E. L.; Alvic, D. R.; and Church, R. L. 1988. Build-1: A disaggregate model of the U.S. electric industry. *European Journal of Operational Research* 35:30–44.

Hoare, A. 1979. Alternative energies: Alternative geographies? *Progress In Human Geography* 3:506–37.

Hoffman, G. W. 1985. *The European energy challenge: East and West*. Durham, NC: Duke University Press.

Hosier, R. H., ed. 1988. *Energy for rural development in Zimbabwe*. Stockholm, Sweden: The Beijer Institute of the Swedish Academy of Sciences.

Isserman, A. M., and Merrifield, J. D. 1987. Quasi-experimental control group methods for regional analysis. *Economic Geography* 63:3–19.

Jackson, E. L. 1988. Public preferences for energy resource options. *Canadian Geographer* 32:162–65.

Janson, B. N. et al. 1987. Network performance evaluation model for HOV facilities. *Journal of Transportation Engineering* 113:381–401.

Johnson, J. H. 1985. Planning for nuclear power plant accidents. In *Geographical dimensions of energy,* eds. F. J. Calzonetti and B. D. Solomon, pp. 123–54. Dordrecht, Holland: D. Reidel.

———, and Zeigler, D. J. 1983. Distinguishing human responses to radiological emergencies. *Economic Geography* 59:386–402.

Jones, D. W. 1988. Some simple economics of improved cookstove programs in developing countries. *Resources and Energy* 10. In press.

———, and Krummel, J. R. 1985. Location and the development of energy supplies from biomass sources. In *Geographical dimensions of energy,* eds. F. J. Calzonetti and B. D. Solomon, pp. 79–99. Dordrecht, Holland: D. Reidel.

——— et al. 1986. Production functions and tract rents in western U.S. surface coal mining. *Resources and Energy* 8:35–61.

Kasperson, R. E. 1983. Social issues in radioactive waste management. In *Equity issues in radioactive waste management,* ed. R. E. Kasperson, pp. 24–65. Cambridge, MA: Oelgeschlager, Gunn & Hain.

——— et al. 1980. Public opposition to nuclear energy. *Science, Technology, and Human Values* 5:11–23.

———; Derr, P. G.; and Kates, R. W. 1983. Confronting equity in radioactive waste management. In *Equity issues in radioactive waste management,* ed. R. E. Kasperson, pp. 331–68. Cambridge, MA: Oelgeschlager, Gunn & Hain.

Keyes, D. L. 1980. The influence of energy on future patterns of urban development. In *The prospective city,* ed. A. P. Solomon, pp. 309–25. Cambridge: MIT Press.

Lakshmanan, T. R. 1981. Regional growth and energy determinants. *Energy Journal* 2(2):1–24.

———. 1983 A multiregional model of the economy, environment, and energy demand in the United States. *Economic Geography* 59: 296–320.

———, and Nijkamp, P., eds. 1983. *Systems and models for energy and environmental analysis.* Hampshire, England: Gower.

———; Anderson, W.; and Jourabchi, M. 1984. Regional dimensions of factor and fuel substitution in U.S. manufacturing. *Regional Science and Urban Economics* 14:381–98.

———, and Bolton, R. 1986. Regional energy and environmental analysis. In *Handbook of regional and urban economics,* ed. P. Nijkamp, pp. 581–628. Amsterdam: North Holland.

Lins, H. F. 1979. Energy developments at Kenai, Alaska. *Annals of the Association of American Geographers* 69:289–303.

Lonergan, S. C. 1985a. Evaluating the regional impact of Canadian energy megaprojects. In *Large-scale energy projects: Assessment of regional consequences,* eds. T. R. Lakshmanan and B. Johansson, pp. 41–61. Amsterdam: North Holland.

———, 1985b. Robustness as a goal in energy models and policies. *Energy* 10:1225–35.

Macey, S. M., and Brown, M. A. 1983. Residential energy conservation through repetitive household behaviors. *Environment and Behavior* 15:123–41.

Manners, G. M. 1981. *Coal in Britain.* London: Allen & Unwin.

———. 1984. North Sea oil. *Geoforum* 15:15–31.

Manners, I. R. 1982. *North Sea oil and environmental planning.* Austin, TX: University of Texas Press.

Marsh, B. 1987. Continuity and decline in the anthracite towns of Pennsylvania. *Annals of the Association of American Geographers* 77:337–52.

Odell, P. R. 1986. *Oil and world power.* 8th ed. Hammondsworth: Penguin Books.

Openshaw, S. 1986. *Nuclear power: Siting and safety.* London: Routledge & Kegan Paul.

O'Riordan, T. 1984. The Sizewell B inquiry and a national energy strategy. *Geographical Journal* 150:172–82.

Osleeb, J. P., and Ratick, S. J. 1983. The impacts of coal conversions on the ports of New England. *Economic Geography* 59:35–51.

Owens, S. 1986. *Energy, planning and urban form.* London: Pion.

Pasqualetti, M. J. 1980. Geothermal energy and the environment: The global experience. *Energy* 5:111–65.

———. 1986a. The dissemination of geographic findings on nuclear energy. *Transactions of the Institute of British Geographers* n.s., 11:325–36.

———. 1986b. Planning for the development of site-specific resources: The example of geothermal energy. *Professional Geographer* 38:82–87.

———. 1988. Decommissioning at ground level. *Land Use Policy* 5:45–61.

———, and Miller, B. A. 1984. Land requirements for the solar and coal options. *Geographical Journal* 150:190–212.

———, and Pijawka, K. D., eds. 1984. *Nuclear power: Assessing and managing hazardous technology.* Boulder, CO: Westview Press.

Pijawka, K. D., and Chalmers, J. 1983. Impacts of nuclear generating plants on local areas. *Economic Geography* 59:66–80.

Pryde, P. R. 1979. Nuclear power. In *The Soviet energy system,* eds. L. Dienes and T. Shabad, pp. 151–70. Silver Spring, MD: Winston & Sons.

———. 1983. *Nonconventional energy resources.* New York: Wiley.

———. 1985. Land requirements for solar electricity alternatives. In *Geographical dimensions of energy,* eds. F. J. Calzonetti and B. D. Solomon, pp. 255–75. Dordrecht, Holland: D. Reidel.

Ratick, S., and Lakshmanan, T. R. 1983. An overview of the strategic environmental assessment system. In *Systems and models for energy and environmental analysis,* eds. T. R. Lakshmanan and P. Nijkamp, pp. 126–52. Hampshire, England: Gower.

Sagers, M. J., and Green, M. B. 1986a. The transportation of refined petroleum products and the efficient location of refineries in the USSR. *Applied Geography* 6:339–57.

———, and ———. 1986b. *The Transportation of Soviet energy resources.* Totowa, NJ: Rowman & Littlefield.

———, and Tretyakova, A. 1986. Constraints in gas for oil substitution in the USSR. *Soviet Economy* 2:72–94.

Sawyer, S. W. 1986. *Renewable energy: Progress, prospects.* Washington: Association of American Geographers.

———, with Friedlander, S. C. 1983. State renewable energy tax incentives. *Energy Policy* 11:272–77.

———, and Wirtshafter, R. M. 1984. Market stimulation by renewable energy tax incentives. *Energy* 9:1017–22.

———; Sorrentino, A.; and Wirtshafter, R. M. 1984. Solar energy adoption patterns in the United States. *Energy Systems and Policy* 8:162–82.

Science and Public Policy Program. 1981. *Energy from the West.* Norman, OK: University of Oklahoma Press.

Sheskin, I. M. and Osleeb, J. P. 1982. Mexican natural gas. *Energy Policy* 10:27–41.

Smil, V. 1983. *Biomass energies.* New York: Plenum.

———. 1984. On energy and land. *American Scientist* 72:15–21.

———. 1985. Acid rain. *Power Engineering* 89(4):59–63.

———. 1987. *Energy in China's modernization.* Armonk, NY: M. E. Sharpe.

Smith, N. 1981. *Wood: An ancient fuel with a new future.* Washington: Worldwatch Institute.

Solomon, B. D. 1980. Agricultural energy. *Environmental Professional* 2:292–95.

———. 1984. Decommissioning nuclear power plants. In *Nuclear power: Assessing and managing hazardous technology,* eds. M. J. Pasqualetti and K. D. Pijawka, pp. 387–412. Boulder, CO: Westview Press.

———, and Cameron, D. M. 1985. Nuclear waste repository siting. *Energy Policy* 13:564–80.

———, and Pyrdol, J. J. 1986. Delineating coal market regions. *Economic Geography* 62:109–24.

———, and Shelley, F. M. 1988. Siting patterns of nuclear waste repositories. *Journal of Geography* 87:59–71.

Sommers, L. M. 1986. Northern high latitude continental shelf areas. *Nordia* 20:71–76.

Sorensen, J. H. 1984. Evaluating the effectiveness of warning systems for nuclear power plant emergencies. In *Nuclear power: Assessing and managing hazardous technology,* eds. M. J. Pasqualetti and K. D. Pijawka, pp. 259–77. Boulder, CO: Westview Press.

——— et al. 1987. *Impacts of hazardous technology.* Albany: SUNY Press.

Southworth, F., and Janson, B. N. 1982. Energy use and emissions impact measurement in TSM. *Journal of Transportation Engineering* 108:328–42.

———; Papathanassopoulos, E.; and Zavattero, D. 1981. Direct and indirect energy consumption in Chicago's urban trucking industry. *Transportation Research Record* 834:20–27.

———, and Westbrook, F. 1985. High-occupancy vehicle lanes. *Transportation Research Record* 1081:31–39.

———, and Peterson, B. E. 1989. Disaggregation within national highway vehicle miles of travel and fuel use forecasts in the United States. In *Spatial energy analysis,* eds. L. Lundqvist, L. G. Mattsson, and E. A. Eriksson, pp. 199–224. Aldershot, England: Gower.

Spooner, D. 1981. *Mining and regional development.* Oxford: Oxford University Press.

———, and Calzonetti, F. J. 1984. Geography and the coal revival. *Progress in Human Geography* 8:1–25.

Sternberg, R. 1983. Hydroelectric energy, repressed demand and economic change in Amazonia. *Acta Amazonica* 13:371–91.

White, A. L. 1985. House prices and house buyers: Does energy matter? In *Geographical dimensions of energy,* eds. F. J. Calzonetti and B. D. Solomon, pp. 325–52. Dordrecht, Holland: D. Reidel.

Whitney, J. B. R.; Dufournaud, C. M.; and Murck, B. W. 1987. An examination of alternatives to traditional fuelwood use in the Sudan. *Journal of Environmental Management* 25:319–46.

Wilbanks, T. J. 1981. *Building a consensus about energy technologies.* Oak Ridge, TN: Oak Ridge National Laboratory, ORNL-5784, September.

———. 1982. Location and energy policy. In *Applied geography,* ed. J. W. Frazier, pp. 219–32. Englewood Cliffs, NJ: Prentice-Hall.

———. 1987a. Lessons from the national energy planning experience in developing countries. *Energy Journal* 8(2):169–82.

———. 1987b. The prospect of synthetic fuels in the United States. In *The unfulfilled promise of synthetic fuels,* eds. E. J. Yanarella and W. C. Green, pp. 193–211, Westport, CT: Greenwood Press.

Zeigler, D. J.; Johnson, J. H.; and Brunn, S. D. 1981. Evacuation from a nuclear technological disaster. *Geographical Review* 71:1–16.

Zelinsky, W., and Sly, D. F. 1984. Personal gasoline consumption, population patterns, and metropolitan structure. *Annals of the Association of American Geographers* 74:257–78.

Water Resources

Graham A. Tobin | Duane D. Baumann | John E. Damron |
Jacque L. Emel | Katherine K. Hirschboeck |
Olen P. Matthews | Burrell E. Montz

The ubiquitous nature of water problems has generated wide research interests throughout the geographic discipline, from the traditional concerns associated with hydrology, climatology, and geomorphology to the socioeconomic analyses and behavioral models of human geography. At the same time, applied geographic work has treated water as a resource, employing policy studies and decision-making criteria in research programs. Many American geographers, within the subdiscipline of water resources, have also been concerned with the advancement of theoretical concepts and/or the practical application of research findings to specific societal problems. Indeed, geographers working in water resources have a prolific publication record, and have often been leaders in their respective fields, providing new models and sound theoretical frameworks for subsequent research in the subdiscipline. The work of White et al. (1958) in developing flood-hazard research and Wolman and Miller (1960) in hydrogeomorphology come immediately to mind.

This chapter outlines some of the important water-related research with which geographers—primarily American—have been concerned over the last decade. The survey is certainly not comprehensive, and no effort has been made to incorporate all references on particular topics. Every attempt has been made, however, to identify important themes and particular issues that merit discussion, to point out where water resource research is weakest, and to suggest future directions.

Current research in water resources is examined critically, ranging from physical studies of hydroclimatology and hydrogeomorphology to investigations of water policy and legislation. Physical geography provides the underlying theme for much of the chapter, while an explicit emphasis is maintained on traditional human-land relationships as they pertain to water resources. The chapter is divided into six research components: (1) Hydrologic, (2) Water Quality, (3) Water Management: Urban Supply and Demand, (4) Flood Hazard, (5) Groundwater Resources, and (6) Law and Water Resources. Conclusions follow which attempt to pull some of these ideas together and indicate future directions for geographically based water-resources research.

HYDROLOGIC RESEARCH

Hydrology has been defined as that branch of physical geography which is concerned with the origin, distribution, and properties of waters of the Earth (Linsley, Kohler, and Paulhus 1958). Thus hydrologic work represents the fundamental physical basis for much of the research undertaken in water resources. To examine the role of contemporary American physical geographers in hydrologic research, these questions must be addressed:

- What are the major research themes within the science of hydrology, and how have American physical geographers contributed to them?
- What critical unanswered research questions exist in hydrology today, and what can a geographic perspective bring to them?

These questions are discussed below with reference to water resources; they are examined in a much wider context in the chapters on climatology and geomorphology.

The four major themes of basic hydrologic research in the 1970s and 1980s have been

1. hydroclimatology and hydrometeorology,
2. hydrologic modeling of surface and groundwater processes,
3. flood analysis, and
4. hydrogeomorphology.

Scientists from a variety of disciplines have examined these topics, using both deterministic and stochastic methodological approaches. American physical geographers have made contributions in all major research areas of hydrology, particularly hydroclimatology and hydrogeomorphology, although much research has emanated from the subdisciplines of climatology and geomorphology.

Hydroclimatology and Hydrometeorology

Rooted in the basic principles of the hydrologic cycle, the water balance is central to all hydrologic processes. Physical geographers have made significant contributions in this area, beginning with the seminal water-balance paper of Thornthwaite (1948), followed by Thornthwaite and Mather (1955). Numerous studies have since focused upon individual components of the hydrologic cycle, the system as a whole, and the system's utility in water-resources management (Mather 1978, 1981; Miller 1977; Muller 1976, 1987; Shelton 1981; Willmott, Rowe, and Mintz 1985). Other research that bridges hydrology and climatology from an energy and water-balance perspective includes snow hydrology (Aguado 1985a, 1985b; Barry 1983), the analysis of lake levels and net lake-basin supplies (Brinkmann 1985), and drought analysis (Easterling and Changnon, 1987; Shelton 1982; Steila 1987). Water-balance studies have been the focus of major investigations of system-wide processes that involve biologic as well as physical systems.

Hydrometeorologic studies have examined the link between the atmosphere and the hydrosphere by focusing upon individual rainfall events and runoff responses. Physical geographers working in this area have defined the synoptic atmospheric circulation patterns that generate extreme rainfall events (Winkler 1987, 1988), catastrophic floods (Hirschboeck 1987a), and regional peak flow and monthly streamflow variations (Hirschboeck 1987b; Keables 1988).

Hydrologic Modeling of Surface and Groundwater Processes

Modeling has been a common theme in all aspects of hydrology, and geographers have contributed from both hydroclimatic and hydrogeomorphic perspectives. Shelton's spatial disaggregation model of the Deschutes River Basin (1985) is exemplary of a comprehensive hydroclimatic approach to surface-water modeling. Other hydroclimatic models have been derived or calibrated for specific applications, such as defining the hydrologic response to timber harvesting (Mahacek-King and Shelton 1987), or simulating streamflow in low-relief, groundwater-dominated coastal watersheds (Sun and Brook 1988).

Surface-water modeling, from a geographic perspective, has revolved around hydrologic responses to changes in the geomorphic fluvial system (Phillips 1985, 1988; Rhoads 1988; Weirich 1985). Groundwater modeling, however, has been dominated by engineers and hydrogeologists, although in the area of groundwater flow in karst terrain geomorphologists have made significant contributions (Crawford 1984; Drake 1984; Ford 1980, 1984). The basis of all modeling must be rooted in field observation and calibration, and here geographers have also been active. Some examples are rainfall-simulation experiments to analyze the spatial and temporal variations in runoff and sediment yield (Abrahams, Parsons, and Luk 1988; Luk, Abrahams, and Parsons 1986) and soil-moisture field measurements (Panciera, Walsh, and Vitek 1986).

Flood Analysis

The integral link among flood processes, streamflow, and river-channel response has prompted most flood-related research by physical geographers to fall within the realm of fluvial geomorphology (see chapter on geomorphology). A notable exception is the body of work on flood-probability analysis by Waylen (1985a, 1985b), Waylen and Woo (1982, 1983), and Woo and Waylen (1984, 1986), which emphasized the analysis of partial-duration flood series and the statistical evaluation of mixed processes in flood series. In a related study on mixed flood distribution, the atmospheric-circulation mechanisms that generate floods were used to separate partial-duration flood series into hydroclimatologically homogeneous components (Hirschboeck 1987b).

Hydrogeomorphology

While hydrology is concerned primarily with water flowing in river channels and across land surfaces, fluvial geomorphology focuses upon the channels and surfaces themselves, and the depositional and erosional processes that produce adjustments in the fluvial system. The two areas are so closely related that a clear distinction between them cannot be defined, and much of what might be discussed here appears in the chapter on geomorphology. Some themes in hydrogeomorphic studies that are closely interrelated with hydrology are:

> Stream channel and basin responses to land use and climatic changes (Brook and Luft 1987; Graf 1977; Knox 1977, 1987; Magilligan 1985; Martin and Johnson 1987)

Sediment budgets and sediment movement within fluvial systems (Irvine and Drake 1987; Phillips 1986; Renwick and Ashley 1984; Trimble 1983; Weirich 1986)

Hydrologic adjustments to geomorphic controls (Graf 1983; Phillips 1988; Rhoads 1987)

Critical Issues in Hydrology

Emerging from the four major themes of research discussed above are some critical issues that are challenging the field of hydrology today and are shaping the direction that hydrology will take in the future. These issues include the role of extreme events in hydrologic processes, the physical basis for traditional stochastic models in hydrology, the importance of scale in the conceptualization and modeling of hydrologic processes, hydrologic regionalization, the impact of climatic change on hydrology, and the emergence of global hydrology. The integrated Earth-science backgrounds of physical geographers, coupled with an inherently geographic perspective of space and scale, qualify them to make unique contributions in each of these critical areas.

The landmark paper by Wolman and Miller (1960) on magnitude and frequency of geomorphic processes provided a framework for evaluating the importance of moderate and extreme events in hydrogeomorphic processes. Since that paper, more evidence has accumulated to support the overriding geomorphic effectiveness of extreme rainfall events and catastrophic flooding in certain environments (Brookes 1987; Clarke et al. 1987; Gupta and Fox 1974; Wolman and Gerson 1978). In flood hydrology, estimation of the probabilities of extreme floods has become a critical issue (National Research Council, 1988); the answer appears to lie in a better understanding of the flood process and its physical controls (Potter 1987), rather than in statistical models. Toward this end, geographers have made significant inroads in defining the atmospheric origins of extreme rainfall and flooding events by applying information on the physical causes of floods to traditional stochastic methods of flood-frequency determination (Waylen and Caviedes 1987).

Geographers have an inherent interest in scale, and this perspective is becoming important in current hydrologic research, especially in the modeling of the dynamics of hydrologic responses to storm events of different sizes (American Geophysical Union, 1987). Physical geographers have analyzed spatial and temporal scale factors in hydrology from the hydroclimatic and hydrogeomorphic traditions (Bartlein, 1982; Church and Mark 1980; Hirschboeck 1988; Wolman and Gerson 1978), but a real need exists for additional input from geographers to this new focus in hydrology.

A related area of rapidly growing interest in hydrology is that of "hydrologic regionalization" (Greis and Wood 1981; Tasker 1982, 1987). In this context, the concept of a region—so central to traditional geographic thought—has become somewhat distorted by a purely statistical approach to defining homogeneous hydrologic regions for the purpose of improving flood-frequency estimates. In contrast, traditional regional hydrology, as described by Walling (1985), focuses upon the spatial variability of hydrologic processes in different areas of the globe. American geographers have been most active in this latter approach, as evidenced by articles on the hydrological processes associated with polar regions (Church 1988; Woo 1983), tropical regions (Granger 1983; Gupta 1988), arid regions (Graf 1988), and humid regions (Irvine and Drake 1987; Knox 1987). Through studies of the climatic and geomorphic factors that determine spatial and regional hydrologic variability, geographers are in a position to make significant contributions to purely stochastic regionalization techniques, particularly in providing better explanations of the physical factors that control statistical distributions of floods (Potter 1987).

The emergence of Earth-systems science and global change as new interdisciplinary paradigms has highlighted the impact of climatic change on water resources and global hydrology. Previously, the impact of climatic change or fluctuations upon hydrologic variability was ignored or disputed in mainstream traditions of hydrology (Yevjevich 1968), while physical geographers, geologists, and paleohydrologists steadily built up an impressive body of knowledge on the subject (Aguado, 1982; Kay and Johnson 1981; Knox 1984, 1985). Now that climatic change is at the forefront of future directions in hydrologic research, physical geographers have every opportunity to use their interdisciplinary strengths to continue making significant contributions, specifically to studies of hydroclimatic variability through modeling of paleohydrologic responses to Pleistocene and Holocene synoptic circulation patterns, simulations of the global water-balance response to an enhanced global warming effect, paleobotanical reconstructions of wet and dry episodes, or the estimation of past stream discharges through geomorphic and botanical evidence.

The same is true for the emergence of global hydrology as a new focus in hydrologic research (Eagleson 1986). Although this geographic-sounding concept did not have its origins within the discipline of geography, it has emphasized many of the principles of geographic-hydrologic research, such as the global water balance (Willmott, Rowe, and Mintz 1985). And, it is technologically based on remote-sensing techniques and Geographic Information Systems (see chapters on those topics). Geographers in general, and physical geographers in particular, should view the emergence of this global approach to water-resources research as an occasion for greater scientific cooperation, both within the discipline of geography, where hydrologic research serves as a natural bridge between climatology and geomorphology, and outside the discipline, where the integrated Earth-science background and global perspective of physical geographers can have a unique impact. In addition, such research will prove invaluable to studies of water-resources policies and management.

WATER-QUALITY RESEARCH

Concerns about the deteriorating quality of water have spurred an increasing level of research activity in recent years. Geographers have been involved in water-quality research and have made important contributions to this topic, both from a physical-geography perspective in understanding the spatial distribution and transmission of pollutants through the hydrologic system, and from a human-geography perspective in understanding the impacts of pollution on society and in the development of policy and management objectives. Wolman has presented two syntheses of the state of water quality in American rivers (Wolman 1971, 1987). Other studies have focused upon the implications of water pollution on water management and planning, including policy perspectives that are both regional (Kromm 1985) and environmental (Walker and Williams 1982).

In addition to water management and planning, geographers have a long tradition of service in investigating the effects of human alteration of the environment, including human effects on water quality. Studies of the physical, biological, and chemical sources of pollution, both human-caused and natural, have been important themes in geographers' research. Graf (1985a, 1985b), in his regional studies of the Colorado River basin, analyzed the water-quality implications of increased salinity, and the presence of heavy metals and radioactive materials in the watershed. Other geographers have focused upon particular water-quality parameters, such as a study in medical geography that reviewed the geographic effects of a water-borne disease organism (Weil and Kvale 1985) or the assessment of copper dispersal in ephemeral streams by Marcus (1987). Similarly, Eney and Petzold (1988) have demonstrated the spatial variability of acid rain. The alkalinity of surface waters, shown on a thematic map prepared by Omernik and Powers (1983) is another indication of the work done by geographers to organize and display a portion of the enormous array of water-quality data now available to researchers.

Another important focus for geographers has been work on water-quality problems in other nations. Contributions have included work in developed as well as developing nations, and have ranged from the regional environmental implications of river pollution in Norway (Sommers and Cullen 1982) to the problems of supplying good-quality water to the developing world (Wolman and Wolman 1986).

In the late 1950s, White et al. (1958) pioneered research by geographers into the behavioral aspects of humans in their environment. Geographers have expanded research on this topic in recent years, especially in identifying the acceptable level of environmental risk to different constituencies (Whyte 1986; O'Riordan 1986) and to groundwater (Waterstone 1983). This approach can also apply to water-quality issues in the process of moving from assumptions of water purity to establishing levels of acceptable risk for different water uses.

Water quality will be an increasingly important focus in coming years due to calls for more stringent regulations and the increased effort necessary to maintain water quality because of an accelerated water-supply demand. As a result, geographers will have greatly expanded opportunities in water-quality management and research, primarily in four areas: water-quality planning, new water-quality standards, increased regulation, and improved data collection and retrieval. These topics are discussed below.

Water-Quality Planning

Geographers have an opportunity to play a role in determining what constitutes acceptable water quality and in dealing with conflicts in perception among water users, government officials, and communities regarding water quality. Work has begun on assessing the public perception of acceptable water-quality standards (Baumann 1986). A need now exists for the development of research methodologies to enable water planners and managers to meet the perceived water-quality needs, as well as those water-quality standards required by law. Indeed, further consideration should be given to the efficiency of water-quality monitoring strategies, to enable planners to assess problems accurately (Rajagopal 1986).

In addition to water basin-management research in developing countries, geographers have made valuable contributions toward meeting the goal of a safe water supply for developing countries for many years. A great deal remains to be done in this area, because only one-third of the developing world now has adequate drinking-water supplies (Wolman and Wolman 1986). A successful approach will require sensitivity to the values and needs of the local culture, as well as the technical judgment to make the program appropriate to the resources of the communities involved. Geographers possess the skills that are necessary to integrate successfully the social/cultural and the physical/technical aspects of problems.

New Water-Quality Standards

The second area of opportunity for geographers is in planning for new water-quality standards. Recent amendments to the Safe Drinking Water Act (1986) and the Clean Water Act (1988) require significant adjustments by municipal and private water-supply systems as individuals and governments work to comply with the new legislation. Basin-wide planning for emergency situations, such as degraded water quality resulting from drought and flood events, is a research need that draws on geographers' strengths.

Groundwater quality is deteriorating across North America, and as a result geographers will have continued opportunity to assist communities and regions that depend on groundwater for domestic water supplies. In the mid-continent region, for example, contamination from pesticides is an increasing problem (Rajagopal 1984). In coastal regions, saltwater intrusion into freshwater aquifers is a cause for concern. Radon is also a problem, and, in addition to uranium, the Safe Drinking Water Act sets standards for radon in groundwater for the first time. Smaller water-supply systems will be particularly affected by these new regulations, accelerating the need to establish costs for achieving various levels of water quality.

Increased Regulation—Atmospheric, Riverine, Estuarine, Marine

Inevitably, concerns over acidic rain and its effects on water quality in riverine and estuarine systems will finally result in the adoption of standards by the federal government. Also, while instream-flow requirements for riverine systems have focused upon assuring adequate water quantity, we can expect that water-quality standards will be added to instream-flow regulations, especially for fish and wildlife values. In fact, the U.S. Water Resources Council bases instream-flow needs on the dominant instream use, and noted that this invariably incorporates fish and wildlife requirements (Wolman and Wolman 1986).

Estuarine water quality is another current topic of concern to water researchers. For example, one of the primary purposes of the U.S. Environmental Protection Agency's (EPA) Chesapeake Bay Program was to collect data on the degradation of water quality in the estuary. Research results from this program served as a basis for the 1987 Chesapeake Bay Agreement signed by Virginia, Maryland, Pennsylvania, the District of Columbia, and the EPA to assure the improvement of water quality by regulating the release of pollutants into the Bay. Geographers can play an important role in assessing the spatial impacts of pollution overloading in estuarine environments. Water-quality research opportunities in riverine and estuarine systems also extend to marine areas and the work of geographers should be better represented in marine-related research. Marine water-quality concerns, such as the closing of New Jersey beaches in Summer 1988 due to medical-waste contamination, will certainly increase. Geographers are trained to examine the relations among the riverine, estuarine, and marine systems, and can make valuable contributions by giving more attention to this area.

Improved Data Collection and Retrieval

Geographers have led the way in developing and applying remote-sensing techniques and Geographic Information Systems (GIS) to water-resource management issues and problems of water-quality. Remote sensing has provided important water-quality information on the distributions of physical phenomena, such as sediment zones (Scheibe, Harrington, and Ritchie 1987); biological phenomena, such as the distribution of aquatic resources (Welch, Remillard, and Slack 1988); and land-based resources (Nellis 1985). For further details see the chapter on remote sensing.

The development of Geographic Information Systems (GIS) software has enabled geographers to link water-quality data with graphic-output capabilities, providing a new approach and methodology for geographic analysis of water quality (for details see the chapter on GIS). The EPA, for example, has developed a stream-cataloging system (called The Reach File), enabling it to link extensive water-quality data by stream reach and to display graphically streams, lakes, reservoirs, estuaries, and other surface-water features. This system was adapted and enhanced by geographers in the Columbia River basin to permit water-resource planners to designate stream reaches that will be protected from future hydropower development, based on fish and wildlife values (Damron and Anderson 1987). The application of GIS technology to water-quality issues gives geographers a powerful tool to do better what we have been doing all along: interpreting detailed spatial information within a larger framework and explaining what it means. Such understanding leads to direct benefits for water-resources research that is focused upon management activities.

WATER-MANAGEMENT RESEARCH: URBAN SUPPLY AND DEMAND

Water management has become an increasingly difficult problem for society, particularly as urban centers have grown, per capita water use has risen, and water pollution levels have increased. Water availability is no longer dictated purely by measures of quantity but must also be determined by the ultimate quality of supplies; water availability is not merely a simple supply-and-demand issue. While water itself may not be a scarce resource, supplying it to urban areas has become more costly. Rising capital, labor, energy, and chemical-treatment costs have translated into increased expenditures for municipal water suppliers; this cost has in turn been passed on as increased rates to consumers. In addition, parameters other than those of hydrology and economics play a role in water management, including policy practices, legislative concerns, and even behavioral aspects of groups and individuals concerned (Grima 1972; Tobin and Montz 1983). A further assessment of strategies in American water management is required (White 1969), with a reexamination of the need for a national water policy (Thomas 1983).

Many major water utilities are now considering demand management strategies in order to cope with rising costs of supply. While such strategies can often postpone large-scale investments and curb rising water-supply costs, a full understanding of consumer behavior and attitudes is needed before effective water-management policies can be implemented. Geographers have made substantial inroads into some of the problems that face urban water managers. This section looks at several of these, including water conservation, water-demand forecasting, and drought management, all from the perspective of the urban community.

Water Conservation

Typically, urban water-supply planning has consisted of acquiring new water supplies or augmenting existing supplies, while options to reduce water use have generally been ignored. However, several factors—such as increasing costs of water-supply development, the quest for environmental protection, and other institutional and political barriers and regulations—have forced planners to consider water-conservation practices in balancing water supply and demand. Geographers have been active in this area. Baumann, Boland, and Sims (1980, 1984), for instance, have developed a procedures manual for evaluating water-conservation measures for municipal water-supply systems. Water conservation was given a conceptual and practical definition, and classified into three broad categories: regulation, management, and education. The basic premise was that the potential for consumer adoption and the probable effectiveness of the measures had to be understood fully, before efficient conservation management is possible. The conservation literature is filled with studies, mainly by civil engineers, that have measured the effectiveness of individual technological solutions, such as toilet inserts and low-flow showerheads. However, these studies have failed to examine consumers' propensity to adopt conservation practices. Thus, real effectiveness can be ascertained only through an understanding of consumer behavior and adoption of water-conservation techniques.

Other research has explored the relation between water use and socioeconomic traits of residential populations. These have found that variables such as income, education, family size, house size, and attitude toward the environment explain considerable variation in water use between different neighborhoods. However, these studies generally share two major weaknesses. First, the representativeness of the sample is often highly restricted—respondents may be drawn from a single geographic area and/or from a narrow range of dimensions, such as gender, age, and education. Because these traits are themselves significantly related to water conservation, it is difficult to apply results from data gathered in humid Illinois to explain water use in arid Arizona, or to generalize results based on data from affluent suburban neighborhoods to poor inner-city areas. Second, many of these studies have not discussed fully the meaning of revealed statistically significant relationships between water conservation and the independent/explanatory variables, and hence they have suffered from uniqueness attributable to the case-study approach.

Recently, several geographers have looked for reasons why consumers adopt water-conservation practices. For example, current research emphasis is on exploration of relations between residential adoption of water-conservation measures and the range of independent variables that had previously been used to explain variations in water use. It is suggested that behavioral water-conservation measures—such as reducing lawn-watering and using full loads in clothes washers—have great potential for saving water, and are likely to be adopted by consumers. However, by and large, behavioral measures have been unrelated to actual reductions in water use, and technological controls remain more effective.

Geographers still face many problems in fully understanding consumer behavior and water use, and specifically what promotes adoption of water-conservation measures at the individual level. Appropriate study designs, state-of-the-art analyses, and careful interpretation of findings can lead to identification of important independent variables. Therefore, geographers, with their many skills, should be at the forefront of this critical water resource-management issue.

Water-Demand Forecasting

For the efficient short-term running of water-resource systems, managers require accurate forecasts of water demand. These forecasts have usually been based on previous levels of water use, and have influenced the day-to-day operation of the supply systems. Long-term forecasts of water use are necessary for establishing sound water-supply plans and for determining the effectiveness of water-conservation measures. In addition, predictions of future water use are essential for planning major investments in new supply facilities, especially for establishing the appropriate scale of any engineering project. Forecasting water use is a complex procedure that should involve economic, environmental, and engineering considerations. Water-resource geographers, again, should play an increasing role in this process.

Traditionally, the most common and widely used forecasting method has been the per capita approach, whereby historical trends of water use are extrapolated to a future date (Baumann and Dworkin 1978). Population growth is then projected for the same period, and multiplied by the estimated per capita use to arrive at a predicted future water use for a particular urban area. Failure to take into account disaggregations in water-use sectors, such as changes in income, housing stock, industrial mix, and price of water, are the most critical shortcomings of this method. The per capita approach can seriously overestimate demand for water, thereby resulting in unnecessary and costly investments.

Increased scarcity of readily available, high-quality water and the rising cost of providing suitable supplies have brought considerable attention to improving forecasting procedures. Many geographers (Baumann and Dworkin 1978; Boland et al. 1984; Grima 1973) have been strong proponents of disaggregated water-use forecasts; these take into account differences in the socioeconomic characteristics of the residential population. Of the many independent variables that have been used to explain variations in water use (see above), the price of water is particularly important because it is the only explanatory variable that a water utility has the power to change. Knowledge of consumer behavior and attitudes toward changes in the price of water, therefore, is necessary to determine the impact of future price fluctuations on municipal demand. While numerous studies deal with price and its effect on water demand, few studies examine these behavioral and attitudinal aspects.

One area of demand forecasting where further research is needed is in the determination of factors that influence water demand in nonresidential (industrial and commercial) sectors. Currently used are simple coefficients, based on historical water use and factors such as employment. Demand equations that explain the variation within and between different nonresidential water users would result in more-accurate forecasts, efficient planning, and improved conservation programs. A better understanding is necessary of the determinants of demand in all sectors of the municipality, particularly in consumer behavior towards price. Until this is accomplished, water-use forecasting will be plagued with uncertainty. Thus, while geographers, working with economists, engineers, and other environmental scientists, have already achieved a great deal in improving water-demand forecasting, with further research they can lead the way in understanding the consumer attitudes and behavior that affect water demand.

Drought Management

The physical attributes and possible causes of droughts have been examined by several geographers, building on the work of Thornthwaite (1948) and Thornthwaite and Mather (1955). For example, Skaggs (1975) employed the Palmer Index to compare the spatial and temporal variability in intensity of the 1930s drought within the U.S. Shear and Steila (1973) discussed the validity of a new drought-intensity index, and Dey (1982) looked at the nature and possible causes of droughts in the Canadian Prairies, using synoptic models of the climate. Through studies such as these, knowledge related to the physical aspects of drought has been built. However, the concerns now expressed over climatic change will undoubtedly promote further research into drought, and consequently the development of more-sophisticated climatological models. Concerns have also been expressed regarding the impact of drought hazard on society. Sawyer (1983) examined the attitudes of water managers in drought conditions, while Beiswenger (1983) has considered water-management plans under arid conditions. Nevertheless, much still needs to be done, and White and Haas (1975), in their assessment of research on natural hazards, called for more broad-based research into both the physical and social aspects of droughts.

The advances made in urban water conservation and demand forecasting have also contributed to more effective drought-management policies. Drought management requires balancing the costs of capacity expansion against the expected damages and costs of a supply shortage. This requires the use of demand forecasting to predict future supply deficits, and a thorough understanding of the costs and effectiveness of supply-augmentation and demand-reduction measures. In addition, a methodology to estimate expected damages resulting from the projected deficits in supply is necessary. This will permit a water manager to place a value on the reliability of the supply, and to determine the optimal strategy to adopt when a shortage occurs.

Russell, Arey, and Kates (1970) pioneered the geographic research in this area. Using economic data from the 1966 Massachusetts drought, they determined procedures for timing and sizing increments in the safe yield of a system to minimize expected drought losses. More recently, based on this and subsequent studies, a procedure known as the Drought Optimization Procedures (DROPS) was developed by Dziegielewski et al. (1986). DROPS uses probabilistic forecasts of supply, combined with disaggregated forecasts of demand, to determine future deficits. Compensation for any water deficit is made up from feasible supply-augmentation options and demand-reduction measures that best minimize economic losses. A shortcoming of the model is that it requires prior knowledge of expected deficits in order for the optimal package of alternatives to be chosen.

Geographic-based research can enhance drought planning in several ways:

First, a better understanding of the physical processes, especially trigger mechanisms such as low reservoir levels which might indicate the onset of a drought, would enable water managers to prepare for and implement strategy options to conserve water.

Second, the development of a model to predict expected drought intensity would greatly improve the efficiency of current operational techniques.

Third, research concerning the social-psychological aspects of water conservation during drought would be useful in increasing the effectiveness of emergency drought campaigns to conserve water.

One of the more difficult aspects of drought management is to place an economic value on water-supply reliability. The uncertainty of damage estimates accruing from potential water shortages and the problems of determining acceptable costs for given levels of reliability have complicated water-management research. Carson and Mitchell (1987) attempted to address one aspect of the problem. They employed an economic technique known as "contingent valuation" to assess the maximum dollar amount Southern Californians were willing to pay for a reliable water supply. Because of the apparent success of this and similar studies, the contingent-valuation method has gradually gained wide acceptance as a tool for measuring benefits associated with water supplies. Mitchell and Carson argued (1987) that basic research on this method is still necessary, particularly to determine the way in which willingness-to-pay values are influenced by a perceived uncertainty of supply, and by perceived property rights.

Thus, water management draws from several disciplines, including economics, statistics, hydrology, and the behavioral sciences. The interdisciplinary nature of geography permits the merging of these disciplines, enabling geographers to continue their contributions to research in this and other areas of water-resources management.

FLOOD-HAZARD RESEARCH

Geographic research into the flood hazard has traditionally focused upon the interaction between the human-use system and the physical processes operating in riverine and coastal environments. Early research activity was concerned primarily with:

1. explanations of why people occupied hazardous areas, and how they adjusted to flooding (White 1945),
2. discussions concerning the extent of development in floodplains (White et al. 1958), and
3. examinations of the choice processes that floodplain residents and managers undergo in coping with the hazard (Kates 1962; White 1964).

This body of work has formed the basis for much of the flood-hazard research subsequently undertaken around the world (Parker and Penning-Rowsell 1983). Criticisms can be, and have been, made of this early work, especially with its methodological emphases on case study analysis and applied geography. However, it remains the basic building block of hazards research from which more sophisticated, theoretically oriented work has grown. Many of the questions and methods originally directed toward flood research have since been applied to other natural and technological hazards, so the impact of this research has extended beyond the concern of water resources (e.g., see the chapter on hazards).

Geographers are uniquely placed to examine the flood hazard and the interaction of the human-use system with the physical world. Many geographers have been active in hydrologic research applicable to floods. (E.g., see the discussion of the work of Hirschboeck 1987b; Waylen 1985a, 1985b; Waylen and Woo 1982, 1983; and Woo and Waylen 1984, 1986.) There are also many additional facets that must be considered in conjunction with the physical parameters, including the social traits of the community, prevailing economic pressures, and political influences at all levels of government. Geographers have brought these many diverse issues together under the realm of flood-hazard research. However, two themes continue to dominate such research activities in the U.S. The first, a more historical one, deals with human adjustments to the flood hazard. The second, a more recent emphasis, focuses upon the impact of flooding on human settlements. These themes, which will serve as the framework for the following discussion, are characterized by both applied and theoretical questions that have not been fully addressed.

Human Adjustment to Flooding

Questions dealing with how humans, both individually and collectively, adjust to the flood hazard have been the center of research activities for almost 50 years. Recently, research has dealt with the effect of management policies on floodplain land use, and on residential perception of the flood hazard. Once it was established that flood-control structures provided a false sense of security for floodplain residents, resulting in the "levee effect" (Segoe 1937; White et al. 1958), geographers began to look at land-use controls as viable solutions to flood problems. This has led to a body of research focused upon the role of governments in determining land uses to alleviate flooding (Platt 1987; Platt and Nechamen 1984). The combination of structural and nonstructural adjustments has also been of concern to geographers, particularly in light of the land-use implications of each measure (Galloway 1980; Platt 1979). While there appears to be general agreement that structural projects alone cannot be successful, what needs to be established is the appropriate mix of structural and nonstructural measures. Therefore, a goal of further hazards research should be the development of decision-making models to aid floodplain managers in determining effective flood-alleviation programs for particular communities. If this is not possible, then research will just continue on a case-by-case basis, a characteristic of hazards research that has been severely criticized.

The structural/nonstructural debate has been satisfactory for slow-rise, riverine flooding. However, a topic that has received increased attention from geographers is flash flooding and the efficacy of warning systems (Gruntfest 1986). Following the 1976 Big Thompson Flood in Colorado, geographers became concerned with developing a range of potential adjustment strategies for flash flooding. Further research has continued on issues associated with the efficient dissemination of the warning message and the determination of appropriate public and institutional responses. As populations continue to expand in areas subject to flash flooding, the need for information about these questions will grow.

One criticism associated with flood-hazard research is the apparent lack of a unifying theoretical base. Some theoretical models have been proposed (Kates 1971), but there has been no development of an all-encompassing research paradigm for flood-hazard research, and few general research frameworks have been forthcoming. However, many hazards researchers are now attempting to answer these criticisms. Indicative of this is the work by Palm (in press) which discusses the feasibility of an integrative theory for all natural-hazards research.

In part, some of the delay in establishing a theoretical base is due to the complexity of the system that must incorporate an understanding of individual perceptions and actual behavior, government policies and requirements, and aspects of the physical environment as well. The delay is also due to the applied nature of hazard research and the overwhelming emphasis on specific problems. Furthermore, because human adjustment to flooding involves more than simple river or land-use control projects, much has been learned from other disciplines, including psychology and the decision sciences. Consequently, behavioral and decision-making models employed in hazards research often incorporate elements from these other disciplines. Further research is necessary to develop spatially oriented models and other geographically based research frameworks through which the floodplain manager can promote sound policy decisions. Some work has been accomplished in this area; e.g., see the research on flooding and land use by Platt (1979, 1986) and disaster planning by Foster (1980). But much more needs to be done.

Impact of Flooding on Human Settlement

Geographers have also focused attention upon the impact of flooding on human settlements. Most research in this area has been economically oriented, although a few studies have been concerned with the impact of flooding on the structure and health of society. The basic economic premise has been that floods would have a negative impact on the value of property, while flood-alleviation projects, such as dams and levees, would have a positive impact. For the most part, research has focused upon the consequences of implementing flood-mitigation programs on adjacent land values, particularly the effect of designating land parcels as flood-prone through the National Flood Insurance Program and other governmental policies (Babcock and Mitchell 1980; Montz 1987; Muckleston 1983). None of the studies to date has reached more than tentative conclusions, although most have suggested that floodplain regulations have had little effect. Again, it is probably the lack of an underlying theoretical base, and the use of an unintegrated case-study approach, that have made more-definitive findings difficult. The establishment of a conceptual framework around which subsequent research can develop has been called for, and geographic research continues to that end (Tobin and Montz 1988; Tobin and Newton 1986).

Critical Issues in Flood Hazard

Historically, flood-hazard research has been a major component of water-resources research in geography, and it continues to be important today. Indeed, many early questions remain, and new questions arise. The applied approach of the past will likely continue because the flood hazard is a real problem for communities. However, this must be combined with the development of a theoretical base, so that an integrated body of knowledge can be established to address specifically the spatial and temporal aspects of the flood hazard.

An area of continuing importance to geographers is the appropriate mix of structural and nonstructural measures. Despite a decreased Federal emphasis on flood control, communities continue to rely on dams and levees for flood protection (Montz and Gruntfest 1986). However, many structures are nearing the end of their design lives, and the protection they offer may be diminishing. How, then, do communities combine nonstructural measures to insure an adequate level of protection? This is a question that will likely be of concern to geographers in the next decade. Many areas subject to flash flooding and hurricanes are experiencing rapid population increases, often of older persons, and many of the early questions relating to human adjustment are still pertinent in these locations. Thus, geographers continue to have a great deal to offer by utilizing an applied, case-study approach.

Flood-hazard researchers must continue to develop a theoretical base upon which future research can build. The previous lack of theory has led to replication of research methods in different settings, and to tentative and sometimes contradictory findings. To integrate flood-hazard research, and to build on the case study approach, a conceptual framework must be sought.

GROUNDWATER-RESOURCES RESEARCH

Because of the importance of groundwater to irrigated agriculture and as a source of drinking water, geographers and other resource analysts have been interested in the individuals and institutions that manage its allocation, development, use, and protection. For instance, between 40% and 50% of the American population depends upon groundwater as a primary source of drinking water, and approximately 75% of cities obtain part or all of their supply from groundwater (Pye, Patrick, and Quarles 1983). Groundwater is also the source of 38% of irrigation water in the western states (U.S. Geological Survey 1984). In several areas of the country, notably the midwestern and western states, significant water level (or pressure) declines have occurred. Furthermore, contamination of the resource exists in most parts of the country, and will continue to occur due to careless chemical disposal and other environmentally detrimental land-use practices.

An examination of geographic research undertaken in the past 10 or 15 years on the topic of groundwater shows three major foci: (1) institutions, policies, and implementation, (2) political intervention, and (3) land use and water quality. (These topics are discussed below.) The type of work undertaken by geographers in groundwater-resource management is similar to that of surface-water and coastal-water management, where much of the emphasis is on practical questions of policy (Platt 1985, 1987). As Wescoat remarked (1987) on water-resources geography in general, little formal commitment to theory, interpretation, or explanation has been attempted. However, there are some signs of change, with "why" questions beginning to receive more attention, and more geographers considering the political and social aspects of groundwater-resource use (Emel and Brooks 1988; Roberts and Gros 1987). Taken as a whole, these geographic studies of groundwater reflect an emphasis on the contextual aspects of human-resource problems. They do not separate the actors from their socioeconomic (and biophysical) landscapes, as do the neoclassical economic treatments of groundwater problems; and they evince a commitment to the wise use of resources (the ecological perspective), which is often missing from political scientists' examinations of decision processes.

Institutions, Policies, and Implementation

Groundwater is a common-pool resource. Initially, this resource could be tapped by anyone who could finance well development and pay pumping costs. Land ownership was generally the only legal requirement for entry into the common-property arrangement that held sway over the resource. As groundwater use increased, congestion and conflict arose due to the fact that pumping and land-use effects are transmitted in aquifers through space. One user could increase the pumping costs of another; many users operating atomistically could unintentionally deplete (or contaminate) a resource important to existing and future supplies. The predominant solution to these stressed-resource problems is institutional intervention.

The general neoclassical economic approach to solving the common-pool problem is to define property rights carefully (Anderson 1983). Indeed, regulation of groundwater in the U.S. to a large extent means the definition and enforcement of property rights associated with the resource (Emel and Brooks 1988). The primary goals of agencies involved in groundwater management, therefore, have been to protect property rights in groundwater, and to manage aquifer development according to public-interest goals like planned depletion, safe yield, and public health and safety. To attain these goals, agencies have used a traditional set of management options. These decisions usually concern the quantity of pumpage for existing wells and the location, number, and quantity of pumpage of new wells (Emel and Maddock 1986). In effect, political and economic goals are translated into hydrologic-response objectives.

Geographers have been particularly interested in examining the spatial distribution of management objectives and frameworks, as well as the comparative effectiveness and equity of these various management strategies. Throughout the U.S. the variety in management approaches and normative patterns is substantial. In the western states (which, but for Florida, are the most dependent upon groundwater), all but four try to manage the resource through statutes on water levels (Emel and Maddock 1986). The four exceptions are Texas, California, Nebraska, and Oklahoma. In some California basins, rates of withdrawal are managed to ensure a sustainable yield. Nebraska, in addition to Arizona, Kansas, New Mexico, and Oklahoma, assigns water duties to manage water demand within at least one of its management districts (Emel and Yitayew 1987). Nearly all states concerned with groundwater depletion encourage or require some form of well spacing to prevent undue lowering of water levels from pumping wells in close proximity. In some cases, notably the Texas High Plains Underground Water Conservation District No. 1 and the Upper Big Blue Natural Resources District in Nebraska, well spacing is the only management control exercised (Emel and Maddock 1986; Templer 1985).

Texas, one of the states most dependent upon groundwater, has the least-comprehensive management approach, and is characteristically held up as a bad case with regard to wise use. Templer has written numerous pieces on the evolution of Texas groundwater law, from a descriptive and evaluative stance (Templer 1978a, 1985). He has been critical of the piecemeal approach, which depends largely upon the voluntary formation of districts that may then choose their management strategies, opting for no formal controls in some cases. His view is that this will "do little to solve the problem of groundwater depletion."

On the other hand, Templer has pointed out (1985) the important services that the local underground-water conservation districts on the Texas High Plains have provided in advocating voluntary water conservation among irrigators and the general public. The Texas High Plains Underground Water Conservation District No. 1 has undoubtedly improved irrigation efficiency within (and probably without) its political boundaries, by promoting technology, conducting and funding research, furnishing irrigators with irrigation scheduling information, testing pumping-equipment efficiency, and publishing a wealth of material on conservation and the importance of saving water. However, as Templer (1985) argued, it is difficult to measure the success of the district in promoting water conservation and slowing the decline of the High Plains Aquifer; researchers have yet to make convincing arguments on this subject.

In an examination of western groundwater, Emel tried to evaluate the potential effectiveness and equity of various management approaches by considering the hydrologic effects of different policies (Emel 1987; Emel and Maddock 1986). These studies concluded that many management approaches would be neither effective nor equitable because of failure to control pumping and nonpumping water levels. The question, then, is why do states or districts adopt strategies which will not ensure attainment of these objectives?

The Politics of Groundwater Management

Following in the White tradition (1969) of examining how people make their choices in managing water, Kromm and White (1984, 1985) have developed a rich empirical database on preferences for adjustments to reduce depletion of the High Plains Aquifer among populations of the High Plains states. Nearly a thousand people were asked questions about the seriousness of the depletion problem, about their preferences for reducing depletion, and about what type of institution they would prefer to implement adjustment strategies. In general, Kromm and White found that most irrigators do not like the idea of financial incentives or disincentives as means to reduce water demand, and preferred somewhat ambiguous measures, such as "improving irrigation efficiency" and "employing conservation tillage practices." Other popular adjustments were to "build reservoirs to hold water," and "encourage conservation laws" (Kromm and White 1985).

Irrigators were least supportive of policies designed to reduce water use or allocate water, whereas public-service professionals and ranchers tended to be supportive. A return to dryland farming was not preferred by farmers, and least liked by irrigators. Preferences for levels of institutional involvement (ranging from local to federal) varied among the states, with Texas respondents preferring "no agency" more frequently than those in other states. By and large, respondents seemed to prefer the status quo. In their more detailed study of respondents from southwestern Kansas, Kromm and White (1984) found that irrigators favored districts or local management (overseen by an elected board of directors), while those outside agriculture tended to prefer state enforcement.

In general, the theme is that irrigators try to protect their rights from further encroachment by the state and by other water users. These other users can be new waves of irrigators, including local interests who are trying to enlarge their enterprises, or municipal users looking to protect future supplies. The type of institutional reform that may occur when a resource is under perceived pressure, whether it be a redefinition of property rights, additional controls on withdrawals, or something largely symbolic, depends upon the historical context, the distribution of power, the magnitude and distribution of costs and benefits, the system in place, and the nature of the threat to the existing right holders. Roberts and Gros (1987), studying the politics of groundwater-management reform in Oklahoma, concluded that High Plains irrigators were able to fight off the reduced allocations proposed by urban legislators because of asymmetries in the distribution of benefits and costs of managing the groundwater. They argued that the advisory role adopted by the Oklahoma Water Resources Board was effective, because it would gradually build the voluntary cooperation needed to implement the existing groundwater-management legislation in a climate of hostility to regulation and severely constrained agency resources.

Emel and Brooks (1988) argued that reforms in groundwater management which occurred in the 1970s in response to increased pressure upon land and resources for agricultural expansion did not significantly alter the normative basis of laws governing water rights. What did change was the form of intervention. Rules promoting uniformity, regularity, rigidity, and increased certainty of expectation supplanted standards that promoted individualized case-specific justice, flexibility, and discretion. Freedom to act was constrained to enhance security. The management systems were rationalized. Some of the products of the change may be very positive in terms of education, professional staffing, monitoring or data collection, and greater security. On the other hand, rules tend to promote imprecision in goal satisfaction, inequity, and individualism.

All the work undertaken so far by geographers in attempting to answer why groundwater-management reform occurs, why the management patterns exist as they do, and what the meaning of reform is to society, points to a very rich future research agenda in water-resource geography.

Land Use and Groundwater Quality

Various geographic studies have examined the relation between land use and groundwater quality. These research efforts have promoted the greater consideration of quality groundwater protection programs under realistic funding, personnel, technology, and knowledge constraints. They emphasized the important relations that exist between land use and water-resource protection. Roberts and Butler (1984), for instance, stated that mapping land and water resources may, in many cases, be a more effective approach to identifying problems than more-expensive and traditional methods of monitoring and modeling.

Waterstone (1987) pointed out that understanding and comparing land/water relationships is not enough to make good risk assessments for cleanup or siting. Also needed is an understanding of the community's capacity to deal with the threat of water pollution in a timely manner. He believed that the objective risks posed by land/ water relations and response capability allowed for the evaluation of net risk. Rajagopal and Tobin (1988) focused upon groundwater experts. They elicited opinions from these experts regarding specific groundwater problems and potential management options, to see if this would lead to a more pragmatic approach or realistic strategy for dealing with groundwater-contamination problems. Stephenson and Lemmon (1983) took a comprehensive view and advocated state involvement in land-use controls to complement existing permitting strategies for enhanced groundwater protection.

Clearly, all the geographic contributions to groundwater issues are attempts to understand, explain, and improve the policy-making and management frameworks that have to do with the resource. With some exceptions, the research is rich in empirical material, rooted in context, and concerned with broadening the human and historical view of human-resource relationships. In this regard, the geographic contribution significantly and importantly departs from the neoclassical-economics and political-science approaches to resource management.

LAW AND WATER RESOURCES

Geographers concerned with water resources have long been entangled with water law and the development of water policy. Where and how water is used is not always controlled by socioeconomic factors, or even by physical/hydrologic variables that are traditionally evaluated by geographers. Water law may limit where and how water is used, leading to an overlap between law and geography. Geographers have had to consider water law when explaining water use, and have contributed to new law through the geographic evaluation of water problems.

U.S. water law evolved at different scales, with individual property rights being established by common law in the East, and the statutory appropriation doctrine being applied in the West. Because water is a mobile, reusable commodity, these individual rights were not exclusive, and a second level of law evolved to protect a wider public interest or to resolve problems that are beyond the scope of an individual property owner. Federal development of flood-control measures and large irrigation projects provided a fruitful field for geographic research, and much early work concentrated on these federal policies. Today the overlap between law and geography can be classified into three different research approaches: (1) legal-impact analysis, (2) legal-system analysis, and (3) legal research methodology for geographic issues. The first two approaches use geographic-research methodology, but all three require knowledge of the legal system and existing laws. Overlaps among these may occur in a single research project.

Legal-Impact Analysis

Legal-impact analysis can be divided into prelaw and postlaw. The prelaw component evaluates a water-resource issue and proposes a legal solution, while the postlaw component asks what impact a specific law has upon water resources. Prelaw policy analysis has a long tradition with geographers.

Geographers frequently study problems that require a legislative solution. These efforts have had an impact upon flood control, and upon other major water problems (Platt 1986). Geographers can provide important information that policy-makers can use in this respect (Lloyd 1982; Roberts and Butler 1984; Templer 1980).

Recent examples of postlaw analysis include an evaluation of the Colorado River Basin Salinity Control Act of 1974 (Wescoat 1986), and the Texas stream adjudication statute (Templer 1981). Galloway (1983) looked at environmental legislation and its impact on water-resource project development. Geographic evaluations of existing laws can determine whether goals of efficiency and equity are reached, and lead to recommendations for revision.

Legal-System Analysis

The second major approach, analyzing the legal system, has in the past consisted mostly of evaluating the responsibilities of institutions or agencies concerned. The way water-management institutions function is critical to understanding the way in which water is used, and studies such as Kromm's (1985) should be encouraged. But the legal system consists of more than institutions. Courts, legislatures, voters, and other elements of the legal system are subject to geographic analysis as well. To some extent these other elements have been incorporated in several excellent case studies (Kirn and Marts 1986; Sewell 1987). Emel and Brooks (1988) have gone a step further by comparing court-controlled water-resource administration with legislative agencies. More research needs to be done in this area.

Legal-Research Methodology for Geographic Issues

The third approach is to use standard legal-research methodologies to solve geographic problems. Geographers frequently use research techniques from other disciplines, and have, to a limited extent, begun to use legal research methods. Part of this comes from a small group of geographers with legal training who have an interest in water resources (Dworkin 1988; Matthews 1984; Platt 1985; and Templer 1978b). But geographers in general can do legal research (Wescoat 1985), and should do much more. Geographers are concerned with attributes of water quality, public access to water for recreation, conflicts between state and federal water policies, transporting water over long distances, and individual rights in water. A full understanding of these geographic issues cannot be obtained without legal research. Legal research is basically another form of library research, where the goal is to find "the law" and/or commentary about the law.

The strongest contributions geographers have made in the past are through legal-impact analysis and evaluating the institutional aspect of the legal system. This type of research will continue, because it is central to understanding water use. Geographers must also develop a better ability to undertake legal research and incorporate it into their problem-solving. Geographers can contribute in new ways by evaluating noninstitutional elements of the legal system and by evaluating the basis of legal theories. Noninstitutional elements of the legal system include court procedures and structure, the process of changing or creating legislation and administrative rules, and the intergovernmental conflicts resulting from overlapping jurisdiction.

Geographers can also combine geographic and legal theories. For example, Emel (1987) has proposed that the definition of an individual's water right be changed to reflect geographic reality. In some ways this is similar to legal-impact analysis, but it goes further in proposing a change in fundamental legal concepts. This redefinition is a matter for geography and law.

Geographers must also be aware of trends in the law as opportunities for new research. At present, one of the major unresolved legal conflicts is jurisdictional. Should the federal government or the states have control over water? Also of concern in the legal community are public rights to water for recreational, aesthetic, environmental, and health reasons. These public rights are often in conflict with state-granted individual rights. Because these issues have a geographic element, the solution should not be a legal one developed in isolation, but a combination of legal and geographic efforts. These contemporary legal issues provide an ideal opportunity to integrate law and geography in a new way.

FUTURE DIRECTIONS IN WATER-RESOURCES RESEARCH

Research undertaken by geographers has not only advanced theoretical concepts in hydrology and water resources, but has also led the way toward sound water-management practices. There is still much to be done, however, ranging from basic research into facets of the hydrological system to detailed analyses of water policies and management practices. Geographers, with their traditional skills in human-land relationships, are well-placed to enhance further research activities in water resources. In addition, geographers' emphases on spatial and temporal concerns will further contribute to water-resources research.

As detailed in this chapter, the future needs of water-resources research are twofold. First, geographers must continue to build a sound theoretical base, and to establish predictive and explanatory models. Second, applied research must continue problem-solving with a goal toward effective policy-making and water-management practices. In following these goals, water-resource researchers must be prepared to adopt new strategies that incorporate latest research techniques.

Theory Development and Model Formulation

Regardless of specialty within water resources, this chapter has argued for the development of theoretical bases for all research agenda. While physical geography has already established structured research frameworks, further, more sophisticated models are called for. Two aspects stand out:

> First, current research has demonstrated the need to consider scale, especially spatial aspects, in any consideration of hydrologic processes. This is pertinent to our understanding of stream discharges and sediment loads and the importance of extreme events in geomorphology.
>
> Second, a call is made to maintain theoretical relevance in the establishment of models. For instance, stochastic and other statistically based models have been developed which often provide a high level of predictive power, but they are abstract and operate at very low levels of explanation. Thus, a goal should be to reorient some research toward incorporating hydrologic theory, to enhance knowledge of hydrologic processes.

In addition, the need for deeper understanding of the human dimensions of water-resources research calls for the development of theory and explanatory models. In particular, management problems related to supply and demand, flood hazard, and groundwater require attention, and lend themselves to the construction of more-sophisticated theorizing. For example:

- In light of perceived water shortages, it is important to develop models of consumer behavior under different supply-and-demand conditions, incorporating models of commercial and industrial water use and models to forecast water demand by different sectors of society.
- In flood research, a common criticism has been the lack of unifying theories. Several scholars are now addressing this, and already various research frameworks designed to enhance hazards research have been proposed.
- Like so many aspects of water-resources research, groundwater research must also combine models that incorporate physical parameters with those of human behavior and decision-making. In the area of water quality, there is a need to develop models that will predict the spatial and temporal distribution of pollutant dispersal, as well as determine potential impacts on society.

Thus, while not losing sight of the specific nature of water resource problems, and the continued importance of the case-study approach, geographers should be developing and testing general principles and models that can be applied widely to water problems. Hydrologists and water-resource researchers must strive for further theoretical developments, looking for fundamental laws and relations to explain accurately the workings of the human/water environment.

Applied Problem-Solving and Policy Recommendations

Following logically from the first research direction—theory development and model formulation—is the need to apply knowledge to the development of sound water-management techniques. Once again, it is important to maintain a balance between the physical and human dimensions of research. Obviously, basic concepts of water quantity and quality are essential for effective management of the resource. At the same time, a clear grasp of social, economic, political, and legal factors is required before any policy recommendation can be seriously considered as a solution to water problems.

Of particular concern here is the impact of climatic change. This pervades the whole realm of water-resources research, since any climate change will have far-reaching impacts, not only of a physical nature but also in terms of impact upon society. Investigations into synoptic climatology, global hydrology, and models of drought and flooding will enhance society's capability to deal with these problems. Thus, specific case-study research into particular problems is warranted, since it will contribute to the accumulating body of knowledge on water resources, and will gradually lead to improved problem-solving capabilities. These studies should include water supply-and-demand management in rural and urban settings, flood-hazard studies, and groundwater management. Similarly, policy formulation and management practices will be greatly enhanced by further research into legal aspects of water resources.

In conclusion, water-resource geographers have maintained a full and active research agenda, incorporating the latest methodological approaches and many of the latest technological developments. Research has focused upon a broad range of spatial and temporal problems, but many more should be investigated. For instance, little has been said of research into international affairs, water-resource problems in developing countries, or water-resource concerns under different political systems. Also, it should be remembered that research by North American geographers does not necessarily represent the epitome of research in this area. There is equally excellent geographic work undertaken elsewhere. Water-resource geographers will benefit greatly from increased awareness of the work of other geographers, and from the research of scientists in other disciplines. The interdisciplinary nature of water research means that water-resource geographers must not be isolated from the rest of the academic world. The challenge is there for water-resource researchers to forge new bridges among disciplines, to develop greater scientific cooperation, and to establish strong, rigorous geographic research programs.

REFERENCES

Abrahams, A. D.; Parsons, A. J.; and Luk, S. 1988. Hydrologic and sediment responses to simulated rainfall on desert hillslopes in southern Arizona. *Catena* 15:103–17.

Aguado, E. 1982. A time series analysis of the Nile River low flows. *Annals of the Association of American Geographers* 72:109–19.

———. 1985a. Radiation balances of melting snow covers at an open site in the central Sierra Nevada, California. *Water Resources Research* 21:1649–54.

———. 1985b. Snowmelt energy budgets in southern and east-central Wisconsin. *Annals of the Association of American Geographers* 75:203–11.

American Geophysical Union. 1987. Rainfall fields: Estimation, analysis, and prediction. Reprinted from *Journal of Geophysical Research* 92:9551–714.

Anderson, T. L., ed. 1983. *Water rights: Scarce resource allocation, bureaucracy, and the environment*. San Francisco: Pacific Institute for Public Policy Research.

Babcock, M., and Mitchell, B. 1980. Impact of flood hazard on residential property values in Galt (Cambridge), Ontario. *Water Resources Bulletin* 16:532–37.

Barry, R. G. 1983. Research on snow and ice. *Reviews of Geophysics, Space Physics* 21:765–76.

Bartlein, P. J. 1982. Streamflow anomaly patterns in the U.S.A. and southern Canada—1951–1970. *Journal of Hydrology* 57:49–63.

Baumann, D. 1986. *Public perception of drinking water quality*. Discussion Paper, Symposium on National Drinking Water. Annapolis, MD: American Water Works Association.

———, and Dworkin, D. 1978. Water resources for our cities. *Resource Paper for College Geography* No. 78-2. Washington: Association of American Geographers.

———; Boland, J. J.; and Sims, J. H. 1980. *The evaluation of water conservation for municipal and industrial water supply—Procedures Manual*. Fort Belvoir, VA: U.S. Army Corps of Engineers, Institute for Water Resources.

———; Boland, J. J.; and Sims, J. H. 1984. Water conservation: The struggle over definition. *Water Resources Research* 20:428–34.

Beiswenger, R. E. 1983. Water management in the arid west: The Cheyenne water project. *Environmental Professional* 5:84–97.

Boland, J. J.; Dziegielewski, B.; Baumann, D. D.; and Optiz, E. M. 1984. *Influence of price and rate structures on municipal and industrial water use*. Fort Belvoir, VA: U.S. Army Corps of Engineers, Institute for Water Resources.

Brinkmann, W. A. R. 1985. Association between summer temperature and precipitation over the Great Lakes Region and water supplies to the Great Lakes. *Journal of Climatology* 5:161–73.

Brook, G. A., and Luft, E. R. 1987. Channel pattern changes along the lower Oconee River, Georgia, 1895/7 to 1949. *Physical Geography* 8:191–209.

Brookes, I. A. 1987. A medieval catastrophic flood in central west Iran. In *Catastrophic flooding*, eds. L. Mayer and D. Nash, pp. 225–46. Boston: Allen & Unwin.

Carson, R. T., and Mitchell R. C. 1987. *Economic value of reliable water supplies for residential water users in the state water project area for the Metropolitan Water District of Southern California*. Palo Alto, CA: QED Research, Inc.

Church, M. 1988. Floods in cold climates. In *Flood geomorphology,* eds. V. Baker, R. Kochel, and P. Patton, pp. 205–29. New York: Wiley.

———, and Mark, D. M. 1980. On size and scale in geomorphology. *Progress in Physical Geography* 4:342–90.

Clarke, G. M.; Jacobson, R. B.; Kite, J. S. and Linton, R. C. 1987. Storm-induced catastrophic flooding in Virginia and West Virginia, November, 1985. In *Catastrophic flooding,* eds. L. Mayer and D. Nash, pp. 355–79. Boston: Allen & Unwin.

Crawford, N. C. 1984. Karst development along the Cumberland Plateau escarpment of Tennessee. In *Groundwater as a geomorphic agent,* ed. R. G. LaFleur, pp. 294–339. Boston: Allen & Unwin.

Damron, J. E. and Anderson, D. A. 1987. Adaptation of EPA's Reach File for use in a regional anadromous fish data base. Paper presented at the annual meeting of the Association of American Geographers, Portland, Oregon.

Dey, B. 1982. Nature and possible causes of droughts on the Canadian Prairies—case studies. *Journal of Climatology* 2:233–49.

Drake, J. J. 1984. Theory and model for global carbonate solution by groundwater. In *Groundwater as a geomorphic agent,* ed. R. G. LaFleur, pp. 210–26. Boston: Allen & Unwin.

Dworkin, J. M. 1988. A critique of geopolitical change resulting from enactment of the Arizona Management Act. Paper presented at the annual meeting of The Association of American Geographers, Phoenix, Arizona.

Dziegielewski, B.; Optiz, E. M.; Baumann, D. D.; Davis, W. Y.; and Padin, C. M. 1986. *Optimal drought plans*. Volume 1: Protypal application of the "Drops" method. Reston, VA: United States Department of Interior, Geological Survey.

Eagleson, P. S. 1986. Global-scale hydrology. *Water Resources Research* 22:6s–14s.

Easterling, W. E., and Changnon, S. A., Jr. 1987. Climatology of precipitation droughts in Illinois based on water supply problems. *Physical Geography* 8:362–77.

Emel, J. L. 1987. Groundwater rights: definition and transfer. *Natural Resources Journal* 27: 653–73.

———, and Maddock, T. 1986. Effectiveness and equity of groundwater management methods in western United States. *Environmental Professional* 8:225–36.

———, and Yitayew, M. 1987. Water duties: Arizona's groundwater management approach. *Journal of Water Resources Planning and Management* 113:82–94.

———, and Brooks, E. 1988. Changes in form and function of property rights institutions under threatened resource scarcity. *Annals of the Association of American Geographers* 78:241–52.

Eney, A. B., and Petzold, D. E. 1988. The spatial variability of rainfall pH in the Washington D.C. metropolitan area. *Professional Geographer* 40:315–26.

Ford, D. C. 1980. Threshold and limit effects in karst geomorphology. In *Thresholds in geomorphology,* eds. D. Coates and J. Vitek, pp. 345–62. Boston: Allen & Unwin.

———. 1984. Karst groundwater activity and landform genesis in modern permafrost regions in Canada. In *Groundwater as a geomorphic agent,* ed. R. G. LaFleur, pp. 340–50. Boston: Allen & Unwin.

Foster, H. D. 1980. *Disaster planning: The preservation of life and property*. New York: Springer-Verlag.

Galloway, G. E. 1980. *Nonstructural measures in flood damage reduction activities*. Washington: U.S. Water Resources Council.

———. 1983. Delays in Federal water resources project development: The environment is not the problem. *Environmental Professional* 5:11–15.

Graf, W. L. 1977. The rate law in geomorphology. *American Journal of Science* 277:178–91.

———. 1983. Downstream changes in stream power in the Henry Mountains, Utah. *Annals of the Association of American Geographers* 73:372–87.

———. 1985a. The Colorado River: instability and basin management. *Resource Publications in Geography*. Washington: Association of American Geographers.

———. 1985b. Mercury transport in stream sediments of the Colorado Plateau. *Annals of the Association of American Geographers* 75:552–65.

———. 1988. Definition of floodplains along arid-region rivers. In *Flood geomorphology,* eds. V. Baker, R. Kochel, and P. Patton, pp. 231–42. New York: Wiley.

Granger, O. E. 1983. The hydroclimatonomy of a developing tropical island: a water resources perspective. *Annals of the Association of American Geographers* 73:183–205.

Greis, N. P., and Wood, E. F. 1981. Regional flood frequency estimation and network design. *Water Resources Research* 17:1167–77.

Grima, A. P. L. 1972. *Residential water demand: Alternative choices for management.* Toronto: University of Toronto Press.

———. 1973. The impact of policy variables on residential water demand and related investment requirements. *Water Resources Bulletin* 9:703–10.

Gruntfest, E. C., ed. 1986. What we have learned since the Big Thompson Flood. *Special Publication* No. 16. Boulder, CO: Natural Hazards Research and Applications Information Center.

Gupta, A. 1988. Large floods as geomorphic agents in humid tropics. In *Flood geomorphology,* eds. V. Baker, R. Kochel, and P. Patton, pp. 301–15. New York: Allen & Unwin.

———, and Fox, H. 1974. Effects of high-magnitude floods on channel form: A case study in Maryland piedmont. *Water Resources Research* 10:499–509.

Hirschboeck, K. K. 1987a. Catastrophic flooding and atmospheric circulation anomalies. In *Catastrophic flooding,* eds. L. Mayer and D. Nash, pp. 23–56. Boston: Allen & Unwin.

———. 1987b. Hydroclimatically defined mixed distributions in partial-duration flood series. In *Hydrologic frequency modeling,* ed. V. P. Singh, pp. 199–212. Dordrecht, Holland: D. Reidel.

———. 1988. Flood hydroclimatology. In *Flood geomorphology,* eds. V. Baker, R. Kochel, and P. Patton, pp. 27–49. New York: Wiley.

Irvine, K. N., and Drake, J. J. 1987. Process-oriented estimation of suspended sediment concentration. *Water Resources Bulletin* 23:1017–25.

Kates, R. W. 1962. Hazard and choice perception in flood plain management. University of Chicago Department of Geography Research Paper No. 78. Chicago: University of Chicago.

———. 1971. Natural hazard in human ecological perspective: Hypotheses and models. *Economic Geography* 47:438–51.

Kay, P. A., and Johnson, D. L. 1981. Estimation of Tigris-Euphrates streamflow from regional paleoenvironmental proxy data. *Climatic Change.* 3:251–63.

Keables, M. J. 1988. Spatial associations of midtropospheric circulation and Upper Mississippi River Basin hydrology. *Annals of the Association of American Geographers* 78:74–92.

Kirn, J. K., and Marts, M. 1986. The Skagit High Ross Dam controversy: Negotiations and settlement. *Natural Resources Journal* 26:261–89.

Knox, J. C. 1977. Human impacts on Wisconsin stream channels. *Annals of the Association of American Geographers* 67:323–42.

———. 1984. Fluvial responses to small scale climatic changes. In *Developments and applications of geomorphology,* eds. J. E. Costa and P. J. Fleisher, pp. 318–42. Berlin: Springer-Verlag.

———. 1985. Responses of floods to Holocene climatic change in the Upper Mississippi Valley. *Quaternary Research* 23:287–300.

———. 1987. Historical valley floor sedimentation in the Upper Mississippi Valley. *Annals of the Association of American Geographers* 77:224–44.

Kromm, D. E. 1985. Regional water management and assessment of institutions in England and Wales. *Professional Geographer* 37:183–91.

———, and White, S. 1984. Adjustment preferences to groundwater depletion in the American High Plains. *Geoforum.* 15:271–84.

———, and ———. 1985. *Conserving the Ogallalas: What's next?* Manhattan, KS: Kansas State University.

Linsley, R. K.; Kohler, M. A.; and Paulhus, L. J. 1958. *Hydrology for engineers.* New York: McGraw Hill.

Lloyd, W. J. 1982. Growth of the municipal water system in Ciudad Juarez, Mexico. *Natural Resources Journal* 22:944–71.

Luk, S; Abrahams, A. D.; and Parsons, A. J. 1986. A simple rainfall simulator and trickle system for hydrogeomorphological experiments. *Physical Geography* 7:344–56.

Magilligan, F. J. 1985. Historical floodplain sedimentation in the Galena River Basin, Wisconsin and Illinois. *Annals of the Association of American Geographers* 75:583–94.

Mahacek-King, V. L., and Shelton, M. L. 1987. Timber harvesting and the hydrologic response of Redwood Creek, California. *Physical Geography* 8:241–56.

Marcus, W. A. 1987. Copper dispersion in ephemeral stream sediments. *Earth Surface Processes* 12:217–28.

Martin, C. W., and Johnson, W. C. 1987. Historical narrowing and riparian vegetation expansion in the Medicine Lodge River Basin, Kansas, 1871–1983. *Annals of the Association of American Geographers* 77:436–49.

Mather, J. R. 1978. *The climatic water budget in environmental analysis*. Lexington, MA: Lexington Books.

———. 1981. Using computed stream flow in watershed analysis. *Water Resources Bulletin* 17:474–82.

Matthews, O. P. 1984. *Water resources, geography and law*. Washington: Association of American Geographers.

Miller, D. H. 1977. *Water at the surface of the earth: An introduction to ecosystem hydrodynamics*. New York: Academic Press.

Mitchell, R. C., and Carson, R. T. 1987. How far along the learning curve is the contingent valuation method? In *The role of social and behavioral sciences in water resources planning and management,* eds. D. D. Baumann and Y. Y. Haimes, pp. 65–98. New York: American Society of Civil Engineers.

Montz, B. E. 1987. Floodplain delineation and housing submarkets. *Professional Geographer* 39:59–61.

———, and Gruntfest, E. C. 1986. Changes in American floodplain occupancy since 1958: the experiences of nine cities. *Applied Geography* 6: 325–38.

Muckleston, K. W. 1983. The impact of floodplain regulations on residential land values in Oregon. *Water Resources Bulletin* 19:1–7.

Muller, R. A. 1976. Comparative climatic analyses of lower Mississippi River floods: 1927, 1973, 1975. *Water Resources Bulletin* 12:1141–50.

———. 1987. Water budget analysis. In *The encyclopedia of climatology,* eds. J. E. Oliver and R. W. Fairbridge, pp. 914–21. New York: Van Nostrand Reinhold.

National Research Council. 1988. *Estimating probabilities of extreme floods*. Washington: National Academy Press.

Nellis, M. D. 1985. Remote sensing techniques for identifying regional agrohydrologic parameters. *Papers and Proceedings of the Applied Geography Conferences*. 8:117–24.

Omernik, J. M., and Powers, C. F. 1983. Total alkalinity of surface waters—A national map. *Annals of the Association of American Geographers* 73:133–36.

O'Riordan, T. 1986. Coping with environmental hazards. In *Geography, Resources and Environment,* eds. R. W. Kates and I. Burton, pp. 272–309. Chicago: University of Chicago Press.

Palm, R. In press. *An integrative theory for natural hazards*. Baltimore: Johns Hopkins University Press.

Panciera, S. E.; Walsh, S. J.; and Vitek, J. D. 1986. Spatial and temporal variations of soil moisture in west-central Oklahoma. *Physical Geography* 7:258–74.

Parker, D. J., and Penning-Rowsell, E. C. 1983. Flood hazard research in Britain. *Progress in Human Geography* 7:182–202.

Phillips, J. D. 1985. Stability of artificially-drained lowlands: A theoretical assessment. *Ecological Modelling* 27:69–79.

———. 1986. The utility of the sediment budget concept in sediment pollution control. *Professional Geographer* 38:246–52.

———. 1988. Incorporating fluvial change in hydrologic simulations: A case study in coastal North Carolina. *Applied Geography* 8:25–36.

Platt, R. H. 1979. Options to improve federal nonstructural response to floods. Prepared for The Subcommittee on Science and Technology. Washington: U.S. Water Resources Council.

———. 1985. Congress and the Coast. *Environment* 27:12–17; 34–40.

———. 1986. Floods and man: A geographer's agenda. *Themes from the work of Gilbert F. White*. Vol. 2 of *Geography, resources and environment,* eds. R. W. Kates and I. Burton. Chicago: University of Chicago Press.

———, ed. 1987. Regional management of metropolitan floodplains: Experience in the United States and abroad. *Program on Environment and Behavior*. Monograph No. 45. Boulder, CO: Institute of Behavioral Science, University of Colorado.

———, and Nechamen, W. 1984. Flood loss reduction through interstate compacts: An under-utilized approach. Amherst, MA: Water Resources Research Center, University of Massachusetts.

Potter, K. W. 1987. Research on flood frequency analysis: 1983–1986. *Reviews of Geophysics* 25:113–18.

Pye, V.; Patrick, R.; and Quarles, J. 1983. *Groundwater contamination in the United States*. Philadelphia: University of Pennsylvania Press.

Rajagopal, R. 1984. Groundwater quality assessment for public policy in Iowa. *First Annual Report*. Chicago: The Joyce Foundation.

———. 1986. Conceptual design for a groundwater quality monitoring strategy. *Environmental Professional* 8:244–64.

———, and Tobin, G. A. 1988. Case studies in groundwater quality protection. *Project Report 1986–1988*. Chicago: The Joyce Foundation.

Renwick, W. H., and Ashley, G. M. 1984. Sources, storages, and sinks of fine grained sediments in a fluvial-estuarine system. *Geological Society of America Bulletin* 95:1343–48.

Rhoads, B. L. 1987. Changes in stream channel characteristics at tributary junctions. *Physical Geography* 8:346–61.

———. 1988. Mutual adjustments between process and form in a desert mountain fluvial system. *Annals of the Association of American Geographers* 78:271–87.

Roberts, R. S., and Butler, L. M. 1984. Information for state groundwater quality policy making. *Natural Resources Journal* 24:1015–41.

———, and Gros, S. 1987. Information for state groundwater policy making. *Natural Resources Journal* 24:1015–42.

Russell, C. S.; Arey, D. G.; and Kates, R. W. 1970. *Drought and water supply*. Baltimore: Johns Hopkins Press.

Sawyer, S. W. 1983. Water conservation: conflicting attitudes of planners and utility managers. *Environmental Professional* 5:124–33.

Scheibe, F.; Harrington, J.; and Ritchie, J. 1987. Remote sensing of suspended sediments of Lake Chicot, Arkansas. *Proceedings of the Sixth Corps of Engineers' Symposium*.

Segoe, L. 1937. Flood control and the cities. *American City* 52:55–56.

Sewell, W. R. D. 1987. The politics of hydro-megaprojects: damming with faint praise in Australia, New Zealand and British Columbia. *Natural Resources Journal* 27:497–532.

Shear, J. A., and Steila, D. 1973. The assessment of drought intensity by a new index. *Southeastern Geographer* 13:12–29.

Shelton, M. L. 1981. Runoff and land use in the Deschutes Basin. *Annals of the Association of American Geographers* 71:11–27.

———. 1982. Spatial characteristics of severe drought in the Sacramento River Basin. In *International Symposium on Hydrometeorology*, eds. A. I. Johnson and R. A. Clark, pp. 355–60. Minneapolis: American Water Resources Association.

———. 1985. Modeling hydroclimatic processes in large watersheds. *Annals of the Association of American Geographers* 75:185–202.

Skaggs, R. H. 1975. Drought in the United States. *Annals of the Association of American Geographers* 65:391–402.

Sommers, L. M., and Cullen, B. T. 1982. River water pollution in Norway: some regional environmental policy implications. *Professional Geographer* 34:208–19.

Steila, D. 1987. Drought. In *The encyclopedia of climatology,* eds. J. E. Oliver and R. W. Fairbridge, pp. 388–95. New York: Van Nostrand Reinhold.

Stephenson, L. K., and Lemmon, J. J. 1983. Land use controls to protect groundwater quality in the arid southwest. *Environmental Professional* 5:98–105.

Sun, C., and Brook, G. A. 1988. A hydrologic model for lower coastal plain watersheds, southeast United States. *Physical Geography* 9:15–34.

Tasker, G. D. 1982. Comparing methods of hydrologic regionalization. *Water Resources Bulletin* 18:965–70.

———. 1987. Regional analysis of flood frequencies. In *Regional Flood Frequency Analysis,* ed. V. P. Singh, pp.1–9. Dordrecht: D. Reidel.

Templer, O. W. 1978a. Texas groundwater law: inflexible institutions and resource realities. *Ecumene* 10: 6–15.

———. 1978b. Texas surface water law: the legacy of the past and its impact on water resources management. *Journal of Historical Geography* 4:11–20.

———. 1980. Conjunctive management of water resources in the context of Texas water law. *Water Resources Bulletin* 16:305–11.

———. 1981. The evolution of Texas water law and the impact of adjudication. *Water Resources Bulletin* 17:789–98.

———. 1985. Water conservation in a semi-arid agricultural region: The Texas High Plains. In *Forum of the Association for Arid Lands Studies,* vol. 1, ed. O. W. Templer, pp. 31–38. Lubbock, TX: Texas Tech University.

Thomas, F. H. 1983. Searching for national water policy: more to come. *Environmental Professional* 5:6–10.

Thornthwaite, C. W. 1948. An approach toward a rational classification of climate. *Geographical Review* 38:55–94.

———, and Mather, J. R. 1955. The water balance. *Publications in Climatology* 8:1–86.

Tobin, G. A., and Montz, B. E., eds. 1983. Water resources management: water resource modeling, institutional framework and policy strategies, and attitudinal implications. *Environmental Professional* 5:1–140.

———, and ———. 1988. Catastrophic flooding and the response of the real estate market. *Social Science Journal* 25:167–77.

———, and Newton, T. G. 1986. A theoretical framework of flood induced changes in urban land values. *Water Resources Bulletin* 22:67–71.

Trimble, S. W. 1983. A sediment budget for Coon Creek, the Driftless Area, Wisconsin, 1853–1977. *American Journal of Science* 283:454–74.

U.S. Geological Survey. 1984. National water summary 1983. *Water Supply Paper No. 2250*. Washington: U.S. Government Printing Office.

Walker, R. A., and Williams, M. J. 1982. Water from power: Water power and regional growth in the Santa Clara Valley. *Economic Geography* 58:95–119.

Walling, D. E. 1985. Physical hydrology. *Progress in Physical Geography* 9:97–103.

Waterstone, M. 1983. Toxics and groundwater: The development and application of net risk analysis. *Environmental Professional* 5:46–56.

———. 1987. Reducing groundwater pollution by toxic substances: Procedures and policies. *Environmental Management* 11:793–804.

Waylen, P. R. 1985a. A method of predicting daily flow peaks in the high-flow season. *Journal of Hydrology* 77:89–105.

———. 1985b. Stochastic flood analysis in a region of mixed generating processes. *Transactions of the Institute of British Geographers* 10:95–108.

———, and Woo, M. K. 1982. Prediction of annual floods generated by mixed processes. *Water Resources Research* 18:1283–86.

———. 1983. Annual floods in southwestern British Columbia, Canada. *Journal of Hydrology* 62:95–105.

———, and Caviedes, C. N. 1987. El Niño and annual floods in coastal Peru. In *Catastrophic flooding,* eds. L. Mayer and D. Nash, pp. 57–77. Boston: Allen & Unwin.

Weil, C., and Kvale, K. M. 1985. Current research on geographic aspects of schistosomiasis. *Geographical Review* 75:186–216.

Weirich, F. H. 1985. Sediment budget for a high energy glacial lake. *Geografiska Annaler* 67:83–99.

———. 1986. A study of the nature and incidence of density currents in a shallow glacial lake. *Annals of the Association of American Geographers* 76:396–413.

Welch, R.; Remillard, M.; and Slack, R. 1988. Remote sensing and geographic information system techniques for aquatic resource evaluation. *Photogrammetric Engineering and Remote Sensing* 54:177–85.

Wescoat, J. L. Jr. 1985. On water conservation and reform of the prior appropriation doctrine in Colorado. *Economic Geography* 61:3–24.

———. 1986. Impacts of federal salinity control on water rights allocation patterns in the Colorado River Basin. *Annals of the Association of American Geographers* 76:157–74.

———. 1987. The practical range of choice in water resources geography. *Progress in Human Geography* 11:41–59.

White, G. F. 1945. Human adjustment to floods. University of Chicago Department of Geography Research Paper No. 29. Chicago: University of Chicago.

———. 1964. Choice of adjustment to floods. University of Chicago Department of Geography Research Paper No. 93. Chicago: University of Chicago.

———. 1969. *Strategies of American water management*. Ann Arbor, MI: University of Michigan Press.

———, et al. 1958. Changes in urban occupance of flood plains in the United States. University of Chicago Department of Geography Research Paper No. 57. Chicago: University of Chicago.

———, and Haas, J. E. 1975. *Assessment of research on natural hazards*. Cambridge: MIT Press.

Whyte, A. V. T. 1986. From hazard perception to human ecology. In *Geography, resources and environment,* eds. R. W. Kates and I. Burton, pp. 240–71. Chicago: University of Chicago Press.

Willmott, C. J.; Rowe, C. M.; and Mintz, Y. 1985. Climatology of the terrestrial seasonal water cycle. *Journal of Climatology* 5:589–606.

Winkler, J. A. 1987. Diurnal variations of summertime very heavy precipitation in the eastern and central United States. *Physical Geography* 8:210–24.

———. 1988. Climatological characteristics of summertime extreme rainstorms in Minnesota. *Annals of the Association of American Geographers* 78:57–73.

Wolman, M. G. 1971. The nation's rivers. *Science* 174:905–18.

———. 1987. Water-quality trends in the nation's rivers. *Science* 235:1607–15.

———, and Miller, J. P. 1960. Magnitude and frequency of forces in geomorphic processes. *Journal of Geology* 68:54–74.

———, and Gerson, R. 1978. Relative scales of time and effectiveness of climate in watershed geomorphology. *Earth Surface Processes* 3:189–208.

———, and Wolman, A. 1986. Water supply: persistent myths and recurring issues. In *Geography, resources and environment,* eds. R. W. Kates and I. Burton, pp. 1–27. Chicago: University of Chicago Press.

Woo, M. K. 1983. Hydrology of a drainage basin in the Canadian high Arctic. *Annals of the Association of American Geographers* 73:577–96.

———, and Waylen, P. R. 1984. Areal prediction of annual floods generated by two distinct processes. *Hydrological Sciences Journal* 29:75–88.

———, and ———. 1986. Probability studies of floods. *Applied Geography,* 6:185–95.

Yevjevich, V. M. 1968. Misconceptions in hydrology and their consequences. *Water Resources Research* 4:225–32.

Coastal and Marine Geography

Niels West

Professional interest in issues related to coastal and marine resources has been stimulated by two related developments. First, the ability to exploit offshore resources including fish, oil, and gas provided research opportunities for geographically trained resource managers. Second, the environmental movement of the 1960s, initiated by population and industrial growth in the coastal region, contributed to professional interest on the part of several disciplines, including geography. Although geographers by no means were the only ones who seized this opportunity, our discipline has made significant contributions to the understanding of the problems encountered in this area. In large part, this is due to geography's spatial and holistic approaches to the study of environmental problems.

The initial attempt to create a marine component in geography probably began in 1954 when the Office of Naval Research (ONR) convened what is believed to have been the first Coastal Geography Conference. This meeting took place nearly 20 years before the idea of a marine geography subdiscipline materialized in 1974 (ONR 1954).

The development of coastal and marine geography as an established subdiscipline within geography is unique. By definition, this subject field should be treated as a regional study, however ill-defined; yet, scholarly contributions have been made almost exclusively by systematic geographers. Coastal and marine geographers can analyze their research problems holistically, identifying a region as a dynamic system. The relationship among climate, nearshore oceanography, and beach morphology is a case in point. Other holistic approaches include impacts on physical and biological systems that are the product of human-made modifications. In the coastal and marine

---

Great appreciation is extended to those who kindly reviewed and commented extensively on earlier drafts: James Allen, Lewis M. Alexander, Donald Davis, Gerald Krausse, Lisle Mitchell, Bruce Marti, and Karl Nordstrom. Appreciation is also extended to the anonymous reviewers, all of whom made many valuable suggestions. Thanks also are extended to Norbert Psuty, Jesse H. Walker, and Donald Davis who willingly shared their extensive marine geographical files. Any omissions or errors are, of course, the responsibility of the author.

COASTAL AND MARINE GEOGRAPHY

The Evolution of a Subdiscipline

To assess the present status of coastal and marine geography, it is useful to briefly review the origins of the subdiscipline. Occasional papers on the coastal and marine environment appeared in the *Annals of the Association of American Geographers* and *Geographical Review* long before the official establishment of the Coastal and Marine Geographers Specialty Group. Most of these early papers were written by systematic geographers. A small sample includes articles on marine transportation (Weigend 1958), fisheries (Cohen 1957), estuarine processes (Walker, Arnborg, and Peippo 1962; Russell and Howe 1935), and maritime jurisdictions (Alexander 1967).

The previously cited ONR sponsorship of geographic coastal research (Pruitt 1979) raises an important question, one that has implications for research currently underway. It appears that the initial emphasis in developing a specialty in marine geography (later to be expanded to include coastal geography) resulted from a need to resolve applied problems, as opposed to development of coastal and marine research methods and concepts. The 1954 ONR *Proceedings* addressed topics that are as much in vogue today as they were during the early 1950s. These concerns include beach and wetland processes, coastal classification, remote sensing (at that time limited to air-photo interpretation), and coastal and marine archaeology.

Current Status

The following analysis is based in part on a brief survey that was mailed to all members of the Coastal and Marine Geographers Specialty Group (hereafter referred to as COMAGS) during the Spring of 1988. In addition, analysis was undertaken of the 577 abstracts submitted to the annual meetings of the Association of American Geographers since 1977. The questionnaire sought information on scholarly and applied research, teaching, and service orientation. The abstract review was done to identify the most important themes that coastal and marine geographers have worked on during the past dozen years (Table 1).

Four themes are analyzed here: coastal geomorphology, marine recreation, marine boundaries, and marine transportation.[1] These represent the research interests of the majority of the coastal and marine geographers, as evidenced by papers presented at the AAG annual meetings, those appearing in refereed literature, and those presented before professional organizations that commonly include coastal and marine geographers.

---

[1] Although coastal management was identified as the second most important area (Table 1), this subject is discussed in other chapters (e.g., Water Resources), and thus is not covered here.

Table 1

Coastal and marine subfields based on submitted AAG annual meetings abstracts

Subfield	Abstracts
Coastal geomorphology	156
Coastal zone management	64
Marine recreation and tourism	64
Port, shipping, and urban waterfront	79
Marine boundaries, law of the sea, sea use	45
International coastal and marine geography	22
Renewable resources	20
Coastal hazards	20
Oil and gas	20
Cultural resource management	18
Coastal land use	14
Wetlands	13
Environmental quality	7
Coastal and marine climate	6
Dredging	6
Remote sensing	6
Tidal power	5
Mangrove/coral management	4
Estuarine research	2
Mariculture/aquaculture	2
Navy	2
Oceanography	2

Note: No attempt was made to eliminate papers that were withdrawn prior to the meeting, or no-shows.

This organization clearly does not fulfill the intent of the original marine geography interest group, which was primarily to analyze applied coastal problems. The successor organization, COMAGS, has sought to develop and apply geographic methods to the analysis of coastal problems, summarized by McArthur (pers. com. 1972) as a "spatial sorting of systematic information, wherein the spatial entity—is the ocean."

COASTAL GEOMORPHOLOGY

In this country, geographically trained geomorphologists represent a distinct minority compared to those from engineering and geology; however, the geomorphologist with a background in geography has several attributes not commonly exhibited by other coastal scientists. The geography-trained geomorphologist is more likely to study and interpret coastal forms and processes from a spatial point of view. Comparative (spatial) analyses and spatial classifications of coastal features are almost exclusively the domain of geographers (Alexander 1966). Perhaps a more-important attribute is that they often include social and behavioral aspects of the coastal geomorphological system. This expansion of the subject may complicate the study of coastal geomorphology, but it is especially important because a significant portion of the population has chosen to live, work, and play along the nation's four coastlines.

Two evolutionary phases can be discerned within coastal geomorphology. The first originated with R. J. Russell's work in Louisiana, while the second extends this tradition quantitatively.

The first coastal geomorphological phase was primarily descriptive of coastal erosional and depositional processes, with emphasis on Louisiana's coastline and nearshore. The philosophical impact of Russell is difficult to gauge. It is clear, however, that the traditions formed under his tutelage continue to influence the current generation of geomorphologists. The topics of recent refereed articles and abstracts presented at AAG annual meetings do not deviate significantly from earlier works. What is different is the methods by which the problems are being addressed.

Russell's work was concerned primarily with coastal problems within the U.S. Many of his students, colleagues, and followers addressed coastal geomorphological issues in other countries. These environments often were vastly different climatologically and morphologically from Louisiana's coastal regions (Walker and McIntire 1967; Psuty 1965; West 1965; Alexander 1969). In common with almost all geographical research up to the late 1950s, these studies were predominantly descriptive, interpretive, and classificatory (regional). Findings and conclusions were based on detailed field observations and mapping, from which spatial relationships and processes were deduced. The studies commonly inferred the evolution of coastal landscapes from present coastal landforms, based on fluvial, eolian, and nearshore processes (Orme and Tchakerian 1986). Research completed during this period was less concerned with the relations between energy inputs and how these were transformed into transportation systems affecting terrestrial and nearshore environments (Petnick 1984).

This descriptive tradition continues to the present. Many well-documented descriptive, nonquantitative studies that address domestic and international coastal geomorphological problems have been presented at recent AAG annual meetings (Stephenson 1986) and in domestic and international journals (Walker 1987).

The second phase follows several directions. One is concerned with human environmental problems, such as flood damage; another addresses the need to set aside ecologically sensitive areas while still ensuring economic growth for coastal states and communities. Much of this work was initiated by several pieces of national legislation. Three of the most significant are the Coastal Zone Management Act (CZMA), the Coastal Barriers Resources Act (COBRA), and the National Environmental Policy Act (NEPA). Several of these acts included planning and/or implementation programs which have provided funding and research opportunities for coastal geomorphologists and other coastal scientists.

Besides ONR's early support for physical and social marine sciences, the Office of Ocean and Coastal Resources Management (OCRM) and the National Park Service (NPS) have supported scientific research along the nation's coasts. From a geomorphological perspective, the NPS has been particularly supportive of erosion and sediment-transport studies along many national parks' barrier systems. Considerable research has been carried out, some by geographically trained geomorphologists working on Cape Hatteras, Sandy Hook, and Cape Cod.

Another direction taken within the later phase of coastal geomorphological research is the acceptance of humans as geomorphological agents. This recognition has resulted in more work which focuses upon social problems in an applied, interdisciplinary coastal-geomorphological context. The scale of these investigations runs the spectrum from answering questions about the location and protection of individual buildings, to comprehensive models for management of state and national resources. Geographical geomorphologists who conduct this work retain the principles of landform evolution as the basis of their evaluations, but often make extensive use of geographic principles from other branches of the discipline.

The specific question that has been addressed is the extent to which geomorphological processes can be attributed to human activity. In many ways, these studies have been made possible by new instrumentation which uses sonar and portions of the electromagnetic spectrum, including optics, to measure the wave environment and sediment transport in the nearshore marine environment (Allen 1988). These and other developments have provided coastal geomorphologists with increased ability to measure the behavior of coastal systems under any number of different environmental constraints (Allen 1981).

Historically, coastal geomorphologists have favored analyzing the landforms that change most rapidly. This stems from the early work at Louisiana State University (LSU), which was concerned with the Mississippi River Delta and associated wetlands—two systems that are subject to dramatic landform changes over brief periods. The emphasis changed during the 1960s. Several geographers from LSU established highly reputable research and educational centers. Rutgers' Center for Coastal and Environmental Studies has undertaken many research projects on Sandy Hook and the New Jersey and New York barrier beaches. More recently, the Center has participated in the NPS research program within the Cape Cod National Seashore and Gateway National Recreational Area.

Another example is the Department of Environmental Science at the University of Virginia, where scientists have been active in quantifying shoreline changes on the dynamic barrier islands of the mid-Atlantic coast (Dolan et al. 1977). Using a quantitative approach, the Department of Geography and Geology at Scarborough College (University of Toronto) has investigated the rapidly eroding shores of the Great Lakes. Sherman and his colleagues at the University of Southern California in Los Angeles have continued this quantitative tradition (Sherman, Greenwood, and Bauer 1988). Another regional geomorphological center is the University of California–Los Angeles, where considerable work has been conducted on the interpretation of past landscapes.

In what direction will the subdiscipline move in the remainder of the twentieth century? The increased dependency on high technology will no doubt continue, enabling coastal geomorphologists to extend their studies of the various sediment-transportation systems that impact coastal landforms. At the present time, most coastal geomorphologists tend to analyze the impact of one system on the coastal environment. There clearly is a need to better understand the relations among the various factors that affect the physiography of the local beach environment.

The reluctance of many nongeography coastal scientists to include socioenvironmental factors in the landform-evolution equation represents a unique opportunity for the geographically trained geomorphologist. Management schemes are being formulated for several of our state shorelines and national seashores, historical monuments and wildlife-management areas. These areas provide future opportunities for coastal geomorphologists willing to accept these challenges.

Members of this subdiscipline have been reluctant to move into a forecasting mode (Sherman, Greenwood, and Bauer 1988) in which the objective is to project coastal geomorphological events in time and space. While this is the most difficult scientific stage to enter, forecasting—whether social, physical, or biological—represents the ultimate attainment by science. Limited efforts of hindcasting have been made by Stow and Chang (1987) and Allen and Nordstrom (1978). It appears that forecasting represents an important challenge to the geomorphologists of the twenty-first century.

MARINE RECREATION

Marine recreation and tourism include leisure-time activities that occur on, or are enhanced by, a coastal location. The marine-recreational subdiscipline has been left behind by geographers who have specialized interests in urban and other recreational issues. The majority of outdoor recreational research done by geographers in the Recreation, Tourism, and Sports Specialty Group has had a distinct landbased tradition, perhaps due to the interior distribution of geography departments that are interested in recreational research and education. Similarly, very little urban-waterfront recreation research has been undertaken by the urban, coastal, or recreation groups.

Within the past few years, some of these problems appear to have been partially alleviated. Several sessions, jointly sponsored by COMAGS and the Recreation, Tourism, and Sports Specialty Group, have provided marine-recreational researchers with an important outlet for interim research results. In addition, traditional geographical and nongeographical journals, such as *Coastal Management Journal, Tourism Management, Journal of Leisure Research,* and *Environmental Behavior,* have increasingly published the results of marine-recreational research.

This subject represents a field to which disciplines such as resource economics, community planning, and sociology can contribute. The geographer's most important contribution to the study of outdoor recreation is the ability to analyze coastal recreational activities using spatial methods and techniques. The majority of the marine-recreational geographical research has been descriptive and systematic, and for analytical purposes may be broken into three groups—historical, structural or systematic, and environmental.

Historical Research

Most of the historical research undertaken by recreational geographers has analyzed passive recreational activities in the context of land-use change. Several geographers, among them Meyer-Arendt (1986) and Demars (1988), have studied the evolution of coastal resort settlements and camps. These range from facilities open to the general public (Davis and Detro 1975) to recent tourist/recreational communities in North Carolina (Spyrou 1988). Other authors have analyzed the unique religious camps that prevailed during the latter part of the nineteenth and early part of the twentieth centuries.

Still other studies have emphasized the socioenvironmental impacts which such activities have had on the preexisting land-use pattern. Several of these are historical geographical vignettes. They describe the recreational land-use evolution in response to changing cultural-social mores (Gerlach 1987). Others have analyzed the impact of a changing transportation system (Stansfield 1980). One of the few exceptions is Fournier's study (1984) in which he sought to use variations of the gravity model to account for recreational use in rural and urban environments. Meyer-Arendt (1988) developed a "model" to account for changing land uses encountered in numerous waterfront communities as the primary means of public transportation changed from rail to automobile.

Structural Research

Most recreational researchers analyze marine recreation from a sectoral or topical approach. This is based on the assumption that use of the marine environment varies significantly from one activity to another. The overwhelming number of these sectoral studies are descriptive, and few of them have been summarized. In addition, much of it is "fugitive" or published as "gray literature" which is not readily available to the scientific community. Another characteristic of the sectoral research is the timing and emphasis on terrestrial uses, as opposed to nearshore or marine-resource uses.

Marine-recreational research has been "clustered," i.e., studies focus on a specific topic or activity. There have been few attempts to study marine recreational systems in which several user groups may occupy a given recreational environment. What are the potential conflicts, and in what way can different uses be accommodated within a given site? Such studies often require the involvement of several disciplines—including geography. Before discussing the more important of these clusters, it is important to keep in mind that each cluster is made up of a relatively small number of geographers who have tended to develop their own research approaches with findings and conclusions unique to specific activities. There is an obvious danger to this approach, as findings, methods, and conclusions are often not tested rigorously.

At the present time, only a limited number of clusters exists, including beach recreational research (Fabbri 1987; McCloy 1987), boating (West 1987), and fishing (Roundy et al. 1985). Although beach activities are among the most popular outdoor recreational activities, many of the problems encountered within this use category can be found among other coastal user groups. One approach to the study of a given recreational site or activity is to identify those factors that appear to inhibit usage. For convenience, these may be broken down into two factors: external (e.g., accessibility[2]) and internal (e.g., social and physical conditions of the site itself). External factors that influence urban beach use have been studied by Heatwole and West (1980).

Another aspect of coastal recreation concerns the users' perception. How do different user groups perceive the coastal and marine environment, and how do those

---

[2]Access and accessibility in this context refer to two separate but related concepts. Access means the physical strip of land required to enable users to get onto the beach and water. Accessibility means those factors that may either encourage or constrain beach use, and include availability of public transportation, parking, discriminatory pricing practices, etc.

perceptions affect the user's decision-making process? A study that incorporates both internal factors affecting beach use and perception is Hecock's (1966) functional segregation study (as opposed to ethnic or social segregation). In this work, he shows how teenagers and young families favor different beaches, thereby avoiding conflict between the two groups.

There is clearly a need to integrate the physical characteristics of the recreational environment with the perceptions of that environment by the different users. For instance, is the perceived beach quality a more important factor influencing how the beach is used than its official water-quality class? Answers to such questions could make more effective use of the coastal resource, and might enhance the relationship between users and the federal agencies that serve their needs. The U.S. Coast Guard is especially important to recreational users, as this agency provides navigational aid, monitors vessel-source pollution, and provides search and rescue (SAR) services to recreational and commercial boaters. Similarly, many of the functions of the U.S. Army Corps of Engineers affect the recreational public, as the Corps is the agency in charge of dredging navigational channels. Finally, NOAA provides a number of services, including weather forecasting, coastal-zone management, and the Sea Grant program, all of which are important to the marine recreational community. The relationships between the users and the various services represents a fertile field for geographical research.

Environmental Research

Another research area is the nearshore marine environment. This has been almost completely neglected by geographers. From an outdoor recreational point of view, this represents a growth area, as more people engage in an ever-increasing number of recreational activities along the nation's shores. A few studies (Mann 1976) have sought to map the recreational use of the nearshore—a difficult problem, considering the spatial multidimensionality of that environment. For example, most boaters use the ocean only temporarily. Some use only the water surface, while others (commercial and recreational fishermen) use a portion of the water column as well. Still others use the water surface, column, and bottom. Diving is another example of a marine recreational activity that uses all three environmental components. These problems are spatial in nature, and require a spatial (geographical) approach to their solution.

The subdiscipline is young, and so it is not surprising that the majority of studies are descriptive. They emphasize historical, structural, or environmental aspects of coastal and marine recreation. Comparative analyses in time and space are few, no doubt because the research community feels it is important to describe the phenomenon before attempting classification and forecasting. As more is learned about marine recreation, one would expect researchers to address problems associated with the total recreational-resource system.

In the past, the pressure on recreational environments was less intense than today; consequently, a sectoral or activity-based analysis of marine recreation made sense. However, in many areas, especially those close to cities and towns, this no longer appears to be justified. The principal contribution of geographers to the study of marine recreation may be in showing how a limited resource base can be managed without impacting the environment, and with a minimum of conflict among competing activities.

Another category of marine recreational research deals with the potential economic, social, and physical environmental impacts that the recreational activity may have, both domestically and in other countries. Marine environmental impacts are both the least studied and the most fertile area in which geographers may make a major contribution.

Americans, and increasingly citizens from other developed countries, have long sought tourist and recreational opportunities in countries other than their own. As our standard of living has increased, our tourist destinations have extended and impacted international destinations not previously affected. Many areas that received few international visitors a generation ago are now adversely affected environmentally, economically, and socially. Matley (1976) described the implications of tourism developments in tropical and subtropical regions, where the environment is capable of supporting only the local population. This author mentions the potential contributions that geographers could make in allocating often scarce resources between tourist and resident populations.

Marine environmental impacts that occur domestically correlate highly with growth in coastal recreational activities. Capital improvements such as marinas, boat ramps, mooring areas, and beach-front condominiums often impact wetlands, dunes, and shellfishing areas. Some geographers have collaborated on assessing impacts caused by such construction activities (Nordstrom 1987), but there is still greater potential for geographers to affect future recreational-investment decisions.

Nonconstruction recreational activities have also adversely affected the physical environment. Boating-generated human waste (West, Heatwole, and Smith 1982), while regulated, represents one such impact. Geographers have undertaken studies addressing the environmental impacts caused by outdoor recreational vehicles (ORV) and pedestrian traffic on dunes (Raup 1978).

Geographers have undertaken promising work in marine environmental studies, but quantitative studies are still few in number. With the exception of contributions by historical marine recreational geographers, previously cited, there is a need to formulate and test hypotheses that have a wider applicability than the specific cases being addressed.

MARINE BOUNDARIES

Professional interest in marine regions and boundaries preceded numerous attempts by the United Nations to address marine resource-allocation problems. A majority of these issues resulted from unilateral extensions of territorial sea by several South American nations during the 1950s and 1960s, a practice that was later adopted by developing nations in Africa, Asia, and elsewhere. The expanded jurisdictions claimed by several Pacific and Atlantic nations did not conflict with other nations' maritime claims. The seaward extensions "only" infringed upon the maritime space held in common by the "family of nations." However, as more and more nations extended their seaward boundaries, conflicts arose in several enclosed or semienclosed seas. Several examples of these conflicts are addressed in Melamid (1986).

Most seafaring nations, including the United States, United Kingdom, and to some extent the Soviet Union, are not generally in favor of such developments, viewing any limitations on marine navigation seaward of the territorial sea as a restriction of "freedom of navigation" principles. Such restrictions are especially problematic for the U.S. and its allies, because maritime accessibility is of paramount importance in cases of conflict.

During the years immediately following the Second World War, up to the passage of the United Nation's Convention on the Law of the Sea (UNCLOS III) in 1982, geographers played an important role in defining the various zones now recognized. It was natural that geographers were called upon to participate in this research, as much of it refers to the definition of a marine region. In fact, several geographers, among them L. Alexander (1973) and Glassner (1986), have analyzed marine boundaries from a regional perspective. Much of this work was conceptual, but helped to identify issues and principles that were eventually adopted at the UNCLOS (Alexander 1959).

Maritime boundary research can be divided into three categories:

> The first concerns seaward extension of the coastal jurisdiction. Many coastal nations extended their territorial sea 50, 100, and in a few instances 200 miles from their coast, while others have exerted jurisdiction over some or all of the resources located seaward of their traditional territorial-sea boundary.
> The second research category concerns the location of a coastal boundary that extends perpendicular from a shoreline separating the nearshore marine environment between adjacent coastal states. An example of such a dispute is determination of the Rhode Island/Connecticut marine boundary.
> The third research category concerns primarily the effects that a boundary has upon control and management of a particular resource or maritime activity (Morris 1986).

There are two considerations common to the three groups. First, the definition of a "boundary" is based primarily on resource considerations and the perceived need for protection against "foreign" intrusion. Second, to be effective, the coastal state must be able to exert its jurisdiction over the claimed marine territory to prevent unauthorized resource extraction and/or access.

Resolution of boundary disputes between adjacent states, while often influenced by resource considerations, was sometimes achieved through bilateral negotiation (Easterly 1982). However, many more issues remain unresolved. A few countries have agreed to final arbitration through the World Court; however, most conflicts exist with no immediate resolution in sight.

Many geographers have analyzed the boundary issues as they pertain to a specific resource. Besides oil and gas exploration and development, where the determination of a boundary can add significantly to a coastal state's income (Maier 1984), other resource conflicts include fish (Alexander 1960), shellfish (Easterly and Keyser 1984), the right to conduct oceanographic research inside a coastal nation's 200 mile territorial sea (Alexander 1970), and shipping (both commercial and military).

What are the most important problems and opportunities that confront geographers interested in boundary issues? Research addressing deep-ocean issues has largely been undertaken by a different group of geographers, compared to those working in the coastal region. There are some indications, largely initiated from outside geography, that the separation is diminishing between those investigating the nearshore geographical problems and those primarily interested in offshore research. One reason is that most nearshore-boundary issues can be resolved by applying existing international legal principles. Nearshore and offshore boundaries represent a continuum with no distinct change at the territorial sea, exclusive economic zone, or other national or international maritime division. Consequently, it is quite possible that researchers who in the past have emphasized problems associated with one region or the other will increasingly treat the marine environment as a system. As we learn more about the physical, social, and economic potential of the deep ocean, there is every likelihood that the distinction between coastal and marine geographers will become increasingly blurred.

MARINE TRANSPORTATION

For organizational purposes, marine-transportation geographers may be divided into two categories: those concerned with port location and operation (Hayuth 1977), and those concerned with shipping (Leving 1978).

The conversion to containerization has been one of the fundamental changes affecting North American ports during the 1950s and 1960s. Research has been conducted on the impacts of containerization (Smith 1982) and associated technical developments (Jaworski and Raphael 1976) and socioeconomic developments (Goodwin 1988) along the urban waterfront. The evolution of the traditional general cargo/bulk port of the 1950s and 1960s into highly specialized ports and port facilities has resulted in a corresponding increase in research efforts to analyze the morphology and socioenvironmental impacts on the port's hinterland (Boehm 1980). Studies have dealt with load centers (Marti 1988a), superports (Schwartz 1978), and offshore oil ports (McIntire 1978; Reed 1978). Other researchers have studied cargo flows, including coal (Osleeb, Ratick, and Lewis 1981; Osleeb 1982), liquid bulk cargoes (Stanley 1978), and cruise shipping (Blick 1979; Marti 1988b).

As in most of the previously discussed marine study areas, research dealing with ports and shipping has been, with few exceptions, largely descriptive and ideographic.

THE FUTURE

The vast majority of coastal and marine geographical research conducted during the past generation has been descriptive, with a few quantitative studies being undertaken, especially by coastal geomorphologists. Nearly all of these research efforts have been topical. Individual researchers have attempted to incorporate aspects of other subdisciplines into their work, but such efforts have been few. Most research has been site-specific, with few efforts made to draw conclusions that might be applicable in other areas or under variant circumstances. A number of comparative studies have been undertaken, but their findings generally have not been tested in other situations.

Where, then, should coastal and marine geography head as we approach the twenty-first century? Many coastal and marine geographers perceive a collective need to move our research from descriptive and classification exercises toward the development of forecasting models. There also is a need to integrate our own research interests with those of geographers who work in corollary areas. These include, but are not limited to, transportation, remote sensing, wetlands, urban, geomorphology, resource management, and perception, to mention just a few of the more-obvious specialties with which we share common interests. This will be difficult for many of us, but it is paramount if the subdiscipline is to grow and mature. The challenge is there; the only question is: Are we sufficiently committed to accept these challenges in the coming years?

REFERENCES

Alexander, C. S. 1966. A method of descriptive shore classification and mapping as applied to the north east coast of Tanganyika. *Annals of the Association of American Geographers* 56:128–40.

———. 1969. Beach ridges in north eastern Tanzania. *Geographical Review* 59:104–22.

Alexander, L. M. 1959. The expanding territorial sea. *Professional Geographer* 11(4):6–8.

———. 1960. Offshore boundaries & fisheries in Northwest Europe. *The British Yearbook of World Affairs* 14:236–59.

———. 1967. Offshore claims of the world. In *The law of the sea: Offshore boundaries and zones,* pp. 71–84. Columbus, OH: Ohio State University Press.

———. 1970. Alternative methods for delimiting the outer boundary of the Continental Shelf. Washington: Office of External Research, U.S. Department of State.

———. 1973. Coastal state competence to regulate traffic in straits and other areas near the coasts vs. world community trends to maximize vessel mobility. In *Hazards of Marine Transit,* eds. T. A. Clingan and L. M. Alexander, pp. 19–28. Cambridge: Ballinger.

Allen, J. R. 1981. Beach erosion as a function of variations in the sediment budget, Sandy Hook, New Jersey, U.S.A. *Earth Surface Processes* 6:139–50.

———. 1988. Nearshore sediment transport. *Geographical Review* 78:(2)148–56.

———, and Nordstrom, K. F. 1978. A simulation model of shoreline dynamics, Sandy Hook, New Jersey. Paper presented at the annual meeting of the Association of American Geographers, New Orleans, Louisiana.

Blick, J. D. 1979. The cruise as a major touristic activity. Paper presented at the annual meeting of the Association of American Geographers, Philadelphia, Pennsylvania.

Boehm, R. G. 1980. The changing concept of Port Hinterland. Paper presented at the annual meeting of the Association of American Geographers, Louisville, Kentucky.

Cohen, S. 1957. Israel's fishing industry. *Geographical Review* 47 (1):66–85.

Davis, D., and Detro, R. A. 1975. Louisiana's marsh as a recreational resource. In *Geoscience and Man, Research Techniques in Coastal Environments* 18:311–20.

Demars, S. E. 1988. Worship by the sea camp meeting and seaside resorts in 19th century. Paper presented at the annual meeting of the Association of American Geographers, Phoenix, Arizona.

Dolan, R. et al. 1977. Shoreline forms and shoreline dynamics. *Science* 197:49–51.

Easterly, E. 1982. Principles of marine boundary delimitation and their relation to exploration. Paper presented at the annual meeting of the Association of American Geographers, San Antonio, Texas.

———, and Keyser, G. L. 1984. Coastal boundaries shell dredging and reef restoration. Paper presented at the annual meeting of the Association of American Geographers, Washington, DC.

Fabbri, P., 1987. The organizational structure of recreational use along the beaches. Paper presented at the annual meeting of the Association of American Geographers, Portland, Oregon.

Fournier, J. P., 1984. Recreational Hinterland and the recreational gravity model as applied to Grand Isle, Louisiana. Paper presented at the annual meeting of the Association of American Geographers, Washington, DC.

Gerlach, J. 1987. Spring break at Padre Island: A new type of tourism. Paper presented at the annual meeting of the Association of American Geographers, Portland, Oregon.

Glassner, M. I. 1986. The new political geography of the sea. *Political Geography,* 5 (1):6–8.

Goodwin, R. F. 1988. Industrial change, water dependency and the character of urban waterfronts. Paper presented at the annual meeting of the Association of American Geographers, Phoenix, Arizona.

Hayuth, Y. 1977. The effects of coastal management on port development. Paper presented at the annual meeting of the Association of American Geographers, Salt Lake City, Utah.

Heatwole, C. A., and West, N. 1980. Mass transit and beach access in New York City. *Geographical Review* 70 (2):210–17.

Hecock, R. D. 1966. Public beach recreation opportunities and patterns of consumption on Cape Cod. Ph.D. diss., Graduate School of Geography, Clark University, Worcester.

Jaworski, E., and Raphael, N. 1976. The cooperative dredge spoil disposal program in the Great Lakes. *Coastal Zone Management Journal* 3 (1):91–96.

Leving, G. L. 1978. Network structure of the coastal shipping system in Thailand. Paper presented at the annual meeting of the Association of American Geographers, New Orleans, Louisiana.

McArthur, 1972. Letter to H. J. Walker, June 27.

McCloy, J. 1987. Risk management of public open water recreational beaches in the U.S. Paper presented at the annual meeting of the Association of American Geographers, Portland, Oregon.

McIntire, W. 1978. Environmental requirements for construction and operation of LOOP. Paper presented at the annual meeting of the Association of American Geographers, New Orleans, Louisiana.

Maier, E. 1984. The maritime boundary in the Gulf of Maine between the U.S. and Canada. Paper presented at the annual meeting of the Association of American Geographers, Washington, DC.

Mann, Roy Associates, Inc. 1976. Recreational boating on the Tidal Waters of Maryland: A management study. Maryland Department of Natural Resources, Annapolis, Maryland.

Marti, B. 1988a. The evolution of Pacific Basin load centers. *Marketing Policy and Management* 15 (1):57–66.

———. 1988b. Geography and the cruise ship port selection process. Paper presented at the annual meeting of the Association of American Geographers, Phoenix, Arizona.

Matley, I. M. 1976. The geography of international tourism. *AAG Resource Paper* No. 76–1, Washington, DC.

Melamid, A. 1986. The division of narrow seas. *Political Geography* 5 (1):39–42.

Meyer-Arendt, K. J., 1986. Coastal resort evolution: The case of Estero Island, Florida. Paper presented at the annual meeting of the Association of American Geographers, Minneapolis, Minnesota.

———. 1988. Morphologic patterns of resort evolution along the Gulf of Mexico. Paper presented at the annual meeting of the Association of American Geographers, Phoenix, Arizona.

Morris, M. A. 1986. Maritime geopolitics in Latin America. *Political Geography Quarterly* 5 (1):43–55.

Nordstrom, K. 1987. Management of tidal inlets on barrier island shorelines. *Journal of Shoreline Management* 3 (3):169–90.

ONR. 1954. *Proceedings of the Coastal Geography Conference*. February 18. Washington, DC.

Orme, A. R., and Tchakerian, V. P. 1986. Quaternary dunes of the Pacific Coast of the Californias. In *Aeolian Geomorphology,* ed. W. G. Nickling, pp. 149–75. New York: Allen & Unwin.

Osleeb, J. B. 1982. Implications of worldwide increased demand for coal on U.S. Ports. Paper presented at the annual meeting of the Association of American Geographers, San Antonio, Texas.

———; Ratick, S. J.; Lewis, G. K. 1981. Aspects of a Programming Model to Analyze Coal Handling in New England Ports. Paper presented at the annual meeting of the Association of American Geographers, Los Angeles, California.

Petnick, John. 1984. *An introduction to coastal geomorphology*. London: Edward Arnold Publishers.

Pruitt, E. L. 1979. The Office of Naval Research and Geography. *Annals of the Association of American Geographers* 69 (1):103–8.

Psuty, N. P. 1965. Beach ridge development in Tabasco, Mexico. *Annals of the Association of American Geographers* 55:112–24.

Raup, H. A. 1978. Patterns of beach vehicle use at Ocracoke, Cape Hatteras National Seashore, North Carolina. Paper presented at the annual meeting of the Association of American Geographers, New Orleans, Louisiana.

Reed, W. 1978. Facilities and operation of Louisiana Offshore Oil Port, Inc. Paper presented at the annual meeting of the Association of American Geographers, New Orleans, Louisiana.

Roundy, R. W.; Tucker, R.; Weinstein, N.; Whelan H. 1985. Perception and mitigation of toxic risk by New Jersey fishermen. Paper presented at the annual meeting of the Association of American Geographers, Detroit, Michigan.

Russell, R. J., and Howe, H. V. 1935. Cheniers of Southwestern Louisiana. *Geographical Review* 25 (3):449–61.

Schwartz, A. R. 1978. Supersport concept overview. Paper presented at the annual meeting of the Association of American Geographers, New Orleans, Louisiana.

Sherman, D. J. 1988. Coastal Geomorphology. *Geographical Review* 78 (2):116–18.

———; Greenwood, B.; and Bauer, B. O. 1988. Longshore current variability and nearshore sediment transport. Paper presented at the annual meeting of the Association of American Geographers, Phoenix, Arizona.

Smith, D. A. 1982. Containerizaton and river and lake ports. Paper presented at the annual meeting of the Association of American Geographers, San Antonio, Texas.

Spyrou, M. 1988. Land development and landuse evolution at Wrightsville Beach, NC. Paper presented at the annual meeting of the Association of American Geographers, Phoenix, Arizona.

Stanley, W. K. 1978. The liquid natural gas trade. Paper presented at the annual meeting of the Association of American Geographers, New Orleans, Louisiana.

Stansfield, C. 1980. The railroads and resorts. South Jersey's rail net and the development of the Jersey Shore. Paper presented at the annual meeting of the Association of American Geographers, Louisville, Kentucky.

Stephenson, R. A. 1986. The shapes of capes. Paper presented at the annual meeting of the Association of American Geographers, Minneapolis, Minnesota.

Stow, D. A., and Chang, H. H. 1987. Numerical simulation of a coastal entrance channel process in Southern California. Paper presented at the annual meeting of the Association of American Geographers, Portland, Oregon.

Walker, H. J. 1987. Riverbank erosion in the Colville River Delta. *Geografiska Annaler* A69 (1):61–70.

———; Arnborg, L.; Peippo, J. 1962. Suspended load in the Colville River, Alaska. *Geografiska Annaler* A4:195–210.

———, and McIntire, W. G. 1967. Tropical cyclones and coastal morphology, Mauritius. *Annals of the Association of American Geographers* 57:582–96.

Weigend, G. 1958. Some elements in the study of port geography. *Geographical Review* 48 (2):185–200.

West, N. 1987. Boat usage on Narragansett Bay. Paper presented at the annual meeting of the Association of American Geographers, Portland, Oregon.

———; Heatwole, C. A.; Smith, L. 1982. Environmental improvement on Narragansett Bay as a result of Section 312 implementation of the Federal Water Pollution Control Act. *Coastal Zone Management Journal* 10 (1-2):125–40.

West, R. C. 1965. Mangrove swamps of the Pacific Coast of Colombia. *Annals of the Association of American Geographers* 46:98–121.

HISTORICAL AND CULTURAL CONTRIBUTIONS TO GEOGRAPHIC UNDERSTANDING

Historical Geography

Carville Earle | Lary Dilsaver |
David Ward | Peter J. Hugill |
Robert D. Mitchell | Cole Harris | Robert C. Ostergren |
James Lemon | Sean Hartnett | Peter Goheen |
William Wyckoff | J. H. Galloway | Ronald Grim |
Stanley W. Trimble | Ronald U. Cooke |
Howard Botts | Jeanne Kay | Craig E. Colten

The mid-1980s are not the mid-1950s. Neither geography nor historical geography has a single, overriding paradigm within which inquiry is conducted. Instead, a profusion of paradigms jockeys for preeminence.

How different this is from the placid days of the early fifties, when geography took its last thorough inventory of the field. Geography seemingly had achieved a consensus in areal differentiation, in the individuality and character of regions. The Hartshornian view of the world, though extremely helpful for inventorying American geographical research, was nonetheless illusory. The consensus was cracking; reformers propositioned geography as science, as logical positivism, as spatial relations. The fight was on: wherein lay the "true geography"—in the abstract geometry of maps, or in the richness of regional individuality?

Today, by contrast, there is no consensus, nor is there much debate on the need for staking out the "true geography." We are instead in a phase of remarkably creative experimentation, with the accent on philosophies and methods of inquiry, rather than geography's essential definition. And instead of one, or even two paradigms, we choose among many: logical positivism, Marxism, structurationism, behavioral geography, institutional perspectives, and phenomenological ones. All more or less are represented in the discipline of historical geography.

As we grope for the paradigm that yields consensus, it is clear that no single individual, or any group of individuals, is capable of capturing the range of experimentation, let alone the likely directions it will take. For these reasons, the Historical Geography Specialty Group very early decided upon multiauthorship. While numbers alone do not confer limited liability, at least they minimize the extent of each author's potential error and embarrassment. The risk, of course, is that a series of small essays may appear more as a pastiche, than as a patterned and coherent mosaic. We trust not. In this context of intellectual fluidity, then, let us look briefly at what our inventory of historical geography yields.

What follows is a status report on historical geography, as practiced by the 300 or so members of the Historical Geography Specialty Group of the Association of American Geographers (AAG). The essay inventories current scholarship, while forecasting research themes. It accents Specialty Group scholarship, while acknowledging intellectual influences in geography, history, and social-science history. And it offers a revealing self-assessment, while posting geographers on the contributions of North American historical geographers.

Some of these contributions are impressive. We now know a vast amount about the historical geography of North America: its regional development, migrations, urban evolution, and rural changes. We are beginning to fit that knowledge into a macrogeographical context, into Meinig's Atlantic World or Wallerstein's world system. We also know a fair amount about Latin America, thanks to the scholarly legacy of Carl Sauer. And we know a little bit about wringing data from a variety of intractable sources in archive and field (Norton 1984c).

But historical geography is not simply a success story. Our inventory points up issues, problems, and regions that need to be addressed by us, and by the hundreds of latent historical geographers in geography. Heading the list are more sophisticated and sensitive treatments of minorities, of women, of human impacts on the physical environment, and of historical geography beyond the Western Hemisphere.

Ours is not an essay in the conventional sense; rather it is a series of desperately compressed sketches on the varied facets of historical geography. Each sketch reduces a massive and complex literature into a few pages, sometimes less. The sketches are arranged into five parts: (1) Theory and Paradigms, (2) North American Perspectives, (3) Perspectives Abroad, (4) Tools of the Trade, and (5) Needs and Opportunities. We trust that the whole suggests something of the accomplishment and the promise of the field.

THEORY AND PARADIGMS

Historical Geography and Geography: Divergence and Convergence

In the U.S., historical geography has long been viewed as peripheral to the broader field of human geography, to which it is nevertheless inextricably connected. This peripheral position has encouraged rich connections with anthropology, history, and more recently sociology. But correspondingly, relations with geography have been decidedly loose—a condition that is happily on the mend. For almost two generations, historical geography and geography displayed divergent tendencies, despite often parallel developments of their research agenda. These divergent tendencies were evident nearly a half century ago, when Carl Sauer presented his presidential address to the Association of American Geographers (1941). Questioning the emphasis on "areal differentiation" espoused in Richard Hartshorne's 1939 history of geographic thought, Sauer attacked a growing preoccupation—a "lingering sickness"—with a narrowly conceived disciplinary autonomy.

HISTORICAL AND CULTURAL CONTRIBUTIONS TO GEOGRAPHIC UNDERSTANDING

In his substantive work, Sauer demonstrated the ways in which geographers might contribute to broad cross-disciplinary themes related to the cultural transformations of natural landscapes. In Sauer's view, this agenda was much broader and intellectually more appealing than that which Hartshorne had derived from historical examination of geographic thought (Hartshorne 1939). Although Hartshorne's regional perspective acknowledged antecedent conditions, his regions and places fit within a world view that stressed the functional relationships of modern world economy derived from recent rather than remote processes. In contrast, Sauer stressed long-term cultural transformations of the natural landscape, and he placed higher value on the environmental adaptations of the "traditional" world than those of the modern world.

As a student of Sauer's and a colleague of Hartshorne's, Andrew Clark knew firsthand the antipathies of their agendas. In his monograph *Three Centuries and The Island* (1959), Clark mediated between the genetic cultural-historical interests of the Berkeley School and the presentist functionalism of Hartshorne's human geography. He sought a culturally sensitive historical geography of modernity, rather than a celebration of the moral and ecological virtues of the traditional world. His methodological premises, however, were unambiguously geographic, and his thematic maps were emblematic of his disciplinary focus. Influenced also by the work of H. C. Darby, Clark combined a series of "past geographies" into a synoptic vision of geographical change in which the geography of change itself became the dependent variable.

Clark spoke of the geography of change, and changing geographies. He combined retrospective geographies of the recent past with meticulous quantitative examination of the kinds and rates of change in multiple socioeconomic variables. Like Ralph Brown (1943), he advocated using archival evidence and local records, but he insisted that local records were wasted on the historical geography of individual settlements or communities; geographic interpretation demanded a mosaic of local studies that collectively revealed the variability of regional change, or in modern parlance, uneven development.

In his monograph on Prince Edward Island, Clark (1959) put all this into practice. He compiled evidence from many local sources and presented it in over 200 maps of geographical change in population, economy, and society. Using a classically inductive exposition, Clark let the evidence (much of it quantitative) speak for itself. Had his data been analyzed more thoroughly, the book might have been recognized as a precursor of the quantitative revolution, but its inductive cast obscured its relation to the "new geography" with its commitment to statistics in developing spatial laws and theories. This search for spatial laws rather than statistics was the root cause of divergence between historical and human geography during the sixties and seventies (Ward 1975a).

In this context, historians seemed much more receptive to the works of historical geographers. Clark concluded that it would be more profitable for historical geographers to examine the environmental and regional implications of historiographic controversies. Using primary sources and exploring historiographic issues, some historical geographers entered into an explicitly bidisciplinary enterprise; indeed, their research is now embedded within historical discourse on the colonial economy, transatlantic migration, and the industrialization of urban life.

This "incorporation" into historical studies dramatically extended historical geography's impact—yet at the same time it raised questions about our relationship to human geography at large.

These anxieties produced two methodological reactions. Some historical geographers attempted to create an explicitly "geographical" historical geography, predicated on changes in material landscape and macroregional patterns. In a series of regional historical geographies, and more recently in the first of three volumes on *The Shaping of America,* Donald Meinig has provided an exemplary model of this methodology (1986). His is a work that only a geographer could have written, and it sets a standard for this genre. At its best, the genre elaborates the relationships between process and pattern, between identity and place; but at its worst it deteriorates into an uncritical preoccupation with simple patterns, local forms, and disembodied geographical change.

The sterility of the latter has prompted a "humanistic" corrective. Drawing upon idealist epistemologies, humanistic geography reconsidered historical changes in the symbolic meaning of landscape and varying ideological contexts of the built-environment (Guelke 1982; Cosgrove and Jackson 1987). Although humanistic geography developed independently of historical geography, their common interests are a source of convergence in human geography.

A second way that historical geographers have forged closer connections with human geography is through retrospective applications of contemporary theory and method in economic and urban geography. Analyses of past cities and city systems have demonstrated the virtues of historical cases and quantitative historical sources. Although this genre too may deteriorate into mindless verifications of contemporary theory, on balance urban historical geography has been sensitive to the inferential naivete about historical processes that is embedded in contemporary theories of spatial organization. History did not always happen as it should have; hence, presentist theory usually requires redefinition and revision when applied to the uneven trajectories of industrialization, modernization, and the rise of capitalism.

In a word, historical *context* matters a great deal. Context has long been one of historical geography's basic presuppositions, and in recent years human geography has acknowledged the importance of institutional contexts for locational and ecological behavior. Convergence is most emphatically revealed in efforts to identify geographic changes in the structure of capitalism, and the degree to which the spatial organization of society may be expressed as part of the uneven regional development of the world economy. Some structural approaches have viewed the mode of production as the predominant restraint upon human agency; others, presuming the contingent nature of these restraints, have explored relations between capitalism and culture, class and community, and nation and region. On a few occasions, historical geographers have wed these approaches into a comprehensive structuralist geographical interpretation (Pred 1986).

After several decades of marginality, historical geography has opened promising lines of convergence with human geography. Experimentation flourishes: approaches range from Marxist to humanist, logical positivist to structurationist, imperialist macroregional modeling to world systems. The recently published anthology edited by Mitchell and Groves reveals that innovation and tradition have been directed toward

HISTORICAL AND CULTURAL CONTRIBUTIONS TO GEOGRAPHIC UNDERSTANDING

a common purpose (1987). The stage is set for a broader arena of discourse within human geography.

Historical Geography and World-System Analysis

The arena of discourse first expanded with the 1974 publication of Immanuel Wallerstein's *The Modern World-System*: macroscale historical geography received new impetus (1974, 1980, 1984). World-system theory promised to unify interdisciplinary approaches around venerable problems in historical and cultural geography. To wit: What happens when markedly dissimilar cultures come into sustained contact? What explains the remarkable success of expansionist Europe after A.D. 1400?

The first of these themes, of course, had been defined by Carl Sauer and the Berkeley School of cultural geography, and the second by the inquiries of Andrew H. Clark and his students at Wisconsin. But in neither case were these provocative themes elaborated into overarching theories of acculturation, diffusion, and expansion. The best work of the Berkeley School was ethnographic, often at the microscale. That of the Wisconsin School tended to be at the regional level. Both tended to be empirical rather than theoretical. Only in some of Sauer's "thought experiments," notably his theory of agricultural origins (1952), were these biases laid aside. Chastened perhaps by an earlier generation's futile search for the environmental determinants of human behavior, cultural and historical geographers shied away from theoretical generalizations.

The empiricist tradition eroded in the late 1960s, as cultural and historical geography came under attack from economic geography—a subdiscipline vigorously making the transition from empiricism to normative social science. The critique stimulated a restoration of theoretical discourse in cultural and historical geography that had been largely abandoned by the generation following Huntington and Semple. In this experimental atmosphere, world-system theory proved exciting for historical geographers.

Wallerstein's spatial model defined three regions in nonstate terms—core, semi-periphery, and periphery. The theory was dynamic: states move in and out of the three categories in a nonevolutionary way. The driving force in European capitalist expansion after A.D. 1400 was, and is, a competitive jockeying for global economic hegemony by core nations—European at first, but later, European-settler colonies, such as the U.S., and states that adopted European economic and technical systems, such as Japan.

Wallerstein's central theme, the expansion of Europe into the Atlantic, has long fascinated historical geographers, although the subtheme of capitalist development has not. Andrew Clark's students have written extensively on Europe overseas, but at the scale of cultural regions rather than the world system. Geography's most ambitious statement is Donald W. Meinig's *The Shaping of America: Atlantic America, 1492–1800* (1986); it is the latest of his programmatic statements on European expansion and culture contact. His ample theoretical framework expands on Wallerstein's, without losing sight of empirical evidence or the spatial core of our discipline.

But unlike most empirical studies, Meinig's transcends political boundaries in producing the first balanced account of Atlantic expansion. His account recognizes the systemic nature of the Atlantic economy: from the northern fishery, to the sugar "carbohydrate frontier" of the Afro-Caribbean Atlantic, to the trade and settler colonies between the Carolinas and southern Maine. But unlike Wallerstein, Meinig insistently focuses on political power, its application, its spatial constraints (e.g., distance), and the remarkable variation in local and regional cultures.

If Meinig surpasses Wallerstein in his empirical and theoretical accounting of imperial expansion and contraction, matters are reversed in certain other spheres. Wallerstein's Marxian analysis is more sensitive to social structures, wealth distribution, and capital provisioning on the "carbohydrate frontier." In this respect, Wallerstein is the intellectual legatee of Fernand Braudel, himself the heir of Marc Bloch, who in turn owed much to French regional geographers such as Paul Vidal de la Blache.

Wallerstein's world-system model has been provocative, so much so that ensuing discourse has exposed its shortcomings as an interpretation of modern history. It bypasses both Weberian idealism (Weber 1958) and Weber's later idea on the rise of the rational bureaucratic state (Weber 1968, vol. 2, chap. 9). Explicitly seeking an alternative to Marxist-Leninist explanations of empire, Weber saw imperialism as "tied to the power position of the state in the international arena" (Collins 1986, 145). Meinig's Atlantic World thus represents an empirical Weberian rebuttal to Marxist historicism and Leninist imperialism. By grounding Weber in the real world, Meinig demonstrates the vitality of Weberian theory as an alternative (or perhaps a complement) to a Wallersteinian theory of world systems.

Geographical theories of power superior to those of Marx and Wallerstein have also been proposed by Anthony Giddens (1981). Borrowing freely the ideas of geographer Torsten Hagerstrand, Giddens derives his central notion of time-space distanciation (1981, 91). Power in world-empires, he argues, assumes a very different form than it does in a capitalist world-economy, and the difference is because of capitalism's efficient spatial expression of power along the "time-space paths" represented by the core, the semiperiphery, and the periphery. Hagerstrand, meanwhile, has expanded the concept of time-space distanciation by adding Goldenweiser's principle of limited possibilities (1988)—a principle in which historical choices narrow the range of ensuing choices among time-space paths.

Wallerstein's work has refreshened social-science discourse, enriching the debate over scale-of-inquiry as well as theory. Moreover, world-systems thinking transcended disciplinary boundaries at a time of disenchantment with logical positivism and historical uses of quantitative methods.

An aggressively interdisciplinary historical social science now confronts the problem of capitalist origins. Ironically, historians have contributed modestly to the new macrohistory, with honorable exceptions such as William H. McNeill (1963, 1982). Its authors are historical social scientists who have been compelled by their "quantitative" colleagues to confront theory in a serious way. Examples abound: The anthropologist Eric Wolf (1982) pays close attention to colonial labor systems, and thus ends up taking a more Marxist position than does Wallerstein. The economist-cum-geographer E. L. Jones focuses on the interactions of environment, technology, and demography in shaping economies and geopolitics (1981). His revised environmentalism demonstrates how configurations of land, water, and topography affected human interaction under conditions of rudimentary transportation systems—a perspective anticipated by Sauer's work on the Caribbean Basin (1962).

Historical geography has many possible futures, but one of these surely lies at the intersection of macroscale theory and thick empirical description. Meinig's *Atlan-*

tic America offers a glimpse of the possibilities, but our macrotheories of world systems must push onward to Marx's great question: "What is the locomotive of history?" While historical geographers are too aggressively multicausal to rest interpretation on only one factor, such as Marx's class struggle, we nonetheless tend to favor one or two causes above others—demography, climatic change, technology, environmental degradation, religion, warfare, family structure; the list goes on.

The task of a satisfactory theory of world systems, of course, is to reconcile these interpretations and integrate them with macrohistorical rhythms such as the Kondratieff long wave in capitalist economies (Mensch 1979). Does the long wave apply to capitalism? Does it account for hegemonic shifts among core states? And, lastly, what drives this relentless rhythm?

World-system theory has been healthy for historical geography and human geography. Its rubric encompasses the two schools of American human geography—cultural-historical and economic—under one roof. It has identified common ground for the convergence of geographical inquiry: Peter Hall (1981) and Brian Berry (Berry, Conkling, and Ray 1987) with Kondratieff; Hagerstrand with "time-space paths" (1978); and Meinig with spatial forms of power (1986).

Such macroperspectives promise geographers a prominent role in a reconstructed historical social science that is preeminently spatial. The reconstruction has already begun: for example, Stuart Corbridge's critique of the "radical development geography" of Wallerstein and others (1986); John Agnew's (1987) revitalization of Anglo-American regional geography from the perspective of world-system theory and historical change; and Peter Taylor's (1985) revised *Political Geography* as informed by Wallersteinian theory. Historical geographers should welcome and nurture these convergent trends. Let us not lose the opportunity.

The repercussions of Wallerstein's and Meinig's macrogeographies have resonated throughout historical geography. While their full impact remains to be assimilated, discernible shifts in nuance, perspective, theory, and scale of inquiry are already underway. A refined sensitivity to the connectedness of peoples, places, and events in macroscale and macrohistorical systems is nowhere more evident than among historical geographers of the North American scene.

NORTH AMERICAN PERSPECTIVES

Colonial Models and Legacies from the Atlantic Seaboard of North America

The study of Anglo-American colonial origins and development during the last generation has been transformed from descriptive regional study, based on secondary sources, to systematic and regional works in which analytical procedures are applied to primary documents and artifacts. The result has been a search for a comparative, process-oriented synthesis that challenges traditional approaches to the colonial era (Meinig 1957–58; Harris 1978; Mitchell 1987; Lemon 1987).

A recurring theme in colonial research has been the continuity and change between British heritage and colonial foundations. Traditional models of American exceptionalism, epitomized by Turner's frontier thesis of declining European influence with distance from the Atlantic Seaboard, have been replaced by models of continuity between metropolis and colonies. Some scholars employ a gemeinschaft-gesellschaft model in describing the "modernization" of late-colonial America (Harris 1977; Henretta 1978). They have argued, based primarily on New England experiences, that an initial process of pioneer simplification gave way to a pattern of intensely local, family-centered life characterized by easy access to land, long-run financial security, and limited availability of markets. The resultant agrarian society—egalitarian, yeoman-based, and primarily subsistent—is presumed to have characterized most of the Northern colonies until the emergence of large-scale commodity markets after 1750.

An alternative thesis has argued for a more interconnected, commercial, and individualistic social system in both Northern yeoman and Southern planter societies. These differentiated societies were based on easy access to land, profitable ventures in land and trade, and upward social mobility (Lemon 1972, 1980; Mitchell 1977). The debate has been fueled by numerous case studies (Merrens 1964; Meinig 1966; Earle 1975; Mitchell 1983; Harris 1985), but the models to date have not addressed adequately the impact of British social and economic influences on the colonies, the variable pace of change, nor the complexity of spatial variations in societal change over time.

Another model connecting the European heritage with colonial origins is based on the culture-hearth thesis. It postulates a mosaic of culture areas, derived in turn from the source areas of British immigration (Zelinsky 1973; Meinig 1978). Integrated, but regionally discrete, ways of life were presumed to have diffused, from the core areas of southern New England, southeastern Pennsylvania, the southern Chesapeake, and the South Carolina low country, to the expanding settlement frontier. Proponents of the culture hearth-diffusion thesis have generated a number of studies of internal migration and culture regions. One approach traces the diffusion of culture complexes from initial hearths to the interior (Kniffen 1965; Pillsbury 1970; Zelinsky 1973). Another focuses on the geographic structure of key seventeenth-century settlements and their reconfiguration on eighteenth-century frontiers (Newton 1974; Mitchell 1978, 1983; Kulikoff 1979; Meinig 1986). These diffusion models, however, have not resolved questions about hearth formation or differential patterns of diffusion and cultural reformation in the interior.

Another recurrent theme in colonial studies has been the relationship between population growth and economic development. Both the population and the occupied area of colonial America increased tenfold between 1700 and 1775. Traditional models of land-use and settlement emphasizing ethnic heritage and Old World tradition have been replaced by dynamic models based on supply and demand. Thus, the surge in population growth, and its southerly expansion after 1720, have focused attention on dramatic increases in colonial consumption patterns and transatlantic markets, which gave rise to a burgeoning wheat and meat trade. One consequence was the emergence of a backcountry specialization in wheat and range livestock, stretching possibly from the central Hudson valley to the South Carolina piedmont (Merrens 1964; Lemon 1972; Mitchell 1977).

A supply-oriented staple thesis has been preferred by scholars interested in explaining the fluctuating fortunes of Southern staples of tobacco, rice, indigo, and naval stores, and the consequent impacts on land, labor (slavery), and capital requirements (Earle 1975; Earle and Hoffman 1976; McCusker and Menard 1985). For example, in the Chesapeake tobacco colonies, traditional concerns with soil deterioration and planter migration have been replaced with the timing of business cycles and their impacts on crop diversification and the transition to commercial grain production.

What has emerged, in short, is a reinterpretation of economic regionalism. It transcends conventional sectional distinctions among New England, Middle, and Southern colonial economies, and it integrates the Southern backcountry firmly into the expanding commercialization of the North.

This research, in turn, raises new questions about colonial-settlement organization. Discoveries of the extent of trade decentralization (especially in the Southern colonies) and the ubiquity of dispersed settlement have raised new questions about colonial urbanism. On one level, debate has focused on the relationship between morphology and function in the creation of town and country landscapes throughout the colonies. A number of case studies have unravelled settlement foundations and dispersions at local and regional scales (Merrens 1964; Lemon 1972; Ernst and Merrens 1973; Earle 1975, 1978; Wacker 1975; Wood 1982; O'Mara 1982). On another scale, research has centered on the relationship between settlement systems and emerging regional economies within the larger Atlantic world. Such studies, in turn, have renewed interest in the evolution of colonial urbanism (Earle and Hoffman 1976; Barton 1977; O'Mara 1983).

The synthesis emerging from these discussions suggests an urban evolution, based initially on an imperial model of colonial port cities operating within a mercantilist-dominated trading system imposed by the metropolis (Price 1974; Meinig 1986). The expansion of urbanism into the eighteenth-century interior seems to fit best a mercantile model of long-distance trade, controlled by coastal ports operating through smaller market centers (or "unraveling points") at key locations in the emerging transportation systems of the late-colonial period (Vance 1970). Nesting within these systems are central-place systems of local and regional retailing, often based on centers of local government (e.g., Lemon 1972; Mitchell 1977). These settlement models, however, do not yet account for differential town growth and system development, or the impact of widespread manufacturing after 1750. Geographical study of late-colonial industrialization and its relationship to economic and urban growth remains in its infancy (but see Earle and Hoffman 1980).

The progress that has been made in understanding the geography of colonial America has been impressive. But rich themes remain neglected, such as Indian-European relations, the ecology of colonial expansion, and the geographical dimensions of slavery. And if we have made contributions to the debate about continuity and change between metropolis and colonies, we have yet to establish a framework for examining the political transformation from colonies to nation, from colonists to Americans.

Canada Before 1800

The outreach of European commercial capital to northern North America, beginning early in the sixteenth century, created transatlantic spatial economies, work camps perched in the wilderness, and eventually a few small towns at break-of-bulk points. European economies in native territories introduced trade goods and diseases, and changed patterns of native trade, warfare, and settlement. Apart from the European staple trades, a few colonists began to clear land, transferring a northwestern European crop-livestock complex and peasant ways to rock-bound pockets of arable land far from markets. These different settlements and spatial economies established the geographical pattern of early Canada. They have been much-studied by historical geographers, whose work now constitutes an essential empirical and interpretive component of early Canadian studies.

Work on staple trades reveals European work camps in a vast land inhabited by native North Americans. In these ephemeral, utilitarian settlements, rhythms of work and transportation dominated social relations. These rhythms were adapted to a severe continental climate, divided time, and the hierarchy of command in a staple trade. Distance and forbidding land discouraged alternative economies. Europeans had penetrated a harsh realm—at first seasonally, then for periods of years as hired men overwintered, and eventually, more permanently, as women arrived and families appeared.

The older of these trades, the cod fishery, remained confined to the northeastern corner of North America and made no use of native labor. But, by preempting the coast, this industry destroyed native economies. The fur trade, dependent on an easily depleted resource and on native labor, expanded rapidly inland. It eventually became a transcontinental enterprise that established much of the basis for the present political border between Canada and the U.S.

Throughout the territory of the fur trade, native populations were decimated by European diseases and heightened intertribal warfare (as muskets became available). Remnant peoples were frequently relocated by warfare or trading opportunities. The depletion of game eventually created patterns of dependence that the earlier introduction of European trade goods had not (Ray 1974).

Until the Napoleonic Wars, merchants in European towns controlled the principal fisheries around Newfoundland and the Gulf of St. Lawrence. The fur trade, on the other hand, soon required an urban base in the New World, and towns developed along the lower St. Lawrence in the seventeenth century. Strategically located for trade and defense, they were foci of commercial, military, and administrative power. Occupationally, they were more diverse than other settlements in early Canada. Socially, they were finely stratified. Culturally, they were heterogeneous, reflecting North Atlantic migration fields. Early Canadian towns reproduced on a modest scale the town plans and urban architecture of their day, looking very much like the transplanted French or English ports that they essentially were. As late as 1800, there was no single system or hierarchy of Canadian towns.

The early Canadian countryside was a more original creation. The relative availability of land, scarcity of labor, and weakness of markets tended to encourage vigorous peasant economies. The blend of immigrant backgrounds and the circumstances of pioneering created distinctive local cultures. With opportunity to create new family farms, the rate of natural population increase was high. As populations rose, the limits of agricultural land were often reached. Land values then rose, rural populations became more stratified, and many of the young moved away. Their routes of migration shaped the pattern and extent of regional cultures that, in early Canada, were isolated from each other by rock and distance (Clark 1959).

Work by historical geographers is revealing these geographical patterns, clarifying the character of the distinctive European penetration of northern North America, and providing a wealth of structured insight into the modification of European economies, societies, and cultures overseas (Harris and Matthews 1987).

A Restless People: Migration in Country and City

After the Revolution, a continuation of the westward migrations of the colonial population resulted in a considerable mixing of people drawn from different regional subcultures. Nevertheless, new regional cultures and economies emerged in the South, Old Northwest, and later on the Great Plains and the Far West. After 1840, this process of regional differentiation was also fueled by massive immigration from overseas. Indeed, nearly one in every five Americans was foreign-born throughout the second half of the nineteenth century. Not only were large numbers of Americans recent immigrants, but high levels of geographic mobility also characterized these migrants once they had established themselves here. For a long time these high levels of mobility were assumed to be a source of rapid assimilation. Historical geographers have done much to clarify these processes and patterns of past migrations to and within America, but they have also contributed to interpretations of these experiences which have emphasized the persistence and redefinition of ethnicity in American society.

What makes people migrate is a perennial question that knows no disciplinary bounds. Like most modern students of migration, historical geographers have treated the phenomenon as a behavioral process, concentrating on the mechanisms that influence migratory behavior. Detailed studies of migration fields connecting the British Isles and parts of eastern Canada have emphasized the role of changing political and economic forces in providing incentive to emigrate (Handcock 1976, 1977; Omner 1977). Others have emphasized the interplay of information about opportunity and the individual situation of potential migrants at key moments in time (Rice and Ostergren 1978; Sauressig-Schreuder 1985a, 1985b). Migration to America, as well as within it, is generally seen as a complex process, often involving multiple and sequential moves, a broad range of social and economic objectives, and idiosyncratic behavior of specific cultural groups (Hudson 1973, 1976; Ostergren 1980).

While migration to the American frontier may appear to have been the dominant nineteenth-century flow, it was in fact often balanced by local, regional, and urban migration (Conzen 1974). The destinations of migrants belonging to specific ethnic, religious, and racial groups were often highly selective. Migrants tended to congregate amongst people of similar origins, a tendency reinforced by geographic variations in opportunity that drew people having specific cultural backgrounds and skills to key reception areas. The unevenness in the regional distribution of groups, and their intense local clustering, have generated curiosity about these ethnic settlement patterns (Jordan 1966; Meinig 1972; Zelinsky 1973; McQuillan 1978b; Ward 1987).

While initial settlement patterns often suggest transplantation of old cultures to new environs, most migrants experienced gradual assimilation with the host society, although the speed and completeness of that process varied. This enduring historiographic theme underlies much of what historical geographers have attempted to do with immigrant settlement in America.

Studies of immigrant rural settlement have been particularly sensitive to ethnocultural persistence in concentrated settlement areas, despite the proverbial geographic mobility of Americans. These studies have also gauged the premigration experience as it influenced the success and status of immigrants faced with the democratizing influences and abundant landed wealth of the frontier (Rice 1973; Ostergren 1979, 1988). Some studies have isolated transplanted cultural factors—landholding patterns, inheritance practices, farming preferences, and material culture—that may have contributed to high levels of cultural segregation and persistence, while others have concluded that it was simply a matter of time before the power of the American economic system leveled all differences (Mannion 1974; Rice 1977, 1978; McQuillan 1978a; Baltensperger 1980; Ostergren 1981a, 1981b; Legreid and Ward 1982; Sauressig-Schreuder 1985b; Omner 1986). Resolving these conflicting interpretations of ethnocultural assimilation remains the central issue.

Urban historical geographers have focused on residential differentiation, with special reference to the emergence of ethnic and racial slums and ghettos (Ward 1982). This research has documented the changing composition of immigrant quarters, their institutional completeness, and the relationship of their inhabitants to the adverse environment of the inner city. The initial ethnic communities usually formed an overlapping mosaic of neighborhoods, rather than extensive homogeneous areas typical of the black belts in modern cities (Groves 1974; Ward 1975b; Groves and Muller 1975; Radford 1976; Conzen and Conzen 1979; Ward 1982). The rates and patterns of dispersal of different ethnic communities to new suburban nodes were highly varied, and interpretations of these differences have emphasized the continuing, if diminished role of ethnic community.

In addition, migration has been linked to the uneven regional development of American capitalism, and the varying rates of social and residential mobility have accordingly been attributed to the fluidity or rigidity of the ethnic division of labor. Assimilation was long linked to the frontier and suburbanization, but it is now clear that the ethnicity of migrants complicated the process. As a result, urban and regional patterns of ethnic pluralism serve as the definitive expressions of the relationships between migration and economic development (Ward 1987).

The Rise and Fall of the American "Rural Freehold Empire"

Rural America, once the nation's bulwark, is now in trouble. Foreclosures on small farmers who dwell on the productive lands of North America, exurban complaints of farmyard smells, environmental contamination by chemical fertilizers, water shortage in irrigated regions, marketing boards to stabilize prices, and widespread government controls over agricultural production signal the decline and even demise of the "rural freehold empire."

Once filled with the promise of independence, stability, and ample material well-being, the empire has been nearly emptied of its farm families. From something like three-quarters of all workers, the number of farmers has shrunk to less than 5%. The contours of this panoramic process are outlined in the recent volume edited by Mitchell and Groves (1987), notably in the chapters by Harris, Mitchell, Lemon, Hilliard, Earle, McIlwraith, Wishart, Hornbeck, Wynn, and Lewis.

Viewed from the perspective of nearly 400 years, this is a drama of enormous proportion. Over the long haul, a tide of rural settlement swept across the continent, broke in the last decades of the nineteenth century, and then ebbed sharply away. At the turn of this century, the empire counted more farms than at any time before or since. But be-

HISTORICAL AND CULTURAL CONTRIBUTIONS TO GEOGRAPHIC UNDERSTANDING

hind this westward-flowing empire, farmers were abandoning marginal lands as early as 1800. And by that date, too, the urban share of the population began its rise, later accelerating so that cities, towns, and villages housed half the people by 1900, and three-quarters or more today (with another fifth in rural nonfarming pursuits).

Many factors contributed to the rise and fall of the "rural freehold empire," but the principal cause has been the adoption of machinery to cut labor costs. Replacing labor with machines rendered redundant large numbers of farmers, farm families, tenants, overseers, and laborers. Regional distinctions in settlement and agriculture were first established with colonization, and over time a multitude of local and regional decisions increasingly sorted crop and livestock specializations into their optimal environmentals. By the end of the great expansion, about 1930, the mosaic of American agricultural regions was in place for cataloging by economic geographers.

This grand story of empire has both temporal and spatial elements. First, how can its periods be structured? In contrast to the optimism of the 1950s, when everyone wanted to forget the Great Depression, we are now far more conscious that a similar crash may be impending; that is to say that we are aware of waves in the social economy of the past. Was the empire's growth and recession affected by similar phases of depression and take-off in the past, most notably in the 1740s, 1790s, mid-1840s, mid-1890s, and the early 1940s? What were the effects on land acquisition and abandonment? On specialization of land use? On the application of technologies? On rural investment?

In this series of "Kondratieffs" from 1600 onward, it would seem that American development hinged midway between Rostow's 1840s "take-off" and the turn of this century. American wealth and real income grew gradually up to that point, and then slowed. In both rural and urban North America, the high point of general prosperity was achieved about 1912. One hastens to qualify this generalization by noting that neither the Old South nor industrial workers shared in the largess until after 1940, when military Keynesianism induced, or helped to induce, the so-called fourth Kondratieff. Our forebears thus celebrated growth, believing it would continue. Indeed, most historical geography has been written about the accelerating side of the ledger. Now we and future generations have to face the downside.

Historical geographers have been fascinated by the rise of the freehold empire. They have accented various themes, only three of which are dealt with in the following paragraphs: (1) frontier evolution, (2) the durability of ethnocultural farming practice and landscape, and (3) environmental perception and ecological change (Clark 1972).

Frontier studies suggest a partial model of the evolution of rural settlement and economy. Mitchell (1977) and Muller (1976) envision a three-stage frontier: initial pioneer subsistence, the emergence of commercial staple crops, and regional economic diversification. Their models aptly describe the expanding wheat frontiers of the Shenandoah Valley in the eighteenth century and the Midwest in the ensuing century (Johnson 1969). We are less certain, however, of the model's applicability to other staple frontiers, or to less-expansive times. Moreover, the model is partial, in the sense that it has yet to incorporate frontier variations in seasonal labor migration, land tenure, rural settlement type, crop combinations, and land-use choice, or receptivity to technical innovation (Rice 1977; McIntosh 1975; Hudson 1976; Conzen 1971, 1974; Earle 1987; and Hewes 1973).

A general frontier model must also incorporate the findings of ethnocultural inquiries on persistence and change in agrarian practice, landscape, and diet (Hilliard 1972; Gerlach 1976; Jordan 1966; Baltensperger 1980). Interpretations are sharply divided over the extent and pace of cultural change. Some studies suggest rapid assimilation of immigrant farming into the mainstream of the market economy (Rice 1973, 1977; Ostergren 1979), while others stress the durability of ethnocultural patterns (Jordan 1967; Meyer 1987). In reconciling these interpretations, historical geographers are torn between disjunctive and conjunctive modes of interpretation: the former argues that only one interpretation is correct, but the latter believes that both may be correct when evidence is specified in time and space.

Environmental perception and ecological change are equally important elements in a general settlement model. Several studies show how perception and evaluation of frontier resources channeled the flow of initial frontier settlers (McMannis 1963; Jakle 1977). Others demonstrate how hazard perception affected technical agrarian adaptations, notably among drought-plagued farmers in the Great Plains and the Great Columbian Plain (Bowden 1975; Meinig 1968). Fewer studies, however, comment upon destructive ecological practices on the frontier, save of course in the South where scholars debate the extent of soil erosion and exhaustion, and the culpability of plantation slavery (Earle 1975; Trimble 1987).

But why have historical geographers been preoccupied by these accelerating years of prosperity? In some degree, our inquiries have been driven by an admirable populist impulse. We have studied ordinary rural people, placing them in a positive if not altogether transparent light. Though conscious of, say, the work of land speculators, we have not often talked about the power of individuals or groups to impose solutions on others. Implied in our populism has been an attempt to simplify Europe overseas, and so we have largely avoided narrative reconstruction of the struggles for rural power. These struggles are viewed discretely from a distance.

The problem of power may also be viewed from the standpoint of status. Although European social ranks were simplified in America, that did not dispose of the problem of status, of finding one's standing in society. For many it may have become more difficult. In early Pennsylvania, for example, internal squabbles and divisions within congregations were frequent when churches lacked respected clergymen. Evangelical revivals and awakenings complicated the search for status in a secular society. Mobility, a grand theme in our work, has been as much a quest for status as for economic security.

We have not carefully distinguished among the ordinary people of the freehold empire. They were seemingly mostly men of the middling sort, the ones who sought land on the frontier. Contemporary concerns about race and gender have muddied this picture, and the lives of slaves, free blacks, and farm women deserve compensatory inquiry.

Similarly, historical geographers have slighted environmental degradation. We have emphasized the spectacular success of farm machinery, rather than the negative consequences of rural displacement and destructive occupance. What conditions allowed North America—indeed, all industrial societies—to lose control over technology? And to suffer an agrarian devolution, manifested in rural depopulation, farm abandonment, and the demise of the family farm? All of these subjects are worthy of their historical geographer.

Doubtless, a commitment to money has had something to do with this. While the long-run trend in social economy has been acceleration and then deceleration, our interest in money-making has been relentless. We have reached a point in North America that it has now been detached from production. Use value has been gradually undercut by exchange value, in the pursuit of status value. Those with power pursue the Janus-faced strategy of persuading ordinary people that government should keep hands-off the economy, while manipulating government toward their own ends. Historical geographers might well pay more attention to the intersection of government, money, and rural society over the nearly 400 years of settlement—a span wherein the rural freehold empire was won, and lost.

The Design of North American Cities: Institutions in the Shaping of Internal Structure and Regional Systems

Our interpretations of the American past have been reshaped by the decline of rural America. The city and urbanization, which until recently were mere factual appendages at the fringes of American historical geography, have become concepts central to our reinterpretation of the continent's development.

Meinig, in his major reinterpretation of the American experience, places towns closer to the center than the hem of his strategic geographical polity. His understanding of the role of Quebec City is as the focus of early French colonial designs on the northern continent, and of New York and Philadelphia as cities that "constituted a mobilization for national life" (1986). In Mitchell's and Groves's edited work, *North America* (1987), the various authors dealing with the period from 1860 to 1920, an era of "consolidation," focused on the network-building and integrating power of urban institutions and systems of cities, emphasizing their contribution to the changing social and economic structure of the nation.

The idea of American space as the product of design has appealed to several authors. They have emphasized the strategic role of towns in rationally planned landscapes, and the power of urban institutions to invent and implement their plans. Hudson concentrates on both points in his study of northern Plains railway-promotion schemes and settlement (1985). He assesses the corporate perceptions and private interests of urban services in their plans that organize land occupance. He offers an effective demonstration of the planning rationale, and of the circumstances that limited its realization. Towns were talismans of the success or failure of these strategies.

Plans for orderly settlement on an impressive scale were not, of course, an invention of late nineteenth-century business. Illustrating designs of an earlier period, Wyckoff has examined the Holland Land Company's promotion of settlement in upstate New York (1986a, 1986b). The trail leads backward to Dutch banking houses, and forward to the survey of mill sites, to encourage organized agricultural occupation of the company's tract. Design implies institutional guidance and urban services. The link to cities is close; to the long-established idea of cities on the frontier, these studies add the further concept of a planned environment, reflecting in greater or lesser measure the purposes and control of urban institutions.

The ability of powerful urban institutions to restructure regions long after their initial settlement has been nicely illustrated in an essay on Pennsylvania anthracite towns and the industrialization of a rural environment (Marsh 1987). It is the story of the power that control of natural resources and transportation conferred on a few corporations that were singlemindedly advancing the interests of capital. What they planned and controlled were production and the movement to market of their product. The design of the cultural landscape was incidental, and it evolved in a distinctly uncoordinated and unregulated way. But once in place, urban networks may exhibit a nearly indestructible quality, even in the face of dramatic economic collapse of the structures that created them. Marsh highlights the inertia of an outdated urban system. This is surely not the only urban palimpsest in America, not the only continuing evidence of a past that is not yet totally erased.

Rendering visible the hands that created and managed urban systems has reduced the necessity for recourse to explanations of technological capacity, and has increased our understanding of how business was organized spatially, and how communication networks grew. The most fully developed example is McCann's research on the impact in the Maritimes of growing metropolitan dominance in post-Confederation Canada (1979, 1985). For instance, the capacity of Halifax firms to structure a regional economy around their own interests declined, in the face of the nationwide scale on which businesses in Montreal (and later, Toronto) came to operate.

Recent communications research has emphasized the role of government (the post office) and business (newspaper editors) in structuring and managing information flows at regional, national, and international scales (Goheen 1985-86, 1987). Pred's books, on the relations between communications and urban growth, highlight the production of urban networks as concrete historical and geographical structures, rather than as mere abstractions or byproducts of a general process of urbanization (Pred 1973, 1980).

That cities sheltered immigrant labor forces has been one of the central themes in American urban historical geography since Ward published his declaration of the field, *Cities and Immigrants* (1971). Emphasis was initially on the floating population as it arrived, stayed, or departed. The concept was largely demographic. Recently, interest has focused on the environment in which the new urban masses found themselves. What was the nature of housing tenure, and how did it change with explosive urban population growth or industrialization (Hertzog and Lewis 1986; Harris, Levine, and Osborne 1981)? What influence can be traced to the activities of promoters, builders, realtors, and landlords as actors whose decisions influenced the physical fabric and social geography of the city (Dennis 1987; Hanna and Olson 1983; Doucet 1982; Doucet and Weaver 1984; Richard Harris 1987)? In these studies, the urban environment is no passive backdrop for the ebb and flow of population.

City-building is a vital process in its own right, and when understood it yields clues to the economic and social texture of urban life (Olson 1979). Thus is opening the possibility of regarding urban residents as conscious decision-makers who are faced with known ranges of choice. A more behavioral approach to the residential structure of nineteenth-century cities begins to seem feasible (Miller 1982).

PERSPECTIVES ABROAD

Comparative Historical Geography: The Frontier

Historical geographers have never been content with an exclusively North American focus. Long before the macrogeographical perspectives of Meinig and Wallerstein made it de rigueur, comparative inquiry in historical geography had described the global variability of spatial and ecological processes. Indeed, since the late 1950s, historians and geographers have used comparative methods to assess North American, South African, Australian, Latin American, and Asian frontier experiences. Often, these comparisons elaborate upon historian Frederick Jackson Turner's "frontier thesis" (Gerhard 1959; Gully 1959; Mikesell 1960; Turner 1920; Wyman and Kroeber 1957), or examine Isaiah Bowman's "pioneer fringe" and its evolution (Bowman 1931; Joerg 1932; Lonsdale and Holmes 1981; Meinig 1959; Thompson 1975; Vanderhill 1982).

Recently, geographers have offered several conceptual frameworks of interest. Frederick Jackson Turner's emphasis on frontier isolation in shaping and simplifying frontier societies has been revived by Harris (1977) and Harris and Guelke (1977). And in various frontier settings, others have explored issues such as the market and its effects on subsistence and commercial agriculture (Norton 1982, 1984a, 1984b); the evolution of frontier urban systems, as viewed through central place and mercantile models (Winters 1981); the effects of evolving systems, of technology and institution upon the development of urban-transport infrastructures (Taaffe, Morrill, and Gould 1963; Meinig 1962; Earle 1977); and migration processes (Hudson in Miller and Steffen 1977, 11–31).

Even more ambitious and provocative are a handful of studies that position frontier experiences in global context. Webb's "Great Frontier" (1964) and Wallerstein's "world-system" (1974) acknowledge the common origins of many frontier settlement systems: their geographies the creation of powerful European states bent on economic, political, and cultural domination. Meinig's (1982) "imperialism" and Osborne and Rogerson's (1978) unequal power relations in frontier settings argue for a political focus on conflict and coaptation. Engerman (1983, 1986) assesses how such power was manifested in exploitive labor systems, while Knapp and Hauptman (1980; Hauptman and Knapp 1977) show how European and Japanese colonial administrative policies in East Asia produced parallel results. Christopher's (1984) description of the shaping of the African landscape by colonial institutions offers another regional example.

Future work on comparative frontiers necessarily will draw on cognate fields. Anthropological perspectives offer possibilities for cross-fertilization. Wolf (1982) and Kirkby (1984) have transfigured world-system abstractions of capitalist global dominance into human manifestations in regional settings—e.g. the changes in native culture and frontier geography on contact with the new global order. Mitchell (1979) and Palmer (1976) have echoed that perspective, advocating a comparative cultural ecology that evaluates colonial impacts on indigenous ecosystems.

Comparative historical geography is a modest but necessary step in the progression toward a truly systemic global interpretation of the past. But if our global interpretations are to be more than gratuitous, they will require considerably larger investments of research energies abroad. Fortunately, a path has already been blazed by historical geographers of Latin America. Their impressive studies en masse reveal the rewards, the possibilities, and the frustrations of inquiry in the "third world periphery."

The Historical Geography of Latin America

The historical geography of Latin America has a long tradition in the AAG, maintained over the decades by some of the Association's most-distinguished members. This interest group, however, has always been a small one. Why this should be so is difficult to know, although our aversion to foreign languages, the difficulty of working abroad, and the fact that regional interests are somewhat out-of-fashion in geography are part of an explanation.

Those who, in spite of these obstacles, ventured south across the Rio Grande have found an intellectually stimulating scene. Latin America is vast, its history long, and necessarily there has been a certain amount of specialization. Historical geographers have turned more to Mexico and the Caribbean than to South America, more to "contact" and colonial times than to the nineteenth century. Demographic and land-use studies still predominate, but thematic changes have taken place. Regional cultural geography has diminished, while historical economic geography has grown. New interpretive themes have arisen. Agricultural change, for example, is explained by external forces of imperialism and capitalism, which are often linked to Wallerstein's world-system or to a Braudelian world view. Similarly, urban and environmental geography are providing new themes for geographical inquiry. This synopsis, though desperately compressed, offers a framework for the discussion that follows.

Historical geographers have long been intrigued by precontact societies, especially their demography and agricultural geography. The two are linked, since agrarian regional productivity provides clues to the numbers of people who were present. Evidence increasingly shows that intensive forms of land use were widespread (Wilken 1987). Yet controversy persists over the size of the population supported by indigenous agriculture. The estimate of 60 million in 1492 has a good deal of support, but higher and lower figures also have their advocates (Denevan 1976: 3, 289–92; Bethell 1984–86, vol. 1). Indian demography continues to prompt the most debate in the region's historical geography.

Discussion of the Indian's postcontact fate goes back to the time of the conquistadores. Although the story's broad outline is well known—the disappearance of the Indians from the Caribbean, and a great decline in numbers elsewhere—modern scholarship is filling in the details in New Spain (Gerhard 1972, 1979), Guatemala (Lovell 1985), Honduras and Nicaragua (Newson 1985, 1987), and Brazil (Hemming 1978, 1987).

The three-century-long colonial period offers ample scope for research. But aside from Peter Boyd-Bowman's classic work on the locational and social origins of sixteenth-century Spanish emigrants and their destinations in the Americas (1976), we know too little about the Iberian colonizers. Nothing comparable exists for Portuguese emigration to colonial Brazil.

Somewhat greater attention has been paid to mixed Indian, European, and African populations (Robinson, 1981). A nice variety of colonial scholarship appears in the important new monographic series, Dellplain Latin American Studies, edited by David Robinson of Syracuse University. Several of these explore the social geography of colonial cities, while others analyze regional mining and agricultural economies. Historical geographers have also taken an active interest in the plantation and the hacienda, the latter having come to be seen as a commercial institution rather than a "feudal" one. The old dichotomy between these settlements is now drawn less sharply than before. Caribbean plantation societies continue to attract attention: Watts (1987) and Pulsipher (1986) emphasized the environmental impact of these societies, and Higman (1987) detailed the spatial economy of Jamaican sugar plantations, using eighteenth- and nineteenth-century maps.

HISTORICAL AND CULTURAL CONTRIBUTIONS TO GEOGRAPHIC UNDERSTANDING

Nineteenth-century Latin America has fared less well than the colonial period, a neglect that is difficult to explain, given the abundance of interesting topics to study. Good works on Central America and the Caribbean do come to mind: Dozier (1985) on Nicaragua's Mosquito Shore, and Richardson (1983, 1985) on the struggle for economic freedom by former slaves and their descendants in the British West Indies. But elsewhere, American geographers are not making significant contributions to, for instance, the study of nineteenth-century economic transformation in Argentina, Uruguay, Chile, or southern Brazil. Similarly, urban geographies of nineteenth-century cities are rare; the principal contribution in recent years, on Buenos Aires, came from the historian James Scobie (1974).

The study of Latin America's historical geography is, and will remain, interdisciplinary. The geographer thus enters into a reciprocal exchange with historians, economists, anthropologists, and archaeologists. All draw from each other. A recent history text, Lockhart and Schwartz (1983), borrows concepts freely from geography and the social sciences; similarly, *The Cambridge History of Latin America* (Bethell 1984–86) is required reading for historical geographers. Latin American scholars share in this collaboration. Levi Marrero y Artiles (1971–1985), geographer and historian, is contributing a multivolume work of fundamental importance on Cuba. Brazilian geographers and historians have begun an analysis of Rio de Janeiro in the late nineteenth and early twentieth centuries (Naro 1987). These interdisciplinary and international collaborations are just two reasons why the study of Latin American historical geography is so enjoyable. There is room for more participants.

TOOLS OF THE TRADE

Among the paradoxes of the "information age" is the odd fact that we often know more about the past than the present. With patience, a bit of specialized skill, and some imagination, historical sources yield unparalleled insights on individual spatial and ecological behavior. The trick is to build upward: gathering names from documents, record-linking the names into real people, and space-linking them into real neighborhoods, communities, and regions. Archival documents, in turn, are collated with field reconstructions of cultural and environmental landscapes. Literary sources help round out the historical-geographic scene. Computers make the whole job faster and more efficient.

These skills are not merely arcane; the wit and imagination they demand have considerable value for contemporary geographical inquiry, particularly in those cases when important geographical changes, e.g. plant closures and deindustrialization, run far ahead of census tabulations. In these times, the tools of our trade assume added value in their transferability to geography at-large.

Documents and Archives

Much of historical geography's research is conducted in archival and manuscript repositories. During the past 20–25 years, the basis of historical geographical research has been the verification of historical geographical data and their spatial distribution through the critical analysis of sources contemporary to the period under study. Although many types of primary sources are potentially useful, certain types of documents, textual and graphic, have been utilized more successfully than others (Ehrenberg 1975; Grim 1982).

Textual documents, whether manuscript or printed, provide the basic source for qualitative and quantitative data. Although governmental archives at the national, state, and local levels normally document matters of a political or legal nature, they often contain information relevant to geography. National and state records, for instance, describe the establishment of boundaries, the regulation of commerce and trade, the founding of cities and towns, the use of natural resources, and the construction of transportation systems. Population counts can be derived from censuses, tax lists, or militia lists; real and movable property data reside in local records of land patent, deed, and probate. Business records, such as merchant and plantation accounts, document the economic activity of individual enterprise, while newspaper advertisements and business directories scan a wider range of economic activity. Exploration accounts and travel diaries usefully document geographical knowledge and initial perceptions of frontier regions.

Graphic sources—maps, aerial photographs, landscape sketches, and photographs—provide spatial and visual perspective for historical studies. Often used only as illustrations, these documents, when critically appraised, sometimes stand alone as primary sources. For example, rural landownership could be studied through a succession of survey plats and county landownership maps; urban land use could be studied through successive fire-insurance maps.

Obviously, primary source material available for any particular problem depends on the recordkeeping practices of the creating agency, and the archival practices of successive holders. Although certain types of records have been used more successfully than others, historical geographers should continually examine repository guides, bibliographic literature, and footnote references as they seek new sources and innovative ways of interpreting them.

The Landscape as Archive

The notion of landscape has a venerable geographical history, deeply rooted in Schluter's "landschaft," Hoskins' *The Making of the English Landscape,* Sauer's "The Morphology of Landscape," and Kniffen's cultural landscape. Although eclipsed somewhat during the "quantitative revolution," this traditional theme has contributed immensely to our understandings of culture groups, regions, and processes. In recent years, something of a revival in landscape and material culture research has discretely gained momentum (Lewis 1985; Ford 1984; Schlereth 1985).

As a tool of historical geography, cultural landscape has been used in three ways: (1) as an indicator of historical process, (2) as an indicator of landscape origin and development, and (3) as a medium of historical preservation.

The first of these approaches looks at landscape elements—buildings, fences, field patterns, roads, cemeteries, gardens, and more—as markers of historical diffusion and culture change. Kniffen's landmark works of 1965 and 1966 (with Glassie), as well as those of Lewis (1975a) and Wacker (1973), show how landscape elements convey information on migration and diffusion routes, on ethnic variation in frontier settlement, and on cultural and technological relations. Landscape relics offer clues to changing land use, culture, and attitudes toward land. Urban landscape interpretations suggest that architecture and morphology are good indicators of a city's evolving functions and social classes (Conzen 1978; Rubin 1977).

The second and perhaps most-popular approach to landscape focuses on the landscape itself—its origins and interplay of culture and environment. Excellent examples in this tradition include studies of log houses (Jordan 1978; Newton and Pulliam-DiNapoli 1977), cemeteries (Jordan 1982), house forms (Pillsbury and Kardos 1970; Lewis 1975a; Glassie 1975), barns (Wacker 1973), and town squares (Price 1968). A few have explored the Old World antecedents of New World landscape forms (Jordan 1980; Mannion 1974). Others have examined the touchy issue of environmental influence, as mediated through the perception of builders (Meinig 1979) or through construction materials (Gritzner 1974) and climate (Francaviglia 1972). Needed are more studies of landscape expressions of class, religious belief, and political ideology.

A third focus for students of cultural landscape is historic preservation. Its advocates, though few in number, have been remarkably eloquent in appealing for landscape identification and preservation (Lewis 1975b; Datel and Dingemans 1980). Here, historical geographers could make valuable contributions to the burgeoning public and private interest in landscape renovation, rehabilitation, and preservation.

Material culture and landscape constitute, as it were, libraries on the land. We must care for these sources, help catalogue and map them, and preserve them for the undeciphered information they contain. Historical geography's vigilance in protecting genuine landscapes is no less vital than in protecting documentary archives from replacement by myth and historical fiction.

Sources for the Study of Environmental Change

Field and archival research are mutually reinforcing in environmental research. Although this mutuality dates from the time of George Perkins Marsh, only in the past two decades have physical geographers fully appreciated the utility of historical records for reconstructing past physical landscapes. Physical geographers use many of the techniques and sources of historical geography. These include travel accounts, early governmental surveys, newspapers and journals, congressional serials, old maps and atlases, aerial and ground photographs, and agricultural statistics. Other sources more particular to physical phenomena are navigation and water-power surveys; stream and sediment discharge records; coastal charts; road, rail, canal, and bridge surveys; reservoir surveys; soil inventories; flood-control reports; drainage and irrigation records; and climatological records.

Fieldwork for the physical geographer then involves remeasuring environments described in old documents, so that processes and rates may be established. The use of similar approaches, techniques, and sources suggests the fruitfulness of increasing the interaction and synergism between historical and physical geography.

Computers in Historical Geography

The laboriousness of historical geographic research is legendary, and to some intimidating. Not every temperament is suited for collecting, linking, aggregating, and mapping a melange of documentary and field data. But thanks to the personal computer, the tedium of historical research is in full retreat. Attitudes began to change in the

1970s when large-scale, heavily financed projects (such as the Philadelphia Social History Project) made historical geographers aware of the labor-saving potential of computers (Hirshberg 1981). But widespread computer applications were few in number, owing to the cost of mainframe computers and the skills required of programmers. The diffusion of the personal computer, however, has enabled historical geographers to automate their databases, analyses, and graphic displays.

This new technology offers powerful solutions for managing and linking varied historical record sets. With the personal computer and well-known sources of archival information, historical geographers can extract and link data. Using commercially available software, data may be matched (by individual name, or some geographic unit such as a street address or township), sorted, and statistically tested. Integrated mapping packages and databases are changing the way we do historical geography. Mapping software, such as *Atlas Graphics,* when combined with predigitized boundary files such as the "U.S. County Historical Boundary Files, 1850–1970" (Earle and Young 1988), offers powerful tools for data analysis.

Even more powerful is Geographic Information Systems software, such as *PC Arc/Info* and *Autocad,* which automates the mapping and linkage of varied spatial data sets (Hornbeck and Botts 1988). The use of GIS in historical research, however, is hardly new. Indeed, some 30 ago, Andrew Clark's work on Prince Edward Island, with its over 200 maps of population, economy, and society, constituted an early, nonautomated version of GIS (Hartnett 1988). As so often happens, innovation has come full circle back to tradition—but now the task takes half the time.

NEEDS AND OPPORTUNITIES

Historical geography's recent achievements are impressive. The discipline has amassed a remarkably varied literature. Drawing upon archival and landscape sources, historical geographers have formulated valuable historical interpretations of regions, particularly in North and South America; explored ecological and locational processes through time; and commenced the exciting task of integrating our knowledge into macroscale models, such as the Atlantic world or the world system.

But the vibrancy of historical geography also illuminates great gaps in our research agenda. We could be more attentive, as the following sketches point out, to the historical geography of minorities, to intersections with physical geography, and to applied historical geography. These are just a few of the enormous opportunities awaiting us, as we attempt to integrate these ideas into the prevailing paradigms and models of historical geography.

Native American Historical Geography

Geography's quest for relevance during the 1960s and 1970s enriched the discipline through the addition of environmental issues and radical interpretations of social problems. But the crusade for social awareness sometimes fell short. Few research careers in historical geography, for example, have been built on studies of racial minorities or women. Nor have mainstream historical geographers effectively incorporated the rich literature on these groups. Eurocentric motifs tenaciously endure.

HISTORICAL AND CULTURAL CONTRIBUTIONS TO GEOGRAPHIC UNDERSTANDING

To illustrate: None of us could conceive of interpreting the British and French colonization of North America as a vanished "stage," of interest primarily because of the kind of food they ate or the battles they lost. Nor would we consider studies that emphasized information and paradigms from a quarter of a century ago "innovative". Yet, American Indians routinely receive this kind of treatment in American historical geography. Of our principal minorities, the historical geography of Hispanic Americans is most firmly established, thanks to the efforts of Carl O. Sauer, Donald W. Meinig, and Richard Nostrand (Nostrand 1980). Conversely, a literature on the historical geographies of Black Americans, Asian Americans, and women scarcely exists.

Historical geography's treatment of American Indians lies somewhere in between these extremes, and thus it nicely illustrates our ambiguities toward, and uneven treatment of, minorities in general. On one hand, the historical geography of American Indians can claim many high-quality publications and strong links to North American anthropology, ethnohistory, and Indian history. This fact is attested by our cognate's use of geography's spatial, environmental, and regional perspectives in a critique of traditional paradigms (Albers 1987; Clifton 1979; Horsman 1982; White 1984).

On the other hand, mainstream historical geography often overlooks these interpretations of Native Americans. Some syntheses scarcely mention indigenous cultures, even in their discussions of ecology and frontiers (Norton 1984a; Butlin 1987). Three recent historical geographies of the United States and Canada (Ward 1979; Meinig 1986; Mitchell and Groves 1987) incorporate some Indian material, but the coverage is uneven, varying with author, region, and period.

Our paradigms for interpreting Native Americans are outdated. Too frequently, the paradigm harkens back a quarter of a century or more to the "stomach geography" school of Indian cultural ecology, to the material culture catalogs popularized by anthropologist Harold Driver (1961) and his predecessors, and to the theme of Indian wars and frontier politics. Studies of European colonization present Indians either as an early stage of sequent occupance or (in 1930s-style regional typology) as a "natural phenomenon," i.e., wildlife, vegetation, and Indians. These narrow analyses ignore Indian-white cooperation in the fur trade; the implications of intermarriage, as with the Metis; certain comparatively long periods of peaceful coexistence; and the persistence of viable Indian populations long after white conquest.

Early revisionists commendably attempted to correct these biases, but often they did so by portraying a romantic "Noble Savage," stoically dispossessed (Jacobs 1973 is the most frequently cited example). A more positive image of Indians in American historiography remains a major priority, but such work must offer more than an updated morality play.

Indian historical geography also has its ironies. Its practitioners have been criticized by contemporary geographers who claim that they have forgotten the modern reservation and the urban Indian (Higgens 1982). Doubtless, historical geography could make valuable contributions to contemporary problems, e.g., to Native claims cases, by reconstructing historic patterns of occupance and resource use. Presently, however, applied research of this kind is conducted by historians and anthropologists.

These criticisms notwithstanding, the historical geography of American Indians has made substantial progress. Two research traditions stand out from the rest: the tribal-history approach, represented by Heidenreich's (1981) encyclopedic *Huronia*, and the regional-ecological approach, exemplified by Ray's (1974) *Indians in the Fur Trade*. The tribal perspective has been applied to the Navajo (Jett and Spencer 1981;

Goodman 1982), the Iroquois (Konrad 1981), and the Choctaw (McKee and Schlenker 1980). The regional-ecological perspective is most prominent in fur-trade studies on, for example, the trading practices of Hudson's Bay Company (Ray and Freeman 1978), the Far West (Wishart 1979), and the western Great Lakes area (Kay 1984), and on Indian contributions to mapping (Lewis 1980).

Further themes in Native American historical geography are explored in the *Journal of Cultural Geography* (Spring/Summer 1982), *A Cultural Geography of North American Indians* (Ross and Moore 1987), and Goodman's essay on "The Native American" (1985). Future research might examine Indians' environmental attitudes, tribal land, and the occupance and resource bases of Native claims cases.

The prospect for Native American historical geography is as rich and diverse as the variety of tribes, regions, and the questions about them posed by geography and cognate disciplines. Native American research needs and opportunities increase geometrically in the case of other minorities and women in the American past. But if the opportunities for minority study are vast, the lingering ambiguities that surround this inquiry serve as a constant reminder of Eurocentrism's enduring and insidious appeal.

Environmental Impacts in the American Past

Although the study of human impacts on the natural environment of Anglo-America is a tradition almost as old as American geography, its effects on geography and historical geography are less potent than one might suspect, particularly given the restlessness of North Americans and their frontiers.

Physical environment was, at the turn of the century, an important component of American geography, but these early professional geographers discounted the role of human agency in environmental processes. The situation deteriorated even further during the 1920s and 1930s when geographers largely abandoned the study of human impacts on the Earth. Lamenting this development, Sauer (1941) enthusiastically advocated the study of "man as an agent of physical geography." Ironically, the federal government (USDA Soil Conservation Service) was at that time sponsoring precisely the kinds of research called for by Sauer, but most of it was undertaken by engineers and geologists.

Sauer's vigil was finally rewarded during the mid-1950s in the mammoth interdisciplinary symposium on "Man's Role in Changing the Face of the Earth" (Thomas 1956). This scholarly volume made many contributions, though few of them focused on North America. Not until the environmental crusade of the 1970s did this continent become a major region of inquiry. Indeed, as late as the 1960s, studies of North American denudation rarely considered the role of humankind.

Since 1970, the study of human-induced environmental change in North America has increased significantly. In 1974, a volume sponsored by the AAG inventoried past and contemporary research (Manners and Mikesell 1974). And in 1986, Goudie's *The Human Impact on the Natural Environment* could summarize the impressive results of such studies in North America and the world. Nonetheless, no single volume chronicles North America alone. The need is pressing, because even now, human impacts on large areas of the continent remain little known.

Although there is now considerable ferment in the field of human impact on past environments, the impetus has come principally from physical geographers, using the methods and techniques of historical geography. Historical geographers meanwhile show signs of another "Great Retreat." At the International Conference of Historical Geographers in 1986, for example, just one paper touched on a physical theme.

Whatever its other merits, Donald Meinig's recent book, *The Shaping of America* (1986), requires no prior knowledge of physical geography whatsoever. The volume barely hints at the impacts which Americans have had upon their environs, and vice versa. Indeed, Meinig seems to have gone out of his way to ignore the physical diversity of North America and its sophisticated interpretation. His is a great book, but how much better it could have been had geography's full arsenal been brought to bear. Contrast it, for example, with the deep understanding of natural landscape consistently displayed in Ralph Brown's classic historical geography in 1948. Is this situation desirable? If good geography synthesizes human and physical elements of landscape (Marcus 1979), should not historical geography do the same?

The time may be right for rapprochement. Surveying their field in 1980, four geomorphologists (Graf, et al. 1980) stated:

> We believe that no other group engaged in geomorphic research is as well qualified to grapple with the human factor [of geomorphic change] as are geographers. Our broad training in both physical and cultural systems and our appreciation of landscape change in the natural and human senses give us perspectives and insight that are rarely found in other disciplines. Geographers need to work closely with engineers and geologists in order to share with them such wide-ranging concepts as spatial analysis and emphasis on the man-land interface. These concepts are endemic to geography but they may be quite foreign to other workers.

Geomorphologists, they conclude, should have training in archaeology, anthropology, and historical geography. Ronald Cooke, in his keynote address to the First International Conference on Geomorphology in Manchester (1985), echoed these sentiments in calling for historical perspectives on physical geography.

We require an increment of physical geography, not just for environmental impact analysis, but for opening up the ample possibilities of geographic research—which include the effects of environment on people. American agriculture offers a case in point. Soil erosion has been a major force in American historical geography (Sauer 1941). But where has it occurred, and how bad has it been? What synergism of physical and cultural elements caused it—or prevented it? We are reminded of Sauer's old question (1941): "Were the Virginians great colonizers because they were notable soil wasters?" If geography lies along a continuum between physical and human systems, should not historical geographers be working all parts of the continuum? A vital and lively historical geography depends on it.

Applied Historical Geography and Careers in Historical Geography

The vitality of historical geography depends on many things, not the least being the recruitment of outstanding minds. This, in turn, depends on the practical matter of career opportunities. Applied historical geography, therefore, has a strategic role to play in a vital, dynamic, and expanding field.

The status of applied research has not improved dramatically since Andrew Clark remarked that "a large part of research in historical geography might be called pure research in that it is not directly focused on contemporary applications" (Clark 1954).

The Journal of Historical Geography has yet to publish an unequivocally applied paper. A recent survey of over 1000 applied geographers identified only 2% as having historical specialties (Russell 1983; Janiskee 1980). Nonetheless, cognate historical disciplines have found geographic theory and methods useful in public scholarship programs (Johnson 1977; Lewis 1984; O'Brien 1984), and this bodes well for increasing involvement by historical geographers.

The boundaries separating academic and applied scholarship are becoming less distinct, and this is particularly true in historical geography. Curricula in public history and widespread participation in public archaeology have made applied scholarship respectable among academic social scientists. At the very least, academic and applied scholarship are complementary; one supplies the theoretical framework, the other provides tests of theory, or collects and preserves valuable information for academic inquiry.

The greatest opportunities for applied historical geography lie in the interaction among the university, the public sector, and interdisciplinary projects (Figure 1). Planning and preservation are obvious examples (Ford 1979; Conzen and Conzen 1979; Konrad 1982). But we have also contributed to adjudication of legal questions (De Vorsey 1973, on historical boundaries), and the understanding of natural resource management—e.g., Trimble's (1974) reconstruction of historical soil erosion, and Dilsaver's (1986) examination of National Park Service policy. Furthermore, historical and geographical analyses of human mobility, industrial waste disposal, and human responses to climatic change have enriched hazards research (Bowden 1982; Colten 1986; McQuillan 1982).

Students prospecting for jobs will find opportunities in historic preservation, collecting and managing databases; in museums and historic districts, developing educational and interpretive programs; and in public agencies, conducting research on natural resources and relict hazards, e.g. toxic industrial wastes. While there is no

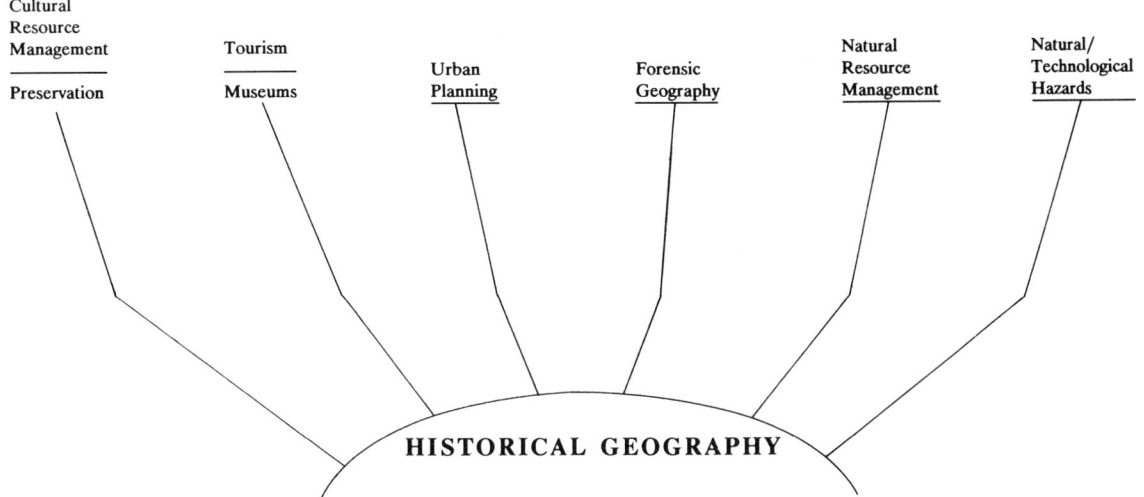

Figure 1
Applied historical geography

HISTORICAL AND CULTURAL CONTRIBUTIONS TO GEOGRAPHIC UNDERSTANDING

equivalent to a mother lode in historical geography's job markets, employment seekers can follow a well-marked trail of interdisciplinary scholarship to rich veins of career possibilities.

It is possible to exaggerate the usefulness of historical geography. But, as one public historian recently mused, "the lesson here is not that planning bureaus should hire public historians, but rather that there should be more history in a planner's training" (O'Donnell 1986, 242). The same can be said for historical geography. Museum directors and lawyers may seldom seek out geographers, but if they are attuned to what we do—and more importantly, if we are familiar with their jobs—opportunities to interact and collaborate will be realized.

* * *

Let us not close without remarking upon the mysterious enchantment of historical geography. For most of us, it is pure fun; indeed, for a few it is a loving avocation, supplemental to our principal geographical callings. Each of us enjoys the challenge of making sense out of past geographies, of puzzling over them relatively unconstrained by doctrinal methodologies or philosophies. Let us trust that, even as we discipline ourselves to the imperatives of grand theory and the practice of applied geography, our liberating eclecticism will endure.

REFERENCES

Agnew, John. 1987. *The United States in the world economy: A regional geography.* New York: Cambridge University Press.

Albers, P. C. 1987. New directions in scholarship on American Indians. *Reviews in Anthropology* 14:221–35.

Baltensperger, Bradley H. 1980. Agricultural change among Nebraska immigrants, 1880–1920. In *Ethnicity on the Great Plains,* ed. Frederick C. Luebke, pp. 170–89. Lincoln, NE: University of Nebraska Press.

Berry, Brian J. L.; Conkling, Edgar C.; and Ray, D. Michael. 1987. *Economic geography: Resource use, locational choices, and regional specialization in the global economy.* Englewood Cliffs, NJ: Prentice-Hall.

Bethell, Leslie, ed. 1984–86. *The Cambridge history of Latin America.* 5 vols. Cambridge: Cambridge University Press.

Bowden, Martin. 1975. Desert wheat belt, plains corn belt: Environmental cognition and behavior of settlers in the plains margin, 1850–99. In *Images of the plains: The role of human nature in settlement,* eds. Brian Blouet and Merlin P. Lawson. Lincoln, NE: University of Nebraska Press.

———. 1982. Geographical changes in cities following disaster. In *Period and place,* eds. A. R. H. Baker and M. Billinge, pp. 114–126. Cambridge: Cambridge University Press.

Bowman, I. 1931. *The pioneer fringe.* New York: American Geographical Society.

Boyd-Bowman, Peter. 1976. Patterns of Spanish emigration to the Indies until 1600. *Hispanic American Historical Review* 56:580–604.

Brown, Ralph E. 1943. *Mirror for Americans: Likenesses of the eastern seaboard, 1810.* New York: American Geographical Society.

Butlin, R. A. 1987. Theory and methodology in historical geography. In *Historical geography: Progress and prospect,* ed. M. Pacione, p. 35. Wolfboro, NH: Croom Helm.

Christopher, A. J. 1984. *Colonial Africa.* London: Croom Helm.

Clark, Andrew H. 1954. Historical geography. In *American geography: Inventory and prospect,* eds. P. James and C. Jones, pp. 71–105. Syracuse: Syracuse University Press.

———. 1959. *Three centuries and the island, a historical geography of settlement and agriculture in*

Prince Edward Island, Canada. Toronto: University of Toronto Press.

———. 1972. Suggestions for the geographical study of agricultural change in the United States, 1790–1840. *Agricultural History* 46:155–72.

Clifton, J. A. 1979. The tribal history: An obsolete paradigm. *American Indian Culture and Research Journal* 3:81–100.

Collins, Randall. 1986. *Weberian sociological theory*. New York: Cambridge University Press.

Colten, Craig E. 1986. Industrial wastes in southeast Chicago: Production and disposal, 1870–1970. *Environmental Review* 10:93–106.

Conzen, Michael P. 1971. *Frontier farming in an urban shadow: The influence of Madison's proximity on the agricultural development of Blooming Grove, Wisconsin*. Madison, WI: State Historical Society of Wisconsin.

———. 1974. Local migration systems in nineteenth century Iowa. *Geographical Review,* 64:339–61.

———. 1978. Analytical approaches to the urban landscape. In *Dimensions of human geography,* ed. Karl Butzer, pp. 128–65. University of Chicago Department of Geography Research Paper No. 186. Chicago: University of Chicago.

———, and Conzen, Kathleen Neils. 1979. Geographical structure in nineteenth-century urban retailing: Milwaukee, 1836–90. *Journal of Historical Geography* 5:45–66.

Corbridge, Stuart. 1986. *Capitalist world development: A critique of radical development geography*. Totowa, NJ: Rowman & Littlefield.

Cosgrove, Denis E., and Jackson, Peter. 1987. New directions in cultural geography. *Area* 19:95–101.

Datel, Robin E., and Dingemans, Dennis. 1980. Historic preservation and urban change. *Urban Geography* 1:229–53.

Denevan, William M., ed. 1976. *The native population of the Americas in 1492*. Madison, WI: University of Wisconsin Press.

Dennis, Richard. 1987. *Landlords and rented housing in Toronto, 1885–1914*. Centre for Urban and Community Studies Research Paper No. 162. Toronto: University of Toronto.

De Vorsey, Louis. 1973. Florida's seaward boundary: A problem in applied historical geography. *Professional Geographer* 25:214–20.

Dilsaver, Lary M. 1986. Land-use conflict in the Kings River Canyons. *The California Geographer* 26:59–80.

Doucet, M. J. 1982. Urban land development in nineteenth-century North America. *Journal of Urban History* 8:299–342.

———, and Weaver, John. 1984. The North American shelter business, 1860–1920: A study of a Canadian real estate and property management agency. *Business History Review* 58(1984):234–62.

Dozier, Craig L. 1985. *Nicaragua's Mosquito shore: The years of British and American presence*. University, AL: University of Alabama Press.

Driver, Harold E. 1961. *Indians of North America*. Chicago: University of Chicago Press.

Earle, Carville V. 1975. *The evolution of a tidewater settlement system: All Hallow's Parish, Maryland, 1650–1783*. University of Chicago Department of Geography Research Paper No. 170. Chicago: University of Chicago.

———. 1977. The first English towns of North America. *Geographical Review* 67:34–50.

———. 1978. A staple interpretation of slavery and free labor. *Geographical Review* 68:51–65.

———. 1987. Regional economic development west of the Appalachians, 1815–1860. In *North America: The historical geography of a changing continent,* eds. R.D. Mitchell and P.A. Groves, pp. 172–97. Totowa, NJ: Rowman & Littlefield.

———, and Hoffman, Ronald. 1976. Staple crops and urban development in the eighteenth-century South. *Perspectives in American history* 10:5–78.

———, and ———. 1980. The foundation of the modern economy: Agriculture and the costs of labor in the United States and England, 1800–60. *American Historical Review* 85:1055–94.

———, and Young, Cyrus. 1988. U.S. historical county boundary files 1850–1970 for the PC, Oxford, Ohio. Oxford, OH: Department of Geography, Miami University.

Ehrenberg, Ralph E., ed. 1975. *Pattern and process: Research in historical geography*. Washington: Howard University Press.

Engerman, Stanley. 1983. Contract labor, sugar, and technology in the nineteenth century. *Journal of Economic History* 43:635–59.

———. 1986. Slavery and emancipation in comparative perspective: A look at some recent debates. *Journal of Economic History* 46:317–39.

Ernst, Joseph A., and Merrens, H. R. 1973. Camden's turrets pierce the skies! The urban process in the southern colonies during the eighteenth century. *William and Mary Quarterly,* 3d ser. 30:549–74.

Ford, Larry R. 1979. Urban preservation and the geography of the city in the USA. *Progress in Historical Geography* 3:211–38.

———. 1984. Architecture and geography: Toward a mutual concern for space and place. *Yearbook of the Association of Pacific Coast Geographers* 46:7–33.

Francaviglia, Richard V. 1972. Western American barns: Architectural form and climatic consideration. *Yearbook of the Association of Pacific Coast Geographers* 34:153–60.

Gerhard, D. 1959. The frontier in comparative view. *Comparative Studies in Society* 1:205–29.

Gerhard, Peter. 1972. *A guide to the historical geography of New Spain.* Cambridge: Cambridge University Press.

———. 1979. *The southeast frontier of New Spain.* Princeton: Princeton University Press.

Gerlach, Russell. 1976. *Immigrants in the Ozarks.* Columbia, MO: University of Missouri Press.

Giddens, Anthony. 1981. *Power, property and the state.* Vol. 1 of *A contemporary critique of historical materialism.* Berkeley: University of California Press.

Glassie, Henry. 1975. *Folk housing in middle Virginia: A structural analysis of historic artifacts.* Knoxville: University of Tennessee Press.

Goheen, Peter. 1985–86. Communications and urban systems in mid-nineteenth-century Canada. *Urban History Review* 14:235–45.

———. 1987. Canadian communications circa 1845. *Geographical Review* 77:35–51.

Goodman, J. J. 1982. *The Navajo atlas: Environments, resources, people and history of the Dine Bikeyah.* Norman, OK: University of Oklahoma Press.

Goodman, J. M. 1985. The Native American. In *Ethnicity in contemporary America: A geographical appraisal,* ed. Jesse O. McKee, pp. 31–54. Dubuque, IA: Kendall/Hunt Publishing Company.

Goudie, Andrew. 1986. *The human impact on the natural environment.* Cambridge, MA: MIT Press.

Graf, W. L.; Trimble, S. W.; Toy, T. J.; and Costa, J. E. 1980. Geographical geomorphology in the eighties. *Professional Geographer* 32:279–84.

Grim, Ronald E. 1982. *Historical geography of the United States: A guide to information sources.* Detroit: Gale Research.

Gritzner, Charles F. 1974. Construction materials in a folk housing tradition: Considerations governing their selection in New Mexico. *Pioneer America* 6:25–39.

Groves, Paul A. 1974. The "hidden" population: Washington alley dwellers in the late nineteenth century. *Professional Geographer* 3:270–6.

———, and Muller, Edward K. 1975. The evolution of black residential areas in late-nineteenth century cities. *Journal of Historical Geography* 1:169–91.

Guelke, Leonard. 1982. *Historical understanding in geography: An idealist approach.* Cambridge: Cambridge University Press.

Gully, J. L. M. 1959. The Turnerian frontier: A study in the migration of ideas. *Tijdschrift voor Economische en Sociale Geografie* 50:65–72, 81–91.

Hagerstrand, Torsten. 1978. Survival and arena. In *Timing space and spacing time,* vol. 2 of *Human activity and time geography,* eds. Tommy Carlstein, Don Parkes, and Nigel Thrift, pp. 122–45. London: Edward Arnold.

———. 1988. Some unexplored problems in the modeling of culture transfer and transformation. In *The transfer and transformation of ideas and material culture,* eds. Peter J. Hugill and D. Bruce Dickson, pp. 217–32. College Station, TX: Texas A & M University Press.

Hall, Peter. 1981. The geography of the fifth Kondratieff cycle. *New Society* March 26:535–37.

Handcock, W. G. 1976. Spatial patterns in a trans-Atlantic migration field: The British Isles and Newfoundland during the eighteenth and nineteenth centuries. In *The settlement of Canada: Origins and transfer,* ed. Brian S. Osborne, pp. 13–40. Kingston, Ontario: Queen's University.

———. 1977. English migration to Newfoundland. In *The peopling of Newfoundland: Essays in historical geography,* ed. J. J. Mannion, pp. 15–48. St. Johns: Memorial University of Newfoundland.

Hanna, David, and Olson, Sherry. 1983. Metiers, logers et bouts de rues: l'armature de la societe montrealaise 1881 a 1901. *Cahiers de Geographie du Quebec* 27:255–75.

Harris, Richard; Levine, Gregory; and Osborne, Brian. 1981. Housing tenure and social class in Kingston,

Ontario, 1881–1901. *Journal of Historical Geography* 7:271–89.

Harris, R. Cole. 1977. The simplification of Europe overseas, *Annals of the Association of American Geographers.* 67:469–83.

———. 1978. The historical geography of North American regions. *American Behavioral Scientist* 22:115–30.

———. 1985. European beginnings in the Northwest Atlantic: A comparative view. In *Seventeenth-century New England,* eds. D. G. Allen and D. D. Hall, pp. 119–52. Charlottesville, VA: University Press of Virginia.

———. 1987. France in North America. In *North America: The historical geography of a changing continent,* eds. R. D. Mitchell and P. A. Groves, pp. 65–92. Totowa, NJ: Rowman & Littlefield.

———. 1987. The growth of home ownership in Toronto, 1899–1913. Paper read before a Housing Tenure Workshop at the Centre for Urban and Community Studies, University of Toronto, Toronto, Ontario, February.

———, ed., and Matthews, Geoffrey J., cartographer. 1987. *From the beginning to 1800.* Vol. 1 of *Historical atlas of Canada.* Toronto: University of Toronto Press.

———, and Guelke, L. 1977. Land and society in Early Canada and South Africa. *Journal of Historical Geography* 3:135–53.

Hartnett, Sean. 1988. The use of personal computers in historical geographic research. Paper presented at the annual meeting of the Association of American Geographers, Phoenix, Arizona.

Hartshorne, Richard. 1939. *The nature of geography, a critical survey of current thought in the light of the past.* Lancaster, PA: Association of American Geographers.

Hauptman, L. M., and Knapp, R. G. 1977. Dutch-Aboriginal interaction in New Netherland and Formosa: An historical geography of empire. *Proceedings of the American Philosophical Society* 121:166–82.

Heidenreich, C. E. 1981. *Huronia: A history and geography of the Huron Indians, 1600–50.* Toronto: McClelland & Stewart.

Hemming, John. 1978. *Red gold: The conquest of the Brazilian Indians.* London: Macmillan.

———. 1987. *Amazon frontier: The defeat of the Brazilian Indians.* London: Macmillan.

Henretta, James. 1978. Families and farms: Mentalite in preindustrial America. *William and Mary Quarterly,* 3d ser. 35:3–32.

Hertzog, Stephen, and Lewis, Robert D. 1986. A city of tenants: Homeownership and social class in Montreal, 1847–1881. *Canadian Geographer* 30:316–23.

Hewes, Leslie. 1973. *The suitcase farming frontier.* Lincoln, NE: University of Nebraska Press.

Higgens, B. 1982. Urban Indians: Patterns and transformations. *Journal of Cultural Geography* 2:110–18.

Higman, B. W. 1987. The spatial economy of Jamaican sugar plantations: Cartographic evidence from the eighteenth and nineteenth centuries. *Journal of Historical Geography* 13:17–39.

Hilliard, Sam B. 1972. *Hog meat and hoecake: Food supply in the Old South.* Carbondale, IL: Southern Illinois University Press.

Hirshberg, Theodore W., ed. 1981. *Philadelphia work, space, family, and group experience in the nineteenth century.* New York: Oxford University Press.

Hornbeck, David, and Botts, Howard. 1988. Seven Oaks dam project water systems. Los Angeles: U.S. Army Corps of Engineers.

Horsman, R. 1982. Well trodden paths and fresh byways: Recent writing on Native American history. *Reviews in American history* 10:234–44.

Hudson, John C. 1973. Two Dakota homestead frontiers. *Annals of the Association of American Geographers* 63:442–62.

———. 1976. Migration to an American frontier. *Annals of the Association of American Geographers* 66:242–65.

———. 1985. *Plains country towns.* Minneapolis: University of Minnesota Press.

Jacobs, W. R. 1973. The Indian and the frontier in American history—A need for revision. *Western Historical Quarterly* 4:43–56.

Jakle, John. 1977. *Images of the Ohio Valley.* New York: Oxford University Press.

Janiskee, Robert. 1980. Socially and ecologically responsible historical geography. *Environmental Review* 4:35–40.

Jett, S. C., and Spencer, V. E. 1981. *Navajo architecture: Forms, history, distributions.* Tucson: University of Arizona Press.

Joerg, W. L. G. 1932. *Pioneer settlement.* New York: American Geographical Society.

Johnson, G. 1977. Aspects of regional analysis in archaeology. *Annual Review of Anthropology* 6:479–508.

Johnson, Hildegard B. 1969. King wheat in Southeastern Minnesota: A case study in frontier agriculture. *Annals of the Association of American Geographers* 59:348–64.

Jones, E. L. 1981. *The European miracle: Environments, economies and geopolitics in the history of Europe and Asia.* Cambridge: Cambridge University Press.

Jordan, Terry G. 1966. *German seed in Texas soil.* Austin, TX: University of Texas Press.

———. 1967. The imprint of the Upper and Lower South on mid-nineteenth-century Texas. *Annals of the Association of American Geographers* 57:667–90.

———. 1978. *Texas log buildings: A folk architecture.* Austin, TX: University of Texas Press.

———. 1980. Alpine, Alemannic, and American log architecture. *Annals of the Association of American Geographers* 70:154–80.

———. 1982. *Texas graveyards, A cultural legacy.* Austin, TX: University of Texas Press.

Kay, Jeanne. 1984. The fur trade and Indian population growth. *Ethnohistory* 31:265–87.

Kirkby, D. 1984. Colonial policy and native depopulation in California and New South Wales, 1770–1840. *Ethnohistory* 31:1–16.

Knapp, R. G., and Hauptman, L. M. 1980. Civilization over savagery: The Japanese, the Formosan frontier and the United States Indian policy, 1895–1915. *Pacific Historical Review* 49:645–52.

Kniffen, Fred. 1965. Folk housing: Key to diffusion. *Annals of the Association of American Geographers* 55:549–77.

———, and Glassie, Henry. 1966. Building in wood in the Eastern United States. *Geographical Review* 56:40–66.

Konrad, V. 1981. An Iroquois frontier: The north shore of Lake Ontario during the late seventeenth century. *Journal of Historical Geography* 7:129–244.

———. 1982. Historical artifacts as recreation resources. In *Recreational land use: Perspectives on its evolution in Canada,* eds. G. Wall and J. Marsh, pp. 391–416. Toronto: Oxford University Press.

Kulikoff, Allan. 1979. The colonial Chesapeake: Seed bed of antebellum southern culture? *Journal of Southern History* 45:513–40.

Legreid, Anne Marie, and Ward, David. 1982. Religious schism and the development of rural immigrant communities: Norwegian Lutherans in Western Wisconsin, 1880–1905. *Upper Midwest History* 2:13–29.

Lemon, James T. 1972. *The best poor man's country: A geographical study of early southeastern Pennsylvania.* Baltimore: Johns Hopkins University Press.

———. 1980. Early Americans and their social environment. *Journal of Historical Geography* 6:115–31.

———. 1987. Colonial America in the eighteenth century. In *North America: The historical geography of a changing continent,* eds. R. D. Mitchell and P. A. Groves, pp. 121–46. Totowa, NJ: Rowman & Littlefield.

Lewis, G. M. 1980. Indian Maps. In *Old trails and new directions: Papers of the third North American fur trade conference,* eds. C. M. Judd and A. J. Ray, pp. 9–23. Toronto: University of Toronto Press.

Lewis, K. E. 1984. *The American frontier: An archaeological study of settlement pattern and process.* New York: Academic Press.

Lewis, Peirce F. 1975a. Common houses, cultural spoor. *Landscape* 19:1–22.

———. 1975b. The future of our past: Our clouded vision of historic preservation. *Pioneer America* 7:1–20.

———. 1985. Learning from looking: Geographic and other writing about the American landscape. In *Material culture: A research guide,* ed. Thomas Schlereth, pp. 35–56. Lawrence, KS: University of Kansas Press.

Lockhart, James, and Schwartz, Stuart B. 1983. *Early Latin America: A history of colonial Spanish America and Brazil.* Cambridge: Cambridge University Press.

Lonsdale, R. E., and Holmes, J. H., eds. 1981. *Settlement systems in sparsely populated regions: The United States and Australia.* New York: Pergamon Press.

Lovell, George W. 1985. *Conquest and survival in colonial Guatemala: A historical geography of the Cuchumatan Highlands 1500–1821.* Kingston: McGill–Queen's University Press.

McCann, L. D. 1985. Metropolitanism and branch business in the Maritimes, 1881–1931. In *Atlantic Canada after confederation: The Acadiensis reader* eds. P. A. Buckner and David Frank, pp. 202–15. Fredericton, N.B.: Acadiensis Press.

McCusker, John J., and Menard, Russell R. 1985. *The economy of British America 1607–1789*. Chapel Hill: University of North Carolina Press.

McIntosh, C. B. 1975. Use and abuse of the timber culture act. *Annals of the Association of American Geographers* 65:347–62.

McKee, J. O., and Schlenker, J. A. 1980. *The Choctaw: Cultural evolution of the Native American tribe*. Jackson, MS: University of Mississippi Press.

McManis, Douglas. 1963. *The initial evaluation and utilization of the Illinois prairies, 1815–1840*. University of Chicago Department of Geography Research Paper No. 94. Chicago: University of Chicago.

McNeill, William H. 1963. *The rise of the West*. Chicago: University of Chicago Press.

———. 1982. *The pursuit of power. Technology, armed force, and society since 1000 A.D.* Chicago: University of Chicago Press.

McQuillan, D. Aidan. 1978a. Farm size and work ethic: Measuring the success of immigrant farmers on the American grasslands, 1875–1925. *Journal of Historical Geography* 4:57–76.

———. 1978b. Territory and ethnic identity: Some measures of an old theme in the cultural geography of the United States. In *European settlement and development in North America: Essays in honour and memory of Andrew Hill Clark,* ed. James R. Gibson, pp. 136–69. Toronto: University of Toronto Press.

———. 1982. The interface of physical and historical geography. In *Period and place,* eds. A.R.H. Baker and M. Billinge, pp. 136–44. Cambridge: Cambridge University Press.

Mannion, John J. 1974. *Irish settlements in eastern Canada: A study of cultural transfer and adaptation*. Toronto: University of Toronto Press.

Manners, I. R., and Mikesell, M. W., eds. 1974. *Perspectives on environment*. Washington: Association of American Geographers.

Marcus, Melvin G. 1979. Coming full circle: Physical geography in the twentieth century. *Annals of the Association of American Geographers* 69:521–32.

Marrero y Artiles, Levi. 1971–1985. *Cuba: Economia y sociedad*. 12 vols. Rio Piedras: Editorial San Juan, Vol. 1 Madrid: Editorial Playor, vols. 2–12.

Marsh, Ben. 1987. Continuity and decline in the anthracite coal towns of Pennsylvania. *Annals of the Association of American Geographers* 77:337–53.

Marsh, George P. 1864. *Man and nature*. New York: Scribner.

Meinig, Donald W. 1957–58. The American colonial era A geographic commentary. *Proceedings of the Royal Geographic Society of Australia, South Australian Branch* 59:1–22.

———. 1959. Colonization of wheatlands: Some Australian and American comparisons. *Australian Geographer* 7:205–13.

———. 1962. A comparative historical geography of two railnets: Columbia Basin and South Australia. *Annals of the Association of American Geographers* 52:394–413.

———. 1966. The colonial period, 1609–1775. In *Geography of New York State,* ed. J.H. Thompson, pp. 121–39. Syracuse: Syracuse University Press.

———. 1968. *The great Columbian plain: A historical geography 1805–1910*. Seattle: University of Washington Press.

———. 1972. American wests: Preface to a geographical interpretation. *Annals of the Association of American Geographers* 62:159–85.

———. 1978. The continuous shaping of America: A prospectus for geographers and historians. *American Historical Review* 83:1186–1205.

———, ed. 1979. *The interpretation of ordinary landscapes*. New York: Oxford University Press.

———. 1982. Geographical analysis of imperial expansion. In *Period and place: Research methods in historical geography,* eds. A. R. H. Baker and M. Billinge, pp. 71–78. Cambridge: Cambridge University Press.

———. 1986. *Atlantic America, 1492–1800*. Vol. 1 of *The shaping of America: A geographical perspective on 500 years of history*. New Haven: Yale University Press.

Mensch, Gerhard. 1979. *Stalemate in technology: Innovations overcome the depression*. Cambridge, MA: Ballinger.

Merrens, H. Roy. 1964. *Colonial North Carolina in the eighteenth century: A study in historical geography*. Chapel Hill: University of North Carolina Press.

Meyer, David. 1987. The national integration of regional economies, 1860–1920. In *North America: The historical geography of a changing continent,* eds. R. D. Mitchell and P. A. Groves, pp. 321–45. Totowa, NJ: Rowman & Littlefield.

Mikesell, Marvin. 1960. Comparative studies in frontier history. *Annals of the Association of American Geographers* 50:62–74.

Miller, D. H., and Steffen, J. O. 1977. *The frontier: Comparative studies*. Norman, OK: University of Oklahoma Press.

Miller, Roger. 1982. Household activity patterns in nineteenth-century suburbs: A time-geographic exploration. *Annals of the Association of American Geographers* 72:355–71.

Mitchell, Robert D. 1977. *Commercialism and frontier: Perspectives on the early Shenandoah Valley*. Charlottesville: The University Press of Virginia.

———. 1978. The formation of early American cultural regions: An interpretation. In *European settlement and development in North America,* ed. J.R. Gibson, pp. 66–90. Toronto: University of Toronto Press.

———. 1979. Perspectives on the colonial frontier: Culture, society, and environment in the Chesapeake. *Comparative Frontier Studies* 13:4.

———. 1983. American origins and regional institutions: The seventeenth-century Chesapeake. *Annals of the Association of American Geographers* 73:404–20.

———. 1987. The colonial origins of Anglo-America. In *North America: The historical geography of a changing continent* eds. R. D. Mitchell and P. A. Groves, pp. 93–120. Totowa, NJ: Rowman & Littlefield.

———, and Groves, Paul A., eds. 1987. *North America: The historical geography of a changing continent*. Totowa, NJ: Rowman & Littlefield.

Muller, Edward K. 1976. Selective urban growth in the Middle Ohio Valley, 1800–1860. *Geographical Review* 66:178–99.

Naro, Nancy P. Smith. 1987. Rio studies Rio: Ongoing research on the First Republic in Rio de Janeiro. *The Americas* 43:429–40.

Newson, Linda A. 1985. *The cost of conquest: Indian decline in Honduras under Spanish rule*. Dellplain Latin American Studies, vol. 20. Boulder, CO: Westview Press.

———. 1987. *Indian survival in colonial Nicaragua*. Norman, OK: University of Oklahoma Press.

Newton, Milton B. 1974. Cultural preadaptation and the Upland South. *Geoscience and Man* 5:143–54.

———, and Pulliam-DiNapoli, Linda. 1977. Log houses as public occasions: A historical theory. *Annals of the Association of American Geographers* 67:360–83.

Norton, William. 1982. Some comments on late nineteenth-century agriculture in areas of European overseas expansion. *Ontario History* 74:113–17.

———. 1984a. Agricultural evolution on the frontier. *Professional Geographer* 36:18–27.

———. 1984b. A comparative analysis of frontier settlement in the Cape Province, South Africa and Southern Ontario, Canada. *South African Geographer* 12:43–55.

———. 1984c. *Historical analysis in geography*. New York: Longman.

Nostrand, Richard L. 1980. The Hispano homeland in 1900. *Annals of the Association of American Geographers* 70:382–96.

O'Brien, M. J. 1984. *Grassland, forest, and historical settlement*. Lincoln, NE: University of Nebraska Press.

O'Donnell, T. 1986. Pitfalls along the path of public history. In *Presenting the Past,* eds. S. P. Bensen, S. Brier, and R. Rosenzweig, pp. 239–44. Philadelphia: Temple University Press.

Olson, Sherry H. 1979. Baltimore imitates the spider. *Annals of the Association of American Geographers* 69:557–74.

O'Mara, James. 1982. Town founding in seventeenth-century North America: Jamestown in Virginia. *Journal of Historical Geography* 8:1–11.

———. 1983. *An historical geography of urban system development: Tidewater Virginia in the 18th century*. Geographical Monographs, vol. 13. Downsview, Ontario: Atkinson College, Department of Geography, York University.

Omner, Rosemary E. 1977. Highland Scots migration to Southwestern Newfoundland: A study in kinship. In *The peopling of Newfoundland: Essays in historical geography,* ed. John J. Mannion, pp. 212–33. St. John's: Memorial University of Newfoundland.

———. 1986. Primitive accumulation and the Scottish clan in the old world and the new. *Journal of Historical Geography* 12:121–41.

Osborne, B. S., and Rogerson, C. M. 1978. Conceptualizing the frontier settlement process: Development or dependency? *Comparative Frontier Studies* 11:1–3.

Ostergren, Robert C. 1979. A Community transplanted: The formative experience of a Swedish immigrant community in the Upper Midwest. *Journal of Historical Geography* 5:189–212.

———. 1980. Prairie bound: Migration patterns to a Swedish settlement on the Dakota Frontier. In *Ethnicity on the Great Plains,* ed. Frederick C. Luebke, pp. 73–91. Lincoln, NE: University of Nebraska Press.

———. 1981a. The immigrant church as a symbol of community and place on the landscape of the American Upper Midwest. *Great Plains Quarterly* 1:224–38.

———. 1981b. Land and family in rural immigrant communities. *Annals of the Association of American Geographers* 71:400–11.

———. 1988. *A community transplanted: The transatlantic experience of a Swedish immigrant settlement in the Upper Midwest, 1835–1915.* Madison, WI: University of Wisconsin Press.

Palmer, G. B. 1976. The ecology of frontier communities: African settlement schemes. *Comparative Frontier Studies* 4:1–3.

Pillsbury, Richard R. 1970. The urban street pattern as a culture indicator: Pennsylvania, 1682–1815. *Annals of the Association of American Geographers* 60:428–46.

———, and Kardos, Andrew. 1970. *A field guide to the folk architecture of the Northeastern United States.* Geographic Publications at Dartmouth No. 8. Hanover, NH: Dartmouth College.

Pred, Allan R. 1973. *Urban growth and the circulation of information: The United States system of cities, 1790–1840.* Cambridge: Harvard University Press.

———. 1980. *Urban growth and city-systems in the United States, 1840–1860.* Cambridge: Harvard University Press.

———. 1986. *Place, practice, and structure: Social and spatial transformation in Southern Sweden, 1750–1850.* Totowa, NJ: Barnes & Noble.

Price, Edward T. 1968. The central courthouse square in the American county seat. *Geographical Review* 58:29–60.

Price, Jacob M. 1974. Economic function and the growth of American port towns in the eighteenth century. *Perspectives in American History* 8:121–86.

Pulsipher, Lydia. 1986. *Seventeenth-century Monserrat: An environmental assessment.* IBG Historical Geography Research Group: Research Series No. 17. Norwich, England: Geo Books.

Radford, John P. 1976. Race, residence and ideology: Charleston, South Carolina in the mid-nineteenth century. *Journal of Historical Geography* 2:329–46.

Ray, A. J. 1974. *Indians in the fur trade: Their role as hunters, trappers, and middlemen in the lands southwest of Hudson Bay, 1660–1870.* Toronto: University of Toronto Press.

———, and Freeman, D. B. 1978. *Give us good measure: An economic analysis of relations between the Indians and the Hudson's Bay Company before 1763.* Toronto: University of Toronto Press.

Rice, John G. 1973. Patterns of ethnicity in a Minnesota county, 1880–1905. *Geographical Reports 4.* Umea, Sweden: Department of Geography, University of Umea.

———. 1977. The role of culture and community in frontier prairie farming. *Journal of Historical Geography* 3:155–75.

———. 1978. The effect of land alienation on settlement. *Annals of the Association of American Geographers* 68:61–72.

———, and Ostergren, Robert C. 1978. The decision to emigrate: A study in diffusion. *Geografiska Annaler* B60:1–15.

Richardson, Bonham C. 1983. *Caribbean migrants: Environment and human survival on St. Kitts and Nevis.* Knoxville: University of Tennessee Press.

———. 1985. *Panama money in Barbados, 1900–1920.* Knoxville: University of Tennessee Press.

Robinson, David J., ed. 1981. *Studies in Spanish American population history.* Dellplain Latin American Studies, vol. 8. Boulder, CO: Westview Press.

Ross, T. E., and Moore, T. G., eds. 1987. *A cultural geography of North American Indians.* Boulder, CO: Westview Press.

Rubin, Barbara. 1977. A chronology of architecture in Los Angeles. *Annals of the Association of American Geographers* 67:521–37.

Russell, J. A. 1983. Specialty fields of applied geographers. *Professional Geographer* 35:471–75.

Sauer, Carl O. 1941. Foreword to historical geography. *Annals of the Association of American Geographers* 31:1–24.

———. 1952. *Agricultural origins and dispersals.* New York: American Geographical Society.

———. 1962. Middle America as a culture historical location. In *Readings in cultural geography,* eds. Philip L. Wagner and Marvin W. Mikesell, pp. 195–201. Chicago: University of Chicago Press.

Sauressig-Schreuder, Yda. 1985a. Dutch Catholic emigration in the mid-nineteenth century Noord Brabant, 1847–1871. *Journal of Historical Geography* 11:48–69.

———. 1985b. Dutch Catholic settlement in Wisconsin. In *The Dutch in America: Immigration, settlement, and cultural change,* ed. Robert P. Swrerenga. New Brunswick, NJ: Rutgers University Press.

Schlereth, Thomas L., ed. 1985. *Material culture: A research guide*. Lawrence, KS: University of Kansas Press.

Scobie, James. 1974. *Buenos Aires: Plaza to suburb, 1870–1910*. New York: Oxford University Press.

Taaffe, Edward J.; Morrill, R. L.; and Gould, P. R. 1963. Transport expansion in underdeveloped countries. *Geographical Review* 53:503–29.

Taylor, Peter. 1985. *Political geography: World-economy, nation-state and locality*. New York: Longman.

Thomas, W. L., ed. 1956. *Man's role in changing the face of the earth*. Chicago: University of Chicago Press.

Thompson, S. I. 1975. The contemporary Latin American frontier. *Comparative Frontier Studies* 1:2–3.

Trimble, Stanley W. 1974. *Man-induced soil erosion on the Southern Piedmont, 1700–1970*. Ames, IA: Soil Conservation Society of America.

———. 1987. Perspectives on the history of soil erosion control in the Eastern United States. *Agricultural History* 59:162–80.

Turner, Frederick J. 1920. *The frontier in American history*. New York: Henry Holt and Co.

Vance, James E., Jr. 1970. *The merchant's world: The geography of wholesaling*. Englewood Cliffs, NJ: Prentice-Hall.

Vanderhill, B. G. 1982. The passing of the pioneer fringe in Western Canada. *Geographical Review* 72:200–17.

Wacker, Peter O. 1973. Folk architecture as an indicator of cultural areas and culture diffusion: Dutch barns and barracks in New Jersey. *Pioneer America* 5:37–47.

———. 1975. *Land and people: A cultural geography of preindustrial New Jersey*. New Brunswick, NJ: Rutgers University Press.

Wallerstein, Immanuel. 1974. *The modern world-system: Capitalist agriculture and the origins of the European world-economy in the sixteenth century*. New York: Academic Press.

———. 1980. *The modern world-system II. Mercantilism and the consolidation of the European world-economy, 1600–1750*. New York: Academic Press.

———. 1984. *The politics of the world-economy. The states, the movements and the civilizations*. Cambridge: Cambridge University Press.

Ward, David. 1971. *Cities and immigrants: A geography of change in nineteenth century America*. New York: Oxford University Press.

———. 1975a. The debate about alternative methods in historical geography. *Historical Methods Newsletter* 8:82–87.

———. 1975b. Victorian cities: How modern? *Journal of Historical Geography* 1:135–51.

———, ed. 1979. *Geographic perspectives on America's past: Readings on the historical geography of the United States*. New York: Oxford University Press.

———. 1982. The ethnic ghetto in North American cities: Past and present. *Transactions, Institute of British Geographers* 7:257–75.

———. 1987. Population growth, migration and urbanization, 1860–1920. In *North America: The historical geography of a changing continent,* eds. R. D. Mitchell and P. A. Groves, pp. 299–320. Totowa, NJ: Rowman & Littlefield.

Watts, David. 1987. *The new West Indies: Patterns of development, culture and environmental change since 1492*. Cambridge: Cambridge University Press.

Webb, Walter P. 1964. *The great frontier*. Austin, TX: University of Texas Press.

Weber, Max. [1904-5] 1958. *The Protestant ethic and the spirit of capitalism*. New York: Scribner's.

———. [1922] 1968. *Economy and society: An outline of interpretive sociology*. 3 vols. New York: Bedminster.

White, R. 1984. Native Americans and the environment. In *Scholars and the Indian experience: Critical reviews of recent writing in the social sciences,* ed. W. R. Swagerty, pp. 179–204. Bloomington, IN: Indiana University Press.

Wilken, Gene. 1987. *Good farmers: Traditional agricultural resource management in Mexico and Central America*. Berkeley: University of California Press.

Winters, C. 1981. The urban systems of medieval Mali. *Journal of Historical Geography* 7:341–55.

Wishart, David. 1979. *The fur trade of the American West, 1807–1840: A geographical synthesis*. Lincoln, NE: University of Nebraska Press.

Wolf, Eric R. 1982. *Europe and the people without history*. Berkeley: University of California Press.

Wood, Joseph. 1982. Village and community in early colonial New England. *Journal of Historical Geography* 8:333–46.

Wyckoff, William. 1986a. Frontier milling in western New York. *Geographical Review*, 76:73–93.

———. 1986b. Land subdivision on the Holland purchase in Western New York State, 1797–1820. *Journal of Historical Geography* 12:142–61.

Wyman, W. D., and Kroeber, C. B., eds. 1957. *The frontier in perspective*. Madison, WI: University of Wisconsin Press.

Zelinsky, Wilbur. 1973. *The cultural geography of the United States*. Englewood Cliffs, NJ: Prentice-Hall.

Cultural Ecology

Karl W. Butzer

Cultural ecology has represented an explicit research perspective within geography since the 1960s. A Specialty Group by that name was organized within the Association of American Geographers in 1980, and its membership has trebled during the last five years, to 196 in March 1988. Cultural ecology was first categorized as a Topical Proficiency in 1987, but only 129 American geographers identify themselves under this heading. Perhaps it is still perceived more as a research perspective than as a separate subfield.

Relatively few geographers have a clear appreciation of what cultural ecologists aspire to do, and so the purpose of this chapter is explanatory and constructive, rather than critical. It represents a personal interpretation of the spirit and the logical structures of cultural ecology, as practiced by geographers in North America.[1] Academic research can rarely be organized in simple subcategories, and cultural ecology is patently in a state of ferment and rapid growth. This chapter therefore attempts to represent both the unifying themes as well as the diversity of what cultural ecologists do. A complementary view of the subfield is given by Turner (1989).

UNIFYING THREADS AND THEMES

Cultural ecology draws upon interdisciplinary roots within geography and anthropology in seeking to understand the interrelationships between people, resources, and space. It focuses upon how people live, doing what, how well, for how long, and with what environmental and social constraints. It emphasizes that human behavior has a cognitive dimension and is dependent on information flow, values, and goals. Finally, cultural ecologists recognize that actions are conceived and taken by individuals, but that such actions must be examined and approved by the community, in the light of

---

[1] Initially, specific responses to a set of questions were solicited from a committee consisting of W. M. Denevan, L. Grossman, P. W. Porter, B. L. Turner, and M. Watts. Their replies provided insights derived from different experiences. Successive drafts also profited with suggestions from my Texas colleagues, W. E. Doolittle, G. Knapp, and K. E. Foote.

tradition and the prevailing patterns of institutions and power, before decisions can be implemented.

These general statements are best followed by more-specific comments on how cultural ecologists formulate their problems, proceed in their analysis, and present their conclusions:

1. Society and nature are seen as intimately interconnected, bound by complex, systemic interrelationships. Within that unified framework, particular attention is given to how people manage resources via a range of strategies in regard to diet, technology, reproduction, settlement, and system maintenance. The variability of the biophysical environment in space and time is an integral component of all such discussions, as is the role of environmental constraints.
2. Cultural behavior is explicitly considered in its functional role, and with respect to material culture as well as the tangible reflections of nonmaterial culture. This is normally achieved by in-depth field studies to gain a comprehensive understanding of how energy-flows and information-flows operate, how alternative options are developed and selected, and how process and form are interrelated. Empirical detail is crucial to such research, as is the connectivity between data and conclusions.
3. Food production is a fundamental theme, especially in regard to demographic variables and sustainability. Most studies in cultural ecology are in fact directed toward rural and agricultural societies—with a Third World bias—and they generally exhibit a specific interest in understanding change.

As a corollary, it follows that cultural ecologists are concerned with the role of people and the manipulation of resources within ecosystems, rather than the delineation or simulation of such systems as a whole. Normative inferences are drawn from intensive, empirical studies. Culture is not treated as a superorganic "black box," but is increasingly presented as a processual context, amenable to analysis. Finally, cultural ecologists are interested in behavioral diversity, alternative outcomes, and feedback loops—far more than they are in causation or prediction.

CULTURAL ECOLOGY AS A NEW PARADIGM

Distant roots for cultural ecology can be sought in German efforts to integrate physical and human research in geography (Butzer 1989a), as well as in Marsh's articulation (1864) of human influences on the environment. However, these are indirect roots, not unique to cultural ecology. Two more proximal traditions are of greater interest, namely Chicago and Berkeley, representing "Midwestern" and "Western" geography (Porter 1978), and espousing different, if mainly implicit, notions of ecology.

The concept of human ecology was formulated at Chicago by J. P. Goode about 1907 (Martin 1987) and brought to wider attention by Barrows (1923). The latter presidential address proposed to shift geography from analytical to applied research, focused upon human economic adjustment to environment. Barrows' influence was directly reflected in the research of White and his students[2] on perception of, adjust-

---

[2] W. D. Pattison (Chicago) kindly clarified the linkages between Goode, Barrows, and White, including the evolution of the Chicago geography program during the 1920s and 1930s.

ment to, and management procedures for environmental hazards (Burton, Kates, and White 1978).

Initially representing an acultural, technocratic approach, this "hazards tradition" had little philosophical impact on the emergence of *cultural* ecology, although several concepts, such as hazard perception, proved to have considerable utility. Barrows's human ecology also contributed indirectly to the evolution of 1960s-style human geography, with its socioeconomic and Western, materialistic orientation. Somewhat ironical, and probably not unrelated, was the parallel development of another Chicago school of human ecology, directed towards urban sociology.[3] It emphasized collective life as an adaptive process, reflecting the interactions of environment, population, and organization (Hawley 1986), with a thrust that is distinctly Western and applied.

The backgrounds of these two Chicago schools explain why the "ecology" of North American human geography tends to deemphasize cultural in favor of sociological processes or behavioral psychology, generally examined in contemporary, industrialized societies. From the geographer's perspective, cultural ecology and human ecology are therefore quite different, although Clarkson (1970), Butzer (1982), and Kates (1987) suggest that the two approaches can be usefully reconciled.

The impact of Sauer's Berkeley school on cultural ecology is more obvious, but by no means unproblematical. Sauer did not explicitly espouse ecology, and some recent interpretations of his work conclude that he was a humanist at heart (see Leighly 1987). But in the context of German geographers such as Hahn, Ratzel, Meitzen, and Gradmann, Sauer (1) saw cultural landscapes as historically informative in their own right, as the product of successive cultural transformations of an original "natural" landscape; (2) recognized the dynamic role of technology and human institutions; (3) communicated with anthropologists, whose sophistication in matters of culture he appreciated; and (4) had a predilection for nonurban and non-Western societies (Sauer 1925, 1927, 1941). It is therefore not surprising that his Berkeley students, and their students in turn, were well equipped and predisposed to participate in the crystallization of cultural ecology as a viable approach in geography since the mid-1960s.

But cultural ecology was not a predictable outgrowth of the Berkeley school, with its indifference to theory and analytical specialization. Only a very few of its graduates have made the decisive shift from a preeminent concern with human impacts, landscape history, or cultural morphology to a direct study of how cultural processes affect adaptive strategies. Equally pertinent is that a majority of cultural ecologists based in North America are not linked to the Berkeley school, even though most of them share a deep appreciation for Sauer. Finally, Denevan, who directly or indirectly supervised the training of most Berkeley-influenced cultural ecologists, acknowledges the strong influence of Brookfield and the Australian school on himself and his students (Denevan, pers. com.).

Two consistent background traits of cultural ecologists in North America can be singled out: (1) considerable training in the Earth sciences or biological ecology, and (2) extensive coursework or long-term association with anthropologists. Brookfield and other early members of the Australian school shared a long interdisciplin-

---

[3]The pioneers of this group—R. E. Park, E. W. Burgess, and R. D. McKenzie—formulated their ideas in the same years as Barrows. Although both the Chicago sociologists and Barrows took pains to disassociate themselves from one another (W. D. Pattison, pers. com.), one must suspect some degree of initial contact.

ary association with American anthropologists such as Brown, Rappaport and Vayda in a New Guinea field project. The influence of Brookfield and his students on Denevan, Nietschmann, Turner, Waddell, Watts, and others was enhanced by faculty appointments at, or degrees from, North American universities. As for other North American cultural ecologists, Porter collaborated extensively with anthropologists in East Africa, while Butzer, Carr, and Kirkby were intimately associated with archaeological projects. Butzer additionally held an anthropology appointment at Chicago. Younger cultural ecologists who first published after 1975 are equally strongly grounded in anthropology.

This analysis shows that, while cultural ecology draws from geographical tradition, it also represents a significant break with that tradition. Both the methods and the theoretical framework—the paradigm, if you will—are different. Cultural processes have become a theme of primary attention, and the scale of analysis has shifted from extensive research on "culture areas" to intensive study of smaller social groupings, with long periods of fieldwork and emphasis on detailed observation or measurement. Primary attention is no longer devoted to the impact of people on the environment or visible features of the cultural landscape, but to food production, demography, and ecological sustainability. This quantum change represents a new set of goals that required a methodology not provided by cultural geography.

Two catalytic agents can be identified in this paradigm shift. One is the impact of the "scientific" methodology widely espoused by the social sciences since the 1960s, derived from ecological, systems, and cybernetics or information theory. The other is the application of analytical modes developed in anthropology. The first of these greatly facilitated the examination and didactic presentation of complex interrelationships and transformations. The second allowed greater understanding of sociocultural processes, switching culture from a "black box" to a set of tangible variables, amenable to direct study. These characteristics distinguish cultural ecology from cultural geography and other nature-society approaches.

COMPLEX INTERRELATIONSHIPS

Ecology, systems theory, and cybernetics or information theory were tapped more or less simultaneously by the social sciences, although with variable degrees of enthusiasm, and more often implicitly rather than explicitly. Their use had tended to accompany rather than follow programmatic statements, such as those offered by several British geographers (e.g., Stoddart 1965; Chorley 1973). Cultural ecologists have been cognizant of these conceptual frameworks, but, like many other social scientists, have been reluctant to cast their research in the terminology of other sciences. The reasons for this restraint become apparent on closer inspection.

Ecology is a biological concept, primarily concerned with energy and organisms. It deals with:

1. organic productivity,
2. the roles of different organisms with distinct econiches as they compete with other organisms of similar feeding-habits (trophic levels), and
3. the food chains (energy pathways) that link groups such as photosynthetic producers, herbivores, and carnivores at successive levels of the food chain.

Both the advantages and problems of transferring a biological paradigm to the social sciences are fairly evident. Ecology allows a structured organization of unlike variables, emphasizes function and hence interchanges between component parts, and is amenable to systematic and nondeterministic study of interrelationships within an organic whole. Much less satisfactory is that it deals with plants and animals, and offers no obvious niche for the role of culture and human cognition. Placing people at the top of the trophic pyramid as ecological dominants only deepens the problem by implying analogies between human and animal behavior.

Systems theory enhances the ecological framework and facilitates the understanding and even simulation of complex interrelationships. Above all it has great heuristic value by emphasizing the degree to which all interactions are interdependent. Changes within one population or variable can affect some, or many, or all of the other components of the ecosystem. Such change is channeled through a chain of interlinked structures that ultimately impinge on the original variable—negative or positive feedback loops that serve to either suppress or amplify change. Such change may be reflected in long-term, net trends (dynamic equilibrium), or in abrupt shifts (metastable equilibrium).

The systems perspective has been particularly helpful in projecting long-term environmental impacts, as well as in explaining and anticipating thresholds, ecological "simplification," or "catastrophic" readjustment. The limitations are equally apparent. Simulation is very difficult and quantification rarely possible, while the approach as such is too mechanistic and prone to overemphasize functional and materialistic attributes (e.g., Ellen 1982).

Cybernetics can be drawn upon to illuminate the peculiarly human role in the ecosystem. Culture represents encoded information, and individual as well as group behavior is regulated and implemented in the context of information. Decisions are made with respect to alternative information, within a social system characterized by established energy and information pathways, complicated by cooperation as well as competition at each "trophic" level, and screened by the experience and deeper values embodied in culture. Finally, technology and social organization in the broadest sense reflect information in varying degrees. Adaptive choices and cultural variety represent critical variables in such an information system, which at the highest level is operated, if not controlled, by human cognition.

Societies can therefore be viewed as interlocking human ecosystems. They operate on the basis of individual initiatives and actions, embodied in aggregate, community behavior and institutional structures. At the individual level, built-in goal conflicts and human unpredictability represent potentially powerful variables for *change,* while at the several community and institutional levels a wide range of negative feedbacks favor *stability*. Prediction, whether for long-term evolutionary change or rapid modification, is therefore difficult, even in probabilistic terms. The system involved is simply too complex to simulate effectively, as exemplified in the difficulty of economic or social prognoses. Even in historical perspective, societal behavior is difficult to analyze and explain satisfactorily.

Given these difficulties of normative study, cultural ecologists follow the precedent of biologists in focusing upon a limited range of variables to gain understanding of component processes. They have also increasingly found that particularistic case studies provide realistic experience with variability, resilience, stability, and change.

These theoretical tools are therefore no more than one means to an end, where cumulative experience, cultural sensitivity, and even intuition are of paramount importance in the drawing of inference.

INTERDISCIPLINARY CONNECTIONS

A widespread impression obtains that geographical cultural ecology is based on the anthropological work of Steward (1955). But both Steward and Sauer were influenced by the British geographer Forde (1934), who emphasized good case studies of subsistence economies. Also, Steward was at least indirectly indebted to Sauer for his emphasis on the environment. Nonetheless, Steward played a catalytic role for both geography and anthropology: he effectively made the point that nature and society are interlinked by cultural *adaptation,* i.e., strategies for ecological success (Denevan 1983).

Steward saw cultural ecology as the study of adaptive processes, whereby cultures adjusted to an environment through their subsistence activities. His "method" was (1) to establish the interrelationships between environment and exploitative technology, (2) to examine the patterns of behavior followed in appropriating specific technologies in that environment, and (3) to assess the degree to which behavioral patterns affected other aspects of culture (Steward 1955, 40–41). He envisioned ecological relationships as part of a network of cultural adjustments and adaptations that, collectively and incrementally, set in train a multilinear process of cultural evolution that incorporated alternative techno-environmental patterns and social behavior. He sought to explain the functional relationship between agricultural technology and output, population density, settlement patterns, and social organization. His unit of analysis was a "culture core," linked to a subregional environment (akin to the "culture area"), and he attempted to show by crosscultural studies that similar functional interrelationships recurred in different areas having different historical trajectories.

A second important contribution was made in the same period by another anthropologist, Barth (1956). In emphasizing the complementary lifeways of farmers and herders in Pakistan, Barth showed that two groups can achieve a symbiotic relationship within a single environment by exploiting different *econiches.*

A third influential study was that of another anthropologist, Geertz (1963). He compared two alternative agrosystems in Indonesia with the structure, productivity, energy flows, and stability of the tropical rainforest that they replaced. Shifting cultivation and wet-rice cultivation were found to be strikingly dissimilar systems with respect to diversity, nutrient cycling, type of equilibrium, and ability to absorb population increase through "involution," i.e., internal elaboration without fundamental change. Geertz demonstrated the utility of comparing *ecosystems,* drew attention to productivity and nutrient flows as empirical processes, and placed agricultural systems in a broader historical context of European overseas expansion and its consequences for local cultures.

A fourth group of productive concepts relates to population. Carneiro (1960), another anthropologist, elaborated the concept of *carrying capacity,* the maximum population that can theoretically be supported by a particular environment and with a particular technology. Greater utility for this measure of resource produc-

HISTORICAL AND CULTURAL CONTRIBUTIONS TO GEOGRAPHIC UNDERSTANDING

tivity awaited a bold proposition from Boserup, an agricultural economist; she argued (1965) that population growth would stimulate technological innovation and agricultural "intensification," thus increasing carrying capacity (see also Brookfield and Brown 1963). Although it is now recognized that population growth and agricultural transformation tend to covary, and are very difficult to separate, Boserup drew attention to the relationships between labor input and productivity in agroecology. As a result, *intensification* has become a major theme of cultural ecologists.

A fifth strand of ideas was integrated by the sociologist Buckley (1967). He proposed that human societies are "complex adaptive systems." Adaptive strategies were defined as sets of behaviors that reflect cognitive mapping of the environment and by which such systems adjust to both external and internal changes. Buckley singled out the value of a pool of adaptive variability in identifying new and more detailed varieties and constraints within the environment, allowing a society to incorporate such information. Adaptive variability implicitly allowed a role for cultural evolution through *cultural selection*. By emphasizing *cognition, decision-making,* and *perception,* Buckley anticipated the utility of identifying alternative adaptive solutions to environmental constraints. The "culture as information" approach was also simultaneously developed by the archaeologist D. Clarke (1968), who linked resources, technology, and culture in a scaled hierarchy. Within this same train of thought, the anthropologist Bennett (1969) outlined a first regional case study of competing "adaptive strategies" within a common ecological and economic environment, focusing upon events and the constant human potential for the emergence of innovative arrangements.

Finally, we can identify a sixth perspective, most effectively promoted by the anthropologist Rappaport (1968). He studied a small New Guinea group as an ecological subsystem, emphasizing the functional role of ritual in daily life as well as in relationships with competitive and reciprocating groups. The mass of quantitative caloric data collected was subsequently applied to quantify the energy cycle of this New Guinea group (Rappaport 1971), setting in train a vigorous *energetics* school. Emphasis was placed on negative feedbacks to maintain a homeostatic equilibrium.

This selection of key themes that emerged in the first decade or so of pioneer research in cultural ecology demonstrates that anthropology and other social sciences shared the parallel, theoretical revolution experienced by geography during the 1960s. It is indeed legitimate to speak of an interdisciplinary ferment, in which priority for ideas and conceptual elaborations is often difficult to assign. The first geographers engaged in cultural ecology, such as Brookfield (1962, 1968, 1969), Butzer (1964), Porter (1965), Denevan (1966), Harris (1969), Clarkson (1970), and Mikesell (1970), were an integral part of that process. As reticent as some geographers are to explicate systemic or normative views in cultural ecology, their key concepts and methods unambiguously derive from this era of logical positivism.

Indeed, as surprising as the conclusion may be to some of us, our particular mode of cultural ecology is very much a product of the theoretical revolution. Decidedly low-keyed, in contrast to the flamboyance of an emergent spatial geography, the more incremental crystallization of cultural ecology represented another fundamental break with established methodologies in geography.

Culture ecological research within geography can be roughly subdivided into two categories: synchronic and diachronic. On the one hand, the synchronic or "contemporary" approach began as a series of local case studies that served to develop a methodology, with successive examples offering new thematic insights at higher levels of generalization. Such work has increasingly been applied to a new view of Third World development. On the other hand, the diachronic or "historical" approach has used local studies to examine technological and related demographic changes over longer periods of time so as to understand the dynamics of cultural adaptation and change. Such historical experience provides a different perspective on equilibrium properties and helps identify alternative scenarios relevant to contemporary problems. Although fundamentally different, these synchronic and diachronic methods are complementary.

From Local Studies to Lessons for Development

Perhaps the key unifying thread in "contemporary" cultural ecology is a preoccupation with traditional farming. Within that context, an evolution of methodology and applicability can be traced from local studies of seemingly isolated groups to complex case studies in which smaller groups form exemplary parts of regional or even global networks.

The New Guinea Tradition. The starting model was the standard anthropological case study, with the goal of intensive and comprehensive understanding of a single community. But, whereas earlier anthropologists tended to select "autonomous" microcultural systems in order to identify cultural processes, Brookfield (1962) immediately redefined the ground rules in his first landmark study. He proposed extensive field observations over a wide area so as to recognize patterns and problems that would then be followed up by detailed local study; integration of the extensive and intensive observations would subsequently generate fresh interpretive insights. The resulting monographic study of the Chimbu of New Guinea by Brookfield and Brown (1963) explored a wide range of concepts, such as carrying capacity, as to how agricultural resources are evaluated, used, and allocated in a densely settled area.

Brookfield (1964) next questioned the premises of Berkeley cultural geography, which disclaimed the need to examine "the inner workings of culture" (Wagner and Mikesell 1962, 5), to argue that understanding of society-environment interrelationships was next to impossible without analyzing values, beliefs, and social organization. Subsequently, Brookfield (1969) explored the potential of perception studies to understand resource utilization, the role of new information, decision-making, and change in a traditional society.

Several dissertations on small New Guinea groups were subsequently completed and published by Americans or Canadians under Brookfield's guidance or influence. W. Clarke (1971) introduced time-scheduling, developed the theme of labor inputs versus yields, examined systemic stability, and argued for the possibility of progressive internal change in the process of steady-state adjustment (in other words, dynamic equilibrium).

Waddell (1972) identified three levels at which a small society operates. The base is provided by the biological resources and the environmental constraints that limit them; in the middle are the adaptive strategies employed to maximize productivity and minimize risk; and at the top is the individualistic manipulation of key human actors. Waddell suggests that intensification may be (1) a direct response to environmental variables that justify specialized techniques to increase productivity, (2) an involuntary response to arrest declining output, or (3) an accommodation to variations in population size, density or growth.

Grossmann (1984b, 1984c) completes this evolution by linking the local case study to a higher-order market economy, in which boom and bust cycles exert powerful feedback influences on social relationships, resource use, and ecological harmony. Grossman challenged the assumption that "subsistence affluence" was an enduring trait of traditional agriculture, by demonstrating that food production was quite variable from year to year, with repeated shortfalls. He showed that contemporary commodity production can conflict with and undermine subsistence agriculture, even when surplus land and labor are available, and that such conflicts can make subsistence systems more vulnerable to environmental problems. His work on time-allocation studies (Grossman 1984a) has broad applicability (see also Bergman 1980).

African Case Studies. In East Africa, Porter (1965, 1978, 1979) worked in association with a team of anthropologists, and was assigned the task of articulating land use and environment for a set of different socioeconomic groups. He employed an energy-water budget approach to relate environmental variety to crops and pasture grasses. Subsequently he moved from a Western categorization of soils to an indigenous one, in terms of terminology, criteria, and taxonomy. Only in this way could he understand indigenous agricultural practices, soil assessment, and management. He concluded by showing that indigenous land appraisal and sociopolitical organization interact to produce a livelihood system that exploits several environments at different elevations; the solution increased productivity, reduced subsistence risks, and smoothed out labor schedules.

Knight (1974), as a student of Porter, developed a comprehensive local study in which agroecology was complemented by systematic investigation of the rationale of indigenous "ethnoscience" (see also Newman 1970). The local study was then integrated into the larger political, social, and economic matrices in which the evolution of the study group was embedded—specifically, a plural society involved in rural modernization, guided and limited by national policies, and dependent upon a global politico-economic system. Knight's conclusions cast doubt on the efficacy of Western innovations and on the validity of the Western scientific system in unfamiliar environments.

Another exemplary African study, by Carr (1977), examined a pastoral group on the Ethiopian border, artificially restricted from using part of its traditional lands inside adjacent countries. Initially working with Butzer in the field, Carr extended her attention from pastoral ecology and supplementary riverine agriculture to an examination of how the social system facilitates the exploitation of multiple and fluctuating resources, and how range deterioration caused instability that has subsequently erupted into intertribal warfare. Johnson (1978) provided a more general culture ecological rationale for nomadic pastoralists, and conflicts between pastoralists and agriculturalists in West Africa were elucidated by Vermeer (1981) and Bassett (1986).

New Approaches in Latin America. In Latin America, the Miskito Indians of Nicaragua were studied by Nietschmann (1973, 1979), a student of Denevan. In an exemplary application of the energetics approach, he first established the energy flows of their traditional economy, spread over four different biotopes, and then examined the impact of economic development on resource deterioration and impoverishment. For the Shipibo of the Amazon, Bergman (1980), also a Denevan student, made daily observations on time inputs into subsistence activities for the full annual cycle.

The second theme ("political ecology") has been developed by Hecht (1982) in regard to soil and forest destruction in the eastern Amazon Basin. The world cattle market and Brazil's strategy with respect to it create incentives and constraints that induce frontier farmers to abandon sound management procedures. Cattle exports in Costa Rica have had similar deleterious effects on environmental resources, food production, and traditional social organization (Place 1985).

In a very different genre is Kirkby's study (1973) of ecology and allocation of farm land and irrigation water in Oaxaca. Although ultimately focused upon prehistoric settlement, this is a model for sophistication in applying the geographer's art to integrate ecological variables into a synthetic whole. A microstudy by Doolittle (1984), a student of Turner, in the Rio Sonora valley demonstrates how the conversion of a tributary channel and its floodplain into a rationally exploited agricultural system is the cumulative result of innumerable ad hoc decisions by individual cultivators. The thrust is that agricultural intensification is, and presumably has often been, an incremental process.

Another case study, by Knapp (1984), a Denevan student, in highland Ecuador, examined the relationships among altitude, climate, slope, soil fertility, labor input, and crop yields for various Andean cultigens, to explain patterns of altitudinal zonation as they changed over time. The study concluded that soil fertility is the greatest single limiting factor and challenge for Andean farmers, and that historical changes of fertility-management technology have been associated with dramatic shifts in niche use and settlement. The resulting appreciation of indigenous adaptive strategies is of equal interest for an understanding of the present-day vulnerability of marginal farmers to climate change (Parry, Knapp, and Cañadas 1988).

Several more general studies can be singled out in this broader context. Turner and Doolittle (1978), for example, devised quantitative measures to assess degrees of agricultural intensification. Wilken (1987) took a didactic and nontheoretical approach in systematically outlining traditional agricultural procedures in Mesoamerica, creating the first approximation of a textbook to facilitate training of the next generation of field-grounded cultural ecologists. Another practical study by Denevan and Padoch (1987) provided recommendations for managed agroforestry based on intensive study of slash-and-burn "abandoned fields" in the Peruvian Amazon. Also at a general level is the collaborative work of a geographer and an anthropologist, Turner and Brush (1987), on comparative farming systems. It specifically addressed agricultural change in different physical, social, economic, and cultural environments.

Common Ground. This selection of synchronic studies in cultural ecology identifies a distinctively geographical approach, illuminating a broad sphere of interaction with respect to resources and the spatial matrix of the cultural and biophysical environment. The trend has also been to link the local group into the larger economic system of which they are part, a more realistic, open system and nonhomeostatic perspective.

By contrast, anthropological cultural ecology has tended to over-refine its theoretical constructs, while limiting empirical work to the processes and structures whereby relatively simple human groups match resources with their needs and incorporate them into their cultural behavior. The differences are striking, yet logical.

The thrust of synchronic cultural ecology within geography has found its primary application in a fresh look at Third World development. Cultural ecologists are firmly opposed to mindless modernization according to Western standards. They argue that traditional agriculture reflects much trial-and-error; minimizes risk; is more often than not based on intuitively good ecological decisions, if not sound evaluation; and that it is intimately interwoven with cultural values and perceptions. A common stance is that Westerners should first learn from indigenous groups before prescribing change, and that any changes should incorporate and emphasize the best components of the traditional system.

Cultural ecologists have also become active participants at international conferences and in national or international agencies which are evaluating development schemes. The demand for such expertise is high, so that the number of properly trained students barely matches the potential demand of the applied sector.

Political Economy. A major new arena for contemporary cultural ecology is the set of problems as to how integration into regional, state, and world economies affects the management of resources. A good example here is Watts (1983), a student of Nietschmann, who examined food production and periodic famine in northern Nigeria, and then integrated institutions and international structures into the explanation of a nature-society problem. Fitting in a similar context is the study of Blaikie and Brookfield (1987) on land degradation and society; in dealing with ecological issues, it again emphasizes institutions and political structures that set a matrix of limits, constraints, and possibilities for resource management.

Such perspectives from political economy move cultural ecology from the context of a small, closed society into a broad, hierarchical system. The focus is upon three basic issues (Watts 1983, 1987):

- ☐ the nested levels of system integration, including social, economic, and power relations beyond the individual or household unit of analysis,
- ☐ the constraints and possibilities imposed on the men and women who manage resources (as individuals and as households), and the degree to which systems of access to and control over resources "marginalize" certain social groups, and
- ☐ the historical processes of integration into the market, state, and world economies. Structuration theory offers further avenues for exploration.

All in all, this research in synchronic cultural ecology represents a very broad canvas, notable for its rapid elaboration and diversification. Each author has a sophisticated grasp of ecological problems and strives to chart new intellectual ground in the understanding of contemporary problems. It is this sense of excitement and commitment that probably explains the rapid growth of the Specialty Group, as a community of scholars intensely concerned with food, population, and the sociocultural mechanisms that link them.

Change and the Historical Perspective

While synchronic investigation can better identify short-term system maintenance, historical or diachronic research seeks to recognize changing configurations and to understand the responsible processes. The "contemporary" cultural ecologist has little access to the time depth necessary to evaluate the nature of systemic change, or more importantly, to study how alternative options to internal or external stress are chosen, and whether in the long run these are successful or not.

The development of historical cultural ecology shows some parallels to its more contemporary counterpart. Much of the work is local and intensive, with the goal of generating hard data as well as larger hypotheses for subsequent, more comprehensive investigation. There also is a similar trend to consider higher-order systemic interactions.

Abandoned Agricultural Landforms. One convenient category of research includes agricultural landforms, such as terraced, raised, channelized, and sunken fields, now mainly found in thinly inhabited parts of Latin America. The reason that these features are interesting is that they demonstrate intensified prehistoric agriculture and, by implication, higher populations in the past. This in turn raises questions about the agrosystems themselves, and the strategies by which they were devised. Equally intriguing are the factors responsible for abandonment, as well as the lessons such features provide for potential future increases in agricultural productivity.

The groundwork for examining past agricultural landforms as one central theme in historical cultural ecology is represented by the wide-ranging investigation of Spencer and Hale (1961) in Southeast Asia. Such features were subsequently found along the Amazonian margins of the Andes (Denevan 1970, 1982), and in the highlands of Ecuador and Peru (Farrington 1985; Denevan, Mathewson, and Knapp 1987). For the Colca Valley of highland Peru, Denevan (1987) directed an interdisciplinary project that exhaustively studied a terraced landscape, now partially abandoned, the origins of which go back well beyond the Inca past. Population data are first available from 1530, allowing wide-ranging inferences on demographic change, labor inputs, and productivity.

The Maya lowlands provide another case in point, but one where the initial question was to identify an adequate subsistence base to support the large populations verified archaeologically. This led to several interdisciplinary studies that identified alternative agrosystems and ultimately demonstrated a critical role for artificial landforms such as raised and channelized fields (Harrison and Turner 1978; Turner and Harrison 1983; Turner 1983). Team study served to disentangle the record of Maya wetland cultivation in terms of surface preparation, biotic associations, hydraulic agriculture, and settlement. The implications for high Maya population densities are surprising (Whitmore et al. 1989), requiring a total reevaluation of the potential productivity of lowland tropical environments, as well as an interpretation via demand-based models.

Early Irrigation. A similar cluster of studies is linked to prehistoric and historic irrigation systems. The origin of complex societies has commonly been linked to a positive feedback system in which resource stress and population growth are thought to be instrumental in the development of irrigation agriculture, thus requiring a managerial bureaucracy, and ultimately, sociopolitical growth. In the case of the Egyptian Nile Valley (Butzer 1976), it can instead be argued that the emergence of an irrigation agrosystem was an incremental process and that it was, and continued to be, managed locally; in other words, intensification was not the stimulus for administrative centralization or social stratification.

In the Rio Sonora valley of northern Mexico, Doolittle (1988) reconstructed an irrigation-based sequence of prehistoric occupance, spanning about a millennium. He demonstrated changing orders of settlement hierarchies that reflected substantial changes in population and links to different exchange networks. The Sonoran example illustrates the flexibility of marginal environments in supporting larger populations, depending on the effort invested to improve productivity, as well as the broader rationale for such inputs within the context of a larger, open system.

Population Cycles. Demography provides the third theme for more comprehensive applications of the historical approach. Population growth is rarely possible without improved technology, social access to resources, or a combination of the two; decline points to fundamental social or environmental problems. Growth, stability, or decline also suggest different questions about the quality of life. Examples of such investigations include the linkages between progressive intensification and systemic breakdowns in ancient Egypt (Butzer 1976, 1980, 1984), several global analyses of demographic "millennial long waves" (Whitmore et al. 1989), and the catastrophic New World population loss due to the introduction of European epidemic diseases (Denevan 1976).

These case studies illustrate that, in the long-term view, populations may not only grow but may experience catastrophic collapse. They show that sociopolitical and socioeconomic variables are tightly interlocked, and that simplification is possible in one or the other, or both domains. Historical studies of this kind can be implemented at intermediate and small scales, to derive more-detailed understanding of the mechanisms of change, of the human costs involved, and of the decisions that communities or larger social groupings make when confronted with crisis.

A case study in the Sierra de Espadan of eastern Spain (Butzer 1989b) illustrates how population expansion and increasing resource scarcity since A.D. 1700 led to a series of temporary adaptive choices among different options, to forestall more fundamental changes that involved cultural values (such as family-size limitation) and ultimately emigration. It serves to show how social groups attempt to manage mounting crises, and that they deliberately weigh options with different sociocultural impacts. Some of the choices made are unpredictable and unexpected. Another such historical evaluation can be cited for European-Indian contacts in New England (Cronon 1983).

Discussion. Historical cultural ecology provides a powerful methodology to examine and understand change in a larger, systemic context. In this we can recognize parallels with the recent trend of synchronic cultural ecology to move from examination of agricultural production to examination of higher-order, interlinked structures. Study need not be confined to non-Western societies, a current predilection that relates more to the history of our endeavor than to the suitability of the materials or the promise of insights complementary to those obtained by other efforts of human geography. Ecological sophistication and cultural expertise can be applied with equal profit in the First World, as in the Third.[4]

By way of general conclusion, cultural ecology currently forms a center of intellectual activity, and the subfield can be expected to continue to evolve and mature over the next decade or two. It is likely that our perspectives and innovative methodologies in integrating the two domains of environment and society will attract many new converts in the process.

---

[4] We could, for example, profit from cultural ecological studies of Midwestern family farming communities, of West Texas cattle ranches, of lobster fishing towns on the Maine coast, or of intensified agriculture in the Salinas Valley of California (W. E. Doolittle, pers. com.) (see also Turner and Brush 1987).

REFERENCES

Barrows, H. H. 1923. Geography as human ecology. *Annals of the Association of American Geographers* 13:1–14.

Barth, F. 1956. Ecologic relationships of ethnic groups in Swat, North Pakistan. *American Anthropologist* 58:1079–89.

Bassett, T. J. 1986. Fulani land movements. *Geographical Review* 76:233–48.

Bennett, J. W. 1969. *Northern plainsmen: Adaptive strategy and agrarian life*. Chicago: Aldine.

Bergman, R. W. 1980. *Amazon economics: The simplicity of Shipibo Indian wealth*. Dellplain Latin American Studies, no. 6. Ann Arbor, MI: University Microfilms International.

Blaikie, P., and Brookfield, H. C. 1987. *Land degradation and society*. London: Methuen.

Boserup, E. 1965. *The conditions of agricultural growth*. Chicago: Aldine.

Brookfield, H. C. 1962. Local study and comparative method: An example from Central New Guinea. *Annals of the Association of American Geographers* 52:242–54.

———. 1964. Questions on the human frontiers of geography. *Economic Geography* 40:283–303.

———. 1968. New directions in the study of agricultural systems in tropical areas. In *Evolution and environment,* ed. E. T. Drake, pp. 413–39. New Haven: Yale University Press.

———. 1969. On the environment as perceived. *Progress in Human Geography* 1:51–80.

———, and Brown, P. 1963. *Struggle for land: Agriculture and group territories among the Chimbu of the New Guinea Highlands*. Cambridge: Cambridge University Press.

Buckley, W. 1967. *Sociology and modern systems theory*. Englewood Cliffs, NJ: Prentice-Hall.

Burton, I; Kates, R. W.; and White, G. F. 1978. *The environment as hazard*. New York: Oxford University Press.

Butzer, K. W. [1964] 1971. *Environment and archeology: An introduction to pleistocene geography*. Chicago: Aldine.

———. 1976. *Early hydraulic civilization in Egypt: A study in cultural ecology*. Chicago: University of Chicago Press.

———. 1980. Civilization: Organisms or systems? *American Scientist* 68:517–23.

———. 1982. *Archaeology as human ecology.* New York: Cambridge University Press.

———. 1984. Long-term Nile flood variation and political discontinuities in Pharaonic Egypt. In *From hunters to farmers,* eds. J. D. Clark and S. A. Brandt, pp. 102–12. Berkeley: University of California Press.

———. 1989a. Hartshorne, Hettner, and *The nature of geography. Annals of the Association of American Geographers.* In press.

———. 1989b. The realm of cultural ecology: Adaptation and change in historical perspective. In *The earth as transformed by human action,* ed. B. L. Turner II. New York: Cambridge University Press. In press.

Carneiro, R. 1960. Slash and burn agriculture: A closer look at its implications for settlement patterns. In *Men and cultures,* ed. A. F. C. Wallace, pp. 229–34. Philadelphia: University of Pennsylvania Press.

Carr, C. J. 1977. *Pastoralism in crisis: The Dasanetch and their Ethiopian lands.* University of Chicago Department of Geography Research Paper 180.

Chorley, R. J. 1973. Geography as human ecology. In *Directions in geography,* ed. R. J. Chorley, pp. 155–69. London: Methuen.

Clarke, D. L. 1968. *Analytical archaeology.* London: Methuen.

Clarke, W. C. 1971. *Place and people: An ecology of a New Guinean community.* Berkeley: University of California Press.

Clarkson, J. D. 1970. Ecology and spatial analysis. *Annals of the Association of American Geographers* 60:700–16.

Cronon, W. J. 1983. *Changes in the land: Indians, colonists, and the ecology of New England.* New York: Hill & Wang.

Denevan, W. M. 1966. A cultural-ecological view of the former aboriginal settlement in the Amazon Basin. *Professional Geographer* 18:346–51.

———. 1970. Aboriginal drained-field cultivation in the Americas. *Science* 169:647–54.

———. ed. 1976. *The native population of the Americas in 1492.* Madison, WI: University of Wisconsin Press.

———. 1982. Hydraulic agriculture in the American tropics: Forms, measures, and recent research. In *Maya subsistence,* ed. K. Flannery, pp. 181–203. New York: Academic Press.

———. 1983. Adaptation, variation, and cultural geography. *Professional Geographer* 35:399–407.

———. 1987. Terrace abandonment in the Colca Valley, Peru. In *Pre-Hispanic agricultural fields in the Andean region,* eds. W. M. Denevan, K. Mathewson, and G. Knapp, pp. 1–43. Oxford: British Archaeological Reports, International Series, 359.

———; Mathewson, K.; and Knapp, G.; eds. 1987. *Pre-Hispanic agricultural fields in the Andean region.* Oxford: British Archaeological Reports, International Series, 359.

———, and Padoch, C., eds., 1987. Swidden-fallow Agroforesty in the Peruvian Amazon. *Advances in Economic Botany* 5:1–107.

Doolittle, W. E. 1984. Agricultural change as an incremental process. *Annals of the Association of American Geographers* 74:124–37.

———. 1988. *Pre-Hispanic occupance in the Valley of Sonora, Mexico: Archaeological confirmation of early Spanish reports.* Tucson: University of Arizona Anthropological Paper No. 48.

Ellen, R. 1982. *Environment, subsistence and system: The ecology of small-scale social formations.* New York: Cambridge University Press.

Farrington, I. S., ed., 1985. *Prehistoric intensive agriculture in the tropics.* Oxford: British Archaeological Reports, International Series, no. 232.

Forde, C. D. 1934. *Habitat, economy, and society.* New York: Harcourt Brace Jovanovich.

Geertz, C. 1963. *Agricultural involution: The processes of ecological change in Indonesia.* Berkeley: University of California Press.

Grossman, L. 1984a. Collecting time–use data in Third World rural communities. *Professional Geographer* 36:444–54.

———. 1984b. *Peasants, subsistence ecology, and development in the highlands of Papua New Guinea.* Princeton: Princeton University Press.

———. 1984c. Sheep, ceremonial exchange, and coffee prices in Papua New Guinea. *Geographical Review* 74:315–30.

Harris, D. R. 1969. The ecology of agricultural systems. In *Trends in geography,* eds. R. U. Cooke and J. H. Johnson, pp. 133–42. New York: Pergamon.

Harrison, P. D., and Turner, B. L., II., eds. 1978. *Pre-Hispanic Maya agriculture.* Albuquerque: University of New Mexico Press.

Hawley, A. H. 1986. *Human ecology: A theoretical essay*. Chicago: University of Chicago Press.

Hecht, S. 1982. *Cattle ranching development in the Eastern Amazon: Evaluation of a development policy*. Ph.D diss., Department of Geography, University of California, Berkeley.

Johnson, D. L. 1978. Nomadic organization of space: Reflections on patterns and process. In *Dimensions of human geography,* ed. K. W. Butzer, pp. 25–47. University of Chicago Department of Geography Research Paper No. 186.

Kates, R. W. 1987. The human environment: The road not taken, the road still beckoning. *Annals of the Association of American Geographers* 77:525–34.

Kirkby, A. V. T. 1973. *The use of land and water resources in the past and present valley of Oaxaca, Mexico*. Memoirs of the Museum of Anthropology, no. 6. Ann Arbor: Museum of Anthropology, University of Michigan.

Knapp, G. 1984. *Soil, slope, and water in the equatorial Andes: A study of prehistoric agricultural adaptation*. Ph.D. diss., Department of Geography, University of Wisconsin, Madison.

Knight, C. G. 1974. *Ecology and change: Rural modernization in an African community*. New York: Academic Press.

Leighly, J. 1987. Ecology as metaphor: Carl Sauer and human ecology. *Professional Geographer* 39:405–16. (With comments by K. Mathewson, P. W. Porter, and B. L. Turner II.)

Marsh, G. P. [1864] 1965. *Man and nature: Physical geography as modified by human action*. Reprinted and edited by D. Lowenthal. Cambridge: Harvard University Press.

Martin, G. J. 1987. The ecological transition in American geography. *Canadian Geographer* 31:74–77.

Mikesell, M. W. 1970. Cultural ecology. In *Focus on geography: Key concepts and teaching strategies,* ed. P. Bacon, pp. 39–61. Washington: National Council for the Social Studies.

Newman, J. L. 1970. The ecological basis for subsistence change among the Sandawe of Tanzania. Washington: National Academy of Sciences.

Nietschmann, B. Q. 1973. *Between land and water: The subsistence ecology of the Miskito Indians, Eastern Nicaragua*. New York: Seminar Press.

———. 1979. Ecological change, inflation and migration in the Far West Caribbean. *Geographical Review* 69:1–24.

Parry, M.; Knapp, G.; and Cañadas, L.; eds. 1988. *The effect of climatic variations on agriculture in the Central Sierra of Ecuador*. Dordrecht: D. Reidel.

Place, S. E. 1985. Export beef production and development contradictions in Costa Rica. Tijdschrift voor Economische en Sociale Geografie 76:288–97.

Porter, P. W. 1965. Environmental potentials and economic opportunities: A background for cultural adaptation. *American Anthropologist* 67:409–20.

———. 1978. Geography as human ecology. *American Behavioral Scientist* 22:15–39.

———. 1979. *Food and development in the semi-arid zone of East Africa*. Foreign and Comparative Studies: African Series, no. 32. Syracuse: Syracuse University.

Rappaport, R. A. [1968] 1984. *Pigs for the ancestors: Ritual in the ecology of a New Guinea People*. New Haven: Yale University Press.

———. 1971. The flow of energy in an agricultural society. *Scientific American* 224(3):116–32.

Sauer, C. O. [1925] 1967. The morphology of landscape. Reprinted in *Land and life,* ed. J. Leighly, pp. 315–50. Berkeley: University of California Press.

———. 1927. Recent developments in cultural geography. In *Recent developments in the social sciences,* ed. E. C. Hayes, pp. 154–212. Philadelphia: Lippincott.

———. [1941] 1967. Foreword to historical geography. Reprinted in *Land and life,* ed. J. Leighly, pp. 351–79. Berkeley: University of California Press.

Spencer, J. E., and Hale, G. A. 1961. The origin, nature, and distribution of agricultural terracing. *Pacific Viewpoint* 2:1–40.

Stoddart, D. R. 1965. Geography and the ecological approach. *Geography* 50:242–51.

Steward, J. H. 1955. *The theory of culture change*. Urbana, IL: University of Illinois Press.

Turner, B. L., II. 1983. *Once beneath the forest: Prehistoric terracing in the Rio Bec region of the Maya Lowlands*. Boulder, CO: Westview.

———. 1989. The specialist-synthesis approach to the revival of geography: The case of cultural ecology. *Annals of the Association of American Geographers* 79. In press.

———, and Brush, S. eds., 1987. *Comparative farming techniques*. New York: Guilford Press.

———, and Doolittle, W. E. 1978. The concept and measure of agricultural intensity. *Professional Geographer* 30:297–301.

———, and Harrison, P. D., eds., 1983. *Pulltrouser Swamp: Ancient Maya habitat, agriculture, and settlement in northern Belize*. Austin, TX: University of Texas Press.

Vermeer, D. E. 1981. Collision of climate, cattle, and culture in Mauritania during the 1970s. *Geographical Review* 71:281–97.

Waddell, E. 1972. *The mound builders: Agricultural practices, environment, and society in the Central Highlands of New Guinea*. Seattle: University of Washington Press.

Wagner, P. L., and Mikesell, M. W. 1962. *Readings in cultural geography*. Chicago: University of Chicago Press.

Watts, M. 1983. Silent violence: Food, famine, and peasantry in Northern Nigeria. Berkeley: University of California Press.

———. 1987. Powers of production—geographers among the peasants. *Environment and Planning D: Society and Space* 5:215–30.

Whitmore, T. M.; Turner, B. L., II; Johnson, D. L.; Kates, R. W.; and Gottschang, T. R. 1989. What goes up comes down: Long-term population change and environmental transformations. In *The earth as transformed by human action,* eds. B. L. Turner II. New York: Cambridge University Press. In press.

Wilken, G. C. 1987. Good farmers: Traditional agricultural resource management in Mexico and Central America. Berkeley: University of California Press.

Cultural Geography

Lester B. Rowntree | Kenneth E. Foote | Mona Domosh

Cultural geography, while often recognized as one of the oldest components of North American human geography, is also one of the newest specialty groups within the AAG, attaining that status only in 1988. This apparent contradiction between longstanding historical roots and recent specialization exemplifies the current status of the subfield in two ways.

First, it reminds us that cultural geography, despite its central role in human geography, has been treated as an intellectual ambient or background out of which have come more focused subfields and specialty groups, such as cultural ecology and environmental perception, thereby leaving a diffuse and pluralistic residual. Second, because of this, many of those responsible for this new specialty group in cultural geography feel a certain discomfort with this unfocused, highly tolerant "anything goes" pluralism that has characterized so much of the discipline in the last decade, and seek a forum for discussion of method and theory that is congruent with the increased interest in culture, space, and landscape shown throughout the social sciences.

Because cultural geography has traditionally been an individualistic, humanistic, and atheoretical endeavor, and because current epistemological and ontological revision in the social sciences is linked closely to and driven by larger currents of social theory, there is a tendency to dichotomize cultural geography into two contrasting camps, often thought of as the "old" and "new": traditional descriptive, particularistic work, and theory-informed, explanatory new directions (e.g., see Cosgrove and Jackson 1987). This dichotomy is false and unworthy, for three reasons.

First, at this point the "new" cultural geography is more talked about than done (Kofman 1988), which is not to deny the existence or vitality or activity in that direction, but to simply state a fact. Second, constructing such a dichotomy is an ill-founded strategy that privileges and reifies one segment over the other without the necessary critique and interactive discourse between disparate elements, and consequently, is unfitting for a critical and reflexive social science.

Last, as social science moves away from the rigid methodologies imposed by natural-science positivism, there is increased emphasis on the intellectual content and epistemological flexibility of traditional humanism and, by extension, of humanistic cultural geography. The subdiscipline of behavioral geography, as an example, integrates humanistic dimensions into its "new" positivism (e.g., see Aitken and Bjorklund 1988; Seamon 1987), and postpositivistic archaeology (referred to as "postprocessual" in that discipline) converges strongly with phenomenology in its search for symbolic meaning in material culture (Hodder 1987). This reminds us that "new" social theory builds upon the "old," and that cultural geography could be crippled by severing ties with our traditional core if we reinforce a "new/old" dichotomy. Put differently, some of the solutions for a new cultural geography may come from synthesis of extant theory (see Norton 1987; Rowntree 1988).

What follows, then, is a critique of contemporary North American cultural geography that is shaped and framed by a sensitivity to emerging themes and issues of the "new," without denying or severing ties to the core of traditional cultural geography. Yet, at the same time, because it is often more interesting to ponder where we might be going instead of simply celebrating where we have been, the weight of this chapter is with a reasoned discussion of current issues and future paths.

AN OVERVIEW OF ISSUES IN CULTURAL GEOGRAPHY

At one level, the overriding issue today is the perceived tension between a traditional atheoretical cultural geography and one that is more "theory-informed." Yet, as discussed, this is somewhat of a bogus controversy. Our historical antipathy toward theory is usually ascribed to Carl Sauer's prominent role in forging a humanistic cultural geography. Although Sauer's skepticism toward social- and natural-science theory was overt and well documented (Solot 1986; Leighly 1987; Speth 1987), his influence on cultural geography unquestioned (Kenzer 1987), and continuation of this bias in Berkeley School research often articulate (e.g., see Parsons 1986), the present debate over the explicitness and appropriateness of theory is far more complicated than a simple reaction to Sauer.

To begin with, a critical analyst would argue that Sauer's humanistic geography was theory-laden because any strong and explicit ideological subscription involves implicit acceptance and use of theory (Johnston 1986, 26). Further, promoting a view of cultural geography's intellectual heritage as "atheoretical" is in itself a theory-informed construction (Hanson 1985). So, rather than argue a counterfeit issue, it is more productive to probe and elucidate the constituent parts of cultural geography, and work toward assessing the implications of synthesizing "old" and "new" theory.

For convenience, let us examine five categories:

1. The epistemological spectrum embraced by cultural geography
2. The centrality and conceptualization of culture
3. The interaction among humans, culture, and landscape
4. Expanding problematic inquiry
5. Studying everyday life and landscapes

These topics are not concerns unique to cultural geography, but are common to most of the social sciences, as they interrogate and interact with larger currents of social theory. Perspective on this matter is important, and tracking similar issues in an allied social science such as anthropology can be instructive (e.g., see Ortner 1984; Hodder 1986; Leone 1986).

The Epistemological Spectrum

Before fueling discussion on method and theory in cultural geography, some definitions are in order. An "epistemology," or theory of knowledge, is central to the way work is conducted within a discipline, and is obviously influenced by "ontology," a theory of existence or of what can be known (Johnston 1986, 5). Separating the two terms is not germane to the following; consequently we will use "epistemology" to encompass both. Epistemologies guide "theory," which is simply an integrated group of fundamental principles that offer a plan or scheme which conditions what is worth thinking about; and theory influences "methodology," the gathering and testing of data. Because all research is guided by some sense of worth and purpose (Johnston 1986, 26), the notion of "atheoretical" cultural geography seems in need of revision.

A step in that direction might be to qualify and articulate the implicitness or explicitness of the theoretical structure: is the research explicitly linked to or driven by an extant theoretical structure, or are there implicit and subtle linkages to unrevealed assumptions and ideologies? The term "common sense" often describes the latter end of that spectrum, and some critics argue that the development of science is the substitution of theory for common sense. This entails the replacement of an ethnocentric and self-correcting system that lacks clear evaluative or testing criteria, with a more explicit and systematic body of knowledge that contains clear and agreed upon methodologies for collecting data and establishing correctness (Dunnell 1982). Only by this replacement, some theorists argue, will knowledge be cumulative instead of particularistic and idiosyncratic. Others resist the rigidity of this scientific canon and seek alternative methodologies through humanism. Cultural geography has offered a tolerant home for both ends of the spectrum.

This epistemological pluralism contains four clear categories: humanism, positivism, structuralism, and poststructuralism. While more detail on each category can be found in works on geographic thought (e.g., Johnston 1986, 1987), a brief review with overt linkages to cultural geography will set the scene for further discussion.

Humanism, with its emphasis on interpretive understanding, is usually regarded as a critique of, and reaction to, positivism (Entrikin 1976). This position was adopted early by Sauer, reaffirmed in response to geography's quantitative revolution in the 1960s, and reinforced in the late 1980s by postpositivism in the social sciences. Examples are found in such highly visible research as Meinig's historical geography of North America (1986), Jordan's regional text on Europe (1987), Butzer's cultural ecology (1988), and Lewis's landscape interpretations (1983). Phenomenology, an important subcategory, with its emphasis on human meaning and interaction with landscape, is experiencing revitalization in environment and design research (Seamon 1987). The fact that humanistic geography is essentially a personal project, one that revolves around individual style and efficacy, makes the assessment of its products difficult, which takes us back to the tension between those who argue that explicit theory is needed for evaluation and accumulation, and those who resist the canon.

Cultural geography's links with *positivism* are more tenuous, because those who subscribe to this philosophy tend to think of themselves as belonging to other subdisciplines within geography. An example is cultural ecology, where there are calls for a more specialized, positivistic method and theory (Turner 1987) than the traditional research path set by Sauer and found within cultural geography. When the focus is upon individual or group interaction with landscapes and environment, arguably a key component of culture, behavioral geography often subsumes research (see chapter on environmental perception and behavioral geography).

Structuralism, with its emphasis on the linkages between the world of appearance and the autonomous structures (social, economic, psychological, cognitive, functional, etc.) that lie beneath and control individual and group action, pervades the social sciences and has two expressions in cultural geography. (We place structural-Marxism, the dominant expression in the discipline of geography, outside of cultural geography for this discussion.)

The first expression, *semiotics* or *semiology,* argues that signifying systems (such as landscape) are analogous to language, and thus can be examined within the precepts of linguistic structuralism (Foote 1985; Duncan 1987). The second expression is usually called *structural-functionalism;* it assumes that there are regularities in the behavior of groups of people, even if those structures are not accounted for by the conscious regularities of the individuals making up that group. These regularities seem to be an adaptive system that serves to maintain the group as a whole (Duncan 1985, 176). Examples are Jackson's scheme (1985, 147–57) for regularities in the American landscape, and Rowntree and Conkey's model of landscape symbolism (1980). Criticism of structuralism focuses upon the denial of individual intention and attitudes by the privileging of superorganic, autonomous existences (Duncan 1985).

Poststructuralism attempts to construct workable theories of action that avoid both the determinism of the structural view and the hyper-individualism of humanistic geographers (Duncan 1985, 178). This new accord is being worked out in different ways, most prominently with the structuration framework promoted by Giddens (1984), and by the convergence of literary and social theory in the rethinking of landscape-as-texts (Duncan and Duncan in press).

The Centrality and Conceptualization of Culture

An overarching issue in the subfield is whether to be content with a passive, superorganic conceptualization of culture that essentially negates individual action, yet is adequate for larger-scale generalizations, or to adopt a more active, processual view sensitive to individuals and small groups and their interaction with space, place, and landscape. Duncan (1980) traced superorganic thought in American cultural geography, spoke to its limitations, and in a later work (1985) promoted an alternative. However, mainstream cultural geography seems satisfied with the superorganic.

For example, this conceptualization seems appropriate and efficacious in Meinig's historical geography of the Eastern colonies (1986) and Jordan's study of log cabins (1985), where the discussion of the inner workings of culture could be distracting. Given the fairly broad scale of inquiry of these studies, a notion of culture that focuses upon the everyday interactions of people would seem somewhat inappropriate. More detailed studies have been able to incorporate a different view of culture that incorporates notions of conflict and change by emphasizing the relation between a particular landscape and the people who produced it (e.g., see studies by Anderson 1987; Cosgrove 1984, 1985; Domosh 1988; and Steinetz in press).

Interaction Among Humans, Culture, and Landscape

While traditional conceptualization of the cultural landscape emphasized unilinearity—humans, working through the agency of culture, modify environments and create cultural landscapes—there are current works celebrating the recursiveness or interaction among humans, culture, and landscape, by emphasizing the role cultural landscapes play in constructing, reinforcing, and reproducing social structure. This emphasis has applicability in both contemporary and historical studies: the Duncans' study of the reinforcing role of landscape for elite social classes in North America (1984) illustrates the former, while both Upton (1985) and Steinitz (forthcoming) discuss how residential architecture and morphology constructed and reinforced social boundaries in eighteenth-century Virginia (Upton) and Massachusetts (Steinitz). As well, Cosgrove's work (1984) tracing the evolution of the landscape idea in European thought is linked to changing uses and perceptions of land and social class during the development of capitalism.

Behind much of this recent research into the meaning of landscape lies the concept of communication. Interest in this concept among cultural geographers stems from a traditional concern for landscape symbolism and an awareness of the prominent role this concept has played in other disciplines. Geographers have reached out to these other disciplines in their attempts to establish a firmer foundation for their generalizations about environmental symbolism. These explorations have led cultural geographers to affirm their ties with anthropology, where the long-held view of "culture as communication" continues to find expression in contemporary linguistic and semiotic models that have attracted proponents from other disciplines (Broadbent 1980; Krampen 1979; Preziosi 1979; Singer 1984).

Contact with sociology, psychology, and related fields has stressed insights to be gained from theories of symbolic interaction and nonverbal communication (Ankerle 1981; Csikszentmihalyi and Rochberg-Halton 1981; Rapoport 1982). Finally, value has been found in theories of symbolism derived from critical literary theory and textual criticism (Gottdiener and Lagopoulos 1986). Together, these explorations stand behind recent attempts to analyze landscape as "text," to relate environmental symbolism to theories of nonverbal communication, and to recast semiotic models for landscape studies.

None of these excursions has been altogether successful in lending insight into environmental and landscape symbolism. This is not to say that communication-based models are inappropriate for cultural geography, but only that available theories require considerable modification to address the concerns of geographers. Linguistic analogies have, for example, proved popular in landscape studies because they suggest the value of searching for grammatical rules in the "text" of landscape. Yet to use these comparisons as anything but suggestive analogies risks accepting many of the basic deficiencies of the linguistic theories themselves (Foote 1983, 16–23).

Today, the most encompassing communication-based models are those of semiotics, a branch of science concerned with the signs and symbols of social life. The difficulty of adapting these models to landscape studies can be traced to the intellectual roots of semiotics in linguistics and logic. As semiotics emerged from these fields, its practitioners were little concerned with the subject matter of geography. When, in time, semiotic theorists turned to the related issues of meaning in architecture and the symbolism of material culture, they often sought only to demonstrate that these symbol systems expressed language-like properties. This research strategy proved of limited value to geographers, because the aspects of landscape symbolism that are of the most interest to them—particularly the quality of temporal durability—are precisely those properties that are furthest removed from the character of verbal behavior (Foote 1985, 160–69; Foote 1988). Indeed, the tendency of many semioticians to conceptualize every mode of human action as a separate system of meaning leads them to overlook issues of great import to geographers—the complex symbiotic relationships among verbal behavior, nonverbal action, and the situational variables and material objects that define landscape.

Thus, for geographers interested in semiotic and communication-based theories of environmental meaning, the basic problem is that only a limited number of insights can be borrowed from previous research in other disciplines. Semiotics, for instance, still has a great distance to go before it will realize the full value of viewing culture as a system of communication, and society as a web of symbolic action. However, by participating directly in these pioneering attempts to realize the potential of these claims, cultural geographers can contribute as much as they can gain. This is because geographers alone are most closely attuned to one of the most-important questions arising from semiotic and communication-based research: how do social and cultural systems express themselves symbolically in the material world of landscapes and environments?

Expanding Problematic Inquiry

Because cultural geography has preferred diachronic depth as central to its methodology, this has allowed and nurtured a perceived and sometimes real distinction between itself and a contemporary, often explicitly problem-oriented social geography. As a result, criticism has been raised over the innocence of traditional topics within cultural geography. These charges depend, of course, on how one chooses to define and see problems in the world, and negates the central role played by cultural geographers in studying human modification of the environment (Kates 1987). Sauer's influence in shaping the landmark "Man's Role in Changing the Face of the Earth" is well known and recently documented (Williams 1987). Nonetheless, as the social sciences emphasize the problematical dimensions of culture, due largely, but not exclusively, to influences of neo-Marxist thought, an emergent theme in the "new" cultural geography focuses overtly upon power relations and social structure (Cosgrove 1983; Warf 1986; Dear 1987).

There seem to be two interrelated foci to this problematic: the recursiveness of place and landscape in the construction and reinforcement of social categories (see "Interaction Among Humans, Culture, and Landscape" above), and unmasking the power of ideologies in cultural landscapes. An excellent example of the first thrust is Anderson's study (1987) of the interaction among Vancouver's Chinatown (from 1880 to the 1920s), social structure, and political practice, working from the premise that racial categories are cultural ascriptions that are consciously constructed and transmitted. The Duncans (1988) explicate the ideological "naturalizing" of social realities once value structures and political assumptions have become concretized in the landscape, where they become a structuring agent that makes certain social processes seem natural and unquestionable. Ley (1987) uses the ideological constellations of postmodernism and neoconservatism as reference points for interrogating the emerging landscape of Vancouver, and Domosh (1988) examines how the differing structures of elite classes and their ideologies shaped urban form in nineteenth-century New York and Boston.

Gender as a category of explanation has been introduced to cultural geography, particularly in analyses of urban morphology (Saegert 1980; Wekerle et al. 1980) and in an examination of the changing ideals of domesticity and housing (Seager 1988; Loyd 1982). Much of this work emphasizes the recursiveness of the relation among socioeconomic structure, the particular historical role of women, and built form. In addition, investigations of how the interpretation of landscapes reflects gender differences is a vital research area in the examination of the cultural landscape (see the chapter on environmental perception).

Studying Everyday Life and Landscape

Although traditional cultural geography has preferred topics with historical depth, there is increasing interest and emphasis on the study of everyday life and landscapes that is complemented by theoretical currents within the social sciences. This theme covers an incredibly broad spectrum. At one end of that spectrum is British contemporary cultural studies, which theorize strategies of resistance that subordinate groups employ to contest the hegemony of those in power (Cosgrove and Jackson 1987, 88). Anchoring the other end of the spectrum is *Landscape* magazine, linked still to J.B. Jackson's French *Annales* agenda celebrating quotidian environments, and reinforced by the postmodern and poststructural emphasis on individualism and locality. Whether the British trajectory becomes an integral part of North American cultural geography remains to be seen. However, there is little controversy over the central role played by *Landscape* in promoting the study of everyday landscapes in elegant prose and photos unburdened by explicitly theoretical jargon.

Three recent articles exemplify the coverage given to everyday landscapes. Zelinsky (1988) examines welcoming signs for towns along America's highways as icons through which mundane settlements construct uniqueness that counteracts environmental blandness. Parsons (1988) collects, maps, and describes another dimension of elevated locality, the giant letters that dot western hillsides and are cultural signatures traceable in time and space, from the "Big C" constructed in 1905 above the University of California at Berkeley, to more modest hillside monograms tended by service clubs and high-school students. And last, O'Brien (1988) moves away from the usual assumptions of unfettered voluntarism central to humanistic geography by examining the way the British Travel Authority monitors and structures how visitors experience public images and are denied access to other realities of contemporary life.

REFERENCES

Aitken, S. C., and Bjorklund, E. 1988. Transactional and transformational theories in behavioral geography. *Professional Geographer* 40:54–64.

Anderson, K. J. 1987. The idea of Chinatown: The power of place and institutional practice in the making of a racial category. *Annals of the Association of American Geographers* 77:580–98.

Ankerl, G. 1981. *An experimental sociology of architecture: A guide to theory, research, and literature*. The Hague: Mouton.

Broadbent, G.; Bunt, R.; and Llorens, T. 1980. *Meaning and behavior in the built environment*. New York: Wiley Interscience.

Butzer, K. W. 1988. Cattle and sheep from old to new Spain: Historical antecedents. *Annals of the Association of American Geographers* 78:29–56.

Cosgrove, D. E. 1983. Towards a radical cultural geography. *Antipode* 15:1–11.

———. 1984. *Social formation and symbolic landscape*. London: Croom Helm.

———. 1985. Prospect, perspective, and the evolution of the landscape idea. *Transactions of the Institute of British Geographers,* n.s. 10:45–62.

———, and Jackson, P. 1987. New directions in cultural geography. *Area* 19:95–101.

Csikszentmihalyi, M., and Rochberg-Halton, E. 1981. *The meaning of things: Domestic symbols and the self*. New York: Cambridge University Press.

Dear, M. J. 1987. Editorial: Society, politics, and social theory. *Environment and Planning D: Society and Space* 5:363–66.

Domosh, M. 1988. Shaping the commercial city: A comparative study of the retail districts in nineteenth-century New York and Boston. Paper presented at the AAG meetings, Phoenix, April 6–10.

Duncan, J. S. 1980. The superorganic in American cultural geography. *Annals of the Association of American Geographers* 70:181–98.

———. 1985. Individual action and political power: A structuration perspective. In *The future of geography,* ed. R. J. Johnston, pp. 174–89. London: Methuen.

———. 1987. Review of urban imagery: Urban semiotics. *Urban Geography* 8:473–83.

———, and Duncan, N. G. 1984. A cultural analysis of urban residential landscapes in North America: The case of the anglophile elite. In *The city in cultural context,* eds. J. A. Agnew, J. Mercer, and D. E. Sopher, pp. 255–76. Boston: Allen & Unwin.

———, and Duncan, N. G. 1988. [Re]reading the landscape. *Environment and Planning D: Society and Space* 6:117–26.

Dunnell, R. 1982. Science, social science, and common sense: The agonizing dilemma of modern archaeology. *Journal of Anthropological Research* 38:1–25.

Entrikin, J. N. 1976. Contemporary humanism in geography. *Annals of the Association of American Geographers* 66:615–32.

Foote, K. E. 1983. Color in public spaces: Toward a communication-based theory of the urban built environment. University of Chicago Department of Geography Research Paper No. 205. Chicago: University of Chicago Press.

———. 1985. Space, territory, and landscape: The borderlands of geography and semiotics. *Recherches Semiotiques/Semiotic Inquiry* 5:158–75.

———. 1988. Object as memory: The material foundations of human semiosis. *Semiotica* 69:243–68.

Giddens, A. 1984. *The constitution of society*. Cambridge, MA: Polity Press.

Gottdiener, M., and Lagopoulos, A. 1986. *The city and the sign: An introduction to urban semiotics*. New York: Columbia University Press.

Hanson, N. R. 1985. *Patterns of discovery*. Cambridge: Cambridge University Press.

Hodder, I. 1986. *Reading the past*. Cambridge: Cambridge University Press.

———. 1987. Converging traditions: The search for symbolic meaning in archaeology and geography. In *Landscape and culture: Geographical and archaeological perspectives,* ed. J. M. Wagstaff, pp. 135–45. Oxford: Blackwell.

Jackson, J. B. 1985. *Discovering the vernacular landscape*. New Haven: Yale University Press.

Johnston, R. J. 1986. *Philosophy and human geography*. 2d ed. London: Edward Arnold.

———. 1987. *Geography and geographers: Anglo-American human geography since 1945*. 3d ed. London: Edward Arnold.

Jordan, T. G. 1985. *American log buildings: An old world heritage*. Chapel Hill, NC: University of North Carolina Press.

———. 1988. *The European culture area: A systematic geography*. 2d ed. New York: Harper & Row.

Kates, R. W. 1987. The human environment: The road not taken, the road still beckoning. *Annals of the Association of American Geographers* 77:525–34.

Kenzer, M., ed. 1987. *Carl O. Sauer: A tribute*. Corvallis, OR: Oregon State University Press.

Kofman, E. 1988. Is there a cultural geography beyond the fragments? *Area* 20:85–87.

Krampen, M. 1979. *Meaning in the urban environment*. London: Pion.

Leighly, J. 1987. Ecology as a metaphor: Carl Sauer and human ecology. *Professional Geographer* 39:405–12.

Leone, M. 1986. Symbolic, structural, and critical archaeology. In *American archaeology past and future,* eds. D. Meltzer, D. Fowler, and J. Sabloff, pp. 415–38. Washington: Smithsonian Institute Press.

Lewis, P. 1983. Learning from looking: Geographic and other writing about the American cultural landscape. *American Quarterly* 35:242–61.

Ley, D. 1987. Styles of the times: Liberal and neo-conservative landscapes in inner Vancouver, 1968-86. *Journal of Historical Geography* 13:40–56.

Loyd, B. 1982. Women, home and status. In *Housing and identity*, ed. James Duncan, pp. 181–97. New York: Holmes & Meier.

Meinig, D. W. 1986. *Atlantic America, 1492–1800*. Vol. 1 of *The shaping of America*. New Haven, CT: Yale University Press.

Norton, W. 1987. Humans, land, and landscape: A proposal for cultural geography. *Canadian Geographer* 31:21–33.

O'Brien, D. W. 1988. Figures against the ground: The imagery of British sight-seeing. *Landscape* 30:33–40.

Ortner, S. B. 1984. Theory in anthropology since the sixties. *Journal of Comparative Study of Society and History* 26:126–66.

Parsons, J. J. 1986. A geographer looks at the San Joaquin Valley. *Geographical Review* 76:371–89.

———. 1988. Hillside letters in the western landscape. *Landscape* 30:15–23.

Preziosi, D. 1979. *The semiotics of the built environment*. Bloomington, IN: Indiana University Press.

Rapoport, A. 1982. *The meaning of the built environment: A non-verbal communication approach*. Beverly Hills: Sage Publications.

Rowntree, L. B. 1988. Orthodoxy and new directions: Cultural/humanistic geography. *Progress in Human Geography* 12:575–86.

———, and Conkey, M. W. 1980. Symbolism and the cultural landscape. *Annals of the Association of American Geographers* 70:459–74.

Saegert, S. 1980. Masculine cities, feminine suburbs: Polarized ideas and contradictory realities. *Signs* 5:96–111.

Seager, J. 1988. Father's chair: Domestic reform and housing change in the progressive era. Ph.D. diss., Clark University.

Seamon, D. 1987. Phenomenology and environment-behavior research. In *Advances in environment, behavior, and design*, eds. G. T. Moore and E. Zube, pp. 3–26. New York: Plenum.

Singer, M. 1984. *Man's glassy essence: Explorations in semiotic anthropology*. Bloomington, IN: Indiana University Press.

Solot, M. 1986. Carl Sauer and cultural evolution. *Annals of the Association of American Geographers* 76:508–20.

Speth, W. 1987. Historicism: The disciplinary world view of Carl Sauer. In *Carl O. Sauer: A tribute*, ed. M. Kenzer, pp. 11–39. Corvallis, OR: Oregon State University Press.

Steinitz, M. Rethinking geographic approaches to the common house: The evidence from eighteenth century Massachusetts. In *Perspectives in vernacular architecture III*, eds. Bernard Herman and Thomas Carter. Columbia, MO: University of Missouri Press. In press.

Turner, B. L. 1987. Comment on Leighly. *Professional Geographer* 39:415–16.

Upton, D. 1985. White and black landscapes in eighteenth-century Virginia. *Places* 2:59–72.

Warf, B. 1986. Ideology, everyday life, and emancipatory phenomenology. *Antipode* 18:268–83.

Wekerle, G.; Peterson, R.; and Morley, D. 1980. *New space for women*. Boulder, CO: Westview Press.

Williams, M. 1987. Sauer and "Man's role in changing the face of the earth." *Geographical Review* 77:218–31.

Zelinsky, W. 1988. "Where every town is above average." Welcoming signs along America's highway. *Landscape* 30:1–10.

Environmental Perception and Behavioral Geography

Stuart C. Aitken | Susan L. Cutter | Kenneth E. Foote | James L. Sell

The underlying rationale for environmental perception and behavioral research lies in the assertion that understanding the geography of space and place requires knowledge of the way in which people experience, perceive, organize, and ascribe meaning to information about the environment, as well as how people act upon this information. Often this research seeks to identify general patterns of perception, cognition, and action common to all humans, or to members of particular social groups.

Environmental perception and behavioral research has maintained a unique perspective on geographical phenomena, by focusing upon the individual and postulating a dynamic interdependence between people and places. It also embraces a wide range of geographic scales and strives toward an interdisciplinary understanding of the interface between persons and environments.

Concern for perception and behavior can be traced quite far into geography's past. Early sources of influence are to be found in the writings of three American geographers—Sauer, Wright, and White—and the British geographer Kirk (Gold 1980, 34). The message propagated by these early researchers was that the subjective component of human lived experience was equally as important as the objective component. But it was not until the 1960s that environmental perception and behavioral geography became a distinctive subfield, inspired by the publication of a number of seminal books and articles by both geographers and nongeographers. Perhaps the most cited is Lynch's *The Image of the City* (1960), although it was followed almost immediately by key works written or edited by Saarinen, Lowenthal, Downs, Stea, Rushton, Golledge, Gould, Sonnenfeld, Blaut, and many others. So influential was this work that its insights diffused quickly throughout the discipline. In fact, many geographers readily integrated perception and behavioral approaches into their research

---

The authors would like to extend their thanks to Martin Cadwallader, Jim Duncan, Reg Golledge, Tom Saarinen, and Joe Sonnenfeld for comments they made on earlier drafts of this chapter, and for their support over the last year.

interests, without ever claiming primary allegiance to the subfield itself. Indeed, when measured by widespread use, the concepts of environmental perception and behavioral geography have framed some very important theoretical and applied advances in the discipline over the last 20 years.

Even though environmental perception and behavior studies have come of age, they have retained a youthful vigor. Perhaps more than many subfields, this one has taken to heart geography's contemporary claim to intellectual pluralism. Philosophical, theoretical, and methodological debate has flourished as scholars have pursued innovative research programs. In the last decade, this research has been the subject of several excellent reviews which provide a strong foundation for the present discussion. Foremost among these are progress reports by Saarinen and Golledge (Saarinen and Sell 1980, 1981; Saarinen, Sell, and Husband 1982; Golledge and Rayner 1982; Golledge and Rushton 1984; Golledge 1987). General textbooks have also been published, along with collections of essays and commentaries on specific subtopics (Ley and Samuels 1978; Gold 1980; Cox and Golledge 1981; Golledge and Timmermans 1988; Walmsley 1988).

Particularly noteworthy are three works that have appeared very recently:

> *Environmental Perception and Behavior: An Inventory and Prospect* (Saarinen, Seamon, and Sell 1984) was an outgrowth of special sessions organized by the Environmental Perception Specialty Group for the 1982 Annual Meeting of the Association of American Geographers.
> *Analytical Behavioral Geography* (Golledge and Stimson 1987) codifies the major themes of one important branch of perception geography.
> *Handbook of Environmental Psychology* (Stokols and Altman 1987) demonstrates the close interdisciplinary linkages maintained by geographers with colleagues in other fields.

This chapter does not attempt to replicate these works, but rather to focus on four themes within geography that have been particularly influenced by environmental perception and behavioral research (Figure 1) (Ford 1984). The first theme is that of spatial cognition and human behavior, or what Golledge and his colleagues have recently termed "analytical behavioral geography" (Couclelis and Golledge 1983; Golledge and Stimson 1987). The root of this theme is spatial analysis and location theory. The second theme focuses on the ecological dimensions of person-environment relationships, one of the original bases of environmental perception research as it grew from study of natural hazards. This theme has remained closely tied to issues of public policy and environmental decision-making.

A third theme is landscape perception and experience, a well-developed research tradition in geography which has served as the focus of recent attempts to tie perception research to cultural and historical themes, often with humanistic methods. The final theme derives from comparative research involving varied social and cultural groups. Although sometimes difficult to classify, these studies provide insight into the social and cultural relativity of perception and behavioral research. These four themes are neither completely exhaustive nor mutually exclusive, but each one serves to highlight important developments within the subfield.

Figure 1
Four themes of contemporary environmental perception and behavioral geography, and their interrelationships

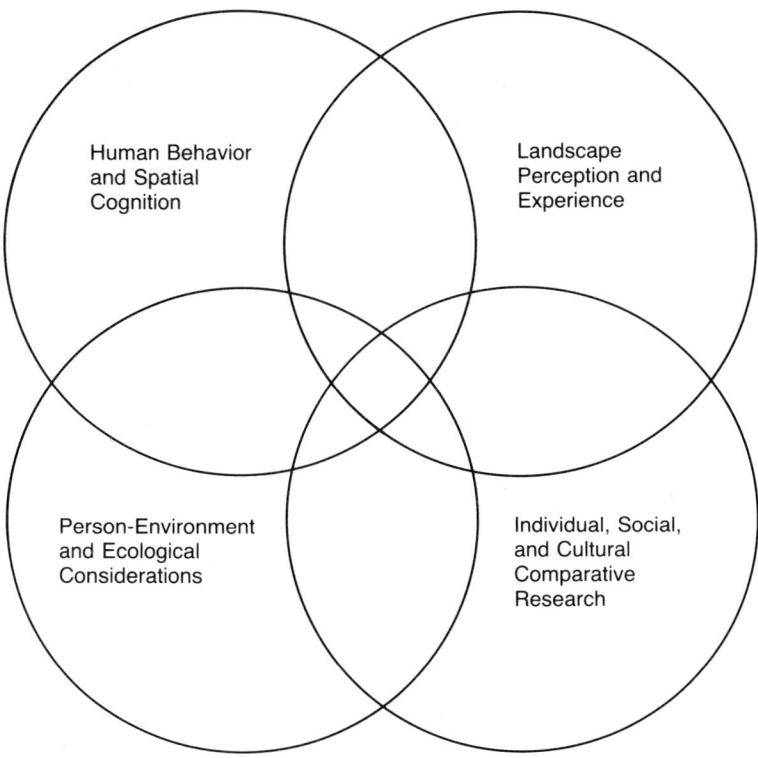

HUMAN BEHAVIOR AND SPATIAL COGNITION

Much of the work in environmental perception and behavioral geography has derived from the view of geography as "spatial science." Although this spatial-scientific approach was initially associated with overly reductive and deterministic models of human behavior, these have been superseded by more-sophisticated analyses in which the individual is seen as an active agent, setting criteria and making decisions based upon information available within particular contexts.

Contemporary analytical behavioral geography is characterized by five major developments:

1. The central tenets of positivism that traditionally guided research have been "softened" to encompass more-realistic assumptions about the separation of interdependent phenomena, such as subjects and objects, values and facts, and observers and those who are observed (Golledge and Rushton 1984).
2. There has been a related broadening of the concept of "cognition" to encompass the full breadth of human consciousness (Couclelis and Golledge 1983, 333).
3. The measurement techniques, statistical analyses and models employed by behavioral geographers have reached new levels of sophistication and rigor.
4. An increasing theoretical coherence has provided a more-stable platform for guiding empirical research (Golledge and Stimson 1987, 9).

5. Some progress has been made toward understanding aggregation problems at both a theoretical and an empirical level (Golledge 1988).

Empirical Research Directions

During its formative years in the 1960s and 1970s, analytical behavioral geography drew many of its concepts from psychology, sociology, and economics. Although Goodey and Gold suggested (1985) that British geographers have, for the most part, abandoned this interdisciplinary exploration, geographers in the U.S. have strengthened these links as a means of developing better research tools and models. And, in sustaining these interdisciplinary contacts, geographers have contributed as much as they have borrowed. Geographers are now recognized outside their own discipline for work in mental mapping and spatial cognition, environmental learning, proximity and preference modeling, discrete-choice modeling, space-time geography, and contextual analysis.

In reviewing a few examples of this interdisciplinary cross-fertilization, it is prudent to turn first to cognitive images and mental maps, or internal representations of the world that people use to guide their activity. In this context, the term "map" is solely a suggestive metaphor for describing complex cognitive processes, which involve gathering, sorting, storing, and using environmental information. In this research area, several issues remain in debate (Downs 1981). First, it is unclear whether these representations are best conceived as "stable images" or "dynamic schemata" (Tuan 1975; Gold 1980; Lloyd 1982; Aitken 1987b). Second, the methods commonly used to gain evidence about representations, such as sketch maps and rating scales, present problems of internal validity, aggregation, and replication. The empirical work investigating these issues revolves, at present, around the topics of distance and direction bias in cognitive imaging (Cadwallader 1979; Staplin and Sadalla 1981; Baird, Wagner, and Noma 1982), and how urban image distortions affect behaviors such as shopping and residential choice (Cadwallader 1981; Recker and Schuler 1981; Preston 1982).

Among the techniques used to explore these empirical issues are the multidimensional scaling (MDS) and repertory-grid methods. MDS is a technique that is designed to construct a "map" showing the relationships among a number of objects, given only a matrix of "distances" between them. This technique has been used extensively to analyze dissimilarity judgments and judgments of preference pertaining to questions of landscape evaluation and assessment (Taylor, Zube, and Sell 1987), distortions in cognitive mapping (Cadwallader 1979; Gale 1982), shopping-center preference (Spencer 1980), and housing choice (Preston 1982). In using MDS, geographers have had to resolve data-scale problems (Golledge and Rayner 1982), and had to introduce probabilistic elaborations (MacKay 1983; MacKay and Zinnes 1988).

The repertory grid, developed initially in clinical psychology to investigate person-person relationships, has been adapted by geographers interested in person-place relationships, primarily in retail and residential environments (Preston and Taylor 1981; Timmermans, der Heijden, and Westerveld 1982; Gill and Smith 1985; Aitken 1987a). Like MDS, the repertory grid attempts to elicit underlying cognitive categories used by individuals to give meaning to environments, without prespecifying these categories for the respondent. An individual's repertoire of images or schemata are classified into patterns of interdependency by techniques such as principal-components analysis or cluster analysis.

The question of how these images develop hierarchically and through time is the subject of anchorpoint theory (Golledge 1978; Clark and Burt 1980; Huff 1986). This theory suggests that some primary locations—such as home and work—serve to anchor, and thereby give order to the spatial information gathered by individuals, eventually conditioning the search for more information between the anchorpoints. This research has led geographers to consider related issues of environmental learning, such as skill acquisition and wayfinding (Sonnenfeld 1982, 1988; Golledge et al. 1985).

A more recent development in the study of spatial cognition has been the application of artificial intelligence (AI) techniques (Smith, Pellegrino, and Golledge 1982; Smith 1984; Couclelis 1986a). These techniques attempt to simulate human mental processes with computer programs. Although they may prove to be insightful, these techniques do not necessarily explain the processes that they aspire to simulate (Nystuen 1984). To date, they have only been used to model simple processes, such as navigation in an airport (Couclelis 1986b) and the wayfinding skill of children in a suburban neighborhood (Golledge et al. 1985).

Geographers have also been recognized for their work with discrete-choice modeling and spatial utility maximization. Although these models have firm roots in microeconomics and psychology, they have been developed and extended by behavioral geographers (Longley 1984). In discrete-choice models, the individual is characterized as an active decision-maker who processes available information, evaluates it against criteria, and makes choices among alternatives. The method has been applied to a wide range of phenomena, including residential choice (Hensher and Taylor 1983; Onaka 1983), shopping behavior (Hanson 1980; Timmermans 1984, 1988), and travel analysis (Hensher 1983).

Discrete-choice modeling has built upon the concepts of activity and action spaces. The related theorems of space-time geography, although developed in Sweden, have also been adapted to transportation and access problems in U.S. cities (Hanson 1982; Aitken and Fik 1988). This research has led logically to a consideration of the social, spatial, and temporal constraints over which individuals have little control. These constraints and contextual effects often complicate the relationship between attitudes and preferences on the one hand, and overt volitional behavior on the other (Burnett 1980; Desbarats 1983; Eagle 1988).

Contextual effects on spatial behavior have been explored in relationship to the search for employment opportunities (Gill and Smith 1985), evaluations of neighborhood attractiveness (Cutter, 1982; Preston, 1986), the perception of relative accessibility (Eagle 1988), and knowledge of housing vacancy information (Mackett and Johnson 1985; Palm 1986b). Pred's work on activity patterns and time geography suggests another avenue for understanding societal and institutional constraints, as well as a means for aggregating information about individual behavior (Pred 1981, 1984).

The Bounds of Theory

Taken together, these empirical works give some indication of the complexity of the cognitive processes underlying spatial decision-making. However, most of the studies cited above do not address at once the full sequence of decision-making, but focus instead on specific aspects of the decision-making process, such as search behavior or spatial choice. It seems clear that research in the 1990s must come to terms with spatial search, evaluation, choice, and behavior as elements of a unified process. In pursuing this goal, analytical behavioral geography will maintain a strong empirical orientation, so much so that empirical description of case studies may continue to outrun conceptualization and theory-building. Indeed, despite earlier optimism (Cox and Golledge 1969), no single theory has been developed to address the full range of person-environment relationships.

Alternative perspectives, such as transactionalism, offer to refine theory building (Couclelis and Golledge 1983, 333). Transactionalism emphasizes the dynamism of person-environment relationships by framing hypotheses specific to behaviors that are confined to particular contexts and sequences of action. Transactionalism recognizes that these contexts and sequences do not necessarily yield to convenient generalizations.

To some reviewers, such developments indicate abandonment of traditional high-level theory construction in behavioral geography (Cox and Golledge 1981, xix), but others see value in these departures. These latter writers advocate development of meta-theories that transcend traditional scientific explanation (Couclelis 1986b; Aitken and Bjorklund 1988). These meta-theories would serve to "outline the bounds of the possible in any given area, help organize and integrate piecemeal empirical research, help clarify the meaning of what is known, suggest which issues are researchable, which problems are in principle solvable, which are the questions worth asking" (Couclelis 1986b, 96).

PERSON-ENVIRONMENT AND ECOLOGICAL CONSIDERATIONS

An original focus of perception and behavioral research, and one of the discipline's oldest traditions, was the study of person-environment relationships. Studies of human perception of, and response to, natural hazards provided an ideal way for geographers to integrate their knowledge of natural Earth processes and human spatial behavior (see the chapter on hazards). Indeed, the development of hazard studies closely parallels the advances in analytical behavioral geography described above.

Yet, in a recent review of this research area, Mitchell noted that hazard-perception studies are presently "handicapped by lack of a clear cut definition of content and purpose" (1984, 58). He suggested a more comprehensive research strategy, focusing upon the processes whereby human perceptions are linked to hazard adjustments and decision-making. In a more recent review, Fischhoff, Svenson, and Slovic (1987) suggest that, although there is a growing literature on behavioral responses to environmental hazards, this research is piecemeal and lacking in either a theoretical or policy-oriented framework.

It seems clear that the resolution of these issues revolves around four fundamental questions:

1. What methods must be used to describe individual perceptions of the environment, environmental quality, and environmental risk?
2. How do people cope with environmental risk?

3. What is the relationship between the perception of environmental quality and risk, and responses to those perceptions?
4. How should person-environment studies inform public policy decision-making?

Environmental Risk Perception

Decision theory and heuristics provide the means of understanding the bases of the perception of environmental risks, including biases in judgment of risk, factors influencing attitudes toward risk, estimates of risk, and the relationship between risk perception and overt behavior (Tversky and Kahneman 1981; Kahneman, Slovic, and Tversky 1982; Covello 1983; Slovic, Fischhoff, and Lichtenstein 1984; Slovic 1987). In particular, psychometric scaling and multivariate analyses have produced quantitative assessments of perceptions and attitudes (Slovic 1985), and have advanced the understanding of perceived environmental risk.

Yet, few geographers have employed these techniques, despite their relevance to human-environment interactions. Noteworthy exceptions are studies of risks associated with nuclear power and power-plant accidents (Cutter 1984a, 1984b; Soderstrom et al. 1984; Sorenson et al. 1987), and noise pollution (Taylor 1982). Perception of social hazards such as crime has also attracted attention (Rengert and Wasilchick 1985; Westover 1985).

Behavioral Responses to Environmental Threats

The vast majority of recent hazards studies by geographers have focused upon individual and public responses to actual or perceived environmental threats which are manifest in both natural and technological systems. They include individual adjustments to natural hazards (Montz 1982; Sims and Baumann 1983; Palm 1986a), and place-specific comparisons of community adjustments to both natural and technological hazards (Preston, Taylor, and Hodge 1983; Kates and Hohenemser 1985).

At a larger scale, geographers have been heavily involved with programs of research on public responses to global environmental problems. Particularly noteworthy is the United Nations Environment Program (UNEP) assessment of *The World Environment, 1972–1982* (Holdgate, Kassas, and White 1982), because it was written by geographers and contains a chapter devoted to environmental education and public understanding. The work of Kates (1980) on perceptions of global climatic change also has reached a large international audience. Much of the research dealing specifically with behavioral responses to technological hazards in the last five years has focused on nuclear power-plant accidents (Cutter and Barnes 1982; Johnson and Zeigler 1983) and chemical disasters (Liverman and Wilson 1981; Bowonder, Kasperson, and Kasperson 1985).

Congruence Between Attitudes and Behavior

In recent years, philosophical debate has heightened over the connection between attitudes and overt behavior. Perception studies have been criticized for not offering empirical evidence to link individual perceptions of risk to overt behavior (Mitchell 1984). Early studies could only offer explanations based on concepts such as cognitive dissonance. Now, however, geographers find it possible to draw on modified versions of Fishbein and Ajzen's theory of reasoned action (Ajzen and Fishbein 1980). This theory examines the processes by which beliefs, attitudes, and behavioral intentions (and ultimately overt behavior) are linked. It has been applied to community attitudes towards pollution (Cutter 1981; Dunlap and Van Liere 1984), energy conservation (Brown and Macey 1982; Macey and Brown 1983), and evacuation behavior in response to nuclear power plant accidents (Zeigler and Johnson 1984; Johnson 1986). Geographers have also made recent breakthroughs in tying affect and cognitive processes to attitudes (Amedeo and York 1988).

Despite these successes, the ability to predict behavior remains limited in the absence of a better understanding of the relationship between attitudes, intent, and behavior, on the one hand, and the role of environmental, spatial, and additional contextual variables on the other.

Behavioral Bases for Environmental Management

Behavioral and perceptual research has a bearing on a wide range of issues relating to environmental management, such as those posed by cultural ecologists in their studies of traditional agricultural systems. Yet, beyond this important work, one of the most-significant advances in the area of person-environment research within the last decade has been its application to public-policy decisions in developed nations. Behavioral studies of responses to the Mount St. Helen's eruption and the accident at Three Mile Island are notable for initiating changes in emergency-response planning and providing substantial evidence in support of public concerns over workable evacuation plans (Cutter 1984b; Saarinen and Sell 1985).

Geographers have strongly advocated the behavioral basis for such evacuation planning (Johnson 1984; Cutter 1984a) and the development of behaviorally based warning systems (Sorenson 1984). Perception and behavior elements are now as regularly incorporated into natural-hazards research as are the physical characteristics of the hazard itself. And perception and behavior studies are now incorporated routinely into public policy decision-making.

LANDSCAPE PERCEPTION AND EXPERIENCE

Landscape has long been one of geography's central themes, and it was natural for perception researchers to turn toward this theme quickly after their early success with spatial analysis and hazard perception in the 1960s and 1970s. But, even though "landscape" is one of the most popular terms in geography's vocabulary, it is one of the most difficult to define. This is true even in perception studies, where "landscape" and "environment" are sometimes used synonymously (Sell, Taylor, and Zube 1984).

For many researchers, however, the terms "environmental perception" and "landscape perception" differ in connotation. Both can refer to natural and humanly created objects, but "environment" is the more general of the two terms. "Landscape perception," as one domain of "environmental perception," tends to focus on tangible visual elements of environments, especially features that express social and cultural traits. Some geographers have sought to broaden the use of the term "landscape" to encompass "soundscapes" and "smellscapes," but all express a tendency to refer back to the visual world (Porteous 1985; Porteous and Mastin 1985). A major contribution of geographers to the interdisciplinary field of landscape perception has been to set "landscape" within its appropriate natural and human environmental context, and to show how landscape and human experience are shaped and changed through time.

Types of Landscape Experience and Their Assessment

No single convenient taxonomy can be applied to the range of modified environments that are open to study. Rural landscapes, natural landscapes, and wilderness areas have long been of research interest. However, many researchers feel uncomfortable applying the term landscape to highly modified urban environments and complex architectural spaces. As a consequence, the term "urban built environment" has been coined as a means of characterizing cityscapes as special types of cultural landscape.

Methods that have been developed to investigate landscape perception differ in some ways from those used in analytical behavioral geography, in large part because research in the two areas strives for different goals. Landscape-perception studies focus generally on questions of the meaning and experience of landscape. Research in analytical behavioral geography is concerned more explicitly with empirical testing of general principles governing cognition of, and action within, environment. The major exception to this rule is work in visual-landscape assessment as applied to natural environments. In this area, researchers have developed a set of empirical research tools, including MDS, designed to elicit people's perceptions and reactions to the visual environment. Apart from important theoretical questions, this work in visual-landscape assessment has a distinct practical dimension, in that it can be applied to answer questions concerning natural-resource management (Zube 1984).

Cultural and Historical Themes in Landscape Studies

Perhaps a more important reason for the divergence of landscape-perception studies from the literature of environmental cognition and behavior stems from its more manifestly humanistic outlook. As alluded to earlier, recent research in environmental perception and behavioral geography has tended, more than in many other fields, to draw from both positivistic and humanistic philosophies. Indeed, studies of landscape experience provided one of the key channels through which humanistic philosophies entered geography.

Early studies of the human values that shape landscape experience and meaning, by scholars such as Lowenthal, have served as the foundation for a still-growing literature. This work has not been confined to rural or so-called "natural" landscapes. There has been a recent surge of interest in these questions as they pertain to the urban built environment (Hummon 1985; Relph 1987). The importance of this research lies in its recognition that landscape experience is tied inextricably to cultural values, and the relationship is not solely unidirectional. Cultural values serve to condition the way people view landscape, but people's activities in landscape also serve to alter these values through time, often very gradually. This concern for cultural values in perception studies has the added consequence of tying it to traditional themes of cultural and historical geography (see chapters on those topics in this volume). But this relationship also raises problems of theoretical compatibility, if only because research paradigms of perception research are not so readily adaptable to the study of cultural and historical processes.

This difficulty arises when cultural geographers and perception researchers turn to their common interest in interpreting the symbolic elements of cultural landscapes. Culture, by definition, transcends the individual, and the discourse of cultural geography is predicated, quite properly, on examination of the long-lived traditions and values that underpin community and society. It is not easy, therefore, for the insights of perception research—as derived from individual human beings—to be carried into the realm of cultural studies. Also, it is difficult to compare the environmental attitudes of individuals with those expressed by human institutions and complex social organizations.

Other more manageable problems arise from attempts to incorporate concepts of perception into the literature of historical geography. The methods of perception research do often rely on first-person surveys and interviews, but the insights the field can offer do not rely exclusively on experimental methods that have no counterparts in historical inquiry. Historical geographers have turned instead to other sources of information, which help them understand how individuals have viewed landscape in the past.

However, the overlap of perception research and historical geography remains small, being confined to the work of a few scholars like Lowenthal, who are committed to both subfields (Lowenthal 1985; Doughty 1987). In this context, it should not be forgotten that perception research can focus on human attitudes toward the past, an approach fruitful to studies of historic preservation (Ford 1979; Lowenthal 1979; Datel and Dingemans 1984).

Alternative Methodological and Philosophical Departures

The difficult questions arising from the study of historical processes, cultural values, and landscape have had the added consequence of attracting to the subfield geographers who are interested in a wide range of alternative methodological and philosophical departures. Some departures hinge on the use of novel source materials to explore attitudes toward landscape, and the significance of environmental symbolism. Among these sources are fictional novels and stories (Pocock 1981), art (Arreola 1984), poetry, travelogues and guidebooks, photography, and cinema (Zonn 1984).

More important are departures based on well-grounded humanistic and structural philosophies, such as phenomenology (Buttimer and Seamon 1980). Although the methods of phenomenology are difficult to apply to most geographical questions, they are well suited to issues of landscape perception. By stressing the inward, personal nature of environmental experience, phenomenology touches on the central themes of the perception literature (Rowles 1978; Hill 1985).

Interest in the idea of "reading" landscape has its basis in textual analysis, as derived from critical literary theory. From this point of view, landscape is viewed as "text" and, as such, is amenable to study in terms of discourse analysis. Of course, the idea that landscape can be "read" and seen as human communication is not new (Lewis 1979; Meinig 1979). But geographers have never taken to heart the challenge of this view of landscape. For example, semiotics, which views communication as a key to understanding society and culture, has gained little attention in geography, even though it can have a bearing on landscape interpretation (Foote 1985). Similarly, symbolic interactionism and ethnomethodology have been applied by researchers in environmental perception, but have not attracted large constituencies.

HISTORICAL AND CULTURAL CONTRIBUTIONS TO GEOGRAPHIC UNDERSTANDING

Marxism, as the preeminent structural philosophy in contemporary human geography, has had an effect on studies of landscape and perception (Duncan and Ley 1982; Blaut 1983; Cosgrove 1984). Alternatively, as Walker suggests in a chapter elsewhere in this volume, socialist geography has now broadened to encompass perspectives on environmental perception. It is also noteworthy that Marxist scholars, such as Harvey, have been turning, of late, to the topics of landscape, built environment, and, indirectly, perception (see Harvey 1985 and the chapter on geography from the left in this volume).

The Concept of Place

The term "place" implies a location and an integration of society, culture, and nature. Profound psychological and emotional links develop between people and the places they experience. In the last decade, a number of authors have discussed "sense of place" as a concept capable of sustaining intradisciplinary contact among geographers, both within and without the subfield of environmental perception. Sense of place may not fulfill these expectations, but it is one of the few concepts that seems to find its way into most methodological and philosophical discussions within the subfield.

No matter how divergent their outlook, perception and behavioral researchers find that "sense of place" crosscuts the most important issues of the field—perception, cognition, symbolism, meaning, historical change, and cultural process. There is a chance, then, that studies of place will provide a means for landscape-perception studies to gain a still broader constituency within the discipline. Of course, the value of such studies is derived from the fact that places differ, and all people express special characteristics that lead them to interact with places in different ways. The last section of this review explores some of the research that celebrates the uniqueness of person-place interactions.

INDIVIDUAL, SOCIAL, AND CULTURAL COMPARATIVE RESEARCH

In the realm of perception research, comparative study serves to stress the interdependence of person-place interactions and such individual characteristics as age, gender, class, ethnicity, value orientation, social group affiliation, and physical and mental ability. Comparative study cautions against assuming that results valid for one group can be applied to other, dissimilar groups, indicating that generalizations must be posed with care and informed by detailed knowledge about particular peoples and environments. For researchers to ignore the unique characteristics of both people and environment is to risk more than being wrong—it is to fail to realize the richness of human potential that is manifest in a plural society.

Environmental-perception researchers continue to demonstrate increased awareness of individual, social, and cultural characteristics. Cross-cultural research continues to build on international contributions, as well as on studies undertaken in North America. Social aspects of perception have also attracted attention, especially class differences, gender, and professional socialization. Insofar as individual characteristics are concerned, much research has centered on women, children, the elderly, and the disabled.

Cultural Attributes

Questions of cultural values arose earlier in this chapter, in relation to issues of landscape perception. Apart from that work, cultural geography and perception studies overlap in their concern for differences in attitudes toward environment, expressed by varying cultural groups. Cross-cultural research in North America has been particularly attentive to ethnic values as they define environmental experience, particularly in regions of cultural conflict or locales experiencing rapid change (Gordon 1984; Schoeller 1984; Desbarats 1986).

Studies outside North America have focused on three themes. The first addresses the perceptual bases of tourism and the problems entailed in the forced relocation of population (Churchill and Hutchinson 1984; Seamon 1984). The second theme involves the emergence of regional cultural symbols (Cline-Cole 1984; Gade 1984; Karan 1984) and how they may be interpreted by outsiders. The third theme is urban cognition and the perception of environmental quality stemming from the fact that many cross-cultural studies center on cities. The articles collected in *The City in Cultural Context* (Agnew, Mercer, and Sopher 1984) provide a good measure of the degree to which cultural geographers have become concerned with questions of urban perception. Other writings have appeared in this area, presenting comparative studies of neighborhood environmental quality, social networks, and residential planning (Hourihan 1984; Walmsley 1988).

Social Groups

Many studies document the strength of socioeconomic status and class affiliation in defining the character of environmental perception. These have focused primarily on feelings of personal attachment to neighborhood areas (Cohen and Shinar 1985; R. L. Smith 1985), commuter behavior (Kipnis and Mansfeld 1986), tourism (Searle and Jackson 1985), class conflict and gentrification (Cox 1984; Ley 1986), and historic preservation (Datel 1985; Gilbert 1985). The topic of "professional socialization" has continued to interest geographers. Studies have helped to show how professional groups differ in their environmental preferences and perceptions (Howard 1985; Rengert and Wasilchick 1985).

The underlying contention of this work is that spatial order can only be properly understood against a background of the underlying dimensions of social organization, as well as human behavior.

Individual Differences

As individuals, people differ by gender, age, ability, and personal preferences. And, in recent years, geographers have addressed the relation between these personal characteristics and environmental perception. Although gender issues are dealt with more fully in the chapter of this volume on women, it is important to note that attention to gender has had a twofold effect on behavioral geography. First, it has gone far to uncover male-female differences in spatial movement and the experience of place (Brooker-Gross and Maraffa 1985), and second, it has led to a heightened awareness of human diversity generally, regardless of gender differences (Monk 1984a).

The feminist perspective in perception and behavior research has been most pronounced in urban studies (Holcomb 1984; Wekerle 1984). Holcomb (1986) has emphasized the ambivalence manifest in this urban focus, by pointing out that women's perception of cities is held in tension between the potential dangers of this male-dominated urban environment and the appeal of its stimulating and liberating opportunities. Monk (1984b) has felt it important to stress that women-environment research should focus on the home and the sense of landscape affiliation it engenders. In accepting this challenge, feminist geography has positioned the theme of women and space as a key research interest. Monk and Norwood have, for example, edited one set of essays that explored the female sense of place and landscape value in the desert Southwest (Norwood and Monk 1987).

Behavioral researchers are now concerned with the sometimes subtle influences of life-cycle characteristics on human needs, environmental and community participation, and spatial-activity patterns. Although some behavioral geographers have sought to make a systematic attack on the relationship of life-cycle and environment, the only existing life-cycle studies have involved residential mobility (Davies and Pickles 1985) and landscape perception (Zube, Pitt, and Evans 1983). To date, however, most life-cycle research has concentrated on children and the elderly, without attending in detail to the period between childhood and late adulthood. Nonetheless, studies of children and the elderly have been very productive.

Interdisciplinary research on children's environments has been particularly insightful (Hart 1984; Moore 1986; Weinstein and David 1987). Geographers, for their part, have focused upon the following:

- Geographic learning (Morrill 1985)
- Children's perceptions of play and the importance of their "doing nothing" (Wood 1985)
- The development of cognitive representations of space (Downs 1985; Golledge et al. 1985)
- Mental maps
- The relationship between neighborhood quality and child development (Sell 1985)

Of methodological importance is the development of the Children's Environmental Response Inventory (CERI), a measurement instrument for assessing environmental dispositions ("personality") among children (Bunting and Cousins 1985).

In reviewing the issues of children and gender, note must be made of recent interest in the demand for and provision of child care. Research has barely begun, but it already seems apparent that child care will emerge as an issue of tremendous importance in coming years (Cromley 1987). New patterns of family composition that result in declining access to economic resources on the part of many women and children is likely to present important new research priorities for geographers.

The chapter in this volume on geography of the aging and aged details much of the research that has been devoted to the relationship of age to environmental perception and spatial behavior. Research on the perceptions and behaviors of the elderly has been concerned with:

- the relative importance of inertia and personal resources in explaining mobility decisions (Preston 1984)
- the limitations of mobility and accessibility in shopping decisions (G. C. Smith 1985)

- nighttime activities in the community (Golant 1984a)
- the experience and meaning of residential environments (Rowles 1978; Warnes 1982; Golant 1984b, 1986)
- the role of recreation-leisure activities in fulfilling life satisfaction needs among elderly people (Romsa, Bondy, and Blenman 1985)

In an essay concerned with evaluation of housing quality for the elderly, Golant (1986) has noted the discrepancies between so-called "objective" and "expert" evaluations and the qualities as perceived by the residents themselves.

Perhaps the greatest value of these studies of children and the elderly is that they suggest the importance of a more complete life-cycle analysis of environmental perception and behavior. Much can be learned by partitioning the lifespan into distinct age groups. But, ultimately, these studies must converge on a more dynamic, developmental approach which is capable of linking the stages of the life cycle into a comprehensive sequence. The pursuit of such a goal will address a major need felt across the full range of perception research—a temporal account of the development of environmental perception and spatial behavior.

In the last decade, attention has turned to assessing the environmental experience of people who are faced with various disabilities, such as mobility impairment and blindness (Hill 1985; Foote 1986). Of particular note is Ulrich's often-cited study (1984) of how exposure to natural views from windows has a restorative effect upon patients recovering from surgery. It is too often the case that the disabled are viewed as exceptions to general rules of perception and behavior. Yet, at any given moment, almost 10% of any population will be faced with some sort of acute or chronic mobility impairment, and the incidence of such impairment increases with age.

Through the creation of a life-span model, these factors may be set in the proper perspective as a natural part of human development. Such a perspective has the advantage of enfranchising other populations that are often overlooked when researchers focus on discrete groups, rather than on continuous processes of development and perceptual differentiation (Lauria and Knapp 1985). In this way, a life-span perspective will do more than produce a profile of human environmental development. It will serve to frame knowledge of environmental perception in ways which account for the richness of human experience at the cultural, social, and individual levels.

CONCLUSIONS

In the decades preceding 1980, geography was sadly lacking in coherent conceptions of person-environment relations. Indeed, perhaps as a result of the discipline's ostracization from academia in the early part of this century, geographers were wary of making statements about how the environment influences people and vice versa. The continued vitality of perception and behavioral geography may be assessed against this backdrop.

The subfield has generated conceptual structures that emphasize the bidirectional, reticulate nature of person-environment relations. Integrating frameworks have emerged to direct the rampant empiricism that characterized early behavioral research. The purview of the subfield has expanded to encompass provocative and challenging new topics. Looking back over the last decade, it can be asserted with confidence that the subfield of perception and behavioral research has not only contributed to the pluralism of contemporary geography, but has supplied the discipline with many of its most innovative concepts as well. Perception and behavioral researchers in the U.S. have demonstrated exceptional willingness to experiment with new ideas developed within geography and other cognate disciplines. Our intellectual associations with psychology, anthropology, sociology, and various planning and design disciplines have been extensively nurtured.

It is also evident that geographers' work on spatial behavior and person-place relationships is now widely accepted and used in other disciplines. Such sharing has incited vigorous internal debate over philosophy, theory, and method. Indeed, many proponents of environmental perception and behavioral geography have sought actively to encourage theoretical eclecticism, trusting, perhaps, that divergent perspectives may eventually converge at a higher level of abstraction.

Environmental perception and behavioral geography is likely to remain a heterogeneous research enterprise. Analytical behavioral geography has emerged as almost a subfield within a subfield. It has developed for itself a vital research agenda, based on years of systematic and rigorous research. Natural-hazards research has emerged from perception studies as a separate specialty group. Its empirical advances in relating cognition to behavior are complemented by a strong and well recognized public-policy component. Studies of landscape perception and experience are likely to remain an important arena for theoretical debate and empirical experimentation as research in this area continues to address humanistic, cultural, and historical themes. It is likely to make major contributions to the recent resurgence of historical and cultural geography.

Finally, comparative research stressing individual, social, and cultural attributes promises to generate a powerful life-span perspective on perceptual development that is capable of lending insight into the subtleties of the human condition. Environmental perception and behavioral geography has most certainly "come of age" as a mature research enterprise. Yet it maintains a youthful vitality capable of sustaining an active position within the discipline of geography.

REFERENCES

Agnew, J. A.; Mercer, J.; and Sopher, D. E., eds. 1984. *The city in cultural context*. Winchester, MA: Allen & Unwin.

Aitken, S. C. 1987a. Evaluative criteria and social distinctions in renters' residential search procedures. *Canadian Geographer* 31:114–26.

———. 1987b. Households moving within the rental sector: Mental schemata and search spaces. *Environment and Planning A* 19:369–83.

———, and Bjorklund, E. M. 1988. Transactional and transformational theories in behavioral geography. *The Professional Geographer* 40:54–64.

———, and Fik, T. J. 1988. The daily journey to work and choice of residence. *The Social Science Journal* 25:463–75.

Ajzen, I., and Fishbein, M. 1980. *Understanding attitudes and predicting social behavior*. Englewood Cliffs, NJ: Prentice-Hall.

Amedeo, D. and York, R. A. 1988. Affective states in cognitively-oriented person-environment-behavior frameworks. In *People's needs/planet management: Paths to co-existence*, eds. D. Lawrence, R. Habe, A. Hacker, and D. Sherrod. Proceedings of the Environmental Design Research Association 19, Washington D.C., pp. 203–11.

Arreola, D. D. 1984. Mexican American exterior murals. *Geographical Review* 74:409–24.

Baird, J.; Wagner, M.; and Noma, E. 1982. Impossible cognitive spaces. *Geographical Analysis* 14:204–26.

Blaut, J. 1983. Assimilation versus ghettoization. *Antipode: A Radical Journal of Geography* 15(1):35–41.

Bowonder, B.; Kasperson, J. X.; and Kasperson, R. E. 1985. Avoiding future Bhopals. *Environment* 27:6–13,31–37.

Brooker-Gross, S. R., and Maraffa, T. A. 1985. Commuting distance and gender among nonmetropolitan university employees. *The Professional Geographer* 37:303–10.

Brown, M. A., and Macey, S. M. 1982. Understanding residential energy conservation through attitudes and beliefs. *Environment and Planning A* 12:175–86.

Bunting, T. E., and Cousins, L. R. 1985. Environmental dispositions among school-age children: A preliminary investigation. *Environment and Behavior* 17:725–68.

Burnett, P. 1980. Spatial constraints-oriented approaches to movement, microeconomic theory, and urban policy-conceptual issues. *Urban Geography* 1:53–67.

Buttimer, A., and Seamon, D., eds. 1980. *The human experience of space and place*. London: Croom Helm.

Cadwallader, M. T. 1979. Problems in cognitive distance: Implications for cognitive mapping. *Environment and Behavior* 11:559–76.

———. 1981. Towards a cognitive gravity model: The case of consumer spatial behavior. *Regional Studies* 15:275–84.

Churchill, R. A., and Hutchinson, D. M. 1984. Flood hazard in Ratnapura, Sri Lanka: Individual attitudes vs. collective action. *Geoforum* 15:517–24.

Clark, W. A. V., and Burt, J. 1980. The impact of workplace on residential relocation. *Annals of the Association of American Geographers* 70:59–67.

Cline-Cole, R. A. 1984. Towards an understanding of man-firewood relations in Freetown (Sierra Leone). *Geoforum* 15:583–94.

Cohen, Y. S., and Shinar, A. 1985. *Neighborhood and friendship networks: A study of three residential neighborhoods in Jerusalem*. University of Chicago Department of Geography Research Paper No. 215. Chicago: University of Chicago Press.

Cosgrove, D. 1984. *Social formation and symbolic landscape*. London: Croom Helm.

Couclelis, H. 1986a. Artificial intelligence in geography: Conjectures on the shape of things to come. *Professional Geographer* 38:1–10.

———. 1986b. A theoretical framework for alternative models of spatial decision and behavior. *Annals of the Association of American Geographers* 76:95–113.

———, and Golledge, R. G. 1983. Analytic research, positivism, and behavioral geography. *Annals of the Association of American Geographers* 73:331–39.

Covello, V. T. 1983. The perception of technological risks: A literature review. *Technological Forecasting and Social Change* 23:285–97.

Cox, K. R. 1984. Neighborhood conflict and urban social movements: Questions of historicity, class and social change. *Urban Geography* 5:343–55.

———, and Golledge, R. G., eds. 1969. *Behavioral problems in geography*. New York: Methuen.

———, and ———, eds. 1981. *Behavioral problems in geography revisited*. New York: Methuen.

Cromley, E. K. 1987. Locational problems and preferences in preschool child care. *Professional Geographer* 39:309–17.

Cutter, S. 1981. Community concern for pollution: Social and environmental influences. *Environment and Behavior* 13:105–24.

———. 1982. Residential satisfaction and the suburban homeowner. *Urban Geography* 3:315–27.

———. 1984a. Emergency preparedness and planning for nuclear power plant accidents. *Journal of Applied Geography* 4:235–45.

———. 1984b. Risk cognition and the public: The case of Three Mile Island. *Environmental Management* 8:15–20.

———, and Barnes, K. 1982. Evacuation behavior and Three Mile Island. *Disasters* 6:116–24.

Datel, R. E. 1985. Preservation and a sense of orientation for American cities. *Geographical Review* 75:125–41.

———, and Dingemans, D. J. 1984. Environmental perception, historic preservation, and sense of place. In *Environmental perception and behavior: An inventory and prospect,* eds. T. F. Saarinen, D. Seamon, and J. L. Sell, pp. 131–44. University of Chicago Department of Geography Research Paper No. 209. Chicago: University of Chicago Press.

Davies, R. B., and Pickles, A. R. 1985. A panel study of life-cycle effects in residential mobility. *Geographical Analysis* 7:199–216.

Desbarats, J. 1983. Spatial choice and constraints on behavior. *Annals of the Association of American Geographers* 73:340–57.

———. 1986. Ethnic differences in adaptation: Sino-Vietnamese refugees in the United States. *International Migration Review* 20:405–27.

Doughty, R. 1987. *At home in Texas: Early views of the land*. College Station, TX: Texas A & M University Press.

Downs, R. M. 1981. Maps and metaphors. *Professional Geographer* 33:287–93.

———. 1985. The representation of space: Its development in children and in cartography. In *The development of spatial cognition,* ed. R. Cohen, pp. 323–45. Hillsdale, NJ: Erlbaum.

Duncan, J., and Ley, D. 1982. Structural marxism and human geography: a critical assessment. *Annals of the Association of American Geographers* 72:30–59.

Dunlap, R. E., and Van Liere, K. D. 1984. Commitment to the dominant social paradigm and concern for environmental quality. *The Social Science Quarterly* 65:1013–28.

Eagle, T. C. 1988. Contextual effects in consumer behavior. In *Behavioral modeling in geography and planning,* eds. R. G. Golledge and H. Timmermans, pp. 299–324. New York: Croom Helm.

Fischhoff, B.; Svenson, O.; and Slovic, P. 1987. Active responses to environmental hazards: Perceptions and decision-making. In *Handbook of environmental psychology,* vol. 2, eds. D. Stokols and I. Altman, pp. 1089–1133. New York: Wiley.

Foote, K. E. 1985. Space, territory, and landscape: The borderlands of geography and semiotics. *Recherche Semiotique/Semiotic Inquiry* 5:158–75.

———. 1986. Mobility impairment and pharmacy accessibility: Conflict in a commercial built environment. *Environment and Behavior* 18:571–603.

Ford, L. R. 1979. Urban preservation and the geography of the city in the U.S.A. *Progress in Historical Geography* 3:211–38.

———. 1984. Where do we go from here? A commentary. In *Environmental perception and behavior: An inventory and prospect,* eds. T. F. Saarinen, D. Seamon, and J. L. Sell, pp. 145–48. University of Chicago Department of Geography Research Paper No. 209. Chicago: University of Chicago Press.

Gade, D. W. 1984. Redolence and land use on Nosy Be, Madagascar. *Journal of Cultural Geography* 4:29–40.

Gale, N. 1982. Some applications of computer cartography to the study of cognitive configurations. *Professional Geographer* 34:313–21.

Gilbert, A. 1985. Et si les geographes s'interessaient aux ideologies moins officielles. *Cahiers de Geographie du Quebec* 29:217–24.

Gill, A., and Smith, G. 1985. Residents' evaluative structures of northern Manitoba mining communities. *Canadian Geographer* 29:17–30.

Golant, S. M. 1984a. Factors influencing the nighttime activity of old persons in their community. *Journal of Geography* 39:485–91.

———. 1984b. *A place to grow old*. New York: Columbia University Press.

———. 1986. Subjective housing assessment by the elderly: A critical information source for planning and program evaluation. *Gerontologist* 26:122–27.

Gold J. R. 1980. *An introduction to behavioral geography*. New York: Oxford University Press.

Golledge, R. G. 1978. Learning about urban environments. In *Making sense of time,* eds. T. Carlstein, D. Parkes, and N. Thrift, pp. 76–98. London: Edward Arnold.

———. 1987. Environmental cognition. In *Handbook of environmental psychology,* vol. 1, eds. D. Stokols and I. Altman, pp.131–74. New York: Wiley.

———. 1988. Science and humanism in geography: Multiple languages in multiple realities. In *A ground for common search,* eds. R. G. Golledge, H. Couclelis, and P. Gould, pp. 63–72. Goleta, CA: Santa Barbara Geographical Press.

———, and Rayner, J. N. 1982. *Proximity and preference: Problems in the multidimensional analysis of large data sets*. Minneapolis: University of Minnesota Press.

———, and Rushton, G. 1984. A review of analytical behavioral geography. In *Geography and the urban environment: Progress in research and application,* vol. 6, eds. D. T. Herbert and R. J. Johnston, pp. 1–44. Chichester: Wiley.

———; Smith, T. R.; Pellegrino, J. W.; Doherty, S.; and Marshall, S. P. 1985. A conceptual model and empirical analysis of children's acquisition of spatial knowledge. *Journal of Environmental Psychology* 5:125–52.

———, and Stimson, R. J. 1987. *Analytical behavioral geography*. London: Croom Helm.

———, and Timmermans, H. 1988. *Behavioral modeling in geography and planning*. New York: Croom Helm.

Goodey, B., and Gold, J. R. 1985. Behavioral and perceptual geography: From retrospect to prospect. *Progress in Historical Geography* 9:585–95.

Gordon, J. J. 1984. Onondaga Iroquois place-names: An approach to historical and contemporary Indian landscape perception. *Names* 32:218–33.

Hanson, S. 1980. The importance of multipurpose journey to work in urban travel behavior. *Transportation* 9:229–48.

———. 1982. The determinants of daily travel-activity patterns: Relative relocation and socio-demographic factors. *Urban Geography* 3:179–202.

Hart, R. 1984. The geography of children and children's geographies. In *Environmental perception and behavior: An inventory and prospect,* eds. T. F. Saarinen, D. Seamon, and J. L. Sell, pp. 99–129. University of Chicago Department of Geography Research Paper No. 209. Chicago: University of Chicago Press.

Harvey, D. 1985. *Consciousness and the urban experience*. Oxford: Blackwell.

Hensher, D. A. 1983. A sequential attribute dominance model of probabilistic choice. *Transportation Research A* 17:215–18.

———, and Taylor, A. K. 1983. Intraurban residential relocation choices for students: An empirical inquiry. *Environment and Planning A* 15:815–30.

Hill, M. 1985. Bound to the environment: Towards a phenomenology of sightlessness. In *Dwelling, place, and environment: Toward a phenomenology of person and world,* eds. D. Seamon and R. Mugerauer, 99–111. Dordrecht: Martinus Nijhoff.

Holcomb, B. 1984. Women in the rebuilt urban environment: The United States experience. *Built Environment* 10:18–24.

———. 1986. Geography and urban women. *Urban Geography* 7:448–56.

Holdgate, M. W.; Kassas, M.; and White, G. F., eds. 1982. *The world environment, 1972–1982: A report by the United Nations Environment Program*. Dublin: Tycooly International Publishing.

Hourihan, K. 1984. Context-dependent models of residential satisfaction. *Environment and Behavior* 16:369–93.

Howard, P. 1985. Painters' preferred places. *Journal of Historical Geography* 11:138–54.

Huff, J. O. 1986. Geographic regularities in residential search behavior. *Annals of the Association of American Geographers* 76:208–27.

Hummon, D. M. 1985. Urban ideology as a cultural system. *Journal of Cultural Geography* 5:1–15.

Johnson, J. H. 1984. Planning for spontaneous evacuation during a radiological emergency. *Nuclear Safety* 25:186–94.

———. 1986. A model of evacuation decision-making in a nuclear reactor emergency. *Geographical Review* 76:405–18.

———, and Zeigler, D. J. 1983. Distinguishing human responses to radiological emergencies. *Economic Geography* 59:386–402.

Kahneman, D.; Slovic, P.; and Tversky, A., eds. 1982. *Judgement under uncertainty: Heuristics and biases*. New York: Cambridge University Press.

Karan, P. P. 1984. Landscape, religion and folk art in Mithilia: An Indian cultural region. *Journal of Cultural Geography* 5:85–101.

Kates, R. 1980. Climate and society: Lessons from recent events. *Weather* 35:17–25.

———, and Hohenemser, H. 1985. *Managing the risks of technology*. Chicago: University of Chicago Press.

Kipnis, B. A., and Mansfeld, Y. 1986. Work-place utilities and commuting patterns: Are they class or place differentiated? *Professional Geographer* 38:160–69.

Lauria, M., and Knapp, L. 1985. Toward an analysis of the role of gay communities in the urban renaissance. *Urban Geography* 6:152–69.

Lewis, P. F. 1979. Axioms for reading the landscape. In *The interpretation of ordinary landscapes: Geographical essays,* ed. D. W. Meinig, pp. 195–244. New York: Oxford University Press.

Ley, D. 1986. Alternative explanations for inner-city gentrification: A Canadian assessment. *Annals of the Association of American Geographers* 76:521–35.

———, and Samuels, M. S. 1978. *Humanistic geography: Prospects and problems*. London: Croom Helm.

Liverman, D., and Wilson. J. P. 1981. The Mississauga train derailment and evacuation, 10–16 November 1979. *The Canadian Geographer* 25:365–75.

Lloyd, R. 1982. A look at images. *Annals of the Association of American Geographers* 72:532–48.

Longley, P. 1984. Discrete choice modeling and complex spatial choice: An overview. In *Recent developments in spatial data analysis,* eds. S. G. Bahrenberg, M. M. Fischer, and P. Nijkamp, pp. 375–93. Aldershot: Gower.

Lowenthal, D. 1979. Environmental perception: Preserving the past. *Progress in Historical Geography* 3:549–59.

———. 1985. *The past is a foreign country*. New York: Cambridge University Press.

Lynch, K. 1960. *The image of the city*. Cambridge: MIT Press.

Macey, S. M., and Brown, M. A. 1983. Residential energy conservation: The role of past experience in repetitive household behavior. *Environment and Behavior* 15:123–42.

Mackay, D. B. 1983. Alternative probabilistic scaling models for spatial data. *Geographical Analysis* 15:173–86.

———, and Zinnes, J. L. 1988. Probabilistic multidimensional scaling of spatial preferences. In *Behavioral modeling in geography and planning,* eds. R. G. Golledge and H. Timmermans, pp. 198–222. New York: Croom Helm.

Mackett, R. L., and Johnson, I. 1985. Residential search behavior: The implication of survey and analytical design. *Tijdschrift voor Economische en Sociale Geografie* 76:173–79.

Meinig, D. W. 1979. Reading the landscape. In *The interpretation of ordinary landscapes: Geographical essays,* ed. D. W. Meinig, pp. 195–244. New York: Oxford University Press.

Mitchell, J. K. 1984. Hazard perception studies: Convergent concerns and divergent approaches during the past decade. In *Environmental perception and behavior: An inventory and prospect,* eds. T. F. Saarinen, D. Seamon, and J. L. Sell, pp. 33–59. University of Chicago Department of Geography Research Paper No. 209. Chicago: University of Chicago Press.

Monk, J. 1984a. Approaches to the study of women and landscape. *Environmental Review* 8:23–33.

———. 1984b. Human diversity and perceptions of place. In *Perception of people and places through media,* vol. 1, ed. H. Haubrich, pp. 45–67. Freiburg, Federal Republic of Germany: Pädagogische Hochschule.

Montz, B. E. 1982. The effect of location on the adoption of hazard mitigation measures. *Professional Geographer* 34:416–23.

Moore, R. C. 1986. *Childhood's domain: Play and place in child development*. London: Croom Helm.

Morrill, R. W. 1985. Childhood geographies. *Journal of Geography* 84:123–25.

Norwood, V., and Monk, J. eds. 1987. *The desert is no lady*. New Haven: Yale University Press.

Nystuen, J. D. 1984. Artificial intelligence and geographical problem solving: Commentary and reply. *Professional Geographer* 36:358–60.

Onaka, J. L. 1983. A multiple-attribute housing disequilibrium model of residential mobility. *Environment and Planning A* 15:751–66.

Palm, R. 1986a. Coming home. *Annals of the Association of American Geographers* 76:469–79.

———. 1986b. Racial and ethnic influences on real estate agent practices. *Social Science Journal* 23:43–54.

Pocock, D., ed. 1981. *Humanistic geography and literature*. London: Croom Helm.

Porteous, J. D. 1985. Smellscape. *Progress in Historical Geography* 9:356–78.

———, and Mastin, J. F. 1985. Soundscape. *Journal of Architectural and Planning Research* 2:169–86.

Pred, A. 1981. Social reproduction and the time-geography of everyday life. *Geografiska Annaler* B63:5–22.

———. 1984. Structuration and the time-geography of becoming places. *Annals of the Association of American Geographers* 74:279–97.

Preston, V. A. 1982. A multidimensional scaling analysis of individual differences in residential area analysis. *Geografiska Annaler* B64:17–26.

———. 1984. A path model of residential stress and inertia among older people. *Urban Geography* 5:146–64.

———. 1986. A case study of context effects and residential area evaluation in Hamilton, Canada. *Environment and Planning A* 18:41–52.

———, and Taylor, S. M. 1981. Personal construct theory and residential choice. *Annals of the Association of American Geographers* 71:437–51.

———; ———; and Hodge, D. C. 1983. Adjustment to natural and technological hazards: A study of an urban residential community. *Environment and Behavior* 15:143–64.

Recker, W. W., and Schuler, H. J. 1981. Destination choice and processing spatial information: Some empirical tests with alternative constructs. *Economic Geography* 57:373–83.

Relph, E. 1987. *The modern urban landscape*. London: Croom Helm.

Rengert, G., and Wasilchick, J. 1985. *Suburban burglary: A time and place for everything*. Springfield, IL: Charles C. Thomas.

Romsa, G.; Bondy, P.; and Blenman, M. 1985. Modeling retirees' life satisfaction levels: The role of recreational, life cycle, and socio-environmental elements. *Journal of Leisure* 17:29–39.

Rowles, G. D. 1978. *Prisoners of space? Exploring the geographical experience of older people*. Boulder, CO: Westview Press.

Saarinen, T. F., and Sell, J. L. 1980. Environmental perception. *Progress in Historical Geography* 4:525–48.

———, and ———. 1981. Environmental perception. *Progress in Historical Geography* 5:525–47.

———, and ———. 1985. *Warning and response to the Mount St. Helen's eruption*. Albany: State University of New York Press.

———; ———; and Husband, E. 1982. Environmental perception: International efforts. *Progress in Historical Geography* 6:515–46.

Saarinen, T. F.; Seamon, D.; and Sell, J. L. 1984. *Environmental perception and behavior: An inventory and prospect*. University of Chicago Department of Geography Research Paper No. 209. Chicago: University of Chicago Press.

Schoeller, P. 1984. Urban values: A review of Japanese and German attitudes. *Urban Geography* 5:43–48.

Seamon, D. 1984. Heidegger's notion of dwelling and one concrete interpretation as indicated by Hassan Fathy's architecture for the poor. *Geoscience and Man* 24:43–53.

Searle, M. S., and Jackson, E. L. 1985. Socioeconomic variations in perceived barriers to recreation participation among would-be participants. *Leisure Sciences* 7:227–49.

Sell, J. L. 1985. Children and neighborhood environmental quality. *Childrens' Environments Quarterly* 2:41–48.

———; Taylor, J. G.; and Zube, E. H. 1984. Toward a theoretical framework for landscape perception. In *Environmental perception and behavior: An inventory and prospect,* eds. T. F. Saarinen, D. Seamon, and J. L. Sell, pp. 61–83. University of Chicago Department of Geography Research Paper No. 209. Chicago: University of Chicago Press.

Sims, J. H., and Baumann, D. D. 1983. Educational programs and human response to natural hazards. *Environment and Behavior* 15:165–89.

Slovic, P. 1985. Characterizing perceived risk. In *Perilous progress: Managing the hazards of technology,* eds. R. W. Kates, C. Hohenemser, and J. X. Kasperson, pp. 91–125. Boulder, CO: Westview Press.

———. 1987. Perception of risk. *Science* 236:280–85.

———; Fischhoff, B.; and Lichtenstein, S. 1984. Behavioral decision theory perspectives on risk and safety. *Acta Psychologica* 56:183–203.

Smith, G. C. 1985. Shopping perceptions of the inner city elderly. *Geoforum* 16:319–31.

Smith, R. L. 1985. Activism and social status as determinants of neighborhood identity. *Professional Geographer* 37:421–32.

Smith, T. R. 1984. Artificial intelligence and its applicability to geographical problem solving. *Professional Geographer* 36:147–58.

———; Pellegrino, J.; and Golledge, R. G. 1982. Computational process modeling of spatial cognition and behavior. *Geographical Analysis* 14:305–25.

Soderstrom, E. J.; Sorenson, J. H.; Copenhaver, E. D.; and Carnes, S. A. 1984. Risk perception in an interest group context: An examination of the TMI restart issue. *Risk Analysis* 4:231–44.

Sonnenfeld, J. 1982. Egocentric perspectives on geographic orientation. *Annals of the Association of American Geographers* 72:68–76.

———. 1988. Abilities, skills, competence: A search for alternatives to diffusion. In *The transfer and transformation of ideas and material culture,* eds. P. J. Hugill and D. B. Dickson. College Station, TX: Texas A & M University Press.

Sorenson, J. H. 1984. Evaluating the effectiveness of warning systems for nuclear power plant emergencies: Criteria and application. In *Nuclear power: Assessing and managing hazardous technology,* eds. M. J. Pasqualetti and K. D. Pijawka, pp. 248–77. Boulder, CO: Westview Press.

———; Soderstrom, E.; Copenhaver, E.; Carnes, S.; and Bolin, R. 1987. *Impacts of hazardous technology: The psycho-social effects of restarting TMI-1*. Albany: State University of New York Press.

Spencer, A. H. 1980. Cognitive and shopping choice: A multidimensional scaling approach. *Environment and Planning A* 12:1235–51.

Staplin, L. J., and Sadalla, E. K. 1981. Distance cognition in urban environments. *Professional Geographer* 33:302–10.

Stokols, D., and Altman, I., eds. 1987. *Handbook of environmental psychology*. New York: Wiley.

Taylor, J. G.; Zube, E. H.: and Sell, J. L. 1987. Landscape assessment and perception research methods. In *Methods in environmental and behavioral research,* eds. R. Bechtel, R. Marans, and W. Michelson, pp. 361–93. New York: Van Nostrand Reinhold.

Taylor, S. M. 1982. A comparison of models to predict annoyance reactions to noise from mixed sources. *Journal of Sound and Vibration* 81 (1):123–38.

Timmermans, H. 1984. Discrete choice vs. decompositional multiattribute preference models: A comparative analysis of model performance in the context of spatial shopping behavior. In *Discrete spatial choice models,* ed. D. A. Pitfield, pp. 8–102. London: Pion.

———. 1988. Multipurpose trips and individual choice behavior: An analysis using experimental design data. In *Behavioral modeling in geography and planning,* eds, R. G. Golledge and H. Timmermans, pp. 356–67. New York: Croom Helm.

———; der Heijden, V.; and Westerveld, H. 1982. Cognition of urban retailing structures. *Tijdschrift voor Economische en Sociale Geografie* 73:2–12.

Tuan, Y. F. 1975. Images and mental maps. *Annals of the Association of American Geographers* 65:205–13.

Tversky, A., and Kahneman, D. 1981. The framing of decisions and the psychology of choice. *Science* 211:1453–58.

Ulrich, R. S. 1984. View through a window may influence recovery from surgery. *Science* 224: 420–21.

Walmsley, D. J. 1988. *Urban living: The individual in the city*. New York: Wiley.

Warnes, A. M. ed. 1982. *Geographical perspectives on the elderly*. New York: Wiley.

Weinstein, C. S., and David, T. G., eds. 1987. *Space for children: The built environment and child development*. New York: Plenum.

Wekerle, G. R. 1984. A woman's place is in the city. *Antipode: A Radical Journal of Geography* 16:11–19.

Westover, T. N. 1985. Perception of crime and safety in three midwestern parks. *The Professional Geographer* 37: 610–20.

Wood, D. 1985. Nothing doing. *Children's Environments Quarterly* 2:14–25.

Zeigler, D. J., and J. H. Johnson. 1984. Evacuation behavior in response to nuclear power plant accidents. *Professional Geographer* 36:207–15.

Zonn, L. 1984. Landscape depiction and perception: A transactional approach. *Landscape Journal* 3:144–50.

Zube, E. H. 1984. Themes in landscape assessment. *Landscape Journal* 3:104–10.

———; Pitt, D. G.; and Evans, G. W. 1983. A lifespan developmental study of landscape assessment. *Journal of Environmental Psychology* 3:115–28.

Geographic Research on Native Americans

Dick G. Winchell | James M. Goodman |
Stephen C. Jett | Martha L. Henderson

Geographic research on Native Americans is as diverse as the discipline itself. Each tribal community, both contemporary and historical, represents a complex and unique cultural and spatial system, making geographic research on Native Americans hard to catalog, and often linked more closely to regional issues than to general theory. Geographers of all subdisciplines are involved in Native American research, from physical geographers doing contract research for tribal governments, to historical geographers examining society-environment relations in pre-European contact tribal societies.

There are two major research emphases within recent geographic literature, both essential to the topic. One is cultural/historical, with a focus upon historical documents and the historical contexts of tribal cultures. The second is contemporary, which promotes tribal sovereignty through applied research and direct services to tribal governments and Indian organizations. This chapter identifies recent trends in geographic research on Native Americans, and discusses research opportunities within the field.

NATIVE AMERICAN GEOGRAPHIC RESEARCH

Native American geographic research has great diversity and complexity. Over 400 tribes are recognized by the Bureau of Indian Affairs (BIA), and 278 are identified by the Bureau of the Census (U.S. Department of Commerce 1984, 3). Although Native Americans represented only 0.6% of the total U.S. population in 1980, tribes con-

---

The authors would like to thank the members of the Native American Specialty Group for their contributions to the development of this paper, especially Barbara Jacquay of Arizona State University, Douglas Richardson of GeoResearch, Inc., Eliott McIntire of California State University–Northridge, and George Van Otten of Northern Arizona University. We would also like to thank Cort Willmott, Gary Gaile, and two reviewers for their comments, suggestions, and support.

HISTORICAL AND CULTURAL CONTRIBUTIONS TO GEOGRAPHIC UNDERSTANDING

trolled 2.5% of the nation's land, including significant energy resources (Goodman 1985). The land base of these reservations ranges from less than one acre for California Rancheria reservations to the Navajo Nation, with over 25,000 square miles.

Land tenure on these reservations also varies greatly. Some tribal lands have status under fee-simple ownership, but most is comprised of either "allotted" lands which are tribally controlled but individually owned by tribal members, or tribally owned land held in trust by the BIA. Because of these differences in land tenure, the percentage of American Indians on reservations also varies greatly. Although residents in Arizona's reservations are predominantly American Indian (88%), many reservations have more non-Indian than Indian residents. California, for example, has only 30.3% American Indian residents on reservations within the state, while Washington has 20.8% and Wyoming has only 18.0% (U.S. Department of Commerce 1984, 21–25).

Tribal traditions, language, religion, and culture are linked to aboriginal lands and create unique "meaning systems" for land and land tenure within tribal cultures. Many reservations are on or near aboriginal lands, but even where tribes have been relocated, the concept of land and the meaning of landscape elements are much different for Indians than non-Indians (e.g., Weightman 1976).

The issues of identity and political power have also been critical to contemporary and historic native settlements. Although concepts of tribal sovereignty existed throughout the history of European contact with Native Americans, political autonomy and power is only a recent experience for Indian communities.

The historical evolution of Indian communities is critical in establishing land claims and unique status for tribal governments. Indian tribal governments are semi-sovereign nations which generally do not come under state or county law but have their own legal codes and powers to govern through treaty or constitutions under the Indian Reorganization Act of 1934. Their status as "dependent domestic nations" generally includes powers of independent local governments, including the ability to institute land-use controls, business and economic development efforts, limited taxation, and the unique task of preserving tribal culture and community.

In addition to land tenure, relations to aboriginal lands, and unique political status, the size of Native American communities also varies greatly from tribe to tribe. The Navajo Nation has more than 150,000 Native Americans living on-reservation, while several California Rancherias have no permanent residents (U.S. Department of Commerce 1984). Native American place of residence varies from the most urban to the most rural setting, with rapid population growth in all settings (Winchell 1987).

NATIVE AMERICAN GEOGRAPHIC LITERATURE

Native American communities and cultures have been a continuing topic of research in the development of the discipline. Historical and cultural approaches represent longstanding traditions within the field. Applied geographic research on Native American communities has also emerged as a major theme during the last two decades. Applied research can be categorized under land tenure and land use, environmental issues, housing, demography and migration, poverty, social geography, and Native

American education. A review of geographic research on Native Americans under these major headings constitutes the body of this chapter.

Historical Geography of Native Americans

Historical geography is a central theme within Native American geographical studies. Three foci of research, as with all cultural geography, have evolved: regional, ecological, and spatial. Each focus is supported with analysis of a number of native groups, but rarely have all three foci been fully researched for any one native group, nor have all native groups in Canada and the U.S. been studied. While an interest in Native American geographic research is evident in both countries, few historical geographers have applied their skills to this inviting task.

Discovering past regions, past ecological systems, and past spatial constructions of land, resources, and culture are historical themes as old as historical geography. Sauer and the "Berkeley School" challenged geographers to identify and analyze past patterns and processes of native groups. The native groups most often studied by these geographers were of Mexican or Middle American locale. Perhaps foreign field work was more attractive, or perhaps the land relationships of American Indians involved more complex social and political inter- and intra-group structures. Nevertheless, the past-landscape topics of culture area, social interaction, economic relations, land occupation and use patterns, and the processes that created these landscapes all invite explanation.

A few historical geographers have responded to this invitation, and their research contributes to the knowledge base within geography. Historical geographers of the current generation are focusing upon the U.S. and Canadian native groups. Recent research is varied, but can be assigned to the three foci:

Regional studies, either in cartographic or written form, contribute to the knowledge of group absolute and relative location and regional dynamics at various times in history (Allen and Turner 1988; Donley et al. 1979; Goodman 1982a; Greer et al. 1981; Hecht and Reeve 1981; Jett and Spencer 1981; Konrad 1981, 1987; Lewis 1987; McKee 1987; McKee and Schelenken 1980; Mitchell and Groves 1987; Pillsbury 1983; Ray and Heidenrich 1976; Rooney, Zelinsky, and Louder 1982; Williams and McAllister 1979).

Ecological studies have established the relations among land and resources, changes in environmental conditions, and associated changes in Indian subsistence patterns and the impact of American economies on Indian livelihood and culture (Aschmann 1974; Ballas 1985; Carlson 1975; Kay 1979, 1982, 1985a, 1985b; Ray 1972; Ray and Freeman 1978; Wishart 1979). Some of these studies have addressed continental-scale conditions involving more than one Indian group, while others have contributed to an in-depth knowledge of one specific Indian group.

Spatial studies which emphasize Indian groups' organization of space to meet cultural preferences are also found in Native American historical research. Past patterns of spatial interaction have been the focus of books by Ross and Moore (1987) and Sutton (1975), and journal articles by individuals (Ballas 1966, 1973, 1974, 1987; Goodwin 1977; Janke 1980, 1982, 1987; Sack 1986). Overall, these spatial investigations contribute to a growth in awareness of Indian landscapes, cultural adaptation and changing land-use patterns.

Cultural Geographic Research on Native North Americans

The modern literature on the cultural geography of Native North Americans is also closely allied with the earlier work of Sauer and his immediate intellectual offspring. Although Sauer and his protegees were particularly interested in Middle and South America, Sauer himself, as well as certain of his students—Homer Aschmann, Donald D. Brand, George F. Carter, Clinton R. Edwards, Fred B. Kniffin, Frederick J. Simoons, and Phillip L. Wagner—have made contributions to the geographical study of Native Americans north of Mexico, both pre- and post-Columbian. Many works have also been written over the years by scholars from other traditions (see Carlson 1972). There have also always been strong links between cultural geographers and anthropologists working on Native American topics—e.g., Franz Boas and his students, including Alfred L. Kroeber, Robert H. Lowie, and Clark Wissler earlier in the century (see Duncan 1980)—as well as with historians, sociologists, botanists, geologists, and other social and environmental scientists.

The growth of interest in Native North Americans on the part of cultural geographers has been manifested in recent years by the devotion of an entire issue of the *Journal of Cultural Geography* (Carlson 1982) to articles on U.S. and Canadian tribes, plus the appearance of the compendium text, *A Cultural Geography of North American Indians* (Ross and Moore 1987).

Culture, cultural origins and dispersals, and culture area are all themes integral to cultural-geographic studies of American Indians. Stephen Jett's work (1970) on the blowgun is an example of an origins-and-dispersal study involving not only North America but the world as a whole (see also Jett 1971, 1983). Jerry N. McDonald (1982) wrote of the "La Jicarrilla" culture area in early New Mexico, and David Hornbeck's (1982) article on Californian Indians epitomizes a regional study (see also Hornbeck and Fuller 1983). Regional interests of geographic research are currently concentrated on the greater Southwest, the Great Plains, the Southeast, and the northern forests and tundra.

The Native American cultural landscape, particularly with respect to settlement geography, has been an important object of geographic research. Traditional dwelling forms and their contrast with house types derived from Euroamerican culture have intrigued geographers, as have native rural-settlement patterns in broader terms. In recent decades research has included studies of the types, origins, evolutions, and diffusions of Navajo dwellings (Jett 1987; Jett and Spencer 1981; Spencer and Jett 1971); Navajo socioterritorial groups (Jett 1978a); Navajo seasonal migration patterns (Jett 1978b), and Navajo homestead situation and site (Jett 1980). Jett (Bohn and Jett 1977) has also studied prehistoric Puebloan architecture.

Eliot McIntire (1971) examined changing housing materials and spatial patterns of the Hopi villages, while settlement morphology of South Dakota reservations has received the attention of Ingolf Vogeler and Terry Simmons (1975). Phillip Wagner (1972) discussed the persistence of native settlement in coastal British Columbia, while Louis De Vorsey, Jr. (1966, 1971) used early maps to help reconstruct historical Southern Indian boundaries and landscapes. Within these studies one can detect not only the influence of the "Berkeley School" but also that of contemporary historical geographers and noncultural settlement geographers.

Nonsettlement aspects of native material culture have also received attention in the geographic literature. Topics include textiles, basketry, twig figurines, pottery, rupestrian art, and the blowgun complex (e.g., Curry-Roper 1982; Jett 1976, 1982, 1983, 1984, 1986, 1987; Jett and Moyle 1986).

Cultural ecology, particularly with respect to traditional subsistence and more recent changes in livelihood, is another focus. The ecological theme derives from the Sauerian man-and-the-land tradition and from the works of anthropologists, geomorphologists, botanists, and ethnobotanists. Among contemporary geographers Ballas (1962, 1973, 1985) has published on land use and subsistence among the Eastern Cherokees and the Sioux; Martha Henderson (1987) described 1930s land use on the Mescalero Apache Reservation; Jett (1977, 1979) dealt with Navajo fruit-tree-raising; and Moodie (1978, 1987a, 1987b) studied Canadian Indian subsistence, especially as related to agriculture and the fur trade.

Frederick Simoons (Johnson et al. 1977, 1978) has explored possible relations between culture and the geography of adult lactose malabsorption for tribes of the Southwest and the Great Basin, as well as the avoidance of fish as food in the Southwest (Simoons, Shoennfeld-Leber and Issel 1979). Jeanne Kay's writings (1979, 1982, 1985a, 1985b; and Albers and Kay 1987) include studies of Wisconsin Indian hunting and gathering patterns, and the impacts of the fur trade upon wildlife and upon Native American population. Arthur J. Ray has written (Ray and Freeman 1978) mainly about the historical Canadian fur trade, but has also dealt with native environmental adaptations (Ray 1972) and disease problems (Ray 1975, 1976).

Culture change and its geographic implications are additional emphases of cultural geography. McKee and Schelenken (1980) described the cultural evolution of the Choctaw, while Goodwin (1977) examined the changing culture and environment of pre-1776 Cherokee, and Pillsbury (1983) examined the acculturation history of the Cherokee in preremoval times. Joseph Manzo (1982a, 1982b, 1984, 1987) described the impacts of Indian removal, and Hewes (1978) discussed the occupation of the Cherokee country of Oklahoma. Cultural change among the Houma was reported by Curry-Roper (1987), and culture history and acculturation is a theme of much of Jett's opus (1971, 1978b, 1980, 1987; Jett and Spencer 1981; Spencer and Jett 1971). Donald Meinig's (1986) work on Atlantic America includes the colonial-period clash of Native American and European cultures and the results, continuing a tradition exemplified by his earlier books (1968, 1971).

Land Tenure and Land Use

Although cultural and historical research have been major foci of Native American geographic research that are directly related to tribal land issues and land tenure, specific land-use topics are presented separately. Sutton (1975), in his extensive study of Indian land tenure, stated:

> Past and present writings in the Indian field underscore an acknowledged fact: land continues to be the crux of the conflict between Indian and white in this country (p. x).

The complex issue of land tenure remains a critical point in understanding Native American issues, as seen in Sutton's more recent work (1985), and as demonstrated by Sam Hilliard (1971, 1972) and Richard Weil (1987).

Historically based research on land tenure has directly contributed to the more applied topics of land-use development and land-use planning on reservations. Several geographers have examined tribal planning issues, or have worked for tribes on land-use issues (Gilbert and Taylor 1966; Graf 1986; Goodman 1982b; Goodman and Thompson 1975; Gribb and Czerniak 1987; Henkel 1985; McKee and Murray 1986; Steiner 1966; Sutton 1967, 1970, 1975, 1976, 1982, 1985; Richardson 1979, 1980a, 1980b, 1983; Van Otten 1985, 1987; Weil 1984; and Winchell 1983, 1985). Early VISTA programs targeted reservations, and the training and development parts of these programs, which had an emphasis on agricultural self-sufficiency, was led by geographer Mayland Parker (1966, 1970). During the 1970s most tribes participated in the Department of Housing and Urban Development (HUD) 701 Comprehensive Planning Programs, and geographers were active in tribal planning (e.g., Richardson 1979, 1980a, 1980b; Winchell 1980).

More recently, tribal planning has emphasized the need for positive intergovernmental relations, recognizing tribal sovereignty and status as autonomous local governments (Knight 1985, 1988; Jose, Stark, and Winchell, in press). Solomon (1987) has actively pursued solutions to the problems of land tenure in the state of Washington through cooperative approaches to development of tribal lands in conjunction with non-Indian local governments.

Environmental Issues

Many tribes have timber, mining, or other resources which require extensive environmental information to develop and operate, as well as claims to water rights and operation of water and utility companies. Environmental-impact assessments, development of geographic databases, and ongoing research and monitoring of environmental quality and natural resources are rapidly expanding areas for tribal government which hold great potential for geographers.

Native American tribal governments are faced with the dilemmas of development decisions for large reserves of coal and other energy resources. The Council of Energy Resource Tribes was formed to assist tribes in these decisions. One major inventory work on coal resources (Richardson 1979) demonstrates an application of geographic database techniques to this issue, and was awarded the first Applied Geographers Specialty Group Award for excellence in applied research. Tribes having energy resources need increased revenues, but wish to avoid the disruption and environmental degradation of their lands. Extensive research in this area by Richardson (1979, 1980a, 1980b, 1983, 1984, and 1985) has directly assisted tribes in renegotiation of leases, development of regulations for control and development of resources, and protection of tribal sovereign powers. Policy decisions on development involve a broader range of social and economic impacts for tribal government (Goodman 1982b; White 1985; Winchell and Becker 1987), and these must be part of the tribal decision-making process.

Many reservation communities have only limited environmental base data or any kind of information on maps. A major contribution to tribal development are tribal-specific atlases such as Goodman's *The Navajo Atlas* (1982a) and Richardson's inventory (1979). Such works serve as integrative informational resources for tribes, and contribute to the potential for understanding appropriate development for tribes.

Housing

Geographic research on Native American housing has emphasized human-land relations and dwellings as cultural artifacts, as discussed previously under cultural research, along with the provision of housing in contemporary Native American communities. The need for adequate shelter and assessments of efforts by the Department of Housing and Urban Development and Indian communities to meet these needs have been the focus of this research (Snyder, Sadalla, and Stea 1976; Spencer 1985; Stanton 1977; Stea 1980, 1981, 1982, 1983a, 1983b; Stea and Buge 1980; Stea, Snyder, and Sadalla 1979, Stea and Wisner 1984a, 1984b; Winchell 1983). There is a need for "appropriate" housing solutions that are sensitive to the environment, to specific tribal cultures, and to the needs of individual homeowners (Winchell 1983, 808–818).

Demography and Migration

Geographers have published surprisingly little that is specifically on Native American demography and migration. Elaine Neils' study (1971) of urban migration is an important foundation for such research, describing Native American migration patterns based upon 1970 census data and records from the Bureau of Indian Affairs' urban-relocation program. There has been only limited follow-up on the impacts of Native American migration (Higgins 1982; Lazewski 1982), and even publication of descriptive analysis of U.S. Census data has been left to general articles (Goodman 1985). Bohland (1982) offers one of the few applications of urban models to Native Americans in his study of Indian residential segregation.

Recent works on demographics include studies by McIntire (1982, 1987) on Hopi population, Ross (1987) on Lumbee population growth, and Winchell (1987) on Indian population change in Arizona. William Denevan's text (1976) covers New World populations as a whole at the time of Columbus.

Poverty

A number of geographical articles have examined poverty on reservations and the political exploitation of Native Americans (Cummings and Harrison, 1972; Foraie and Dear, 1978; Flad 1972; Goodey 1970; Hickcox 1982; Jojola 1984; Stea and Wisner 1984a, 1984b; Stillwaggon 1984). While contributing to our understanding of contemporary communities, the philosophical debates that have developed over tribal or "fourth world" issues in relation to Marxist and neo-Marxist perspectives have not been enjoined (Churchill 1982; Churchill and La Duke 1985; Jorgensen 1971, 1972, 1978).

Social Geography: Philosophical and Methodological Approaches

Much of the nongeographic literature on Native Americans has examined the meaning of events within the unique cultural systems of Native Americans. Ethnicity, identity, and symbolic actions form important components of research that, although explored by geographers (Clarke, Ley, and Peach 1984; Ley 1974), has had few applications to Native American settings. David Saile's humanistic description (1985) of Tewa space and place, Barbara Weightman's assessment (1976) of Musquem settlement, and Dick Winchell's conceptual development (1982) of Yavapai space and place indicate that an examination of the unique meaning frameworks of spatial activity within a tribal culture offer important contexts for this type of research. Studies of urban Indians (Higgins 1982; Lazewski 1982) emphasize social as well as spatial contexts of migration.

Behavioral research is also limited, but Sonnenfeld's research (1982) on the spatial perception of Eskimos and Natives in Alaska demonstrates the need for further exploration of the topic. Special sensitivity to Native American communities is necessary for any research in the field, and includes special or appropriate research techniques (Geisler et al. 1982; Guyette 1983).

Native American Education

American Indian studies programs were formed in many colleges and universities in the 1960s and 1970s. These programs were generally different from Black studies, Chicano studies, or women's studies programs in that they were based in part on the unique political relations of tribes and tribal culture within the U.S. These programs (Heth and Guyette 1984) offer special courses, curricula, and degrees, and provide counseling support for Native American Indian students on campuses. They also help coordinate Native American interest and research across campuses.

Geographers have formed linkages to Native American studies programs through the development of special courses and programs that serve Native Americans and through the development of resource materials (Lipka 1987; McKee 1985; Goodman 1981, 1985), training workshops, and college courses (Van Otten and Swarts 1978). Several geographers have been active in the broader issue of Indian education (Van Otten and Narcho 1980; Winchell, Saffron, and Porter 1980, 1981) and its historic context (Wishart 1982).

THE FUTURE FOCUS OF NATIVE AMERICAN RESEARCH

The future and vitality of a subdiscipline can be gauged by its identification of important research problems. Comparison of the relatively small number of American geographers who conduct investigations pertaining to Native Americans to the large number of consequential problems that could be addressed indicates the magnitude of challenges to the Native American Specialty Group.

The purpose of the Native American Specialty Group of the Association of American Geographers is to "encourage and disseminate research contributing to an understanding of the problems, concerns, and issues of Native Americans (American Indians, Eskimos, Hawaiians)." From its origins at a special paper session at the Atlanta annual meeting (Lazewski 1973, 1977), the organization has increased in membership and level of activity, sponsoring over 20 paper sessions, field trips, publications, and a newsletter between 1977 and 1988. Early development of the specialty group included publication of a collection of papers (Lazewski and McDonald 1980) and a debate about which perspective should dominate, the cultural/historical or the contemporary (McIntire 1979, 1980; Lazewski 1980a, 1980b, 1980c). Presently, the importance of both efforts are recognized, and the specialty group seeks to encourage and promote all research in the field.

The following four broad areas of research as practiced by geographers have significance for the future: (1) cultural and historical interpretations, (2) contemporary Native American communities, (3) application of current and new technology, and (4) formulation of research methodologies.

Cultural and Historical Interpretations

Cultural and historical interpretations of Native American peoples have provided numerous topics for geographic research, yet the possibilities for additional work are virtually unlimited. There will always be a strong demand for the geographic perspective in both historical and cultural investigations of Native American questions. The challenge to correctly identify Indian patterns, geographic processes, and relations with the dominant culture is there for all historical and cultural geographers. Geographic interpretations supplement and complement research perspectives offered by other disciplines—such as anthropology, history, and sociology—that are more traditionally identified with Native American studies.

Geographers may also make important contributions to the maintenance of Indian self-awareness and the spatial component of ethnicity. An understanding of the spatial realm within which the tribe has formerly operated, and the relations of their culture and older lifeways to this contemporary community setting, becomes essential to the preservation of tribal heritage.

Contemporary Native American Communities

The study of contemporary Native American communities promises to be rewarding to geographers who are oriented toward planning endeavors and public-policy questions. Service to contemporary Native American communities is a primary focus of applied research conducted by Native American Specialty Group members. Native American tribes as political entities constantly face a number of complex and critical issues, such as economic development for the welfare of the tribe as a common group and for individual tribal members; the status of sovereignty and the struggle for self-determination; the preparation and administration of claims cases; and the monitoring of demographic conditions. Geographers can work directly for tribes or Indian organizations on these critical issues.

Successful tribal economic development is contingent upon a number of factors, especially the formulation of strategies that allow a tribe to manage its own affairs. Each tribe needs to develop information on its resources and potentials for development, and to train its members to direct the development of their resources through planning. Geographers can assist with both the collection and analysis of environmental data and the training of Native American personnel to perform these tasks. Geographic expertise has already been employed in the development and interpretation of laws, plans, and actions pertaining to resource development, but further contributions are needed.

Geographers can also contribute to the improvement of management strategies and the training of tribal personnel. Geographers need to design research that is sensitive to native attitudes, so that resource development and the modification of environment and native lifeways can be carried out in an appropriate manner as defined by the tribes.

Sovereignty of Native American tribes over their lands is a major topic for geographic investigations. Spatial autonomy and government varies from tribe to tribe. The unraveling of these patterns is essential to an understanding of the geographic qualities of Native American landscapes. Tribal administration is also a matter of management style that varies from traditional to progressive, but in all cases tribal governments have a much broader mission than local non-Indian governments. Tribal practices of government and intergovernmental relations constitute interesting points of inquiry for political geographers. The characteristics of Native American self-determination within the context of state and local interests and laws are also intriguing, and are integral to the basic aspects of economic development.

Geographers have been requested to assist in the adjudication of Native American claims for land, water, resources, and political autonomy against the federal and other levels of government. Since many of these claims are based on improper guidance from the federal government, the preparation for these actions constitutes a means of self-study and improvement of tribal decision-making. Corrections for past improprieties call for policy analysis and planning, which are major extensions of any geographic investigation.

Research on migration and population change is also important for tribes. Shifts in settlement patterns during the current century, especially the past several decades, reveal high population-growth rates and unique population characteristics related to health, welfare, and lifeway. Identifying the spatial patterns of these changes is a major task which has not been completed.

Application of Current and New Technology

The application of geotechniques, remote sensing, geographic information systems, and computer cartography to Native American resources and environments has been barely touched, despite the fact that many tribes have access to funding for such efforts. The development of electronic atlases for Native American land holdings on conventional GIS databases would be a first step in this process. Discipline-wide conventions and standardization of data systems would allow the interchange of data between researchers and Indian communities.

An untouched challenge to the Native American Specialty Group is the recruitment of Native American people to the ranks of geography. The development of special academic programs leading to professional degrees that utilize the skills and expertise of geography would be a major contribution to the discipline, as well as to the needs of the Native American. A task force headed by Jim Goodman has been established to begin work in this area. Especially important are educational programs in community planning, resource assessment, and resource management.

Formulation of Research Methodologies

Finally, perhaps the single greatest challenge to geographers is the formulation of appropriate research methodologies and paradigms within the context of tribal cultures, perceptions, and language structures. Since most traditional Indian languages are spoken and have no written format, basic collection of data must often be through oral interviews with non-English speakers. Nuances of translation may create problems on many subtle or even major points, and variation in thought and meaning patterns presented in the dialogue may not be properly examined using interviewers.

Special tools for adapting spatial concepts of Native Americans to geographic explanation also provide distinct challenges to the geographer who seeks to view the spatial realm of the Native American without the biases of the non-Indian world. Interdisciplinary efforts among anthropologists, sociologists, psychologists, and linguists can contribute to appropriate research designs.

CONCLUSION

There has been extensive research on Native American lands and people, but much remains to be done. The Native American Specialty Group represents a microcosm of the discipline, and our future depends to a great extent on attracting others to the challenges of historical and cultural research and to the research and educational needs of Native American communities. If our efforts are productive, the value of geography to these special issues will be demonstrated, along with development of a clearer understanding of spatial systems within unique and complex tribal cultures and communities.

REFERENCES

Albers, Patricia, and Kay, Jeanne. 1987. Sharing the land: A study in American Indian territoriality. In *A cultural geography of North American Indians,* eds. T. E. Ross and T. G. Moore, pp. 47–91. Boulder, CO: Westview Press.

Allen, James Paul, and Turner, Eugene James. 1988. *We the people: An atlas of America's ethnic diversity.* New York: Macmillan.

Aschmann, Homer. 1974. Environment and ecology in the "Northern Tonto" claim area. In *A study of Western Apache Indians, 1846–1886: Apache Indian V,* pp. 167–232. American Indian Ethnohistory Series. New York: Garland Publishing.

Ballas, Donald J. 1962. The livelihood of the Eastern Cherokees. *Journal of Geography* 61:342–50.

———. 1966. Geography and the American Indian. *Journal of Geography* 26:94–104.

———. 1973. Early agriculture and livestock raising among the Teton Dakota Indians. *Bulletin of the Illinois Geographical Society.* December: 53–62.

———. 1974. The land of the Rosebud Sioux. *Places* (July):43–46.

———. 1985. Changing ecology and land-use among the Teton Dakota Indians, 1680–1900. *Bulletin of the Illinois Geographical Society* 27:35–47.

———. 1987. Historical geography and American Indian development. In *A cultural geography of North American Indians,* eds. T. E. Ross and T. G. Moore, pp. 15–31. Boulder, CO: Westview Press.

Bohland, J. R. 1982. Indian residential segregation in the urban Southwest: 1970 and 1980. *Social Science Quarterly.* 63:749–61.

Bohn, Dave, and Jett, Stephen C. 1977. *House of three turkeys: Anasazi reboubt.* Santa Barbara, CA: Copra Press.

Carlson, A. 1972. A bibliography of the geographical literature on the American Indian, 1920–1971. *Professional Geographer* 24:258–63.

———. 1975. Spanish American acquisition of cropland within the Northern Pueblo Indian grants, New Mexico. *Ethnohistory* 22:95–110.

———, ed. 1982. *Journal of Cultural Geography* 2(2):1–134.

Churchill, Ward, ed. 1982. *Marxism and Native Americans.* Boston: South End Press.

———, and La Duke, Winoma. 1985. Radioactive colonization and the Native American. *Socialist Review* 81:95–119.

Clarke, C.; Ley, D; and Peach, C., eds. 1984. *Geography and ethnic pluralism.* London: Allen & Unwin.

Cummings, Harry, and Harrison, Melzetta. 1972. The American Indian: The poverty of assimilation. *Antipode* 2:55–60.

Curry-Roper, Janel M. 1982. Houma blow guns and baskets in the Mississippi River delta. *Journal of Cultural Geography* 2(2):13–23.

———. 1987. Cultural change and the Houma Indians:

A historical and ecological examination. In *A cultural geography of North American Indians,* eds. T. E. Ross and T. G. Moore, pp. 227–41. Boulder, CO: Westview Press.

De Vorsey, Louis, Jr. 1966. *The Indian boundary in the Southern Colonies, 1763–1775.* Chapel Hill, NC: University of North Carolina Press.

———. 1971. Early maps as a source in the reconstruction of Southern Indian landscapes. In *Symposium on Indians in the Old South: Red, white, and black,* ed. C. M. Hudson, pp. 12–30. Athens, GA: Southern Anthropological Society.

Denevan, William M., ed. 1976. *The native population of the Americas in 1492.* Madison, WI: University of Wisconsin Press.

Donley, Michael W. et al. 1979. *Atlas of California.* Culver City, CA: Pacific Book Center.

Duncan, James S. 1980. The superorganic in American cultural geography. *Annals of the Association of American Geographers.* 70:181–98.

Foraie, J., and Dear, M. 1978. The politics of discontent among Canadian Indians. *Antipode* 10(1):34–45.

Flad, Harvey. 1972. The urban Indians of Syracuse, New York: Human exploration of urban ethnic space. *Antipode* 4(2):88–99.

Geisler, Charles C. et al, eds. 1982. *Indian SIA: The social impact assessment of rapid resource development on native peoples.* Monograph No. 3. Ann Arbor, MI: School of Natural Resources, University of Michigan.

Gilbert, William H., and Taylor, John L. 1966. Indian land questions. *Arizona Law Review* 8:102–31.

Goodey, Brian. 1970. The role of the Indian in North Dakota's geography: Some propositions. *Antipode* 2:11–24.

Goodman, James M. 1981. The Native American Specialty Group, 1980–1981. *Transition* 11(2):34–35.

———. 1982a. *The Navajo atlas: Environments, resources, people, and history of the Dine Bikeyah.* Norman, OK: University of Oklahoma Press.

———. 1982b. Resource development and its significance to the future of the Navajo. *Journal of Cultural Geography* 2(2):101–10.

———. 1985. The Native American. In *Ethnicity in America,* ed. Jesse O. McKee, pp. 31–55. Dubuque, IA: Kendall/Hunt.

Goodman, James, and Thompson, Gary L. 1975. The Hopi-Navaho land dispute. *American Indian Law Review* 3:397–415.

Goodwin, G. C. 1977. *Cherokees in transition: A study of changing culture and environment prior to 1775.* University of Chicago Department of Geography Research Paper No. 181. Chicago: University of Chicago Press.

Graf, W. L. 1986. Fluvial erosion and federal public policy in the Navajo Nation. *Physical Geography* 7(2):97–115.

Greer, D. C., et al. 1981. *Atlas of Utah.* Provo, UT: Weber State College and Brigham Young University Press.

Gribb, W. J. and Czerniak, R. J. 1987. Land development decisions based on integrating the San Juan County Comprehensive Plan and computer mapping system. In *Papers and proceedings of Applied Geography Conferences, vol. 10,* eds. J. W. Frazier, et. al., pp. 62–68. Binghamton, NY: Department of Geography, Applied Geography Conferences.

Guyette, S. 1983. *Community based research: A handbook for Native Americans.* Los Angeles, CA: American Indian Studies Center, University of California, Los Angeles.

Hecht, Melvin E., and Reeve, Richard W. 1981. *The Arizona atlas.* Tucson, AZ: Office of Arid Land Studies, University of Arizona.

Henderson, Martha L. 1987. 1936 land use inventory of the Mescalero Apache Reservation: Prospectus for modern land uses. In *Papers and proceedings of the Applied Geography Conferences, vol. 10,* eds. J. W. Frazier, et. al., pp. 210–17. Binghamton, NY: Department of Geography, Applied Geography Conferences.

Henkel, R. 1985. Policy implications of non-Indian population growth on Indian land and resources in Arizona. In *Papers and proceedings of Applied Geography Conferences, vol. 8,* eds. J. W. Frazier, B. J. Epstein, and F. A. Schoolmaster, pp. 298–307. Binghamton, NY: Department of Geography, Applied Geography Conferences.

Heth, C., and Guyette, S. 1984. *Issues for the future of American Indian studies.* Los Angeles, CA: University of California, Los Angeles.

Hewes, Leslie. 1978. *Occupying the Cherokee country of Oklahoma.* Lincoln, NE: University of Nebraska Studies.

Hickcox, D. H. 1982. The Tongue River National Sacrifice Area. *Transition* 12(1):2–4.

Higgins, Brian. 1982. Urban Indians: Patterns and transformations. *Journal of Cultural Geography* 2(2):110–18.

Hilliard, S. B. 1971. Indian land cessions west of the Mississippi. *Journal of the West* 10:493–510.

———. 1972. Map Supplement Number 16: Indian land cessations. *Annals of the Association of American Geographers* 62(2):374.

Hornbeck, David. 1982. The California Indian before European contact. *Journal of Cultural Geography* 2(2):23–39.

———, and Fuller, David L. 1983. *California patterns: A geographical and historical atlas*. Palo Alto, CA: Mayfield Publishing.

Janke, R. A. 1980. Prehistoric origins of the Chippewa Indians. *Geographical Bulletin* 19:37–43.

———. 1982. Chippewa land losses. *Journal of Cultural Geography* 2(2):84–100.

———. 1987. The loss of Indian lands in Wisconsin, Montana and Arizona. In *A cultural geography of North American Indians,* eds. T. E. Ross and T. G. Moore, pp. 127–48. Boulder, CO: Westview Press.

Jett, Stephen C. 1970. The development and distribution of the blowgun. *Annals of the Association of American Geographers* 60:662–88.

———. 1971. Diffusion versus independent development: The bases of controversy. In *Man across the sea: Problems of pre-Columbian contacts,* eds. Carroll L. Riley et al, pp. 5–53. Austin, TX: University of Texas Press.

———. 1976. Pleasures afield on the Colorado Plateau. *Places* 3(2):8–10.

———. 1977. History of fruit tree raising among the Navajo. *Agricultural History* 51:681–701.

———. 1978a. Navajo seasonal migration patterns. *Kiva* 44(1):65–75.

———. 1978b. The origins of Navajo settlement patterns. *Annals of the Association of American Geographers* 68:351–62.

———. 1979. Peach cultivation and use among the Canyon de Chelly Navajo. *Economic Botany* 3:298–310.

———. 1980. The Navajo homestead: Situation and site. *Yearbook of the Association of Pacific Coast Geographers* 42:101–18.

———. 1982. War dogs in the Spanish expedition mural, Canyon del Muerto, Arizona. *Kiva* 46(4):273–80.

———. 1983. Commentary, "Houma blowguns and baskets: Further observations." *Journal of Cultural Geography* 4:126–28.

———. 1984. Making the Stars of Navajo Planeteria. *Kiva* 50(1):25–40.

———. 1986. Observations regarding Chartkoff's California Rock Feature Complex. *American Antiquity* 51:615–17.

———. 1987. Cultural fusion in Native-American architecture: The Navajo hogan. In *A cultural geography of North American Indians,* eds. T. E. Ross and T. G. Moore, pp. 243–58. Boulder, CO: Westview Press.

———, and Moyle, Peter B. 1986. The exotic origins of fishes depicted on prehistoric pottery from New Mexico. *American Antiquity* 51:688–720.

———, and Spencer, Virginia E. 1981. *Navajo architecture: Forms, history, distribution*. Tucson, AZ: University of Arizona Press.

Johnson, John D. et. al. 1977. Lactose malabsorption among the Pima Indians of Arizona. *Gastroenterology* 73(6):1299–1304.

———. 1978. Lactose metabsorption among adult Indians of the Great Basin and American Southwest. *American Journal of Clinical Nutrition* 31:381–87.

Jojola, T. S. 1984. The conflicting role of national governments in the tribal development process: Two case studies. *Antipode* 16(2):19–26.

Jorgensen, J. G. 1971. Indians and the metropolis. In *The American Indian in urban society,* eds. J. Waddell and O. M. Watson, pp. 67–113. Boston: Little, Brown.

———. 1972. *The sun dance religion: Power for the powerless*. Chicago: University of Chicago Press.

———. 1978. Energy, agriculture, and social science in the American West. In *Native Americans and energy development,* eds. J. G. Jorgensen et. al., pp. 3–16. Cambridge, MA: Anthropology Resource Center.

Jose, Cecil; Stark, Dan; and Winchell, Dick G. *Tribal planning and local intergovernmental relations*. Cheney, WA: Eastern Washington University Press. In press.

Kay, J. 1979. Wisconsin Indian hunting patterns, 1634–1836. *Annals of the Association of American Geographers* 69:402–18.

———. 1982. The ecological basis of Menominee ethnobotany. *Journal of Cultural Geography* 2(2):1–12.

———. 1985a. The fur trade and Native American population growth. *Ethnohistory* 31:265–87.

———. 1985b. Native Americans in the fur trade and

wildlife depletion. *Environmental Review* 9:118–30.

Knight, David B. 1985. Minorities and self-determination. In *Our geographic mosaic,* ed. David B. Knight, pp. 139–47. Ottawa: Carleton University Press.

———. 1988. Self-determination for indigenous people: The context for change. In *Nationalism, self-determination and political geography,* eds. David Knight and Eleonore Kofman, pp. 117–34. London: Croom Helm.

Konrad, V. A. 1981. An Iroquois frontier: The north shore of Lake Ontario during the late 17th century. *Journal of Historical Geography* 7:129–44.

———. 1987. The Iroquois return to their homeland: Military retreat or cultural adjustment. In *A cultural geography of North American Indians,* eds. T. E. Ross and T. G. Moore, pp. 191–211. Boulder, CO: Westview Press.

Lazewski, Tony. 1973. Geographic research and Native Americans. Unpublished paper. *Native Americans in contemporary society: Problems and issues special session.* Annual Meeting of the Association of American Geographers. Atlanta, Georgia, April 16.

———. 1977. Report of the committee on Native Americans. *Transition* 7(2):24–25.

———. 1980a. Native Americans and geography: Another view. *Transition* 10(1):23.

———. 1980b. News of the Native American Specialty Group. *Transition* 10(2):26.

———. 1980c. Report on the Native American Specialty Group. *Transition* 9(4):18.

———. 1982. American Indian, Puerto Rican and Black urbanization. *Journal of Cultural Geography* 2(2):119–34.

———, and McDonald, Jerry. 1980. *Geographical perspectives on Native Americans: Topics and resources.* Publication No. 1. Milwaukee, WI: Native American Specialty Group, Association of American Geographers.

Lewis, G. Malcolm. 1987. Indian delimitations of primary biogeographic regions. In *A cultural geography of North American Indians,* eds. T. E. Ross and T. G. Moore, pp. 93–104. Boulder, CO: Westview Press.

Ley, D. 1974. *The Black inner city as frontier outpost: Images and behavior of a Philadelphia neighborhood.* Monograph Series No. 18. Washington: Association of American Geographers.

Lipka, Jerry. 1987. The Alaska Native Claims Settlement Act: A land selection simulation. *Journal of Geography* 86:174–77.

Manzo, Joseph T. 1982a. The Indian pre-removal network. *Journal of Cultural Geography* 2(2):72–83.

———. 1982b. Native Americans, Euro-Americans: Some shared attitudes toward life in the prairies. *American Studies* 23:39–49.

———. 1984. Economic aspects of Indian removal. *Southeastern Geographer* 24(2):115–25.

———. 1987. Women in Indian removal. In *A cultural geography of North American Indians,* eds. T. E. Ross and T. G. Moore, pp. 213–26. Boulder, CO: Westview Press.

McDonald, Jerry N. 1982. La Jicarilla. *Journal of Cultural Geography* 2(2):40–57.

McIntire, E. G. 1971. Changing patterns of Hopi Indian settlement. *Annals of the Association of American Geographers* 61:510–21.

———. 1979. Native Americans and geography. *Transition* 9(3):20–22.

———. 1980. And a response (to Native Americans and geography). *Transition* 10(1):24.

———. 1982. First Mesa Hopi in 1900: A demographic reconstruction. *Journal of Cultural Geography* 2:58–71.

———. 1987. Early twentieth century Hopi population. In *A cultural geography of North American Indians,* eds. T. E. Ross and T. G. Moore, pp. 275–96. Boulder, CO: Westview Press.

McKee, Jesse O. 1985. *Ethnicity in contemporary America: A Geographical appraisal.* Dubuque, IA: Kendall/Hunt.

———. 1987. The Choctaw: Self-determination and socioeconomic development. In *A cultural geography of North American Indians,* eds. T. E. Ross and T. G. Moore, pp. 173–87. Boulder, CO: Westview Press.

———, and Schelenken, J. A. 1980. *The Choctaw: Cultural evolution of a Native American tribe.* Jackson, MS: University of Mississippi Press.

———, and Murray, Steve. 1986. Economic progress and development in the Mississippi band of Choctaw since 1945. In *After the removal: The Choctaw in Mississippi,* eds. Samuel J. Wells and Roseana Tubby, pp. 122–36. Jackson, MS: University of Mississippi Press.

Meinig, Donald W. 1968. *The great Columbia Plain: A*

historical geography, 1805–1910. Seattle, WA: University of Washington Press.

———. 1971. *Southwest: Three peoples in geographical change, 1600–1970*. New York: Oxford University Press.

———. 1986. *Atlantic America 1492–1800*. Vol. 1 of *The shaping of America: A geographic perspective on 500 years of history*. New Haven: Yale University Press.

Mitchell, Robert D., and Groves, Paul A. 1987. *North America: The historical geography of a changing continent*. Totowa, NJ: Rowman & Littlefield.

Moodie, D. Wayne. 1978. Agriculture and the fur trade. In *Old trails and new directions,* eds. Arthur J. Ray and Carol Judd, pp. 272–92. Toronto: University of Toronto Press.

———. 1987a. Indian Maps. Plate 59. *From the beginning to 1800*. Vol. 1 of *Historical atlas of Canada,* ed. R. Cole Harris. Toronto: University of Toronto Press.

———. 1987b. The trading post settlement of the Canadian Northwest, 1774–1821. *Journal of Historical Geography* 13:360–74.

Neils, Elaine. 1971. *Reservation to city, Indian migration and federal relocation*. University of Chicago Department of Geography Research Paper No. 131. Chicago: University of Chicago Press.

Parker, L. Mayland. 1966. *Training VISTAs for Indian reservations*. Program Grant, Department of Geography, Arizona State University, 1964–1966.

———. 1970. *Indian Community Action Program (ICAP) Training and Technical Assistance Program*. Department of Geography, Arizona State University, 1966–1970.

Pillsbury, Richard. 1983. The Europeanization of the Cherokee settlement landscape prior to removal: A Georgia case study. *Geoscience and man*. Vol. 23 of *Historical archaeology of the eastern United States: Papers from the R. J. Russell Symposium,* ed. R. W. Neuman, pp. 59–69. Baton Rouge, LA: School of Geoscience, Louisiana State University.

Ray, Arthur J. 1972. Indian adaptations to the forest-grassland boundary of Manitoba and Saskatchewan, 1650–1821. *Canadian Geographer* 9(2):85–98.

———. 1975. Smallpox: The epidemic of 1837–38. *Beaver* 306:8–13.

———. 1976. Diffusion of diseases in the western interior of Canada, 1830–1850. *Geographical Review* 66(2):139–57.

———, and Heidenrich, Conrad. 1976. *The early fur trades: A study in cultural interaction*. Toronto: McClelland & Stewart.

———, and Freeman, D. B. 1978. *Give us good measure: An economic analysis of relations between the Indians and the Hudson's Bay Company before 1763*. Toronto: University of Toronto Press.

Richardson, Douglas B. 1979. *The control and reclamation of surface mining on Indian lands*. Washington: Council of Energy Resource Tribes.

———. 1980a. The regulation of surface mining rights on American Indian lands: Some key policy issues. In *Proceedings of Applied Geography Conferences, vol. 3.* eds. John W. Frazier and Bart J. Epstein. pp. 127–41. Binghamton, NY: Department of Geography, Applied Geography Conferences.

———. 1980b. What happens after the lease is signed? *American Indian Journal of the Institute for the Development of Indian Law* 6(2):11–17.

———. 1983. *Environmental data inventory and assessment, coal bearing region of the Crow Indian Reservation and Ceded Area, Montana*. Billings, MT: GeoResearch, Inc.

———. 1984. *Complex mineral ownership patterns, An annotated bibliography*. Washington: GeoResearch, Inc.

———. 1985. *Wind energy and solar energy potential on the Northern Cheyenne Reservation*. Billings, MT: GeoResearch, Inc.

Rooney, John G.; Zelinsky, Wilbur; and Louder, Dean R., eds. 1982. *This remarkable continent: An atlas of United States and Canadian society and culture*. College Station, TX: Texas A & M University Press.

Ross, Thomas E. 1987. The Lumbees: Population growth of a non-reservation Indian tribe. In *A cultural geography of North American Indians,* eds. T. E. Ross and T. G. Moore, pp. 297–309. Boulder, CO: Westview Press.

———, and Moore, T. G., eds. 1987. *A cultural geography of North American Indians*. Boulder, CO: Westview Press.

Sack, Robert David. 1986. *Human territoriality: Its theory and history*. New York: Cambridge University Press.

Saile, David G. 1985. Many dwellings: Views of a Pueblo world. In *Dwelling, place and environment: Towards a phenomenology of person and world,* eds. David Seamon and Robert Mugerauer, pp. 159–81. Dordrecht, Holland: Martinus Nijhoff.

Simoons, Frederick J.; Schönfeld-Leber, Bärbel; and Issel, Helen L. 1979. Cultural deterrents to use of fish as human food. *Oceanus* 22:66–71.

Snyder, Peter Z.; Sadalla, Edward K.; and Stea, David. 1976. Socio-cultural modifications and user needs in Navajo housing. *Journal of Architectural Research* 5(3):4–9.

Solomon, Shirley. 1987. *Indian land tenure and economic development project: Phase I*. Seattle, WA: Northwest Renewable Resources Center.

Sonnenfeld, J. 1982. Egocentric perspectives on geographic orientation. *Annals of the Association of American Geographers* 72:68–76.

Spencer, Virginia E. 1985. *Population, poverty and housing condition of American Indians, Eskimos and Aleuts: 1980*. Seattle, WA: Housing Assistance Council.

———, and Jett, Stephen C. 1971. Navajo dwellings of Rural Black Creek Valley, Arizona-New Mexico. *Plateau* 43(4):133–42.

Stanton, T. H. 1977. *Trail of broken promises: An assessment of HUD's Indian housing programs*. Washington: Center for the Study of Responsive Law.

Stea, David. 1982. *Native American reservation housing*. Fourth World Studies in Planning. Los Angeles: University of California.

———. 1980. *Native American energy development and human settlement forms*. Fourth World Studies in Planning, No. 2. Los Angeles: University of California.

———. 1981. *Indian reservation housing: A brief summary of recent events, with bibliography*. Fourth World Studies in Planning, No. 14. Los Angeles: University of California.

———. 1983a. Indian reservation housing: Progress since the Stanton Report? In *An inquiry into critical perspectives in planning: Proceedings from the first regional conference on new perspectives on planning in the West,* ed. Joochul Kim, pp. 783–801. Tempe, AZ: College of Architecture and Environmental Design, Arizona State University.

———. 1983b. The sacred and the profane, again. *Transition* 13(3):16–19.

———, and Buge, C. 1980. *Cultural impact assessment on Native American reservations*. Fourth World Studies in Planning, No. 4. Los Angeles: University of California.

———; Snyder, P. Z.; and Sadalla, E. K. 1979. *Three papers on Navajo housing*. Fourth World Studies in Planning, No. 6. Los Angeles: University of California.

———, and Wisner, B. 1984a. Introduction. *Antipode* 16(2):3–12.

———, and ———, eds. 1984b. The Fourth World: A geography of indigenous struggles. *Antipode* 16(2).

Steiner, Rodney. 1966. Reserved lands and the supply of space for the Southern California metropolis. *Geographical Review* 56:344–62.

Stillwaggon, E. M. 1984. Anti-Indian agitation and economic interests. *Antipode* 16(2):13–18.

Sutton, Imre. 1967. Private property in land among reservation Indians in Southern California. *Yearbook of the Association of Pacific Coast Geographers* 29:69–89.

———. 1970. Dams and the environment. *Geographical Review* 60(1):128–29.

———. 1975. *Indian land tenure: Bibliographical essays and a guide to the literature*. New York: Clearwater Publishing.

———. 1976. Sovereign states and the changing definition of the Indian reservation. *Geographical Review* 66:281–95.

———. 1982. Indian land rights and the Sagebrush Rebellion. *Geographical Review* 72:357–59.

———. 1985. *Irredeemable America: The Indians' estate and land claims*. Albuquerque, NM: University of New Mexico Press.

U.S. Department of Commerce, Bureau of the Census. 1984. *Census of population: American Indian areas and Alaska native villages: 1980. Supplementary Report PC80-S1-13*. Washington: U.S. Government Printing Office.

Van Otten, George A. 1985. A geographer's perception of land use planning in Arizona's Native American reservations. In *Papers and proceedings of Applied Geography Conferences,* vol. 8, eds. J. W. Frazier, B. J. Epstein, and F. A. Schoolmaster, pp. 307–12. Binghamton, NY: Department of Geography, Applied Geography Conferences.

———. 1987. Economic development and land use planning for the San Lucy District of the Tohono O'odham Nation. In *Papers and proceedings of Applied Geography Conferences,* vol. 10, eds. J. W. Frazier et. al, pp. 169–174. Binghamton, NY: Department of Geography, Applied Geography Conferences.

———, and Narcho, Ruth J. 1980. Adult education and land use planning. *Journal of American Indian Education* 19(3):5–7.

———, and Swarts, S. W. 1978. Effective adult education in applied geography on the Navajo reservation. *Journal of Geography* 78:277–79.

Vogeler, Ingolf, and Simmons, T. 1975. Settlement morphography of the Indian reservations of South Dakota. *Yearbook of the Association of Pacific Coast Geographers* 37:91–108.

Wagner, Philip L. 1972. The persistence of native settlement in coastal British Columbia. In *Peoples of the living land: Geography of cultural diversity in British Columbia,* ed. Julian Minghi, pp. 13–27. Vancouver: University of British Columbia Geography Series.

Weightman, B. A. 1976. Indian social space: A case study of the Musquem band of Vancouver, British Columbia. *Canadian Geographer* 20:171–86.

Weil, Richard. 1984. Fragmented delivery systems: Law enforcement on two Indian reservations. *Association of North Dakota Geographers Bulletin* 34:32–43.

———. 1987. The loss of lands inside Indian reservations. In *A cultural geography of North American Indians,* eds. T. E. Ross and T. G. Moore, pp. 149–71. Boulder, CO: Westview Press.

White, Richard. 1985. Introduction to American Indians and the environment. *Environmental Review* 9:101–3.

Williams, Jerry L., and McAllister, Paul E. 1979. *New Mexico in maps*. Albuquerque, NM: University of New Mexico Press.

Winchell, D.G. 1980. *The Fort McDowell 701 comprehensive plan*. Fort McDowell, AZ: Fort McDowell Indian Community.

———. 1982. *Space and place of the Yavapai*. Unpublished Ph.D. diss. Department of Geography, Arizona State University.

———. 1983. HUD housing and appropriate housing at Fort McDowell. In *An inquiry into critical perspectives in planning: Proceedings from the first regional conference on new perspectives on planning in the West,* ed. Joochul Kim, pp. 802–21. Tempe, AZ: College of Architecture and Environmental Design, Arizona State University.

———. 1985. The cultural impacts of urban pressures on Indian lands: A Fort McDowell case study. In *Papers and proceedings of Applied Geography Conferences, vol. 8,* eds. John W. Frazier, B. J. Epstein, and F. A. Schoolmaster, pp. 313–19. Binghamton, NY: Department of Geography, Applied Geography Conferences.

———. 1987. American Indian population change in Arizona: An analysis of recent census data. *Journal of the Arizona-Nevada Academy of Science* 21:45–51.

———, and Becker, R. James. 1987. A basic needs analysis of Navajo energy development. In *Papers and proceedings of the Applied Geography Conferences, vol. 10,* eds. J. W. Frazier, et. al., pp. 368–76. Binghamton, NY: Department of Geography, Applied Geography Conferences.

———; Saffron, S.; and Porter, Robert N. 1980. Indian self-determination and the community college. *Community College Review* 19(3):17–23.

———; ———; and ———. 1981. Tribal management: A response to the vocational needs of Native American communities. *Journal of American Indian Education* 8(4):46–49.

Wishart, David J. 1979. The dispossession of the Pawnee. *Annals of the Association of American Geographers* 69:382–401.

———. 1982. Education, geography, and Indian assimilation, 1887-1933. *Journal of Geography* 81:204–10.

ANALYSIS AND MANAGEMENT OF SOCIETAL GROWTH AND CHANGE

Population Geography

Stephen E. White | Lawrence A. Brown |
William A. V. Clark | Patricia Gober | Richard Jones |
Kevin McHugh | Richard L. Morrill

Population geography emerged as a recognizable, systematic branch of geography in the early 1950s (Jones 1981). In his 1953 presidential address to the Association of American Geographers, Glenn Trewartha noted the virtual absence of population studies by American geographers and argued that a focus upon population should be essential to an understanding of areal differentiation. He observed that "Population is the point of reference from which all the other elements are observed and from which they all, singly and collectively, derive significance and meaning" (Trewartha 1953, 83). Shortly thereafter, studies of population geography appeared which focused upon distribution, growth, composition, and migration within the context of areal differentiation.

Many recent studies continue to provide descriptive assessments of regional variation for different population components. However, with increasing frequency, population geography researchers have employed quantitative methodologies designed to seek broad generalizations, test hypotheses, and model the complex relationships that explain population change. The emphasis upon areal differentiation has reached a plateau, while the focus upon spatial interaction has increased. In addition, the emphasis of behavioral perspectives upon population change has increased (Golledge 1980; White 1980a). Unlike demography, which has emphasized the empirical, statistical, and mathematical study of fertility, mortality, and migration, geographers have targeted migration as their primary topical interest. Of secondary frequency are studies that examine density and distribution, followed by those that assess the spatial variation of specific population characteristics. Even less work focuses upon mortality or fertility, despite the fact that these topics are always covered in population-geography textbooks.

A review of feature-length articles that appeared in the *Annals of the Association of American Geographers,* the *Geographical Review,* and *The Professional Geographer*

The authors wish to thank Professors Tom Kontuly and Curt Roseman and the editors for their excellent comments and suggestions.

over a 10-year period beginning January, 1978 provides several insights about the recent scope of population geography in the U.S. Whether or not works in these journals reflect a truly representative cross-section of recent population-geography research is certainly debatable. However, it is clear that Trewartha's hope that population would take a more central role in geographic understanding has been realized. Of the 769 articles surveyed, 77 (10%) focused upon population-geography topics. Of these, 53.3% emphasized migration, 30.6% examined changing patterns of population distribution, density, or urban growth, while 14.7% dealt with patterns of specific population characteristics. Seven articles (9.1%) were cast in an historic-settlement mode, while 10.4% discussed the relationships between population change and economic development. The areas of fertility, mortality, and population-estimating procedures were each represented by just one article.

Although the frequency with which geographers have employed inferential statistics and quantitative expression to model spatial processes has increased, slightly more than half of the articles were essentially descriptive or review, and used no statistics or mathematics beyond the simplest measures of central tendency or dispersion. However, those authors who used quantitative techniques rarely opted for basic descriptive statistics, preferring more high-powered analytical methods.

Forty-five studies (58.4%) were set in the U.S., while 41.6% focused upon population issues in other countries. About 65% of the articles focused upon population topics within one specific areal unit. Although 30.7% of the research was cast within an intraurban framework, the preferred scale of study was an areal unit larger than a city. The remaining studies examined linkages between areal units; 17.3% at the interregional scale, 8% on rural-urban change, and 5.3% on interurban movement.

The literature survey reflects a conservative estimate of the degree to which population is an important topic in geographic research. Studies that examined urban, political, and other topics, and which may have had population implications, were omitted if the primary focus was not population. In several studies, it was difficult to judge the authors' intention concerning the significance of population relative to other issues. Ironically, if Trewartha was right and population is a point of reference from which other elements derive significance and meaning, then it makes little sense to suggest that "population geography" is a well-defined subfield. Population's importance to geography may be that it is a prerequisite ingredient to meaningful synthesis in geography, rather than an isolated entity of geographic study. This may be even more true in the future if geographers begin to examine the impact of population dynamics on social change and public policy in greater detail.

Recognizing the difficulty of summarizing the key facets of population research in a limited space, the remainder of this chapter highlights six population themes that appear frequently in recent geographic research. The themes are organized by scale. Two sections review research that has been completed primarily at the intraurban level: *residential mobility* and *urban housing*. The third section examines the process of *counterurbanization,* while the following section is concerned with long-distance migration between cities and regions within a nation (*internal migration*). The last two sections have an international scope. Emphasis is placed upon geographic research that focuses upon *migration between countries* and the relationship between *population change and economic development.*

RESIDENTIAL MOBILITY

Geographical research on mobility has experienced many changes since Rossi's *Why Families Move* was published (1955). The research literature has been enriched by studies of residential decision-making, population flows within cities, and studies of the intersection of residential mobility and public policy. That residential mobility is an important issue in demography generally, and population geography specifically, is reflected in the more than 700 articles that have been published in the past decade (Clark 1982; Clark and Van Lierop 1986).

Residential mobility is a topic that has been dominated by the research of geographers. Although there have been important contributions by sociologist-demographers, and to a lesser extent by economists, most of the research contributions reflect activity by geographers whose research in residential mobility was stimulated by the behavioral-geography paradigm of the 1970s (Clark 1981).

In this part of the chapter, it is only possible to comment on some of the major articles and themes within residential mobility and to highlight the research directions that seem to be of greatest concern in future geographic research. A more comprehensive literature review is available in Clark (1982). For simplicity, a distinction is made between papers that are focused upon understanding and modeling of the actual decision-making process that underlies the residential-mobility process, and research that is focused upon the outcomes and context of residential mobility.

Theory and Decision-Making

Wolpert's initial research (1965) on stress and individual decision-making created a body of work which focused attention upon the dynamics of the mobility process. Likewise, Simmons' review (1968) brought a large number of sociological studies to the attention of geographers. The translation of Wolpert's ideas by Brown and Moore (1970) provided a basis for a series of model formations of stress and utility concepts. Although papers by Brummel (1979) and Huff and Clark (1978) took somewhat different directions and had different emphases in their attention to the decision-making process, they were concerned with evaluating that process. (This literature is reviewed in Clark 1983.) Most recently, research on the stress-inertia trade-offs has been enriched with work by Phillips and Carter (1984), who have provided specific parameter estimates for the stress-inertia models.

A parallel approach to explaining the decision to move focuses upon housing-disequilibrium models. These economic models (Quigley and Weinberg 1977; Hanushek and Quigley 1978) use a housing-costs approach to the stimulus to move. In an attempt to bridge the economic and more-general stress approaches to residential decision-making mobility, Brummell (1981) has incorporated housing-disequilibrium notions into utility trade-off models.

The notion of disequilibrium that actually underpins both stress and housing-costs models also has influenced the development of housing-search models. Drawing on the early work by Flowerdew (1976), a number of geographers have attempted to enlarge our understanding of housing search, and at the same time provide a theoretical understanding of search in relation to the relocation process. That research literature varies from individuals who have focused upon the specific search behavior of minorities (Lake 1981), to models of the search process (Smith et al. 1979), to specific models of spatial search (Huff 1986). Extensions of the work on residential search have suggested that the expected utility-theory notions used in the housing-disequilibrium models can be translated, with modification, to understand aspects of housing-market choice and selection (Clark and Smith 1982). However, an alternative approach has emphasized the nonnormative nature of search behavior (Meyer 1980).

Attempts to link mobility with neighborhood choice led to work in the area of sequential decision-making and residential selection (Clark and Van Lierop 1986). While data limitations have restricted the extent to which it has been possible to evaluate the usefulness of sequential-choice models for understanding residential mobility, neighborhood choice, and dwelling selection, they do seem to be an appropriate way of providing a context for the residential-selection process (Onaka and Clark 1983). At the same time, the focus upon such models has led to a series of research endeavors which have provided a methodology for simplifying the categorical models to provide what are termed "robust" models of residential mobility and tenure choice (Deurloo, Dieleman, and Clark 1987).

Residential Mobility and Urban Contexts

Although the research on residential decision-making has stimulated a rich literature and increased our understanding of the decision-making process, considerable dissatisfaction arose about the lack of a link between the decision-making and the wider social and political context within which that decision-making was set (Moore and Harris 1979; Moore and Clark 1980). This concern, coupled with a research tradition that has increasingly emphasized social processes, produced a series of papers that attempted to set mobility within the context of urban structural change.

In part, the development of this research tradition reflected the interest in residential mobility by British geographers, who argued convincingly that the context or the constraints on residential mobility had powerful impacts on the kinds of mobility that occurred. As Murie noted, "while the family's life-cycle indicates needs it does not imply that the housing system will distribute resources according to need" (Murie 1974, 14). The issue of how people have access to housing, and their ability to move within the housing-market context, is not just a function of characteristics of the household, but characteristics of how society provides housing to society. The British geographers emphasized the need to consider the residential-mobility process and its links with the housing market as an interrelated set of institutional arrangements. This emphasis raised or reiterated a debate between choice and constraint, which had been an element of the behavioral discussions of the late 1970s (Clark 1981).

The literature that arose in response to the emphasis on institutional constraints (explicit or implicit) is broad and wide-ranging. It includes the following:

> Emphases on the role of real-estate agents and financial agents as they influence residential mobility (Palm 1976b)
> Studies of the relative role of economic preferences and discrimination as influences on the mobility of minority households (Clark 1986)
> Studies of the role of specific institutional contexts (Murie 1974)
> The role of accessibility in influencing residential mobility (Clark and Burt 1980)
> Analyses of housing-market change (Moore 1978)
> Studies of residential conflict (Oliver and Johnson 1986)

Examinations of suburbanization (Gober 1986b)

Specific studies of elderly mobility (Wiseman and Roseman 1979; Golant 1980)

Upward mobility and residential mobility are inextricably meshed in views of societal change. Residential mobility is a way of achieving upward mobility, and increases in status produce residential mobility. Indeed, this view of a restless American society persists today, even though mobility is thought to be declining from its historically high levels (Long 1978). The notion of the disruptive impacts of rapid immigration is still debated in the sociological literature (South 1987), but the geographical research on residential mobility suggests that these negative views are misplaced. Work by Michelson (1977) and Fischer (1976) suggests that mobility is the natural outcome of the changing needs of households, and social consequences are not clearly negative as a result of mobility. Furthermore, the attachment to local environments may well remain strong even though mobility is relatively high. Certainly, there are no simplistic links between mobility and public outcomes.

Within the analyses of contextual impacts, the studies of policy influences are fewer and less straightforward in their results. While context (the amount of rental housing, the nature of the central city) does influence mobility, the policy impacts are less clear. Indeed, the major study of the relationship between mobility and housing—carried out by the Department of Housing and Urban Development in the early 1970s, mostly by economists and sociologists—concluded that there was only a weak relationship between mobility and housing-market interventions (the provision of additions to rent or subsidized housing), and that the relationships tend to be dominated by the more traditional sociodemographic variables. At the same time, a specific study of mobility in relationship to rent stabilization provides evidence that mobility does decline when there are direct housing controls that lead to substantial differentials between controlled and uncontrolled units (Clark and Heskin 1978).

Broad studies of mobility and public policy, encompassed in studies of neighborhood change, white flight, and analyses of household behavior in the light of social intervention (specifically via the intervention in school desegregation and busing programs), have provided a rich literature which establishes quite clearly that households do respond to these particular interventions. Lord and Catau (1976) were able to show that, in addition to a general suburbanization trend, a large number of relocation decisions of households were in response to the desegregation program.

Continuing Research Themes in Residential Mobility

The proliferation of research in residential mobility has generated two broad research areas—studies of decision-making and analyses of migration in urban contexts. With the increase in the availability of large data sets, it is likely that both of these themes will continue to dominate research in residential mobility. It is likely, too, that specific studies of minorities, women, the elderly, and their access to housing will be important ways in which both the studies of decision-making and the links of decision-making to the larger urban context will be developed. It is this latter issue—the links between the microscale decision-making studies and the macroscale studies of the changing built environment—that still waits to be developed.

URBAN HOUSING AND HOUSEHOLDS

An emerging area of concern in population geography is the way in which people organize themselves into households, and the way those households are distributed within urban areas. The study of households represents an important intersection of population and housing research. Household structure influences the demand for housing having particular space, location, tenure, and price attributes. The availability of housing in turn affects household size and composition, as for example when scarce and expensive housing forces people to take on roommates or remain with parents, rather than to maintain individual living units.

The nature of households nationwide underwent a dramatic transformation in recent decades. This transformation involved a decline in average household size, and a shift from households consisting of married couples with children to less-traditional types of families and nonfamilies. The number of persons living alone grew rapidly, both in absolute terms and as a relative proportion of all households. Demographers documented the phenomenon itself, as well as its demographic, social, and economic causes. Demographer/planner Dowell Myers is currently editing a state-of-the-art volume on the status of what he calls "housing demography" dealing with trends in household formation and composition, the relationships between household structure and housing choices, and the intraurban patterns of these relationships (Myers 1989).

Geographers have long-recognized the role of household structure on the propensity for and directionality of residential mobility. Less well-studied are the territorial outcomes of household change and the mobility process. These issues have emerged recently in the form of two strongly geographic research themes in housing: (1) how households of various sizes and compositions are distributed in urban space, and (2) how mobility changes (or maintains) the demographic structure of neighborhoods and communities.

Household Size and Composition

Studies of the distribution of intraurban households have been strongly influenced by the life-cycle view of neighborhood change. Hoover and Vernon (1959) first described the evolution of neighborhoods from single-family residences inhabited by young families with children, to a transition stage when apartments and other multiple-family quarters are subdivided to accommodate an increasingly poor and minority population, to a thinning-out stage in which density falls because household size declines. Cross-sectionally, this process creates a landscape in which household size increases with distance from the city center. Old housing near the city is occupied by small, nonfamily households, while large households with married couples gravitate toward the suburbs.

Until recently, empirical evidence tended to substantiate this view of the urban-household structure. Using data from the 1970 Census, Myers (1978) found a strong correlation between the ages of housing and population in the San Francisco Bay area. The resulting spatial pattern was one where large young households occupied new housing at the periphery, while smaller, older, immobile households concentrated in older housing near the city center. Gober (1980) found a very similar pattern in metropolitan Phoenix in 1970.

More recent evidence shows greater complexity in this pattern (Gober 1987). In Phoenix, the largest declines in average household size between 1970 and 1980 occurred at the periphery in suburban communities, pushing several suburbs below the central city in terms of average household size. Large suburban declines resulted from profound shifts in household structure toward childless couples, singles, and cohabitation. These shifts were related to the evolution of suburban business cores and apartment districts. Small, less family-oriented households were superimposed on areas heretofore dominated by married couples with children. This trend was consistent with Stapleton's national study (1982) in which primary individuals were more likely to live in suburbs in 1979 than in 1971.

Newton and Johnston (1976), Palm (1976a), and Gober (1986a) examined the pattern of household diversity and changes therein within urban areas. Newton and Johnston (1976) found high intra-area heterogeneity in household-structure correlates such as age and marital status of neighborhood residents in Christchurch, New Zealand. In Minneapolis and Des Moines, Palm (1976a) found older, well-established neighborhoods to contain a large mixture of households, whereas young suburban and inner-city neighborhoods house homogeneous household structures—young families with children in the first, and nonfamilies in the second.

The implication is that neighborhoods begin with a high degree of homogeneity, pass through a transitional period of relative heterogeneity as original residents are gradually replaced by a mix of households, and ultimately return to homogeneity—this time dominated by persons living alone and with unrelated individuals. Gober (1986a) examined the diversity of households at the census-tract level across 20 U.S. cities and found that households in 1980 were more likely to reside in proximity to those having different compositions. In contrast to Palm's earlier findings, the most genuinely diverse census tracts in 1980 were those having recently constructed single-family housing.

Mobility and Household Change

A related research theme involves the processes underlying neighborhood population change. Geographers and demographers have investigated the composition of migration streams in an attempt to assess whether and how intraurban mobility affects areal population characteristics. R. B. White (1982), for example, examined the role of intraurban mobility in altering household sizes between central cities and suburbs for the intervals 1965–70, 1970–75, and 1975–79. Families that migrated from central cities to suburban areas were larger in mean size than families that comprised the respective counterstreams. White concluded that central-city decline and suburban growth were accentuated by these opposing stream differentials.

Morrow-Jones (1988) reported similar results for the period 1978–83. Suburbs sent a higher percentage of divorced heads of households to the central city than they got back, and received a higher percentage of married couples and never-married persons than they sent. In general, the suburbs sent smaller households to the central city than they received in return.

Focusing upon a single suburb, Behr (1986) tried to assess the effect of residential mobility on population characteristics by comparing out-migrating households with subsequent residents of the same housing units. She found that mobility brought about little change in household composition in new neighborhoods that had large, expensive housing. In older neighborhoods containing small and relatively inexpensive housing, in-migrating households were younger, less well-to-do, and less likely to have children than out-movers.

While there can be no question that mobility alters the population characteristics of some neighborhoods, Gober (1986b) demonstrated that mobility can also serve to maintain an area's household structure. In Phoenix, little household change under conditions of high turnover occurred in older, multiple-family housing. Stability in household structure was achieved by what Moore (1972) calls "high through-put" of neighborhood residents in areas having inflexible, small apartments requiring high levels of mobility to maintain a population of young singles or childless marrieds.

Intense concern with the process of residential mobility and its spatial outcomes has caused us to undervalue the role of immobility on changing the household structure of an area. The characterization of this process as "aging in place" gives the false impression of a slow and gradual process, in contrast to rapid population change resulting from mobility. In a recent analysis of household change using Panel Study of Income Dynamics (PSID) data, Koo (1985) investigated the instability of households over a 13-month period beginning in 1979, and found that 19% were unstable. This figure is comparable in magnitude to the 20% of the population which change residences annually, and suggests that in-place household restructuring is at least as important as mobility in altering small-scale population characteristics.

Conclusions on Urban Housing and Households

The sheer magnitude and pace of small-scale population change has awakened geographers to the significance of societal shifts in household size and composition for urban areas. Moreover, significant spatial variation in household composition and changes therein portend monumental geographic ramifications in the urban demographic landscape. To understand contemporary changes in this landscape, we must learn to gain a more-accurate picture of how people organize themselves into residential units and the way that organization is linked to residential location.

Also needed are more innovative models of the interrelationships among households, housing, and location. The life cycle of the family and of neighborhoods is an increasingly outmoded view of the way households age and make residential decisions (Stapleton 1980). Population geographers must attach spatial references to various life-course changes, to capture the effects of changing household composition on residential location.

In addition to its theoretical contributions, geographical studies of housing have many practical applications. Understanding the precise linkages between households and housing enables real-estate analysts to forecast more accurately the demand for particular types of housing and enables planners to better predict small-area population change and school enrollments. Housing studies offer population geographers the opportunity to tackle an inherently geographic phenomenon having direct linkages to the study of urban housing, while at the same time making a meaningful contribution to the applied side of our discipline.

COUNTERURBANIZATION

A fundamental question of all geographic inquiry is the degree of concentration or dispersion of a phenomenon. Human settlement is arguably the key phenomenon of human geographic analysis. Thus the relative concentration or dispersion of population in settlements is the most obvious outcome of human behavior that population geographic theory should be able to explain.

Empirically, a survey of the evolution of human settlement reveals a rather continuous and inevitable concentration of population—first in the form of the burgeoning population of agricultural civilizations having favored physical environments and effective organizational structures; and second in the form of never-ending urbanization and metropolitanization (e.g., Berry 1981).

On the theoretical side, the metaprinciple is simply that, whether humans are pursuing goals of greater production/consumption or social interaction, the larger the concentration, the greater the value of output and interaction. These ideas have been operationalized through such specific concepts as returns to scale and economies of agglomeration. But the tyranny of space—the fact that the inherent productive potential of the environment to support people is incredibly dispersed—has always prevented the realization of that optimum "world city."

Much human ingenuity over the millennia has been aimed at overcoming space, and thus the amazing improvements in transportation and communications have seemed continually to "shrink the world" and to make possible ever-greater concentrations. In this unidirectional evolution, we perhaps have ignored the existence of other possible principles of human behavior, including the positive valuation of nature, and the positive valuation of people to the setting from which their very being evolved.

In any event, a major theme of modern population geography has been urbanization cum metropolitanization (Gottman 1961; Gibbs 1963; Williams 1983; Ledent 1985). People by the hundreds of millions in the last few centuries have shifted from rural to urban settings, in by far the largest human movement in history. While other social systems may have engendered a similar process of concentration, it seems that the evolution of capitalism, via an ever-greater social division of labor, almost inevitably required this geographic expression: the accumulation of capital was most efficient when the concentrations of labor became greater.

In much of the world—including the second, "newly industrializing" world, as well as the third world—the dominant processes continue to be urbanization, its concomitant rural-to-urban migration, and even the tendency toward extreme concentration in the "primate" or largest cities. Thus much population-geography research is and must be devoted to explaining urbanization, including the role of social, political, and economic structures, local-to-international (Rondinelli 1983).

But the title of this section is "counterurbanization." The very term seems to defy the pattern of evolution of human settlement generally, and under capitalism in particular. The term is recent and ill-defined, and not important in itself. Its meaning derives from the observation that population settlement seems not, after all, to be unidirectional toward ever greater concentrations. Again empirically, the nature and magnitude of suburbanization may be viewed as a limited form of deconcentration (Muller 1981). At a national scale, suburbanization is but a particular geographic expression of continuing metropolitan concentration. But at the labor-market scale, suburbanization is—in fact has long been—a form of settlement deconcentration, to which a variety of theoretical frameworks may apply.

Optimizing behavior of the firm in the face of increasing costs at the core, and taking advantage of changes in transportation, have encouraged local decentralization of some activities. Obviously, housing had to extend territorially as population grew. Its characterization as "suburban" depends upon the political/class/racial fragmentation of the metropolis, which is an outcome of competitive behavior under highly unequal power. Falling average population densities and the spillover of urban workers into the rural, exurban hinterland is another expression of deconcentration, one engendered by changes in transport technology and provision, and by government housing policy.

At the same time, suburbanization—in the sense of low-density, single-family settlements, and in the form of separate small jurisdictions—seemingly permitted the realization of goals of a more "natural" environment and a sense of control over and protection of one's property. It allowed some to escape the social and economic externalities of the central city, while still responding to the ultimate power of the firm's placement of jobs.

Just as there are bases for forms of recentralization at the national scale, so are there firms and social groups for which reclaiming and redeveloping inner urban areas may be optimal. Again, this probably has been enabled by transfers of real income from the less affluent to the more affluent in recent years.

Beginning even before 1970 in several of the more advanced urban countries, a degree of large-scale decentralization took place, characterized by the simultaneous absolute and relative decline of many large metropolitan areas and by a resurgence of growth in many rural small-town areas which had experienced long-term decline. So dramatic an apparent reversal of virtual "geographic laws of concentration" of course has been a major theme of contemporary research (e.g., Fielding 1982; Long and DeAre 1980; Vining and Kontuly 1978; Vining, Pallone, and Plane 1981; Kontuly and Vogelsang 1988; Zelinsky 1978).

Population deconcentration, as reflected in net movement from metropolitan to nonmetropolitan areas in the U.S. in the 1970s was first identified by Beale (1975). The nonmetropolitan turnaround received much attention because it ran counter to the prevailing expectation of continued metropolitan growth and expansion. In fact, some remained skeptical that the turnaround actually represented a "clean break" with past trends, arguing that it could be explained by metropolitan spillover (Gordon 1979). While much nonmetropolitan growth was in countries adjacent to metropolitan areas, it is clear that remote, rural counties in the U.S. registered substantial population gains during the 1970s (McCarthy and Morrison 1979).

Both economic and noneconomic explanations for the nonmetropolitan turnaround in the U.S. have been advanced. The broad conclusion from this research is that both employment growth in nonmetropolitan America as well as preferences for smaller places in attractive environmental settings underlie the turnaround (see Fuguitt 1985 for an excellent review).

Just as counterurbanization became accepted and population deconcentration trends were being extrapolated into the future, the patterns shifted back in the early 1980s, with metropolitan areas once again growing faster than nonmetropolitan areas!

This has stimulated another round of studies (Beale and Fuguitt 1985; Long and DeAre 1988; Frey 1988; Elo and Beale 1988), and has reinforced Zelinsky's observation (1977, 177) that we are "forever sprinting to catch up with unanticipated events we can never quite explain."

Much of the counterurbanization research may be characterized as pseudoscientific, since many came to view such decentralization as "good" or "bad" or "real" or "unreal," and tried to prove their prejudices. For example, the current dogma is that recentralization has reclaimed its rightful place, while others deny it. The theoretician of human behavior, who can peer beneath surface sentiment, should not be surprised at the changes or at cycles of concentration or deconcentration. The very economic geographic theory of concentration via returns to scale and agglomeration always implied the far end of the curve—that of diseconomies from costs of congestion, excess travel, generation of unavoidable externalities, and the like. Remember that in probably all societies, and certainly in capitalist ones, firms make the basic location decisions. Thus under certain social, economic, and political conditions, it simply became more profitable to shift some kinds of activities to regions and countries having lower costs and offering higher returns. Under other conditions, core areas become more profitable.

But at the same time, significant numbers of people were able or determined to ignore the controlling power firms, by retiring or accepting less-than-maximum wages to live in preferred environments or settings, perhaps where they grew up. As we began to better-understand human behavior, we often found that such nonmetropolitan locations really did maximize the utility and well-being of some people. And, the simplest ideas of comparative advantage, expressed through prices and costs of exploiting particular mineral or recreational resources, played an obvious if short-term role in other forms of rural and small-town resurgence.

Similarly, the "recentralization" of the 1980s, exemplified by the growth of New York and Los Angeles, should not be viewed as a return to some empirical law or even as capitalist inevitability, but should be analyzed via theory of human behavior. While more firms found it more profitable to leave than to enter many large cities in the 1970s, the balance reversed in the 1980s. Perhaps the shift reflected changes in industrial and occupational structure, which have in turn accompanied the large-scale shift in real income to the more affluent. And perhaps the shift reflected deregulation, and/or a change in paying for the external costs of greater urban concentration from firms to workers and governments. It is to these processes and their stability that we must look to assess future changes in the balance of centralization and dispersion of people.

Around the world, as across the landscape of most countries, one can see evidence of the human processes of firm location, relocation, and restructuring, and of household migration, refereed by public policies. It is evidenced as settlement change, which at certain times and at certain scales looks like continuing concentration, but at other times or scales suggests counterurbanization. Extrapolation of short-term trends may be popular, but it is irresponsible: the population geographer can and should go behind and beneath such outcomes to examine the processes that lead to these results.

INTERNAL MIGRATION

Internal migration has received increasing attention over the last few decades because it now represents the principal mechanism by which population and labor are redistributed within most developed countries of the world. The following is a brief overview of major themes in contemporary migration research by geographers, although some works by nongeographers are cited. Emphasis is on macrolevel and behavioral approaches to internal migration research in the U.S.

Macrolevel Approaches

Macrolevel studies seek to describe and explain patterns of migration for population aggregates and geographic areas. Perhaps the most distinguishing characteristic of work by geographers is a concern for describing and explaining place-to-place migration flows. Much of the work on migration flows is descriptive, with recent research focusing upon interregional migration from core to peripheral regions, and migration between metropolitan and nonmetropolitan areas.

The U.S. was the first developed nation to experience net migration from the traditional core region (Northeast and industrial Midwest) to peripheral regions (South and West). Net flows from core to periphery greatly accelerated in the 1970s, leading to (1) work that documented patterns of interregional migration in the U.S. (Berry and Dahmann 1977; Rogerson and Plane 1985), (2) discussion of the underlying causal forces, such as changing patterns of industrial location and environmental amenities (Vining 1982; Plane 1984), and (3) examination of core-periphery migration trends in other developed nations (Vining and Pallone 1982).

Geographers have also examined patterns of interstate migration in the context of regional restructuring and core-periphery relations (Plane and Isserman 1983; Plane 1984; Morrill 1988). The overall conclusion of these studies is that patterns of population redistribution within the U.S. are more complex than the simple core-periphery dichotomy advanced by Vining. Morrill (1988), for example, identifies eight migration regions based upon a cluster analysis of the demographic effectiveness of 2450 net migration exchanges among the 50 states.

Over the last two decades, geographers have increasingly turned to analytic modeling of migration flows. This work largely falls within two traditions (Rogerson 1984). The first is a demographic accounting tradition, in which the focus is upon forecasting future migration patterns on the basis of previous patterns. This ranges from simple Markov models to sophisticated multiregional demographic analyses advanced by Rogers and his associates (Rogers and Willekens 1986). These approaches are informative for descriptive purposes and provide reasonably good predictions in the short run, but they do not "explain" observed migration patterns (Rogerson 1984).

A second modeling tradition seeks to explain migration flows based upon origin and destination characteristics and relational measures of separation, typically distance. Seeking to identify the determinants of migration, this tradition includes spatial-interaction models (Wilson 1971; Alonso 1978; Haynes and Fotheringham 1984), and migration models derived from economic theory (Borts and Stein 1964; Sjaastad 1962).

A recent contribution in spatial-interaction modeling is the work by Fotheringham (1983, 1986) on the theory of competing destinations. He shows that gravity models are misspecified because they do not include a measure of the relationship between interaction and competition between destinations. Distance parameters for a given origin are biased downward (less negative) if destinations are highly accessible, and biased upward (more negative) if destinations are inaccessible. Including a measure of competing destinations in the model (i.e., destination accessibility) controls for this spatial structure effect and produces unbiased distance parameters.

There have also been recent advances in our understanding of the economic determinants of migration. A paper by Clark and Ballard (1980) has been influential. Using migration from Appalachia (1958–75) as a case study, they develop and test a two-stage model that separates the decision to leave (Appalachian region) and the decision of where to move (destination selection). Their primary contribution is that in separating the migration process into two stages, they show that economic conditions at origin are significant predictors of out-migration. Previous aggregate-level studies, following the work of Lowry (1966) and others, had found economic factors at origin to be unimportant in explaining migration flows.

A prominent research direction is in modeling temporal change in migration. Plane and Rogerson have advanced several methods for modeling temporal change in migration-flow matrices, including information theoretic principles, lag structures, and a causative-matrix approach (Rogerson and Plane 1984; Plane and Rogerson 1986). In addition, Plane (1987) has adapted shift-share analysis to the examination of change in migration systems. Overall, this line of research highlights the dynamic nature of interregional migration and the importance of modeling origin-destination interdependencies.

Modeling temporal change is also evident in work by Rogerson and Plane on developing nonstationary forecasting models in which migration flows evolve through time as economic conditions change (Plane and Rogerson 1982; Rogerson 1984). The heart of this research is in explicating the role of past migration flows in current migration, and processes by which flows change over time. This work holds considerable potential because it synthesizes Markovian and economic approaches in migration.

The role of information flows and vacancy chains has also been emphasized in modeling interregional migration. Rogerson and MacKinnon (1982), for example, specify a model in which migrants react to the relative attractiveness of destinations, based upon information about job vacancies transmitted through both employers-agencies and personal communication. Some interesting implications of the model are demonstrated via a simulation exercise, but model verification is not possible because requisite data are not available.

Behavioral Studies

Two characteristics distinguish behavioral studies in migration. First, the focus is upon the behavior of individuals and households, rather than population aggregates. A second characteristic is a concern for decision-making aspects of migration, such as the role of dissatisfaction in stimulating moving considerations, information gathering and search, evaluation of potential destinations, and so on.

There is considerable overlap and cross-fertilization between behavioral studies in intraurban mobility and those addressing migration. Rossi (1955), Wolpert (1965), and Brown and Moore (1970) developed early frameworks incorporating stress and utility concepts; these seminal works have been influential in research on intraurban mobility and migration.

One of the thrusts in behavioral research has been to turn more explicitly to psychological theory, methodology, and measurement (Lieber 1978; White 1980b, 1981; DeJong and Fawcett 1981; Haberkorn 1981; Sell and DeJong 1983; Desbarats 1983; McHugh 1984). The stereotypical view of an individual becoming dissatisfied with the community, evaluating alternatives, formulating migration plans, and subsequently moving is only one of many paths in the migration decision process.

Behavioral models have also examined the role of situational and contextual factors in shaping migration decisions and circumscribing behavior (DeJong and Fawcett 1981; Desbarats 1983; Sell 1983; Landale and Guest 1985). Much more work needs to be done in this arena, but results to date have shown that socioeconomic characteristics of migrants and the spatial structure of opportunities influence migration indirectly by shaping the formation of migration desires and expectations. Situational and contextual variables may also influence behavior directly by blocking anticipated moves, facilitating unanticipated moves, and forcing relocations.

A third area of behavioral research is examining the role of place ties in migration. It has long been recognized that social and economic ties in a community, often measured by duration of residence, hold people in place (Lansing and Mueller 1967; Speare, Kobrin, and Kingkade 1982). Recent studies have examined the role of ties elsewhere in the migration process, including ties established via previous residence, family and friends, repeated vacations, and business connections (McHugh 1984; Roseman and Oldakowski 1984). Results show that place ties developed over the life course serve to define and limit potential destinations. Most migrants consider only one or two destinations, often places where they have established ties.

Linkages between place ties and return migration have received attention. Using a national sample of migrants, DaVanzo and Morrison (1981) show that the propensity to return declines with time, reflecting the depreciation of place ties over time. They also found that the propensity for return is higher if the former residence is where the person grew up, presumably because people tend to have stronger ties to the place of their upbringing.

Place ties and returning "home" are the focus in White's examination (1983, 1987) of the recent upsurge in migration to eastern Kentucky. He argues that the stem-family migration system is still operating, with return to Appalachia being attributable to strong sociocultural ties. Place ties and return migration are also central in studies examining the South's shift to net in-migration of blacks after almost a century of net black movement to other regions (Johnson and Brunn 1980; McHugh 1987; Cromartie and Stack 1987).

Research Directions in Internal Migration

The macrolevel and behavioral traditions in migration are well established; both will continue to contribute to our understanding of migration patterns and processes. Yet, we are cognizant of our limited ability to explain, let alone to predict

. . . how many of what sorts of people will go where, in what patterns of flow, and when, for what reasons and with what effects upon places of origin and destination, and upon the entire social system of which they form a part. (Zelinsky 1983, 20)

There are three research directions that will help accomplish this goal. First, greater effort in linking macro and micro approaches in migration studies is warranted. This includes incorporating more-realistic behavioral assumptions and processes in aggregate-level models, and couching behavioral studies within the larger socioeconomic and geographic context. Such linkages have proven difficult, but the potential rewards are great.

Second, sweeping demographic changes in our society point to the need of more migration research on population subgroups, such as the elderly and other age cohorts, women, family and household types, minorities, and ethnic groups. To what degree do the patterns and determinants of migration vary across segments of the population? And what are the impacts and consequences of migration occurring in conjunction with demographic shifts?

Third, we should broaden our definition of migration, and rely less upon conventional government migration statistics. Greater use of longitudinal and specialized surveys indicates a trend in this direction, but we have only begun to scratch the surface in identifying and explaining human movement at various dimensions of space and time.

INTERNATIONAL MIGRATION

International migration is one of the more pressing and salient social phenomena of our time. In comparison to the "immigrant century," international migration today is, on the average, shorter-distance and more temporary; it cuts across steeper income gradients between origin and host countries; and it has a much larger component of political refugees. Since 1980, the general world recession has sharpened international economic differentials, making emigration more compelling for sending societies, and its impacts more acute for host societies. Huge new international refugee populations—Afghans, Salvadorans, Angolans—have changed the world migration map and created major problems for new host societies, many of which are themselves underdeveloped.

Several major theoretical statements have furthered our understanding of international migration. These have included explanations of the economic inevitability of international wage-labor migration (Piore 1979) and the long-run economic and social consequences it entails (Bohning 1981; Kritz 1983; Kritz and Keely 1981). The involvement of American geographers in these trends has been relatively slight, but the situation is changing. A sample of three recent works (Kritz and Keely 1981; Kritz 1983; *International Migration Review* 1986) reveals only two geographers out of 58 contributors. In this sample, demographers and sociologists dominate, followed by political scientists and economists.

Nevertheless, the spatial approach has stimulated considerable interest within the other disciplines. In a recent volume on undocumented migration within the U.S. (Jones 1984), only three of eleven authors were geographers, but the topics were all spatial and comparative. Many sociologists, political scientists, and economists recognize that entire countries are too macroscale for understanding the structural forces, timing, and impacts of migration. In addition, the many single-village studies that exist have not provided a comprehensive, comparative picture of international migration streams. Therefore, spatial studies are now being undertaken by nongeographers.

Spatial Patterns of International Migrations

Population geographers have contributed theoretical perspectives as well as empirical studies of both wage-labor and refugee migration. Regarding theoretical perspectives, Zelinsky (1971), Brown and Sanders (1981), and Conway (1986) have provided the most notable contributions to our understanding of the migration process. Zelinksy, Brown, and Sanders argue that migration patterns change as a country passes through its mobility transition:

> In the early-transitional period, internal migration is primarily rural-to-urban and emigration is prominent, both guided by kinship ties and push factors at the origin.
> In the late-transitional society, urban-to-urban migration equals rural-to-urban, push-and-pull factors are in equilibrium, and emigration is declining.
> In the advanced society, urban-to-urban and intraurban migration dominate, destination-pull factors are pronounced, and emigration is slight.

The study of Jones and Brown (1985) provides empirical proof for the Brown/Sanders model, across a set of developing countries. Conway (1986) applies the concept of circulation (short-term, repetitious moves with a lack of declared intention of permanent change) to distinguish Caribbean emigration from more permanent flows. The circulation concept has also been presented by Roseman (1985), in a topology of multiple residence applied to internal migration in developed societies.

Wage-labor migration studies with a spatial or comparative perspective have recently been conducted by European geographers (e.g., Salt 1981), as well as by nongeographers such as Kritz (1981) on the Caribbean; Ling (1984) and Birks, Seccombe, and Sinclair (1986) on the Middle East; Stahl and Arnold (1986) on Asia; and Bustamante (1977), Reichert and Massey (1979), and Mines (1982) on Mexico. These studies are characterized by a concern for differences between and within sending countries in the magnitude of migrant remittances, the local impacts of these remittances, repercussions of emigration in general, and the microscale factors that generate the migration.

American geographers have just begun to contribute to international wage-labor migration studies. Research on Mexico-U.S. migration has uncovered (1) a high degree of spatial concentration and flow channelization (Jones 1982a, 1982b, 1988; Dagodag 1975; Guitterrez 1984); (2) increasing spatial dispersion, both of origins and destinations over time (Jones 1982b; Jones, Harris, and Valdez 1984); (3) differences in sending forces, timing, and distance of migration for Mexican interior-versus-border sending areas (Jones and Murray 1986; Jones 1988); and (4) intraurban concentration of undocumented migrations in traditional downtown manufacturing/warehousing/low-rent districts (Valdez and Jones 1984; Dagodag 1984). This research suggests, indeed, a complex pattern of flows and motives, rather than a consistent, generalizable one.

Refugee-migration studies by geographers are even more recent. They tend to be more empirical and persuasively oriented than the wage-labor studies, in keeping with their immediate social relevance. Spatial refugee patterns have been the focus of several as yet unpublished pieces. Sechrist (1987) described the diffusion of the Sanctuary movement in the U.S., from its origin in 1985 in Arizona, to California, Illinois, and New York in 1986, and hence to the late-adopter states such as Alabama, Georgia, and North Carolina in 1987. Jones (1987) attempts to sort out the economic and political motivations of Salvadoran refugees in the U.S. by correlating the spatial incidence of political violence, economic setbacks, and departmental origins of Salvadorans apprehended by the U.S. Immigration Service over the period 1980–85, finding that the U.S. migration is more closely related to economic setbacks than to political violence. McElroy's analysis (1987) of refugee migration flows out of El Salvador reveals social and political barriers in Guatemala which channel many migrants northward to Belize.

Impacts on Sending Areas

This research area has interested anthropologists, sociologists, and economists for the last decade. Much of this research has sprung from a dependency-theory perspective, focusing upon negative impacts such as local disinvestment, inequality, and the deterioration of traditional social institutions (e.g., Swanson 1979; Brana-Shute and Brana-Shute 1982; Fergany 1982; Cobbe 1982). Geographers writing within this perspective have perceived similar problems. Shrestha (1985), for example, notes that underdevelopment in modern Nepal stems partly from the British practice of recruiting the most able young men for the British Army.

Conway's research (1985) presents a different view, in the case of the British Caribbean. He grants that external dependence and decline of the traditional economy (sugar cane) do indeed result from emigration to the U.S. and other regions. However, these are more than offset by increases in foreign-exchange earnings vital for capital goods imports, by better quality of life at the family level, and by the enabling of a portion of the population to leave dead-end, often hopeless jobs in agriculture. Conway's findings for the Caribbean are supported in a study in rural Zacatecas (Jones 1988). Data from a sample of 302 households in Villanueva municipio indicate that after 1–4 years of U.S. migration, migrant household incomes exceed those of resident nonmigrants, a trend even more accentuated after 5 years of migrant experience. Peasant families with little chance of securing local urban jobs are able with U.S. migration to significantly improve not only their homes and household possessions, but also to invest in agricultural inputs and in human capital such as medical care and their children's education.

Impacts on Migrants and Host Societies

In the last decade this topic has been of great interest to economists, and to a lesser degree to sociologists and anthropologists (Smith and Newman 1977; Grossman 1984; Strand 1984; Bean, Lowell, and Taylor 1986). Geographers working on this topic include Desbarats (1986), Symanski (1986), and Jones (1985). Desbarats, in a number of recent articles (e.g., 1986), argues for greater attention by policymakers to the plight of Indochinese refugees. She finds that Laotians and Cambodians (in contrast to Vietnamese) in the U.S. have faced a decade of underclass status only partly explicable in terms of their more deprived demographic and economic backgrounds. A deeper explanation is found in their identification with a more "passive" form of Buddhism and in the physical and psychological effects of their privations vis-à-vis the Communist takeover and subsequent programs of mass ruralization and expulsion.

Symanski (1986) traces the privations of Salvadorans in bordering Mexico and Texas, through the combined techniques of interviews and participant observation. As does Desbarats, he raises questions of how U.S. policy treats the lowest rung of migrant minorities. How are their rights preserved, their opportunities increased? Finally, Jones (1985), in an analysis of monthly time-series data on INS apprehensions, unemployment, and wage data for a sample of southwestern U.S. cities, finds no consistent evidence that recent undocumented migration increases unemployment or depresses wages in this region.

Challenges in the Study of International Migration

Two trends in geographic research on international migration have occurred in the last decade: (1) a strong spatial focus centering upon models of circulation, flow patterns, and changing types of migration over time, and (2) an emerging concern with consequences to migrants, to origin societies, and to destination societies. One challenge facing population geographers is to become more involved in the pressing social questions facing developing as well as developed societies. In the context of international migration, several questions may be mentioned:

- ☐ What is the impact of government immigration policy on specific origins and destinations in sending and receiving countries?
- ☐ How does emigration fit into a portfolio of income-generating options available to governments in less-developed countries?
- ☐ Why are some ethnic immigrant groups assimilating much slower than others, and how does this phenomenon vary over space?
- ☐ How are cyclic trends in climate, international debt, recession, political unrest, and wars related to patterns of international migration over longer periods of time?

Despite the lack of adequate models, adequate data, and adequate funding, we should be addressing these problems.

POPULATION AND DEVELOPMENT

A mainstay of introductory courses in human geography is demographic-transition theory (e.g., Jackson and Hudman 1986, ch. 3). Based on the economic history of Europe and North America, this hypothesizes that fertility and mortality both decline with development, but at different rates. A four-phase framework is common: high fertility and mortality, identified with traditional societies; high fertility and declining mortality, identified with low-income developing nations; declining fertility and low mortality, identified with middle-income developing nations; and low fertility and mortality, identified with high-income developed nations.

An ancillary issue is the age distribution of a population. A high percentage at ages 15 or less (45% is not uncommon!) typifies developing nations, rendering a triangular-shaped population pyramid, while the population of developed nations is more evenly distributed among age cohorts, rendering a more bullet-shaped population pyramid.

National variations in fertility, mortality, and age distribution also are mirrored within a country, according to development level and/or rural-urban distinctions.

Because pronounced spatial variation typifies the *natural increase* portion of the demographic equation (Shryock and Siegel 1976), this is an obvious topic for geographic research. But geographers have largely confined themselves to descriptive, atlas-type studies. Virtually ignored have been the processes underlying spatial variation in fertility and mortality, between and within nations and at the individual and aggregate levels of analysis. This is, then, an important topic for future research.

A third component of the demographic equation, in addition to fertility and mortality, is *migration,* the primary concern of geographers studying population and development. It is discussed here as three topics: (1) development-migration interrelationships in previous research, (2) development paradigms of migration, and (3) beyond conventional thinking.

Development-Migration Interrelationships in Previous Research

Population geographers and other social scientists frequently assume that migration is governed by laws that apply across geographic settings and points in time, a tradition dating at least to Ravenstein (1885, 1889). The search for universality is most-recently exemplified by statistical formulations, rooted in neoclassical economics, that account for aggregate migration flows through variables pertaining to wage rates, job opportunities, amenity levels, migration costs, and information flows. This model was originally articulated for developed-world settings such as the U.S. (Lowry 1966), but applied with minimal modification to areas as diverse as Costa Rica, Peru, Colombia, Mexico, Venezuela, Ghana, Kenya, Egypt, India, and Taiwan.

Although conventional modeling emphasizes spatial variations in wage levels and job opportunities, an equally important component is information flows between places, which communicate economic differentials to prospective migrants. Gravity-model variables or distance between origin and destination are commonly used as surrogates. Given concern for development influences, however, particular note should be made of the *migration chain* effect, i.e., that current migrants tend to follow the paths of relatives, friends, and acquaintances who have moved earlier. The strong role of migration chains in Third World settings reflects a paucity of formal communication mechanisms; significance of one's community, which enhances the perceived reliability of informal communications sent through migration chains; and the role of friends, relatives, and acquaintances in a destination who ameliorate adjustment should migration occur.

The importance of migration chains also can be understood in terms of the active and passive migrant distinction of Hagerstrand (1957). Actives systematically seek a new locale that promises future prosperity. Passive migrants, who are considerably more numerous, follow and are dependent upon impulses or information emanating from their active counterparts.

Development is an element of conventional migration models, although not explicitly so. This is illustrated by the central role ascribed to market conditions, which vary from place to place in reflection of economic growth-and-decline experiences at national, regional, and local levels. An explicit statement of the development-migration interface is provided by dual-economy conceptualizations of Third World development. In these, movements from traditional to modern locales are motivated by wage and job opportunity differentials between them; migration persists until the modern sector develops sufficiently to transform the traditional and/or to absorb its excess labor; meanwhile, economic growth proceeds by drawing on extensive human resources provided by traditional areas.

This dynamic is further elaborated in spatial forms of the dual-economy model, notably core-periphery and growth-center constructs. These distinguish between polarization or backwash effects, whereby economic growth impulses move toward the core or growth center(s), and spread or trickle-down effects, whereby growth impulses are directed to places in the periphery (Gaile 1980). The balance between these effects determines whether economic conditions, represented in conventional migration models by wage and job opportunity variables, are more attractive in core or periphery locales, and migration flows would shift accordingly.

Complementing dual-economy, core-periphery, and growth-center constructs is the human-resource perspective, one segment of which addresses migration effects on development (Brown and Kodras 1987; Brown and Lawson 1988; Gober-Meyers 1978a; Lipton 1976, 1980). Because migration is selective, origin places are drained of quality human capital to the benefit of destinations, thus altering development prospects of each locale. To put this in a broader framework, migration occurs in response to place-specific economic differentials between core and periphery (growth center and hinterland, modern and traditional sectors), which then are exacerbated by human-resource effects on development trajectories, leading to further migration, and so on, until polarization reversal occurs (i.e., the switch away from core dominance).

Neoclassical research, then, suggests a highly interrelated system wherein migration and the development context within which it occurs have significant effects upon each other. Nevertheless, studies of aggregate migrations (by geographers as well as others) have tended to neglect place characteristics related to development, much less to take account of their effects.

By contrast, development is a primary concern of historical-structural perspectives, which include dependency and Marxist approaches. These seek generalizations concerning forces dictating the political, economic, social, and geographic organization of society, and migration is viewed as resulting from those forces interacting with local conditions (Bach and Schraml 1982; Wood 1982). Accordingly, conclusions concerning migration pattern and process tend to be stated in terms of the locale being studied, not as universals, and migration may even be regarded as a purely derived phenomenon. For example, Friedmann and Wulff observe (1976, 26–27):

> . . . migration to cities reflects merely a demographic adjustment to changes in the spatial structure of economic and social opportunities . . . (it) is a derived phenomenon, a symptom of urbanization and not the thing itself. . . . Demographers and others insisted on treating migration as a major policy variable when it was, in fact, dependent on the major structural features of the economy.

One manifestation of development to which historical-structural research has drawn attention is the role of origin factors in fomenting migration. Land availability and income-producing opportunities in rural areas, for example, are affected by development-related phenomena such as high rates of population increase, technology diffusion, urban bias in the location of manufacturing, and prevailing social-political structures (Brown and Lawson 1985b). Illustrative in this regard is Costa Rica's promotion of commercial cattle production (Taylor 1980). Because such production is markedly less labor-intensive than alternative agricultural enterprises, and, in Costa Rica, it involved land-reform measures that disrupted traditional subsistence agriculture, income-producing opportunities for the peasant were reduced and out-migration increased. This exemplifies, more generally, Brown and Lawson's assertion (1985b, 417) that:

> . . . structural change or disequilibrium of any sort, including cataclysmic events . . . may be the single most important progeniture of population movements in Third World settings.

Further, Taylor notes (1980, 86) that the importance of origin factors:

> . . . contradict[s] the traditional view that urban attraction is at the root of the migration process . . . [Data] show that most peasants leaving Guanacaste have favored destinations within the periphery over the urban center. At the same time . . . the restriction of employment and land owning opportunities within the periphery make it increasingly difficult for the peasant to avoid a final move into the city. It seems that urban "pull" factors are playing a relatively minor role in the Costa Rican rural-urban migration process.

Also relevant to Taylor's point is evidence of sizable migrations to rural destinations in a number of Third World countries (Brown and Lawson 1985b), which is counterintuitive to neoclassical, pull-oriented models of development.

Although research has tended to take either a neoclassical or historical-structural stance, the complementarity of these views suggests a need for integrating within one framework structural-institutional forces and individual or household responses to such factors. The issue is addressed in the next section, which concerns two "development paradigms of migration." These meld macroeconomic structure and individual decision-making by emphasizing the importance of regional character or development-milieu effects.

Structural forces and individual decision-making also have been integrated by using simultaneous equations or two-stage least-squares formulations that explicitly treat the migration-development link (Gober-Meyers 1978a, 1978b; Greenwood 1975a, 1975b; Salvatore 1981). However, this approach has rarely been applied to Third World settings, Greenwood's study (1978) of Mexico being one exception. More important, by estimating parameters on a countrywide basis, simultaneous-equation models embodying migration-development links have assumed that similar relationships govern all locales, whereas Brown and coauthors (Brown and Lawson 1985a, 1985b; Brown and Sanders 1981) argue that the role of migration process components may differ considerably across development milieus.

Development Paradigms of Migration

This section addresses two conceptual frameworks based on the observation that economic growth involves structural changes in society which, in turn, alter the role of factors influencing migration. Work by Mabogunje (1970), Zelinsky (1971), Brown and Sanders (1981), and Brown (1987, ch. 3) argues that conditions pertinent to human movements are affected by the development-dependent mix of social and economic conditions, government policies, infrastructure, technological achievement, and other aspects of regional systems. Accordingly, modern-sector wage rates and job opportunities may play a dominant role in advanced settings, whereas migration-chain or rural-push effects may dominate under less-advanced conditions. Different development milieus give rise, therefore, to different "processes" of migration. Further, since development varies both temporally and from place to place, these differences may be found for a nation at different times in its evolution, or alternatively in comparing nations or subnational regions at a given point in time.

Given this, ambiguities in the findings of previous research may be attributed to differences in the level or nature of development among the locales studied (Jones and Brown 1985) or to differences in the (historical) developmental processes characterizing those locales. That incorporating the development dimension is critical to understanding Third World migration processes also has been noted by Goldstein (1981) and Urzua (1981).

Initial attempts to link development and migration were primarily in terms of movement patterns, not processes. For example, Connell et al. (1976, 201) conclude that:

> Patterns of migration from a rural community may well change in "stages," following integration (into the national urban system) and development of that community. Circular migration usually comes early . . . succeeded by directed migration, but still relatively little differentiated by socioeconomic group. . . . Subsequent integration often differentiates migrant streams . . . both by status and by age, sex, and destination. . . . The process also often involves a shift from personal to household migration.

Similarly, Zelinsky's hypothesis (1971) of the mobility transition posits rate and pattern changes in terms of five development phases: (1) a premodern traditional society, (2) an early-transitional society, (3) a late-transitional society, (4) an advanced society, and (5) a future superadvanced society.

Skeldon (1977) applies Zelinsky's framework to an area of highland Peru (Cuzco) for its period of change from a premodern traditional society to an early-transitional society, finding that pattern and rate shifts diffuse down the settlement hierarchy and from richer to poorer social groups within each settlement. Underlying this pattern is the accessibility (or distance) of each settlement to major urban centers and its resource endowment, and similarly for social groups.

Forbes (1981) notes the parallel between his findings for Indonesia and Zelinsky's hypotheses, but argues that societal structures and their articulation with local conditions differ from place to place in a manner that defies generalization, essentially taking a historical-structural position.

In a shift of focus, Brown (1987, ch. 3; Brown and Sanders 1981) puts forth a development paradigm of migration that is concerned with process elements more than with pattern. To establish continuity with earlier research, it is articulated in terms of conventional model variables and Zelinsky's development phases. The outcome of Brown's reasoning is summarized:

1. Early migrations, occurring in Zelinsky's early-transitional society or—using Brown's terminology—the move toward modernization, are hypothesized to

be highly chain in nature, origin-pushed, and oriented toward opportunities of the informal, small-scale enterprise labor market. Rural-to-rural migration streams would be as likely as rural-to-urban ones.
2. As development proceeds, entering later phases of the move toward modernization or Zelinsky's late-transitional society, migration by better-off social classes would be pulled by modern-sector employment and educational opportunities, but retain a significant chain dimension owing to transportation and communication systems that are somewhat rudimentary. At the same time, migration by poorer social classes would maintain its origin push motivation, orientation toward informal and small-scale enterprise labor markets, and chain characteristics. Rural-to-urban flows should increase.
3. Finally, as development reaches a relatively advanced level, entering modernization or Zelinsky's advanced society, migration of all social classes would be oriented toward formal, modern-sector employment; and formal communication channels would be the primary sources of information, thus reducing, and in many instances eliminating, the chain dimension. Further, the dominant pattern of migration should be urban-to-urban, rather than rural-to-urban.

Three further observations are appropriate. First, an integral element of the Zelinsky and Brown frameworks is that places (as origins/destinations of aggregate migration streams) represent different development milieus, which may be characterized by different migration processes. Second, the Zelinsky and Brown frameworks are directed toward aggregate migration flows (paralleling conventional models) and the effect on those of place characteristics. Finally, one outcome of development is a form of migration known as circulation: short-term movements, either seasonal or sporadic, for temporary employment. Geographic studies of circulation include Brea (1986); Brown, Brea, and Goetz (1988); Chapman and Prothero (1985); Forbes (1981); Hugo (1982, 1985); and Prothero and Chapman (1985).

Beyond Conventional Thinking

Demonstrating that a single (conventional) framework applies to many settings was an important task at one time, but that has been accomplished. It is now appropriate to take the next step, by incorporating place and process specificity into our knowledge of Third World migrations.

Conventional approaches to studying migration (reviewed above) revolve around variables such as job opportunities, wage rates, and information flows. When these represent the range of one's analytical framework, other factors are masked, factors that add richness to our understanding of human movements in Third World settings. Situations also differ in the degree to which conventional modeling is applicable. For example, rural areas and nations with more-agrarian economies are less suitable for this approach. Finally, conventional approaches to migration embody a conventional view of development that has been increasingly questioned (Brown 1988).

Incorporating place and process specificity into our knowledge of Third World migrations is unlikely to provide the level of generality associated with conventional modeling. Critical, therefore, are procedures (or a methodology) which provide specificity but avoid inundation by idiosyncratic events. This requires a research protocol that raises and addresses questions about broad relationships, an important element of which is a sense of place and/or geographic procedures for obtaining it.

Gains from broadening the range of inquiry in this manner are illustrated by Brown and Lawson's study (1985b) of rural migration in Costa Rica for the period 1968–73. One set of findings pertains to conventional wisdom that rural-to-urban flows dominate Third World settings. Rather than being universally applicable, it appears that significant rural-directed migration is typical of agrarian Third World economies such as Costa Rica or Ecuador; that conventional expectations are typical of more industrialized countries such as Brazil, Mexico, or Venezuela; and that rural-to-urban transfers of population in both settings result from factors fomenting rural outmigration, as readily as from destination pulls.

While factors underlying rural-directed migration pertain to economic opportunities in general, the greater level of specificity deepens our understanding of development as it actually occurs and migration responses thereto. Especially interesting in Brown and Lawson's account are the cyclical, rather than developmental, underpinnings of several factors fomenting migration, and the important role of world economic events, donor-nation actions, and government policies.

FUTURE DIRECTIONS

Recent population-geography research has been dominated by migration studies, while fertility and mortality topics have received very little attention. Geographers have continued to examine migration at a variety of scales for both individual and aggregate behavior, while employing methodologies that range from simple description to very sophisticated analytical model-building procedures. Current research activities are too numerous and varied in context and methodology to detail in just one chapter. Instead of a comprehensive review, summaries of research in six topical areas, progressing in scale from intraurban to international, have been offered to illustrate the breadth of mainstream activity in population geography.

Several research directions are worthy of attention, based on the assessment of recent work. One important challenge that emerges is the need to frame population-geography research in such a way as to increase its emphasis on social relevance. Regardless of scale, there is a widespread recognition that while population distribution and relocation are products of the socioeconomic differences between places, they also in turn can influence social, economic, and political changes. Population geography can have substantial public-policy ramifications and more studies should examine the human impacts resulting from population change.

Population geographers are becoming increasingly aware of the need to conceptually bridge microlevel studies, which focus upon individual behavior, with those at a macroscale that are more concerned with aggregate behavior. Also, while it is often operationally convenient to study human migration at a specific scale, more synthesis is required to make generalizations about the complex factors that influence human movement over a range of migration distances.

Attempts to conceptually bridge scales of migration will undoubtedly cloud definitions of human mobility and migration even further. For example, the distinctions between "circulation" and "migration" have already been debated. Yet a bridging of scale requires that population geographers sort out a complex hierarchy of short-distance, short-term movements that are influential precursors of longer-distance, longer-term movements. In short, we need to understand the role of one sequence of movements on another, within the context of time, space, life cycle, and cultural differences.

There is a continued need for more studies that (1) identify the role of regional differences in migration behavior, (2) assess the role of both time and space in human mobility, and (3) examine the movement of specific groups of people. The demand for comparative studies between regions, longitudinal studies that permit documentation of temporal changes in spatial interaction, and detailed research of specific minority groups is not likely to decline in the near future.

Geographers might disagree about the most significant topics within population geography, the most appropriate research questions, and certainly the most appropriate research methodologies. However, it is clear that despite these honest differences, population-geography research has grown rapidly since 1953, and shows no signs of slowing. Undoubtedly, population geography is pivotal to an understanding of the cultural landscape.

REFERENCES

Alonso, W. 1978. A theory of movements. In *Human settlement systems,* ed. N. Hansen, pp. 197–212. Cambridge, MA: Ballinger Publishing.

Bach, R. L., and Schraml, L. A. 1982. Migration, crisis, and theoretical conflict. *International Migration Review* 16:320–41.

Beale, C. L. 1975. *The revival of population growth in nonmetropolitan America.* Washington: USDA, Economic Research Service.

———, and Fuguitt, G. V. 1985. *Metropolitan and nonmetropolitan growth differentials in the United States since 1980.* University of Wisconsin, Center for Demography and Ecology, Working Paper 85–6.

Bean, F. D.; Lowell, B. L.; and Taylor, L. 1986. The impact of undocumented migration on the earnings of other groups in metropolitan labor markets in the U.S. Austin, TX: Texas Population Research Center, University of Texas-Austin, Paper 8.007.

Behr, M. 1986. Residential mobility and neighborhood change in a suburban context. Unpublished Ph.D. diss., Department of Geography, Arizona State University.

Berry, B. J. L. 1981. *Comparative urbanization.* New York: St. Martin's Press.

———, and Dahmann, D. C. 1977. Population redistribution in the United States in the 1970s. *Population and Development Review* 3:443–71.

Birks, J. S.; Seccombe, I. J.; and Sinclair, C. A. 1986. Migrant workers in the Arab gulf: The impact of declining oil revenues. *International Migration Review* 20:799–814.

Bohning, W. R. 1981. Elements of a theory of international economic migration to industrial nation states. In *Global trends in migration: Theory and research on international population movements,* eds. M. M. Kritz and C. B. Keely, pp. 28–43. New York: Center for Migration Studies.

Borts, G., and Stein, J. 1964. *Economic growth in a free market.* New York: Columbia University Press.

Brana-Shute, R., and Brana-Shute, G. 1982. The magnitude and impact of remittances in the eastern Caribbean: A research note. In *Return migration and remittances developing a Caribbean perspective,* ed. W. F. Stinner, et al., pp. 267–89. Washington: Smithsonian Institution.

Brea, J. A. 1986. Effects of structural characteristics and personal attributes on labor mobility in Ecuador. Ph.D. diss. Department of Geography, Ohio State University.

Brown, L. A. 1987. Development, geography, and societal processes: With particular reference to migration, labor force experiences, and regional change in Latin America. Studies on the Interrelationships Between Migration and Development in Third World Settings. Discussion Paper 35, Department of Geography, Ohio State University.

———. 1988. Third World development as the local articulation of world economic conditions, donor nation actions, and government policies: A rationale and research implications. Studies on the Interrelationships Between Migration and Development in Third World Settings. Discussion Paper 35, Department of Geography, Ohio State University.

———, and Moore, E. G. 1970. The intraurban migration process: A perspective. *Geografiska Annaler* 60B:1–13.

———, and Sanders, R. L. 1981. Toward a development paradigm of migration: With particular reference to third world settings. In *Migration decision making: Multidisciplinary approaches to micro-level studies in developed and developing countries,* eds. G. F. DeJong and R. W. Gardner, pp. 149–85. New York: Pergamon Press.

———, and Lawson, V. A. 1985a. Migration in third world settings, uneven development, and conventional modelling: A case study of Costa Rica. *Annals of the Association of American Geographers* 75:29–47.

———, and ———. 1985b. Rural-destined migration in Third World settings: A neglected phenomenon. *Regional Studies* 19:415–32.

———, and Kodras, J. E. 1987. Migration, human resource transfers, and development contexts: A logit analysis of Venezuelan data. *Geographical Analysis* 19:243–63.

———, and ———. 1988. Polarization reversal, migration related shifts in human resource profiles, and spatial growth policies: A Venezuelan study. *International Regional Science Review* 12.

———; Brea, J. A.; and Goetz, A. R. 1988. Policy aspects of development and individual mobility: Migration and circulation from Ecuador's rural sierra. Submitted for publication.

Brummell, A. C. 1979. A model of intraurban mobility. *Economic Geography* 55:338–52.

———. 1981. A method of measuring residential stress. *Geographical Analysis* 13:248–61.

Bustamante, J. 1977. Undocumented immigration from Mexico: Research report. *International Migration Review* 2:149–77.

Chapman, M., and Prothero, R. M., eds. 1985. *Circulation in population movement: Substance and concepts from the Melanesian case.* London: Routledge & Kegan Paul.

Clark, G. L., and Ballard, K. P. 1980. Modeling out-migration from depressed regions: The significance of origin and destination characteristics. *Environment and Planning A* 12:799–812.

Clark, W. A. V. 1981. Residential mobility and behavioral geography: Parallelism or interdependence. In *Behavioral problems in geography revisited,* eds. K. Cox and R. G. Golledge, pp. 182–208. London: Methuen.

———. 1982. Recent research on migration and mobility: A review and interpretation. *Progress in Planning* 18:7–56.

———. 1983. Structures for research on the dynamics of residential mobility. In *Evolving geographical structures,* eds. D. A. Griffith and A. C. Lea, pp. 372–97. The Hague: Martinus Nijhoff.

———. 1986. Residential segregation in American cities: A review and interpretation. *Population Research and Policy Review* 5:95–127.

———, and Heskin, A. 1978. The impact of rent control on tenure discounts and residential mobility. *Land Economics* 58:109–17.

———, and Burt, J. 1980. The impact of workplace on residential location. *Annals of the Association of American Geographers* 70:59–67.

———, and Smith, T. 1982. Housing market search behavior and expected utility theory II: The process of search. *Environment and Planning A* 14:717–37.

———, and Van Lierop, W. F. J. 1986. Residential mobility and household location modelling. *Handbook of Regional and Urban Economics* 1:97–132.

Cobbe, J. 1982. Emigration and development in southern Africa, with special reference to Lesotho. *International Migration Review* 16:837–68.

Connell, J. et al. 1976. *Migration from rural areas: Evidence from village studies.* Oxford: Oxford University Press.

Conway, D. 1985. Remittance impacts on development in the eastern Caribbean. *Bulletin of Eastern Caribbean Affairs* 11:31–40.

———. 1986. Caribbean migration as international circulation. Paper presented at the 82nd annual meeting of the Association of American Geographers, Twin Cities, May 1986.

Cromartie, J. B., and Stack, C. B. 1987. Who counts? Black homeplace migration to the South, 1975–1980. Unpublished paper.

Dagodag, W. T. 1975. Source regions and composition of illegal Mexican immigration to California. *International Migration Review* 9:499–511.

———. 1984. Illegal Mexican aliens in Los Angeles: Locational characteristics. In *Patterns of undocumented migration: Mexico and the United States,* ed. R. C. Jones, pp. 199–217. Totowa, NJ: Rowman & Allanheld.

DaVanzo, J., and Morrison, P. A. 1981. Return and other sequences of migration in the United States. *Demography* 18:85–102.

DeJong, G. G., and Fawcett, J. T. 1981. Motivations for migrational assessment and a value-expectancy research model. In *Migration decision making,* eds. G. F. DeJong and R. W. Gardner. New York: Pergamon Press.

Desbarats, J. 1983. Spatial choice and constraints on behavior. *Annals of the Association of American Geographers* 73:340–57.

———. 1986. Cambodian and Laotian refugees: An underclass. Paper presented at the annual meeting of the Population Association of America, San Francisco, April 1986.

Deurloo, M. C.; Dieleman, F. M.; and Clark, W. A. V. 1987. Tenure choice in the Dutch housing market. *Environment and Planning A* 19:763–81.

Elo, I. T., and Beale, C. L. 1988. The decline in counterurbanization in the 1980s. Paper presented at the annual meeting of the Association of American Geographers, Phoenix, Arizona, April 6–10.

Fergany, N. 1982. The impact of emigration on national development in the Arab region: The case of the Yemen Arab Republic. *International Migration Review* 16:757–80.

Fielding, A. 1982. Counter urbanization in western Europe. *Progress in Planning* 17:1–52.

Fischer, C. S. 1976. *The urban experience.* New York: Harcourt Brace.

Flowerdew, R. 1976. Search strategies and stopping rules in residential mobility. *Transactions of the Institute of British Geographers* 1:47–57.

Forbes, D. 1981. Mobility and uneven development in Indonesia: A critique of explanations of migration and circular migration. In *Population mobility and development: Southeast Asia and the Pacific,* eds. G. W. Jones and H. V. Richter, pp. 51–70. Canberra: Development Studies Centre, Monograph 27, Australian National University.

Fotheringham, A. S. 1983. A new set of spatial interaction models: The theory of competing destinations. *Environment and Planning A* 15:15–36.

———. 1986. Modelling hierarchial destination choice. *Environment and Planning A* 18:401–18.

Frey, W. H. 1988. Counterurbanization and the migration process: A study of developed countries. Paper presented at the annual meeting of the Association of American Geographers, Phoenix, Arizona, April 6–10.

Friedmann, J., and Wulff, R. 1976. The urban transition: Comparative studies of newly industrializing societies. In *Progress in geography,* vol. 8, eds. C. Board, R. J. Chorley, P. Haggett, and D. R. Stoddart, pp. 1–93. New York: St. Martin's Press.

Fuguitt, G. V. 1985. The nonmetropolitan population turnaround. *Annual Review of Sociology* 11:259–80.

Gaile, G. L. 1980. The spread-backwash concept. *Regional Studies* 14:15–25.

Gibbs, J. 1963. The evolution of population concentration. *Economic Geography* 39:119–29.

Gober, P. 1980. Shrinking household size and its effect on urban population density patterns: A case study in Phoenix, Arizona. *Professional Geographer* 32:55–62.

———. 1986a. Homogeneity versus heterogeneity in household structure: The recent experience of 20 U.S. cities. *Environment and Planning A* 18:715–27.

———. 1986b. How and why Phoenix households changed: 1970–1980. *Annals of the Association of American Geographers* 76:536–49.

———. 1987. Geographic ramifications of the changing American household. Paper presented to the annual meeting of the Association of American Geographers in Portland, Oregon, April 1987.

Gober-Meyers, P. 1978a. Employment motivated migration and economic growth in post-industrial market economies. *Progress in Human Geography* 2:207–29.

———. 1978b. Interstate migration and economic growth: A simultaneous equations approach. *Environment and Planning A* 10:1241–52.

Golant, S. 1980. Future directions for elderly migration research. *Research on Aging* 2:271–80.

Goldstein, S. 1981. Research priorities and data needs for establishing and evaluating population redistribution policies. In *Population Distribution Policies in Development Planning,* eds. G. Demko and R. Fuchs, pp. 183–203. New York: United Nations,

Department of International Economic and Social Affairs, Population Studies 75.

Golledge, R. G. 1980. A behavioral view of mobility and migration research. *Professional Geographer* 32:14–21.

Gordon, P. 1979. Deconcentration without a 'clean break'. *Environment and Planning A* 11:281–90.

Gottman, J. 1961. *Megalopolis*. Cambridge: MIT Press.

Greenwood, M. J. 1975a. Research on internal migration in the United States: A survey. *Journal of Economic Literature* 8:397–483.

———. 1975b. Simultaneity bias in migration models: An examination. *Demography* 12:519–36.

———. 1978. An economic model of internal migration and regional economic growth in Mexico. *Journal of Regional Science* 18:17–31.

Grossman, J. B. 1984. Illegal immigrants and domestic employment. *Industrial and Labor Relations Review* 37:240–51.

Guitterrez, P. R. 1984. The channelization of Mexican nationals to the San Luis Valley of Colorado. In *Patterns of undocumented migration: Mexico and the United States,* ed. R. C. Jones, pp. 183–98. Totowa, NJ: Rowman & Allanheld.

Haberkorn, G. 1981. The migration decision-making process: Some social psychological considerations. In *Migration decision making,* eds. G. F. DeJong and R. W. Gardner. New York: Pergamon Press.

Hagerstrand, T. 1957. Migration and area. In *Migration in Sweden: A symposium,* eds. D. Hannerberg, T. Hagerstrand, and B. Odeving, pp. 27–158. Lund Studies in Geography 13. Lund, Sweden: Gleerup.

Hanushek, R., and Quigley, J. 1978. Housing market disequilibrium and residential mobility. In *Population mobility and residential change,* eds. W. A. V. Clark and E. G. Moore, pp. 51–98. Evanston, IL: Northwestern University, Studies in Geography, No. 25.

Haynes, K. E., and Fotheringham, A. S. 1984. *Gravity and spatial interaction models*. Beverly Hills: Sage Publications.

Hoover, E. M., and Vernon, R. 1959. *Anatomy of a metropolis*. Cambridge: Harvard University Press.

Huff, J. O. 1986. Geographic regularities in residential search behavior. *Annals of the Association of American Geographers* 76:208–27.

———, and Clark, W. A. V. 1978. Cumulative stress and cumulative inertia: A behavioral model of the decision to move. *Environment and Planning A* 10:1101–19.

Hugo, G. J. 1982. Circular migration in Indonesia. *Population and Development Review* 8:59–83.

———. 1985. Structural change and labour mobility in rural Java. In *Labour circulation and the labour process,* ed. G. Standing, pp. 46–88. London: Croom Helm.

International Migration Review. 1986. Special issue, Temporary worker programs: Mechanisms, conditions, consequences, 20.

Jackson, R. H., and Hudman, L. E. 1986. *World regional geography: Issues for today*. 2d ed. New York: Wiley.

Johnson, J. H., and Brunn, S. D. 1980. Spatial and behavioral aspects of the counterstream migration of blacks to the south. In *The American metropolitan system: Present and future,* eds. S. D. Brunn and J. O. Wheeler. New York: Wiley.

Jones, H. R. 1981. *A population geography*. London: Harper & Row.

Jones, R. C. 1982a. Channelization of undocumented Mexican migrants to the U.S. *Economic Geography* 58:156–75.

———. 1982b. Undocumented migration from Mexico: Some geographical questions. *Annals of the Association of American Geographers* 72:77–87.

———. 1984. *Patterns of undocumented migration: Mexico and the United States*. Totowa, NJ: Rowman & Allanheld.

———. 1985. Economic impacts of undocumented migration on the U.S. Southwest. Paper presented at the conference, The Sunbelt: A Region and Regionalism in the Making? Miami, Florida.

———. 1987. Salvadoran migrants to the U.S.: Political or economic refugees? Paper delivered at the 83rd annual meeting, of the Association of American Geographers, Portland, Oregon, April 1987.

———. 1988. The role of U.S. migration in rural Zacatecas: Preliminary conclusions. Talk delivered at the University of Monterrey, Mexico.

———, and Brown, L. A. 1985. Cross-national tests of a Third World development-migration paradigm, with particular attention to Venezuela. *Socio-Economic Planning Sciences* 19:357–61.

———; Harris, R. J.; and Valdez, A. 1984. Occupational and spatial mobility of undocumented migrants from Dalores Hidalgo, Griangnato. In *Patterns of undocumented migration: Mexico and the United States,* ed. R. C. Jones, pp. 159–82. Totowa, NJ: Rowman & Allanheld.

———, and Murray, W. B. 1986. Occupational and spatial mobility of temporary Mexican migrants to the U.S.: A comparative analysis. *International Migration Review* 20:973–85.

Kontuly, T., and Vogelsang, R. 1988. Intensification of counterurbanization in the FRG. *Professional Geographer* 40:42–53.

Koo, H. P. 1985. Short-term change in household and family structure. Unpublished paper of the Research Triangle Institute.

Kritz, M. M., ed. 1981. *U.S. immigration and refugee policy: Global and domestic issues*. Lexington, MA: Lexington Books.

———. 1983. International migration patterns in the Caribbean basin: An overview. In *Global trends in migration: Theory and research on international population movements,* eds. M. M. Kritz and C. B. Keely, pp. 208–33. New York: Center for Migration Studies.

———, and Keely, C. B., eds. 1981. *Global trends in migration: Theory and research on international population movements*. New York: Center for Migration Studies.

Lake, R. W. 1981. *The new suburbanites: Race and housing in the suburbs*. New Brunswick, NJ: Center for Urban Policy Research, Rutgers University.

Landale, N. S., and Guest, A. 1985. Constraints, satisfaction, and residential mobility: Speare's model reconsidered. *Demography* 22:199–222.

Lansing, J. B., and Mueller, E. 1967. *The geographic mobility of labor*. Ann Arbor: Institute for Social Research, University of Michigan.

Ledent, J. 1985. The urbanization process. *Urban Geography* 6:69–82.

Lieber, S. 1978. Place utility and migration. *Geografiska Annaler* 60B:16–27.

Ling, L. H. M. 1984. East Asian migration to the Middle East: Courses, consequences, and considerations. *International Migration Review* 18:19–36.

Lipton, M. 1976. *Why poor people stay poor: A study of urban bias in world development*. Cambridge: Harvard University Press.

———. 1980. Migration from rural areas of poor countries: The impact on rural productivity and income distribution. *World Development* 8:1–24. Also in *Migration and the labor market in development countries,* ed. R. Sabot, pp. 191–228. Boulder, CO: Westview Press.

Long, L. 1978. The geographical mobility of Americans: An international comparison. *U.S. Department of Commerce Bureau of the Census Population*. Reprint 64.

———, and DeAre, D., 1980. Migration to nonmetropolitan areas. *CDS 80-2,* U.S. Bureau of Census.

———, and ———. 1988. Measures of population concentration and deconcentration in the U.S.: 1900–1986. Paper presented at the annual meeting of the Association of American Geographers, Phoenix, Arizona, April 6–10.

Lord, J., and Catau, J. 1976. School desegregation, busing and suburban migration. *Urban Education* 11:275–94.

Lowry, I. A. 1966. *Migration and metropolitan growth: Two analytical models*. San Francisco: Chandler.

Mabogunje, A. L. 1970. Systems approach to a theory of rural-urban migration. *Geographical Analysis* 2:1–17.

McCarthy, K. F., and Morrison, P. A. 1979. *The changing demographic and economic structure of nonmetropolitan areas in the 1970s*. Santa Monica, CA: The Rand Corporation.

McElroy, C. 1987. El Salvadoran refugee patterns of migration to Belize. Paper presented at the 83rd annual meeting of the Association of American Geographers, Portland, Oregon, April 1987.

McHugh, K. E. 1984. Explaining migration intentions and destination selection. *Professional Geographer* 36:315–25.

———. 1987. Black migration reversal in the United States. *Geographical Review* 77:171–82.

Meyer, R. 1980. A descriptive model of constrained residential search. *Geographical Analysis* 12:21–32.

Michelson, W. 1977. *Environmental choice, human behavior and residential satisfaction*. New York: Oxford University Press.

Mines, R. 1982. Migration to the United States and Mexican rural development: A case study. *American Journal of Agricultural Economics* 64:444–54.

Moore, E. G. 1972. *Residential mobility in the city*. Commission on College Geography, Resource Paper No. 13. Washington: Association of American Geographers.

———. 1978. The impact of residential mobility on population characteristics at the neighborhood level. In *Population mobility and residential change,* eds. W. A. V. Clark and E. G. Moore. Evanston, IL: Northwestern University Press.

———, and Clark, W. A. V. 1980. The policy context for mobility research. In *Residential mobility and public policy,* eds. W. A. V. Clark and E. G. Moore, pp. 10–28. Beverly Hills: Sage Publications.

———, and Harris, R. 1979. Residential mobility and public policy. *Geographical Analysis* 11:175–83.

Morrill, R. L. 1988. Migration regions and population redistribution. *Growth and Change* 19:43–60.

Morrow-Jones, H. A. 1988. Housing tenure change in American suburbs. Paper presented to the annual meeting of the Population Association of America in New Orleans, Louisiana, April 1988.

Muller, P. 1981. *Contemporary suburban America.* Englewood Cliffs, NJ: Prentice-Hall.

Murie, A. 1974. *Household movement and housing choice.* Birmingham: Centre for Urban and Regional Studies, University of Birmingham.

Myers, D. 1978. Aging of population and housing: A new perspective on planning for more balanced metropolitan growth. *Growth and Change* 9:8–13.

———. 1989. *Urban demography.* Madison, WI: University of Wisconsin Press. In press.

Newton, P. W., and Johnston, R. J. 1976. Residential area characteristics and residential area homogeneity: Further thoughts on extensions to the factorial ecology method. *Environment and Planning A* 8:543–52.

Oliver, M., and Johnson, J. 1986. Inter-ethnic conflict in an urban ghetto: The case of blacks and Latinos. In *Research, social movements, conflict and change,* ed. R. L. Racliffe. New York: JAI.

Onaka, J., and Clark, W. A. V. 1983. A disaggregate model of residential mobility and housing choice. *Geographical Analysis* 15:287–304.

Palm, R. 1976a. An index of household diversity. *Tijdschrift voor Economische en Sociale Geographie* 67:194–201.

———. 1976b. Real estate agents and geographical information. *Geographical Review* 66:267–80.

Phillips, A. G., and Carter, J. E. 1984. An individual-level analysis of the stress-resistance model of household mobility. *Geographical Analysis* 15(2):176–89.

Pickles, A. 1980. Models of movement: A review of alternative methods. *Environment and Planning A* 12:1383–1404.

Piore, M. 1979. *Birds of passage: Migrant labor and industrial societies.* Cambridge: Cambridge University Press.

Plane, D. 1984. A systematic demographic efficiency analysis of U.S. interstate population exchange, 1935–1980. *Economic Geography* 60:294–312.

———. 1987. The geographic components of change in a migration system. *Geographical Analysis* 19:283–99.

———, and Rogerson, P. A. 1982. Spatial economic-demographic modelling of interregional migration with limited data. Paper presented at the International Conference on Forecasting Regional Population Change, Airlie, Virginia.

———, and Isserman, A. M. 1983. U.S. interstate labor force migration: An analysis of trends, net exchanges, and migration subsystems. *Socioeconomic Planning Sciences* 17:251–66.

———, and ———. 1986. Dynamic flow modelling with interregional dependency effects: An application to structural change in the U.S. migration system. *Demography* 23:91–104.

Prothero, R. M., and Chapman, M., eds. 1985. *Circulation in Third World countries.* London: Routledge & Kegan Paul.

Quigley, J., and Weinberg, D. 1977. Intraurban residential mobility: A review and synthesis. *International Regional Science Review* 2:41–66.

Ravenstein, E. G. 1885. The laws of migration. *Journal of the Royal Statistical Society* 48:167–235.

———. 1889. The laws of migration. *Journal of the Royal Statistical Society* 52:241–305.

Reichert, J. S., and Massey, D. S. 1979. Patterns of U.S. migration from Mexican sending community: A comparison of legal and illegal migrants. *International Migration Review* 13:599–623.

Rogers, A., and Willekens, F. J. 1986. *Migration and settlement: A multiregional comparative study.* Dordrecht, Holland: D. Reidel.

Rogerson, P. A. 1984. New directions in the modelling of interregional migration. *Economic Geography* 60:111–21.

———, and MacKinnon, R. D. 1982. Interregional migration models with source and interaction information. *Environment and Planning A* 14:445–54.

———, and Plane, D. A. 1984. Modelling temporal change in flow matrices. *Papers of the Regional Science Association* 54:147–64.

———, and ———. 1985. Monitoring migration trends. *American Demographics* 7:26–29; 47.

Rondinelli, D. 1983. Dynamics of growth of secondary cities of developing countries. *Geographical Review* 73:42–57.

Roseman, C. C. 1985. A typology of multiple residence. Paper presented at the 81st annual meeting of the Association of American Geographers, Detroit, Michigan, April 1985.

———, and Oldakowski, R. K. 1984. Place ties and migration expectations of a central city population. *Urban Geography* 5:95–110.

Rossi, P. 1955. *Why families move*. Glencoe, IL: Free Press.

Salt, J. 1981. International labor migration in western Europe: A geographical review. In *Global trends in migration: Theory and research on international population movements,* eds. M. M. Kritz and C. B. Keely, pp. 133–57. New York: Center for Migration Studies.

Salvatore, D. 1981. *Internal migration and economic development: A theoretical and empirical study*. Washington: University Press of America.

Sechrist, R. P. 1987. The sanctuary movement: Spatial and temporal growth trends. Paper delivered at the 83rd annual meeting of the Association of American Geographers, Portland, Oregon, April 1987.

Sell, R. R. 1983. Analyzing migration decisions: The first step–whose decisions? *Demography* 29:29–311.

———, and DeJong, G. F. 1983. Deciding whether to move: Mobility, wishful thinking, and adjustment. *Sociology and Social Research* 67:146–65.

Shrestha, N. R. 1985. The political economy of economic underdevelopment and external migration in Nepal. *Political Geography Quarterly*.

Shryock, H. S., and Siegel, J. S. 1976. *The methods and materials of demography*. New York: Academic Press.

Simmons, J. 1968. Changing residence in the city: A review of intraurban mobility. *Geographical Review* 53:622–51.

Sjaastad, L. 1962. The costs and returns of human migration. *Journal of Political Economy* 70:80–93.

Skeldon, R. 1977. The evolution of migration patterns during urbanization in Peru. *Geographical Review* 67:394–411.

Smith, B., and Newman, R. 1977. Depressed wages along the U.S.-Mexico border: An empirical analysis. *Economic Inquiry* 15: 51–66.

Smith, T. R.; Clark, W. A. V.; Huff, J. O.; and Shapiro, P. 1979. A decision making and search model for intraurban migration. *Geographical Analysis* 11:1–22.

South, S. 1987. Metropolitan migration and social problems. *Social Science Quarterly* 68:3–18.

Speare, A.; Kobrin, F.; and Kingkade, W. 1982. The influence of socioeconomic bonds and satisfaction on interstate migration. *Social Forces* 61:551–74.

Stahl, C. W., and Arnold, F. 1986. Overseas workers' remittances in Asian development. *International Migration Review* 20:899–925.

Stapleton, C. M. 1980. Reformulation of the family lifecycle concept: Implications for residential mobility. *Environment and Planning A* 12:1103–18.

———. 1982. Spatial distribution of primary individuals. *Professional Geographer* 34:167–77.

Strand, P. J. 1984. Employment predictors among Indochinese refugees. *International Migration Review* 18:50–64.

Swanson, J. C. 1979. The consequences of emigration for economic development: A review of the literature. *Papers in Anthropology* 20:39–56.

Symanski, R. 1986. Along the Salvadoran pipeline. *Focus* 36(4):3–11.

Taylor, J. E. 1980. Peripheral capitalism and rural-urban migration: A study of population movements in Costa Rica. *Latin American Perspectives* 26:75–90.

Trewartha, G. 1953. A case for population geography. *Annals of the Association of American Geographers* 43:71–97.

Urzua, R. 1981. Population redistribution mechanisms as related to various forms of development. In *Population distribution policies in development planning,* eds. G. J. Demko and R. J. Fuchs, pp. 53–69. New York: United Nations, Department of Economic and Social Affairs, Population Studies 75.

Valdez, A., and Jones, R. C. 1984. Geographical patterns of undocumented Mexicans and Chicanos in San Antonio, Texas: 1970 and 1980. In *Patterns of undocumented migration: Mexico and the U.S.,* ed. R. C. Jones, pp. 218–35. Totowa, NJ: Rowman & Allanheld.

Vining, D. R. 1982. Migration between the core and the periphery. *Scientific American* 247:44–53.

———, and Kontuly, T. 1978. Population dispersal from major metropolitan regions. *International Regional Science Review* 3:49–74.

———; Pallone, R.; and Plane, D. 1981. Recent migration patterns in the developed world. *Environment and Planning A* 13:243–50.

———, and Pallone, R. 1982. Migration between core and peripheral regions: A description and tentative explanation of the patterns in 22 countries. *Geoforum* 13:339–410.

White, R. B. 1982. Family size composition differentials between central city-suburb and metropolitan-nonmetropolitan migration streams. *Demography* 19:29–51.

White, S. E. 1980a. Awareness, preference, and interurban migration. *Regional Science Perspectives* 10:71–86.

———. 1980b. A philosophical dichotomy in migration research. *Professional Geographer* 32:6–13.

———. 1981. The influence of urban residential preferences on spatial behavior. *Geographical Review* 71:176–87.

———. 1983. Return migration to appalachian Kentucky: An atypical case of nonmetropolitan migration reversal. *Rural Sociology* 48:471–91.

———. 1987. Return migration to eastern Kentucky and the stem family concept. *Growth and Change* 18:38–52.

Williams, L. 1983. The urbanization process. *Urban Geography* 4:122–37.

Wilson, A. 1971. A family of spatial interaction models and associated developments. *Environment and Planning A* 3:1–32.

Wiseman, R. F., and Roseman, C. C. 1979. A typology of elderly migration based on the decision-making process. *Economic Geography* 55:324–37.

Wolpert, J. 1965. Behavioral aspects of the decision to migrate. *Papers and Proceedings of the Regional Science Association* 15:159–69.

Wood, C. H. 1982. Equilibrium and historical-structural perspectives on migration. *International Migration Review* 16:298—319.

Zelinsky, W. 1971. The hypothesis of the mobility transition. *Geographical Review* 61:219–49.

———. 1977. Coping with the migration turnaround: The theoretical challenge. *International Regional Science Review* 2:175–78.

———. 1978. Are nonmetropolitan areas repopulating? *Demography* 15:13–39.

———. 1983. The impasse in migration theory: A sketch map for potential escapees. In *Population movements: Their forms and functions in urbanizations and development,* ed. P. A. Morrison. Brussels: International Union for the Scientific Study of Population, Orina Editions.

Industrial Geography

William B. Beyers | Susan Christopherson |
Rodney A. Erickson | Lay J. Gibson | Geoffrey G. D. Hewings |
Edward J. Malecki | James E. McConnell | John Rees

Industrial geography is clearly one of the most dynamic subfields of economic geography. Industrial geographers have provided aggressive intellectual leadership in development and adaptation of theories and technical approaches to enhance the understanding of the space economy. The outputs of scholars working in this area are distinctive, yet well integrated into a larger body of work produced by economists, planners, and researchers from several business disciplines.

Industrial geographers are known both for their basic research and for their applied research. Frequently the line between these two orientations is blurred, and basic research themes are carried through to their ultimate conclusion as policy statements. Few specialties in geography have more potential to contribute to public policy and private-sector decision-making than industrial geography.

In many ways, what this chapter reveals is evidence for that time-honored dictum that economics is too important to be left to economists. If it were not for the research of industrial geographers, we would know very little about the locational dimension of industrial change. In this regard, David Smith's *Industrial Location* (1981) still stands as one of the best manifestations of the locational dimension to economic theory. He integrates neo-Weberian cost models with demand factors over time, but an explicit concern with public policy is relegated to the last section of the book.

One theme that more recently has unified the various research traditions within industrial geography has been concern over the limitations of orthodox neoclassical economics. However, this dissatisfaction has manifest itself in a number of different ways, and geographers have not provided any real theoretical alternative to it. One explicit reaction to the neoclassical orthodoxy is Ian Hamilton's (1974) volume, *Spatial Perspectives on Industrial Organization and Decision Making*. It stands out as a symbol of the behavioralists' concern with how business-location decisions are actually made, how such decision-making fits the broader context of investment policy and company growth, and how companies adapt to a changing economic environment through their location decisions.

Hamilton's book also represents a milestone of a different kind. It launched a series of books on spatial analysis, industry, and the industrial environment as a select outlet for the work of the international community of industrial geographers, specifically the International Geographical Union (IGU) Commission on Industrial Systems.[1] This group spawned a number of study groups in different countries, including the IGU/Association of American Geographers (AAG) Specialty Group in Industrial Geography. It has now met for 10 years and is the oldest specialty group within the AAG.

Industrial geography is certainly a focused specialty, but it is not narrow. Given the variety of subjects covered, theoretical perspectives offered, and technical approaches employed, an overview of the field benefits from a specialist approach.

The number of specific topics and perspectives that might be offered in a review of industrial geography is indeed large. For example, a section on telecommunication and spatial restructuring and a section on corporate behavior might have been offered, had space permitted. Instead, these and other potential themes have been integrated into the seven general discussions that are provided. No claim is made that this chapter presents an exhaustive review of industrial geography. But every effort has been made to present a comprehensive and balanced picture of industrial geography's historical development, topics of current interest, potential changes in emphasis and new directions for research, and the potential of industrial geography's influence on formulation of public policies and private decision-making.

LOCATIONAL SHIFTS IN MANUFACTURING

The spatial distribution of manufacturing activities has long been a topic of interest to industrial geographers. Locational change is inherent in these distributions, and its pace has quickened in recent decades. These locational shifts have often signaled important repercussions of growth and decline in area economies, with important effects on labor-market adjustment. The economic turmoil that has been felt in many communities and regions across the U.S. over the past two decades bears evidence of the wide-ranging problems of adaptation that must occur in the wake of industrial restructuring. The expertise of industrial geographers has contributed greatly to a better understanding of the causes of change, the prospects for adaptation, and the policies for industrial revitalization.

Traditions of Industrial Location

The microeconomics of a firm's production function, and the resulting least-cost location, have always had a strong tradition in industrial geography. This paradigm has underpinned many case studies in which the locational patterns of industries have been analyzed. Transportation, labor, land, tax, energy, and other cost differences have played a particularly prominent role in the analysis of regional manufacturing-location

---

[1] The first collective manifestation of an "American" industrial geography of recent vintage is represented by Rees, Hewings, and Stafford's (1981) collection on *Industrial Location and Regional Systems,* where many papers were drawn from the Specialty Group meeting in Philadelphia in 1979. Two major themes emerged in that volume and have dominated our research since: (1) the need to critically examine our theoretical perspectives, and (2) an emerging concern with the interface between industrial location and public policy.

patterns and shifts. Likewise, agglomeration and urbanization economies also represent a major explanatory theme in industrial location. The location-factor approach has also characterized many analyses of industrial decentralization from core areas to the suburban fringe, where aggregate costs of production are often lower than in core areas.

Over the past quarter of a century, industrial-location research has been greatly influenced by the "geography of enterprise." Here, the focus of attention shifted from the analysis of aggregate industrial sectors to the firm as an organization, especially the corporation (Krumme 1969). In this conceptualization, locational shifts result from the decision-making of corporate managers, who pursue goals for the firm and themselves. Only one of these goals may be profit maximization.

Location choice for firms is influenced by a complex array of elements, both internal and external to the firm (Townroe 1969). The application of theories of risk and uncertainty plays a critical role in such decision analysis. In these, the firm is seen as responding to opportunities and constraints in its environmental milieu (Hanink and Cromley 1987a).

Structural Change and Current Approaches

The focus on the interplay of firms' decision-making has led to a reinterpretation of regional and metropolitan patterns of industrial shifts, under the rubric of "structural change." Structural change reflects a complex set of alterations in a regional economy that derive from the cumulative investment decisions of firms over time. Structural change may involve alterations in the levels of capital invested, and the mix of regional activities as investment decisions are made. There are obvious effects on the work force that may include the substitution of capital for labor, and resultant deskilling of jobs as well as redundancy. Employment relations frequently change, and nonlocal ownership of capital has also accompanied the restructuring process (Clark, Gertler, and Whiteman 1986).

Industrial geographers increasingly view locational shifts within a structural-change framework in which industrial-location patterns themselves, while interesting, are only a manifestation of processes of investment decision-making in the capitalist mode of production. This orientation is reflected in geographers' use of components-of-change approaches, in which regional manufacturing change has been disaggregated into establishment births, deaths, and migration (Lloyd 1979; Mason 1980). Likewise, corresponding employment changes have been ascribed to establishment births and deaths, in situ expansions and contractions, and relocations of plants into and out of the region. This approach has been used both in broad regional analysis and in intrametropolitan analysis.

Such components-of-change approaches have provided much useful information concerning the broad patterns of regional investment and disinvestment in manufacturing. Nonetheless, the processes underlying locational shifts remain rather illusive in this "change accounting" framework. Searches for richer explanatory frameworks have led geographers further into theories of industrial organization, technological change, and social theory.

One of the more prominent paradigms of the past decade has been the product cycle or profit cycle (Norton and Rees 1979). In the product cycle, the relative importance of various production factors (such as capital, labor, technology, external economies, and management during the life cycle of a product) leads firms to different locational investments during subsequent phases of the cycle. The location of branch plants in smaller metropolitan and nonmetropolitan areas, which has characterized industrial restructuring during much of the past two decades, has often been portrayed in this explanatory framework (Erickson and Leinbach 1979). It has also been used to account for differences in the rates of new firm formation among various regional geographies.

In a similar vein, the profit cycle—while focusing more explicitly on profitability and market structure (oligopoly) during the life cycle—has emphasized the broad regional shifts that have occurred in the location of manufacturing employment over rather long periods of U.S. economic history (Markusen 1985).

The relationships between capital and labor have been further elucidated under the rubric of the employment relation and the spatial division of labor (Clark 1981). Here, the conflict between labor and capital over control of the production process leads management to adopt strategies of spatial decentralization of its plants and facilities. In this decentralized production mode, depressed and less-organized labor markets often become the sites of standardized production, where lower wage rates prevail (Peet 1983).

Alternatively, firms that require highly skilled workers with specific technical expertise are more dependent upon the locational preferences of these latter labor pools. The same general themes relating technology of production and labor requirements underpin the labor theory of production. The labor process has clearly become more central to industrial location and shifts, as rapid technological change has impacted much of contemporary manufacturing (Storper and Walker 1983).

Another related paradigm, applied particularly to industrial restructuring at the intrametropolitan scale, has been the comparative roles of capital and labor under a Heckscher-Ohlin construct. Wage rates have been shown to vary systematically across large metropolitan areas. Industrial decentralization is seen as a response to the comparative advantage of labor-intensive production in older core areas and capital-intensive production in newer suburban areas (Scott 1981 and 1982). Thus, capital deepening and attendant industrial restructuring are continually changing the industrial geography of the metropolis.

Future Directions and Emerging Issues in Industrial Location

Rapid technological change has created an urgent need to focus even more deeply on the role of both labor and capital in manufacturing location and its shifts. The traditional treatment of labor as a homogeneous and passive input to the production process has been largely discarded by contemporary industrial geographers. The preeminence of the corporation as the kingpin of regional industrial change also has been challenged, by studies that stress the role of new firm formation and small enterprises as the principal generators of net new jobs in the American industrial economy. Here again, labor has been shown to be a crucial element in the growth and localization of such establishments.

The difficulties of acquiring data on capital have long hampered the efforts of geographers to assess its specific role in industrial location and change. However, recent research indicates that manufacturing investment contributes a major share of private capital accumulation and consistently leads investment in other sectors of metropolitan economies. This implies that technological change has involved increasing capital substitution for labor (Gertler 1986). Other current research involves the analysis of productivity differences over space. These differences are related to capital input as an explanatory mechanism for industrial shifts (Casetti 1984).

Capital movements are clearly a crucial determinant of industrial location and restructuring at all geographical scales. Contemporary industrial geographers realize that the flows of capital take place over long distances and multinational space. Investment decisions of multinational corporations and global capital interests reach down into virtually every region, and it is no longer sufficient to study industrial location, even at the level of the metropolis, without due concern for the potential influences of global capital movements (Susman and Schutz 1983). Accordingly, studies of both foreign direct investment in manufacturing and global capital flows are aspects of research in industrial geography that will surely intensify (McConnell 1980; O'hUallachain 1984).

Contemporary international and interregional industrial restructuring have also signaled a shift toward greater interest in competitiveness and firm competitive strategy among industrial geographers (Schoenberger 1986). Some studies have shunned the more deterministic paradigms of industrial change in favor of an approach that stresses management responses to alternative production techniques, industrial organization, and the actions of competitor firms (Harrington 1985).

Finally, the broad institutional setting for industrial location has become a topic of increasing interest to geographers. We are seeing a resurgence of interest in the role of government in industrial location and restructuring. This has included specific studies of the influences of taxation, zoning, environmental regulation, and government development policies, among others, in contributing to the contemporary locational shifts in manufacturing activities (Stafford 1985). Thus, it is not surprising to find industrial geographers at the heart of the dialogue concerning industrial policy and public efforts to alleviate the human consequences of locational shifts.

RESEARCH ON THE PRODUCER SERVICES

Within the last decade, an increasing amount of research by industrial geographers has focused on the "producer services." Outputs of the producer services are used as inputs in the production process by manufacturers, governments, and service industries, as well as by households. This segment of the economy includes financial, insurance, real estate, legal, business, and professional services, and central administrative office (CAO) functions. Interest in the producer services stems from their rapid growth rate and emergence as an important component in the economic base of many cities. Consider recent North American work on the producer services from a conceptual and empirical perspective.

Conceptual Approaches in Producer Services

It would be accurate to say, at this point in time, that geographers have not developed a distinctive body of theory to explain the location decisions of establishments/firms in the producer services. Existing conceptualizations (such as that offered by Pred 1977) view the development of the services as a part of a multiregional system, evolving over time as a function of changes in technology, scale, external market relations, and corporate organizational structure. "City-systems" of the type conceptualized by Pred have hierarchical structures, presumably mirrored in the organizational structures of firms embedded in these hierarchies. Locational decision-making is framed in market-oriented models of either a central place or Weberian theoretic nature (Pred 1977; Gottmann 1983; Daniels 1985). However, it should be noted that Pred's empirical research showed poor correspondence to a hierarchical model in the organizational structures of individual firms.

Agglomeration economies have been emphasized in explanations of the pattern of headquarters and, presumably, services linked to the support of headquarters activity (Borchert 1978; Semple 1985; Green 1983; Noyelle and Stanback 1983). Some scholars have questioned whether changes in telecommunications and information-processing technology will allow agglomerations to disperse (Gottmann 1983; Sternlieb and Hughes 1983; Hepworth 1986; Blazar 1985). Arguments have been made in support of the dispersal option. Arguments have also been posed in support of the many personal/human forces that tend to keep agglomerations together as centers of power for the producer services (Gottmann 1983; Daniels 1986). However, little progress appears to have been made in developing measures of agglomeration economies.

Walker (1985) has cogently argued that much of the apparent growth of producer services is a rearticulation of work previously done in other industrial settings, a change which he argues is to be expected in the process of development in a capitalist society. Walker's emphasis on the continued division of labor—and consequent formation of new industrial categories—has yet to be verified by case studies that would allow us to confirm, reject, or assess the relative importance of his arguments.

Empirical Approaches to Producer Services

The bulk of American work on the producer services is of an empirical nature, either at the national level or in regional case studies. Noyelle and Stanback (1983) analyzed change within the largest standard metropolitan statistical areas (SMSAs) in the U.S. from 1959 through 1976; their work emphasized the growing role of the producer services and the "complex of corporate activities" in the development of the nation's city-system. They found that the urban areas of the country could be divided into four groups: diversified service centers, specialized service centers, production centers, and consumer-oriented centers.

Noyelle and Stanback also found that older, large cities, which were important centers of manufacturing as well as service activity, and which have now developed a healthy, export-oriented producer-services sector, have tended toward relatively rapid growth. Cities that have not displayed this type of structural change have often become stagnant. Specialized service centers, such as state capitals, entertainment centers, and university towns—all places trading or exporting their services—have also grown rapidly. Cities whose service structures are just oriented to their hinterlands have tended to grow more slowly.

Kirn (1987) analyzed service-sector growth in the U.S. over the 1958–1977 time period. Kirn's sample of places ranged from small nonmetropolitan counties to the nation's largest cities. Kirn found evidence of relatively rapid growth in nonmetropolitan areas in some producer services, notably in accounting; finance, insurance, and real estate; advertising; and management, consulting, and public relations. On the other hand, legal services and engineering and architectural services tended to grow more rapidly in larger places.

Another theme in national analyses of the producer services has involved studies of the location of corporate headquarters functions. Borchert's (1978) work on "control points," identified as the location of corporate headquarters for firms in different lines of business activity, serves as a good example. Similar efforts have been undertaken by Semple (1985) and Green (1983). Wheeler and Brown (1985) undertook a similar study in the southern U.S. and demonstrated the weakened relationship between the corporate and population hierarchies in recent years, indicating a dispersal of the pattern of corporate control.

Other studies have focused on linkages or market relationships. Polese and Coffey (1987) and Polese (1982) explored the sources of producer services used by various Canadian firms, from the standpoint of their internal acquisition within the firm, or their contractual purchase externally. They found that intrafirm trade possibilities were rich and significant with respect to the functions of the producer services. They also discovered that long-distance intrafirm and external service trade was common and is becoming more common as service markets are internationalized and as information-transmission costs (related to distance) decline.

Beyers and Alvine (1985) reported strong, and growing, degrees of external market activity in a sample of producer-service firms in the Puget Sound region. They found that those firms involved in relatively specialized "market niches" tended to be the most externalized. They also found, as did Polese and Coffey (1987), that the organizational structure of the firm influenced its propensity to be involved with export markets. Locally headquartered firms with branch offices in other regions tended to be relatively export-oriented. But as Polese and Coffey point out, there are wide variations in the mode by which these service functions are actually performed. Ley and Hutton (1987) offer similar evidence for a sample of Vancouver producer-service firms.

Other themes explored in studies of the producer services include the impact of information technologies on the organization of industries, the impact of the changing nature of the regulatory environment, and intrametropolitan locational trends. Hepworth's (1986) research on Canadian multilocational firms shows how new telecommunications technologies have allowed the rationalization of production systems within the services and manufacturing industry. Blazar (1985) has shown how nonmetropolitan regions can capitalize upon these new technical opportunities to diversify their economic base.

Many producer services have been regulated by states and the federal government; Holly (1987) has provided important insights into the responses emerging within the U.S. banking industry to new organizational possibilities in the current climate of deregulation. Daniels (1986) has contrasted the development of international banking in London and New York. Wheeler and Dillon (1985) have focused on the development of banking and urban structure, and found meaningful relationships between the level of commercial banking activity in a city and the significance of local corporate-headquarters activity, as well as the size of the market area.

Daniels (1982), Rees (1978), and Erickson (1983) provide case studies of suburban office development. Daniels studied the development of suburban offices in Seattle. He found most growth was by firms indigenous to the suburbs, with little migration of firms from central city locations. Erickson found suburban service growth to be relatively rapid, in comparison to overall metropolitan-area services growth. Rees found that Dallas had a relatively decentralized pattern of central administrative office (CAO) functions, but suggested that the importance of these sectors in newer "postindustrial" cities is likely to be less than is the case for older "industrial" cities having a strong complement of CAO activity in their economic base.

While conceptual and empirical work on the producer services has advanced our understanding of the causes of this sector's development, much work remains to be done. Extension and integration of bodies of location and development theory, articulated for segments of the economy that flowered in an earlier era, now need to be undertaken for the producer services. National-level analyses of trends in producer-services development for more recent time periods must be completed.

Also, field research on causes of development within this sector must be undertaken, to a much greater extent than has been accomplished. For example, there is little systematic knowledge on the relative importance of vertical-disintegration processes within manufacturing, or other sectors of the economy, in leading to the explosive growth of employment that is counted as producer services. This is in contrast to the research on technology-driven entrepreneurship that is autonomously innovating new approaches to service-industry delivery.

Industrial geographers need to work to develop a systematic research agenda on this segment of the economy. This agenda must be linked conceptually and empirically to research on other key sectors. It also must be linked to a needed concern for the social implications brought about by the transformation of our economy into one in which information—its generation, transmission, storage, and use—has become an ever-more critical basis for the competitiveness of the national economy in an increasingly internationalized world economy.

HIGH TECHNOLOGY AND INDUSTRIAL INNOVATION

The topics of high technology and industrial innovation are important ones within industrial geography. They are among the topics where geographers contribute to the formulation of policy as well as theory, because many, if not most, regions are actively pursuing economic development by attempting to enhance the technological level of their firms and to nurture new local or indigenous firms. Industrial innovation activities, especially R&D, contribute to the innovative potential of regions, by providing new products and processes for economic activity, and by employing the highly skilled professional and technical workers who are among a region's most probable entrepreneurs.

The Legacy

The history of research on high-technology industries and industrial innovation is actually quite recent. It is part of the major resurgence of work in economic geography that began in the 1970s. Research on innovation and technology grew out of concern within regional-development research over the varying structures and prospects of regions in an era of rapid technological change (Thomas 1975). To some extent, the research also had origins in studies of innovation diffusion, which was also beginning to focus on firms and their multifaceted role in technological change (Thomas 1975). But it also grew out of research on individual firms and their role in regional economies. For a review of this early work, see Malecki (1983).

Early Approaches

The shift to a broader focus on the role of firms in regional economic change grew largely out of Thomas' identification of product-cycle notions as critical to understanding the nature of regional economic change. But before dealing with critiques of this model and its applications, a brief discussion should be helpful.

In the product cycle, the activities of research and development, innovation, and other nonroutine functions are the primary focus of the first stage, or *innovation phase*. Skilled labor, product innovation, small-volume production, and minimal automation are the hallmarks of this phase. Although some products—especially custom and small-batch items—never progress beyond this first stage, many products move into the *growth phase*. This second stage in the cycle is a period of capital investment and process innovation, usually involving automated production, associated with increases in production volume.

By the third stage, or *mature phase,* production is accomplished in a routine fashion by either highly automated equipment or unskilled labor, and little further innovation takes place. In this stage, manufacturing plants can be located virtually anywhere in the world. Manufacturing activity of this sort has dispersed widely in search of low-cost labor in such industries as motor vehicles and electronics.

The product-cycle model has had several useful applications. First, it allows economic activities, or even regions, to be distinguished on the basis of the skills required of the labor force, rather than only on the basis of their industrial sector, as was common previously. The separation within firms of different activities—some utilizing highly skilled workers, others with predominantly low-skilled labor—also gave rise to the spatial division of labor and the regional specialization of activities as key characteristics of modern firms.

Secondly, the product-cycle model sets industrial innovation in the context of the broader set of firms' activities, which include both nonproduction and production activities. Commitment to innovation, or reaction to it, is necessary for virtually all firms, even those whose primary concern is to copy the innovations of others and produce them at lower cost.

The regional economic implications of the product cycle are extensive. Around the world, firms are observed to concentrate their innovation-phase activities, such as R&D laboratories and pilot plants, in a relatively small number of locations. These locations, then, also are concentrations of skilled and technical workers, who will continually be working on new innovations and improvements to older products. By contrast, routine production of standardized products is more widespread because more plants are needed to produce the larger output. But the plants producing these products are predominantly locations of unskilled workers who are unlikely to spawn new innovations as time goes on.

The product cycle is not a general model for all industries and firms because it imposes a sequence of phases and therefore a somewhat artificial view of regional life cycles that mirror the cycles for a product. Similarly, others emphasize that in some industries, especially those with relatively small production volumes and small firms, innovation is a continual process, rather than one which tapers off over time. The differences among high-tech industries largely result from the variations in volume, and the degree to which standardization is critical to competition. Recent advances in computer-integrated manufacturing and "flexible automation" appear to render obsolete the standardization found in the product cycle, as firms are able to concentrate on custom and small-batch items.

Geographical research on industrial innovation, and its importance in an increasingly international marketplace, has focused on research and development (R&D) and innovation as critical elements for regional success, just as they are thought to be related to the success of firms and of nations. The interrelationship among these scales and the search for regional and local policies have been the impetus behind many of the recent research efforts concerning high-technology firms, industries, and regions (Hall and Markusen 1985; Rees 1986).

Current Approaches

While the research described above has been rather broadly concerned with firms and their regional economic impacts, a second area of research has attempted to learn more about the geographical dimensions of a central element of high technology, corporate R&D. This research has noted the integral nature of R&D with the organization of the rest of the firm. Firms place great importance on links internal to the firm, and external geographic or regional characteristics are secondary (Malecki 1980). Close proximity between R&D and corporate headquarters assures contact with marketing and other key corporate functions.

In addition to such internal organizational priorities, firms also tend to locate R&D where it can be close to information networks for the industry and close to pools of research workers. This is a secondary priority for large firms, but it is critically important for small firms, which rely more on external sources of information and people. Both of these factors serve to concentrate industrial R&D in large urban regions, where such agglomeration economies are present. These areas are the sites of corporate headquarters as well as major labor pools, and they are attractive areas for the recruitment and retention of highly mobile professional workers.

As a result, the geographic pattern of R&D is far more concentrated than that of manufacturing. R&D remains near major headquarters cities, and is unlikely to disperse to the same degree as manufacturing. High-tech manufacturing, by contrast, is quite widespread (Markusen, Hall, and Glasmeier 1986). The potential for dispersion and the emergence of new centers of high-tech activity have raised the hopes of local and state leaders that high-tech regions can be created and fostered away from high-tech core regions such as Silicon Valley and Boston (Hall and Markusen 1985; Rees 1986).

Case Studies of High-tech Regions

A great deal of the current research on high-technology firms and industries has taken the form of case studies of particular firms or regions. The journals *Environment and*

Planning A and *Regional Studies* are favored outlets for these studies; others are collected in Hall and Markusen (1985) and Rees (1986). While much of the research looks only at a single region, Oakey (1984) has completed a comparative study of three high-tech regions—the San Francisco Bay area of the U.S., also known as Silicon Valley; the area of southern Scotland called Silicon Glen; and the southeast region of England centered on London. His results show that the Silicon Valley region is clearly the most innovative, mainly because of the degree to which information, workers, markets, and investment capital are concentrated. He thus suggests that agglomeration economies—with their advantages to firms from clustering, especially in large urban regions—are a feature present to a greater degree in California than in England or Scotland.

Several others have focused on the Silicon Valley and California as a whole. They illustrate the spatial division of labor, the increasing concentration in high-technology industries as they attain large-volume production, and the distinctions made by firms between professional and unskilled workers (see the studies collected in Hall and Markusen 1985). Scott (1983a, 1986) has examined high-tech development in southern California, another growing agglomeration of high-technology industry.

The role of government policy in the development of high technology includes that of the federal government, and military spending in particular. Much of the pattern of high-tech industry derives directly from defense-industry locations, both for manufacturing and for R&D (Markusen, Hall, and Glasmeier 1986; Rees 1986). A growing area of interest is the arena of state and local policy, and the effort to build or create the conditions for high-technology-based regional development. Applying the results of theoretical and empirical studies to specific regions and places is a challenge that continues to attract the attention of industrial geographers (Rees 1986).

Future Directions and Emerging Issues in High Technology

Interest in high technology has also begun to develop a broader concern for understanding of its place in the wider process of urbanization and regional development. Although high-tech industries are not the only focus of study, they provide excellent laboratories for understanding the dynamics of entrepreneurial opportunity, the development of interfirm linkages, and the development of specialized industrial complexes wherein the division of labor within and among firms is able to be efficiently organized. Recently, Scott has shown this in several contexts, among them the printed-circuits industry of greater Los Angeles (Scott 1983b), the development of a high-technology complex in Orange County (Scott 1986), and in the internationalization of the semiconductor industry (Scott 1987).

The integration of the behavior of firms with their territorial context thus allows industrialization in a high-tech era to be analyzed as a counterpart to earlier eras of industrial activity. We are likely to see less focus on high technology than on the more widely applicable processes of interdependent industrial and urban-regional change.

A second growing focus for future research is the issue of firm formation, or entrepreneurship. None of the previous approaches adequately addresses the foundation of new firms, a process which takes place at a higher rate in some locales than in others. Research by geographers on this topic is perhaps more advanced in Europe, where Silicon Valleys are less abundant and their creation is an object of public policy.

Thomas (1987) provides a conceptualization that distinguishes the "regional factor" from factors internal to the firm, and from global pressures on firms. Indeed, much of the emphasis of policy at local and regional levels is intended to generate the regional conditions that would promote regional economic growth and development. The focus on entrepreneurship contrasts with much of the previous work in industrial geography, in that it deemphasizes the global environment and corporate strategy and structure issues of multilocation firms, and stresses the regional attributes that affect economic activity.

Entrepreneurial regions appear to contrast with branch-plant regions, where little interaction may take place within the regions in which they are located. Newly formed firms require more than large firms from their local environment, including capital, information, labor, and linkages. These elements must be available locally, or a new firm will not survive or thrive. But these factors are not the usual products of public policies, which have emphasized tax breaks and subsidies rather than information flows, linkages, and contacts. There is much to learn empirically about entrepreneurs, and there are challenges to the developed theory base which has been centered around the role of large firms. Future regional growth will be a combination of the effects of large and small firms and of the interactions between them. Technological change will continue to affect the economic structure of regions and the nature of economic activity.

INTERNATIONAL TRADE AND FOREIGN INVESTMENT

International trade and foreign investment bridge the increasingly artificial boundary between the world economy and national and regional space economies. They provide the focus for investigating the interrelationships between the restructuring taking place within the international business environment and the dynamics of industrial operations occurring within and between countries and regions. Associated with the movement of trade and investments across national frontiers are important policy issues. These include the growth and development of nation-states and regions, the transborder flow of information and technology, the international division of labor, economic justice, the nation-state vs. transnational corporations, industry and corporate growth and decline, and the crossborder transmission of business cycles and culture. In short, the topics of international trade and foreign investment provide geographers with a rich research agenda involving theory building, empirical analyses, and policy evaluation.

Historical Roots

The historical roots of geographers' interest in the global patterns of trade, investment, and industrial production can be traced to the early writings on commercial geography (Chisholm 1889; Smith 1913). Through the years, close ties have been recognized among international trade, capital investment flows, and the location of industrial production. Ohlin, for example, argues that the theory of international trade is part of a general theory of localization (1933, vii). Later, Isard notes that "trade theory can be broadly conceived as synonymous with the general theory of location and space-economy" (1956, 54). And more recently, Johnston suggests that "the history of location theory during the last two centuries is the history of tentative answers" to "the spatial determinants of [capital] investment" in productive activities (1986, 268).

Prominent Research Themes

Although the interrelationships among international trade, capital investment flows, and the location of industrial activity have long been recognized by geographers, we have been less than enthusiastic about exploring such linkages. Even more puzzling has been our relative lack of concern for the interrelationships between the changing patterns of trade and investment that are taking place in the world economy, and the evolving characteristics of national and regional space economies.

Instead, a review of the research by American geographers on international trade and foreign investment over the past decade or so suggests the existence of two general approaches, a traditional nonintegrated one, and one focusing upon interrelationships.

Traditional Approach. The first approach is a more-traditional perspective, in which the changing patterns of international trade and foreign investment are analyzed without much attention to integrating the two topics or to relating them to patterns of national and regional growth and development.

This work has been focused upon trade as transportation, country-to-country commodity flows, and a measure of regional economic integration (McConnell 1986). Studies of trade as transportation describe the development and changing functions of seaports and their hinterlands and forelands, the morphology of ocean trade routes, and the influence of transport rate structures. Most studies of trade as commodity flow provide an econometric analysis of the various factors presumed to influence the direction and intensity of trade movements. Studies of trade and regional economic integration assess the influence of various international trade agreements on the creation and diversion of commerce.

Although several exceptions exist, for the most part these analyses are static in nature. They emphasize the macro-aspects of international business and ignore variations across industry sectors and the role of corporate decision-makers. Also, they do not make explicit the interconnections among trade, the location of industrial production, and policy issues associated with national and regional growth and development.

In like manner, the more-traditional studies of foreign direct investment are focused upon what Sayer (1982) has termed "descriptive monitoring." In general, the authors of these works describe the spatial distribution of foreign direct investments, and they measure the interrelationships between changing patterns of foreign direct investments and other associated spatial and nonspatial factors, and they advance suppositions about the processes surrounding foreign direct-investment decisions and regional growth and development (Gaile and Hanink 1984).

More specifically, as noted by O'hUallachain (1986), most of these studies are designed (1) to examine the location and spatial distribution of foreign-owned subsidiaries within a national or regional setting; (2) to measure and assess the impacts these investment decisions are having upon national and regional economies; and (3) to relate the outcome of the direct-investment decision to the technological structure of the host region and to the organizational and external linkage characteristics of the investing corporations.

In most instances, however, these studies lack the longitudinal, firm-level data sets that are necessary to examine the investment-location process in a systematic manner. They are linked in only a superficial manner to the evolving theory of investment behavior, and, therefore, little has been done in the area of synthesis. Also, there is a need for further studies by American geographers on corporate-investment strategy, or the impact of direct investments upon space economies (Hanink and Cromley, 1987b).

Interrelationships Approach. The second approach focuses upon the interrelationships between (1) shifting patterns of global trade and investment flows and the restructuring of industrial production, and (2) the changing patterns of economic growth and development occurring within and between nation-states and regions.

Two examples of such work are noteworthy. Harrington, Burns, and Cheung (1986) use the market-imperfections model (from the theory of foreign direct investments) to test hypotheses about the sector, activity, size mix, and regional-development policy impact of Canadian direct investments in the western region of New York State. At a much-broader geographic scale, Scott (1987) analyzes the location and international division of labor associated with the evolution and spatial organization of the semiconductor industry in Southeast-Asia.

Both of these studies emphasize the interrelationships between the trade and investment decisions of foreign-owned enterprises and the industrial structure, technological capabilities, and economic growth and development performances of national or regional host economies. Moreover, both studies rely upon survey data from individual enterprises. And both help us begin to develop an appropriate synthesis from which effective and appropriate policy formulation can be put forth for national and regional planners and policymakers.

Emerging Issues and Future Directions in International Trade and Foreign Investment

Considering the future, several developments already unfolding throughout the global environment are likely to influence greatly the directions of research over the next decade or so. One of the most apparent of these developments is the increasing internationalization of production and service activities, and the growing interdependence and rivalry of nation-states and transnational corporations. Emerging from this environment are new industries and enterprises that are generating services as their principal output. They are relying upon knowledge-based assets rather than resource-intensive strengths for their competitive advantage. They are global in orientation from the very beginning and are often footloose with respect to their choice of international locations.

For industrial geographers, these shifts suggest the need for research that is focused upon several areas:

- ☐ transnational trading companies and the processes associated with intrafirm transmissions of information and services across international frontiers.
- ☐ the impact that locations of these new industries and enterprises is likely to have on the rise and decline of indigenous firms and labor markets in particular nation-states and regions.

ANALYSIS AND
MANAGEMENT OF
SOCIETAL
GROWTH AND
CHANGE

- ☐ changing national and international patterns of organizational linkages, intrafirm trade, and crossborder investment decisions.
- ☐ the policy implications that the transborder flow of knowledge and services has for nation-states and corporations.

Another dominant trend in the world economy is the shift that is occurring in the geographical center of gravity for international trade and investment, from the countries of the Atlantic community to those rimming the Pacific Ocean. This reorientation has been accompanied by the rationalization and restructuring of multinational enterprises. It has resulted in a new international division of labor, and a reallocation of resources in developed countries from capital-intensive traditional industries to knowledge-based enterprises. The reorientation has also prompted the concurrent emergence of strongly competitive transnational corporations in east Asia. And it has affected the spatial disaggregation of the internal functions of the firm and the relocation of these functions within and between nation-states.

For geographers, this new axis of global business raises important questions about:

- ☐ the impact of this restructuring upon traditional global movements of capital, commodity, and service.
- ☐ the transformations that are occurring in national and regional labor markets, industry mix, and corporate linkages.
- ☐ the changes that are taking place in the processes associated with urban and regional growth, as corporations and nation-states shift their global orientation.
- ☐ the impact these international shifts in competitive advantage are having upon national and regional policies on allocating government subsidies to particular geographic areas and industry sectors, fostering internationally competitive domestic industries, formulating trade and investment agreements with other nation-states, and resolving the continuing conflict between national control and the countervailing power of global corporations.

LABOR AND LABOR MARKETS

From a rather narrow conceptual base that defined the topic until the mid-1970s, there has been a profusion of work dealing with the subject of labor in economic and industrial geography. This section briefly reviews the work on labor that preceded the recent expansion of interest in this subject. It then traces some of the major lines of inquiry in recent work on labor and labor markets, and suggests how geographers are contributing to the analysis of the changing U.S. labor force and new sources of differentiation among regional labor markets.

Historical Background

Prior to 1950, economic geography was dominated by the study of primary-sector industries, trade routes, and commodities. Because of this resource orientation, labor did not appear as a significant variable. In the 1950s, however, developments in the subject matter and method of geography began to alter this orientation:

1. Economic geographers became more interested in geographic patterns and in the composition of regional economies.

2. With the expansion of urban geography and its intersection with economic geography, geographic analysis became "less oriented toward the natural world and more oriented toward the social world" (McNee 1959). In particular, work on labor mobility and the journey-to-work stimulated interest in workplace location and in the dynamics of urban and regional development.

A few geographers expressed the need to move beyond the description of land-use patterns to explanations of areal economic patterns, with labor as a central component. According to Vance, "An economic geography without labor and capital is one without understanding" (Vance 1960). Methodologically, a dynamic analysis of the reasons for the location of economic activities that included labor was more characteristic of urban historical geography and of analysis of intraurban patterns.

For most economic geographers during this period, labor was simply a factor of production. Variations among cities and regions were measured with respect to differences in the quantity of employment in the various sectors (Nelson 1955; Alexander 1952). Underlying these regional differences was a mechanical process through which industries appear in places as a consequence of the relative cost of factors, including labor (cf. Alexander 1952).

The Radical Urban Critique

In the 1960s and early 1970s, the conception of labor as employment and a factor of production equivalent to other factors continued to prevail in economic geography. With the development of a radical critique within urban and economic geography, however, there was more emphasis on the dynamics of the capital-labor relationship in shaping spatial patterns (see chapters on the urban problematic and geography from the left in this volume). This analysis was either urban-oriented (Harvey 1973), or examined the emerging division of labor at the international scale. One of the lasting contributions of this work was to infuse a range of geographic studies at all scales with a concern for the construction of social as well as spatial inequality (Peet 1977).

New Explanations for Labor Market Differences

In the 1970s, Allan Pred developed an analysis of economic development of city systems that provided a more complex basis for differentiating among labor markets. Pred's historical and contemporary work on the evolution of city systems suggested that expanded industrial activity gave rise to specialized labor requirements, which could only be met in a large city with a substantial number of skilled workers. Pred's work contradicted the tenets of central place and growth-pole theory in arguing for the asymmetry of interurban linkages. This asymmetry is demonstrated by the way in which major multilocational firms distribute jobs among major urban centers (Pred 1977). This work began a focus on the firm as an actor, creating and destroying employment opportunities in labor markets.

Two factors brought labor more centrally into the analysis of industrial location and regional development in the late 1970s and early 1980s. First, geographers participated in the development of a radical regional analysis in the U.S. and the United Kingdom. Secondly, economic geographers from a wide range of perspectives recognized that transport costs were no longer the central variable in industrial location.

The general reconsideration of location theory led to a major reevaluation of the role of labor in firm industrial-location decisions (Storper and Walker 1983). Much of the early analysis in this vein built on concepts about divisions within the work force, such as dual labor-market theory, or labor-market segmentation models or constructs (e.g. the international division of labor).

A Proliferation of Approaches to the Labor Question

The 1980s have been a period of expansion in the approaches taken to labor—an intersection of traditional regional questions, with the more complex understanding of labor demand. Much of this work has been strongly influenced by the "restructuring" debate. Urban and economic geographers have joined a more general process in the social sciences, to reconsider assumptions about the direction of development in capitalist economies (Scott and Storper 1986).

Among this work, some particular strands stand out. A significant amount of work has been done in the area of industrial organization, considering its implications for labor demand and industrial location. A series of studies by Allen Scott was particularly significant in demonstrating how production organization in different industries is related to the formation of segmented labor markets (Scott 1983a, 1983b, 1984a, 1984b). Recent research on transactions-intensive and information-laden production activities also has contributed an understanding of the processes that are shaping contemporary labor markets, particularly those in the highest-order urban centers. With agglomeration has come a reconcentration of the labor engaged in those production activities (Storper and Christopherson 1987). Related research has considered the effects of new technologies on industrial organization and labor demand (Walker 1985), and the ways in which the gender composition of the work force enters into firm location decisions (Nelson 1986).

Much of the new work on labor, industrial location, and labor markets has grown out of the traditional questions of economic geography. Gordon Clark, however, has opened up a new area of inquiry—the geography of labor relations—examining the role of regulatory structures on the spatial pattern of labor organization. This spatial pattern, in turn, influences how crises in employment relations are resolved, nationally and within communities (Clark 1981, 1986; Clark and Gertler 1983).

In contrast with the work that preceded it, much of the recent attention to labor questions concentrates on the dynamics operating in already-developed economies. Finally, recent changes in the economy, such as labor flexibility and the increasing separation of the firm from sectorally specific production activities, have raised a new set of issues that geographers are just beginning to address. These include the gender division of labor, the limits of sectoral analysis, and how to account for differences among labor markets in an economy in which 70% of workers are service workers.

MODELS FOR REGIONAL INDUSTRIAL ANALYSIS

The processes of industrial change have been described and analyzed by industrial geographers using a variety of techniques. However, for the most part, modeling has not been a prominent feature of the many analytical tools that have been utilized. In this section, an attempt will be made to review the current stable of models.

More attention will be focused on the prospects for the future use of analytical models. It is here that some of the major challenges emerge, because the restructuring processes that have characterized so many economies have created new demands on models, both in terms of their sophistication and their conceptual underpinnings.

There has been an unfortunate tendency in the literature to view regional analytical models in a competitive manner, rather than as alternative perspectives on the way in which an economy is assumed to function. Since many of these models have been derived from a common base, it is just as important to stress this commonality, as it is the differing paths along which the models have evolved (see Hewings and Jensen 1986, 1988; Polenske 1988).

Historical Development

In the initiation of modeling of urban and regional economies, few scholars would dispute the role played by international trade and development theory, the concept of the foreign-trade multiplier, and the role of the nascent field of regional science. The economic base model provided the first stage in an attempt to describe and interpret the growth and development of subnational economic entities. This model essentially allocates to exports a major role in conditioning the future path along which an economy will grow. Since the model was developed in an era in which growth was the dominant feature of economic systems, little attention was focused on the problems of decline. The economic base model has often become more like an equation econometric model (see Ledent 1978), or an input-output model (see Romanoff 1974), as scholars attempted to increase its utility and flexibility.

While several scholars raised the problems of interpretation of the mechanisms of change as an economy grew, there has never been any clear resolution of the issues. The economic base model remains, today, a system that is still in widespread use and is subject to voluminous criticism (see Richardson 1985). However, the base model does have many problems. These often can negate the substantial advantages that its simplicity provides to the analyst who is interested in a rapid estimation of the impact of a new industry on a local economy.

Greater elaboration and specification of the interactions within the regional economy is provided by the input-output model. This model proceeds to describe an economy comprised of interactions among many sectors driven by final demand, including household expenditures, state and local government, and capital expenditures, as well as exports. In the simple economic base model, a change in exports in any sector would create the same impact upon the regional economy. But in the input-output formulation, the effect may be different, depending upon the sector or sectors in which the change takes place. Furthermore, this model provides the capability to examine impacts on a detailed level.

The input-output model is not without its many detractors. Some important limitations to the uncritical use of the model are problems of fixed-coefficient production functions, absence of consideration of scale economies, capacity limitations, and the age and productivity of capital stock, together with issues of the commodity composition of output. These conceptual issues have often paled in importance, compared to the problems of constructing input-output models at the regional level. For a number of years, debate raged between analysts who claimed that nonsurvey methods could be found to approximate the structure of an economy and those who argued for the use of extensive survey methods. Jensen's (1980) compromise seems to have settled many of the issues and allowed attention to be focused on some of the problems discussed here.

A final set of models might be generically referred to as "econometric models." These range from very simple formulations of an economy (the shift and share approaches), to those requiring an elaborate specification of time-series interdependence (see Glickman 1977). The issues raised here center on the specification of the models—issues that move the focus of attention very rapidly into the mainstream of the more important debates in the econometric literature. Issues of model specification, serial and spatial autocorrelation, and estimation present a compendium of problems that are often as intractable as those facing the input-output analyst.

Most of the models have been used for a variety of impact analyses—new firm locations, fiscal incidence, defense spending—as well as some attempt to explore market potential. Links have been made with policy questions through procedures designed to reveal the existence of key sectors within the regional economy.

The Current State of Research in Regional Industrial Analysis

The litany of problems associated with almost any model used in regional analysis might suggest that the prudent scholar avoid their use altogether. However, the picture is not that bleak. In fact, the problems created by the use of these models have often resulted in important new insights into the functioning of an economy, as analysts attempted to find ways to solve the original problems. Several common threads can now be detected.

First, there has been an increasing realization that none of the current models can, by itself, provide the necessary flexibility to ensure a wide range of uses at urban and regional levels. Accordingly, there has been a marked interest in promoting the development of more-comprehensive models at the subnational level. These efforts have taken a variety of forms, of which these two seem to have dominated:

1. Models that essentially link together two or more independent models (for example, linking input-output and econometric models, or the multiple-module approach of Isard and Anselin 1982).
2. Models that provide an integration of activities normally handled by two or more independent models (the linkage between demographic and economic activities by Batey and Madden 1983; extensions include the development of social accounting systems along the lines of Stone 1977; Miyazawa 1976; and van Dijk and Oosterhaven 1986).

Second, there has been a general concern for more-rigorous specification of the models and greater attention devoted to issues of error and error analysis (see Anselin 1988). In fact, the field of spatial econometric analysis is gradually gaining ground and influencing a wide array of analytical endeavors within the broader area of spatial analysis.

While the Isard and Anselin model (1982) represents an ambitious proposal, the work completed in the last five years suggests that the evolution of current trends might result in more movement in this direction. One of the major byproducts from this research has been the increasing realization that the notion of analytical importance of parameters—the degree to which correct estimation of individual elements within a model affects the accuracy of results obtained with the model—provides a useful vehicle to narrow the sense for accuracy to a subset of a model's parameters (see Hewings 1986).

Emerging Trends and Future Directions in Regional Industrial Analysis

For the most part, the current developments in regional analytical modeling have been true evolutions of previous paradigms. As the structure of regional economies in developed nations continues to undergo the rather dramatic metamorphoses noted in this chapter (in the sections on the producer services industry, policy analysis, and high-tech and R&D activities), regional analysts are presented with a new "layer" of challenges.

First, the way activities have been classified in the past is beginning to make less sense in the latter part of the twentieth century. The distinction between manufacturing and service activity, for example, is often very blurred. In this regard, the work of Karunaratne (1986) provides an example of some possible explorations of the ways in which a traditional model (input-output) can be modified to identify the extent and nature of the "information economy."

Second, the conception of linkages and interactions within regional and interregional systems has been changing and expanding. Third, the development of what might be referred to as more general equilibrium and nonequilibrium models of regional economies is no longer limited by computer technology.

Hence, the major directions would appear to be in terms of the evolution of regional models, to accommodate new dimensions of importance in the structure and functioning of regional systems such as aspects of control, ownership, and conflict. It is surprising to realize that there does not exist a generally accepted theory about the way in which one would expect the structure of a regional economy to evolve over time. Very little comparative analysis has been undertaken at the regional level. Little is known about the expected evolutionary path of the myriad linkages that characterize economies at the subnational level. While Jensen et al. (1988) have attempted to provide the rationale for a taxonomy of economies, and have proposed a working hypothesis for evolution, there is a great deal of empirical and theoretical work yet to be done.

In addition, the notion that economies interact with each other in a hierarchical fashion (nation with nation, region with the nation of which it is a part, subregion with region, etc.) has been seriously undermined, as pointed out in the section of this chapter on trade and foreign investment. The hegemony displayed by multinational and multiregional corporate ownership has completely distorted the hierarchical structure. There are now some intense regional-national linkages that are often stronger than the nation-nation linkage. Though the existence of these linkages has been known for some time, the linkage process has created some new tensions in the development of policy priorities, in terms of international trade agreements and effective protection arguments.

In essence, the movements of modeling toward greater integration will continue. In the process, some received theory will have to be critically reexamined, and some models will have to be substantially revised or even discarded. However, the general need to provide information for both ex ante and ex post evaluation, for impact analyses, and for project selection will continue to require models in the spirit of those currently in use.

INDUSTRIAL GEOGRAPHY AND POLICY ANALYSIS

Many will agree that the field of economic geography, including industrial geography, has been one of the most active in the U.S. in recent years, both in terms of theory development and empirical analysis. Policy analysis, however, has been relatively neglected by both economic and industrial geographers in this country until recently. Despite the achievements reported in this chapter for American industrial geography as a whole, much more can be achieved in terms of policy analysis by geographers, as well as by geographers having more input in the policy area.

In this section, various research agenda and their evolution in industrial geography are examined, with particular attention to how they deal with government policy and the role of the state. While the progress of American geography as a whole may be better judged by its great articles, rather than its great books, certain books will be highlighted here as milestones along an evolutionary trajectory.

Present and Future Directions for Policy Research in Industrial Geography

To date, three policy eras can be identified in the U.S. over the last 30 years. First, the 1960s saw the heyday of *regional policy* in this country, typified by the creation of the Economic Development Administration (EDA) and the Appalachian Regional Commission (ARC). The name of the most pervasive piece of legislation, the *Public Works and Economic Development Act of 1965,* tells us that the approach to regional policy was by investment in public works, and not by any direct subsidy to industry.

Second, the 1970s was the decade of *urban policy,* where a more explicit targeting of federal aid to cities and to particular parts of cities was the order of the day. The Community Development Block Grant Program (CDBG) and the Urban Development Action Grant (UDAG) are examples of this geographical targeting.

Third, the 1980s to date can be typified as an era of *de facto industrial policy and indirect regional policy*. "Reaganomics" has meant a mounting federal deficit, where increased defense spending can be interpreted as an indirect form of regional industrial policy. If major cuts in defense spending were to take place, the high-tech-oriented military-industrial complexes of California, Texas, Connecticut, and other states would quickly go into recession. But not enough is known about the indirect impact of various government policies on industrial change.

As industrial geographers, there is much to do to assess the impact of government policy on industrial location. The only book that takes a geographical view of "United States public policy" is the collection edited by the late John House (1983), and only one chapter (Rees and Weinstein) deals with government policy and industrial location. One important consequence of the dearth of research on these topics is the conspicuous absence of geographers in policy research on Capitol Hill. Exceptions to this include Rees' work for the Joint Economic Committee of Congress (Rees 1980;

Rees, Briggs, and Hicks 1984), and the work of others for the (Congressional) Office of Technology Assessment (1984). In this regard it is appropriate to recall Brian Berry's words of 1970: "If we, as geographers, fail to perform in policy relevant terms, we will cease to be called on to perform at all" (p. 22).

The need for further policy-oriented research by industrial geographers appears insatiable. Two important areas for future research come to mind; these are discussed below.

Impact of Macroeconomic Policies. The first topic deals with the impact of various macroeconomic policies on industrial change (Rees 1987). In addition to the impact of the federal deficit on regional economies, one major topic in need of research is the impact of tax policy and recent changes in that policy on industrial location. To elucidate: One of the most pervasive pieces of legislation during the Reagan era was the Economic Recovery Tax Act of 1981 (ERTA), when the Accelerated Cost Recovery System (ACRS) introduced rapid write-offs for investment in new plants, equipment, and R&D. To date we know very little about the differential regional impact of this legislation.

More recently, the Tax Reform Act of 1986 resulted in the largest structural change in the U.S. tax code in 40 years. The overall impact of this new tax system on the U.S. economy as a whole is uncertain, because intersectoral shifts in investment patterns are bound to result, and these will have differential regional impacts. As a consequence, each state will have to reexamine its own tax code, relative to changes in the federal structure. This round of changes is destined to have a differential impact on various localities in each state.

In addition, changes in monetary policy from the high interest rates of the early 1980s to the lower interest rates of today are already having a major impact on particular sectors. High interest rates in the early 1980s contributed to the "revival of the northeast" (particularly the large financial centers like New York and Boston), because they caused a large influx of foreign investment in response to higher rates of return. On the other hand, the decline of exports from the U.S. in the early 1980s has already reversed itself. The decline in the dollar has made the textile industries of the Southeast much more competitive internationally than they were just five years ago, while the Japanese companies (Honda in particular) have shown that they are not averse to exporting cars from the U.S. back to Japan.

Impact of State Policies. A second research topic involves state policy, or what Goldstein (1987) calls state industrial policy. The budget cutbacks of the Reagan era have spawned a de facto form of New Federalism, wherein states and localities have initiated numerous new economic-development initiatives around the country. The (Congressional) Office of Technology Assessment cataloged over 150 new programs by 1984, including new university-industry relationships, incentives to generate new-technology-based companies in various states, and programs to upgrade the existing technology base of a particular state via industrial extension services and the like (Rees and Bradley 1988). All of these experiments at the state and local level are in need of assessment as to how they have impacted, and will impact, the destiny of industry in particular localities.

In sum, the policy-oriented research agenda for industrial geographers is filled with opportunities. These are opportunities not only to assess the impact of policy, but also to influence and direct appropriate policies from the highest levels of government.

REFERENCES

Alexander, J. W. 1952. Industrial expansion in the United States, 1939–47. *Economic Geography* 28:128–42.

Anselin, L. 1988. *Spatial econometrics: Methods and models*. Dordrecht: Martinus Nijhoff.

Batey, P. W. J., and Madden, M. 1983. The modelling of demographic-economic change within the context of regional decline: Analytical procedures and empirical results. *Socio-economic Planning Sciences* 17:315–28.

Berry, B. J. L. 1970. The geography of the United States in the year 2000. *Transactions, Institute of British Geographers* 51:21–53.

Beyers, W. B., and Alvine, M. J. 1985. Export services in post-industrial society. *Papers of the Regional Science Association*. 57:33–45.

Blazar, W. 1985. Telecommunications: Harnessing it for development. *Economic Development Commentary* (Fall 1985):8–11.

Borchert, J. 1978. Major control points in American economic geography. *Annals of the Association of American Geographers* 62:214–32.

Casetti, E. 1984. Manufacturing productivity and Snowbelt-Sunbelt shifts. *Economic Geography* 60:313–24.

Chisholm, G. G. 1889. *Handbook of commercial geography*. London: Longmans, Green and Company.

Clark, G. 1981. The employment relation and spatial division of labor: A hypothesis. *Annals of the Association of American Geographers* 71(3):412–24.

———. 1986. The crisis of the Midwest auto industry. In *Production, work, territory: The geographical anatomy of industrial capitalism,* eds. A. J. Scott and M. Storper, pp. 127–48. Boston: Allen & Unwin.

———, and Gertler, M. 1983. Local labor markets: Theories and policies in the U.S. during the 1970s. *Professional Geographer* 35:274–85.

———; ———; and Whiteman, J. 1986. *Regional dynamics: Studies in adjustment theory*. Boston: Allen & Unwin.

Daniels, P. W. 1982. An exploratory study of office location behavior in greater Seattle. *Urban Geography* 3:58–78.

———. 1985. *Service industries: A geographical appraisal*. Andover: Methuen.

———. 1986. Foreign banks and metropolitan development: A comparison of London and New York. *Tijdschrift voor Economische en Sociale Geografie* 78:269–88.

Erickson, R. A. 1983. The evolution of the suburban space economy. *Urban Geography* 4:95–121.

———, and Leinbach, R. 1979. Characteristics of branch plants attracted to nonmetropolitan areas. In *Nonmetropolitan industrialization,* eds. R. E. Lonsdale and H. L. Seyler, pp. 57–78. Washington: Winston/Wiley.

Gaile, G. L., and Hanink, D. M. 1984. Basic issues in the geography of foreign direct investment. *Geography Research Forum* 7:261–73.

Gertler, M. 1986. Regional dynamics of manufacturing and non-manufacturing investment in Canada. *Regional Studies,* 20:523–34.

Glickman, N. 1977. *Econometric analysis of regional systems*. New York: Academic Press.

Goldstein, H. A., ed. 1987. *The state and local industrial policy question*. Chicago: APA Press.

Gottman, J. 1983. *The coming of the transactional city*. College Park, MD: University of Maryland Institute of Urban Studies.

Green, M. 1983. The interurban corporate interlocking directorate network of Canada and the United States: A spatial perspective. *Urban Geography* 4:338–54.

Hall, P., and Markusen, A., eds. 1985. *Silicon landscapes*. Boston: Allen & Unwin.

Hamilton, F. E. I., ed. 1974. *Spatial perspectives on industrial organization and decision making*. London: Wiley.

Hanink, D. M., and Cromley, R. G. 1987a. A risk-return model for multiregion and multiproduct diversification of the firm. *Environment and Planning A* 19:81–92.

———, and ———. 1987b. Minimizing the geographical risk of foreign direct investment. *Geoforum* 18:247–56.

Harrington, J. W., Jr. 1985. Corporate strategy, business strategy, and activity location. *Geoforum* 16:349–56.

———; Burns, K.; and Cheung, M. 1986. Market-oriented foreign investment and regional development: Canadian companies in western New York. *Economic Geography* 62:155–66.

Harvey, D. 1973. *Social Justice and the City*. Baltimore: The Johns Hopkins University Press.

Hepworth, M. 1986. The geography of technological change in the information economy. *Regional Studies* 20:407–24.

Hewings, G. J. D. 1986. Problems of integration in the modelling of regional systems. In *Integrated analysis of regional systems*, eds. P. W. J. Batey and M. Madden, pp. 37–53. London: Pion.

———, and Jensen, R. C. 1986. Regional, interregional and multiregional input-output analysis. *Regional economics*. Vol. 1 of *Handbook of regional and urban economics*. ed. P. Nijkamp, pp. 295–355. Amsterdam: North Holland.

———, and ———. 1988. Emerging challenges in regional input-output analysis. *Annals of Regional Science* 22:43–53.

Holly, B. P. 1987. Regulation, competition, and technology: The restructuring of the U.S. commercial banking system. *Environment and Planning A* 19:633–52.

House, J. W., ed. 1983. *United States public policy: A geographical view*. Oxford: Clarendon Press.

Isard, W. 1956. *Location and space-economy*. Cambridge: MIT Press.

———, and Anselin, L. 1982. Integration of multiregional models for policy analysis. *Environment and Planning A* 14:359–76.

Jensen, R. C. 1980. The concept of accuracy in input-output. *International Regional Science Review* 5:139–54.

———; Hewings, G. J. D.; Sonis, M.; and West, G. R. 1988. On a taxonomy of economies. *Australian Journal of Regional Studies* 1. In press.

Johnston, R. J. 1986. The state, the region, and the division of labor. In *Production, work, territory: The geographical anatomy of industrial capitalism*, eds. A. J. Scott and M. Storper, pp. 265–80. Boston: Allen & Unwin.

Karunaratne, N. D. 1986. An input-output approach to the measurement of the information economy. *Economics of Planning* 20:87–103.

Kirn, T. J. 1987. Growth and change in the service sector of the U.S. economy: A spatial perspective. *Annals of the Association of American Geographers* 77:353–72.

Krumme, G. 1969. Toward a geography of enterprise. *Economic Geography* 45:30–40.

Ledent, J. 1978. Regional multiplier analysis: A demometric approach. *Environment and Planning A* 10:537–60.

Ley, D., and Hutton, T. 1987. Vancouver's corporate complex and producer services sector: Linkages and divergence within a provincial staple economy. *Regional Studies* 21:413–24.

Lloyd, P. E. 1979. The components of industrial change for Merseyside inner area: 1966–1975. *Urban Studies* 16:45–60.

McConnell, J. E. 1980. Foreign direct investment in the United States. *Annals of the Association of American Geographers* 70:259–70.

———. 1986. Geography of international trade. *Progress in Human Geography* 10:471–83.

McNee, R. 1959. The changing relationships of economics and economic geography. *Economic Geography* 35(3):189–98.

Malecki, E. J. 1980. Corporate organization of R and D and the location of technological activities. *Regional Studies* 14:219–34.

———. 1983. Technology and regional development: A survey. *International Regional Science Review* 8:89–125.

Markusen, A. 1985. *Profit cycles, oligopoly and regional development*. Cambridge: MIT Press.

———; Hall, P.; and Glasmeier, A. 1986. *High-tech America: The what, how, where, and why of the Sunrise industries*. Boston: Allen & Unwin.

Mason, C. M. 1980. Industrial decline in greater Manchester, 1966–1975: A components of change approach. *Urban Studies* 17:173–84.

Miyazawa, K. 1976. *Input-output analysis and the structure of income distribution.* Berlin: Springer-Verlag.

Nelson, H. 1955. A service classification of American cities. *Economic Geography* 31:189–210.

Nelson, K. 1986. Labor demand, labor supply and the suburbanization of the low wage office worker. In *Production, work, territory: The geographical anatomy of industrial capitalism,* eds. A. J. Scott and M. Storper, pp. 149–71. Boston: Allen & Unwin.

Norton, R. D., and Rees, J. 1979. The product cycle and the spatial decentralization of American manufacturing. *Regional Studies* 13:141–51.

Noyelle, T. J., and Stanback, T. M., Jr. 1983. *The economic transformation of American cities.* Totowa, NJ: Rowman & Allenheld.

Oakey, R. P. 1984. *High technology small firms.* London: Frances Pinter.

Office of Technology Assessment. 1984. *Technology, innovation and regional economic development.* OTA-STI-238.

Ohlin, B. 1933. *Interregional and international trade.* Cambridge: Harvard University Press.

O'hUallachain, B. 1984. Linkages and foreign direct investment in the United States. *Economic Geography* 60:238–53.

———. 1986. The role of foreign direct investment in the development of regional industrial systems: Current knowledge and suggestions for a future American research agenda. *Regional Studies* 20:151–62.

Peet, R. 1977. *Radical geography: Alternative viewpoints on contemporary issues.* Chicago: Maaroufa Press.

———. 1983. Relations of production and the relocation of United States manufacturing industry since 1960. *Economic Geography* 59:112–43.

Polenske, K. R. 1988. Historical and new international perspectives on input-output accounts. In *Frontiers of input-output analysis,* eds. R. E. Miller, K. R. Polenske, and A. Z. Rose. New York: Oxford. In press.

Polese, M. 1982. Regional demand for business services and inter-regional service flows in a small Canadian region. *Papers of the Regional Science Association* 50:151–63.

———, and Coffey, W. J. 1987. Trade and location of producer services: A Canadian perspective. *Environment and Planning* A 19:597–611.

Pred, A. 1977. *City systems in advanced economies.* New York: Halsted Press.

Rees, J. 1978. Manufacturing headquarters in a post-industrial context. *Economic Geography* 54:337–54.

———. 1980. Government policy and industrial location in the United States. In *Special study on economic changes, vol. 7.* Prepared for the Joint Economic Committee, Congress of the United States.

———. 1987. What happened to macroeconomics? *Environment and Planning* A 19:139–41.

———, ed. 1986. *Technology, regions, and policy.* Totowa, NJ: Rowman & Littlefield.

———; Hewings, G.; and Stafford, H. 1981. *Industrial location and regional systems.* New York: Bergin Publishers.

———; Briggs, R.; and Hicks, D. 1984. *New technology in the American machinery industry: Trends and implications.* Prepared for Joint Economic Committee, Congress of the United States.

———, and Bradley, R. 1988. State science policy and economic development in the United States: A critical perspective. *Environment and Planning* A. In press.

Richardson, H. W. 1985. Input-output and economic base multipliers: Looking backward and forward. *Journal of Regional Science* 25:608–61.

Romanoff, E. 1974. The economic base model: A very special case of input-output analysis. *Journal of Regional Science* 14:121–29.

Sayer, A. 1982. Explaining manufacturing shift: A reply to Keeble. *Environment and Planning* A 14:119–25.

Schoenberger, E. 1986. Competition, competitive strategy, and industrial change: The case of electronic components. *Economic Geography* 62:321–33.

Scott, A. J. 1981. The spatial structure of metropolitan labor markets and the theory of intra-urban plant location. *Urban Geography* 2:1–30.

———. 1982. Locational patterns and dynamics of industrial activity in the modern metropolis. *Urban Studies* 19:111–42.

———. 1983a. Industrial organization and the logic of intra-metropolitan location II: Theoretical considerations. *Economic Geography* 59:233–50.

———. 1983b. Industrial organization and the logic of intra-metropolitan location II: A case study of the printed circuits industry in the Greater Los Angeles area. *Economic Geography* 59:343–57.

———. 1984a. Industrial organization and the logic of intra-metropolitan location III: A case study of the women's dress industry in the Greater Los Angeles region. *Economic Geography* 60:3–27.

———. 1984b Territorial reproduction and transformation in a local labor market: The animated film workers of Los Angeles. *Environment and Planning* D 2:277–307.

———. 1986. High technology industry and territorial development: The rise of the Orange County complex, 1955–1984. *Urban Geography* 7:3–45.

———. 1987. The semiconductor industry in South-East Asia: Organization, location and the international division of labor. *Regional Studies* 21:143–60.

———, and Storper, M., eds. 1986. *Production, work, territory: The geographical anatomy of industrial capitalism.* Boston: Allen & Unwin.

Semple, R. K. 1985. Towards a quaternary place theory. *Urban Geography* 6:285–96.

Smith, D. M. 1981. *Industrial location,* 2d ed. New York: Wiley.

Smith, J. R. 1913. *Industrial and commercial geography.* New York: Henry Holt and Co.

Stafford, H. A. 1985. Environmental protection and industrial location. *Annals of the Association of American Geographers* 75:227–40.

Sternlieb, G., and Hughes, J. 1983. The uncertain future of the central city. *Urban Affairs Quarterly* 18:455–72.

Stone, J. R. N. 1977. Foreword. In *Social accounting for development planning,* eds. G. Pyatt and A. R. Roe, pp. xvi–xxxi. London: Cambridge University Press.

Storper, M., and Christopherson, S. 1987. Flexible specialization and regional industrial agglomerations. *Annals of the Association of American Geographers* 77(1):104–17.

———, and Walker, R. 1983. The theory of labour and the theory of location. *International Journal of Urban and Regional Research* 7:1–41.

Susman, P., and Schutz, E. 1983. Monopoly and competitive firm relations and regional development in global capitalism. *Economic Geography* 59:161–77.

Thomas, M. D. 1975. Growth pole theory, technological change, and regional economic growth. *Papers of the Regional Science Association* 34:3–25.

———. 1987. The innovation factor in the process of microeconomic industrial change: Conceptual explorations. In *New technology and regional development,* eds. E. Wever and B. van der Knaap, pp. 21–44. London: Croom Helm.

Townroe, P. M. 1969. Locational choice and the individual firm. *Regional Studies* 3:15–24.

Vance, J. 1960. Labor-shed, employment field, and dynamic analysis in urban geography. *Economic Geography* 36 (3):189–220.

Van Dijk, J., and Oosterhaven, J. 1986. Regional impact of migrants' expenditure: An input-output/vacancy-chain approach. In *Integrated analysis of regional systems,* eds. P. W. J. Batey and M. Madden, pp. 122–47. London: Pion.

Walker, R. 1985. Is there a service economy? The changing capitalist division of labor. *Science and Society* 39:42–83.

Wheeler, J. O., and Brown, C. L. 1985. The metropolitan corporate hierarchy in the U.S. South, 1960–1980. *Economic Geography* 61:66–78.

———, and Dillon, C. L. 1985. The wealth of the nation: Spatial dimensions of U.S. metropolitan commercial banking. *Urban Geography* 6:297–315.

Transportation Geography

William R. Black

Transportation geography began in the writings of political economists and civil engineers of the last century. Nevertheless, most geographers would begin a review of the field with Mark Jefferson's 1928 "Civilizing Rails," a very popular early paper in transportation geography. Others would suggest that the field began with the publication of Ullman and Mayer's 1954 review of transportation geography in *American Geography: Inventory and Prospect*. However, Ullman and Mayer's transportation geography had only a small body of literature and a virtual absence of theory. It was a field without data, where transport facilities were examined because of the flows they implied, and a field where the only model known was the unconstrained form of the gravity model. It was viewed then by many as a subfield of economic or political geography, although Ullman and Mayer thought that it was more.

Although one does not like to dwell on definitions, it is useful to at least define the discipline, so that others can understand the boundaries that delimit the field. Numerous authors have provided definitions or structures, inter alia Wheeler (1971), Hay (1973), Eliot Hurst (1974), Lowe and Moryadas (1975), and Rimmer (1978). Probably none of them is widely accepted. For the purposes of this chapter, transportation geography is *the study of the spatial aspects of transport networks and the interaction or movement of goods and people (except migration)*.

The spatial aspects of networks include (1) location, (2) structure (including connectivity), (3) environment, and (4) development. Within this context, one could include studies of location theory for routes, optimal network configurations, network accessibility and connectivity, impacts of transport routes and networks on the environment or the economy, and network development or change.

The spatial aspects of interaction include studies of trade flows and human spatial interaction. Trade flows may involve commodities (including energy flows). Human interaction would include studies of travel, such as the journey to work, shopping trips, choice of mode or route, trip-chaining, activity patterns, and travel time. Most of the so-called optimization problems that are studied in transport geography are actually concerned with optimal flows (MacKinnon and Barber 1977).

Agreement on content, however, is no longer a reflection of homogeneity, as there now are competing philosophical perspectives. Individuals view content through the variegated eyes of a logical positivist, a humanist, a phenomenologist, or a Marxist, to name a few of the current philosophical views of the field.

PARADIGMS SINCE 1954

Once again, transport geography barely existed at the time of *American Geography: Inventory and Prospect* (i.e. before Ullman and Mayer's 1954 paper). Recall that this period also marked the end of the descriptive, qualitative, or regional era of geography. Ullman and Mayer's paper was of fundamental importance in that it identified the major problems that were holding back the development of transportation geography: too few practitioners and a lack of data on flows. The authors also suggested areas that seemed to hold future promise.

The network-analysis research tradition (Laudan 1977) was born in the following decade. The catalyst for much of this work was a U.S. Army contract with Northwestern University. Works by Garrison and Marble (1965) and Kansky (1963) on network structure formed the basis for numerous early studies. Berry's studies (1966) of the spatial structure of commodity flows also were carried out at this time. Studies by Werner (1968) on route location, and Boyce (1963) on network development, stemmed from this same effort. Chorley and Haggett (1970) later validated the paradigm with their *Network Analysis in Geography*. Although there was considerable "research" that followed on the heels of these early efforts, the network-analysis paradigm practically had been discarded by the 1970s. Most of the research questions it sought to answer were left unanswered.

Various factors caused the movement away from the network-analysis paradigm. One might like to say that it no longer solved contemporary problems, or that a more-powerful paradigm took its place. But, this would not be true. Some believe the paradigm was too abstract and that it yielded only descriptive results. But, this explanation also is questionable. Others attribute the shift to the rapid growth of funding opportunities in urban and social transport planning or to students' cries for relevance. But, none of these explanations completely account for the shift, and it was doubtless a combination of all of them. Be that as it may, the field did not so much set aside the paradigm as much as it cast it out.

The decade from 1965 to 1975 saw the field of transport geography concerned with social issues, in response to federal funding. Nearly every transport geographer was involved in at least one UMTA (Urban Mass Transportation Administration) planning, research, or demonstration project. Transit systems were planned, transport needs of the young and elderly were examined, questions of spatial equity were raised, and mobility and accessibility were defined for groups in different urban areas.

Rimmer (1978), among others, sought to redirect the field to what he called a humanistic transport geography. The concern with human transport and a shift toward behavioral research in the field of geography led to studies of choice of mode, choice of route, and so forth. By and large, this decade of application (as opposed to research) made us all feel relevant. However, it did little to advance the field, since it tended to ignore theory development. The mass of behavioral research produced during this period seems to have added very little to the field.

Philosophers of science may say that paradigm development in a mature science should not be stimulated by external social factors, as transportation geographers obviously were during this decade. However, the current flurry of research on cancer and AIDS and on the components of the strategic defense initiative (SDI) would argue against this. There are many other examples. In other words, external influences, notably the availability of research funding, can and does lead to paradigm formation.

By 1975, most of this social-behavioral research paradigm had begun to wither. Federal interests had shifted to regional transport policy issues. Examples include railroad reorganization and abandonment, intercity bus deregulation, airline passenger transport deregulation, motor carrier deregulation, and energy and environmental problems related to transportation.

Geographers moved into these areas as well (see O'Sullivan, Holtzclaw, and Barber 1979; O'Sullivan 1980; and Black 1986). Urban transport geography, which flourished during the 1960s and 70s, also slipped from the limelight in response to budget cuts, as well as to concern for major regional transport questions. It would have disappeared except that enough established transport geographers continued to work in the field.

Disaggregate choice modeling became a major focus within urban transport geography, because of its relevance to those energy and environmental problems that are related to modal-choice decisions. As a result, the research of the last decade has tended to have multiple foci. These include some network analysis, more flow analysis, a continuation of interest in urban transport and modal-choice studies, and analyses of regional transport policy issues.

CONTENT OF THE FIELD

Some familiar with the field might suggest that our content is adequately outlined in the table of contents contained in the recent urban-transportation geography reader edited by Hanson (1986). However, although Hanson's anthology was written largely by geographers, some of the chapters are devoid of even a single citation of research by a geographer. Witness the chapters on urban-transport planning, aggregate characteristics of travel, residential location and mode of travel to work, transportation and energy, and environmental impacts. The book represents a vision of what urban-transportation geography should be, and is not necessarily a status report. It also should be remembered that urban-transportation geography is simply an application of transport geography to an urban environment. It should be possible, then, to define a field where urban and regional applications are equally possible.

Areas to Include in a Holistic Transport Geography

Some possible areas that should be included in such a holistic transport geography would be the following:

1. Transport routes and network location
2. The structure of transportation networks
3. The growth and development of transport networks—spatial and temporal
4. Travel patterns and their spatial variation

5. Spatial interaction of goods and people
6. Reciprocal impacts of transport networks or flows and the economic, social, and physical environment

These six topics define much of the work that has been done, or should be done, in transportation geography. The major research thrust has been in studies of spatial interaction (e.g., see Haynes and Fotheringham 1984; Werner 1985). A discussion of each topic follows.

Transport Routes. Transport location has not been handled very well. Losch (1954) and Warntz (1957) have noted that transportation routes tend to bend or refract as they seek a least-cost route across different construction-cost surfaces. This "law of refraction," as it is called, has been extended beyond the two-surface case by Werner (1968). Although an interesting anecdote in this area, it should not be the sole focus of theoretical route-location work. Major transport-location writings of the 1800s (e.g., Parnell 1838; Ellet 1839; Gillespie 1871; Wellington 1876) are unknown to many transport geographers, yet this early work is rich in relevant content. There continues to be a genuine need for historically based empirical research in this area.

Transportation Networks. Studies of network structure include the literature on connectivity, nodal dominance, and accessibility of urban and regional systems. Interesting problems include evaluations (and hopefully reduction in the number) of network measures, ways of increasing network accessibility, and the relationships between network structure and accessibility. Questions of spatial equity and mobility also are pertinent.

Growth of Transport Networks. Little is known about the growth of transport networks. The ideal sequence of transport development, according to Taaffe, Morrill, and Gould (1963), consists of the development of a series of coastal ports, followed by lines of penetration and development of feeder lines, the beginnings of interconnection, complete interconnection, and the development of routes of dominance. This process may be applicable to numerous underdeveloped areas. A more recent major work in this area is Vance's *Capturing the Horizon* (1986); it should form the basis for a considerable quantity of research.

Research questions needing examination are:

How do transport networks develop?
How do networks contract?
Is the process of network expansion the opposite of network contraction?
Is network growth really a diffusion process?

Such network-development questions, unfortunately, seem to be more commonly addressed by geographers using operations-research techniques, where optimality is the main consideration. Optimal systems are the orderly exception in this field, which is generally quite chaotic but subject to generalization.

Travel Patterns. Travel patterns and their variation represent another area that has not been examined in much detail. All cities have distinctive patterns of travel. Geographers tend to generalize these into morning and evening peaks, but to what other factors are the observed variations related? The daily pattern of travel in rural areas also reveals seasonal and weekly peaks, but these have not been examined heretofore. Similarities in commodity flows also should be evaluated. Are complexities in flows due to complexity of the phenomenon being examined or to complexity of the network? Also, very little is known about how activity spaces—those areas within which individuals do most of their daily travel—vary for different cultural groups, and this should be examined.

Spatial Interaction. Without doubt, the bulk of the literature in transport geography is concerned with spatial interaction. Beginning with Ullman's "bases for interaction" (1956), this area includes the gravity model, entropy-maximization models, logit choice models, and the recent trip-chaining models. A major flaw with such modeling is that the focus has been on the model, as opposed to the spatial interaction. As a result, we often know more about the models than we do about the interaction. This literature may be divided into three general areas: aggregate spatial-interaction modeling, disaggregate-choice modeling, and trip-stop chaining models. Let us examine these briefly before making an overall observation on modeling in transport geography.

Aggregate spatial-interaction modeling has been concerned with improving our understanding of the gravity model in its constrained forms. Questions about how parameters vary and the role of the geographical distribution of origins and destinations have been examined, but more needs to be done. Even now, our understanding is meager of why the parameters of spatial-interaction models (primarily gravity-type) vary across phenomena, space, and time. Existing generalizations are couched in terms of morphology, and some models are so abstract that they are difficult to equate with reality. Progress has been made only where researchers have carried out a long series of studies on the same data set. Most studies examine a single set of flows for one point in time, and as a result, not a great deal has been learned.

Models seeking to replicate modal or spatial choice have been abundant in the literature. Early work by Burnett (1980) and Louviere (1981), and more recent work by Horowitz (1985) and Pipkin (1986), are representative of this area. Some of the more promising research questions in this area are related to geographical and temporal stability of the models, or alternatively to their transferability, i.e., the extent to which models derived in one region can be used in another region.

Heavy emphasis on methodology also is evident in recent trip-chaining research (e.g., O'Kelly 1981; Southworth 1985), which has its origins in early work of Marble (1967), Nystuen (1967), and Marble and Bowlby (1968). The early and more recent work seeks to determine the sequence of shopping or other trips in urban travel. Of necessity, this research must focus on the individual, and it is reasonable to ask if one can generalize the findings. Many are skeptical.

Modeling is desirable in each of the three areas of spatial interaction, but at the same time, it creates new and difficult questions. Quality databases also are needed to examine research questions so that frequently cited caveats about data deficiencies can be minimized.

Reciprocal Impacts. Studies of the impacts of transportation on the economic, social, or physical environment, as well as studies of the impact of these environments on transportation, are few in geography. Research by Horowitz (1982) represents an important exception to this statement.

THE LAST DECADE

Let us now examine the research contributions of American transportation geographers during the past decade. "American transportation geographers" are defined as members of the AAG who are writing in the field of transportation, members of the AAG Transportation Specialty Group, or transportation-geography faculty in U.S. and Canadian colleges and universities.

Thirteen major journals covering the period from 1977 through 1987 were examined, and a list of transport articles by American geographers (as defined above) was compiled. Books and monographs by these same transport geographers also were compiled, resulting in a comprehensive bibliography of 333 items. References then were categorized into one of 22 different classes (Table 1).

It should be recognized that there is considerable subjectivity in this process. In addition, it should be apparent that the approach taken in this section has as its focus transport *geographers,* not transport geography. (For example, there have been numerous papers in the environmental area of transportation.) But, the goal here was to summarize the research undertaken by transport geographers. Put another way, research in the topical areas by nongeographers is excluded here.

If specific geographers have several publications in an area, or perhaps a single paper that is viewed as very significant in one of the 22 areas, this is noted in Table 1.

GENERAL FINDINGS

During the time period from 1977 to 1987, the emphasis has been on flow modeling (Table 1). Approximately one-third of all articles evaluated either a choice model or an interaction model. By contrast, traditional areas of geography are nearly absent (e.g., environmental studies, growth and impact studies, and location studies). In addition, several of the papers written by transport geographers and included in the compilation are not traditional transport geography, e.g., Black (1977, 1979). Because of this, the content of the field was identified before the research was reviewed. To define this subarea only in terms of the research that has been done would result in a "geography is what geographers do" view of the discipline. The author is unwilling to accept such a definition.

Transportation Geography Research 1977–1987

The following subsections discuss each topical area in Table 1.

Abstract Flow Analysis. Abstract flow analysis focuses upon methods, more than upon the items being transported. Tobler (1981) perhaps has made the major contribution of the decade. In his 1981 paper, he developed a potential model that enables researchers to estimate net flow of different phenomena. The model is also discussed

Table 1
Transportation geography research 1977–1987

Topical Area	Number of Papers	Dominant Geographers
Abstract flow analysis	8	Waldo Tobler
Activity and travel patterns	17	Susan Hanson
		Perry Hanson
		James O. Huff
Data	8	Ira M. Sheskin
Energy and transportation flows	35	David L. Greene
Environmental aspects of transport	7	Joel Horowitz
Flow analysis—general	7	Daniel C. Knudsen
General transportation	10	Donald G. Janelle
Impact analysis	6	
Location analysis and models	4	
Network growth and change	4	Gerald M. Barber
		Bruce Ralston
Network structure and accessibility	12	Eliahu Stern
Optimization methods	7	Patrick T. Harker
Regional transport policy, studies, and planning	22	William R. Black
		Patrick O'Sullivan
Routing	7	Frank Southworth
Spatial behavior and choice models	61	K. P. Burnett
		Joel Horowitz
		Jordan J. Louviere
		John S. Pipkin
Spatial-interaction models	50	Gordon O. Ewing
		A. S. Fotheringham
		Eric S. Sheppard
		Frank Southworth
Trade theory and studies	6	
Transport history	5	James E. Vance, Jr.
Transport in developing countries	6	Thomas R. Leinbach
Trip chaining	15	Gordon F. Mulligan
		Morton O'Kelly
		Frank Southworth
Urban transport policy and planning	25	Gordon J. Fielding
		Genevieve Giuliano
Urban transport and travel—general	11	

elsewhere (Tobler 1983, 1985). Tobler has called for more research on it, but the response has been disappointing. This model continues to be worthy of our attention.

Activity and Travel Patterns. Research concerned with the analysis of activity spaces and travel patterns had its origins in the federally funded transportation studies of the 1960s. Trip-generation models, which attempted to estimate the traffic produced by and attracted to traffic zones, activities, or households, also were developed in those studies. The earliest generalizations in this area were presented by Mitchell and Rapkin (1954) and Oi and Shuldiner (1962), with the most recent knowledge presented by Sheppard (1986) and Hanson and Schwab (1986). Although there have been some refinements in the models developed, the basic knowledge is not significantly different than it was two decades ago.

Travel generated by residential traffic zones in urban areas is functionally related to income, labor-force participation, auto ownership, and household size. Travel generated by activities or zones is related to the number of jobs (for industries), selling space (for retail activities), enrollments (for schools), and so forth.

Disaggregate travel studies have examined traffic generated by the household and the individual. These studies have found tripmaking to be related to income, occupation, automobile availability, education, employment status, age, sex, race, and other variables of less importance. The general findings are not a great deal different from the results of aggregate studies, but the possible inferences are quite different.

Data. Some of the fundamental problems of doing empirical research are related to data. Data has been a problem since the time of the Ullman and Mayer review, and as will be noted later, it is a problem of the future as well. The few papers that fell into this category concerned either the evaluation of data quality (Black and Robbins 1986), the development of data-collection methods for urban-transit data (Sheskin and Stopher 1982a, 1982b), the development of an automobile classification system based on characteristics data (Dubin, Greene, and Begovich 1979), or use of archeological data for measuring the flow of goods (Clark 1979). Primary data collection presents the researcher with numerous problems that must be resolved, but those who use secondary data sources are faced with decreases in quality and increases in the price of these data.

Energy and Transportation Flows. Research concerned with the flow or use of energy has been abundant over the last decade, in response to the awareness of limitations in fuel resources during the energy crises of the 1970s. This area ranks third as a research-publication area (Table 1), with a majority of these publications being assisted by the newly formed U.S. Department of Energy. Work by Greene (1979, 1980) and his associates at Oak Ridge (e.g., Greene and Chen 1983) has spatial components that are of more than passing interest.

Environmental Aspects of Transport. There has been little research in the areas of environmental impacts of transportation. A notable exception is the volume by Horowitz (1982) on air-quality analysis. As noted, this does not mean that there has been an absence of research in these areas; it simply means that geographers have not been doing it.

Flow Analysis—General. There were seven papers in this category, and they were concerned primarily with the analysis of actual flows and methods of comparing flows. Of the papers in this category, the efforts of Rogerson and Plane (1984) and Knudsen (1985) perhaps are most notable. Knudsen examined changes in the pattern of rail commodity flows in the U.S. from 1972 to 1981. He found that the substantial change observed was not due to changes in the deterrence function, but rather to annual changes in aggregate supply and demand for the commodities of interest.

ANALYSIS AND MANAGEMENT OF SOCIETAL GROWTH AND CHANGE

General Transportation. Few papers appeared in this category, but they should not be overlooked. For example, the work of Janelle (1986) on communications-transportation trade-offs, and of Janelle and Goodchild (1983) on space-time autonomy, are especially innovative. This category also includes several review papers written during the decade.

Impact Analysis. Nearly 20 years ago Werner (1970) wrote an insightful paper on impact analysis that led several geographers to undertake research in this area. Nevertheless, the amount of impact research in the last decade is significantly less than it once was. This may be due to the fact that it is extremely difficult to comprehensively evaluate all the pertinent, interacting systems in an impact situation.

More research is needed in this area because we are witnessing changes in economic space brought on by deregulation of the major transport modes. Heretofore, one could only imagine what would happen to travel if the cost of movement was changed substantially. Also, it was difficult to consider the impact on industrial concentrations if trucking and rail costs were to be significantly reduced. These events have now occurred, but the research challenge created by deregulation—the most significant change in the U.S. space economy in a century—has not been met.

Location Analysis and Models. Although not nearly as popular an area as it once was, this topic did see some activity as O'Kelly (1986) began pursuing the locational attributes of hub facilities, and Weaver and Church (1986, 1987) examined optimal location problems on a network.

Network Growth and Change. Little is understood about the manner in which transport networks grow and develop. The few substantive papers published in this area during the last decade are well represented by the work of Barber (1978), Barber and Ralston (1980), and Ralston and Barber (1982). While innovative, such research seems to place the problem in a planning context, and it does not provide much of an explanation for existing patterns.

Network Structure and Accessibility. In the past, network analysis undoubtedly had network structure as its centerpiece. While research continues to be undertaken on connectivity of different systems, accessibility, and such, network structure is no longer the main interest of network analysts. For instance, research has examined the positive relationship between evolving network structure and neocolonialism (Gaile and Hanink 1984) and the role of changing network structure in avoiding South African sanctions (Gaile 1988). Some of the research by Stern (1979, 1980) is of interest for the manner in which it examines interdependencies between network structure and flow.

Optimization Methods. Optimization methods were brought into geography primarily by transportation geographers, but there is very little work in this area by geographers today. The primary reasons for this are that (1) the methods long have been recognized as inappropriate for the analysis of human flows, and (2) they are unstable in most real-world network situations, i.e., the results are too easily changed by minor system changes. These statements are somewhat oversimplified, but they are correct.

Regional Transport Policy, Studies, and Planning. Consistent with the funding realities of the last decade, there has been a lot of research into regional transport policy and planning. These areas have been examined from the perspective of the practicing planner and policy-maker, and much of this work has been summarized (at least in the rail case) by Black (1986). O'Sullivan (1980) and O'Sullivan, Holtzclaw, and Barber (1979) also have done significant work in this area. What appears to be missing is a treatment of exactly what geography has to offer transport-policy analysis. Few geographers, and even fewer policy analysts, fully appreciate that the majority of transport and related policy differences across a country such as the U.S. are due in large part to geographical variation in other phenomena. This area deserves more attention. The research to date has been too focused upon individual cases to be of much value to transportation geography.

Routing. Although routing has never been a very active research area, it did see considerably more interest than usual during the last decade, as researchers became very interested in the problem of transporting hazardous materials (e.g., Pijawka, Foote, and Soesilo 1985; Radwan et al. 1986; Robbins 1981; Stough and Hoffman 1986). The other routing research was concerned with locating truck terminals in urban areas.

Spatial Behavior and Choice Models. Spatial behavior represents an area of empirical research that provides a foundation for numerous choice models. Psychologists such as Lewin (1936) ventured into the area more than a half century ago, but they realized that spatial behavior might have very little to do with stimulus-response, which was the dominant paradigm within psychology at the time. Spatial behavior was a response, but it might not be a logical response; it might be a *manifestation* of the response and, therefore, difficult to link to the stimulus. Spatial behavior, in other words, was influenced by far more factors than psychologists wanted to tackle.

The behavioral movement in transportation geography had its origins in the late 1960s and was associated with some of the market-area research being undertaken at the time. Desbarats (1983) has noted the influence of constraints on behavior and called for their recognition. Our knowledge of individual travel behavior, however, remains meager. It is possible to speak in general terms about spatial behavior. The variables that influence mode and route choice were known in the 1960s, and our understanding of their role has not changed that much in spite of the significant volume of research undertaken following the work of Domencich and McFadden (1975). Once again, the reason for this appears to be too great a focus on the models and not enough on the flows being examined.

Spatial-Interaction Models. If spatial-interaction modeling was not the leading research area of the last decade, it would have to be second. This research area in transportation geography dates from the 1950s and research on the initial unconstrained gravity model. Most of the research of the past decade took as its point of departure the research of Britain's A. G. Wilson in the latter 1960s and early 1970s. Wilson's initial work in this area (1967) and his later voluminous writings introduced the notion of a family of unconstrained and constrained gravity models. Research by transport geographers and others over the last decade has sought to improve our understanding of factors influencing the behavior of the model. It has been determined that the distribution of origins and destinations influence the models' parameters, and that adding variables (e.g., accessibility) to the models improves their performance.

Trade Theory and Studies and Transport History. These two areas saw little activity during the decade, but an interesting theoretical paper on "spatial price equilibria" did appear (Sheppard and Curry 1982). As for historical studies, the work of Vance (1986) has been cited previously as a major contribution. Thompson's study (1987) of the influence of railroad management from 1920 to 1941 on passenger decline is also worthy of review.

Transport in Developing Countries. Regional transport studies in less-developed countries were few during the decade. Southeast Asia was the focus for the research that was found, and the bulk of this was by Leinbach (1983, 1985). Leinbach's research focuses upon Indonesia and concerns everything from the role of transport in development to urban-transport problems.

Trip Chaining. The value of trip-chaining research lies primarily in its utility for retail-location planning. It is believed that, if it is possible to model the sequence of trip stops, then this information will assist in identifying those retail activities that should be located together. One of the major barriers to accomplishing this goal is the need for very large databases on individual shopping behavior. Existing databases are rarely detailed enough for research in this area, and new data-collection efforts would probably be of considerable value.

However, in order for the findings of this area to have utility in the retail-location area, the findings obtained must be generalized so that they are applicable to any urban area. Results thus far appear to be heavily influenced by the local urban geography of the city examined. Nevertheless, this research area may be worthy of further research, to follow the early work of O'Kelly (1981, 1983, 1984) and Southworth (1985), and the more recent research of McLafferty and Ghosh (1986) and Mulligan (1987).

Urban Transport Policy and Planning. As is true of much of the work in regional transport policy and planning, work in this area comes largely from academic practitioners. The major geographic contributions of Fielding, Babitsky, and Brenner (1985), and Fielding, Brenner, and Faust (1985), have been in defining the elements of transit and developing a classification which displays contrast from place to place. Giuliano has also been active in this area and has developed a method for transport-investment planning (Giuliano 1985), and (in conjunction with Berechmann) has researched such areas as the cost structure of transit firms (Berechmann and Giuliano 1984) and economies of scale in bus transit (Berechmann and Giuliano 1985). Research by transportation geographers in the urban-transit economics field enables that field to benefit from a geographic perspective that would otherwise be missing.

Urban Transport and Travel—General. The journey to work, commuting patterns, the relationship of land use to transportation, and similar notions represent the last set of papers examined. Most of these were done in a case-study format involving a single city, and the extent to which the generalizations derived prevail awaits further research.

PERSPECTIVES ON RESEARCH IN THE FIELD

In spite of all this research, there has been very little new added to the field in recent years. Little more is known today than a quarter-century ago about network growth, route location, differences in commodity flows, interregional and international trade, and impact analysis.

Part of the failure to examine these fundamental research areas is due to the fact that most transportation geographers are not involved in research, which literally means "to look again." As a result, a problem area is looked at once or perhaps twice, and another area is moved to. This is not the way that substantive contributions are made to any body of knowledge. It is common for transportation geographers to make the comment that a given individual is "still working on the gravity model" or to hear statements such as "not another paper on energy flows." On the contrary, the only way the field will progress is by researchers having such a focus.

Many university faculty also do not encourage focus. How many times has proposed doctoral research been criticized because the general subject has been examined with a similar method before? Once again, such critics have very little understanding of the purpose of research. This perceived need to continually come up with so-called "new problems" will help make the field of transport geography an intellectual potpourri.

Our journals also do not encourage focus; rather they try to serve the pluralistic as opposed to the scientific nature of the field. Journals of applied geography, urban geography, and several new journals in physical geography, for instance, have sprung up during the past decade in response to a need to move ahead. Transportation geographers may also consider the creation of their own journal, even though our numbers may be too small to support such an effort. Geography's major journals will not be of much help in developing the science of transportation geography.

THE FUTURE

Equilibrium models represent pattern, but rarely, if ever, the observed pattern. Optimal transport networks can be designed. Optimal flow systems can be identified. But, with the change of a few units of supply or demand, they are no longer optimal. Do we want to *understand* what Lösch called "our sorry reality," or do we want to become normative planners? This author prefers the former. Other transport geographers, however, must share this perspective before the field moves too far in the planning direction.

At the same time, case studies that dwell on detail will not advance the field unless they are part of a series of such studies concerned with generalizations.

Is there nothing new on the horizon to capture our imagination? In addition to some of the fascinating and yet unanswered questions of the past, there is some significant value in new approaches to some old problems.

In the area of transport-network growth, nonlinear dynamic systems models hold considerable promise (e.g., Wilson 1981). In their simplest form, these deterministic models yield smooth graphs for the traffic growth of some short regional railroad systems, which are followed by irregular fluctuations that may lead to bankruptcy, the economic equivalent of extinction. These graphs are surprisingly similar to historical accounts of numerous small railroads, and the area appears worthy of investigation. Do not expect too much, because we presently lack the theoretical content that would support use of these models. As Smale (1978) has noted ". . . good mathematical models are not generated by mathematicians throwing models to sociologists, biologists, etc. for the latter to pick up and develop . . . Good mathematical models don't start with the mathematics, but with a deep study of certain natural phenomena." Far more in-depth study of transportation networks and spatial interaction is needed before some of these models should even be considered for application in our field.

While most transport geographers have examined transport problems from the perspective of a regulatory structure concerned with economic efficiency, there is a second school of thought. This school views transport problems from the perspective of a regulatory structure based on maximizing political support (see Keeler 1983, 66–67). The view has considerable merit in areas of urban transport and regional transport service abandonment, and it deserves investigation.

Perhaps there may also be a decrease in the logical positivist approach to research in transport geography. This should occur in response to the increasing ease of handling large data sets. In effect, hypothesis testing and statistical theory of this area never anticipated the ability to analyze 100,000 data records in microseconds. Statistical methods yield results that are (nearly) always significant with very large data sets. Assuming massive data sets tell us something: the standard statistical approach for examining them should be set aside. The alternative is to draw samples from these data sets and retain the use of classical methods.

More research that is done with old data and new methods is needed. Any transport geographer who has a good data set should hold onto it. In an era when agency budgets are cut, as is occurring, data collection begins to suffer. The annual railroad carload waybill sample, for example, is threatened with termination; its cost to researchers and planners has increased from $100 to over $1,000 in the last decade. The size of the National Personal Travel Survey also has shrunk from 18,000 home interviews in 1977 to 6,500 in 1983, a reduction of nearly two-thirds. Current plans call for the 1988–89 survey to be conducted by telephone, as opposed to personal interview. In addition to these, the Census of Transportation also is threatened with termination. Ironically, the current deregulated (free-market driven) transport environment requires access to such information, if it is to function properly.

A natural outgrowth of this data-shortage problem is the need for a center for spatial interaction studies that could serve as a depository for transport-flow databases. Databases then could be available to researchers throughout the world, and comparative evaluations of many different models and research findings would be possible.

CONCLUSIONS

Transportation geographers must continue to participate in all types of transportation research, even though it may not yield a direct return. Our perspective and solutions to today's transport problems are useful and vital. We can give something back to transport geography while we educate others about the value of the field.

A recent book review typifies the ambient ignorance about our field. In an otherwise glowing review of the recent urban-transport geography volume edited by Hanson (1986), Wachs (1987), a planner, comments: "I am not sure what distinguishes a geographer's view of urban transportation from the perspective of a planner or engineer. . . ."

The transportation geographer's view should be a holistic one, guided by a body of theory and generalizations regarding spatial interaction, and a knowledge of urban and regional variations. We should not lose sight of these perspectives, theory, and knowledge. Nevertheless, we must improve the field, or one day we will find we no longer have anything special to offer.

REFERENCES

Barber, G. M. 1978. Regional transport investment planning. *Annals of the Association of American Geographers* 68:384–95.

———, and Ralston, B. A. 1980. The elementary dynamics of road development. *Geographical Analysis* 12:258–62.

Berechmann, J., and Giuliano, G. 1984. Analysis of the cost structure of an urban bus transit property. *Transportation Research B,* 18B:273–87.

———, and ———. 1985. Economies of scale in bus transit: A review of concepts and evidence. *Transportation* 12:313–32.

Berry, B. J. L. 1966. *Essays on commodity flows and the spatial structure of the Indian economy.* University of Chicago Department of Geography Research Paper No. 111. Chicago: University of Chicago Press.

Black, W. R. 1977. Negotiations for local rail service continuation: The major issues. *Traffic Quarterly* 31:455–69.

———. 1979. On the development of management fees for subsidized rail service. *Transportation Journal* 18:20–27.

———. 1986. *Railroads for rent: The local rail service assistance program.* Bloomington, IN: Indiana University Press.

———, and Robbins, J. C. 1986. An assessment of the geographical accuracy of the carload waybill sample for state rail planning. *Transportation Research Record* 1038:85–89.

Boyce, D. 1963. *The generation of synthetic transportation networks.* Submitted under contract DA-44-177-TC-685, U.S. Army Transportation Research Command, Fort Eustis, VA.

Burnett, K. P. 1980. Spatial constraints oriented modeling: Empirical analysis. *Urban Geography* 1:153–66.

Chorley, R. J., and Haggett, P. 1970. *Network analysis in geography.* New York: St. Martin's Press.

Clark, J. R. 1979. Measuring the flow of goods with archaeological data. *Economic Geography* 55:1–17.

Desbarats, J. 1983. Spatial choice and constraints on behavior. *Annals of the Association of American Geographers* 73:340–57.

Domencich, T., and McFadden, D. 1975. *Urban travel demand: A behavioral analysis.* Amsterdam: North-Holland.

Dubin, R.; Greene, D. L.; and Begovich, C. 1979. Multivariate classification of automobiles by use of Automobile Characteristics Data Base. *Transportation Research Record* 726:29–37.

Eliot Hurst, M. E. 1974. The geographic study of transportation, its definition, growth and scope. In *Transportation geography: Comments and readings,* ed. M. E. Eliot Hurst, pp. 1–15. New York: McGraw-Hill.

Ellet, C. 1839. *An essay in the laws of trade in reference to the works of internal improvement in the United States.* Richmond: P. D. Bernard, printer.

Fielding, G. J.; Babitsky, T. T.; and Brenner, M. E. 1985. Performance evaluation for bus transit. *Transportation Research A,* 19A:73–82.

———; Brenner, M. E.; and Faust, K. 1985. Typology for bus transit. *Transportation Research A,* 19A:269–78.

Gaile, G. L. 1988. African airline connectivity: South African sanctions, neocolonialism, and development. *African Urban Quarterly.* In press.

———, and Hanink, D. M. 1984. Caribbean airline connectivity and development. *Caribbean Geography* 1:272–83.

Garrison, W. L., and Marble, D. F. 1965. *A prolegomenon to the forecasting of transportation development.* Evanston, IL: The Transportation Center at Northwestern University.

Gillespie, W. M. 1871. *A manual of the principles and practice of roadmaking.* New York: A. S. Barnes.

Giuliano, G. 1985. A multicriteria method for transportation investment planning. *Transportation Research A,* 19A:29–41.

Greene, D. L. 1979. State differences in the demand for gasoline: An econometric analysis. *Energy Systems and Policy* 3:191–212.

———. 1980. The spatial dimension of gasoline demand: An econometric analysis of state gasoline use. *Geographical Survey* 9(2):19–28.

———, and Chen, C. K. 1983. A time series analysis of state gasoline demand. *Professional Geographer* 35:40–51.

Hanson, S., ed. 1986. *The geography of urban transportation.* New York: Guilford Press.

———, and Schwab, M. (1986). Describing disaggregate flows: Individual and household activity patterns. In *Geography of urban transportation,* ed. S. Hanson, pp. 193–219. New York: Guilford Press.

Hay, A. 1973. *Transport for the space economy.* Seattle: University of Washington Press.

Haynes, K. E., and Fotheringham, A. S. 1984. *Gravity and spatial interaction models.* Beverly Hills, CA: Sage Publications, Scientific Geography Series, 3.

Horowitz, J. L. 1982. *Air quality analysis for urban transportation planning.* Cambridge: MIT Press.

———. 1985. Travel and location behavior: State of the art and research opportunities. *Transportation Research A,* 19A:441–53.

Janelle, D. G. 1986. Metropolitan expansion and the communication-transportation trade-off. In *The geography of urban transportation,* ed. S. Hanson, pp. 357–85. New York: Guilford Press.

———, and Goodchild, M. C. 1983. Transportation indicators of space-time autonomy. *Urban Geography* 4:317–37.

Jefferson, M. 1928. The civilizing rails. *Economic Geography* 4:217–31.

Kansky, K. J. 1963. *Structure of transportation networks.* University of Chicago Department of Geography Research Paper No. 84. Chicago: University of Chicago Press.

Keeler, T. E. 1983. *Railroads, freight, and public policy.* Washington: The Brookings Institution.

Knudsen, D. C. 1985. Exploring flow system change: U.S. rail freight flows, 1972–1981. *Annals of the Association of American Geographers* 75:539–51.

Laudan, L. 1977. *Progress and its problems: Towards a theory of scientific growth.* Berkeley: University of California Press.

Leinbach, T. R. 1983. Transport evaluation in rural development: An Indonesian case study. *Third World Planning Review* 5:23–35.

———. 1985. Commuting and circulating characteristics in the intermediate-sized city: The example of Medan, Indonesia. *Singapore Journal of Tropical Geography* 6:35–47.

Lewin, K. 1936. *Principles of topological psychology.* New York: McGraw-Hill.

Lösch, A. 1954. *The economics of location.* New Haven: Yale University Press.

Louviere, J. J. 1981. A conceptual and analytical framework for understanding spatial and temporal choices. *Economic Geography* 57:304–14.

Lowe, J. C., and Moryadas, S. 1975. *The geography of movement.* Boston: Houghton Mifflin.

MacKinnon, R. D., and Barber, G. M. 1977. Optimization models of transportation network improvement. *Progress in Historical Geography* 3:387–412.

McLafferty, S. L., and Ghosh, A. 1986. Multipurpose shopping and the location of retail firms. *Geographical Analysis* 18:215–26.

Marble, D. F. 1967. A theoretical exploration of individual travel behavior. In *Economic and cultural topics,* Part 1 of *Quantitative geography,* eds. W. L. Garrison and D. F. Marble, pp. 33–53, Northwestern University Studies in Geography, 13.

———, and Bowlby, S. 1968. Shopping alternatives and recurrent travel patterns. In *Geographic studies of urban transportation and network analysis,* ed. F. E. Horton, pp. 42–75, Northwestern University Studies in Geography, 16.

Mitchell, R. B., and Rapkin, C. 1954. *Urban traffic, a function of land use.* New York: Columbia University Press.

Mulligan, G. F. 1987. Consumer travel behavior: Extensions of a multipurpose shopping model. *Geographical Analysis* 19:364–75.

Nystuen, J. D. 1967. A theory and simulation of intraurban travel. In *Economic and cultural topics,* Part 1 of *Quantitative geography,* eds. W. L. Garrison and D. F. Marble, pp. 54–83, Northwestern University Studies in Geography, 13.

Oi, W. Y., and Shuldiner, P. 1962. *An analysis of urban travel demands.* Evanston, IL: The Transportation Center at Northwestern University.

O'Kelly, M. 1981. A model of the demand for retail facilities incorporating multistop, multipurpose trips. *Geographical Analysis* 13:134–48.

———. 1983. Impacts of multistop, multipurpose trips on retail distributions. *Urban Geography* 4:173–90.

———. 1984. Multipurpose shopping trips and the size of retail facilities. *Annals of the Association of American Geographers* 74:231–39.

———. 1986. The location of interacting hub facilities. *Transportation Science* 20:92–106.

O'Sullivan, P. 1980. *Transport policy: Geographic, economic and planning aspects.* Totowa, NJ: Barnes & Noble.

———; Holtzclaw, G. D.; and Barber, G. M. 1979. *Transport network planning.* London: Croom Helm.

Parnell, H. 1838. *A treatise on roads.* London: Longman, Orme, Brown, Green, and Longmans.

Pijawka, K. D.; Foote, S.; and Soesilo, A. 1985. Risk assessment of transporting hazardous material: Route analysis and hazard management. *Transportation Research Record* 1020:1–6.

Pipkin, J. S. 1986. Disaggregate travel models. In *The geography of urban transportation,* ed. S. Hanson, pp. 179–206. New York: Guilford Press.

Radwan, A. E.; Pijawka, K. D.; Soesilo, A.; and Shieh, F-Y. 1986. Transportation of hazardous wastes in Arizona: Development of a data base management system for basic analysis. *Transportation Research Record* 1063:1–7.

Ralston, B., and Barber, G. M. 1982. A theoretical model of road development dynamics. *Annals of the Association of American Geographers* 72:201–10.

Rimmer, P. 1978. Redirection in transport geography. *Progress in Historical Geography* 2:6–100.

Robbins, J. C. 1981. *Routing hazardous materials shipments.* Unpublished Ph.D. diss. Department of Geography, Indiana University.

Rogerson, P. A., and Plane, D. A. 1984. Modeling temporal change in flow matrices. *Papers of the Regional Science Association* 54:147–64.

Sheppard, E. 1986. Modeling and predicting aggregate flows. In *The geography of urban transportation,* ed. Susan Hanson, pp. 91–118. New York: Guilford Press.

———, and Curry, L. 1982. Spatial price equilibria. *Geographical Analysis* 14:279–304.

Sheskin, I. M., and Stopher, P. 1982a. Pilot testing of alternative administrative procedures and survey instruments. *Transportation Research Record* 886:8–22.

———, and ———. 1982b. Surveillance and monitoring of a bus system. *Transportation Research Record* 862:9–15.

Smale, S. 1978. Review of "Catastrophe theory: Selected papers, 1972–1977, by E. C. Zeeman." *Bulletin of the American Mathematical Society* 84(6):1360–68.

Southworth, F. 1985. Simulation of trip chaining potential in urban land use patterns. *Modeling and Simulation* 16:331–35.

Stern, E. 1979. Inter-nodal association of bus services. *GeoJournal* 3:89–96.

———. 1980. A cognitive scalar for the public transport connection matrix. *Professional Geographer* 32:326–34.

Stough, R. R., and Hoffman, J. 1986. Assessing the risk of hazardous materials flows: Implications for incidence response and enforcement training. *Transportation Research Record* 1063:27–32.

Taaffe, E. J.; Morrill, R.; and Gould, P. 1963. Transport expansion in underdeveloped countries: A comparative analysis. *Geographical Review* 53:503–29.

Thompson, G. L. 1987. Management's role in U.S. rail passenger decline, 1920–1941. *Transportation Research A,* 21A:95–108.

Tobler, W. 1981. A model of geographical movement. *Geographical Analysis* 13:1–20.

———. 1983. An alternative formulation for spatial inter-action modeling. *Environment and Planning A* 15:693–703.

———. 1985. Derivation of a spatially continuous transportation model. *Transportation Research A* 19A:169–72.

Ullman, E. L. 1956. The role of transportation and the bases for interaction. In *Man's role in changing the face of the earth,* ed. W. Thomas, pp. 862–80. Chicago: University of Chicago Press.

———, and Mayer, H. M. 1954. Transportation geography. In *American geography: Inventory and prospect,* eds. P. E. James and C. L. Jones, pp. 311–32. Syracuse: Syracuse University Press.

Vance, Jr., J. E. 1986. *Capturing the horizon: The historical geography of transportation*. New York: Harper & Row.

Wachs, M. 1987. Review of "The geography of urban transportation," ed. S. Hanson. *Transportation Research A* 21A:477–78.

Warntz, W. 1957. Transportation, social physics, and the law of refraction. *Professional Geographer* 9:2–7.

Weaver, J. R., and Church, R. 1986. A location model based on multiple metrics and multiple facility assignments. *Transportation Research B* 20B:283–96.

———, and ———. 1987. The formal and computational relationships of the supporting median problem to the p-median problem. *Transportation Research B* 21B:323–29.

Wellington, A. M. 1876. *The economic theory of the location of railways*. New York: Wiley.

Werner, C. 1968. The law of refraction in transportation geography: Its multivariate extension. *Canadian Geographer* 12:28–40.

———. 1970. Formal problems of transportation impact research. *Annals of Regional Science* 4:134–47.

———. 1985. *Spatial transportation modeling*. Beverly Hills, CA: Sage Publications, Scientific Geography Series, 5.

Wheeler, J. O. 1971. An overview of research in transportation geography. *East Lakes Geographer* 7:3–12.

Wilson, A. G. 1967. A statistical theory of spatial distribution models. *Transportation Research A,* 1A:253–69.

———. 1981. *Catastrophe theory and bifurcation*. Berkeley: University of California Press.

Contemporary Agriculture and Rural Land Use

Darrell Napton

The Contemporary Agriculture and Rural Land Use Specialty Group (CARLU) focuses upon agriculture and rural land use in developed nations. CARLU members exhibit interests in a variety of agricultural, rural, and natural-resource issues. Because of this diversity, CARLU is a heterogeneous organization that shares interests with several other AAG specialty groups. In turn, this has led to fruitful joint-paper sessions with other specialty groups since CARLU was organized in 1985. Among the specialty groups with which CARLU has close ties are Rural Development, Hazards, Water Resources, and Political Geography.

Rural and agricultural geographers in North America are pursuing a host of research programs. Nonetheless, some major themes, minor themes, and evidence of new research directions emerge from the literature. The focus of this chapter is upon a few of the salient research themes that have been pursued since 1975, with a discussion of some of the possible topics for future research.

Over the past 15 years, scholarly interest in rural-research themes has increased dramatically, so much so that in 1980 the AAG sponsored a conference on nonmetropolitan America to identify emerging land-use trends and issues (Platt and Macinko 1983). The resulting book provides a useful counterpoint to this essay.

CONTEXT

Rural and agricultural geographers have been trying to understand the processes and relations that underlie cultural and natural distributions outside of cities. Rural regions cannot be understood without examining the values and economic conditions of the larger society and the events and forces that have recently influenced rural land-use decisions. These have often been forces of instability to which rural land-use decision-makers and policy-makers have had to adapt.

Rural geographical research today must be viewed within the context of these recent events and trends, which may have long-term implications. Farmers have had to contend with instability and rapidly changing conditions as they have faced high interest rates, inflation, fluctuating energy prices, inflating land prices, deflating land prices, changing federal policies, and increasing international demand for grain, followed by decreasing demand. Rural communities are in the midst of an on-again, off-again "farm crisis" that has been addressed, and sometimes complicated, by successive waves of federal programs. The current crisis in the American agricultural economy—marked by low prices and low demand for American farm products—may be a consequence of the internationalization of the food system. This internationalization has resulted in decreased international demands for U.S. food and fiber, and decreased land prices that have reduced farm equity at a time of high farm indebtedness (Hart 1986a).

The long-term impacts of this internationalization pose fundamental challenges to American farmers:

Will they be able to compete in the world market?
If not, how will American farming adjust to meet only domestic demands?
If, as some suggest, exports become increasingly important, how will the spatial structure of farming and other rural land uses adjust?

Perhaps a more fundamental force affecting rural America will be the union of agriculture and biotechnology, a change that some are comparing to the Industrial Revolution (Hite 1987). This latest stage in the application of the scientific method may mean food and fiber surpluses for the foreseeable future. This could lead to less demand for farmland in developed nations and more farmland being converted to nonfarm uses.

CURRENT THEMES IN CONTEMPORARY RURAL AND AGRICULTURAL RESEARCH

Farmland Conversion, Protection, and Adequacy

The loss of farmland has been the most popular rural research topic of the last decade in both the U.S. and Canada (Berry and Plaut 1978; Lapping 1980; Bryant and Russwurm 1979, 1981; Pierce 1981). Most concern within the commuting field of cities was directed toward the impacts of urban growth, because periurban areas are zones of land-use conflict. In these areas, urban influences upon farmland use and ownership are sometimes the paramount factors influencing individual decision-making (Furuseth and Pierce 1982). Farmers within the development frontier have to contend with nonfarm development distorting not only the land market and land assessment, but also tax rates and local political decisions. Urbanization also influences land tenure near the city, where hobby farming and part-time farming are more common (Layton 1981). Many farmland owners have also faced uncertainty about the locations of proposed highways, airports, sewers, and waste sites. Concern has perhaps been greatest in California, the nation's premier agricultural state and the most populous (Pryde 1982), but virtually every region of the U.S. and Canada that experienced urban growth had areas of concern (Furuseth and Pierce 1982; Godfrey and Ward 1981; Pease and Jackson 1979; Platt 1977). There were fears that prime soil, unique farming areas, and farmland that had agricultural value because of its accessibility to cities were all threatened, though some urban fringe counties have shown recent increases in farm productivity (Gregor pers. com. 1988). Urban farming may becoming polarized, with some types of farming withering and others flourishing in urban areas (Lawrence 1988).

Farmland conversion first emerged as a salient issue during the rapid urban growth and transportation restructuring of the 1950s and 1960s. Geographers concluded both then (Harris 1956) and more recently (Hart 1976) that urban growth did not pose a threat to U.S. food supplies, but that severe competition between urban and agricultural land uses could occur in some areas, particularly near cities. Some geographers were less sanguine (see Furuseth and Pierce 1982 for debate summary).

Nonfarm development has not been confined to the edges of cities, but may occur at highway intersections (Moon 1987), at the edges of smaller communities, or far from the city limits (Pyle 1985). The distinction between rural and urban has blurred with the low-density settlement pattern and long-distance commutes that cars and limited-access highways have allowed (Lewis 1983). Ribbons of settlement follow high-speed highways deep into the countryside, and many farmers and farm spouses today have part-time or full-time jobs in a town or city (Hart 1980; Pyle 1985). Additional rural nonfarm growth has been supported by rural industrialization and footloose retirement pensions (Hart 1980, 1984).

Even if urbanization and other land uses that compete with agriculture pose little threat to the long-term production of North American food and fiber, these alternative land uses reduce future options, because all land is not equal. Some land is better able to economically meet food and fiber demands (Chapman, Smit, and Smith 1984). In Canada, for example, land in the warmer areas and with higher fertility is particularly important, especially for vegetables and fruits (Smit et al. 1983).

The role of geographers has not been limited to theoretical research. Much geographic research has evaluated and helped design more-effective methods of protecting urban farmland, or has focused upon determining to what degree the nation's or a region's farmland was threatened. Recent research has begun to model the loss of farmland (Cocklin, Gray, and Smit 1983; Furuseth 1983), and to provide policy-makers with relevant information to determine which farmland could or should be saved (Pease et al. 1987).

The long-term capacity of agricultural resources to produce adequate food and fiber at both the national and international scale is a research topic intertwined with farmland conversion. Geographers have been active in resource assessment and capability classification, with most of these efforts devoted to assessing the adequacy of agricultural land (Pierce and Furuseth 1983), pinpointing constraints to increased food production (Pierce and Furuseth 1986), or optimizing production options (Chapman, Smit, and Smith 1984). Environmental and resource constraints affecting the long-term productive capacity of North American agriculture include land degradation, competition for agricultural land, energy, water availability and cost, and climatic change (Pierce and Furuseth 1986). These constraints have both natural and social origins that are often interconnected.

North America has relatively little potential for low-cost cropland expansion, with pasture conversion and increased double-cropping the most likely sources of new cropland. The costs of increased cropland will likely be reflected in higher food prices and in a reduced flexibility in the agricultural system as resource limits are approached and environmental and resource constraints become more critical (Pierce and Furuseth 1983).

Changing Structure of North American Agriculture

The changing structure of agriculture in developed nations has attracted the interest of an increasing number of geographers. Public policy, industry, science, and economies of scale have allowed and encouraged farms to become larger, more mechanized (Oshiro 1985), and more specialized, while the nonfarm part of the food system has increased in importance. Today less land is needed to meet national food and fiber demands, a few large farms produce an increasingly disproportionate share of the nation's agricultural output, and middle-sized farms are becoming increasingly scarce and vulnerable, while there has been an increase in small, part-time and semi-commercial farmers whose primary livelihood is an off-farm job (Layton 1981). More research is needed to assess the role of this latter group.

The increase in farm size and decrease in number of farms has been one of the most visible outcomes of the changing farm structure (Baltensperger 1987; Gregor 1982a). The shifting scale of agricultural production has had numerous economic, social, and environmental consequences that have been the focus of research by many geographers. For example, the concentration of land ownership has been a concern, particularly because of its threats upon the rural social structure (Gregor 1982b), though the long-term outcome and implications are not yet clear, partly because data are difficult to find (Steiner 1982). One result of larger farm operating units has been the emergence of fragmented farming, where farms consist of scattered parcels of land because adjacent land was not available (Smith 1975). The extent to which farmers expand by renting or leasing land versus purchasing it is still being examined.

The size of farm or ranch may also be a legacy from the past (Hart 1986b), and may influence management strategies. In the Kansas Flint Hills, owners of small ranches were less likely to burn pastures, a method associated with increased production and pasture improvement, and much more likely to overgraze their pastures than large landowners (Wilds and Nellis 1988).

America's largest farms and ranches include giant corporations, specialized firms, and farms and ranches that have diversified into nonagricultural activities (Smith 1980). These firms consume an increasing amount of energy and capital. They can be expected to expand as long as the economy rewards large, specialized firms (Steiner 1982).

Also associated with the larger farm sizes has been the increasingly important role of capital. Capital is normally used to purchase technology, with larger farms favored over small farms in their ability to use specialized and mechanized technologies and labor (Gregor 1982b). Perhaps the most noticeable new technology has been center-pivot irrigation, which has spread across the nation from its Great Plains origin. Large investments, such as irrigation, must be carefully planned and located to take maximum advantage of the physical environment, especially soils and climate (McKnight 1983).

Large-scale, capital-intensive farming has often increased the desirability of inherently high-quality land and better climate. Modern farming techniques may exacerbate the differences in productivity among farming areas, because the inherently better soils have greater yield increases and are better able to respond to fertilizers and modern management techniques than marginal soils. This has often resulted in marginal lands being abandoned (Hart 1976), crops becoming concentrated on the best land (Hart 1978), and regional shifts in farming (Hart 1977). Market shifts and improvements in soil management, however, may negate the advantages of inherently fertile soil (Gersmehl and Brown 1986).

Research has shown that changes in ownership generally precede changes in land use. Recently many rural areas have witnessed an influx of nonfarmers who purchased farmland for consumption rather than production uses. Exurban population increases are associated with parcelization, or the dividing of large parcels into smaller tracts (Healy and Short 1981). Buyers, sellers, and local governments operate to produce the new settlement pattern. Many new rural residences are scattered among parcels that remain in agriculture. The resulting pattern cannot be predicted with land-use models, and is apparently associated as much with local government policies and personal inclination, as with distance and expectations of maximizing profit (Pyle 1985).

Changing Crop and Livestock Patterns

Another persistent theme among rural geographers is to describe and interpret the changing distributions of crops and livestock. This continuing research is necessary because of the emergence of new crops and crop regions (Nellis 1984a), the decline or new distributional patterns of old crops (Hart 1977, 1978; Lewis 1980; Rumney 1984), and the actions of farmers responding to problems of available resources, changed markets, or altered policies (De Blij 1981; Granger 1980; Keddie 1983; Lewthwaite 1975; Manners 1979; Peters 1984). This avenue of research may become more important as geographers grapple with the new crop and livestock distributions that may emerge from farmers' attempts to remain competitive by diversifying output (Texas Department of Agriculture 1986), or adapting to changed resource conditions (Nellis 1987b).

Fewer geographers have investigated animals and animal-product businesses, though harvesting animals is a basic industry in some rural areas (Wallach 1981; Mealor and Prunty 1976). Some of these industries are old and are trying to adapt to modern economic and environmental changes (Alford 1975), while others are new or have grown in popularity. Private hunting preserves, for instance, provide a "rural experience" for pay (Kouba 1976). Increased leisure time and income, coupled with the paucity of hunting areas near cities and the increased posting of land, have caused the rapid growth of this rural industry. Most lessors hunt, but some pay simply to have access to a rural environment.

Adapting to a World Market

In today's global economy, American farmers are no longer relatively isolated production units. Developing nations have established agricultural-research programs, increased domestic crop and animal production, and are becoming exporters of agricultural commodities. Farming regions can no longer achieve economic prosperity by possessing comparative advantage within the American market. Today, a region must compete in the global economy (Drucker 1986); "to be successful it must have comparative advantage in world markets" (Hite 1987).

Increasingly, international markets and policies will have national, regional, and local implications. The transition of the New Zealand dairy industry from British market to world market provides one example (Lewthwaite 1980). Between 1954, when Great Britain purchased 88% of New Zealand's dairy exports, and 1973, when Britain entered the European Common Market and access to the market became limited, the share of New Zealand's dairy exports to Britain had declined to less than half. During those years, changing economic conditions accelerated the transformation of the dairy industry. Marginal farmers withdrew; remaining farmers enlarged their holdings, specialized, and adopted labor-saving technology. Dairy companies and cooperatives merged and diversified their product. The elected National Dairy Board representing the cooperatives actively shaped the pattern of production, sought to guide milk flow into the most-profitable areas, and oversaw the export of dairy products. By 1973, New Zealand dairy marketing spanned the globe.

With nonfarm influences pervading the rural countryside, evidence that American farmers regularly produce more food and fiber than can be consumed, and indications that farmland and farmland ownership will continue concentrating in selected regions, geographers who do traditional agricultural research may find their efforts focused in regions favored for agricultural concentration by climate, soils, or local or international market conditions. The trend toward research programs that attempt to understand farming by evaluating the agribusiness system and local, state, and national policy will continue, and geographers will focus more of their efforts upon traditionally nonagricultural aspects of the rural environment, including amenity resources, reserved lands, forests, and rural nonfarm settlement (Foresta 1984).

Natural Resources and Rural America

Rural areas provide many of the natural resources needed to sustain and increase our quality of life. Research in this area by CARLU members overlaps with the Biogeography, Energy, and Water Resources Specialty Groups. Rural geographers have focused most of their attention upon one of four resources—forests, soils, water, or energy—particularly in the aftermath of the 1970s energy-price increases (Lins 1979).

Most efforts, however, have been devoted to water and soil resources. As population increases, limited water resources must be managed more effectively (Pigram 1986a). Large-scale water management has progressed further in California than anywhere else in the nation (Cantor 1980). There the world's largest water-transfer project, the California State Water Project, transports water from the moist northern third of the state to the populated, industrial, agricultural, dry south. In the arid and semiarid American West, competition among consumers for scarce water is growing (Georgianna and Haynes 1981). On the High Plains, where farmers have been mining the Ogallala aquifer, geographers have used remote sensing to measure, interpret, predict, and model water use (Nellis and Briggs 1987; Nellis 1987a, 1987c, 1984c). As the growing population and increased agricultural uses place greater demands upon the nation's water supply, problems of water deficits will become common in other parts of the nation.

Interest groups have played an active role in developing water law and management policies. As water needs change, these groups may impede the revision of the policies because their favored status may be threatened (Gros and Roberts 1985). They may also be slow to acknowledge that new water supplies will not materialize from the ether. Recognizing that water deficits may be a problem is only a first step, albeit a significant one. Solutions to problems of water availability fall into two categories that are often promoted by antagonistic groups. The first solution is technological conservation or water transfer projects. The second emphasizes reduced demands. Water development to achieve economic growth and protection, coupled with decreased demand, may be incompatible.

Although farmers are aware of impending water shortages, they are more likely to postpone economic shortages by converting to more water-efficient crops while continuing to mine groundwater, than they are likely to abandon irrigation and return to dryland farming (Nellis 1987b). The magnitude of irrigation-equipment investments may make it economically advantageous for the irrigator to maximize exploitation for short-term benefit (Nellis 1984b).

Historically, land and water use have too often been viewed as separate resource issues. Fortunately, research has resulted in an increased awareness that they are integrated (Carlyle 1984). Half the U.S. population relies on groundwater as its primary source of drinking water. The quality of these supplies is increasingly compromised by land use-related pollution (Roberts and Butler 1984). All too often, agricultural productivity gains have been acquired at the expense of degrading soil and water resources (Pigram 1986b). Traditional monitoring and modeling of groundwater is expensive and inadequate, and is more successful in pinpointing an existing problem than in preventing one. If groundwater were reconceptualized from a physical resource to a resource use, then groundwater uses rather than groundwater could be protected, and mapped land-use information could become a significant component in determining the location of groundwater demands and stresses (Roberts and Butler 1984). This resource-use approach is inexpensive, flexible, and could use much available data. It also has the potential for protecting groundwater quality, while minimizing the tendency of policy organizations to overregulate or underregulate.

When marginal or unnecessary farmland is taken from production, some of it may be converted to forests (Hart 1978). But reforested land in the Southeast has resulted in a reduced streamflow, because of increased evapotranspiration and water consumption by the trees (Trimble and Weirich 1987). The impact downstream may become significant if there is increased demand for water, hydroelectric power, navigation, or pollution dilution.

A related issue is the land resource needed for energy production. Energy needs may compete with farming for farmland and water. Biomass from forest plantations is one alternative to produce renewable energy from rural areas (Cocklin, Lonergan, and Clarke 1985). Forests may provide an important element in some energy programs even during unfavorable economic times, because the environmental impacts may be less than they would be for developing some traditional energy resources, while providing jobs at the same time.

Past research on soils geography has been physically oriented. A handful of geographers, however, have used soil to integrate physical and human geography. No-till farming has been advocated as one answer to the high energy prices of the 1970s.

No-till farming produced crops without plowing or other major soil manipulation. Economic and psychological barriers have inhibited Corn Belt farmers from adopting this innovation, while physical barriers of low soil temperature to the north, variable rainfall to the west, and weed invasion to the south will likely limit the geographic spread of this revolutionary crop-production technique (Gersmehl 1978). Seminal research by Trimble has demonstrated the relationship of the agricultural system to past soil erosion (Trimble 1971), and has shown that appropriate conservation measures do reduce erosion and sedimentation (Trimble and Land 1982).

Geographers are also concerned about rural-resource regions that are particularly vulnerable to changing economic conditions. These areas may have weak economies because of dependency upon one product, such as forestry or the mining of a mineral. They may face problems of little investment capital, limited and costly services, environmental degradation, or isolation (Frederic 1981; Wallach 1980, 1981).

The public will have to develop allocation schemes that balance society's needs for water, energy, and agriculture with other needs including recreation, fish, wildlife, open space, and human and industrial consumption (Georgianna and Haynes 1981). Geographers can help clarify the problems and develop tools that will enable resource-management schemes to achieve their goals without losing sight of the trade-offs.

Government and Policy Research

Direct and Indirect Policy Impacts. The interaction of government policies and rural resources is a major research theme for CARLU members (Borchert 1985). Normally, geographers have examined the impact of direct-action policies where government programs were designed to elicit a change in public behavior. In agricultural geography, these efforts have been directed toward government policies that are designed to decrease U.S. grain surpluses or maintain farm income. The U.S. Payment-in-Kind (PIK) program, for instance, was designed to reduce agricultural surpluses by "paying" farmers with grain so that they would not plant additional crops (Furuseth 1984). Future activities are likely to be directed toward soil conservation and nonpoint-pollution policies. The National Resources Inventories conducted by the U.S. Department of Agriculture pinpointed soil erosion as the major conservation problem in over half the nation's cropland (U.S. Soil Conservation Service 1979).

Evaluating policies that have an indirect impact is becoming increasingly important (Roberts 1987; Moon 1987). Sometimes these side effects are unexpected consequences that may be at cross-purposes with other policies or with the long-term goals of the nation. For example, an unexpected outcome of policies that have a cost-share component may be to discourage innovations by their rigid guidelines, thus freezing technology in time and place (Napton 1985). Some geographers have investigated opposing policies, sometimes within the same federal agency, that may waste tax dollars, add to the complexity of an issue, and exacerbate one problem while trying to alleviate another. The USDA Farmers Home Administration (FmHA), for instance, contributed to farmland conversion at the same time that other divisions of the agency were alarmed about farmland conversion into urban uses (Ward 1979). The FmHA provided loans for farmland purchase and house construction in farm areas for non-farmers who commuted to cities to work. The resulting exurban growth sometimes resulted in local governments borrowing money from FmHA to construct rural water and sewer systems, thus exacerbating the problem.

Other examples of conflicting policy-making can be found in the USDA soil-conservation programs. Current agricultural policies are in part designed to fallow erodible soils that price-support programs of the 1970s encouraged farmers to till (Heimlich 1986; Pierce and Furuseth 1986). There is preliminary evidence that these programs are being farmed for profit by many farmers who do not fulfill the national expectation that they adhere to good farming practices (Gersmehl pers. com. 1988). It also appears that requiring compliance with soil-conservation standards to obtain program benefits (cross-compliance) is not scale-neutral; it favors large farms over the small farms that are in most need of better conservation practices (Green and Heffernan 1986). Indirect and unexpected impacts may be either positive or negative, but it is increasingly important that they be accounted for when developing, modifying, or implementing policy.

Regional and Local Responses to National Policies. Future research will likely compare regional variations in response to national policies. There is an increasing awareness that rural policies which may be effective in one part of the country may be ineffective in another (Hart 1986b; U.S. Congress 1986). For instance, the Conservation Reserve Program, designed to take eroding land from production, has been more popular in the High Plains than in more erosion-prone areas of the Midwest (Sloggett and Dickason 1986). Of the first 43 counties to reach their Conservation Reserve Program limit of withdrawing land from production, 28 are in the High Plains where Ogallala groundwater is falling fastest.

Rural Settlement Patterns. Geographers have also investigated the impact of policies upon the rural-settlement pattern (Lewis and Roberts 1985). In the American South, government loans helped the black population redistribute itself from dispersed to nucleated settlements, while becoming disassociated from agriculture (Aiken 1985). Today, blacks are more likely to live in small towns or new rural hamlets than in scattered houses associated with tenant cropping. Though this spatial pattern resembles the one of the slave era, it represents a new freedom by being disassociated from the agrarian objects of white domination and control. Unfortunately, black rural poverty continues to be pervasive.

Focus on Spatially Limited Zones of Competition. Rural land use in the U.S. exhibits a general ordering, despite the absence of national or large-scale regional land-use planning (Borchert 1983). This broad pattern has been relatively stable in 90% of the nation for the past half-century. Although there is a general abundance of land, there has been an intensification of land use in some limited regions associated with urban development, agriculture, mining, and water resources. These areas comprise a relatively small amount of the nation's land, but they have the greatest potential for land-use competition and conflict, and the greatest need for resource regulation and monitoring.

Rural and agricultural geographers focus many of their efforts upon these zones of tension where there is competition for the best use of land (Mason 1988; Georgianna and Haynes 1981; Cocklin, Smit, and Johnston 1987). These areas normally are near cities, or sometimes are in more distant regions where urban-resource or amenity demands pose serious environmental questions. Any land-use issue associated with these tension zones is normally one of three:

1. Competition between multiple land uses
2. Scale problems, where different communities have disassociated the prerogatives of decision-making from the responsibility for paying
3. Externalities, where a cost has been imposed on the public without its consent, and performance standards have not been enacted

In the future, geographers will try to obtain an increased understanding of the physical, economic, and cultural characteristics of regional conflicts and link them to the larger setting of social and political forces.

THEORY, METHODOLOGY, AND TECHNOLOGY

Theory

Agricultural-location theory rests upon the work of Heinrich von Thünen and the concept of bid rent (Hall 1966). Most recent theoretical research has critiqued or expanded upon his work. Jones demonstrated (1978) that von Thünen's land rent model is equivalent to Ricardo's theory of rent. He applied the general von Thünen model of the shifting of transportation costs into prices to farm tenure. This linked the model with spatial components of social organization (Jones 1982). Jones concluded that different organizations and different individual behavior at varying locations may reveal more about different incentives than about underlying motivations. Jones also found empirical support for the von Thünen assertion that risk increases with distance from market (Jones 1983). With all other conditions held steady, income from distant farms was more risky than for farms near market, and farmers organized their activities and resources differently to cope with these distance-associated risks.

Visser demonstrated that the von Thünen model is relevant in explaining the spatial structure of local agriculture (Visser 1980). At the regional level, however, environmentally biased types of farming diminished the influence of distance to market when farmers adopted capital-using technologies. These farmers were able to expand in area by replacing costly labor with capital in regions where agricultural intensity should have decreased because of distance. He found that capital-using technological change was specific to irrigated, pastoral, and urban agriculture; locations of these farming types were not a function of access to the global market.

Kellerman (1978) found that rent from farmland surrounding cities may be interpreted using a von Thünen model, while Lawrence (1988) observed urban agricultural changes that did not support the model. Much research is still necessary to determine the usefulness of the model to the relations among urban fringe processes and agricultural products (Kellerman 1978).

The von Thünen explanation of rent at the national scale, however, seems weak, casting doubts on the validity of the model in a nationwide context (Kellerman 1977). Visser stated that the von Thünen model cannot provide the basis of a general theory of agricultural location, because it is not independent of time and culture, and because of the role of the market and environment in relation to land-use type (Visser 1982). Recently, agricultural-location theory has been recast within a spatial-interaction framework (Wilson and Birkin 1987). This approach allows models to accommodate landscape evolution, but it has yet to be empirically tested.

Crossley (1976) was able to apply the industrial marginal rent principle to explain the location of beef-processing centers. He found that cattle were moved to market as industrial products, rather than agricultural. This approach might be applicable to any agricultural products that are treated as raw materials by several industries.

Recently, Marxist political economy has been used to evaluate the penetration of capitalism into agriculture and changes in agriculture. This view emerged outside of North America to evaluate uneven development as it affects agriculture, to conceptualize the family farm under capitalism, and to integrate government policy into agricultural production models (Bowler 1987). Recent essays and presentations indicate that some American geographers are beginning to adopt this approach (Fitzsimmons 1985; Vogeler 1982). Others are attempting to apply the humanistic approach to rural areas (Harper 1987).

Methodology and Technology

Social-survey tools are becoming increasingly popular with rural geographers (Pyle 1985). Both mail and telephone questionnaires can provide information about how and why individuals make land or policy decisions. The Delphi method has also been used to obtain data on physical or economic characteristics of resource areas, and to determine trends in resource use when developing and implementing policy (Pease 1984).

Optimization methods provide one framework for capturing the diversity inherent in natural-resource planning. Single-objective linear programming has wide applicability. The technique is normally used to optimize solutions, but geographers have extended it to identify strategic areas and to gauge the impacts of different land uses and management strategies (Smit et al 1984).

Natural-resource policy-making increasingly demands the recognition of multiple goals. Goal programming incorporates all objectives and minimizes deviations from them to produce its solution (Cocklin, Lonergan, and Smit 1986). This model may be useful as both a descriptive and prescriptive tool to evaluate alternative development priorities and resource options (Cocklin, Lonergan, and Smit 1986). While goal programming models cannot substitute for resource analysts or decision-makers, they are particularly useful in identifying goal conflicts, alternative resource priorities, and assessing the regional consequences of different resource-management alternatives. Input-output analysis may also be used to define the interactions between the environment and the economy (Lonergan and Cocklin 1985). If appropriate data are available, input-output analysis can describe the magnitude of relations and identify the impacts of policy alternatives.

Computers are increasingly being used to store databases, or geographic-information systems (GIS), and to manipulate or combine data files rapidly to produce new maps. The data files, unfortunately, may provide misleading maps and conclusions if they are used for purposes incompatible with the rules of sampling and data storage for the particular variable. One of the most-common problems in using computer databases is to assume that natural resources behave the same throughout their range, and that natural-resource interpretations can be safely transferred from one location to another (Napton and Luther 1981). This assumption is often false, and calls into question the validity of state and regional computer-derived resource interpretations. There should be a local recalibration of natural-resource variable behavior and relations with other variables (Gersmehl and Napton 1982).

As remote-sensing technology has become refined, it has emerged as an increasingly important tool for monitoring rural land-use change (Nellis and Briggs 1987). Remote sensing, coupled with computer cartography and geographic-information systems, may allow geographers unprecedented insight and monitoring, and predictive capabilities concerning rural land use.

The map has been the central tool of the geographer, and this will undoubtedly continue. The revolution of harnessing computers to geographic databases, the development of computer-mapping technology, plus the refinement of remote sensing, all provide geographers with powerful mapping tools that promise to reveal much about the patterns, relations, and processes underlying rural land use.

Application of Agricultural and Rural Geography

Agricultural and rural land-use geography have always had an applied component and a planning component (Vogeler 1982). This will continue and will grow. Increasingly, problems may be tackled by problem-oriented institutions or interdisciplinary groups such as the Rural and Small Town Research and Studies Programme of Mount Allison University's Geography Department, or the Land Evaluation Group at the University of Guelph (Bowler 1984). The Guelph Land Evaluation Group is an interdisciplinary team, including a large number of geographers, that has developed analytical techniques to assess current and future land capabilities (Brklacich pers. com. 1988). They have focused upon such topics as soil erosion (Smit et al. 1988), urban growth (Smit et al. 1983), acid precipitation (Ludlow and Smit 1987), and climate change and agriculture (Brklacich and Smit 1988). The impacts and implications of global warming, for instance, cannot be fully understood by any one discipline. Geographers, with their propensity for linking human and physical systems and knowledge of scale and map use, can significantly contribute to society's understanding of, and adapting to, this problem.

As state and local resource plans and protection zones become more common, an important thrust of applied research will be to help planners and policy-makers link databases to policy. This will make policy formation and implementation more objective, and legally and politically defensible (Pease et al. 1987). Associated with this trend are needs to develop better local land-use information and a better understanding of how best to interpret this information and portray it to a policy-making audience (Macinko 1983; Napton and Luther 1981).

Future Directions

Rural geographers are grappling with how to identify and understand general processes and their interrelations, to interpret how they manifest themselves locally, and to explain how the local variation of processes and their interrelations, in turn, influences the wider physical and social systems of which they are a part (Massey 1984).

They are striving to portray how regions and places manifest themselves as variations on a theme, exhibiting tremendous diversity while adhering to the behavioral limits of the general processes.

Rural geographers will continue to be concerned with distributions to gain understanding of how social and natural processes affect the character of places and regions, and how, in turn, the economic, political, cultural, and environmental attributes of particular places, regions, and interconnecting flows and routes shape the manner in which social and natural processes operate. Increasingly, geographers will demonstrate that geographic differentiation and spatial distributions are not only the result of social and natural processes, but that they also affect how those processes work (Massey 1984). Spatial is both outcome and part of the explanation. The goal is to understand the interrelations of social and natural relations between town and country, city and farm, and region and nation, without losing sight of the distinctive history and character that different places bring to bear upon the impact of those processes.

CONCLUSION

Rural America is undergoing a major transformation. A world surplus of grain, continued increases in farm productivity, increasing costs of production, and federal disinclination to maintain artificial market prices have driven American farmers from the international market. Meanwhile, domestic demand is not high enough to provide many farmers with a profit. One result is a surplus of farmland and many impoverished farmers.

Our relationship with the natural environment is also undergoing a profound change. The potential impact of biotechnology and the computer upon agriculture cannot be overestimated. Biotechnology may assure continued farm surpluses for the foreseeable future. Together, biotechnology and computers may reduce the amount of farmland necessary to meet national food and fiber needs, reinforce the trend toward larger farms, and may favor regions that can use computer-driven irrigation systems or respond most effectively to fluctuations in the world market (Hite 1987). Human activities have altered the biosphere, and growing evidence indicates that a global warming is one result. Modern technology can allow farms to flourish in the desert, and resource development to take place under the sea and in the polar regions.

Rural geographers must develop new tools, models, and paradigms to evaluate these changes and their impacts. The new thrust has already started. For instance, mountain land use has normally been described with biogeographic analogies or models. The construction of new roads and the growth of the new settlements that they stimulate behave as catalysts, transforming highland land uses and integrating mountain habitats into lowland cultures. Now, because of this increased accessibility into mountains, these models have to be replaced by human-oriented models (Allan 1986).

Many factors are altering the world and the role of farmer, agriculture, and rural land within it: the increase in population; growth of cities; transportation and communication links that have made almost all parts of the world accessible; internationalization of agriculture; widespread use of computers; changes in the Earth's atmosphere; and the incipient biotechnological revolution. These changes raise a set of questions that geographers are well equipped to address:

How much land should remain in farming, and where should it be located?
What are the "best" alternatives for rural land and resource use?
How will anticipated changes alter the rural settlement pattern?
How will the comparative advantage of regions change, and how will the affected regions adjust?
What are the impacts of national and state policies in local areas?
How do rural areas affect state and national policy and the quality of life for city folk?
To what extent might changes in biophysical and socioeconomic conditions impinge upon or enhance the opportunities for sustainable development of the biosphere?
How can our resources be managed to maintain and enhance the quality of life for future generations?

The research frontier for rural and agricultural geographers is ripe with promise. Geographers will be able to continue exploring many traditional research themes, and will have the opportunity to increase their value to society by helping solve many emerging problems that have a geographic content.

REFERENCES

Aiken, C. 1985. The new settlement pattern of rural blacks in the American South. *Geographical Review* 75 (4):383–404.

Alford, J. J. 1975. The Chesapeake oyster fishery. *Annals of the Association of American Geographers* 65 (2):229–39.

Allan, N. J. R. 1986. Accessibility and altitudinal zonation models of mountains. *Mountain Research and Development* 6 (3):185–94.

Baltensperger, B. 1987. Farm consolidation on the Northern and Central Great Plains. *Great Plains Quarterly* (Fall): 256–65.

Batie, S. S., and Healy, R. G. 1983. The future of American agriculture. *Scientific American* 248 (2) February:45–53.

Berry, D., and Plaut, T. 1978. Retaining agricultural activities under urban pressures: A review of land use conflicts and policies. *Policy Sciences* (9):153–78.

Borchert, J. R. 1983. American land use in a national perspective. In American-German International Seminar. *Geography and regional policy: Resource management by complex political systems*. Heidelberger Geographische Arbeiten Heft 78 Göttingen 1983. pp. 21–37.

———. 1985. Geography and state-local public policy. *Annals of the Association of American Geographers* 75(1):1–4.

Bowler, I. R. 1984. Agricultural geography. *Progress in Historical Geography* 8(2):255–62.

———. 1987. Agricultural Geography. *Progress in Historical Geography* 11(3):425–32.

Brklacich, M., and Smit, B. 1988. Effects of climatic change on agricultural land resource potential. *Perspectives on land modelling: Workshop proceedings Toronto, Ontario, November 17–20, 1986.* Guelph, Ontario: Occasional Papers in Geography No. 13.

Bryant, C. R., and Russwurm, L. H. 1979. The impact of non-farm development on agriculture: A synthesis. *Plan Canada* 19:122–39.

———. 1981. Agriculture in the urban field: Canada, 1941 to 1971. In *The rural-urban fringe: Canadian perspectives,* eds. K. B. Beasley and L. H. Russwurm, pp. 34–52. Downsview, Ontario: Department of Geography, Atkinson College, York University.

Cantor, T. M. 1980. The California state water project: A reassessment. *Journal of Geography* 79(4):133–40.

Carlyle, W. J. 1984. Water in the Red River Valley of the North. *Geographical Review* 74(3):331–58.

Chapman, B. R.; Smit, B. R.; and Smith, W. R. 1984. Flexibility and criticality in resource use assessment. *Geographical Analysis* 16(1):52–64.

Cocklin, C.; Gray, E.; and Smit, B. 1983. Future urban growth and agricultural land in Ontario. *Applied Geography* 3:91–104.

———; Lonergan, S. C.; and Clarke, D. W. 1985. Forest energy plantations: An international perspective. *Geoforum* 16(3):257–64.

———; ———; and Smit, B. 1986. Assessing options in resource use for renewable energy through multiobjective goal programming. *Environment and Planning A* 18(10):1323–38.

Cocklin, C.; Smit, B.; and Johnston, T., eds. 1987. *Demands on rural lands: Planning for resource use*. Boulder, CO: Westview Press.

Crossley, J. C. 1976. The location of beef processing. *Annals of the Association of American Geographers* 66(1):60–75.

de Blij, H. J. 1981. *Geography of viticulture*. Miami: Miami Geographical Society, University of Miami.

Drucker, P. F. 1986. The changed world economy. *Foreign Affairs* 64(4):768–91.

Fitzsimmons, M. 1985. Hidden philosophies: How geographic thought has been limited by its theoretical models. *Geoforum* 16(2):139–49.

Foresta, R. A. 1984. *America's national parks and their keepers*. Washington: Resources for the Future.

Frederic, P. B. 1981. Trees to cattle feed: A joint Canadian-U.S. effort to improve Maine's dairy and beef industry. *Proceedings: New England–St. Lawrence Valley Geographical Society* 21:59–61.

Furuseth, O. J. 1983. A model for rural land conversion. *Rural Systems* 1(3):177–87.

———. 1984. Payment-in-kind (PIK), a new U.S. agricultural policy. *GeoJournal* 8(2):185–87.

———, and Pierce, J. T. 1982. *Agricultural land in an urban society*. Resource publications in geography. Washington: Association of American Geographers.

Georgianna, T. D., and Haynes, K. E. 1981. Competition for water resources: Coal and agriculture in the Yellowstone Basin. *Economic Geographer* 57(3):225–37.

Gersmehl, P. J. 1978. No till farming: The regional applicability of a revolutionary agricultural technology. *Geographical Review* 68(1):66–79.

———, and Napton, D. 1982. Interpretation of resource data: Problems of scale and spatial transferability. In *Practical applications of computers in government. Papers from the annual conference of the Urban and Regional Information Systems Association, August 22–25, 1982,* eds. R. R. Schmitt and H. J. Smolin, pp. 402–5.

———, and Brown, D. A. 1986. The diminishing advantage of the Muscatine Silt Loam and the Dixification of the Midwest. *Journal of Geography* 85(5):212–17.

Godfrey, M. A., and Ward, R. M. 1981. Failure of Michigan's subdivision control act to preserve agricultural land. *Geographic Perspectives* 48:13–21.

Granger, O. E. 1980. Climatic variations and the California raisin industry. *Geographical Review* 70(3):300–13.

Green, G., and Heffernan, W. 1986. Government programmes for soil conservation: Progressive or regressive effects. *Environment and Behavior* 18(3):369–84.

Gregor, H. 1982a. *Industrialization of U.S. agriculture—An interpretive atlas*. Westview Special Studies in Agriculture/Aquaculture Science and Policy. Boulder, CO: Westview Press.

———. 1982b. Large-scale farming as a cultural dilemma in U.S. rural development—The role of capital. *Geoforum* 13(1):1–10.

Gros, S. L., and Roberts, R. S. 1985. Groundwater management reform in Oklahoma: The farmers verdict. *Papers and proceedings of applied geography conferences* 8:1–11.

Hall, P. 1966. *Von Thünen's isolated state*. Translated by C. M. Wartenburg. Oxford: Pergamon.

Harper, S. 1987. A humanistic approach to the study of rural populations. *Journal of Rural Studies* 3(4):309–19.

Harris, C. 1956. The pressure of residential-industrial land use. In *Man's role in changing the face of the earth,* ed. W. L. Thomas, pp. 881–95. Chicago: University of Chicago Press.

Hart, J. F. 1976. Urban encroachment on rural areas. *Geographical Review* 66(1) January:1–17.

———. 1977. The demise of King Cotton. *Annals of the Association of American Geographers* 67(3):307–22.

———. 1978. Cropland concentrations in the South. *Annals of the Association of American Geographers* 68(4):505–34.

———. 1980. Land use change in a Piedmont County. *Annals of the Association of American Geographers* 67(3):492–527.

———. 1984. Population change in the Upper Lake States. *Annals of the Association of American Geographers* 74(2):221–43.

———. 1986a. Change in the Corn Belt. *Geographical Review* 76(1):51–72.

———. 1986b. Facets of the geography of population in the Midwest. *Journal of Geography* 85(5):201–11.

Healy, R. G., and Short, J. L. 1981. Rural land: Market trends and planning implications. *Journal of the American Planning Association* 45:305–17.

Heimlich, R. E. 1986. Agricultural programs and cropland conversions. *Land Economics* 62(2):174–81.

Hite, J. C. 1987. Natural and rural community resources and the environment and their users. In *Proceedings of phase I workshop*. Neill Schaller, comp., pp. 153–77. In Social science agricultural agenda project, Spring Hill Conference Center, Minneapolis, Minnesota, June 9–11, 1987. Copies can be obtained from Resources and Technology Division, Economic Research Service, U.S. Department of Agriculture.

Jones, D. W. 1978. Rent in an equilibrium model of land use. *Annals of the Association of American Geographers* 68(2):205–13.

———. 1982. Location and land tenure. *Annals of the Association of American Geographers* 72(3):314–31.

———. 1983. Location, agricultural risk, and farm income diversification. *Geographical Analysis* 15(3):231–46.

Keddie, P. D. 1983. The renewed viability of grain corn production in Southern Ontario, 1961–1971, and changes in harvesting and storage methods. *Canadian Geographer* 27(3):223–39.

Kellerman, A. 1977. The pertinence of the Macro-Thünen analysis. *Economic Geographer* 53(3):255–64.

———. 1978. Determinants of rent from agricultural land around metropolitan areas. *Geographical Analysis* 10(1):1–12.

Kouba, L. J. 1976. The evolution of shooting preserves in the United States. *Professional Geographer* 28(2):142–46.

Lapping, M. B. 1980. Agricultural land retention: Responses, American and foreign. In *The farm and the city: Rivals or allies?*, pp. 145–78. American Assembly, Columbia University, Englewood Cliffs, NJ: Prentice-Hall.

Lawrence, H. W. 1988. Changes in agricultural production in metropolitan areas. *Professional Geographer* 40(2):159–75.

Layton, R. I. 1981. Attitudes of hobby and commercial farmers in the rural urban fringe of London, Ontario. *Cambria* 8:33–44.

Lewis, M. E., and Roberts, R. S. 1985. Would an effective cropland retirement program threaten rural communities?: Some preliminary results. *Papers and Proceedings of Applied Geography Conferences* 8:30–41.

Lewis, P. 1983. The galactic metropolis. In *Beyond the urban fringe: Land use issues of nonmetropolitan America*, eds. R. H. Platt and G. Macinko, pp. 23–49. Minneapolis: University of Minnesota Press.

Lewis, T. R. 1980. Declining cigar tobacco production in Southern New England. *Journal of Geography* 79(3):108–11.

Lewthwaite, G. R. 1975. Australian milk and its markets: A dairy system in transition. *Queensland Geographical Journal*, 3d. ser. 3:1–22.

———. 1980. New Zealand milk on the map. *Annals of the Association of American Geographers* 70(4):475–91.

Lins, H. 1979. Energy development at Kenai, Alaska. *Annals of the Association of American Geographers* 69(2):289–303.

Lonergan, S. C., and Cocklin, C. R. 1985. The use of input-output analysis in environmental management. *Journal of Environmental Management* 20(2):129–47.

Ludlow, L., and Smit, B. 1987. Assessing the implications of environmental change for agricultural production: The case of acid rain in Ontario, Canada. *Journal of Environmental Management* 25(1):27–44.

Macinko, G. 1983. Recapitulation. In *Beyond the urban fringe: Land use issues of nonmetropolitan America*, eds. R. Platt and G. Macinko, pp. 385–89. Minneapolis: University of Minnesota.

McKnight, T. L. 1983. Center pivot irrigation in California. *Geographical Review* 73(1):1–14.

Manners, I. R. 1979. The persistent problem of the boll weevil: Pest control in principle and practice. *Geographical Review* 69(1):25–42.

Mason, R. J. 1988. *Implementing the New Jersey pinelands plan: A study of environmental conflict and accommodation*. Philadelphia: Temple University Press.

Massey, D. 1984. Introduction: Geography matters. In *Geography Matters,* eds. D. Massey and J. Allen, pp. 1–11. Cambridge: Cambridge University Press.

Mealor, W. T., Jr., and Prunty, M. C. 1976. Open-range ranching in Southern Florida. *Annals of the Association of American Geographers* 66(3):360–76.

Moon, H. E. Jr. 1987. Interstate highway interchanges reshape rural areas. *Rural Development Perspectives.*

Napton, D. E. 1985. Incentives to the landowner: Unexpected results of farm policy. *Geographical Perspectives* 13:65–68.

———, and Luther, J. 1981. Transferring resource interpretations: Limitations and safeguards. In *Remote sensing: An input to geographic information systems in the 1980s. Proceedings of the Pecora VII Symposium*. pp. 175–86.

Nellis, D. M. 1984a. The emergence of an American oilseed sunflower region. *Geographical Perspectives* 54(3):47–53.

———. 1984b. Land use related adjustments to aquifer depletion in South-western Kansas. *The Geographical Bulletin* 26:10–18.

———. 1984c. A remote sensing approach for modeling water resource use. *Water Resources Bulletin* 20(5):789–93.

———. 1987a. Assessing groundwater resource dynamics in Northwest Kansas. *Kansas Applied Remote Sensing* 15(1):4–5.

———. 1987b. Land use adjustments to aquifer depletion in Western Kansas. In *Demands upon rural lands: Planning for resource use,* eds. C. Cocklin, B. Smit, and T. Johnston, pp. 71–83. Boulder, CO: Westview Press.

———. 1987c. A remote sensing approach for predicting water demand in irrigated areas of Western Kansas. *Pecora XI: Satellite Land Remote Sensing* 440–46. (Annual symposium).

Nellis, D. M., and Briggs, J. 1987. Micro-based Landsat TM data processing for tall-grass prairie monitoring in the Konza National Area, Kansas. *Papers and proceedings of the Applied Geography Conferences* 10:76–80.

Oshiro, K. K. 1985. Mechanization of rice production in Japan. *Economic Geographer* 61(4):323–31.

Pease, J. R. 1984. Collecting land use data. *Journal of Soil and Water Conservation* 39(6):361–64.

———, and Jackson, P. 1979. Farmland preservation in Oregon. *GeoJournal* 6(6):547–53.

———; Huddleston, J. H.; Forrest, W. G.; Hickerson, H. J.; and Langridge, R. W. 1987. The use of agricultural land evaluation and site assessment in Linn County, Oregon. *Environmental Management* 11(3):389–405.

Peters, G. L. 1984. Trends in California viticulture. *Geographical Review* 74(4):455–67.

Pierce, J. T. 1981. Conversion of rural land to urban: A Canadian profile. *Professional Geographer* 33:163–73.

———. 1983. Assessing the adequacy of North American agricultural land resources. *Geoforum* 14(4):413–25.

———, and Furuseth, O. J. 1986. Constraints to expanded food production: A North American perspective. *Natural Resources Journal* 26(1):17–39.

Pigram, J. J. 1986a. *Issues in the management of Australia's water resources*. London: Longman.

———. 1986b. Salinity and basin management in Southeastern Australia. *Geographical Review* 76(3):249–64.

Platt, R. H. 1977. The loss of farmland: Evolution of public response. *Geographical Review* 67(1):93–101.

———, and Macinko, G. eds. 1983. *Beyond the urban fringe: Land use issues of nonmetropolitan America*. Minneapolis: University of Minnesota Press.

Pryde, P. R. 1982. Is there any hope for agriculture in California's rapid growth areas? *GeoJournal* 6:433–42.

Pyle, L. A. 1985. The land market beyond the urban fringe. *Geographical Review* 75(1):32–43.

Roberts, R. S. 1987. Rural population loss and cropland change in the Southern plains: Implication for cropland retirement policy. *Professional Geographer* 39(3):275–87.

———, and Butler, L. M. 1984. Information for state groundwater quality policymaking. *Natural Resources Journal* 24(4):1015–41.

Rumney, T. 1984. A look at world soybean production: Distribution, uses, and dilemmas. *Geographical Perspectives* 53:64–67.

Sloggett, G., and Dickason, C. 1986. *Ground-water mining in the United States*. Economic Research Service, U.S. Department of Agriculture. Washington: Government Printing Office.

Smit, B.; Rodd, S. R.; Bond, D.; Brklacich, M.; Cocklin, C.; and Dyer, A. 1983. Implications for food production potential of future urban expansion in Ontario. *Socio-Economic Planning Science* 17(3):109–19.

———; Brklacich, M.; Dumanski, T.; MacDonald, K. B.; and Miller, M. H. 1984. Integral land evaluation and its application to policy. *Canadian Journal of Soil Science* 64: 467–79.

———; Brklacich, M.; McBride, R.; Yongyuan, Y.; and Bond, D. 1988. Assessing implications of soil erosion for future food production: A Canadian example. *Geoforum* 19(2).

Smith, E. G., Jr. 1975. Fragmented farms in the United States. *Annals of the Association of American Geographers* 65(1):58–70.

———. 1980. America's richest farms and ranches. *Annals of the Association of American Geographers* 70(4):528–41.

Steiner, R. 1982. Large private landholdings in California. *Geographical Review* 72:315–26.

Texas Department of Agriculture. 1986. *Economic growth through agricultural development: A blueprint for action*. Austin, TX: State of Texas.

Trimble, S. W. 1971. *Man-induced erosion on the Southern Piedmont, 1700–1970*. Ankeny, IA: Soil Conservation Society of America.

———, and Land, S. W. 1982. *Soil conservation and the reduction of erosion and sedimentation in the Coon Creek Basin, Wisconsin*. U.S. Geological Survey Professional Paper No. 1234.

———, and Weirich, F. H. 1987. Reforestation reduces streamflow in the Southeast. *Journal of Soil and Water Conservation* 42:274–76.

U.S. Congress. 1986. *Technology, public policy, and the changing structure of American agriculture*. Office of Technology Assessment, OTS-F-285. Washington: U.S. Government Printing Office.

U.S. Soil Conservation Service. 1979. *National summaries of the 1977 national resource inventories*. Washington: U.S. Government Printing Office.

Visser, S. 1980. Technological change and the spatial structure of agriculture. *Economic Geographer* 56(4):311–19.

———. 1982. On agricultural location theory. *Geographical Analysis* 14(1):167–76.

Vogeler, I. 1982. Applied rural geography: Choices and opportunities. In *Applied geography,* ed. John W. Frazier, pp. 283–304. Englewood Cliffs, NJ: Prentice-Hall.

Wallach, B. 1980. Logging in Maine's empty quarter. *Annals of the Association of American Geographers* 70(4):542–52.

———. 1981. Sheep ranching in the dry corner of Wyoming. *Geographical Review* 71(1):51–63.

Ward, R. M. 1979. Resolution of conflicting land use policy in the U.S.D.A.: FmHA initiative in Michigan. *Journal of Soil and Water Conservation* 34(5):240–42.

Wilds, S., and Nellis, D. M. 1988. Land tenure and range management practices in the northern Flint Hills. *The Geographical Bulletin* 30(1):41–50.

Wilson, A. G., and Birkin, M. 1987. Dynamic models of agricultural location in a spatial interaction framework. *Geographical Analysis* 19(1):31–56.

Regional Development and Planning

Ray Bromley | Peter Hall | Ashok Dutt |
Debnath Mookherjee | John E. Benhart

This chapter reviews some of the seminal work done by geographers and by scholars from related disciplines on the process of regional development and on the role of regional planning as a mechanism intended to guide that development. Regional development and planning take place, to some degree, in every country of the world, and they result in changes in the spatial distribution of human activity within national territories. The term "development" is variously used to refer to the process of change, to economic growth, to general socioeconomic progress, to the intensification of land-use and natural-resource exploitation, and to an improvement in the living conditions of the poorest groups in society. When such "development" is guided by purposive human action responding to a carefully conceived strategy, then the term "planning" is appropriate. "Conceived in its broadest terms, planning involves designing the future of a community over time, thus giving it some rational, meaningful pattern, and the shaping of its history to the extent that control over environmental factors permits" (Friedmann 1973a, 6).

Regional planning can be defined as "planning intended to bring about desired changes in the spatial distribution of wealth, population, economic activity and service provision, and in the relationships between different levels of the administrative system and different parts of the national territory." Its focus is generally "subnational" and "supralocal"; in other words, it deals with intermediate-scale units within nation-states. In general, these units, most commonly called "regions," include numerous different forms of land use, a substantial variety of natural resources, and various conurbations and cities separated by rural areas. Thus, regional planning is at a lower scale than national planning, though in many cases it may include the disaggregation of a national development strategy into packages of programs and projects for specific regions. In turn, regional planning is at a broader scale than urban or neighborhood planning, generally involving the coordination of various local authorities and the reconciliation of urban and rural interests. Regional planning generally involves, but is not necessarily limited to, the fields of such sectoral spheres of planning as land use, environment, built form, agriculture, industry, education, health, and recreation.

Regional planning can be viewed as a form of mesoscale coordination, disaggregating national schemes and adapting them to regional needs, aggregating and coordinating the proposals and activities of local governments and units, assigning regional resources to local areas, and interrelating the various economic and social sectors at the regional level. Of course, not all bodies conducting regional planning perform all of these activities. A very helpful two-fold distinction, derived from P. Hall (1970, 1982b), is between national/regional planning (the assignment of resources to regions by national governments) and regional/local planning (the work of specific regional-development agencies concerned with the assignment of resources and the coordination of activities within their specific regions). A further general category of interregional planning may be established when two or more neighboring regional agencies work together to bring about changes in an area of overlap between their territories.

The size of regions varies enormously for four major reasons:

1. Countries vary greatly in size and population, so that major regions of large countries like the U.S. are generally bigger and more populous than most of the countries of the world, while the regions of small countries are often no larger than mere "localities" or "local-government areas" in the biggest nations.
2. There is no hard-and-fast rule as to how many regions a country should be divided into, though most nations establish something between three and twenty.
3. A whole hierarchy of regions can be established within specific countries, beginning with macroregions, which in turn may be divided into regions, which as a further stage may be divided into microregions or subregions.
4. Numerous different regional divisions may exist within the same country, each division reflecting the different objectives and criteria of different analysts and organizations.

There is of course a massive literature on the definition of regions and the process of regionalization, much of it generated by geographers (cf. Berry 1961a, 1967b; Haggett, Cliff, and Frey 1977, 450–90; Massam 1980, 33–58), but for reasons of brevity we will not give much attention to this issue here. To say the least, however, the regional concept is very much "in the public domain," and numerous scholars other than geographers have made substantial contributions (cf. Odum 1936; Jensen 1951; Bogue and Beale 1961; Isard and Cumberland 1961; C. A. Smith 1976; Markusen 1985, 1987). The range of approaches to the definition of regions is summarized in one of the most recent textbooks, Melville C. Branch's *Regional Planning: Introduction and Explanation* (1988). Branch also demonstrates very clearly that regional planning is not only practiced by civilian governmental agencies, but also by the military, by business corporations, and by a wide range of international, private, voluntary, and political organizations. In effect, he broadens the concept of regional planning from the conventional image of governmental decision-making in specialized regional-planning agencies, to the assignment of resources and coordination of activities within functional territories, irrespective of the nature of the entities involved.

The analysis of regional development and the conduct of regional planning require many different disciplinary skills, including engineering, environmental science, management science, public administration, economics, sociology, political science, geography, and planning. Both scholars and practitioners come from a wide range of disciplinary backgrounds, and geographers have no proprietary claim to the field. Nevertheless, geographers have played a major role as both scholars and practitioners specializing in regional development and planning, and it is rare to find an important team or school of thought that does not include at least a few geographers among their number.

Geographers have provided many of the seminal contributions in the emergence of location theory (von Thünen [1826, 1875] 1966; Christaller [1933] 1966) and the environmentalist movement (Marsh 1864; Thomas 1956). Furthermore, some of the greatest interdisciplinary scholars who have contributed to this field, most notably Patrick Geddes (1915; cf. Stalley 1972), Benton MacKaye, and Lewis Mumford, had at least as much sympathy with geography as with any other academic discipline. Even though MacKaye's academic background was in forestry, for example, his book *The New Exploration: A Philosophy of Regional Planning* (MacKaye 1928), and his collected essays, *From Geography to Geotechnics* (Bryant 1968), are seminal contributions to the emergence of regional planning, acknowledging the inspiration of William Morris Davis, Patrick Geddes, and Isaiah Bowman.

MacKaye and Mumford provided the main environmentalist and regionalist inputs for the Regional Planning Association of America (RPAA), a group of inspirational thinkers centered in New York City who, between 1923 and 1933, greatly influenced American thinking on regional issues (see Sussman 1976). The regionalist perspectives of the RPAA are well synthesized in the "Regional Plan" issue of *Survey Graphic* (Mumford 1925) and in Mumford's 1938 classic, *The Culture of Cities*. Some of the RPAA's principles were shared with the Regional Plan Association of New York (RPA-NY), a more business-oriented and much larger group, also formed in New York City in the 1920s, but in general the RPAA concentrated more on rural, environmental, and neighborhood issues, while the RPA-NY focused upon metropolitan management problems and transportation infrastructure.

During the 1920s the RPA-NY concentrated on producing the famous *Regional Plan of New York and Its Environs,* published between 1929 and 1931 (Committee on the Regional Plan of New York 1929; Mumford 1932; Adams 1932), a monumental work proposing a long-term development strategy for the New York metropolitan region. Despite the fact that the plan came from a nongovernmental pressure group, many elements of it were eventually implemented, most notably when those elements won the support of Robert Moses, one of the most powerful nonelected officials in the history of the U.S. (cf. Caro 1974). Because of the backing the RPA-NY received from the Russell Sage Foundation and from many of the corporate and governmental elites of New York City, however, it has continued to be active as a regional-development lobbying body, right to the present day (Hays 1965; Heiman 1988).

GEOGRAPHY AND PLANNING: AN EVOLVING INTERFACE

The academic discipline of geography is much older than that of planning, and geographers like George Perkins Marsh and Sir Halford Mackinder played key roles in public-policy debates long before planning was taught at any U.S. university. Nevertheless, social reformers and city elites were far more instrumental than geographers in establishing urban and regional planning as an activity in the U.S., and the most direct academic roots for the emergence of planning in U.S. universities are in departments of architecture and landscape architecture (cf. Boyer 1983; Foglesong 1986). The first city-planning course taught in a U.S. university was not given until 1909, and regional planning did not emerge as a focus for specialized courses until well after the first planning program had been established at Harvard in 1923 for the degree of Master of Landscape Architecture in City Planning (Perloff 1956). Most planning programs and departments in U.S. universities were established during or after the Second World War, and as an academic discipline planning is still relatively small in comparison with geography. Some major universities have both disciplines in separate departments, and often in separate colleges, while others have only one of the two, or even neither.

A striking trend in recent years has been the modification of geography departments with the introduction of planning courses and programs, sometimes also with a change of departmental name to "Geography and Planning" or some similar title to reflect a new bidisciplinary identity. This extensive interlinkage with planning has not been totally one-sided. Some publishers have begun to produce joint catalogs for the two disciplines, and this seems to reflect the very real overlap between the topics studied, rather than any search for the aggrandizement of geography. In recent years some major planning departments, notably at University of California–Los Angeles and Cornell University, have added several geographers to their faculty. Urban and regional planning has become one of the principal fields of professional and applied geography, and it may well be that geography is the best possible disciplinary background for practicing planners because it combines extensive study of location and analytical methods with a broad understanding of the relations between human activity and the surrounding environment.

Numerous geographers have argued that geography has a central role in planning. Thus, for example, J. Lewis Robinson (1956, 6) suggests that:

> . . . both geography and planning have many principles and methods in common. No successful plan can be based on preconceived ideas of what is best for the region. It must be a plan which fits its physical setting, historical development, and adjusts to the present patterns. Planning is regional geography projected into the future, with the hope of guiding the region into desirable patterns in harmony with its environment.

Similarly, in his pioneering book entitled *Geography and Planning,* T. W. Freeman (1958, 13) argues that:

> Planning has an inescapable geographical basis. . . . Geography . . . is essential to the planner's work, for the planner must understand the existing landscape before he tries to reform it, both in town and country.

Peter Hall (1982b, 9) takes a more balanced and somewhat less-optimistic view:

> One can argue . . . that spatial planning, or urban and regional planning, is essentially human geography . . . harnessed or applied to the positive task of action to achieve a specific objective. Many teachers in planning schools would hotly deny this. They would argue that planning, as they teach it, necessarily includes many aspects which are not commonly taught in geography curricula. . . . The law relating to the land is one of these; civil engineering is another; civic design is another. . . . What does seem true is that the central

body of social sciences which relate to geography—economics, sociology, politics and psychology—does form the core of the subject matter of urban and regional planning. By "subject matter" I mean that which is actually planned. It is, however, arguable that there is another important element in planning education not covered in this body of social science: that is the study of the *process of planning* itself, the way men assume control over physical and human matter and process it to serve their defined ends.

In the New Deal period, there was a rapid expansion of federal and state government employment throughout the U.S., including the creation of many job opportunities for enterprising individuals who claimed expertise in urban and regional planning. The field of urban planning also underwent substantial growth in local governments throughout the country, and this expansion has continued to the present day. There is widespread support for local planning controls to preserve historic areas, to maintain the character of suburban residential neighborhoods, to limit new "developments," and to stimulate the local economy, and urban planning has gradually been institutionalized throughout the country. Regional planning, however, has followed a more-difficult trajectory, declining in the 1940s with the gradual eclipse of the New Deal programs, regaining a small part of its old momentum in the 1960s and 1970s, and then falling back again after the abolition of most federal financial support to regional-development agencies by the Reagan Administration in 1981 (cf. Lim 1983). Nevertheless, if we add urban planning and regional planning together, and if we count federal, state, local, and consultant jobs, it is probably true that the total of planning-related, nonacademic career opportunities in the U.S. has expanded continuously since the New Deal.

The emergence of a substantial jobs-and-consultancy market in planning has prompted many geographers to ponder the relations between geography and planning, and there is at least the beginning of a dialogue between the two disciplines (cf. Inskeep 1962; Hodge 1965, 1966; Griffin 1965; Cooper 1966; Harris 1967; J. N. Jackson 1967; Marshall 1966; Harrison and Larsen 1977; Christensen 1977). In spite of apparent differences in the philosophy, values, and goals of geography and planning, the relationship between the two fields continues to evolve. In recent years, planning agencies have shown great interest in acquiring hardware, software, and expertise for computer cartography, remote sensing, and geographic information systems, and the role of geography as an applied professional discipline that contributes to planning has now been firmly established. The much quoted phrase ". . . planning is the art of which geography is the science . . ." remains worthy of consideration.

The traditional pluralistic approach and the integrative skills of the geographer to draw upon diverse elements from both the physical and social sciences are considered by many to be directly applicable to urban, regional, and developmental planning studies. The issue is: how can one link the diversified facets of the two fields of geography and planning in a coordinated fashion? The rationale for the systematic development of an "applied interface" with planning, whereby geography could become more "marketable" outside academia while retaining its integrity and vitality as a discipline, has clearly been articulated (Harrison and Larsen 1977, 144–45). The fullest development of such an interface will require a significant proportion of both geographers and planners to undertake formal study in the other's discipline, and the encouragement of mixed academic and practitioner careers in which new generations of scholars undertake substantial periods of work in government and business before returning to teach and research in university posts. The applied interface must be an evolving exchange, guided on one hand by the research and theory of university-based geography and planning, and on the other hand by the changing needs of government, nongovernmental organizations, and the business community.

FIVE KEY INTERDISCIPLINARY CAREERS

The interdisciplinary character of regional development and planning is well illustrated by the careers of Rexford Tugwell, Harvey Perloff, Walter Isard, John Friedmann, and Brian Berry, probably the five greatest American contributors to the theory and practice of regional development and planning since the New Deal. The seminal work and massive output of these five scholars provides a benchmark for the field as a whole, and much of the intellectual history of regional planning derives from their interactions and from the inspiration they have given to other scholars.

Of the five, only the youngest, Brian Berry, is a full-blooded geographer. All five, however, have worked with geographers, have inspired geographers, and have contributed to the published output of several academic disciplines, including geography. Their careers emphasize the crucial importance of interdisciplinary work in regional development and planning, and the importance of an effective blend of academic and practical experience in order to fruitfully link scholarship with an understanding of major public-policy issues.

A striking common feature of the careers of all five is that they have emphasized quantitative economic analysis and location theory for at least the first two decades of their careers. In all cases, however, as mature scholars they have paid increasing attention to the context within which policy is formulated and implemented, recognizing the limitations of pure economic analysis and paying more attention to political variables and issues.

Rexford Tugwell

An economist trained at the University of Pennsylvania, Tugwell was one of the three members of Franklin Roosevelt's "Brain Trust"—the other two being Raymond Moley and Harold Ickes—who were responsible for advising the President on how to bring the U.S. out of the catastrophic depression that had begun in 1929. Tugwell was the most active practitioner of regional planning in the New Deal period, notably as Under Secretary of Agriculture, Director of the Resettlement Administration, and then Chairman of the New York City Planning Commission. Subsequently, as Governor of Puerto Rico from 1941 to 1946, he laid the groundwork for the classic "Operation Bootstrap" regional economic-development program, which reached its height under Governor Luis Muñoz Marín in the 1950s and early 1960s.

In 1947 he moved to the University of Chicago as a Professor of Political Science, and there he was responsible for the establishment of the Planning Program, involving Harvey Perloff as the first core faculty member. After retirement in 1957, Tugwell held visiting appointments at various universities and developed proposals for major changes in the U.S. Constitution that would establish 20 regional "republics" to be carved from the 50 states. His voluminous writings are relatively little-known to most geographers, yet they provide crucial insights into the relations among national economic change, political circumstances, and regional-development policies (e.g., Tugwell 1934, 1958, 1970; cf. Padilla 1975).

Harvey Perloff

As a student of economics at the University of Pennsylvania during Roosevelt's "first hundred days," and as Tugwell's intellectual heir at the University of Chicago, Perloff shared the same missionary spirit for planning and the same belief that regional planning for economic development and resource management was far more important than its more localized cousin, urban planning. He had first met Tugwell on a planning assignment in Puerto Rico in the 1940s, and while there he worked extensively with the famous Puerto Rican geographer Rafael Pico, who played a major role in the island's economic-development programs over the following decades. The planning program that Perloff came to direct at the University of Chicago had strong links with such geographers as Edward Ackerman, Harold Mayer, and later Brian Berry, as well as with many of the University's leading sociologists and economists.

In a period of financial crisis in the early 1950s, however, the University decreed the phased closure of the program, and in 1955 Perloff moved on to Resources for the Future, a Washington research foundation, where he worked until 1968. There he conducted surveys of U.S. planning education and the work of universities in regional studies, and he specifically noted the preeminence of planning and geography (vis-à-vis other disciplines such as economics and sociology) in regional studies (Perloff 1957a, 1957b; cf. Burns and Friedmann 1985, 350).

Most importantly, however, together with three other economists he produced what many would argue is the finest study of regional development in the U.S. (Perloff et al. 1960). Subsequently, both in Washington and at the University of California–Los Angeles after 1968, he shifted his attention to city planning and metropolitan-development issues, contributing to the emergence of a national urban policy in the U.S. in the 1960s and 1970s (Perloff et al. 1975; Perloff 1980).

Walter Isard

An economist by training, Walter Isard found economics so limiting as a discipline for his regional interests that he laid the groundwork for the emergence of a new interdisciplinary field which he chose to call "Regional Science." He founded the Regional Science Association in 1954, and it has grown to a worldwide membership of over 2500 scholars, headquartered at the University of Illinois. It has members in over 60 countries and includes 20 national sections. Most of the RSA's members have geography, economics, or planning as their disciplinary background, but there are also members with backgrounds in other disciplines such as business, sociology, public administration, and mathematics. *The Papers of the Regional Science Association,* the *International Regional Science Review,* the *Journal of Regional Science,* and other regional-science periodicals are major outlets for geographers publishing on regional development and planning. If we define geography as "locational analysis," then many of the works of nongeographers in regional science are clearly geographical in character.

Working from his main base at the University of Pennsylvania, where he was a full-time faculty member from 1956 to 1979, Isard has published a long sequence of classic books, most importantly *Location and Space Economy* (1956), *Industrial Complex Analysis and Regional Development* (1959), *Methods of Regional Analysis* (1960), *General Theory: Social, Political, Economic and Regional* (1969), and *Introduction to Regional Science* (1975). Several of these books involve collaboration with junior authors, notably Tony Smith, Gerald Carruthers, John Cumberland, and Michael Dacey, and the whole series develops and extends location theory, economic analysis, and operations research in their application to regional issues.

Regional Science has interacted continuously with urban and regional economics, and with quantitative geography, and it has developed a very extensive literature. Its level of quantification has become increasingly sophisticated, making it a very dynamic research area for the most mathematically inclined geographers. With the new processing capability of supercomputers, we are even offered the specter of a unified location theory, which would enable us to understand all location patterns simultaneously instead of having to work with partial theories—one for agriculture, one for industry, one for retailing, etc. Regrettably, however, the mathematical sophistication of much of the work published in regional-science journals and in such kindred geographic journals as *Geographical Analysis* and *Environment and Planning, Series A and B* has tended to isolate this work from the mainstream of regional development and planning (Rodwin 1987).

Most practicing planners cannot understand the work of regional scientists, and many others argue that the quality and availability of real-world data are not adequate for such sophisticated analyses. Some of the more strident critics of regional science go two steps further. They first argue that the crucial issues and problems in regional development are political in character and are not susceptible to the forms of mathematical models used by regional scientists. They secondly argue that distance—a crucial variable in regional science and in geography as a whole—is no longer a major factor in many locational decisions.

A closer examination of Isard's whole work offers a potential avenue out of the impasse that seems to face regional science. As well as giving us an interdisciplinary framework with which to link quantitative geography, economics, and regional planning, his broad-ranging intellect has focused upon decision-making, militarism, and environmental issues. He has played a leading role in the development of Peace Science, and in the study of mechanisms for conflict resolution (cf. Isard and Smith 1982, Isard 1988). Hopefully, his political analyses will lead regional scientists to a fuller appreciation of administrative structures, bureaucratic behavior, and conflicts among rival interest groups in the regional-development process.

John Friedmann

Friedmann studied geography as a second field for his Ph.D. in Planning at the University of Chicago, and his doctoral dissertation on the Tennessee Valley Authority was published in the Chicago Research Papers in Geography (Friedmann 1955). Nevertheless, Friedmann, like his mentor Perloff, is generally viewed as a specialist in the discipline of planning, and his work has been extraordinarily influential in attracting new recruits to the field of regional planning and in provoking periodic major rethinking of that field. His career and intellectual output can be clearly divided into three major stages, the crucial separations being in the late 1950s and in 1973 (the year of the brutal military coup against the elected civilian government of Chile).

Until the late 1950s, he worked primarily in the U.S. and particularly on the TVA, following the intellectual paths of such pioneers as MacKaye, Odum, Mumford, and Tugwell, and interacting with a wide range of social scientists including geographer Edward Ackerman. His work followed the tradition of regional planning as natural-resource management, and as the preparation of special development plans for problem areas and for zones of unusual natural-resource potential.

During the second stage of his career, from the late 1950s until 1973, he became increasingly involved in regional science and international development activities, notably in Brazil, Korea, Venezuela, and Chile. Together with William Alonso, the first Ph.D. from Walter Isard's new doctoral program in regional science at the University of Pennsylvania, he edited the classic reader *Regional Development and Planning* (Friedmann and Alonso 1964), which brought together a broad range of essays by economists, geographers, planners, and others to provide a panorama of the subject, and which became the text for a generation of regional planners.

Then, involved in a major technical-assistance program coordinated by Lloyd Rodwin for the Guayana region of Venezuela, he wrote his classic book, *Regional Development Policy: A Case Study of Venezuela* (Friedmann 1966). Its first 100 pages must be the most frequently photocopied essay on regional planning. In that first section, he outlined a general theory of regional development based on core-periphery relations, growth-pole theory, and the gradual increase and subsequent decrease of interregional inequalities. His work on Venezuela was clearly influenced by emerging paradigms in international development studies, particularly the classic writings on regional inequalities by Myrdal (1957, 13–26), Hirschman (1958, 183–94), and Williamson (1965).

In the late 1960s he became increasingly interested in innovation diffusion and in the importance of major cities as centers of socioeconomic activity, developing an emphatically pro-urban "strategy of deliberate urbanization" and a "general theory of polarized development" (Friedmann 1968, 1972, 1973b). He also prepared a major second edition of his reader with Alonso (Friedmann and Alonso 1975), changing the title and a substantial proportion of the essays and including a brilliant final review essay on the progress made in the field over the preceding decade (Friedmann 1975). In his assessment of progress he refers extensively to the work of quantitative geographers, notably on spatial organization (Morrill 1970; Abler, Adams, and Gould 1972), innovation diffusion (Hagerstrand 1967; Pedersen 1974), the diffusion of modernization (Soja 1968; Gould 1970), the structure and growth of urban systems (Whebell 1969; Berry 1971), and growth-pole policies (Kuklinski and Petrella 1972; Kuklinski 1972).

By the time that this second edition had come out, however, Friedmann had entered his new, radical and activist phase—which continues to the present day—signaled by the publication of *Retracking America* (Friedmann 1973a), a call for a highly participatory and decentralized democracy in the U.S. with numerous social-learning mechanisms to facilitate effective grass-roots planning. His major changes of view were both summarized and depersonalized in *Territory and Function* (Friedmann and Weaver 1979), a historical essay written with one of his former students, whereby the three-stage evolution of his views was generalized to represent the general evolution of thinking among specialists in regional planning. He interpreted this evolution as a shift from territorial policies that emphasized community needs and solutions, to functional solutions that emphasized the global workings of the capitalist system, and back again to territorial solutions that emphasized "selective territorial closure" as the peoples of peripheral regions and localities seek to break away from an exploitive global system.

The first shift, from territorial to functional, probably does reflect a major change in thinking about regional planning in the 1950s and 1960s, with the emergence of regional science and the rapid growth of research and consultancy relating to Third World economic development. The second shift, however, is not so much a description of what has definitely happened in the study and practice of regional planning, as it is an outline of what Friedmann would like to see happening throughout the world. He outlines a model of "agropolitan development" (cf. Lo and Salih 1981) to illustrate how selective territorial closure could operate in the densely populated and predominantly rural areas of Asia. Regrettably, however, there is little detailed consideration of how such changes might take place, and in the real world little progress toward agropolitan models seems to have been made in the decade since *Territory and Function* was written.

Clyde Weaver, who obtained his bachelor's and master's degrees in geography before specializing in planning, has followed up his work with Friedmann with *Regional Development and the Local Community* (1984), a comprehensive historical survey of regional planning. He dedicates this book to Friedmann, Perloff, and Francois Perroux, and he provides a very detailed analysis of the interplay among geography, economics, planning, the other social sciences and a wide range of social-reform movements in the emergence and evaluation of regional planning.

Meanwhile, Friedmann has massively extended *Retracking America* as his new classic, *Planning in the Public Domain* (1987), and most recently he has brought together many of his recent essays on urban, rural, and community development issues in Third World countries (1988). Unlike most scholars, he has shown a strong propensity to rethink his ideas and to change paradigms on various occasions during his career, and he may well soon have new brainstorms to inspire or provoke the community of specialists in regional development and planning.

Brian Berry

There can be little doubt that the most obvious role model for young geographers who seek to specialize in regional development and planning is Brian Berry. He was born in Britain, earned his B.Sc. in Economics at University College London, and initially came to the U.S. in 1955 to earn his master's and Ph.D. in geography at the University of Washington. At that time, Washington was the emerging powerhouse of quantitative geography, with a community of faculty and graduate students that included the seasoned economic geographer Edward Ullman and such rising stars as William Garrison, Duane Marble, William Bunge, Richard Morrill, John Nystuen, Waldo Tobler, and Michael Dacey. In 1958 Berry took a faculty position in geography at the University of Chicago, and that was his prime appointment until he moved on to Harvard University in 1976, to Carnegie-Mellon University in 1981, and to the University of Texas at Dallas in 1986.

From 1966–1976, in addition to his post at Chicago, he was a member of the faculty of the Advanced Study Program in Urban Policy at the Brookings Institution in Washington, and since 1976 his professorships have been in City and Regional Planning, in Urban Studies and Public Policy, and in Social Science. From an early stage he interacted strongly with Walter Isard's regional-science group at the University of Pennsylvania. At Chicago he worked with scholars from various other departments, notably sociology and economics, and he was Director of the Center for Urban Studies from 1974–76. He is a full member of the American Institute of Certified Planners, and he has been very active in professional associations for planning, geography, and regional science. His career has been punctuated by frequent involvements in governmental projects and consultancy work, both in the U.S. and in a wide range of Third World countries, most notably India and Indonesia.

Berry's classic text *Geography of Market Centers and Retail Distribution* (1967a), has recently been revised and expanded (Berry and Parr 1988), and several of his other books have contributed to the geographical education of a generation of scholars (Berry and Marble 1968; Berry and Horton 1970; Berry 1973, 1776; Berry, Conkling, and Ray 1987). Many would recommend his best articles as the key initial readings for all researchers who plan to work on city size distributions, central-place systems, suburbanization and commuting, inner-city revitalization, and counterurbanization (cf. Berry 1961b, 1964, 1970, 1979, 1980; Berry and Barnum 1964). Some of his most seminal papers have recently been compiled by Clark and Wrigley (1989).

Less well known, but equally important in establishing his credentials as one of the foremost specialists in regional development and planning, are his major policy reports to governments and international organizations, notably on Indian regional policies, on Indonesian regional development, and on U.S. urban policy (cf. Berry 1966, 1974; Berry and Hanson 1983). The quality and sheer quantity of Berry's work is remarkable, and he has played a central role in the consolidation of urban geography, quantitative geography, locational analysis, regional science, central-place studies, and international-development studies as major areas of specialization within the geographical profession.

PLANNING THEORY: CONTRIBUTIONS BY GEOGRAPHERS

As stressed by a number of writers (Harris 1967; Faludi 1973; Los 1981), there are two kinds of theory relating to any kind of planning: theory *in* planning (theory about the content of what is planned), and theory *of* planning (theory about the nature and content of the planning process itself). American geographers, in the last decade as earlier this century, have made particularly distinguished contributions to theory *in* regional planning, and notable contributions to a theory *of* planning as well. A look at a longer time horizon demonstrates a general rule: by powerfully helping to shape planners' notions of appropriate planning content, geographers have also strongly influenced their concepts of the planning process.

Thus, in the 1930s—the heyday of the region and regionalism in Anglo-American geographical thought (Johnston 1983, 42)—geographers like George T. Renner persuasively argued for a concept of regional planning that was based upon natural-resource conservation. They shared this philosophy with other influential individuals, like Benton MacKaye and Lewis Mumford—who had derived it from Patrick Geddes (who has in turn derived it from French geographers like Vidal de la Blache and Elisée Reclus)—and like Howard Odum, the southern regionalist sociologist who built his school at the University of North Carolina (Odum and Moore 1938; Odom and Jocher 1945).

All these groups and individuals were highly influential during the New Deal in establishing the National Resources Planning Board (later the Natural Resources Committee). Its 1935 report, *Regional Factors in National Planning and Development*, contained a plan for a regionalization of America for planning purposes drawn up by Renner, who was one of its members (U. S. National Resources Committee 1935; cf. Clawson 1981). Renner's approach, based on the concept of human ecology—which he described more fully in 1936, and then used as the basis for a successful textbook in that same year—was heavily derived from the French geographers' idea of organic natural regions, which formed a central object of criticism in Hartshorne's 1939 intellectual polemic *The Nature of Geography* (Renner 1936; White and Renner 1936; Hartshorne 1939). Despite the devastating nature of this critique, Renner's writings on conservation and regional geography continued to be highly influential during the 1940s (Renner 1942; White and Renner 1948).

With the so-called geographical revolution of the mid-1950s, and its base in logical positivism, theory-testing, and model-building, geographical and regional-planning theory interfused in a new way. The school that clustered around William Garrison at the University of Washington was centrally concerned with the distribution of urban centers in a hierarchical system, and with the rules governing flows of people, goods, and information within such a system (Garrison 1959/60; Berry 1964). Logically, following the pathbreaking study of Mitchell and Rapkin of the relation between land use and traffic generation (Mitchell and Rapkin 1954), it came to be used immediately in the new generation of land-use/transportation planning models, which then began to appear in huge numbers as the basis for the metropolitan component of the Interstate Highway System during this, the heroic age of freeway construction. The new geography thus provided a vital basis for theory-in-planning at the metropolitan regional scale during the heroic age of American suburban growth. But it also provided an implicit theory of planning: "a view of the city as a piece of optimizing economic machinery, in which individual decision-makers located their factories, offices, homes, schools, parks and trip patterns according to pure principles of economic advantage as expressed in economic space" (P. Hall 1984a, 29).

In the intellectual turmoil of the late 1960s came an inevitable countermovement against this mechanistic, black-box view of the urban world. It expressed itself in a number of ways: in a demand for greater citizen participation in the planning process, in a new concern for the distribution of urban costs and benefits, and in a rediscovery of the concerns of the 1930s for natural-resource conservation. Then, in the early 1970s, it crystallized as a new radical view of human geography. In the celebrated distinction of David Harvey (1973), it came in two versions, "liberal" and "Marxist." Both found their natural subject matter in the city, particularly the deprived inner city. Both stressed the geography of distribution: the study of who got what, and where. But the Marxists, who provided a more coherent account of the underlying processes, came to dominate the field in the late 1970s and early 1980s. And, like their systems-analytical predecessors of the 1960s, in providing a new framework of theory-in-planning, they also provided a powerful new concept—generally implicit, but sometimes quite explicit—of a theory of planning.

The first is more properly analyzed elsewhere in this book, under geography from the left, for it amounted to a new theory of regional and urban-development processes. These could be properly understood as results of changes in the structure of late capitalism. Increasingly global in extent, corporations could readily switch production from city to city, region to region, or country to country. Production was increasingly divorced from control; the meaningful division of labor was no longer by product, but by process. *"The Anatomy of Job Loss"* (Massey and Meegan 1982) and *"The Deindustrialization of America"* (Bluestone and Harrison 1982) could be properly explained only through these new *"Spatial Divisions of Labor"* (Massey 1984).

The significant work in this field was done by a close group of workers, some American, some British, some geographers, some economists: Scott, Storper, Clark, Markusen, Bluestone, and Harrison in America; Massey, Harloe, Pickvance, and Sayer in Britain. The clear conclusion that emerged from these studies was that global capitalism was now almost beyond the range of control of any national government, let alone regional planning authority. Consequently their policy implications were strangely quietist; these studies in geography were without clear suggestions for planning intervention.

But at the same time, geographers made a powerful—even dominant—contribution to the development of a new theory of planning. The first signs of this were clearly seen in the contribution of David Harvey (1978) to a symposium on *Planning Theory in the 1980s* (Burchell and Sternlieb 1978). In another contribution, Britton Harris confessed that "this conference as a whole may turn out to have a certain old-fashioned flavor, since it involves many of the Old Guard of planning and planning theory" (Harris 1978, 254). There is indeed a striking intellectual gap between this symposium and another that appeared a mere three years later, *Urbanization and Urban Planning in Capitalist Society* (Dear and Scott 1981). In the first, only 2 contributions out of 21 represent the then-new Marxist approach; in the second, all 29 do. But, equally significant, in the first only two contributors are geographers: Harvey's ideological onslaught *On Planning the Ideology of Planning* is neatly balanced by Brian Berry's bitterly ironic *Notes on an Expedition to Planland*. The latter describes an imaginary island where priests worship the God Plan, but the priests have divided into competing sects or clans. The members of one sect, "the lefties,"

> . . . are easily picked out in any crowd because they wear long hair, go bearded and unwashed, and rally around their banner, the bar sinister, as they shout a leftist antithesis to every rightist thesis, find a contradiction in every contribution, and relish in their favorite blood sport, character assassination. (Berry 1978, 202)

No such spirited controversy is found in the Dear and Scott symposium. Here, 10 of the 21 contributors are geographers, and all but one American; the rest are a mixture of economists, sociologists, political scientists, and other social scientists. But it is difficult to discern the disciplinary origins of the authors from their writings, and the internal debates are highly scholastic ones, informed by French and German theory, about the precise role of the State in contemporary capitalist society. Insofar as geographers (and others of similar persuasion) tackled planning, they interpreted it as merely a reaction to the crisis of the capitalist state:

"neither urbanization in general, nor urban planning in particular, constitute independent, self-determinant occurrences . . ." (Dear and Scott 1984, 4)

"A viable theory of the capitalist State is one which . . . derives the form of the State from its relationship to the wider structural relations of the capitalist society . . ." (Clark and Dear 1981, 49)

". . . planning is a historically-specific and socially-necessary response to the self-disorganizing tendencies of privatized capitalist social and property relations as these appear in urban space . . ." (Dear and Scott 1981, 13)

Planning, in this view, is one of a number of State functions necessary to facilitate capital accumulation; to assist the reproduction of the labor force through provision of education and social services; to maintain a balance between labor and capital, and thus avert social disintegration; and to legitimate social and property relations. In particular, planning seeks to supply certain public goods and services, to regulate and facilitate market operations, to intervene to pursue certain social policy objectives, and to arbitrate between social groups and classes (Clark and Dear 1981, 49; cf. Harvey 1978, 222). As Harvey argued in his pathbreaking contribution to the 1978 symposium:

> . . . the planner's task is to contribute to the processes of social reproduction, maintenance and management . . . in so doing the planner is equipped with powers . . . which permit him or her to stabilize, to create the conditions for "balanced growth," to contain civil strife and factional struggles by repression, cooptation or integration. (Harvey 1978, 223)

But there is a paradox: since capitalists constantly want to escape outside interference, they oppose the very structures and institutions that they indirectly bring into being. Thus, planning solves one problem only to create a new one (Dear and Scott 1981, 14–15); the inherent contradiction between private accumulation and collective action remains (Scott and Roweis 1977, 1107). And—an important gloss—capitalists by no means form a single homogenous group: different factions may have contradictory interests, and they may form complex alliances in consequence. Because of this, some of the best Marxist work of the 1980s (cf. Mollenkopf 1983) has come close to being pluralist. As Dear and Scott (1981, 16) put it:

> The more the State intervenes in the urban system, the greater is the likelihood that different social groups and factions will contest the legitimacy of its decisions. Urban life as a whole becomes progressively invaded by political controversies and dilemmas.

It is generally recognized that planning is often an unstable decision-making process which is affected by forces beyond the dictates of the planning rationale or strategy. P. Hall (1982a), for example, points out that there are three principal actors that form a concert which formulates a modern-day planning project. The first is the massive and well-established bureaucracy which is concerned with policy and program maintenance. The second is the activist pressure groups, usually involving small but vocal minorities. And the third is a group of politicians, always motivated by the intensity of popular opinion because they are interested in votes for future elections. On occasions, grave mistakes have occurred, aborting vital projects, implementing unnecessary ones, and selecting unduly expensive alternatives, because of the decision-making process involving the three groups. Nevertheless, genuine Marxist analyses are distinguished from pure pluralist approaches, such as the one Peter Hall (1982a) used in his analysis of *Great Planning Disasters,* by their strong structural element.

Traditional planning theory has avoided the true nature of capitalist planning, so this argument runs; and because of this it is meaningless. It seeks to define an ideal kind of planning, without economic and social context. Worse, it serves the function of depoliticizing planning, thus legitimating it (Scott and Roweis 1977, 1098). The reality is the opposite: planning theory is simply a creation of the social forces that bring planning into existence (Scott and Roweis 1977, 1099). This is a powerful body of criticism; the only problem is where it leaves the theory of planning. Clearly, if planning is only this, then it is not very interesting and not worth doing. A true theory of planning must provide both a guide to what planners should do, and a justification for it. On this crucial point, the Marxist school of planning theory divides. A minority draw the logical conclusion: that the job of the planner is to change the system. Thus, Harvey suggests:

> . . . we might even come to see that it is the commitment to an alien ideology which chains our thought and understanding in order to legitimate a social process that preserves, in a deep sense, the domination of capital over labor. Should we reach that conclusion, then we would surely witness a markedly different reconstruction of the planner's world view than we are currently seeing. We might even begin to plan the reconstruction of society, instead of merely planning the ideology of planning. (Harvey 1978, 231)

Yet, equally remarkable about even the Dear-Scott symposium is that—in contradistinction to Harvey in 1978—very few contributions contain much suggestion of a commitment to action. The most explicit is the nongeographer Shoukry Roweis, who ends his paper:

> The question now is this: will planners and prospective planners act to realize the potentials of urban planning? And the other question is: what if they do not act? (Roweis 1981, 177)

Here he reflects the conclusion of an earlier joint paper with Allen Scott:

> . . . a viable theory or urban planning should not only tell us what planning is, but also what we can, and must, do as progressive planners. (Scott and Roweis 1977, 1099)

Even this, of course, leaves the whole question open. Harvey is at least specific: Marxist planners should work for the achievement of a socialist state, without which true planning is impossible. For the others, the right course of action is far from clear.

Thus, since the mid-1970s, planning theorists (among whom American geographers have played a major role) have produced a new and persuasive theory of planning. Yet it tells planners almost nothing about what they should do in their day-to-day activities. On the contrary, it powerfully suggests that they are engaged in a fundamentally unworthy activity. It has left planning in a kind of theoretical impasse from which, at the end of the 1980s, there appears as yet no escape. Many practicing planners have responded to such views of planning theory by dismissing academics as impractical and subversive, and gatherings of professional planners and consultants are often characterized by a strong anti-intellectual bias.

The division of the American planning profession into separate tribes of "scholars" and "practitioners" is sharply visible when a comparison is made between the content of the annual congress of the Association of Collegiate Schools of Planning (ACSP) and that of the American Planning Association (APA). For career academics anxious to specialize in regional development and planning, there is a continual conflict between the need to acquire hands-on planning experience in government, business corporations, and voluntary bodies, and the pressures generated by their universities for research grants and publications. This conflict is made a great deal more difficult to handle when the academics also have qualms about the political complexion, morality, or objectives of their "real-world" employers.

RESEARCH FOCUSING UPON THE UNITED STATES

It is paradoxical that the U.S. has produced much—perhaps even most—of the world's finest scholarship on regional development and planning, while national/regional and regional/local planning have been of relatively little importance during most of the country's history. The brief flowering of regional planning that occurred in the 1920s and 1930s with the work of the RPAA and the RPA-NY, and then the proliferation of New Deal programs, was rapidly squashed in the 1940s by a coalition of state and local governments and of pork-barrel politicians protesting at the intrusion of technocratic planning agencies into their areas of power, responsibility, and patronage. By the time that Senator Joseph McCarthy reached the apex of his influence around 1950, "planning" had effectively been limited to local governments and the work of urban planners in such fields as zoning, subdivision ordinances, and capital budgeting. The suggestion that the country as a whole should be "planned" was often branded as the easy road to communism, and essential federal government activities involving long-term resource allocations to particular sectors and regions were usually described using substitute words instead of "planning" and "plan": for example, allocation, management, administration, policy, strategy, program, project, and scheme. Many of the pioneers of the 1930s followed Tugwell's example by going to work outside the continental United States, or by finding an academic position well away from the federal government.

Despite the reaction against planning in the 1940s and 1950s, the federal government continued to make major allocations across the national territory, most notably in creating the Interstate Highway System, in aerospace and defense contracting, and in water-resource management. These decisions should not be viewed as simply the product of pork-barrel politics. Even pork-barreling has its geographic rationale, but many of the locational guidelines and criteria were governed by other rationales—technical, economic, environmental, and military. Many of these massive investments were made rationally and with great foresight, and they had major affects on the spatial organization of the country, most notably through accelerating suburbanization, supporting the rise of the Sunbelt, and accelerating the development of the West (Rose 1979; Markusen 1986; Worster 1985). Regional planning was taking place in everything but name, and these investment programs merit much greater attention from specialists in regional development and planning. In the 1960s and 1970s, there was even a move toward developing a National Urban Policy—just as much a national/regional plan as the river-basin proposals of the 1930s (cf. Alonso 1972; Ehrlichman 1972; U.S. National Resources Committee 1935, 106–11)—but again, the term "regional planning" was not applied to this activity.

The paradox that the U.S. seems to simultaneously have a great deal of scholarship on regional planning and only a modest degree of activity that is formally called regional planning is well illustrated by the range of organizations that have performed regional-planning functions in recent years. These organizations are diverse and only weakly coordinated with one another, and they have very different origins and institutional trajectories. The three most important and best known are:

- The Port Authority of New York and New Jersey, established in 1921, which concentrates on transportation infrastructure and regional economic development in the New York metropolitan area (Warf 1988)
- The Tennessee Valley Authority (TVA), founded in 1933 for the integral development of the Tennessee and Cumberland basins, but now heavily focused upon power generation from coal and nuclear sources to supply consumers far outside those basins (Hargrove 1983)
- The Appalachian Regional Commission (ARC), founded in 1965 and gradually dismantled after 1981 (Newman 1972; Branch 1988, 1989)

In addition, there are numerous smaller agencies that perform more specialized or localized functions (cf. Lim 1983 and Walker 1987), coordinate neighboring local governments, provide public services, or tackle specific environmental problems. Lim (1983, 9–11) summarizes these as follows:

> At the state level, various departments and commissions are engaged in some type of regional planning. Examples include coastal zone management, land use and housing planning, energy planning, health planning, and flood plain control. . . .In addition, some states have developed a comprehensive system of regional planning councils. At the substate level, several approaches to regional planning can be identified. The first is the consolidation of city and county government. . .a second approach has been to strengthen existing urban counties by expanding the service responsibilities of county governments. . . .A third. . .is a system of two-tier government. . . .The fourth approach is the creation of a regional government (there are only two in the United States). . . .The fifth and most popular is the Council of Governments (COG). The number of COGs increased from 23 in 1950 to 670 in 1980. . . .Their functions are limited to information exchange, cooperation, coordination, technical assistance, and provision of some services.

The 1977 Census of Governments found 1932 regional planning organizations, spending more than $1.3 billion, employing more than 148,000 workers, and involving more than 43,000 citizens on their governing bodies (So, Hand, and McDowell 1986, xiv). It is important to recognize, however, that most of the expenditures and resources were concentrated in a handful of organizations, and that the overwhelming majority were small, weak, and severely limited in their functions. Since then the number of organizations has diminished substantially because of the abolition of most forms of federal funding in 1981, and many of the surviving organizations have had to reduce their personnel and activities. A major anthology on these organizations recently concluded that:

> Almost all regional organizations face major difficulties in implementing their plans. They could use greater authority, more stable organizational structures and financial resources, stronger linkages with other regional organizations, more consistent support from a wide variety of federal aid programs, and strong support from and linkages to state governments. But all of these are hard to come by. (So, Hand, and McDowell 1986, 164)

The demise of many regional-planning organizations in the 1980s is highlighted by spectacular cases like the ARC and the Tri-State Regional Planning Commission for the New York Metropolitan Area, which was liquidated in 1981 after 20 years of operation, initially as a Transportation Committee and later as a Metropolitan Planning Organization (Danielson and Doig 1982). This demise, however, is not just the result of the Reagan Administration's abolition of many forms of federal financial support. It also reflects the entrenched vested interests of federal, state, and local governments and of elected political representatives, the national emphasis on individualism, choice and support for business interests, and the continuing proliferation of local governmental units based on the desire of suburban communities to maintain separate identities, services, tax bases, and school systems from those of the major cities (cf. J. K. Jackson 1985). While suburban communities have generally embraced local-level urban planning as a means to preserve and enhance their local environments and their political, class, and ethnic identities, they have generally resisted subordination to larger regional bodies. The strength of these *"Politics of Exclusion"* (Danielson 1976) is well illustrated by Wood's 1961 classic on the New York Metropolitan Region, *1400 Governments,* and by Miller's *Cities by Contract* (1981), which describes why and how the number of municipalities in Los Angeles County rose from 45 to 81 between 1954 and 1981.

Recent geographical research on regional development and planning in the U.S. has focused upon three major themes: regional inequalities and national spatial organization, regions and regionalism, and land-use patterns and environmental issues.

Regional Inequalities and National Spatial Organization

During the last 25 years there has been a massive restructuring of the U.S. space economy, featuring the decline of most traditional industries, the rise of the service sector, and major shifts of population and economic activity away from the traditional "core" of the Northeast toward the Sunbelt and the West Coast. The country's foreign-trade links have gradually been reoriented away from Europe and toward the Far East—an area of the world which we should rename the Far West, given that the normal way to get there is across the Pacific! U.S. corporations have shifted many of their manufacturing activities to subsidiaries and subcontractors outside the country, U.S. banks have lent heavily to Third World governments and corporations, and during the 1980s the country's mushrooming fiscal and trade deficits have fueled large-scale foreign investment in U.S. stocks, bonds, corporations, and real estate.

Major national overviews have focused upon such topics as the territorial evolution of the country as a whole (Meinig 1986), the historical development of the urban system (Pred 1966, 1980), the contemporary urban system (Pred 1977; Berry and Hanson 1983; Johnston 1982), the regional structure and performance of the economy (Clark, Gertler, and Whiteman 1986), social well-being and patterns of poverty (Morrill and Wohlenberg 1971; D. M. Smith 1973), and congressional voting patterns and the distribution of federal investment (Johnston 1980). The general pattern of metropolitan development is a structured outgrowth of the dynamics of the production system (Green 1965; Scott 1982), and the regional metropolis is gradually decentralizing and dispersing (cf. Berry 1970, Borchert 1972; Gottdiener 1977; Patel 1980; Griffith 1981; Rogerson 1984; Zelinsky and Sly 1984).

Studies of regional population change have shown tendencies toward nonmetropolitan growth in older industrial states and rapid metropolitanization in newly industrializing states (Morrill 1979), with the appearance of pockets of high unemployment and economic distress in areas of rapid deindustrialization (Clark 1980). Regional changes have been linked to economic restructuring, both in traditional industries (cf. Markusen 1985), and through the rise of new high-technology industries (cf. Rees 1979, 1986; Hall and Markusen 1985; Markusen, Hall, and Glasmeier 1986). Numerous geographical studies have analyzed the gradual shift of population and jobs from the Snowbelt to the Sunbelt (cf. Beyers 1979; Casetti 1981; Clark 1982; Kirn 1987; Plane 1984), and various authors have examined the relations between the rise of the Sunbelt and the spatial allocation of aerospace and defense contracts (cf. Rees 1981; Malecki 1984; and O'hUallacháin 1987).

Regions and Regionalism

American geographers continue to produce fine theme studies, monographs, and atlases that focus upon individual states and regions. Examples are California (Vance 1972), the Colorado Plateau (Durrenberger 1972), the Great Plains (Mather 1972), the Middle West (Hart 1972), Minnesota (Borchert and Gustafson 1980), the Northern Heartland (Borchert 1987), the New York Metropolitan Region (Bergman and Pohl 1975), and the Piedmont Cotton Region (Prunty and Aiken 1972). Such works follow in the noblest traditions of regional geography, and they play a very important role in relating geography departments to their surrounding regional communities. They demonstrate the geographer's very special concern for the character of places (cf. Tuan 1974), and they provide an empirical basis for relating perceptions to scale and for continuously reconsidering how we should define regions (Laity 1984).

A further line of research examines how the conceptualization of places and regions relates to the field of time-geography (Hagerstrand 1978; Hudson 1979, Pred 1984), and such work is greatly assisted by detailed historical studies of specific localities in their broader regional context (cf. P. F. Lewis 1972). Regional-scale population changes have also been subjected to detailed analyses (cf. Hart 1984; G. K. Lewis 1972; B. Marsh 1987), and such work is of direct interest to most state and local planning agencies.

Identity and territory are defined as major considerations in geographical perspectives on nationalism and regionalism (Knight 1982), and in turn, regionalism is a central issue in defining regions and in providing a motivation for regional planning. In recent years, there has been a revival of interest in regionalism, both in North America and in other areas of the world, and there is ample scope for the development of a major field of collaboration among political science, political geography, and regional development and planning.

Shortridge (1984, 1987) has developed ingenious research methods to examine perceived regional identities and to define such vague entities as the East and the Midwest, and the literature on behavioral geography and environmental planning (cf. Gould and White 1974; Saarinen 1976) provides many useful pointers to the relations between perceptions and decision-making. Such work is given added significance when it is related to the content of three recent books, written by nongeographers but tremendously geographical in character: Gibbins (1982), Price (1982), and Markusen (1987). These provide the basis for a political economy of regionalism, analyzing regional identities and pressure groups in terms of coalitions of interest, and impacting directly on the patterns of economic growth and decline in the country as a whole.

Land-Use Patterns and Environmental Issues

The American landscape is being remodeled by vast regional land-use changes in both metropolitan and nonmetropolitan areas (Berry and Gillard 1977; Platt and Macinko 1983). The relative abundance of the American land resource when compared to most other areas of the world should not obscure the fact that it is still finite (R. H. Jackson 1981). Land-use and land-tenure studies can play a central role in monitoring regional development and planning (Healy 1976; D. W. Jones 1978, 1982), and geographers can play a key role in devising operational models for land management (Charnes et al. 1975; Merrifield 1983). The U.S. National Environmental Policy Act of 1969 (NEPA) and the ensuing environmental-impact legislation have recognized the need for studies of the environmental impacts a project will have on a region before construction begins, and these studies have generated many new opportunities for geographers to become involved in public-policy debates and research (Greenberg, Anderson, and Page 1978; Porter 1985).

Urban encroachment upon valuable rural lands, and the need to reconcile the conflicting goals of preservation of agricultural and wilderness land and the expansion of urban space, have generated innumerable local controversies (Clawson 1972). Special difficulties have arisen in areas having outstanding tourist potential, notably along the coasts and close to resort areas and historic sites (cf. Heiman 1986). Environmental-planning issues associated with the disposal of wastes and the location of unpopular facilities have also been very contentious, and geographers have conducted numerous studies of garbage and toxic-waste disposal, sewer-line construction, open-pit mining, and the siting of power plants and mental facilities (cf. Seley 1983; Benhart and McCartney 1984).

INTERNATIONAL AND COMPARATIVE STUDIES

Even when we concentrate on the work of U.S. geographers on regional development and planning outside the U.S., it is extraordinarily difficult to present a representative picture of research progress. The sheer diversity of countries, cultures, and environmental conditions, combined with the enormous variety of publication outlets in a wide range of languages, mean that our selection is somewhat arbitrary, to say the least. We concentrate here on three major themes in geographical research: regional inequalities and national/regional planning; the management of large metropolitan areas—the world's megacities; and marketing and central-place systems.

Regional Inequalities and National/Regional Planning

Ever since the publication of the U.S. National Resources Committee's 1935 study of *Regional Factors in National Planning and Development,* and the U.K. Barlow Commission's 1940 *Report of the Royal Commission on the Distribution of Industrial Population,* geographers, planners, and economists have taken a keen interest in devising means to identify backward or lagging regions and to formulate strategies for balancing regional development. The key debate has been among (1) advocates of public policies to canalize investment, employment, and services toward depressed and peripheral regions ("jobs to people" strategies); (2) advocates of laissez-faire approaches having no explicit regional policies; and (3) advocates of policies to encourage people to migrate out of depressed and peripheral regions to more dynamic areas ("people to jobs" strategies). "Jobs to people" strategies tend to be territorially focused, and therefore focus upon "place prosperity," paying relatively little attention to intraregional distributions of income and wealth, while "people to jobs" strategies tend to directly focus upon "people prosperity," ignoring territorial units and focusing upon national distributions, totals, and averages.

Though national/regional planning has not been formally conducted in the U.S. since the Second World War, it is a major activity of national-planning ministries and other central government agencies in many other countries. Various geographers have examined national/regional planning and its implications for patterns of regional development in specific countries, for example: Dutt and Costa (1980, 1985) on China and the Netherlands, Bromley (1977) on Ecuador, Mabogunje (1978) on Nigeria, and Misra and Natraj (1981) on India. A particularly interesting case is Britain, where a very elaborate system of regional-development incentives to support the Northwest of the country was gradually built up from 1929 till 1979, and since has been almost totally dismantled by the Thatcher government (cf. Martin and Hodge 1983; Chisholm 1985, 1987).

The study of spatial inequalities is a central theme in modern geography (Coates, Johnston, and Knox 1977), and many authors have examined uneven regional-development patterns, for example: Dutt (1968, 1969/70, 1970) for Britain and the Netherlands; Berentsen (1978, 1979, 1981a, 1981b) for Austria and East Germany; Mookherjee (1974), Mookherjee and Morrill (1973), and Costa and Dutt (1978) for India; and C. Hall (1984) for Costa Rica. Some of the processes of regional change underlying these inequalities are well illustrated by Soja (1968), Gould (1970), and Pedersen's classic studies (1974) of the impact of the "diffusion of modernization" on regional inequalities. Further, spatiotemporal analyses of this type are very much needed, though it is vital that new analysts escape from the ethnocentric and deterministic mold which characterized many of the early studies of modernization.

Other national case studies relate regional inequalities to major socioeconomic-development issues. Yapa (1979), for example, shows how the green revolution in India has accelerated regional economic growth, while intensifying the polarization between the rich and the poor within most regions and localities. In analyzing the role of politics in regional inequalities in Israel, Gradus (1983) notes that certain areas dominate the decision-making process of the country. He argues that the centralized, unitary character of government reinforces both authority-dependency relationships between core and periphery, and the underdevelopment of certain peripheral regions.

In her analysis of the locational patterns of five multinational corporations in Brazil, as well as small locally owned firms, Cunningham (1981) found that the multinationals were predominantly concentrated in the metropolitan areas of the Southeast, while the smaller locally owned firms were more widely spread across the country. Also for Brazil, Enders (1981) evaluated the national government's policy of industrial dispersion during the 1960s and 1970s, and found it to have been relatively successful. He attributed an above-average growth in a large number of peripheral states to the "trickle down mechanism" or "spiral effect" from the developed cores, but he noted the need for more vigorous governmental programs and economic stimulation in three states having locations and circumstances that were too unfavorable to draw industrial activities.

Regional economic and social inequalities in the socialist countries have been analyzed by Hoffman (1972), Karaska (1975), Fuchs and Demko (1979), and Lakshmanan and Hua (1987). Though the governments of the USSR and East Europe are officially committed to eliminate such disparities, the studies indicate that they have not succeeded. According to Fuchs and Demko, the main reason for the persistence of the inequalities lies in the socialist decision-making system, which places "a higher value on growth, productivity and efficiency than on equity, whether spatial or structural." Thus, there is a significant gap between their avowed "balanced regional development" policies and the practice of fostering development where it is more economically efficient (Fuchs and Demko 1979, 317). Similarly, no effective regional policy has been established to minimize the regional disparities in China, where the difference between the richest area (Shanghai) and the poorest area (Tibet) stands at a ratio of 16:1 (Lakshmanan and Hua 1987).

Management of Large Metropolitan Areas

United Nations estimates (1987, 25–26) suggest that by the year 2000 the world will have three urban agglomerations with over 20 million inhabitants each (Mexico City, São Paulo, and Tokyo/Yokohama), and 45 more with over five million inhabitants. Both in rich and poor countries, the growth of such giant urban agglomerations poses major metropolitan planning problems which must be solved in a broad, regional context. Food, fuel, electric power, and drinking water must be brought in from considerable distances, and massive quantities of waste materials must be disposed of. The giant metropolis needs very sophisticated transportation systems if it is to function effectively, and city dwellers may have to travel considerable distances to recreational facilities and surviving farm and wilderness areas. Most major cities also have some special problems resulting from the peculiarities of their sites, for example subsidence, earthquakes, and air and groundwater pollution buildups in Mexico City, and flooding and typhoon damage in Dacca.

A major intellectual debate, still unresolved, has raged between (1) proponents of the idea that metropolitan growth can continue and urban primacy can accentuate almost ad infinitum because the economies of scale virtually always outweigh the diseconomies of urban growth (cf. Alonso 1968; Richardson 1972, 1976; Mera 1973), and (2) proponents of balanced urban hierarchies with severe controls on the major metropoli and deliberate stimulation of secondary cities and rural growth centers (cf. Gilbert 1976, 1977; Johnson 1970; Rondinelli and Ruddle 1978; Rondinelli 1983). Further complexities are added to the discussion on urban hierarchies and optimum urban size when political decentralization and grass-roots development strategies are recommended, and most proponents of such strategies (cf. Stöhr and Taylor 1981; Rondinelli 1981; Cheema and Rondinelli 1983; Friedmann 1988) favor regional-development policies designed to correct "urban bias" and to divert geographical transfers of value away from the major cities.

Geographers have played a major role in case-study research on the growth and management problems of major metropoli, concentrating particularly on land-use planning issues on the metropolitan fringe. Some books have focused upon individual cities and countries, for example P. Hall's *London 2000* (1969) and *The Containment of Urban England* (1973); others have gathered essays on a variety of major cities (Pacione 1981a, 1981b; P. Hall 1984b; Ewers, Goddard, and Matzerath 1986; Dogan and Kasarda 1988); while still others have compared urban growth management and the evolution of urban systems in different countries (Clawson and P. Hall 1973; P. Hall and Hay 1980). A new burst of publication is imminent with the book series of individual city monographs on major world cities edited by Ron Johnston and Paul Knox. Hopefully, this new wave of case studies will be accompanied by new thematic contributions on megacities, focusing upon such issues as optimum urban size, residential density and neighborhood organization, metropolitan government, and the choice of appropriate technologies for transportation and utilities.

Marketing and Central-Place Systems

A very fruitful area of geographical and anthropological research in the 1970s focused upon the spatiotemporal organization of systems of periodic and daily markets (cf. R. H. T. Smith 1971, 1978, 1979/80; C. A. Smith 1976; Bromley, Symanski, and Good 1975; Bromley 1976, 1980), examining the rationale of periodic markets, the activity patterns of mobile traders, the factors determining market periodicity, and the factors underlying changes in the structure of regional central-place systems. This work linked in with studies of marketing chains, commodity flows, and interregional trade in Third World countries (cf. Berry 1966; R. H. T. Smith and Hay 1969; Hay and R. H. T. Smith 1970; Bromley 1974), and with the emergence of spatial-planning strategies in India which are based on a mixture of industrial growth poles, rural service centers, and regulated markets (cf. Johnson 1970; Roy and Patil 1977; Misra and Sundaram 1978). Since then, case-study research has been done in a wide range of countries (cf. Eighmy 1972; Good 1975; Park 1981), and a number of rural growth center experiments have been tried, most notably in Kenya (Obudho and Waller 1976; Gaile 1979, 1989; B. G. Jones 1986).

In the late 1970s, a fully fledged regional development strategy for the enhancement of central-place systems was proposed by Rondinelli and Ruddle (1978) under the title "Urban Functions in Rural Development" (UFRD), linking and extending the earlier pioneering studies of Berry (1967a), Mosher (1969), and Johnson (1970). The URFD strategy set out to improve conditions in rural areas of the Third World through the installation of public services and the promotion of commercial activities and agroindustries in strategically located "key market towns." The methodology has now been more fully elaborated (Rondinelli 1985a; Bromley 1984), and it has been integrated into broader proposals for multifaceted rural-development strategies (Bar-El, Bendavid-Val, and Karaska 1987).

Nevertheless, some significant criticisms have been made (cf. Bromley 1983; Gore 1984, 149–56), most notably on the grounds that the strategy does not directly attack the traditional mechanisms by which town-based merchants, landlords, and moneylenders have exploited the peasantry over large areas of the Third World—mechanisms which are embodied in the concepts of "internal colonialism" (Gonzalez Casanova 1964) and "urban bias" (Lipton 1977). The continuing poverty of most of rural India, despite some 20 years of rural growth center policies, demonstrates the limitations of UFRD and similar spatial strategies, though more long-term field research in a variety of countries will be needed before a full assessment of their potentials and limitations can be made.

FUTURE AGENDA

Geographers have made substantial contributions both to the analysis of regional development and to the theory and practice of regional planning. One important question remains: what future agenda will further enhance the regional development and planning aspects of our discipline? We see three research themes as being particularly crucial, and yet not receiving adequate attention: nation-building and territorial remolding; the application of scale criteria to development theories; and environmental change, most notably in humid tropical and semiarid areas.

Nation-Building and Territorial Remolding

Charles Gore's general survey (1984) of regional-development theory, *Regions in Question,* is perhaps the best "state of the art review" for the 1980s. It raises serious doubts about every major theoretical underpinning of our subject, and it accuses most regional planners of spatial fetishism and inappropriate spatial separatism. To say the least, however, Gore's overwhelmingly critical tone (cf. Rondinelli 1985b) tends to divert our attention from two crucial realities: first, every day and in every country, governments and corporations are busily distributing investments and collecting taxes and profits across the national territory, and these classic geographic processes are of great intrinsic interest; and second, some public policies in some countries produce dramatic and permanent changes in national spatial organization. The fact that our current theoretical tool kit is inadequate should not induce us to abandon the field, but rather to redouble our efforts to find new theories and models "grounded" (Glaser and Strauss 1968) in real-world examples of national transformation.

It is particularly crucial to understand *when, why, and how* major spatial policies emerge, and *how much and in what ways* they eventually influence spatial organization. Key cases are easy to find: for example, the massive spatial reorganization of Russia embarked upon by Peter the Great (Massie 1980), the remolding of Nigerian federalism since the Civil War (Panter-Brick 1978; Moore 1985), the Brazilian "assaults" on the Mato Grosso and the Amazon (Katzman 1977; Bourne 1978), the continuing search for viable territorial and social policies to hold India together as a nation-state (Majeed 1984; Akbar 1985), and such crucial elements of U.S. history as westward expansion, the search for an *Empire on the Pacific* (Graebner 1982), and the construction of the Interstate Highway System.

To date, such themes have received more attention in political geography and in the emerging geography of public finance (Bennett 1980) than from specialists in regional development and planning, but their full integration into our field is crucial for us to adequately envisage the impact of current policies and to appreciate the range of alternatives that are available. Regional planners should be tackling the great national unity and disunity problems of our age—Lebanon, Northern Ireland, Sri Lanka, etc.—and a key element here is to more fully develop the concept of "nation-building" (cf. Rokkan 1973; Rokkan and Unwin 1983; Anderson 1983) so as to analyze the range of policies that national governments can employ to encompass diversity within unity.

Scale Criteria for Development Theories

Gore's *Regions in Question* (1984) is also a very good exploration of the relations between regional-development theory and national/international development theory. All major development concepts and theories—core-periphery, agglomeration, dominance-dependence, growth poles, urban bias, leading-sector strategies, import-substitution strategies, selective territorial closure, etc.—can be applied at different scales and in different contexts. Because of their mesoscale role between national and local, specialists in regional development and planning should be exploring the significance of scale and context, and the limitations of particular theories and approaches at subnational levels. The emerging field of development geography (cf. Brookfield 1975; Mabogunje 1980; Forbes 1984; Corbridge 1986) is important here, and there is a particular need for research on the geographical transfer of value at all scales from global to local (Forbes and Rimmer 1984).

Perhaps the most useful of all contributions, however, is by the British economist and politician Stuart Holland (1976a, 1976b). He argues that most conventional national/regional policies, designed to provide incentives for the private sector to invest more in peripheral regions, are now worthless because the globalization of capital and the behavior of multinational corporations implies that the crucial locational decisions are between countries and not between the regions of any specific country. If development processes are now essentially global in scope, what hope is there for meaningful local economic development strategies? What can the mayors of crisis-ridden cities or the governors of troubled states in the U.S. Rustbelt do to generate growth in the local economy, and how much is what they do just part of a zero-sum game where they are snatching investment from competing cities and states (Goodman 1979)?

Environmental Change

The Earth is currently under greater human-generated environmental pressure than ever before, and massive problems are emerging associated with pollution, the depletion of nonrenewable resources, the loss of genetic diversity, deforestation, erosion-provoking agricultural and livestock-raising practices, and the essentially permanent loss of farmland and wilderness to urban development and infrastructure provision. Specialists in regional planning have a special responsibility to prescribe environmentally sound development strategies, and there is a great need for more applied research in this area.

Regional development and planning should emerge as a rallying point for geographers who "take the high ground" by interrelating physical and human conditions in the analysis of major environmental problems (Stoddart 1987), and the regional perspective is particularly crucial here to link local studies and pilot projects with national policies. There are some commendable examples of relevant recent work by geographers (cf. Bernard 1985; Blaikie 1985; Hemming 1985; Turner 1989), but these should be viewed as new beginnings, rather than as final statements.

A Concluding Comment

The future growth of the field of regional development and planning is conditional on effective collaborative links with both scholars in cognate disciplines and practitioners in government, business, and voluntary organizations. This will involve substantial participation by geographers in multidisciplinary teams and projects, the development of new career patterns to mix scholarship with practice, and the emergence of new generations of interdisciplinary scholar-practitioners to emulate such great pioneers as Marsh, Geddes, MacKaye, Mumford, Tugwell, Perloff, Isard, Friedmann, and Berry. In his classic Presidential Address to the AAG, Brian Berry (1980, 453) provides an eloquent justification for such efforts:

> Applied geography is not something to be contrasted with, and set beneath, an academic geography that is somehow "pure," . . . for there is an ineluctable interdependence of speculation, application, and theoretical imagery that only those fearful of emerging into sunlight from their cloisters would deny.

He illustrates this by quoting Mao Zedong:

> If you want knowledge, you must take part in the practice of changing reality. If you want to know the taste of a pear, you must change the pear by eating it. We must expand rather than deny our commitment to and reliance upon the world of practice—what others more felicitously term the world of the nonteaching professional—for this world must be to the explorers of future geography what field observation was to geographical exploration in times past.

Berry's final plea is for us to:

> . . . join together in new and vigorous efforts to create *future geographies* that are consistent with our enduring values, *environments* that expand freedom and choice and opportunity, *organised fantasies* that contain new delights for the scholarly practitioners and the practical scholars who will succeed us. (Berry 1980, 458)

REFERENCES

Abler, R.; Adams, J. S.; and Gould, P. 1972. *Spatial organization: The geographer's view of the world*. Englewood Cliffs, NJ: Prentice-Hall.

Adams, T. 1932. A communication: In defense of the regional plan. *New Republic* 71(6 July):207–10.

Akbar, M. J. 1985. *India: The siege within*. Harmondsworth: Penguin.

Alonso, W. 1968. Urban and regional imbalances in economic development. *Economic Development and Cultural Change* 17:1–14.

———. 1972. Problems, purposes, and implicit policies for a national strategy of urbanization. In *Population distribution and policy,* ed. S. M. Mazie. Washington: U.S. Government Printing Office, U.S. Commission on Population Growth and the American Future, Research Reports, vol. 5.

Anderson, B. 1983. *Imagined communities: Reflections on the origin and spread of nationalism*. London: Verso.

Bar-El, R.; Bendavid-Val, A.; and Karaska, G. J., eds. 1987. *Patterns of change in developing rural regions*. Boulder, CO: Westview Press.

Benhart, J. E., and McCartney, D. M. 1984. Solid waste management in Southcentral Pennsylvania. *Pennsylvania Geographer* 22:1–7.

Bennett, R. J. 1980. *The geography of public finance*. London: Methuen.

Berentsen, W. H. 1978. Austrian regional development policy. *Economic Geography* 54:115–34.

———. 1979. Regional planning in the German Democratic Republic. *International Regional Science Review* 4:137–54.

———. 1981a. Conflicts between national and regional planning objectives: Austria and East Germany. *Papers of the Regional Science Association* 48:135–48.

———. 1981b. Regional change in the German Democratic Republic. *Annals of the Association of American Geographers* 71:50–66.

Bergman, E. F., and Pohl, T. W. 1975. *A geography of the New York metropolitan region*. Dubuque, IA: Kendall/Hunt.

Bernard, F. E. 1985. Planning and environmental risk in Kenyan dryland. *Geographical Review* 75:58–70.

Berry, B. J. L. 1961a. City size distributions and economic development. *Economic Development and Cultural Change* 9:573–88.

———. 1961b. A method of defining multi-factor uniform regions. *Przeglad Geograficzny* 33:263–82.

———. 1964. Cities as systems within systems of cities. *Papers of the Regional Science Association* 13:147–63.

———. 1966. *Essays on commodity flows and the spatial structure of the Indian economy*. University of Chicago Department of Geography Research Paper No. 111. Chicago: University of Chicago Press.

———. 1967a. *Geography of market centers and retail distribution*. Englewood Cliffs, NJ: Prentice-Hall.

———. 1967b. Grouping and regionalizing. In *Quantitative geography,* eds. W. L. Garrison and D. F. Marble, pp. 219–51. Evanston, IL: Northwestern University Studies in Geography.

———. 1970. The geography of the United States in the year 2000. *Transactions of the Institute of British Geographers,* o.s. 51:21–53.

———. 1971. City size and economic development. In *South and Southeast Asia urban affairs annual.* Vol. 1 of *Urbanization and national development,* eds. L. Jacobson and V. Prakash, pp. 111–56. Beverly Hills: Sage Publications.

———. 1973. *The human consequences of urbanization*. New York: St. Martin's Press.

———. 1974. *A framework for regional planning in Indonesia*. 3 vols. Washington: World Bank.

———, ed. 1976. *Urbanization and counter-urbanization*. Urban Affairs Annual Review, No. 22. Beverly Hills: Sage Publications.

———. 1978. Notes on an expedition to Planland. In *Planning theory in the 1980s,* eds. R. W. Burchell and G. Sternlieb, pp. 201–208. New Brunswick, NJ: Center for Urban Policy Research, Rutgers University.

———. 1979. Inner city futures: An American dilemma revisited. *Transactions of the Institute of British Geographers,* NS 5:1–28.

———. 1980. Creating future geographies. *Annals of the Association of American Geographers* 70:449–58.

———, and Barnum, H. G. 1964. Aggregate relations and elemental components of central place systems. *Journal of Regional Science* 4:35–68.

———, and Marble, D. F. 1968. *Spatial analysis: A reader in statistical geography*. Englewood Cliffs, NJ: Prentice-Hall.

———, and Horton, F. 1970. *Geographic perspectives on urban systems*. Englewood Cliffs, NJ: Prentice-Hall.

———, and Gillard, Q. 1977. *The changing shape of metropolitan America*. Cambridge, MA: Ballinger.

———, and Hanson, R. 1983. *Rethinking urban policy*. Washington: National Academy Press.

———; Conkling, E.; and Ray, M. 1987. *Economic geography: Problems, approaches, basic theory*. Englewood Cliffs, NJ: Prentice-Hall.

———, and Parr, J. B. 1988. *Market centers and retail location: Theory and applications*. Englewood Cliffs, NJ: Prentice-Hall.

Beyers, W. H. 1979. Contemporary trends in the regional economic development of the United States. *Professional Geographer* 31:34–44.

Blaikie, P. 1985. *The political economy of soil erosion in developing countries*. London: Longman.

Bluestone, B., and Harrison, B. 1982. *The deindustrialization of America*. New York: Basic Books.

Bogue, D. J., and Beale, C. L. 1961. *Economic areas of the United States*. Glencoe, IL: Free Press.

Borchert, J. R. 1972. America's changing metropolitan regions. *Annals of the Association of American Geographers* 62:352–73.

———. 1987. *America's northern heartland*. Minneapolis, MN: University of Minnesota Press.

———, and Gustafson, N. C. 1980. *Atlas of Minnesota resources and settlement*. Minneapolis, MN: Center for Urban and Regional Affairs, University of Minnesota.

Bourne, R. 1978. *Assault on the Amazon*. London: Victor Gollancz.

Boyer, M. C. 1983. *Dreaming the rational city: The myth of American city planning*. Cambridge: MIT Press.

Branch, M. C. 1988. *Regional planning: Introduction and explanation*. New York: Praeger.

Bromley, R. 1974. Interregional marketing and alternative reform strategies in Ecuador. *European Journal of Marketing* 8:245–64.

———. 1976. Contemporary market periodicity in highland Ecuador. In *Economic systems,* vol. 1 of *Regional analysis,* ed. C. A. Smith, pp. 91–122. New York: Academic Press.

———. 1977. *Development and planning in Ecuador.* London: Grant & Cutler.

———. 1980. Trader mobility in systems of periodic and daily markets. In *Geography and the urban environment, vol. 3,* eds. R. J. Johnston and D. T. Herbert, pp. 133–74. Chichester: Wiley.

———. 1983. The urban road to rural development: Reflections on USAID's 'urban functions' approach. *Environment and Planning A* 15:429–34.

———. 1984. Market centre analysis in the urban functions in rural development approach. In *Equity with growth? Planning perspectives for small towns in developing countries,* eds. H. D. Kammeier and P. J. Swan, pp. 294–340. Bangkok: Asian Institute of Technology.

———; Symanski, R.; and Good, C. M. 1975. The rationale of periodic markets. *Annals of the Association of American Geographers* 65:530–37.

Brookfield, H. C. 1975. *Interdependent development.* London: Methuen.

Bryant, P. T., ed. 1968. *From geography to geotechnics, by Benton MacKaye.* Urbana, IL: University of Illinois Press.

Burchell, R. W., and Sternlieb, G., eds. 1978. *Planning theory in the 1980s: A search for future directions.* New Brunswick, NJ: Center for Urban Policy Research, Rutgers University.

Burns, L. S., and Friedmann, J., eds. 1985. *The art of planning: Selected essays of Harvey S. Perloff.* New York: Plenum.

Caro, R. A. 1974. *The power broker: Robert Moses and the fall of New York.* New York: Knopf.

Casetti, E. 1981. A catastrophe model of regional dynamics. *Annals of the Association of American Geographers* 71:572–9.

Charnes, A. et al. 1975. A hierarchical goal programming approach to environmental land use management. *Geographical Analysis* 7:121–30.

Cheema, G. S., and Rondinelli, D. A., eds. 1983. *Decentralization and development.* Beverly Hills: Sage Publications.

Chisholm, M. 1985. De-industrialization and British regional policy. *Regional Studies* 19:301–13.

———. 1987. Regional development: The Reagan-Thàtcher legacy. *Environment and Planning C: Government and Policy* 5:197–218.

Christaller, W. 1933. *Die zentralen Orte in Süddeutschland.* Jena. Translated in 1966 as *Central places in southern Germany.* Englewood Cliffs, NJ: Prentice-Hall.

Christensen, D. E. 1977. Geography and planning: Some perspectives. *Professional Geographer* 19:148–52.

Clark, G. L. 1980. Capitalism and regional inequality. *Annals of the Association of American Geographers* 70:226–37.

———. 1982. Dynamics of interstate labor migration. *Annals of the Association of American Geographers* 72:297–313.

———, and Dear, M. 1981. The state in capitalism and the capitalist state. In *Urbanization and urban planning in capitalist society,* eds. M. Dear and A. J. Scott, pp. 45–61. London: Methuen.

———; Gertler, M. S.; and Whiteman, J. 1986. *Regional dynamics: Studies in adjustment theory.* Boston: Allen & Unwin.

———, and Wrigley, N., eds. 1989. *Cities and regions: Structure and change. Selected essays by Brian J. L. Berry.* London: Routledge & Kegan Paul.

Clawson, M. 1972. *America's land and its uses.* Baltimore: Johns Hopkins University Press.

———. 1981. *New Deal planning: The National Resources Planning Board.* Baltimore: Johns Hopkins University Press.

———, and Hall, P. 1973. *Planning and urban growth: An Anglo-American comparison.* Baltimore: Johns Hopkins University Press.

Coates, B. E.; Johnston, R. J.; and Knox, P. L. 1977. *Geography and inequality.* Oxford: Oxford University Press.

Committee on the Regional Plan of New York. 1929. *Regional plan of New York and its environs.* 2 vols. and appendices. New York: Committee on the Regional Plan of New York and Its Environs.

Cooper, S. H. 1966. Theoretical geography, applied geography, and planning. *Professional Geographer* 18:1–2.

Corbridge, S. 1986. *Capitalist world development: A critique of radical development geography.* Totowa, NJ: Rowman & Littlefield.

Costa, F. J., and Dutt, A. K. 1978. Ideological orientations of the national planning process in India. In *Indian urbanization and planning: Vehicles of modernization,* eds. A. G. Noble and A. K. Dutt. New Delhi: Tata McGraw-Hill.

Cunningham, S. M. 1981. Multinational enterprises in Brazil. *Professional Geographer* 33:48–62.

Danielson, M. N. 1976. *The politics of exclusion.* New York: Columbia University Press.

———, and Doig, J. W. 1982. *New York: The politics of urban regional development.* Berkeley: University of California Press.

Dear, M. S., and Scott, A. J., eds. 1981. *Urbanization and urban planning in capitalist society.* London: Methuen.

Dogan, M., and Kasarda, J. D., eds. 1988. *The metropolis era.* 2 vols. Beverly Hills: Sage Publications.

Durrenberger, R. 1972. The Colorado Plateau. *Annals of the Association of American Geographers* 62:211–36.

Dutt, A. K. 1968. Levels of planning in the Netherlands, with particular reference to regional planning. *Annals of the Association of American Geographers* 54:670–85.

———. 1969/70. Regional planning in England and Wales: A critical evaluation. *Plan: Journal of the T.P.I.C.* 10:7–23; 59–71.

———. 1970. A comparative study of regional planning in Britain and the Netherlands. *Ohio Journal of Science* 70:321–35.

———, and Costa F. J. 1980. An evaluation of national economic planning in the People's Republic of China. *Geoforum* 11:1–15.

———, and ———, eds. 1985. *Public planning in the Netherlands: Perspectives and changes since the World War.* Oxford: Oxford University Press.

Ehrlichman, J. D. 1972. National growth policy. *Congressional Record.* 92nd Congress, 2nd Session, Vol. 118, No. 153, Sept. 28.

Eighmy, T. H. 1972. Rural periodic markets and the extension of an urban system: A Western Nigeria example. *Economic Geography* 48:299–315.

Enders, W. T. 1981. Regional disparities in industrial growth in Brazil. *Economic Geography* 56:300–310.

Ewers, H. J.; Goddard, J. B.; and Matzerath, H., eds. 1986. *The future of the metropolis.* Berlin: Walter de Gruyter.

Faludi, A. 1973. *Planning theory.* Oxford: Pergamon.

Foglesong, R. E. 1986. *Planning the capitalist city: The colonial era to the 1920s.* Princeton: Princeton University Press.

Forbes, D. K. 1984. *The geography of underdevelopment.* London: Croom Helm.

———, and Rimmer, P. J., eds. 1984. *Uneven development and the geographical transfer of value.* Human Geography Monograph 16. Canberra: Australian National University.

Freeman, T. W. 1958. *Geography and planning.* London: Hutchinson.

Friedmann, J. 1955. *The spatial structure of economic development in the Tennessee Valley.* University of Chicago Department of Geography Research Paper No. 39, and Program of Education and Research in Planning Research Paper No. 1. Chicago: University of Chicago Press.

———. 1966. *Regional development policy: A case study of Venezuela.* Cambridge: MIT Press.

———. 1968. A strategy of deliberate urbanization. *Journal of the American Institute of Planners* 34(6):364–73.

———. 1972. A general theory of polarized development. In *Growth centers in regional economic development.* ed. N. Hansen, pp. 82–107. New York: Free Press.

———. 1973a. *Retracking America: A theory of transactive planning.* Garden City, NY: Anchor/Doubleday.

———. 1973b. *Urbanization, planning, and national development.* Beverly Hills: Sage Publications.

———. 1975. Regional development planning: The progress of a decade. In *Regional policy: Readings in theory and applications,* eds. J. Friedmann and W. Alonso, pp. 791–808. Cambridge: MIT Press.

———. 1987. *Planning in the public domain: From knowledge to action.* Princeton: Princeton University Press.

———. 1988. *Life space and economic space: Essays in Third World planning.* New Brunswick, NJ: Transaction Books.

———, and Alonso, W., eds. 1964. *Regional development and planning: A reader.* Cambridge: MIT Press.

———, and ———, eds. 1975. *Regional policy: Readings in theory and applications.* Cambridge: MIT Press.

———, and Weaver, C. 1979. *Territory and function: The evolution of regional planning.* London: Edward Arnold.

Fuchs, R. J., and Demko, G. J. 1979. Geographic inequality under socialism. *Annals of the Association of American Geographers* 69:304–18.

Gaile, G. L. 1979. Distance and development in Kenya. In *The spatial structure of development,* eds. R. A. Obudho and D. R. F. Taylor, pp. 201–22. Boulder, CO: Westview Press.

———. 1989. Choosing locations for small town development to enable market and employment expansion: The case of Kenya. *Economic Geography* 65. In press.

Garrison, W. 1959/60. Spatial structure of the economy. *Annals of the Association of American Geographers* 49:328–39, 471–82; 50:357–73.

Geddes, P. 1915. *Cities in evolution*. London: Williams & Norgate.

Gibbins, R. 1982. *Regionalism: Territorial politics in Canada and the United States*. Toronto: Butterworths.

Gilbert, A. 1976. The arguments for very large cities reconsidered. *Urban Studies* 13:27–34.

———. 1977. The arguments for very large cities reconsidered: A reply. *Urban Studies* 14:225–27.

Glaser, B. G., and Strauss, A. L. 1968. *The discovery of grounded theory*. London: Weidenfeld.

Gonzalez Casanova, P. 1964. Internal colonialism and national development. *Studies in Comparative International Development* 1:27–37.

Good, C. M. 1975. Periodic markets and traveling traders in Uganda. *Geographical Review* 65:49–72.

Goodman, R. 1979. *The last entrepreneurs: America's regional wars for jobs and dollars*. Boston: South End Press.

Gore, C. 1984. *Regions in question: Space, development theory and regional policy*. London: Methuen.

Gottdiener, M. 1977. *Planned sprawl*. Beverly Hills: Sage Publications.

Gould, P. R. 1970. Tanzania 1920-63: The spatial impress of the modernization process. *World Politics* 22(2):149–70.

———, and White, R. 1974. *Mental maps*. Harmondsworth: Penguin.

Gradus, Y. 1983. The role of politics in regional inequality: The Israeli case. *Annals of the Association of American Geographers* 73:388–403.

Graebner, N. A. 1982. *Empire on the Pacific: A study in American continental expansion*. 2d ed. Santa Barbara, CA: ABC-Clio.

Green, J. L. 1965. *Metropolitan economic republics*. Athens, GA: University of Georgia Press.

Greenberg, M. R.; Anderson R.; and Page, G. W. 1978. *Environmental impact statements*. Resource Paper No. 78-3. Washington: Association of American Geographers.

Griffin, D. W. 1965. Some comments on urban planning and the geographer. *Professional Geographer* 17:4–7.

Griffith, D. A. 1981. Evaluating the transformation from a monocentric to a polycentric city. *Professional Geographer* 33:189–96.

Hagerstrand, T. 1967. *Innovation diffusion as a spatial process*. Chicago: University of Chicago Press.

———, 1978. Survival and arena: On the life history of individuals in relation to their geographical environment. In *Human activity and time-geography*. Vol. 2 of *Timing space and spacing time,* eds. T. Carlstein, D. Parkes, and N. Thrift, pp. 122–45. London: Edward Arnold.

Haggett, P.; Cliff, A. D.; and Frey, A. 1977. *Locational analysis in human geography*. 2d ed. London: Edward Arnold.

Hall, C. 1984. Regional inequalities in well-being in Costa Rica. *Geographical Review* 74:48–62.

Hall, P. 1969. *London 2000*. 2d ed. New York: Praeger.

———. 1970. *Theory and practice of regional planning*. London: Pemberton Books.

———. 1973. *The containment of urban England*. 2 vols. Beverly Hills: Sage Publications.

———. 1982a. *Great planning disasters*. Berkeley: University of California Press.

———. 1982b. *Urban and regional planning*. 2d ed. London: Allen & Unwin.

———. 1984a. Geography: Descriptive, scientific, subjective and radical images of the city. In *Cities of the mind: Images and themes of the city in the social sciences,* eds. R. Hollister and L. Rodwin, pp. 95–115. New York: Plenum Press.

———. 1984b. *The world cities*. 3d ed. London: Weidenfeld.

———, and Hay. D. 1980. *Growth centers in the European urban system*. Berkeley: University of California Press.

———, and Markusen, A., eds. 1985. *Silicon landscapes*. Boston: Allen & Unwin.

Hargrove, E. C. 1983. *TVA, fifty years of grass-roots bureaucracy*. Urbana, IL: University of Illinois Press.

Harris, B. 1967. The limits of science and humanism in planning. *Journal of the American Institute of Planners* 33:324–35.

———. 1978. The comprehensive planning of location. In *Planning theory in the 1980s,* eds. R. W. Burchell and G. Sternlieb, pp. 255–68. New Brunswick, NJ: Center for Urban Policy Research, Rutgers University.

Harrison, J. D., and Larsen, R. D. 1977. Geography and planning: The need for an applied interface. *Professional Geographer* 19:139–47.

Hart, J. F. 1972. The Middle West. *Annals of the Association of American Geographers* 62:258–82.

———. 1984. Population change in the upper lake states. *Annals of the Association of American Geographers* 74:221–43.

Hartshorne, R. 1939. *The nature of geography*. Lancaster, PA: Association of American Geographers.

Harvey, D. 1973. *Social justice and the city*. London: Edward Arnold.

———. 1978. On planning the ideology of planning. In *Planning theory in the 1980s,* eds. R. W. Burchell and G. Sternlieb, pp. 213–33. New Brunswick, NJ: Center for Urban Policy Research, Rutgers University.

Hay, A. M., and Smith, R. H. T. 1970. *Interregional trade and money flows in Nigeria*. Ibadan, Nigeria: Oxford University Press.

Hays, F. B. 1965. *Community leadership: The Regional Plan Association of New York*. New York: Columbia University Press.

Healy, R. G. 1976. *Land use and the states*. Baltimore: Johns Hopkins University Press.

Heiman, M. K. 1986. *Coastal recreation in California: Policy, management, access*. Berkeley: Institute of Governmental Studies, University of California at Berkeley.

———. 1988. *The quiet revolution: Power, planning and profits in New York State*. New York: Praeger.

Hemming, John, ed. 1985. *Change in the Amazon Basin*. 2 vols. Manchester: Manchester University Press.

Hirschman, A. O. 1958. *The strategy of economic development*. New Haven: Yale University Press.

Hodge, G. 1965. The new vista for regional planning: What role for the geographer? *Canadian Geographer* 9:122–27.

———. 1966. Geography, space, and planning. A reply. *Canadian Geographer* 10:49–53.

Hoffman, G. W. 1972. *Regional development strategy in southeast Europe*. New York: Praeger.

Holland, S. 1976a. *Capital versus the regions*. London: Macmillan.

———. 1976b. *The regional problem*. London: Macmillan.

Hudson, R. 1979. Space, place, and placelessness. *Progress in Human Geography* 3:169–73.

Inskeep, E. L. 1962. The geographer in planning. *Professional Geographer* 14:22–24.

Isard, W. 1956. *Location and space economy*. New York: Wiley.

———. 1959. *Industrial complex analysis and regional development*. New York: Wiley.

———. 1960. *Methods of regional analysis*. New York: Wiley.

———. 1969. *General theory: Social, political, economic and regional*. Cambridge: MIT Press.

———. 1975. *Introduction to regional science*. Englewood Cliffs, NJ: Prentice-Hall.

———. 1988. *Arms races, arms control and conflict analysis*. New York: Cambridge University Press.

———, and Cumberland, J. H., eds. 1961. *Regional economic planning: Techniques of analysis for less-developed areas*. Paris: Organization for Economic Cooperation and Development.

———, and Smith, C. 1982. *Conflict management analysis and practical conflict management procedures*. Cambridge, MA: Ballinger.

Jackson, J. K. 1985. *Crabgrass frontier: The suburbanization of the United States*. New York: Oxford University Press.

Jackson, J. N. 1967. Geography and planning: Two subjects or one? *Canadian Geographer* 11:357–65.

Jackson, R. H. 1981. *Land use in America*. New York: Wiley.

Jensen, M., ed. 1951. *Regionalism in America*. Madison, WI: University of Wisconsin Press.

Johnson, E. A. J. 1970. *The organization of space in developing countries*. Cambridge: Harvard University Press.

Johnston, R. J. 1980. *The geography of federal spending in the United States of America*. Chichester: Research Studies Press.

———. 1982. *The American urban system: A geographical perspective*. New York: St. Martin's Press.

———. 1983. *Geography and geographers: Anglo-American human geography since 1945.* 2d ed. New York: Wiley.

Jones, B. G. 1986. Urban support for rural development in Kenya. *Economic Geography* 62:201–14.

Jones, D. W. 1978. Rent in an equilibrium model of land use. *Annals of the Association of American Geographers* 68:205–13.

———. 1982. Location and land tenure. *Annals of the Association of American Geographers* 72:314–31.

Karaska, G. 1975. Perspectives on less developed regions in Poland. In Development regions in the Soviet Union, Eastern Europe and Canada, ed. A. F. Burghardt, pp. 43–64. New York: Praeger.

Katzman, M. T. 1977. *Cities and frontiers in Brazil: Regional dimensions of economic development.* Cambridge: Harvard University Press.

Kirby, A. M. 1988. High-level nuclear waste transportation: Political implications of the weakest link in the nuclear fuel cycle. *Environment and Planning C: Government and Policy* 6:311–22.

Kirn, T. J. 1987. Growth and change in the service sector of the U.S.: A spatial perspective. *Annals of the Association of American Geographers* 77:353–72.

Knight, D. B. 1982. Identity and territory: Geographical perspectives on nationalism and regionalism. *Annals of the Association of American Geographers* 72:514–31.

Kuklinski, A., ed. 1972. *Growth poles and growth centers in regional planning.* The Hague: Mouton.

———, and Petrella, R., eds. 1972. *Growth poles and regional policies.* The Hague: Mouton.

Laity, A. L. 1984. Perceiving regions as scattered objects. *Professional Geographer* 36:285–92.

Lakshmanan, T. R., and Hua, C. 1987. Regional disparities in China. *International Regional Science Review* 11:97–104.

Lewis, G. K. 1972. Population change in Northern New England. *Annals of the Association of American Geographers* 62:307–22.

Lewis, P. F. 1972. Small towns in Pennsylvania. *Annals of the Association of American Geographers* 62:323–51.

Lim, G. C. 1983. Regional planning in transition. In *Regional planning: Evolution, crisis and prospects,* ed. G. C. Lim, pp. 3–14. Totowa, NJ: Allanheld Osmun.

Lipton, M. 1977. *Why poor people stay poor: A study of urban bias in world development.* London: Temple Smith.

Lo, F., and Salih, K. 1981. Growth poles, agropolitan development, and polarization reversal. In *Development from above or below?* eds. W. Stöhr and D. R. F. Taylor, pp. 123–52. Chichester: Wiley.

Los, M. 1981. Some reflections on epistemology, design and planning theory. In *Urbanization and urban planning in capitalist society,* eds. M. Dear and A. J. Scott, pp. 63–88. London: Methuen.

Mabogunje, A. L. 1978. Growth poles and growth centres in the regional development of Nigeria. In *Regional policies in Nigeria, India, and Brazil,* ed. A. Kuklinski, pp. 1–93. The Hague: Mouton.

———. 1980. *The development process: A spatial perspective.* London: Hutchinson.

MacKaye, B. 1928. *The new exploration: A philosophy of regional planning.* New York: Harcourt Brace.

Majeed, A., ed. 1984. *Regionalism: Developmental tensions in India.* New Delhi: Cosmo Publications.

Malecki, E. J. 1984. Military spending and the U.S. defense industry: Regional patterns of military contracts and subcontracts. *Environment and Planning C: Government and Policy* 2:31–44.

Markusen, A. R. 1985. *Profit cycles, oligopoly, and regional development.* Cambridge: MIT Press.

———. 1986. Defense spending: A successful industrial policy? *International Journal of Urban and Regional Research* 10:105–22.

———. 1987. *Regions: the economics and politics of territory.* Totowa, NJ: Rowman & Littlefield.

———; Hall, P.; and Glasmeier, A. 1986. *High tech America.* Boston: Allen & Unwin.

Marsh, B. 1987. Continuity and decline in the anthracite towns of Pennsylvania. *Annals of the Association of American Geographers* 77:337–52.

Marsh, G. P. 1864. *Man and nature: Or, physical geography as modified by human action.* New York: Charles Scribner.

Marshall, J. U. 1966. Geography, space and planning: A reply. *Canadian Geographer* 10:49–53.

Martin, R. L., and Hodge, J. S. C. 1983. The reconstruction of British regional policy. *Environment and Planning C: Government and Policy* 1:133–52 and 317–40.

Massam, B. H. 1980. *Spatial search: Applications to planning problems in the public sector.* Oxford: Pergamon.

Massey, D. 1984. *Spatial divisions of labor: Spatial structures and the geography of production*. London: Macmillan.

———, and Meegan, R. 1982. *The anatomy of job loss: The how, why and where of employment decline*. London: Methuen.

Massie, R. K. 1980. *Peter the Great: His life and work*. New York: Knopf.

Mather, E. C. 1972. The American Great Plains. *Annals of the Association of American Geographers* 62:237–57.

Meinig, D. W. 1986. *Atlantic America, 1492–1800*. Vol. 1 of *The shaping of America: A geographical perspective on 500 years of history*. New Haven: Yale University Press.

Mera, K. 1973. On the urban agglomeration and economic efficiency. *Economic Development and Cultural Change* 21:309–24.

Merrifield, J. 1983. Using analog regions to assess the economic impact of federal land management policies. *Professional Geographer* 35:298–302.

Miller, G. J. 1981. *Cities by contract: The politics of municipal incorporation*. Cambridge: MIT Press.

Misra, R. P., and Sundaram, K. V. 1978. Growth foci as instruments of modernization in India. In *Regional policies in Nigeria, India, and Brazil,* ed. A. Kulinkski, pp. 95–185. The Hague: Mouton.

———, and Natraj, V. K. 1981. India: Blending central and grass roots planning. In *Development from above or below? The dialectics of regional planning in developing countries,* eds. W. B. Stöhr and D. R. F. Taylor, pp. 259–79. Chichester: Wiley.

Mitchell, B. R., and Rapkin, C. 1954. *Urban traffic, a function of land use*. New York: Columbia University Press.

Mollenkopf, J. 1983. *The contested city*. Princeton: Princeton University Press.

Mookherjee, D., and Morrill, R. L. 1973. *Urbanization in a developing economy: Indian perspectives and patterns*. Sage Professional Paper in International Studies 2, 02-018. Beverly Hills: Sage Publications.

——— and ———. 1974. Dimensions of Indian urban centers: Some policy implications. *Review of Regional Studies* 4:88–97.

Moore, J. 1985. The political history of Nigeria's new capital. *Journal of Modern African Studies* 22:167–75.

Morrill, R. L. 1970. *The spatial organization of society*. Belmont, CA: Wadsworth.

———. 1979. Stages in patterns of population concentration and dispersion. *Professional Geographer* 31:55–65.

———, and Wohlenberg, E. H. 1971. *The geography of poverty in the United States*. New York: McGraw-Hill.

Mosher, A. T. 1969. *Creating a progressive rural structure*. New York: Agricultural Development Council.

Mumford, L., ed. 1925. Regional plan. *Survey Graphic* 54(7):128–206. Special theme issue.

———. 1932. The plan of New York. *New Republic* 71(15 June):121–26; 72 (22 June):146–54.

———. 1938. *The culture of cities*. New York: Harcourt Brace.

Myrdal, G. 1957. *Economic theory and underdeveloped regions*. London: Duckworth.

Newman, M. 1972. *The political economy of Appalachia: A case study in regional integration*. Lexington, MA: Lexington Books.

Obudho, R. A., and Waller, P. W. 1976. *Periodic markets, urbanization and regional planning: A case study from Western Kenya*. Westport, CT: Greenwood Press.

Odum, H. W. 1936. *Southern regions of the United States*. Chapel Hill, NC: University of North Carolina Press.

———, and Moore, H. E. 1938. *American regionalism: A cultural-historical approach to national integration*. New York: Henry Holt.

———, and Jocher, K., ed. 1945. *In search of the regional balance of America*. Chapel Hill, NC: University of North Carolina Press.

O'hUalacháin, B. 1987. Regional and technological implications of the recent buildup in American defense spending. *Annals of the Association of American Geographers* 77:208–23.

Pacione, M., ed. 1981a. *Problems and planning in Third World cities*. London: Croom Helm.

———, ed. 1981b. *Urban problems and planning in the developed world*. London: Croom Helm.

Padilla, S. M., ed. 1975. *Tugwell's thoughts on planning*. Rio Piedras: University of Puerto Rico Press.

Panter-Brick, K., ed. 1978. *Soldiers and oil: The political transformation of Nigeria*. London: Frank Cass.

Park, S. 1981. Rural development in Korea. *Economic Geography* 57:113–26.

Patel, D. I. 1980. *Exurbs: Urban residential developments in the countryside*. Washington: University Press of America.

Pedersen, P. O. 1974. *Urban-regional development in South America: A process of diffusion and integration*. The Hague: Mouton.

Perloff, H. S. 1956. Education of city planners: Past, present, and future. *Journal of the American Institute of Planners* 23(Fall):186–217.

———. 1957a. *Education for planning: City, state and regional*. Baltimore: Johns Hopkins University Press.

———. 1957b. *Regional studies at U.S. universities*. Washington: Resources for the Future.

———. 1980. *Planning the post-industrial city*. Chicago: American Planning Association.

——— et al. 1960. *Regions, resources, and economic growth*. Baltimore: John Hopkins University Press.

——— et al. 1975. *Modernizing the central city: New towns intown . . . and beyond*. Cambridge, MA: Ballinger.

Plane, D. A. 1984. Migration space: Doubly constrained gravity model mapping of relative interstate separation. *Annals of the Association of American Geographers* 74:244–56.

Platt, R. H., and Macinko, G., eds. 1983. *Beyond the urban fringe: Land use issues in nonmetropolitan America*. Minneapolis, MN: University of Minnesota Press.

Porter, C. F. 1985. *Environmental impact assessment*. New York: University of Queensland Press.

Pred, A. R. 1966. *Spatial dynamics of US urban-industrial growth 1800-1914*. Cambridge: MIT Press.

———. 1977. *City systems in advanced economies*. New York: Wiley.

———. 1980. *Urban growth and city systems in the United States, 1840-1860*. Cambridge: Harvard University Press.

———. 1984. Place as historically contingent process: Structuration and the time-geography of becoming places. *Annals of the Association of American Geographers* 74:279–97.

Price, K. A., ed. 1982. *Regional conflict and national policy*. Washington: Resources for the Future.

Prunty, M. C., and Aiken, C. S. 1972. The demise of the Piedmont cotton region. *Annals of the Association of American Geographers* 62:283–306.

Rees, J. 1979. Technological change and regional shifts in American manufacturing. *Professional Geographer* 31:45–54.

———. 1981. The impact of defense spending on regional industrial change in the United States. In *Federalism and regional development,* ed. G. W. Hoffman, pp. 193–222. Austin, TX: University of Texas Press.

———. 1986. *Technology, regions and policy*. Totowa, NJ: Rowman & Littlefield.

Renner, G. T. 1936. National regional planning in resource use. In *Our national resources and their conservation,* eds. A. E. Parkins and J. R. Whitaker, pp. 592–615. New York: Wiley.

———. 1942. *Conservation of national resources: An educational approach to the problem*. New York: Wiley.

Richardson, H. W. 1972. Optimality in city size, systems of cities and urban policy: A skeptic's view. *Urban Studies* 9:29–48.

———. 1976. The arguments for very large cities reconsidered: A comment. *Urban Studies* 13:307–10.

Robinson, J. L. 1956. Geography and regional planning. *Canadian Geographer* 8:1–8.

Rodwin, L. 1987. On the education of urban and regional specialists. *Papers of the Regional Science Association* 62:1–11.

Rogerson, P. A. 1984. The demographic consequences of metropolitan population deconcentration in the U.S. *Professional Geographer* 36:307–14.

Rokkan, S. 1973. *Nation-building*. The Hague: Mouton.

———, and Unwin, D. W. 1983. *Economy, territory, identity: Politics of West European peripheries*. Beverly Hills: Sage Publications.

Rondinelli, D. A. 1981. Government decentralization in comparative perspective. *International Review of Administrative Sciences* 47:133–45.

———. 1983. *Secondary cities in developing countries: Policies for diffusing urbanization*. Beverly Hills: Sage Publications.

———. 1985a. *Applied methods of regional analysis: The spatial dimensions of development policy*. Boulder, CO: Westview Press.

———. 1985b. Regions in question: A critical review. *Third World Planning Review* 7:263–68.

———, and Ruddle, K. 1978. *Urbanization and rural development: A spatial policy for equitable growth*. New York: Praeger.

Rose, M. H. 1979. *Interstate: Express highway politics 1941–1956*. Lawrence, KS: Regents Press of Kansas.

Roweis, S. 1981. Urban planning in early and late capitalist societies: A theoretical perspective. In *Urbanization and urban planning in capitalist society,* eds. M. Dear and A. J. Scott, pp. 159–77. London: Methuen.

Roy, P., and Patil, B. R. 1977. *Manual for block level planning.* New Delhi: Macmillan.

Saarinen, T. F. 1976. *Environmental planning: Perception and behavior.* Boston: Houghton Mifflin.

Scott, A. J. 1982. Production system dynamics and metropolitan development. *Annals of the Association of American Geographers* 72:185–200.

———, and Roweis, S. T. 1977. Urban planning in theory and practice: An appraisal. *Environment and Planning A* 9:1097–1119.

Seley, J. E. 1983. *The politics of public facility planning.* Lexington, MA: Lexington Books.

Shortridge, J. R. 1984. The emergence of the "Middle West" as an American regional label. *Annals of the Association of American Geographers* 74:209–20.

———. 1987. Changing usage of four American regional labels. *Annals of the Association of American Geographers* 77:325–36.

Smith, C. A., ed. 1976. *Regional analysis.* 2 vols. New York: Academic Press.

Smith, D. M. 1973. *The geography of social well-being in the United States.* New York: McGraw-Hill.

Smith, R. H. T. 1971. West African market-places: Temporal periodicity and locational spacing. In *The development of indigenous trade and markets in West Africa,* ed. C. Meillassoux, pp. 319–46. London: Oxford University Press.

———, ed. 1978. *Market-place trade: Periodic markets, hawkers and traders in Africa, Asia, and Latin America.* Vancouver: Centre for Transportation Studies, University of British Columbia.

———. 1979/80. Periodic market-places and periodic marketing. *Progress in Human Geography* 3:471–505; 4:1–31.

———, and Hay, A. M. 1969. A theory of spatial structure of internal trade in underdeveloped countries. *Geographical Analysis* 1:121–36.

So, F. S.; Hand, I.; and McDowell, B. D. 1986. *The practice of state and regional planning.* Chicago: American Planning Association.

Soja, E. 1968. *The geography of modernization in Kenya.* Syracuse, NY: Syracuse University Press.

Stalley, M., ed. 1972. *Patrick Geddes: Spokesman for man and the environment.* New Brunswick, NJ: Rutgers University Press.

Stoddart, D. R. 1977. To claim the high ground: Geography for the end of the century. *Transactions of the Institute of British Geographers,* n.s. 12:327–36.

Stöhr, W. B., and Taylor, D. R. F., eds. 1981. *Development from above or below? The dialectics of regional planning in developing countries.* Chichester: Wiley.

Sussman, C., ed. 1976. *Planning the fourth migration: The neglected vision of the Regional Planning Association of America.* Cambridge: MIT Press.

Thomas, W. L., Jr., ed. 1956. *Man's role in changing the face of the earth.* Chicago: University of Chicago Press.

Tuan, Y-F. 1974. *Topophilia: A study of environmental perception, attitudes, and values.* Englewood Cliffs, NJ: Prentice-Hall.

Tugwell, R. 1934. *Our economic society and its problems.* New York: Harcourt.

———. 1958. *The place of planning in society.* San Juan: Government Service Office, Printing Division.

———. 1970. *A model constitution for a United Republics of America.* Santa Barbara, CA: Center for the Study of Democratic Institutions.

Turner, B. L., II, ed. 1989. *The earth transformed by human action.* Cambridge: Cambridge University Press.

U.K. Barlow Commission. 1940. *Report of the Royal Commission on the Distribution of Industrial Population.* London: Her Majesty's Stationery Office.

United Nations. 1987. The prospects of world urbanization revised as of 1984-85. New York: United Nations.

U.S. National Resources Committee. 1935. *Regional factors in national planning and development.* Washington: Government Printing Office.

Vance, J. E., Jr. 1972. California and the search for the ideal. *Annals of the Association of American Geographers* 62:185–210.

Von Thünen, J. H. 1826; 1875. *Der Isolierte Staat in Beziehung auf Landwirtschaft und Nationalökonomie.* Hamburg. Translated in 1966 as *Von Thünen's isolated state,* ed. P. Hall. Oxford: Pergamon.

Walker, D. B. 1987. Snow White and the 17 dwarfs: From metro cooperation to governance. *National Civic Review* 76(1):14–28.

Warf, B. 1988. The Port Authority of New York and New Jersey. *Professional Geographer* 40:289–97.

Weaver, C. 1984. *Regional development and the local community: Planning, politics, and social context*. Chichester: Wiley.

Whebell, C. F. J. 1969. Corridors: A theory of urban systems. *Annals of the Association of American Geographers* 59:1–26.

White, C. L., and Renner, G. T. 1936. *Geography: An introduction to human ecology*. New York: Appleton-Century.

———, and ———. 1948. *Human geography: An ecological study of society*. New York: Appleton-Century-Crofts.

Williamson, J. G. 1965. Regional inequality and the process of national development. *Economic Development and Cultural Change* 13 (4, pt. 2):3–45.

Wood, R. C. 1961. *1400 governments: The political economy of the New York metropolitan region*. Cambridge: Harvard University Press.

Worster, D. 1985. *Rivers of empire: Water, aridity and the growth of the American West*. New York: Pantheon.

Yapa, L. S. 1979. Ecopolitical economy of revolution. *Professional Geographer* 31:371–76.

Zelinsky, W., and Sly, D. F. 1984. Personal gasoline consumption, population patterns, and metropolitan structure: The United States, 1960-1970. *Annals of the Association of American Geographers* 74:257–78.

The Geography of Recreation, Tourism, and Sport

Lisle S. Mitchell | Richard V. Smith

The birth of recreation geography as a legitimate research topic has been traced to 1930 (Mitchell 1981). The subdiscipline will, therefore, be 60 years old at the end of 1989. It is not, however, the purpose of this chapter to trace the historical development of subject matter; rather it is to provide a research perspective of the subspecialties of recreation, tourism, and sport, with an emphasis on the past 25 years.

The chapter opens with a brief statement of the evolution of recreation geography in the U.S. and Canada, and identifies some fundamental concepts. Then, 10 research themes and contributions will be discussed, followed by an assessment of the state of the specialty. Finally, some of the major challenges and future prospects that face recreation geographers will be detailed.

HISTORICAL BACKGROUND AND FUNDAMENTAL CONCEPTS

Recreation geography was born during the Great Depression, with the publication of an *Annals* article by McMurry (1930) on the topic of land use for recreation. Additional publications and the presentation of papers at AAG meetings increased slowly from 1930 until the mid 1970s, when a quantum leap occurred in the volume of scholarly productivity on the topics of recreation, tourism, and sport (Mitchell 1981; Mitchell and Smith 1985).

The contribution of Canadian recreation geographers should not be underestimated, because their level of productivity has increased dramatically, as evidenced by the large number of recent recreation publications in the *Canadian Geographer* (Mitchell and Smith 1985; Janiskee and Mitchell forthcoming). Recreation geography is also flourishing in Europe, as noted by Wolfe (1964), Matley (1974), and Coppock (1982). It has been estimated that over three quarters of the literature on international tourism is in German, French, Italian, Russian, or East European languages (Matley 1976, 4–5).

The term *recreation geography* can be traced to a four-page chapter fragment written by McMurry (1954). Recreation geography has historically been interpreted to include all phenomena related to the experiences and processes involved in the use of leisure time. No individual or group has defined the three general topics under which recreation geographers function—recreation, tourism, and sport.

However, specific distinctions are recognized. Recreation and tourism differ more in degree than in kind, because both types of activity are pursued for pleasure, and because both imply the five stages of a recreation experience: (1) planning, (2) travel to the experience, (3) participation in the activity itself, (4) travel back home, and (5) recall of the experience (Clawson and Knetsch 1966). Recreation tends to take place after work or school, during the week and during weekends, and often occurs in close proximity to one's primary residence. Tourism, on the other hand, happens much less frequently, usually during holiday or summer vacation periods, and involves travel of some considerable distance and/or time. Individual definitions differ a great deal on the specifics of these distinctions, but the duration and distance of movement are the critical geographic factors. Debate about the definitions of these terms has appeared in the *Canadian Geographer* (Britton 1979; Mieczkowski 1981; Chadwick 1981; Britton 1981).

Sport is seen as a specific group of activities, facilities, programs, and institutions that can be classified as either recreation or tourism, or both. For example, Little League Baseball is primarily recreation in nature, while Major League Baseball has a major tourism component. Likewise, participation in sport activities is seen as recreation, while the viewing of sporting events for entertainment is seen as tourism. Sport research usually concentrates attention on a hierarchy of sport phenomena, with professional activities at the pinnacle, and unorganized experiences at the base. Major emphasis is placed on highly organized sports programs that play an important role in popular culture.

RESEARCH THEMES AND CONTRIBUTIONS

The identification of major research themes pursued by geographers in the areas of recreation, tourism, and sport is exceedingly difficult. The variety of contributions is remarkable, and that variety is particularly complex because many papers might be classified in more than one category. For the present assessment, a combination of objective and subjective criteria were employed. Articles published in 10 major journals, and papers presented at annual Association of American Geographers meetings, were placed into relatively homogeneous groups. Of the resulting 10 research themes, 4 pertain to recreation, 4 to tourism, 1 to a combination of recreation and tourism, and 1 to sport. The 10 themes are:

Recreation perception
Recreation participation
Urban recreation
Studies of places and areas
Tourism development
Resorts and the resort cycle
Tourism travel
Tourism impacts
Recreation and tourism planning
Sport

Recreation Perception

The study of perception is taking on increasing importance in recreation geography. As elsewhere in the discipline, this is occurring because the role of communications, values, information available to the individual, and other factors are clearly of increasing importance in influencing the kinds of recreational decisions made. Perception, in the context of recreation, is very much centered on the notion of *available opportunities*.

The distance-decay concept which is so familiar to geographers is of special importance in this situation. Some activities are commonly engaged in close to home, and others may be sought at great distances. In such cases, the distance-decay concept is very much complemented by the notion of range, as it is thought of in terms of central place. Thus, downhill skiing is perceived very differently than tennis, in terms of where it will be done and at what cost.

Perception studies by recreation geographers are quite diverse. Ewing and Kulka (1979) investigated ski-resort attractiveness and provided an in-depth analysis of a specific type of facility. Dilley (1986), on the other hand, produced an extensive analysis of tourist brochures and recorded the responses of individuals to the ones they read. Jackson and Schinkel (1981) considered different perceptions of activity preferences between resident and tourist campers in a specific area.

Barker (1980) studied the relationship among national parks, conservation, and agrarian reform in Peru, a study which examines the responses of people in a particular context in a developing country. Ford and Griffin (1981) examined a Chicano park and reported the specific cultural feelings produced during the creation of a special recreation facility. Finally, a variety of works by Lucas (1964) and Lime (1975) on diverse aspects of river recreation incorporate significant attributes of perceptions.

There is a long list of published studies about spatial perceptions, conducted by recreation researchers from other disciplines. Their work lays an increasingly solid base upon which geographers can build. Since the fundamental issues ultimately relate to available recreation opportunities, this topic has an increasing importance, both for the privileged and the less-privileged in American society. The ultimate goal is to assist in providing better and fuller information to larger parts of society, so that more-informed choices can be made from the available opportunities.

Recreation Participation

Relatively few formal studies on participation have been published by geographers, although unpublished research by students is relatively common. Participation investigations typically examine activities, including analysis of where, why, when, how often, and how the quality of the experience was rated. Recreational travel is a critical element in examining participation and is perhaps more fully studied than other aspects. Other studies might focus on one or more of the stages in the typical recreational experience, the nature of involvement in terms of active/passive roles, and a basic consideration of how many people actually participate in any activity.

Nearly all of the formal and detailed participation studies have been major components of state recreation plans. Two of the most visible contributions by geographers have emanated from their work on such plans. Young and Smith (1979) produced a model which dealt with the scale issues involved in studying participation. Such issues have had a particular relevance to the state planning process, given the desire in most studies of this nature to satisfy the planning and information needs of officials at both state and local levels. Fesenmaier and Lieber, in their state planning work in Illinois and Oklahoma, developed comprehensive participation studies. One product of their studies helps in determining the value of models (Fesenmaier and Lieber 1985).

In Canada, recreation geographers have been involved in participation studies at the provincial level, especially in Ontario (Tourism and Outdoor Recreation Planning Study 1977). At the national level, geographers have added significantly to *Canadian Outdoor Recreation Demand Studies* (Avedon 1976).

In a more general context, many researchers have looked at a particular situation and studied the participation patterns. Butler and Smith (1986), as an example, analyzed the behavior of a group of onshore and offshore oil-industry employees in Newfoundland. And yet another example of a useful study includes Ryan's (1984), which provides a perspective on the relationship between clusters of recreation activities.

Behavioral analyses are increasingly important in studying participation. Combined with that tendency is a realization that a relatively small part of any population participates in any one activity, and that a substantial part of a population rarely participates in outdoor recreation at all. People are motivated to participate, or prevented from participating, by a large number of factors. The providers of facilities are prominent figures in influencing participation. Further study of these factors from the geographer's spatial perspective will help deepen understandings of this important dimension of recreation.

Urban Recreation

Interest in urban-recreation research has had two primary stimuli. One emanated from the substantial increase in urban and community studies, fostered by the reports of the Outdoor Recreation Resources Review Commission (1962). The second was the great increase in urban research in geography and related disciplines during the decade of the 1960s. The study of urban land use, economic activity, environmental impacts, and social and cultural patterns inevitably involved consideration of the lands and facilities devoted to recreation activity. Studies of the resource base and user preferences became important. The number of researchers concentrating on the study of recreation topics increased slowly at first and then expanded in the late 1970s and 1980s.

The literature on urban recreation has appeared in a wide array of journals, monographs, books, and agency reports. The presentation of papers at national and regional meetings has grown steadily in volume. As Mitchell and Smith reported in 1985, 10 articles appeared in a group of 10 major geography and leisure journals between 1976 and 1982, as compared to just one in the preceding six years. Similarly, 28 urban-recreation papers were presented at AAG meetings in the 1976–1982 period, as compared to seven in the earlier period. Far more research results appeared in the other outlets—agency reports, local and regional journals, and books and monographs.

The literature is extremely diverse in character. The majority of studies have been place-specific or activity-specific. A few have made contributions to improved methodologies. Several prominent researchers have included major treatments of urban themes in books. Especially notable are the contributions of Chubb and Chubb in *One Third of Our Time?* (1981); VanDoren, Priddle, and Lewis in their edited volume *Land Use and Leisure* (1979); Lieber and Fesenmaier in an edited manuscript *Recreation Planning and Management* (1983); and Murphy in his *Tourism: A Community Approach* (1985b).

Relatively few recreation geographers have turned their attention to the development of a theory of urban recreation. Mitchell (1969) was the first to attempt such a statement. It begins with a series of assumptions that underlie four propositions fundamental to the study. The resulting propositions are analogous to those involved in central-place theory and refer to hierarchies, spacing, distributions, hinterland sizes and shapes, and threshold populations. The second part of the article involves modification of certain assumptions to more closely approximate reality.

Smith (1983a) points out several strengths in Mitchell's theory, while also observing difficulties that result from ignoring a number of real-world variables such as land prices, budget limitations, political influence, and others. In more recent papers, Mitchell and Lovingood (1978) explored other issues relevant to an urban theory.

Other contributions to the urban-recreation literature have been more focused. For example, hiking has attracted the attention of Lieber and his associates. Fesenmaier, Goodchild, and Lieber (1980) studied day-hiking travel. Lieber and Allton (1983) produced a model of trail evaluations in metropolitan Chicago. Fesenmaier and Lieber have continued to explore these and related topics, with a strong focus on modeling recreation behavior in urban areas (1985). Heatwole and West (1980, 1981, 1982) have explored a number of aspects of beach access, fishing marinas, and other dimensions of user access to water-based recreational activities in the metropolitan New York City area.

Smith has explored several dimensions of urban recreation in articles pertaining to intervening opportunity (1980) and the spatial aspects of restaurant locations (1983b). His contributions are marked by a major effort to extend theory and to apply familiar methods in compelling ways. Wall has made innumerable contributions to an understanding of urban cultural facilities. His work includes numerous studies of Toronto facilities (1983), as well as articles with Sinnott about cultural facilities as tourism attractions (1980). Murphy (1980a, 1980b) has conducted several studies emphasizing urban facilities as tourism attractions, as well as studying visitor characteristics and behavior.

Much work remains to be accomplished on urban topics. Theories need fuller development and testing. The developing literature needs careful scrutiny to bring relatively inaccessible contributions to a larger audience. For example, review articles on the urban sections of state and national recreation plans would make visible useful empirical research results and sizable data sets. These might provide basic information for further theoretical observations and development.

Studies of Places and Areas

A large proportion of the studies conducted by recreation geographers are concerned with particular types of places and areas. For example, the coastal and nearshore environment, national parks, the backyards of homes, and other readily identified locales are the focus of many studies. Such studies may deal with user patterns; resource characteristics; management problems; environmental, economic, or social impacts; and other individual questions. Or, the study may be a comprehensive examination of a place or area.

West (1988) has provided a comprehensive summary on "Marine Recreation in America" in a manuscript prepared as a working paper for the IGU Commission on the Coastal Environment. In it, he notes the great and growing importance of the coastal and nearshore environments for active and passive recreation. Factors that combine to enhance the attractiveness and accessibility of coastal recreation include the great concentration of population along the coasts, more leisure time, affluence, and new and improved equipment.

Fishing, boating, diving, and beach-related activities are the primary traditional forms of participation. New marine-recreation pursuits constantly emerge, often combining two or more of the traditional activities. The constantly growing pressure of both the traditional and new forms of activity on scarce and sensitive coastal resources is a major cause of concern. Enlightened managers and planners are essential if the resource base is to be maintained and if positive recreation experiences are to be enjoyed.

Similar concerns are being investigated for other important components of the nation's recreation resource base, such as the national parks, national forests, state parks, and other public lands. At another scale, backyard and neighborhood recreation is starting to receive closer examination. Place-specific and area-specific studies have a promising future, by themselves and as contributions to the larger literature, where they can help lay the basis for further conceptual advances.

Tourism Development

Geographic research on the topic of tourism development is not extensive, but is of sufficient quantity to provide useful insights into the process. A genetic approach to tourism development has been presented by Pearce (1980). This article demonstrates the significance of viewing development from a temporal perspective by viewing tourism change and growth in New Zealand. The historic approach aids in understanding the dynamic, evaluative aspect of tourism development, by providing analytical impressions through time.

Murphy (1985b) views tourism development from the perspective of the destination community. This text examines the environmental and accessibility context of the host region and relates them to economic and cultural factors. Planning strategies are then considered that increase the probabilities of success, and minimize the negative aspects of tourism development.

Tourism as a tool of economic development has been examined by a number of researchers (e.g., Britton 1977; Clarke 1981; Loukissas 1982; Choy 1984). All of these articles consider the attempts by governments first to rationalize tourism as a method or tool of economic diversity and improvement, and then to analyze the success or failure of these efforts. In most cases, the benefits of tourism development have been less rewarding than anticipated by government agencies.

The effect of scale on tourism development has also been assessed (e.g., Rodenburg 1980; Loukissas 1982; Choy 1984). It has been discovered that large-scale developments tend to be appropriate in regions that have a previous history of economic and/or tourism development, and where existing institutions can accommodate rapid modification. Such developments also seem to achieve their goals if the area being impacted is large enough and sufficiently diverse in natural and cultural resources to accept drastic change. Obviously, a large number of places have neither the physical size nor the ideological structure required to adjust to the economic, social, and political impacts brought about by tourism development.

A number of case studies have been provided that explain why tourism projects have either been slow to development, or have not achieved original objectives. The lack of tourism infrastructure has been a limiting factor in the development of Tamil Nadu, India in spite of an abundance of natural and cultural resources (Hyma and Wall 1979). Conflicts between the cultural values of Samoans and the western world have resulted in limited tourism growth beyond the initial stage in American Samoa (Choy 1984). Victor Teye (1986) has illustrated how liberation wars have hampered the development of tourism in Zambia.

Development of existing or "virgin" locations for tourism facilities and activities is in need of further investigation. The examination of the static and dynamic aspects of such growth, using temporal, systematic, or regional approaches, could lead to a better understanding of the process. Study of tourism as a means of improving local, regional, or national economies is in great demand, and case studies pertaining to the rationalization of growth, scale of development, and social and political effects need to be added to the existing literature.

Tourism Resorts and Their Development

One of the most fundamental contributions of recreation geography is the notion of a sequential process of growth and change, from some initial attraction to a culminating stage of development or deterioration. This conceptual process was first described by Christaller (1963), within the context of a larger study pertaining to a peripheral theory of tourism location. The idea of a location theory which is based on tourist attractions that are positioned on the edges of countries and/or continents has not developed, but the idea of a resort cycle has persisted. The thought that resorts may be observed to change over time, in a series of steps or stages, was reintroduced to the literature in a study pertaining to Atlantic City (Stanfield 1978). This study did not systematize the resort cycle, but the concept was noted and used in a historical description of the evolution of Atlantic City as a middle-class resort.

A theoretic framework for the concept was first postulated in the *Canadian Geographer,* where Butler (1980) stated that the frame of reference was based on the idea of the product cycle. He divided the temporal process into six possible stages: (1) exploration, (2) involvement, (3) development, (4) consolidation, (5) stagnation, and (6) decline or rejuvenation. This model was interpreted flexibly and provided a high degree of explanatory power.

Since the concept was formally postulated, it has been widely used to examine a host of evolutionary patterns (e.g., Hovinen 1981, 1982; Meyer-Arendt 1985; Murphy 1985a; Stough 1985). Keller (1987) used the model to discover whether an ideal strategy exists for avoiding a center-periphery conflict.

Beyond the study of resort process is the investigation of resorts as morphological and tourism phenomena. The concept of "recreation business district" was first presented in 1970 (Stanfield and Rickert). This notion introduced the idea of a specialized district having linear land-use patterns and specific spatial interrelationships among and between retail establishments. Pearce (1978) described the form and function of French resorts.

As has been noted in other disciplines, the morphology of resorts tends to reflect the tourism activities on which they are based. Resorts, however, are not homogeneous with regard to form; rather, they vary due to differences in evolution processes, site characteristics, situational relationships, and economic status. The development, planning, and construction of Cancún as a resort in Mexico was analyzed in terms of site and situation strategies (Collins 1979). Analysis of the physical and cultural character of potential destinations in Mexico, in conjunction with the cost and convenience of transportation between competing locations, resulted in the decision to position the resort at Cancún. An examination of leisure activities and the resulting activity spaces in a Caribbean resort demonstrated greatly different patterns between residents and tourists (Husbands 1986).

The influence of European (especially British) experience on resorts has been examined by a number of geographers. The contributions of the English resort experience to the values, activities, facilities, and programs of American resorts was studied by Demars (1979). Traditions of southern U.S. spas, such as their leisurely paced but passive activities in elegant surroundings, were investigated by Lawrence (1983) and found to be based on European models.

A comparison of spa experiences among Canada, the U.S., and Europe (Weightman and Wall 1985) suggested a large number of similarities and differences. Canadian and American resorts were found to be essentially identical, with differences being more of degree than of kind. Differences between North American and European resorts were primarily based on the period of historical development, degree of government involvement, infrastructure, social structure, belief in medical benefits of spa waters, and leisure philosophies.

Tourist resorts remain a fertile area of investigation, due to a limited number of published studies. The resort-cycle concept needs to be examined in numerous physical and cultural contexts, in order to establish its validity. In addition, the morphological nature of resorts need to be studied, with regard to differing economic, social, political, environmental, and location elements.

Tourism Travel

Travel between places of demand or origins, and places of supply or destinations, has long been a significant research theme. Patterns of tourism travel are probably the single topic in tourism most researched by geographers. An article by Williams and Zelinsky (1970) was the forerunner of this type of investigation. They examined international tourist flows among the 14 countries having the largest number of tourists, using unique graphic flow-assignment models. These models illustrated the flow of tourists between origination and destination countries, and suggested eight explanations. Unfortunately, this methodology has never been replicated, and the findings have yet to be verified.

A more recent study of international tourism, with Canada as the focus, was conducted by Bailie (1980). This study primarily dealt with the factors attributed to tourism that impacted Canada's foreign-trade deficit. The major cause was a decrease of U.S. tourists. Population shifts in the states were suggested as a principal reason for this change, but other factors were considered, such as the exchange rate and the fuel crisis. A second important cause was an increase of Canadians traveling to the U.S. This shift was explained by exchange rates, a larger and more affluent middle class, cheaper airfares to the U.S., social changes in Canada, and the long Canadian winter with associated warmer destinations to the south.

An even more recent study of international tourism in Europe by Husbands (1983) concluded that distance plays a less-important role in tourism travel than in day-to-day behavior. This finding was based on a study of 11 European countries, and focused on the center/periphery model of Christaller and the definition of attraction. He found that, the greater was a country's attraction, the further was its location from the center. He argued that attractiveness is relative and could be inferred from the spatial choices of tourists between competing destinations. He also strongly suggested that the core/periphery relationship was a relationship of exploitation, with the tourist from the origin (i.e., center) traveling to some destination (i.e., periphery) to take advantage of the contrast.

Smith (1985) replicated the statistical aspect of Husbands' study, but came to some different conclusions. First, he noted that Husbands' interpretation was flawed, because Christaller's model was based on the location of tourist facilities, and not destination countries. In addition, Christaller, in his original statement, thought of tourism as an acceptable manner by which peripheral locations could improve the economic imbalance between differing locations, and not that the locations on the edge could be economically exploited by individuals from the center.

Secondly, Smith found that, when corrections were made for population and distance between origins and destinations, differences between center and periphery countries were greatly diminished. This finding, in turn, tends to invalidate Husbands' contention that distance was of lesser importance for tourism travel. Smith also noted that tourists seek destinations on the basis of benefits received, and these benefits in and of themselves do not constitute exploitation. Although international tourism is an important theme, it has not produced a large volume of research by geographers from the U.S.

Investigation of patterns of domestic tourism make up a significant percentage of travel literature and presentations. Three studies, one at a microscale (Murphy and Rosenblood 1974) and the other two at a macroscale (Smith and Brown 1981; Smale and Butler 1985), dealt with tourism travel as spatial search process and directional bias. The microstudy pertained to first-time travellers to Victoria, British Columbia. Four distinct clusters of tourists were identified. It was discovered that search patterns were related primarily to prior mental images of the destination, and to personal motivations.

Smith and Brown found a definite east-to-west orientation of tourist travel at the provincial level (i.e., macro) in Canada for 1968 to 1978. This directional bias was consistent over time, with few anomalies, and it was suggested that the statistical procedures followed could be used to accurately model interprovince tourism travel. Smale and Butler discovered above-average numbers of person-trips to the Quebec City–Windsor corridor. This directional bias was attributed to the density of population and the concentration of attractions in the corridor.

Two studies, both published in 1985, pertained to gross tourist travel patterns between states. Smith (1985) noted a great deal of state-to-state variation in vacation travel and that most patterns adhered to the principle of distance decay. He also found that three variables were important shifters of demand curves: the number of state parks, auto ownership, and a willingness to travel long distances. He concluded that geographic and environmental variables were more significant than generally thought.

Purdue and Gustke (1985) reported that, for interstate tourist travel, the southern states and mountain-west states were the primary recipients of tourists, and that the Great Lakes and eastern-gateway states were the most important dispatchers of tourists. It was noted that travel to visit friends and relatives was the most dominant reason, and that a significant association between leisure travel and trip purpose existed.

Another contribution to tourism studies relates to the location of accommodations. An article by Liu and Var (1980) presented an analysis of the lodging industry in Victoria, British Columbia, using standard accounting procedures. They found that categorizing hotels and motels into groups according to their ratio of food and beverages sales to room sales provided useful information to tourism planners, as well as to the management of individual facilities.

A spatial analysis of the growth points of the U.S. lodging industry between 1963 and 1977 (VanDoren and Gustke 1982) identified the Sunbelt as the region of growth. Coastal development in the regions was dramatic, but the predominant trend was growth in mid-sized interior cities.

A paper that calls attention to urban places as tourist destinations, and the importance of lodging establishments as a functional land use, was published by Wall, Dudycha, and Hutchinson (1985). This research used three different point-pattern methods for analyzing travel accommodations in Toronto. In general, they concluded that tourist resting places tend to be concentrated in particular locations, such as the downtown area and a highway strip near the airport. Location factors such as high-density transport routes, agglomeration near competitors, and proximity of tourist facilities and amenities were noted.

Another theme of tourism travel is automobile touring, and portions of a book and two articles best address this topic. Chronologically, the first was a manuscript on an individual's automobile tour to the west coast in 1932 (Jakle 1981). The trip was seen as a regional adventure, and travel inconveniences were seen as a rudimentary part of the whole. It was concluded that the individual's sense of history and routes selected worked, in an almost deterministic fashion, to provide only stereotypical perspectives of the American West. An extensive investigation of automobile touring before, between, and after the two world wars was presented in *The Tourist* (Jakle 1985). The evolution of vehicles, routeways, available services, costs, and encapsulated environments was traced in three chapters.

Hugill (1985) expanded on some of Jakle's ideas in an article on the rediscovery of America by elite automobile tourers. During the first decade of this century, a small, wealthy elite took to touring as an adventure and, in the process, rediscovered the romantic Northeast as depicted by artists. During the second decade, the automobile trickled down from the elite to the upper, middle, and lower classes. As has happened in many instances, mass followed class, and automobile touring as a leisure pastime was transformed forever.

International tourism travel has not been a popular research topic. But recent global economic changes make this a theme of great importance, and ought to lead to new investigative efforts. Domestic tourism, on the other hand, has been studied to a much larger extent. Nevertheless, there is a real need to provide better understanding of the patterns and process of interstate and intrastate travel, location characteristics of accommodations, attractiveness of destination regions, and transportation modes.

Tourism Impacts

As one might suspect, from the voluminous literature on the impacts of tourism on both developed and undeveloped landscapes, this type of study has been a major theme and contribution to the study of recreation geography. The single best review of this research is the book *Tourism: Economic, Physical and Social Impacts* (Mathieson and Wall 1982). Impact studies vary from positive to negative to balanced. However, negative assessments dominate the literature. According to the conclusions of Mathieson and Wall, economic influences tend to be positive, while physical and social effects are generally negative.

Hudman (1978) considered the need for regional planning as a way of minimizing harmful impacts. Environmental problems associated with large-scale recreation subdivisions were listed and described by Stroud (1983). Problems related to water, soil, drainage, and scenic erosion were detailed, and it was suggested that such development should conform to the needs of the region and to the environmental constraints of a particular area.

Economic impacts were discussed in articles by Var and Quayson (1985) and Mescon and Vozikis (1985). The former article determined the economic-multiplier effect of four different types of tourists to Okanagan Basin in British Columbia. The latter analyzed the economic impact of the cruise-ship industry on Dade County, Florida and found that the industry contributed 264 million dollars to the gross regional product in 1982.

Social conflicts resulting from tourism have been described by Farrell (1979) and Keogh (1982). Farrell noted that conflicts between hosts and guests were as much psychological as physical manifestations of behavior, and were fundamentally based on differences in values. This notion was confirmed in Keogh's study, although the hosts of New Brunswick held a more favorable view of visitors that seemed to be partially based on the shortness of the summer tourist season.

Although studies of tourism impacts constitute one of the largest bodies of literature, much remains to be done. The pioneering review by Mathieson and Wall is almost 10 years old, and recent additions need to be collected, cataloged, classified, and evaluated. In addition, a more systematic analysis of the positive and negative impacts of tourism on physical and cultural environments needs to be initiated.

Recreation and Tourism Planning

Research in recreation planning has taken two forms: One is the critiquing of existing plans, and the other is participating directly in the planning process. The first group contains numerous studies. Examples include Lundgren (1980); Murphy (1981); and Nelson, Cordes, and Masyk (1972). These are evaluative studies of particular plans, conducted by academic geographers. Such critiques have much merit, as they provide independent, empirical analyses of current public-policy issues.

Direct involvement of geographers in planning is exemplified by a number of studies. A large set of planning studies is that completed by geographers who are employed by federal agencies. The much-discussed concept of carrying capacity has seen numerous contributions from geographers like Lime and Stankey (1975). Lime (1975) has conducted a wide range of river-recreation studies. To note the extent of one individual's contributions, his work with the USDA Forest Service resulted, in part, in developing the Wilderness Permit System, developing interpretive and visitor information services, establishing criteria for locating and designing wilderness campsites, and establishing party-size limits in wilderness areas.

A number of academic geographers have become involved in state recreation planning programs. Examples are Hook and Jencks in Indiana, Smith and Young in Ohio, Lieber in Illinois, and Fesenmaier in Oklahoma and Texas. In all five states, large surveys of the user population were conducted, along with numerous special surveys and studies. In each instance, substantial portions of the state's planning reports were completed by these individuals. Each study involved a modeling of the state's participation and future needs. Such applied projects can provide the basis for further development of concepts and theories.

The text *Recreation Planning and Management* provided an ". . . integrated perspective on the value of leisure participation and outdoor recreation and some of the major components of the planning process" (Lieber and Fesenmaier 1983, viii). This text contains perhaps the single largest concentration of geographic research on the topic of planning in existence: 9 of 26 chapters were authored by geographers. Nevertheless, recreation and tourism planning is a seldom-cultivated research area that needs extensive exploitation.

Sports

The very beginning of geographic investigation of sport may be traced to an article appearing in the *Journal of Geography* (Shaw 1963). However, the publication that initiated sport geography's development into a viable research specialty was authored by Rooney (1969).

The conceptual framework followed by most sport geographers was formulated by Rooney in two books, a book chapter, and an atlas chapter. The book *A Geography of American Sport* (Rooney 1974) established sport geography as an important and recognized element of recreation geography, even before there was a substantial body of literature to support its contentions. This initial conceptualization was elaborated upon and diffused to a larger audience in a book chapter (Rooney 1975). *The Recruiting Game* (Rooney 1980, 1987) added detail to the evolution and spatial nature of recruiting intercollegiate athletes. In addition, he proposed practical alternatives to the present method of organizing collegiate sports programs and recruiting athletes. An atlas chapter coauthored with McDonald (1982) presented 39 maps and associated text that expanded upon his ideas and provided concrete examples.

Since 1969, the publication of articles and presentation of papers on sport geography has increased at a rapid rate. Most of these research endeavors build on the foundation laid by Rooney. John Bale, who has probably completed more survey articles on sport geography than any other individual, noted in a monograph chapter (1983) that the conceptual framework of his efforts was taken directly from Rooney's publications. The pioneering nature of Rooney's contributions can be seen in the following descriptions, because all of these research efforts were anticipated by Rooney.

Five topics will be analyzed: (1) origins and diffusion of sport, (2) sport regions, (3) spatial variations in sport, (4) sport landscapes, and (5) women in sport.

Origins and Diffusion of Sport. Three articles on the origins and diffusion of sport have examined social and demographic characteristics of U.S. football player demand, supply, and migration (Yetman and Eitzen 1973; McConnell 1983, 1984). This was done by identifying the productive surplus regions and then tracing the movement of student-athletes from their origins to their destinations. Bale (1987) illustrated the migratory patterns of blue-chip athletes from foreign origins to American university destinations.

In a related but different context, Sage and Loy (1978) studied the mobility patterns of college coaches. The sport of volksmarching (i.e., people-walking) was analyzed by Wepfer (1984). She located the diffusion centers of the sport in the U.S., and then traced the spatial aspects of the dispersal and the social mechanisms that explained the process.

Sport Regions. Research on sport regions is exemplified by baseball and bowling regions. Shelley and Cartin (1984) and Shelley (1987) have investigated fan regions for professional baseball and found them to be coextensive with local media coverage. Commuting distance to the ballpark was discovered to be a major variable, as a majority of loyalists support the team in proximity to their residence. However, individuals who migrate in their adult years have a tendency to retain childhood favorites.

Harmon (1985) noted that the pattern of bowling regions in North America is complex and dynamic. Five different types of indoor bowling exist in Anglo-America, and their regional character is undergoing change because of a host of cultural and technological factors.

Spatial Variations in Sport. The spatial variation of sport is a popular research theme, because of a wealth of both sports and intellectual approaches. The spatial distribution of stock-car races and drivers was studied by Pillsbury (1974). The agglomeration of hockey activities, facilities, and players along the political border separating the U.S. and Canada was investigated in 1975 (Ojala and Kureth). A microscale examination by Gould and Gatrell (1979) of basketball and soccer games revealed significant differences in their spatial variations. Manzo (1987) explored areal distinctions between various levels of soccer competition in the U.S. and West Virginia.

Cockfighting, as a sport, was examined by Hawley (1987) to demonstrate the differences between this and other sports, and to illustrate the cultural and "macho" nature of the activity. Eyton (1987) analyzed the distance and slope characteristics of a Rocky Mountain road race, using both quantitative and graphic methods.

Sport Landscapes. Sport-landscape studies were given a relatively early start by Oriard (1976) when he delved into the character of sport places. More recently, Neilson (1984) examined the evolution of the modern baseball park, with all of its technological equipment, as a sport place within the context of urban growth and expansion.

Adams and Rooney (1984, 1985) traced the development of golf courses from traditional linear, coastal locations to the current sprawling, residential designs. The integration of golf and home construction has been fostered for two reasons: conservation of scarce and expensive land, and enhancement of the economic viability of the sport establishment and the residential development.

The conceptualization of sport landscapes as ensembles by Raitz (1987) is one of the most innovative and useful perspectives in the cultural-environmental tradition of recreation geography. Providing the specialty with an imaginative and valuable intellectual tool are the four aspects of sport ensembles: proper context, experience and gratification, aesthetic dimension, and symbols of higher values.

Women in Sport. The most-recent series of sport studies to develop and attract widespread media attention has been written by Ojala (1987) and Ojala and Edmondson (1987). These articles, appearing in two very different journals, deal with the recent and rapid growth of women's sport programs and participation levels, due to relatively recent federal legislation. Interest in the "war between the sexes," changing concepts of sex roles, and national women's organizations have focused much needed attention on sport geography, in particular, and recreation geography, in general.

Sport geography, as a research subspecialty of recreation geography, has the smallest number of active participants. The volume of articles published and papers presented is also small (Mitchell and Smith 1985). Nevertheless, interest in sport within and beyond recreation geography is increasing, and the need to expand both the quantity and quality of research is high. The research topics surveyed here, and others still in the formative stage, are rich opportunities for individuals to make a significant contribution to sport geography. The publication of a new journal, *Sports Place* (Pillsbury 1987), is an indication of the health and vitality of this important component of recreation geography.

THE STATE OF THE SPECIALTY

The status of recreation geography and recreation geographers in 1989 needs to be considered in the broadest possible context. Recreation geography and recreation geographers make up a small proportion of the total number of disciplines and individuals making contributions. Recreation research is an interdisciplinary field that encompasses a large number of professional perspectives. The three leading interdisciplinary journals—*Annals of Tourism Research, Journal of Leisure Research, Leisure Sciences*—contain a broad range of articles from a multitude of viewpoints. Between 1979 and 1987, the *Annals of Tourism Research* produced 18 special issues which demonstrate the breadth of interest and approaches to the study of tourism. In addition, the National Recreation and Parks Association and the Travel and Tourism Research Association encompass a broad range of individuals and groups, from academia to the public and private sectors of the economy.

It should also be noted that other geographic specialties and specialists touch upon recreation, tourism, and sport topics during the course of their investigations. The interests of the recreation geographer are not exclusive, and therefore, other specialists consider the ideologies, activities, facilities, and institutions that are central to recreation functioning. For example:

 □ Cultural geographers examine the importance of recreational lifestyles to societal processes.

- Economic geographers study the significance of income, taxes, and employment generated by recreation and tourism.
- Urban geographers evaluate recreation land-use patterns in cities.
- Transportation specialists study recreation traffic flows between origins and destinations.
- Coastal marine geographers are concerned about the negative impacts of resort development along coastlines.

Any analysis of the written and oral research of recreation geographers will illustrate a wide range of topics, although there are numerous areas of common concern. Geography as a discipline is pluralistic in nature. Recreation geography as a subdiscipline is no different. The number of research topics in recreation geography is huge. In many cases, there are not enough studies on a particular topic to form the critical mass necessary to advance comprehension. A host of one-of-a-kind studies dominates research endeavors, and these unique investigations do not always add a great deal to understanding static or dynamic recreation patterns. The morphology of the research frontier has been compared to a bird's foot (Mitchell 1985), and this comparison implies that subject-matter diversity hinders the focusing of attention on a few areas of common concern.

Descriptions of recreation landscapes, activities, facilities, and institutions are absolutely necessary to any research endeavor because they are the first step in the comprehension process. Nevertheless, no matter how important accurate data collection and description are to fundamental research efforts, this information has to be synthesized to produce generalizations and/or principles in order to attain a higher level of understanding. A large proportion of recreation research is factually rich, and adds greatly to the existing stockpile of readily available insights. However, the development of concepts, understandings, explanations, and predictions cannot be solely based on investigations that place emphasis on description and information accumulation.

A fundamental cause of subject-matter diversity and description is the limited volume of easily accessible databases. (A notable exception is the contribution of Clawson and VanDoren, 1984.) Information collected at local, sectional, state, regional, and national levels is either highly specific, confined in scope, or difficult to obtain.

Such restrictions are a major barrier to research in recreation geography. Due to this lack, most researchers have to collect their data on an individual basis to meet specific research objectives. Therefore, a large majority of all investigations are unique or ideographic. Examination of the unique case has great value in highlighting the atypical situation, but this kind of study does not add significantly to a better understanding of the static and dynamic patterns of recreation landscapes. The call for more readily available databases dates to the early years of the specialty, and is repeated in almost all publications that present a general overview of recreation geography.

In addition, the academic community is either unaware of, or is unable to obtain, a large volume of data collected by the private sector in a host of recreation, tourism, and sport-related enterprises. And the lack of reliable and consistent data in a large number of different places makes it difficult, if not impossible, to replicate sound spatial research. Related to replication is the problem of a lack of temporal studies. The examination of recreation phenomena through time is almost impossible because time-series data are seldom collected and made available to interested researchers.

ANALYSIS AND MANAGEMENT OF SOCIETAL GROWTH AND CHANGE

By almost any measure, recreation geography is flourishing. Membership in the Specialty Group is about 200, which is near its all-time high. The number of papers (48), special sessions (8) and panel discussions (3) presented at the 1988 meeting of the AAG in Phoenix is also close to the record level. Articles being published in geographic and leisure journals continue to increase at a slow-but-steady pace. Books and monographs on recreation, tourism, and sport constantly expand in number. The number of theses and dissertations being completed on an annual basis has declined since the late 1970s (Mitchell 1981), but research at the graduate level remains constant and at a fairly high level of production (*Guide to Departments of Geography in the United States and Canada 1987–1988*). Courses on recreation, tourism, sport, travel, parks (national, state, local), and leisure are slowly increasing in number, as indicated in the *Schwendeman's Directory of College Geography of the United States* (1988).

Geographers are more frequently called upon to apply their expertise to recreation planning, policy, and decision-making. In addition, recreation geographers are participating in a wide variety of geographic organizations (e.g., Applied Geography Conference and Society for the North American Cultural Survey) and leisure organizations (e.g., Travel and Tourism Research Association and National Recreation and Parks Association). The *Newsletter* of the Specialty Group is published on a regular basis, and the content and format reflect a high level of commitment and involvement of the membership. All of these indicators demonstrate that the vitality of the Specialty Group and individual recreation researchers is high.

Although a multitude of problems faces the individual who conducts research on topics pertaining to recreation, tourism, and sport, there is also a sense of excitement and enthusiasm. This sense appears to motivate the small number of individuals to participate in scholarly pursuits at an extremely high level of production.

In summary, the Recreation, Tourism, and Sport Specialty Group is in a robust state of health. Recent changes in political, social, and economic structures, due to the shift from a manufacturing to an information society, indicate a bright future for those who study leisure-time phenomena. Note that tourism is destined to become the most-significant item in international trade in the near future; the wellness and preventative-health revolutions of the 1970s and 1980s will increase the importance of active recreation pastimes; and the high interest in sports will continue. These are all omens that the future of recreation geography and recreation geographers will be full of exciting possibilities.

CHALLENGES

The number of geographers studying recreation, tourism, and sport topics has increased substantially in the past decade. An extensive literature has become visible, the presentation of papers at professional meetings has expanded notably, and communication among researchers is much more common. Synthesizing works have begun to appear, and the needed groundwork for conceptual and theoretical advances is increasingly in place. The three specialties identified in this chapter are receiving increasing levels of attention as their role becomes evident in the quality of life, economic development and redevelopment processes, local and regional land-use patterns, and social and cultural fabric of communities.

Recreation, tourism, and sport geographers use familiar geographic concepts and models effectively—e.g., distance decay functions, the central-place concept, economic rent, and economic base. Increasingly, recreation researchers are asking comparable questions and are establishing common areas of concern. However, recreation geography has not yet achieved its full potential, because the efforts of geographers continue to be characterized by remarkable diversity in fundamental philosophies, methods, and the kinds of issues under study.

The lack of readily available quality information is a continuing concern in the search for better understanding of recreation processes and patterns. High quality databases are essential if basic principles of human spatial behavior in pursuit of recreation, tourism, and sport are to be identified and tested. An example of the kind of database that is needed is exemplified by that developed since 1976 in the National River Recreation Study, where standardized data-collection instruments were used to describe characteristics, perceptions, and preferences of river floaters for a variety of environments nationwide (Lime, Knopf, and Peterson 1981). As that long-term study illustrated, the questions used in generating sizable data files must be critically and sensitively developed if data pertinent to the concerns of geographers are to be provided.

Most studies to date have depended on small surveys conducted by the researcher, or on large-scale surveys conducted for state or national recreation and/or tourism development plans. In the latter case, the surveys are normally designed to provide data suitable to the pragmatic questions confronting the agency, and rarely probe for comprehensive user-behavior and resource-base data sets. In addition, many states prepare new plans and conduct large-scale surveys every four or five years, giving little attention to the need for consistency with earlier surveys. Comparative data for a discrete area is rarely generated. Little communication among states occurs as they develop comparable planning studies. Therefore, the possibility is limited for finding data sets that provide information suitable for comparative analyses.

Clearly, the solution to some of the difficulties faced in generating central questions, developing hypotheses, and in acquiring usable data depends on cooperation with public and private officials. National and state agencies need to be encouraged to develop better data. Similarly, fuller communication with leisure specialists from other disciplines is another important step that many geographers need to take. However, it must also be demonstrated that what geographers are studying is of basic importance to society. The impact of recreation, tourism, and sport on total land use, the prominence of tourism as a major economic sector, the significance of positive uses of leisure as a central element in the quality of life, and the impact of sport on large segments of the population require documentation and communication to the academic and technical audiences, and to the larger public as well. Proof of the relevance of intellectual understandings to public needs of recreation geography will assist in developing support for the creation of the kinds of data sets and other information that are required. At that point, the identification of central questions, and the development of underlying theory, will have a far better chance of being realized.

CONCLUSION

This chapter has summarized what geographers have accomplished in recent years in studying recreation, tourism, and sport. It has also identified a number of areas of study that need greater attention in the years immediately ahead. Significant components of a future research agenda include these needs:

1. Strengthen the linkages between pure and applied research, and to ensure that each sustains the other.
2. Communicate research findings to a broader audience, through more and better articles in leading geographic and nongeographic journals.
3. Emphasize issue-driven research as a top priority.
4. More thoroughly examine the relationships between researchers and the users of applied research.
5. Enter recreation and tourism concerns into regional-development models.
6. Apply traditional geographic concepts and models more fully and more often to research efforts.
7. Further develop basic understandings through an increased emphasis on patterns and processes.
8. Encourage more research on environmental and cultural resources and facilities.
9. Increase the number of investigations that duplicate and/or replicate important research findings.
10. Focus on the dynamic nature of recreation experiences through the study of new or fad activities, and to note their resource base and management implications.
11. Examine the implications for recreation, tourism, and sport on changing demographics and economic structure.
12. Provide greater spatial insights into environmental, social, and economic impact studies.
13. Study the needs of special groups in the population, such as the elderly and the physically or mentally handicapped.
14. Expand research into the critical issue of public access, because providing reasonable access for the total population is a continuing need.
15. Stimulate additional study into coastal and nearshore environments.
16. Include conflict and risk analysis in appropriate research situations.
17. Encourage much fuller study of residentially based recreation, because it is the location of the majority of leisure activities.
18. Examine whether the study of recreation, tourism, and sport should be moved more visibly under the broader umbrella of leisure studies.

In conclusion, the future of the geographic investigation of recreation, tourism, and sport is bright. Megatrends of the late 1980s and the growing need for a comprehensive, internally consistent, and logical understanding of recreation processes and patterns is an indicator that the golden age of recreation geography is at hand.

REFERENCES

Adams, R. A., and Rooney, J. F. 1984. Condo Canyon: An examination of emerging golf landscapes in America. *North American Culture* 1:65–76.

———. 1985. Evolution of American golf facilities. *Geographical Review* 75:419–38.

Avedon, E. M., ed. 1976. *Canadian outdoor recreation demand study*. Federal-Provincial Parks Conference Proceedings. 3 vols. Waterloo: Ontario Research Council on Leisure.

Bailie, J. G. 1980. Recent international travel trends in Canada. *Canadian Geographer* 24:13–21.

Bale, J. 1983. Directions in the geographical study of sport. In *Geographical Perspectives on Sport*, eds. J. Bale and Charles Jenkins. Keele, England: Department of Education, University of Keele.

———. 1987. The muscle drain: Foreign student athletes in American universities. *Sports Place* 1:3–15.

Barker, M. L. 1980. National park conservation and agrarian reform in Peru. *Geographical Review* 70:276–82.

Britton, R. A. 1977. Making tourism more supportive of small state development. *Annals of Tourism Research* 4:268–78.

———. 1979. Some notes on the geography of tourism. *Canadian Geographer* 23:276–82.

———. 1981. Some notes on the geography of tourism: A reply. *Canadian Geographer* 25:197–99.

Butler, R. W. 1980. The concept of a tourism area cycle of evolution: Implications for management of resources. *Canadian Geographer* 24:5–12.

———, and Smith, D. C. 1986. Recreational behavior of onshore and offshore oil industry employees in Newfoundland, Canada. *Leisure Sciences* 8:297–318.

Choy, D. 1984. Tourist development: The case of American Samoa. *Annals of Tourism Research* 11:573–90.

Chadwick, R. A. 1981. Some notes on the geography of tourism: A comment. *Canadian Geographer* 25:191–97.

Christaller, W. 1963. Some considerations of tourism locations in Europe: The peripheral region, underdeveloped countries, recreation areas. *Regional Science Association Papers.* 12:95–105.

Chubb, M., and Chubb, H. R. 1981. *One third of our time?* New York: Wiley.

Clarke, A. 1981. Coastal development in France: Tourism as a tool for regional development. *Annals of Tourism Research* 8:447–61.

Clawson, M., and Knetsch, J. L. 1966. *Economics of outdoor recreation*. Baltimore: Johns Hopkins University Press.

———, and VanDoren, C. S. 1984. *Statistics on outdoor recreation*. Washington: Resources for the Future, Inc.

Collins, C. 1979. Site and situation strategy in tourism planning: A Mexican case study. *Annals of Tourism Research* 6:351–66.

Coppock, J. T. 1982. Geographical contributions to the study of leisure. *Leisure Studies* 1:1–27.

Demars, S. 1979. British contributions to American seaside resorts. *Annals of Tourism Research* 6:285–93.

Dilley, R. S. 1986. Tourist brochures and tourist images. *Canadian Geographer* 30:59–65.

Ewing, G. D., and Kulka, T. 1979. Revealed and stated preference analysis of ski resort attractiveness. *Leisure Sciences* 2:249–75.

Eyton, J. R. 1987. Morphometrics of the Chasquis invitational Rocky Mountain road race. *Sports Place* 1:39–47.

Farrell, B. H. 1979. Tourism's human conflicts: Cases from the Pacific. *Annals of Tourism Research* 6:122–26.

Fesenmaier, D. R.; Goodchild, M. F.; and Lieber, S. R. 1980. Correlates of day-hiking travel: The effects of aggregation. *Journal of Leisure Research* 12:213–28.

———, and Lieber, S. R. 1985. Evaluating the stability of outdoor recreation participation models. *Professional Geographer* 37:15–21.

Ford, L. R., and Griffin, E. 1981. Chicano Park: Personalizing an institutional landscape. *Landscape* 25:42–48.

Gould, P., and Gatrell, A. 1979. A micro-geography of team games: Graphical explorations of structural relations. *Area* 11:275–78.

Guide to departments of geography in the United States and Canada: 1987–1988. Washington: Association of American Geographers.

Harmon, J. E. 1985. Bowling regions in North America. *Journal of Cultural Geography* 6:109–24.

Hawley, F. 1987. Cockfighting in the Pine Woods: Gameness in the New South. *Sports Place* 1:18–26.

Heatwole, C. A., and West, N. C. 1980. Mass transit and beach access in New York City. *Geographical Review* 70:210–17.

———. 1981. *Perspectives on marina development in New York City*. New York: New York Sea Grant Institute.

———. 1982. Recreational-boating patterns and water surface zoning. *Geographical Review* 72:304–11.

Hovinen, G. R. 1981. A tourist cycle in Lancaster County, Pennsylvania. *Canadian Geographer* 25:283–86.

———. 1982. Visitor cycles: Outlook for tourism in Lancaster County. *Annals of Tourism Research* 9:565–83.

Hudman, L. E. 1978. Tourist impacts: The need for regional planning. *Annals of Tourism Research* 5:112–26.

Hugill, P. 1985. The rediscovery of America: Elite automobile touring. *Annals of Tourism Research* 12:435–47.

Husbands, W. C. 1983. Tourist space and touristic attraction: An analysis of the destination choices of European travellers. *Leisure Sciences* 5:289–308.

———. 1986. Leisure activity resources and activity space formation in periphery resorts: The response of tourists and residents in Barbados. *Canadian Geographer* 30:243–49.

Hyma, B., and Wall, G. 1979. Tourism in a developing area: The case of Tamil Nadu, India. *Annals of Tourism Research* 6:338–50.

Jackson, E. L., and Schinkel, D. R. 1981. Recreational activity preferences of resident and tourist campers in the Yellowknife region. *Canadian Geographer* 25:350–64.

Jakle, J. A. 1981. Touring by automobile in 1932: The American West as stereotype. *Annals of Tourism Research* 8:534–49.

———. 1985. *The tourist: Travel in twentieth-century North America*. Lincoln, NE: University of Nebraska Press.

Janiskee, R. L., and Mitchell, L. S. 1989. *Applied recreation geography*. In press.

Keller, C. P. 1987. Stages of peripheral tourism development—Canada's Northwest Territories. *Tourism Management* 8:20–32.

Keogh, B. 1982. L'Impact social du tourisme: Le cas de Shediac, Nouveau Brunswick. *Canadian Geographer* 26:318–31.

Lawrence, H. 1983. Southern spas: Source of the American resort tradition. *Landscape* 27:1–12.

Lieber, S. R., and Allton, D. 1983. Modeling trail area evaluations in metropolitan Chicago. *Journal of Leisure Research* 15:184–202.

———, and Fesenmaier, D. R. 1983. *Recreation planning and management*. State College, PA: Venture Publishing.

Lime, D.W. 1975. Sources of congestion and visitor dissatisfaction in the Boundary Waters canoe area. *Proceedings of the Boundary Waters Canoe Area Conference*. Minneapolis: Quetico-Superior Foundation, 68–82.

———, and Stankey, G. H. 1975. Carrying capacity: Maintaining outdoor recreation quality. *Proceedings of Forest Recreation Symposium*. Upper Darby, PA: Northeast Experiment Station, 174–84.

———; Knopf, R. C.; and Peterson, G. L. 1981. *The National River recreation study*. In *Some Recent Products of River Recreation Research*. St. Paul, MN: North Central Forest Experiment Station.

Liu, J. C., and Var, T. 1980. The use of lodging ratios in tourism. *Annals of Tourism Research* 7:406–27.

Loukissas, P. J. 1982. Tourism's regional development impacts: A comparative analysis of the Greek Islands. *Annals of Tourism Research* 9:523–41.

Lucas, R. C. 1964. Wilderness perception and use: The example of Boundary Waters canoe area. *Natural Resources Journal* 3:394–411.

Lundgren, J. O. J. 1980. The land component in national recreational planning: The Swedish case and its Canadian applications. *Canadian Geographer* 24:22–31.

Manzo, J. T. 1987. A ball in the grass: An exploratory look at soccer in the United States. *Sports Place* 1:30–38.

Mathieson, A., and Wall, G. 1982. *Tourism: Economic, social and physical impacts*. London: Longman.

Matley, I. 1976. *The geography of international tourism*. Washington: Association of American Geographers.

McConnell, H. 1983. Southern major college football: Supply, demand and migration of players. *Southeastern Geographer* 23:78–106.

———. 1984. Recruiting patterns in midwest major college football. *Geographical Perspectives* 53:27–43.

McMurry, K. C. 1930. The use of land for recreation. *Annals of the Association of American Geographers* 20:7–20.

———. 1954. Recreational geography. In *American geography: Inventory and prospect,* eds. P. James and C. Jones, pp. 251–57. Syracuse: Syracuse University Press.

Mescon, T., and Vozikis, V. 1985. The economic impact of tourism at the port of Miami. *Annals of Tourism Research* 12:515–28.

Meyer-Arendt, K. 1985. The Grand Isle, Louisiana resort cycle. *Annals of Tourism Research* 12:449–65.

Mieczkowski, S. T. 1981. Some notes on the geography of tourism: A comment. *Canadian Geographer* 25:186–90.

Mitchell, L. S. 1969. Toward a theory of public urban recreation. *Proceedings of the Association of American Geographers*. Washington: Association of American Geographers.

———. 1981. Unpublished paper presented at the annual meeting of the AAG.

———. 1985. The Tao of recreational geography. *Journal of Cultural Geography* 6:141–51.

———, and Lovingood, P. E., Jr. 1978. The structure of public and private recreation systems: Columbia, South Carolina. *Journal of Leisure Research* 10:21–36.

———, and Smith, R. V. 1985. Recreation geography: Inventory and prospect. *Professional Geographer* 37:6–14.

Murphy, P. E. 1980a. Perceptions and preferences of decision-making groups in tourist centers: A guide to planning strategy? In *Tourism planning and development issues.* Washington: George Washington University Press.

———. 1980b. Tourism management using land-use planning and landscape design: The Victoria experience. *Canadian Geographer* 24:60–71.

———. 1981. Tourism management using land use planning and landscape design: The Victoria experience. *Canadian Geographer* 24:60–71.

———. 1985a. Geography's potential as a catalyst for tourism theory. Unpublished paper presented at the Conference on Scientific Geography.

———. 1985b. *Tourism: A community approach*. New York: Methuen Inc.

———, and Rosenblood, L. 1974. Tourism: An exercise in spatial search. *Canadian Geographer* 18:201–10.

Neilson, B. 1984. Dialogue with the city: The evolution of the baseball park. *Landscape* 29:39–47.

Nelson, J. G.; Cordes, L. S.; and Masyk, J. 1972. The proposed master plans for Banff National Park: Some criticisms and an alternative. *Canadian Geographer* 16:29–49.

Ojala, C. F. 1987. A geography of major interscholastic women's sports in the United States. *Geographical Bulletin* 29:24–43.

———, and Kureth, E. 1975. From Saskatoon to Perry Sound, a geography of skates and sticks in North America. *Geographical Survey* 4:177–98.

———, and Edmondson, B. 1987. Good sports. *American Demographics* 9:34–37.

Oriard, M. V. 1976. Sport and place. *Landscape* 21:32–40.

Outdoor Recreation Resources Review Commission. 1962. The future of outdoor recreation in metropolitan regions of the United States. Washington: *Outdoor Recreation Resources Review Commission* 21:1–560.

Pearce, D. G. 1978. Form and function in French resorts. *Annals of Tourism Research* 5:142–56.

———. 1980. Tourism and regional development: A genetic approach. *Annals of Tourism Research* 7:69–82.

Pillsbury, R. 1974. Carolina thunder: A geography of southern stock car racing. *Journal of Geography* 73:39–47.

———, ed. 1987. *Sports Place*. Stillwater, OK: Black Oak Press.

Purdue, R., and Gustke, L. 1985. Spatial patterns of leisure travel by trip purpose. *Annals of Tourism Research* 12:167–80.

Raitz, K. 1987. Perception of sports landscapes and gratification in the sport experience. *Sports Place* 1:4–19.

Rodenburg, E. E. 1980. The effects of scale in economic development: Tourism in Bali. *Annals of Tourism Research* 7:177–96.

Rooney, J. F. 1969. Up from the mines and out from the prairies—Some geographical implications of football in the U.S. *Geographical Review* 59:471–92.

———. 1974. *A geography of American sport*. Reading, MA: Addison-Wesley.

———. 1975. Sports from a geographic perspective. In *Sport and social order: Contributions to the sociology of sport,* eds. Donald W. Ball and John W. Loy. Reading, MA: Addison-Wesley.

———. 1975. Sports from a geographic perspective. In *Sport and social order: Contributions to the sociology of sport,* eds. Donald W. Ball and John W. Loy. Reading, MA: Addison-Wesley.

———. 1987. *The recruiting game.* 2d ed. Lincoln, NE: University of Nebraska Press.

———, and McDonald, D. B. 1982. Sports and Games. In *This Remarkable Continent.* College Station, TX: Texas A & M University Press.

Ryan, B. 1984. Activity clustering in urban recreation. *North American Culture* 1:3–34.

Sage, G., and Loy, J. 1978. Geographical mobility patterns of college coaches. *Urban Life* 7:252–74.

Schwendeman's directory of college geography of the United States. 1988. Richmond, KY: Southeastern Division of the Association of American Geographers.

Shaw, E. B. 1963. Geography and baseball. *Journal of Geography* 42:74–76.

Shelley, F. 1987. Geographic factors in fan support of major league baseball teams. *North American Culture* 3:30–36.

Shelley, F. M., and Cartin, K. F. 1984. The geography of baseball fan support in the United States. *North American Culture* 1:77–95.

Smale, B. J. A., and Butler, R. W. 1985. Domestic tourism in Canada: Regional and provincial patterns. *Ontario Geography* 26:37–56.

Smith, S. L. J. 1980. Intervening opportunity and travel to urban recreation centers. *Journal of Leisure Research* 12:296–308.

———. 1983a. *Recreation geography.* New York: Longman.

———. l983b. Restaurants and dining out: Geography of a tourist business. *Annals of Tourism Research* 10:515–49.

———. 1985. Tourist space and tourist attraction. *Leisure Sciences* 7:65–71.

———, and Brown, B. 1981. Directional bias in vacation travel. *Annals of Tourism Research* 8:257–70.

Stanfield, C. 1978. Atlantic City and the resort cycle: Background to the legalization of gambling. *Annals of Tourism Research* 5:238–51.

Stanfield, C. A., and Rickert, J. E. 1970. The recreational business District. *Journal of Leisure Research* 2:213–25.

Stough, R. R. 1985. A risk assessment model for tourism and resort investment decisions. Unpublished paper presented at the Conference on Scientific Geography.

Stroud, H. 1983. Environmental problems associated with large recreation subdivisions. *Professional Geographer* 35:303–13.

Teye, V. B. 1986. Liberation wars and tourism development in Africa: The case of Zambia. *Annals of Tourism Research* 13:589–608.

Tourism and outdoor recreation planning study. 1977. Toronto: Government of Ontario. Seven parts.

VanDoren, C. S.; Priddle, G. B.; and Lewis, J. E. 1979. *Land and Leisure.* 2d ed. Chicago: Maaroufa Press, Inc.

———, and Gustke, L. D. 1982. *Spatial analysis of the U.S. lodging industry, 1963–1977.*

Var, T., and Quayson, T. 1985. The multiplier impact of tourism in the Okanagan. *Annals of Tourism Research* 12:497–514.

Wall, G. 1983. The economic value of cultural facilities: Tourism in Toronto. In *Recreation planning and management.* State College, PA: Venture Publishing.

———, and Sinnott, J. 1980. Urban recreational and cultural facilities, and tourist attractions. *Canadian Geographer* 24:50–59.

———; Dudycha, D.; and Hutchinson, J. 1985. Point pattern analyses of accommodation in Toronto. *Annals of Tourism Research* 12:603–18.

Weightman, D., and Wall, G. 1985. The spa experience at Radium Springs. *Annals of Tourism Research* 12:393–416.

Wepfer, A. 1984. The diffusion of volksmarching in the United States. *Geographical Bulletin* 26:9–28.

West, N. C. 1988. Marine recreation in North America. Working paper contribution on recreational uses of coastal areas. Report to the Commission on the Coastal Environment. International Geographical Union biannual meeting: Sydney, Australia.

Williams, A., and Zelinsky, W. 1970. On some patterns in international tourism flows. *Economic Geography* 46:549–67.

Wolfe, Roy I. 1964. Perspective on outdoor recreation: A bibliographical survey. *Geographical Review* 54:203–38.

Yetman, N. R., and Eitzen, D. S. 1973. Some social and demographic correlates of football productivity. *Geographical Review* 63:553–57.

Young, C. W., and Smith, R. V. 1979. Aggregated and disaggregated outdoor participation models. *Leisure Sciences* 2:143–54.

ASSESSMENT AND MANAGEMENT OF HAZARDS AND INFIRMITY

Hazards Research

James K. Mitchell

In a recent *New York Times* column, Russell Baker provided an alphabetized list of the 125 most publicized things that could "get you" (*New York Times* November 11, 1986). It is an epigram of contemporary hazards. As well as generally acknowledged threats like tornadoes, earthquakes, and nuclear power stations, the list includes a wide variety of other real or imagined hazards, such as botulism, casual sex, children in satanic possession, dental plaque, exhausted air-traffic controllers, fire ants, lack of calcium, midlife crises, ozone breakdowns, phone calls in thunderstorms, runaway electric carving knives, sharks, steroids, whiskey, and worn-out windshield wipers. In a sense, that list—or something like it—constitutes the potential subject matter of hazards research. It is a range of natural events, manufactured systems, and people that threaten our lives and life-support systems, our emotional security, our property, and the functioning of our societies. When these threats materialize and overwhelm our coping capabilities, they are known as disasters.

Over a period of 40 years, geographers who study hazards have made intellectual and practical contributions that are significantly out of proportion to the small number of workers in the field. Approximately 5% (about 120) of the 2400 college faculty who are members of the Association of American Geographers identify "hazards" as one of their primary fields of specialization. However, two of the eight geographers who are members of the National Academy of Sciences—Gilbert F. White and Robert W. Kates—are leaders in hazards research.[1] Four other NAS members who are geographers have also made significant additions to the literature on natural, technological, or social hazards—John R. Borchert, Walter Isard, M. Gordon Wolman, and Julian Wolpert.

Several reviews of hazards research have been published in recent years: Hewitt (1983), Kates (1978), Kasperson and Pijawka (1985), Mitchell (1984), Mitchell (forthcoming–a), O'Riordan (1986), Whyte (1986), and Whyte and Burton (1980). This chap-

---

[1]Geographers who are members of the National Academy of Sciences, with dates of their induction, include Gilbert F. White (1973), Brian J. L. Berry (1975), Robert W. Kates (1975), John R. Borchert (1976), Julian Wolpert (1977), Waldo R. Tobler (1982), Walter Isard (1985), and M. Gordon Wolman (1988).

ter will not attempt to duplicate that material. Instead, it will focus attention upon several broad questions which are likely to be of interest both to specialists and nonspecialists in the field:

- Why are geographers attracted to this area of scholarship?
- What is distinctive about their approaches to hazards research?
- How have interpretations of hazards changed, as we have come to know more about them?
- And finally, what are likely to be some of the significant issues and developments in the years ahead?

As you consider these questions and the answers provided in the following pages, it is hoped that you will come to share the opinion that hazards research is one of the most intellectually challenging and socially relevant branches of modern geography.

WHY DO GEOGRAPHERS STUDY HAZARDS?

Hazards touch our emotions. They involve dramatic situations that excite curiosity, threaten safety, and mobilize humanitarian feelings. However, the main reasons for studying hazards are pragmatic and idealistic, rather than emotional. Hazards are major public-policy problems, and researchers seek to lighten the burden of hazard on society by reducing or preventing losses. Research on hazards also provides insights about one of the great enduring themes of scholarship—the relationships of people and their physical environments.

There has long been a close connection between hazards research and public-policy applications. Research advances often are stimulated by major disasters that reveal the inadequacies of existing hazard-management systems. This was clearly displayed in the classic flood-hazard studies carried out by Gilbert White and his students at the University of Chicago (Platt 1986a). That work followed in the wake of widespread disastrous floods during the 1930s. Its point of departure was the finding that previous heavy expenditures on flood protection—mainly in the form of structural engineering works—had not curbed flood losses in the U.S.

In the search for better flood-management policies, researchers focused attention upon alternative concepts of flood hazard, particularly on the changing relation between physical risks of floods and the human use of floodplains. This led to the development of a "human ecological" model of flood hazard, and the identification of additional points of intervention for the application of public policies. Not only were new theoretical insights about hazards gained; more alternatives for flood-loss reduction were uncovered, and the range of policy choices was substantially broadened and diversified. As this example illustrates, in the course of addressing pressing human problems, hazards research also engages important intellectual issues (White 1988c).

Moreover, the practice of framing research questions in ways that are relevant to decision-makers has helped to shape the methodology of hazards research. Formulation of "human ecological" models of hazard, identification of opportunities for intervention, assessment of alternative courses of action, and expansion of the range of choice that is available to decision-makers—all have become important features of

ASSESSMENT AND
MANAGEMENT OF
HAZARDS AND
INFIRMITY

subsequent research on a wide variety of natural and technological hazards (Burton, Kates, and White 1978; Kates, Hohenemser, and Kasperson 1984; Ricketts 1986; White 1974; White and Haas 1975).

Despite the obvious links with public-policy issues, hazards research is also a fundamental field of inquiry in the oldest tradition of geographical scholarship: the study of relationships between society and environment (Pattison 1964). Connections with this theme occur on several levels. On the level of grand theory, hazard research informs broad philosophical debates about the relative significance of human and nonhuman factors in shaping the Earth and its inhabitants (Turner et al. forthcoming). For example, studies of hazards illuminate the role of catastrophe in the evolution of the Earth's physical systems and in the economic and sociopolitical development of its peoples. Information about hazards also serves to temper the view that the Earth is a storehouse of unconstrained resources and unlimited opportunities for human betterment. The very existence of hazards challenges concepts about order, balance, and continuity in Nature that underlie scientific theories of Earth processes and religious and philosophical explanations of existence (Glacken 1973). At another level, hazard research also illuminates a variety of more specific topics, such as the patterns of global occupance, perceptions and decisions under conditions of uncertainty, responses of human and engineered systems to stress, and the role of human agency in the modification and transformation of natural systems.

THE STATUS OF HAZARDS RESEARCH

Today the field of hazards research is composed of a loosely grouped cluster of subfields that includes *disaster research, natural-hazards research,* and *risk analysis* (Mitchell forthcoming–b). In this analysis, the contributions of geographers are highlighted, but it is important to note that scholars and professionals from many disciplines and professions have also shaped the development of all three subfields. Interdisciplinary perspectives and collaboration between researchers and users of research findings are hallmarks of hazards research. Decision-makers, managers, affected publics, and other groups are all regarded as members of the hazards-research community. Theory and praxis are firmly linked. Typically, theory arises as a consequence of addressing practical problems that involve hazards; findings are directed toward informing public policy-makers and the management of the human environment.

Hazards geographers have worked primarily in the subfields of natural-hazards research and risk analysis. Natural-hazards research focuses upon human responses to geological, meteorological, and hydrological risks including earthquakes, tornadoes, and floods. Risk analysis is mainly concerned with the assessment of technological or quasitechnological risks (White 1988b). Early geographical work focused upon dramatic, short term, localized, natural extremes in the U.S., especially floods. Later, a wide range of natural hazards throughout the world were canvassed (White 1974). In the 1970s, investigators began to address hazards of technology, hazards that are international or global in scope, and hazards that occur on time scales of decades to centuries. New topics that have come under scrutiny in recent years include nuclear-energy risks (Pasqualetti and Pijawka 1984); nuclear detonations in the atmosphere (Harwell and Hutchinson 1985); soil erosion, deforestation, and other processes of

Table 1
Publication trends in natural-hazards research

Period	North American Geographers	Other Geographers	Non-Geographers	Totals
1960–1969	5	3	1	9
1970–1979	44	15	39	98
1980–1983	38	45	102	185
1960–1983 totals	87	63	142	292

Source: Mitchell, J. K. 1987. Borba so stikhiynymi bedatviyami (Human responses to natural hazards). In Geograficheskie aspekty vzaimodeystviya khozyaystva i okruzhayuschev sredy. Moscow: Akademiya nauk SSSR, pp. 178–87.

gradual environmental deterioration (Blaikie 1985; Blaikie and Brookfield 1987); and insidious global atmospheric changes that are likely to occur during the next 100 years (Kates, Ausubel, and Berberian 1985).

Since World War II, the volume of literature in this field has grown rapidly. Table 1 illustrates trends in the number of articles about *natural* hazards that appeared in 86 major English-language journals in the natural, social, and policy sciences during the period between 1960 and 1983 (Mitchell 1987a). Of the 292 articles on hazards published during those years, only 9 appeared during the 1960s, compared with 98 in the 1970s, and 185 in the first four years of the 1980s.

Natural and technological hazards have also become popular topics for thesis and dissertation research (Table 2). In North American universities and colleges, at least 44 Ph.D. dissertations and 126 master's theses on hazards topics were completed during the six years from 1981 to 1986. Sixty percent of both dissertations and theses focused upon natural hazards; a quarter of the dissertations and 20% of the theses focused upon technological hazards. Floods and droughts dominated the list of topics, followed by other sudden-onset geological or meteorological hazards. Hazardous industrial facilities and nuclear wastes were the most commonly studied technological hazards.[2]

Important changes in the theoretical basis of hazards research have accompanied expansion of the research agenda. Early work employed rational theories of individual decision-making and emphasized opportunities for broadening the range of choice; later studies took more explicit account of the cognitive and behavioral factors that modify decisions about hazards (Mitchell 1984; Kasperson et al. 1988). Recent research has examined circumstances where responses to hazard are few and are tightly constrained. Researchers have begun to explore the utility of various theoretical perspectives, including conflict theory (Solomon et al. 1987), catastrophe theory (Olson 1987), structuralist-materialist viewpoints (Marston 1983; Susman, O'Keefe, and Wisner 1983), and humanistic explanations (Tuan 1979).

---

[2] Behavioral aspects of hazards, energy hazards, and geomorphological dimensions of hazards are addressed in other chapters of this volume on environmental perception, energy geography, and geomorphology.

Table 2
Hazards dissertations and theses, 1981–1986

Year	Natural Hazard	Technological Hazard	Other Hazard	Total
Ph.D. dissertations				
1981	2	1	2	5
1982	3	0	1	4
1983	6	3	2	11
1984	7	4	0	11
1985	4	1	1	6
1986	4	2	1	7
1981–86 totals	26	11	7	44
Master's theses				
1981	11	4	4	19
1982	12	5	3	20
1983	19	4	6	29
1984	20	4	3	27
1985	8	6	6	20
1986	6	2	3	11
1981–86 totals	76	25	25	126

These explorations have focused upon the social dimensions of hazards, but physical geographers also have added important new insights about synergisms between physical processes and management responses (Cooke 1984; Alexander 1985; Penning-Rowsell, Parker, and Harding 1985). Various new interdisciplinary perspectives have also been pioneered (Kates 1985; Clark and Munn 1986). As a result of these developments, there is increasing theoretical diversity within hazards research. For a period in the late 1970s and early 1980s, exponents of the mainstream behavioral decision-making approach engaged in spirited debate with exponents of a structuralist political-economy approach (Marston 1983; Mitchell 1984). More recently, there has been some convergence among alternative approaches as ideas from different groups are shared with and incorporated into the work of other groups (Cooke 1984; Kirby 1986; Mitchell forthcoming-a; Handmer 1987).

Hazards research now resembles a rapidly growing tree. The roots are spreading out to draw upon an increasingly large number of disciplines for inputs that enable the trunk to send out many new branches in the form of specialized research institutions and specialized fields of inquiry. Geographers provide leadership in innovative hazards-research units throughout North America—e.g. Natural Hazards Research and Applications Center, University of Colorado; Center for Technology, Environment, and Development, Clark University; World Hunger Center, Brown University; and International Federation of Institutes of Advanced Study, Toronto.

Geographers also make major contributions to specialized branches of inquiry that are concerned with the assessment, communication, and management of hazards. Assessment research focuses upon measurement and mapping of physical risks and their impacts, upon awareness of risk, upon the evolution of vulnerability, and upon other factors that influence human behavior in the face of risk (Foster 1980, 1984; Greenberg 1987; Kates, Ausubel, and Berberian 1985; Kotlyakov et al. 1988; Mitchell 1984; Sternberg 1987; Tobin 1985; Wisner 1986; Zeigler, Johnson, and Brunn 1983).

Communication research addresses the dissemination and interpretation of information about risks, including education programs and public participation in the making of decisions about hazards (Kasperson 1986; Sims and Baumann 1983). Management research engages issues that affect the organizations and programs having responsibility for protecting society against hazards (Arnell, Clark, and Gurnell 1984; Kasperson and Kasperson 1988; Liverman 1987; Mitchell 1985, 1986; Platt 1985, 1987; Platt, Pelczarski, and Burbank 1987; Sorensen, Vogt, and Mileti 1987).

In summary, a great many studies of hazard have been undertaken in different parts of the world by a wide variety of investigators who use different analytical techniques and different systems of explanation. Obviously, much has been learned about hazards as a result of these activities. How has that information affected interpretations of hazards?

CHANGES IN THE INTERPRETATION OF HAZARDS

Important changes in interpretation of hazards have occurred during the last four decades. In the past, storms, landslides, droughts, and other "natural" hazards were commonly regarded as "acts of God"—external events that impinged on unsuspecting people. This interpretation has largely been replaced by the view that natural hazards are dynamic interactive phenomena that involve people as contributors and modifiers, as well as victims. Concepts of technological hazards have undergone more recent but similar changes. Although it has been long known that humans play major roles in technological hazards, most pre-1970 analyses focused upon failures of the material components of technologies—not the human elements. Now, technological hazards are increasingly interpreted as failures of people-machine-environment systems, just as natural hazards are regarded as expressions of inadequate adjustments in people-environment systems. In other words, the concept of *interaction* is firmly entrenched at the heart of hazards research, and the main elements of hazard have been uncovered by investigators who have elaborated this concept.

From the outset, geographers stressed the responsive capabilities of people. At first, the focus was upon responses to risks that arise in physical systems. Somewhat later, factors that affect the exposure of human populations to risk were built into explanations of hazard. More recently, the role of differing potentials for loss and recovery among affected populations has been highlighted. Scholars have arrived at a finding that hazards are a function of four sets of interacting variables: risk, exposure, vulnerability, and response (Mitchell, forthcoming–b):

- ☐ Risk is the probability of extreme events.
- ☐ Exposure is a measure of the population at risk.

- Vulnerability is the potential for loss.
- Response is the choice of measures to reduce, avoid, or prevent loss.

Acceptance of the concept of hazards as interactive phenomena has had several important consequences. It has encouraged the development of interdisciplinary approaches to hazards research. This is because the basic model can be applied to many different kinds of hazards, and no one discipline provides the skills or experience to address all of its components, even for a single hazard. When interactive models of hazard replaced models that emphasized physical causation, new ways of managing hazards were suggested. Policies that emphasize the modification of physical risks have been augmented by policies that address the modification of exposure and vulnerability. As the complex "causes" of hazard have been elaborated, it has become possible to formulate preparedness, mitigation, and prevention programs, instead of relying exclusively on post-disaster relief and reconstruction schemes. Both the range of potential adjustments to hazards and the number of permutations among different adjustments has increased.

Now another phase in the evolution of hazards research has begun. Changes are occurring on two levels. First, the nature of hazard is changing, as new types of risks emerge and new hazard-management systems are created. Second, the concept of hazard is again being modified. It is being extended to include not just hazards themselves, but also the contexts in which they are embedded.

Increased Hazards

In the past, natural hazards, wars, famines, and epidemic diseases posed a multitude of immediate local threats. These traditional hazards have declined in some countries, but they are intensifying and spreading in many other parts of the world (National Research Council 1987). The mix of hazards is also changing. Everywhere, new and increasingly potent hazards are emerging (Karan, Bladen, and Wilson 1986). Some, like AIDS, resemble existing hazards, but are potentially more virulent. Others, such as nuclear-weapons detonations in the atmosphere and the destruction of stratospheric ozone, are entirely unprecedented (White 1985; Kates 1985).

As the spreading rash of hazardous-materials spills illustrates, many of the new hazards are site-specific. Others, like increased atmospheric levels of carbon dioxide, are global in scope. The combined effects of potent new hazards and worsening old ones, such as deforestation, soil degradation, and desertification, threaten the long-term viability of global life-support systems (Allen and Barnes 1985; Karan and Mather 1985). This represents the most profound of all the physical changes that have affected hazards (White 1985). It stems from the accelerating capacity of people to modify the biosphere, both deliberately and inadvertently. Humans are increasingly seen as the preeminent factor affecting natural processes.

In the parlance of a recent international conference on the subject, the Earth has been "transformed" by human agency, and thresholds of environmental stability are being tested and crossed in increasing numbers (Turner et al. forthcoming). The interpretation of hazards reflects this transition. There is now genuine concern about the consequences of global atmospheric warming, sea-level rise, genetic engineering, and other processes of environmental modification that seemed fanciful only a few decades ago.

Institutional and Policy Changes

Changes in the number and types of hazards make it likely that their management will be significantly different. Moreover, other factors have already brought about changes in hazard-management institutions and policies. For example, partly in response to the results of geographical research, during the last two decades there has been a well-marked shift of emphasis in public policy, from remedial strategies to anticipatory strategies. Although emergency-management actions and post-disaster relief are still important adjustments to hazard, significantly greater emphasis is now placed on preparedness measures such as warning systems and mitigation, or prevention measures like land-use and development controls (Mitchell 1985).

The shift toward mitigation and prevention is most marked in developed countries, but it is also underway in developing states. It will probably continue unless increasing uncertainty about new global hazards encourages a return to policies dominated by contingency planning and disaster-response strategies.

Hazards management has taken on an increasingly international and global aspect as new hazards, such as the Chernobyl nuclear reactor fire, produce troubling impacts far from their sources. This trend has been aided by the growing areal impacts of more familiar hazards, such as African drought. As a result, new international institutions and scientific initiatives have begun to provide information about hazards, and to address the task of coordinating hazards policies. These include nongovernmental organizations that lobby for hazards reduction, satellite-based environmental-monitoring systems, government-sponsored regional hazards-management information and training centers, international quick-response disaster teams, and various collaborative global data-gathering and research programs, such as the International Geosphere-Biosphere Program (IGBP) and the International Decade for Natural Disaster Reduction (IDNDR) (see next section).

NEW DEPARTURES IN HAZARDS RESEARCH

Changes that are affecting the nature of hazards and hazards management are of large magnitude and wide scope. These changes have increased the complexity of the task that faces hazards researchers (Mitchell forthcoming–a). As a result, not only is the hazards-research agenda being redirected to address more complex questions, but strategies of inquiry are also beginning to change. Two new developments stand out. The first is awareness of the need to increase the use of knowledge about hazards by hazards managers. The second is a search for broader theoretical explanations of hazard. Both of these developments are reflected in ambitious new research and management programs that are likely to play important roles in shaping knowledge about hazards and societal coping capabilities during the next decade.

Improving Utilization of Information About Hazards

Much more is known than is put into practice about ways to reduce potential losses from most natural hazards and most conventional technological hazards (National Research Council 1987; Mitchell 1988). Some of the reasons that individuals and institutions fail to make optimal use of existing knowledge are known. Among others, they include lack of awareness of relevant information, lack of mechanisms and resources for applying knowledge, and unwillingness to deviate from routine procedures. For example, it is known that there is relatively little exchange of experience and insights between students of natural hazards and students of technological hazards (White, 1988c). Also, hazard-management organizations exhibit significant reluctance to undertake searching postaudits of policies, programs, and other hazard-management activities that might provide a means of learning from experience (White 1988b). These and many similar findings underscore the importance of learning about factors that impair the use of hazards research.

Some geographers have already begun to address this task through searching evaluations of contemporary public programs. Such studies often point out inconsistencies and deficiencies that stem from failure to apply existing knowledge. Among others, they include analyses of changes in the occupancy of hazard zones that have been subject to development regulations (Montz and Gruntfest 1986), postaudits of major facilities and programs (White 1988a; Mitchell 1987b), assessments of projected environmental impacts of new technologies (Harwell and Hutchinson 1985), analyses of nuclear preparedness and evacuation plans (Cutter 1988; Platt 1986b), and evaluations of programs for managing hazardous lands (Platt, Pelczarski, and Burbank 1987). More efforts of this kind are clearly warranted.

But it will also be necessary for researchers to go well beyond the subject of risk communication. Among others, future studies might evaluate prospects for the adoption of new adjustments to hazard, including the role of alternative incentives and the development of new methodologies and procedures for incorporating hazards information into investment, planning, and development decisions.

Increased use of available information about hazards is a central objective of the forthcoming International Decade for Natural Disaster Reduction, which the United Nations has declared for 1990–2000. It is planned that many nations will collaborate to reduce losses from sudden-onset disasters, such as earthquakes and flash floods, in part by improving the application of mitigation and preparedness measures that have proven their worth elsewhere (Mitchell 1988). Hazards geographers helped to shape the Decade's early planning, and they have provided critiques of its subsequent development (Oliver 1988; Parker 1988). It is too early to determine the prospects for success, but this initiative bears watching as a test of efforts to improve the use of available information about hazards.

Contexts of Hazard

The bulk of previous research has been concerned with single hazards, but several researchers have extended the research paradigm to include complex threats such as hunger (Currey and Graeme 1984; Herkstra and Liverman 1986), warfare (Cutter 1988; Hewitt 1987; Harwell and Hutchinson 1985), and global environmental change (Burton 1987; Kates, Ausubel, and Berberian 1985; Chen, Boulding, and Schneider 1983; Kates 1987; Parry 1986). In the light of these and other studies, some observers have concluded that geographers lack an adequate theoretical basis for analyzing megahazards like hunger and nuclear war (Cutter 1988; Mitchell forthcoming–a). Because of the need to develop broader analytic methods to accommodate such complex phenomena, a new concept of hazards is emerging.

Hazards geographers are beginning to rework and extend the concept of interaction to take account of a range of different *contexts* which color interpretation and make difficult the grasping of commonalities among different theoretical explanations. Although there are many types of context, two are particularly important.

First are contexts that stem from the fact that "real world" problems are not easily bounded. Hazards tend to overlap and interpenetrate with other problems. Natural and technological hazards are members of a set of "interlocking crises" that make up a giant global problematique (Kates and Burton 1986). Among others, this interactive set includes environmental mismanagement, resource depletion, hunger, geopolitical tensions, the burden of international debt, the widening gap between rich and poor nations, and the threat of nuclear war. Because such problems individually and collectively threaten the sustainability of life on Earth, hazard has become a metaphor for the age.

The second context is the dichotomy between a unitary global perspective and fragmented local perspectives. This is especially interesting to geographers, because it touches on issues of scale and spatial organization. Thus, we inhabit "One Earth" whose physical systems are interdependent, but we live in "Many Worlds" that are fragmented into competitive groups by divisions of religion, language, class, ethnicity, and historical experience. It is evident that hazards must be considered in the context of several hierarchical spatial mosaics of socioenvironmental conditions.

Contextual research on hazards has been under way for many years, but until recently each context has been treated mainly as a kind of backdrop to the general interactive model of hazard. Contextual study has not yet been elevated to a general principle of hazard research, but several researchers are at work on this theme (e.g., Palm forthcoming; Wisner 1986; White 1987). Though hazards research is moving in this direction, society still awaits a broad, integrated theory of hazard which draws together work from the differing contexts, that sets hazards in relation to other problems, and that builds bridges across the ancient philosophical divide that separates those who believe life is primarily about the avoidance of pain and those who see it as the pursuit of pleasure.

Contextual issues are being addressed in a series of collaborative international activities and programs that focus upon the theme of global environmental change. Some of these are initiatives by geographers, and others are multidisciplinary in scope. For example, geographers from the United States and the Soviet Union are addressing hazard issues in a joint project on global change (Kotlyakov et al. 1988). Two committees of the International Geographical Union are also undertaking work on hazards in the context of long-term and short-term global environmental changes (i.e., Commission on the Coastal Environment; Working Group on Global Change). Organizations that have developed multidisciplinary perspectives on global change include the International Council of Scientific Unions (Scientific Committee on Problems of the Environment), the U.S. National Research Council (Committee on Global Change) (Clark 1988), the Human Response to Global Change Program (sponsored by the International Federation of Institutes of Advanced Study, United Nations University and the International Social Science Council) (Burton 1987), and the International Association of Applied Systems Analysis (Clark and Munn 1986).

The largest and most ambitious of the multidisciplinary initiatives is a natural-science research program—the International Geosphere-Biosphere Program. It seeks to encourage the emergence of a new science of global change, based on the integration of large-scale physical and biological models of environmental processes (Malone 1986). In part, because of an emphasis on global-modeling methods, IGBP fosters a "top-down" perspective on global change. Unfortunately, the role of human agency has been largely neglected, and little attempt has been made to address the task of providing information that might be useful for managing the risks of long-term global change (Kates 1987). Hazards geographers have moved quickly to plug this gap by helping to lay out a program that focuses upon human responses to global change (Burton 1987), and by bringing together American and Russian geographers—Kotlyakov, Mather, Sdasyuk, and White—in the first publication to address the topic of human agency in global change (Kotlyakov et al. 1988). What is particularly notable about both the cooperative USSR-U.S. project and the Human Response to Global Change program is that they are both "bottom-up" perspectives which focus upon the direct experience that people have had with environmental hazards in diverse local settings.

The necessity of coping with environmental risks and hazards is also a key aspect of activities that are designed to define and operationalize the concept of "sustainable development" (Brown et al. 1987). Since formulation of the World Conservation Strategy in 1980 an increasing number of governments, intergovernmental organizations, and other international institutions (e.g., World Bank) have acknowledged that future resource-development schemes should incorporate measures for anticipating, avoiding, or mitigating short- and long-term environmental hazards.

SUMMARY AND CONCLUSIONS

Relations between people and their environments are affected by many constraints and uncertainties. Hazards research addresses the subset of constraints and uncertainties that threaten society with significant losses. It is a subset of increasingly complex and increasingly serious problems that range from catastrophic floods in Bangladesh and industrial emergencies in many countries to alteration of the chemical composition of the global atmosphere. A growing stream of evidence about ongoing disasters and emerging threats is reported in the mass media and the scientific literature.

Geographers have laid the foundations of several subfields of hazards research. By calling attention to neglected human dimensions of hazard—and to the dynamic interaction of societies and environments—they have also helped to shape improved institutional responses to hazard in the U.S. and other countries. In the course of this work, concepts of hazards and approaches to the study of hazard have gradually become broader and more integrative. New developments in hazards research are now occurring in response to a combination of growing and spreading hazards, flawed hazard-management systems, and incomplete analytic methods. In particular, there is widespread recognition of the need for more effective use of knowledge about hazards and the adoption of more effective hazard-management systems.

Similarly, it is becoming evident that hazards cannot adequately be addressed in isolation from other problems and processes. The field is going through a period of intellectual exploration that centers on the search for alternative methods that will make it possible to analyze complex hazards and the contexts in which they are embedded. In the next decades, efforts to improve the use of knowledge about hazards and to improve understanding of hazard contexts are likely to have major impacts on research and public policy programs in the U.S. and throughout the world. Innovations in collaborative international research and management programs such as the IGBP and the IDNDR are particularly worth watching for clues about broader developments in hazards research.

It should be abundantly clear that hazards research is a thriving, challenging, and socially useful endeavor which spills over the boundaries of geography to nourish a variety of other fields. After four decades of growth, it continues to evolve at a rapid pace, moving in the direction of ever-higher levels of theoretical and methodological integration among its diverse components. In this regard, it will stand as a model for the development of other geography subfields into the twenty-first century.

REFERENCES

Alexander, D. 1985. Culture and environment in Italy. *Environmental Management* 9:121–34.

Allen, J. C., and Barnes, D. F. 1985. The causes of deforestation in developing countries. *Annals of the Association of American Geographers* 75:163–84.

Arnell, N. W.; Clark, M. J.; and Gurnell, A. M. 1984. Flood insurance and extreme events: The role of crisis in prompting changes in British institutional response to flood hazard. *Applied Geography* 4:167–82.

Blaikie, P. 1985. *The political economy of soil erosion in developing countries*. London: Longman.

———, and Brookfield, H. 1987. *Land degradation and society*. London: Methuen.

Brown, B. J. et al. 1987. Global sustainability: Towards definition. *Environmental Management* 11:713–19.

Burton, I. 1987. Our common future: The World Commission on Environment and Development. *Environment* 29:25–29.

———; Kates, R. W.; and White, G. F. 1978. *The environment as hazard*. New York: Oxford University Press.

Chen, R. S.; E. Boulding; and Schneider, S. H., eds. 1983. *Social science research and climate change*. Dordrecht: D. Reidel.

Clark, W. C. 1988. The human dimensions of global change. Report prepared for the U.S. National Research Council's Committee on Global Change. Cambridge: John F. Kennedy School of Government, Harvard University.

———, and Munn, R. E., eds. 1986. *Sustainable development of the biosphere*. Cambridge: Cambridge University Press.

Cooke, R. U. 1984. *Geomorphological hazards in Los Angeles*. London: Allen & Unwin.

Currey, B., and Graeme, H. 1984. *Famine as a geographical phenomenon*. Dordrecht: D. Reidel.

Cutter, S. L. 1988. Geographers and nuclear war: Why we lack influence on public policy. *Annals of the Association of American Geographers* 78:132–43.

Foster, H. D. 1980. *Disaster planning: The preservation of life and property*. New York: Springer Verlag.

———. 1984. Reducing vulnerability to natural hazards. *The Geneva Papers on Risk and Insurance* 9.

Glacken, C. J. 1973. *Traces on the Rhodian shore*. Berkeley: University of California Press.

Greenberg, M. 1987. Urban/rural differences in behavioral risk factors for chronic diseases. *Urban Geography* 8:146–51.

Handmer, J., ed. 1987. *Flood hazard management: British and international perspectives*. Norwich: Geo Books.

Harwell, M. A., and Hutchinson, T. C., eds. 1985. *Ecological and Agricultural Effects*. Vol. 2 of *Environmental consequences of nuclear war*. Chichester: Wiley.

Hewitt, K. 1987. The social space of terror: Towards a civil interpretation of total war. *Environment and Planning D: Society and Space* 5:445–74.

———, ed. 1983. *Interpretations of calamity*. Boston: Allen & Unwin.

Herkstra, G. P., and Liverman, D. M. 1986. Global food futures and desertification. *Climatic Change* 9:59–66.

Karan, P. P., and Mather, C. 1985. Environmental stress in the Himalaya. *Geographical Review* 75:71–92.

———; Bladen, W. A.; and Wilson, J. A. 1986. Technological hazards in the Third World. *Geographical Review* 76:195–208.

Kasperson, R. E. 1986. Six propositions on public participation and their relevance for risk communication. *Risk Analysis* 6:275–82.

———, and Pijawka, K. D. 1985. Societal response to hazards and major hazard events: Comparing natural and technological hazards. *Public Administration Review* Special Issue.

———, and Kasperson, J. X., eds. 1988. *Nuclear risk analysis in comparative perspective*. London: Allen & Unwin.

——— et al. 1988. The social amplification of risk: A conceptual framework. *Risk Analysis* 8:177–85.

Kates, R. W. 1978. *Risk assessment of environmental hazard*. SCOPE Report No. 8. New York: Wiley.

———. 1985. Success, strain and surprise: An introduction. *Issues in Science and Technology* 2:46–58.

———. 1987. The human environment: The road not taken, the road still beckoning. *Annals of the Association of American Geographers* 77:525–34.

———; Hohenemser, C.; and Kasperson, J. X. 1984. *Perilous progress: Technology as hazard*. Boulder, CO: Westview Press.

———; Ausubel, J. H.; and Berberian, M., eds. 1985. *Climate impact assessment*. SCOPE Report No. 27, p. 33. New York: Wiley.

———, and Burton, I. 1986. The great climacteric, 1798–2048: The transition to a just and sustainable human environment. In *Geography, resources, and environment*. 2 vols. Chicago: University of Chicago Press.

Kirby, A. 1986. Technological risks in urban areas. *Cities*:137–41.

Kotlyakov, V. M. et al. 1988. Global change: Geographical approaches (A review). *Proceedings of the U.S. National Academy of Sciences* 85:5986–91.

Liverman, D. M. 1987. Forecasting the impact of climate on food systems: Model testing and model linkage. *Climatic Change* 11:267–85.

Malone, T. F. 1986. Integrating studies of global change. *Environment* 28:6–11; 39–42.

Marston, S. A. 1983. Natural hazards research: Towards a political economy perspective. *Political Geography Quarterly* 2:339–48.

Mitchell, J. K. 1984. Hazard perception studies: Convergent concerns and divergent approaches during the past decade. In *Environmental perception and behavior: An inventory and prospect,* eds. Thomas F. Sarinnen, David R. Seamon, and James L. Sell. University of Chicago Department of Geography Research Paper No. 209. Chicago: University of Chicago Press.

———. 1985. Post-disaster prospects for improved hurricane protection on oceanic islands: Hawaii after hurricane Iwa. *Disasters* 9:286–94.

———. 1986. Coastal management since 1980: The U.S. experience and its relevance for other nations. *Ocean Yearbook 6,* eds. Elizabeth Mann Borgese and Norton Ginsburg, pp. 319–45. Chicago: University of Chicago Press.

———. 1987a. Borba so stikhiynymi bedatviyami (Human responses to natural hazards). In *Geograficheskie aspekty vzaimodeystviya khozyaystva i okruzhayuschev sredy,* pp. 178–87. Moscow: Akademiya nauk SSSR.

———. 1987b. National Academy of Sciences/National Research Council post-disaster surveys: Their applicability for mitigation purposes. *Proceedings of the International Symposium on Housing and Urban Redevelopment After Natural Disasters,* Bar Harbour, Florida, October 23–26, 1985. Washington: American Bar Association.

———. 1988. Confronting natural disasters. *Environment* 30:25–29.

———. Forthcoming–a. Complexity, disparity, and the search for guidance in hazards research. Berkeley: University of California Press. In press.

———. Forthcoming–b. Risk assessment. In *Global change: A geographical approach*. Moscow: Progress Press. In press.

Montz, B., and Gruntfest, E. C. 1986. Changes in American urban floodplain occupancy since 1958: The experience of nine cities. *Applied Geography* 6:325–38.

National Research Council. 1987. *Confronting natural disasters: An international decade for natural hazard reduction*. Washington: National Academy Press.

Oliver, J. 1988. Commentary–Natural hazard reduction. *Environment* 6:2–3.

Olson, S. 1987. Red densities: The landscape of environmental risk in Madagasgar. *Human Ecology* 15:67–89.

O'Riordan, T. 1986. Coping with environmental hazards. *Themes from the work of Gilbert F. White*. Vol. 2 of *Geography, resources, and environment*, eds. Robert W. Kates and Ian Burton, pp. 272–309. Chicago: University of Chicago Press.

Palm, R. I., n.d. *An integrative theory for natural hazards research*. Baltimore: Johns Hopkins University Press. In press.

Parker, D. J. 1988. Commentary. *Environment* 7:2–3.

Parry, M. L. 1986. Some implications of climate change for human development. In *Sustainable development of the biosphere,* eds. W. C. Clark and R. E. Munn, pp. 378–406. Cambridge: Cambridge University Press.

Pasqualetti, M. J., and Pijawka, K. D., eds. 1984. *Nuclear power: Assessing and managing hazardous technology*. Boulder, CO: Westview Press.

Pattison, W. D. 1964. The four traditions of geography. *Journal of Geography* 63:211–16.

Penning-Rowsell, E. C.; Parker, D. J.; and Harding, D. M. 1985. *Floods and drainage: British policies for hazard reduction, agricultural improvement and wetland conservation*. London: Allen & Unwin.

Platt, R. H. 1986a. Floods and man: A geographer's agenda. *Themes from the work of Gilbert F. White*. Vol. 2 of *Geography, resources, and environment*, eds. R. W. Kates and I. Burton, pp. 28–68. Chicago: University of Chicago Press.

———. 1986b. Nuclear crisis relocation: Issues for a host community–the case of Greenfield, Massachusetts, USA. *Environmental Management* 10:189–98.

———. 1987. Coastal wetland management: The advance designation approach. *Environment* 29:16–20; 38–43.

———, ed. 1985. *Regional management of metropolitan floodplains*. Program on Environment and Behavior Monography 45. Boulder, CO: Institute of Behavioral Science, University of Colorado.

———; Pelczarski, S. G.; and Burbank, B. K. R., eds. 1987. *Cities on the beach: Management issues of developed coastal barriers*. University of Chicago Department of Geography Research Paper No. 224. Chicago: University of Chicago Press.

Ricketts, P. J. 1986. National policy and management responses to the hazard of coastal erosion in Britain and the United States. *Applied Geography* 6:197–222.

Sims, J. H., and Baumann, D. D. 1983. Educational programs and human response to natural hazards. *Environment and Behavior* 15:165–90.

Solomon, B. D. et al. 1987. Radioactive waste management policies in seven industrialized democracies. *Geoforum* 18:415–31.

Sorensen, J. R.; Vogt, B. M.; and Mileti, D. S. 1987. *Evacuation: An assessment of planning and research*. ONRL-6376. Oak Ridge, TN: Oak Ridge National Laboratory.

Sternberg, H. O'R. 1987. Aggravation of floods in the Amazon River as a consequence of deforestation? *Geografiska Annaler: Series A–Physical Geography* 69A:201–19.

Susman, P.; O'Keefe, P.; and Wisner, B. 1983. Global disasters, A radical interpretation. In *Interpretations of Calamity,* ed. K. Hewitt. pp. 263–83. Boston: Allen & Unwin.

Tobin, G. A. 1985. Environmental ethics and geography: Some thoughts. *Geographical Perspectives* 55:6–14.

Tuan, Y. F. 1979. *Landscapes of Fear*. Minneapolis, MN: University of Minnesota Press.

Turner, B. L. et al., n.d. *The earth as transformed by human action*. New York: Cambridge University Press. In press.

White, G. F. 1985. Geographers in a perilously changing world. *Annals of the Association of American Geographers* 75:10–16.

———. 1987. Greenhouse gases, Nile snails, and human choice. Unpublished lecture, Distinguished Lecture Series on Behavioral Science 1985–1986. Boulder, CO: Institute of Behavioral Science, University of Colorado.

———. 1988a. The environmental effects of the high dam at Aswan. *Environment* 30:5–11; 34–40.

———. 1988b. Paths to risk analysis. *Risk Analysis*. 8:171–75.

———. 1988c. When may a post-audit teach lessons? *The flood control challenge: Past, present, and future*. Proceedings of a National Symposium, New Orleans, Louisiana, September 26, 1986, eds. H. Rosen and M. Reuss, pp. 53–63. Washington: Public Works Association.

———, ed. 1974. *Natural hazards: Local, national, global*. New York: Oxford University Press.

———, and Haas, J. E. 1975. *Assessment of research on natural hazards*. Cambridge: MIT Press.

Whyte, A. V. 1986. From hazard perception to human ecology. *Themes from the work of Gilbert F. White*. Vol. 2 of *Geography, resources, and environment,* eds. Robert W. Kates and Ian Burton, pp. 240–71. Chicago: University of Chicago Press.

———, and Burton, I., eds. 1980. *Environmental risk assessment*. SCOPE Report No. 15. New York: Wiley.

Wisner, B. 1986. Land Management in Lesotho: Tragedy of the commons, or suburbanization of a labor reserve. In *Human ecology and geography,* eds. D. Steiner, and B. Wisner, pp. 87–105. Zuricher Geographische Scriften No. 28. Zurich: Geographisches Institut Eidgenossische Techische Hochule Zurich.

Wolpert, J. 1980. The dignity of risk. *Transactions of the Institute of British Geographers* 5:391–401.

Zeigler, D. J.; Johnson, J. H.; and Brunn, S. D. 1983. *Technological Hazards*. Washington: Association of American Geographers.

Medical Geography

Robert J. Earickson | Michael R. Greenberg |
Nancy D. Lewis | Melinda S. Meade | S. Martin Taylor

Since the mid-1970s, medical and health-related research by at least 65 American geographers has proceeded worldwide, in both developed and developing nations, in rural and urban settings, and from the microscale of a neighborhood or residential compound to the global scale. This chapter, based on articles and books numbering over 200 titles, seeks to portray the nature of this research. In so doing, it will reveal some of the strengths of this subdiscipline, as well as research that needs to be addressed. Space limitations prohibit identification of all but a few individual contributions; however, an attempt has been made to include them all in three tables and the list of references.

Medical geographers tend to divide their research between two major foci. The first can be loosely termed *disease ecology,* and the second broad category, medical and health services location and utilization, is usually abridged to *health services research* (Meade, Florin, and Gesler 1988; Pyle 1979). It will become apparent in this review that both of these abstractions mask increasing concerns about the geographic aspects of related matters such as aging, malnutrition, economic development policy, and planning at all scales.

DISEASE ECOLOGY AND MALNUTRITION

Diseases and malnutrition occur not only in populations, but in places. Whether wellness and illness are considered by class or age or nationality, there is a spatial dimension. The geographic patterns of morbidity and mortality are the result of the interaction of social, cultural, behavioral, environmental, and biological processes. As Meade puts it (1977, 379), disease is the culmination of "maladaptive reactions among the familiar triad of population, environment and culture." Medical geography seeks to understand as holistically as possible the correlates of health and disease in order to advance knowledge of disease etiology and to promote adaptable population/behavior/environment interactions. Although the relationship between generative processes and spatial form is intuitively obvious, its connections are often ambiguous and confusing. "As in good mystery novels, the exact nature of the process-form (motive—opportunity—murder) relationship may become apparent only *ex post facto*," notes Mayer (1983b, 1213).

Judging from the literature (Hunter and Thomas 1984; Hunter, Rey, and Scott 1982; Stock 1986), the health consequences of abortive attempts to improve on nature for economic reasons—particularly in Third World nations—have, like other process-form relationships, also become apparent only after the expenditure of much energy and large sums of money. Stock argues that, in light of these consequences, there has been a fundamental shift in the social sciences away from developmental theories (modernization and neomodernization) of socioeconomic change in the Third World to dependency and Marxist theories of underdevelopment. Other than the writings of Hunter and a few others, however, "the medical geographical literature on health and development/underdevelopment of the past 15 years has only weakly reflected these trends" (Stock 1986, 689).

Research on disease ecology is conveniently subdivided into investigations of:

1. Infectious and arthropod-borne disease
2. Infant mortality, famine, and malnutrition
3. Noninfectious maladies such as cancer, heart disease, alcoholism, etc.
4. Morbidity and mortality associated with environmental toxins

The latter two categories are collapsed here into a section entitled "chronic disease."

Infectious and Arthropod-Borne Diseases

The study of infectious and arthropod-borne diseases, rooted in the disease-ecology tradition, has continued to be on the research agenda of many American geographers (Table 1). Most infectious-disease research is based in the tropical developing countries. The most significant exception to this involves AIDS.

Medical geographers address infectious disease from several viewpoints. They usually research the distributions and endemism of afflictions that have high priority on the world or national public-health agendas. Schistosomiasis, for instance, is the most widespread trematode infection in the world, affecting an estimated 200 million people (Kloos et al. 1987; Kvale 1981). Malaria, although considerably reduced since World War II, is still far from eradicated in the tropics and subtropics. Given the physical possibility for malarial occurrence in the physically most susceptible regions, in an area where poverty, underdevelopment, and the possibility of contact with mosquitoes are still great, malaria eradication may well be impossible (Dutta and Dutt 1978).

Geographers identify meteorological, biological, and cultural phenomena that are associated with a disease (Lewis 1984; Meade 1978), and most researchers investigate avenues for amelioration of illness, often revealing political and economic barriers to positive change (Stock 1986). A prevailing thesis from disease ecology is that cultural behavior and environmental distributions are complementary processes. Thus, changes in spatial infrastructure, population distribution, and agricultural technology can affect elements of disease systems and have profound impact on the health status of the population (Meade 1977). Medical geographers provide useful maps that either

Table 1
Citations since 1976 referring to infectious diseases and hypothesized ecological correlates

Causes	Citations
Schistosomiasis	Haddock (1981); Hunter (1981b); Kvale (1981); Kloos et al. (1983, 1987); Roundy (1985b); Weil and Kvale (1985)
Onchocerciasis	Edungbola, Asaolu, and Watts (1986); Hunter (1980, 1981a)
Dracunculiasis	Edungbola and Watts (1984); Watts (1986, 1987)
Trypanosomiasis	Matzke (1979, 1983)
Dengue hemorrhagic fever	Meade (1976b)
Relapsing fever	Good (1978)
Malaria	Dutt, Akhtar, and Dutta (1980); Dutta and Dutt (1978); Hyma and Ramesh (1984); Hyma, Ramesh, and Chakrapani (1983)
Cholera	Stock (1976)
Influenza	Patterson and Pyle (1983); Pyle (1985, 1986); Pyle and Patterson (1984)
AIDS	Dutt (1987); Wood (1988)
Canine heartworm disease	Haddock (1987)
Tuberculosis	Hunter and Arbona (1984, 1985)
Infant mortality, famine, and malnutrition	Bailey (1988); Bhardwaj and Paul (1986); Currey and Hugo (1984); Hunter (1984); Kloos (1982); Knox (1981a, 1985); Newman (1980); Taylor et al. (1986)

Ecological Factors	Citations
Environmental and socioeconomic	Gesler (1979c); Gesler and Webb (1983); Lenz (1988); Lewis (1984); Meade (1978); Roundy (1985a); Weil (1979)
Economic development	Haddock (1979); Hunter and Thomas (1984); Hunter, Rey, and Scott (1982); Lewis (1984, 1986); Meade (1976a); Stock (1986)
Population movement/migration	Mayer (1980a); Meade (1977); Roundy (1976, 1978)

identify areas of endemicity, or show associative factors. Figure 1, for example, is from research done by Hunter and Thomas (1984, 35) on the association of urbanization with leprosy and tuberculosis in Africa. It illustrates one way in which economic development results in migration corridors that may lead to increased levels of disease transmission.

Other investigators document and provide original research to illustrate the association of infections with poverty and underdevelopment. In a study of guinea-worm prevalence, Watts (1986) demonstrated how diseases of this kind occur most widely in rural areas that lack safe, year-around potable water supplies. Still other geographers provide additional evidence that public works projects, particularly water impoundments, aggravate health risks in the tropical world (Hunter, Rey, and Scott 1982).

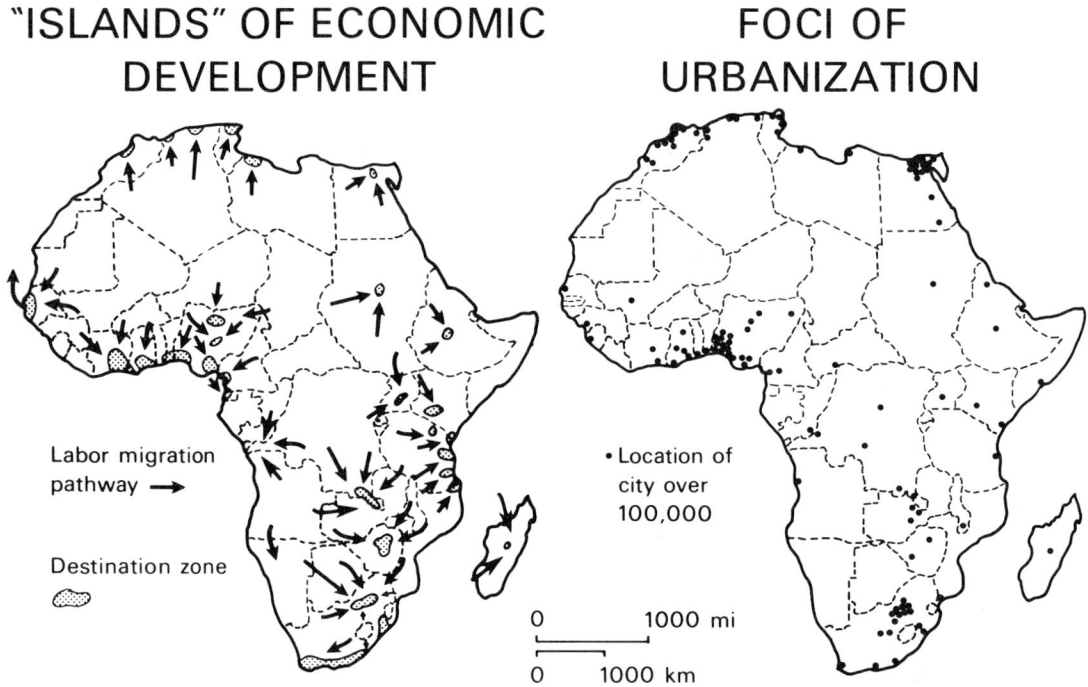

Figure 1
Africa's "islands of economic development" generate corridors that can lead to increased levels of disease transmission. In Africa, rapid urbanization seems to play the role of propagating diseases like tuberculosis. (Reprinted with permission from *Social Science and Medicine,* Volume 19:35, J. M. Hunter and M. Thomas, "Hypothesis of Leprosy, Tuberculosis and Urbanization in Africa," © 1984 Pergamon Press plc.)

Among infectious diseases, acquired immunodeficiency syndrome (AIDS) is doubtless the most important infectious disease of this century. It is the first known instance of a communicable disease that directly attacks the body's defenses, leaving its victims vulnerable to a wide variety of unusual and in some cases untreatable infections and malignancies. To medicine, AIDS has posed many questions, including: What causes it? Where did it come from? How can it be controlled? Certainly it has influenced monogamy, divorce, open practice and the scale of homosexual activities, use of blood transfusions, and the prevalence of drug dependence among the populations affected. To the medical geographer, the paramount questions are: How and where does AIDS spread? How does it affect human behavior in different places?

Geographers are just beginning to plan research on this pandemic, but to date they have had little if any reliable data on which to base their work. The first exploratory publications on the geography of AIDS have just appeared (Dutt 1987; Wood 1988).

Infant Mortality, Famine, and Malnutrition

The continuing search for data with which to tie indicators of health to economic development has led many to employ measures of infant mortality, child survival, and malnutrition (Kloos 1982). Acknowledged as imperfect, in part due to data limitations, infant mortality is a commonly used index of the health status of populations.

North American medical geographers have not generally engaged in research using this index for regional or historical comparisons. Perhaps more surprisingly, as the plight of children in the developing world (14.5 million children under the age of five die annually in the less-developed world) has drawn the attention of social scientists, geographers have not been particularly well represented. There are some notable exceptions, with the work of Knox (1981a, 1985), Taylor et al. (1986), and Bhardwaj and Paul (1986) in rural Bangladesh. This and other (unpublished) work is quite recent, and in response to heightened awareness of the magnitude of the problem.

Bailey (1988) has investigated child morbidity in Jamaica, demonstrating that in countries with high rates of unemployment, the young, uneducated working mothers have offspring with the highest rates of illness. She points out that these women "often have the sole responsibility of providing for themselves and others, [thus] predictably, they raise extremely vulnerable families" (Bailey 1988, 1124).

Malnutrition and the food crisis in Third World development is another topic of importance on which medical geographers have only scarcely written, and which cries out for geographical analysis. The most relevant work is found in the literature of cultural ecology, nutrition, and anthropology. We note the earlier work of Kloos (1982), Newman (1980), and Currey and Hugo (1984) among medical geographers' work on malnutrition.

Chronic Disease

In contrast to contagious diseases, the susceptibility of individuals to chronic disease is usually a direct response to their lifestyles or to environmental factors. Rooted in disease ecology, environmental health, and human/urban ecology traditions, medical geographers are working with other health scientists in unprecedented efforts to characterize the potential adverse health effects of modern lifestyles and exposure to human-made and natural chemical, biological, and physical agents (Table 2).

Cardiovascular Disease. The leading cause of death in the U.S., as well as in many other developed countries, is heart disease. Given the importance of understanding cardiovascular diseases, medical geographers have contributed to the epidemiological literature. Foggin and Godon (1986), for instance, examined relationships between employment-specific cardiovascular mortality and certain spatially based potential risk factors in Quebec. They concluded that there is an ongoing need to monitor certain areas of employment that may be creating unnecessary risks to health, especially in the case of female workers. This thesis was expanded and generalized in Foggin and Thouez (1987).

Geographers have traditionally looked for areas having abnormally high disease rates and for associated explanatory variables. For example, Meade (1983) studied the coastal plain of the southeastern U.S., the so-called "stroke belt." In Savannah, Georgia, she found hypertension to be associated with old city housing, the municipal water supply, high room densities, wharves and warehouses, and industrial surroundings. Lower-density, suburban, and open areas had lower rates. Following up on this research, Gesler and Meade (1988) found that urban ecologic structure in metropoli-

Table 2

Citations since 1976 referring to noninfectious mortality and morbidity and hypothesized ecological correlates

Causes	Citations
Cancers	Armstrong (1980); Armstrong and Armstrong (1983); Armstrong, et al. (1978, 1979, 1983); Foster (1986); Glick (1979a, 1979b, 1980, 1982); Greenberg, Preuss, and Anderson (1980); Lam (1986); Thouez, Beauchamp, and Simard (1981)
Cardiovascular diseases	Foggin and Godon (1986); Foggin and Thouez (1987); Meade (1983); Thouez (1979)
Suicide and homicide	Greenberg, Carey, and Popper (1987)
Multiple sclerosis	Mayer (1981); Laborde, Dando, and Teetzen (1988)
General public health	Greenberg (1986, 1987a, 1987c, 1987d); Shannon (1977)

Ecological Factors	Citations
Hunger and poverty	Currey and Hugo (1984); Kloos (1982)
Environmental toxins	Cutter (1987); Earickson and Billick (1988); Greenberg (1983, 1987b); Greenberg et al. (1982); Greenland and Yorty (1985); Hunter (1976, 1977, 1978)
Sociodemographic	Mayer (1983b); Weil (1981)
Diet and nutrition	Griffith and Innes (1983); Newman (1977, 1980); Weil (1979)

tan Savannah played a more important role in health care-seeking behavior than did the personal characteristics of individuals.

In the inner city, where mortality and morbidity rates are often excessive, geographers also find community characteristics that are correlated with illness. Gesler (1980) and Gesler and Cromartie (1985) found statistical associations between morbidity and poverty and race. Earickson and Billick (1988) also documented the association between high levels of air pollutants and elevated levels of lead in children in the poor inner sections of industrial cities.

Cancer. The second leading cause of death in industrial nations—and growing increasingly serious—is the various forms of cancer. An agent is classified as a hazard if laboratory tests with animals, chemical molecular analyses, and observations of exposed humans show discernible risk that is judged to be unacceptable. Comparison of exposed and unexposed people is one method. For example, the Armstrongs and their associates found that Malaysians who smoke, eat salted fish, consume large quantities of alcohol, use nasal medications, and are exposed to workplace dust and smoke had much higher rates of nasopharyngeal cancer than those without these exposures (Armstrong, Kutty, and Armstrong 1978).

Glick (1980) demonstrated systematic patterns of cancer mortality in the U.S. that suggest possible environmental associations. Kennedy's investigation (1988) of male and female lung-cancer mortality rates in the southeastern U.S. suggests that lung cancer in males there may be occupational in origin. Lam's investigation (1986) of cancer mortality in China suggested that certain kinds of cancer could be associated with diet, while others are possibly associated with air, water, and noise pollution produced by intensive automobile use, and a high density of population in industrialized areas.

Thouez, Beauchamp, and Simard (1981) attempted to account statistically for some of the spatial variation of cancers in Quebec. They, like many other disease ecologists, found it difficult to pin cancer incidence on any specific chemical causes, lamenting that the data did not allow them to ascertain the degree of exposure of the population to contaminants: "Unfortunately, we do not possess data banks on the quality of drinking water (or even of air) which are precise enough temporally and spatially" (Thouez, Beauchamp, and Simard 1981, 221).

Substance Abuse. Alcoholism and drug abuse are other high-priority chronic medical problems in North America. Geographic studies of alcohol abuse are of relatively recent origin. Three major themes are summarized and illustrated by Smith and Hanham (1982). The first is primarily descriptive, identifying regional patterns and variations in alcohol consumption with consideration given to possible cultural determinants. A second includes epidemiological studies which relate prevalence measures to various personal and environmental factors, based on the public-health model. The third theme emphasizes planning and policy implications, building on the findings of research under the first two headings. Since then, the authors have examined the effect of urban lifestyles on the use and abuse of alcohol (Smith and Hanham 1985), and regional variations of problem-drinking in the U.S. (Smith and Hanham 1984).

A series of studies using individual-level survey data by Snow and colleagues (1985, 1986, 1987) examined the spatial correlates of drunk driving. The findings show that drunk drivers are a heterogeneous group, differentiated in terms of sociodemographic and personality characteristics, reasons for drinking, and drinking location. Rooney and Butt (1978) have contributed to the research on the spatial extent and social consequences of nationwide alcohol use and abuse.

From a planning and policy perspective, Smith (1983a) has examined the provision of alcohol-treatment facilities in the wake of the U.S. Uniform Alcoholism and Intoxication Treatment Act of 1971. Using county-level data for Oklahoma, he reveals evidence of the "inverse care law," i.e., the overrepresentation of services in areas of lesser need. More generally, Smith and Hanham (1982) argued for the need to go beyond the limits of the medical model of alcohol use and abuse in developing intervention strategies.

Accidents and Violence. Closely associated with alcohol and drug abuse, and therefore more grist for the medical geographers' mill, is the subject of accidents, injury, and violence. In a very recent paper, Greenberg, Carey, and Popper (1987) wrote on the rapidly increasing youth death rate due to suicide and violence. The authors used mortality data compiled by the National Center for Health Statistics to illustrate death rates at the state level. They associate violence mortality with a number of social, economic, and regional factors. This subject, however, is virtually untouched in the geographic literature, again probably due to a paucity of data at relatively local scales.

Risk Assessment. Another critical area of study is that of risk assessment, particularly the risk associated with environmental pollution, a hazard in both the developed and developing world. At high concentrations, air pollutants can exacerbate respiratory and heart conditions, cause central nervous system damage, and some scientists think they contribute to carcinogenesis. Cutter (1987), Muschett (1981), and Greenland and Yorty (1985) all have documented the geographical distribution of airborne pollutants in different cities.

Studies of people who relocate are often a first step toward identifying hazards. For example, Mayer (1981) became intrigued by the fact that the incidence, prevalence, and mortality rates of multiple sclerosis are lowest at the equator and increase regularly with latitude. Young people who move from high- to low-risk areas typically have low rates, but adult migrants retain the high risk of their original location. No one knows why, but this finding can perhaps lead to fruitful comparative studies. Further research on this disease is under way at the University of North Dakota, in a collaborative effort that includes geographer William Dando.

Most scientists believe that exposure—i.e., number of exposed people—is the weakest link in risk assessment. Medical geographers are helping to fill this void. Hunter (1976) demonstrated widespread exposure of children to roadside lead. This and other papers were part of the evidence used by the U.S. Environmental Protection Agency as justification for phasing out lead from gasoline.

Environmental Toxins. Personal monitoring (e. g., people wearing monitors) is a recent development in air-pollution exposure assessment. Ziegenfus (1987) analyzed the primary data for the U. S. Environmental Protection Agency and found that high exposure is more likely to result from indoor exposure (e.g., dry cleaners; working with someone who smokes) than from breathing outdoor air.

Medical geographers have also contributed to water and land contamination exposure assessments. For example, Greenberg et al. (1982) found that, contrary to widespread belief, high concentrations of toxins are much more likely to be found in groundwater than in surface water. Furthermore, one can usually predict the type of groundwater contamination by knowing nearby land uses. Greenberg (1987b) used hexagonal, square, and rectangular grids and random-search methods to find toxins spreading through soil and groundwater into water supplies. Kloos (1983), working in an area of sharp competition between agriculturalists and environmentalists in California, analyzed spatial and temporal patterns of dichlorobromopropane pesticide contamination of well water in eastern Fresno County, and Thouez (1979) tested hypotheses about the association of water pollution and cancer in Quebec.

Problem Magnitude. Estimating the *magnitude* of a public-health problem is the final and most important step in risk assessment, because it leads to policy formation. Medical geographers have made contributions in developing and Western nations. Meade (1978) found that changing rain forest to modern agriculture decreases enteric infections, but increases the risk of scrub typhus and malaria, as well as the risk of motor-vehicle accidents. Lewis (1986) found that ciguatera—a form of fish poisoning—was a controllable problem among subsistence fishermen of the Pacific Islands, but has become a more serious problem as the islands have urbanized. Hunter, Rey, and Scott (1982) point to human-made lakes as sources of serious diseases in developing nations. Development of Asia, Africa, and Latin America has also increased the burden of some preventable chronic diseases by increasing access to cigarettes and alcohol (Armstrong 1986).

Because of their ability to manipulate large spatial data sets, medical geographers have made important contributions to our understanding of the spread of infectious and chronic diseases in the U.S. Pyle (1986) analyzed the diffusion of pandemics and less-deadly influenza epidemics in the U.S. His research shows why influenza is so difficult to control: strains vary, outbreaks occur in different places under a wide variety of circumstances, and government's response is inconsistent. Greenberg (1987c) found a trend toward homogenization of chronic-disease mortality rates across the U.S., and an association between this trend and the distribution of smoking, hypertension, alcohol abuse, and other behavioral habits.

The reliability of data is always questionable in the study of any disease ecology, and one of the problems, even in the U.S., is faulty reporting of mortality and morbidity (Greenberg 1986). This problem is especially critical in the developing world, as was brought to our attention by Hunter and Arbona (1985) working in Puerto Rico and by the research of Lenz (1988) in Indonesia.

Medical geographers' involvement in risk assessment is an exciting trend. It not only contributes to solving important problems, enables other researchers to recognize the important contributions that geography can make, and requires that we integrate physical and human geography; it also provides opportunities to learn by working with epidemiologists, toxicologists, environmental scientists, and environmental engineers.

MEDICAL AND HEALTH SERVICES LOCATION AND UTILIZATION

The second major focus of medical geography is the provision and location of health services and patient spatial behavior. Although this concern developed later than that of disease ecology, it has enjoyed a fruitful outpouring of literature in the past decade (Table 3). The geographical approach has been rooted deeply in location theory, theories of public-service provision, and transportation geography, including optimization models (Mayer 1983a; Knox 1982). Most of this research has centered around issues of equity, efficiency, social well-being, and community conflict surrounding the location of particular facilities and amenities.

The interface between disease ecology and health services, at least at the intraurban level, was earlier explicated by Shannon (1977). In this research, the author explored the linkages among space, time, neighborhoods, communities with varying risk of disease or injury, the relative concentration of medical practitioners, and patient travel patterns to health services. However, Knox, Bohland, and Shumsky (1984) have more recently addressed the broader questions of the role of social institutions and urbanization on the medical-care system in the U.S., shedding light on the present spatial organization of physician offices and dispensaries. They demonstrated that disparities in accessibility to primary medical care tended to compound many other patterns of urban socioeconomic disparity in both the United Kingdom and the U.S.

Table 3

Citations since 1976 referring to location and utilization of health and medical facilities

Topic	Citations
Developing world	Annis (1981); Armstrong (1984, 1986); Enders (1981); Freeman et al. (1983); Gesler (1979a, 1979b, 1984); Gesler and Gage (1984); Rushton (1984, 1987); Scarpaci (1985, 1987, 1988b); Stock (1983, 1985, 1987)
Traditional and alternative medicine	Bhardwaj (1980); Bhardwaj and Paul (1986); Good (1977, 1980, 1987a, 1987b); Good et al. (1979); Hyma and Ramesh (1985); Ramesh and Hyma (1981)
Emergencies and accidents	Bianchi and Church (1988); Brodsky (1984); Brodsky and Hakkert (1985); Mayer (1979 1980b)
Pharmacies	Knox (1981b); Selya (1988)
Physicians and clinics	Gesler (1988); Gober and Gordon (1980); Shumsky, Bohland, and Knox (1986); Henry (1978)
Hospitals	Bohland (1984); Gesler and Cromartie (1985); Ingram, Clarke, and Murdie (1978); Joseph (1986); McLafferty (1982, 1986, 1988); Mayer (1983a); Mayer, et al. (1987); Scarpaci (1988a)
Health maintenance organizations	Cromley and Shannon (1983)
Mental health and substance abuse	Dear (1977a, 1977b, 1977c, 1978, 1981, 1984); Dear and Taylor (1982); Dear and Wolch (1979, 1987); Dear, Clark, and Clark (1979); Hall (1988); Hall and Taylor (1983); Hall et al. (1987); Hunter (1987); Hunter, Shannon, and Sambrook (1986); Joseph (1979); Joseph and Broeckh (1981); Kearns, Taylor, and Dear (1987); Miller, Dear, and Streiner (1986); Rooney and Butt (1978); Smith (1976, 1983a, 1983b, 1986); Smith and Hanham (1981a, 1981b, 1982, 1984, 1985); Snow and Cunningham (1985); Snow and Landrum (1986); Snow and Wells-Parker (1986); Snow and Anderson (1987); Taylor and Dear (1980); Taylor et al. (1984)
Elderly health	Bohland and Frech (1982); Cromley and Shannon (1986)
Ecological factors in utilization	Gesler (1986); Gesler and Meade (1988); Gordon (1987); Gordon and Stephenson (1980); Gordon et al. (1987); Joseph (1982, 1987); Joseph and Bantock (1984); Joseph and Phillips (1984); Kennedy (1988); Knox (1979a); Pyle (1985); Schneider (1986); Shannon and Spurlock (1976); Shannon, Bashshur, and Spurlock (1978)
History, planning, and policy	Bohland and Shumsky (1983); Knox (1978, 1979b, 1980 1982); Knox, Bohland, and Shumsky (1984); Mayer (1986); Meade (1986); Rosenberg (1988)

Rosenberg (1988) argues that an "intellectual cul-de-sac" has been reached because of the lack of linkage in methodologies used to analyze health-care delivery systems that explicitly recognize the sociocultural and political-economic influences in the environment. To illustrate his thesis, Rosenberg uses the example of abortion services in Canada in general, and Ontario specifically.

Geographers who investigate facilities location and travel behavior may focus on one or more of the major actors in the system: the patient, or the practitioner, or the institution. This research usually involves statistical relationships or mathematical modeling of the patient-provider system. Examples of this approach are provided in the writings of Bohland (1984), Hall (1988), Joseph (1987), McLafferty (1988), Mayer (1983b), and Rushton (1984).

For convenience, this material will be disaggregated into sections that examine the research on (1) primary-care provider location and hospital characteristics and location, and (2) mental-health facilities location and community response.

Primary and Secondary Medical Care

Primary providers include physicians, nurses, native doctors, and in some cases, hospital emergency rooms. Primary health care is not only fundamental to the operation of effective medical-care systems; it is also arguably the most cost-efficient component of that system (Joseph and Phillips 1984). Secondary care includes hospitals and outpatient surgeries, and in developing countries may include clinics that provide both traditional and biomedical treatment.

General practitioners are the key element in rural health-care delivery in developed nations, through their provision of basic care and their control over referral to higher levels of care. However, these providers have a tendency to concentrate in larger settlements, putting the dispersed rural population at a distinct and lasting disadvantage (Joseph and Bantock 1984).

In rural Third World settings, where most patients travel on foot and few have access to transport, distance significantly affects the utilization of health-care services (Gesler 1984). Figure 2 illustrates proximity to health facilities by the populace of two departments (provinces) in rural Guatemala. The author depicts the average range of patients from facilities with the use of central-place hexagons. In rural Nigeria, patients living some distance from health-care facilities tend to delay using their services, usually preferring to use some alternate form of treatment, including self-treatment with traditional or patent medicines (Stock 1983). In both rural Nigeria and rural Guatemala, quality of service appears to be at least as important a determinant of facility utilization.

Despite local notions of disease etiology and perceptions of medical alternatives, the basic fact remains that the bottom line for the patient is to *get cured*. This may be an enormously complex concept culturally, but it is usually relatively clear-cut biologically. As Annis (1981, 521) puts it, "In the long run, those alternatives with the best curing track record (in the 'now it doesn't hurt' sense) will probably be the ones that the patient returns to, particularly if these alternatives are also inexpensive and socially dignified."

Multiple types of medical care and disease prevention exist everywhere in the Third World, and people choose among them according to prevailing notions of efficacy. These multiple systems are to a large degree the result of diffusion—e.g., the spread of Islam and its Greek-based systems, the introduction of traditional Chinese medical beliefs with Chinese migration to Southeast Asia, and the introduction of African techniques and ideas to the Americas—or by cultural imperialism and con-

ASSESSMENT AND MANAGEMENT OF HAZARDS AND INFIRMITY

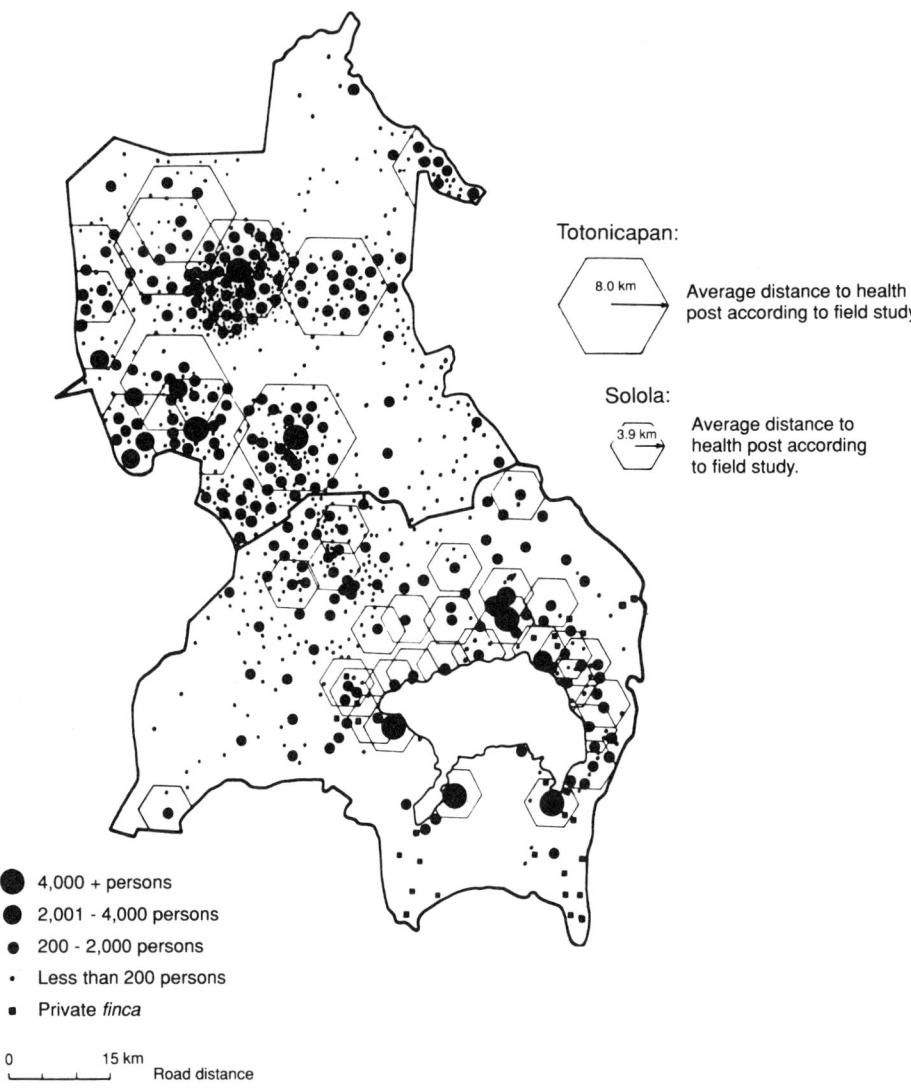

Figure 2
Population falling within 3.9 km and 8.0 km radii of existing Ministry of Health facilities in two departments (provinces), Totonicapan and Solola, Guatemala (Reprinted with permission from *Social Science and Medicine*, Volume 15D, S. Annis, "Physical Access to Utilization of Health Services in Rural Guatemala," © 1981 Pergamon Press plc.)

quest, as with the introduction of European medicine to African peoples or to Native Americans.

The use of traditional Chinese medicine, as well as Western (or scientific) medicine, has greatly lowered mortality in China and has had an immense influence on international attitudes toward the legitimacy of traditional medicine. There has been a shift from the perspective that it is quackery and anathema to one of credibility, although it can be harmful if it promotes neglect of scientific care.

Geographic research focuses upon the transference of medical systems through space and upon their competitive uses when rural people are urbanized or when people in less developed nations (with their more holistic concepts) encounter practitioners of scientific Western medicine (which is highly specialized). The latter is also focused upon organs, while the former highlights relationships, stress, and behavior. Most of the important work on alternative medicine in the social sciences emanates from anthropology. This is not to say that geographers have neglected this aspect; many of the works mentioned in this section and under disease ecology relate to alternative medicine. The major recent publication is Good's book (1987a). Major contributors other than those already mentioned include Hyma (1985), Weil (1981), and Ramesh and Hyma (1981).

In the medical-care systems of developed countries, the urban physician office location plays an important role. Shumsky, Bohland, and Knox (1986) trace the changing locations of doctors' offices in San Francisco between 1881 and 1941 to illustrate how the urban transformations that occurred late in nineteenth-century America affected medical practitioners. In most U.S. cities, people who reside near downtown are more likely to be found near the bottom of the socioeconomic ladder. Although there are hospitals in these communities, physicians often have fled to the suburbs, creating a "medical wasteland" for residents, forcing them to seek primary care in hospital emergency rooms (Knox 1980). This means that restricting primary care in emergency rooms or closing inner-city hospitals and shifting hospital functions to suburban locations could, in the absence of other alternatives, seriously reduce the ability of residents from lower-status neighborhoods to secure medical care (Bohland 1984).

Physicians tend to locate their offices where it is both profitable and convenient for them, if not for their patients. This results in an inequitable distribution of private-practice physicians in many metropolitan areas (Gober and Gordon 1980). Originally, it was the Health Maintenance Organization (HMO) that was supposed to partially fill this vacuum in the inner city and other medically underserved areas. However, HMOs must be economically viable to survive; hence they tend to be found most often in the suburbs or at the fringe of metropolitan areas. Cromley and Shannon (1983) conclude that HMO development, nationally, has been concentrated in medical resource-rich areas.

Some medical geographers have investigated the potential role of primary-care providers who are not presently licensed to practice biomedicine. Chiropractors and pharmacies have been suggested as alternatives. Gesler (1988) examined the distribution of chiropractors in North Carolina, finding that they tended to locate in small towns and rural areas having white, low-income populations, particularly where certain religions were prevalent. Selya (1988) explored the feasibility of adding another actor to the expanded network of medical personnel. He argued that if the distribution of pharmacies is more equal than the distribution of doctors, then pharmacies might play an effective role in providing more comprehensive medical care, especially in medically underserved areas.

Geographers have for decades explored spatial patterns of patient travel for hospital care, as well as the location of hospitals themselves. Within our time frame, Mayer's research (1983a) found that the distance patients travel to hospitals in Rhode Island varies depending on their particular diagnoses. Gesler and Cromartie (1985) used data from New York City to demonstrate the distance-decay effect of hospital utilization.

A larger issue in the recent hospital literature is that of closure of institutions that are not economically viable. McLafferty (1982, 1986, 1988) has probably done more in this area than any other geographer. Hospitals in New York City that have been forced to close are most often those in low-income, high-infant-mortality neighborhoods. This is a serious loss in these communities because low-income people have higher rates of emergency-room use for primary care than the more affluent population.

In many cases, it is no less serious for a rural hospital to close. For-profit hospitals are much more likely to close in rural areas than are community hospitals, as there is little incentive for the owners to keep unprofitable institutions open. "This reflects a fundamental philosophical difference between community and for-profit hospitals. While the services which both provide may be similar or identical, for-profit hospitals may view their services as economic entities, while community hospitals, which are still concerned with financial viability, are viewed as providing public services" (Mayer et al. 1987, 333). On the other hand, it is not clear that closure of a hospital is necessarily detrimental to a community or to the health of its residents. If the quality of the referent hospital is inferior, Mayer argues, the community might be better off to lose that institution.

Mental-Health Care

Deinstitutionalization, the move from hospital to community-based treatment of the chronically mentally disabled (CMD), has been the catalyst for a range of studies dealing with the geography of mental-health care. Four main issues have been addressed:

- Public attitudes towards the CMD and the facilities and services introduced to meet their needs
- The effects of community settings on the coping ability of clients
- Patterns of utilization of psychiatric services
- The broader issue of the implementation and impacts of community mental-health care policies

Geographers have been in the forefront in investigating the sets of factors that together affect public attitudes at both the individual and neighborhood scale. At the individual level, there is strong evidence of widespread sympathy in principle for the needs of the CMD (Dear and Taylor 1982), but it remains unclear to what extent this translates into support in practice. Several analyses have explored the determinants of location of facilities and attitudes towards clients and facilities, e.g., using path analysis and regression to estimate the effects of a wide range of personal and situational factors (Smith and Hanham 1981a; Hall and Taylor 1983). The effect of geographic propinquity to facilities has received special attention with the consistent finding that the presence of a facility engenders more positive attitudes; however, residents in closest proximity are the most likely to demonstrate opposition (Dear and Taylor 1982).

At the neighborhood scale, spatial variations in the prevalence of opposition to clients and facilities are more apparent. In simplest terms, there is a central city–suburban gradient in tolerance (Dear and Wolch 1987). The profiles of accepting and rejecting neighborhoods reflect this spatial variation, with stronger opposition associated with areas of social stability and higher socioeconomic and family status (Taylor et al. 1984).

Studies of the effects of the community environment on the coping ability of clients are complementary to the research just described. Smith (1976) identified general ecological characteristics that enhanced clients' chances for community involvement, and thereby their recuperation potential. Dear and Wolch (1987) listed housing, employment, finances, and social and psychiatric support as fundamental dimensions of coping. The effects of these factors on hospital readmission and the client's subjective experience in the community have been the focus of recent detailed investigation (Kearns, Taylor, and Dear 1987). Housing has been identified as a particularly important factor, and is now prompting applied research by geographers on the development of community-housing programs (Hall, Nelson, and Smith-Fowler 1987).

Research on the utilization of mental-health care facilities has largely followed in the tradition of the seminal work by Jarvis, seeking to determine the effects of distance on both hospital admissions and, more recently, the use of small-scale community facilities. Historical analyses point to the spatial and temporal persistence of distance-decay effects in asylum admissions in North America through the nineteenth century (Hunter, Shannon, and Sambrook 1986). Analysis of current data at the regional and urban scale shows that distance-decay effects persist, but are modified by various ecological factors (Miller, Dear, and Streiner 1986; Hall 1988) and individual-level factors, such as referral procedures (Dear 1977b; Joseph 1979) and types of diagnosis (Joseph and Boeckh 1981).

Growing interest among geographers in the interrelations between society and space has been reflected in the mental-health field by analysis of broader policy issues. Good examples are Smith's evaluation (1983b) of community mental-health policy in the U.S. between 1965 and 1980, or the focus of Dear (1981) on policy from a social-theory perspective and the recent analysis of the relations between chronic mental disability and homelessness (Dear and Wolch 1987).

Research Gaps and Priorities for the Future

Arguably the most important immediate area of concern today is the contagious and—so far—incurable disease AIDS. Apart from the certainty of death that accompanies AIDS, the means of its transmission, the spatial patterns of its diffusion, and the groups at highest risk of infection vary considerably among world regions. Active AIDS cases are only the tip of the iceberg; a much larger number of people are estimated to be infected with the Human Immunodeficiency Virus (HIV) that causes AIDS, though this "hidden" AIDS population shows no symptoms, and again, no one knows what proportion will ultimately develop the disease.

All fields of inquiry are frustrated with the genuine paucity of data on this epidemic. Little is known about the statistics of AIDS prevalence, but without these data, health providers will be hard-pressed to determine how AIDS diffuses through the population and how to concentrate resources to fight it. A national random survey of some 50,000 persons, to be conducted by the National Center for Health Statistics in 1989 (using questionnaires and taking blood samples), will be a positive, but limited, step in the right direction.

The NCHS survey has already proven controversial. AIDS is a very sensitive issue everywhere, because of its potential to expose individuals to alienation and/or loss of livelihood. From the geographer's point of view, NCHS survey data are problematic because they provide only a "snapshot" in time and are disaggregated only to a four-region level. To model diffusion successfully, cases of AIDS reported by county over several years are required. These data do exist in Centers for Disease Control data banks, but they are not, as of this writing, available to anyone outside CDC. If the epidemic continues unabated, however, research on AIDS diffusion should become less constrained and better funded. Again, multidisciplinary research will be required to analyze the unique dynamics of AIDS and the policies and programs aimed at curbing its spread.

Geographers interested in health problems in the developing world must work simultaneously on three interrelated fronts, according to Stock (1986):

> First, they must familiarize themselves with advances in the literature on underdevelopment theory.
> Second, there is a large agenda for empirical research involving not only the reassessment of popular "disease and development" topics, such as the consequences of water-development projects and their health implications, but also the investigation of diverse topics on the health-underdevelopment interface about which geographers have only recently shown interest (Table 4).
> Third, Stock calls for the development of a unified theory of the relationship between underdevelopment and health in the Third World. This theory would advance the understanding of the varying health manifestations of underdevelopment in different regions and different social classes.

Another subject that requires additional attention is that of child health and nutrition, in both developed and developing countries. The data that exist in developed nations are at either the state level or some other gross administrative level. The National Institute of Health has the only major nutritional survey with information that could be of use, but its geographic resolution is to only four regions of the U.S.! Nutritional information is extremely expensive to collect (e. g., blood analyses and dietary studies with individual subjects for days, which is typical in anthropological field work). And, it is available only for very small and select populations in, say, one village, neighborhood, or health center. If medical geographers are to become involved with nutrition beyond the field study/cultural ecology of diets and crops developed by Jacques May, they must become involved in creating data sets through cooperation with nutritionists.

Recently there has been a call for a national nutritional monitoring program in the U.S. If such a program is forthcoming, it would be most helpful if a finer resolution of geographic boundaries could be imposed at the outset. As methodologies to investigate the spatial aspects of child survival continue to be developed, geographers might do well to take note of the charge set by the Bellagio conference, which resulted in the volume *Child Survival: Strategies for Research* (Mosley and Chen 1984), and to develop conceptual frameworks for bridging the gap between biomedical and social-science approaches to the study of mortality in human populations.

Another area of research that would naturally fit into the realm of medical geography is crime and violence. Law-enforcement data have been used to demonstrate

Table 4
Selected themes for research on health and underdevelopment

Aspect	Research Theme
Colonialism	1. Health and underdevelopment under colonial rule.
	2. The impact of health hazards, both tangible and perceived, on the pattern of colonial development.
	3. The sanitation syndrome: health considerations in the structuring of colonial cities.
Social conditions	1. Apartheid and health in South Africa.
	2. Health status and social status of women.
	3. Class: who gets what, where, and why.
Nutrition	1. Commerciogenic malnutrition: infant formula and other processed foods.
	2. Health consequences of agricultural modernization (green revolution; agricultural development projects).
	3. Export-oriented agriculture and the underdevelopment of health.
Technology	1. The export of noxious industries to the Third World.
	2. Occupational safety and health: miners, industrial workers, and agricultural laborers.
	3. Pesticides and herbicides in the Third World.
Health care	1. Iatrogenic health care in underdeveloped countries.
	2. Health consequences of gross inequalities in access to health services.
	3. Pharmaceutical dumping.
Global politics	1. The militarization of the Third World: health linkages.
	2. Evaluating aid policies and their ultimate health impacts.
	3. Strategies for liberation.

Source: Stock (1986, 697)

the locations of criminal and violent activity, but data showing the impact of such activity upon health and health facilities are more difficult to find, if they exist at all. Surgeon General Koop, the National Institute of Mental Health, and the Centers for Disease Control have all shown an active interest in the effects of violence.

Various perspectives incorporated into the public-health approach are consistent with paradigms in medical geography. First, public-health researchers are prone to make regional and international comparisons in order to gauge the relative status of cities or states. Second, analyses tend to express comparisons in terms of "relative risk," a concept that is a simple yet effective expository device, and one that lends itself to geographic applications. Before substantial progress can be made, however, standardized databases must be obtained from sources such as emergency rooms, school systems, social services, and medical examiners.

There seem to be three directions in which research on the geography of mental health will continue:

- ☐ The urgent need to better understand the complex of factors that affect coping and quality of life for the chronically mentally disabled in the community will generate further studies with both basic and applied goals in view.
- ☐ The growing emphasis on the relations between society and space will likely continue to find an empirical footing in the mental-health field, not least be-

cause some of the most obvious sociospatial inequalities in the contemporary city involve disabled groups such as the mentally disabled.

- A strongly applied focus can be anticipated with attention given to such pressing issues as housing the CMD in a shrinking low-income housing market.

The avenues of inquiry pursued by geographers and allied researchers in the context of alcoholism are equally applicable to other aspects of substance abuse, including drug and tobacco consumption. Given that the burden of illness in North America is heavily weighted towards lifestyle-related chronic morbidities, it is reasonable to anticipate an extension of geographic research on alcohol and substance abuse. In this context, geographers are likely to be participants in multidisciplinary research of prevalence, etiology, and interventions founded on a socioecological model of health and health policy.

Finally, medical geographers will be researching a host of other important subject areas. These include, but are not limited to:

- Use of traditional and alternative practitioners in primary health care (e.g., midwives, chiropractors, etc.)
- Bridging the gap between disease studies and medical care
- Environmental hazards, including the effects of industrial pollution on health, and the export of banned chemicals
- Effects of tobacco and drug production and marketing practices on health in less-developed nations
- Disease and famine as natural hazards
- Diseases of unknown etiology, such as Parkinsonism and multiple sclerosis, which are linked to unidentified but strongly suspected environmental toxins or have marked geographical patterns
- Diffusion of health innovations, including family planning

Medical geographers are, of course, involved in issues that literally are of life-or-death importance. Unfortunately, as they take very small steps forward, they often have an uncomfortable sense of futility. It was there for all to see on international television, for example, how well-intentioned efforts to bring massive supplies of food and medicine to the famine-ridden Sahel became mired in greed, selfishness, bureaucratic muddle, and ignorance.

New patterns of living, increased crowding and environmental contamination, an aging population, and the continued social and geographic mobility of the population in the industrialized world contribute to variants of ill health and shifts in demand for health care. But, growth, change, and uncertainty present opportunities for medical geography to play a more active role in policy formation and in the maintenance (or creation) of a healthful environment.

REFERENCES

Annis, S. 1981. Physical access to utilization of health services in rural Guatemala. *Social Science and Medicine* 15D:515–23.

Armstrong, R. W. 1980. Geographical aspects of cancer incidence in Southeast Asia. *Social Science and Medicine* 14D:299–306.

———. 1984. Health status and differential use of medical resources: A survey sample in Kuala Lumpur. *Medical Journal of Malaysia* 39:257.

———. 1986. Tobacco and alcohol use among urban Malaysians in 1980. *International Journal of the Addictions* 20:1803–08.

———, and Armstrong, M. J. 1983. Environmental risk factors and Nasopharyngeal Carcinoma in Selangor, Malaysia: A cross-ethnic perspective. *Ecology and Disease* 2:185–98.

———; Kutty, M. K; and Armstrong, M. J. 1978. Self-specific environment associated with Nasopharyngeal Carcinoma in Selangor, Malaysia. *Social Science and Medicine* 12:149–56.

———, et al. 1979. Incidence of Nasopharyngeal Carcinoma in Malaysia, 1968–1977. *British Journal of Cancer* 40:557–80.

———, et al. 1983. Salted fish and inhalants as risk factors for Nasopharyngeal Carcinoma in Malaysian Chinese. *Cancer Research* 43:2967–70.

Bailey, W. 1988. Child morbidity in the Kingston metropolitan area, Jamaica 1983. *Social Science and Medicine* 26:1117–24.

Bhardwaj, S. M. 1980. Medical pluralism and homeopathy: A geographic perspective. *Social Science and Medicine* 14B:209–16.

———, and Paul, B. K. 1986. Medical pluralism and infant mortality in a rural area of Bangladesh. *Social Science and Medicine* 23:1003–10.

Bianchi, G., and Church, R. L. 1988. A hybrid FLEET model for emergency medical service system design. *Social Science and Medicine* 26:163–72.

Bohland, J. 1984. Neighborhood variations in the use of hospital emergency rooms for primary care. *Social Science and Medicine* 19:1217–26.

———, and Frech, P. 1982. Spatial aspects of primary health care for the elderly. In *Geographical perspectives on the elderly,* ed. A. M. Warnes, pp. 339–54. New York: Wiley.

———, and Shumsky, L. 1983. The urban transition and the evolution of the medical care delivery system in America. *Social Science and Medicine* 17:34–43.

Brodsky, H. 1984. The bystander in highway injury accidents. *Social Science and Medicine* 11:1212–16.

———, and Hakkert, A. S. 1985. Accessibility and bystander response in an emergency. *Transactions of the Institute of British Geographers* 10:303–16.

Cromley, E. K., and Shannon, G. W. 1983. The establishment of health maintenance organizations: A geographical analysis. *American Journal of Public Health* 73:184–87.

———, and ———. 1986. Locating ambulatory medical care facilities for the elderly. *Health Services Research* 21:499–514.

Currey, B., and Hugo, G. 1984. *Famine as a geographical phenomenon. Geo-journal library 1*. Dordrecht: D. Reidel.

Cutter, S. 1987. Airborne toxic releases: Are communities prepared? *Environment* 29:12–17.

Dear, M. J. 1977a. Impact of mental health facilities upon property values. *Community Mental Health Journal* 13:150–57.

———. 1977b. Locational factors in the demand for mental health care. *Economic Geographer* 53:223–40.

———. 1977c. Psychiatric patients and the inner city. *Annals of the Association of American Geographers* 67:588–94.

———. 1978. Planning for mental health care: A reconsideration of public facility location theory. *International Regional Science Review* 3:93–112.

———. 1981. Social and spatial reproduction of the mentally ill. In *Urbanization and urban planning in capitalistic society,* eds. M. J. Dear and A. J. Scott, pp. 481–90. London: Methuen.

———. 1984. Health services planning: Searching for solutions in well-defined places. In *Planning and analysis in health care systems,* ed. M. Clarke, pp. 7–21. London: Pion.

———, and Wolch, J. 1979. The optimal assignment of human service clients to treatment settings. In *Location and environment of elderly population,* ed. S. Golant, pp. 197–210. Washington: V. H. Winston.

———, and ———. 1987. *Landscapes of despair: From deinstitutionalization to homelessness*. Princeton: Princeton University Press.

———; Clark, S.; and Clark, G. 1979. Economic cycles and mental health care policy: An examination of the macro-context for social service planning. *Social Science and Medicine* 13D:43–53.

———, and Taylor, S. M. 1982. *Not on our street: Community attitudes to mental health care*. London: Pion.

Dutt, A. K. 1987. Geographical patterns of AIDS in the United States. *Geographical Review* 77:468–83.

———; Akhtar, R.; and Dutta, H. M. 1980. Malaria in India with particular reference to two west-central states. *Social Science and Medicine* 14D:317–30.

Dutta, H. M., and Dutt, A. K. 1978. Malarial ecology: A global perspective. *Social Science and Medicine* 12:69–84.

Earickson, R. J., and Billick, I. H. 1988. The areal association of urban air pollutants and residential characteristics: Louisville and Detroit. *Applied Geography* 8:5–23.

Edungbola, L., and Watts, S. J. 1984. An outbreak of dracunculiasis in a peri-urban community of Ilorin, Kwara State, Nigeria. *Acta Tropica* 41:155–64.

———; Asaolu, A.; and Watts, S. 1986. The status of human onchocerciasis in the Kainji Reservoir Basin areas 20 years after the impoundment of the lake. *Tropical and Geographical Medicine* 38:226–32.

Enders, W. T. 1981. Subjective evaluation and utilization by low-income residents in Porto Alegre, Brazil. *Social Science and Medicine* 15D:525–36.

Foggin, P., and Godon, D. 1986. Cardiovascular mortality as it relates to the geographic distribution of employment in non-metropolitan Quebec. *Social Science and Medicine* 22:559–69.

———, and Thouez, J-P. 1987. The social ecology of cardiovascular disease and mortality. In *Perspectives in urban geography,* ed. C. S. Yadev. New Delhi: K. K. Agencies Publishers.

Foster, H. D. 1986 *Reducing cancer mortality: A geographical perspective*. Western Geographical Series No. 23. Victoria, B. C.: University of Victoria Department of Geography.

Freeman D. H., et al. 1983. A categorical data analysis of contacts with the family health clinic, Calabar, Nigeria. *Social Science and Medicine* 17:571–78.

Gesler, W. M. 1979a. Barriers between people and health practitioners in Calabar, Nigeria. *Southeastern Geographer* 19:27–41.

———. 1979b. Illness and health practitioner use in Calabar, Nigeria. *Social Science and Medicine* 13:23–30.

———. 1979c. Measurement of morbidity in household surveys in developing areas. *Social Science and Medicine* 13D:223–26.

———. 1980. Spatial variations in morbidity and their relationship with community characteristics in central Harlem health district. *Social Science and Medicine* 14D:387–96.

———. 1984. *Health care in developing countries*. Washington: Association of American Geographers.

———. 1986. The uses of spatial analysis in medical geography: A review. *Social Science and Medicine* 23:963–74.

———. 1988. The place of chiropractors in health care delivery: A case study of North Carolina. *Social Science and Medicine* 26:785–92.

———, and Webb, J. L. 1983. Patterns of mortality in Freetown, Sierra Leone. *Singapore Journal of Tropical Geography* 4:99–118.

———, and Gage, G. 1984. Health care delivery for under-fives in rural Sierra Leone. In *Health and disease in tropical Africa: Geographical and medical viewpoints,* ed. R. Akhtar, pp. 427–68. London: Harwood Academic Publishers.

———, and Cromartie, J. B. 1985. Patterns of illness and hospital use in Central Harlem health district. *Journal of Geography* 84:211–16.

———, and Meade, M. 1988. Locational and population factors in health care seeking behavior in Savannah, Georgia. *Health Services Research* 23:443–62.

Glick, B. 1979a. The spatial autocorrelation of cancer mortality. *Social Science and Medicine* 13D:123–30.

———. 1979b. Distance relationships in theoretical models of carcinogenesis. *Social Science and Medicine* 13D:253–70.

———. 1980. The geographic analysis of cancer occurrence: Past progress and future directions. In *Conceptual and methodological issues in medical geography,* ed. M. S. Meade, pp. 170–93. Chapel Hill: University of North Carolina Department of Geography.

———. 1982. The spatial organization of cancer mortality. *Annals of the Association of American Geographers* 72:471–81.

Gober, P., and Gordon, R. J. 1980. Intraurban physician location: A case study of Phoenix. *Social Science and Medicine* 14D:407–17.

Good, C. M. 1977. Traditional medicine: An agenda for medical geography. *Social Science and Medicine* 11:705–13.

———. 1978. Man, milieu and the disease factor: Tick-borne relapsing fever in East Africa. In *Disease in African history,* eds. G. Hartwick and K. Patterson. Durham, NC: Duke University Press.

———. 1980. Ethnomedical systems in Africa and the LDCs: Key issues for the geographer. In *Concep-*

tual and methodological issues in medical geography, ed. M. S. Meade, pp. 93–116. Chapel Hill: University of North Carolina Department of Geography.

———. 1987a. *Ethnomedical systems in Africa*. New York: Guilford Press.

———. 1987b. Community health in tropical Africa: Is medical pluralism a hindrance or a resource? In *Health and disease in tropical Africa: Geographical and medical viewpoints*, ed. R. Akhtar. London: Harwood Academic Publishers.

———, et al. 1979. The Interface of dual systems of health care in the developing world: Toward health policy initiatives in Africa. *Social Science and Medicine* 13D:141–54.

Gordon, R. J. 1987. *Arizona rural health provider atlas*. Tucson, AZ: University of Arizona College of Medicine.

———, and Stephenson, L. K. 1980. Medical deprivation and area deprivation: The case of Phoenix 1960 to 1970. *Proceedings of applied geography conference* 3:286–97.

———, et al. 1987. The effect of malpractice liability on the delivery of rural obstetrical care. *Journal of Rural Health* 3:7–13.

Greenberg, M. R. 1983. Environmental toxicology in the United States. In *Geographical aspects of health*, eds. N. D. McGlashan and J. R. Blunden, pp. 157–74. London: Academic Press.

———. 1986. Disease competition as a factor in ecological studies of mortality: The case of urban centers. *Social Science and Medicine* 23:929–34.

———. 1987a. *Public health and the environment*. New York: The Guilford Press.

———. 1987b. Sampling strategies for finding contaminated land. *Applied Geography* 7:197–202.

———. 1987c. The changing geography of major causes of death among middle age White Americans, 1939–1981. *Socio Economic Planning Sciences* 21:223–28.

———. 1987d. Urban-rural differences in behavioral risk factors for chronic diseases. *Urban Geography* 8:146–51.

———; Preuss, P. W.; and Anderson, R. 1980. Clues for case control studies of cancer in the northeast urban corridor. *Social Science and Medicine* 14D:37–43.

———, et al. 1982. Empirical test of the association between gross contamination of wells with toxic substances and surrounding land use. *Environmental Science and Technology* 16:14–19.

Greenberg, M.; Carey, G. W.; and Popper, F. J. 1987. Violent death, violent states, and American youth. *Public Interest* 87:38–48.

Greenland, D., and Yorty, R. 1985. The spatial distribution of particulate concentrations in the Denver metropolitan area. *Annals of the Association of American Geographers* 75:69–82.

Griffith, P., and Innes, F. 1983. The relationship of socio-economic factors to the use of vitamin supplements in the city of Windsor. *Nutrition Research* 3:445–55.

Haddock, K. 1979. Disease and development in the tropics: A review of Chagas' disease. *Social Science and Medicine* 13D:53–60.

———. 1981. Control of schistosomiasis: The Puerto Rican experience. *Social Science and Medicine* 15D:501–14.

———. 1987. Canine heartworm disease: A review and pilot study. *Social Science and Medicine* 24:225–46.

Hall, G. B. 1988. Monitoring and predicting community mental health centre utilization in Auckland, New Zealand. *Social Science and Medicine*. 26:55–70.

———, and Taylor, S. M. 1983. A causal model of attitudes toward mental health facilities. *Environment and Planning A* 15:525–42.

———; Nelson, G.; and Smith-Fowler, H. 1987. Housing for the chronically mentally disabled: Part 1. Conceptual framework and social context. *Canadian Journal of Community Mental Health* 6:65–78.

Henry, N. F. 1978. The diffusion of abortion facilities in the northeastern United States, 1970–1976. *Social Science and Medicine* 12:7–15.

Hunter, J. M. 1976. Aerosol and roadside lead as environmental hazard. *Economic Geography* 52:147–60.

———. 1977. The summer disease: An integrative model of the seasonality aspects of childhood lead poisoning. *Social Science and Medicine* 11:691–703.

———. 1978. The summer disease: Some field evidence on seasonality in childhood lead poisoning. *Social Science and Medicine* 12:85–94.

———. 1980. Strategies for the control of river blindness. In *Conceptual and methodological issues in medical geography*, ed. M. S. Meade, pp. 38–76.

Chapel Hill: University of North Carolina Department of Geography.

———. 1981a. Progress and concerns in the World Health Organization onchocerciasis control program in Africa. *Social Science and Medicine* 15D:261–75.

———. 1981b. Past explosion and future threat: Exacerbation of red water disease (schistosomiasis haematobium) in the Upper Volta region of Ghana. *Geojournal* 5:305–13.

———. 1984. Insect clay geophagy in Sierra Leone. *Journal of Cultural Geography* 4:2–13.

———. 1987. Need and demand for mental health care: Massachusetts 1854. *Geographical Review* 77:139–56.

———; Rey, L.; and Scott, D. 1982. Man-made lakes and man-made diseases: Toward a policy resolution. *Social Science and Medicine* 16:1127–45.

———, and Arbona, S. 1984. Disease rate as an artifact of the public health care system: Tuberculosis in Puerto Rico. *Social Science and Medicine* 19:997–1008.

———, and ———. 1985. Field testing along a disease gradient: Some geographical dimensions of tuberculosis in Puerto Rico. *Social Science and Medicine* 21:1023–42.

———, and Thomas, M. 1984. Hypothesis of leprosy, tuberculosis and urbanization in Africa. *Social Science and Medicine* 19:27–57.

———; Shannon, G. W.; and Sambrook, S. L. 1986. Rings of madness: Service areas of 19th century asylums in North America. *Social Science and Medicine* 23:1033–50.

Hyma, B., and Ramesh, A. 1980. The reappearance of malaria in Sathanur Reservoir and environs: Tamilnadu. *Social Science and Medicine* 14D:337–44.

———, and ———. 1985. Siddha system of health care: A profile of users in Madras, India. *Transactions of the Institute of Indian Geographers* 7:17.

———; Ramesh, A.; and Chakrapani, K. 1983. Urban malaria control situation and environmental issues: Madras City, India. *Ecology of Disease* 2:321–35.

Ingram, D. R.; Clarke, D. R.; and Murdie, R. A. 1978. Distance and the decision to visit an emergency department. *Social Science and Medicine* 12:55–62.

Joseph, A. E. 1979. The referral system as a modifier of distance decay effects in the utilisation of mental health care services. *Canadian Geographer* 23:159–69.

———. 1982. On the interpretation of the coefficient of localization. *Professional Geographer* 34:443–46.

———. 1986. The accessibility of public hospital services in Auckland: A geographic study. *New Zealand Geographer* 42:11–17.

———. 1987. Measuring access to health services: An experiment with spatial indices. *East Lakes Geographer* 22:10–20.

———, and Boeckh, J. L. 1981. Locational variation in mental health care utilization dependent upon diagnosis: A Canadian example. *Social Science and Medicine* 15D:395–404.

———, and Bantock, P. R. 1984. Rural accessibility of general practitioners: The case of Bruce and Grey Counties, Ontario, 1901–1981. *Canadian Geographer* 28:226–39.

———, and Phillips, D. R. 1984. *Accessibility and utilization: Geographical perspectives on health care delivery*. London: Harper & Row.

Kearns, R. A.; Taylor, S. M.; and Dear, M. 1987. Coping and satisfaction among the chronically mentally disabled. *Canadian Journal of Community Mental Health* 6:13–24.

Kennedy, S. 1988. A geographic regression model for medical statistics. *Social Science and Medicine* 26:119–30.

Kloos, H. 1982. Drought and famine in the Awash Valley of Ethiopia. *African Studies Review* 25:21–48.

———. 1983. DBCP pesticide in drinking water wells in Fresno and other communities in the central valley of California. *Ecology of Disease* 2:353–67.

——— et al. 1983. Schistosomiasis and human behavior in an upper Egyptian village. *Social Science and Medicine* 17:545–62.

———, et al. 1987. Coping with intestinal illness among the Kamba in Machakos, Kenya, and aspects of schistosomiasis control. *Social Science and Medicine* 24:383–94.

Knox, P. L. 1978. The intra-urban ecology of primary medical care: Patterns of accessibility and their policy implications. *Environment and Planning C: Government and Policy* 10:415–35.

———. 1979a. The accessibility of primary care to urban patients: A geographical analysis. *Journal of the Royal College of General Practitioners* 29:160–68.

———. 1979b. Medical deprivation, area deprivation and public policy: A review. *Social Science and Medicine* 13D:111–21.

———. 1980. Urban deprivation and health care provision. *Medicine in Society* 5:54–59.

———. 1981a. Convergence and divergence in regional patterns of infant mortality in the United Kingdom, 1949–51 to 1970–72. *Social Science and Medicine* 15D:323–28.

———. 1981b. Retail geography and social well-being: A note on the changing distribution of pharmacies in Scotland. *Geoforum* 12:255–64.

———. 1982. The geography of medical care: An historical perspective. *Geoforum* 13:245–51.

———. 1985. Regional socio-economic change in Western Europe since 1930: The evidence of infant mortality rates. *L'Espace Geographique* 14:227–34.

———; Bohland, J.; and Shumsky, L. 1984. Urban development and the geography of personal services: The example of medical care in the United States. In *Public Service provision and urban development,* eds. A. Kirby, P. Knox, and S. Pinch, pp. 152–75. London: Croom Helm.

Kvale, K. M. 1981. Schistosomiasis in Brazil: Preliminary results from a case study of a new focus. *Social Science and Medicine* 15D:489–500.

Laborde, J. M.; Dando, W. A.; and Teetzen, M. L. 1988. Climate, diffused solar radiation and multiple sclerosis. *Social Science and Medicine* 27:231–38.

Lam, N. S-N. 1986. Geographical patterns of cancer mortality in China. *Social Science and Medicine* 23:241–47.

Lenz, R. 1988. Jakarta Kampong morbidity variations: Some policy implications. *Social Science and Medicine* 26:641–50.

Lewis, N. D. 1984. Ciguatera: Parameters of a tropical health problem. *Human Ecology* 12:253–73.

———. 1986. Disease and development: Ciguatera fish poisoning. *Social Science and Medicine* 23:983–94.

McLafferty, S. 1982. Neighborhood characteristics and hospital closures. *Social Science and Medicine* 16:1667–74.

———. 1986. The geographical restructuring of urban hospitals: Spatial dimensions of corporate strategy. *Social Science and Medicine* 23:1079–86.

———. 1988. Predicting the effect of hospital closure on hospital utilization patterns. *Social Science and Medicine* 27:255–62.

Matzke, G. 1979. Settlement and sleeping sickness control: A dual threshold model of colonial and traditional methods in East Africa. *Social Science and Medicine* 13D:209–14.

———. 1983. A reassessment of the expected development consequences of Tsetse control efforts in Africa. *Social Science and Medicine* 17:531–38.

Mayer, J. D. 1979. Paramedic response time and survival from cardiac arrest. *Social Science and Medicine* 13D:267–72.

———. 1980a. Migrant studies and medical geography: Conceptual problems and methodological issues. In *Conceptual and methodological issues in medical geography,* ed. M. S. Meade, pp. 136–54. Chapel Hill: University of North Carolina Department of Geography.

———. 1980b. Response time and its significance in medical emergencies. *Geographical Review* 70:79–87.

———. 1981. Geographical clues about multiple sclerosis. *Annals of the Association of American Geographers* 71:28–39.

———. 1983a. The distance behavior of hospital patients: A disaggregated analysis. *Social Science and Medicine* 17:819–27.

———. 1983b. The role of spatial analysis and geographic data in the detection of disease causation. *Social Science and Medicine* 17:1212–21.

———. 1986. International perspectives on the health care crisis in the United States. *Social Science and Medicine* 23:1059–66.

———, et al. 1987. Patterns of rural hospital closure in the United States. *Social Science and Medicine* 24:327–34.

Meade, M. S. 1976a. Land development and human health in West Malaysia. *Annals of the Association of American Geographers* 66:428–39.

———. 1976b. A new disease in Southeast Asia: Man's creation of Dengue hemorrhagic fever. *Pacific Viewpoint* 17:133–46.

———. 1977. Medical geography as human ecology: The dimensions of population movement. *Geographical Review* 67:379–93.

———. 1978. Community health and changing hazards in a voluntary agricultural resettlement. *Social Science and Medicine* 12D:95–102.

———. 1983. Cardiovascular disease in Savannah, Georgia. In *Geographical aspects of health,* eds.

N. D. McGlashan and J. R. Blunden, pp. 175–96. London: Academic Press.

———. 1986. Geographic analysis of disease and care. *Annual Review of Public Health* 7:313–35.

———; Florin, J.; and Gesler, W. 1988. *Medical geography*. New York: Guilford Press.

Miller, G. H.; Dear, M. J.; and Streiner, D. L. 1986. A model for predicting utilization of psychiatric facilities. *Canadian Journal of Psychology* 31:424–30.

Mosley, W. H., and Chen, L. C. 1984. *Child survival: Strategies for research. Population and Development Review, Supplement 10*. Cambridge, MA: Cambridge University Press.

Muschett, F. 1981. Spatial distribution of urban atmospheric particulate concentrations. *Annals of the Association of American Geographers* 71:552–65.

Newman, J. L. 1977. Some considerations in the field measurement of diet. *Professional Geographer* 29:171–76.

———. 1980. Dietary behavior and protein-energy malnutrition in Africa south of the Sahara: Some themes for medical geography. In *Conceptual and methodological issues in medical geography*, ed. M. S. Meade, pp. 77–92. Chapel Hill: University of North Carolina Department of Geography.

Patterson, K. D., and Pyle, G. F. 1983. The diffusion of influenza in Sub-Saharan Africa during the 1918–1919 pandemic. *Social Science and Medicine* 17:1299–1307.

Pyle, G. F. 1979. *Applied medical geography*. New York: Wiley.

———. 1985. Spatial perspectives on influenza innoculation acceptance and policy. *Economic Geography* 60:273–93.

———. 1986. *The diffusion of influenza: Patterns and paradigms*. Totowa, NJ: Rowman & Littlefield.

———, and Patterson, D. 1984. Influenza diffusion in European history: Patterns and paradigms. *Ecology of Disease* 2:173–84.

Ramesh, A., and Hyma, B. 1981. Traditional Indian medicine in practice in an Indian metropolitan city. *Social Science and Medicine* 15D:69–81.

Rooney, J. F., and Butt, P. L. 1978. Beer, bourbon and Boone's Farm: A geographical examination of alcoholic drink in the United States. *Journal of Popular Culture* 11:832–56.

Rosenberg, M. W. 1988. Linking the geographical, the medical and the political in analysing health care delivery systems. *Social Science and Medicine* 26:179–86.

Roundy, R. 1976. Altitudinal mobility and disease hazards for Ethiopian populations. *Economic Geography* 52:103–15.

———. 1978. A model for combining human behavior and disease ecology to assess disease hazard in a community: Ethiopia as a model. *Social Science and Medicine* 12:121–30.

———. 1985a. Clean water provision in rural areas of less developed countries. *Social Science and Medicine* 20:293–300.

———. 1985b. Schistosomiasis assessment: Agricultural development projects in Lofa and Bong Counties, Liberia. *Rural Africana* 22:63–72.

Rushton, G. 1984. Use of location-allocation models for improving the geographical accessibility of rural services in developing countries. *International Regional Science Review* 9:217–40.

———. 1987. Meeting the need for services in developing rural regions. In *Patterns of change in developing rural regions,* eds. R. Bar-El, A. Bendavid-Val, and G. J. Karaska, pp. 63–77. London: Westview Press.

Scarpaci, J. 1985. Restructuring health care financing in Chile. *Social Science and Medicine* 21:415–31.

———. 1987. HMO promotion and the privatization of health care in Chile. *Journal of Health Politics, Policy and Law* 12:551–67.

———. 1988b. *Primary medical care in Chile: Accessibility under military rule*. Pittsburgh: University of Pittsburgh Press.

———. 1988a. DRG calculation and utilization patterns: A review of method and policy. *Social Science and Medicine* 26:111–18.

Schneider, D. 1986. Planned out-of-hospital births, New Jersey, 1978–1980. *Social Science and Medicine* 23:1011–16.

Selya, R. M. 1988. Pharmacies as alternative sources of medical care: The case of Cincinnati. *Social Science and Medicine* 26:409–16.

Shannon, G. W. 1977. Space, time and illness behaviour. *Social Science and Medicine* 11D:683–89.

———, and Spurlock, C. W. 1976. Urban ecological containers, environmental risk cells, and the use of medical services. *Economic Geography* 52:171–80.

———; Bashshur, R. L.; and Spurlock, C. W. 1978. The search for medical care: An exploration of urban black behavior. *International Journal of Health Services* 8:519–30.

Shumsky, N. L.; Bohland, J.; and Knox, P. 1986. Separating doctors' homes and doctors' offices: San Francisco, 1881–1941. *Social Science and Medicine* 23:1051–58.

Smith, C. J. 1976. Residential neighbourhoods as humane environments. *Environment and Planning A* 8:311–26.

———. 1983a. Innovation in mental health policy: Community mental health in the United States of America, 1965–1980. *Environment and Planning D: Society and Space* 1:447–68.

———. 1983b. Locating alcoholism treatment facilities. *Economic Geography* 59:368–85.

———. 1986. Equity in the distribution of health and welfare services: Can we rely on the state to reverse the inverse care law? *Social Science and Medicine* 23:1067–78.

———, and Hanham, R. Q. 1981a. Any place but here! Mental health facilities as noxious neighbors. *Professional Geographer* 33:326–34.

———, and Hanham, R. Q. 1981b. Proximity and the formation of public attitudes towards mental illness. *Environment and Planning A* 13:147–65.

———, and ———. 1982. *Alcohol abuse: Geographical perspectives*. Washington: Association of American Geographers.

———, and ———. 1984. Regional change and problem drinking in the United States: 1970–1978. *Regional Studies* 19:149–62.

———, and ———. 1985. What drives people to drink? Interpreting the effect of urban living on the use and abuse of alcohol. *Urban Ecology* 9:195–213.

Snow, R. W., and Cunningham, O. R. 1985. Age, machismo, and the drinking locations of drunken drivers: A research note. *Deviant Behavior* 6:57–66.

———, and Landrum, J. W. 1986. Drinking locations and frequency of drunkenness among Mississippi DUI offenders. *American Journal of Drug and Alcohol Abuse* 12:405–18.

———, and Wells-Parker, E. 1986. Drinking reasons, alcohol consumption levels, and drinking locations among drunken drivers. *International Journal of the Addictions* 21:671–89.

———, and Anderson, B. J. 1987 Drinking place selection factors among drunk drivers. *British Journal of Addiction* 82:85–95.

Stock, R. 1976. *Cholera in Africa. African environment report 3*. London: International African Institute.

———. 1983. Distance and the utilization of health facilities in rural Nigeria. *Social Science and Medicine* 17:563–70.

———. 1985. Health care for some: A Nigerian study of who gets what, where and why. *International Journal of Health Services* 15:469–84.

———. 1986. Disease and 'Development' or the 'Underdevelopment' of health: A critical review of geographical perspectives on African health problems. *Social Science and Medicine* 23:689–700.

———. 1987. Understanding health care behavior: A model, together with evidence from Nigeria. In *Health and disease in tropical africa: Geographical and medical viewpoints,* ed. R. Akhtar, pp. 427–68. London: Harwood Academic Publishers.

Taylor, S. M., and Dear, M. J. 1980. Scaling community attitudes toward the mentally ill. *Schizophrenia Bulletin* 7:225–40.

——— et al. 1984. Predicting community reaction to mental health facilities. *American Planning Association Journal* 50:36–47.

——— et al. 1986. Modelling the incidence of childhood diarrhea. *Social Sciences and Medicine* 23:995–1002.

Thouez, J-P. 1979. Caracteristiques physico-chimiques de l'Eau potable et la mortalite ischemique du coeur: Application aux municipalites des Cantons de l'Est (Quebec). *Canadian Geographer* 23:308–21.

———, Beauchamp, Y., and Simard, A. 1981. Cancer and the physicochemical quality of drinking water in Quebec. *Social Science and Medicine* 15D:213–23.

Watts, S. J. 1986. The comparative study of patterns of Guinea worm prevalence as a guide to control strategies. *Social Science and Medicine* 23:975–82.

———. 1987. Dracunculiasis in Africa in 1986: Its geographic extent, incidence and at-risk population. *American Journal of Tropical Medicine and Hygiene* 37:119–25.

Weil, C. 1979. Morbidity, mortality and diet as indicators of physical and economic adaptation among Bolivian migrants. *Social Science and Medicine* 13D:215–22.

———. 1981. Health problems associated with agricultural colonization in Latin America. *Social Science and Medicine* 15D:449–61.

———, and Kvale, K. 1985. Current research on geographical aspects of schistosomiasis. *Geographical Review* 75:186–216.

Wood, W. B. 1988. AIDS North and South: Diffusion patterns of a global epidemic and a research agenda for geographers. *Professional Geographer* 40:266–79.

Ziegenfus, R. 1987. Air quality and health. In *Public health and the environment,* ed. M. Greenberg, pp. 139–72. New York: Guilford Press.

Aging and the Aged

Stephen M. Golant | Graham D. Rowles | Judith W. Meyer

Unprecedented growth of the population over age 65 in the U.S. and other developed countries, and in particular the rapidly increasing size of the old-old population (persons over 75 years of age), has given rise to pressing societal concerns. Demographic trends assume even greater significance, because these persons clearly are not transitory. The future growth of the aged is assured because of the inevitable aging of large cohorts of younger populations and the promise of longer life expectancy.

In response, a large and complex infrastructure of special-interest groups and social programs emerged to address the needs of older adults. Social and behavioral scientists from various disciplines—collectively identified as *gerontologists*—began to cogently argue for their research agendas to include age, developmental, stage-in-life, and generational issues. Correlative manifestations of this interest included the greater availability of research funding, a proliferation of publication outlets, and the increased influence of the gerontological academic community in policy development.

Within this very supportive academic climate, there developed a number of intellectual and practical issues of direct relevance to geographers. First, it became apparent that elderly people might occupy, utilize, and experience environments in ways distinctively linked to the aging process. As a corollary to this, it appeared likely that older people would respond or cope differently with environmental stresses than younger people. Such a perspective implied that the locations or environments of older people might be modified or manipulated to relieve age-related stresses.

A second set of issues emanated from the proposition that the social problems, injustices, and inequalities that afflict older populations could be linked to their locations and environments. It followed that addressing such concerns might require age-specific solutions and policies focused on changes in location, or modification of existing environments.

Finally, interest developed in the distinctive social, political, economic, and aesthetic features of human settlements that contain large elderly population concentra-

ASSESSMENT AND MANAGEMENT OF HAZARDS AND INFIRMITY

tions, and in the impact of such concentrations of older people on the quality of life for younger residents and on surrounding communities.

Motivated by such concerns, geographers—primarily in North America, France, Australia, New Zealand, and the United Kingdom—have produced a voluminous literature. These works have been exhaustively documented in earlier reviews and book collections (Cribier 1980; Golant 1979a, 1980, 1984d, 1986b, 1988; Rowles 1986b, 1987; Rudzitis 1984; Wiseman 1978; Warnes 1981, 1982). Thus, a comprehensive review is neither necessary nor appropriate. Rather, our goal is to highlight representative and influential contributions, with a bias towards the work of North American geographers. Because of the goals of this volume, our discussion will restrict itself to the writings of geographers. However, readers should recognize that many of the themes examined here have been addressed by researchers in other disciplines (e.g., Altman, Lawton, and Wohlwill 1984; Flynn et al. 1985; Hancock 1987; Newcomer, Lawton, and Byerts 1986; Serow 1987; Lawton 1985; Scheidt and Windley 1985).

THE GERONTOLOGICAL LITERATURE OF GEOGRAPHERS—AN OVERVIEW

Contributions by geographers can be subsumed under two broad categories: (1) the patterns, antecedents, and consequences of older people's residential locations and residential migration/mobility behavior, and (2) the utilization (activity patterns), meanings (perceptions and interpretations), and impact of older people's residential environments.

Epistemologically, this literature has been diverse and represents a literal microcosm of the philosophical and methodological approaches found in the geographic literature. Empirical investigations have ranged from simple descriptive analyses to elaborate mathematical modeling formulations. Philosophical underpinnings have been provided by both positivist and humanistic thinking. Environments of older people have been conceptualized and measured in both objective and subjective terms. Analyses have variously focused on the situation or context occupied and utilized by older persons, on individually defined environments, and on individuals and their environments viewed as inseparable entities, engaged in ongoing, reciprocal, constantly changing, and redefining transactions. Consistent with this conceptual eclecticism, data sources for the gerontological inquiries of geographers have included census materials, structured interview and questionnaire surveys, and ethnographic studies employing participant observation.

THE RESIDENTIAL LOCATIONS AND MIGRATION/MOBILITY BEHAVIOR OF THE ELDERLY—PATTERNS, ANTECEDENTS, IMPLICATIONS

Residential Location Patterns of the Elderly

Where older people live, as a group and relative to younger populations, has been investigated at various levels—national, regional (areas within countries), state, and local (county, community, census tract, neighborhood). The locations occupied by older people have been distinguished according to their urban (city and suburban) and rural qualities. These residential patterns have been studied from historical, contemporary, and future perspectives. At least four atlases have been produced, displaying the locations and qualities of older people's residential situations (Birdsall, Hallman, and Kopec 1979; Howe, Newton, and Sharwood 1987; Rowles, Frey-McClung, and Pratt 1987; Rowles et al. 1988).

Earlier studies in the U.S. documented residential concentrations of older people in inner or central cities, in rural and economically depressed counties, in states experiencing out-migration of the young, and in a number of retirement-oriented, amenity-rich states (Golant 1972; Graff and Wiseman 1978; Smith and Hiltner 1975; Wiseman 1978).

Later studies (Hanham, Rowles, and Bohland, 1984; Gober 1985; Golant 1975, 1979b, 1984e; Rudzitis 1982) have emphasized:

- Suburbanization of the elderly
- Growth of selected nonmetropolitan retirement-oriented residential enclaves
- Tendency of older people to live in more dispersed locations
- Appearance of residential concentrations outside of the traditional retirement/amenity destinations of the South
- Reduced number of counties experiencing large increases in their percentage of elderly residents
- Decline of elderly populations in areas (such as Appalachia) that are experiencing the dying-in-place of an older generation whose children had left earlier to seek employment opportunities
- Growth of elderly concentrations as a result of the development of planned retirement housing projects and communities

Comparable themes were investigated in the United Kingdom by Warnes and Law (1984) and in Australia by Hugo (1984a) and Hugo, Rudd, and Downie (1984).

The physical (human-made) and social characteristics of urban neighborhoods that are dominated by large elderly population concentrations were investigated by Golant (1972, 1984e), Rudzitis (1982), and Smith and Hiltner (1975). These studies identified a variety of population and housing characteristics that were more prevalent in old-age-homogeneous than in age-heterogeneous neighborhoods. Rowles (1986a) contrasted the environmental qualities of three different subregions of Appalachia that were occupied by large concentrations of elderly.

The social, economic, and political impact of elderly population concentrations resulting from planned age-segregated retirement housing has been considered by several geographers. These increasingly available residential alternatives were shown to have both advantages and disadvantages, depending on whose perspective was emphasized. On the one hand, these accommodations represented very satisfying environments to the older people who selected them. But on the other hand, they worried some local politicians, who feared that these residents would make excessive demands on their community's social and medical services (Gober 1985; Golant 1979b, 1985a; Phillips, Vincent, and Blackwell 1987; Zonn and Zube 1987).

Migration/Mobility Patterns of Older People

The less-frequent and shorter-distance residential moves of older people have been universally reported (Golant 1977–1978; Rogers 1988; Warnes 1983a, 1983b). For those who do make long-distance moves, the highest propensities to migrate tend to be around retirement for men, and somewhat earlier for women. For moves within the same county, there is an upturn in mobility rates in very old age (a pattern not displayed by intercounty or interstate movers) (Rogers 1988; Warnes 1983a, 1986). These age-specific migration rates have been mathematically summarized via "model migration schedules" for various developed counties (Rogers 1988).

The regional origins and destinations of old people's longer-distance moves have been studied in the U.S. and Great Britain. A relatively spatially concentrated set of destinations is a distinguishable feature (Bohland and Treps 1982; Law and Warnes 1980; Wiseman 1978, 1986). Recent research has also indicated that the relative role of elderly migration in relation to nonelderly migration in explaining changes in elderly population concentration has increased in recent years (Bohland and Rowles 1988). In addition, mathematical models of multiregional demography have been constructed to project the interregional migration patterns of the U.S. elderly through 2020 (Rogers and Watkins 1987).

More focused investigations have analyzed the migration behavior of older people in specific regions of the U.S.—New England (Meyer 1987a, 1987b) and California (Ormrod 1986)—distinguishing counties having consistently positive or negative net migration rates.

Analyses of intrametropolitan moves by the elderly in the U.S. reveal the dominance of moves within the central cities and suburbs of metropolitan areas. When older people do move outside these territories, the central city-to-suburb flow is considerably stronger. Whereas cities experience net migration losses of older movers, suburbs experience large net gains (Golant 1987b; Golant, Rudzitis, and Daiches 1978; Wiseman and Virden 1977). Geographers have also documented the net migration gains of older people in U.S. nonmetropolitan areas since the mid-1960s (Hanham, Rowles, and Bohland 1984). But Golant (1987b) cautions that these gains result from only a very small *percentage* of older metropolitan residents relocating to nonmetropolitan places. Most elderly metropolitan residents who move either remain in the same metropolitan area or relocate to another.

Antecedents of Older People's Residential Locations and Moves

Older people's immobility, or their aging-in-place, was early identified by Golant (1972, 1975) as an important demographic antecedent of their current and future population concentrations. Various social and psychological factors have contributed to older people's residential inertia, including their emotional attachments to their homes; their reluctance to give up homeownership; the attractiveness of maintaining the status quo; the ties of nearby friends, neighbors, and family; the security of living in a familiar place; and their usually high level of satisfaction with their present surroundings (Golant 1984a, 1984e; Rowles 1978, 1980, 1983a, 1983c, 1987).

On the other hand, immobility may reflect the obvious constraints associated with low income, poor health, and perceptions that alternative, more-suitable housing options are unavailable (Golant 1984e; Rowles 1984). The extent to which elderly population growth in a place can be explained by aging-in-place and net migration processes has been mathematically established by Rogers and Woodward (1988) for the states of Florida, Arizona, California, Illinois, and New York. As a cautionary note,

Hugo's (1984b) research in Australia revealed the statistical pitfalls of population projections that underestimate the rate of international in-migration, and especially the downward shift in the mortality rate of future generations of elderly.

Various research efforts throughout the developed world have identified the individual and household characteristics of older people that account for their local and longer-distance moving patterns. Along with income, gender, employment status, marital status, ethnic status, race, and social status, influential factors include family ties (especially the nearness to children), place of birth, amount of previous mobility experience, the need for assistance (because of poor functional health), the need to reduce living costs, previous residential ties, and the location of former friends (Bohland and Treps 1982; Concord 1984; Cribier 1982; Gober and Zonn 1983; Golant 1972, 1984e; Karn 1977; Law and Warnes 1980, 1982; Liaw and Kanaroglou 1986; Meyer and Speare 1985; Oldakowski and Roseman 1986; Ring 1986; Warnes, Howes, and Took 1985b).

Moving plans also are more likely for older homeowners, who feel that caring for their houses is too expensive, takes too much time, and is too tiring and for renters who feel that their buildings are not kept clean and in good repair, are bothered by insects or rodents, and who do not find the janitor or superintendent in their building to be very helpful. Fear of crime, unattractive weather, difficulties with neighbors, and dissatisfaction with a community's facilities also play a part (Golant 1984e, Wiseman 1986).

Typologies that generalize the major types and motivations of older people's relocation behavior have been developed by Wiseman (1980), Wiseman and Roseman (1979), Meyer and Speare (1985), and Speare and Meyer (1988).

National and regional analyses have identified the social and economic characteristics of places that are statistically associated with the in- and out-migrations of older people. Older people have been attracted to residential environments distinguished for their retirement-oriented amenities (e.g., presence of ocean, lakes); retirement housing accommodations; higher economic status; warmer climates; lower housing rents; suburban settings; and large existing concentrations of older people.

Places losing, rather than gaining, older migrants have included older urban cores and rural, economically depressed and isolated areas (Cribier 1982; Gober and Zonn 1983; Hall, Roseman, and Joseph 1986; Karn 1977; Liaw and Kanaroglou 1986; Meyer 1987a, 1987b; Ormrod 1986; Warnes and Law 1984; Wiseman 1986).

THE RESIDENTIAL ENVIRONMENTS OF OLDER PEOPLE

Activity Patterns and the Utilization of Environments

The activity patterns of older persons have been conceptualized as a major mechanism by which they purposively and selectively utilize different parts of their physical (human-made) settings, in order to satisfy everyday needs and goals. Testable propositions have related old people's activity patterns to their environmental experiences (Golant 1984a, 1984e; Rowles 1978). For example, Golant (1984a, 264) proposes that "the smaller the space-time locus of activity behavior, the more likely that people's environmental experiences derive from memories and fantasies of earlier environmental transactions." This proposition has been given empirical support by Golant (1984e) and Rowles (1978, 1983c).

Older people's activities have been studied in diverse contexts, both place (central/inner city, suburbs, rural) and temporal (day, night). The findings caution against making simple generalizations about the frequency, regularity, purposes, or locational context of their activities. The variable ways in which older people utilize their environment reflect where they live and personal characteristics such as age, gender, employment status, marital status, ethnicity, race, religion, health, personality, life-style, and economic status (Bohland and Frech 1982; Golant 1972, 1976, 1984b, 1984c; Hanson 1977; Meyer 1981a; Peace 1982; Robson 1982; Hiltner, Smith, and Sullivan 1986; Smith 1987; Shannon, Cromley, and Fink 1985).

An important issue is the extent to which these activity patterns result from individual preference, or are imposed involuntarily on older persons because of some combination of impaired behavioral and mental functioning (Golant and McCaslin 1979) and environmental barriers and constraints. On the one hand, a sought-after change from a working to a leisure life-style, a welcomed reduction in family responsibilities, or a preference for more passive recreational activities (e.g., reading, home-centered hobbies) will translate to less-frequent outside-the-dwelling activities and smaller activity spaces. On the other hand, the same observable activity patterns may reflect the constraints imposed by lack of an automobile and the absence of public transportation, the burden of excessive transportation costs, the presence of mobility-limiting health problems, the inaccessibility of community services, a well-founded fear of crime, the special difficulties of nighttime activity, or lost social opportunities resulting from the death of a spouse or the moving away of friends.

While isolating the most important influences is sometimes clear-cut, older people's environmental utilization patterns frequently reflect complex adaptations to their life-styles and personal resources, on the one hand, and environmental opportunities and constraints, on the other (Golant 1984e; Peace 1982; Robson 1982; Rowles 1984).

Understanding the Meaning and Impact of Older People's Environments

Major research efforts have focused on understanding and explaining older people's perceptions and interpretations of their residential settings, and the impact of these settings on their lives. Several assumptions and propositions—supported to varying degrees by empirical evidence—have stimulated these inquiries:

1. Where people grow old matters. "That it is better, more enjoyable, easier, and less adaptationally costly to grow old in some places than in others" (Golant 1984e, 2; Rowles and Ohta 1983).
2. Old age is accompanied by physical and mental deficits that make the impact of the everyday environment greater on older than younger people (Lawton 1985).
3. Because of individually defined motives and constraints, older persons display considerable variability in how they perceive and experience the qualities and consequences of their residential settings (dwelling, neighborhood, community, region). Simply put, the same residential setting is likely to differently

affect (cognitively, emotionally, overtly) individual older persons (Golant 1984e; Rowles 1978).
4. Objective assessments of old people's housing situations (i.e., standardized environmental assessments performed by detached, uninvolved professionals who employ predefined standards of quality) often do not agree with the personal assessments and feelings expressed by the older occupants of these environments (Golant 1982, 1984e, 1986a; Rowles 1978, 1980, 1983c, 1984, 1987). Older people's subjective responses are generally more favorable than indicated by these objective ratings.

The patterns and antecedents of older people's environmental experience have been investigated by three major research efforts: in an inner-city neighborhood, an older suburb, and a rural setting. While their methodological approaches differed, the results of these studies were found to complement and reinforce each other; they are described below.

Adopting a humanistic, ethnographic perspective, Rowles (1978) personally investigated the geographical experience (environmental values, meanings, and intentionalities) of older people living in an inner-city neighborhood of a northeastern U.S. city. He provided insights into how their environmental actions (physical movement in space), orientation (cognitive differentiation of space), feelings (emotional involvements in place) and fantasies (vicarious involvements in environments displaced in space and/or time) changed as they coped with old age and their activity became more physically restricted.

In a second study, focused on a small and declining rural Appalachian community, Rowles (1980, 1981, 1983a, 1983b, 1983c) expanded on these themes in his exploration of its older population's place satisfaction and residential inertia. Based on their own cognitive designations, he distinguished how various "zones" of the residential setting (home, surveillance zone, vicinity, community, subregion, region, and nation) assumed new importance as sources of social support and material assistance as individuals aged. He also examined how attachments to these places shifted, and how memories and fantasies often assumed greater importance as components of residential satisfaction. Rowles emphasized how these rural elderly experienced a physical, social, and psychological "insidedness" within their environment. This was reflected by strong familiarity with their physical surroundings, by a strong integration within a local social network of neighbors, friends, and family with whom they shared similar norms of behavior and values, and by a strong historical sense of place—an "autobiographical insideness."

Golant's study (1982, 1984e, 1985b) of older people living in a midwestern urban community similarly investigated the qualities of the residential setting from the perspective of its occupants. His emphasis was on describing the variability of old persons' environmental assessments and experienced outcomes, and on establishing how these were influenced by their personal attributes and behaviors. In contrast to Rowles' approach, the data for this study were obtained through structured surveys, administered by professional interviewers.

Golant theorized that the environmental outcomes experienced by older people are influenced by their "behavioral relationships," as indicated by how they utilize their residential settings (their activity patterns) and by the length of time they have lived in a place as an owner or renter (their residential behaviors). As a result of these behavioral relationships (with their environments), older people are variously exposed to different opportunities (and constraints) that facilitate or impede the realization of their needs or demands.

Golant also theorized that their experienced outcomes are influenced by how they know their environments. That is, older people are differently aware of their environment's resources, have different salient needs, associate different amounts of benefits and harms with their residential settings, have different expectations and aspirations for achieving their needs, have different abilities or competencies for achieving their demands, vary as to their psychological well-being (influencing their overall view of their surroundings), and have different abilities and needs to recall earlier experienced environmental outcomes. Thus, when older people evaluate their residential situations, they are not just telling us about the objective attributes of their environs. They also are revealing much about themselves—about their needs, expectations, activities, life-styles, competence, and their overall sense of well-being.

The above studies also reported a variety of specific findings about the attributes of home, neighborhood, community, and region that impinged both positively and negatively on older people's well-being. For example, Golant (1985b) found that older people who more-frequently enjoyed having *new* community experiences were also more likely to report higher levels of psychological well-being. Rowles (1986c) discussed how the efforts of the church in a rural community could complement older people's social support networks and contribute to their improved overall well-being.

Other studies have also provided valuable insights about the significant aspects of older people's residential settings: the importance of pharmacy accessibility (Shannon, Cromley, and Fink 1985); the participation of alone, poor, and frail elderly in congregate meal programs (Meyer 1981b); the availability of shopping opportunities (Smith 1987); the social qualities of the neighborhood, particularly its friendliness (Bohland and Herbert 1983); and the adoption of energy-conservation measures (Macey 1985).

FUTURE DIRECTIONS

Almost two decades of research have clearly established the study of aging and the aged as a legitimate domain for geographical inquiry. Some of this research has been pioneering work: it has involved forays into unexplored domains, albeit sometimes with a degree of naivete. In other cases, geographers have been able to build upon well-developed themes in gerontology, demography, and other allied disciplines. A certain eclecticism and diversity is an inevitable concomitant of the emergence of a new field. It is now appropriate to consolidate existing research themes and to expand our horizons in new directions.

Research on the residential locations and migration/mobility behavior of the elderly in developed societies has established a baseline of data and insight. This has facilitated a transition from simple descriptive distributional studies to more sophisticated research that seeks to monitor change over time and to project into the future. There are also encouraging current trends toward more in-depth analyses of local demographic trends on a sub-national scale, and toward more refined consideration of particular subcomponents of the elderly population. Examples include the distinguishing of the behaviors of the young-old, old-old, and oldest-old; acknowledging the existence of populations having special needs; and differentiating between different types of residential environments and residential histories.

More research is needed on the characteristics, patterns of relocation, and motivations of both elderly nonmigrants and migrants:

> Why do the majority of the elderly elect not to move?
> Is the emergence of multiple regional retirement destinations that cater to local elderly migrants and those from nearby states an ephemeral phenomenon? Or does it portend significant future reorientation of elderly population concentrations?
> In determining propensity to migrate, what is the relative contribution of migration history, the presence of kin, the availability of medical care, the existence of recreational amenities, and the environmental quality of potential destinations?
> How significant is the phenomenon of "snowbirding," the seasonal migration of elderly populations to the Sun Belt?
> What are the geographical implications of the emergence of "full-timers," a population of older people who have liquidated their assets and live year-round in campers?

Finally, in the context of major anticipated growth of elderly populations in developing nations, it is important to complement research on the location and migration of the elderly in developed societies with studies adopting a cross-cultural and more global perspective (Rowles 1986b).

With regard to studies of the residential environments of older people, geographers have barely skimmed the surface. There is a critical need for much more research on the characteristics and lifestyles of the elderly, in a much wider range of environments, in order to facilitate comparative analyses. What is it like to grow old in an inner-city neighborhood, in a single-room occupancy (S.R.O.) hotel district, in a suburban development, in a rural environment, in a large-scale retirement community amidst thousands of age peers, in the age-segregated environment of a congregate housing project, in a nursing home, or in a life-care facility?

Currently, for most of these settings, there are no more than one or two studies that adopt a geographical perspective. With such limited empirical grounding, it is difficult to stimulate appropriate critical comparative discourse. Future studies might profitably focus on some of the themes identified in the few existing studies: on activity patterns; on mental maps and the manner in which older people cognitively differentiate their environments; on the way in which individual locations, including home and "special" environments, become imbued with meanings; and on the way in which memories and vicarious involvements in temporally and spatially displaced environments are manifest within the lifeworlds of individual older people.

Unfortunately, the number of geographers involved in research on aging and the aged is far too small. In order to adequately address the wide range of issues arising out of extant studies, it will be necessary to enlist many more geographers. Such an infusion of talent would not only provide the possibility of completing an existing agenda. It also would enable the field to progress in several new directions that offer

ASSESSMENT AND MANAGEMENT OF HAZARDS AND INFIRMITY

both significant intellectual challenge and the prospect of major contributions to public policy.

Investigation of the *impacts* of increasing concentrations of the elderly should become a major focus, particularly in Sun Belt retirement states and in recently emergent regional retirement areas such as western North Carolina and lakeshore areas of northern Michigan. What is the impact upon local economies of the in-migration of significant numbers of young-old, bringing their retirement income and spending potential with them? What are the costs to communities when, as this population ages, it develops increasing needs for medical and social services? How are local economies affected by the seasonal arrival and departure of snowbirds (Happel, Hogan, and Sullivan 1983)?

There is also a need for research focusing on the social and cultural impacts of elderly population concentrations. Do elderly population concentrations jeopardize service provision for the young by creating a powerful constituency able to vote down school bonds or, through their political influence, effectively veto community programs that are not of direct benefit to its members? To what extent does the presence of large numbers of elderly people in a community result in the development of a distinctive sociocultural milieu pervaded by norms of behavior and value systems that imbue the place with a gerontological personality?

A second worthwhile focus lies in efforts to contribute to what might be termed a *geography of dependency*. Most people remain healthy and active during their young-old years, and a significant number maintain such status until shortly before their death. However, many older people, especially as they enter their eighth and ninth decades, become increasingly frail, are restricted by chronic illness, and experience increased need for assistance with activities of daily living. In addition, significant numbers suffer strokes, are afflicted with Parkinson's disease, or fall victim to Alzheimer's disease and other maladies that are more prevalent among the elderly.

By forging an alliance with medical geography, it may be possible to enhance understanding of the etiology of such diseases, their prevalence, and their spatial distribution. It may also be possible to utilize the perspective of humanistic geography in deepening our understanding of experiential and environmental context-related aspects of such diseases. For example, to what extent is the progressive disorientation and separation from the world associated with the progress of Alzheimer's disease, a separation from place? Is there a way in which the debilitating effects of the disease can be slowed or limited through environmental modifications or therapies designed to facilitate continued orientation to place (Shannon and Rowles 1986)?

Focusing on broader family and societally related aspects of chronic illness and increasing dependency, it is important to build upon valuable recent research on geographical aspects of elder/family relationships as they relate to caregiving (Warnes 1987; Warnes, Howes, and Took 1985a, 1985b).

> To what extent is the substitution effect (among children, siblings, other relatives, friends, and neighbors, with respect to the assumption of a caregiving role) intimately intertwined with the geographical distance of the elderly person from each of these resources (Warnes 1987)?
> What is the role of neighbors within the surveillance zone (space within the visual field of home) in the everyday monitoring and practical support of older people as they become more vulnerable (Rowles 1981)?

To what extent is the geography of elder/family relationships something that evolves over a lifetime as individuals relocate or, perhaps even more significantly, choose not to relocate, as a result of anticipating their future needs for proximity to sources of practical and social support?

A third focus is to emphasize the geographical aspects of service delivery and public policy for the elderly population. Some geographers have already made contributions in this domain (Golant and McCaslin 1979; Golant 1987a; Bohland and Frech 1982; Cromley and Shannon 1986; Hiltner, Smith, and Sullivan 1986; Meyer 1981a, 1981b; Rowles 1986c; Macey 1985). However, these efforts have been fragmentary and diverse in emphasis, and the number of geographers involved has been too few to establish a clear paradigm in this area.

Yet, there is rich potential for contributions on several levels. First, as Warnes (1988) has noted, there is a critical need for basic information on the status of the elderly in different geographical areas. The development of informational atlases (Birdsall, Hallman, and Kopec 1979; Rowles, Frey-McClung, and Pratt 1987; Rowles et al. 1988; Howe, Newton, and Sharwood 1987) and the conduct of basic needs surveys that include the elderly as a significant component (Sheskin 1987) is not a glamorous undertaking. But it provides spatial information that may be invaluable to policy makers.

Second, geographers can contribute by undertaking contract research and serving as consultants to policy-making bodies (Meyer 1981b; Swartz 1979). Third, there is a need for increased research on the patterns of service delivery and the barriers (both spatial and social) that limit access for elderly subpopulations. Finally, there is considerable potential to make both theoretical and philosophical contributions in this domain. A recent commentary by Golant (1987a, 16) serves to illustrate this potential, with regard to alternative approaches for providing nutrition services:

> . . .specially prepared meals could be provided at a nutrition site, home-delivered to the older person's dwelling, or be prepared and served in a congregate apartment building's kitchen facilities. One obvious issue that arises is the substitutability or complementarity of these different categories of services—and in particular their comparable ease and flexibility of delivery, their relative costs, and their relative acceptability by the elderly consumer.

Research on the characteristic lifestyles and values of older people in segregated environments can assist in the ongoing debate on the implications of the proliferation of such environments. Does spatial segregation—especially when it is institutionalized, as in the case of elderly overlay zoning—increase the potential for intergenerational tensions? Reflecting increasing societal concern with ethical issues, geographers may contribute to the debate on spatial aspects of dependency. To what extent is society beholden to maintain those of its older citizens who choose to live, or remain, within geographically remote but familiar and preferred environments, regardless of the cost of supplying needed services? At what point does the older person assume total responsibility for the consequences of his or her locational choices?

As research by geographers on aging and the aged becomes more sophisticated at its core and reaches out in new directions at its periphery, it is important to remember the academic and intellectual roots of such inquiry. While individual studies contribute to a growing corpus of knowledge, their value is enhanced if they can be related to an emerging body of theory. There are encouraging signs that such theory may be evolving in two domains: first, in the analysis of the dynamics of older people's transactions with their environments, and second, in the specification of the antecedents and consequences of elderly residential mobility patterns.

But too few studies use theoretical formulations as a focus for ongoing research. Moreover, North American studies are almost exclusively framed within positivist or phenomenological paradigms. Contributions are conspicuously absent toward developing macrosocietal or structural theories, and toward developing a political-economy perspective on older people and the social construction of old age (Warnes 1988). As with many subspecialties within geography, a reluctance to frame research within a more global context exacerbates the danger of accumulating a mass of unrelated studies that provide useful information, but limited insight.

The next 20 years offer exciting prospects for developing a geography of aging and the aged in a society that by 2010 will be on the threshold of a "senior boom" as the postwar baby-boom generation moves into old age. Narrowness of focus and limited vision must not be allowed to undermine the potential for making a significant contribution to understanding the milieu in which this generation will grow old.

REFERENCES

Altman, I.; Lawton, M. P.; and Wohlwill, J. F., eds. 1984. *Elderly people and the environment*. New York: Plenum Press.

Birdsall, S. S.; Hallman, S. P.; and Kopec, R. J. 1979. *North Carolina: Atlas of the elderly*. Chapel Hill: Department of Geography, University of North Carolina.

Bohland, J. R., and Frech, P. 1982. Spatial aspects of primary health care for the elderly. In *Geographical Perspectives on the Elderly*, ed. A. M. Warnes, pp. 339–54. New York: Wiley.

———, and Treps, L. 1982. County patterns of elderly migration in the United States. In *Geographical Perspectives on the Elderly*, ed. A. M. Warnes, pp. 139–58. New York: Wiley.

———, and Herbert, D. T. 1983. Neighborhood and health effects on elderly morale. *Environment and Planning* 15:929–44.

———, and Rowles, G. D. 1988. The significance of elderly migration to changes in elderly population concentration in the United States: 1960–1980. *Journal of Gerontology* 43:S145–52.

Concord, C. M. S. 1984. Intraurban residential mobility of the aged. *Geografiska Annaler* 66B:99–109.

Cribier, F. 1980. A European assessment of aged migration. *Research on Aging* 2:255–70.

———. 1982. Aspects of retired migration from Paris: An essay in social and cultural geography. In *Geographical Perspectives on the Elderly*, ed. A. M. Warnes, pp. 111–38. New York: Wiley.

Cromley, E. K., and Shannon, G. W. 1986. Locating ambulatory medical facilities for the elderly. *Health Sciences Review* 21:499–514.

Flynn, C. B.; Longino, Jr., C. F.; Wiseman, R.; and Biggar, J. C. 1985. The redistribution of America's older population: Major national migration patterns for three census decades, 1960–1980. *Gerontologist* 25:292–96.

Gober, P. 1985. The retirement community as a geographical phenomenon: The case of Sun City, Arizona. *Journal of Geography* 84:189–98.

———, and Zonn, L. E. 1983. Kin and elderly amenity migration. *Gerontologist* 23:288–94.

Golant, S. M. 1972. *The residential location and spatial behavior of the elderly*. University of Chicago Department of Geography Research Paper No. 143. Chicago: University of Chicago.

———. 1975. Residential concentrations of the future elderly. *Gerontologist* 15:16–23.

———. 1976. Intraurban transportation needs and problems of the elderly. In *Community Planning for an Aged Society*, eds. M. P. Lawton, R. J. Newcomer, and T. Byerts, pp. 282–316. Stroudsburg, PA: Dowden, Hutchinson, and Ross.

———. 1977–1978. The spatial context of residential moves by elderly persons. *International Journal of Aging and Human Development* 8:279–89.

———, ed. 1979a. *Location and environment of elderly population*. Washington: V. H. Winston & Sons.

———. 1979b. Locational-environmental perspectives on old-age segregated residential areas in the United States. In *Geography and the Urban Environment,* eds. R. J. Johnston and D. T. Herbert, pp. 257–94. London: Wiley.

———. 1980. Future directions for elderly migration research. *Research on Aging* 2:271–80.

———. 1982. Individual differences underlying the dwelling satisfaction of the elderly. *Journal of Social Issues* 38:121–33.

———. 1984a. The effects of residential and activity behaviors on old people's environmental experiences. In *The Elderly and the Environment,* eds. I. Altman, J. Wohlwill, and M. P. Lawton, pp. 239–78. New York: Plenum.

———. 1984b. Factors influencing the locational context of old people's activities. *Research on Aging* 6:528–48.

———. 1984c. Factors influencing the nighttime activity of old persons in the community. *Journal of Gerontology* 39:485–91.

———. 1984d. The geographic literature on aging and old age: An introduction. *Urban Geography* 5:262–72.

———. 1984e. *A place to grow old: The meaning of environment in old age*. New York: Columbia University Press.

———. 1985a. In defense of age-segregated housing for the elderly. *Aging* 345:22–26.

———. 1985b. The influence of the experienced residential environment on old people's life satisfaction. *Journal of Housing for the Elderly* 3:23–49.

———. 1986a. Subjective housing assessments by the elderly: A critical information source for planning and program evaluation. *Gerontologist* 26:122–27.

———. 1986b. The suitability of old people's residential environments: Insights from the geographic literature. *Urban Geography* 7:437–47.

———. 1987a. *Is there anything new under the sun: Housing packages for our older population*. Issues in Aging No. 3. Chicago: Center for Applied Gerontology.

———. 1987b. Residential moves by elderly persons to U.S. central cities, suburbs, and rural areas. *Journal of Gerontology* 42:534–39.

———, n.d. The residential moves, housing locations, and travel behavior of older people. *Urban Geography*. In press.

———; Rudzitis, G.; and Daiches, S. 1978. Migration of the elderly from U.S. central cities. *Growth and Change* 9:20–35.

———, and McCaslin, R. 1979. A functional classification of services for older people. *Journal of Gerontological Social Work* 1:187–209.

Graff, T. O., and Wiseman, R. 1978. Changing concentrations of older Americans. *Geographical Review* 68:379–93.

Hall, G. B.; Roseman, C.; and Joseph, A. E. 1986. The changing geography of the elderly in metropolitan Auckland: Process and policy implications. *New Zealand Geographer* 42:46–56.

Hancock, J. A. 1987. *Housing for the elderly*. New Brunswick, N.J.: Center for Urban Policy Research, Rutgers University.

Hanham, R. Q.; Rowles, G. D.; and Bohland, J. R. 1984. The changing geography of the elderly in the United States. Paper presented at the annual meeting of the Association of American Geographers, Washington, April 1984.

Hanson, P. 1977. The activity patterns of elderly households. *Geografiska Annaler* B59:109–24.

Happel, S. K.; Hogan, T. D.; and Sullivan, D. 1983. The social and economic impact of Phoenix area winter residents. *Arizona Business* 30:3–10.

Hiltner, J.; Smith, B. W.; and Sullivan, J. A. 1986. The utilization of social and recreational services by the elderly: A case study of northwestern Ohio. *Economic Geography* 62:232–40.

Howe, A. L.; Newton, P.; and Sharwood, P. 1987. *Aging in Victoria: An Electronic Social Atlas*. National Research Institute of Gerontology and Geriatric Medicine. Melbourne: University of Melbourne.

Hugo, G. J. 1984a. *The aging of ethnic populations in Australia*. Occasional paper in Gerontology, No. 6, National Research Institute of Gerontology and Geriatric Medicine. Melbourne: University of Melbourne.

———. 1984b. Projecting Australia's aged population: Problems and implications. *Journal of the Australian Population Association* 1:42–56.

———; Rudd, D. M.; and Downie, M. C. 1984. Adelaide's aged population: Changing spatial patterns and their policy implications. *Urban Policy and Research* 2:17–25.

Karn, V. 1977. *Retiring to the seaside*. London: Routledge & Kegan Paul.

Law, C. M., and Warnes, A. M. 1980. The characteristics of retired migrants. Vol. 3 of *Geography and the Urban Environment,* eds. D. T. Herbert and R. J. Johnston, pp. 175–222. New York: Wiley.

———, and ———. 1982. The destination decision in retirement migration. In *Geographical Perspectives on the Elderly,* ed. A. M. Warnes, pp. 53–81. New York: Wiley.

Lawton, M. P. 1985. Housing and living environments of older people. In *Handbook of Aging and the Social Sciences,* second edition, eds. R. Binstock and E. Shanas, pp. 450–78. New York: Van Nostrand Reinhold.

Liaw, K., and Kanaroglou, P. 1986. Metropolitan elderly out-migration in Canada, 1971–1976. *Research on Aging* 8:201–31.

Macey, S. M. 1985. Residential energy conservation among the elderly. In *Geographical Dimensions of Energy,* eds. F. J. Calzonetti and B. D. Solomon, pp. 353–71. Dordrecht, Holland: Reidel Publishing.

Meyer, J. W. 1981a. Elderly activity patterns and demand for transportation in a small city setting. *Socio-Economic Planning Sciences* 15:9–17.

———. 1981b. Equitable nutrition services for the elderly in Connecticut. *Geographical Review* 71:311–22.

———. 1987a. County characteristics and elderly net migration rates: A three-decade regional analysis. *Research on Aging* 9:441–52.

———. 1987b. A regional scale temporal analysis of the net migration patterns of elderly persons over time. *Journal of Gerontology* 42:366–75.

———, and Speare, Jr., A. 1985. Distinctively elderly mobility: Types and determinants. *Economic Geography* 61:79–88.

Newcomer, R.; Lawton, M. P.; and Byerts, T. O., eds. 1986. *Housing an aging society*. New York: Van Nostrand Reinhold.

Oldakowski, R. K., and Roseman, C. C. 1986. The development of migration expectations throughout the lifecourse. *Journal of Gerontology* 41:290–95.

Ormrod, R. K. 1986. Recent intrastate net migration flows of the elderly in California. *The California Geographer* 26:45–57.

Peace, S. 1982. The activity patterns of elderly people in Swansea, South Wales, and southeast England. In *Geographical Perspectives on the Elderly,* ed. A. M. Warnes, pp. 281–301. New York: Wiley.

Phillips, D. R.; Vincent, J. A.; and Blackwell, S. 1987. Spatial concentration of residential homes for the elderly: Planning responses and dilemmas. *Transactions of the Institute of British Geographers* 12:73–83.

Ring, M. L. 1986. *Elderly migration: A cross-national comparative analysis*. Working Paper 86-3. Boulder, CO: Population Program, Institute of Behavioral Science.

Robson, P. 1982. Patterns of activity and mobility among the elderly. In *Geographical Perspectives on the Elderly,* ed. A. M. Warnes, pp. 265–80. New York: Wiley.

Rogers, A., n.d. Age patterns of elderly migration: An international comparison. *Demography*. In press.

———, and Watkins, J. 1987. General versus elderly interstate migration and population redistribution in the United States. *Research on Aging* 9:483–529.

———, and Woodward, J. 1988. The sources of regional elderly population growth: Migration and aging-in-place. *Professional Geographer* 40:450–59.

Rowles, G. D. 1978. *Prisoners of space? Exploring the geographical experience of older people*. Boulder, CO: Westview Press.

———. 1980. Growing old "inside:" Aging and attachment to place in an Appalachian community. In *Transitions of Aging,* eds. N. Datan and N. Lohmann, pp. 153–70. New York: Academic Press.

———. 1981. The surveillance zone as meaningful space for the aged. *Gerontologist* 21:304–11.

———. 1983a. Between worlds: A relocation dilemma for the Appalachian elderly. *International Journal of Aging and Human Development* 17:301–14.

———. 1983b. Geographical Dimensions of Social Support in Rural Appalachia. In *Aging and Milieu: Environmental Perspectives on Growing Old,* eds. G. D. Rowles and R. J. Ohta, pp. 111–30. New York: Academic Press.

———. 1983c. Place and personal identity in old age: Observations from Appalachia. *Journal of Environmental Psychology* 3:299–313.

———. 1984. Aging in rural environments. In *Elderly People and the Environment,* eds. I. Altman, M. P. Lawton, and J. F. Wohlwill, pp. 129–57. New York: Plenum Press.

———. 1986a. *The elderly of Appalachia*. Appalachian Data Bank Report 3. Lexington, KY: Appalachian Center, University of Kentucky.

———. 1986b. The geography of aging and the aged: Toward an integrated perspective. *Progress in Human Geography* 10:511–39.

———. 1986c. The rural elderly and the church. *Journal of Religion and Aging* 2:79–98.

———. 1987. A place to call home. In *Handbook of Clinical Gerontology,* eds. L. Carstensen and B. Edelstein, pp. 335–53. New York: Pergamon Press.

———, and Ohta, R. J., eds. 1983. *Aging and milieu: Environmental perspectives on growing old*. New York: Academic Press.

———; Frey-McClung, V.; and Pratt, S. G. 1987. *West Virginia atlas of the elderly*. Morgantown, West Virginia: West Virginia University Gerontology Center.

———; Watkins, J. F.; Ilvento, T.; Raitz, K.; Danner, D.; and Nathalang, M. 1988. *Kentucky Atlas of the Elderly*. Lexington, Kentucky: Sanders-Brown Center on Aging, University of Kentucky.

Rudzitis, G. 1982. *Residential location determinants of the older population*. University of Chicago Department of Geography Research Paper No. 202. Chicago: University of Chicago.

———. 1984. Geographical research and gerontology: An overview. *Gerontologist* 24:536–42.

Scheidt, R. J., and Windley, P. 1985. The ecology of aging. In *Handbook of the Psychology of Aging,* 2d ed., eds. J. E. Birren and K. W. Schaie, pp. 245–58. New York: Van Nostrand Reinhold.

Serow, W. J. 1987. Determinants of interstate migration: Differences between elderly and non-elderly movers. *Journal of Gerontology* 42:95–100.

Shannon, G. W.; Cromley, E. K.; and Fink, J. L., III. 1985. Pharmacy patronage among the elderly: Selected racial and geographical patterns. *Social Science and Medicine* 30:473–78.

———, and Rowles, G. D. 1986. Alzheimer's Disease: Is medical geography relevant? Paper presented at the annual meeting of the Association of American Geographers, Minneapolis, Minnesota, May 1986.

Sheskin, I. M. 1987. The Jewish Federation of Palm Beach County Demographic Study. West Palm Beach, FL: Jewish Federation of Palm Beach County.

Smith, B. W., and Hiltner, J. 1975. Intraurban location of the Elderly. *Journal of Gerontology* 30:473–78.

Smith, G. C., n.d. The spatial shopping behavior of the urban elderly: A review of the literature. *Geoforum*. In press.

Speare, A., and Meyer, J. W. 1988. Types of elderly residential mobility and their determinants. *Journal of Gerontology* 43:574–81.

Swartz, R. D. 1979. Providing retail goods and services for residents of government assisted senior housing. *The East Lakes Geographer* 14:43–49.

Warnes, A. M. 1981. Towards a geographical contribution to gerontology. *Progress in Human Geography* 5:317–41.

———. 1982. Geographical perspectives on aging. In *Geographical Perspectives on the Elderly,* ed. A. M. Warnes, pp. 1–31. New York: Wiley.

———. 1983a. Migration in late working age and early retirement. *Socio-Economic Planning Sciences* 17:291–302.

———. 1983b. Variations in the propensity among older persons to migrate: Evidence and implications. *Journal of Applied Gerontology* 17:20–27.

———, and Law, C. M. 1984. The elderly population of Great Britain: Locational trends and policy implications. *Transactions of the Institute of British Geographers* 9:37–59.

———; Howes, D. R.; and Took, L. 1985a. Intimacy at a distance under the microscope. In *Aging: Recent Advances and Current Responses,* ed. A. Butler. London: Croom Helm.

———; ———; and ———. 1985b. Residential locations and inter-generational visiting in retirement. *The Quarterly Journal of Social Affairs* 1:231–47.

———. 1986. The residential mobility histories of parents and children and relationship to present proximity and social integration. *Environment and Planning* 18:1581–94.

———. 1987. Microlocational issues in housing for the elderly. In *Aging: The Universal Experience,* eds. G. L. Maddox and E. W. Busse, pp. 534–54. New York: Springer Publishing Co.

———. 1988. Geographers' contributions to gerontology: Needed directions for research. Paper presented at the annual meeting of the Association of American Geographers, Phoenix, Arizona, April 1988.

Wiseman, R. 1978. *Spatial aspects of aging*. Resource Paper for College Geography No. 78-4. Washington: Association of American Geographers.

———. 1980. Why older people move: Theoretical issues. *Research on Aging* 2:141–54.

———. 1986. Concentration and migration of older Americans. In *Housing an aging society: Issues, alternatives, and policy,* eds. R. J. Newcomer, M. P. Lawton, and T. O. Byerts, pp. 69–82. New York: Van Nostrand Reinhold.

———, and Virden, M. 1977. Spatial and social dimensions of intraurban elderly migration. *Economic Geography* 53:1–13.

———, and Roseman, C. C. 1979. A typology of elderly migration based on the decision making process. *Economic Geography* 55:324–37.

Zonn, L. E., and Zube, E. 1987. Sun City as suburban landscape. *Landscape Research* 12:19–25.

INTERNATIONAL UNDERSTANDING THROUGH REGIONAL SYNTHESIS

Perspectives on Africa in the 1980s

Thomas J. Bassett

The 1980s have been a decade of profound social, political, and economic change in Africa. Much has been written on the "crisis" nature of these changes, such as the "food crisis," the "debt crisis," the "environmental crisis," the "population crisis," and the "energy crisis." Amplified by television newscasts, food aid benefits, and popular books like Lloyd Timberlake's *Africa in Crisis* (1985), the two words *Africa* and *crisis* have come to be synonymous. Although usage of both the word "crisis" and some of the data are debatable (Johnston and Taylor 1986, 2–5; Lawrence 1986; Sender and Smith 1986), there is general agreement that the deepening economic crisis of the 1980s in Africa is qualitatively different from the boom-bust growth of the industrialized capitalist economies. Persistent negative economic growth rates, declining per capita food production, deteriorating terms of trade, and chronic balance-of-payments deficits are just a few indicators of the secular economic decline experienced by many sub-Saharan African countries (Sutcliffe 1986; World Bank 1984).

As will be evident in this review of Africanist research by North American geographers in the 1980s,[1] human geographers have devoted considerable attention to Africa's economic malaise. This work can be broadly divided into two research areas, "reformist development geography" and the "geography of underdevelopment."[2]

Reformist development geography refers to the research conducted in the postmodernization era, which is distinguished by its neopopulist approaches to

---

Thanks to Maarten de Witte who provided research assistance in the preparation of this review, and to Philip Porter, C. Gregory Knight, J. Barry Riddell, and to the editors of this book for their helpful comments on an earlier version of this chapter.

[1]This review concentrates on peer-reviewed articles and books, and consequently neglects significant contributions made by Africanist geographers in other outlets. This is particularly true in the case of development geographers who are producing research reports for international aid organizations and development programs.

[2]These terms are used by Rimmer and Forbes (1982) in their discussion of the major theoretical orientations represented in geographical analyses of Third World development and underdevelopment.

spatial planning and resource management. Although still primarily concerned with macrospatial organization, these studies are now directed toward rural-development planning akin to the World Bank's basic needs orthodoxy and integrated rural-development policies of the 1970s. The populist focus of this research is highlighted by its concerns with equity, agriculture, employment, appropriate technology, decentralization, pragmatism, local participation in development planning, and appropriate spatial scales. A frequent recourse to sophisticated economic and statistical analyses and the elaboration of models proposing to integrate rural areas and inhabitants more closely with national economies distinguishes this work as neopopulist (see Kitching 1982). The overall emphasis on spatial planning and incremental change to address the needs of the rural poor has prompted Rimmer and Forbes (1982) and Riddell (1987) to label the work undertaken in this area as "neomodernization" studies.

The *geography of underdevelopment* perspective, in contrast, draws inspiration from the various political-economy approaches in analyzing the spatial and human-environmental dimensions of African development and underdevelopment. The focus of political economy upon the dynamics of the world economic crisis, class analysis, capital accumulation, the role of the state, and the production process has been attractive to geographers interested in the relations between world-capitalist development and traditional human geographical concerns like peasant production, resource management, and population mobility. Recent theoretical trends in this area range from deterministic fundamentalist Marxism and articulation of modes of production theory (Forbes 1984) to conjunctural approaches that emphasize the interactive effects of many endogenous and exogenous forces.

The major distinction between the populist and political-economy approaches lies in their very different analyses of the causes and constraints producing the patterns under investigation. The populist view of the food crisis, for example, stresses the "neglect" of food-crop research and development by colonial and postcolonial governments, and urges development planners to build upon the strengths of indigenous agricultural systems to increase output (Knight 1980; Vermeer 1983). Political-economy approaches, on the other hand, emphasize the position of peasant production in the accumulation strategies of specific social groups to determine the causes of stagnation. Thus what might appear to be neglect to some observers appears to others as the deliberate outcome of agricultural policies within a historically changing regime of accumulation (M. Watts and Bassett 1986). In sum, the political-economy perspective focuses upon the processes and agents involved in restructuring production systems and spatial organizations, while the populist approach tends to accept the results of these processes as given, and by advocating certain policy prescriptions, seeks to rectify past "mistakes."

Table 1 presents an outline of the following review within a framework of the major research themes, substantive foci, and theoretical orientations representative of this work. Only those themes and foci for which a significant literature exists are reviewed in this essay. Moreover, some work is only partially reviewed, because of substantial overlap with another chapter in this volume (medical geography). The brief focus upon physical geographic research reflects the scant work currently undertaken by North American geographers in this field.

Table 1
Research themes, foci, and theoretical orientation of Africanist geographical research

Research Themes	Substantive Foci	Theoretical Orientation
Human Geography		
Population and resources	Environmental degradation	Boserupian; political ecology; eco-demographic
	Changing agrarian systems	Human ecology; political ecology
	Population and disease control	Neo-Malthusian; articulation of modes of production; behavioral
Planning African development	Appropriate rural development	Populist
	Energy planning	Neomodernization
	Integrated rural (under)development	Neomodernization; fundamentalist Marxism; political economy
Sociospatial patterns and processes	Regional inequalities	Regional political economy
	Rural-urban relations	Neomodernization
	Migration studies	Articulation of modes of production
	Urban history	Descriptive; sociospatial
	Informal-sector studies	Neomodernization
Physical Geography		
Climatology	Rainfall patterns; agroclimatology	
Environments and resources	Environmental resource management	Ecosystems; neomodernization

POPULATION AND RESOURCES

Recent research in this area has concentrated on three substantive issues: environmental degradation, changing agrarian systems, and population and disease control. The theoretical orientations range from single-hypothesis approaches (e.g., Boserupian) to conjunctural explanations that attempt to interpret local patterns of resource use and distribution within the context of broader political and economic processes (e.g., agrarian capitalism; internationalization of capital).

The Paradox of Degradation

The issue of environmental degradation has received considerable attention in the African crisis literature (Glantz 1987; Timberlake 1985; Harrison 1987). Yet what is striking about this work is the absence of reliable data to support the often sensational claims of inexorable desert encroachment in the semiarid regions of the West African Sahel and East Africa. Leaving aside the measurement problem (see Stocking 1987), geographers generally accept that land degradation is occurring, and seek to uncover its underlying causes.

One theme that stands out in this current research is the apparent maladaptiveness of indigenous production systems to specific environments. M. Watts (1987b) views the land-degradation issue as a paradox in the light of recent human ecological research emphasizing the adaptiveness of African agricultural systems to tropical environments (Richards 1985, 1986; M. Watts 1983b). How does one explain, therefore, the apparent failure of these proven resource-management strategies? Two perspectives are offered in the recent geographical literature: Boserupian and political ecology.

Thom and Martin (1983) focus upon population growth and inappropriate indigenous technology as the main causes of severe soil erosion in the Baringo-Kerio Valley of western Kenya. They argue that peasant agronomic techniques and pastoral herd-management strategies have failed to adjust to increased population pressure. In short, the Boserupian agricultural evolution, in which population pressure on resources induces agricultural innovation and intensification, has failed to materialize. To avert further environmental degradation, they recommend the establishment of formal soil conservation and range-management programs.

Bernard and Thom (1981) present a methodology for calculating the human carrying capacity of semiarid areas of Kenya. Their model considers human ecological factors such as edaphic and ecoclimatic information, crop yields, and land-use patterns to determine the number of households that can fully or partially subsist in a region. In this model, land degradation, migration, and hunger are the manifestations of excessive population pressure on an area's human carrying capacity (Bernard 1982).

Campbell and Riddell (1984) similarly discuss the relations between demographic pressures, land-use patterns and land degradation in the Mandara Mountains of northern Cameroon. They argue that population pressure in the past led to the development of intensive, environmentally conservative farming practices on mountain terraces (Riddell and Campbell 1986). Population growth and extraregional employment have led, however, to terrace abandonment and the colonization and partial degradation of lowland areas. What is puzzling to these authors is the nontransfer to the lowlands of intensive agricultural techniques which are practiced in the mountains. The absence of population pressure is presented as the major reason for degradation.

In contrast to these Boserupian approaches, in which land use and production are reduced to agrotechnology questions and population pressure is viewed as a key mechanism of agrarian change, M. Watts (1987b) presents a "social theory of dryland ecology" to explain the paradox of degradation. His approach combines the concerns of human ecology and political economy to identify the key social, political, economic, and environmental processes that result in environmental destruction. The thrust of Watts's argument is that some peasants and pastoralists who are dependent on commodity production for their survival frequently become entrapped in a "simple reproduction squeeze" which forces them to intensify production. Resource-poor households are particularly vulnerable to production shortfalls and adverse terms of trade which force them to exhibit seemingly irrational behavior, such as crop mortgaging and mining the soil. The point Watts makes is that such responses are rational at certain conjunctures.

Blaikie (1985, 117) similarly argues that "marginal" peasants and pastoralists are often forced to commit "ecocide" under adverse socioeconomic and environmental conditions, despite their known repertoire of land-management skills. This common focus upon the political and economic dimensions of human ecological issues like hunger and famine is what characterizes the political ecology approach to human-environmental problems (Blaikie and Brookfield 1987, 17–19).

Changing Agrarian Systems

The theme of crisis and change in African agrarian systems dominates the recent geographical literature on Africa. Considerable attention is devoted to analyzing the famines of 1973–74 and 1984–85, with emphasis on the historically changing responses of peasants and pastoralists to natural hazards like drought. A common thread tying together this work is the general consensus that the severity of drought has been a function of social, political, and economic factors as much as it has been the result of adverse environmental conditions. Most of these studies draw some inspiration from the political-ecology approach. Yet, despite a common interest in examining the political and economic contexts of specific human-ecological issues, what distinguishes the various uses of this model are the different interpretations of, and analytical approaches to, its very political, economic, and environmental dimensions.

Watts's studies of famine in northern Nigeria stress the need to examine the interactive effects among society, political economy, and environment to understand the periodic collapse of rural production and distribution systems in the Sudano-Sahelian region (M. Watts 1983a, 1983b). From this perspective, he considers various mechanisms of surplus extraction (taxation, export commodity production, and unequal exchange), different forms of state intervention in the agricultural sector, and processes of socioeconomic differentiation as central to understanding the vulnerability to drought of poor rural households. In a comparative study of Nigeria and Ivory Coast, M. Watts and Bassett (1986) underscore the complex origins and dynamics of the African food crisis by contrasting the different political economies (regimes of accumulation) of declining per capita food production in these countries.

Kates (1981) analyzes the comparative impact of drought in the Sahel on different "livelihood systems" during two periods, 1910–14 and 1968–74. On the basis of admittedly weak data sets, he concludes that morbidity and mortality rates were lower during the 1968–74 drought. He cites food relief as being the single most important factor behind this reduced vulnerability to famine. However, the "shifting fortune" of some groups has made them more vulnerable to drought. Nomadic pastoralists, for example, are believed to have fared better during the earlier drought, when they enjoyed more political and economic power in the region. Kates (1985) and Downing (1987) present a range of methodological approaches to the study of climate-impact assessment. A major challenge to model-building within this paradigm is to show the linkages between social processes and differential responses to natural hazards. For a discussion of some of the theoretical problems involved, see M. Watts (1982).

Campbell (1981, 1984) and Bassett (1986, 1988d) focus upon recent changes occurring in the pastoral sectors of East and West Africa as a result of drought, land-use competition, and migration. Campbell discusses the deteriorating conditions of Maasai pastoralism as farmers and the Kenyan government alienate rangelands crucial to nomadic land-use patterns. He shows how reduced access to different ecological zones has increased the Maasai's vulnerability to drought, and argues for land-use zoning as a precondition for sustainable agricultural systems in Kenya's semiarid areas. Bassett examines the nature of Fulani pastoralism and peasant-herder conflicts in northern Ivory Coast from a political-ecology perspective. He argues that one must look at the intersection of Ivorian political economy and the human ecology of agricultural and pastoral systems in the savanna region to grasp the dynamics of current land-use conflicts.

The relation between access to and control of productive resources and agricultural performance is examined in a number of studies. Weiner et al. (1985) show that land concentration in Zimbabwe is marked by the underutilization of resources. Given the dramatic increase in maize output by peasant producers in the former tribal trustlands since independence, it is argued that land redistribution will lead to more efficient and productive land utilization.

Carney (1988a, 1988b), Mackenzie (1986, 1987) and Bassett (1988a, 1988b) discuss some of the sociocultural changes associated with the restructuring of production systems under agrarian capitalism and the implications of these changes for agricultural performance. In the context of a contract-farming scheme involving the expansion of irrigated rice in The Gambia, Carney shows how intrahousehold struggles over land rights and labor have depressed rice yields. The failure to intensify production stems from the refusal of women to work on land they formerly controlled, but which men have succeeded in usurping by reclassifying it as compound versus individual land, with the acquiescence of project authorities. Mackenzie examines tensions over access to productive resources by gender in Central Province, Kenya. She shows how women are individually and collectively dealing with the pressures of increased economic insecurity resulting from their weakened control over land and their own labor power.

In a historical study of cotton development in Ivory Coast, Bassett (1988b) argues that a necessary precondition for the contemporary expansion of cotton there was the decomposition throughout the colonial period of large, lineage-based units of production into smaller, social units with new economic needs. In a related study, he shows how peasant efforts to manage labor bottlenecks in a cotton-centered farming system have led to the emergence of a group of technologically innovative, commercially oriented male farmers whose existence is being nurtured by foreign agribusiness and international aid agencies (Bassett 1988a). Porter (1987) focuses upon more humanistic aspects of sociocultural change in his lament of the death of precapitalist lifeworlds. He asks how the "nonexploitive, kin-based" features of societies based on an economy of affection might be preserved in the transition to industrial capitalism in the Third World.

The question of labor control in agrarian systems is taken up by Crush (1985b, 1986, 1987) in his examination of labor-recruitment schemes in Swaziland and South Africa during the early colonial period. Contrary to the notion that land alienation and taxation created sufficient conditions to compel the dispossessed to form a migrant-labor stream to the mines and white-settler farms, Crush argues that Africans exercised their options and were not so easily manipulated into becoming proletarians (Crush 1985a).

Population and Disease Control

The confluence of famine and rapid population-growth rates in Africa over the past two decades has elicited a neo-Malthusian response in international-aid circles that family planning is urgent if African states hope to break out of vicious cycles of poverty. While some geographical analyses reflect this view (Rogge 1982a), other studies seek to understand the dynamics of demographic regimes (Newman and Lura 1983, M. Watts 1987a). As Rogge (1982a) reveals in his sampling of a range of pronatal and antinatal population policies, considerable diversity exists among African governments in their interpretation of the population problem. His own view is that "Africa's problem is less a matter of too many people but rather one of excessive growth in too short a time frame." Exacerbating internal population growth is the special problem of Africa's refugees—"the least fortunate migrant" (Rogge 1982b). Rogge discusses the causes and patterns of forced migration and evaluates attempts of African governments to integrate refugees into different national economies (Rogge 1986).

Newman and Lura (1983) focus upon the issue of fertility control in an examination of "culturally variable demographic regimes" among the pastoral Kipsigis and Fulani. They argue that social, political, and economic factors play a greater role in regulating fertility than do deliberate efforts to control population growth. In the case of the Kipsigis of Kenya, low fertility rates during the precolonial period were related to culturally prescribed practices of late marriage and child spacing that tended to reinforce the political-economic authority of elders over juniors. Ironically, with colonization and modernization, fertility rates have increased, due to the weakening of the most important mechanism that determines low fertility levels—the elders' authority to regulate (late) marriages.

M. Watts (1987a) similarly focuses upon material and sociocultural conditions underlying high fertility regimes in a discussion of risk, household security, and fertility. He gives greatest weight to the utility-of-children argument, emphasizing the importance of children as "insurance against systematic and widespread sources of risk." If we are to understand why a premium is placed upon children and by whom, Watts argues that geographers must examine the origins and nature of risk at different spatial and institutional levels.

Straddling the subfields of medical geography and human ecology, the work by geographers on disease control has sought to illuminate the relations between disease ecologies and human behavior. Roundy (1980) provides an overview of disease patterns involving human contact with, and modification of, vegetative habitats. S. Watts (1986, 1987) focuses upon the influence of different types of population mobility on the transmission of guinea worm (*dracunculiasis*) in Nigeria.

Matzke (1983) strikes a cautious note in his discussion of the anticipated benefits of tsetse fly eradication. Recent estimates suggest that 10 million square kilometers (a third of Africa's land area) is off-limits to cattle due to the tsetse fly. Harrison (1987, 234) argues that, if livestock-raising and mixed farming could be extended into this area, the annual value of agricultural output could increase by $50 billion. In Matzke's view, the tsetse fly is a convenient political scapegoat for other causes and agents of underdevelopment. Rather than replicating the fiascos of previous fly-eradication programs, he argues that a reduction in fly habitat through the extension and intensification of agriculture into uninhabited areas ("agricultural prophylaxis") will have the greatest impact on limiting the tsetse's range.

In contrast to the dominant developmentalist view in African medical geography, which poses ill-health as an unfortunate side-effect of development processes, Stock (1986) and Wisner (1980) suggest that the underdevelopment of health is a structural feature of capitalist development. Wisner examines the nutritional consequences of uneven regional development in Kenya from the perspective of an articulation of modes of production. He utilizes the mesolevel concepts of marginalization and functional dualism to theorize the processes of increasing vulnerability of specific social groups to hunger and malnutrition. Stock contrasts the approaches and contributions of medical geographers working in the developmentalist and underdevelopment of health paradigms. He calls for a "fully developed holistic" approach to African health problems that would combine the strengths of political economy and human ecology in the analysis of morbidity and health policies "in the interests of common people and states attempting to liberate themselves" (Stock 1986, 689).[3]

PLANNING AFRICAN DEVELOPMENT

Africanist geographers have been active in the development-planning process, and have sought to contribute to policy formulation. This applied dimension of the discipline has been most visible in the areas of energy planning and rural development. It is difficult to categorize the work in this section by theoretical orientation, because of the common use of keywords (i.e., "the state," "uneven development," "context") by representatives of both the populist and political-economy schools. Generally speaking, those perspectives focusing upon bottom up and basic needs strategies are representative of the populist approach, while those stressing state intervention and production relations fall within the political-economy school.

Appropriate Rural Development

Geographers have been at the forefront of populist research on alternative paths to agricultural development and rural change (Knight 1980; Lewis and Berry 1988; Porter 1979; Scott 1984, 1985; Vermeer 1983). Consonant with the human-environmental tradition of human geography, the themes most often stressed in these studies center upon the adaptiveness of indigenous agricultural systems to the environment. The strength and appeal of this work lies in its sophisticated yet simple conclusion that agricultural research and development in sub-Saharan Africa should be firmly based on peasant agro-ecological knowledge or ethnoscience, rather than on inappropriate models issuing from distant experiment stations.

Uniting all this work is a "development from within" approach that abandons the trickle down strategies of conventional development economics by its emphasis on the importance of local knowledge, participation, and control of the development process. It is assumed, for example, that development planners and experiment-station personnel have much to learn from peasant farmers, and that their R & D agendas should be reoriented to enhance indigenous agricultural systems rather than to replace them.

Knight provides a clear rationale for taking the ethnoscientific approach to planning African agricultural development. He views ethnoscience as a pragmatic way of opening lines of communication between peasants and planners, to "squelch misdirected regulations and plans" (Knight 1980, 228). Porter (1979) and Vermeer (1983)

---

[3] For a more comprehensive review of the contributions of Africanist geographers to the field of medical geography, see the chapter on that topic in this volume.

similarly argue that agricultural-development planners might also find technical solutions to African food-supply problems by looking more closely at indigenous agricultural systems. Porter argues that the combined effects of state policies and population pressure in semiarid areas of East Africa have created stresses in food production and procurement systems which demand "ethno-agronomic" and socio-political solutions. This focus upon both the human ecological and political aspects of underdevelopment represents one of the earliest contributions to the political-ecology approach to African development geography.

Working within the spatial-organization tradition, Taylor (1981, 1985) adopts a development from within approach to identify appropriate spatial strategies for African rural development. He argues that popular participation in and control over the development process requires "territorially based community self-reliance" which will come about only through a devolution of power to local communities. Taylor recognizes that such shifts in power relations will entail conflicts between local groups and the state. He underestimates, however, the range of conflicts within these local communities by assuming relatively homogenous social groups occupying indigenously defined "people's spaces" (Taylor 1985, 41–42). The work of Carney (1988b) and Mackenzie (1987), for example, reveals considerable local socioeconomic differentiation and inter- and intra-household conflicts over the control of this space. The contribution of Taylor's work is its emphasis on the politics of organizing "traditional" development space, which represents a clear break with, and critique of, the spatial-analysis tradition of the modernization school (see below).

Energy Planning

Africa's energy needs, supplies, and alternatives became of increasing concern to governments and foreign donors with the oil-price hikes of the 1970s. With some countries spending more than half of their foreign-exchange earnings on oil imports, the energy crisis focused attention upon Africa's consumptive patterns and future supplies. The overwhelming importance of wood as an energy source, even in oil-rich Nigeria where fuelwood accounts for 80% of total energy consumption, has focused attention upon traditional as well as conventional fuel needs and supplies. A recurring theme in this literature is that planning inevitably entails conflicts over the distribution of resources.

Hosier et al. (1982) critique the dual-economy approach to development planning, and argue for a (populist) restructuring of energy planning, based on local participation. They recognize the inherent political obstacles to this bottom-up approach in which "'the plan' becomes the terrain of struggle over the power to allocate and consume." In contrast to energy-sector studies which advocate conservation and the diffusion of more fuel-efficient stoves in Africa, the authors debunk these notions while accentuating the need for political solutions that address distributional issues. These equity concerns receive less attention in a case study of Zimbabwe's energy needs in which Hosier (1986) promotes a sectorally integrated end-use approach to projecting future energy needs over the traditional supply-oriented models.

Wisner (1985) accentuates the political when he argues that the development process should be viewed as "a struggle over resources, not an exercise in utilitarian rationality." He traces the fate of the basic-needs approach and shows how it became transformed into a "de-politicized shopping list" conducive to "dis-integrated" (versus integrated) rural development. He advances the basic-needs position in another study on the effects of the woodfuel crisis on poor women in Kenya and Lesotho (Wisner 1987). Wisner argues that planners might implement more appropriate development policies to meet the needs of poor women if they were to focus upon the specific causes of rural poverty, rather than upon their common symptoms.

Integrated Rural (Under)Development

The effect of government policies on increasing the vulnerability of poor households to climatic hazards is the subject of two studies by Silberfein (1984) and Bernard (1985). Silberfein delineates the history of uneven development in Machakos District, Kenya, in which she shows how colonial state interventions in the agricultural sector were linked to a policy of "subregional favoritism." This process of regional differentiation was accompanied by a widening of socioeconomic inequalities among the Kamba. A core-periphery structure gradually evolved, characterized by labor migration and unequal exchange. It is within this context of regional and socioeconomic disparities that Silberfein shows how the capacity of lowlanders to adjust to environmental stress had become severely undermined by the end of the colonial period.

Bernard (1985) examines postcolonial development policies in Kenya's arid lands, and concludes that, rather than reducing risks, Kenya's integrated rural-development projects have inadvertently weakened the ability of peasants and pastoralists to adjust to environmental hazards. A checklist of impediments is presented to explain the failure of development planning in Kenya's arid lands.

Bassett (1988c) critiques the World Bank's food-crop development policy for sub-Saharan Africa by testing its food crop/cash crop complementarity thesis against the results of a Bank-funded project in northern Ivory Coast. He shows that the complementarity between food crops and cotton is weak in the study area, due to a number of institutional and technical constraints. He concludes by arguing that the food/export crop complementarity/competition debate can only be advanced through analyses of different agrarian systems in their political-economic contexts in which the repercussions of agricultural policies can be examined.

The crisis of peasant production and government planning in socialist states is examined by Samatar (1985, 1989), Franke (1984), and Kruks and Wisner (1984). Samatar suggests that Somalian state structures and representatives present a much greater obstacle to economic development than unfavorable environmental conditions. He argues that, despite the rhetoric of scientific socialism and decentralized development planning, Somalian administrative structures ("administrative deconcentration") impede peasant participation in rural-development planning. Samatar (1987) also critiques the World Bank's free-market prescriptions for Africa's economic crisis, by arguing that it is not state control but the inherent contradictions of unfettered merchant capital that are largely responsible for the crisis in the Somali livestock sector.

Kruks and Wisner (1984) similarly critique the socialist structures of "democratic centralism" in Mozambique for excluding women, the main food producers, from rural development planning. They note that "the crisis of peasant production is in large part a crisis of women's production," and argue that only when domestic labor relations are seriously examined and transformed will women be in the position to improve agricultural output.

Franke (1984) is more optimistic about the role of the (socialist) state in rural development, in his study of five experiments among the Tuareg. He argues that the persistence of feudal-class structures remains the greatest obstacle to egalitarian development, and that the state has the power to restructure these societies. The destitution surrounding the 1968–74 drought tended to reduce class differences between the Tuareg nobility and vassal and slave classes. Some Sahelian development programs, such as the Niger Livestock and Range Project, are seen as rebuilding these former structures, while others, notably in socialist Algeria, are viewed as leveling such differences to the benefit of the Tuareg majority.

SOCIOSPATIAL PATTERNS AND PROCESSES

The spatial-organization tradition in Africanist geography has experienced some profound permutations in the 1980s. Although some geographers still abide by the "spatial fetish rules" of modernization theory (Logan 1985; Jones 1986), much current research reflects a radical break with the diffusionist paradigm. Riddell's work represents the most trenchant (self-)critique of the modernization approach (Riddell 1981b, 1985a). He calls on geographers to forsake their former fixation on diffusion surfaces, and to examine the underlying causes producing specific spatial patterns. This new sociospatial analysis, much of it influenced by the infusion of political-economy paradigms in the social sciences, promises to catapult the study of spatial organization out of its "spacious cul-de-sac" (Riddell 1981b). The research themes currently dominating this area are (1) regional inequalities, (2) rural-urban relations, (3) migration, (4) urban historical studies, and (5) informal-sector studies.

Regional Inequalities

The basic research question raised in this literature is: what are the underlying causes producing the patterns of regional and social inequality in Africa today? In a case study of Uganda, Ede (1981) takes a historical approach and argues that regional inequalities evolved from the interplay of internal and external forces during the colonial period. He chronicles the emergence of three historically distinct spatial divisions of labor, and argues that if development planners hope to rectify these inequalities, it is imperative that they develop strategies to address these specific conditions.

Riddell (1985a, 1985b) examines regional disparities in Sierra Leone through the theory of articulation of modes of production. The genesis of these inequalities is related to the interaction of endogenous and exogenous forces at four interlocking "levels of spatial resolution:" the global, continental, national, and local. In Riddell's model, the global context conditions but does not determine internal developments (Riddell 1985b). In a second study, Riddell (1985a) illuminates these processes more concretely by examining the role of the state as a "redistributive mechanism" to show how economic surpluses are extracted and distributed to the benefit of some groups and regions over others.

Rural-Urban Relations

A major focus of the geography-of-modernization studies concentrated on the development gap between rural (traditional) and urban (modern) areas (sectors). Spatial planning was to narrow this gap by facilitating the diffusion of institutions, goods, and services through various networks and hierarchies. An objective of these studies was to provide logistical support to modernization planners, through an exercise that Riddell (1983) refers to as "spatial engineering." The question of how town and countryside can be better linked to accelerate economic growth continues to be posed by geographers. The neomodernization studies of Gaile (1989), Jones (1986), and Silberfein and Kessler (1988) focus upon the role of small towns in the process of innovation diffusion and rural development in Kenya and Sierra Leone. The emphasis in these studies is on identifying the ideal characteristics and locations of towns that will favor the extension of development activities (extension services, marketing places). Once these spatial-logistical problems are solved, it is expected that increased production and prosperity will follow (McNulty 1987).

Freeman (1985) and Riddell (1985c) are not so sanguine about the prospects of prosperity for the mass of rural producers through new innovation-diffusion networks. Freeman (1985) introduces the notion of preemption rents, which he defines as monopoly profits captured and maintained by early adopters of innovations, through political means such as legislation restricting the area under coffee cultivation or the location of milk-processing plants.

Distortions in the innovation-diffusion process are similarly noted by Riddell (1985c), who explains the nonadoption of intensive rice cultivation by peasants in Sierra Leone in terms of urban bias. He argues that rice policies and export-oriented rural-development projects tend to benefit urban residents, the state, and the beneficiaries of state investments in the urban and agro-industrial sectors, at the expense of the rural poor. In sum, these studies suggest that small-town development can open up both new economic opportunities for rural residents as well as new territory to be exploited by a predatory state and its support classes. Riddell (1985c) calls for a consideration of the political economy of rural-urban relations, to determine who is to benefit from linking town and countryside (see Armstrong and McGee 1985).

Migration Studies

There are surprisingly few Africanist geographers doing research on population mobility. Yet, the role of migration in the phenomenal expansion of Africa's primate cities is believed to account for 25–50% of recent urban-growth rates. Of the few studies that have been devoted to this theme, a major focus has been upon the processes underlying migration patterns.

Riddell (1981a) makes a theoretical contribution to the field by offering an alternative to the standard push-pull model, in which population movements are presented as the outcome of rational individual decisions. His migration model emphasizes the processes of capitalist penetration and proletarianization as major forces directing migration. Capitalist development has conditioned migration, Riddell argues, by its "restructuring of the space-economy" (development of mines, plantations, and transportation), and the "manipulation of migration" (taxation and creation of labor-reserve economies) by colonial administrators and settlers. Riddell (1981a) places greatest emphasis on coercion, and migrants are thus largely viewed as passive victims of external forces.

Crush's study on Swazi migrant workers to the Witwatersrand mines advances this discussion, by showing how the Swazi were not so easily manipulated by mining interests and the state, despite a range of coercive measures (Crush 1986, 1987). He stresses the unevenness of Swazi incorporation into the migrant labor system by revealing how they vacillated between the mines, commodity production, and various forms of urban/rural informal-sector activities. This attention to the internal dynamics of Swazi social history is in contrast to the more-deterministic views suggesting a smooth and unambiguous response to admittedly coercive external forces—a perspective that Crush criticizes because "it fails to accord the Swazi the common courtesy of being present at the making of their own past."

Urban Historical Studies

The literature on African urbanization is primarily occupied with the questions of city form or morphology and social geography. Identifying and classifying the "colonial city" and other ideal types such as the "African city" and "apartheid city" dominate the discussion (Winters 1982; Western 1985, 1986; S. Watts and Watts 1986). Simon (1984) has argued that such studies have contributed to our understanding of certain structural features of colonial cities, but largely fail to relate these sociospatial forms to the broader political economic contexts in which they are embedded. Although his own analysis falls short of specifying the relations between urban form and political economy, his theoretical contribution brings a dynamism to African urban studies that has been lacking.

Winters' study of urban morphogenesis in francophone Africa is representative of the descriptive culture-contact models which emphasize the influence of culture in shaping urban structures (Winters 1982). The colonial city, for example, is described as a dual and often segregated city, comprised of a European city and an African city. The European city resembles the order of post-Haussmann continental cities in its buildings, residential neighborhoods, and business districts, which are integrated by a grid pattern. In contrast, the African city has an organic form composed of spontaneous settlements, ethnic concentrations, and strong rural ties. In the end, it is unclear what is specifically African or cultural about squatter settlements. Recourse to superorganic notions of culture and a neglect of production relations result in such simplistic findings that the colonizer and colonized tended to live in separate sections of the city.

Frenkel and Western (1988) illuminate more fully the contradiction between the ideology of the British "civilizing mission" and the practice of racial segregation in colonial Sierra Leone. Despite a plethora of official rationalizations, particularly the health benefits (to whites) of constructing a hill station, the authors argue that racist impulses were the primary motivation behind this new urban form at the turn of the twentieth century.

Western's work on South African urbanization focuses upon the relations among society, economy, and "politico-spatial" urban planning (Western 1981, 1984). His examination of the 1950 Group Areas Act on the (d)evolution of Cape Town into an apartheid city is located in the context of both apartheid policies and the lives of displaced people. One senses the tensions creating these cities, and the gravity of the conflicts that might ultimately tear them apart (Western 1982, 1986).

White (1985) evaluates the poor results of a World Bank-financed sites and services housing project in Senegal, in which less than one-quarter of the final occupants matched the original low-income profile. He shows how conflicting perceptions of the project by both the Bank and Senegalese government combined with adverse external economic conditions to open the doors of this housing project to wealthier families.

Informal-Sector Studies

Reflecting the neomodernization concern with small-scale, labor-intensive employment in rural areas, geographers have sought to contribute to the study of the informal sector. Hosier (1987) addresses the issue of spatial and temporal variation in the informal sector of Kenya. His findings indicate a correlation among city size and employment characteristics, competitiveness, profit rates, and the number of informal-sector establishments.

In contrast to the urban informal sector, Freeman and Norcliffe (1985) provide a detailed analysis of the rural nonfarm sector in Kenya. Employing eight times the number of people involved in the urban informal sector, the rural nonfarm sector is shown as fulfilling a vital economic role in the rural economy. The authors urge rural-development planners to support this sector, and recommend specific policy measures to encourage its growth.

Although not explicitly linked to informal-sector studies, Watts's work on small-scale women traders of foodstuffs falls within this domain (S. Watts 1984). Her study reveals that women's involvement in the processing and trading of a "maize-meal snack" (*eko*) in Ilorin, Nigeria, is characterized by arduous work, low profits, and little upward mobility in the trading hierarchy. Contrary to prevailing notions of economically independent African market women, S. Watts notes that the only reason women continue to engage in this low-earning trade is because of their access to the labor power of young girls and the lack of other income-earning opportunities.

Freeman (1980) contrasts the marketing patterns of mobile traders in Kenya to non-African and West African periodic-market models. The major difference observed is the existence of part-time mobile merchants who engage in trade for "target saving" motives. The point of his study is that the profit-maximizing assumptions transferred from central-place and industrial-location theories to periodic-market research are inappropriate in the East African context.

CLIMATOLOGY

Similar to the human geographical literature focusing upon African development and underdevelopment, physical geographers have focused upon issues pertinent to contemporary development geography. However, this literature is much smaller, and largely concentrates on climatological issues. Some of the research reviewed here is representative of the human-environment tradition, and could have been reviewed above as well (Vermeer 1981; Porter 1983). But the greater emphasis in these studies on physical geographical processes versus sociocultural, political, and economic processes led to their inclusion in this section.

Rainfall Patterns

Successive droughts and below-average rainfall since the late 1960s throughout sub-Saharan Africa have generated much debate among climatologists about the patterns and processes of rainfall regimes. Nicholson (1980, 1981) has made some important contributions in this area with her work on rainfall fluctuations in West Africa. Her work challenges some of the simplistic notions about coherence in rainfall variation and the importance of the Inter-Tropical Convergence Zone (ITCZ) in determining rainfall changes throughout the region. She identifies a number of Sahelian drought types, each produced by different mechanisms, and suggests that dry years may be related to the intensity of the Hadley circulation, rather than to the relative displacement of the ITCZ.

Vermeer's (1981) analysis of rainfall data for Mauritania shows a drought cycle characterized by a 30-year interval that most severely affects the west-central part of the country. He argues that development planners need to be sensitive to the temporal and spatial dimensions of drought, if they seek to promote both livestock-raising and environmental conservation.

Agroclimatology

Both de Blij (1985) and Porter (1983) focus upon agrometeorological attributes of African farming systems. De Blij's study of the agricultural geography of wine-grape varieties in South Africa indicates that the heat-summation system is of little predictive value, since cool-climate varietals are found in warm environments. Porter's research in Kenya examines the relation between energy-water balance and crop yields in a semiarid zone. His study shows that planting densities and dry-season soil-management techniques play critical roles in maintaining adequate soil moisture for reliable crop yields in these drought-prone environments.

ENVIRONMENTAL-RESOURCE MANAGEMENT

Complementing the literature in human geography on environmental degradation, recent research by Lewis and Berry (1988) examines the diversity of African environments and the major environmental-resource management problems facing Africa today. Their study shows that Africa's natural and human-modified environments range widely in their physical geographical characteristics, potentials, and proneness to degradation. Development planners, they argue, must consider these basic differences to ensure proper resource use and conservation.

CONCLUSION

The primary objective of this study has been to provide an overview of the current research interests and findings of North American Africanist geographers writing in the 1980s. I have suggested that the human geographical studies can be divided into two distinctive schools, based on the general theoretical orientation of the author. While it has been a fairly straightforward exercise to identify the basic research questions and contributions, it has not been easy to place these studies within these schools. This is the result, in part, of the general absence of theory in much of this work, and partly due to the widespread use of loosely defined concepts and keywords in the literature.

One of these keywords is "context." Despite its common usage, there is very little agreement over this word's meaning. For example, geographers focusing upon land degradation have very different interpretations of the political and economic contexts of their studies. Some are content with citing political, economic, and social *factors,* while others attempt to theorize the relations among these factors in explaining degradation. What exactly constitutes a contextual approach is a complicated philosophical and methodological issue that is of primary concern to human geographers (see Forbes 1984,126–29; Johnston 1986). The fact that many Africanist geographers are grappling with these questions in social theory represents significant intellectual progress in the field since the 1970s (Doherty 1986; Porter and De Souza 1974). Indeed, if one can point to a major advance in Africanist geographical scholarship in the 1980s, it is the ability to move from micro- to macro-level analysis via clearly articulated conceptual frameworks.

Among the themes and foci reviewed in this essay, a number of substantive areas appear insufficiently developed, if not totally neglected by Africanist geographers. The most obvious gaps are in physical geography (geomorphology, remote sensing, biogeography), historical cartography (although see Kirchheer 1982, 1987), political and cultural geography, and regional geography. Above all, there is a clear need for a comprehensive regional geography of Africa, for teaching undergraduates as well as for demonstrating the strengths of geographical analysis to our colleagues in African studies. Recent advances in "new wave" regional geography (Gilbert 1988; Pudup 1988) and African historiography (Crush and Rogerson 1983) suggest the potential for a rejuvenated African regional geography that will transcend the largely descriptive studies that have dominated the field (Best and de Blij 1977; de Blij and Martin 1981; Weigend 1985a, 1985b). The theoretically informed and empirically rich studies of development geographers reviewed by Riddell (1988) demonstrate the capacity of geographers to break new ground in this area.

Another substantive area in which Africanist geographers can make important contributions is regional political ecology. The human-environment tradition is a hallmark of human geography, and has maintained its predominance in Africanist research since the beginning of this century. The infusion of political-economy approaches into the field during the 1970s has shifted the focus of inquiry to society-environment relations, reflecting the greater attention given to social processes that condition human-environmental interactions (Doherty 1986, 258). Geographers working within this paradigm are now faced with the challenge of combining human and physical geographic approaches toward the analysis of a wide range of resource use and access problems (Blaikie and Brookfield 1987). Such a "fully developed holism" is still in gestation, but it should come forth in the 1990s to advance our understanding of the processes and patterns of African development and underdevelopment.

REFERENCES

Armstrong, W., and McGee, T. 1985. *Theatres of accumulation: Studies in Asian and Latin American urbanization.* London: Methuen.

Bassett, T. J. 1986. Fulani herd movements. *Geographical Review* 76(3):233–48.

———. 1988a. Breaking up the bottlenecks in food crop and cotton cultivation in northern Ivory Coast. *Africa* 58(2):147–74.

———. 1988b. The development of cotton in northern Ivory Coast, 1910–1965. *Journal of African History* 29(2):267–84.

———. 1988c. Development theory and reality: The World Bank in northern Ivory Coast. *Review of African Political Economy* 41:45–59.

———. 1988d. The political ecology of peasant-herder conflicts in the northern Ivory Coast. *Annals of the Association of American Geographers* 78(3):453–72.

Bernard, F. 1982. Population pressure and population redistribution in Kenya. In *Population redistribution in Africa,* eds. J. Clarke and L. Kosinski, pp. 150–56. London: Heinemann.

———. 1985. Planning and environmental risk in Kenyan drylands. *Geographical Review* 75:58–70.

———, and Thom, D. 1981. Population pressure and human carrying capacity in selected locations in Machakos and Kitui Districts. *Journal of Developing Areas* 15(2):381–406.

Best, A., and de Blij, H. J. 1977. *African Survey.* New York: Wiley.

Blaikie, P. 1985. *The political economy of soil erosion in developing countries.* London: Longman.

———, and Brookfield, H. 1987. *Land degradation and society.* London: Methuen.

Campbell, D. 1981. Land use competition at the margins of the rangelands: An issue in development strategies for semi-arid areas. In *Planning African development,* eds. G. Norcliffe and T. Pinfold, pp. 39–61. Boulder, CO: Westview Press.

———. 1984. Response to drought among farmers and herders in southern Kajiado District, Kenya. *Human Ecology* 12(1):35–64.

———, and Riddell, J. 1984. Social and economic change and the intensity of land use in the Mandara mountains region of north Cameroon. *Tijdschrift voor Economische en Sociale Geografie* 75(5):335–43.

Carney, J. 1988a. Struggle over crop rights and labour within contract farming households in a Gambian irrigated rice project. *Journal of Peasant Societies* 15(3):334–49.

———. 1988b. Struggle over land and crop rights in the Gambia: Conflict and accumulation in the household. In *Agriculture, women and land: The African experience,* ed. J. Davison, pp. 59–78. Boulder, CO: Westview Press.

Crush, J. 1985a. Colonial coercion and the Swazi tax revolt of 1903–1907. *Political Geography Quarterly* 4(3):179–90.

———. 1985b. Landlords, tenants and colonial social engineers: The farm labor question in early colonial Swaziland. *Journal of Southern African Studies* 11:235–57.

———. 1986. Swazi migrant workers and the Witwatersrand gold mines 1886–1920. *Journal of historical geography* 12(1):27–40.

———. 1987. *The struggle for Swazi labour, 1890–1920.* Kingston, Ontario: McGill-Queen's University Press.

———, and Rogerson, C. 1983. *New wave African historiography and African historical geography* 7(2):203–31.

de Blij, H. J. 1985. Heat summation regions and cultivar distribution at the cape, South Africa. *National Geographic Society Research Reports* 20:117–21.

———, and Martin, E. B., eds. 1981. *African perspectives: An exchange of essays on the economic geography of nine African states.* New York: Methuen.

Doherty, J. 1986. Social geography and development in sub-Saharan Africa. In *Social geography in international perspective,* ed. J. Eyles, pp. 251–74. London: Croom Helm.

Downing, T. 1987. Climate impact assessment in central and eastern Kenya: Notes on methodology. In *Planning for drought: Toward a reduction in societal vulnerability,* eds. D. A. Wilhite, W. E. Easterling, and D. A. Wood. Boulder, CO: Westview Press.

Ede, K. 1981. An analysis of regional inequality in Uganda. *Tijdschrift voor Economische en Sociale Geografie* 72(5):296–303.

Forbes, D. K. 1984. *The geography of underdevelopment.* Baltimore: Johns Hopkins University Press.

Franke, R. W. 1984. Tuareg of West Africa: Five experiments in fourth world development. *Antipode* 16(2):45–53.

Freeman, D. 1980. Mobile enterprises and markets in Central Province, Kenya. *Geographical Review* 70:36–49.

———. 1985. The importance of being first: Preemption by early adopters of farming innovations in Kenya. *Annals of the Association of American Geographers* 75(1):17–28.

———, and Norcliffe, G. B. 1985. *Rural enterprise in Kenya: Development and spatial organization of the nonfarm sector*. University of Chicago Department of Geography Research Paper No. 214. Chicago: University of Chicago Press.

Frenkel, S., and Western, J. 1988. Pretext or prophylaxis? Racial segregation and malarial mosquitos in a British tropical colony: Sierra Leone. *Annals of the Association of American Geographers* 78(2):211–28.

Gaile, G. L. 1989. Choosing locations for small town development to improve market and labor expansion: The case of Kenya. *Economic Geography* 65. In press.

Gilbert, A. 1988. The new regional geography in English and French-speaking countries. *Progress in Human Geography* 12(2):208–28.

Glantz, M., ed. 1987. *Drought and hunger in Africa*. Cambridge: Cambridge University Press.

Harrison, P. 1987. *The greening of Africa: Breaking through in the battle for land and food*. London: Penguin.

Hosier, R. 1986. Energy planning in Zimbabwe: An integrated approach. *Ambio* 15(2):90–96.

———. 1987. The informal sector in Kenya: Spatial variation and development alternatives. *Journal of Developing Areas* 21(4):383–402.

———, et al. 1982. Energy planning in developing countries: Blunt axe in a forest of problems. *Ambio* 11:180–87.

Johnston, R. J. 1986. *On human geography*. Oxford: Blackwell.

———, and Taylor, P. J., eds. 1986. *A world in crisis: Geographical perspectives*. Oxford: Blackwell.

Jones, B. G. 1986. Urban support for rural development in Kenya. *Economic Geography* 62(3):201–14.

Kates, R. 1981. Drought in the Sahel: Competing views as to what really happened in 1910–14 and 1968–74. *Mazingira* 5(2):72–83.

———. 1985. The interaction of climate and society. In *Climate impact assessment: Studies of the interaction of climate and society*, eds. R. Kates, J. Ausubel, and M. Berberian, pp. 3–36. New York: Wiley.

Kirchheer, E. C. 1982. Towards an explanation of the transitionary nature of the political map of Africa in the period 1950–1978. *Africa Quarterly* 21(2-4):5–22.

———. 1987. *Place names of Africa: A political gazetteer*. Metuchen, NJ: Scarecrow Press.

Kitching, G. 1982. *Development and underdevelopment in historical perspective*. London: Methuen.

Knight, C. G. 1980. Ethnoscience and the African farmer: Rationale and strategy. In *Indigenous knowledge systems and development*, eds. D. Brokensha, D. Warren, and O. Werner, pp. 205–31. Lanham, MD: University Press of America.

Kruks, S., and Wisner, B. 1984. The state, the party and the female peasantry in Mozambique. *Journal of Southern African Studies* 11(1):106–27.

Lawrence, P., ed. 1986. *World recession and the food crisis in Africa*. Boulder, CO: Westview Press.

Lewis, L. A., and Berry, L. 1988. *African environments and resources*. Boston: Unwin & Hyman.

Logan, B. 1985. Evaluating public policy costs in rural development planning: The example of health care in Sierra Leone. *Economic Geography* 61(2):144–57.

Mackenzie, F. 1986. Local initiatives and national policy: Gender and agricultural change in Murang'a District, Kenya. *Canadian Journal of African Studies* 20(3):337–401.

———. 1987. Local organization: Confronting contradiction in a smallholding district of Kenya. *Cahiers Geographie Quebec* 31(83):273–86.

McNulty, M. 1987. Urban/urban relations and rural regional development. In *Patterns of change in developing rural regions*, eds. R. Bar-EL, A. Ben David-val, and G. Karaska, pp. 33–39. Boulder, CO: Westview Press.

Matzke, G. 1983. A reassessment of the expected development consequences of tsetse control efforts in Africa. *Social Science and Medicine* 17(9):531–37.

———, and Lura, R. 1983. Fertility control in Africa. *Geographical Review* 73:396–406.

Nicholson, S. 1980. The nature of rainfall fluctuations in subtropical West Africa. *Monthly Weather Review* 108:473–87.

———. 1981. Rainfall and atmospheric circulation during drought periods and wetter years in West Africa. *Monthly Weather Review* 109:2191–2208.

Porter, P. 1979 [1984]. *Food and development in the semi-arid zone of East Africa*. Third printing with new preface. Foreign and Comparative Studies/African Series 32. Syracuse, NY: Maxwell School of Citizenship and Public Affairs, Syracuse University.

———. 1983. Problems of agrometeorological modeling in Kenya. In *Agroclimate information for de-

velopment: Reviving the green revolution, ed. D. Cusack, pp. 276–90. Boulder, CO: Westview Press.

———. 1987. Wholes and fragments: Reflections on the economy of affection, capitalism, and the human cost of development. *Geografiska Annaler* 69B(1):1–14.

———, and De Souza, A. 1974. *The underdevelopment and modernization of the Third World.* Resource Paper No. 2. Commission on College Geography, Association of American Geographers.

Pudup, M. B. 1988. Arguments within regional geography. *Progress in Human Geography* 12(3):369–90.

Richards, P. 1985. *Indigenous agricultural revolution.* Boulder, CO: Westview Press.

———. 1986. *Coping with hunger: Hazard and experiment in an African rice-farming system.* London: Allen & Unwin.

Riddell, J. B. 1981a. Beyond the description of spatial pattern: The process of proletarianization as a factor in population migration in West Africa. *Progress in Human Geography* 5(3):371–92.

———. 1981b. The geography of modernization in Africa: A re-examination. *Canadian Geographer* 25(3):290–99.

———. 1983. Spatial fetish rules: O.K.? *Canadian Journal of African Studies* 17(1):119–23.

———. 1985a. Beyond the geography of modernization: The state as a redistributive mechanism in independent Sierra Leone. *Canadian Journal of African Studies* 19(3):529–45.

———. 1985b. Internal and external forces acting upon disparities in Sierra Leone. *Journal of Modern African Studies* 23(3):389–406.

———. 1985c. Urban bias in underdevelopment: Appropriation from the countryside in post-colonial Sierra Leone. *Tijdschrift voor Economische en Sociale Geografie* 76(5):374–83.

———. 1987. Geography and the study of Third World underdevelopment. *Progress in Human Geography* 11(2):264–74.

———. 1988. Geography and the study of Third World underdevelopment revisited. *Progress in Human Geography* 12(1):116–25.

Riddell, J. C., and Campbell, D. 1986. Agricultural intensification and rural development: The Mandara mountains of North Cameroon. *African Studies Review* 29(3):89–106.

Rimmer, P. J., and Forbes, D. K. 1982. Underdevelopment theory: A geographical review. *Australian Geographer* 15(4):197–211.

Rogge, J. 1982a. Population trends and the status of population policy in Africa. *Journal of Geography* 81(5):164–73.

———. 1982b. Refugee migration and resettlement. In *Redistribution of population in Africa,* eds. J. Clarke and L. Kosinski, pp. 39–43. London: Heinemann.

———. 1986. *Too many, too long: Sudan's twenty-year refugee dilemma.* Totowa, NJ: Rowman & Allanheld.

Roundy, R. 1980. The influence of vegetational changes on disease patterns. In *Conceptual and methodological issues in medical geography,* ed. M. Meade, pp. 16–37. Studies in Geography 15. Chapel Hill, NC: Department of Geography, University of North Carolina.

Samatar, A. 1985. The predatory state and the peasantry: Reflections on rural development policy in Somalia. *Africa Today* 3:41–56.

———. 1987. Merchant capital, international livestock trade and pastoral development in Somalia. *Canadian Journal of African Studies* 21(3):301–15.

———. 1989. *The state and rural transformation in Somalia: 1884–1986.* Madison, WI: University of Wisconsin Press.

Scott, E., ed. 1984. *Life before the drought.* Winchester, MA: Allen & Unwin.

———. 1985. Development through self-reliance in Zambia. *Journal of Geography* 84(6):282–90.

Sender, J., and Smith, S. 1986. *The development of capitalism in Africa.* New York: Methuen.

Silberfein, M. 1984. Differential development in Machakos District, Kenya. In *Life before the drought,* ed. Earl Scott, pp. 101–23. Winchester, MA: Allen & Unwin.

———, and Kessler, S. 1988. The role of a small town in rural development: A Sierra Leone case study. *Studies in Comparative International Development* 23(1):85–101.

Simon, D. 1984. Third World colonial cities in context: Conceptual and theoretical approaches with particular reference to Africa. *Progress in Human Geography* 8(4):93–514.

Stock, R. 1986. 'Disease and development' or 'the underdevelopment of health': A critical review of geographical perspectives on African health problems. *Social Science and Medicine* 23(7):689–700.

Stocking, M. 1987. Measuring land degradation. In *Land degradation and society,* eds. P. Blaikie and H. Brookfield, pp. 49–63. London: Methuen.

Sutcliffe, B. 1986. Africa and the world economic crisis. In *World recession and the food crisis in Africa,* ed. P. Lawrence, pp. 18–28. Boulder, CO: Westview Press.

Taylor, D. R. F. 1981. Conceptualizing development space in Africa. *Geografiska Annaler* 63B(2):87–93.

———. 1985. Rural development space in Africa. In *Our geographic mosaic: Research essays in honour of G. C. Merrill,* ed. D. B. Knight, pp. 33–45. Ottawa: Carleton University Press.

Thom, D., and Martin, N. 1983. Ecology and production in Baringo-Kerio Valley, Kenya. *Geographical Review* 73(1):15–29.

Timberlake, L. 1985. *Africa in crisis: The causes, the cures of environmental bankruptcy.* Washington: Earthscan.

Vermeer, D. 1981. Collision of climate, cattle and culture in Mauritania during the 1970s. *Geographical Review* 71(3):281–97.

———. 1983. Food sufficiency and farming in the future of West Africa: Resurgence of traditional agriculture? *Journal of African Studies* 10(3):74–83.

Watts, M. 1982. On the poverty of theory: Natural hazards research in context. In *Interpreting calamities,* ed. K. Hewitt, pp. 231–62. London: Allen & Unwin.

———. 1983a. Hazards and crises: A political economy of drought and famine in northern Nigeria. *Antipode* 15(1):24–34.

———. 1983b. *Silent violence: Food, famine and peasantry in northern Nigeria.* Berkeley: University of California Press.

———. 1987a. Conjunctures and crises: Food, ecology and population, and the internationalization of capital. *Journal of Geography* 86(6):292–99.

———. 1987b. Drought, environment and food security: Some reflections on peasants, pastoralists and commoditization in dryland West Africa. In *Drought and hunger in Africa,* ed. M. Glantz, pp. 171–211. Cambridge: Cambridge University Press.

———, and Bassett, T. 1986. Politics, the state and agrarian development: A comparative study of Nigeria and the Ivory Coast. *Political Geography Quarterly* 5(2):103–25.

Watts, S. 1984. Rural women as food processors and traders: *Eko* making in the Ilorin area of Nigeria. *Journal of Developing Areas* 19:71–82.

———. 1986. Human behavior and the transmission of *dracunculiasis*: A case study from the Ilorin area of Nigeria. *International Journal of Epidemiology* 15(2):252–56.

———. 1987. Population mobility and disease transmission: The example of guinea worm. *Social Science and Medicine* 25(10):1073–81.

———, and Watts, S. 1986. Morphology, planning and cultural values: A case study of Ilorin, Nigeria. *Third World Planning Review* 8(3):237–49.

Weigend, G. 1985a. Economic activity patterns in White Namibia. *Geographical Review* 75(4):462–81.

———. 1985b. German settlement patterns in Namibia. *Geographical Review* 75(2):156–69.

Weiner, D. et al. 1985. Land use and agricultural productivity in Zimbabwe. *Journal of Modern African Studies* 23(2):251–85.

Western, J. 1981. *Outcast Cape Town.* London: Allen & Unwin.

———. 1982. The geography of urban social control: Group areas and the 1976 and 1980 civil unrest in Cape Town. In *Living under apartheid,* ed. D. Smith, pp. 217–29. London: Allen & Unwin.

———. 1984. Autonomous and directed cultural change: South African urbanization. In *The city in cultural context,* eds. J. Agnew, J. Mercer, and D. Sopher, pp. 205–36. Boston: Allen & Unwin.

———. 1985. Undoing the colonial city. *Geographical Review* 75(3):335–57.

———. 1986. South African cities: A social geography. *Journal of Geography* 85(6):249–55.

White, R. R. 1985. The impact of policy conflict on the implementation of a government-assisted housing project in Senegal. *Canadian Journal of African Studies* 19:505–28.

Winters, C. 1982. Urban morphogenesis in francophone black Africa. *Geographical Review* 72(2):139–54.

Wisner, B. 1980. Nutritional consequences of the articulation of capitalist and non-capitalist modes of production in eastern Kenya. *Rural Africana* 8-9:99–132.

———. 1985. Making ends meet: Food, fuel and water need conflicts in rural development perspective. *Rural Systems* 3(2):105–20.

———. 1987. Rural energy and poverty in Kenya and Lesotho: All roads lead to ruin. *Institute for Development Studies Bulletin* 18(1):23–29.

World Bank. 1984. *Towards sustained development in sub-Saharan Africa.* Washington: World Bank.

Latin America

David Robinson

It is ironic that, in a decade that has witnessed the increasing impact of Latin America on the public's perception—be it via drugs, debts, droughts, democracy, or civil-political disorder—progress in the study of its geography by North American Latinamericanists should have been relatively desultory. Why this has been so is not difficult to understand.

First, one should note the decreasing number of geographers specializing in the study of Latin America. In contrast to the halcyon days of the late sixties—when research-funding agencies, ever-anxious to encourage academics to find out what our southern neighbors were like, poured money into projects that nowadays may seem far from the frontier of "relevant" issues—Latinamericanists now confront an ever-decreasing supply of funds for travel and research. As Augelli (1980a) correctly predicted, American geographers have increasingly sacrificed the potential delights and risks of a Latin Americanist's career for the safer paths of U.S. specialization.

Second, one should also note that one of the most serious impacts of the rise in popularity of the quantitative-systematic side of the discipline in the sixties was to steer many of the brightest of the new generation away from any regional interest, except of course those regions that could be generated via factor scores or other data-manipulation techniques. To know a region—its culture, its people, its places, its history, its symbolism—became unfashionable: better to know FORTRAN and PASCAL than Spanish or Portuguese (Swearingen 1984). The investment of time and energy to live in a foreign land, to learn its languages and the customs of its folk, was viewed as not worth the effort, especially since the only important geography that appeared to exist was that written about the U.S., or at least the developed world! If data were so much more easily collected at home, and if many more colleagues would read one's research findings if they were published in English, why bother learning about the Latins who share our hemisphere? Fewer and fewer graduate students have ventured south of the border during the last decade (Lonsdale 1986; Fuller et al. 1984), and when they have, they have gravitated mostly toward the same popular research areas (Davidson 1980).

Third, in addition to this lack of new blood entering the field, one must also note the increasing loss of prior expertise caused by death, retirement, and redefinitions of interest (Hausladen and Wyckoff 1985). The last two decades have witnessed the death of several notable geographers who for decades had exemplified the skills and excellence of pioneer Latinamericanists (Meigs, Sauer, James, Innes, and others). We have also "lost" through retirement yet another class of Latinamericanists who for long provided intellectual leadership (Parsons, West, Stanislawski, Johannessen, Pennington, and H. Sternberg), although, fortunately, most appear to be still publishing, more than ever before. Yet another group of geographers abandoned strictly Latin American studies to devote their time and energies to global or systematic contexts, only occasionally using Latin American examples when the subject matter demanded (Morrill and Angulo 1979).

These personnel losses have had a serious impact on the field. The departure of senior scholars means fewer sympathetic ears on grant-selection committees, and a lack of weighty argument in favor of Latin America as opposed to other regions and specialties. One may note the nonreplacement of Latinamericanists in several major departments.

Yet, having allowed regional specialties to wither during the great retreat into positivistic pseudoscientific geography, the discipline is now once again rediscovering the region, as both concept and analytical tool. It is now highly fashionable to demonstrate one's interest in "contingent" conditions; processes operating over considerable historical time are now thought to be relevant to an analysis of contemporary patterns; and the fact that such patterns vary by culture is admitted as significant.

Yet, born-again regionalists argue that their new regions will not be like those of old. The re-regionalizations and re-regionalisms are now to be considered in the context of global processes of the world economy, or polity. It is not enough to know of Mexico, or Peru, or Caribbea, for example; now these have to be embedded within a grander scheme of things. Explanation now demands multilevel and multicultural analysis. Structures have given way to structurations, and each season seems to produce yet another social theoretic stance.

Yet anyone reading many of the recent pronouncements will note that almost all are the products of geographers who have virtually no experience in cultures other than their own, and who prefer preaching to practice. Not a few demonstrate the arrogance of ignorance, rejecting what has been done in the past without even reading it. And most of the new gurus read almost nothing that is not written in English (Gade 1983a). The "closing" of the North American geographical mind is a sad fact of life. It will be interesting indeed to see what happens when the new regionalists confront the real developing portions of the world.

Equally significant for Latinamericanist geographers is the fact that North Americans in general are also showing considerably more interest in things Latin. It may have been the Contras, or General Noriega, or the Colombian cartels, or Alan García, or geographer Pinochet, or Amazonian deforestation, or Cuban emigration, but whatever combination of factors one may offer, the fact is that the public is now much more aware of "them." Indeed, the fact is that Hispanics are now the dominant social

INTERNATIONAL UNDERSTANDING THROUGH REGIONAL SYNTHESIS

(and soon to be political?) group in several sections of the U.S.; some observant commentators have quite properly begun to ask just how far north Latin America extends. The Latin Americanist geographer will soon not have to travel outside of the U.S. to undertake research!

The most positive note to be sounded concerning the study of Latin American geography during the last decade is the continued healthy state of an organization that has come to represent the majority of Latinamericanists in North America—the Conference of Latin Americanist Geographers (CLAG). In its publications, and at its annual meetings, one can find abundant evidence of a steady stream of significant and highly diverse research. CLAG, perhaps more than any other AAG specialty group, has fostered a group consciousness and camaraderie that has nurtured neophytes, and permitted the sharing of research interests and ideas in congenial social settings. It has also made significant and eminently successful efforts to integrate the work of its members with those of geographers from within Latin America.

Another important item on the scene is the *Dellplain Latin American Studies* series, published out of Syracuse. It has provided an outlet for some 24 volumes during the last decade, many of them by geographers, an important fact when one notes the relative demise of older series such as *Iberoamericana*.

Clearly, in the brief space allotted, one can do no more than highlight themes of research, assess the advances that have been made during the last decade (as well as point to gaps in the record), and suggest additions to the research agenda. The bibliographical items upon which this essay is based do not include as separate items the articles that appeared in the CLAG Yearbooks, Proceedings, and Special Publications (Denevan 1978; Martinson and Elbow 1981; Lentnek 1983; Kvale 1984; Clawson 1986; Works 1987b; Horst 1981, 1982). To have done so would have doubled the listing.

Readers are encouraged to read for themselves the rich diversity of studies, dealing with bees, black-boned chickens, coconuts, crime, geopolitics, lobsters, urban renewal, recreational landscapes, cattle-ranching diffusion, settlement systems, migration, subsistence, and much more. The bibliography should provide an indication of the range of topics under investigation during the last decade, as well as an indication of the diversity of publishing outlets used by Latinamericanists. It is worth noting, however, that of the 400 or so items consulted in the preparation of this essay, 60% relate to contemporary issues, 30% to historical matters, a mere 8% to physical topics, and no more than 2% to methods or techniques. There is clear evidence that Latinamericanists have expanded the range and improved the quality of their work during the last decade.

HISTORICAL ANTECEDENTS

Although the problems of contemporary Latin America are as numerous and as grave as one can find in any other part of the Third World, a continuing and significant preoccupation with historical aspects of Latin America characterizes the studies of Latinamericanist geographers in North America. This is perhaps best explained by the enduring influence of a handful of historically minded geographers who are training new students. Students often assimilate the interests of their advisors, and thus perpetuate their historical leanings.

It would appear that relatively few scholars who were and are interested in contemporary issues have produced many new graduates to form the next generation of academics. This means that, although the range of interests may have expanded during the last decade, it is hard to document much of an intensification of research on contemporary topics. Few major projects have been initiated on such urgent topics as urban growth, capital flows, decentralization, etc. Thus, for example, much more attention has been paid to pre-Hispanic agriculture in the past decade than to the persistent problems associated with current agrarian-reform issues.

While some have criticized what they regard as an overemphasis on matters that are not directly "relevant" to contemporary Latin America, there can be little doubt of the excellence of the historical studies (see the chapter on historical geography), and their impact on adjacent academic disciplines. Indeed, given the historical continuities in Latin American development, such historical tendencies should perhaps be encouraged: the past has much to teach us of and for the present.

Aboriginal Relics and Reminders

Notable studies have focused upon artifacts of Amerindian agricultural systems, allowing us to learn of the sophistication of the cultures that produced a multiplicity of cultural landforms, as well as to reappraise the population-carrying capacity of specific ecotones in the historic past. Terraces have now been mapped and measured, from within the forests of the Maya region (Turner 1983; Turner and Harrison 1978, 1983), to the depths of the Colca valley of southern Peru (Denevan et al. 1987). The hunt for raised fields (*eras, camellones*), has uncovered a veritable continentwide distribution (Siemens 1983; Parsons 1985; Knapp 1982; Pozorski et al. 1983; Denevan et al. 1987). This is forcing an appreciation of the complex techniques used by ancient cultures to modify their habitats. The dramatic impact of what was previously, and somewhat pejoratively, called "traditional agriculture" on the changing environment is now better understood.

Similarly, drainage channels, sunken fields, irrigation systems, and the like have received considerable attention (Knapp 1986; Doolittle 1980, 1984; Denevan 1982). Eidt (1984) has provided us with new methods of identifying abandoned settlement sites on the ground, while many of the studies have benefited from remotely sensed data from satellites (Mower et al. 1983). Such analyses have resulted in reassessments of population densities in the pre-Hispanic period (Driever and Hoy 1982), as well as a renewed interest in looking beneath the forest (Turner 1983) and grass for subtle indicators of prior land uses.

Yet another group of studies has focused upon the varieties and complexities of contemporary traditional agriculture (Wilken 1987; Mathewson 1984; Denevan et al. 1984) that clearly reflect aboriginal practices of conservation, land management, and a nurturing of resources that contrasts markedly with the characteristics of European systems. Whether it comes from the Brazilian Amazon (H. Sternberg 1987a, 1987b), or *Sertão* (Johannessen 1983), the *selva* of eastern Peru (Bergman 1980), or the Caribbean rim (Nietschmann 1979a), the evidence is clear: if one only takes time to look and learn, there is still much to benefit from an ecological experience that has extended over millennia.

Colonial Considerations

Whereas in the 1960s and 1970s the focus of research was predominantly upon population change in the early contact phase, and upon historical regional studies, the last decade has seen a notable extension of investigation. The analysis of Indian population decline now includes Central America (Newson 1985, 1986, 1987; Lovell 1983b, 1985; Veblen 1983a) and Peru (Cook 1981). The relocation of Indians in new Spanish-designed villages has also been monitored (Gade and Escobar 1982; Lovell 1983a, 1986), as well as the creation of what might be called "Hispanicized" landscapes (Licate 1981; Stanislawski 1983; Delson and Dickenson 1984).

E. Barrett (1981, 1987) has completed her studies of the Mexican copper industry, paralleling that of Ewald (1985) on that nation's salt industry. Both reveal how little is known of the incipient industrial development of colonial Latin America. Much needed are similar studies of the colonial textile, food, and construction industries.

Agriculture has not received the attention it deserves by geographers, especially when one considers the flood of recent studies on the secular hacienda in Mexico, and the Jesuit estates in South America. Galloway (1985) has demonstrated the benefits of a macroscale perspective on sugar production and the intricacies of agricultural reform in colonial Brazil (1979), and the publication of his monograph on the historical geography of sugar is eagerly awaited. Murphy (1986) has elegantly demonstrated the nature of colonial irrigation agriculture, and Ormrod (1979) the evolution of soil-management systems. W. Barrett (1979) and Chardon (1980) have documented the complexity of measurements in colonial Spanish America, a matter of no little import when one considers how little is known of the landed estates, commodity trade, and transportation. Colonial metrics are too important to be left to economic historians.

Relatively new to the colonial scene is the series of studies that have examined the ecology of the colonial city (Swann 1982; Robinson 1979a), and the nature and extent of urban systems (Robinson 1979b). Similarly, the last decade has seen significant advances in the study of the changing nature of late colonial population, especially racial mixing, household structure, and regional population change (Robinson 1981). The recent contributions of geographers to the analysis of colonial population migration has established as an empirical fact the continual flux of Indians and Hispanics alike (Robinson and McGovern 1980; Parsons 1983b; Robinson 1989).

An interesting new development has been the greater attention paid by geographers to colonial sources. Recent studies include that of Mexican colonial parish registers (Robinson 1980a), which should promote more demographic research; an analysis of the early colonial surveys of the Crown (Edwards 1980); an in-depth study of an eighteenth-century provincial governor's report (Robinson 1988); and a detailed guide to documents relating to mortgages on property in colonial New Galicia (Robinson and Greenow 1986). Pulsipher (1987) stands alone in her attempts to use cartographic sources in historical reconstructions of colonial geographies, and no geographer has yet matched the erudition and insight of Gerhard's trilogy of studies (1972, 1979, 1982) on the administrative historical geography of New Spain.

Of considerable significance, especially when viewed in light of the debate over the nature of mercantilist or capitalist penetration in colonial Latin America, are Greenow's studies (1983, 1984) of credit structures and processes. Since similar data exist for many other locations, it is to be hoped that this type of study will be extended to permit better measure of the relations among economic sectors, social classes, and ecological zones.

With the exception of the magisterial survey of the Caribbean culture area by Watts (1988), relatively few regional syntheses have appeared during the last decade. Pulsipher has examined seventeenth-century Montserrat (1986), Hall (1985) has placed contemporary Costa Rica in historical perspective, and Lovell (1985) has provided an elegant account of a Guatemalan region. But of the rapidly developing regionalism of late colonial Latin America, relatively little is known.

The Nineteenth-Century Vacuum

Although the nineteenth century witnessed major political, economic, and social developments in Latin America, studies by geographers have been extremely limited in number. To the formation of the nation states, or to the shift from Hispanic to Northern European and U.S. hegemony in the economic and political orders, we have paid little attention. Earlier interests in agricultural and ethnic frontiers, in the rapidly growing ports and capitals, in the immigration streams, in the mining ventures, in transportation evolution—all of these and more have been excluded from our research agenda. Of course, it is often the fact that data are difficult to extract from the archives, regional level documentation being particularly thin in comparison with late colonial sources. But when one considers the fine studies that social and economic historians have produced over the last decade, a renewed interest in this forgotten century is perhaps in order. Arreola (1980, 1982) has shown how landscapes can be reconstructed; Craig (1979), how governmental inquiries may be used; Dunbar (1988), the significance of French interest in Mexico; and Mathewson (1986), the relevance of Humboldt's interpretations. Yet, much more remains to be done.

CONTEMPORARY COMPLEXITIES

In numerical terms, during the last decade the majority of Latinamericanists in North America have been attracted to the analysis of contemporary patterns and processes, involving an extremely diverse range of topics. It is also worth noting that several geographers have moved into professional consulting positions, either temporarily (e.g., Bromley, Robinson, Elbow, Rushton) or permanently (e.g., Dickinson), in which they have provided advice to Latin American governments or to international agencies. Even though the number is still quite restricted, this trend demonstrates that geographical expertise, be it ecological, economic, administrative, or methodological, can have a significant direct impact on the development process. The challenge of "doing" development, rather than studying it, provides a humbling experience that hopefully more will enjoy (for some of the problems, see Robinson 1984).

The Difficulties of Regional and Urban Analysis

Several notable advances have taken place in the study of contemporary urban Latin America. Griffin and Ford's model (1980) of urban structure has provided an elegant argument that has now been successfully tested beyond Mexico (Elbow 1983), and awaits further extensions into the full range of urban settlements. Everitt (1984) has diagnosed the new capital of Belize, and one looks forward to the publication of Bromley's systematic analysis of new capital cities in Latin America, reflecting as they do the changing political and socioeconomic structure of the respective nation states. Planned residential segregation of urban population has been analyzed by Scarpaci et al. (1988), and Tata and his coauthors (1985, 1986) have provided excellent studies of urban differentiation in Venezuela and Trinidad.

What has not been seen is the type of in-depth and penetrating urban analysis provided by British geographers (Gilbert and Ward 1985; Ward 1986; Bromley 1981b, 1981c, 1982; Bromley and Birkbeck 1984; Clarke 1986). There have been no major funded projects on Latin American cities by North American Latinamericanists, a strange development after the promising research of the 1970s. Geographers appear to have abandoned the field to the urban anthropologists and sociologists.

With respect to regional development, the situation is somewhat better, especially with the arrival of Ray Bromley in America. His publications on USAID's "urban functions approach" (1983b), on the problems of development planning (1983a), and on more general issues of urban industrial change (1984) raise the productivity of this area of research considerably. Antonini and his coauthors (1979a, 1979b, 1982), have demonstrated the potential for regional integration, as well as the role that a university might play in offering assistance. The impact on development of the connectivity of the air-transport network in the Caribbean region in colonial and postcolonial times has been demonstrated by Gaile and Hanink (1984). Clawson (1984) has shown that religion and other cultural factors are ignored at peril, and Dickinson has elegantly outlined models of ecodevelopment for western Venezuela (1982a) and Honduras (1982b). The environmental risks attached to major developmental projects have been treated in Guatemala (Hoy and Belisle 1984), and Brazil (R. Sternberg 1983; Hecht 1985).

What is still lacking is an integrated account of the geography of environmental degradation or risk management at the continental level, or at least a major collection of essays that demonstrate the significance of what geographers have been saying amongst themselves for years.

The financing and administration of regional development has not yet attracted many geographers, though one can cite the valuable case studies of Kent (1983; Kent and Sandoval 1984) and Jones (1982). Rondinelli (1978, 1983, 1985) appears to have monopolized what could have been a geographic corner of the development market with his prolific output of studies utilizing spatial concepts in development.

Smith's studies of Amazonian regional development (1982) and resource use (1981) provide excellent examples of what geographers can offer to complex subjects. But it is an economist (Bunker 1985) who has most cogently articulated the contemporary predicament of Amazonian development, both from an internal-resource perspective, as well as within the context of transnational capitalism.

Notwithstanding the deleterious impact of the economic crisis of the early 1980s on Latin Americans, few geographers have attempted to analyze the patterns of poverty (Robinson and Ortíz 1984b; Tata 1982). Considerably greater attention has been given to migration, which has long been a subject favored by Latinamericanist geographers (Alexander 1979; Brown and Lawson 1985; Lawson and Brown 1987; Richardson 1983, 1985; Thomas and Wittick 1981; Weil 1983; Dawsey 1983; Godfrey 1982; Jones 1978; Kirchner 1980; Nietschmann 1979b; Wilkie and Wilkie 1980). With the publication of valuable resource books such as that of Wilkie (1984), it is likely that migration analyses will proliferate even more. Again, however, perhaps the time has come for an overview of the relationships among the many types of migration, (of which there are now countless cases), and the economic and political factors at the several spatial levels.

Agricultural Issues: Ecology or Economics?

In spite of the significant number of studies published on agriculture during the last decade, it is difficult to isolate any strikingly new themes. Geographers are still primarily interested in the ecological basis of traditional agriculture (Denevan 1984; Hiraoka 1985; Kent 1979; Works 1987a), and show relative neglect of market-oriented production, or commercial estates and collective-farming organizations. The flash of interest in agrarian reform that so illuminated the 1960s and early 1970s has disappeared, except for isolated studies (Barker 1980), probably reflecting both the difficulty of saying something new, and the relative insignificance of continuing reform in most countries. This inattention, however, does not mean that most Latin American countries do not need more reform than ever.

Publications have included an interesting investigation concerning the rejection of the "Green Revolution" in Mexico (Clawson and Hoy 1979), the problems of extending the agricultural frontier under distinctive environmental conditions (Harnapp 1985; Doolittle 1983; Hiraoka 1980; Hiraoka and Yamamoto 1980; Augelli 1987; Bromley 1981a), the role of government policy on agricultural change (Lawson 1988), the distinctive tropical land systems (Posner et al. 1983), and the cattle industry of Costa Rica (Parsons 1983a). Yet, there still is no agricultural geography of Latin America, and in spite of Gonzalez's (1985) attempt to situate agriculture within the context of underdevelopment and trade relations, there still is no major study on that topic.

One might indeed raise the question as to why geographers have not confronted the problems of commercial agriculture, especially since so many Latin American countries are now so poor and indebted that they can barely afford to import foodstuffs. This condition may be a consequence of governmental neglect of traditional agriculture, but faced with the long-term problems of attempting to change food habits (i.e. replace wheat flour with manioc or quinoa), most governments would welcome studies of the potential for improving agricultural production.

Then again, it may not be agriculture itself that requires investigation, but rather a field almost completely neglected in Latin American geography: the exchange and trade in commodities at the local, regional, and national levels. One looks in vain for studies of the role of middlemen, of the truckers who form the vanguard of capitalist penetration (often rolling over roads paved by international agencies!), of the gatekeepers to urban wholesale markets, and the like. Given the spatial dimensions of the patterns and processes of trade, it is remarkable that so few have studied these critical phenomena (Carroll et al. 1984).

Political Dimensions of Development

No one who studies Latin America can fail to recognize the significance of political factors in the development process. Indeed, one might argue that Latin American development is, and always has been, at root a political process. At all levels—from presidential or congressional power, to the influence of local authorities and status groups and individuals—what has happened, where, and at what pace is a product of politics. Economic geographers would do better to focus upon questions of political economy, rather than the economics of development; it is power that affects the generation and maintenance of most of the processes that are included under the term "development."

Yet, the past decade has seen relatively few major studies of political actions in Latin America. Perhaps most significant, and overtly supportive of subnational groups' rights against the dominant state mechanisms, are those of Nietschmann (1984, 1985). He, like Ballard (1984), has become involved in one of the most politically sensitive areas within Latin America, and is to be applauded for his moral stance and the skill he has demonstrated in applying his ecological expertise to the Miskito nation's cause. Others who have examined political issues include Augelli (1980b), who provides an excellent interpretation of the Dominican borderlands problems, Caviedes (1984b), who synthesizes a great deal of material concerning the authoritarian state of the southern cone, Glassner (1983), who discusses the problems of landlocked states, Johnson (1982), who analyzes peasant struggles in Mexico, and R. Sternberg (1985), who demonstrates the politics of hydroelectric developments in Brazil.

However, for an overview of the geopolitics of the region at large, one must turn to the excellent analysis of political-scientist Child (1985), and if one examines the *Political Geography Quarterly,* one finds only English contributions (Momsen 1984; M. Morris 1986). Given the dramatic state of the political crises that face many countries, one would expect more interest to be shown by geographers. Two studies that attempt to provide a national perspective on development deserve mention: Tata (1982) and Hall (1985).

Peoples and Landscapes of the Present

In the haste to analyze, it is right to pause and celebrate those who have provided rich, insightful interpretations of the peoples, cultures, and landscapes of Latin America. It is a tradition that should be valued, and one that will no doubt be reappraised by the "new" cultural geography school. The past decade has seen major ethnographic studies from Pennington (1979, 1980, 1983) on various northern Mexican groups. Dickenson (1982) has provided a model of landscape appreciation that should be required reading for all geographers. Porteous (1986) has penetrated within the structure of Lowry's novels to enrich our understanding of landscapes of the mind. Johannessen (1981) has argued cogently for Chinese traits in Guatemala, and is now deep into a study of the presence of maize in India well before the arrival of the Spanish in the New World. Hunter and De Kleine (1984) have recently described the surprisingly widespread occurrence of geophagy in Central America.

These and other studies (e.g., Gade 1983b) provide abundant evidence of the enduring nature of cultural traits, absorbing, rejecting, and diverting the process of modernization that a few decades ago was thought to be all-powerful. We should rejoice in such diversity and differentiation: for many of us to sit on a hill, and soak it in, is more than adequate recompense for the hours of plodding investigation.

Physical Conditions and Ecological Change

The Latinamericanists involved in research into physical geography remain small in number, but they have produced a distinguished record of scholarship. The two most prolific scholars are Veblen (Veblen et al. 1980, 1981; Veblen 1983a, 1983b, 1985), who has produced a series of superb studies of forest structures in central-southern Chile; and H. Sternberg (1987a, 1987c), whose knowledge of the physical parameters of Amazonia is probably unequalled. Others have examined paleoclimates (Alexander 1982), doline morphology (Day 1983), the El Niño phenomenon (Caviedes 1984a), Caribbean climates (Granger 1985), Andean environmental change (Parsons 1980, 1982), and the vegetation of northeastern Brazil (Johannessen 1983). It is clear that environmental hazards and the ecological basis of agriculture have diverted the attention of many others from strictly physical-geography topics.

Texts and Interpretations

In this brief survey of the field during the last decade, mention should be made of the major texts that are available for university students interested in Latin America (Williams 1980). Some of the old stalwarts have clearly seen better days (James and Minkel 1986), based as they are upon outdated paradigms. Another (West and Augelli 1988), refurbished and extended, presents a superb synthesis of Middle America, but lacks a companion volume on South America. The Bromleys (1982) and Morris (1981) present concise and highly generalized accounts of Latin American development, and Preston (1987) does the same via multiple authors. The Blouets (1982) and Blakemore and Smith (1983) provide a mixed collection of essays from distinctive specialist views.

It is clear that there is a need for new texts that incorporate the very distinctive perspectives of Latinamericanists in North America: the ecological, the political economy, the historical, etc. Perhaps one can no longer hope (nor should one expect) for a single-volume synthesis of the entire continent.

A GLIMPSE AHEAD

The last few years have seen the publication of several festschriften, and other salutes to distinguished Latinamericanist scholars (Davidson and Parsons 1980; Robinson 1980b; Denevan 1988), each of which allows a most valuable glimpse into the minds and around the boots of some of our most distinguished predecessors. The opportunity to record for posterity all that can be found should not be missed, and the lesser scholars also probably have just as much to say as those who have commanded the heights.

In looking ahead, one cannot but mention the year 1992, the celebration of the "encounter of the two worlds," surely a golden opportunity for Latinamericanists to make major contributions to the literature. One such project that will undoubtedly make its mark is the study organized by Denevan on aboriginal agriculture (with National Endowment For the Humanities funding). Yet, it is sad to note the paucity of such research proposals.

Presented here are a few of the tasks offered as worthy of study, if necessary by teams of researchers under the coordination of CLAG:

- First, there is the matter of our absolute lack of good atlases. In comparison with the superb products on Anglo-America (National Geographic Society 1988), and Canada (Harris and Matthews 1987), there exists only a thin historical product (Lombardi and Lombardi 1983). One hopes that a cartographic specialist would consider at least a modern atlas of the subcontinent, if not a historical companion volume.
- Second, it is time to fully assess the state of geography in Latin America, its curriculum, its geographical societies, its resources, and the ways in which geographers could assist in its development. Perhaps after centuries of extracting geographical resources, it is time to repay our disciplinary debts, and not in expensive English-language articles and books! Given the recent innovations in microcomputers, in GIS, and other "hot" technologies (Harnapp 1979), it might be opportune to provide our colleagues south of the border with new tools and training.
- Third, there is great need for a geography of colonialism in Latin America. Historians are more than ever interested in regions and spatial structures, and geographers should help them in their quest for an understanding of this formative phase. If only someone would write sweeping syntheses like those of Sauer (1966), Spate (1979), or Meinig (1986)!
- Fourth, Latin America provides an excellent arena in which to test Wallerstein's world-systems theory. Given the early entry of capitalism, the differential response to it in the Latin periphery, and the second stage of late nineteenth-century expansion of the world economic system via Britain and the U.S., the exercise would be most rewarding.
- Fifth, the formative years of the nineteenth century cry out for analysis by geographers. The rise of the nation state and recent challenges to its hegemony (Slater 1985, 1986) make it a perfect target for investigation. The making of the new mining, agricultural, commercial, and urban landscapes demands study, and there are photographs, maps, and documents of all types awaiting the enterprising geographer.
- Sixth, in the near future the opportunity should be seized to integrate the interpretations of our more distant neighbors with the Hispanics within the U.S.: what better opportunity than the encounter of 1992?
- Seventh, since CLAG will reach the ripe old age of 20 in 1990, the time has come to reassess its structure and function. While it serves admirably as a focus of interest, it has not proven to be a very useful generator of ideas or research proposals. Perhaps a joint meeting with our European colleagues (cf. Kleinpenning and Hoetink 1988) to discuss such matters would prove stimulating.
- Eighth, in view of the dispersion of Latinamericanist publications, it would be in the interest of the specialty group to consider the advantages of a "Hispanic American Geographical Review." One reason for our relative lack of identity and visibility is that there is no one place to which one can turn to find the best of our scholars' publications.
- Ninth, there is a need to address the issues of the "new" cultural geography, and the "new" regional geography (Pudup 1988). If our work is worthy of

consideration by our non-Latinamericanist colleagues, then we must confront these new perspectives with confidence. Our cause will not be served by burying our heads in the empiricism of the past.

☐ Tenth, perhaps the most profitable way to ensure that our Latinamericanist interests are appreciated by others would be to organize joint sessions with other groups, as is done on a regular basis among members of the specialty groups of the Institute of British Geographers. If we had sessions with our African or Anglo-American colleagues on poverty, indebtedness, capital flight, regional decline, and the like, then perhaps we would all realize that cultural relativism requires multicultural perspectives, and we should not limit ourselves to just two. A good dose of outbreeding might be just what we all need.

REFERENCES

Alexander, C. 1979. Modernization and rural population movements in western Puerto Rico. *Journal of Interamerican Studies and World Affairs* 23:523–50.

———. 1982. A comparative study of modern and ancient beach morphologies: Insights into the paleoclimate of Margarita Island, Venezuela. *Journal of Geology* 90:663–78.

Antonini, G. A., and Boswell, T. D. 1979a. El uso de censos en la regionalización agro-socioeconómico en un país latinoamericano. *Revista Geográfica* 90:9–31.

———, and York, M. A. 1979b. Integrated regional development and the role of the university in the Caribbean: The case of the Plan Sierra, Dominican Republic. *Revista Geográfica* 90:97–113.

——— et al. 1982. Chambo: Evolución del paisaje y su relación con el potencial productivo agrícola. *Revista Geográfica* 96:5–24.

Arreola, D. D. 1980. Landscapes of nineteenth-century Veracruz. *Landscape* 24:27–31.

———. 1982. Nineteenth-century townscapes of Eastern Mexico. *Geographical Review* 72:1–19.

Augelli, J. P. 1980a. Latin American geography in the seventies: Inventory and prospects. In *Geographic research on Latin America: Benchmark 1980*, eds. T. Martinson and G. S. Elbow, pp. 467–74. Muncie, IN: Conference of Latin Americanist Geographers.

———. 1980b. Nationalization of Dominican borderlands. *Geographical Review* 70:19–35.

———. 1987. Costa Rica's frontier legacy. *Geographical Review* 77:1–16.

Ballard, P. L. 1984. Toward indigenous liberation: The Sandinistas and the Miskito of Northeast Nicaragua. *Antipode* 16:54–64.

Barker, M. L. 1980. National parks, conservation and agrarian reform in Peru. *Geographical Review* 70:1–18.

Barrett, E. 1981. The king's copper mine, Inguaran in New Spain. *The Americas* 38:1–29.

———. 1987. *The Mexican copper industry*. Albuquerque, NM: University of New Mexico Press.

Barrett, W. 1979. Jugerum and caballería in New Spain. *Agricultural History* 53:423–37.

Bergman, R. W. 1980. *Amazon economics: The simplicity of Shipibo wealth*. Dellplain Latin American Studies, No. 6. Ann Arbor, MI: University Microfilms International.

Blakemore, H., and Smith, C. T. 1983. *Latin America: Geographical perspectives*. 2d ed. New York: Methuen.

Blouet, B. W., and Blouet, E. M., eds. 1982. *Latin America: An introductory survey*. New York: Wiley.

Bromley, R. 1981a. The colonization of humid tropical areas in Ecuador. *Singapore Journal of Tropical Geography* 2:15–26.

———. 1981b. From calvary to white elephant: A Colombian case of urban renewal and marketing reform. *Third World Planning Review* 2:205–32.

———. 1981c. Market centers and market places in highland Ecuador: A study of organization, regulation and ethnic discrimination. In *Cultural transformations and ethnicity in modern Ecuador,* ed. N. E. Whitten, pp. 233–59. Urbana, IL: University of Ilinois Press.

———. 1982. Working in the streets: survival strategy, necessity or unavoidable evil? In *Urbanization in contemporary Latin America,* eds. A. Gilbert, J. E.

Hardoy, and R. Ramírez, pp. 59–77. New York: Wiley.

———. 1983a. Development planning in adversity. *Yearbook of World Affairs* 37:187–203.

———. 1983b. The urban road to rural development: Reflections on USAID's 'urban functions' approach. *Environment and Planning A* 15:429–32.

———, ed. 1984. *Planning for small enterprises in Third World cities*. Oxford: Pergamon Press.

———, and Bromley, R. D. F. 1982. *South American development: A geographical introduction*. Cambridge: Cambridge University Press.

———, and Birkbeck, C. 1984. Researching street occupations of Cali: The rationale and methods of what many would call an "informal sector" study. *Regional Development Dialogue* 5:184–203.

Brown, L. A., and Lawson, V. A. 1985. Migration in Third World settings: Uneven development and conventional modeling: A case study of Costa Rica. *Annals of the Association of American Geographers* 75:29–47.

Bunker, S. G. 1985. *Underdeveloping the Amazon: Extraction, unequal exchange, and the failure of the modern state*. Urbana, IL: University of Illinois Press.

Carroll, T. et al. 1984. Exploration of rural-urban linkages and market centers in highland Ecuador. *Regional Development Dialogue* 5:22–69.

Caviedes, C. 1984a. El Niño 1982–1983. *Geographical Review* 74:267–90.

———. 1984b. The southern cone: Realities of the authoritarian state. Totowa, NJ: Rowman & Allanheld.

Chardon, R. 1980. A quantitative determination of a second linear league used in New Spain. *Professional Geographer* 32:462–66.

Child, J. 1985. *Geopolitics and conflict in South America: Quarrels and neighbors*. New York: Praeger.

Clarke, G. C. 1986. *East Indians in a West Indian town: San Fernando, Trinidad*. London: Allen & Unwin.

Clawson, D. L. 1984. Religious allegiance and economic development in rural Latin America. *Journal of Interamerican Studies and World Affairs* 26:499–524.

———, ed. 1986. *CLAG 1986 Yearbook*. Muncie, IN: CLAG Publication No. 12.

———, and Hoy, D. R. 1979. Nealticán, Mexico: A peasant community that rejected the 'green revolution,' *American Journal of Economics and Society* 38:371–87.

Cook, N. D. 1981. *Demographic collapse: Indian Peru, 1520–1620*. Cambridge: Cambridge University Press.

Craig, A. K. 1979. Exploration of Eastern Peru by the Junta de Vias Fluviales. *Revista Geográfica* 90:199–212.

Davidson, W. V., compiler. 1980. *Geographical research on Latin America: A cartographic guide and bibliography of theses and dissertations*. Muncie, IN: CLAG Occasional Publication No. 2.

———, and Parsons, J. J., eds. 1980. *Historical geography of Latin America*. Vol. 21 of *Geoscience and Man*. Baton Rouge, LA: Louisiana State University Press.

Dawsey, C. B. 1983. Push factors and pre-1970 migration to southwest Paraná, Brazil. *Revista Geográfica* 98:54–57.

Day, M. 1983. Doline morphology and development in Barbados. *Annals of the Association of American Geographers* 73:206–19.

Delson, R. M., and Dickenson, J. P. 1984. Perspectives on landscape change in Brazil. *Journal of Latin American Studies* 16:101–25.

Denevan, W. M., ed. 1978. *The role of geographical research in Latin America*. Muncie, IN: CLAG Publication No. 7a.

———. 1982. Hydraulic agriculture in the American Tropics: Forms, measures, and recent research. In *Maya subsistence: Studies in memory of Dennis E. Puleston,* ed. K. V. Flannery, pp. 181–203. New York: Academic Press.

———. 1984. Ecological heterogeneity and horizontal zonation in the Amazon floodplain. In *Frontier expansion in Amazonia,* eds. M. Schmink and C. H. Wood, pp. 311–36. Gainesville, FL: University of Florida Press.

——— et al. 1984. Indigenous agroforestry in the Peruvian Amazon: Bora Indian management of swidden fallows. *Interciencia* 9:346–57.

———, Mattewson, K., and Knapp, G., eds. 1987. *Pre-Hispanic agricultural fields in the Andean region*. International Series No. 359. Oxford: B.A.R.

———, ed. 1988. *Hispanic lands and peoples: Selected writings of James J. Parsons*. Dellplain Latin American Studies, No. 23. Boulder, CO: Westview Press.

Dickenson, J. P. 1982. *Brazil*. Essex: Longman.

Dickinson, J. C. 1982a. Development planning at the interface of mountain and plain: A Venezuelan example. *Mountain Research and Development* 2:32–48.

———. 1982b. Honduras country environmental profile. In *AID Honduras Country Profile*. Washington: USAID.

Doolittle, W. E. 1980. Aboriginal agricultural development in the valley of Sonora, Mexico. *Geographical Review* 70:328–42.

———. 1983. Agricultural expansion in a marginal area of Mexico. *Geographical Review* 73:301–13.

———. 1984. Settlements and the development of statelets in Sonora, Mexico. *Journal of Field Archaeology* 11:13–24.

Driever, S. L., and Hoy, D. R. 1982. Población potencial de los mayas durante el período clásico. *Revista Geográfica* 96:25–37.

Dunbar, G. S. 1988. The compass follows the flag: The French scientific mission to Mexico, 1864–1867. *Annals of the Association of American Geographers* 78:229–40.

Edwards, C. 1980. Geographical coverage of the sixteenth century *Relaciones de Indias* from South America. *Geoscience and Man* 21:75–82.

Eidt, R. C. 1984. *Advances in abandoned settlement analysis: Application to prehistoric anthrosols in Colombia*. Milwaukee, WI: University of Wisconsin Press.

Elbow, G. S. 1983. Determinants of landuse change in Guatemalan urban centers. *Professional Geographer* 35:57–65.

Everitt, J. 1984. Belmopán, dream and reality: A study of the other planned capital in Latin America. *Revista Geográfica* 99:121–33.

Ewald, U. 1985. *The Mexican salt industry, 1560–1980*. Stuttgart: Gustav Fischer.

Fuller, G.; Murton, B.; and Lewis, N. D. 1984. Crisis in graduate student foreign area field study. *Geographical Perspectives* 53:15–26.

Gade, D. 1983a. Foreign languages and American geography. *Professional Geographer* 35:261–6.

———. 1983b. Lightning in the folk life and religion of the Central Andes. *Anthropos* 78:770–88.

———, and Escobar, M. 1982. Village settlement and the colonial legacy in Southern Peru. *Geographical Review* 72:430–49.

Gaile, G. L., and Hanink, D. M. 1984. Caribbean airline connectivity and development. *Caribbean Geography* 1:272–83.

Galloway, J. H. 1979. Agricultural reform and the enlightenment in late colonial Brazil. *Agricultural History* 53:763–79.

———. 1985. Tradition and innovation in the American sugar industry, c.1500–1800: An explanation. *Annals of the Association of American Geographers* 75:334–51.

Gerhard, P. 1972. *A guide to the historical geography of New Spain*. Cambridge: Cambridge University Press.

———. 1979. *The southeast frontier of New Spain*. Princeton: Princeton University Press.

———. 1982. *The north frontier of New Spain*. Princeton: Princeton University Press.

Gilbert, A. G., and Ward, P. 1985. *Housing, the state and the poor: Policy and practice in Latin American cities*. Cambridge: Cambridge University Press.

Glassner, M. I. 1983. The transit problems of land-locked states: The cases of Bolivia and Paraguay. In *Ocean Yearbook 4,* eds. E. Mann and N. Ginsburg, pp. 366–89. Chicago: University of Chicago Press.

Godfrey, B. J. 1982. Xingu Junction: Rural migration and land conflict in the Brazilian Amazon. *Proceedings of the Pacific Coast Council on Latin American Studies* 9:71–82.

Gonzalez, A. 1985. Latin America: Population, food supply and agricultural dependency. *Revista Geográfica* 101:91–96.

Granger, O. 195. Caribbean climates. *Progress in Physical Geography* 9:16–43.

Greenow, L. L. 1983. *Credit and socioeconomic change in colonial Mexico: Loans and mortgages in Guadalajara, 1720–1820*. Dellplain Latin American Studies, No. 12. Boulder, CO: Westview Press.

———. 1984. City and region in the credit market of late-colonial Guadalajara, Mexico. *Journal of Historical Geography* 10:263–78.

Griffin, E., and Ford, L. 1980. A model of Latin American city structure. *Geographical Review* 70:397–422.

Hall, C. 1985. *Costa Rica: A geographical interpretation in historical perspective*. Dellplain Latin American Studies, No. 17. Boulder, CO: Westview Press.

Harnapp, V. R. 1979. Landsat as a land evaluation tool in Gulf Coast Mexico. *Revista Geográfica* 90:151–6.

———. 1985. Agricultural development and environmental change: Expanding ranching in the Huasteca Region of Mexico. In *Latin America: Case studies,* pp. 143–51. Dubuque, IA: Kendall Hunt.

Harris, R. C., and Matthews, G. J., eds. 1987. *From the Beginning to 1800*. Vol. 1 of *Historical atlas of Canada*. Toronto: University of Toronto Press.

Hausladen, G., and Wyckoff, W. 1985. Our discipline's demographic futures: Retirements, vacancies and appointment priorities. *Professional Geographer* 37:339–43.

Hecht, S. 1985. Environmental development and politics: Capital accummulation in the livestock sector in Eastern Amazonia. *World Development* 13:663–84.

Hiraoka, M. 1980. Settlement and development of the Upper Amazon: The East Bolivian example. *Journal of Developing Areas* 14:327–47.

———. 1985. Mestizo subsistence in riparian Amazonia. *National Geographic Research* 1:236–46.

———, and Yamamoto, S. 1980. Agricultural development in the Upper Amazon of Ecuador. *Geographical Review* 70:423–45.

Horst, O. H., ed. 1981. *Papers in Latin American geography in honor of Lucia C. Harrison*. Muncie, IN: CLAG Special Publication Vol. 1.

———, ed. 1982. *New themes in instruction for Latin American geography*. Muncie, IN: CLAG Special Publication Vol. 2.

Hoy, D. R., and Belisle, F. J. 1984. Environmental protection and economic development in Guatemala's Western Highlands. *Journal of Developing Areas* 18:161–76.

Hunter, J. M., and De Kleine, R. 1984. Geophagy in Central America. *Geographical Review* 74:157–69.

James, P. E., and Minkel, C. W. 1986. *Latin America*. 5th ed. New York: Wiley.

Johannessen, C. L. 1981. Melanotic chicken use and Chinese traits in Guatemala. *Revista de Historia de Amárica* 93:73–89.

———. 1983. Maximización de la utilidad largo plazo de la vegetación del Sertao, nordeste del Brasil. *Síntesis geográfica*, Caracas, Univèrsidad Central de Venezuela 6:36–43.

Johnson, J. 1982. Peasant struggles in contemporary Mexico. *Antipode* 14:39–50.

Jones, R. C. 1978. Myth maps and migration in Venezuela. *Economic Geography* 54:75–91.

———. 1982. Regional income inequalities and government investment in Venezuela. *Journal of Developing Areas* 16:373–90.

Kent, R. B. 1979. Diversidad ecológica y las regiones apícolas de Costa Rica. *Revista Geográfica* 90:65–96.

———. 1983. The Municipal Development Institute and local institution building: Recent Bolivian experience. *International Review of Administrative Sciences* 49:279–87.

———, and Sandoval, A. 1984. Evolución y provisión de servicios públicos en la Sierra Central del Perú. *Revista Geográfica* 99:35–56.

Kirchner, J. A. 1980. *Sugar and seasonal labor migration: The case of Tucumán, Argentina*. University of Chicago Department of Geography Research Paper No. 192. Chicago: University of Chicago Press.

Kleinpenning, J. M. G., and Hoetink, H., eds. 1988. Latin American studies in the Netherlands, 1970–1987. *Boletín de Estudios Latinoamericanos y del Caribe*, 44. Special issue.

Knapp, G. W. 1982. Prehistoric flood management on the Peruvian coast: Reinterpreting the "sunken fields" of Chilca. *American Antiquity* 47:144–54.

———. 1986. Una perspectiva de la irrigación en los Andes del norte. *América Indígena* 46:349–56.

Kvale, K. M. 1984. *CLAG 1984 yearbook*. Muncie, IN: CLAG Publication No. 10.

Lawson, V. A. 1988. Government policy biases and Ecuadorian agricultural change. *Annals of the Association of American Geographers* 78:433–52.

———, and Brown, L. A. 1987. Structural tension, migration and development: A case study of Venezuela. *Professional Geographer* 39:179–88.

Lentnek, B., ed. 1983. *Contemporary issues in Latin American geography*. Muncie, IN: CLAG Publication No. 9.

Licate, J. A. 1981. *Creation of a Mexican landscape: Territorial organization and settlement in the Eastern Puebla Basin, 1520–1605*. University of Chicago Department of Geography Research Paper No. 201. Chicago: University of Chicago Press.

Lombardi, C. L., and Lombardi, J. V. 1983. *Latin American history: A teaching atlas*. Madison, WI: University of Wisconsin Press.

Lonsdale, R. E. 1986. The decline of foreign-area specialization in geography doctoral work. *Journal of Geography* 85:4–7.

Lovell, W. G. 1983a. Settlement change in Spanish America: The dynamics of congregación in the Cuchumatán Highlands of Guatemala. *Canadian Geographer* 27:163–74.

———. 1983b. To submit or to serve: Forced native labour in the Cuchamatán Highlands of Guatemala, 1525–1821. *Journal of Historical Geography* 9:127–44.

———. 1985. *Conquest and survival in colonial Guatemala*. Kingston, Ontario: McGill—Queen's University Press.

———. 1986. Rethinking conquest: The colonial experience in Latin America. *Journal of Historical Geography* 12:310–17.

Martinson, T., and Elbow, G. S., eds. 1981. Geographical research in Latin America: Benchmark 1980. Muncie, IN: Conference of Latin Americanist Geographers.

Mathewson, K. 1984. *Irrigation horticulture in Highland Guatemala: The tablón system of Panajachel*. Boulder, CO: Westview Press.

———. 1986. Alexander von Humboldt and the origins of landscape archaeology. *Journal of Geography* 85:50–56.

Meinig, D. W. 1986. *Atlantic America, 1492-1800*. Vol. 1 of *The shaping of America: A geographical perspective on 500 years of history*. New Haven, CT: Yale University Press.

Momsen, J. 1984. Caribbean conflict: Cold war in the sun. *Political Geography Quarterly* 3:145–51.

Morrill, R. L., and Angulo, J. J. 1979. Spatial aspects of a smallpox epidemic in a small Brazilian city. *Geographical Review* 69:319–30.

Morris, A. 1981. *Latin America: Economic development and regional differentiation*. London: Hutchinson.

Morris, M. A. 1986. Maritime geopolitics in Latin America. *Political Geography Quarterly* 5:43–55.

Mower, R. D. et al. 1983. *Remote sensing in Colombia*. Grand Forks, ND: Copy Cat Press.

Murphy, M. E. 1986. *Irrigation in the Bajío Region of Colonial Mexico*. Dellplain Latin American Studies, No. 19. Boulder, CO: Westview Press.

National Geographic Society. 1988. *Historical atlas of the United States: Centennial Edition*. Washington: National Geographic Society.

Newson, L. A. 1985. Indian population patterns in colonial Spanish America. *Latin American Research Review* 20:41–74.

———. 1986. *The cost of conquest: Indian decline in Honduras under Spanish rule*. Dellplain Latin American Studies, No. 20. Boulder, CO: Westview Press.

———. 1987. *Indian survival in colonial Nicaragua*. Norman, OK: University of Oklahoma Press.

Nietschmann, B. Q. 1979a. *Caribbean Edge*. New York: Bobbs-Merrill.

———. 1979b. Ecological change, inflation, and migration in the far western Caribbean. *Geographical Review* 69:1–24.

———. 1984. Misurasata/Sandinista Negotiations. *Cultural Survival Quarterly* 9:59–61.

———. 1985. Miskito and Kuna struggle for nation. In *Tribal peoples and development issues: A global overview*, ed. J. Bodley, pp. 271–80. Mountain View, CA: Mayfield Press.

Ormrod, R. K. 1979. The evolution of soil management practices in early Jamaican sugar planting. *Journal of Historical Geography* 5:157–70.

Parsons, J. J. 1980. Europeanization of the savanna lands of northern South America. In *Human ecology of savanna environments*, ed. D. R. Harris, pp. 267–89. London: Academic Press.

———. 1982. The northern Andean environment. *Mountain Research and Development* 2:253–62.

———. 1983a. Beef cattle. In *Costa Rican natural history*, ed. D. Janzen, pp. 77–79. Chicago: University of Chicago Press.

———. 1983b. The migration of Canary Islanders to the Americas: An unbroken current since Columbus. *The Americas* 39:447–81.

———. 1985. Raised field farmers as pre-Columbian landscape engineers: Looking north from the San Jorge, Colombia. In *Prehistoric intensive agriculture in the tropics*, ed. I. Farrington, pp. 149–65. International Series No. 232. Oxford: B.A.R.

Pennington, C. W. 1979. *Arte y vocabulario de la lengua Dohema, Have o Eudeva*. Mexico: Universidad Nacional Autónoma de México.

———. 1980. *Material Culture*. Vol. 1 of *The Pima Bajo of Central Sonora, Mexico*. Salt Lake City: University of Utah Press.

———. 1983. Tarahumara and Northern Tepehuán. In *Handbook of North American Indians*, ed. W. C. Sturtevant, pp. 276–89 and 306–14. Washington: Smithsonian Institute.

Porteous, J. D. 1986. Inscape: Landscapes of the mind in the Canadian and Mexican novels of Malcolm Lowry. *Canadian Geographer* 30:123–31.

Posner, J. L. et al. 1983. Land systems of hill and highland tropical America. *Revista Geográfica* 98:5–22.

Pozorski, T. et al. 1983. Pre-Hispanic ridged fields of the Casma Valley, Peru. *Geographical Review* 73:407–16.

Preston, D. A., ed. 1987. *Latin American development: Geographical perspectives*. Essex: Longman.

Pudup, M. B. 1988. Arguments within regional geography, *Progress in Human Geography* 12:369–90.

Pulsipher, L. M. 1986. *Seventeenth century Montserrat: An environmental impact statement*. Norwich: Geo Books.

———. 1987. Assessing the usefulness of a cartographic curiosity: The 1673 map of a sugar island. *Annals of the Association of American Geographers* 77:408–22.

Richardson, B. C. 1983. *Caribbean migrants: Environment and human survival on St. Kitts and Nevis*. Knoxville, TN: University of Tennessee Press.

———. 1985. *Panama money in Barbados, 1900–1920*. Knoxville, TN: University of Tennessee Press.

Robinson, D. J. 1979a. Córdoba en 1779: Ciudad y campana. *Gaea* 17:279–312.

———, ed. 1979b. *Social fabric and spatial structure in colonial Latin America*. Dellplain Latin American Studies, No. 1. Ann Arbor, MI: University Microfilms International.

———. 1980a. *Research inventory of the Mexican collection of colonial parish registers*. Salt Lake City, UT: University of Utah Press.

———, ed. 1980b. *Studying Latin America: Essays in honor of Preston E. James*. Dellplain Latin American Studies, No. 4. Ann Arbor, MI: University Microfilms International.

———, ed. 1981. *Studies in Spanish American population history*. Boulder, CO: Westview Press.

———. 1984. *Syracuse University technical assistance to the Integrated Regional Development Project, Peru: Final Report*. Syracuse Metropolitan Studies International Series, Monograph No. 15. Syracuse, NC: Syracuse University Press.

———. 1988. *Relación de la Provincia de Antioquia*. Medellín: Gobierno de Antioquia.

———, ed. 1989. *Migration in colonial Spanish America*. Cambridge: Cambridge University Press.

———, and McGovern, C. 1980. La migración yucateca en la epoca colonial. *Historia Mexicana* 30:99–125.

———, and Ortíz, A. 1984. La pobreza en Ayacucho. *Socialismo y Participación* 28:15–33.

———, and Greenow, L. L. 1986. *Catálogo del Archivo del Registro Público de la Propiedad de Guadalajara: Libros de Hipotecas, 1566–1820*. Guadalajara: Secretaría del Gobierno de Jalisco.

Rondinelli, D. A. 1978. *Urbanization and rural development: A spatial policy for equitable growth*. New York: Praeger.

———. 1983. *Secondary cities in developing countries: Policies for diffusing urbanization*. Beverly Hills: Sage Publications.

———. 1985. *Applied methods of regional analysis*. Boulder, CO: Westview Press.

Sauer, C. O. 1966. *The early Spanish Main*. Berkeley: University of California Press.

Scarpaci, J. L. et al. 1988. Planning residential segregation: The case of Santiago, Chile. *Urban Geography* 9:19–36.

Siemens, A. H. 1983. Wetland agriculture in pre-Hispanic Mesoamerica. *Geographical Review* 73:166–81.

Slater, D. 1985. *New social movements and the state in Latin America*. Dordrecht, Holland: Foris Publications.

———. 1986. Socialism, democracy and the territorial imperative: Elements for a comparison of the Cuban and Nicaraguan experiences. *Antipode* 18:155–85.

Smith, J. H. 1981. *Man, fishes and the Amazon*. New York: Columbia University Press.

———. 1982. *Rainforest corridors: The Transamazon colonization scheme*. Berkeley: University of California Press.

Spate, O. H. K. 1979. *The Spanish lake*. Minneapolis, MN: University of Minnesota Press.

Stanislawski, D. 1983. *The transformation of Nicaragua: 1519–1548*. Ibero-Americana, No. 54. Berkeley: University of California Press.

Sternberg, H. O'R. 1987a. Aggravation of floods in the Amazon River as a consequence of deforestation? *Geografiska Annaler* 69A:201–19.

———. 1987b. Biogeografía i desenvolvupament economic als tropics. *Revista Catalana de Geografía* 3:5–22.

———. 1987c. Life sciences and economic development in the tropics: A holistic perspective. In *The open research problems in the life sciences under tropical conditions*, eds. D. O. Hall et al., pp. 133–47. Rotterdam: A. Balkema.

Sternberg, R. 1983. Hydroelectric energy, repressed demand and economic change in Amazonia. *Acta Amazónica* 13:371–91.

———. 1985. Large scale hydroelectric projects and Brazilian politics. *Revista Geográfica* 101:29–44.

Swann, M. M. 1982. *Tierra Adentro: Settlement and society in colonial Durango*. Dellplain Latin American Studies, No. 10. Boulder, CO: Westview Press.

Swearingen, W. D. 1984. Foreign languages and terrae incognitae. *Professional Geographer* 36:73–75.

Tata, R. J. 1982. *Haiti: Land of poverty*. Washington: University Press of America.

———, and Campbell, M. I. 1985. La variabilidad de los barrios de Caracas. *Revista Geográfica* 102:81–92.

———, and Evans, A. S. 1986. Racial separation versus social cohesion: The case of Trinidad-Tobago. *Revista Geográfica* 104:23–32.

Thomas, R. N., and Wittick, R. I. 1981. Migrant flows to La Paz, Bolivia, as related to the internal structure of the city: A methodological treatment. *Revista Geográfica* 94:41–52.

Turner, B. L. 1983. *Once beneath the forest: Prehistoric terracing in the Río Bec Region of the Maya lowlands*. Dellplain Latin American Studies, No. 13. Boulder, CO: Westview Press.

———, and Harrison, P. D., eds. 1978. *Pre-Hispanic Maya agriculture*. Albuquerque, NM: University of New Mexico Press.

———, and ———, eds. 1983. *Pulltrouser Swamp: Ancient Maya habitat, agriculture, and settlement in Northern Belize*. Austin, TX: University of Texas Press.

Veblen, T. T. 1983a. Relación de los caciques y número de Yndios que hay en Guatemala, 21 de abril de 1572. *Mesoamerica* 4:212–35.

———. 1983b. Temperate broad-leaved evergreen forests of South America. In *Temperate broad-leaved evergreen forests,* ed. J. D. Ovington, pp. 5–31. Amsterdam: Elsevier.

———. 1985. Stand dynamics in Chilean nothofagus forests. In *The ecology of disturbances and patch dynamics,* eds. S. T. A. Pickett and P. S. White, pp. 35–51. New York: Academic Press.

——— et al. 1980. Structure and dynamics of old-growth nothofagus forests in the Valdivian Andes, Chile. *Journal of Ecology* 68:1–31.

——— et al. 1981. Forest dynamics in South Central Chile. *Journal of Biogeography* 8:211–47.

Ward, P. 1986. *Welfare politics in Mexico: Papering over the cracks*. London: Allen & Unwin.

Watts, D. 1988. *The New West Indies: Patterns of development, culture and environmental change since 1492*. Cambridge: Cambridge University Press.

Weil, C. 1983. Migration among landholdings by Bolivian campesinos. *Geographical Review* 73:182–97.

West, R. C., and Augelli, J. P. 1988. *Middle America: Its lands and peoples*. 3d ed. Englewood Cliffs, NJ: Prentice-Hall.

Wilken, G. C. 1987. *Good farmers: Traditional agriculture and resource management in Mexico and Central America*. Berkeley: University of California Press.

Wilkie, R. 1984. *Latin American population and urbanization analysis: Maps and statistics, 1950–1982*. Los Angeles: UCLA Latin American Center Publications, No. 8.

———, and Wilkie, J. R. 1980. *Migration and an Argentine rural community in transition*. Amherst, MA: Program in Latin American Studies, Publication No. 12.

Williams, L. S. 1980. Regional texts on Latin America: A review. *Journal of Geography* 79:32–38.

Works, M. A. 1987a. Aguaruna agriculture in eastern Peru. *Geographical Review* 77:343–58.

———. 1987b. *CLAG 1987 yearbook*. Muncie, IN: CLAG Publication No. 13.

Asia

P. P. Karan | Nanda Shrestha | David G. Dickason |
James A. Hafner | Kenji K. Oshiro | Shakib Al-Khameri

In an area extending from the Pacific and Indian Oceans to the shores of the eastern Mediterranean and reaching out to the vast wedge of central plateau, desert basins, and mountain ranges lies a great land mass of enormous diversity. This is Asia—home to two-thirds of the world's population—with a cultural tapestry of almost infinite variety. Long subjected to powerful alien forces, Asia has emerged as a vital element in the world's economic and political patterns. For the geographer interested in place and space relations, Asia presents both research challenges and opportunities, ranging from a search for broad patterns and generalizations to the investigation of a subtle regional mosaic.

This chapter, which reviews the work of American geographers on Asia, is organized in the four common divisions: East Asia, Southeast Asia, South Asia, and Southwest Asia. An introduction briefly traces the development of Asian geographic studies in North America.

ASIAN STUDIES DEVELOPMENT IN NORTH AMERICA

The foundation for serious research and field studies in Asia by American geographers was laid in the early 1900s by a small group of scholars that includes Huntington (1907), Jones (1921, 1929), Semple (1912, 1921), Orchard (1928), Trewartha (1928, 1930a, 1930b, 1934a, 1934b), Hall (1931, 1932, 1934), Davis (1934a, 1934b, 1934c), Cressey (1934), Spencer (1935), and Lattimore (1932). The research of these geographers was largely concentrated on Japan and China.

Between 1903 and 1906, Huntington made extensive field observations in Asia—the first American geographer to do so. He concluded that the historic outpourings

---

This chapter was prepared by P. P. Karan, based on materials submitted by N. Shrestha, David G. Dickason (South Asia), James A. Hafner (Southeast Asia), Kenji K. Oshiro (Japan, Korea, and Taiwan), and Shakib Al-Khameri (Southwest Asia). Additional materials on China and Japan were provided by Laurence J. C. Ma, John D. Eyre, and Allan G. Noble. The assistance of William Noble and other geographers, too numerous to mention here, is also acknowledged.

of nomadic peoples from Central Asia, which led to the Mongol conquest of India and China, and the invasions of eastern Europe in the thirteenth century, were caused by the drying up of pastures on which the nomads were dependent. This thesis was presented in his book, *The Pulse of Asia* (Huntington 1907), which made him famous as a student of climatic influences on human society. The vivid descriptions of places in his book are among the most effective examples of geographical writing.

Regional geography's popularity in the 1930s and 1940s and the need for organized knowledge about Asian countries for strategic intelligence during and after World War II led to the further expansion of research on Southeast Asia (Pelzer 1945; Broek 1942; Ginsburg and Roberts 1957; Pendleton 1943), Japan (Trewartha 1945), and Korea (McCune 1956). In particular, during the post-World War II period, the emergence of independent nations in South Asia and the establishment of the state of Israel stimulated geographic interest in South and Southwest Asia (Brush 1949; Mayfield 1955; Karan 1952; Cohen 1957; Crary 1951).

Cressey's *Asia's Lands and Peoples* (1951) provided the most complete descriptive reference source on the continent at that time. Spencer's *Asia, East By South* (1954) portrayed the historical background and cultural geography of Monsoon Asia. *The Pattern of Asia*, edited by Ginsburg (1958), presented Asian political and economic problems from a geographical standpoint. The regional chapters were authored by Brush, McCune, Randall, and Wiens, each expert in his area. This book contributed to a basic understanding of the processes of change that were radically transforming the Asian landscape.

Spencer (1973) provided an excellent introductory textbook for Asian geography courses. It presented insights into the dilemmas and problems facing the two-thirds of humanity who live in Asia. He stressed the crises created by population growth, needed agrarian reform, uneven capital accumulation, and varying government systems.

Geographic research by Murphey (1977) offered an understanding of the political-economic expansionist activities of the West in Asia. Murphey offered historical analyses of the ways in which the affected Asian societies have responded and reacted, and an assessment of the interplay between political-economic pressures by the West and the reaction patterns of the two largest nations of Asia, India and China. Murphey investigated the colonial port city and the treaty-port theme. The founding of international trading centers at port cities was aimed at transforming the traditional inward-facing economies that were dominated by profit-seeking European technologies, facilities, and energies. As Murphey saw it, the colonial port dynamics did transform the Indian administrative and economic systems. The experience in India led the West to try the same process in China. The earliest treaty-port concessions followed the Indian pattern, but control over Chinese affairs and business practices eluded European grasp. Murphey's conclusion, that India was brought into the international orbit whereas China remained largely outside of it and in control of her own economy, is well argued.

Huke (1982a) spent many years researching the human geography of Asia; his agroclimatic and dry-season maps of South, Southeast, and East Asia have contributed to the understanding of Asian environments. Three dry-season maps portrayed the length, time, and intensity of the water-deficit period for rice-producing areas from Pakistan through Korea. Each map set presented a climate-based regional division of

Table 1
Annals articles on Asia, Africa, and Latin America, 1955–87

Years	Total Articles	Asia		Africa		Latin America	
		Articles	Percent	Articles	Percent	Articles	Percent
1955–60	141	6	4.3	5	3.5	6	4.3
1961–65	160	13	8.1	5	3.1	12	7.5
1966–70	216	23	10.6	11	5.1	7	3.2
1971–75	195	12	6.2	8	4.1	11	5.6
1976–80	164	8	4.9	2	1.2	4	2.4
1981–87	220	9	4.1	3	1.4	3	1.4
Total	1096	71	6.5%	34	3.1%	43	3.9%

Source: *Annals of the Association of American Geographers* (1955–87), compiled by Nanda Shrestha. Commentaries not included.

the areas that produce the vast majority of the world's rice. Huke also studied (1982b) the human geography of rice production in southern and eastern Asia, where rice is by far the leading food, and where production must increase by at least eight million tons per year just to maintain the current level of per capita consumption. Huke's research represents the first attempt to standardize the database for Asia's rice-producing areas.

Turning to geographic instruction about Asia in North American colleges and universities, it is notable that during 1955–65 the number of geography departments offering courses on Asian geography jumped from 96 to 243, and enrollment in these courses quadrupled from 1571 in 1955 to 6159 in 1965.[1] However, between 1965 and 1985 greater emphasis on topical and systematic geography and a reduction of regional geography courses in many departments contributed to a decline of enrollment in Asian geography courses. Only 2931 students were enrolled in such courses in 1985. Recent enrollment data indicate that this declining trend is reversing, with an enrollment of 4138 in 1986 and 4476 in 1987.

The relative importance of Asian research may be judged from research articles on Asia, Africa, and Latin America in the *Annals of the Association of American Geographers* (Table 1) and in the *Geographical Review* (Table 2). Between 1955 and 1987, research papers on Asian regions and topics exceeded those on Africa and Latin America in these two major American geographical journals.

Further evidence of the relative importance of Asian geography in America can be gleaned from the number of Ph.D. dissertations on Asian topics. Doctoral dissertations completed in American universities during 1977–87 totalled 1319.[2] Only 102 dealt with Asia: 29 on East Asia, 26 on Southeast Asia, 23 on South Asia, and 24 on Southwest Asia. The relatively small number of doctoral dissertations on Asia is not surprising if one considers the number of American Ph.D. programs in geography.

---

[1] All enrollment figures in this section were calculated by Nanda Shrestha from 34 volumes (1955–88) of *Schwendeman's Directory of College Geography of the United States* (Richmond, KY: Eastern Kentucky University). The *Directory* was published annually at the University of Kentucky between 1949 and 1966, and has been published at Eastern Kentucky University since 1967.

[2] Figures on dissertations were compiled by Nanda Shrestha from various issues of *The Professional Geographer* (1977–79) and *Guide to Graduate Departments of Geography in the United States and Canada* (1984–87).

Table 2

Geographical Review articles on Asia, Africa, and Latin America, 1955–87

Years	Total Articles	Asia Articles	Asia Percent	Africa Articles	Africa Percent	Latin America Articles	Latin America Percent
1955–60	175	20	11.4	16	9.1	14	8.0
1961–65	140	19	13.6	10	7.1	13	9.3
1966–70	124	12	9.7	12	9.7	9	7.3
1971–75	119	12	10.1	10	8.4	20	16.8
1976–80	137	8	5.8	7	5.1	17	12.4
1981–87	167	27	16.2	19	11.4	11	6.6
Total	862	98	11.4%	74	8.6%	84	9.7%

Source: *Geographical Review* (1955–87), compiled by Nanda Shrestha. Commentaries not included.

According to the 1987–88 *Guide to Departments of Geography,* 49 departments award doctoral degrees in geography. Of 828 active faculty members, only 62 list Asia or some region of Asia as their research interest. Even more revealing is the fact that only 15 of these 49 doctoral programs mention Asia as one of the foci of their graduate programs. Many departments have no regional emphasis. Graduate departments at the Universities of Akron, Hawaii, and Kentucky have a major concentration of faculty with active research programs in various regions of Asia.

In 1976, with a view toward promoting geographic research and studies in Asia, geographers at the University of Akron initiated publication of the *Bulletin of Asian Geography,* edited by Laurence J. S. Ma (1976–78), P. P. Karan (1979–83), M. Rahman (1984–85), and B. L. Sukhwal (1986–present). The *Bulletin* serves as an important vehicle for the exchange of research news and activities among American geographers having Asian interests. Akron initiated the first research contacts between geographers in the U. S. and in the People's Republic of China. Akron also hosted the first Asian Urbanization Conference in 1985. Selected papers presented at this conference have been edited and published by Costa et al. (1986).

EAST ASIA

East Asia has attracted the research attention of leading American geographers. As noted above, in the early 1900s Huntington, Semple, Orchard, Trewartha, Hall, Davis, Cressey, Spencer, and Lattimore published many classic geographic studies on different regions of East Asia, based on their intensive field research and wide travels. Both Cressey (1934) and Trewartha (1945) published their classic regional studies of China and Japan more than four decades ago. McCune's geography of Korea (1956) was published several years after the Korean War ended.

During 1949–76, there were no formal contacts between geographers in the U.S. and in the People's Republic of China. American geographers having scholarly interest in China—such as Wiens (1954) and Tuan (1969)—published studies that were based on archival research. Field research was limited to the Republic of China in Taiwan.

In 1977, formal contact between Chinese and American geographers was initiated by the visit of a delegation of 10 American geographers led by Professors Allen G.

INTERNATIONAL UNDERSTANDING THROUGH REGIONAL SYNTHESIS

Noble and Laurence J. C. Ma (under the sponsorship of the Ohio Academy of Sciences). Based on their visit, Ma and Noble (1979) provided an excellent assessment of developments in Chinese geography. A 10-member delegation of Chinese geographers then visited the U.S. in 1978. Their American program included a symposium in which the two groups of geographers met and discussed their varied concerns.

Papers presented at the symposium were edited by Ma and Noble (1981). The contributions—which focused upon agricultural development, settlement, and the environment; the effects of water on the environment; urban development and the environment; and environmental monitoring—offer a good sampling of the state of geography in the People's Republic of China and the direction and practical nature the discipline has taken in recent years.

Since the establishment of formal contact with geographers in the People's Republic, several American geographers have been actively pursuing research in China. Among them are Clifton Pannell and C. P. Lo (University of Georgia), Laurence J. C. Ma and Allen G. Noble (University of Akron), Mei-Ling Hsu (University of Minnesota), and Sen-dou Chang (University of Hawaii). Their major works (noted below) represent important American contributions on the geography of contemporary China. The research emphasis has been on development and modernization, population and urbanization, water issues, energy, environment, agricultural land use, and historical cartography.

Pannell (1983) has edited a set of essays on East Asia. Following his brief introduction, Thomas gives a factual survey of physical geography. Kessler then sketches early Chinese history, and McColl contributes an essay on "How the Chinese View Themselves." Beers follows with "Dimensions of Revolution: China and Japan in the 17th and 20th Centuries." Kornhauser gives a brief history of Japanese urbanization, followed by Davis's essay on "China and Japan on the Eve of Modernization." A brief factual survey of population growth and distribution in each East Asian country is given by Wang. Williams analyzes "Patterns of Economic Development," contrasting Japan, Taiwan, and China. Knapp reviews the changing American attitudes toward China since 1949. Greer offers a short survey of the Yellow River water problem. Salter's essay on "China's March on Nature" focuses upon Tachai, a theme emphasized by Salter (1972, 1978) in earlier papers. McCune provides an essay on the political geography of the Korean Peninsula. In the last chapter Pannell reiterates East Asia's economic and political importance. Williams adds a useful survey of the cartographic and remote sensing coverage of East Asia.

China and Taiwan

Pannell and Ma (1983) have pioneered American geographic studies on China during the post-Nixon era. They provide an up-to-date geography of the development and modernization of the People's Republic. Human-environment issues, spatial organization, problems of modernization, and economic development are emphasized. The strength of this study lies in the fact that some topics that are aspatial but significant in understanding today's China are emphasized, including the central-planning process, the communist ideology, and the frequent changes of policy.

Population and urban studies dominate American geographic research on China. Ma (1983) and Hsu (1985) have made important contributions on population geography. Ma compares the 1982 census and earlier population statistics on China. New changes in urban definitions and designations since 1983 have significantly undermined the consistency and relevance of the "total population of cities and towns" statistics in representing the post-1982 urban population numbers. Ma and Cui (1987) identify the major types of China's urban population and relate their relations to areal units. This research clarifies "urban" definitional complexities.

Thirty-three years ago, when Cressey published his revised edition of the geography of China (1955), he gave it the title *Land of the 500 Million*. Today, the population of the country is 1076 million and still growing. In recent years the Chinese government, in an attempt to curb population growth, has established the policy of limiting each family to one child. This method of population control has been more successful in cities than in rural areas. Using the 1982 Chinese census and data from a national sampling on fertility, Hsu (1985) examined the urban-rural contrasts in demographic and marital behavior, rates of growth, and implementation of the one-child policy. Hsu probed the reasons for rural couples wanting larger families and found that the traditional concerns of old age and family propagation are more important to them than are economic factors.

Chang, Hsu, Lo, Ma, Noble, and Pannell have made significant contributions to our understanding of urban population growth and urbanization trends in China. A special issue of *Urban Geography*—1986, 7(4)—contains papers authored by these geographers. Ma and Noble identify a number of important issues related to China's cities and the urbanization processes. Most of the issues discussed are new, stemming from the reforms that moved ahead with full speed in the early 1980s, but a few older areas of inquiry that have not been pursued by scholars are also included. Ten major research themes for study are identified:

1. Level of urbanization
2. The role of cities in stimulating rural economic growth
3. The changing status of counties
4. Internal migration
5. City size and urban-growth policy
6. Policy changes and spatial patterns of growth
7. The Chinese model of urban structure
8. Internal structure and functioning of the urban economy
9. The Chinese approach to urban and regional planning
10. Urban development in Hong Kong and Taiwan

China's modernization program has significantly altered traditional urban systems. Chang (1981) has examined China's economic development and national building activities and their contribution to the emergence of a more integrated national urban system. Lo, Pannell, and Welch (1977) revealed that recent land-use changes in two Chinese cities have led urban growth and development to a distinctive urban spatial structure associated with communist planning and philosophy. Data on Shenyang and Canton, derived from historical sources, maps, aerial photographs, and Landsat-1 images, were used to describe changing patterns of land use and urban structure. The authors assert that, despite their different functional roles and land-use zones, the spatial structure of the two cities appears to be evolving toward socialist goals of classlessness and uniformity. Tan (1986a) has examined the socioeconomic conditions that contributed to the decline of small towns in China between 1949 and 1979 and their revitalization since 1979. In a subsequent paper, Tan (1986b) investigated the effect of this transformation on rapid urbanization, and discussed the problems and prospects of small towns as designated growth poles in China's urban geography.

Among other studies on Chinese urbanization, mention must be made of the rural-urban migration research by Chan (1988). Chan constructs a series of annual net rural-urban migration estimates for post-1949 China, and examines the temporal patterns of rural-urban migration and urbanization and their relation with economic growth. Chan shows that the volume of net rural-urban migration since 1949, accounting for about half the urban population increase, is by no means small, and is certainly more than was previously realized. He concludes that periods of high net migration generally coincide with periods of better economic performance.

Murphey (1975) authored an interesting essay on a Maoist model for urban development. The main elements of this model include limits on the growth of existing big cities, the allocation of new urban and industrial investments—primarily to previously neglected or underdeveloped parts of the country—and the fostering of regional and local self-sufficiency so that urban and industrial functions need not be so heavily concentrated in a few giant centers. In a later publication (1980), Murphey argued a strong case for city-based industrialization.

For students of Chinese city planning, a bibliography by Ma (1980) is of great value. Ma and Hanten (1981) have edited a volume of essays on the geographical, political, and economic aspects of the Chinese city, beginning with urbanization in the late nineteenth and early twentieth centuries and emphasizing the contemporary urban scene in the People's Republic. Ma writes a fine introductory essay on the city in modern China, followed by an account of the evolution of the urban administration in Shanghai, China's largest metropolis. Chan's essay on the central-place system in Guangdong province is based on 1912–49 data derived from primary Chinese sources. Using central functions as the key indicator of centrality, he classifies the central places of Guangdong into seven levels. Chan's conclusion that the size of a central place is positively related to its number of central functions is not surprising to geographers who are familiar with the earlier works of Skinner (1964) and others. The value of Chan's contribution lies in the fact that the specific types of central functions at various levels of the central-place hierarchy are clearly identified. Pannell analyzes the growth and change in China's urban system since 1840. The patterns of China's urban growth, especially the growth of its larger cities, are analyzed through rank-size relations of the Chinese cities for the years 1843, 1937, 1953, and 1970. Pannell's essay is a valuable contribution to the scant existing literature on China's urban system and the major forces shaping that system. Buck's essay discusses the factors favoring the growth of medium-sized and small urban centers, and Kwok surveys the major trends of urban planning and development. Kwok's discussion on industrial location, community facilities, suburban development, satellite towns, urban housing, neighborhood design, land use, urban renewal, and urban-rural relations is valuable. Fung critically examines the factors contributing to uncontrolled urban expansion and the wasteful use of urban and suburban land. Ma, in the last essay of this collection, deals with urban housing stock, neighborhood structures, and urban population density.

Land use, population, and economic base are the "Holy Trinity" of the American planner. In all three areas there have been enormous panoramic changes in China. Most of the planning publications in China look to the year 2000 as a kind of watershed. Per capita income is expected to increase to $800–1000. Development will proceed from the coastal provinces inland, and on a very large scale will begin to reach the central provinces and spill over into the underdeveloped West. Currently, however, there is still a relatively impressive degree of interregional spatial equality of incomes (Gaile 1983).

Lee and Schmidt (1986) analyzed the quantity and quality of the knowledge of the urban environment possessed by the residents of Guangzhou. Sketch maps drawn by 146 first-year college geography majors were categorized by the authors on the basis of accuracy and dominant elements. Factors that contributed to the frequency and type of elements included on the maps were considered. Relations among map coverage, map detail, urban environment, and respondent characteristics are discussed. The authors conclude that the absence of detailed, distinct information, distance-decay effects, and the fragmented character of the maps reflect the early stages of information processing by new urban arrivals and the restricted travel fields of the respondents. Leung and Ginsburg (1980) have edited a volume of essays on urbanization and national development.

American geographers have not been very active in examining recent changes in the urban geography of Hong Kong and Taiwan. The population dynamics of the colony have been researched by Lo (1986). Because sovereignty over Hong Kong will return to China on July 1, 1997, geographic research on the current and future spatial relations between Hong Kong and China should be fruitful.

Pannell's monograph (1973) on Taichung is a significant comprehensive study of the city's structure and function, including its history, location, administration, land use, journey-to-work of the residents, and urban planning. More than one-third of the monograph is devoted to a description and analysis of land use. Pannell's study reveals that declines in land values and in the intensities of land use are similar to the patterns in Western cities, a function of distance from the city center. This conclusion is reached mainly through the use of a correlation analysis and is explained in part by the theory of land rent. Pannell suggests that there are probably universal economic factors as well as unique cultural reasons that affect urban development in Taiwan. Pannell's work is the most exhaustive and detailed study yet of a Taiwanese city.

Geographers have examined urban environmental issues and land-use patterns. Karan and Chao (1986) have critically examined the perception of environmental pollution in Taipei. Their results are compared with similar studies of an Indian city to provide some generalizations on the perception of environmental pollution in non-Western cities. Their comparative analyses of environmental perception within the non-Western cities help to refine theories of environmental cognition. Selya (1977) has written a fine essay on urban agriculture. Williams, Sutherland, and Chang (1988) have produced an excellent map of land use in the Taipei Basin.

The development of and uses of water and energy resources in China have been studied by Greer (1979) and Smil (1979). The development of the water resources of the Yellow River (Huang Ho) is the subject of Greer's book. He provides a detailed discussion of the river's behavior, uneven precipitation and discharges, silt runoff, and silt content. He describes the long history of Chinese water management in the basin, and reviews Western interest in the Yellow River problem between the 1920s and late 1940s, the emergence of basinwide development schemes, and the rise of modern Chinese hydraulic engineering. The progress of the complex utilization of the Huang Ho, involving the construction of 46 dams on the river, is discussed by Smil (1979). Liu and Ma (1983) consider the need for water transfer from the Chang Jiang (Yangtze) to the basins in northern China, which have experienced water shortages. The authors analyze the potential environmental effects of these water-transfer projects.

China has made remarkable progress in developing its economy and resources since 1949. Where does the country get the energy resources to serve as the basis of economic growth? Smil (1976) has carefully sifted through information from various sources to provide a general picture of China's energy problems and prospects.

American geographers have investigated the problems of environmental degradation in China. The best work on this topic is by Smil (1984), who deals with four types of environmental problems: land, air, water, and vegetation. In each case, Smil cites policies that have led to a worsening environmental situation. Reforestation programs have resulted in net decreases in forested areas and in increased soil erosion. The expansion of grain cultivation under a "grain first" policy has resulted in a net loss of arable land and the spread of desertification. Flood-control projects have increased the risks of floods. Industry, which was designed to improve the quality of life, has produced pollution, overcrowding, and health hazards.

Smil concludes that environmental degradation is severe and worsening in China, and lists several contributing factors:

- Repeated policy failures
- Poverty and the rising expectation of improved standards of living
- Indifferent and irresponsible behavior on the part of overlapping, competing, understaffed, and underfunded bureaucracies
- Past tendencies under Mao to put abstract ideology ahead of simple economic realities
- Inescapable ecological imperatives
- Unrealistic production goals and inappropriate technology

Chang (1984) discusses urban environmental quality in China, and Selya (1975) reviews Taiwan's water and air pollution.

A microscale observation in southeastern Qinghai-Xizang (Tibet) by Ives (1981) led him to speculate that in the more-distant past this high-plateau area had considerable forest cover. Subsequent reconnaissance in western Sichuan and northwestern Yunnan by Messerli and Ives (1984) led to the conclusion that extensive mountain deforestation occurred long before 1930.

Agriculture and land-use patterns in China have not been as thoroughly researched by American geographers as have urbanization patterns. Pannell (1985) examined China's changing agricultural policies during the past three decades, reflecting governmental efforts to improve production and increase yields. The most recent policy—the responsibility system—allows farmers to participate in cropping and marketing decisions. Regional variations in yields and per capita output also were analyzed by Pannell. Chang (1982) provided an appraisal of China's agricultural landscapes. Welch, Lo, and Pannell (1979) have written a very informative essay on mapping China's new agricultural land. Welch, Pannell, and Lo (1975) have mapped the land use in Northeast China from satellite imagery.

A major focus of research by American geographers is the thematic mapping of contemporary China. Hsieh and Salter (1973) produced a fine *Atlas of China* characterized by a clarity and simplicity of maps. The topical organization and special portrayal of regional features provide a useful framework for students of Chinese geography. Major economic and political developments in China make some of the maps dated now, but the *Atlas* remains a useful publication. More specific cartographic research is represented by the work of Hsu (1978), who discusses early Chinese cartography and identifies an earlier beginning of the analytical tradition in Chinese cartography. Williams (1974) has authored a very useful carto-bibliography of China.

Cultural geographic research in China has not attracted as much attention as has urban/economic geography. Knapp's work (1986) represents a major contribution to the cultural theme, with his studies on the traditional Chinese peasant dwellings. The principles of architecture and the cosmic symbolism of Chinese cities, palace buildings, large temples, and gentry mansions are emphasized. Detailed accounts of the different types of rural houses and their construction are given, for both the Chinese mainland and Taiwan. Much of the information presented is based on Knapp's extensive field experience.

Another volume on Taiwan edited by Knapp (1980) examines the theme of settlement patterns and economic development from the perspective of historical geography. Hsieh provides examples of the linguistic origins of Taiwanese placenames. And Hsu, Pannell, and Wheeler analyze the development and structure of transportation networks on Taiwan between 1600 and 1972, using graph theory to study connectivity and accessibility. Williams wrote an essay on the sugar industry for this volume.

Sinkiang, Mongolia, and Tibet

During the last 40 years American geographers have not contributed to research in Sinkiang, Mongolia, and Tibet because of the difficulty of fieldwork and travel in these regions. The works of Lattimore (1962), Wiens (1951, 1966) and Jackson (1962) on Central Asia are among the most important.

The impact of Chinese ideology on Tibet's landscape has been analyzed by Karan (1976). In 1950 Chinese forces marched into Tibet, forcing the capitulation of her newly installed god-king, the fourteenth Dalai Lama. The years since have witnessed dramatic changes in this once isolated plateau. Chinese rule has brought about the construction of roads, airfields, modern communication networks, the collectivization and modernization of agriculture, public education, and the first public-health facilities in Tibet's history—all creating a vastly changed material and cultural landscape. The impact of the Han colonization and the extensive changes wrought in the Tibetan economy are examined by Karan.

Japan and Korea

Few countries have changed so much within so short a time as Japan. In recent years American geographers such as Hall, Kornhauser, Eyre, Harris, Wheatley, and Oshiro have made contributions to our understanding of the changing geographic patterns in Japan. The emphasis of Hall's 1976 book is on the changes taking place in Japan. Much of "traditional" Japan remains in customs, ways of thought, art, and agriculture, but the focus of Hall's work is upon Japan's fast-developing, modern, urban-industrial society.

Kornhauser's book on Japan (1982) represents a notable geographic contribution. He has also authored a valuable bibliography of Japan (1984) which is of special interest to geographers. Eyre, who has long studied Japan, has made numerous scholarly contributions. A recent one (1982), which deals with a detailed study of Nagoya, is among the few comprehensive studies on individual cities or urban agglomerations in Japan. It is based on Eyre's extensive fieldwork and repeated visits during a 20-year period, on consultations with local officials and scholars, on substantial Japanese literature on the history, geography, and economy of the city and region, and on official city and statistical sources. Eyre provides an excellent case study of Japanese urbanization.

Harris (1982) examines aspects of the urban and industrial transformations that have swept over Japan. Attention is directed to the transformation of the country as a whole: the growth in urban population, the associated demographic transition, the interrelated changes of the occupational structure, and the remarkable industrial development of Japan. Harris also discusses the areal expression of this transformation in the internal patterns of Japan:

- The massive migration streams from peripheral rural regions to metropolitan and industrial clusters
- The Pacific Industrial Belt in which most large-scale industry is localized
- The areal concentration of urban population in the Tokaido Megalopolis
- The distribution of central places
- The localized pattern of the largest cities
- The large metropolitan areas with their commuting flows from the suburbs and satellites to central cities
- The rise of new towns on the expanding peripheries

Wheatley and Sea (1978) have made a major contribution to understanding Japan's urban tradition. Japan is examined as an example of a "secondary" urban generation since influences from outside the islands are clearly evident in the rise of the Japanese urban tradition. As Wheatley (1971) did in his earlier work, the authors adopt an ecological perspective. Geographers interested in the origin and growth of cities cannot afford to ignore these two excellent volumes.

Gigantic commercial centers exist at principal transportation nodes in several large Japanese cities. Cybriwsky (1988) developed a model which generalizes the characteristic components of these Japanese commercial centers: a transportation hub; a staging area; zones for department stores and office towers; an area for restaurants, bars, and coffee shops; specialized shopping streets; an amusement quarter; and a love-hotel zone. These places attract huge crowds, and the author uses the Shibuya center in Tokyo as an example to illustrate the model.

Oshiro has made several contributions to the agricultural and rural aspects of Japan. The migration of seasonal labor from the agricultural areas of Japan is one of the major themes of Oshiro's research (1975a, 1978, 1984). Seasonal labor migration has been practiced in Japan since the Meiji period. Recently, the number of seasonal labor migrants from the agricultural regions of Hokkaido, Tohoku, Hokuriku, Kyushu,

Shikoku, and Chugoku has increased. Oshiro asserts that seasonal migration is an adaptive mechanism to overcome the problem of insufficient household income from farming caused by the inadequate size of the farm. Oshiro concludes that in recent years a large number of farm wives have entered the seasonal migration stream to work in the industrialized and urbanized areas of Japan, and that rural households have come to accept the seasonal separation of family members as a necessary part of their yearly cycle of survival.

Several factors led to a decline in the number of seasonal migrants in the 1980s (Oshiro 1984). Some farmers who had engaged in seasonal migration had become weary of their double life as farmer and migrant factory worker. The preference of employers for younger workers, as well as the expanded opportunities for employment in rural areas away from the factories or public-work projects, reduced the number of seasonal migrants. Oshiro found that improved modes of transportation and increased car ownership extended the commuting distance that a farmer could travel from his farm home to factory workplace.

Oshiro (1985, 1987) authored papers on Japan's rice and wheat production. Though the consumption of wheat products increased in Japan, wheat production declined steadily to its lowest level in 1973. A major shift in agricultural policy has led to a reversal of this trend. The increased level of food self-sufficiency and the conversion of paddy land to wheat production under government guidance have contributed to the increase in wheat acreage. Oshiro (1975b) also researched the diffusion of dairy cows in a post-war Japanese settlement in Tohoku. McCune (1975) has authored an excellent monograph on agricultural change in the Ryukyu Islands.

The Korean Peninsula served as a "bridge" between China and Japan, and much of Japan's civilization and culture, which derive largely from China, came through Korea. The geographic study of Korea has generally been neglected by Americans, despite the strategic and economic importance of the peninsula. McCune's (1966) study remains the major work on this area. Most significant among recent research on this region is the work by Corey (1980, 1981, 1984). He has articulated the application of the transactional metropolitan planning paradigm for Seoul, Korea and the use of qualitative methodology such as the Program Planning Model for South Korea.

SOUTHEAST ASIA

Geographic research on Southeast Asia by Americans is relatively recent, dating from the post-World War II period. American geographers have made important recent contributions to the study of various aspects of this area, a composite region lying in the contact zone of the Indian and Chinese civilizations. Fryer, Dutt, Ulack, and Pauer provide systematic geographic assessment of the entire region. Fryer's (1979) book, based on many years of observation and research in the area, is the first comprehensive study of the region by an American geographer.

The volume on Southeast Asia edited by Dutt (1985) provides an excellent introduction to the region. Eleven authors have contributed essays on systematic and regional geography. Political, environmental, cultural, economic, urban, and planning aspects of Southeast Asia are covered in the first half of the book. Bladen and Karan have authored an essay on the geopolitical base of the region that provides the frame-

INTERNATIONAL UNDERSTANDING THROUGH REGIONAL SYNTHESIS

work for understanding the economic and political developments in the area. Noble describes the physical environment, Ma writes on culture, Barton on agriculture, Withington on the economic infrastructure and rural-urban change, and Dutt discusses planning in Southeast Asia. Regional geography is assessed in essays by Withington on Indonesia, Dickason on Burma, Cutshall on the Philippines, Lo on Malaysia, Dutt on Indochina, and Dutt and Liversedge on Singapore, Thailand, and Brunei.

Ulack and Pauer's *Atlas of Southeast Asia* (1988) is an admirable reference work. An overview is provided in the first section, and country-by-country surveys in maps and text comprise the second section of the *Atlas*. Special maps, such as those on Southeast Asian refugees and guerrilla activities, highlight this fine work.

Topical and regional research by American geographers in Southeast Asia is largely concentrated in Indonesia, Singapore, Malaysia, Thailand, and the Philippines. Other areas such as Burma, Laos, Kampuchea, and Vietnam have not been studied because of political difficulties related to field research and travel in these countries.

Among the most significant studies are contributions on the development and environment in peninsular Malaysia by Aiken et al. (1982); Fryer and Jackson's study of Indonesia (1977); and Wernstedt and Spencer's work on the Philippines (1967). Leinbach (1982, 1983, 1986a, 1986b) reviews the role of transportation in development and modernization. This research examines the issues of network structure, the social and economic impacts of accessibility improvements, and systems planning for rural and urban development in Indonesia and West Malaysia. A recent work by Leinbach and Sien (1988) provides an assessment of the transport dimension of the development process in this region. A related theme, which has linked accessibility and the evolving systems of urban places, is represented in studies of the geography of modernization. A notable example of research in this area is Leinbach's analysis (1982) of the relations between road development and indices of modernization in West Malaysia.

Interest in the form, structure, and process of urbanization and in urban/regional planning has figured prominently in the work of American geographers on this region. Reflected here is a broad spectrum of historical, structural, and planning perspectives. The city as a center of social, economic, and political change in a historical and colonial context has been examined by Doeppers (1976, 1984), Cobban (1976), and Ginsberg (1955), among others. Reed's study (1976b, 1978) is the most authoritative work on Southeast Asian urbanization before the Industrial Revolution. He deftly traces the beginnings of Philippine urbanism and interprets the establishment, early growth, and character of Manila. Reed discusses the establishment of Manila in 1571, the layout of a major colonial capital, a conspicuous morphological symbolism, and the heyday of the early growth of Manila as a cosmopolitan entrepôt in the exchange of New World silver for Old World handicrafts. He notes that, in contrast to other contemporary European ventures in Asia, the Spanish alone sought to control and to convert an extensive territory and populace. Reed observes that the Spanish used the establishment of urban centers and quasi-urban mission villages as strategies for control and conversion.

Yeung and Lo (1976) have edited a book on Southeast Asian cities. The structure and problems of the contemporary primate city landscapes in the Philippines, Thailand, West Malaysia, Singapore, and Indonesia have been profiled by Leinbach and Ulack (1983), Lew (1984), and Krausse (1985).

While the primate city has remained a research focus (McGee 1967), whole systems of cities and intermediate cities have received considerable attention. Withington (1985) studied Medan in northern Sumatra, and identified the "Cities of Sumatra," using urban functions and accessibility elements as indicators. More recently, Wood (1986) studied Samarinda and Balikpapan in eastern Kalimantan.

The growth of intermediate-sized cities has been evaluated for Indonesia, West Malaysia, the Philippines, and Thailand in works by Costello, Leinbach, and Ulack (1987), Ulack (1975, 1979), Withington (1988), and Leinbach (1987). Cobban (1985) has explored questions of land-use planning and planned development in Jakarta. A related concern with small-scale, decentralized rural development planning as an alternative to earlier strategies of top-down development is exemplified by Hafner's (1986) study of small-scale participatory water-resource development projects in northeast Thailand. It is also important to note the work of Drake (1981a, 1981b) on the pattern and problems of national integration and development in Indonesia.

A historical perspective on selectivity in Chinese migration to Philippine cities in the nineteenth century has been provided by Doeppers (1986). A number of geographers have explored questions of circular migration among intermediate-sized cities, rural-urban migration, urban squatter settlements, remittances and migration, and employment in the Philippines (Ulack, Costello, and Costello 1985), and in Indonesia, Thailand, and Malaysia (Ulack 1978, 1986; Ulack and Leinbach 1985; Leinbach and Suwarno 1985; McGee and Yeung 1977).

Most important among these is the work of Ulack, Costello, and Costello (1985), examining the characteristics and effects of circular migration in the Cebu and Misamis Oriental provinces in the Philippines. The authors indicate that individuals who participate in this type of migration are young, relatively well-educated, single, and economically motivated. Circulation is conceptualized by the authors as an alternative to permanent migration.

The lack of accurate data on squatters and squatter settlements in Asia is a perennial problem. Robert (1981) has provided an excellent account of squatters in Kuala Lumpur. Ulack (1978), in a study of squatter settlements in the southern Philippines, concluded that old squatter settlements had higher socioeconomic status and better locations with respect to employment and amenities, and played a more positive role in the life of the city, than did new settlements.

Chapman (1981) and Gosling and Lim (1979) have considered more general policy and planning issues attendant upon population redistribution processes, especially with respect to Southeast Asia. Of particular interest is the discussion of spontaneous rural migration and government intervention in the form of land settlements and forced resettlement, issues not widely explored in the general literature on population mobility in Southeast Asia (Gosling et al. 1978; Gosling 1979). More narrowly focused is Hafner's speculative assessment (1980) of the prospects for generating employment in intermediate-sized regional centers in northeast Thailand in the face of spontaneous and planned rural resettlement.

A number of problems consistent with the dynamics of the human-environment equation in the region are also represented in American geographic research. The study by Gosling and Lim (1983) on the Chinese in Southeast Asia considers this minority in their contemporary national settings and is distinctly interdisciplinary, although several papers by geographers are included. Huke (1985) investigated the impact of the Green Revolution's hybrid rice technology in the Philippines. Clawson's studies (1985) of small-scale polyculture suggested alternative development models for tropical agriculture. Hafner (1979) considered the dynamics of the changing human ecology in central Luzon in the eighteenth and nineteenth centuries and the emergence of Chinese market gardening as an ethnic monopoly in central Thailand in the nineteenth and twentieth centuries.

The rise in the number of Asians migrating to the U.S. is well known, but studies on the geographical origin of the new residents have attracted less attention. Desbarats (1979) analyzes the Thai migration stream to the U.S., the factors underlying migration decisions, and the destination choices of migrants and their distribution within the Los Angeles area.

Southeast Asia is a region of great complexity, which is mirrored in its marine environment with the seas, gulf, and straits, its islands and archipelagos, its marine resources, and its problems of offshore control. Morgan and Valencia (1984) have edited a handsome atlas with over 100 maps devoted to the natural environmental setting, resources, maritime jurisdictions and boundaries, fisheries, shipping, oil and gas, and pollution sources.

Hill stations, upland resorts, and tourism have provided a cluster for research and publication. Aiken (1987) most recently contributed a historical study on the "Early Penang Hill Station." Earlier, Reed (1976a, 1979) reviewed hill stations, using both Bogor (Buitenzorg) and Baguio as case studies. Spencer and Thomas (1948) provided an early postwar statement about hill stations and summer resorts in the Orient. Withington (1961) provided an addendum and update for Indonesian hill stations.

Over large areas of Southeast Asia, the dominant form of land use is shifting cultivation, which is known by a variety of regional names in Burma, Thailand, Indonesia, Malaya, the Philippines, and Vietnam. Spencer (1976) has made a major geographic contribution on this topic, concluding that shifting agriculture, when not under pressure, is really a long-range soil-and-crop-rotation system. Governments in various countries discourage shifting agriculture because it makes census-taking and taxation difficult, destroys timber, interferes with lowland water supplies, and induces violent soil erosion when carried out on slopes that are too steep or cleared too thoroughly and too frequently. Crooker (1988) has analyzed the forces of change among the opium-growing householders living in the hills of northeast Thailand.

SOUTH ASIA

The emergence of India and Pakistan from colonialism was a product of one of the twentieth-century's great nationalist movements. The old and sundered land became young nations trying to create order from turmoil, unity from divergence, decency from degradation, and progress from stagnation. American social scientists, including geographers, were challenged by the opportunity to study the geography of this absorbing, problem-rich, and truly unique subcontinent. The immense physical, cultural, social, and economic variety of South Asia presented great rewards in sorting out the facts spatially and explaining and interpreting them meaningfully. Karan (1966), Sopher (1973), and Schwartzberg (1983) earlier reviewed the geographic contributions on South Asia, and Sukhwal (1974) assembled an excellent bibliography. The emphasis here is on the major areas and topics of contemporary research by American geographers.

American geographers have carried on research in nearly all areas of South Asia (Table 3) in recent years, but India has been the major focus of their research. The majority of research on South Asia reflects the focus of American geography upon human aspects of the discipline (Table 4). A large share of the published work is on agricultural and cultural geography, urban and settlement patterns, and economic geography, including location analysis, population, and applied geography. Relatively less attention has been paid to human-environment issues, historical geography, geographic methodology, and physical and behavioral geography.

In the agricultural-geography research cluster, American geographers have examined the spatial aspects of the Green Revolution in India, including its social and economic impact, patterns of Indian grain sufficiency, problems of rural development and change in Pakistan and Sri Lanka, and India's livestock problems. Problems of irrigation in the Indus Valley, food production in Bangladesh, and the concept of land-support units have also received attention.

South Asia is the site of one of the world's most successful agricultural experiments. By its Green Revolution, India has demonstrated that high-tech agriculture, relying on hybrid seeds, intensive irrigation, and fertilization, can produce dramatically higher yields in a variety of crops. Starvation, known previously, is now rare. In 20 years India has changed from a net importer of wheat to a net exporter.

The food-grain sufficiency pattern in India has been mapped by Chakravarti (1970). Blynn (1983), Chakravarti (1973, 1986), and Wanmali and Bayliss-Smith (1984) have edited a volume of essays analyzing the spatial pattern of the Green Revolution. Among other things the essays point out the major hurdles facing the further spread

Table 3
Number of published South Asian research reports by country, subregion, and state

Number	Country, Subregion, or State	Number	Country, Subregion, or State
38	South Asia (entire region)	4	Haryana
9	Afghanistan	4	Karnataka
12	Bangladesh	5	Kerala
5	Bhutan	5	Maharashtra
246	India(entire nation)	1	Manipur
2	"North India"	1	Meghalaya
1	"Northeast India"	12	Punjab
28	"South India"	11	Rajasthan
1	"West India"	21	Tamil Nadu
6	Andhra Pradesh	35	Nepal
1	Assam	56	Pakistan
2	Bihar	1	Baluchistan
3	Bengal	23	Sri Lanka
1	Goa	14	(Bibliographies)
1	Gujarat		

Source: A.A.G. membership survey, 1987, compiled by David G. Dickason.

Table 4
Number of South Asian studies by topical area

Number	Topical Area
152	Agricultural geography
146	Cultural geography
126	Urban and settlement geography
111	Economic geography and location analysis
98	Population geography
94	Policy and applied geography
50	Human-environment issues
45	Historical geography
45	Geographic methodology
35	Physical geography
32	Behavioral geography

Source: A.A.G. membership survey, 1987, compiled by David G. Dickason.

of the new varieties of crops. In a rigorous analysis, Chakravarti (1976a) investigated the major impact of a high-yielding varieties program on food-grain production in India. Stoddard and Weerasinghe (1987) have identified agents of change in rural Sri Lanka, and Joshi (1976) has mapped agricultural patterns in far-western Nepal.

Using the Green Revolution as an example, Yapa (1979) has shown the costly errors that arise from efforts to improve productive forces by focusing exclusively upon the diffusion of technology. The adoption of high-yielding varieties of wheat and rice requires fertilizer, pesticides, and irrigation. Yapa points out that in rural areas of India small cultivators face discriminatory prices for scarce agricultural inputs when they compete with landlords and cultivators who have large holdings. The consequence of the uneven acceptance of HYVs has been an increase in income inequalities.

India supports a large cattle population which, including buffaloes, comprises an estimated 20% of the world's total. Chakravarti (1984, 1985a, 1985b) mapped the distribution of cattle and the cattle-development problems in India. His research (1987) shows that good cattle are found in areas where plentiful fodder crops are grown. The scarcity of livestock feed and the poor genetic potential of the existing stock are among the principal obstacles to improving the productivity of Indian livestock. The increasing pressure of human population on the cultivated land does not allow much space for growing fodder crops.

N. E. Shrestha (1985) has examined the political economy of the underdevelopment in rural areas, and has related it to external migration. N. E. Shrestha and Conway (1985) have discussed the population pressure, land resettlement, and development issues in Nepal. Shrestha's two papers enhance our understanding of the problems of Nepal's rural development.

Rural-development problems of Pakistan have been analyzed by Rahman (1983, 1988). Agriculture remains the mainstay of Pakistan's economy, and agricultural areas coincide with irrigable flood plains. The solution of the agricultural problem lies not so much in an increase of the agricultural area, but in the improvement of crop yields per acre—the major weakness that characterizes South Asia's rural economy.

One of the most serious economic problems posed by the partition of British India concerned water management of the Indus rivers system. Pakistan has the largest integrated irrigation plain in the world, yet the headwaters of its rivers and many important dam sites and barrages fell on the Indian side of the partition line. Much of the early relationship between India and Pakistan turned on the Indus axis, and yet, until Michel's superb study (1967), much of it was obscure. Michel's contribution represents genuine geographic scholarship: original, humane, painstaking, and, in the best sense, intellectual. The understanding of the Indus River basin presents several problems; only a holistic geographic approach can fathom the relations among soil type, climate, surface and ground water, crop cycles, institutional effects on human ecology, provincial politics, the effects of foreign aid, recent history, and contrasts in strategy and style between India and Pakistan. Without oversimplifying complex problems, Michel makes them understandable, fits them into perspectives, and carefully weighs alternatives. His book reveals the dynamic relationship between rivers and people, their institutions, and their politics.

More than 110 million people reside in an area the size of Wisconsin: 85% rural, 80% illiterate, mostly malnourished, nearly all poor! Bangladesh, the "international basketcase" as it was described in 1971, continues to fascinate academics especially, because it has negotiated nearly two decades of independence. Noble, Dutt, and Rahman (1980) have analyzed food production in Bangladesh, where only about 15% of the rice raised is of high-yielding varieties. Rice acreage and population trends have been examined, using a time-series analysis. The authors compared projected rice production with that of the estimated population to produce maps showing the regional food balance in Bangladesh. The study reveals that the areas of food deficit will increase considerably, despite the additional land under high-yielding varieties of rice. It is quite evident that even with the Green Revolution, Bangladesh is fighting a losing battle against the population explosion.

Mather and Karan (1978) have advanced the concept of land-support units needed to sustain one average human being in countries at varying stages of development and capital endowment, farm technology and skill, quality of land, and living standards. The land-support unit involves both animal and crop production: livestock is converted into equivalent cropland and incorporated therewith as a statistical measure of human support. Based on field studies and interviews, Mather and Karan found that for Bhutan Himalaya one head of livestock was equal to 0.19 acre of cropland. They converted the livestock into acres of cropland, which were then added to the acreage of cropland to give the total cropland equivalent. The total cropland-equivalent data for various regions in Bhutan were divided into the population figures to derive the land-support units. By measuring actual land-support units it is possible to show, for any developing country, how far the land required to support one human being in a given area falls short of the standard for that country. Land-support units provide a measure of the regional patterns of farming efficiency and the degree of population pressure on land resources.

Cultural geographic studies constitute the next most-frequent category of study. In general, these studies deal with a wide range of topics. Two important collections of essays on India's cultural geography were edited by Sopher (1980) and A. G. Noble and Dutt (1982). Sopher's collection, which contains essays by his former students, attempted to elucidate the spatial dimensions of Indian culture and its underlying geographical order. Sopher, who was born in Shanghai, China, came to the U.S. after World War II. Before his recent death he did field research in South Asia on a variety of cultural geographic problems; his major study was on the Hindu pilgrim circulation in the state of Gujarat (Sopher 1968).

A. G. Noble and Dutt (1982) have edited and contributed to a stimulating book containing 17 substantive essays on a variety of themes on Indian cultural geography. Among the more significant essays in this volume are those on types of Indian pastoralism by Palmieri, factors in agrarian spatial change by Murton, and religious diversity by Dutt and Davgun. Three essays focus upon items of the cultural landscape: rural house types and village patterns by Mitra, urban Brahmin houses in two southern states by Hirt, and the architecture of Kerala-style Hindu temples by W. A. Noble. Four essays deal with diet, health, and disease: cultural aspects of food use in India by Chakravarti, patterns of health and disease by Simoons, the sacred-cow concept and its impact on the economic life of India and its borderlands by Simoons, and the practice of *ahimsa* (nonviolence) applied to the protection of animal life in Western India by Lodrick. Mather and Karan discuss regional patterns of folk art in India. Sircar and Manjusri Chaki-Sircar focus upon traditional regional dance patterns. Ethnic neighborhoods in two Bihar towns are portrayed by Noble and Dhussa, and the perception of the environment by residents of Calcutta is the topic of an excellent contribution by Mookherjee.

A. G. Noble, Dutt, and Singh (1979) have made an important contribution to the geography of crime in India. Using district-level data, they have mapped crime's spatial incidence and have identified a zone of violent crime extending from Punjab to West Bengal. A. G. Noble, Dutt, and Sharma (1985) have analyzed the geographical pattern of crime in the Indian city of Ajmer, and Dutt and Venugopal (1983) have studied the regional variations of crime among Indian cities.

Dutt and Davgun (1977) have also studied the diffusion of Sikhism and the recent migration patterns of the Sikhs in India. They found that the Sikhs are largely rural in Punjab, but outside Punjab the Sikh population is overwhelmingly urban due to recent migrations. In another study, Dutt and A. G. Noble (1985) have analyzed the geographical pattern of religions in Rajasthan and have measured the diversity of those patterns in comparison to India as a whole.

Although the caste system has been a pervasive feature of the cultural geography of India, few recent studies have examined the geographic aspects of the Hindu caste system to complement Schwartzberg's census-based analyses (1965, 1968). Shamim and Chakravarti (1981) provide us with an interesting study on the caste system among India's Muslims. They examine regional variations in the caste system and relate these variations to the acculturation process between the Hindu and Muslim caste systems in different parts of India.

Dutt, Khan, and Sangwan (1985) have analyzed the spatial pattern of languages in India. They have identified several core areas of individual languages which are surrounded by frontier zones. Schwartzberg (1985) has reviewed linguistic factors in restructuring Indian states. A. G. Noble and Dhussa (1981) have analyzed the language patterns in an Indian town.

Dutt and Dhussa (1976) have made an important contribution to the literary geography of India, using literature as a source of information about the landscape, regions, and human attitudes toward the environment. Their discussion of the attitudes toward the urban landscape of Calcutta is based on several novels. A. G. Noble (1976) has analyzed the townscape of Malgudi, a town created by novelist R. K. Narayan as a microcosm of urban India.

A. G. Noble, Dutt, and Davgun (1981) have studied the socioeconomic factors affecting the marriage distance in two Sikh villages of Punjab. They found that illiterate farmers marry at short distances, whereas the artisan-laborer illiterates marry at greater distances. College graduates and members of a particular caste demonstrate a propensity for long-distance marriages; accessibility and exposure to modernization tend to expand the marriage field in India. Libbee (1980) has analyzed the spatial structure of marriage in rural India, using Indian census data.

The cultural and religious diversity of the Indian subcontinent has resulted in regional patterns of food preferences. Chakravarti (1974) authored an excellent paper on the regional patterns of food habits in India. The cultural geography of the houses in India has been studied by W. A. Noble (1987). He traced the origin of centered-courtyard houses in India and found that they are associated with terrestrial-celestial models. Based on the evidence of archaeological remains, he concluded that such models played a major role in the development of India's traditional architecture (W. A. Noble 1985).

Mather and Karan's contribution (1976) on geography and art in the Himalaya demonstrates that vernacular art mirrors historical and geographical forces and provides insight into social aspirations. In another contribution, Karan (1984a) examined important aspects of the cultural geography of Mithila as illustrated through its folk-paintings and songs.

The study of travel to holy places in India has attracted the attention of cultural geographers. Bhardwaj (1973) dealt with the phenomenon as it occurs in Hinduism. He examined pilgrim traffic to sacred places in the Himalayan foothills, and noted that contemporary pilgrimage patterns sustain a long, rich tradition. He mapped pilgrim fields on festival occasions, and considered the role of sacred places in integrating and diffusing religious beliefs. The inhibitions on social interaction among pilgrims who speak different languages and belong to different castes are discussed. Bhardwaj also notes the critical role of the religious movement in the spatial integration of India. Stoddard (1988) has written a paper on the characteristics of Buddhist pilgrimage in Sri Lanka. In another paper (1979–80), he discusses perceptions about the geography of religious sites in the Kathmandu Valley.

Ethnic divisions between the majority-Sinhalese and minority-Tamils in Sri Lanka have heightened during the last five years. Until recently, these two main ethnic groups lived in peace, and intermarriages between the Sinhalese and Tamils have been common. But since Sri Lanka gained independence from Britain in 1948, the Hindu Tamils have protested against political and economic discrimination by the dominant, mostly Buddhist, Sinhalese. Manogaran (1987) provides an important study of this ethnic conflict in Sri Lanka. Political groups in the country have exploited the ethnic divisions. Many Tamils have been successful in professional fields and in private business, and the Sinhalese have deeply rooted apprehensions of domination by the Tamils, whose ethnic kin in India's southern Tamil Nadu state number over 50 million.

The complex relations that South Asian societies have forged with the animal world around them form a distinct research focus which is exemplified in the work of Simoons, Lodrick, and Palmieri. The processes of animal domestication, the antiquity and geographical distribution of dairying, the use and avoidance of food having an animal origin, the cultural determinants of food habits, and the place of domesticated animals in their culture, rituals, and religion—all have received the close attention of these scholars.

Simoons (1968) studied the domestication of a little-known bovine, the mithan, among the hill tribes that live along the eastern margins of the Indian subcontinent. Simoons' study, based on archival research, mapped the geographic distribution of the mithan and speculated on the means and motives behind their domestication. Simoons concluded that cattle were domesticated for sacrificial reasons. Simoons and Lodrick have also examined the sacred-cow complex of Hindu India, its origins, and its manifestations (Simoons 1979; Lodrick and Simoons 1981). Their view of the sacred-cow complex as one that likely developed out of religious controversy in ancient India sparked a debate with cultural materialists, who posited a technoenvironmental rationale both for the origin of the cow's sanctity in Hinduism and for its position in Hindu India today (Chakravarti 1979).

The concept of food prejudice as a culturally defined and spatially expressed phenomenon has intensely interested Simoons (1974). He has published studies on the avoidance of fish in India, and the cultural-historical geography of dairying in Southern Asia (Simoons 1970). He has mapped the traditional limits of dairying in India and in Southeast Asia.

The cultural geography of cattle festivals and cattle rites in India has been examined by Lodrick (1984, 1987). He recorded the cow's symbolic meaning, and found that the festivals and rites mirrored regional patterns of culture and historical tradition in India. Festivals assume a range of social functions and form part of the process by which people are exposed to their cultural heritage. Festivals dedicated to cattle and to the cowherd-god Krishna reinforce traditional Hindu concepts concerning the sanctity of the cow. Lodrick's study (1981) of animal shelters is another example of research on India's rich cultural geography.

Hoffpauir and Palmieri have also advanced our knowledge of South Asian bovines. Focusing upon what he calls India's "other" bovine, Hoffpauir (1982) investigated the major economic importance of the water buffalo to Indian life. Palmieri (1980) studied the parallel importance of the domesticated yak and yak-cattle hybrids to the people living in the highlands of Nepal and Tibet.

During the past few years a number of publications on South Asian urbanization have appeared. This concern of geographers has been caused by the unprecedented growth of towns and cities in South Asia as a result of accelerated modernization and industrialization. Brush authored several important studies on the urban structure of Indian cities; his study (1968) of the intraurban population distribution in Indian cities pointed out that the pattern was too complex to conform to a single model for the entire country. In the traditional cities—the group least influenced by colonial forces—high population densities within a compact center in or adjacent to the bazaar were identified by Dutt and Amin (1986). Signs of decentralization have recently been revealed in these cities, reflecting the outward migration of centrally located high-status groups to new suburbs.

A second pattern has been noted in the British-built port cities—Calcutta, Bombay, and Madras—in which, due to commercially developed centers, relatively low population densities predominate, until a mile or so out from the center density levels increase dramatically, with multistoried, closely packed dwellings. Farther out, a gradual decline in density levels is observed (Kosambi and Brush 1988; Dutt, Barai, and Sami 1984).

A third group comprises modern planned cities, such as Jamshedpur and Chandigarh. They show relatively low population densities throughout, with no significant correlation of density with distance from the city center. A fourth pattern of population densities is revealed in South Asia's binuclear cities, where both the indigenous city and the British appendage may have similarly high densities while presenting distinctly separate nuclei. Dutt (1983) has summarized the principal aspects of the density gradients and morphology of South Asian cities.

Census data have been employed in numerous studies—such as Berry and Spodek (1971) and Dickason (1986)—to discern urban social areas. One conclusion that can be drawn from these studies, especially those that attempt to create a structural model, is that Indian urban sociospatial structures defy simplistic modeling. Caste associations and cultural and linguistic diversity (A. G. Noble and Dhussa 1981) are significant in the spatial distribution of social groups. Berry and Spodek found central high-status zones in Ahmadabad, Bombay, Madras, Pune, and Sholapur. Such dimensions have also been revealed for Delhi by Dickason, using previously untabulated Census of India data. High-status peripheral suburban developments have arisen, based upon socioeconomic class dimensions, occupation, and income. Such a trend, contrasting with central high-status locations, has been revealed by Brush's recent research on Delhi and Bombay.

A. G. Noble, Dutt, and Venugopal (1979) analyzed relations between the land value and land use in Madras, and found that commercial use played the most significant role in determining the land value in Indian urban areas. They found, however, that the rate of decline in land values with increasing distance from the central business district was not uniform in all directions.

Geographers have described and analyzed urbanization trends in India (Mookherjee 1978), Sri Lanka (Panditharatna and A. G. Noble 1988), and Nepal (N. E. Shrestha and Conway 1980). Rural-urban correlates of Indian urbanization have been examined by Dutt, Monroe, and Vakamudi (1986). Using 1981 census data for 388 districts, they found that variables which correlated strongly with urban districts included the female/male ratio, population density, female literacy, and number of household workers.

Among the numerous problems confronting urbanization in developing areas, a major one is the massive stream of low-income and generally low-skilled migrant workers directed toward the large cities and metropolitan centers. The slums and squatter settlements that are inhabited by these poor migrants form a striking feature of the urban environment. Lall (1987) studied the geographic origins of in-migrants to the slums around Chandigarh. He found that most of the migrants were young people coming from the pockets of rural economic stagnation and overpopulation in eastern Uttar Pradesh, Rajasthan, Southern Haryana, and Tamil Nadu. Distance was not a deterrent, and stepwise migration was a common strategy to overcome barriers of distance and culture. Mookherjee (1982) has described the slums of Calcutta.

In a volume of essays edited by Tietze (1988), American and foreign scholars have described and analyzed various aspects of South Asia's urban geography: the early

European suburbanization in Indo-British port cities, planned cities, urban primacy, sex-ratio variation in cities, and the slums and morphology of temple towns. Nepal's migration patterns are examined by M. N. Shrestha (1982a).

Studies in economic geography are oriented toward development problems (Tirtha 1980; Sukhwal 1987). Few studies of South Asia's industrialization have been undertaken. Siddiqi (1979a) analyzed the manufacturing pattern in Pakistan, and found that, despite the large concentrations of manufacturing in the Karachi, Lahore, Lyallpur, Hyderabad, and Multan areas, a few new areas of industrial activity are emerging in the agricultural region of the Indus plains, due to public planning policies. The public sector in Pakistan is locating industries in new areas, taking advantage of the availability of raw materials and infrastructure.

Siddiqi (1981) evaluated the impact of Pakistan's development policies on the spatial structure of economic development. He found wide spatial variations in levels of development; such factors as structural rigidities, natural calamities, nondevelopment expenditures, and population growth were responsible for regional imbalances. Siddiqi notes that one of the problems in Pakistan's "depressed" regions is the lack of interaction and transportation through which growth impulses could be transmitted.

Using 25 variables, Bladen and Karan (1974, 1975a) mapped the regional patterns of socioeconomic and environmental degradation in India. Distinct areal groups of districts emerged in the Middle Ganges Valley and Tamil Nadu, with relatively low levels of socioeconomic well-being. In general, they found that social well-being improved nearer the major metropolitan areas. This result lends support to the hypothesis that, as a developing country begins to move from a backward subsistence economy through a transitional commercial and industrial development phase and toward a modern complex socioeconomic system, spatial disparities in well-being tend to change from small in the subsistence economy phase to large in the transitional phase, and again to small in a mature, well-developed economic system.

Geographers have studied various aspects of population geography, such as the regional patterns of human fertility and acceptance of family planning in India, and population pressure in Pakistan. Four areas of relatively higher fertility and three areas of lower fertility were identified in India by Karan (1973). He found that indicators of economic development, urbanization, literacy, density of road mileage, and religious affiliation greatly influence the spatial patterns of fertility. His findings lend support to the demographic-transition theory. While economic improvements tend to stimulate fertility directly, indirectly the entire set of social changes associated with socioeconomic development—particularly manufacturing activity, urbanization, and education—tends to lower fertility levels.

Using a regression model, Siddiqi (1979b) mapped the population pressure in Pakistan in terms of the carrying capacity of agricultural land as well as the overall population-resource relation. Chakravarti (1976b) studied population changes in India and identified major population-change typologies. Population patterns and related development issues in the Himalaya have been analyzed by Karan (1987b). He found that forces involved in the fertility decline—such as improvements in the status of women, the spread of education, improved health care, lower child mortality, and employment opportunities—are operating far too slowly to produce any change in the current population growth rate (2.7% annually) in the Himalaya. The encouragement of employment-intensive growth in rural areas, along with improvements in health, education, and social equality, would go far toward solving the population and development problems.

Bladen and Karan (1975c, 1976a) have analyzed areal variations in the acceptance and diffusion of family-planning methods in India, and the relation between acceptance patterns and levels of economic development (Bladen, Karan, and Singh 1978). Their studies reveal that demographic, socioeconomic, and administrative variables explain regional variations in the acceptance of family-planning practices. Further, there is a clear areal association between a higher level of development and a higher rate of adoptions of family-planning methods.

Numerous geographic studies in South Asia have policy implications or potential applications to planning, development, and ecological-resource management. Wanmali (1987) has authored a number of significant works that focus upon government policy and its impact on the service-provision system in rural areas of India. Wanmali is critical of urban-industrial-based models of rural development which failed because the requisite mechanisms for spatial development were missing. His studies of the rural service system in 1968 and 1978 (Wanmali 1975, 1980, 1981, 1983, 1985) provided insights into the application of central-place concepts in India, and revealed the presence of most of the spatial characteristics of the central-place system. Not only spatial but also temporal nestings of service centers and service areas were found, and more of the Christallerian than Löschian hierarchies were observed. The question of the provision of basic needs in rural areas of India was investigated by Kulkarni and Karmarkar (1984).

Although ecological elements enter into many rural-development studies, most explicit nature-society geographic research has focused upon the ecology of mountain areas. Allan (1987) examined the ecological impact of modernization in the mountain areas of Pakistan. Karan and Iijima (1985) found that economic changes and population increases are creating threatening ecological conditions in the Himalaya. Bishop (1978) documented the changing geoecology of the Karnali Zone in Western Nepal Himalaya and found widespread environmental stress. Ives (1987) has presented challenging ideas on the Himalayan degradation problem, which spans the highland cultural, politicoeconomic, and physical systems. Development problems in the mountainous terrain of Sikkim and Bhutan Himalaya were discussed by Karan (1984b, 1987a). The environmental consequences of development in Sikkim are documented, and Bhutan's unique policy of development change amid environmental and cultural preservation has relevance to development planning in other mountain regions. Deforestation in South India has been studied by Wallach (1985b).

The regional and village land-resource issues and environmental conditions in Western Nepal were discussed by Zurick (1988) in the context of the degradation of Nepal's mountain environment. Zurick attributed ecological stress in agriculture to the forest/land ratio, farmland capability, and changes in land use.

The impact of tourism on the environment and economy of various areas in the Himalaya has been examined by Allan (1988), Mather and Karan (1985), and M. N. Shrestha (1982b). These studies document the growth of tourism, which has been drawn by the scenic grandeur of the Himalaya. However, little attention has been given to developing a comprehensive plan for tourism that would preserve the environment and ecological balance of the high mountain regions.

Studies in historical geography have been carried out by Schwartzberg et al. (1978), Murton (1977, 1983), Kelly (1981), Siddiqi (1984), and Wallach (1985a). Schwartzberg's historical atlas of South Asia is the best of its kind and a classic example of historical-geographic scholarship. Murton's research deals with the geographic past of areas in South India. He emphasizes development and change in the late eighteenth century to explain the character of the area. Kelly examines the agricultural changes that have taken place in the Hooghly area during the second half of the nineteenth century, in response to the demands of the Calcutta market. Siddiqi analyzed the agricultural developments in the Punjab during 1850–1900, and Wallach studied the development of an irrigation network in India's Krishna basin. Schwartzberg (1981) has cautioned researchers regarding the difficulties of using census materials effectively in historical studies.

Studies in physical geography are exemplified by the works of Bryson and Campbell (1982) on forecasting the Indian monsoon, and by Dey and Kumar (1983) on the relation between the Himalaya's winter snow cover and India's summer monsoon rainfall. Snead (1978, 1980) investigated the geomorphology of specific areas in Afghanistan and Pakistan, and Siddiqi (1978) researched Pakistan's climate. Trewartha (1982) and Huke and Sardido (1980) published papers on the monsoon and climatic changes in India.

The interest of geographers in the problems of perception and attitude formation is reflected in studies by Bladen and Karan. They investigated the perception of air pollution in six industrial towns of Chotanagpur, and found that differences in perception do exist among various cultural groups (Bladen and Karan 1976b). The importance of cultural traits as they affect perception are exemplified in the study. The perception of urban environment in Kathmandu, Nepal revealed that the perception of size and shape of places declined with increased distance from the city center, and the detailed image of different parts of the city varied directly with a person's mode of travel, travel behavior, and the attractiveness of a particular area (Bladen and Karan 1982b).

Distinct social groups have their own characteristic perception of the urban environment. Bladen, Karan, and Singh (1980) found that slum dwellers' and squatters' perceptions of an Indian city were influenced partly by visual factors, partly by economic and sociocultural reasons, and partly by their own travel behavior within the city. The exactness, areal range, and comprehensiveness of the slum dwellers' and squatters' mental maps were influenced by their level of education and by the distance between their home and workplace. The existence of an inverse relationship between the perception of places and the distance from the city center does not agree with the generalizations of urban perception studies in North America and Western Europe. A cross-cultural comparison of the perception of environmental problems in the coal-mining areas of India and the U.S. (Bladen and Karan 1975b) showed both similarities and differences in perception between developing and developed areas.

Technological hazards are acute in the rapidly industrializing countries of the Third World, where factories are concentrated in densely populated areas. Karan, Bladen, and Wilson (1986) delineated areas of technological hazards in India, and mapped the perceived zone of impact which resulted from an actual technological malfunction at Bhopal in 1984.

Medical geography and the delivery of health services in South Asia was the topic of 1980 special issue of *Social Science and Medicine* (14D, No. 3) which contains contributions by Dutt, Meade, Armstrong, Rahman, and Siddiqi. Singh and Dutta (1981) have mapped the pattern of smallpox in an Indian city. A study by Bladen and Karan (1982a) revealed a marked relation between air pollution and acute respiratory illness in Bombay. N. E. Shrestha (1988) analyzed Nepal's primary health-care system and found that its success depends largely upon the personal relationships between the medical staff of the health-care centers and their target populations.

The problem of urban noise pollution has been studied by A. G. Noble, Dutt, and Venugopal (1985) in Bangalore, and by A. G. Noble (1986) in Kandy, Sri Lanka. Cartography and mapping is another research area; Dutt and Geib (1987) produced an *Atlas of South Asia,* and Karan, Pauer, and Iijima have published large-scale maps of Nepal, Sikkim, and Bhutan which are based on field surveys.

SOUTHWEST ASIA

American geographic research on Southwest Asia began relatively recently (Bonine 1976). The first major comprehensive geography of the area ever published by an American geographer appeared less than 30 years ago (Cressey 1960). Cressey's field studies in Southwest Asia covered every region from Turkey to Afghanistan and south to Arabia. His book *Crossroads* dealt with descriptions of the land, its physical environment and resources, and an analysis of the various peoples who occupied the region, their history, and cultural developments. Cressey included a chapter on oil and its significance in the region's economy as well as to the rest of the world. At the time of its publication, *Crossroads* offered the most complete bibliography on Southwest Asian geography available.

English (1966) conducted a detailed study of city-village relations in Southern Iran, outlining the pattern of urban dominance of the rural hinterlands. His work was based on extensive first-hand knowledge of and field observation in the Kirman basin. English concluded that it is incorrect to regard a Middle Eastern city and its surrounding villages as two separate entities, each following a different line of economic and social conditions; they are closely linked.

In another contribution, Bonine (1979) discussed the basic morphology of Iranian cities, which was created by houses occupying adjacent rectangular fields and orchards. Traditional Iranian cities have an orthogonal network of streets that does not conform to the maze of irregular, twisting lanes postulated for the ideal Islamic city. Irrigation and agricultural practices have played an overriding role in the layout and morphology of Iranian cities that have grown along the existing streets and water channels.

Drysdale and Blake (1985) provide a spatial perspective on political processes in Southwest Asia. Their study is a comprehensive, balanced treatment of the political partitioning of space, as well as the accompanying disputes over territory and resources. A major strength of this study is that it presents theories from political geography and then supplies a number of examples from Southwest Asia to illustrate the concepts.

Cohen (1977) analyzes the dispute between Israel and the Arabs over Jerusalem, using the concepts and terminology of modern political geography. Cohen does not favor the territorial internationalization of Jerusalem; for reasons of internal and external security, he advocates Israeli sovereignty over the whole area. His argument is reinforced by a geopolitical analysis of Israel's development from two core areas: (1) political, centered in Tel Aviv on the coast, and (2) ideological, centered in the Upper Jordan Valley in the northeast. This historical duality is reflected today in a political division, expressed by its parliamentary representation. The political core tends to be willing to make territorial concessions to achieve peace between Israel and the Arab states, whereas the ideological core is significantly less inclined to do so. Cohen suggests restructuring the city of Jerusalem into a five-tier integrated structure of boroughs and neighborhoods, consisting largely of separate national or religious elements. Despite objections to territorial internationalization, Cohen favors international control of holy places as a step in the promotion of peace.

Surrounded by hostile Arab neighbors, Israel has a deep concern for security. Cohen (1983) examines the spatial morphology of Israel and discusses the pattern of defensible borders for the country. In 1967, Israel captured the West Bank and East Jerusalem from Jordan, the Sinai Peninsula and Gaza strip from Egypt, and the Golan Heights from Syria. At first the Israelis wanted to trade some of the conquered territory for peace. Today many Israelis regard the territories as essential to the nation's security. Some 1.3 million Palestinians live under Israeli military rule, including thousands in 30 refugee camps. The Israeli government has encouraged Jewish settlement in occupied territory, and the presence of over 60,000 settlers on the West Bank complicates the issue of exchanging land for peace.

Beeley (1978) has analyzed the Greek-Turkish boundary conflict in the Aegean Sea, one of the serious maritime disputes in the region. Apart from access to possible seabed resources, Turkey does not want the approaches to the Turkish straits to be through Greek territorial waters. Israel's maritime boundaries were studied by Glassner and Unger (1974).

Abudaud and Karan (1988) examined the evolution, functions, and spatial patterns of movement across the international boundaries of Saudi Arabia. In a study of boundary perception and movement, they used both quantitative and qualitative variables, such as the number of people crossing the boundary, distance between the point of origin and destination of journeys, number of previous journeys, travelers' socioeconomic status, and the level of socioeconomic development and political stability within Saudi Arabia.

Drysdale (1987) wrote about Jordanian access to the sea. He also published a useful spatial and social analysis of the political elite of Syria (Drysdale 1981). Soffer and Minghi (1986) have written an excellent account of Israel's security landscapes.

A useful general reference on Saudi Arabia is the planning and development atlas prepared by Stanley (1987). As regards settlement studies, Shamekh (1975) has given us a substantial contribution on Bedouin sedentarization in the Qasim region of Saudi Arabia. The Bedouins are abandoning their nomadic way of life to become sedentary, and are doing so in increasing numbers. Shamekh points out that they are settling mainly in existing rural Bedouin villages and larger urban centers, or are creating new villages. The government, in encouraging the nomads to settle, must consider the impact of sedentarization on the environment as well as the availability of water supplies, the logistics of administration, and health and other related problems which are created when spatial behavior is altered on such a large scale.

Alexander Melamid (1986, 1987) has made many fine geographic contributions on developments in the Persian Gulf states. Bonine (1980) provides a thorough study of the city of Yazd, Iran, which may serve to stimulate a more ambitious program of urban study in Southwest Asia by American geographers. Few better models can be found. A more comprehensive account of the evolution of the Iranian city of Yazd is provided elsewhere by Bonine (1979). Peter Lewis (1982, 1983) also has written two provocative essays on Iranian cities.

Problems of the changing contemporary city in Southwest Asia form the research focus of several American geographers. Bladen and Karan (1983) offer a study of the growth and morphology of Arab cities. The dire condition of the traditional *madina* and the attempts to arrest its deterioration are poignantly examined. The contributions of Costa and A. G. Noble (1986), Moustapha, Costa, and A. G. Noble (1985), and Melamid (1980) stress the urban-planning aspects of Arab cities. The authors point out that much remains to be done before we can comprehend the processes and changes that are occurring in Southwest Asian cities. One of the most valuable contributions of Costa and Noble and others is the great number of significant questions they raise, which identify problems that must be addressed not only by scholars but by planners and policy-makers as well.

Shair and Karan (1979) have contributed to the geography of pilgrimage to Mecca. They employed a multiple-regression model to explain the spatial pattern of pilgrim circulation. The results of the regression program reveal that per capita national income is the most important factor that determines the ratio of pilgrims per thousand Moslem population in a given country. Per capita income influences the decision one must make in order to undertake the pilgrimage. There is a positive correlation between the rise in the number of pilgrims and increased prosperity; the distance to Mecca and the percentage of Moslem population in a given country are secondary factors. Using archaeological data, Clark (1978), whose research interests include migration and city growth in the Middle East, has applied an interaction model to measure changes in the ease of trade in ancient Syria.

The relative lack of interest in physical geography on the part of American geographers is reflected in a paper on precipitation in Iran by Alijani and Harman (1985). Hidore and Albokhair (1982) discuss the process of sand encroachment in Al-Hasa, one of the most productive and largest oases in Saudi Arabia. They point out that encroachment is a natural phenomenon; there is little evidence that human use of the area has accelerated the encroachment.

Kolars (1982) delimited areas in Turkey that are particularly vulnerable to earthquake disasters. He estimated the number of people who live in seismic jeopardy, and describes briefly the socioeconomic conditions of these groups to reveal the potential effects of earthquakes on them.

CONCLUSION

This review reveals serious gaps in American geographic studies on Asia. This is not surprising in a nation that has spent 200 years looking across the Atlantic Ocean, drawn Eurocentrically by ethnic and cultural heritages, political goals, and economic needs. In the eyes of most North Atlantic-oriented Americans, Asia has long been perceived as the remote Orient. But that does not change what is happening in Asia. Almost unnoticed during the past three decades, Asian countries from South Korea and Japan to Saudi Arabia and Israel have assumed global importance as a result of growth in trade, finance, energy resources, and migration. In Japan, South Korea, Hong Kong, Taiwan, and Singapore, America's Calvinist work ethic has met, been fused with, and reinforced by Confucian principles. The result is a region where human aspiration is actually matched by human endeavor. Along with expanding economic growth in India and China, Americans—for the first time in history—have been trading more across the Pacific Ocean than across the Atlantic. That trend is expected to continue because trade across the Pacific accelerates at a faster pace than total world trade.

Despite the heightened economic, political, and demographic importance of Asia, only a relatively small number of American geographers are concerned with the area. There are perhaps 300 professional geographers in North America who are identifiably interested in Asia. Probably fewer than 25 of them have accounted for about half of the published geographic literature. Geographers having expertise in languages that would be useful in field research requiring personal interviews and the recording of oral traditions are still fewer. Almost none (A. G. Noble is an exception) have done field research in all of the major regions of Asia. An increasingly large proportion of the contemporary scholarly American literature on South and East Asia is by geographers native to these regions who now live in North America. This situation contrasts sharply with Southeast Asia or Southwest Asia, where most of the literature is by scholars from outside the region who have adopted these areas as a research focus.

The character of geographic research conducted by Americans in Asia has been influenced by their academic training in the U.S. The sparse contribution by Americans to the study of Asia's physical geography reflects their lesser interest in the physical side of the discipline, a condition that recently emerging environmental-ecological concerns might be changing.

Examinations of contemporary systems of production and transportation, human migration and settlement, and rural and urban landscapes comprise much of the writings of American geographers on Asia. They offer information and insight of considerable intrinsic interest, recording the human scene in its geographical context, with concern for the testing of hypotheses.

There is a virtual absence of cultural geography in the available geographic literature on China. This stands in sharp contrast to other areas of Asia. Detailed field studies by American or other geographers from outside China are rare. Also, research on China is based primarily on statistical data and other sources; only to a limited extent is it based on direct field observation. Geographers who have written on China have visited the country only briefly, and have not produced monographs on specific areas or topics based on sustained field observation, mapping, interviewing, and recorded data comparable to the extensive corpus of such studies for other parts of the world. The work of some scholars reflects more emphasis on theory or on similarities and generalizations rather than differences and uniqueness, along with greater emphasis on quantification and more formalized theory. There are still relatively few substantive works in Asian geography directed toward contemporary development problems, the optimizing of investment or location, economic regionalization, and investment disparities.

The conspicuous focus upon the phenomenon of urbanization by geographers in all regions of Asia is readily explainable. As the organizational center of human society and as the geographical phenomenon of interest in terms of its origin, morphology, functions, growth, external linkages, or its role as a center of innovation and political power, the Asian city offers a multitude of significant geographical themes for research.

Among the more pressing problems in Asia are its population and its rural development, both of which are rapidly expanding fields of contemporary research. Political geography has received relatively little attention in Asia except for studies of boundary problems. The political organization of space is of extraordinary importance in both contemporary as well as historical contexts. Regional separatist movements, ethnic conflicts, disputes over the extension of territorial seas, national integration, and the division of international river waters are but a few of the geographical issues that continue to becloud various regions of Asia. At another level, geographical aspects and the consequences of multinational capital investment and its treatment have special importance in several Asian countries.

In substantive regional works—a major tradition in European geography—American geographers have lagged behind. We also lag in rigorous ecological studies. The maintenance of the nature-society system in some sort of ecological equilibrium is recognized as a major challenge in various parts of Asia. One obvious item on our future research agenda must be the identification and analysis of fragile environments in human terms. Floods, droughts, earthquakes, and landslides in Asian lands have also had serious consequences for humanity. Geographic analyses or even case studies of such phenomena are needed. Geographers have a vital role to play in analyzing and helping to solve the problems resulting from the impact of humans on the physical environment of Asian regions.

REFERENCES

Abudaud, A. S., and Karan, P. P. 1988. *International boundaries of Saudi Arabia*. Lexington, KY: Department of Geography, University of Kentucky.

Aiken, R. S. 1987. Early Penang hill station. *Geographical Review* 77:421–39.

——— et al. 1982. *Development and environment in Peninsular Malaysia*. Singapore: McGraw-Hill.

Alijani, B., and Harman, J. R. 1985. Synoptic climatology of precipitation in Iran. *Annals of the Association of American Geographers* 75:404–16.

Allan, N. J. R. 1987. Ecotechnology and modernization in Pakistan mountain agriculture. In *Western Himalaya,* ed. S. C. Joshi. New Delhi: Sterling.

———. 1988. Highways to the sky: The impact of tourism on South Asian mountain culture. In *Impact of tourism on the mountain environment,* ed. S. C. Singh. New Delhi: Oxford University Press.

Beeley, B. W. 1978. The Greek-Turkish boundary: Conflict at the interface. *Transactions of the Institute of British Geographers* 3:351–66.

Berry, B. J. L., and Spodek, H. 1971. Comparative ecologies of large Indian cities. *Economic Geography* 47:266–85.

Bhardwaj, S. M. 1973. *Hindu places of pilgrimage in India: A study in cultural geography*. Berkeley: University of California Press.

Bishop, B. C. 1978. The changing geoecology of Karnali Zone, Western Nepal Himalaya: A case of stress. *Arctic and Alpine Research* 10:531–48.

Bladen, W. A., and Karan, P. P. 1974. Geography of socioeconomic deprivation in India. *National Geographic* 9:1–6.

———, and ———. 1975a. Inter-regional disparities of income in India. *Geographical Review of India* 37:210–20.

———, and ———. 1975b. Perception of environmental problems in coal mining areas of India and the United States. *National Geographer* 10:1–8.

———, and ———. 1975c. Spatial aspects of the diffusion of family planning methods in India. *National Geographical Journal of India* 21:1–5.

———, and ———. 1976a. Geographical patterns of acceptance of family planning methods in India. *National Geographical Journal of India* 22:25–42.

———, and ———. 1976b. Perception of air pollution in a developing country. *Journal of Air Pollution Control Association* 26:139–41.

———, and ———. 1982a. Air pollution and health in Bombay. *National Geographer* 17:1–4.

———, and ———. 1982b. Perception of the urban environment in a Third World country. *Geographical Review* 72:228–32.

———, and ———. 1983. Arabic cities. *Focus* 33(2):1–8.

———; ———; and Singh, S. 1978. Birth control practices and levels of development in India. *Journal of Geography* 77:229–37.

———; ———; and ———. 1980. Slum dwellers' and squatters' images of the city. *Environment and Behavior* 112:81–100.

Blynn, G. 1983. The Green Revolution revisited. *Economic Development and Cultural Change* 31:705–25.

Bonine, M. E. 1976. Where is the geography of the Middle East? *Professional Geographer* 28:190–93.

———. 1979. Morphogenesis of Iranian cities. *Annals of the Association of American Geographers* 69:208–24.

———. 1980. Yazd and its hinterland: A central place system of dominance in the Central Iranian plateau. *Marburger Geographische Schriften* Heft 83.

Broek, J. O. M. 1942. *Economic development of the Netherlands Indies.* New York: Institute of Pacific Relations.

Brush, J. E. 1949. The distribution of religious communities in India. *Annals of the Association of American Geographers* 39:81–98.

———. 1968. Spatial patterns of population distribution in Indian cities. *Geographical Review* 58:362–91.

Bryson, R. A., and Campbell, W. H. 1982. Year-in-advance forecasting of the Indian monsoon rainfall. *Environmental Conservation* 9:51–56.

Chakravarti, A. K. 1970. Foodgrain sufficiency patterns in India. *Geographical Review* 60:208–28.

———. 1973. Green Revolution in India. *Annals of the Association of American Geographers* 63:319–30.

———. 1974. Regional preference for food: Some aspects of food habit patterns in India. *Canadian Geographer* 18:395–410.

———. 1976a. The impact of high-yielding varieties program on foodgrain production in India. *Canadian Geographer* 20:95–110.

———. 1976b. Population growth types in India, 1961-1971. *Journal of Geography* 75:343–50.

———. 1979. Comment on "questions in the sacred cow controversy." *Current Anthropology* 20:476–77.

———. 1984. Some characteristics of spatial distribution of cattle population in India. *Rural Systems* 2:67–76.

———. 1985a. Cattle development problems and programs in India. *GeoJournal* 10(1):21–45.

———. 1985b. The question of surplus cattle in India: A spatial view. *Geografiska Annaler* 67B:121–30.

———. 1986. Green Revolution in India. In *Contributions to Indian geography: Agricultural geography,* ed. P. S. Tiwari, pp. 227–42. New Delhi: Heritage Publishing.

———. 1987. Availability of cattle fodder in India. *Geographical Review* 77:209–17.

Chan, K. W. 1988. Rural-urban migration in China 1950–1982, estimates and analysis. *Urban Geography* 9:53–84.

Chang, S-d. 1981. Modernization and China's urban development. *Annals of the Association of American Geographers* 71:202–19.

———. 1982. Restructuring China's agricultural landscapes. *Geographical Journal* (Hong Kong) 4:41–66.

———. 1984. Urban environmental quality in China: A luxury or a necessity? *China Geographer* 12:81–99.

Chapman, M. 1981. Policy implications of circulation: Some answers from the grassroots. In *Population mobility and development: Southeast Asia and the Pacific,* pp. 71–92. Development Studies Centre Monograph No. 27. Canberra: Australian National University.

Clark, J. R. 1978. Measuring changes in the ease of trade with archaeological data: An analysis of coins found at Dura Europus in Syria. *Professional Geographer* 30:256–63.

Clawson, D. L. 1985. Small-scale polyculture: An alternative development model. *Philippine Geographical Journal* 29:92–103.

Cobban, J. L. 1976. Geographic notes on the first two centuries of Djarkarta. In *Changing South-East Asian cities: Readings on urbanization,* pp. 45–57. Singapore: Oxford University Press.

———. 1985. The ephemeral historic district in Jakarta. *Geographical Review* 75:300–18.

Cohen, S. B. 1957. Israel's fishing industry. *Geographical Review* 47:66–85.

———. 1977. *Jerusalem: Bridging the four walls; a geographical perspective.* New York: Herzl Press.

———. 1983. Israel's defensible borders: A geo-political map. Paper No. 20. Tel Aviv: Jaffee Center for Strategic Studies.

Corey, K. E. 1980. Transactional forces and the metropolis: Towards a planning strategy for Seoul in the year 2000. *In the year 2000: Urban growth and perspectives for Seoul,* ed. Won Kim, pp. 54–89. Seoul, Korea: Korea Planners Association.

———. 1981. The transactional metropolitan paradigm: An application to planning the metropolitan region of Seoul, Korea. In *Proceedings of the fourth annual applied geography conference,* eds. B. Epstein and J. Frazer, pp. 39–49. Binghamton, NY: State University of New York.

———. 1984. Qualitative planning methodology: An application in development planning research to South Korea and Sri Lanka. *Development Planning Review* 3:1–14.

Costa, F. J., and Noble, A. G. 1986. Planning Arabic towns. *Geographical Review* 76:160–72.

——— et al. 1986. *Urbanization in Asia.* Honolulu: University of Hawaii Press.

Costello, M. A.; Leinbach, T. R.; and Ulack, R. 1987. *Mobility and employment in urban Southeast Asia.* Boulder, CO: Westview Press.

Crary, D. 1951. Recent agricultural developments in Saudi Arabia. *Geographical Review* 41:366–83.

Cressey, G. B. 1934. *China's geographic foundations: A survey of its land and its people.* New York: McGraw-Hill.

———. 1951. *Asia's lands and peoples.* New York: McGraw-Hill.

———. 1955. *Land of the 500 million: A geography of China.* New York: McGraw-Hill.

———. 1960. *Crossroads: Land and life in Southwest Asia.* New York: Lippincott.

Crooker, R. A. 1988. Forces of change in the Thailand opium zone. *Geographical Review* 78:241–56.

Cybriwsky, R. 1988. Shibuya Center, Tokyo. *Geographical Review* 78:48–61.

Davis, D. H. 1934a. Present status of settlement in Hokkaido. *Geographical Review* 24:386–99.

———. 1934b. Some aspects of urbanization in Japan. *Journal of Geography* 33:205–20.

———. 1934c. Type of occupance pattern in Hokkaido. *Annals of the Association of American Geographers* 24:201–23.

Desbarats, J. 1979. Thai migration to Los Angeles. *Geographical Review* 69:302–18.

Dey, B., and Kumar, B. 1983. Himalayan winter snow cover and summer monsoon rainfall over India. *Journal of Geophysical Research* 88:5471–74.

Dickason, D. G. 1986. Socio-demographic areas of Delhi. Paper presented at the annual meeting of the Association of American Geographers, Minneapolis, MN, May 5.

Doeppers, D. F. 1976. The development of Philippine cities before 1900. In *Changing Southeast Asian cities: Readings on urbanization,* eds. Y. M. Leung and C. P. Lo. Singapore: Oxford University Press.

———. 1984. *Manila, 1900–1941: Social change in a late colonial metropolis.* New Haven, CT: Yale University Press.

———. 1986. Destination, selection, and turnover among Chinese migrants to Philippine cities in the nineteenth century. *Journal of Historical Geography* 12:381–401.

Drake, C. 1981a. National integration and public policies in Indonesia. *Studies in Comparative International Development* 15:59–84.

———. 1981b. The spatial pattern of national integration in Indonesia. *Transactions of the Institute of British Geographers,* n.s. 6:471–90.

Drysdale, A. 1981. The Syrian political elite, 1966–1976: A spatial and social analysis. *Middle Eastern Studies* (London) 17:3–30.

———. 1987. Political conflict and Jordanian access to the sea. *Geographical Review* 77:86–102.

———, and Blake, G. H. 1985. *The Middle East and North Africa: A political geography.* New York: Oxford University Press.

Dutt, A. K. 1983. South Asian city. In *Cities of the World,* eds. S. Brunn and J. Williams, pp. 325–70. New York: Harper & Row.

———, ed. 1985. *Southeast Asia: Realm of contrasts*. Boulder, CO: Westview Press.

———, and Dhussa, R. C. 1976. The contrasting image and landscape of Calcutta through literature. *Proceedings of the Association of American Geographers* 8:102–6.

———, and Davgun, S. 1977. Diffusion of Sikhism and recent migration patterns of Sikhs in India. *GeoJournal* 1:81–90.

———, and ———. 1979. Religious pattern of India with a factorial regionalization. *GeoJournal* 3:201–14.

———, and Venugopal, G. 1983. Spatial patterns of crime among Indian cities. *Geoforum* 14:223–33.

———; Barai, D.; and Sami, A. 1984. Changes and characteristics of density gradients of colonial and traditional cities of India. *Asian Geographer* 3:103–9.

———, and Noble, A. G. 1985. Religious diversity patterns of Rajasthan within an Indian framework. *Asian Geographer* 4:137–46.

———; Khan, C. C.; and Sangwan, C. 1985. Spatial pattern of languages in India: A culture-historical analysis. *GeoJournal* 10:51–74.

———; Monroe, C. B.; and Vakamudi, R. 1986. Rural-urban correlates for Indian urbanization. *Geographical Review* 76:173–83.

———, and Amin, R. 1986. Toward a typology of South Asian cities. *National Geographical Journal of India* 32:30–39.

———, and Geib, M. 1987. *Atlas of South Asia*. Boulder, CO: Westview Press.

English, P. W. 1966. *City and village in Iran*. Madison, WI: University of Wisconsin Press.

Eyre, J. D. 1982. *The changing geography of a Japanese regional metropolis*. Chapel Hill, NC: Department of Geography, University of North Carolina.

Fryer, D. W. 1979. *Emerging Southeast Asia*. New York: Wiley.

———, and Jackson, J. C. 1977. *Indonesia*. Boulder, CO: Westview Press.

Gaile, G. L. 1983. Reanalyses of Chinese spatial inequality. *Professional Geographer* 35:281–83.

Ginsburg, N. 1955. The great city in southeast Asia. *American Journal of Sociology* 60:455–62. New York: Praeger.

———, ed. 1958. *The pattern of Asia*. Englewood Cliffs, NJ: Prentice-Hall.

———, and Roberts, C. F. 1957. *Malaya*. Seattle, WA: University of Washington Press.

Glassner, M., and Unger, M. 1974. Israel's maritime boundaries. *Ocean Development and International Law Journal* 1:303–13.

Gosling, L. A. P. et al. 1978. *Pa Mong resettlement*. Ann Arbor, MI: Department of Geography, University of Michigan.

———, ed. 1979. *Population resettlement in the Mekong basin*. Chapel Hill, NC: Department of Geography, University of North Carolina.

———, and Lim, L. Y. C., eds. 1979. *Population redistribution: Patterns, policies and prospects*. New York: United Nations Fund for Population Activities.

———, and ———. 1983. *The Chinese in Southeast Asia: Ethnicity and economic activity*. Singapore: Maruzen-Asia.

Greer, C. 1979. *Water management in the Yellow River Basin of China*. Austin, TX: University of Texas Press.

Hafner, J. A. 1983. Chinese market gardening in Thailand: The origins of an ethnic monopoly. In *The Chinese in Southeast Asia: Ethnicity and economic activity,* eds. L. Y. C. Lim and L. A. P. Gosling, pp. 30–45. Singapore: Maruzen-Asia.

———. 1980. Urban resettlement and migration in Northeast Thailand: The spectre of urban involution. *Journal of Developing Areas* 14:483–500.

———. 1986. View from the village: Participatory rural development in Northeast Thailand. *Community Development Journal* 22:87–97.

Hall, R. B. 1931. Some rural settlement forms in Japan. *Geographical Review* 21:93–123.

———. 1932. The Yamato Basin. *Annals of the Association of American Geographers* 22:243–92.

———. 1934. The cities of Japan: Notes on distribution and inherited forms. *Annals of the Association of American Geographers* 24:175–200.

Hall, R. B., Jr. 1976. *Japan: Industrial power of Asia*. New York: Van Nostrand.

Harris, C. D. 1982. The urban and industrial transformation of Japan. *Geographical Review* 72:50–89.

Hidore, J. H., and Albokhair, Y. 1982. Sand encroachment in al-Hasa Oasis, Saudi Arabia. *Geographical Review* 72:350–56.

Hoffpauir, R. 1982. The water buffalo: India's other bovine. *Anthropos* 77:215–38.

Hsieh, Chiao-min, and Salter, C. I. 1973. *Atlas of China*. New York: McGraw-Hill.

Hsu, Mei-Ling. 1978. The Han maps and early Chinese cartography. *Annals of the Association of American Geographers* 68:45–60.

———. 1985. Growth and control of population in China: The urban rural contrast. *Annals of the Association of American Geographers* 75:241–57.

Huke, R. E. 1982a. *Agroclimatic and dry season maps of South, Southeast, and East Asia*. Manila: International Rice Research Institute.

———. 1982b. *Rice area by type of culture: South, Southeast, and East Asia*. Manila: International Rice Research Institute.

———. 1985. The Green Revolution. *Journal of Geography* 86:248–54.

———, and Sardido, S. 1980. Climate change in India? In *WMO and IRRI, Agrometeorology of the rice crop*, pp. 173–80. Manila: International Rice Research Institute.

Huntington, E. 1907. *The pulse of Asia*. Boston: Houghton Mifflin.

Ives, J. D. 1981. High mountains and plateaus—an excursion to the roof of the world. *Mountain Research and Development* 1:79–83.

———. 1987. The theory of Himalayan environmental degradation: Its validity and application challenged by recent research. *Mountain Research and Development* 7:189–99.

Jackson, W. A. D. 1962. *The Russo-Chinese borderlands*. New York: Van Nostrand.

Jones, W. D. 1921. Hokkaido, the northland of Japan. *Geographical Review* 11:16–30.

———. 1929. An isopleth map of land under crops in India. *Geographical Review* 19:495–96.

Joshi, T. R. 1976. Factorial agricultural patterns of the farwestern region of Nepal. *Himalayan Review* 8:10–19.

Karan, P. P. 1952. Geopolitical structure of India. *Proceedings, 17th International Geographical Congress,* pp. 524–7. Washington: National Academy of Sciences–National Research Council.

———. 1966. Recent contributions to the geography of South Asia. *Cahiers de Geographie* 10:317–32.

———. 1973. Spatial patterns of human fertility behavior in India. *National Geographer* 8:1–13.

———. 1976. *The changing face of Tibet: The impact of Chinese communist ideology on the landscape*. Lexington, KY: University Press of Kentucky.

———. 1984a. Landscape, religion and folk art in Mithila: An Indian cultural region. *Journal of Cultural Geography* 5:85–101.

———. 1984b. *Sikkim Himalaya: Development in mountain environment*. Tokyo: Institute for the Study of Languages and Cultures of Asia and Africa.

———. 1987a. Environment and development in Bhutan. *Geografiska Annaler* 69B:15–26.

———. 1987b. Population characteristics of the Himalayan region. *Mountain Research and Development* 7:271–74.

———, and Iijima, S. 1985. Environmental stress in the Himalaya. *Geographical Review* 75:71–92.

———, and Chao, T. K. 1986. Perception of environmental pollution in a Chinese city: A case study of Taipei. *Journal of Asian and African Studies* 32:65–89.

———; Bladen, W. A.; and Wilson, J. R. 1986. Technological hazards in the Third World. *Geographical Review* 76:195–208.

Kelly, K. 1981. Agricultural change in Hooghly, 1850-1910. *Annals of the Association of American Geographers* 71:237–52.

Knapp, R. G. 1986. *China's traditional rural architecture: A cultural geography of the common house*. Honolulu: University of Hawaii Press.

———, ed. 1980. *China's island frontier: Studies in the historical geography of Taiwan*. Honolulu: University of Hawaii Press.

Kolars, J. 1982. Earthquake-vulnerable populations in modern Turkey. *Geographical Review* 72:20–35.

Kornhauser, D. H. 1982. *Japan*. 2d ed. London: Longman.

———. 1984. *Studies of Japan in western languages of special interest to geographers*. Tokyo: Kokon Shoin.

Kosambi, M., and Brush, J. E. 1988. Three colonial port cities in India. *Geographical Review* 78:32–47.

Krausse, G., ed. 1985. *Urban society in Southeast Asia*. Hong Kong: Asian Research Services.

Kulkarni, G. S., and Karmarkar, P. R. 1984. Provision of basic needs in rural areas of India. In *Rural public services: International comparisons,* eds. R. E. Lonsdale and E. Gyorgy. Boulder, CO: Westview Press.

Lall, A. 1987. Migration behavior of settlers in shanty towns in India and some planning implications: A case study of Chandigarh, India. *Proceedings, 8th International Symposium on Asian Studies,* pp. 1075–89. Hong Kong: International Centre for Asian Studies.

Lattimore, O. 1932. Chinese colonization in Manchuria. *Geographical Review* 22:177–95.

———. 1962. *Inner Asian frontiers of China.* Boston: Beacon Press.

Lee, Y., and Schmidt, C. G. 1986. Urban spatial cognition: A case study of Guangzhou, China. *Urban Geography* 7:397–412.

Leinbach, T. R. 1982. Towards an improved rural transport strategy: The needs and problems of isolated Third World communities. *Asian Profile* 10:15–23.

———. 1983. Rural transport and population mobility in Indonesia. *Journal of Developing Areas* 17:349–64.

———. 1986a. Occupational dynamics and migration: The case of Medan, Indonesia. *Southeast Asian Journal of Social Science* 14:1–15.

———. 1986b. Transport developments in Indonesia: Progress, problems, and policies under the new order. In *Central government and local development in Indonesia,* ed. C. MacAndrews, pp. 190–220. Singapore: Oxford University Press.

———. 1987. Economic growth, development planning and policy: Alternatives in Medan, Indonesia. *Journal of Southeast Asian Studies* 18:118–40.

———, and Ulack, R. 1983. Cities of Southeast Asia. In *Cities of the World,* eds. S. Brunn and J. Williams, pp. 371–407. New York: Harper & Row.

———, and Suwarno, B. 1985. Commuting and circulating characteristics in the intermediate-sized city: The example of Medan, Indonesia. *Singapore Journal of Tropical Geography* 6:35–47.

———, and Sien, C. L. 1988. *Southeast Asia transport: Issues in development.* Kuala Lumpur: Oxford University Press.

Leung, C. K., and Ginsburg, N., eds. 1980. *China: Urbanization and national development.* University of Chicago Department of Geography Research Paper No. 196. Chicago: University of Chicago Press.

Lew, A. A. 1984. Singapore: Some considerations of size and its impact. *Transition, Journal of the Socially and Ecologically Responsible Geographers* 14:25–30.

Lewis, P. G. 1982. The politics of Iranian place-names. *Geographical Review* 72:99–102.

———. 1983. Iranian cities. *Focus* 33(3):12–15.

Libbee, M. J. 1980. Territorial endogamy and the spatial structure of marriage in rural India. In *An exploration of India,* ed. D. E. Sopher, pp. 65–104. Ithaca, NY: Cornell University Press.

Liu, C. , and Ma, L. J. C. 1983. Interbasin water transfer in China. *Geographical Review* 73:253–70.

Lo, C. P. 1986. The evolution of the ecological structure of Hong Kong: Implications for planning and future development. *Urban Geography* 7:311–35.

———; Pannell, C. W.; and Welch, R. 1977. Land use changes and city planning in Shenyang and Canton. *Geographical Review* 67:268–83.

Lodrick, D. O. 1981. *Sacred cows, sacred places: Origins and survivals of animal homes in India.* Berkeley: University of California Press.

———. 1984. A cattle fair in Rajasthan: The Kharwa Mela. *Current Anthropology* 25:218–25.

———. 1987. Gopashtami and Govardhan Puja: Two Krishna festivals of India. *Journal of Cultural Geography* 7:101–16.

———, and Simoons, F. J. 1981. Background to understanding the cattle situation of India: The sacred cow concept in Hindu religion and folk culture. *Zeitschrift fur Ethnologie* 106:27–37.

Ma, L. J. C. 1980. Cities and city planning in the People's Republic of China: An annotated bibliography. HUD USER Bibliography Series, U. S. Department of Housing and Urban Development, Office of Policy Development and Research and Office of International Affairs. Washington: Government Printing Office.

———. 1983. Preliminary results of the 1982 census in China. *Geographical Review* 73:198–210.

———, and Noble, A. G. 1979. Recent developments in Chinese geographical research. *Geographical Review* 69:63–78.

———, and ———, eds. 1981. *The environment: Chinese and American views.* New York: Methuen.

———, and Hanten, E. W., eds. 1981. *Urban development in modern China.* Boulder, CO: Westview Press.

———, and Cui, Gonghao. 1987. Administrative changes and urban population in China. *Annals of the Association of American Geographers* 77:373–95.

McCune, S. B. 1956. *Korea's heritage, a regional and social geography.* Tokyo: Charles Tuttle.

———. 1966. *Korea: Land of broken calm.* Princeton, NJ: Van Nostrand.

———. 1975. *Geographical aspects of agricultural change in the Ryu Kyu islands.* Gainesville, FL: University Presses of Florida.

McGee, T. G. 1967. *The Southeast Asian city.* New York: Praeger.

———, and Yeung, Y. 1977. *Hawkers in Southeast Asian cities.* Ottawa: International Development Research Centre.

Manogaran, C. 1987. *Ethnic conflict and reconciliation in Sri Lanka.* Honolulu: University of Hawaii Press.

Mather, C., and Karan, P. P. 1976. Art and geography: Patterns in the Himalaya. *Annals of the Association of American Geographers* 66:485–515.

———, and ———. 1978. Concept of land support units: Bhutan. *Geografiska Annaler* 60B:28–35.

———, and ———. 1985. Tourism and environment in the Mount Everest region. *Geographical Review* 75:93–95.

Mayfield, R. C. 1955. A geographic study of the Kashmir issue. *Geographical Review* 45:181–96.

Melamid, A. 1980. Urban planning in eastern Arabia. *Geographical Review* 70:473–77.

———. 1986. Interior Oman. *Geographical Review* 76:317–21.

———. 1987. Qatar. *Geographical Review* 77:103–5.

Messerli, B., and Ives, J. D. 1984. Gongga Shan (7,556m) and Yulongxue Shan (5,596m): Geoecological observations in the Hengduan Mountains of Southwestern China. In *Natural environment and man in tropical high mountains,* ed. W. Lauer. Stuttgart: Franz Steiner Verlag Weisbaden.

Michel, A. 1967. *The Indus rivers.* New Haven, CT: Yale University Press.

Mookherjee, D. 1978. Urbanization trends in India. *Geographica Polonica* 38:79–90.

———. 1982. A profile of slums in a Third World city: Calcutta. *Ekistics* 49:476–80.

Morgan, J. R., and Valencia, M. J., eds. 1984. *Atlas for marine policy in Southeast Asian seas.* Berkeley: University of California Press.

Moustapha, A. F.; Costa, F. J.; and Noble, A. G. 1985. Urban development in Saudi Arabia: Building and subdivision codes. *Cities* 2:140–48.

Murphey, R. 1975. Aspects of urbanization in contemporary China: A revolutionary model. *Proceedings of the Association of American Geographers* 7:165–68.

———. 1977. *The outsiders: Westerners in India and China.* Ann Arbor, MI: University of Michigan Press.

———. 1980. *The fading of Maoist vision.* New York: Methuen.

Murton, B. J. 1977. Land and class: Cultural, social and biophysical integration: Tamil Nadu in the late eighteenth century. In *Land tenure and peasant in South Asia,* ed. R. E. Frykenberg, pp. 81–99. New Delhi: Orient Longman.

———. 1983. Towns and markets in the Salem district in the late eighteenth century. In *Perspectives in Indian archeology, art, and culture,* ed. K. V. Raman. Madras: New Era Publications.

Noble, A. G. 1976. The emergence and evolution of Malgudi: An interpretation of South Indian townscapes from the fictional writings of R. K. Narayan. *Proceedings of the Association of American Geographers* 8:106–10.

———. 1986. The geography of noise in the South Asian city: A case study of Kandy, Sri Lanka. In *Spectrum of modern geography: Essays in memory of Professor Mohammed Anas,* eds. M. Shafi and M. Raza, pp. 437–46. New Delhi: Concept Press.

———; Dutt, A. K.; and Singh, S. 1979. Is there a north-central sub-culture of violence in India? *National Geographical Journal of India* 25:101–11.

———; ———; and Venugopal, G. 1979. Land values and land use: Spatial relations in Madras, India. *Proceedings,* East Lakes Division, Association of American Geographers pp. 45–52.

———; ———; and Rahman, A. 1980. Food production and population growth in Bangladesh. *Asian Profile* 8:52–77.

———, and Dhussa, R. 1981. Language patterns in a North Indian town. In *New perspectives in geography,* ed. L. R. Singh, pp. 162–68. Allahabad, India: Thinkers Library.

———; Dutt, A. K.; and Davgun, S. 1981. Socio-economic factors affecting marriage distance in two Sikh villages of Punjab. *Journal of Cultural Geography* 2:13–25.

———, and Dutt, A. K., eds. 1982. *India: Cultural patterns and processes.* Boulder, CO: Westview Press.

———; ———; and Sharma, K. K. 1985. Variation in the spatial patterns of crime in Ajmer, India. *Indian Journal of Criminology* 13:52–72.

———; ———; and Venugopal, G. 1985. Variations in noise generation–Bangalore, India. *Geografiska Annaler* 67B:15–19.

Noble, W. A. 1985. Terrestrial-celestial models and the renaissance of monumental architecture in South Asia. In *India: Culture, society and economy–geographical essays in honor of Professor Asok Mitra,* eds. A. B. Mukherji and A. Ahmad, pp. 117–59. New Delhi: Inter-India Publications.

———. 1987. Houses with centered courtyards in Kerala and elsewhere in India. In *Dimensions of social life: Essays in honor of David G. Mandelbaum,* ed. P. Hockings, pp. 215–62. Berlin: Mouton de Gruyter.

Orchard, J. E. 1928. The pressure of population in Japan. *Geographical Review* 18:374–401.

Oshiro, K. K. 1975a. Intra-regional and inter-regional migration of seasonal labor from the agricultural areas in Japan: Its implications for national food policy. *Proceedings of the Association of American Geographers* 7:174–78.

———. 1975b. The role of group consensus in innovation acceptance: The acceptance of dairy cows in a postwar Japanese settlement. *East Lakes Geographer* 10:78–87.

———. 1978. Female seasonal migration from the rural areas of Japan. *East Lakes Geographer* 13:45–61.

———. 1984. Post-war seasonal migration from rural Japan. *Geographical Review* 74:147–56.

———. 1985. Mechanization of rice production in Japan. *Economic Geography* 61:323–31.

———. 1987. Changes in wheat production in Japan, 1960–1980. *Ohio Geographer* 15:66–77.

Palmieri, R. 1980. Field investigations of yak in the Nepal Himalaya. *National Geographic Society Research Reports* 12:529–34.

Panditharatne, B. L., and Noble, A. G. 1988. Trends in urban development in Sri Lanka. In *Asian urbanization: Problems and processes,* ed. W. Tietze. Berlin: Gebruder Borntraeger.

Pannell, C. W. 1973. *Tai-Chung, Taiwan: Structure and function*. Chicago: University of Chicago Press.

———, ed. 1983. *East Asia: Geographical and historical approaches to foreign area studies.* Dubuque, IA: Kendall/Hunt.

———. 1985. Recent Chinese agriculture. *Geographical Review* 75:170–85.

———, and Ma, L. J. C. 1983. *China: The geography of development and modernization*. New York: Wiley.

Pelzer, K. J. 1945. *Pioneer settlement in the Asiatic tropics*. New York: American Geographical Society.

Pendleton, R. L. 1943. Land use in Northeastern Thailand. *Geographical Review* 33:15–41.

Rahman, M. 1983. *Rural development in Pakistan: Policies and problems*. Hong Kong: Asian Research Service.

———. 1988. *Agriculture in Pakistan*. Budapest: Akademiai Kiado.

Reed, R. R. 1976a. *City of pines: The origins of Baguio as a colonial hill station and regional capital*. Berkeley: Center for South and Southeast Asian Studies, University of California.

———. 1976b. Indigenous urbanism in Southeast Asia. In *Changing South-east Asian cities: Readings on urbanization,* eds. Y. M. Leung and C. P. Lo, pp. 14–27. Singapore: Oxford University Press.

———. 1978. *Colonial Manila: The context of Hispanic urbanism and process of morphogenesis*. Berkeley: University of California Press.

———. 1979. The colonial genesis of hill stations: The genting exception. *Geographical Review* 79(4):463–68.

Robert, A. S. 1981. Squatters and squatter settlements in Kuala Lumpur. *Geographical Review* 71:158–75.

Salter, C. L. 1972. The litany of Tachai and the foolish old man: Agricultural landscape modification in Mainland China. *Professional Geographer* 24:113–17.

———. 1978. The enigma of Tachai: Landscape and love. *China Geographer* 9:43–62.

Schwartzberg, J. E. 1965. The distribution of selected castes in the North Indian Plain. *Geographical Review* 55:477–95, plus map.

———. 1968. Caste regions in the North Indian Plain. In *Structure and change in Indian society,* eds. M. Singer and B. S. Cohn, pp. 81–113. Chicago: Aldine.

———. 1981. On the sources and types of census error. In *The censuses of British India,* ed. G. Barrier, pp. 41–60. New Delhi: Manohar Publications.

———. 1985. Factors in the linguistic reorganization of Indian states. In *Region and nation in India,* ed. P. Wallace, pp. 155–82. New Delhi: Oxford and I.B.H. Publishing Co.

———. et al. 1978. *An historical atlas of South Asia*. Chicago: University of Chicago Press.

———. 1983. The state of South Asian geography. *Progress in Human Geography* 7:69–72.

Selya, R. M. 1975. Water and air pollution in Taiwan. *Journal of Developing Areas* 9:177–202.

———. 1977. Urban agriculture in Taiwan. *China Geographer* 8:15–28.

Semple, E. C. 1912. Influences of geographical conditions upon Japanese agriculture. *Geographical Journal* 40:589–603.

———. 1921. The regional geography of Turkey. *Geographical Review* 11:338–50.

Shair, I., and Karan, P. P. 1979. Geography of the Islamic pilgrimage. *GeoJournal* 3:599–608.

Shamekh, A. 1975. *Spatial patterns of Bedouin settlements in Al-Qasim region, Saudi Arabia*. Lexington, KY: Department of Geography, University of Kentucky.

Shamim, A. S., and Chakravarti, A. K. 1981. Some regional characteristics of Muslim caste systems in India. *GeoJournal* 5:55–60.

Shrestha, M. N. 1982a. Rural migration in Nepal. *East Lakes Geographer* 17:25–31.

———. 1982b. Tourism and economic development: A case study of Nepal. *Ohio Geographer* 10:63–71.

Shrestha, N. E. 1985. The political economy of economic underdevelopment and external migration. *Political Geography Quarterly* 4:289–306.

———. 1988. Human relations and primary health care delivery in rural Nepal: The case of Deurali. *Professional Geographer* 40:202–14.

———, and Conway, D. 1980. Urban growth and urbanization in least-developed countries: The experience of Nepal, 1952-1971. *Asian Profile* 8:477–94.

———, and ———. 1985. Issues in population pressure, land resettlement, and development: The case of Nepal. *Studies in Comparative International Development* 20:55–82.

Siddiqi, A. H. 1978. Spatial patterns of climate and irrigation in Pakistan: A multivariate statistical approach. *ArchivFur Meteorologie Geophysik und Bioklimatologie* 35B:345–57.

———. 1979a. Manufacturing and planning industrial development in Pakistan. *Tijdschrift Voor Economische en Sociale Geografie* 70:194–206.

———. 1979b. Population pressure in Pakistan. *Geoforum* 10:183–94.

———. 1981. Regional inequality in the development of Pakistan. *GeoJournal* 5:17–32.

———. 1984. 19th century agricultural development in Punjab: 1850–1900. *Indian Economic and Social History Review* 21:293–312.

Simoons, F. E. 1968. *A ceremonial ox of India*. Madison, WI: University of Wisconsin Press.

———. 1970. The traditional limits of milking and milk use in southern Asia. *Anthropos* 65:547–93.

———. 1974. Fish as forbidden food: The case of India. *Ecology of Food and Nutrition* 3:185–201.

———. 1979. Questions in the sacred cow controversy. *Current Anthropology* 20:467–93.

Singh, S., and Dutta, H. M. 1981. Smallpox pattern and its correlates: A case study of an Indian city. *GeoJournal* 5:77–78.

Skinner, G. W. 1964. Marketing and social structure in rural China. *Journal of Asian Studies* 24:3–43; 195–228; 363–99.

Smil, Vaclav. 1976. *China's energy: Achievements, problems and prospects*. New York: Praeger.

———. 1979. Controlling the Yellow River. *Geographical Review* 69:253–72.

———. 1984. *The bad earth: Environmental degradation in China*. Armonk, NY: M. E. Sharpe.

Snead, R. E. 1978. Geomorphic history of the Mundigak Valley. *Afghanistan Journal* Jg. 5, Heft 2:59–69.

———. 1980. Destruction and loss of archeological sites along the Makran coast of Pakistan from recent tectonic movements and severe storms. In *Proceedings of the CCE Field Symposium, Coastal Archeology Session,* ed. M. L. Schwartz, pp. 70–81. Shimoda, Japan: International Geographical Union Commission on the Coastal Environment.

Soffer, A., and Minghi, J. V. 1986. Israel's security landscapes: The impact of military considerations on land uses. *Professional Geographer* 38(1):28–41.

Sopher, D. E. 1968. Pilgrim circulation in Gujarat. *Geographical Review* 58:392–425.

———. 1973. Toward a rediscovery of India: Thoughts on some neglected geography. In *Geographers Abroad,* ed. M. W. Mikesell, pp. 110–33. University of Chicago Department of Geography Research Paper No. 152. Chicago: University of Chicago Press.

———. 1980. *Exploration of India: Geographical perspectives in society and culture*. Ithaca, NY: Cornell University Press.

Spencer, J. E. 1935. Salt in China. *Geographical Review* 25:353–66.

———. 1954. *Asia, east by south: A cultural geography.* New York: Wiley.

———. 1973. *Oriental Asia.* Englewood Cliffs, NJ: Prentice-Hall.

———. 1976. *Shifting cultivation in southeastern Asia.* Berkeley: University of California Press.

———, and Thomas, W. L. 1948. The hill stations and summer resorts of the Orient. *Geographical Review* 38:635–51.

Stanley, W. R. 1987. Planning and development atlas of Saudi Arabia. *GeoJournal* 14:479–88.

Stoddard, R. H. 1979–80. Perceptions about the geography of religious sites in the Kathmandu Valley. *Contributions to Nepalese Studies* 7:97–118.

———. 1988. Characteristics of Buddhist pilgrimages in Sri Lanka. *Geographia Religionum* 45–61.

———, and Weerasinghe, G. M. S. 1987. Agents of change in rural Sri Lanka. *Transition: Journal of the Socially and Economically Responsible Geographers* 16:2–5.

Sukwahl, B. L. 1974. *South Asia. A systematic geographic bibliography.* Metuchen, NJ: Scarecrow Press.

———. 1985a. *Modern political geography of India.* New Delhi: Sterling.

———. 1985b. Problems, probable solutions and prospects of canal water supply on the ecosystem of the Rajasthan canal command area, India. In *Resource management in drylands,* Stuttgart Geographische Studien, Band 105, pp. 115–24. Stuttgart: Geographische Institut.

———. 1987. *India: Economic resource base and contemporary political patterns.* New Delhi: Sterling.

Tan, K. C. 1986a. Revitalized small towns in China. *Geographical Review* 76:138–48.

———. 1986b. Small towns in Chinese urbanization. *Geographical Review* 76:265–75.

Tietze, W., ed. 1988. *Asian urbanization: Problems and processes.* Berlin: Gebruder Borntraeger.

Tirtha, R. 1980. *Society and development in contemporary India: Geographical perspectives.* Detroit: MI: Harlow Press.

Trewartha, G. T. 1928. A geographic study in Shizuoka prefecture, Japan. *Annals of the Association of American Geographers* 18:127–259.

———. 1930a. The Iwaki Basin: Reconnaissance field study of a specialized apple district in northern Honshiu, Japan. *Annals of the Association of American Geographers* 20:196–223.

———. 1930b. The Suwa Basin. *Geographical Review* 20:224–44.

———. 1934a. Japanese cities: Distribution and morphology. *Geographical Review* 24:404–17.

———. 1934b. Notes on a physiographic diagram of Japan. *Geographical Review* 24:400–403.

———. 1945. *Japan: A geography.* Madison, WI: University of Wisconsin Press.

———. 1982. Monsoons: With a focus on South Asia and tropical East Africa. *Journal of Geography* 81:4–11.

Tuan, Y-F. 1969. *China.* Chicago: Aldine.

Ulack, R. 1975. The impact of industrialization upon the population characteristics of a medium-sized city in the developing world. *Journal of Developing Areas* 9:203–19.

———. 1978. The role of urban squatter settlements. *Annals of the Association of American Geographers* 68:535–50.

———. 1979. The impact of migration on Iligan City, Mindanao. In *Migration and development in Southeast Asia: A demographic perspective,* ed. R. J. Pryor. Kuala Lumpur: Oxford University Press.

———. 1986. Ties to origin, remittances, and mobility: Evidence from rural and urban areas in the Philippines. *Journal of Developing Areas* 20:330–56.

———; Costello, M. A.; and Costello, M. P. 1985. Circulation in the Philippines. *Geographical Review* 75:439–50.

———, and Leinbach, T. R. 1985. Migration and employment in urban Southeast Asia: Examples from Indonesia and the Philippines. *National Geographic Research* 1:310–31.

———, and Pauer, G. 1988. *Atlas of Southeast Asia.* New York: Macmillan.

Wallach, B. 1985a. British irrigation works in India's Krishna Basin. *Journal of Historical Geography* 11:155–73.

———. 1985b. Deforestation: The view from South India. *Garden* 9:8–15.

Wanmali, S. 1975. Rural service centers in India: Present identification and acceptance of extension. *Area* 7:167–70.

———. 1980. The regulated and periodic markets and rural development in India. *Transactions of the Institute of British Geographers* 5:446–86.

———. 1981. *Periodic markets and rural development in India*. New Delhi: B. R. Publishing Corporation.

———. 1983. *Service provision and rural development in India: A study of Miryalguda Taluka*. Washington: International Food Policy Research Institute.

———. 1985. *Rural household use of services: A study of Miryalguda Taluka, India*. Washington: International Food Policy Research Institute.

———. 1987. *Geography of a rural service system in India*. New Delhi: B. R. Publishing Corporation.

———, and Bayliss-Smith, T., eds. 1984. *Understanding the Green Revolution: Agrarian change and development planning in South Asia*. Cambridge: Cambridge University Press.

Welch, R. A.; Pannell, C. W.; and Lo, C. P. 1975. Land use in Northeast China, 1973–A view from Landsat–1. *Annals of the Association of American Geographers* 65:595–96.

———; Lo, C. P.; and Pannell, C. W. 1979. Mapping China's new agricultural land. *Photogrammetric Engineering and Remote Sensing* 45:1211–28.

Wernstedt, F. L., and Spencer, J. E. 1967. *The Philippine island world*. Berkeley: University of California Press.

Wheatley, P. 1971. *The pivot of the four quarters*. Chicago: Aldine.

———, and Sea, T. 1978. *From court to capital: A tentative interpretation of the origins of the Japanese urban tradition*. Chicago: University of Chicago Press.

Wiens, H. J. 1951. Geographical limitations to food production in the Mongolian People's Republic. *Annals of the Association of American Geographers* 41:348–69.

———. 1954. *China's march towards the tropics*. Hamden, CT: Shoe String Press.

———. 1966. Cultivation, development and expansion in China's colonial realm in central Asia. *Journal of Asian Studies* 26:67–88.

Williams, J. 1974. China in maps, 1890-1960: *A selective and annotated cartobibliography*. East Asian Occasional Paper No. 4. Lansing, MI: Michigan State University.

———; Sutherland, C. L.; and Chang, C. 1988. Land use in the Taipei Basin. Map supplement. *Annals of the Association of American Geographers* 78:358–61.

Withington, W. A. 1961. Uplands resorts and tourism in Indonesia: Some recent trends. *Geographical Review* 51:418–23.

———. 1985. The cities of Sumatra, Indonesia in 1980. *Indonesia Circle* 38:15–30.

———. 1988. The intermediate city concept reviewed and applied to major cities in Sumatra, Indonesia. In *Asian urbanization: Problems and processes*, ed. W. Tietze. Stuttgart: Gebruder Borntraeger.

Wood, W. 1986. Intermediate cities on a resource frontier. *Geographical Review* 76:149–59.

Yapa, L. S. 1979. Ecopolitical economy of the Green Revolution. *Professional Geographer* 31:371–76.

Yeung, Yue-man, and Lo, C. P., eds. 1976. *Changing south-east Asian cities: Readings on urbanization*. New York: Oxford University Press.

Zurick, D. N. 1988. Resource needs and land stress in Rapti zone, Nepal. *Professional Geographer* 40:428–44.

Soviet and East Europe

Philip R. Pryde | William H. Berentsen | Ihor Stebelsky

Geographic research on the Soviet Union dates mainly from the 1940s. From its start in the depths of the "cold war" to the present era of arms agreements and *glasnost,* research on Soviet affairs by geographers has proven to be challenging, timely, and rewarding.

The Soviet and East Europe Specialty Group (SEESG) was formed in 1980. The membership of SEESG has defined its purpose as being to promote excellence in research on geographic topics pertaining to the USSR and the countries of eastern Europe. Major research themes have included climatology, urban and economic development, environmental problems, historical geography, demographics, agriculture, and geographic thought and biography. A second major purpose has been to assist and promote both intellectual and physical contact between professional geographers working in these countries and their North American colleagues within the Association of American Geographers (AAG). The progress to date toward both of these objectives is elaborated in this chapter.

MAJOR RESEARCH THEMES AND PUBLISHED WORKS ON THE SOVIET UNION

This section summarizes the main areas of research, with reference to some of the major works contributed by American geographers specializing in the Soviet Union. Space limitation necessarily prevents inclusiveness; the section primarily covers major works by AAG-affiliated geographers in the period since 1977.

Bibliographies, Directories, and Geographic Thought

A recurring effort of western geographers has been to identify the most prominent Soviet geographers, and to make biographical and bibliographical material on them available to English-speaking readers. No other American geographer has contributed as much to bibliographical research on the Soviet Union as Chauncy Harris. His comprehensive *Guide to Geographical Bibliographies and Reference Works in Russian or on the Soviet Union* (1975) remains the most authoritative compilation of its type available. The late Theodore Shabad also compiled directories of Soviet Geographers and published them in *Soviet Geography* (September 1977). This initial list was updated by Shabad and Harris in the March and April 1988 issues of the same journal. These unique works represent the primary guide in English to the largest national assemblage of geographers in the world.

The methodological ideas of Soviet geographers have been presented to American geographers through a number of translated collections. Among the most important are those edited by Harris (1962) and by Fuchs and Demko (1974). The most recent offering is David Hooson's chapter on continuity and change in Soviet geographical thought, in the volume by Demko and Fuchs (1984). Literature by Soviet geographers on Marxist approaches to the geographical environment was reviewed in works by Matley and others prior to 1975 and more recently by Matley in *Progress in Human Geography* (1982). These works remain the standard references in English on these subjects.

Climate and Agriculture

By far the most detailed studies by an American on the climates of the USSR are those by Lydolph. His early analysis of the threats to agriculture posed by the "sukhovey" phenomenon was followed by his comprehensive study of Soviet climates (Lydolph 1964, 1977a). Lydolph, with Martell and Erickson, also assessed the impact of weather on Soviet agricultural performance (in Demko and Fuchs 1984), as well as on human activity. Lydolph's works on Soviet climates have consistently provided important insights which have been incorporated into the writings of researchers on Soviet agriculture. A very thorough examination of the constraints of climate and related factors on the development of the Soviet north and far east has been prepared by Mote (1983).

The first detailed, analytical North American research on Soviet agriculture was done by W. A. Douglas Jackson. His numerous articles and monographs on Soviet agricultural policy and structure, on the Virgin Lands of Siberia and Kazakhstan, and on other Soviet agricultural regions laid the methodological foundation for much of the Western writings on this topic that followed (Jackson 1962, 1971a). Dando has also contributed significantly to the systematic study of Soviet agriculture, and in addition authored the *Geography of Famine* (Dando 1980; Dando and Schlichting 1987). Soviet agricultural land-management policies, including soil erosion, have been reviewed extensively by Stebelsky, as has the topic of Soviet food-consumption patterns (Stebelsky 1974, 1978, 1987a, 1987b).

The topic of Soviet agriculture was examined in a variety of important perspectives by numerous geographers during the period 1954–70, including at least seven articles in the *Annals of the Association of American Geographers* and the *Geographical Review*. As this period predates the main scope of the present volume, these works will not be detailed in this section. However, these early writings on agriculture by Jackson, Jensen, Field, Lewis, and Matley greatly influenced later work, and are duly noted.

Resource Management and Conservation

Soviet natural-resources research was initially introduced to Americans by Jackson with a volume of translations in 1971, followed a few years later by an edited volume of articles by Western specialists (Jackson 1971b, 1978). Two other major compendiums were edited by a British geographer, and contain many works by AAG members evaluating various Soviet environmental issues (Singleton 1976, 1987). The first single-author monograph on Soviet natural-resource conservation practices was prepared by Pryde (1972), who found a commonality of problems when compared to the West. Pryde's recent articles have included such topics as environmental-impact analysis and a review article in *Science* (Pryde 1983a, 1987b), as well as articles in *Soviet Geography* on energy and nature-preserve topics that will be expanded upon in a forthcoming 1990 volume on Soviet environmental management.

Perhaps the most serious environmental problems in the Soviet Union today are those involving the Aral Sea, and the related issues of the augmentation of water supplies in the Volga Basin and Central Asia. These problems have been studied at length by Micklin, who has produced several authoritative studies on proposed Soviet water transfers (Micklin 1983, 1985, 1986, 1988). The subject of air and water pollution in the USSR has attracted several writers, who have evaluated these topics from both regional and management perspectives (ZumBrunnen 1978, 1984; Mote 1978; Bond 1984), and found them to be very serious.

Research on biotic resources in the USSR has included studies that deal with forests, soils, and wildlife (Barr 1984, 1987; Stebelsky 1974; Braden 1987, 1988; Pryde 1987a). The most significant recent work in this area is a volume analyzing Soviet forestry practices (Barr and Braden 1988) that notes longstanding inefficiencies in the industry.

The most substantial contribution to the thematic research on natural resources in the 1980s has been the AAG-sponsored *Soviet Natural Resources in the World Economy*. This collection groups 29 chapters into those dealing with regional dimensions, those dealing with policy problems and potentials for specific sources of energy and raw materials, and those dealing with the role of raw materials in Soviet foreign trade. Sixteen of the chapters were authored by SEESG members; some of these will be noted in the appropriate topical sections that follow. The overview chapter on natural resources was prepared by Shabad (1983).

Historical and Demographic Studies

Among the principal research topics in Soviet historical geography have been regional migration and colonization, both internally and in North America. Early monographs on these subjects included those by Gibson (1969, 1976) and Demko (1969). A more recent volume is *Studies in Russian Historical Geography* (Bater and French 1983), which contains chapters by Gibson, Stebelsky, and Gohstand, among others. Additional contributions have included studies of the Amur Expedition (Bassin 1983), research on Ukrainian peasant colonization east of the Urals (Stebelsky 1984), and research on Russia's nineteenth-century trade (Gohstand 1976b, 1983). These studies have been important in establishing the historical and geographic backgrounds upon which later Soviet cultural and economic developments have been built.

Among the leading research on demographic topics has been a number of monographs on Soviet population change and redistribution (Clem 1975, 1986; Lewis and Rowland 1979; and Wixman 1984b). These studies made significant contributions to geographers' understanding of the regional and ethnic aspects of Soviet demographic changes. Lewis and Wixman also contributed chapters to the recent volume honoring Chauncy Harris (Demko and Fuchs 1984).

Articles on Soviet demography have explored such subjects as the patterns of population change (Clem 1980), internal migrations (Demko 1969), the labor resources of Siberia (Lewis 1983), mobility and settlement system integration (Fuchs and Demko 1983), analysis of census data (Rowland 1986a, 1986b), and demographic trends in minority areas (Wixman 1984a). A great many other papers have been authored by these same researchers and others, especially in the pre-1980 period, and references to these can be found in the more recent works cited above. An important demographic research task in the early 1990s will be an analysis of the 1989 Soviet census data, and many of the above researchers will be reporting their findings.

Urban Geography

Major research themes dealing with Soviet cities and urban planning have included typologies, differential growth patterns, historical studies, and regional comparisons. Some of the more-important monographs have included Harris's seminal work on the cities of the Soviet Union (1970), subsequent works on the same subject by French and Hamilton (1979) and Bater (1980), and most of the articles in the recent festschrift honoring Harris (Demko and Fuchs 1984).

Similar themes developed in major journal and book articles in recent years include population change and urban growth (Bond and Lyndolph 1979), urban growth in Siberia (Hausladen 1987), and the antimetropolitan syndrome in Soviet urban policy (Jensen 1984). In addition, Gohstand has produced a multitheme map depicting the evolution of Moscow and its internal land uses, which appeared as a supplement to the *Annals* in 1976. A major focus of all these works is upon identification and analysis of the ways in which Soviet urban planning differs from that practiced in America; for example, Soviet efforts to plan the size of cities, to consolidate suburbs, to build new cities in Siberia, etc.

Industrial Geography

The development, siting, resource base, and regional comparisons of Soviet industrial and energy complexes have been longstanding research themes among SEESG members. The early comprehensive work in this field, still widely used as a reference, was *Basic Industrial Resources of the USSR* (Shabad 1969). Among the topics covered by recent major monographs are the Soviet energy supply system and resource base (Dienes and Shabad 1979), the same topic expanded westward to include Europe (Hoffman 1985b), and the Soviet iron and steel industry (ZumBrunnen and Osleeb 1986).

Articles published over the past 10 years or so include studies of small-town industrialization (Lonsdale 1977), the synthetic-rubber industry (Lewis 1979), the distribution of industrial employment (Sagers 1984), fossil-fuel development policy

(Dienes 1983), current prospects for the petrochemical industry (Sagers and Shabad 1987), and one of the earliest detailed assessments of the effects of the Chernobyl disaster (Shabad 1986).

Reference was made earlier to the comprehensive compendium published in 1983 on the natural resources of the USSR and their existing and potential impact on the world economy (Jensen, Shabad, and Wright 1983). A great many American geographers contributed detailed chapters to this volume, dealing with one or more components of the Soviet natural-resource base. Among the resources analyzed were forest resources and timber export (Barr, Braden), iron ore and coking coal (ZumBrunnen and others), chrome and manganese (Jackson), and nickel and platinum (R. B. Adams). This volume should serve as the standard reference work on Soviet natural resources for years to come.

In addition to the articles that have appeared over the past decade, it should be noted that industrial geography was one of the strongest emphases of American specialists on the USSR in the two decades from 1955 to 1975. Among the most prominent of these early researchers were Lonsdale, Dienes, Lydolph, Shimkin, Shabad, and Rodgers.

For the future, a major research theme will be the spatial consequences of Gorbachev's *perestroika* (restructuring) policy, as reflected in forthcoming five-year plans.

Transportation

As with industry, some excellent early work on Soviet transportation predates the scope of this present volume, especially that of Taaffe, Kish, and Guest (e.g., Taaffe 1960). Some of the primary monographs on Soviet transportation topics since the late 1970s include works on transport in Western Siberia from a historical perspective (North 1979), the new Baikal-Amur railroad and its strategic importance (Shabad and Mote 1977), and the transportation of Soviet energy resources (Sagers and Green 1986).

Research themes that appear in articles published in the 1980s include Soviet containerization trade and the trans-Siberian land bridge (Mote 1984), commodity flows and regional economic development in the Soviet Far East (Rodgers 1983), spatial efficiency in Soviet electrical transmission (Sagers and Green 1982), and freight rate structure on Soviet railroads (Sagers and Green 1985). Throughout many of these studies, a recurring theme is the inefficiencies produced by the long average-haul distances that are common in all modes of Soviet freight transportation.

Regionalization and Regional Development

Regional studies on the USSR have included both works that compare regions within the USSR, and those that focus upon one particular region, with Central Asia, Siberia, and the Far East being the most studied among the latter. It is incumbent again to begin this section by noting the many excellent texts and monographs dating from the 1950–70 period. Early texts on the geography of the USSR were prepared by Shabad, Cressey, and Hooson. The most widely used text at present by an American author is that of Lydolph (1977b). Topical atlases of the USSR were prepared by Taaffe and Kingsbury (1965) and Kish (1971). In more-recent years, significant regional studies have included monographs and collections on the Soviet Far East and Siberia

(Shabad and Mote 1977; Holzner and Knapp 1987), on economic development and national-policy choice in Soviet Asia (Dienes 1987), and a new compendium on the future development of the Soviet Far East (Rodgers 1989).

Journal articles of note on regional issues include studies on the economic potential of the Siberian regions (Dienes 1982), on Soviet regional development in the 10th Five Year Plan (Jensen 1978), on regional effects of international trade (North 1983), and on Soviet regional investment strategy (Liebowitz 1987). In addition, the unique arctic city of Norilsk has been studied in a series of articles in *Soviet Geography* (Bond 1984), and Stebelsky contributed a number of entries on the geography of Ukraine in the 1988 edition of the *Encyclopedia of Ukraine*. Several American geographers have also contributed to the volumes on the Soviet economy that are issued periodically by the Joint Economic Committee of Congress (Sagers and Shabad 1987; ZumBrunnen 1987; and others). Significant pre-1970 journal articles involving regional studies were authored by Hooson, Jackson, Jensen, Shabad, and others; these laid the methodological groundwork for most of the regional studies that followed. A probable future research theme will be the regional consequences—especially in Central Asia and Siberia—of Gorbachev's economic restructuring.

RESEARCH THEMES ON EASTERN EUROPE

Only a relatively small number of U.S. and Canadian geographers conduct research on Eastern Europe. Of these, only Hoffman is a true East European specialist, and he remains active in research publication. Most of the other geographical specialists in this region have research interests in topical fields to which they devote considerable effort; a few regularly or occasionally work on both East European and Soviet topics.

Within the discipline, geographers have no journal with a specific focus upon Eastern Europe to which scholarly work can be submitted, though journals in related disciplines devoted to this region do exist. As a result, published works by SEESG members appear in a wide variety of geographic journals and those in cognate disciplines. The most active research areas are regional geography, urban geography, and energy studies.

Regional Geography

Rugg has written the only regional monograph on Eastern Europe in recent years (1985). This volume should serve as a standard reference for a number of years. An introduction to the region is also provided by Hoffman (1983b) in a recent European geography text; a new edition is expected in 1989. Many SEESG members working on Eastern Europe specialize in a particular country or region. These include Fedor and Regulska on Poland, Danta on Hungary, and Berentsen on the German Democratic Republic (GDR). Their contributions are summarized in the ensuing sections.

Urban Geography

This is one of three topical subfields in which East Europeanist geographers have been most active during the 1980s. Fuchs provided an excellent introduction to urban change in Eastern Europe early in the decade (1980). Two special issues of *Urban*

Geography appeared in 1987 on the socialist city, edited by Demko and Regulska. Their various articles provide overviews of urbanization in the GDR (Berentsen 1987c), Hungary (Danta 1987), and Poland (Regulska 1987). Danta has also published widely on urbanization and the urban system in Hungary (e.g., 1983). Other recent publications include articles on the urban landscape in Southeastern Europe (Hoffman 1986), on the cities of the Soviet Union and Eastern Europe (Matley 1983), on the relation of urban policy to political systems (Regulska 1986b), and on inner-city changes in Warsaw (Zaniewski 1986).

Regional Development

Recent excellent overviews on this topic are provided by papers in *Regional Development, Problems, and Policies in Eastern and Western Europe* (Demko 1984), as well as in articles by Demko and Fuchs (1979) and Hoffman (1985a). A volume of papers from a joint U.S.-Hungarian seminar on regional development, forthcoming in 1989, will include papers by a number of geographers (Danta, Berentsen, and Daroozi 1989). In addition, there have been recent studies of aspects of regional development and planning in the GDR by Berentsen (1981, 1987b) and Rugg (1986), and a study on regional planning in Poland (Regulska 1986a).

Geography of Energy

Hoffman (1983a, 1985b) has reported on the topic of energy in Eastern Europe in several venues, including studies of the region's dependency on the Soviet Union. Dienes (1985) has written on energy use and conservation in Eastern Europe as well, comparing the situation there with Western Europe.

Other Research Areas

Relatively little has been written in the area of East European physical geography by AAG members. Some work by American climatologists relates at least implicitly to Eastern Europe. A variety of aspects of social geography have been covered in articles dealing with Poland and East Germany (Fedor 1984; Berentsen 1987a). Short biographical sketches of East European geographers have been done, by Kish on Paul Teleki and by Matley on Simon Mehedinti.

The outlook for continued research on Eastern Europe by AAG members is less promising than might be desired. There are few actively publishing specialists now in the field, and relatively few new students are being trained as specialists on Eastern Europe (Demko and Sagers 1985). Nonetheless, foundation support for graduate training and research on Eastern Europe has increased of late, and a small number of people continue to begin work in the field, often as the result of original training for work on the USSR. The infusion of emigre talent into the field is much less notable than it was in the earlier postwar era.

Ironically, despite the small size of the research cadre working on Eastern Europe, several AAG members hold distinguished positions within the research field, or within academe more generally. Hoffman was the interim director of the newly founded East European program within the Woodrow Wilson International Center for Scholars, and serves or has served on many foundation boards. The latter type of service also holds true for Demko, now Director of the Office of the Geographer in the Department of State, who is also a former AAG president and Executive Secretary of the American Association for the Advancement of Slavic Studies (AAASS). Fuchs now serves as a vice president of the International Geographical Union (IGU), and is vice rector of the UN University, Tokyo.

RESEARCH EXCHANGES WITH EASTERN EUROPE

AAG members have been involved in several forms of research interaction with scholars in Eastern Europe. Notably, there have been several National Science Foundation-sponsored joint conferences in recent years with Hungary (1975, 1978, 1985) and Bulgaria (1987). These meetings have created a modest number of informal contacts among geographers in North America and Eastern Europe, but have yet to create conditions allowing for active interchange between, or joint research by, scholars in the two regions. Encouraging efforts in this direction, however, are being made with Hungary and Bulgaria by the Regional Research Institute (Andrew Isserman, Director) at West Virginia University.

Geographers have also sought research experiences and contacts in Eastern Europe on an individual basis. Support often comes from the International Research and Exchanges Board (IREX), which has also helped to fund travel by East Europeans to the USA. The East European Program at the Wilson Center is beginning to support scholars from abroad in Washington, as well as junior researchers from the USA. Most AAG members going to Eastern Europe on such exchanges have gone to the GDR, Poland, and Hungary. Thus far, little exchange activity exists with other nations, and none at all with Albania.

HISTORY OF CONTACTS AND EXCHANGES WITH THE USSR

A primary purpose of the Soviet and East Europe Specialty Group from its inception has been to promote contacts between American and Soviet geographers (East European exchanges have already been discussed). In order to provide the necessary historical framework, this overview briefly surveys the entire span of these efforts and their accomplishments, even though many took place prior to the formation of the specialty group.

A comprehensive review of the history of cooperation by U.S. and USSR geographers appeared in the journal *Soviet Geography* in 1984 (Annenkov and Demko), and this report forms the basis for much of this section. Annenkov and Demko suggest that the history of these contacts can be divided into three stages: 1945–60, the familiarization stage; 1962–72, the information-gathering stage; and 1973 onward, the problem-oriented, collaborative stage.

During the first period, major American geographic works became available in the USSR, and in 1957 the definitive work, *American Geography: Inventory and Prospect,* was translated into Russian (*Amerikanskaya geographiya*). The first significant post-war contact between Soviet and American geographers occurred in 1956 at the 18th International Geographical Congress in Brazil. The following year, AAG President

INTERNATIONAL UNDERSTANDING THROUGH REGIONAL SYNTHESIS

Chauncy Harris toured various Soviet cities, and subsequently published one of the first American summaries of the state of Soviet geography (Harris 1958).

A major event in the effort to familiarize Americans with the work of Soviet geographers occurred in 1960, when the American Geographical Society began publishing *Soviet Geography: Review and Translation,* under the editorship of Theodore Shabad. Two years later, *Soviet Geography: Accomplishments and Tasks,* which systematically reviewed the status of the discipline in the Soviet Union, was published in English in the U.S., also by the American Geographical Society (Harris 1962).

By the start of the 1960s, American graduate students had begun to conduct part of their education in the Soviet Union. The earliest ones included Taaffe, Demko, Fuchs, and Gohstand, all of whom studied within the Department of Economic Geography at Moscow State University. Harris attended the 3rd All-Union Geographical Congress in Kiev in 1960.

In the summer of 1961, Gerasimov, Davitaya, Kovalevskiy, Krotov, Mavlyanov, and Salishchev became the first Soviet exchange group of geographers to travel to the U.S. The reciprocal exchange group of Harris, James, Russell, Espenshade, McCarty, and Horbaly toured extensively through the Soviet Union a few months later. This exchange was significant in opening the way for increased cooperation between Soviet and American geographers. However, it was to be over a decade before another exchange of geographers could be accomplished. In the interim, Shabad attended the fourth congress of the USSR Geographical Society in 1964, as well as the fifth in 1970.

A marked increase in the scope of geographic contacts occurred during the 1970s. These contacts involved topical seminars, the first one (on the theme of the urban environment) taking place in the USSR in 1975. The seminar papers were published in the Soviet Union in Russian (*Gorodskaya . . . 1977*). The return group visited the United States in 1977, and their report was translated in *Soviet Geography* in 1979 (Preobrazhenskiy, Annenkov, and Lappo). Also during the 1970s, several Soviet geographers spent time at various U.S. universities.

A major opportunity for professional interaction occurred in 1976 when Moscow hosted the 23rd International Geographical Congress. About 200 Americans participated in the numerous paper sessions and field trips, resulting in very beneficial interaction between the Western delegates and their Soviet hosts. Three years later, a delegation from the Institute of Geography of the Soviet Academy of Sciences delivered papers at a special session at the 1979 AAG meeting in Philadelphia.

A number of American geographers have independently engaged in research or teaching in the Soviet Union, in addition to those already mentioned. As examples just from the 1980s, Micklin spent considerable time in the USSR in 1984 and 1987 studying the water problems of West Siberia and central Asia, Demko was visiting professor at Moscow State University in 1980, and Mote was one of the first Westerners to visit and report on the new Baikal-Amur rail line (Mote 1985). Perhaps a dozen American scholars in all have benefited from individual IREX research grants in the Soviet Union, and at least as many have traveled there for research purposes, either independently or with funding from other sources. Much useful research has resulted from these visits, although it must be noted that on occasion American (as well as Soviet) geographers have encountered significant problems with the Soviet bureaucracy regarding travel, access to data, attendance at conferences, etc. It is hoped, and believed, that these problems will be minimized in the future.

The current series of US-USSR exchanges were set in motion in 1977, when Fuchs and Demko succeeded in getting Geography included within the framework of a new exchanges agreement of the American Council of Learned Societies (ACLS) and the Academy of Sciences of the USSR. After a four-year delay, a program of joint research was agreed upon by the Academy of Sciences and the ACLS in 1981, utilizing the joint themes of "the geographical aspects of interaction within the system society-economy-environment," and "the dynamics of urbanization, systems of settlement and environmental modification." The coordinators for the environment subtheme were Lydolph and Kibal'chich, and for the urban subtheme, Demko and Lappo.

This agreement began to be implemented in 1983, when an American delegation arrived in Moscow for a two-week exchange. The urban subgroup consisted of Demko, Fuchs, Wolpert, Adams, and Rose, while the natural environment subgroup was composed of Lydolph, Dienes, ZumBrunnen, Pryde, and Shabad. The full group visited Irkutsk Oblast for a discussion of environmental problems in such areas such as the Irkutsk metropolitan area, the basin of Lake Baikal, and the new city of Ust'-Ilimsk. Upon returning to Moscow, the two subgroups pursued separate itineraries; the urban specialists spent a day in Moscow and then visited the Oka River science town of Pushchino, while the remainder of the delegation travelled to Kursk to observe environmental-management techniques in the Kursk iron ore region and at the Central Chernozem Biosphere Reserve. A full report of the trip appeared in Soviet Geography (Pryde 1983b).

The reciprocal visit by a delegation of 10 Soviet geographers occurred in July and August of 1987. The Soviet delegation was headed by Vladimir Kotlyakov, new head of the Institute of Geography of the Academy of Sciences; the remainder of the delegation consisted of Lappo, Annenkov, Sdasyuk, Katagoshchin, Dreier, Petrov, Karpov, Vorob'yev, and Maksakovsky. The itinerary included stays in New York, Washington, Milwaukee, San Diego, and Seattle.

The protocol of the visit set forth plans for a subsequent exchange of delegations, as well as a series of joint publications. The latter will include a history of Soviet-American geographical contacts and exchanges; bibliographic reviews of the work done in each country in the two subtheme areas; and eventual publication of joint monographs dealing with each of the two subthemes, as well as on other topics such as "the Art and Science of Geography in the US and USSR." All of the works will be published in each country in its respective language. It is significant testimony to the strength of this exchange that discussions designed to translate this volume, *Geography in America,* into Russian and publish it in the Soviet Union had begun while the book was still in the writing stage.

The Soviet and East European Specialty Group looks forward to continuing to help promote this type of international cooperation in the future. Although, as noted, these efforts have not been directly sponsored by the specialty group, its members have been centrally involved in organizing all of the exchanges and joint agreements that have been conducted to date, and almost all of the members of the 1983 American exchange delegation were members of the group.

Another significant joint effort just getting underway at the time of this writing is a volume being prepared on global changes as part of the International Geosphere-Biosphere Program. Resulting from the initiative of Gilbert White and V. M. Kotlyakov, the global change volume will outline geographical approaches and potential future contributions to the IGBP agenda (Borchert 1988). It is the hope of the Soviet and East European Specialty Group that such exchanges and cooperative efforts will multiply in the future.

THE JOURNAL *SOVIET GEOGRAPHY*

Although it is not a publication directly prepared or sponsored by the Soviet and East Europe Specialty Group, failure to mention the periodical *Soviet Geography* would render this chapter quite incomplete. This journal has been perhaps the single most important reference work for North American geographers who seek pertinent information about the Soviet Union, and its editors and Advisory Board are largely SEESG members.

The journal was the inspiration of Theodore Shabad, and the first issue appeared in January of 1960 under its original name *Soviet Geography: Review and Translation*. The title of the journal was shortened to *Soviet Geography* in 1984, and it appears 10 times per year. Dr. Shabad remained as editor of the journal until his most untimely death in 1987.

Soviet Geography has achieved world-wide recognition as the leading non-Soviet source of scholarly information about the geography and economic development of the Soviet Union. Its content typically has five components: translated articles by Soviet authors, articles by western specialists on the Soviet Union, book reviews, news notes concerning the Soviet economy, demography, regional development, etc., and translated tables of contents of Soviet geographical journals.

In addition to its normal content, several special issues have been published. The combined May-June 1965 issue translated the USSR's *Physical Geographic Atlas of the World*; the combined May-June 1967 issue did the same for the *Atlas of the Antarctica*. The September 1967 issue presented the first western directory of Soviet geographers; this was updated in 1977, and an even-more thorough listing appeared in the March and April 1988 issues (Shabad and Harris 1988). In May 1970 a special issue on the cities of the Soviet Union appeared, and the June 1986 issue contained a long analysis of the effects of the Chernobyl accident (Shabad 1986). Additional special thematic issues are being prepared for publication during 1988.

The widespread use and popularity of *Soviet Geography* was evidenced when it was identified as ranking first (by a wide margin) out of 106 surveyed geographical serials in terms of number of citations in leading international geographical bibliographies (Harris 1980).

Following the death of Dr. Shabad, the editorial duties are being shared by Andrew Bond, Leslie Dienes, and Matthew Sagers, under the general editorial supervision of Chauncy Harris. The editorial policies of *Soviet Geography* are expected to remain unchanged, and to continue to reflect the high standards set by Dr. Shabad.

SUMMARY

In keeping with other AAG specialty groups, the Soviet and East European group has as its main purpose the promotion of scholarly research in all topic areas within its regional bounds. However, the group is unique in that the difficult political circumstances that pertain to the countries it studies, and the fact that one of these countries contains the largest concentration of geographers of any nation in the world, have made the creation and improvement of contacts with Soviet and East European geographers an extremely important secondary goal. Hence, exchanges and contacts—and more recently, cooperative efforts—between American and Soviet/East European geographers have been a longstanding emphasis of American specialists on these regions, both prior to the specialty-group era and in the 1980s. We are hopeful that, with the emergence of "new thinking" in the late 1980s, new and much more productive levels of contacts and cooperation can be realized.

The major research themes of the Soviet and East European Specialty Group (SEESG) have largely reflected some of the dominant developmental goals and problems of the various countries concerned. Given the extreme difficulty (bordering on impossibility) of carrying out meaningful field research in the USSR and Eastern Europe, and the relative paucity of materials available from these countries in English, a major task of Soviet and East European specialists has been simply to describe, and to the extent possible analyze, what transpires in the fields of internal academic and governmental research. Thus, the earliest Western research was largely in the realm of economic geography, particularly focusing upon natural resources, industrial, and transportation themes. This research also examined how socialist planners were approaching—and the extent to which they were achieving—their stated developmental goals.

Basic textbooks and atlases on the USSR were also an important early contribution, as were biographical compilations. A longstanding part of western studies of these countries has dealt with analyzing census data, to probe for significant trends and conclusions. Historical studies, too, often have had as their specific foci transportation and demographic themes. In the past two decades, urban and environmental themes have become two of the strongest interest areas, with some of the main findings being that major types of problems in these areas are often similar in both socialist and capitalist societies, although in other cases (e.g., housing) they are quite different. Physical geography, with the exception of climatology, has not been a major research area.

Future research themes are likely to focus strongly upon the implications and consequences of Gorbachev's *perestroika* initiative, especially its presumed renewed emphasis on the western portions of the country at the possible expense of Siberia. Environmental forecasting is another emerging theme. The topic of future geographic themes in the Soviet Union was explored by a panel of American geographers in the June 1987 issue of *Soviet Geography*.

The volume of research on the Soviet Union has increased slowly over the past two decades; that on Eastern Europe has remained at a relatively stable level. It is hoped that, with the increased ease of visiting some of these countries (especially the USSR), a corresponding easing of the constraints on carrying out research within them might also come about. Should this occur, the magnitude of research on the countries of Eastern Europe and the USSR could enjoy a significant increase as American geography enters the twenty-first century.

REFERENCES

Amerikanskaya geografiya. Sovremennoye sostoyaniye i perspektivy. 1957. Moscow: Izdatelsvo inostrannaya literatura (Russian translation of *American geography: Inventory and prospect*. 1954. Syracuse: Syracuse University Press.)

Annenkov, V. V., and Demko, G. J. 1984. Development of relations between geographers of the United States and the USSR from the 1950's to the 1980's. *Soviet Geography* 25:749–57.

Barr, B. M. 1984. The Soviet forest in the 1980s: Changing geographical perspectives. In *Geographical studies on the Soviet Union,* eds. G. Demko and R. Fuchs, pp. 235–56. University of Chicago Department of Geography Research Paper No. 211. Chicago: University of Chicago Press.

———. 1987. Regional alternatives in Soviet timber management. In *Environmental problems in the Soviet Union and Eastern Europe,* ed. F. Singleton, pp. 97–124. Boulder, CO: Lynne Rienner.

———, and Braden, K. E. 1988. *The disappearing Russian forest: A dilemma in Soviet resource management*. Totowa, NJ: Rowman & Littlefield.

Bassin, M. 1983. The Russian geographical society, the "Amur Epoch," and the Great Siberian Expedition, 1855–63. *Annals of the Association of American Geographers* 73:240–56.

Bater, J. 1980. *The Soviet city: Ideal and reality*. Beverly Hills: Sage Publications.

———, and French, A., eds. 1983. *Studies in Russian historical geography*. London: Academic Press.

Berentsen, W. H. 1981. Regional change in the German Democratic Republic. *Annals of the Association of American Geographers* 71:50–66.

———. 1987a. German infant mortality, 1960–1980. *Geographical Review* 77:157–70.

———. 1987b. Relationships between trends in regional development and regional policy goals in Central Europe, 1950–1980. *Environment and Planning C: Government and Policy* 5:105–12.

———. 1987c. Settlement structure and urban development in the GDR, 1950–85. *Urban Geography* 8:405–19.

Bond, A. 1984. Air pollution in Norilsk: A Soviet worst case? *Soviet Geography* 25:665–80.

———, and Lydolph, P. 1979. Soviet population change and city growth, 1970–79. *Soviet Geography: Review and Translation* 20:461–88.

Borchert, J. 1988. Global environment program update. *AAG Newsletter* 23(5):2; 10.

Braden, K. E. 1987. The function of nature reserves in the Soviet Union. In *Environmental problems in the Soviet Union and Eastern Europe,* ed. F. Singleton, pp. 59–70. Boulder, CO: Lynne Rienner.

———. 1988. Environmental issues in Soviet forest management. *Soviet Geography* 29:599–607.

Clem, R. 1975. *The Soviet West: Interplay between nationality and social organization*. New York: Praeger.

———. 1980. Regional patterns of population change in the Soviet Union, 1959-1979. *Geographical Review* 70:137–56.

———, ed. 1986. *Research guide to the Russian and Soviet censuses*. Ithaca, NY: Cornell University Press.

Dando, W. 1980. *Geography of famine*. New York: Wiley.

———, and Schlichting, J. 1987. *Soviet agriculture today: Insights, analyses, and commentary*. Grand Forks, ND: University of North Dakota.

Danta, D. R. 1983. Urban systems development and regional policy in Hungary. *Modeling and Simulation* 14:722–73.

———. 1987. Hungarian urbanization and socialist ideology. *Urban Geography* 8:391–404.

———; Berentsen, W.; and Daroozi, E., eds. 1989. *Regional development processes and policies* (Vol. 2, Regional Research Reports, Centre for Regional Studies). Budapest: Hungarian Academy of Sciences.

Demko, G. J. 1969. *The Russian colonization of Kazakhstan, 1896–1916*. Bloomington, IN: Indiana University Press.

———. 1984. *Regional development, problems, and policies in Eastern and Western Europe*. London: Croom Helm.

———, and Fuchs, R. 1979. Georgraphic inequality under socialism. *Annals of the Association of American Geographers* 69:304–18.

———, and ———. 1984. *Geographical studies on the Soviet Union*. University of Chicago Department of Geography Research Paper No. 211. Chicago: University of Chicago Press.

———, and Sagers, M. 1985. The nonreplacement of senior scholars in the Soviet-East European specialty in geography. *Soviet Geography* 26:199–209.

Dienes, L. 1982. The development of Siberian regions. *Soviet Geography* 23:205–44.

———. 1983. Soviet energy policy and the fossil fuels. In *Soviet natural resources in the world economy,* eds. R. G. Jensen, T. Shabad, and A. Wright, pp. 275–95. Chicago: University of Chicago Press.

———. 1985. Energy use and conservation in Eastern and Western Europe. In *The European energy challenge, east and west,* ed. G. Hoffman, pp. 89–112. Durham, NC: Duke University Press.

———. 1987. *Soviet Asia: Economic development and national policy choices.* Boulder, CO: Westview Press.

———, and Shabad, T. 1979. *The Soviet energy system: Resource use and policies.* New York: Wiley.

Fedor, T. S. 1984. Social inequalities and the Polish crisis. *East Lakes Geographer* 19:28–33.

French, R. A., and Hamilton, F. E., eds. 1979. *The socialist city.* New York: Wiley.

Fuchs, R. 1980. Urban change in Eastern Europe: The limits to planning. *Urban Geography* 1:81–94.

———, and Demko, G. 1974. *Geographical perspectives in the Soviet Union.* Columbus, OH: Ohio State University Press.

———, and ———. 1983. Mobility and settlement system integration in the USSR. *Soviet Geography* 24:547–59.

Gibson, J. 1969. *Feeding the Russian fur trade.* Madison, WI: University of Wisconsin Press.

———. 1976. *Imperial Russia in frontier America.* New York: Oxford University Press.

Gohstand, R. 1976a. Moscow (map supplement). *Annals of the Association of American Geographers* 66(1).

———. 1976b. The shaping of Moscow by 19th century trade. In *The city in Russian history,* ed. M. Hamm, pp. 160–81. Lexington, KY: University of Kentucky Press.

———. 1983. Geography of trade of 19th century Russia. In *Studies in Russian historical geography,* eds. J. Bater and A. French, pp. 329–74. London: Academic Press.

Gorodskaya sreda i yeye optimizatsiya (The urban environment and its optimization). 1977. Moscow: Academy of Sciences.

Harris, C. D. 1958. Geography in the Soviet Union. *Professional Geographer* 10:8.

———. 1962. *Soviet geography: Accomplishments and tasks* (English edition edited by Chauncy D. Harris). New York: American Geographical Society.

———. 1970. *Cities of the Soviet Union.* AAG Monograph No. 5. Chicago: Rand McNally.

———. 1975. *Guide to geographical bibliographies and reference works in Russian or on the Soviet Union.* University of Chicago Department of Geography Research Paper No. 165. Chicago: University of Chicago Press.

———. 1980. *Annotated world list of selected current geographical serials.* 4th ed. Chicago: University of Chicago Press.

Hausladen, G. 1987. Recent trends in Siberian urban growth. *Soviet Geography* 28:71–89.

Hoffman, G. 1983a. Energy dependence and policy options in Eastern Europe. In *Soviet natural resources in the world economy,* eds. R. G. Jensen, T. Shabad, and A. Wright, pp. 659–67. Chicago: University of Chicago Press.

———, ed. 1983b. *A geography of Europe.* New York: Wiley.

———. 1985a. *Eastern Europe: Fifty years of change and constraints.* Occasional Paper No. 1, East European Program and Kennan Institute. Washington: The Wilson Center.

———. 1985b. *The European energy challenge, east and west.* Durham, NC: Duke University Press.

———. 1986. The transformation of the urban landscapes in southeastern Europe. In *World patterns of modern urban change,* ed. M. Conzen, pp. 129–50. Chicago: University of Chicago Press.

Holzner, L., and Knapp, J., eds. 1987. *Soviet geography studies in our time.* Milwaukee, WI: University of Wisconsin.

Hooson, D. 1984. Continuity and change in Soviet geographical thought. In *Geographical studies on the Soviet Union,* eds. G. J. Demko and R. Fuchs, pp. 11–27. University of Chicago Department of Geography Research Paper No. 211. Chicago: University of Chicago Press.

Jackson, W. A. D. 1962. The virgin and idle lands program reappraised. *Annals of the Association of American Geographers* 52:69–79.

———. 1971a. *Agrarian policies and problems in Communist and non-Communist countries.* Seattle: University of Washington Press.

———, ed. 1971b. *Natural resources of the Soviet Union: Their use and renewal.* San Francisco: Freeman.

———, ed. 1978. *Soviet resource management and the environment.* Columbus, OH: American Association for the Advancement of Slavic Studies.

Jensen, R. G. 1978. Soviet regional development policy and the 10th five year plan. *Soviet Geography: Review and Translation* 19:196–201.

———. 1984. The anti-metropolitan syndrome in Soviet urban policy. In *Geographical studies on the Soviet Union,* eds. G. J. Demko and R. Fuchs, pp. 71–92. University of Chicago Department of Geography Research Paper No. 211. Chicago: University of Chicago Press.

———; Shabad, T.; and Wright, A. 1983. *Soviet natural resources in the world economy.* Chicago: University of Chicago Press.

Kish, G. et al. 1971. *Economic atlas of the Soviet Union.* Ann Arbor, MI: University of Michigan Press.

Lewis, R. A. 1979. Innovation in the USSR: The case of synthetic rubber. *Slavic Review* 38:48–59.

———. 1983. Regional manpower resources and resource development in the USSR: 1970–1990. In *Soviet natural resources in the world economy,* eds. R. G. Jensen, T. Shabad, and A. Wright, pp. 72–96. Chicago: University of Chicago Press.

———, and Rowland, R. 1979. *Population redistribution in the USSR: Its impact on society, 1897–1977.* New York: Praeger.

Liebowitz, R. 1987. Soviet investment strategy: A further test of the "equalization hypothesis." *Annals of the Association of American Geographers* 77:396–407.

Lonsdale, R. E. 1977. Regional inequity and Soviet concern for rural and small-town industrialization. *Soviet Geography: Review and Translation* 18:590–602.

Lydolph, P. 1964. The Russian Sukhovey. *Annals of the Association of American Geographers* 54:291–309.

———. 1977a. *Climate of the Soviet Union.* Vol. 7 of *World survey of climatology,* ed. H. Landsberg. Amsterdam: Elsevier.

———. 1977b. *Geography of the USSR.* 3d ed. New York: Wiley.

———; Martell, G.; and Erickson, R. 1984. Recent weather and agriculture in the Soviet Union. In *Geographical studies on the Soviet Union,* eds. G. J. Demko and R. Fuchs, pp. 215–34. University of Chicago Department of Geography Research Paper No. 211. Chicago: University of Chicago Press.

Matley, I. M. 1982. Nature and society: The continuing Soviet debate. *Progress in Human Geography* 6:367–96.

———. 1983. Cities of the Soviet Union and Eastern Europe. In *Cities of the world,* eds. S. D. Brunn and J. F. Williams, pp. 123–61. New York: Harper & Row.

Micklin, P. P. 1983. Water diversion proposals for the European USSR: Status and trends. *Soviet Geography: Review and Translation* 24:479–502.

———. 1985. The vast diversion of Soviet rivers. *Environment* 27(2):12–20; 40–45.

———. 1986. The status of the Soviet Union's north-south water transfer projects before their abandonment in 1985–86. *Soviet Geography* 27:287–330.

———. 1988. Desiccation of the Aral Sea: A water management disaster in the Soviet Union. *Science* 241:1170–6.

Mote, V. L. 1978. Soviet atmospheric resource management. In *Soviet resource management and the environment,* ed. W. A. D. Jackson, pp. 202–214. Columbus, OH: American Association for the Advancement of Slavic Studies.

———. 1983. Environmental constraints to the economic development of Siberia. In *Soviet natural resources in the world economy,* eds. R. G. Jensen, T. Shabad, and A. Wright, pp. 15–71. Chicago: University of Chicago Press.

———. 1984. Containerization and the trans-Siberian land bridge. *Geographical Review* 74:304–14.

———. 1985. A visit to the Baikal-Amur mainline and the new Amur-Yakutsk rail project. *Soviet Geography* 26:691–716.

North, R. N. 1979. *Transport in Western Siberia: Tsarist and Soviet development.* Vancouver: Centre for Transportation Studies, University of British Columbia.

———. 1983. The impact of recent trends in Soviet foreign trade on regional economic development in the USSR. In *Soviet natural resources in the world economy,* eds. R. G. Jensen, T. Shabad, and A. Wright, pp. 97–123. Chicago: University of Chicago Press.

Preobrazhenskiy, V. S.; Annenkov, V. V.; and Lappo, G. M. 1979. Problems of the urban environment as treated by U.S. geographers. *Izvestiya Akademii*

Nauk SSSR, seriya geograficheskaya, 1978 (5):127–37; translated in *Soviet Geography,* 1979, 20:372–81.

Pryde, P. R. 1972. *Conservation in the Soviet Union.* London: Cambridge University Press.

———. 1983a. The "decade of the environment" in the U.S.S.R. *Science* 220:274–9.

———. 1983b. Visit by American geographers to the USSR. *Soviet Geography* 24:771–5.

———. 1987a. The distribution of endangered fauna in the USSR. *Biological Conservation* 42:19–37.

———. 1987b. The Soviet approach to environmental impact analysis. In *Environmental problems in the Soviet Union and Eastern Europe,* ed. F. Singleton, pp. 43–58. Boulder, CO: Lynne Rienner.

Regulska, J. 1986a. Methods of physical plan evaluation under different political systems (in Polish). *Bulletin of the Committee of Regional Science and Space Economy* 130:125–60.

———. 1986b. Urban policy and political systems. *Nijmeegse Planologische Cahiers* 22:5–20.

———. 1987. Urban development under socialism: The Polish experience. *Urban Geography* 8:321–39.

Rodgers, A. L. 1983. Commodity flows, resource potential, and regional economic development. In *Soviet natural resources in the world economy,* eds. R. G. Jensen, T. Shabad, and A. Wright, pp. 188–213. Chicago: University of Chicago Press.

———. 1989. *The Soviet Far East: Development and prospect.* London: Croom Helm.

Rowland, R. 1986a. Changes in the metropolitan and large city populations of the USSR. *Soviet Geography* 27:638–59.

———. 1986b. Regional population redistribution in the USSR. *Soviet Geography* 27:158–82.

Rugg, D. 1985. *Eastern Europe.* New York: Longman.

———. 1986. Land use in the German Democratic Republic. In *International handbook on land use planning,* ed. N. Patricios. Westport, CT: Greenwood.

Sagers, M. J. 1984. Regional distribution of industrial employment in the USSR. *Soviet Geography* 25:166–76.

———, and Green, M. 1982. Spatial efficiency in Soviet electrical transmission. *Geographical Review* 72:291–303.

———, and ———. 1985. The freight rate structure on Soviet railroads. *Economic Geography* 61:305–22.

———, and ———. 1986. The *transportation of Soviet energy resources.* Totowa, NJ: Rowman & Littlefield.

———, and Shabad, T. 1987. The Soviet petrochemical industry. In *Gorbachev's economic plans,* vol. 1, pp. 321–41. Washington: Joint Economic Committee.

Shabad, T. 1969. *Basic industrial resources of the USSR.* New York: Columbia University Press.

———. 1983. The soviet potential in natural resources: An overview. In *Soviet natural resources in the world economy,* eds. R. G. Jensen, T. Shabad, and A. Wright, pp. 251–74. Chicago: University of Chicago Press.

———. 1986. Geographic aspects of the Chernobyl nuclear accident. *Soviet Geography* 27:504–26.

———, and Mote, V. L. 1977. *Gateway to Siberian resources (the BAM).* New York: Halstead Press.

———, and Harris, C. 1988. Directory of Soviet geographers, 1946-1987. *Soviet Geography* 29:153–366.

Singleton, F. 1976. *Environmental misuse in the Soviet Union.* New York: Praeger.

———. 1987. *Environmental problems in the Soviet Union and Eastern Europe.* Boulder, CO: Lynne Rienner.

Stebelsky, I. 1974. Environmental deterioration in the central Russian black-earth region: The case of soil erosion. *Canadian Geographer* 18:232–49.

———. 1978. Soviet agricultural land resource management, policies, and future food supply. In *Soviet resource management and the environment,* ed. W. A. D. Jackson, pp. 171–86. Columbus, OH: American Association for the Advancement of Slavic Studies.

———. 1984. Ukrainian peasant colonization east of the Urals, 1896-1914. *Soviet Geography* 25:681–94.

———. 1987a. Agricultural development and soil degradation in the Soviet Union: Policies, patterns, and trends. In *Environmental problems in the Soviet Union and Eastern Europe,* ed. F. Singleton, pp. 71–96. Boulder, CO: Lynne Rienner.

———. 1987b. Regional food imbalances in the USSR: The example of meat in 1970. *Soviet Geography* 28:34–43.

Taaffe, E. J. 1960. *Rail transportation and the economic development of Soviet Central Asia.* University of Chicago Department of Geography Research Paper No. 64. Chicago: University of Chicago Press.

———, and Kingsbury, R. 1965. *An atlas of Soviet affairs.* New York: Praeger.

Wixman, R. 1980. *Language aspects of ethnic patterns and processes in the North Caucasus.* University of Chicago Department of Geography Research Paper No. 191. Chicago: University of Chicago Press.

———. 1984a. Demographic trends among Soviet Moslems. *Soviet Geography* 25:46–60.

———. 1984b. *The peoples of the USSR: An ethnographic handbook.* New York: M. E. Sharpe.

Zaniewski, K. 1986. Changes in the inner city: The case of Warsaw, Poland. *Geographical Perspectives* 57:19–31.

ZumBrunnen, C. 1978. An estimate of the impact of recent Soviet industrial and urban growth upon surface water quality. In *Soviet resource management and the environment,* ed. W. A. D. Jackson, pp. 83–104. Columbus, OH: American Association for the Advancement of Slavic Studies.

———. 1984. A review of Soviet water quality management: Theory and practice. In *Geographical studies on the Soviet Union,* eds. G. J. Demko and R. Fuchs, pp. 257–94. University of Chicago Department of Geography Research Paper No. 211. Chicago: University of Chicago Press.

———. 1987. Gorbachev, economics, and the environment. In *Gorbachev's economic plans,* vol. 2, pp. 397–424. Washington: Joint Economic Committee.

———, and Osleeb, J. 1986. *The Soviet iron and steel industry.* Totowa, NJ: Rowman & Allanheld.

Study of Canadian Geography

William D. Romey

Almost any issue of the Association of American Geographers' journals contains at least one article, review, or comment about Canada. In 1987, 384 members of the AAG reported areal proficiency in Canada (AAG 1988), and the AAG Canadian Geography Specialty Group (CGSG) had 105 members. Yet Canadian geography has for many years been an invisible, often unnamed partner in the AAG. Sessions devoted exclusively to Canada have been rare, with Canadian topics mainly submerged in other contexts. At the four AAG annual meetings from 1985 through 1988, 158 papers dealing with aspects of Canadian geography were presented, although the name "Canada" did not appear in the indexes of the programs. Of those papers, 127 (80%) were by authors from Canadian institutions.

The purpose of this chapter is to identify major themes in Canadian geography that are currently being reported in the American geographical literature. The term "American geographers" must be broadly interpreted, because such a large percentage of the research reported in American journals and at the AAG meetings originates in Canadian universities. Many students of Canadian geography working in U.S. institutions are also Canadian by birth, if not by citizenship. This chapter does not attempt a worldwide survey of studies of Canada by "American" authors in the strict sense. Rather, it attempts to report on some major themes being studied by geographers who have recently published, mainly in the American geographical literature. It specifically includes work done by members of the CGSG, regardless of their nationalities.

The work mentioned in this chapter was published (or reviewed) mainly between 1982 and mid-1988. Some studies by Canadian authors are included because they have entered the American literature through publication or review in American publications, or because their authors are "American geographers" by virtue of their membership in the AAG. Among geographers interested in Canada, the question of who is "Canadian" and who is "American" is complicated by vagaries of place of birth, place of residence, citizenship, membership in professional societies, and educational histories.

A primary problem in reviewing the geography of a single country is that it must include a wide range of topical and thematic areas. As a matter of convenience, this survey is divided into the following subheadings, with the order established more or less on the basis of the number of papers in each area (given in parentheses) that were presented at AAG meetings over the years 1985–1988, taking this number as a rough index of levels of research activity:

Social and urban geography (45)
Physical geography and biogeography (36)
Industrial and agricultural geography (27)
Cultural and human geography (23)
Perception of place and landscape studies (1)
Environmental geography (15)
Historical geography (12)
Border questions (8)
Regional geography (6)
Other areas (16)

Since much of the activity in the cultural and human geography area is also related to the interests of geographers who study perception of place and landscape, these topical areas are placed adjacent to each other. A section on educational aspects of Canadian geography concludes the treatment by topical areas.

SOCIAL AND URBAN GEOGRAPHY

As defined here, this section most strongly emphasizes urban geography, consonant with the fact that more than three-quarters of all Canadians live in cities. Among specific topics included are some aspects of urban historical geography, gentrification, political and economic themes, residential patterns, migration among cities, residential care, the urban-economic base, and development of the urban fringe. Because of their interconnectedness, some studies that deal with aspects of social geography are also to be found in the sections on industrial and agricultural geography, historical geography, and elsewhere in this review.

Goldberg and Mercer's work (1986) provides a well-documented argument that "continentalist" approaches to the study of Canadian and U.S. urban areas are misleading. Canadian cities must be studied and understood within the contexts of their own special histories and cultural environments (Palm 1987). Stelter and Artibise (1986) edited a collection of essays on Canadian-American urban development; a section on political growth of the economy ranges from public-sector impacts to historical aspects of entrepreneurial elites. In this work, a comparison between public and private initiatives in Saint John, New Brunswick and Portland, Maine shows the differences fostered on opposite sides of the border in New England. Canada must be considered in the larger North American context (Doucet 1987).

Ley (1986) sought explanations for inner-city gentrification in 22 Canadian metropolitan areas between 1971 and 1981 by looking at 35 independent variables and correlations. Economic and urban amenity factors are especially important in the gentrification of Canadian cities, while demographic and housing factors are less important. Housing and labor markets are interrelated. Ley's work deals with the contexts for inner-city revitalization within the Canadian national urban system. The next step is to review detailed studies of inner-city change in particular neighborhoods, supported by this work on the roles of general processes within the overall system.

Pratt (1986) studied economic and political impacts of housing tenure arrangements in Canada, with an outer suburb of Vancouver taken as a case history. An urban national sample supported the idea that political attitudes and housing tenure (including home ownership) are related, but only in certain social classes. Home ownership may weaken class-consciousness, especially among the working classes, perhaps leading to greater conservatism. Zelinsky (1987) complained that Pratt's study has no geographic content and should not have been published in an AAG journal. Although Pratt's work may push the limits of geographical research strongly into the areas of sociology and political science, even Zelinsky conceded that many urban and political geographers will undoubtedly add it to their "bibliographic arsenals."

In a study of residential patterns in Kingston, Ontario, R. Harris (1984) showed that class segregation inhibited the development of local political activity by leading to ignorance of economic differences and then to complacency. Segregation, perhaps inevitable in cities, was successfully used in Kingston to create alliances against landlords. Segregation does not necessarily produce class fragmentation.

R. Harris and Hamnett (1987) contrasted the diffusion of home ownership in Britain and North America. Although the proportion of families owning homes in late-nineteenth-century Canada and the U.S. was much higher than in the United Kingdom, ownership proportions have changed so that levels in Britain and North America are now very similar. However, the homes owned in Britain are not the same in size or household equipment. Regionally, housing markets in the three countries show important variations. Property in London, New York, and Toronto offers better returns than in Birmingham, Pittsburgh, and Calgary. Goldberg and Mercer (1986) show that inner-city properties have often turned out to be poor investments in the U.S. where race is a more-explosive factor than in Canada.

Intermetropolitan migration received its first major systematic analysis in the work of Shaw (1985). In seeking to construct a predictive, explanatory model and evaluating various hypotheses on the nature of migration determinants in evolving Canadian society between 1956 and 1981, it lays the groundwork for future studies, showing a diminishing impact of market variables on migration between cities (Langlois 1987).

In studies of residential-care facilities in Toronto, Joseph and Hall (1985) showed a high degree of localization of adult services, children's services, and psychiatric group residential-care facilities. They suggested their results might be used to help reduce the burden on areas having large numbers of facilities by better targeting of future group homes. Hall and Joseph (1988) also studied the ecology of several of these locations in Toronto, concluding that associations identified for one kind of facility do not necessarily apply for other kinds of facilities. Present areas of concentration of group homes may not be compatible with their host environments.

The incomes of many Canadians from investments and pensions and government allowances, in addition to the usual labor-force data, are important in determining the urban economic base (Forward 1982). Pronounced differences exist between heartland cities and more peripheral urban areas. More work will be needed to explain both the regional differences observed and the common characteristics found in some groups of cities.

Studies of the rural-urban fringe in Canada are taking on considerable importance as Canadians continue to move toward large metropolitan areas, occupying what was previously farmland and small villages (Walker and Beesley 1984). Beesley and Russwurm (1981) addressed some general conceptual questions, such as whether an urban fringe is an area where forces of urbanization are working themselves out, or an ecological system in its own right (Sinclair 1984). Urban-rural social research along these lines represents a significant attempt to identify and quantify social and perceptual considerations emphasizing quality-of-life aspects in the outward growth of Canadian cities.

PHYSICAL GEOGRAPHY AND BIOGEOGRAPHY

Physical geography and biogeography include a wide range of subtopics. Physical aspects of arctic areas are of special importance in studies of Canada, and this area has seen the most activity in the American literature. Other areas that are treated in this section include glaciology, soil formation, climatology, hydrology, geomorphology, and the broad area of biogeography.

In the study of Arctic and Subarctic Canada, consideration of permafrost has been especially important. Nelson (1986) developed a revised permafrost index based on mean monthly air temperature and snow-cover data. It is a simple, easily interpreted, physically based prediction device. His model successfully predicts permafrost in peatlands. Permafrost boundaries are sensitive to soil properties, which must be factored into analyses. The index shows clear latitudinal zonation of permafrost in central Canada.

In a major work on periglacial geomorphology edited by Church and Slaymaker (1985), descriptive chapters are augmented by analyses of underlying physical processes in periglacial environments and by attempts at modeling the ways in which soils freeze and thaw. This work also evokes the difficult conditions of field work in harsh periglacial areas (Andrews 1986).

In an issue of the AAG *Annals* (v. 73(4), 1983) devoted primarily to problems of Arctic geography, Kay and Andrews (1983) developed a model to provide consistent reconstructions of July temperatures back to the mid-Holocene, based on taxa of pollens. Further surface samples of modern pollen in northern Canada are needed to test the stability of transfer functions among subregions in the Arctic and Subarctic.

Elliott-Fisk (1983) reported that the northern tree limit in eastern Canada is in equilibrium. In contrast, the tree limit in western and central Canada is out of equilibrium with present climatic conditions. Tree populations are susceptible to destruction by climatic or human-induced disruption.

The hydrology of a drainage basin in the Canadian high Arctic forms the basis for some of Woo's work (1983). Most previous hydrologic research in permafrost areas has dealt with individual processes rather than with interaction of processes in a single basin. Of the annual precipitation, 80% falls as snow, and 70% leaves the basin as runoff.

Bunting (1983) provided photomicrographic descriptions of soil fabrics and textures from Devon and Axel Heiberg Islands. Microscopy may be the only method that provides data permitting realistic assessment of physical processes in the internal formation of Arctic soils.

A map by Price and Vaucher (1983) illustrates the dimensions of the Canadian Arctic and its relation to the Arctic zones of other countries, showing some of the potential problems for Canada. The complexity of the Canadian archipelago has recently led to important legal and jurisdictional problems with the U.S.

In a more general treatment of the Arctic, Archer and Scivener (1986) included all of the Canadian north as their subject matter. Canada's arctic waters are seen as "a sea of opportunity, a sea of uncertainty" with respect to oil and gas, mineral deposits, and renewable resources. In the same book, Tracy describes Canadian Arctic policy as the art of the possible, in connection with its commitments to NATO (Haglund 1988).

Yarnal (1984) showed that the atmospheric environment controls the extent and behavior of the Sentinel Glacier in southern British Columbia on a scale ranging from days to years. A significant statistical relation exists between the mass balance of the Sentinel Glacier and variability of local small-scale weather patterns. Rock glaciers from the Saint Elias Mountains to the Ruby Range east of Kluane Lake in the Yukon show evidence of retrogressive slope failure, high hydrostatic pressures, and avalanche activity as primary models for their formation (Johnson 1984).

Meentemeyer (1984) developed and validated in the field a model for predicting the geography of litter decomposition rates for Canada and the U.S. By far the dominant factor is climate (annual actual evapotranspiration), and materials with high lignin content are the most difficult to decompose. Several maps show much-slower decomposition rates in the northern parts of Canada.

Nkemdirim (1988) identified a quasiperiodic three-year pattern of above-normal and below-normal precipitation around Calgary through studies of precipitation records over the 86-year period from 1885-1970. Between 1940 and 1970, the number of days with precipitation increased by 45% over the preceding 30 years, probably as a result of effects of the urban environment, although the amount of rain per event decreased.

In a detailed study of the drainage basin of the Red River Valley, Carlyle (1984) found that, although there are important problems of flooding in the drainage basin, periods of drought are a more serious long-term threat. Flood-control measures such as floodways and ring dikes now protect most settled areas. The Garrison diversion project, now postponed, is perceived as posing a special threat in the possibility it holds for allowing unwanted species of fish to enter the Lake Winnipeg–Hudson Bay drainage from the Missouri River basin.

In the U.S., much work on geomorphology appears in geological rather than geographical journals and will not be covered in this review. A sampling of work from *The Canadian Geographer* shows that most issues of this journal contain at least some physical geographical papers on Canada. A regular series entitled "Canadian Landform Examples" began with the Fall 1986 issue. More research devoted to geomorphological aspects of Canadian geography needs to be done by American geographers and published in American journals. A series of the type being produced in *The Canadian Geographer* could provide an opportunity for American geographers of all persuasions to review their geomorphology in the light of the most current ideas about landform-building processes in Canada.

INDUSTRIAL AND AGRICULTURAL GEOGRAPHY

On Canada's industrial geography, aside from some general books, little detailed work has been reported recently in the American literature. In agriculture, studies have included work on Canada's frontier areas, on field and computerized analyses of crop patterns, and on timber resources.

Walker (1980) provided a source of information on the manufacturing geography of Canada, synthesizing geographical research done in this area during the 1960s and 1970s. He included treatment of past and present industrial location patterns, looked at current trends, and reviewed planning and development policies (Ironside 1982). Blackbourn and Putnam (1984) paid special attention to analytical and policy issues, as well as to presenting factual material.

Foreign investment, technological dependence, industrial structure, state intervention, and similar matters are related to industrialization (Cannon 1986). Industrialization has become a significant factor in both U.S. and Canadian agriculture (Troughton 1985). Few of the general public have shown much concern about the disappearance of family farms as large-scale agribusiness increases its dominance.

Commercialization, regionalization, and socialization have increased through time in the agricultural evolution of the Ontario frontier, according to Norton (1984). Southern Ontario is typical of frontier areas that are undergoing rapid growth. Detailed statistical descriptions need to be added to cultural factors in the attempt to develop a valid framework for regional evaluation of agriculture in frontier areas.

Problems of agricultural development of Canada's "Middle North" and the agricultural fringe zone of "bush farming" in western Canada have been studied by Vanderhill (1971, 1979, 1982). Further spread of agriculture into the north seems unlikely, although some evidence exists that land settlement on the western Canadian agricultural fringe is still viable. The Peace River Basin in Alberta is the only really open agricultural frontier. True "pioneering" no longer goes on, and the fringe now looks just like the established farm districts close by.

De Lisle (1982) argued that distance, physical environment, and social and technological change all have significant effects upon cropping patterns on Mennonite farms in Manitoba. Farmers travel distances of up to several miles to their outfields from their farmsteads located in villages, but grow their root crops on infields close to the farmstead. Grains and oilseeds are grown throughout the system. Soil productivity and type are also important in cropping patterns.

Analysis of Landsat multispectral-scanner imagery from the north-central U.S. and Saskatchewan enabled Mohler et al. (1986) to develop a computerized technique for estimating spring small-grains acreage. Results of computer estimates compare favorably with former labor-intensive procedures, with the promise of even higher accuracy once further refinements are made.

Gillis and Roach (1986) provided a detailed history of Canada's forest conservation movement. Chapters on Ontario, Quebec, British Columbia, and New Brunswick show that no monolithic conclusions can be drawn about Canadian forests, resource politics, or industry interests (Sanders 1988).

CULTURAL AND HUMAN GEOGRAPHY

The American geographical literature has included a relatively limited number of items pertaining strictly to cultural and human geography. This section describes reports of work conducted specifically on ethnic groups in Canada—French-Canadians, Scottish, Ukrainian, and Chinese—and on vernacular architecture. Cultural and human geographic aspects of other studies are included in other sections of this review, such as those on social and urban geography, historical geography, and environmental perception.

An atlas of society and culture edited by Rooney, Zelinsky, and Louder (1982) contains 387 maps that cover a wide variety of cultural traits. Study of these maps and cross-correlations among them will undoubtedly suggest many areas for investigation of the geography of Canada.

Canada's ethnic minorities have provided an important topic for cultural geographic studies in the past. LeBlanc (1970, 1983, 1985a, 1985b, 1988) has studied interaction between French-Canadians and Franco-Americans, with his current focus being upon connections between the Franco-American elite and Canada. The "French Fact" in North America is imperiled by the continuing rapid assimilation of francophone groups in both Franco-America and Franco-Canada. If the French language and culture do survive in North America, they will most likely do so in the traditional cultural hearth along the St. Lawrence River.

Anderson (1987) ascribed the ultimate legitimization of the ethnic-ghetto concept, such as Vancouver's Chinatown, to the racial attitudes of that city's municipal authorities—an exercise of white European cultural domination. Studies of the meaning of place have too seldom taken into account the role of the state in creating and perpetuating problems of racism and other social problems. Ethnic patterns in Nova Scotia show increasing diversity in the Halifax-Dartmouth area and decreasing diversity around Sydney and Glace Bay (Millward 1981). Although Millward included no material on the English, Irish, and Scots (Macpherson 1983), Hornsby (1988 and pers. com.) has partially remedied this lack with his studies of Scottish immigrants of Cape Breton Island.

Luciuk of the CGSG, together with Hillmer, Kordan, and Dreisziger, organized a conference on "Ethnicity, the State, and War: Canada and Its Ethnic Minorities, 1939–45" held in September, 1986 at Queen's University at Kingston. A mimeographed summary of the proceedings described sessions on:

1. Bureaucratic approaches to ethnic loyalty
2. Loyalties in question
3. Special problems the government encountered when ethnic identity was also "a matter of faith"

4. A contentious session with papers on the Canadian government's attempts to control and channel Canada's large Ukrainian minority and Japanese-Canadians during the Second World War

In a study of vernacular architecture, Coffey (1988) examined building transformations in Augusta Township, Ontario on a lot-by-lot basis to provide a better understanding of factors involved in the transformation of Ontario from a landscape of log buildings to one of brick, stone, and frame. Noble (1984a, 1984b) addressed early vernacular architecture in northeastern Canada along with his primary emphasis on American structures. The distinctiveness of much Canadian folk architecture is apparent in this work.

PERCEPTION OF PLACE, LANDSCAPE STUDIES, AND ENVIRONMENTAL PERCEPTION

Study of the landscape and of environmental perception remains an important subject for geographers interested in Canada. This section describes studies regarding images of Niagara, symbolism of the landscape, and views of the landscape in literature.

Early in the development of Niagara Falls for hydroelectric power, some engineers and entrepreneurs believed that this area would become the greatest manufacturing center in the world (McGreevy 1987). Novelists saw Niagara as a place that would experience a disastrous future of wars and dangerous inventions. After World War I, more began to see the area as a potential site of great progress, and after World War II this peaceful border point continued to inspire optimism. To these early predictions and perceptions, McGreevy has compared the actual development that occurred at Niagara Falls, showing it to be the center of a vigorously developed landscape. The natural splendor of the falls proved to be a magnet for several kinds of economic and tourist development. Although the disastrous future imagined by some did not develop, the Love Canal, originally an expression of confidence in a progressive future, has rendered that area unsuitable for all future inhabitation instead of leading to the predicted perfect future city.

Konrad (1986) has gathered short essays on symbolism of the landscape by Holdsworth ("Architectural Expressions of the Canadian National State"), Zelinsky ("The Changing Face of Nationalism in the American Landscape"), and himself ("Recurrent Symbols of Nationalism in Canada"). The symbols Canadians use—buildings, signs, works of art, photography, and others—all reflect the distinctiveness of the Canadian identity and record changing perceptions of what it is like to live in Canada.

The portrayal of Canadian landscapes in literature forms a subject of research for several American geographers. Romey (1987) pointed out how the attraction and symbolism of Quebec's physical and cultural landscape persuaded Maria Chapdelaine to remain a frontier woman in the harsh physical conditions around Lac St.-Jean rather than emigrating to a factory town in New England. Bureau (1984) pursued the same theme of utopian sense of territoriality in Quebec (Gade 1986), drawing on history and philosophy as bases for understanding current political and social conditions in Quebec. Mallory and Simpson-Housley (1987) included an essay on geographic dimensions of American and Canadian literature by Kenneth Mitchell. This area of geography provides a fertile field for further work by American geographers.

ENVIRONMENTAL GEOGRAPHY

Although one might expect much research in issue-oriented areas of environmental geography to be reported in American journals, most current work of this type is being done by Canadians and reported in Canadian sources. Nonetheless, some work on acid rain, water supplies, and energy issues in Canada is being reported in the American geographical literature. In a useful environmental atlas produced by Environment Canada, the EPA, and Brock and Northwestern Universities, Botts and Krushilnicki dealt with problems of water quality in the Great Lakes and attempts to solve them (Sanderson 1988). The acid-rain problem continues to be a major source of Canadian irritation toward the U.S. (Schmandt and Roderick 1985). Miles (1987) pointed out the lack of a Canadian perspective throughout the book, although it contains a great deal of information, historical depth, and a comparative approach. A properly geographic perspective on this spatially variable issue is still greatly needed.

Problems of water supplies in the Great Lakes were studied by Cohen (1986). A combination of the effects of population growth and carbon dioxide-induced climatic changes may lead to a significant lowering of lake levels and water shortages in this area, which heretofore has been regarded as a possible exporter of water to rain-deficient areas of the U.S. Important policy decisions regarding interbasin transfer of water resources will arise in forthcoming years.

Churchill (1985) pointed out that huge increases in power production have resulted from using ice booms to stabilize the ice cover on the east end of Lake Erie. Unfortunately, despite all evidence to the contrary, the local population developed a perception that the new ice cover was a partial cause of colder weather in the area. Once an erroneous environmental perception of this type has developed, large problems in the development of policy occur.

Studies of the geography of energy include Sanderson's (1983) study of heating degree days and gas consumption in Alberta. At temperatures below 18°C, there is a high correlation between degree days and use of natural gas. After an increase in gas prices in 1973, however, the use of natural gas fell off sharply.

HISTORICAL GEOGRAPHY

In addition to two new major books on aspects of Canada's historical geography, the past five years have seen work on the history of communications systems, mining, and the regional history of New Brunswick.

Meinig (1986) has produced one of the major historical geographic works to emerge in the recent past. The first volume of his proposed trilogy on 500 years in the shaping of America treats the geographic dimensions of early exploration and settlement of the Atlantic Canadian coastal region. Emphasis is on the nature of the implantations of various cultural and ethnic groups from northwestern Europe into Nova Scotia, Newfoundland, Quebec, New Brunswick, and Prince Edward Island. The initial role of Atlantic Canada as a principal fishery for Europe was later supplanted by its role as a settling ground for various groups from France and Britain. Meinig's description of the colonization of eastern Canada and of the later reorganization of British North America after the American Revolution provides one of the most readable accounts in the literature on the role of geographic factors in the historical development of Canada and the evolution of its early relationship to the newly emerging United States. The role of the loyalists and their importance in the creation of the new Canada is clearly represented.

Another seminal work on the historical geography of Canada is the atlas edited by R. C. Harris (1987). It combines current understanding of physical geography, prehistory, and early settlement of Canada into a series of exceptional maps and diagrams. This volume contributes more new knowledge about the early history of Canada than any existing single source (Konrad 1988).

In a study of the history of Canadian communications systems in the mid-nineteenth century, Goheen (1987) related that, although the Post Office was the only legal long-distance carrier, its control from Britain was diffuse. Newspaper editors, publishers, postmasters, and users of the system had to create their own networks, which often worked in ways different from the inadequate official plan.

Newell (1986) described geographic aspects of the mining industry's evolution in Ontario between 1840 and 1880. She emphasized the role of technological innovation on the frontier and its diffusion just before the development of the Sudbury basin and other important mining camps (Wallace 1987). Wynn (1981) synthesized economic, technological, political, social, and human factors, showing how British capitalism transformed early nineteenth-century New Brunswick (Clarke 1983).

BORDER QUESTIONS

The nature of the 5526-mile U.S.-Canadian border has led to intensified research activity through the activities of the Borderlands Project headquartered at Montana State University's Forty-Ninth Parallel Institute. The steering committee of this project includes a geographer, Victor Konrad. One of the first published geographic studies supported by the Borderlands Project involved a statistical analysis of data on the number of crossings recorded for 25 Great Lakes border-crossing points (Merrett 1988).

The inhibiting function of the border has decreased over time. Free trade negotiations may reflect a changing attitude toward the border, as Canada and the U.S. become more dependent upon each other.

Frederic (1975) studied leisure homes along the international section of the St. Lawrence River. Later, he turned to comparative work on the geography of agriculture in Maine and the neighboring provinces of New Brunswick and Quebec (Frederic 1981, 1983, 1986). Of primary concern have been the Canadian-U.S. lumber trade and use of lumber resources, and attempts to improve Maine's dairy and beef production.

REGIONAL GEOGRAPHY

In the strictest sense, major texts on the geography of Canada might be the only works included in this section. However, additional regionally based work on the St. Lawrence River, the prairies, and the Arctic are also included. Several new textbooks transcend the lack of defined conceptual frameworks that has characterized past regional geographic texts on Canada (Miles 1984). Robinson (1983) offered eight chapters on specific regions in Canada, along with a more general introduction on his concept of Canadian regions and human-land relationships. McCann (1982) provided 13 essays which examined Canada from a heartland-hinterland viewpoint. Birdsall and Florin (1981) included substantial coverage of Canadian regional geography in the second edition of their textbook on North American landscapes.

For September-October, 1984, the *Journal of Geography* published a special issue on Canada which included 11 contributions on several aspects of Canada's regional geography, with a useful bibliography for educators. Among them were these contributions:

- An introduction by de Souza that confronted the issue of Canada's position as a neighbor—which is taken for granted by the U.S. (1984).
- Mather explored images held by Americans as contrasted with some of the realities of Canadian geography (1984).
- The Canadian identity is visible in its urban landscapes, attitudes toward government, and strong regional loyalties, as well as in the way the nation has to adjust to adjacent American realities (Holdsworth 1984).
- Canada has developed its own special ties to the Third World (Burghardt 1984).
- Within its borders, the Canadian state has had to deal with a wide range of regionalisms and nationalisms (Knight 1984).
- Whitney, in a study of Canadian residential preferences, found that his subjects had a general aversion to living in the North. For the most part, they preferred to live in their current home areas; he also noted a special preference for the Vancouver area (1984).
- Further exploring the prospects of the Canadian North, Wonders pointed out the lack of long-range future plans for this area, despite its advantages. The opportunity to produce an innovative, novel approach to Canada's last frontier may have been lost (1984).
- Although many North Americans may find little that is exceptional about Ontario's "ordinary" countryside, McIlwraith showed it to be vibrant and distinctive (1984).
- The special importance in Canada of the Windsor-Quebec axis, where more than half the population of Canada lives, was reviewed by Yeates (1984), whose earlier book (1975) first called attention to this region.
- No treatment of Canada is complete without consideration of its urban landscape, and Spelt used the evolution of Toronto's urban area as a case history (1984).

Lasserre's monograph (1980) on the St. Lawrence River region provided a detailed historical treatment of the use of the river as a primary access route into the North American continent by early explorers and later by immigrants, many of whom used it primarily as a passageway into the U.S. The French Canadians, on the other hand, developed its shores as their cultural hearth. Lasserre created a detailed picture of industrial development along the river, the development of the canal systems that culminated in the St. Lawrence Seaway, and the current status of shipping and the role of the river today (R. C. Harris 1982).

Rogge (1981) showed the prairies to be evolving in directions similar to those in the north-central U.S. The impact of energy resources and their consumption has been a strong regional determinant. With recent slowdowns in oil and gas exploitation, however, agriculture and the wide-open spaces have again become important (Paul 1982).

Sugden's regional geography of the Arctic and Antarctic (1982) devoted considerable attention to the Canadian Arctic. He discussed physical and natural aspects of the Arctic and Antarctic regions, the Inuit peoples, and the impacts of urban systems, which are increasingly affecting Arctic areas, including in Canada. Sugden showed the devastating effect that publicity surrounding the slaughter of baby harp seals off Newfoundland has had on one of the few cash opportunities available to the innocent Inuit (Haglund 1984).

OTHER AREAS OF RESEARCH ON CANADIAN GEOGRAPHY

In addition to the areas included under headings above are other studies reported by members of the CGSG or found during the present survey of literature on Canadian geographic studies in the U.S. This section briefly reports some of this work, in transportation, recreation, and bibliography.

Murphy (1983) provided a collection of essays on several issues of tourism in Canada. Marketing and development issues, travel prospects for the elderly, developmental issues, problems of evaluating the social carrying capacity for tourism, the role of government in tourism, and park management are among the subjects treated. Unfortunately, Atlantic Canada is little discussed, and the treatment of several areas is uneven (Young 1985). Fairweather's studies (1985) of Canadian transportation systems have included an analysis of the development of Air Canada's network.

Murphy and Andressen (1988) examined the willingness of the government of Victoria, British Columbia to share its tourism business with the rest of Vancouver Island, and the expectations they placed on citizens of outlying districts. The attitudes and perceptions of residents who live on other parts of the island, and who were expected to participate in and welcome the development of tourism, were found to differ from what government officials expected.

Harmon's study (1985) of bowling games showed that the game of rubberband duckpins is highly specific to Quebec, and candlepins are played mainly in northern New England and the Maritime Provinces. Restriction of the one game to Quebec provides another example of cultural distinction between francophones and anglophones.

Thomas Rumney has created several bibliographies of Canadian Studies publications for Vance Bibliographies of Monticello, Illinois (1985a, 1985b, 1985c, 1986, 1987a, 1987b). Although these bibliographies were not reviewed in detail for the present essay, they provide a broader listing of research on the geography of Canada than it has been possible to adopt in the present essay with its emphasis on the AAG and the CGSG.

EDUCATION

The increasing numbers of Canadian-studies programs in the U.S., even though they are interdisciplinary groups usually not housed directly in geography departments, provide an incentive for more American educational institutions to include courses on the geography of Canada in their curricula. Special programs partially supported by grants from the government of Canada and from various provincial governments have begun to help more U.S. educators develop courses on Canada.

The SUNY at Plattsburgh Quebec Summer Seminar began this process 10 years ago, and at least two or three geographers attend this seminar each year. In the summer of 1988, the Plattsburgh seminar began a new summer program in Ontario as well. The Atlantic Canada Faculty Institute conducted by the University of Maine at Orono extended this kind of consciousness-raising activity into the Atlantic provinces.

The Center for Canadian and Canadian-American Studies at Western Washington University has conducted "Study Canada" institutes that focus upon the Washington-British Columbia border area. A Pacific Northwest Canadian Studies Consortium was formed in April, 1987 to further Canadian Studies in Alaska, Idaho, Oregon, and Washington; they have now begun to conduct summer faculty institutes. The 49th Parallel Institute at Montana State University deals with Canadian-American border issues.

"Theme" conferences on Canada have been conducted at Plymouth State College, New Hampshire. The University of Maine's Canadian-American Center has held conferences on free trade (Konrad, Morin, and Erb 1986) and other issues.

In 1985, in an effort to promote the development of more courses on Canada, E. J. Miles of the University of Vermont convened a meeting on the geography of Canada for geographers from the northeast. Such efforts are helping American geographers who are interested in Canada to develop the sense of identity and importance of their task that will be necessary to promote geographical research on Canada and the teaching of Canadian geography. Most U.S. universities that have interdisciplinary Canadian Studies programs offer at least one course in some aspect of the geography of Canada. Geographers in some institutions have begun to offer more specialized courses on Canada as well.

Discussions and research results on the geography of education in Canada and on the status of geography in Canadian schools have appeared in the American geographical literature. Wolforth (1986) reviewed the status of geography teaching in Canadian schools; regional approaches dominate, rather than spatial theory. In Ontario, teachers develop distinctive courses under broad guidelines from the Ministry of Education, whereas a more prescriptive approach with ministry-defined objectives is used in Quebec. Geography in London, Ontario's elementary schools includes a unit designed to expose local children to the similarities and differences of life in Quebec and in Ontario, an important attempt to increase understanding of intercultural differences in Canada (Cecil and Mitchell 1985). Rogers (1983) reported strong growth in biogeography offerings in Canadian universities during the 1970s.

Robinson (1986) reviewed the history of the Canadian Association of Geographers (CAG) and the status of geography in Canadian universities. About 80 Americans belonged to the roughly 1500-member CAG in 1985. Employment was not a problem for graduates of Canadian geography departments at the time of Robinson's review.

REFLECTIONS AND CONCLUSIONS

The largest volume of work on Canadian geography done by American geographers, published in the American geographical literature, or presented at AAG meetings over the past few years, has been in the areas of social and urban geography, physical geography, industrial and agricultural geography, cultural and human geography, environmental geography, and historical geography, with a smattering of work from other areas. Areas of relative weakness are rural social geography, economic geography, political geography, gender studies, and others. Wide avenues are open for more activity in the study of many areas of the geography of Canada.

The CGSG has seen its membership increase steadily over the past few years. It now has a newsletter that is normally distributed twice a year. Large numbers of AAG members claim areal proficiency in Canadian geography, and the number of papers on Canada given each year ranges between 30 and over-50 on a wide range of topics. Sessions devoted completely to Canadian geography are rare, however, and are usually organized by specialty groups other than the Canadian Specialty Group, showing a lack of pull toward the "Canadian" part of the label.

Of the papers on Canadian geography at annual AAG meetings during the past four years, 80% were by Canadians and only 16% were presented by Americans. The title of a recent article by Porteous and Dyck (1987)—"How Canadian are Canadian Geographers?"—could be rephrased for the U.S. as "How American are 'American' Geographers of Canada?"

Canadian geography departments commonly have a distinctly British or American flavor, and Canadian undergraduates often feel themselves in an alien academic environment. In tracing the history of development of geography departments in Canada, Robinson (1986) concluded that many of them depended heavily upon American, British, and French geographers during their period of rapid development which began about 1935. Now, however, the flourishing Canadian departments greatly exceed American departments in size and scope. Could it be that all of those Canada-related papers at AAG meetings and in the Canadian geographical literature are actually being given by expatriate Americans? The evidence at hand confirms that by far the greatest number of geographers studying Canada are in fact Canadians.

Although AAG journals carry occasional articles on Canada (most by Canadians), the best journals on Canadian geography are the *The Canadian Geographer* or *The Operational Geographer*. Only rarely do articles by geographers appear in *The American Review of Canadian Studies, The Journal of Canadian Studies, Quebec Studies,* or other generalized Canadian studies periodicals.

Understanding of Canada in the U.S. at a critical time when free trade is being negotiated and developed and when Canada's political and military relations with the U.S. and NATO are being actively reconsidered demands more devotion to the study of Canada by more American geographers. Deeper studies of Canada in U.S. geography departments need to be developed. Actively developing the geography elements of emerging Canadian Studies programs provides one avenue for increasing the interest in Canadian geography.

More U.S. geography departments need to offer specific Canada-related courses, rather than simply inserting a couple of lectures on Canada into general courses on the geography of North America. Canadian geography needs to be introduced in more universities in mid-America and in the southeast and southwest. More active teaching about Canada can provide the background and incentive needed to encourage research on Canada by Americans in American universities.

REFERENCES

Association of American Geographers. 1988. AAG topical and areal proficiencies. *AAG Newsletter*. 23(2):20.

Anderson, K. 1987. The idea of Chinatown: The power of place and institutional practice in the making of a racial category. *Annals of the Association of American Geographers* 77:580–98.

Andrews, J. 1986. Review of *field and theory. Lectures in geocryology* by M. Church and O. Slaymaker. *Annals of the Association of American Geographers* 76:466–67.

Archer, C., and Scivener, D., eds. 1986. *Northern waters*. Totowa, NJ: Barnes & Noble.

Beesley, K. B., and Russwurm, L. H., eds. 1981. *The rural-urban fringe: Canadian perspectives*. Downsview, Ontario: York University, Atkinson College, Geographical Monographs No. 10.

Birdsall, S. S., and Florin, J. W. 1981. *Regional landscapes of the United States and Canada*. New York: Wiley.

Blackbourn, A., and Putnam, R. G. 1984. *The industrial geography of Canada*. New York: St. Martin's Press.

Bunting, B. T. 1983. High Arctic soils through the microscope: Prospect and retrospect. *Annals of the Association of American Geographers* 73:609–16.

Bureau, L. 1984. *Entre l'Eden et l'Utopie*. Montréal: Editions Québec/Amérique.

Burghardt, A. F. 1984. Canada and the Third World. *Journal of Geography* 83:205–11.

Cannon, J. 1986. Review of *The industrial geography of Canada*. *Professional Geographer* 38:106–7.

Carlyle, W. J. 1984. Water in the Red River Valley of the north. *Geographical Review* 74:331–61.

Cecil, R. G., and Mitchell, P. 1985. Teaching the geography of Quebec to Ontario schoolchildren. *Journal of Geography* 84:220–22.

Church, M., and Slaymaker, O., eds. 1985. *Field and theory. Lectures in geocryology*. Vancouver: University of British Columbia Press.

Churchill, R. R. 1985. The Lake Erie-Niagara River ice boom. *Geographical Review* 75:111–24.

Clarke, J. 1983. Review of *Timber Colony: A historical geography of early nineteenth century New Brunswick*. *Professional Geographer* 35:395–96.

Coffey, B. 1988. Building materials in early Ontario: The example of Augusta Township. *Canadian Geographer* 32:150–9.

Cohen, S. J. 1986. Climatic change, population growth, and their effects on Great Lakes water supplies. *Professional Geographer* 38:317–23.

De Lisle, D. de G. 1982. Effects of distance on cropping patterns internal to the farm. *Annals of the Association of American Geographers* 72:88–98.

de Souza, A. 1984. Canada: A neighbor taken for granted. *Journal of Geography* 83:194.

Doucet, M. J. 1987. Review of *Power and place: Canadian urban development in the North American context*. *Professional Geographer* 39:383.

Elliott-Fisk, D. L. 1983. The stability of the northern Canadian tree limit. *Annals of the Association of American Geographers* 73:560–76.

Fairweather, M. 1985. Air Canada: The growth of its domestic route system. *Bulletin of Canadian Studies* 9:87–96.

Forward, C. N. 1982. The importance of nonemployment sources of income in Canadian metropolitan areas. *Professional Geographer* 34:289–96.

Frederic, P. 1975. Leisure homes along the international section of the St. Lawrence River. *Proceedings of the Middle States Division of the Association of American Geographers* 9:7–10.

———. 1981. Trees to cattle feed: A joint Canadian—U.S. effort to improve Maine's dairy and beef industry. *Proceedings of the New England—St. Lawrence Valley Geographical Society* 11:59–61.

———. 1986. The structure of agriculture in the upper St. John Valley: Some Maine–New Brunswick comparisons. In *Resource economics in emerging free trade: Proceedings of a Maine/Canada trade conference,* pp. 99–106. Orono, ME: University of Maine, Canadian–American Center.

———, ed. 1983. *Maine–Quebec lumber trade: A discussion.* Farmington, ME: University of Maine Council for Canadian Studies.

Gade, D. W. 1986. Review of *Entre l'Eden et l'Utopie*. *Professional Geographer* 38:297.

Gillis, R. P., and Roach, T. R. 1986. *Lost initiatives: Canada's forest industries, forest policy and forest conservation*. New York: Greenwood Press.

Goheen, P. G. 1987. Canadian communications circa 1845. *Geographical Review* 77:35–51.

Goldberg, M. A., and Mercer, J. 1986. *The myth of the North American city: Continentalism challenged.* Vancouver: University of British Columbia Press.

Haglund, D. K. 1984. Review of *Arctic and Antarctic: A modern geographical synthesis. Professional Geographer* 36:135–36.

———. 1988. Review of *Northern waters. Professional Geographer* 40:248–49.

Hall, G. B. and Joseph, A. E. 1988. Group home location and host neighborhood attributes: An ecological analysis. *Professional Geographer* 40:297–306.

Harmon, J. 1985. Bowling regions on North America. *Journal of Cultural Geography* 6(1):109–23.

Harris, R. 1984. A political chameleon: Class segregation in Kingston, Ontario 1961–1976. *Annals of the Association of American Geographers* 74:454–76.

———, and Hamnett, C. 1987. The myth of the promised land: The social diffusion of home ownership in Britain and North America. *Annals of the Association of American Geographers* 77:173–90.

Harris, R. C. 1982. *Review of Le Saint-Laurent, grande porte de l'Amérique. Annals of the Association of American Geographers* 72:144–45.

———, ed. 1987. *Historical atlas of Canada: From the beginning to 1800.* Toronto: University of Toronto Press.

Holdsworth, D. 1984. Dependence, diversity, and the Canadian identity. *Journal of Geography* 83:199–204.

Hornsby, S. 1988. Scottish emigration and settlement in early nineteenth-century Cape Breton. In *Cape Breton Historical Essays,* ed. K. Donovan. Sydney, N.S.: University College of Cape Breton Press. In press.

Ironside, R. G. 1982. Review of Canada's industrial space-economy. *Annals of the Association of American Geographers* 72:146–47.

Johnson, P. G. 1984. Rock glacier formation by high-magnitude low-frequency slope processes in the southwest Yukon. *Annals of the Association of American Geographers* 74:408–19.

Joseph, A. E., and Hall, G. B. 1985. The locational concentration of group homes in Toronto. *Professional Geographer* 37:143–55.

Kay, P. A., and Andrews, J. T. 1983. Re-evaluation of pollen-climate transfer functions in Keewatin, northern Canada. *Annals of the Association of American Geographers* 73:550–59.

Knight, D. B. 1984. Regionalisms, nationalisms, and the Canadian state. *Journal of Geography* 83:212–20.

Konrad, V., ed. 1986. Focus: Nationalism in the landscape of Canada and the United States. *Canadian Geographer* 30:167–80.

———. 1988. Review of *Historical atlas of Canada: From the beginning to 1800. Annals of the Association of American Geographers* 78:368–72.

Konrad, V.; Morin, L.; and Erb, R. 1986. *Resource economics in emerging free trade: Proceedings of a Maine/Canada trade conference.* Orono, ME: University of Maine, Canadian American Center.

Langlois, A. 1987. Review of *Intermetropolitan migration in Canada: Changing determinants over three decades. Professional Geographer* 39:264.

Lasserre, J-C. 1980. *Le Saint-Laurent, grande porte de l'Amérique.* Quebec: Editions Hurtubise.

LeBlanc, R. G. 1970. The Acadian migrations. *Canadian Geographical Journal* 81:11–19.

———. 1983. Regional competition for Franco-American repatriates, 1870–1930. *Quebec Studies* 1:110–29.

———. 1985a. Colonisation et rapatriement au Lac Saint-Jean (1895–1905). *Revue d'histoire de l'Amérique Francaise* 38:379–408.

———. 1985b. The Francophone "conquest" of New England: Geopolitical conceptions and imperial ambition of French-Canadian nationalists in the nineteenth century. *American Review of Canadian Studies* 15:288–310.

———. 1988. Emigration, colonisation, rapatriement: The Acadian dimension. *La Société historique Acadienne: Les Cahiers.* In press.

Ley, D. 1986. Alternative explanations for inner-city gentrification: A Canadian assessment. *Annals of the Association of American Geographers* 76:521–35.

McCann, L. D., ed. 1982. *Heartland and hinterland: A geography of Canada.* Scarborough, Ontario: Prentice-Hall.

McGreevy, P. 1987. Imagining the future at Niagara Falls. *Annals of the Association of American Geographers* 77:48–62.

McIlwraith, T. F. 1984. Ontario's ordinary countryside. *Journal of Geography* 83:234–39.

Macpherson, A. G. 1983. Review of *Regional patterns of ethnicity in Nova Scotia: A geographical study. Professional Geographer* 35:126–27.

Mallory, W. E., and Simpson-Housley, P., eds. 1987. *Geography and literature*. Syracuse, NY: Syracuse University Press.

Mather, C. 1984. O Canada! Reality and the image in America. *Journal of Geography* 83:195–98.

Meentemeyer, V. 1984. The geography of organic decomposition rates. *Annals of the Association of American Geographers* 74:551–60.

Meinig, D. 1986. *Atlantic America, 1492–1800*. Vol. 1 of *The shaping of America: A geographical perspective on 500 years of history*. New Haven: Yale University Press.

Merrett, C. 1988. Changing border function of the Canadian-American boundary: 1947–1977. *Proceedings of the New England–St. Lawrence Valley Geographical Society*. In press.

Miles, E. J. 1984. Review of *Concepts and themes in the regional geography of Canada and of heartland and hinterland: A geography of Canada*. *Annals of the Association of American Geographers* 74:501–4.

———. 1987. Review of *Acid rain and friendly neighbors: The policy dispute between Canada and the United States*. *Professional Geographer* 39:512–13.

Millward, H. 1981. *Regional patterns of ethnicity in Nova Scotia: A geographical study*. Halifax: International Education Centre, St. Mary's University.

Mohler, R. R. J. et al. 1986. An automatic procedure for estimating spring wheat and other small grain acreage from Landsat data. *Professional Geographer* 38:375–82.

Murphy, P. E., ed. 1983. *Tourism in Canada: Selected issues and options*. Victoria: University of Victoria.

———, and Andressen, B. 1988. Tourism development on Vancouver Island: An assessment of the core-periphery model. *Professional Geographer* 40:32–42.

Nelson, F. E. 1986. Permafrost distribution in central Canada: Applications of a climate-based predictive model. *Annals of the Association of American Geographers* 76:550–69.

Newell, D. 1986. *Technology on the frontier: Mining in old Ontario*. Vancouver: University of British Columbia Press.

Nkemdirim, L. C., 1988. On the frequency of precipitation-days in Calgary, Canada. *Professional Geographer* 40:65–76.

Noble, A. G. 1984a. *Houses*. Vol. 1 of *Wood, brick, and stone: The North American settlement landscape*. Amherst, MA: University of Massachusetts Press.

———. 1984b. *Barns and farm structures*. Vol. 2 of *Wood, brick, and stone: The North American settlement landscape*. Amherst, MA: University of Massachusetts Press.

Norton, W. 1984. Agricultural evolution on the frontier. *Professional Geographer* 36:18–27.

Palm, R. 1987. Review of *The myth of the American city: Continentalism challenged*. *Annals of the Association of American Geographers* 77:136–37.

Paul, A. H. 1982. Review of *The prairies and plains: Prospects for the 80's*. *Professional Geographer* 34:483–84.

Porteous, J. D., and Dyck, H. 1987. How Canadian are Canadian geographers? *Canadian Geographer* 31:177–79.

Pratt, G. 1986. Housing tenure and social cleavages in urban Canada. *Annals of the Association of American Geographers* 76:360–88.

Price, R. L., and Vaucher, M. E. 1983. Propective maritime jurisdictions in the polar seas. *Annals of the Association of American Geographers* 73:617–18 and map.

Robinson, J. L. 1983. *Concepts and themes in the regional geography of Canada*. Vancouver: Talonbooks.

———. 1986. Geography in Canada. *Professional Geographer* 38:411–17.

Rogers, G. F. 1983. Growth of biogeography in Canadian and U.S. geography departments. *Professional Geographer* 35:219–26.

Rogge, J. R. 1981. *The prairies and plains: Prospects for the 80's*. Winnipeg: University of Manitoba Department of Geography.

Romey, W. D. 1987. Geographic imagery and toponymy of the Lac St.-Jean Area, Quebec, in Louis Hemon's novel, *Maria Chapdelaine*. *Proceedings of the New England–St. Lawrence Valley Geographical Society* 16:42–47.

Rooney, H. F. Jr.; Zelinsky, W.; and Louder, D. R., eds. 1982. *This remarkable continent: An atlas of United States and Canadian society and culture*. College Station, TX: Texas A & M University Press.

Rumney, T. 1985a. *Historical geography of Canada: A selected bibliography*. Monticello, IL: Vance Bibliographies.

———. 1985b. *A selected bibliography of the economic geography of Canada: Agriculture, land use, resources, energy development, recreation and tourism.* Monticello, IL: Vance Bibliographies.

———. 1985c. *A selected bibliography of the economic geography of Canada: Industry, transportation, urban and tertiary systems.* Monticello, IL: Vance Bibliographies.

———. 1986. *The urban geography of Canada: A selected bibliography.* Monticello, IL: Vance Bibliographies.

———. 1987a. *The physical geography of Canada—Climate, ice, and water studies: A selected bibliography.* Monticello, IL: Vance Bibliographies.

———. 1987b. *The physical geography of Canada—Geomorphology: A selected bibliography.* Monticello, IL: Vance Bibliographies.

Sanders, R. A. 1988. Review of *Lost initiatives: Canada's forest industries, forest policy and forest conservation. Professional Geographer* 40:117.

Sanderson, M. 1983. Heating degree day research in Alberta: Residents conserve natural gas. *Professional Geographer* 35:437–40.

———. 1988. Review of *The Great Lakes: An environmental atlas and resource book. Professional Geographer* 40:250–51.

Schmandt, J., and Roderick, H., eds. 1985. *Acid rain and friendly neighbors: The policy dispute between Canada and the United States.* Durham, NC: Duke University Press.

Shaw, R. P. 1985. *Intermetropolitan migration in Canada: Changing determinants over three decades.* Toronto: NC Press.

Sinclair, R. 1984. Review of *The urban-rural fringe: Canadian perspectives. Professional Geographer* 36:106–7.

Spelt, J. 1984. Toronto: The evolution of an urban landscape. *Journal of Geography* 83:250–55.

Stelter, G. A., and Artibise, A. F. J., eds. 1986. *Power and place: Canadian urban development in the North American context.* Vancouver: University of British Columbia Press.

Sugden, D. 1982. *Arctic and Antarctic: A modern geographical synthesis.* Totowa, NJ: Barnes & Noble.

Troughton, M. J. 1985. Industrialization of US and Canadian agriculture. *Journal of Geography* 84:255–63.

Vanderhill, B. G. 1971. The ragged edge: A review of contemporary agricultural settlement along the Canadian northern frontier. *Geografisch Tijdschrift, Nieuwe Reeks V* (2):123–33.

———. 1979. Canada's middle north: Some problems for development. *Southeastern Geographer* 19:13–26.

———. 1982. The passing of the pioneer fringe in western Canada. *Geographical Review* 72:200–17.

Walker, D. 1980. *Canada's industrial space economy.* Toronto: Wiley.

Walker, G., and Beesley, K. B. 1984. Urbanites in the rural-urban fringe: The case of northwest Toronto. *Geographical Perspectives* 53:44–57.

Wallace, I. 1987. Review of *Technology on the frontier: Mining in old Ontario. Professional Geographer* 39:377–78.

Whitney, H. A. 1984. Preferred locations in North America: Canadians, clues, and conjectures. *Journal of Geography* 83:221–25.

Wolforth, J. 1986. School geography—alive and well in Canada? *Annals of the Association of American Geographers* 76:17–24.

Wonders, W. C. 1984. The Canadian North: Its nature and prospects. *Journal of Geography* 83:226–33.

Woo, M. 1983. Hydrology of a drainage basin in the Canadian high Arctic. *Annals of the Association of American Geographers* 73:577–96.

Wynn, G. 1981. *Timber colony: A historical geography of early nineteenth century New Brunswick.* Toronto: University of Toronto.

Yarnal, B. 1984. Synoptic-scale atmospheric circulation over British Columbia in relation to the mass balance of Sentinel Glacier. *Annals of the Association of American Geographers* 74:375–92.

Yeates, M. 1975. *Main Street: Windsor—Quebec City.* Toronto: Macmillan.

———. 1984. The Windsor–Quebec City axis: Basic characteristics. *Journal of Geography* 83:240–49.

Young, B. 1985. Review of *Tourism in Canada: Selected issues and options. Professional Geographer* 37:514–15.

Zelinsky, W. 1987. Commentary on housing tenure and social cleavages in urban Canada. *Annals of the Association of American Geographers* 77:651–52.

EMERGING PERSPECTIVES ON GEOGRAPHIC INQUIRY

Political Geography

David R. Reynolds | David B. Knight

Political geography has changed significantly in the past two decades. By the mid-1980s a "new" political geography had emerged from attempts to develop a more critical, postpositivistic human geography. In particular, there now is a concern for social theory and a readiness to examine afresh such central concepts as state, society, nationalism, space, and place. Political geography has joined mainstream social science in its concern for contributing to the creation of new perspectives through which to comprehend and interpret social reality. Political geography also is no longer isolated from other streams within the discipline; indeed, increasingly it is being recognized as having an important, even central, role to play in human geography.

This chapter identifies some significant aspects of the changes that have occurred, cites and discusses a small sample of the literature, and outlines some possible directions the field may take in the future. To the extent possible, these developments are examined within the particular contexts of ongoing empirical research in political geography. Some of the empirical themes discussed involve intrastate issues, some involve interstate issues, and some issues overlap, for it is not always possible to discuss matters internal to states without also considering international processes, and vice versa. The topics identified for discussion are: changing definitions, urban regional development, political parties and elections, local autonomy and the local state, nation-building and national disintegration, and international behaviors.

CHANGING DEFINITIONS

Neither a generally accepted definition of political geography nor an explicit body of social theory underlay the work of political geographers earlier in this century. The seeming lack of focus and apparent peripherality of much of its writings, relative to what other geographers were doing, led Sauer (1927, 207) to call the subfield the "wayward child" of the geographical sciences! In a 1935 review of the subfield, Hartshorne (1935, 957) defined it as "the study of the state as a characteristic of areas in relation to the other characteristics of areas." In a later, more complete review, Hartshorne (1954, 178) provided a similar definition: "the study of areal differences and similarities in political character as an interrelated part of the total complex of areal differences and similarities."

The areal "functional" approach (Hartshorne 1950; Jones 1954) dominated the textbooks over the next decade, but texts were little more than regional inventories, generally from a human-environment perspective, and often "working the same materials over and over again" (Jackson 1958). Jackson felt that there was "a real need for the development of systematic thinking in political geography."

A more-promising evolution came in the 1960s when the National Academy of Sciences—National Research Council Ad hoc Committee on Geography (1965, 32) defined political geography as "the study of the interaction of geographical area and political process." In a major study, Kasperson and Minghi (1969, xi) defined political geography as "the study of the spatial and areal structures and interactions between political processes and systems, or simply, the spatial analysis of political phenomena." They felt that their definition was "sufficiently broad to comprehend the pluralistic nature of the field and to involve political geography in the mainstream of current social science theory and research" (ibid.).

Despite this conclusion, and the introduction of systems perspectives (Cohen and Rosenthal 1971), political geography still tended to examine "old" themes, such as territory and population; territorial organization and power; the raison d'etre for particular states; core areas; boundaries; the unitary state; the federal state; and capital cities. These themes remain important. But what must be noted is that, while the state was central to discussion, there was little explicit attention directed either to the relevance of contrasting theories of the state or to political processes (Pounds [1963] 1972; de Blij [1967] 1973; Glassner and de Blij 1980). Indeed, all too often there was a political geography without politics. Little genuine progress was made in revitalizing the field until the end of the 1970s.

Paradoxically, numerous geographers taught popular political geography courses to thousands of students in the 1960s and early 1970s, even as there was a general lack of interest by those same geographers in developing the subfield, even if only to "bring it into line" with other subfields of the discipline which were then actively exploring new perspectives and methods. The reluctance to develop political geography at that time has been attributed to a floundering "in the wake of the quantitative revolution in geography" (Taylor et al. 1982, 1).

Whether or not this was the case, change was soon to come as new avenues were explored (Kliot 1982; Hall 1982; Gottmann 1982; Burnett 1984; Archer and Shelley 1986; O'Loughlin 1986a; Knight 1982b, 1986; Brunn and Yanarella 1987). Of special importance, emanating from an international political geography conference in 1980, Burnett and Taylor (1981, 4) developed a definition that was wide in scope: political geography refers to "political studies carried out by geographers using techniques and ideas associated with their spatial perspectives." In this, geographers were understood to be part of the larger body of scholars who are involved with political studies, whether within geography or in cognate disciplines.

Since the start of this decade, there has been a remarkable growth in strength, numbers, and activities among political geographers. Political geographers have organized themselves into national and international organizations;[1] new international journals have come into being—e.g., *Political Geography Quarterly* and *Society and Space;* major international conferences have been held; and the number of books and articles by political geographers is increasing at an accelerating rate. In short, through various activities and organizations, and by a marked increase of publications, considerable life is being displayed by political geographers.

POLITICAL GEOGRAPHY AND SOCIAL THEORY: THEORETICAL DEBATE IN THE "NEW" POLITICAL GEOGRAPHY

The radical critique of positivist human geography in the 1970s provided a clear demonstration that human geography, if it entertained any theory of society, tended to adopt the one implicit in neoclassical economics: as an arbitrary, territorial aggregation of individuals whose behavior was driven by underlying individual utility functions. Utility functions were treated simply as unproblematic "givens" (e.g., see Scott and Roweis 1978). The critique was, of course, not limited to a demonstration of geography's failure to confront any of the central issues of social theory (the structure/agency debate, theoretical realism, theories of the state), or its reduction to an almost exclusive concern with a single variable—distance (Massey and Allen 1984). More fundamentally, the critique forced virtual abandonment of positivism as an appropriate philosophy for human geography (for discussion, see Harvey 1973; Johnston 1979a; Gould and Olsson 1982).

But, if the world of appearances was not a reliable source of empirical generalizations from which to fashion geographic theories and explanations, what alternative philosophies were available? The earlier quantitative revolution had created what amounted to an epistemological crisis in geography: explanation required theory; geography possessed very little indigenous theory, and hence needed to seek it from other disciplines; but, with the reduction of geography to a science of distance, there was little basis upon which effective communication with other social-science disciplines could occur (see Harvey 1969). The radical critique did, however, provide at least the rudiments of an alternative philosophy of science and methodology—historical materialism—which held promise of overcoming the deficiencies of positivism and geography's isolation from social theory.

Having rediscovered society, the new generation of human geographers had by the mid-1970s set out to identify spatial patterns of social deprivation and inequality. They had also begun to develop explanations of how these patterns were produced,

---

[1] The organizations are: Political Geography Specialty Group of the Association of American Geographers (PGSG), Political Geography Study Group of the Institute of British Geographers, and *Groupe de trevail en geographie politique* in France. The PGSG is the largest national group; it has consistently been one of the most active of the specialty groups within the AAG. Internationally, under the rubric of the International Geographical Union, most political geographers are involved with the Study Group on the World Political Map (Taylor and House 1984; Blake 1987; Knight and Davies 1987; Johnston, O'Loughlin, and Taylor 1987; Johnston, Knight, and Kofman 1988). Some also are involved with the Study Group on Geography and Public Administration.

reproduced, and ameliorated, and to explore how such patterns might be transcended or otherwise transformed (Busteed 1983). Perhaps the clearest exposition of this new "welfare-oriented" approach to what might be called a "new" political geography was Cox's *Location and Public Problems* (1979). This study sought to identify the dominant patterns of social inequality existing at various geographic scales and the structures and processes producing them. Figuring centrally in Cox's analysis were various forms of economic competition among various actors, set within the juridical context provided by the state (or system of states), which itself was structured by various forms of political competition.

The implicit social theory underlying Cox's analysis was an admixture of pluralism, neoclassical economics, and liberal political philosophy, onto which some Marxian categories had been grafted. It adopted an epistemological stance in which spatial forms and patterns of material well-being were socially constructed, rather than naturally produced. Despite its eclecticism, it nonetheless was very effective in demonstrating that, if the pluralist theory of politics was actually descriptive of "who got what," it was a pluralism riven with inequality. Although the state as an empirical entity had always been on the agenda of human geography, Cox's attempt to identify a satisfactory framework for its analysis helped point to the need for a deeper, more-critical theorization of politics and the state.

The rise of a distinctive urban political geography in the 1970s that focused upon locational dimensions of urban social conflicts (Cox, Reynolds, and Rokkan 1974; Cox 1979; Cox and Johnston 1982; Janelle 1977; Janelle and Millward 1976; Wolpert et al. 1972) was also influential in leading to recognition of the need for a more critical analysis of politics. This work had tended to view conflicts over aspects of the urban built environment as the unavoidable results of individual conflicts of interest over the distribution of costs and benefits associated with land-use changes. Unanswered in any particularly profound way were such questions as: Why did the state never seem to be playing the role attributed to it in pluralist theory—that of neutral arbiter? Why were such conflicts so obviously on the political scene in the 1960s and 1970s, but not in earlier periods?

Dear (1982, 180), in his critique of *Political Geography Quarterly*'s agenda for political geography (Taylor et al. 1982), called for research to focus on why there is a separate "political sphere" in society. He suggested that a theory of society was needed which contained within it a theory of the state capable of explaining "geographical outcomes as a process of mediation (or conflict) between civil society and the apparatuses of the state." He was not alone in calling for critical analysis of the state. Indeed, all those expressing themselves in print on the issue of theory in political geography between 1980 and 1985 saw the exploration of social theory and theories of the state as the most urgent tasks confronting the field (e.g., see Reynolds 1981; Short 1982; Cox and Johnston 1982; Johnston 1982, 1984a; Taylor 1982, 1983; Hudson 1983; Clark and Dear 1984; Burnett 1984).

There was also widespread agreement on the value of grounding theorizations of the state, at least to some extent, in contemporary Marxist social theory. Marxist theory attempted to address the big questions necessary for understanding how society as a whole operates; clearly identified conflict as the "motor" of history; and was sensitive to the need for explanation over time.

Concern for Theory and the Debate over an Appropriate Theory of the State

Geographers have tended to follow the lead of other social scientists and equate the theorization of the political with the development of theories of the state. In this regard, the pioneering work of Clark and Dear (1978, 1981, 1984), which introduced the debate on the state into human geography, has been particularly influential. Recent, particularly Marxist, work on theories of the state is important because it moves beyond purely abstract analysis, and enters the realm in which the abstract becomes linked with historical contingency in a way that permits the analysis of the concrete—what the state actually does.

The theory of the state that has received the most widespread attention in geography is the materialist variant of the Frankfurt School's "state-derivation" model (Clark and Dear 1984; Holloway and Picciotto 1978). This perspective—unlike other neo-Marxist functionalist perspectives, which view the state as relatively autonomous from other social structures constituting society—is concerned primarily with elucidating how the specific functions of the state have changed with the changing nature of capitalism, and vice versa. Although dialectical, the perspective remains rooted largely in functionalist explanation. State activity is viewed as satisfying two functional imperatives: (1) facilitating expanded capital accumulation and domestic economic growth (the accumulation function), and (2) maintaining social cohesion by sustaining the false impression that capitalist social relations are both inevitable and desirable (the legitimation function).[2] The separation of the state from civil society is viewed as deriving from the general nature of capitalist social relations, while the specific mechanisms (or apparatuses) through which a state's functional requisites are carried out are the products of historical contingency.

Within human geography, debate over social theory and the state has centered on two key issues. First, if theories of the state are to be derived conjointly from social theory and the historical analysis of particular societies, how are the specific "societies" that are analyzed to be bounded? Second, should emphasis be placed on the derivation of the form and functions of the state under capitalism, or on the derivation and persistence of an international system characterized by the multiplicity of states, or on both?

Most geographers have finessed the first question altogether, by implicitly assuming that the societies to be studied were unproblematically the "national" populations of states. Regarding the second question, most have followed the lead of social theory, and have assumed that an approach which derives the specifics of the capitalist state in terms of forms and functions will also account for the multiplicity of states and the absence of political forms "higher" than the state.

Taylor (1982, 1985a, 1986, 1987) has outlined a world-systems approach for political geography (building on the work by Wallerstein 1974, 1979, 1980). Taylor argues that the forms and functions of the state cannot be derived solely from an analysis of social relations contained within its territorial boundaries and, as a corollary, that the concept of society cannot and should not be equated simply with the population of a

---

[2] Other such formulations differ in terms of the number of functional imperatives confronted by the state. Clark and Dear (1984), for example, identify three: securing social consensus, securing the conditions of production, and securing social integration.

state. Instead, he suggests that a basic tenet of the world-systems approach—the existence of a single, capitalist world economy—be accepted, and that this "historical system" be the "society" from which various state forms and functions are derived. From Taylor's perspective, what is needed is not just a "theory of the state," but a "theory of states" in which the existence of a multiplicity of states is one of the key structural properties of the world economy that is in need of explanation. It is in this sense that Taylor feels there is some danger of an excessive concentration on the state in the "new" political geography.

Although this particular debate has not drawn written comment from geographers, other than from the two protagonists directly involved (Taylor 1985b; Dear 1986), it is nonetheless a very important debate. It can be viewed as a special case of the more general debate in realist philosophy over what constitutes the "necessary relations" of capitalist social formations, and how they might be identified. In particular, for the development of a better understanding of political power, it is vitally important to determine whether the general form of the relationship between states and societies—as a system of states separated from one another by the practice of sovereignty, and from their respective civil societies by ideological means—is necessary or contingent (Sayer 1984). At the outset, a tentative determination can only be made on philosophical and epistemological grounds. But over the longer run, the issue will be resolved through a comparison of the depth and specificity of understandings provided by the social theories developed under differing conceptions of what is necessary and what is contingent.

The debate is also laden with methodological implications for the conduct of historical/empirical research on the state and the state system. If the dominant functionalist explanation entailed in the state-derivationist perspective cannot be sustained, then the essentially state-centered analyses proposed by Clark and Dear (1984) will either be unguided or misguided theoretically. Essentially, this is the same point that Taylor (1986) made recently in his review of Wallerstein's *Politics of the World Economy* (1984), where he suggests that the "state" in the various "theories of the state" is a "chaotic conception" (see Sayer 1984).[3]

What is perhaps most surprising about this debate is that it has arisen at all. Although the capitalist state can doubtless be interpreted fruitfully as acting in ways which legitimate capitalist social relations (why else call it "capitalist"?), the legitimation of state power itself derives in part from the concept of sovereignty, which in turn is predicated on a territorial definition of society. In moving from abstract analysis to the concrete, it is surprising that more attention has not been given to analyz-

---

[3]Apparently drawing inspiration from Skocpol (1979), Clark and Dear (1984) distinguish between "state-centered" and "society-centered" approaches to development of a satisfactory theory of the state. A state-centered approach inquires into the actual behavior of the state as a real social institution with its own agenda, organization, and language. A society-centered approach views the state as a condensate of the more important relations that characterize society, whether those relations be capitalist or otherwise. Clark and Dear's distinction between state- and society-centered theory is somewhat confusing. On the one hand, they are attracted to a historical materialist methodology; hence, they come close to arguing that the distinction is another of the dualisms in need of expurgation in social theory. On the other hand, they spy (justifiably) an overemphasis on society-centered theorization, with its attendant problems of historical/empirical corroboration, etc., and wish to rectify this imbalance in the direction of a more "tractable" state-centered analysis. Ultimately, they appear to call for a methodology that is both society-centered and state-centered: society-centered in explaining the general form and functions of the state, and state-centered when it comes to providing the historical and spatial specifics necessary for explaining social change.

ing whether territoriality is a necessary feature of the capitalist state. Indeed, it can be argued that Taylor's (and Wallerstein's) demand that a satisfactory social theory (of the world-economy) explain the multiplicity of states must, at a higher level of abstraction, also entail the essential territoriality of the state system.

Sack's (1986) recent work on territoriality as a general strategy of social control may prove to be an important beginning in such an endeavor, if it can be embedded within existing social theory. Johnston (1986a) has made some progress along these lines, but much more detailed historical work remains to be done if his approach is to avoid a form of crude functionalism.

Human Geography's Entry into the Structure/Agency Debate: Implications for the Appropriate Theorization of Space and the "Political"

In recent years, human geographers have spent far more effort attempting to clarify the nature of space in social theory, than they have in contributing to the development of state theory. In many regards, this work can be viewed as an effort to capitalize on the theoretical insights and methodological strengths of essentially Marxist perspectives on social theory, while avoiding, and perhaps even overcoming, their shortcomings.

Soja (1980, 1985; Soja and Hadjimichalis 1979) argued that what was needed was a new mode of theorizing, one he dubbed the "socio-spatial dialectic." Whereas one consequence of the radical revolution was to reject the importance of the spatial organization of things in the constitution of society (Massey 1985), Soja (1980) argued for a radical geography that not only "Marxified" spatial analysis but also "spatialized" Marxist analysis. His argument in 1980 was that capitalist production and reproduction produced both a social and a territorial division of labor, and that both needed to be analyzed conjointly in concrete Marxist analysis. To do otherwise ran the risk of rendering analysis problematic, because of its neglect of human and social spatiality. This is a theme developed independently and at much greater length in sociology by Giddens (1981) in his critique of historical materialism.

By 1985, the epistemological and conceptual groundwork for a theory of structuration had been laid by theoretical realists (Keat and Urry 1982; Bhaskar 1979; Sayer 1984), by Giddens (1981, 1984), and by Thrift (1983). Hence, Soja (1985) shifted his argument from the traditional Marxian realm of the "concrete" to one calling for the full-scale retheorization of spatiality within Marxist thought.

A number of political geographers (including Agnew 1987a; Cox 1986, 1987; Dear and Moos 1986; Johnston 1986a, 1987) have found realist philosophy and the theory of structuration appealing. Structuration offers the promise of placing human geography at the very center of social theory in overcoming the major methodological deficiencies encountered in extending Marxist social theory. The form of geographical analysis called for, however, is quite different from that heretofore practiced in either radical geography or spatial analysis. Agnew (1987a), drawing heavily on Pred (1984), has described what is called for as a "place-based" analysis of politics. In a similar vein, Harvey (1985a, 1985b) identifies historical-geographical materialism as the appropriate method of analysis, with the historical geography of capitalism as its subject matter.

It is too early to be very specific, yet it appears that central issues in the new political geography will be developing ways of identifying a theoretically and practically meaningful concept of place, and coming to grips with the production and reproduction of real places. Such places include any region on the Earth's surface over which social struggles have produced some politically relevant sense of place in the consciousness of its inhabitants. Since social struggles of one form or another are fundamental in the production and reproduction of places in a capitalist world-economy, there are sure to be some rather distinctive political geographies to be discovered and explained in the years ahead.

In the sections that follow, the more-major developments are reviewed in several substantive areas of politics to which geographers have contributed. In all cases, "developments" are linked to the broader theoretical issues discussed here. The central concern throughout is in analyzing the extent to which research in political geography has become more theoretically coherent over the past several years.

POLITICAL GEOGRAPHY OF URBAN REGIONAL DEVELOPMENT

One of the major achievements of radical geography over the past decade has been to place the "geography of capitalism" and its inherent spatial unevenness prominently on the research agenda of human geography (e.g., see Smith 1984; Peet 1987). Unquestionably, the leader in extending Marx's theory and logic in order to understand the geography of advanced capitalism has been Harvey.

Harvey (1982) reconstructs the logic of Marx's analysis and methodology, and applies them specifically to the forms and trends of capitalist urbanization. Some of Harvey's more-original insights center on how the capitalist search for accumulation of surplus value sets off place-specific devaluations of capital, and how these in turn produce further devaluations in the form of layoffs and decreases in effective local demand for wage goods. These insights are developed at greater length, and with specific attention to politics, in his more recent work (1985a, 1985c). Hence, these will be the focus of attention here.

Harvey's ideas are reviewed in some detail because they lay an appropriate theoretical foundation for the further elaboration of a more place-based political geography. First, Harvey's explanation for the rise of a territorially based and relatively autonomous urban politics is examined. Second, other recent work which might usefully be linked to it is reviewed briefly.

Formation of Regional Class Alliances: Implications for a Local Politics and the Production and Reproduction of Places

Central to Harvey's analysis (1985c) is the contradiction implicit in Marx's observation that capitalism is necessarily characterized by the continual attempt to "annihilate space with time" (Marx 1973, 539). The overcoming of spatial barriers requires the "production of space" in the form of capital investments at specific locations. The contradiction is the requirement of increased mobility of capital on the one hand, and the necessity of spatial fixity on the other.

To this contradiction must be linked the material conditions leading to a distinct place-based (especially urban) politics, and (what can most simply be referred to as) the production and reproduction of places. In a manner consistent with recent work in industrial geography, Harvey (1985a) begins his analysis by suggesting that the places created are more or less bounded by the areal extent of urban labor markets—that area over which exchanges and substitutions of labor power occur routinely. Within labor markets, labor qualities, segmentations, and the extent of surpluses are place-specific and highly idiosyncratic.

Over the short run, these present capitalists with a set of "structured" rigidities to which they must adapt. Over a longer time-horizon, a wider range of possibilities opens up for both capital and labor, but even these are "structured." These entail the options of out-migration, local collective action (to enhance the respective capacities of their classes, or segments thereof, for successful struggle in place), and upgrading the various qualities of labor.

Harvey (1985a, 134) suggests that capital as a class can help ensure that qualities of labor in a place do not run down by putting a floor under their competition, by colluding on an appropriate means for regulating the local labor market, and/or by granting specific concessions to segments of labor. However, these cannot be a permanent solution, because once the surpluses of labor power within an urban labor market are exhausted, capitalists have only four options: move elsewhere, or seek out labor-saving innovations in order to create an industrial reserve where there was none before, or force changes in the conditions of labor-power utilization, or simply mobilize the wholesale importation of labor surpluses from elsewhere.

The options of moving elsewhere or of introducing production processes based on new technologies are not chosen lightly. Firms often have considerable investment tied up in fixed facilities, investments which have not been fully amortized over their useful lifetimes. The longer these lifetimes, the more vulnerable particular production systems become, and the more likely that monopoly control of technology and location appears as a necessary means of guaranteeing conditions favorable for long-term investment. The greater the monopoly power of the larger firms, "the more the geographical landscape of capitalist production tends toward a relatively stationary state" (Harvey 1985a, 138). Monopoly power, however, is never complete, and sooner or later competition will require that monopoly tendencies be broken, thereby forcing a change in either the technology employed in production, or a change in location, or both. This can only be accomplished through the devaluation of both capital and labor in specific places.

Harvey argues that the behavior of relatively place-bound capitalists is contingent on the behavior of capital in other places, and on the extent of class consciousness and class struggle across places. However, the behavior of both place-bound capital and local labor does produce a "structured coherence" in the political economy of the place, characterized by a particular technological mix and a dominant set of social relations in both production and reproduction.

The structured coherence of a place forms the material basis that gives rise to the formation of a cross-class alliance. Place-bound factions of capital and those segments of labor with vested interests in an existing "structured coherence" can be expected to attempt the formation of class alliances with the objective of preserving or enhancing an already existing "structured coherence." Such alliances "loosely bounded within a territory and usually (though not exclusively or uniquely) organized through the state, are a necessary and inevitable response to the need to defend values already embodied and a structured regional coherence already achieved" (Harvey 1985c, 151). Such alliances are critically dependent on the production of a sense of place, which must be built up around the set of class-bound processes of production and consumption that give the place territorial definition.

The class-alliances formed, Harvey argues, are inherently unstable, because of the ever-present danger of defections by key members from either the capitalist or laboring classes, prompted by changes in the nature of a place's structured coherence. The devaluation necessarily associated with either capital flight or with the introduction of new technologies can be expected to heighten class conflict or otherwise change its specific nature, thereby producing a restructured coherence and/or class-based defections from an existing alliance.

Harvey contends that the necessity of class alliance formation, coupled with the inherent instabilities of these alliances, creates a "political space in which a relatively autonomous urban politics can arise" (Harvey 1985a, 152). From an otherwise perpetually shifting set of class alliances, the enterprising politician can forge a relatively permanent, ruling coalition of interests by articulating a sense of place anchored materially and socially in an existing "structured coherence."[4]

The impact of a ruling coalition can be profound. It can engage in various forms of local boosterism to attract additional capital investment, which enhances that already in place and tends to channelize potential class conflict in directions less likely to threaten the support that is given the coalition by key segments of labor. A ruling-class alliance in Harvey's theory is invariably a cross-class alliance involving some (perhaps even a majority) of labor in a place. It is, nonetheless, invariably a procapitalist alliance, because it must accommodate the basic logic of capitalist commodity and social relations if it is to reproduce itself through successful "competition" with other places within the capitalist system. He argues that the relative autonomy and apparently unique place-boundedness of ruling coalitions are "not only compatible with but also vital to the logic of accumulation in geographical space" (1985a, 156). Building on the earlier work of Molotch (1976), Harvey views politics as preceding the economy in the special sense that ruling coalitions speculate on the production of the "preconditions" of accumulation by collectivizing risks through the state and finance capital. Typically (but by no means invariably), a ruling coalition functions as a "growth machine" by using its political and economic power to push the region into an upward spiral of further accumulation.

To enter into such an upward spiral of accumulation, the coalition must "search for some appropriate mix of life-styles, social provision, cultural forms, and politics and administration" (Harvey 1985a, 158) which parallels the technological and organizational dynamism of capitalist production. Here Harvey's concept of "appropriate" is the "local" equivalent of the concept of "regime of accumulation" developed by Aglietta (1979), Lipietz (1987), and others. Since the owners of capital must compete within a global world economy, it is possible for places to be left behind, stagnate,

---

[4]Local government can be an important means through which ruling coalitions are forged, but it is neither the only, nor the most important means: "Local jurisdictions frequently divide rather than unify the urban region, thus emphasizing the segmentations (such as that between city and suburb) rather than the tendency toward structured coherence and class-alliance formation" (Harvey 1985a, 153).

and go bankrupt. But, the degree to which a place remains competitive changes over time, and is dependent on the qualities of its labor power, the nature of its infrastructures (both human and physical), the state of local class struggle or other forms of social conflict, the degree to which its culture is consistent with its regime of accumulation, the rationality of its politics, its location vis-à-vis competing places, and its resource endowments.

A ruling coalition's power need not be limited to its own territory, but can be projected "geopolitically onto other spaces" if it possesses the requisite political-economic power (Harvey 1985a, 159). One unusual feature of Harvey's work in this regard is his implicit assumption that a national politics derives from attempts to form a more distinctly local politics that is capable of protecting and sustaining the viability of investments. This leads him to assert that the prior theorization of the capitalist state is unnecessary for the theoretical comprehension of the historical geography of capitalism. For him, it is sufficient to recognize that the state as a system of social control predates capitalism. The result is that each system of local politics is almost invariably nested within a larger politics, defined by the state.

In this sense, it is clear that Harvey adopts a stated-centered definition of society which is closer to that of Clark and Dear than to that of Wallerstein and Taylor. The political power of any given place will depend primarily on the compatibility and legitimacy of its system of local politics and culture in relation to that of the larger state's (or nation's) politics and culture.

The weakest feature of Harvey's theory of politics is that it fails to identify clearly the mechanisms through which local or regional ruling coalitions coalesce into, or are otherwise related to, national ruling coalitions. It would appear that he views state formation as the result of some sort of spatio-economic struggle for hegemony between various regional class-alliances.

Relevance of Harvey's Theory of Urban Politics to Other Recent Work

Much of the research in the new political geography has explicitly focused upon various forms of conflict and struggle around aspects of the urban built environment. Recently, Cox (1984, 1986) has suggested that this work can be classified into one of two groups, based upon whether emphasis is on spatial relations or social relations. In the first category, he places research focusing on neighborhood or locational conflict. In the second, he places research focusing on urban social movements.

Not surprisingly, these differences in emphasis stem, at least in part, from the research traditions of disciplines in which they originated—locational conflict in urban geography, and urban social movements in urban sociology. Cox (1984) criticizes the two literatures for their specific forms of myopia: the locational-conflict work for its general failure to situate the actual conflicts studied in some broader social theoretical and historical context; and the urban social-movement work for its narrow social-constructionist conception of space.

Harvey's theory of politics may well be defective,[5] but it is certainly not subject to the criticism that it fails to treat space seriously as both a social precondition and

---

[5] To date, the best empirical explication of Harvey's theory of urban politics is his own detailed historical analysis of urban development and politics in Paris in the period 1850–70 (see Harvey 1985b).

as a social product. For this reason alone, Harvey's work deserves to be singled out as a significant recent development in political geography. Future work in geography on urban conflict (and conflicts in "places" in general), whether within the tradition of locational conflict or that of urban social movements, will need to be compared with or linked to Harvey's theory. There is also likely to be considerable value in reinterpreting findings from past empirical work on locational conflict and urban social movements, in light of Harvey's theory of urban politics.

Studies of locational conflict and neighborhood activism, in their attempts at explanation, should consider the possibility that some conflicts represent "politics as usual" without challenging the hegemony of an existing ruling coalition, while others can be linked much more directly to the economic restructuring of a place (Ley 1980; Ley and Mercer 1980; Fincher 1984). In general, geographers have tended to focus only on those conflicts over spatial effects of changes in the built environment (typically conceptualized in terms of spatial externality fields). More attention should be directed at studying conflicts in which a local political status quo is specifically called into question.

From this review of Harvey's theory, it should be clear that the politics entailed within it are a place-based, frequently defensive, politics. This should be particularly evident when the continued existence of a local regime of accumulation (defined by its structured coherence) comes under threat from spatial and technological changes in the larger capitalist economy within which it must compete. In this sense, Harvey's theory is completely consistent with Urry's observations (1981, 1983) on urban social movements in their relation to local politics, and with Agnew's observations (1984; 1987a) on the importance of place in twentieth-century Scottish politics. However, it is causally the converse of Logan's argument (1978) that one major outcome of the process of place-creation is a collective effort within places to pursue local advantages through competition between places.

The place-based nature of urban social movements has important implications for the development of class struggle at "higher" spatial scales, and hence for social theory in general. In this regard, Hudson and Sadler (1986) have argued that the competition they observed between territorially defined groups to defend places from the closures of large plants had the unintended consequence of further reinforcing and legitimating the capital/wage labor relation as the dominant and decisive social relation under capitalism.[6] If Harvey's theory of politics has merit, then there is no need to appeal to "instrumentalist" theories which posit that capital as a class actually sought to further divide labor by creating a political-jurisdictional basis for segmentation (e.g., see Hoch 1979; Newton 1978). This is because the necessary spatiality of labor markets leading to the formation of ruling coalitions in places results in a fragmentation of both labor and capital anyway. This is obviously a very important issue that only additional historically based research in political geography will resolve.

Although this is not the place to engage in a detailed review of the literature outside of geography that bears indirectly on Harvey's theory, this review would be remiss if it failed to mention the recent work of Castells (1983), in which he attempted

---

[6]Hudson and Sadler provide empirical support for this argument by drawing on their earlier case study of the social movement that was organized to fight the government's closure of the Consett steelworks in northeast England (Sadler 1984), and on a review of other such movements in other countries within the European community (Hudson and Sadler 1983).

to specify characteristics common to "successful" social movements. Castells contended that social movements must remain separate from political parties, if they are to be successful in producing social change. In addition, his conception of "successful social change" was transformed from its earlier, more revolutionary connotation (Castells 1977), to now imply the social production of a new and particular "urban meaning" and levels of consciousness. Castells' implicit theoretical framework is remarkably similar to Harvey's. Future work could profit from taking Castells' historical/empirical observations on social movements, and reinterpreting them in light of Harvey's much more explicit theorization of urban politics.

POLITICAL PARTIES AND ELECTIONS

Electoral geography was a major growth area in political geography during the 1970s: "elections are a positivist's dream" (Taylor 1978, 153), for they produce huge amounts of quantitative data in a form readily analyzed via cartographic and statistical methods. These data ostensibly also possess considerable face-validity as reflections of the popular will on a key social issue—the selection of governments. Three issues have been emphasized:

- The geography of voting, which seeks to explain the spatial pattern of voting in terms of some other mappable characteristics
- The geographical influences on voting (where the object is to explain voting, typically the decision-making of individual voters, on the basis of "spatial" contexts)
- The geography of representation (which explores the means through which votes are converted into "seats" in alternative electoral systems) (Taylor and Johnston 1979; Taylor 1984b)

to which Archer and Shelley (1986) add a fourth:

- Electoral dynamics and historical change in the geographies of elections

Long recognized as a subfield of political geography, electoral geography has, however, had little in common either theoretically or methodologically with the rest of the field. By the early 1980s, considerable progress had been made concerning relationships between the spatial distribution of voters and the spatial organization of the electorate (Morrill 1981; O'Loughlin 1982; and O'Loughlin and Taylor 1982). While this work enhanced knowledge of the spatial dimensions of elections, it tended to ignore the social theory underlying the various theories of elections.

Some electoral geographers have recently explored ways of integrating research in electoral geography with the "new" political geography. Impetus for this effort largely can be attributed to the emergence of a more-explicit concern within electoral geography for understanding electoral change and its links to social change. Attempts at integration have proceeded along two fronts: (1) efforts to situate political parties and elections within theories of the state, and (2) efforts which have led to the rediscovery of the politics of locality or place as a means of explaining the persistence of regional and local variations in voting in Western democracies.

Political Parties, the State, and the Two Geographies of Elections

Johnston (1979b, 1980) was the first to attempt an integration of electoral and political geography by developing a variant of the input/output "systems" model of Easton (1965). For Taylor, however, the integration required a much-more critical appraisal of the roles of political parties in different societies, and an explanation of why the holding of regularly scheduled elections occurs only in a minority of countries in the world (Taylor 1984a; Osei-Kwame and Taylor 1984; see also Johnston 1984b). In Taylor's view, the world-systems approach of Wallerstein (1979) offered an appropriate framework for initiating such an integration (Archer and Taylor 1981; Taylor 1981, 1982, 1984a).

One attraction of the world-systems approach is that it is defined by two complementary structures, each of which is necessary for the operation of the world-economy: a capitalist mode of production for organizing economic competition, and an interstate system of sovereign states to organize the distortion or even mystification of that competition. Each state exists to relate its (territorial) fragment of capital to the larger world-economy in such a way as to differentially benefit "national" capital, but the success of a state in this endeavor is constrained by its structural position in the world-economy (Osei-Kwame and Taylor 1984). In this view, the state is a capitalist state in the dual sense that it is part of a larger capitalist system and pursues policies designed to enhance capital accumulation by those capitalists who control its governmental apparatus.

There are some strong parallels between this perspective on "international" politics and Harvey's on "urban" politics. A basic difference is that, in Harvey's theory, ruling coalitions are cross-class alliances, whereas in Taylor's world-systems perspective the dominant coalitions controlling states consist of particular factions of "national" capital. Harvey's theory is specific to advanced capitalist states (those in the "core" of the world-economy), whereas the world-systems framework is meant to elucidate politics in any part of the now-global world economy.

Taylor (1984a, 1985a) introduced the concepts of a politics of power and a politics of support into the lexicon of electoral geography. Although he is vague on this point, it appears that the politics of power refers to competition between factions of capital, while the politics of support is more fundamentally concerned with class conflict and conflicts between segments of labor. In countries in the core, Taylor suggests that political parties operate through these two sets of political processes, promoting "national" capital accumulation in general through the politics of power, and mobilizing mass support through the politics of support. He also suggests that there are two geographies underlying elections: a geography of power and a geography of support. The latter is simply the geographical distribution of "normal" party voting, while the former appears to be a reflection of how a state must disburse public expenditures geographically, in order to reward its major backers from amongst "national" capitals. In this schema, Taylor implicitly assumes a simple functionalist theory, wherein the state acts to facilitate capital accumulation and to maintain the legitimacy of its capitalist social relations. The politics of power is the politics of accumulation, and the politics of support is the politics of legitimation.

Taylor (1984a) suggests that there is no general relationship between the two politics (and hence, no invariant spatial relationship between the two geographies). Instead, he argues that a major task of a reconstituted electoral geography would be to relate these two geographies in particular national contexts. Taylor (1984a) provides a number of historical examples to illustrate the descriptive veracity of his schema. But, it needs much more theoretical and historical elaboration, if it is to provide a satisfactory linking of electoral geography to political geography.

Taylor's analysis becomes particularly insightful theoretically when he attempts to apply these ideas to elections in the Third World. In general, he argues that the structural position of most Third World countries in the world-economy is such that no politics of power pursued by a party can lead to sustained capital accumulation for its supporters. Hence, the politics of support cannot be successful in legitimating the economic policies of a government, or in legitimating the government itself. Instead, if elections are to continue to be the means of selecting governments, parties must continuously mobilize different sections of the electorate (Osei-Kwame and Taylor 1984).

Although an important beginning, Taylor's attempt to integrate electoral geography and political geography through the introduction of world-systems concepts must be adjudged as only a partial success. Part of the reason lies in his implicit treatment of parties as theoretical entities, rather than as historical creations. He views political parties as some sort of complex switching mechanism, engaged in what might be called "double duping." On the one hand, parties are the dupes of various fragments of domestic capital, and on the other they are seen as cynically duping the masses in order to obtain and remain in power. Parties may be masters at organizing some issues into politics and others out of politics (to paraphrase Schattschneider 1960). But they are unlikely to do so solely because of "structural imperatives" (also see Cutter et al. 1987). The more likely reason is to win elections without alienating their traditional supporters.

A problem with Taylor's analysis, even with the introduction of a more historically sensitive treatment of parties, is that by retaining the concept of "politics of support," it continues to bestow ontological privilege to particular "cleavages" in an essentially national electorate, rather than to particular places in a highly regionalized electorate. In other words, Taylor's analysis assumes that national parties, in their attempt to win elections, identify issues that appeal to particular segments of the electorate, wherever they may be found. This assumption is a questionable one in most Western democracies, particularly in the U.S., where recent research clearly indicates the persistence of sectionalism in voting patterns (Shelley 1988). An alternative view of electoral politics (and by implication of electoral geography) assumes that places provide an essential context in any explanation of political behavior.

The Politics of Locality or Place

Throughout the 1960s and 1970s, the standard approach to explaining particular geographies of voting was to relate measures of the social composition in constituencies to some measure of partisan support within them. This is the so-called "social-cleavage" model. Departures from "model" expectations frequently were attributed to unspecified regional "effects," or to underspecified sectional parochialisms. Such regional effects, whatever their cause, were deemed to be relatively minor and soon to be relegated to the "scrap heap of history" as the voting responses of the electorate became "nationalized" (e.g., see Agnew 1987a).

In general, electoral geographers accepted the dominant position of modern political sociology—"that political alignments have crystallized around national social cleavages to produce national patterns of political mobilization and partisanship" (Agnew 1987a, 80). In those particular national contexts in which voting alignments varied regionally, the term "sectional alignment" or "sectionalism" was employed to describe the situation. Only to the extent that sectionalism persisted was it thought that place mattered.

As attention turned to examining changes in voting patterns, it became obvious that nationalization was not occurring in either of the two most studied national contexts—the United Kingdom and the United States—or in any of the other industrial democracies. Sectional alignments, where they existed, were changing, not disappearing (Archer and Taylor 1981; Archer and Shelley 1986; Bensel 1984; Archer et al. 1988). And smaller-scale variations in electoral alignments were not only not disappearing, but in many cases were actually increasing (Agnew 1987a; Johnston 1985; Murauskas, Archer, and Shelley 1988). Elections were becoming volatile affairs, often reflecting decreases in the strength of individual identifications with traditional parties. But, with the exception of the U.S., this occurred without significant (or rapid) change in the spatial structure of existing alignments (Johnston 1987).

These findings forced a reexamination of the nationalization thesis and the traditional approach to explaining the geography of elections. If sectionalisms and local variations in alignments were not disappearing, then they could no longer be ignored as minor aberrations. However, as both Agnew (1987a) and Johnston (1986d) have made clear, these findings, while disturbing to the old orthodoxy, have not yet led to its abandonment, particularly in political science and political sociology.

The social-cleavage model, wherein the geography of elections is accounted for by the geography of social class, continues to be "protected," usually by appending "fixing-accounts" which supply ad hoc explanations of any "residual" local effects. Electoral geography remains loath to abandon its heavy reliance on what Thrift (1983) has called a "compositional" approach to the explanation of human behavior, in favor of a more "contextual" approach. Johnston (1987, 10) has suggested that, in a compositional approach, a person's partisan attitudes are viewed as a function of his or her position within the division of labor, with the result that "knowledge of a society's cleavage structure should be sufficient to provide excellent predictions of voting behavior." Contextual explanation, on the other hand, argues in various ways that place must be accorded a key role in the analysis of voting.

The various "fixing accounts" themselves fall into either compositional or contextual categories. In the compositional category are attempts to account for over- and under-predictions from the social-cleavage model by selective migration (Curtice and Steed 1982), or on the basis of consumption-cleavages (Dunleavy 1979, 1980). Also in this category are (1) attempts to account for local variations from the social-cleavage model on the basis of various forms of local party activities,[7] (2) appeals to uneven spatial development and differentiation of areas into core and periphery (McAllister

---

[7]The various party activities analyzed include agenda setting and local organization (Johnston 1986d, 1986e), local campaign spending (Johnston 1986c), local variations in links to the workplace and/or control of local governments (Johnston 1986b, 1986d), and the activities of "third" parties (Curtice and Steed 1982; Agnew 1987a).

and Rose 1984; Agnew 1987a), and (3) appeals to differences in local "political culture" or local adaptations to a dominant ideology (Butler and Stokes 1969; McAllister and Rose 1984; Archer and Shelley 1986; Johnston 1986a; Agnew 1987a; Savage 1987). Each of these has met with some success in particular cases, but for a compositional approach to be successful, by definition it must be appropriate in all contexts.

There are far fewer types of "fixing accounts" in the contextual category, but the few that exist have attracted much attention. By far, the most frequently employed conceptual devise is the so-called "neighborhood effect" of individual voting. The term describes the supposed tendency for a dominant local opinion in a constituency to either enhance or counteract the "effect" of social class (Ennis 1962; Cox 1968; Butler and Stokes 1969). As such, the neighborhood effect is dependent on a form of socio-spatial "contagion" within constituencies (places), and seems to imply that increasing exposure to locally dominant ideas breeds a consensus around those ideas. This interpretation has been assailed on a number of grounds, the most damning of which is the inability of "contagion" to be interpreted as the causal mechanism underlying the phenomenon (Dunleavy 1979).

Spatial propinquity is undoubtedly necessary for the effect to be perpetuated, but it still begs the question of how locally dominant political attitudes are produced in the first place. In this sense, explanation by appealing to the neighborhood effect appears "contextual," but is little different from any of the more standard "fixing accounts" of the compositional variety. For the neighborhood effect to be a plausible addendum to an otherwise macrosociological explanation based on social-class cleavages, it must be linked to the historical development of politics in actual places.

Agnew (1987a, 1987b) argues for a fully contextualized view of political behavior, in which people who live in different places are led to vote in specific ways by being actively socialized into the different political attitudes, depending on the social structure and historical antagonisms of those places. Similar views have also been expressed by Johnston (1986d, 1987) and, to a lesser extent, by Taylor (1985a).

Two recent attempts by Savage (1987) and by Cox (1987) specify more precisely how such analysis might proceed. Savage theorizes local political processes as a means of understanding political alignments in contemporary Great Britain. He concludes that increases in local variations in voting alignments have been caused neither by changes in the geography of social class, nor by an intensification of local political culture, but are the result of similar "types" of localities becoming increasingly similar in their voting patterns. While not rejecting the causal efficacy of social structure in accounting for voting, he argues that local social structures are not simply local fragments of an otherwise national social structure. The material interests of persons occupying different positions in the social structure are locally based, not nationally. In particular, Savage suggests that voting by persons in different social positions is contingent upon the performance or "trajectory" of local labor and housing markets vis-à-vis other such markets, specifically in terms of employment and wage levels and the local market values of housing.

Although Savage's analysis is clearly specific to Great Britain and is highly tentative, he is able to present empirical evidence in support of his argument. The particulars of his argument are less important than the basic thrust of his analysis. Rather than turning immediately to a historical-geographic analysis to develop some concept of local political ideology or political culture, he turns to a simpler, place-specific mode of analysis. In this mode, voting behavior is caused, not directly by British or capitalist social structure, but by these social structures as they are mediated by materially significant aspects of place. Although Savage makes no attempt to link his analysis to Harvey's theory of politics, the parallels between the two analyses are striking.

Mention should also be made of Cox (1987). Drawing upon Sayer (1984), he criticizes Johnston's (1987) particular interpretation of "compositional" forms of explanation in social science, and its implicit conception of social structure as consisting only of relations between people based on the formal similarity of their enumerable characteristics, rather than on the internal (or necessary) structure of relations between them.[8] Unlike Johnston, Cox sees Harvey's theory of regional class-alliance formation as providing an explanation for instabilities and dealignment in the geography of voting in advanced capitalist societies. But he prematurely limits the theorization of the necessary relations underlying social life in real places under capitalism to the capital-wage labor relation and the space-time contradictions of capitalist accumulation.

Harvey's theory is better interpreted as providing a theoretical understanding for the necessity of a local politics, based solely on the operation of spatially constrained class relations. But, these class relations are insufficiently linked historically with other social relations (including those of gender, ethnicity, and racism) to produce a very satisfactory explanation of the concrete political reality characterizing real places. Harvey's theory is an important beginning, but it is still only a beginning.

A compelling case can be made for the necessity of incorporating the concept of place in explaining the geography of elections. The problem is in specifying more precisely how this might be accomplished, without engaging in ontological pragmatics and/or in theoretical eclecticism. It is certainly true that place or "locale is where people learn their politics—at home and beyond" (Johnston 1986d, 594). But it is clear that a place-based, contextual approach to understanding politics would bear little or no relationship to electoral geography as it is presently practiced. Analysis would necessarily shift from the relatively narrow present emphasis on changes in voting alignments over time, to deeper historical analyses of the exercise of power and social struggles against it in particular places during periods of significant social change. Traditional electoral geography must yield to a more-thoroughly political geography.

LOCAL AUTONOMY AND THE LOCAL STATE

Although local governments have long been on the research agendas of the social sciences, attempts to situate local politics within a historical materialist theory of the state are recent (Cockburn 1977; Saunders 1979).[9] Cockburn appears to have coined the term "local state" to describe such efforts, and for several years her work and that of Saunders attracted a considerable amount of attention in geography. A number of

---

[8] In some respects, Cox's concern here matches Savage's for identifying causal mechanisms. They differ, however, in that Cox appears concerned only with highly abstract causal mechanisms that are necessary, while Savage is more concerned with the causal mechanisms that are concrete and either necessary or contingent.

[9] Saunders' perspective on the local state is more in the Weberian than in the Marxian tradition. Nonetheless, it is heavily dependent on Marxist categories and theory.

writers were outspoken in their calls for the development of theories of the local state (Clark and Dear 1978; Dear 1981; Johnston 1982; Kirby 1982; Short 1982).

Work on the local state in geography suffers from ambiguity in the concept itself. "Very often the term is simply swapped for 'local government' and seems little more than a radical rhetoric used to denote a non-traditional viewpoint" (Duncan and Goodwin 1982a, 77). Indeed, there seem to be only three papers by geographers in which this does not occur—Duncan and Goodwin (1982a, 1982b) and Lauria (1986). Some writers even go so far as to equate the local state with any subnational authority or jurisdiction. Thus, the analysis of local politics frequently requires consideration of a large number of "local states."

Simply put, geographers have tended to treat local states much as they have tended to treat national states: not as a theoretical concept to aid in understanding or interpreting local politics, but as concrete institutions, the behavior of which constitutes local politics. The problem this creates epistemologically and methodologically is one of confusing theoretical abstractions of social objects with their empirical appearances (see Sayer 1984). This is not to say that no valuable insights have been forthcoming from radical perspectives on local government. Nor can it be claimed that more traditional work on local government, drawing theoretical inspiration from public finance theory, is without merit. What is claimed is that an alternative perspective on the local state—one emphasizing processes of change in local social relations—offers greater promise for the development of a more decidedly political and theoretically coherent political geography in the future.

One issue has received more attention in political geography than any other: the extent to which the local state is autonomous from the central (national) state. This is obviously a very important question, both empirically and theoretically. However, its answer is dependent on whether the state and local state are viewed as "things," as processes, or as "structuration in place." Clark (1984, 1985), although clearly reifying the local state by creating it as a legal entity which either possesses or does not possess the powers of initiative and immunity, has shown that local governments in the U.S. possess practically no legal autonomy from state governments.[10] A number of writers (Johnston 1982; Short 1982; Clark and Dear 1984) have also argued forcefully that local government in both North America and Britain is less a bastion of local self-determination, than an apparatus of the state.

On this state- and local-state centric view, there would appear to be no strong impetus for the development of a theory of a specifically local state. The focus instead should be on theorizing the state, in general. This seems to have been precisely what has happened in the "new" political geography. Kirby (1987), for example, has noted that some prominent geographers writing on urban politics appear to have dropped the term "local state" from their vocabularies altogether.

Some writers, following Cockburn (1977), view the local state as an apparatus of the state that facilitates accumulation and capitalist reproduction by providing state services and employment, and by furthering the ideological hegemony of the state through local electoral politics and the administration of state policy locally (Clark and Dear 1984). Others, following Saunders (1979), draw an institutional separation among the production, consumption, and legitimation functions of the state: the local

---

[10]Clark defines initiative as the power of local governments to legislate and regulate the behavior of their residents "without fear of the oversight authority of higher tiers of the state" (Clark 1984, 198).

state functions as a mechanism for mediating conflicts surrounding collective consumption; the national state facilitates capitalist production; and both "levels" of the state possess legitimation functions (Johnston 1982; Kirby 1982, 1983).

There is, however, an alternative perspective which deserves attention, derived from Duncan and Goodwin (1982a, 1982b). It argues that the local state can more fruitfully be theorized as the "form" in which social relations occur locally. The question of local autonomy in this case becomes one of determining the extent to which local social relations derive from, and are indistinguishable from, national social relations. According to Duncan and Goodwin (1982b), the fact that social relations and consciousness are unevenly developed spatially is crucial in explaining why local governments and elections are necessary in capitalist social democracies: such institutions are a response to local class struggle, and are thus a possible means of transforming local class relations into those of the more abstract, individual relations of local citizenship.

But, this is only one possibility. Local governments can also become a "means of realizing and expressing such consciousness" (Duncan and Goodwin 1982b, 168). In this sense, localities, through their local governments, are always potential battlegrounds in the attempts of the dominant classes to restructure social relations. The attempts to restructure social relations always produce some unintended consequences, generative of additional conflicts and contradictions. Under this "social relations" perspective on the local state, it is clear that the local state cannot simply be a local version of the national state. Hence, it requires specific, but not necessarily separate, theorization.

The legitimacy of the central state is also dependent on at least the appearance of local democracy (Reynolds and Shelley 1985). If this appearance cannot be maintained by dint of ideological hegemony, then the lack of local autonomy in a legal sense may well be irrelevant. In crises of legitimacy, the central state will have to accede to local autonomy in practice anyway. In this regard, the U.S. Supreme Court (particularly in the case of public schooling), has rendered opinions predicated on the intrinsic reasonableness of the "local control" as a "legitimate state purpose" (Reynolds and Shelley 1988). Agnew (1987a, 39), following Gramsci (1971), makes a similar point: "The central state itself depends upon attracting support and legitimacy from large sections of the territory over which it holds sway. The successful and continuing integration of the modern territorial state requires a considerable degree of popular consent and participation in political life."

Although Duncan and Goodwin (1982b) do not provide a completely satisfactory beginning for the development of a theory of the local state, it should be clear that their concept of local state bears strong resemblance to what others writing on politics have called "place" (Agnew 1987a). Indeed, with the exception of Harvey's theory of urban politics (1985a), their analysis of how "reformism" in British local government can be viewed as a series of attempts to restructure social relations in the interests of "national" capital accumulation is probably the best example of an attempt to theorize the politics of place to be found anywhere in the literature of the "new" political geography.[11] The task of specifying the necessary relations in the politics of place

---

[11] In this context, Lauria's recent attempt (1986) to develop some theoretical understanding of local state intervention strategies in the specific context of averting a plant closure also deserves mention. Lauria adopts an implicit "social relations" conception of the local state, although he appears to eschew a place-based focus in favor of one emphasizing a modified form of corporatism.

under capitalism would thus appear to be identical with that of specifying the necessary relations underlying the capitalist local state.

NATION-BUILDING AND NATIONAL DISINTEGRATION

Political geographical writings on nation, nationalism, and self-determination were slight prior to the 1970s, and largely derivative theoretically from other fields. However, Hartshorne (1950) and Pounds ([1963] 1972) were among those who sought to formulate a geographical understanding of processes involved in nation-building (see Smith 1979). As with the many developments already discussed, recent advances have resulted from new insights based on theoretical explorations. Underlying much of the theorizing is a sharper awareness of various layers of identity, and of linkages between nationalism and territory.

Layers of Identity

All people undoubtedly are linked to several conceptual levels (sometimes competing, sometimes complementary) of attachment to "place." When taken together, and mediated by class and other social elements, these form "identity" within a particular socioregional setting (Knight 1982a). "Society," bound by a nationalism, in "place," will lead people to "see" themselves differently—a we/they dichotomy is inherent in most group identities, including nationalism.

It cannot be assumed that all people have the same conceptions of individual self: as a generalization, most Africans have communally based identities, whereas for people in the West, identity generally is individually based, moderated by class and perhaps ethnic considerations. But in both instances, it is possible for a person in a particular part of a state to have an affiliation with an identity at a higher level, as a regionalism within the state. Thus, contrasting local, regional, and national affiliations may influence how the state and "nation" are perceived (Burghardt 1980), or how public issues are viewed and argued and can affect legislators' voting records (Brunn 1974, 1975; Knight 1977).

Local and regional attachments may be more important than a national one, and each may be influenced by what van der Wusten (1988) has called "competing ideologies" that impact in different ways upon various levels of attachments and on political processes and structures. Most people give priority to local and national identity before an international one, although few people consciously operate on a daily basis with the national identity to the fore.

The relative importance of competing attachments is not confined to individuals or to groups within states. For instance, the Belgian government may one day be adamant in formulating "national" policy, yet on the next may have to support a NATO policy that cuts against Belgian nationalism and constrains Belgian autonomy. The former ultimately is related to attempts at formulating or safeguarding a state-based identity, while the latter recognizes that states are parts of larger systems, and that the concept of sovereignty should not be thought to imply isolation. This example, of a state having to compromise its sovereignty, may be somewhat akin to the experience of former colonial territories now having to be subservient to international agencies such as the World Bank, and to transnational corporations.

Relatedly, there are suprastate identities. For example, in the Arab world there is a belief in an Arab nation, although economic and social integrative experiences have operated within imperially derived state boundaries. Drysdale (1988, 99) claims that the placement of states' interests above those of the Arab nation as a whole is, in a sense, to accept its partition, to foster subnational allegiances, and to engage in separatist activity. Arab states thus now view themselves in a rather conventional way, and the identities they are cultivating are quasinational ones.

Drysdale's conclusion that terms like nation, nationalism, and national identity have several layers of meaning cannot be used in the context of the Arab world without some qualification and refinement has far wider world-regional significance. More work needs to be done to identify and measure these several layers of identity, and to determine how the different levels of attachment are manipulated ideologically by elites at different societal scales.

Nationalism and Territory

An important level of identity for most people is their nationalism. Is Anderson (1983) correct in positing that nations are "imagined communities"? Certainly, Portugali (1988, 155) concluded that "nationalism is simultaneously a 'plan' for society's geopolitical organization and a false consciousness as it presents social constructs such as territorial homeland, nation and nation-state as 'natural' and 'eternal'." These "social constructs" are just that—constructs—and are not givens; yet they may be potent in their symbolism. A nation refers to a community of people (1) who feel that they belong together, (2) who have a historically derived sense of nation—since the past is thought to produce the present, which in turn leads to the future—and (3) who have, or desire to have, control of a territory for the betterment of the community. The ultimate territory—as homeland—that any nation may seek to control through its nationalism is a state, but political realities of locale within existing states may mitigate against such a goal.

Most nationalisms are grounded on cultural bases (Smith 1979), but to succeed they need to develop a political principle that is linked to practical politics. Thus Linz (1985) has shown how Basque nationalism shifted from its "primordial" elements (i.e., with emphasis on common descent, race, language, cultural traditions, and religion) at the end of the nineteenth century, to become a territorial conception which permitted an appeal to and support from all who lived in the territory, regardless of their origins and cultural traits.

It is such a linking between group identities and territory that led Knight (1982a) to recognize that nationalism is inherently territorial, and relatedly, that self-determination could be defined as the "right of a group with a distinctive political-territorial identity to determine its own destiny." Political geography's major contribution to the now vigorous interdisciplinary debates on nationalism has been the explicit demonstration that nationalism is a form of social and political movement firmly rooted in territory, in place, and space. It both operates territorially and interprets and appropriates space, place, and time, thereby constructing alternative geographies and histories (Augelli 1980; Cohen and Kliot 1981; Kliot 1983; Williams and Smith 1983; Williams 1985b; Meinig 1986; Zelinsky 1984).

Contemporary writings on nationalism in geography are based, in part, on Gottmann's (1973) exploration of the significance of territory to western societies and their territorial expressions, notably the state. Gottmann suggested that the concept of territory is rooted not only in the substantial and the physical, but also derives from the meanings invested in it by people acting as deliberate social beings. Territory, he wrote, is a very substantial, material, measurable, and concrete entity. Yet it is also the product, and indeed the expression of, the psychological features of human groups—territory is a "psychosomatic phenomenon of the community, and as such is replete with inner conflicts and apparent contradictions" (Gottmann 1973, 15). Organized territory offers both opportunity (by emphasizing an outward looking stance) and security (by stressing an inward-looking orientation). Tension exists between these contending orientations, and thus territory implicitly includes conflict.

Gottmann's work opened many doors for other geographers, and much is now known about the various ways in which the state claims, controls, or seeks to control its territory, its resources, and its "people" (Burghardt 1973; Burnett and Taylor 1981; Kliot and Waterman 1983; Paddison 1983; Sack 1986). Of course, a new emphasis on expressions of national territoriality and territorial functions and structures may provide still further insights, including the possible impact on interstate interaction and foreign-policy formulation. Even as new insights are sought, we must heed the warnings of Anderson: national territoriality involves control over territory—rights to it, access to it, and exclusion from it. More subtly, it reifies power relations, making them appear more real and "natural." A focus on territoriality can displace attention from the actual relations of power yielded by dominant groups, as in the neutral sounding "law of the land." And, of special concern, territoriality can involve a "fetishism of space," whereby relations between social groups and classes are obscured by being presented as relations between areas or regions, or as relationships within areas or regions (1988, 25–26).

Nationalism also can be two-edged, but in a slightly different way from territory. It at once directs attention *inward* as dominant elites seek to unify the nation and its constituent territory (held or coveted), while at the same time, it directs attention *outward* as its acceptance tends to divide the nation and territory from all others. Such self-centeredness leads to the formulation of what Agnew (1982) calls "national exceptionalism," with its "pandering to national conceit," which finds expression in justifications for national development and foreign policies (Hoffman, 1982).

There is always the danger that work done at a national scale may become statist in emphasis, blinkered in its analyses, and thus a "servant" of the state—a perspective that is a carryover from the last century (see Hudson 1977; Capel 1981; Taylor 1985c; MacLaughlin 1986). Many political geographers are indeed servants of their states, and thus of their nations. But surely there must also be an openness to permit challenges to conventional thought and "expected" structures, without recrimination.

Nation-Building—But Whose Nation?

The process of nation-building was uncritically taken as a given by many political geographers who adopted the state-building perspective dominant in the period of the 1950s–1970s. Nation-building, it was assumed, occurred in tandem with state development. Little focus was placed directly on the concept of nation, nor, more importantly, was serious attention given to recognizing that, while the two are linked, it is critical to focus on the nation and on the ways in which nationalism can be used to bring about state development.

This realization became most explicit when scholars taking a world-systems approach tackled the issue of state-development, for they recognized the interplay among world, regional, and locality-based processes (especially Taylor 1985a; Agnew 1987c). Becker (1988), for instance, traced the close association between nationalism and state intervention as Brazil first was incorporated into the world economy as a colony and huge resource frontier, and then achieved the position of a "newly industrialized country" in the semiperiphery. She stressed that researchers must become conversant with the ideological underpinnings of nationalism, and not just with the processes of state development. The two go hand in hand.

The state-building focus generally was predicated upon a "modernization" or integration model that assumed all people within a state would sooner or later accept state apparatuses. Recently, work on "peripheral" regions of Old World states, in which exist many examples of "separatist nationalisms" and "internal colonies" (Williams 1980), has called the modernization thesis into question. Equally, however, there is still much to learn about the manner by which both Western (in the past) and former colonial states (today) have attempted to forge "national" identities for their generally plural societies, so that central governments may govern successfully.

Given the geographical realities—physical geography, contrasting distributions and densities of peoples, modes of production, means of circulation, etc.—areal variations in any state's development are to be expected. Why, then, were some researchers surprised to find uneven development, with contrasting "core-periphery" relationships in certain states? Be this as it may, the "findings" and resulting interpretations have challenged modernization models and their assumptions. It was explained that "internal colonialism" caused uneven development, and as a result, ethnic resurgence was the means by which subjected regional peoples within states were able to express their discontent. (On internal colonialism, see Hechter 1975, and Hechter and Levi 1979; on uneven development, see Nairn 1977.)

Williams (1979) challenged Hechter by arguing that Celtic nationalism in the periphery of the United Kingdom was inadequately explained by a core-periphery model, for it did not take into account either the internationalization of capital in suprastate economies, or the attempt by the ethnic intelligentsia to redefine its role vis-à-vis the state bureaucracy. Orridge (1981) challenged Nairn, contending that the uneven-development theory could not accommodate instances of nationalism unaccompanied by great differences in development from their surroundings, or conversely, instances of uneven development without the attendance of pronounced nationalism. A weakness of such literature was that the territorial aspect was only implicitly invoked, or more seriously, territorial relations were reified, and social relations in the core and periphery were obscured (Johnston, Knight, and Kofman 1988, 10).

New theorizing is seeking to understand the territorial consequences of pluralism. Many substate peoples' sense of separate identity fosters national disintegration, as when a regional minority tries to use its own nationalism to generate (1) the desire for some type of autonomy within or linked to a state, or (2) the desire for self-determination (Knight 1984a; Knight and Davies 1987). Of the potential consequences, the first (autonomy) may entail a substate territorial restructuring, as the state attempts to accommodate "ethnic" or some other regionally focused diversity, as in Nigeria (Rogge 1977), Switzerland (Jenkins 1986), or Belgium (Murphy 1988). Or, it may involve the rewriting of a constitution, as in Canada (Knight 1984b). The second case (self-determination) could involve territorial separation, and the creation of at least two states from one (Whebell 1973; Waterman 1987).

Despite attempts by national governments to forge a sense of total national identity, substate peoples often remain alienated, such as with many East Timorese, Basques, Punjabis, indigenous peoples in numerous states (Mercer 1987; Knight 1988), and countless others. Most studies to date have focused on the manner in which the state copes with such substate groups, such as providing for devolution of power via some territorially based means (Dikshit 1975; Paddison 1983) or depoliticizing by privileges of place (Smith 1988). An alternative is to approach the issue from the perspective of the minorities in question, a perspective Williams (1984, 1988) feels must be adopted.

Of concern is Williams' observation (1986, 227) that national leaders (1) may "represent political and economic developments as essentially populist," even when such clearly is not the case, and (2) that when, on behalf of the state, they then seek to strengthen the supposed "solidarity of interest and involvement among the populace," they "often resort to centralist bureaucratic institutions to buttress the state apparatus." Marginalized minorities may then perceive the state apparatus as the monopoly institution of an opposition group that is both unwilling and unable to recognize their legitimate grievances within the framework of state activities. Thus, Williams feels, in seeking to incorporate the masses in the national socialization processes, the bureaucratic state often reemphasizes the very lines of conflict that discriminate against subordinate groups.

Towards a Coherent Theoretical Framework

Recent political geographical writing has demonstrated that theorizing on nationalism has largely been insensitive to place and geographical variations (e.g., Agnew 1984, 1987a, 1987b; Anderson 1988; Knight 1984a; Mikesell 1983; Williams 1985a, 1985b). Nationalism has been accepted as an autonomous force (Blaut 1982), a thing in itself, having causal power. Yet nationalism does not have the same force at all times and places. It waxes and wanes, and since the geography of nationalism does not just consist of the variations of certain strata (because nationalism is a response conditioned by different social environments), it must be the political product of a particular set of circumstances, and hence contingent.

Place can be said to mediate the impact of "global" influences. However, a danger may arise from overemphasis upon contextualization and on place as the mediator, in that a lack of effort may then be given to analyses of the interaction between necessary and contingent relations. To focus on nationalism as "contextually constituted" would prevent the possible development of necessary relations in nationalism. Is the latter possible? Williams (1985b) concluded that "unity of explanation" was impossible, but Anderson (1988, 19) has argued that "a general theory of nationalism, remains important as a goal" for without it there is a danger of retreating into empiricism.

With respect to the interrelationship between the necessary and the contingent, MacLaughlin (1986) finds that both structuralist and autonomous theories of nationalism lack an appreciation of the dialectical relationship of the economic, cultural, and political. He feels (1986, 306–307) that a dialectical approach would lead to an avoidance of "the fallacy of dualism," whereby factors that are inextricably linked in the real world are treated as though they do not interact for purposes of theorizing and model-building, because the approach "recognizes the existence of deep-seated contradictions both in relationships of production and in social-relationships." He argues (313) that we should look to a historical materialist interpretation in which "social interaction and class struggles are structured but not determined by the regional setting."

Johnston, Knight, and Kofman (1988, 12) suggest that a Gramscian analysis may serve best to achieve this: there is need to incorporate the relationship between the state and national and regional civil society; the economic, social, and cultural transformation these societies have undergone; the different responses to social classes and groups to these changes; and finally, the internal geographical division of the territory claimed by the nationalist movements. The resulting rounded understanding will presumably permit the recognition of nationalism as but one among several possible territorial and political responses to a changing world, and not as an inevitable strategy.

Structural Marxists have long debated the "national question" (the right to national self-determination) in terms of "progressive" and "reactionary" forces, and have had to deal with Lenin's belief that nationalism could be reconciled neither with Marxism, nor with the declared right of the working class to consolidate its power over any right to self-determination by a substate regional group. According to Richmond (1987, 9), the major difficulty with such analyses is the relegation of ethnicity (and related categories of race, religion, and language) to an epiphenomenal status in the causal model, wherein economic determination and the primacy of the class struggle are dogmatically asserted as a fundamental premise. The evidence of history, and of contemporary developments in advanced industrial societies, Richmond believes, suggests that, on the contrary, ethnic factors may have a greater influence than class on the development of political and social systems. He concludes that they are more salient at the level of consciousness than class membership, and cannot be dismissed as mere manifestations of a "false consciousness," once the dogma of a narrow "materialist" view of history is rejected.

In practice, nationalism denies or downplays class divisions within a nation in order to promote national unity. Class may be used by, and indeed may be the basis of, some forms of nationalism. But the "national interest" may be defined simply as the interest of the dominant class, who may invoke it as a means for generating popular support by the dominated classes. Blaut (1986, 8–9) notes that there is not a single "ideology of nationalism" but many, with at least two opposing ideologies in each "national struggle": (1) "bourgeois-nationalist ideology," and (2) "an ideology which fights for national liberation and against the bourgeoisie—the direct opposite of bourgeois nationalism."

Anderson (1988, 27–28) challenges Blaut, however, by stating that it does not follow that nationalism "functions as a neutral tool or implement." Nor can nationalism be used equally to further all class interests, for nationalism cannot be neutral in the immediate and obvious sense that a national movement or ideology at any particular time and place has definite class content, with some classes standing to gain more than others. Anderson also feels that it is not enough to assess the class implications of nationalism simply on a case-by-case basis, important as that is. It is also necessary to see it in terms of a world system of nation-states, or would-be nation-states, which are dominated by the capitalist mode of production.

Anderson (1988, 29–30) contends that, although the logic of nationalism is to play down class divisions, in practice the furthering of class interests, and domination by particular classes within implicit and explicit class alliances, are central features of nationalism. But then he asks, "why is nationalism effective in furthering class interests?" To answer this question, he feels that it is necessary to understand (1) that nationalism has a partial basis in the political fact that even subordinate groups may achieve real benefits from independent statehood, and (2) that nationalism has a basis in spatial aspects of economics and culture.

At this point, then, political geographers examining nationalism can come into contact with the theorizing by Harvey as discussed earlier, for the cultural and economic basis for territoriality and class alliances can be approached from scales other than the national. Culture generally is territorially expressive and more or less place-bound, perhaps being roughly coincident with the state or part of it; whereas the economic basis of nationalism involves "state" or "internal" markets with labor power that is essentially spatially immobile, in contrast to other "commodities" which may be internationalized. As a result, competition may arise on an "area-versus-area" basis (Anderson 1988, 31) as, in Harvey's words (1982, 420), "each alliance seeks to capture and contain the benefits to be had from flows of capital and labor power through territories under their effective control."

Cultural, economic, or political categorizations alone are not enough to define the territorial delimitation of a nationalism, because, as Anderson stated (1988, 37), a general theoretical framework has to integrate all of these elements, and cannot afford to be one-sidedly "cultural" or "economic" or "political." Also to be encompassed are nationalism's links with statehood and democratic ideals, its two-faced nature with respect both to time and space, and its use to further different class interests (whether of domination or liberation), as well as its basis in spatial aspects of economics and culture, with territoriality providing a fundamental key. From this, one is led to recognize that nationalism is a territorial ideology.

All too often, writing on nationalism has passed over the significance of territory in nationalist ideologies, politics, and strategies. Two things thus remain to be done:

1. The development of a coherent overall theoretical framework that will permit a general appreciation of nationalism and its associated concept of self-determination (as it applies to decolonization and to national disintegration)
2. The critical development of particular nationalisms (state and substate), so that a new awareness can be developed of such movements and their geographical and historical contexts

Both approaches are needed, for the former will enhance the understanding of the latter, and the latter will enable the overall framework to be constructed more substantially.

INTERNATIONAL BEHAVIORS

From the forgoing sections, it should be apparent that the "old" separation of an internal-to-the-state political geography from an international political geography needs rethinking. Too many issues exist which are linked; unless this is recognized, important elements will be ignored and explanations will be weak. It is with hesitation, therefore, that we include a section under the present title (although not to do so would ignore a large and important literature).

The international system of states has at its base individual states, although many processes (especially economic) operate that override and sometimes subvert states' sovereignty. Taylor (1985a, 6), after Wallerstein (1979, 1984), has argued that states are inappropriate units for studying change, since they are not self-contained systems that developed separately from one another, but are all part of a larger whole. However, the system of states is the basic structure of world political territories, and almost all international organizations are based upon that reality. There is the U.N., wherein representatives of states debate and vote (Brunn and Ingalls 1983). But there are also many regional groupings of states, some based on military alliances, some on treaty relationships, and still others on perceived economic mutual self-interest (as with the European Community and OPEC).

Two points can be added:

1. International law—the code of expected interstate behaviors—again is state-based, although selected international human rights perspectives can be claimed as also pertaining to intrastate behaviors.
2. It is clear that states in one part of the world (the core) can directly affect states in other parts (the periphery), as in the case of Europe in connection with Africa.

Examination of states as they interact is an old theme within political geography, because, as in other fields that focus upon international relations, interstate interaction has geographical dimensions. The intellectual roots and development of international political geography, often referred to as "geopolitics," is only now being critically reevaluated in light of abuses that occurred in the name of geopolitics in the 1930s and 1940s (see many papers in recent issues of *Political Geography Quarterly*). Much theorizing has been accomplished to help explain certain international processes and interactions. This work includes Mackinder's remarkable thesis on power regionalizations (see Parker 1982; Blouet 1987); Cohen's explorations of the development of a "hierarchical integration" in world regional interaction (1973, 1982, 1984); the world-systems perspective of Wallerstein and others, noted above (including Taylor 1985a, 1986, 1987); and Modelski's long-cycle theory, which draws upon general-systems theory (1978, 1987).

Research into violence done in the name of the state (see Johnston, O'Loughlin, and Taylor 1987), usually for the supposed sake of the nation, is generally at two scales, either (1) internal to states—as in the subjugation of peoples by the ruling class, who often are closely aligned to and supported by the military, or who may *be* the military; or (2) international in scale. Other work focuses upon:

- Border conflicts (e.g., Starr and Most 1983; Prescott 1985, 1987)
- Land and sea territorial claims and utilization, and management of contentious territories (Harris 1980; Cohen 1986; Kliot 1986; Samuels 1982; Blake 1987)
- The interaction of world-power competition in local conflicts in the Third World (O'Loughlin 1986c)
- The effects of conflicts between major contending states or groups of states, especially with reference to questions of peace and security (O'Loughlin 1984, 1986b; Brunn and Minghst 1985; Douglas 1985; Pepper and Jenkins 1985; van der Wusten 1985; Cohen 1986; Ó Tuathail 1986; van der Wusten and O'Loughlin 1986; O'Loughlin and van der Wusten 1986)
- Future worlds (Brunn 1981)

Much of the work done on these and other traditional topics remains quite narrowly problem-oriented. However, there is some indication that a theoretical perspective is emerging, even in research on those topics that is linked to the perspective developed earlier in this chapter. In our view, the foundation of this essentially place-based perspective has already been well laid by Harvey, in his attempts to develop theories of both a politics of urban development (1985a) and a geopolitics of capitalism (1985c) in which he derives a domestic and an international politics from the logic of uneven development and the search for a "spatial fix" to crises of capital accumulation. Although it is clear that Harvey's work has been influential in urban political geography, it remains to be seen whether this work will make a similar impact on international political geography.

CONTINUING AGENDAS

Throughout this chapter, it has been argued that a new political geography emerged in the early 1980s, one in which much greater stress is given to the theorization of both the "political" and the "spatial." In this effort, it has been argued that a central substantive focus of research is, and is likely to continue to be, upon the production and reproduction of places—both theoretically and historically—in the capitalist world economy. In this view, "places" refer to any area on the Earth's surface over which significant social struggle occurs, or has occurred, which produces a politically relevant sense of place in the consciousness of its inhabitants. A number of other, more traditional research themes have been examined as they relate to theoretical developments in the field as a whole. It has been noted that some of these focus on intrastate issues, some on interstate issues, and still others overlap.

In the modern world, it is clear that states are real places, as well as significant actors in the world economy and in the international political arena. Like any other place, to remain significant in a practical political sense, and as a complex set of socioterritorial relationships, states must be reproduced over time. The propagation of various ideologies is a central means of this reproduction. French political geographer Lacoste is very much on the mark in reminding us that geography itself can be "a strongly ideological form of knowledge" (Hepple 1986, S31), and that a radical geography has at once to be critique, so as to identify ideological bases of existing theories and policies, and at the same time, must be able to develop alternative theories and policies.

While it can be openly acknowledged and accepted that there are a variety of approaches to political geography, we hold that a principal continuing agenda item in political geography must be the search for theoretical coherence, both internally and in terms of political geography's linkage to human geography and social science in general—a coherence defined in terms of the power and richness of its constituent ideas and concepts, rather than as methodological orthodoxy. The concept of place is likely to be important in this coherence, but so too are such concepts as class, ideology, region, territory, and nationalism.

REFERENCES

Aglietta, M. 1979. *A theory of capitalist regulation*. London: New Left Books.

Agnew, J. A. 1982. An excess of 'National exceptionalism': Towards a new political geography of American foreign policy. *Political Geography Quarterly* 2(2):151–66.

———. 1984. Place and political behaviour: The geography of Scottish nationalism. *Political Geography Quarterly* 3(3):191–206.

———. 1987a. *Place and politics: The geographical mediation of state and society*. London: Unwin & Allen.

———. 1987b. Place anyone?: A comment on the McAllister and Johnston papers. *Political Geography Quarterly* 6(1):39–40.

———. 1987c. *The United States in the world-economy*. Cambridge: Cambridge University Press.

Anderson, B. 1983. *Imagined communities: Reflections on the origin and spread of nationalism*. London: Verso.

Anderson, J. 1988. Nationalist ideology and territory. In *Nationalism, self-determination and political geography,* eds. R. J. Johnston, D. Knight, and E. Kofman, pp. 18–39. London: Croom Helm.

Archer, J. C., and Shelley, F. M. 1986. *American electoral mosaics*. Washington: Association of American Geographers.

———, and Taylor, P. J. 1981. *Section and party*. Chichester: Wiley.

———, et al. 1988. The geography of U.S. Presidential elections. *Scientific American* 259(1):44–51.

Augelli, P. 1980. Nationalization of the Dominican borderlands. *Geographical Review* 70:19–35.

Becker, B. K. 1988. Nation-state building in a 'Newly-Industrialized Country': Reflections on the Brazilian Amazonia case. In *Nationalism, self-determination and political geography,* eds. R. J. Johnston, D. Knight, and E. Kofman, pp. 40–56. London: Croom Helm.

Bensel, R. F. 1984. *Sectionalism and American political development 1880–1980*. Madison, WI: University of Wisconsin Press.

Bhaskar, R. 1979. *The possibility of naturalism*. Hassocks, U.K.: Harvester.

Blake, G., ed. 1987. *Maritime boundaries and ocean frontiers*. London: Croom Helm.

Blaut, J. M. 1982. Nationalism as an autonomous force. *Science and Society* 46(1):1–23.

———. 1986. A theory of nationalism. *Antipode* 18(1):5–10.

Blouet, B. W. 1987. *Halford Mackinder: A biography*. College Station, TX: Texas A&M University Press.

Brunn, S. D. 1974. *Geography and politics in America*. New York: Harper.

———. 1975. Vietnam war defense contracts and the House Armed Services Committee. *East Lakes Geographer* 10:17–32.

———. 1981. Geopolitics in a shrinking world: A political geography of the twenty-first century. In *Political studies from spatial perspectives,* eds. A. D. Burnett and P. J. Taylor, pp. 131–56. Chichester: Wiley.

———, and Ingalls, G. L. 1983. Identifying regional blocs in the United Nations voting. In *Pluralism and political geography,* eds. N. Kliot and S. Waterman, pp. 270–83. London: Croom Helm.

———, and Minghst, K. A. 1985. Geopolitics. In *Progress in political geography,* ed. M. Pacione, pp. 41–76. London: Croom Helm.

———, and Yanarella, E. 1987. Towards a humanistic political geography. *Studies in Comparative International Development* 22(2):3–86.

Burghardt, A. F. 1973. The bases of territorial claims. *Geographical Review* 63:225–45.

———. 1980. Nation, state and territorial unity: A trans-Outaouais view. *Cahiers de geographie du Quebec* 24(61):123–34.

Burnett, A. 1984. The application of alternative theories in political geography: The case of political participation. In *Political geography: Recent advances and future directions,* eds. P. J. Taylor and J. House, pp. 25–49. London: Croom Helm.

———, and Taylor, P. J., eds. 1981. *Political studies from spatial perspectives.* Chichester: Wiley.

Busteed, M. A. 1983. The developing nature of political geography. In *Developments in political geography,* ed. M. Busteed, pp. 1–67. London: Academic Press.

Butler, D., and Stokes, D. 1969. *Political change in Britain: Forces shaping electoral choice.* London: Macmillan.

Capel, H. 1981. Institutionalization of geography and strategies of change. In *Geography, ideology and social concern,* ed. D. R. Stoddard, pp. 37–69. Oxford: Blackwell.

Castells, M. 1977. *The urban question.* London: Edward Arnold.

———. 1983. *The city and the grassroots.* London: Edward Arnold.

Clark, G. L. 1984. A theory of local autonomy. *Annals of the Association of American Geographers* 74(2):195–208.

———. 1985. *Judges and the cities.* Chicago: University of Chicago Press.

———, and Dear, M. 1978. The state and geographic process: A critical review. *Environmental and Planning A* 10:173–83.

———, and ———. 1981. The state in capitalism and the capitalist state. In *Urbanization and urban planning in capitalist society,* eds. M. Dear and A. J. Scott, pp. 45–61. New York: Methuen.

———, and ———. 1984. *State apparatus: Structures and languages of legitimacy.* Winchester, MA: Allen & Unwin.

Cockburn, C. 1977. *The local state.* London: Pluto Press.

Cohen, S. B. 1973. *Geography and politics in a world divided.* 2d ed. New York: Oxford University Press.

———. 1982. A new map of global equilibrium: A developmental approach. *Political Geography Quarterly* 1(3):223–42.

———. 1984. Asymmetrical states and global equilibrium. *School of Advanced International Studies Review* 4(2):193–212.

———. 1986. *The geopolitics of Israel's border question.* Boulder, CO: Westview Press.

Cohen, S. B., and Kliot, N. 1981. Israel's place-names as a reflection of continuity and change in nation-building. *Names* 29(3):227–47.

———, and Rosenthal, L. D. 1971. A geographical model for political systems analysis. *Geographical Review* 61:5–31.

Cox, K. R. 1968. Suburbia and voting behavior in the London metropolitan area. *Annals of the Association of American Geographers* 58:111–17.

———. 1979. *Location and public problems: A political geography.* Chicago: Maaroufa Press.

———. 1984. Neighborhood conflict and urban social movements: Questions of historicity, class, and social change. *Urban Geography* 5(4):343–55.

———. 1986. Urban social movements and neighborhood conflicts: Questions of space. *Urban Geography* 7(6):536–46.

———. 1987. Comments on "Dealignment, volatility, and electoral geography." *Studies in Comparative International Development* 22(4):26–34.

———, and Johnston, R. J., eds. 1982. *Conflict, politics and the urban scene.* New York: St. Martin's Press.

———; Reynolds, D. R.; and Rokkan, S. 1974. *Locational approaches to power and conflict.* New York: Halsted.

Curtice, J., and Steed, M. 1982. Electoral choice and the production of government. *British Journal of Political Science* 12:249–98.

Cutter, S. L. et al. 1987. From grass roots to partisan politics: Nuclear freeze referenda in New Jersey and South Dakota. *Political Geography Quarterly* 6(4):287–300.

Dear, M. J. 1981. A theory of the local state. In *Political studies from spatial perspectives,* eds. A. D. Burnett and P. J. Taylor, pp. 183–200. Chichester: Wiley.

———. 1982. Research agendas in political geography—A minority view. *Political Geography Quarterly* 1(2):179–80.

———. 1986. Editorial comment: Theory and object in political geography. *Political Geography Quarterly* 5(4):295–97.

———, and Moos, A. I. 1986. Structuration theory in urban analysis 2: Empirical application. *Environment and Planning A* 18(3):351–73.

de Blij, H. J. ([1967] 1973). *Systematic political geography*. New York: Wiley.

Dikshit, R. D. 1975. *The political geography of federalism*. New Delhi: Macmillan.

Douglas, J. N. H. 1985. Conflict between states. In *Progress in political geography,* ed. M. Pacione, pp. 77–110. London: Croom Helm.

Drysdale, A. 1988. National integration problems in the Arab world: The case of Syria. In *Nationalism, self-determination and political geography,* eds. R. J. Johnston, D. Knight, and E. Kofman, pp. 87–101. London: Croom Helm.

Duncan, S. S., and Goodwin, M. 1982a. The local state and restructuring social relations. *International Journal of Urban and Regional Research* 6(2):157–86.

———, and ———. 1982b. The local state: Functionalism, autonomy and class relations in Cockburn and Saunders. *Political Geography Quarterly* 1(1):77–96.

Dunleavy, P. 1979. The urban basis of political alignment. *British Journal of Political Science* 9:409–43.

———. 1980. The political implications of sectoral cleavages and the growth of state employment: Part 1. The analysis of production cleavages. Part 2. Cleavage structures and political alignment. *Political Studies* 28(3):364–83; (4):527–49.

Easton, D. 1965. *A systems analysis of political life*. New York: Wiley.

Ennis, P. 1962. The contextual dimension in voting. In *Public opinion and congressional elections,* eds. W. McPhee and W. Glaser. New York: Free Press.

Fincher, R. 1984. Identifying class struggle outside commodity production. *Environment and Planning D: Society and Space* 2:309–27.

Giddens, A. 1981. *A contemporary critique of historical materialism*. London: Macmillan.

———. 1984. *The constitution of society*. Berkeley: University of California Press.

Glassner, M. I., and de Blij, H. J. 1980. *Systematic political geography*. 3d ed. New York: Wiley.

Gottmann, J. 1973. *The significance of territory*. Charlottesville, VA: University Press of Virginia.

———. 1982. The basic problem of political geography: The organization of space and the search for stability. *Tijdschrift voor Economische en Sociale Geografie* 73 (6):340–49.

Gould, P., and Olsson, G., eds. 1982. *Search for common ground*. London: Pion.

Gramsci, A. 1971. *Selections from the prison notebooks*. New York: International.

Hall, P. 1982. The new political geography: Seven years on. *Political Geography Quarterly* 1(1):65–76.

Harris, W. W. 1980. *Taking root: Israeli settlement in the West Bank, the Golan and Gaza-Sinai, 1967–1980*. Chichester: Research Studies Press.

Hartshorne, R. 1935. Recent developments in political geography. *American Political Science Review* 29:785–804;943–66.

———. 1950. The functional approach in political geography. *Annals of the Association of American Geographers* 49:95–130.

———. 1954. Political geography. In *American geography: Inventory and prospect,* eds. P. E. James and C. J. Jones, pp. 167–225. Syracuse: Syracuse University Press and the Association of American Geographers.

Harvey, D. 1969. *Explanation in geography*. New York: St. Martin's Press.

———. 1973. *Social justice and the city*. London: Edward Arnold.

———. 1982. *The limits to capital*. Chicago: University of Chicago Press.

———. 1985a. *Consciousness and the urban experience*. Baltimore: Johns Hopkins University Press.

———. 1985b. The geopolitics of capitalism. In *Social relations and spatial structures,* eds. D. Gregory and J. Urry, pp. 128–63. New York: St. Martin's Press.

———. 1985c. *The urbanization of capital*. Baltimore: Johns Hopkins University Press.

Hechter, M. 1975. *Internal colonialism*. London: Routledge & Kegan Paul.

———, and Levi, M. 1979. The comparative analysis of ethnoregional movements. *Ethnic and Racial Studies* 2:260–74.

Hepple, L. 1986. The revival of geopolitics. *Political Geography Quarterly* 5(4):S21–S36.

Hoch, C. 1979. Social structure and suburban spatio-political conflicts in the United States. *Antipode* 11(3):44–55.

Hoffman, G. W. 1982. Nineteenth-century roots of American world power relations. *Political Geography Quarterly* 1(3):279–92.

Holloway, J., and Picciotto, S. eds. 1978. *State and capital: A Marxist debate*. London: Edward Arnold.

Hudson, B. 1977. The new geography and the new imperialism, 1870–1919. *Antipode* 9(1):12–19.

Hudson, R. 1983. The question of theory in political geography: Outlines for a critical theory approach. In *Pluralism and political geography,* eds. N. Kliot and S. Waterman, pp. 29–35. London: Croom Helm.

———, and Sadler, D. 1983. Region, class and the politics of steel closures in European community. *Environment and Planning D: Society and Space* 1:405–28.

———, and ———. 1986. Contesting works closures in Western Europe's old industrial regions: Defending place or betraying class. In *Production, work, territory,* eds. A. J. Scott and M. Storper, pp. 172–93. Boston: Allen & Unwin.

Jackson, W. A. D. 1958. Whither political geography? *Annals of the Association of American Geographers* 48:178–83.

Janelle, D. G. 1977. Structural dimensions in the geography of locational conflicts. *Canadian Geographer* 21:311–28.

———, and Millward, H. 1976. Locational conflict patterns and urban ecological structure. *Tijdschrift voor Economische en Sociale Geografie* 67:102–13.

Jenkins, J. R. G. 1986. *Jura separatism in Switzerland*. Oxford: Clarendon.

Johnston, R. J. 1979a. *Geography and geographers: Anglo American geography since 1945*. London: Edward Arnold.

———. 1979b. *Political, electoral and spatial system*. London: Oxford University Press.

———. 1980. Electoral geography and political geography. *Australian Geographical Studies* 18:37–50.

———. 1982. *Geography and the state: An essay in political geography*. London: Macmillan.

———. 1984a. Marxist political economy, the state and political geography. *Progress in Historical Geography* 8:473–92.

———. 1984b. The political geography of electoral geography. In *Political geography: Recent advances and future directions,* eds. P. J. Taylor and J. House, pp. 133–48. London: Croom Helm.

———. 1985. *The geography of English politics*. London: Croom Helm.

———. 1986a. The neighborhood effect revisited: Spatial science or political regionalism? *Environment and Planning D: Society and Space* 4(1):40–55.

———. 1986b. Places and votes: The role of location in the creation of political attitudes. *Urban Geography* 7(2):103–17.

———. 1986c. Places, campaigns and votes. *Political Geography Quarterly* 5(4):S105–S117.

———. 1986d. Placing politics. *Political Geography Quarterly* 5(4):S63–S78.

———. 1986e. A space for place (or a place for space) in British psephology. *Environment and Planning A* 18:599–618.

———. 1987. Dealignment, volatility, and electoral geography. *Studies in Comparative International Development* 22(4):3–25.

———; Knight, D.; and Kofman, E., eds. 1988. *Nationalism, self-determination and political geography*. London: Croom Helm.

———; O'Loughlin, J; and Taylor, P. J. 1987. The geography of violence and premature death. In *The quest for peace,* eds. R. Veryrnen et al. Beverly Hills: Sage Publications.

Jones, S. B. 1954. A unified field theory of political geography. *Annals of the Association of American Geographers* 24:111–23.

Kasperson, R., and Minghi, J. 1969. *The structure of political geography*. Chicago: Aldine.

Keat, R., and Urry, J. 1982. *Social theory as social science*. 2d ed. London: Routledge & Kegan Paul.

Kirby, A. 1982. The external relations of the local state in Britain: Some empirical examples. In *Conflict, politics and the urban scene,* eds. K. R. Cox and R. J. Johnston, pp. 88–104. New York: St. Martin's Press.

———. 1983. A public city: Concepts of space and the local state. *Urban Geography* 4(3):191–202.

———. 1987. The local state and urban politics. *Urban Geography* 8:273–79.

Kliot, N. 1982. Recent themes in political geography: A review. *Tijdschrift voor Economische en Sociale Geografie* 73(5):270–79.

———. 1983. Dualism and landscape transformation in Northern Sinai: Some outcomes of the Egypt-Israel peace treaty. In *Pluralism and political geography,* eds. N. Kliot and S. Waterman, pp. 173–86. London: Croom Helm.

———. 1986. Lebanon: A geography of hostages. *Political Geography Quarterly* 5(3):199–220.

———, and Waterman, S., eds. 1983. *Pluralism and political geography: People, territory and state*. London: Croom Helm.

Knight, D. B. 1977. *A capital for Canada: Chicago*. University of Chicago Department of Geography Discussion Paper No. 182. Chicago: University of Chicago Press.

———. 1982a. Continuing agendas for research in political geography. *Political Geography Quarterly* 1(2):174–78.

———. 1982b. Identity and territory: Geographical perspectives on nationalism and regionalism. *Annals of the Association of American Geographers* 72(4):514–31.

———. 1984a. Geographical perspectives on self-determination. In *Political geography: Recent advances and future directions,* eds. P. J. Taylor and J. House, pp. 168–90. London: Croom Helm.

———. 1984b. Regionalisms, nationalisms, and the Canadian state. *Journal of Geography* 83:212–20.

———. 1986. Humanistic political geography? In *Humanism and Geography,* ed. S. Mackenzie. Carleton Geography Discussion Paper 3:22–29. Ottawa: Carleton University.

———. 1988. Self-determination and indigenous people: The context for change. In *Nationalism, self-determination and the world political map,* eds. R. J. Johnston, D. Knight, and E. Kofman, pp. 117–34. London: Croom Helm.

———, and Davies, M. 1987. *Self-determination: An interdisciplinary annotated bibliography*. New York: Garland.

Lauria, M. 1986. Towards a specification of the local state: State intervention strategies in response to a manufacturing closure. *Antipode* 18:39–65.

Ley, D. 1980. Liberal ideology and the post-industrial city. *Annals of the Association of American Geographers* 70:239–58.

———, and Mercer, J. 1980. Locational conflict and the politics of consumption. *Economic Geography* 56(2):89–109.

Linz, J. 1985. From primordialism to nationalism. In *New nationalisms of the developed west,* eds. E. A. Tiryakian and R. Rogowski, pp. 203–53. Boston: Allen & Unwin.

Lipietz, A. 1987. *Mirages and miracles: The crises of global fordism*. London: Verso.

Logan, J. R. 1978. Growth, politics and the stratification of places. *American Journal of Sociology* 84:404–16.

McAllister, I., and Rose, R. 1984. *The nationwide competition for votes: The 1983 British election*. London: Frances Pinter.

MacLaughlin, J. G. 1986. The political geography of 'nation-building' and nationalism in social sciences: Structural vs. dialectical accounts. *Political Geography Quarterly* 5(4):299–329.

Marx, K. 1973. *Grundrise*. Harmondsworth: Penguin.

Massey, D. 1985. New directions in space. In *Social relations and spatial structures,* eds. D. Gregory and J. Urry, pp. 9–19. New York: St. Martin's Press.

———, and Allen, J., eds. 1984. *Geography matters*. Cambridge: Cambridge University Press.

Meinig, D. W. 1986. *Atlantic America 1492-1800*. Vol. 1 of *The shaping of America*. New Haven: Yale University Press.

Mercer, D. 1987. Patterns of protest: Native land rights and claims in Australia. *Political Geography Quarterly* 6(2):171–94.

Mikesell, M. W. 1983. The myth of the nation state. *Journal of Geography* 82:257–60.

Modelski, G. 1978. The long cycle of global politics and the nation-state. *Comparative Studies in Society and History* 20:214–35.

———, ed. 1987. *Exploring long cycles*. Boulder, CO: Lynne Rienner.

Molotch, H. 1976. The city as a growth machine: Toward a political economy of place. *American Journal of Sociology* 82(2):309–32.

Morrill, R. L. 1981. *Political redistricting and geographic theory*. Washington: Association of American Geographers.

Murauskas, G. T.; Archer, J. C.; and Shelley, F. M. 1988. Metropolitan, non-metropolitan and sectional variations in voting behavior in recent presidential elections. *Western Political Quarterly*. In press.

Murphy, A. B. 1988. Evolving regionalism in linguistically divided Belgium. In *Nationalism, self-determination and political geography,* eds. R. J. Johnston, D. Knight, and E. Kofman, pp. 135–50. London: Croom Helm.

Nairn, T. 1977. *The break-up of Britain*. London: New Left Books.

National Academy of Sciences—National Research Council Ad hoc Committee on Geography. 1965. Studies in political geography. In *The science of geography,* pp. 31–44. Washington: National Academy of Sciences—National Research Council.

Newton, K. 1978. Conflict avoidance and conflict suppression: The case of urban politics in the United States. In *Urbanization and conflict in market societies,* ed. K. R. Cox, pp. 76–93. Chicago: Maaroufa Press.

O'Loughlin, J. 1982. The identification of racial gerrymandering. *Annals of the Association of American Geographers* 72:165–84.

———. 1984. Geographic models of international conflicts. In *Political geography: Recent advances and future directions,* eds. P. J. Taylor and J. House, pp. 202–26. London: Croom Helm.

———. 1986a. Political geography: Tilling the fallow field. *Progress in Historical Geography* 10:69–83.

———. 1986b. Spatial models of international conflict: Extending current theories of war behavior. *Annals of the Association of American Geographers* 76:63–80.

———. 1986c. World-power competition and local conflicts in the Third World. In *A world in crisis?,* eds. R. J. Johnston and P. J. Taylor, pp. 231–68. Oxford: Blackwell.

———, and Taylor, A. M. 1982. Choices in redistricting and electoral outcomes: The case of Mobile, Alabama. *Political Geography Quarterly* 1:317–40.

———, and van der Wusten, H. 1986. Geography, war and peace: Notes for a contribution to a revived political geography. *Progress in Historical Geography* 10:484–510.

Orridge, A. W. 1981. Uneven development and nationalism: 2. *Political Studies* 39:181–90.

Osei-Kwame, P., and Taylor, P. J. 1984. A politics of failure: The political geography of Ghanaian elections, 1954–1979. *Annals of the Association of American Geographers* 74(4):574–89.

Ō Tuathail, G. O. 1986. The language and nature of the new geopolitics—The case of US-El Salvador relations. *Political Geography Quarterly* 5(1):73–86.

Paddison, R. 1983. *The fragmented state: The political geography of power*. Oxford: Blackwell.

Parker, W. H. 1982. *Mackinder: Geography as an aid to statecraft*. Oxford: Clarendon.

Peet, R., ed. 1987. *International capitalism and industrial restructuring*. London: Allen & Unwin.

Pepper, D., and Jenkins, A., eds. 1985. *The geography of war and peace*. Oxford: Blackwell.

Portugali, J. 1988. Nationalism, social theory and the Israeli/Palestinian case. In *Nationalism, self-determination and political geography,* eds. R. J. Johnston, D. Knight, and E. Kofman, pp. 151–65. London: Croom Helm.

Pounds, N. J. G. [1963] 1972. *Political geography*. 2d ed. New York: McGraw-Hill.

Pred, A. 1984. Place as historically contingent process: Structuration and the time-geography of becoming places. *Annals of the Association of American Geographers* 74(2):279–97.

Prescott, J. R. V. 1985. *The maritime political boundaries of the world*. London: Methuen.

———. 1987. *Political frontiers and boundaries*. Boston: Allen & Unwin.

Reynolds, D. R. 1981. The geography of social choice. In *Political studies from spatial perspectives,* eds. A. Burnett and P. J. Taylor. Chichester: Wiley.

———, and Shelley, F. M. 1985. Procedural justice and local democracy. *Political Geography Quarterly* 4(4):267–88.

———, and ———. 1988. Local control in American public education: Myth and reality. In *Spatial dimensions of U.S. social policy,* eds. J. P. Jones, D. Whalley, and J. Kodras. London: Edward Arnold.

Richmond, A. H. 1987. Ethnic nationalism: Social science paradigms. *International Social Science Journal* 3:3–18.

Rogge, J. R. 1977. The Balkanization of Nigeria's federal system. *Journal of Geography* 76(4):135–40.

Sack, R. D. 1986. *Human territoriality: Its theory and history*. Cambridge: Cambridge Unversity Press.

Sadler, D. 1984. Works closure at British steel and the nature of the state. *Political Geography Quarterly* 3:297–311.

Samuels, M. S. 1982. *Contest for the South China Sea*. New York: Methuen.

Sauer, C. O. 1927. Recent developments in cultural geography. In *Recent developments in the social sci-*

ences, eds. C. A. Ellwood, C. Wissler, and R. H. Gault. Philadelphia: Lippincott.

Saunders, P. 1979. *Urban politics: A sociological interpretation*. London: Longman.

Savage, M. 1987. Understanding political alignments in contemporary Britain: Do localities matter? *Political Geography Quarterly* 6(1):53–76.

Sayer, A. 1984. *Method in social science: A realist approach*. London: Hutchinson.

Schattschneider, E. E. 1960. *The semi-sovereign people*. New York: Holt, Rinehart & Winston.

Scott, A. J., and Roweis, S. T. 1978. The urban land question. In *Urbanization and conflict in market societies*, ed. K. R. Cox. Chicago: Maaroufa Press.

Shelley, F. M. 1988. Structure, stability and section in American politics. *Political Geography Quarterly* 7(2):153–60.

Short, J. R. 1982. *An introduction to political geography*. Boston: Routledge & Kegan Paul.

Skocpol, T. 1979. *States and social revolutions*. Cambridge: Cambridge University Press.

Smith, G. E. 1979. Political geography and the theoretical study of the East European nation. *Indian Journal of Political Science* 40(2):59–83.

———. 1988. Ethnoregional societies, 'developed socialism' and the Soviet ethnic intelligentsia. In *Nationalism, self-determination and political geography*, eds. R. J. Johnston, D. Knight, and E. Kofman, pp. 166–88. London: Croom Helm.

Smith, N. 1984. *Uneven development*. Oxford: Blackwell.

Soja, E. W. 1980. The socio-spatial dialectic. *Annals of the Association of American Geographers* 70(2):207–25.

———. 1985. The spatiality of social life: Towards a transformative retheorisation. In *Social relations and spatial structures*, eds. D. Gregory and J. Urry, pp. 90–127. New York: St. Martin's Press.

———, and Hadjimichalis, C. 1979. Between geographical materialism and spatial fetishism: Some observations on the development of Marxist spatial analysis. *Antipode* 11(2):3–11.

Starr H., and Most, B. A. 1983. The substance and study of borders in international relations research. *International Studies* 20:581–620.

Taylor, P. J. 1978. Progress report: Political geography. *Progress in Historical Geography* 2:153–62.

———. 1981. Political geography and the world-economy. In *Political studies from spatial perspectives*, eds. A. D. Burnett and P. J. Taylor, pp. 157–74. Chichester: Wiley.

———. 1982. A materialist framework for political geography. *Transactions of the Institute of British Geographers* NS7:15–34.

———. 1983. The question of theory in political geography. In *Pluralism and political geography*, eds. N. Kliot and S. Waterman, pp. 9–18. London: Croom Helm.

———. 1984a. Accumulation, legitimation, and the electoral geographies within liberal democracy. In *Political geography: Recent advances and future directions*, eds. P. J. Taylor and J. House, pp. 117–32. London: Croom Helm.

———. 1984b. The geography of elections. In *Progress in political geography*, ed. M. Pacione. London: Croom Helm.

———. 1985a. *Political geography: World-economy, nation-state, and locality*. New York: Longman.

———. 1985b. Review of Clark, G. L. and M. Dear. State apparatus: Structures and language of legitimacy. *Progress in Historical Geography* 9(3):465–67.

———. 1985c. The value of a geographical perspective. In *The future of geography*, ed. R. J. Johnston, pp. 92–110. New York: Methuen.

———. 1986. Chaotic conceptions, antinomies, dilemmas and dialectics: Who's afraid of the capitalist world-economy, *Political Geography Quarterly* 5(1):87–93.

———. 1987. The poverty of international comparisons: Some methodological lessons from world-systems analysis. *Studies in Comparative International Development* 22(1):12–81.

———, and Johnston, R. J. 1979. *The geography of elections*. Harmondsworth: Penguin.

——— et al. 1982. Editorial essay: Political geography—Research agendas for the nineteen eighties. *Political Geography Quarterly* 1(1):1–17; additional agendas and comments, 1(2):167–89.

———, and House, J. W., eds. 1984. *Political geography: Recent advances and future directions*. London: Croom Helm.

Thrift, N. 1983. On the determination of social action in space and time. *Environment and Planning D: Society and Space* 1(1):23–58.

Urry, J. 1981. Localities, regions and social class. *International Journal of Urban and Regional Research* 5:455–74.

———. 1983. Realism and the analysis of space. *International Journal of Urban and Regional Research* 7:122–27.

van der Wusten, H. 1985. The geography of conflict since 1945. In *The geography of peace and war,* eds. D. Pepper and A. Jenkins, pp. 13–28. Oxford: Blackwell.

———. 1988. The occurrence of successful and unsuccessful nationalisms. In *Nationalism, self-determination and political geography,* eds. R. J. Johnston, D. Knight, and E. Kofman, pp. 189–202. London: Croom Helm.

———, and O'Loughlin, J. 1986. Claiming new territory for a stable peace: How geography can contribute. *Professional Geographer* 38(1):18–28.

Wallerstein, I. 1974. *The modern-world system: Capitalist agriculture and the origins of the European world-economy in the sixteenth century.* New York: Academic Press.

———. 1979. *The capitalist world-economy.* Cambridge: Cambridge University Press.

———. 1980. *The modern world-system II: Mercantilism and the consolidation of the European world-economy 1600–1750.* New York: Academic Press.

———. 1984. *The politics of the world-economy.* New York: Academic Press.

Waterman, S. 1987. Partitioned states. *Political Geography Quarterly* 6(2):151–70.

Whebell, C. F. J. 1973. A model of territorial separatism. *Proceedings of the Association of American Geographers* 5:295–98.

Williams, C. H. 1979. Ethnic resurgence in the periphery. *Area* 11:279–83.

———. 1980. Ethnic separatism in Western Europe. *Tijdschrift voor Economische en Sociale Geografie* 71(3):142–58.

———. 1984. Ideology and the interpretation of minority cultures. *Political Geography Quarterly* 3(2):105–26.

———. 1985a. Conceived in bondage—Called into liberty: Reflections on nationalism. *Progress in Historical Geography* 9:331–55.

———. 1985b. Minority groups in the modern state. In *Progress in political geography,* ed. M. Pacione, pp. 111–51. London: Croom Helm.

———. 1986. The question of national congruence. In *A world in crisis?,* eds. R. J. Johnston and P. J. Taylor, pp. 196–230. Oxford: Blackwell.

———. 1988. Minority nationalist historiography. In *Nationalism, self-determination and political geography,* eds. R. J. Johnston, D. Knight, and E. Kofman, pp. 203–22. London: Croom Helm.

———, and Smith, A. D. 1983. The national construction of social space. *Progress in Historical Geography* 7(4):502–18.

Wolpert, J. et al. 1972. *Metropolitan neighborhoods: Participation and conflict over change.* Resource Paper No. 16. Washington: Association of American Geographers, Commission on College Geography.

Zelinsky, W. 1984. O say, can you see?: Nationalistic emblems in the landscape. *Winterthur Portfolio* 19(4):277–86.

Geography from the Left

Richard Walker

Geography on the left in America has come a long way over the last 20 years. A new generation of scholars has expanded the ranks of left-oriented faculty, bringing the analytic framework and progressive social agenda of Marxism and allied schools of thought into most of the traditional subject areas of the discipline. Only part of this sweep of material can be presented here. While this menu of topics necessitates some overlap with other essays in this book, our purpose is to highlight the special contribution of left theorists and researchers to the development of geographic thought in the 1980s.

Contemporary left geography emerged around 1970, pressed by a mere handful of adherents at the professorial level: Jim Blaut, Dick Peet, David Harvey, and Bill Bunge among them. Its banner was carried by the informally organized Union of Socialist Geographers and its ideas were most prominently featured in the journal *Antipode*; this early history of left geography has been ably told by Peet (1975), so we concentrate here on developments in the 1980s. The last decade has seen a number of new turns. Organizationally, a choice was made to join in the formal subject-area groupings of the Association of American Geographers, as the Socialist Geography Specialty Group (SGSG), and the USG faded away. *Antipode* passed from the editorship of Peet and associates at Clark University to the equally able hands of Eric Sheppard of the University of Minnesota and Joe Doherty of the University St. Andrews (U.K.), and, like rebel publications of the left across the disciplines, became a legitimate, institutionalized journal of the field. It has been joined by another publication of the left, *Society and Space,* whose editorial group has less affinity to Marxism and socialism. At the same time, left geographers now appear regularly in all the established journals of geography (and related fields, such as regional studies, planning, or

---

This chapter was written with the assistance of members of the Socialist Geography Specialty Group: Vera Chouinard, Phil Cooke, Ruth Fincher, Margaret Fitzsimmons, Julie Graham, Michael Heiman, Kevin Cox, Suzanne MacKenzie, Andy Mair, Jeff McCarty, Mary Beth Pudup, Allen Scott, Eric Sheppard, Neil Smith, Michael Watts, and Michael Webber.

urban sociology), feature prominently on editorial boards, and even act as editors for several mainstream journals and publishing-house series.

The leading role of David Harvey and Doreen Massey in the 1970s is not to be gainsayed: Harvey is principally responsible for (re)introducing Marxist theory into geography, and at the same time becoming the leading urbanist in North America. Massey is responsible for the great turn to industrial studies by left geographers in the late 1970s, and for a renewed enthusiasm for the geographic dimension of social research in the 1980s. As might be expected, with growth has come greater diversity of subject matter and approach, and a more diverse corps of lead thinkers, such as Allen Scott, Michael Webber, Allan Pred, Ruth Fincher, Kevin Cox, Eric Sheppard, Neil Smith, and Andrew Sayer. While a recognizable core of people exists in North America and Great Britain, the wider penumbra of younger left scholars, sympathetic thinkers from a broad range of backgrounds, and friends outside the Anglophone world should not be overlooked. In the U.S., the left is heavily concentrated in the SGSG, but by no means exclusively so. As left geography has edged toward the mainstream, both its currents and its eddies are deepened and broadened by the encounter.

The agenda of left research has not only expanded but deepened. Initially, a great deal was to be gained by bringing the classic insights of Marxist theory to a variety of topics. But the momentum of the search for a better geographic social science has propelled left scholars down several roads. Some went on to develop Marxist theory itself more fully, that it might provide a more complete set of conceptual tools (e.g., Harvey 1982). Others took another look at method, welcoming the clarifications that realism, critical theory, and structuration theory might add to the understanding of social processes and how to grasp them (Gregory 1978; Thrift 1983; Sayer 1984; Pickles 1985; Pred 1986). Another thread was the search for more finely tuned "middle level" theories of such things as labor relations or local government (Storper and Walker 1983; Clark and Dear 1984; Clark, Gertler, and Whiteman 1986).

The demand for a more explicitly spatialized theory of capitalist societies became increasingly urgent in the face of massive geographic realignments in the world, painful absences in our explanations for the fate of particular places, and disenchantment with some popular "global" spatial theories of the 1970s such as center-periphery dependency models or the new international division of labor (Massey 1984; Harvey 1985b; Storper and Walker 1989). Finally, the growing number of women geographers brought feminist concerns regarding the oppression of women into an overwhelmingly male discipline—including its left wing (Women and Geography Study Group 1984). Revived militancy around racism in the late 1980s may well be the next crucial challenge to the left political and theoretical agenda.

Left geographers can be proud of their achievements in a discipline that is not always noted for either its explanatory depth or overriding concern with human oppression and liberation. The left can claim a good deal of credit for broadening the intellectual respectability of the geographic enterprise outside the discipline in recent years, and can claim a measure of intellectual leadership and even hegemony within certain geographic subfields. At a time when prospects for the discipline have not always been the brightest, this large dose of energy for new research agendas and commitment to greater theoretical sophistication has been exceedingly helpful in moving geography forward.

Socialist analyses of urban geographic change have focused upon a wide variety of issues, but perhaps the central one has been the attempt to demonstrate the "unnaturalness" of the urban order under capitalism. That is, urban-development patterns and the city form are not the inevitable outcome of natural scarcity, individual consumer desires, transport costs, or the technological genie. They are, rather, deeply etched by rivers of capital investment and carved out by forces of social difference along class, gender, and racial lines. In a word, things could be different, but they are kept as they are by the powers of a social order that is more interested in exploitation and accumulation than in the human contours of urban life. Left urban geography might therefore be termed the study of the politics of urban space (Cox 1984c).

This agenda was set by David Harvey's maverick study, *Social Justice and the City* (1973), which enjoyed widespread influence throughout the social sciences. A flurry of research has fleshed out the critique in several directions (for an overview, see Badcock 1984). Some work emphasizes the role of the financial system in the provision and orchestration of urban living space (Stone 1975; Harvey 1977; Williams 1976, 1978; Florida 1986; Meyerson 1986). Other writing takes on prevailing economic theories of rent and the land market, with their implications of optimal performance and benign outcomes (Barnbrock 1974; Harvey 1974a; Walker, 1974, 1975; Roweis and Scott 1978; Scott 1980). Particular stress has been laid on the active role of land owners, property investors, and developers in the process of urban development (Ambrose and Colenutt 1975; Massey and Catalano 1978; Boddy 1980; Feagin 1983; Haila 1988). One important theme is the way in which the land market, in concert with class and racial divisions, generates persistent patterns of residential segregation and conflict over the control and renewal of urban space (Harvey 1975, 1978; Cox 1978, 1982; Rose 1981; Hoch 1984; Lauria 1984). Another is the process by which the American city took on an increasingly suburban form as it expanded in the nineteenth and twentieth centuries, by offering an outlet for surplus capital and social tensions. (Walker 1978, 1981). By the end of the 1970s, moreover, it became apparent that similar forces were at work reconstructing the inner cities through the process of gentrification (Hamnett 1973; Smith 1979, 1987b; Lauria 1982; Hamnett 1984; Schaffer and Smith 1986; Smith and Williams 1986). An important branch of inquiry led toward land-use regulation and the political control of urban space (Walker and Heiman 1981; Logan and Molotch 1986; Plotkin 1987; Heiman 1988b). Another branch led to the sources of widespread homeownership as a key mode of consumption that modifies class relations and the social production of urban space in crucial ways (Rose 1984; Belec, Holmes, and Rutherford 1987; Harris 1986; Pratt 1986a, 1986b; Harris and Hamnett 1987; Florida and Feldman 1988).

The dynamics of capital flows into the built-environment and class struggles in the consumption realm—which had virtually defined the field of left urban geography in the 1970s—was challenged in the present decade from two directions. An emerging feminist critique forcefully placed questions of social reproduction on the agenda, in part to help explain suburbanization, gentrification, residential form, and the like, but also to insist on the centrality of gender difference and the oppression of women in urbanization (Stimpson et al. 1981; Christopherson 1982; Brownill 1984; Rose 1984;

Mackenzie 1987; Pratt and Hanson 1988). Urban geography has been most thoroughly reoriented, however, by the processes of capitalist change which have been at work on American cities in recent years. Such work has often fallen under the rubric of "urban restructuring" (Soja, Morales, and Wolff 1983; Fainstein et al. 1983; M. Smith and Feagin 1987). Nonetheless, advances have been made in key areas of understanding (e.g., see *Society and Space* 1986). One, growing out of the new industrial geography, is a powerful statement that was previously lacking of the relation of urbanization to capitalist production (Scott 1986b, 1988a; Storper and Walker 1989). A second is a consideration of the rise of new office centers in big cities and the accompanying reconstitution of residential areas through gentrification (Walker and Greenberg 1982; Nelson 1986; Smith and Williams 1986; also Walker 1985a; Urry 1986a, 1987). A third is an expansive reinterpretation of urban fragmentation and flux as part of the experience of postmodernity (Davis 1985; Soja 1986, 1989; Dear 1986; Harvey 1987; Knox 1987).

David Harvey has, of course, continued to lead the way in many areas of inquiry (see the essays collected in Harvey 1985b). To understand better the relation of the urban process to capitalism, he undertook a monumental reconsideration of the Marxist theory of capital, in search of an adequate conceptualization of money, finance capital, land rent, fixed investment, crisis, and spatial expansion, among other things (Harvey 1982). This was followed by a sustained investigation of the development of Paris in the mid-nineteenth century that provides perhaps the most complete integration of the various facets of urbanization yet achieved in a single essay (Harvey 1985a). While there has been some difference of opinion between those who take a more unified cut at the city through capital accumulation and those who stress the jostling of other causal forces—from gender to modernist ideology—a positive development running across the work of left geographers in the 1980s has been an increasingly comprehensive and vibrant picture of the immensely complex phenomenon of contemporary urbanization (e.g., Soja 1986; Harvey 1985a, 1985b; Marston and Kirby 1988).

INDUSTRIAL GEOGRAPHY

In the 1970s Marxist geographers, led by Doreen Massey, began a critique of traditional industrial-location theory, reevaluating everything from its neoclassical roots to its efflorescence in sophisticated quantitative models. In the 1980s, they have been responsible for the emergence of a new industrial geography which is a powerful alternative to the traditional field. Some of the consolidated results of this movement are just now appearing in book form.

The impact of Marxism within the new industrial geography is reflected in an emphasis on production and particularly what have been called "spatial divisions of labor" (Massey 1984; Scott and Storper 1986; Storper and Walker 1989). But there is a wide range in scales of analysis and an evolution in the debate that we may be able to capture through a simple tripartite scheme of *micro-, meso-,* and *macro-geographies* of capitalist production.

Microgeography of Production

The microgeography of industry above all concerns relations between capital and labor, or employment relations, at the plant level (factory or office). Clark (1981),

Walker and Storper (1981), and Peet (1983) provided early schematics for the way in which capitalists could exploit spatial differences in labor markets to their advantage in plant-location decisions. These models were later augmented by more subtle analytics of "employment relations" (Storper and Walker 1983, 1984; Moulaert 1987), the interaction of place and industry (Massey 1984), and regional adjustment (Clark, Gertler, and Whiteman 1986). Important applications of the labor-market approach to the spatial division of labor can be found in Nelson (1986) and Angel (1987). It has been amplified, in these and other works, by increased attention to plant- and firm-specific machine technologies, work organization, management practices, union organizing, and strategies for coping with uncertainty (see also Storper 1982; Massey and Meegan 1982; Sayer 1986b; Clark and Johnston 1987; Morgan and Sayer 1988; Walker 1988c; Clark 1989b). For all the contributions of the spatial-division-of-labor approach, however, it still has a residual flavor of Weberian location theory with a laborist twist, and has therefore had to be supplemented.

Mesogeography of Production

Mesolevel analysis began with a turn away from the attributes of places to those of industries (Massey 1979). This heralded a salutary revival of interest in specific case studies of sectors (Massey and Meegan 1978; Scott 1983b, 1984a, 1984b; Markusen 1985; Markusen, Hall, and Glasmeier 1986; Bradbury 1987; Storper and Christopherson 1987; Morgan and Sayer 1988; Holmes 1988). At first such industry studies served principally as a backdrop for understanding plant closure and job loss in declining sectors, under the rubric of "restructuring" theory (Massey and Meegan 1982). A spatial-division-of-labor model reasserted itself at the mesolevel when it came to assessing the overall pattern of plant location, and this was dominated by notions of spatial hierarchies of skill and corporate functions (e.g., Massey 1984; Taylor and Thrift 1982; Bradbury 1985) that owed more to core-periphery models (Frank 1968) and the geography of enterprise (Hymer 1972; Watts 1981) than to the new industrial geography.

The mesogeography of industry has moved in more original directions in three respects. The first has been to inquire further into the division of labor and how it is organized (Scott 1983a, 1986a, 1988a; Walker 1988b). Attention has been turned to the dynamics of the *social division of labor,* as expressed in patterns of specialization at the level of the individual plant and firm. This work has drawn in particular upon the theories of Coase (1937) and Williamson (1975) who have provided a strong analytical language for thinking about processes of economic and institutional organization. This language builds on the logic of inter-firm transacting to show which particular sets of production activities will be internalized within the firm, and how the boundaries between firms will be determined.

Research by geographers in the Coase-Williamson mode has been devoted to processes of *vertical disintegration,* i.e., the deepening and widening of the social division of labor as the fragmentation of production activities proceeds (Scott 1983a, 1986a; Storper and Christopherson 1987). This fragmentation is often equivalent in organizational terms to an increase in subcontracting (Holmes 1986), product innovation and diversification (Schoenberger 1988a), and overall *flexibility* of the production system, for it enormously increases transacting possibilities for any particular production activity (Scott 1988b). It has also been shown that vertical disintegration or fragmentation engenders strong external economies within the production system.

Since vertical disintegration tends to increase levels of external transactional activity in any system, it is also associated with rising distance-dependent costs. Groups of producers often seek to reduce these costs strategically by clustering together in geographic space. In this manner, external economies (a nonspatial phenomenon) are transformed into and consumed in the form of agglomeration economies. Localized production complexes are therefore organizational-cum-spatial systems which are mediated through locational processes out of the social division of labor. They are further sustained by local labor markets, whose flexibility under conditions of agglomeration is greatly increased—i.e., information exchange, job search, and job-matching processes are all enhanced by the close proximity of many employers (and job seekers) in one place.

The second direction in which the mesography of industry has moved has been to render dynamic the analysis of production, in three ways (Walker 1988a; Storper and Walker 1989):

> One is to incorporate the disequilibrium that constantly besets companies in the form of uncertainty and competitive struggle (Clark, Gertler, and Whiteman 1986; Schoenberger 1987; Scott 1983a).
>
> Another is to view location as a process of the technologically driven growth of whole sectors, rather than as the static allocation of plants having known features, or even as short-term restructuring (Walker 1985b).
>
> Yet another is to focus upon the force of external economies that propel growth across wide segments of industry, as well as individual sectors (Scott and Storper 1986; Scott 1988a).

The third direction of inquiry at the mesolevel joins the dynamics of industries with those of regions in a joint process of territorial industrialization (Scott 1988b; Storper and Walker 1989). This goes beyond agglomeration economies to the way in which whole new territories evolve and are affixed to the existing space-economy of capitalism. It builds on the insights of Harvey (1982) regarding spatial expansion and disequilibrium in capitalist growth, but provides a firmer base in the process of production. It also tries to solve the standing theoretical puzzle of the relation of social process to spatial outcomes.

The new industrial geographers, in reaction to much of the statistical and empirical work of regional science, have frequently cautioned against deducing geographic outcomes from industrial characteristics, and generalizing from particular industries and areas to larger social and spatial processes. Thus, Massey (1984) views spatial structures of production in particular firms and industries as only one element in the complex constitution of "spatial divisions of labor." Researchers have acknowledged that the relation between social processes as technological and organizational change, and spatial outcomes such as decentralization and agglomeration, is "mediated" by a multitude of intervening events and processes that defy easy generalization (e.g., Walker 1985b; Sayer 1986a; Gertler 1988). These cautions have helped to liberate industrial geography from monolithic and unidirectional conceptions of industrial development such as that offered by product-cycle theory, but can lead to a kind of explanatory nihilism if carried too far. While the geography of industry is seen as complexly determined, and while industry and regions are seen as mutually constitutive, the specter of essentialism haunts the literature (Storper 1985a). A Marxist-realist perspective has allowed an uneasy coexistence of the classic theory of capital accumulation and class struggle with new strains of multicausal analysis and fresh empirical study. The solution appears to lie in the way capital accumulation and the constitution of places—what Smith (1984) has called "the production of place"—unfold together through the periodic creation of "new industrial spaces" (Scott 1988b; Storper and Walker 1989).

Macrogeography of Production

The macrogeography of production involves an effort by Marxists to situate industrial geography within a broader social context of the evolution of capitalism. On the level of the economy and society as a whole, many industrial geographers are beginning to view geographic change as most effectively periodized in terms of regimes of accumulation and corresponding modes of social regulation (Storper and Scott 1986; Harvey 1987; Harvey and Scott 1989; Storper and Walker 1989). This work draws inspiration from the French Regulationist School of Marxism, which has always been strongly geographic in its approach (e.g., Lipietz 1987). There is general agreement that the period from about the 1920s to the early 1970s (in North America and Western Europe) can be categorized as dominated by a "Fordist" regime of accumulation, based on assembly-line mass-production sectors which formed large growth poles. The geography of this phenomenon corresponded to large industrial cities concentrated in the Manufacturing Belt of the U.S. and the great industrial region stretching across the North European Plain. These cities experienced a devastating process of deindustrialization and job loss over the 1970s and early 1980s.

There is increasing evidence that since the end of the 1960s or early 1970s a new regime of accumulation has begun to appear, based on *flexible* forms of technology, production organization, and labor markets (whence the designation "regime of flexible accumulation"), although the matter is still fiercely debated (Sayer 1988b; Gertler 1988; Schoenberger 1988b). This new regime is strongly associated with the re-agglomeration of production and the emergence of a series of new industrial spaces in various parts of North America and Western Europe (e.g., Silicon Valley, Orange County, the French Cite Scientific, the Third Italy, and so on). These new industrial spaces in general occur in areas that remained free from intensive Fordist forms of industrialization. They are typically based on flexible patterns of production, above all high-technology industry and craft-specialty production (Storper and Scott 1988; Scott and Angel 1987; Scott 1988b; Florida and Kenney 1989).

Research at all three levels of industrial geography has been associated with a heightened sensibility to the connection between industrial base and changing local political configurations; as a result, a "new" political geography seems to be emerging as a complement to the new industrial and regional geographies. Thus, as part of the post-Fordist regime of accumulation, new flexible manufacturing complexes that have sprung up at various locations over the last few decades have for the most part appeared in places that have had little or no prior history of industrialization and working-class community development. In such places, the capital-labor relation has frequently been reconstituted on new (flexible) foundations that have significantly benefited capital. A new kind of *politics of place* seems to have been ushered in, involving disorganized labor (i.e., nonunionized) and neoconservative community formation (Storper and Walker 1989). At the same time, the reconstituted transactional relations between producers in new flexible-production localities engender a search for new forms of business-community development, ranging from local growth coalitions to just-in-time collectives of producers and dependent subcontractors.

QUANTITATIVE METHODS

Quantitative methods are used for two purposes: for confronting theory with data, and as a language for theory development. Both uses are quite recent in left geography. It has become increasingly accepted that certain aspects of Marx's economic theory are subject to analysis using these tools (Farhi 1973; Barnbrock 1976; Scott 1980). In the realm of theory development, mathematical models have been particularly influential in increasing our understanding of the relation between labor values and prices, of the factors affecting historical tendencies in the rate of profit, and the possibilities and limitations of unequal exchange.

The implications of this work for economic geography and uneven development have been investigated by several writers in recent years. Sheppard (1987) has modeled the structure of production prices and labor values in space-economy. On the basis of this analysis, it has been shown that a capitalist space-economy is inherently unstable (Sheppard 1982); that inter- and intra-class conflict is a logical consequence of the economic social structure of capitalism (Sheppard and Barnes 1986); and that regional class alliances have a material foundation in spatial variations in rates of exploitation and wage levels that develop in a capitalist space-economy (Liossatos 1983; Sheppard 1984). In addition, it has been shown how introduction of space into the mathematical analyses of Marxist economists has called into question some of the theoretical conclusions of those analyses, including debates over the likelihood of falling rates of profit, the existence of comparative advantage and benefits from trade, and the tendency of intersectoral capital flows to equalize the rate of profit (Sheppard and Barnes 1986; Webber 1987c). They have also brought significant clarification to debates on land rent, concerning the role of rents by comparison to profits and wages, the various kinds of differential rent, and the status of monopoly and absolute rent (Scott 1980; Huriot 1981; Barnes and Sheppard 1984; Barnes 1984, 1988). Not all of the quantitative work by left geographers has been so closely related to a reexamination of aspects of Marx's theory. Clark, Gertler, and Whiteman (1986), for example, have developed an approach to analyzing regional dynamics from a political economic perspective, involving heavy use of statistical analysis (also Gertler 1984a, 1984b).

Mathematical approaches to Marxist economic theory have been controversial, particularly in the realm of value theory. While their use has clarified the relation between labor values and prices, it has at the same time been argued that there is no direct way that prices can be read from labor values, leading some authors to conclude that labor values are theoretically useless (Hodgson 1981). On the other hand, others have used these analytical developments to show how labor values and other Marxist categories may be calculated empirically (Webber 1987a, 1987b). Geographers have contributed significantly to this empirical literature, calculating the degree of unequal exchange (Webber and Foot 1984), the relation between labor values and exchange values (Gibson et al. 1986), and historical changes in profit rates (Webber and Rigby 1986; Webber 1987a, 1988). Perhaps surprisingly, these empirical analyses have shown that labor values are in fact very strongly correlated with production prices, and that the rate of profit has indeed shown a secular downward trend.

The use of quantitative and mathematical methods has generated considerable debate. Some have argued that formal, deductive logic can help to clarify Marxian economics, even if the entire project is dialectical in nature (Farhi 1973; Barnbrock 1976; Roemer 1986). It may be objected, however, that supposedly "analytic" models introduce an unwarranted degree of individualism into economic theory (e.g., Roemer 1980, 1986); that their class analysis is more Ricardian than Marxian in inspiration (e.g., Scott 1980; Barnes 1988); that their analysis of competition is neoclassical in spirit (Shaikh 1980); and that they tend toward static formulations where dynamic ones are warranted (Walker 1988a). Similarly, Sayer (1984) argues that extensive statistical analysis is atheoretical and incapable of identifying the casual mechanisms behind observed events, and, equally, statistical work loses the notion of agency and struggle inherent in Marxist political theory.

Despite these criticisms, mathematical and statistical work is becoming more common in left geography. This is not to say that statistical work by Marxist geographers will necessarily be any better than that of their empiricist counterparts (Sheppard 1982; Sayer 1984). However, such work can offer two things to left geography: first, it becomes possible to identify what is happening (is the rate of profit failing?), and secondly, the task of proposing to measure a category forces an exact definition of that category (just what is productive labor?). With appropriate limitations, quantification can contribute to both theory development and empirical analysis within left research programs.

DEVELOPMENT STUDIES

Marxist political economy has provided the central questions, and some of the most vital theory, in the study of Third World development over the past two decades. The first steps were taken by Paul Baran in the early 1950s, from whose work flowed dependency theory in its various guises (Frank, ECLA, Cardoso, Amin—see the excellent review by Palma 1978). Following on the anti-imperialist politics of the New Left in the 1960s came a second return to the classical Marxism of Marx, Lenin, and Luxemburg in the early 1970s; this later became synonymous with what is loosely called "structural Marxism," but particularly the modes-of-production approach (Foster-Carter 1978). A third current in this broad stream was the historical work of Wallerstein (1974) and his world-systems school.

Yet, just as the theory to explain the backward condition of the Third World had been fully articulated, the "end of the Third World" was being proclaimed on the basis of the internal differentiation of the periphery and the dramatic postwar industrialization of the newly industrializing countries (NICs) in Latin America and Southeast Asia. Within the Marxist camp there are three milestones in the reaction against the sweeping claims of the global theories of underdevelopment. One is Robert Brenner's seminal critique (1977, 1986) of Frank and Wallerstein for undue emphasis on market exchange, and his reassertion of class relations and technical change in economic growth. The second is Warren's return (1980) to the classical Marxist assertion of capitalism's progressive qualities and the new realities of capitalist growth in the periphery. The third is Lipietz's application (1987) of the French Regulationist school's theory of regimes of accumulation to the dilemmas of Third World industrialization, in a manner that unites Samir Amin's insights into disarticulated development with the methods of production, particularly "bloody Taylorism," and links the rise of flexible accumulation in the core countries to a move toward peripheral Fordism in the Third World.

Geographers have taken up these same themes. Slater (1973, 1977) brought dependency theory into the study of Third World regional and urban development among geographers; Peet (1987) and Watts (1981) have written on modes of production; Bassett (1988) and Watts (1986) have made use of Lipietz; Slater (1987) has taken on the Warren thesis "economism"; and so on. Generally, however, our impression is that, while development studies have been theoretically upgraded, this has happened in a more systematic way in Europe, and it has been overshadowed by other subspecialties within North American geography. Large parts of foreign area studies in American geography are still wanting in rigor and energy. Nonetheless, there is still a good deal to be enthusiastic about concerning the work of the new generation of left development geographers.

An important part of the debate in left development theory in the 1980s is whether the new realities in the periphery—and by extension the extraordinary internationalization of capital since the early 1970s—can be captured by the theory of capitalism outlined by Marx. A strongly dissenting note has been struck by Corbridge (1986), from within the left, based on a post-Structuralist view of Marxism as an economic determinism and unilinear theory of history. Corbridge argues that radical development geography has failed because it is essentialist, oppositional, and confrontational, and cannot explain the existence and peculiarities of the NICs. This dissent is misguided in several ways, however, not the least of which is having overlooked much of the recent work in the field that does not fall prey to the sins he enunciates (Watts 1988).

There are three main fronts along which left research is moving in the 1980s. The first is the study of industrialization and capital accumulation in the periphery (and the NICs in particular). This embraces the general work on transnational capital by Taylor and Thrift (1982), the research on labor-intensive industries taking advantage of cheap labor in Mexico and Southeast Asia (Christopherson 1982; Browett 1986; Scott 1987), the debate on urbanization and labor control in São Paulo (Storper 1984), studies of spatial segregation in South Africa (Crush 1982; Pickles 1985; Mabin 1988), and work on the informal sector (Bromley 1980; Burgess 1982).

Another major focus, harking back to the Marxist classics (particularly Kautsky and Lenin), is the fate of the peasantry (the "agrarian question") in peripheral capitalism, and by extension the whole question of food as a wage good in a world economy. Probably the most insightful work from an explicitly Marxist perspective in this area has been by deJanvry (1981) on the semiproletarianization of the peasantry and the contributions of Harriet Friedmann (1982, 1987) on the international food order since 1945. This entire body of literature has been characterized by a great internal vitality and debate (Watts 1988). Geographers have made a certain contribution to it (Watts 1987). Hecht (1985) has drawn on deJanvry in her studies of Amazonian development in Brazil; Watts (1983, 1986) and Wisner (1977, 1985) have linked the transformation of the peasantry with the intensification of famine in Africa (also Watts and Bassett 1985); Johnson (1982) has looked at peasant struggles for survival in rural Mexico; and Richards (1985) has made a sustained defense of the environmental knowledge of the peasant farmer.

The third front, and perhaps the most exciting, is the coming together, under Marxist tutelage, of political economy and cultural ecology of the old Berkeley type. That is, people are looking at ecological and resource questions through the prism of the relations of production, class domination, and state intervention. The work of Hecht (1985) and Grossman (1984) falls into this mold. More recently, though not explicitly Marxist, Blaikie (1985) and Blaikie and Brookfield (1986) examine soil erosion and land degradation in terms of the constraints facing local managers at the point of production. A large group of Africanists are examining similar aspects of rural and agrarian development in terms of the articulation of the state with local (household) resource management (Bassett 1986, 1988; Richards 1985; Weiner et al. 1985; Samatar 1985, 1988). And a few are now looking more closely at the key role of women in agrarian systems of production and exchange, and how women's oppression can act as a brake on development (Carney 1986).

Some emerging areas of interest should also be noted. There are the long-overdue beginnings of left scholarship on gender and development (Momsen and Townsend 1987), migration (Crush 1986), nationalism (Blaut 1987), world debt (Corbridge 1987), and the state (Watts 1984).

ENVIRONMENTAL AND RESOURCE GEOGRAPHY

The left contribution to environmental and resource studies has been selective, as this remains much "the forgotten dimension" of left geography that Peet (1975) decried over a decade ago. Fitzsimmons (1989) is still chiding us for this neglect. Nonetheless, the quality of work has often been very high, giving it an influence out of proportion to the scale of production. One thinks immediately of the widespread use of Harvey's essay (1974b) on population as a counter to the resurgent Malthusianism of the time.

One of the earliest and most telling areas of work was the refutation of "natural hazard" research, which blamed nature for social catastrophes and the victims for their vulnerability to events such as drought and flood, owing to their primitive and irrational beliefs and behaviors in the face of nature's furies. The left critique pointed instead to the political economic sources of risk, particularly the exposure of peasant agriculture to the vicissitudes of the world market, the erosion of traditional methods of husbandry and social adjustment, increasing pressure on land and people with new methods of production, and new forms of class exploitation (Wisner 1977; O'Keefe and Wisner 1977, 1983; Marston 1983; Watts 1983, 1986). Much of this work has come out of Africa, perhaps because there the transition to commodity production is the most recent, and the degree of human marginalization is the greatest; certainly, calamities have smitten that beleaguered land with haunting regularity in recent years. These trenchant interventions have moved the whole discussion of famine, soil erosion, and environmental degradation among development geographers visibly to the left (Hewitt 1983; Blaikie 1985; Blaikie and Brookfield 1986).

In the advanced capitalist countries, the greatest hazards are often those posed by industrial pollutants, or what are sometimes called "technological hazards." As the epochal pollution-control legislation of the 1970s took hold in the U.S., it triggered a fierce sequence of political and legal struggles between industry and environmentalists; while the new laws to clean up air, water, or workplace were generally positive, they were often pitifully ineffective (Walker and Storper 1978; Walker, Storper, and Widess 1979). As time has passed, however, attention has turned increasingly to the previously underestimated hazards of toxic substances (Fitzsimmons 1987). Toxic-dump siting has become a favorite pastime for some geographers, but is fervently opposed by most local citizens and activists (Heiman 1988a). Repeatedly, an aroused environmental movement has created obstacles to growth that capitalism must circumvent, typically through the intervention of higher levels of the state (Heiman 1988b).

A considerable source of environmental disruption in the advanced countries—and now spreading rapidly into the Third World—has been the large-scale water project. The theme of hydraulic civilizations has long been a favorite of geographers and their fellow travelers. Yet the leading model for modern capitalist water development is undoubtedly the American West, especially California (Worster 1985). The irrationalities of this model have been sharply criticized, and shown to be the result of the brazen exercise of power by the growers and developers whom it most benefits (Leveen 1979; Storper and Walker 1984; Walker and Williams 1982; Westcoat 1984).

The study of nature transformed by human activity in what may be loosely called "rural environs" has led to a reconsideration of the hoary field of agricultural geography. Geographers have joined an assemblage of left theorists from disparate disciplines in tackling the nature of agricultural systems in the advanced capitalist countries, particularly sociologists such as Friedmann (1982), Buttel and Newby (1980), and Friedland (1982). Fitzsimmons (1986) draws on the ideas of the new industrial geography to analyze the structure of farm production in California, while Vail (1982), Pudup and Watts (1987), and Vogeler (1981) focus upon the crucial distinction between capitalist and petty commodity production in even the most advanced farm sectors. Munton and his coworkers in the U.K. are pursuing similar themes, and the theoretical issues are well aired in Marsden et al. (1986). A few left geographers have looked into primary-resource sectors other than agriculture, exploring the relations between economic conditions and the exploitation of both land and the people involved in mining and forestry (Bradbury 1982, 1984; Warf 1988). Hecht's (1988) work on Amazonia, in a very different context, bears mentioning again for its analysis of the political economic origins of resource despoilation.

A quite different perspective emerges from those writers concerned with the immediate consumptive use of nature in parks, nature preserves, and "naturalized" landscapes formerly in other uses. Such areas reveal the most stark contrasts between an apparent state of nature and the often quite exaggerated social machinations and meanings that protect and illuminate it in the properly defined ways. Because societies so generously read their hopes and fears into the verdant landscape of symbols that vital "Nature" presents, the disposition of natural lands cannot be left to chance, nor can their interpretation. Indeed, naturalized landscapes of the most varied kinds, from English gardens to National Parks, have been created by different societies for quite specific reasons, such as luxury consumption and national identity. Raymond Williams (1973) and John Berger (1973) have undoubtedly provided much of the inspiration for this line of thinking, which has been developed in geography by Olwig and Olwig (1979), Cosgrove (1984), Olwig (1980, 1984), and Heiman (1989). The left thus turns cultural geography's traditional obsession with landscapes on its head, giving political and social substance to the very meaning and definition of that cultural product, "nature's ideological landscape" (Olwig 1984).

While the fragmentation of environmental and resource geography on the left is apparent, Smith (1984) makes a sophisticated effort to integrate the field at the highest levels of historical materialist concepts, in terms of "the social production of nature"—to which Fitzsimmons (1989) has recently appealed once again. In a similar manner, Sayer (1979) injects a useful element of philosophical clarification, in a realist vein, into the debate on the relation between people and nature. On the more active side of left geography's engagement with environmental issues has been the frankly political interest expressed through academic writings. This is apparent with respect to specific issues of the moment, such as hazardous-waste removal (Heiman 1988a), petrochemical plant location (Walker, Storper, and Widess 1979), and Amazonian forest clearance (Hecht 1988). It is taken up more broadly in the reflective essays on the Green movement by Redclift (1984) and the U.S. environmental movement by Fitzsimmons and Gottlieb (1988).

SOCIALIST-FEMINIST GEOGRAPHY

Geographic feminist analysis entered disciplinary discourse at about the same time, motivated by many of the same social issues as historical materialism. Realizing that conventional concepts and methods were inadequate to exposing the condition—and oppression—of women, feminists developed an independent theoretical base that saw gender as a socially constructed category which changed over time and varied by place (for further discussion see Little, Peake, and Richardson 1989; Mackenzie 1989). This led many feminists to adopt (and subsequently adapt) the historical materialist framework for examining the overall development of gender relations.

Much of the initial work on the "geography of women" concentrated on empirical documentation of the spatial constraints facing women, and this remains the focus of mainstream research (for a review see Zelinsky, Monk, and Hanson 1982; see also Stimpson et al. 1981 and Wekerle, Peterson, and Morley 1980). Feminist geographers have contributed heavily to the effort to make women's plight more visible (Bowlby, Foord, and MacKenzie 1981; Hanson and Monk 1982; Women and Geography Study Group 1984). Studies of "women's place" in contemporary society have ranged from the home, to the labor market, to city life in general (e.g., Hayden 1984; Christopherson 1982, 1983; Cooke 1984; England 1986; Nelson 1986; MacKenzie 1987, 1989; Little, Peake, and Richardson 1989). Increasingly, attention has turned to the active role of women in altering gender relations and social practices in response to such things as neighborhood transformation (Holcomb 1981; Brownill 1984; Rose 1984; Breitbart 1985), running single-parent households (Klodawsky and Spector 1985; Klodawsky and Mackenzie 1987), labor-force segmentation (Christopherson 1988), and industrial restructuring (Massey 1984; Mackenzie 1986; Murgatroyd et al. 1985) (see generally Andrew and Moore-Milroy 1989; Bowlby et al. 1989).

This work suggests, however, that integrating gender relations into geographic analysis is not just an empirical question of disclosing the spatial element in women's oppression (McDowell 1983). It requires modification of geographic historical materialism at the most fundamental level. A lively debate has been taking place over the force of patriarchy as opposed to class in history and geography (Foord and Gregson 1986; McDowell 1986; Gier and Walton 1987; Knopp and Lauria 1987; Johnson 1987). Some theorize patriarchal relations independently of class, perhaps as a separate mode of production, while others subordinate patriarchy to capitalist class relations. In either case, it is largely agreed that feminist historical materialism must see gender as constituted simultaneously with class and the development of social production (Gregson and Foord 1987; Lerner 1986). This necessitates a shift in focus to the intersection of production and the reproduction of biological and social beings, and a richer concept of "human nature" as creative, androgynous, and changing. It also brings sexuality, and all it implies for the psychic life of individuals, profoundly into the picture (Hartsock 1987). If a single theme must be extracted from this for the geographer, it is that gender is a fundamental parameter of the appropriation, creation, and alteration of environment, physical or human (Breitbart, Foord, and MacKenzie 1984). Both men and women act on the world, and do so in gender-structured ways; as they do, they change themselves, altering their ways of working, of gendering, of loving, and oppressing one another. These alterations in the patterns of women's and men's activity are a profound and prognostic source of environmental change.

REGIONAL GEOGRAPHY AND LOCALITY STUDIES

The 1980s have witnessed a resurgence of interest in regional geography on the left. Long associated with synthetic description, environmental determinism, and cultural modes of explanation, regional geography was part of the corpus of traditional geography rejected by the spatial-science school in the 1960s. The left went even farther, in the 1970s, laying a pox on both houses for their apparent "fetish of space," or attribution of exaggerated causal powers to distance and place in social affairs (Castells 1977). A lone voice for a socialist reading of spatial relations was Henri Lefebvre (1975, 1987), who influenced Harvey (1973), but was largely overlooked in the rush to learn what other disciplines had to say about social theory (Massey 1984, 52). Soja (1989) would attribute this to the prevailing failure of modern social science to handle the spatial dimension in human affairs, including the impoverished theoretical state of traditional geography that made it incapable of reversing the general tide. Indeed, Soja (1980) was among the first to raise a voice against the relegation of space and place to mere containers or playing fields for social processes, resonating with the roughly contemporaneous call of Gregory (1978, 119) in Britain. While the reaction against spatial science and synthetic description led radicals to underestimate the actual importance of geography, it nonetheless produced a generation more thoroughly steeped in philosophy, the social sciences, and historical materialism than had ever been present in the discipline before. The left was poised, at the beginning of the 1980s, to look out upon the landscape of human affairs with new eyes.

The "reconstruction of regional geography" (Thrift 1983) has come from several directions and claimed various sources of inspiration. One that we have already touched on is the powerful stream of the new industrial geography. After initially calling for a move from regional to industrial studies to understanding the declining fortunes of many areas, Massey came to the conclusion that the particulars of regional history and conflict could not be omitted from a theory of spatial divisions of labor, and that, to put it crisply, "geography matters" (Massey 1979, 1985; Massey and Allen 1984). Industry-restructuring theory came to serve as principal inspiration for regional and local studies (e.g., Hudson et al. 1983; O'Keefe 1984; Peet 1987; Graham et al. 1988; Warde 1988; Warf 1988). Harvey was similarly drawn to the peculiarities of capitalist urbanization in Second Empire Paris (Harvey 1985a). Scott's inquiries into territorial clustering of industry led him toward an increasingly spatialized theory of industrial organization (Scott 1988a, 1988b; also Storper and Christopherson 1987), which has been followed up by Storper and Walker in a theory of territorial industrialization in which the essence of capitalism is its ever-shifting regional foundations (Storper and Scott 1988; Walker 1988b; Storper and Walker 1989). Soja has, of course, been in the thick of this renewed urban-industrial geography, bringing his acute powers of kaleidoscopic spatial analysis to bear on the problem (Soja, Morales, and Wolff 1983; Soja 1986, 1989). Clark and Gertler, among others, have also been reasserting the importance of regionalism in their studies of labor and capital relations in a spatially fragmented division of labor (Clark 1981; Clark, Gertler, and Whiteman 1986; Clark 1989b; Gertler 1984a, 1984b).

A different cut on the regional-development problem has been taken by those working through the optic of time geography. Prominent among this group are Gregory, Pred, and Thrift (Gregory 1981, 1982, 1986; Pred 1981, 1984a, 1986; Thrift 1981, 1983). Having soon outgrown the restrictive initial formulation of Hagerstrand, these theorists took strongly to the work of sociologist Anthony Giddens (1979, 1981, 1984) and his structuration theory of the dialectic of human agency and social structure; Giddens reciprocated by rediscovering the importance of space in human affairs, thanks to association with Gregory and Pred, and a reading of Hagerstrand. Giddens' method appeared as a freshet to those most chary of the structuralist tendencies in Marxist geography and most keen on restoring the human face of daily life to grand theory. Yet Giddens has not been able to transcend Marxism as the principle framework of left inquiry, and the foundation of this group's work remains implicitly, if somewhat uneasily, historical materialist (see also Storper 1985b; Gregson 1986; Watts 1987). Their closest antecedents may be found in the historical work of the British Marxists, particularly E. P. Thompson, and the French *Annales* school, particularly Fernand Braudel. Thompson's populist and activist approach to class formation and Braudel's rich investigation of everyday material life bookend the concerns of the new regionalists in their historical studies of the rise of capitalism. Indeed, the former's spirited engagement with things peculiarly *English* in capitalist development, and the latter's close attention to the geographic ebb and flow of life in Europe—thanks to Vidal's impact on French thought—make them apt models for a geography integrated with other social sciences, and yet aware of its own potential contribution. (For an assessment of Braudel, see Baker 1984; Pred 1984b; Kirby 1986).

Certain geographers and sociologists have vigorously advanced this spirit of collaboration without capitulation of geography to more muscular disciplines (Gregory and Urry 1985). This venture has been joined most strikingly under the rubric of "locality studies" by the CURS initiative (Murgatroyd et al. 1985; Urry 1986b; Boddy, Lovering, and Bassett 1987; Cooke 1988). This project, conceived by Doreen Massey as a way of advancing the study of contemporary regional transformation in Britain (cf. Massey 1984; Cooke 1986), has been criticized for a lack of integrative theory

(Smith 1987a; Harvey and Scott 1989); but the danger may be more one of large, collective research projects losing their focus, than of erroneous conception or political drift. The best of the locality studies can be very good indeed (Warde 1988; Beauregard 1988). On the American side, especially the West Coast, regional transformation has been conceived much more in the dynamic terms of capitalist growth than in industrial restructuring, urban decline, and job loss (Cox and Mair 1988; Scott 1988b; Storper and Walker 1989). This would also be true of Harvey's (1985a) Paris study, in which that city's local transformation is seen as emblematic of the highest form of capitalist development at the time, and the locus of social conflicts, the resolution of which set the political agenda for whole nations during the next half century or more.

An important complement to the prevailing economic and sociological approach to regional transformation and locality studies is a revivified interest in the cultural dimension of urbanization and regional development. This work rejects the uncritical approach of traditional regional geographers toward cultural regions and landscape interpretation by scrutinizing the political-economic foundations of culture and ways of seeing implicated in the conventional concept of landscape itself (Blaut 1979; Cosgrove 1983; Cosgrove and Daniels 1987; Daniels 1988). Cosgrove (1984) examines the historical sweep of landscape interpretation from renaissance Europe to industrial America. Olwig (1984) focuses upon regional landscapes widely believed to be natural, and how that condition is invented during periods of dramatic regional change. Pudup (1987) takes up the "problem of Appalachia" in the dominant American vision and shows the discordance with reality, while emphasizing the distinctive economic and social history of that singular region.

The same is being done for the urban landscape. Harvey (1985b) delivers a vibrant political analysis of conflict over Paris and its monuments, one which resonates strongly with the work of nongeographers Berman (1982) and T. J. Clark (1985) on images of nineteenth-century urban transformations under the onslaught of a rising capitalism. Recent work has emphasized the cultures of urban consumption, now typically clothed in the decorative symbols of postmodern architecture (Harvey 1985b, 1987; Davis 1985; Knox 1987; Soja 1989). However, cheek by jowl with such consumer delights as gallerias, luxury plazas, and new museums are immigrant enclaves that are home to vigorous small-business activity and intensive labor exploitation (Zukin 1982; Sassen 1988). Geographers of an only mildly left stamp have been making some trenchant observations about the stamp of race and racist practices on such immigrant and "deviant" neighborhoods in North American cities (Anderson 1987, 1988; Godfrey 1988).

Explorations of culturalist themes often run parallel between Marxist and humanist geographers, and indications of a general revival of cultural geography are afoot (Cosgrove and Jackson 1987; *Society and Space* 1988). While one should not underestimate the differences between idealist and materialist points of entry into the analysis of culture formation, some nimble theorists have been able to work both sides of the fence, to good effect; but the material-cultural hermeneutic remains a poorly understood and greyish area of analysis (Sayer 1988a).

As a result of these various lines of work, there is now something recognizable as a "new" regional geography to which scholars can lay claim and on which they can formulate their research agendas (Pudup 1988; Gilbert 1988). Virtually everyone in the left of geography has been looking closer to the ground for answers to persistent puzzles of difference in time and space, despite seemingly unifying systems of world markets, capitalist production, central states, national culture, or modern media. All have been concerned, despite important disagreements, with the interaction of structure and agency, place and time, the specific and the abstract. The program of the new regionalism appears to be capturing the middle ground between the immediate and the global as a significant level of social causality.

The first keywords are "regional transformation": regions are formed and transformed by human activity; indeed, they are defined by social function and integration, not physiographic boundaries (Massey 1984, 108). The second is "locale" (Thrift 1983, 40) or "territory" (Scott and Storper 1986): the mesolevel formations at which critical social processes take place—not region as a container for action, but as a force in the production and reproduction of industries, genders, or classes. And the last is, somewhat presciptively, the local becomes the general: the way in which locally incubated changes in social relations or forces of production become decisive for nations, capitalism, or the world at large (Storper and Walker 1989). But enthusiasms for this new trend must be weighed against the loose coalition of interests pursuing it, and their often-disturbing lack of clarity on the scale of analysis or ways in which local studies illuminate critical causal forces in a realist manner; the result is often a false sense of concreteness in which a wealth of empirical detail elides a poverty of theory (Sayer 1988a; Cox and Mair 1989).

POLITICAL GEOGRAPHY AND THE STATE

Geographic questions about the political economy of the capitalist state have changed over the last five years, reflecting broader theoretical debates as well as changing economic and political realities. In place of descriptive or deterministic explanations of the state apparatus, there have emerged conceptions of the state as a complex set of social relations that shape peoples' political experiences and practices in places. The state is seen to be reproduced and contested through an array of sites that span the workplace, residential community, political arena, and family.

Two broad areas of research on the state can be identified. The first is "state-centered" and takes up questions of the nature of political relations within the state apparatus. The second is "society-centered" and deals with interactions between state institutions and the broader social politics within which they are located, including the role of the state in reproducing and transforming social relations in capitalism. Much of this geographic research has concentrated on North American and European states, and focuses upon either the national or local scale, rather than attempting comparisons of state forms at different scales. The local state has received particularly close attention, probably reflecting the rise of interest in the state within subspecialities other than political geography, such as urban studies; and it must be said that the work of North American researchers brings in local government in a more vigorous way, on the whole, than similar British work on the left (Fincher 1987).

State-Centered Research

In the past, state analysts have been concerned with describing the functions performed by different branches of the state apparatus (e.g., local, regional, central), and classifying expenditures as productive or legitimating (Dear 1981). These concerns were encouraged by O'Conner's analysis (1973) of the capitalist state, and by British researchers such as Cockburn (1977). But new themes are emerging, as analysts grapple with a range of theories of the state and changing political circumstances. There is an interest in the politics of interaction between central and local government, and the social relations of bureaucracy within the state. These concerns have arisen in part in reference to the local socialist governments of mid-1980s Britain, like the now-fallen Greater London Council (Boddy and Fudge 1984). They have also been evident in new research on the North American state, which seeks to understand local political relations (Johnston 1983; Cox 1986; Pinch 1985; Lauria 1986; Kirby 1989). G. Clark (1985), for example, discusses the role of judges and courts in defining local government autonomy in the U. S. and Canada.

These new directions raise important issues for further research. One is how different allocations of functions between state apparatuses influence degrees of local government autonomy. Patterns of specialization within the state may permit both variation in local government arrangements, and in possibilities of contesting dominant relations within the state (Kirby 1989). A second issue warranting further attention is how the social practices of government workers, particularly bureaucratic policies, procedures, routines, and ideologies, help to influence state actions (Walker and Williams 1982). Under what conditions may state workers actively contest dominant managerial practices, opening up possibilities for struggles over the relations between state and society?

Another significant line of investigation is into the language or discourse of the state (Clark and Dear 1984; Clark 1988a, 1988b). The state is not only being viewed as an institution with determinate rules and impacts, but as a participant in the ongoing construction of social life. Drawing on poststructuralist interpretive studies of texts and discourse, Clark (1989a) is pursuing research on the narratives that set the terms of debate over local political issues and the proper province of state action. In a similar vein, Johnston (1983) inquires as to texts of political managers and the political organization of space, and Kirby (1988) takes up the interpretative dimension of place and local context in political life.

State-Society Relations

Urban geographers have pursued this slant on the political economy of the state, often focusing upon how changes in state form and policies influence the daily lives of urban residents and workers. There are a number of themes in this work, all proceeding apace and generating new insights. Here we can identify three.

The role of the state in the process of creating the city continues be a central theme. Harvey's recent essays (1985a, 1985b) on urban development and planning, including a detailed historical analysis of Haussman's transformation of nineteenth-century Paris, indicate how state intervention in the built environment is intimately connected to shifting patterns of accumulation, class formation, and political struggle. Analyses of postwar housing policies in the U.S. and Canada have helped to illuminate the role of the state in sustaining an intensive regime of accumulation and mass consumption of suburban housing (Walker 1981; Florida and Feldman 1988; Chouinard 1988a). Work on territorial politics and the formation of local growth coalitions provides important insights into how local social relations shape struggles over state intervention in the built environment and the location of economic activity (Kirby 1982; Kirby, Knox, and Pinch 1984; Kirby 1985; Logan and Molotch 1986; Leitner n.d.; Mair

1987; Cox and Mair 1988). Recent work on gentrification explores the state's role in facilitating the physical and social transformation of urban neighborhoods (Lauria 1982; Smith and Williams 1986).

Social-service provision and the decline of the welfare state is another important area of research. Dear and Wolch (1987) describe and explain the origins of policies of deindustrialization, and their social and spatial effects in contemporary cities. Wolch has also drawn attention to the significance of voluntary workers and organizations in the implementation of social policy in the U.S. and Britain (Geiger and Wolch 1986; Wolch 1987). Chouinard (1988a, 1988b, 1988c) links postwar changes in Canada's assisted housing policies to changing conditions of intensive accumulation and social struggle. Focusing upon cooperative housing, she argues that class capacities to resist privatization and recommodification of housing assistance have been limited by the policies and procedures of the state, and by the ways in which people have contested state regulation of cooperative housing nationally, and within localities.

The impacts of urban social movements on the politics of localities have also received considerable attention, following on the powerful (though post-Marxist) work of Castells (1983) (for reviews see Cox 1984a, 1986, 1988). Cox has long written about urban social movements and neighborhood conflicts, struggles that help to form the political relations of North American cities and their governments (Cox 1978, 1982; Cox and McCarty 1982). Cox has tried to situate such turf politics within the context of postwar urban development (Cox 1984b; Cox and Mair 1988). In their most recent work, Cox and Mair (1988) argue that possibilities for effective struggles to legitimize competitive growth-coalition strategies have increased with economic uncertainty and the penetration of capitalist commodity relations into daily life. Lauria (1986) shows how struggles over responses to plant closings help to shape the apparatuses of the state in localities. Knopp (1987) and Knopp and Lauria (1985) have investigated the role of gay movements in local politics and urban renaissance. Slater (1985) treats the relation of new social movements to state power in Latin America. Marston (1988) looks at the political mobilization of the Irish in nineteenth-century American industrialization.

Work on the general theory of the state and politics also continues, as the political economy tradition responds to changes in world capitalism. Peter Taylor's efforts (1982, 1985, 1987), deeply influenced by Wallerstein's world-systems approach, are notable in this regard, especially his sensitivity to the political problem of geographical scale. North American state theory includes Chouinard and Fincher's (1988) attempt to specify concrete terrains of struggle over state development, where the social relations of the state are explained in reference to precise spatial, economic, and political context. Current social processes of state formation are likely to continue, and this research will surely include closer attention to the social construction of political identities or subjectivities and the role of political experiences in state development.

CONCLUSION

Left geography has clearly cut a broad swath across the discipline. No movement of this breadth can be expected to be unified, and our purpose has been to embrace many divergent tendencies within a wide spectrum of the left, rather than to exclude anyone on the grounds of adherence to Marxism or any other core theory. A few of the authors mentioned might chafe a bit at their inclusion with others of either more radical or more liberal stripe, but the purpose has been to indicate the range of activity and intelligence being applied to virtually the whole field of human geography. This was meant to be a celebratory essay, not a critical position paper, since the internal disagreements on the left have been well aired recently in the pages of *Society and Space* and *Antipode*.

Marxism has for long provided the fulcrum of opposition to conventional theory in geography, but there has been a movement away from Marxism in the 1980s for political and intellectual reasons. The challenges to orthodox Marxism as the flagship of progressive social theory are well known, and appear in various guises under the topical headings considered here (e.g., Smith 1987a versus Cooke 1987; Corbridge 1986 versus Watts 1988). Some geographers on the left, particularly those arrayed around the journal *Society and Space,* have now entered the lists of the new "post-Marxists," who draw succor from such varied sources as Foucault and the poststructuralists, Lyotard and the postmodernists, and feminism (see *Society and Space* 1987). Some of this represents a healthy diversification and extension of left inquiry, and a justifiable suspicion of overblown claims to thoroughgoing knowledge of how society works, where it is headed, and how it ought to be changed (Graham 1988). But very little Marxist work in geography ever adopted such a stance (Beauregard 1988), and much of the criticism lapses into longstanding idealist, Weberian, or individualist responses to Marxist theory that cannot stand up to close scrutiny (e.g., Duncan and Ley 1982; Saunders and Williams 1986; Dear 1986). As a result, historical materialism still holds sway in left geography.

Most of the current disputation involves reasoned quarrels generated by real difficulties of social theory and political strategy that need not portend major ideological schisms within our ranks (*Antipode* 1989). It is important that such concerns not be unfairly inflated into theory-bashing of an unnecessarily contentious sort (Watts 1988). Unity has its virtues when one contemplates the relatively limited numbers of American geographers who remain avowedly left in purpose and outlook, and the enormity of the task before us in expanding and consolidating the considerable achievements of left geography over the last two decades. In that spirit, this chapter is dedicated to John Bradbury, a good socialist and a fine person whose unexpected death leaves us visibly diminished.

REFERENCES

Ambrose, P., and Colenutt, R. 1975. *The property machine*. Harmondsworth: Penguin.

Anderson, K. 1987. Chinatown as an idea: The power of place and institutional practice in the making of a racial category. *Annals of the Association of American Geographers* 77:580–98.

———. 1988. Cultural hegemony and the race-definition process in Chinatown, Vancouver: 1880–1980. *Environment and Planning D: Society and Space* 6(2):127–50.

Andrew, C., and Moore-Milroy, B. 1989. *Life spaces: Gender, household, employment*. Vancouver: University of British Columbia Press. In press.

Angel, D. 1987. Agglomeration and local labor markets in the U.S. semiconductor industry. Unpublished paper, Clark University, Worcester, MA.

Antipode. 1989. Special issue: What's left to do? In press, June.

Badcock, B. 1984. *Unfairly structured cities*. Oxford: Blackwell.

Baker, A. 1984. Reflections on the relations of historical geography and the *Annales* school of history. In *Explorations in historical geography,* eds. A. Baker and D. Gregory, pp. 1–27. Cambridge: Cambridge University Press.

Barnbrock, J. 1974. Prolegomenon to a methodological debate on location theory: The case of Von Thünen. *Antipode* 6(1):59–65.

———. 1976. Marx's model of accumulation. *Antipode* 8(2):12–23.

Barnes, T. 1984. Theories of agricultural rent within the surplus approach. *International Regional Science Review* 9:125–40.

———. 1988. Scarcity and agricultural land rent theory in the light of the capital controversy. *Antipode* 20. In press.

———, and Sheppard, E. 1984. Technical choice and reswitching in space economies. *Regional Science and Urban Economics* 14:345–62.

Bassett, T. 1986. Fulani herd movements. *Geographical Review* 76(3):233–48.

———. 1988. The development of cotton in Northern Ivory Coast, 1910–1965. *Journal of African History* 29(2):267–84.

Beauregard, R. 1988. In the absence of practice: The locality research debate. *Antipode* 20(1):52–59.

Belec, J.; Holmes, J.; and Rutherford, T. 1987. The rise of Fordism and the transformation of consumption norms: Mass consumption and housing in Canada, 1930–45. In *Social class and housing tenure,* eds. R. Harris and G. Pratt, pp. 187–237. Gavle, Sweden: National Swedish Institute for Building Research.

Berger, J. 1973. *Ways of seeing*. New York: Viking.

Berman, M. 1982. *All that is solid melts into air*. New York: Simon & Schuster.

Blaikie, P. 1985. *The political economy of soil erosion in developing countries*. Harlow: Longman.

———, and Brookfield, H., eds. 1986. *Land degradation and society*. London: Methuen.

Blaut, J. 1979. A radical critique of cultural geography. *Antipode* 11:25–29.

———. 1987. *The national question*. Atlantic Highlands, NJ: Zed Press.

Boddy, M. 1980. *The building societies*. London: Macmillan.

———, and Fudge, C., eds. 1984. *Local socialism?* London: Macmillan.

———; Lovering, J.; and Bassett, K. 1987. *Sunbelt city? Economic change in Britain's M4 growth corridor*. Oxford: Oxford University Press.

Bowlby, S. et al. 1989. The geography of gender. In *New models in geography,* eds. R. Peet and N. Thrift. New York: Allen & Unwin. In press.

———; Foord, J.; and Mackenzie, S. 1981. Feminism and geography. *Area* 13(4):711–16.

Bradbury, J. 1982. Regional and international restructuring of the iron ore industry. *Tijdschrift voor Economische en Sociale Geografie* 5:295–306.

———. 1984. The impact of industrial cycles in the mining sector: The case of the Quebec-Labrador region in Canada. *International Journal of Urban and Regional Research* 8(3):311–31.

———. 1985. Regional and industrial restructuring processes in the new international division of labour. *Progress in Human Geography* 9:38–63.

———. 1987. Technical change and the restructuring of the North American steel industry. In *Technical change and industrial policy,* eds. K. Chapman and G. Humphrys, pp. 157–73. Oxford: Blackwell.

Breitbart, M. 1985. Terrains of protest: Women in urban struggle. Paper presented to the Association of American Geographers, Detroit, April 26.

———; Foord, J.; and Mackenzie, S., eds. 1984. Women and the environment. *Antipode* (Special issue) 6(3).

Brenner, R. 1977. The origins of capitalist development: A critique of neo-Smithian Marxism. *New Left Review* 104:25–92.

———. 1986. The social bases of economic growth. In *Analytical Marxism,* ed. J. Roemer, pp. 23–53. New York: Cambridge University Press.

Bromley, R. 1980. *Labor and the casual poor*. New York: Pergamon.

Browett, J. 1986. Industrialization in the global periphery. *Environment and Planning D: Society and Space* 4(4):401–19.

Brownill, S. 1984. From critique to intervention: Socialist-feminist perspectives on urbanization. *Antipode* 16(3):21–34.

Burgess, R. 1982. The politics of urban residence in Latin America. *International Journal of Urban and Regional Research* 6:467–79.

Buttel, F., and Newby, H., eds. 1980. *The rural sociology of the advanced societies*. Montclair, NJ: Allanheld, Osmun.

Carney, J. 1986. Struggles over land and crop rights in the Gambia: Conflict and accumulation in the household. In *Land, women and agriculture in Africa,* ed. J. Davison. Boulder, CO: Westview Press.

Castells, M. 1977. *The urban question: A Marxist approach.* Cambridge: MIT Press.

———. 1983. *The city and the grassroots.* Berkeley: University of California Press.

Chouinard, V. 1988a. Changing forms of the capitalist state: Assisted housing policies in postwar Canada. Unpublished paper, Geography Department, McMaster University, Hamilton, Canada.

———. 1988b. Explaining local experiences of state formation: The case of cooperative housing in Toronto. *Environment and Planning D: Society and Space.* In press.

———. 1988c. Social reproduction and housing alternatives: Cooperative housing in postwar Canada. In *The power of geography: How territory shapes social life,* eds. M. Dear and J. R. Wolch. Allen & Unwin. In press.

———, and Fincher, R. 1988. State formation in capitalist societies: A conjunctural approach. *Antipode* 19(3):329–53.

Christopherson, S. 1982. Family and class in a new industrial city. Ph.D. diss., University of California, Berkeley.

———. 1983. Households and class formation. *Environment and Planning D: Society and Space* 1(3):323–38.

———. 1988. Labor flexibility and new forms of labor segmentation. Paper presented to the Institute of British Geographers, Loughborough, January 8.

Clark, G. 1981. The employment relation and the spatial division of labor. *Annals of the Association of American Geographers* 71:412–24.

———. 1985. *Judges and the cities: Interpreting local autonomy.* Chicago: University of Chicago Press.

———. 1988a. Propaganda and transcendental reason: The National Labor Relations Board's regulation of the language of union representation campaigns. Working Paper No. 88-10. School of Urban and Public Affairs, Carnegie-Mellon University, Pittsburgh.

———. 1988b. A question of integrity: The national labor relations board, collective bargaining and the relocation of work. *Political Geography Quarterly* 6. In press.

———. 1989a. The anxiety of becoming. *Antipode.* In press.

———. 1989b. *Unions and communities under siege: American communities and the crisis of organized labor.* Cambridge: Cambridge University Press.

———, and Dear, M. 1984. *State apparatus: Structures and language of legitimacy.* Boston: Allen & Unwin.

———; Gertler, M.; and Whiteman, J. 1986. *Regional dynamics.* London: Allen & Unwin.

———, and Johnston, K. 1987. The geography of US union elections 1: The crisis of U.S. unions and a critical review of the literature. *Environment and Planning A* 19:33–57.

Clark, T. 1985. *The painting of modern life.* New York: Alfred Knopf.

Coase, R. 1937. The nature of the firm. *Economica* 4:386–405.

Cockburn, C. 1977. *The local state.* London: Pluto Press.

Cooke, P. 1984. Region, class and gender: A European comparison. *Progress in Planning* 22(2):85–146.

———. 1986. The changing urban and regional system in the United Kingdom. *Regional Studies* 20(3):243–51.

———. 1987. Clinical inference and geographic theory. *Antipode* 19(1):407–16.

———, ed. 1988. *Localities.* London: Hutchinson.

Corbridge, S. 1986. *Capitalist world development.* Totowa, NJ: Rowman & Littlefield.

———, ed. 1987. Special issue on world debt. *Geoforum* 9.

Cosgrove, D. 1983. Towards a radical cultural geography. *Antipode* 15:1–11.

———. 1984. *Social formation and symbolic landscape.* London: Croom Helm.

———, and Daniels, S., eds. 1987. *The iconography of landscape.* Cambridge: Cambridge University Press.

———, and Jackson, P. 1987. New directions in cultural geography. *Area* 19(2):95–101.

Cox, K. 1978. Local interests and urban political processes in market societies. In *Urbanization and conflict in market societies,* ed. K. Cox, pp. 94–112. Chicago: Maaroufa Press.

———. 1982. Housing tenure and neighborhood activism. *Urban Affairs Quarterly* 18(1):107–29.

———. 1984a. Neighborhood conflicts and urban social movements: Questions of historicity, class and social change. *Urban Geography* 5(4):343–55.

———. 1984b. Social change, turf politics and concepts of turf politics. In *Public provision and urban development,* eds. A. Kirby, P. Knox, and S. Pinch, pp. 283–315. New York: St. Martin's Press.

———. 1984c. Space and the urban question: A review essay of *Social theory and the urban question,* by P. Saunders. *Political Geography Quarterly* 3(1):77–84.

———. 1986. Urban social movements and neighborhood conflicts: Questions of space. *Urban Geography* 7(6):536–46.

———. 1988. Urban social movements and neighborhood conflicts: Mobilization and structuration. *Urban Geography* 9(4):412–25.

———, and McCarty, J. 1982. Neighborhood activism as a politics of turf. In *Conflict, politics and the urban scene: Case studies in urban political geography,* eds. K. Cox and R. Johnston, pp. 196–219. London: Longman.

———, and Mair, A. 1988. Locality and community in the politics of local economic development. *Annals of the Association of American Geographers* 78(2):307–25.

———, and ———. 1989. Levels of abstraction in locality studies. *Antipode*. In press.

Crush, J. 1982. The southern African regional formation: A geographical perspective. *Tijdschrift voor Economische en Sociale Geografie* 73:200–12.

———. 1986. Swazi migrant workers and the Witwatersrand gold mines, 1886-1920. *Journal of Historical Geography* 12(1):27–40.

Daniels, S. 1988. Marxism, culture and the duplicity of landscape. In *New models in geography,* eds. R. Peet and N. Thrift. In press.

Davis, M. 1985. Urban renaissance and the spirit of post-Modernism. *New Left Review* 151:106–13.

Dear, M. 1981. A theory of the local state. In *Political studies from spatial perspectives: Anglo-American essays on political geography,* eds. A. Burnett and P. J. Taylor, pp. 183–200. New York: Wiley.

———. 1986. Postmodernism and planning. *Environment and Planning D: Society and Space* 4(3):367–84.

———, and Wolch, J. 1987. *Landscapes of despair: From deinstitutionalization to homelessness.* Princeton: Princeton University Press.

deJanvry, A. 1981. *The agrarian question and reformism in Latin America.* Baltimore: Johns Hopkins University Press.

Duncan, J., and Ley, D. 1982. Structural Marxism and human geography: A critical assessment. *Annals of the Association of American Geographers* 72/1:30–59.

England, K. 1986. Spatial variations in the employment of women. Paper presented to the Association of American Geographers, Minneapolis, April 15.

Fainstein, S. et al. 1983. *Restructuring the city.* New York: Longman.

Farhi, A. 1973. Urban economic growth and conflicts: A theoretical approach. *Papers of the Regional Science Association* 31:95–124.

Feagin, J. 1983. *The urban real estate game.* Englewood Cliffs, NJ: Prentice-Hall.

Fincher, R. 1987. Space, class and political processes: The social relations of the local state. *Progress in Human Geography* 11(4):496–516.

Fitzsimmons, M. 1986. The new industrial agriculture: The regional integration of specialty crop production. *Economic Geography* 622(4):334–52.

———. 1987. Review of *Controlling chemicals,* by Brickman et al. *Annals of the Association of American Geographers* 77(4):675–78.

———. 1989. The production of nature. *Antipode*. In press.

———, and Gottlieb, R. 1988. A new environmental politics. *The Year Left* 3:114–32.

Florida, R. 1986. The political economy of financial deregulation and the reorganization of housing finance in the United States. *International Journal of Urban and Regional Research* 10(2):207–31.

———, and Feldman, M. 1988. Housing in U.S. Fordism: The class accord and postwar spatial organization. *International Journal of Urban and Regional Research* 12(2):187–210.

———, and Kenney, M. 1989. *The breakthrough economy.* New York: Basic Books.

Foord, J., and Gregson, N. 1986. Patriarchy: Toward a reconceptualization. *Antipode* 18(2):186–211.

Foster-Carter, A. 1978. The modes of production controversy. *New Left Review* 107:47–77.

Frank, A. 1968. *Capitalism and underdevelopment in Latin America.* New York: Monthly Review Press.

Friedland, W. 1982. The end of rural society and the future of rural sociology. *Rural Sociology* 47:589–608.

Friedmann, H. 1982. The political economy of food. In *Marxist inquiries,* eds. M. Burawoy and T. Skocpol, pp. 248–86. Chicago: University of Chicago Press.

———. 1987. The family farm and international food regimes. In *Peasants and peasant societies,* ed. T. Shannin, pp. 247–59. Oxford: Blackwell.

Geiger, R. K., and Wolch, J. R. 1986. A shadow state? Voluntarism in metropolitan Los Angeles. *Environment and Planning D: Society and Space* 4(3):351–66.

Gertler, M. 1984a. The dynamics of regional capital accumulation. *Economic Geography* 60:150–74.

———. 1984b. Regional capital theory. *Progress in Human Geography* 8(1):50–81.

———. 1988. The limits to flexibility: Comments on the post-Fordist vision of production and its geography. *Transactions of the Institute of British Geographers.* In press.

Gibson, K. et al. 1986. *Toward a Marxist empirics.* Book manuscript.

Giddens, A. 1979. *Central problems in social theory.* Berkeley: University of California Press.

———. 1981. *A contemporary critique of historical materialism.* Berkeley: University of California Press.

———. 1984. *The constitution of society.* Berkeley: University of California Press.

Gier, J., and Walton, J. 1987. Some problems with reconceptualizing patriarchy. *Antipode* 19(1):54–58.

Gilbert, A. 1988. The new regional geography in English and French speaking countries. *Progress in Human Geography* 12(2):208–28.

Godfrey, B. 1988. *Neighborhoods in transition: The making of San Francisco's ethnic and nonconformist communities.* Berkeley: University of California Press.

Graham, J. 1988. Post-modernism and Marxism. *Antipode* 20(1):60–65.

——— et al. 1988. Restructuring in U.S. manufacturing: the decline of monopoly capitalism. *Annals of the Association of American Geographers* 78(3):473–91.

Gregory, D. 1978. *Science, ideology and human geography.* London: Hutchinson.

———. 1981. Human agency and human geography. *Transactions of the Institute of British Geographers* 6:1–18.

———. 1982. *Regional transformation and industrial revolution: A geography of the Yorkshire woolen industry.* London: Macmillan.

———. 1986. Presences and absences: Time-space relations and structuration theory. In *The critical theory of the advanced societies,* eds. D. Held and J. Thompson. Cambridge: Cambridge University Press.

———, and Urry, J., eds. 1985. *Social relations and spatial structures.* Oxford: Polity.

Gregson, N. 1986. On duality and dualism: The case of structuration and time geography. *Progress in Human Geography* 10:184–205.

———, and Foord, J. 1987. Patriarchy: Comments on critics. *Antipode* 19(3):371–75.

Grossman, L. 1984. *Peasants, subsistence ecology and development in Highland Papua New Guinea.* Oxford: Blackwell.

Haila, A. 1988. Land as a financial asset: The theory of urban rent as a mirror of economic transformation. *Antipode* 20(2):79–101.

Hamnett, C. 1973. Improvement grants as an indicator of gentrification in inner London. *Area* 5:252–61.

———. 1984. Gentrification and residential location theory: A review and assessment. In *Geography and the urban environment,* eds. D. Herbert and R. Johnston, pp. 283–319. Chichester: Wiley.

Hanson, S., and Monk, J. 1982. On not excluding half the human in human geography. *Professional Geographer* 34(1):11–23.

Harris, R. 1986. Homeownership and class in modern Canada. *International Journal of Urban and Regional Research* 10(1):67–86.

———, and Hamnett, C. 1987. The myth of the promised land: The social diffusion of home ownership in Britain and North America. *Annals of the Association of American Geographers* 77(2):173–90.

Hartsock, N. 1987. *Money, sex and power: Toward a feminist historical materialism.* Boston: Northeastern.

Harvey, D. 1973. *Social justice and the city.* London: Edward Arnold.

———. 1974a. Class-monopoly rent, finance capital and the urban revolution. *Regional Studies* 8:239–55.

———. 1974b. Population, resources and the ideology of science. *Economic Geography* 50(3):256–77.

———. 1975. Class structure in a capitalist society and the theory of residential differentiation. In *Processes in physical and human geography: Bristol essays,* eds. R. Peet, M. Chisholm, and P. Haggett, pp. 354–69. London: Heinemann.

———. 1977. Government policies, financial institutions and neighbourhood change in U.S. cities. In *Captive cities,* ed. M. Harloe, pp. 123–40. New York: Wiley.

———. 1978. Labor, capital and class struggle around the built environment in advanced capitalist societies. In *Urbanization and conflict in market societies,* ed. K. Cox, pp. 94–112. Chicago: Maaroufa Press.

———. 1982. *The limits to capital.* Oxford: Blackwell.

———. 1985a. *Consciousness and the urban experience.* Baltimore: Johns Hopkins University Press.

———. 1985b. *The urbanization of capital.* Baltimore: Johns Hopkins University Press.

———. 1987. Flexible accumulation through urbanization: Reflections on post-modernism in the American city. *Antipode* 19(3):260–86.

———, and Scott, A. 1989. The practice of human geography: Theory and empirical specificity in the transition from Fordism to flexible accumulation. In *Re-modelling geography,* ed. W. MacMillan. Oxford: Blackwell. In press.

Hayden, D. 1984. *Redesigning the American dream.* New York: Norton.

Hecht, S. 1985. Environment, development and politics: Capital accumulation and the livestock sector in eastern Amazonia. *World Development* 13(2):663–84.

———. 1988. Contemporary dynamics of Amazonian development: The politics of colonist attrition. Unpublished manuscript. School of Architecture and Urban Planning, UCLA.

Heiman, M. 1988a. Hazardous waste facility siting: From not in my backyard to not in anyone's backyard. Paper presented to the Association of Collegiate Schools of Planning, October 31.

———. 1988b. *The quiet evolution.* New York: SUNY Press.

———. 1989. Production confronts consumption: Landscape perception and social conflict in the Hudson Valley. *Environment and Planning D: Society and Space.* In press.

Hewitt, K., ed. 1983. *Interpretations of calamity.* Boston: Allen & Unwin.

Hoch, C. 1984. City limits: Municipal boundary formation and class segregation. In *Marxism and the metropolis,* 2d ed., eds. W. Tabb and L. Sawers, pp. 101–22. New York: Oxford University Press.

Hodgson, G. 1981. *Capitalism, value and exploitation.* Oxford: Oxford University Press.

Holcomb, B. 1981. Women's roles in distressing and revitalizing cities. *Transition* 11(2):1–6.

Holmes, J. 1986. The organizational and locational structure of production subcontracting. In *Work, production, territory,* eds. A. Scott and M. Storper, pp. 80–106. Boston: Allen & Unwin.

———. 1988. Industrial restructuring in a period of crisis: An analysis of the Canadian automobile industry, 1973–83. *Antipode* 20(1):19–51.

Hudson, R. et al. 1983. *Redundant spaces in cities and regions? Studies in industrial decline and social change.* New York: Academic Press.

Huriot, J-M. 1981. Rente fonciere et modele de production. *Environment and Planning A* 123:1125–49.

Hymer, S. 1972. The multinational corporation and the law of uneven development. In *Economics and world order,* ed. J. Bhagwai, pp. 113–40. New York: Free Press.

Johnson, K. 1982. Peasant struggles in contemporary Mexico. *Antipode* 14(3):39–50.

Johnson, L. 1987. (Un)Realist perspective: Patriarchy and feminist challenges in geography. *Antipode* 19(2):210–15.

Johnston, R. 1983. Texts, actors and higher managers: Judges, bureaucrats and the political organization of space. *Political Geography Quarterly* 2(1):4–19.

Kirby, A. 1982. *The politics of location.* New York: Methuen.

———. 1985. Nine fallacies of local economic change. *Urban Affairs Quarterly* 21(2):207–20.

———. 1986. Survey 10: Le monde braudellien. *Environment and Planning D: Society and Space* 4:211–19.

———. 1988. Context, common sense and sense of place. *Journal for Theory of Social Behavior* 18(2):239–50.

———. 1989. Context, spatiality, state and local state. In *The state in comparative and international perspective,* ed. J. Caporaso. Beverly Hills: Sage Publications. In press.

———; Knox, P.; and Pinch, S. 1984. *Public provision and urban development.* New York: St. Martin's Press.

Klodawsky, F., and Spector, A. 1985. Mother-led families and the built environment in Canada. *Women and Environments* 7(2):12–17.

———, and Mackenzie, S. 1987. Gender sensitive theory and the housing needs of mother-led families: Some concepts and some buildings. *Feminist Perspectives* 8. Ottawa: Canadian Research Institute for the Advancement of Women.

Knopp, L. 1987. Social theory, social movements and public policy: Recent accomplishments of the gay and lesbian movements in Minneapolis, MN. *International Journal of Urban and Regional Research* 11:243–61.

———, and Lauria, M. 1987. Gender relations as a particular form of social relations. *Antipode* 19(1):48–53.

Knox, P. 1987. The social production of the built environment: Architects, architecture and the postmodern city. *Progress in Human Geography* 11(3):354–78.

Lauria, M. 1982. Selective urban redevelopment: A political economic perspective. *Urban Geography* 3:224–39.

———. 1984. The implication of Marxian rent theory for community controlled redevelopment strategies. *Journal of Planning Education and Research* 4(1):16–24.

———. 1986. Toward a specification of the local state: State intervention strategies in response to a manufacturing plant closure. *Antipode* 18(1):39–63.

———, and Knopp, L. 1985. Toward an analysis of the role of gay communities in the urban renaissance. *Urban Geography* 6:152–69.

Lefebvre, H. 1974. *La production de l'espace*. Paris: Anthropos.

———. 1987. An interview with Henri Lefebvre. *Environment and Planning D: Society and Space* 5(1):1–118.

Leitner, H. n.d. Pro-growth coalitions, the local state and downtown development: The case of six cities. In *Society-economy-state,* eds. M. M. Fischer and M. Sauberer. Vienna: In press.

Lerner, G. 1986. *The creation of patriarchy*. New York: Oxford University Press.

Leveen, P. 1979. Natural resource development and state policy: Origins and significance of the crisis in reclamation. *Antipode* 11(2):61–80.

Liossatos, P. 1983. Commodity production and interregional transfers of value. In *Regional analysis and the new international division of labor* eds. F. Moulaert and P. Salinas, pp. 57–76. The Hague: Martinus Nijhoff.

Lipietz, A. 1987. *Mirages and miracles*. London: Verso.

Little, J.; Peake, L.; and Richardson, P., eds. 1989. *Women in cities: Gender and the urban environment*. London: Macmillan. In press.

Logan, J., and Molotch, H. 1986. *Urban fortunes: The political economy of place*. Berkeley: University of California Press.

Mabin, A. 1988. The struggle for the city: Urbanization and political strategies of the Southern African state. Unpublished paper. South African Research Program, Yale University.

Mair, A. 1987. Urban growth coalitions in historical perspective. Paper presented to the annual meeting of the Association of American Geographers, Portland, Oregon, April 21 to 26.

Mackenzie, S. 1986. Women's responses to economic restructuring: Changing gender, changing space. In *The politics of diversity: Feminism, Marxism and nationalism,* eds. R. Hamilton and M. Barrett, pp. 81–100. London: Verso.

———. 1987. Neglected spaces in peripheral places: Homeworkers and the creation of a new economic centre. *Cahiers de Geographie du Quebec* 31(83):247–60.

———. 1989. Women in the city. In *New models in geography,* eds. R. Peet and N. Thrift. New York: Allen & Unwin. In press.

Markusen, A. 1985. *Profit cycles, oligopoly, and regional development*. Cambridge: MIT Press.

———; Hall, P.; and Glasmeier, A. 1986. *High tech America*. Boston: Allen & Unwin.

Marsden, T. et al. 1986. Towards a political economy of capitalist agriculture: A British perspective. *International Journal of Urban and Regional Research* 10(4):498–521.

Marston, S. 1983. Towards a political economic approach to natural hazards research. *Political Geography Quarterly* 2.

———. 1988. Neighborhood and politics: Irish ethnicity in nineteenth century Lowell, Massachusetts. *Annals of the Association of American Geographers* 78(3):414–32.

———, and Kirby, A. 1988. Urbanization, industrialization and the social creation of a space-economy: A reconstruction of the historical development of Lowell and Lawrence, Massachusetts. *Urban Geography* 9(4):358–75.

Massey, D. 1979. In what sense a regional problem? *Regional Studies* 13(2):233–43.

———. 1984. *Spatial divisions of labor: Social structures and the geography of production*. London: Macmillan.

———. 1985. New directions in space. In *Social relations and spatial structures*, eds. D. Gregory and J. Urry, pp. 9–19. Oxford: Polity Press.

———, and Catalano, A. 1978. *Capital and land*. London: Edward Arnold.

———, and Meegan, R. 1978. Industrial restructuring versus the cities. *Urban Studies* 15:273–288.

———, and ———. 1982. *The anatomy of job loss*. London: Methuen.

———, and Allen, J. 1984. *Geography matters!* Cambridge: Cambridge University Press.

McDowell, L. 1983. Towards an understanding of the gender division of urban space. *Environment and Planning D: Society and Space* 1(1):59–72.

———. 1986. Beyond patriarchy: A class-based explanation of women's subordination. *Antipode* 18(3):311–21.

Meyerson, A. 1986. The changing structure of housing finance in the United States. *International Journal of Urban and Regional Research* 10(4):465–97.

Momsen, J., and Townsend, J., eds. 1987. *Geography of gender in the Third World*. London: Hutchinson.

Morgan, K., and Sayer, A. 1988. *Micro-circuits of capital*. Oxford: Polity Press.

Moulaert, F. 1987. An institutional revision to the Storper-Walker theory of labour. *International Journal of Urban and Regional Research* 11(3):309–30.

Murgatroyd, L. et al. 1985. *Localities, class and gender*. London: Pion.

Nelson, K. 1986. Labor demand, labor supply and the suburbanization of low-wage office work. In *Work, production, territory*, eds. A. J. Scott and M. Storper, pp. 149–71. Boston: Allen & Unwin.

O'Connor, J. 1973. *The fiscal crisis of the state*. New York: St. Martin's Press.

O'Keefe, P., ed. 1984. *Regional restructuring under advanced capitalism*. London: Croom Helm.

———, and Wisner, B., eds. 1977. *Land-use and African development*. London: International African Institute, Report No. 5.

———, and ———. 1983. Global disasters: A radical interpretation. In *Interpretations of calamity*, ed. K. Hewitt, pp. 263–80. Boston: Allen & Unwin.

Olwig, K. 1980. Historical geography and the society/nature problematic. *Journal of Historical Geography*. 6(1):29–45.

———. 1984. *Nature's ideological landscape*. London: Allen & Unwin.

———, and Olwig, K. 1979. Underdevelopment and the development of natural park ideology. *Antipode* 11(2):17–26.

Palma, G. 1978. Dependency: A formal theory of underdevelopment. *World Development* 6:881–924.

Peet, R. 1975. Editor's introduction. *Radical geography*. Chicago: Maaroufa Press.

———. 1983. Relations of production and the relocation of United States manufacturing since 1960. *Economic Geography* 59(2):112–31.

———, ed. 1987. *International capitalism and industrial restructuring*. Boston: Allen & Unwin.

Pickles, J. 1985. *Phenomenology, science and geography: Spatiality and the human sciences*. New York: Cambridge University Press.

Pinch, S. 1985. *Cities and services: The geography of collective consumption*. London: Routledge and Kegan Paul.

Plotkin, S. 1987. Property, policy and politics: Towards a theory of urban land-use conflict. *International Journal of Urban and Regional Research* 11(3):382–404.

Pratt, G. 1986a. Against reductionism: The relations of consumption as a mode of social structuration. *International Journal of Urban and Regional Research* 10(3):377–400.

———. 1986b. Housing tenure and social cleavages in urban Canada. *Annals of the Association of American Geographers* 76:366–80.

———, and Hanson, S. 1988. Gender, class and space. *Environment and Planning D: Society and Space* 6(1):15–36.

Pred, A. 1981. Production, family and free-time projects: A time-geographic perspective on the industrial and societal changes in 19th century U.S. cities. *Journal of Historical Geography* 7:3–36.

———. 1984a. Place as historically contingent process. *Annals of the Association of American Geographers* 74:279–97.

———. 1984b. Structuration, biography formation and knowledge: Observations on port growth during the late mercantile period. *Environment and Planning D: Society and Space* 2:251–75.

———. 1986. *Place, practice and structure*. Cambridge: Polity Press.

Pudup, M. 1987. Land before coal: Class and regional development in southeastern Kentucky. Unpublished Ph.D. diss., Department of Geography, University of California, Berkeley.

———. 1988. Arguments within regional geography. *Progress in Human Geography* 12:369–90.

———, and Watts, M. 1987. Growing against the grain: Mechanized rice farming in the Sacramento Valley, California. In *Comparative farming systems,* eds. S. Brush and B. Turner, pp. 345–84. New York: Guilford Press.

Redclift, M. 1984. *Development and the environment: Red or green alternatives?* London: Methuen.

Richards, P. 1985. *Indigenous agricultural revolution*. London: Hutchinson.

Roemer, J. 1980. A general equilibrium approach to Marxian economics. *Econometrica* 48(2):505–30.

———, ed. 1986. *Analytical Marxism*. New York: Cambridge University Press.

Rose, D. 1981. Accumulation versus reproduction in the inner city: The recurrent crisis of London revisited. In *Urbanization and urban planning in capitalist society* eds. M. Dear and A. Scott, pp. 339–82. London: Methuen.

———. 1984. Rethinking gentrification: Beyond the uneven development of Marxist urban theory. *Environment and Planning D: Society and Space* 1(2):47–74.

Roweis, S., and Scott, A. 1978. The urban land question. In *Urbanization and conflict in market societies,* ed. K. Cox, pp. 94–112. Chicago: Maaroufa Press.

Samatar, A. 1985. The predatory state and the peasantry: Reflections on rural development policy in Somalia. *Africa Today* 3:41–56.

———. 1988. Merchant capital, international livestock trade and pastoral development in Somalia. *Canadian Journal of African Studies* 21(3):301–15.

Sassen, S. 1988. *The mobility of labor and capital*. New York: Cambridge University Press.

Saunders, P., and Williams, P. 1986. Guest editorial. The new conservatism: Some thoughts on recent and future developments in urban studies. *Environment and Planning D: Society and Space* 4(4):393–400.

Sayer, A. 1979. Epistemology and conceptions of people and nature in geography. *Geoforum* 10:19–43.

———. 1984. *Method in social science: A realist approach*. London: Hutchinson.

———. 1986a. Industrial location on a world scale: The case of the semiconductor industry. In *Work, production, territory,* eds. A. J. Scott and M. Storper, pp. 107–23. Boston: Allen & Unwin.

———. 1986b. New developments in manufacturing: The just-in-time system. *Capital and Class* 30:43–72.

———. 1988a. The 'new' regional geography and problems of narrative. Unpublished paper, Urban and Regional Studies, University of Sussex, UK.

———. 1988b. Post-Fordism in question. Paper presented to the Association of American Geographers, Phoenix, April 6.

Schaffer, R., and Smith, N. 1986. The gentrification of Harlem? *Annals of the Association of American Geographers* 76(3):347–65.

Schoenberger, E. 1987. Technological and organizational change in automobile production: Spatial implications. *Regional Studies* 21(3):199–214.

———. 1988a. From Fordism to flexible accumulation: Technology, competitive strategies and international location. *Environment and Planning D: Society and Space*. In press.

———. 1988b. Thinking about flexibility: A response to Gertler. Unpublished paper, Department of Geography and Environmental Engineering, Johns Hopkins University, Baltimore.

Scott, A. 1980. *The urban land nexus and the state*. London: Pion.

———. 1983a. Industrial organization and the logic of intra-metropolitan location I: Theoretical considerations. *Economic Geography* 59:233–50.

———. 1983b. Industrial organization and the logic of intra-metropolitan location II: A case study of the printed circuits industry in the greater Los Angeles region. *Economic Geography* 59:343–67.

———. 1984a. Industrial organization and the logic of intra-metropolitan location III: A case study of the women's dress industry in the greater Los Angeles region. *Economic Geography* 60:3–27

———. 1984b. Territorial reproduction and transformation in a local labor market: The animated film workers of Los Angeles. *Environment and Planning D: Society and Space* 2:277–307.

———. 1986a. Industrial organization and location: Division of labor, the firm and spatial process. *Economic Geography* 62(3):215–31.

———. 1986b. Industrialization and urbanization: A geographical agenda. *Annals of the Association of American Geographers* 76(1):25–37.

———. 1987. The semiconductor industry in southeast Asia: Organization, location and the international division of labor. *Regional Studies* 21(2):143–60.

———. 1988a. *Metropolis: From the division of labor to urban form*. Los Angeles: University of California Press.

———. 1988b. *New industrial spaces*. London: Pion.

———, and Storper, M., eds. 1986. *Work, production, territory: The geographical anatomy of contemporary capitalism*. Boston: Allen & Unwin.

———, and Angel, D. 1987. The U.S. semiconductor industry: A locational analysis. *Environment and Planning A* 19:875–912.

Shaikh, A. 1980. Marxian competition versus perfect competition. *Cambridge Journal of Economics* 4(1):75–83.

Sheppard, E. 1982. City size distributions and spatial economic change. *International Regional Science Review* 7:127–51.

———. 1984. Value and exploitation in a capitalist space economy. *International Regional Science Review* 9:97–107.

———. 1987. A Marxian model of the geography of production and transportation in urban and regional systems. In *Urban systems: Contemporary approaches to modelling,* eds. C. Bertuglia et al., pp. 189–250. London: Croom Helm.

———, and Barnes, T. 1986. Instabilities in the geography of capitalist production: Collective versus individual profit maximization. *Annals of the Association of American Geographers* 76(4):493–507.

Slater, D. 1973. Geography and underdevelopment, 1. *Antipode* 5:21–53.

———. 1977. Geography and underdevelopment, 2. *Antipode* 9:1–31.

———. 1985. Social movements and a recasting of the political. In *New social movements and the state in Latin America,* ed. D. Slater, pp. 1–25. Cinnaminson, NJ: Foris Publications USA.

———. 1987. On development theory and the Warren thesis: Arguments against the predominance of economism. *Environment and Planning D: Society and Space* 5(3):263–83.

Smith, M., and Feagin, J. 1987. *The capitalist city*. Oxford: Blackwell.

Smith, N. 1979. Toward a theory of gentrification: A back to the city movement by capital not people. *Journal of the American Planning Association*. 45:538–48.

———. 1984. *Uneven development*. Oxford: Blackwell.

———. 1987a. Dangers of the empirical turn: The CURS initiative. *Antipode* 19(1):59–68.

———. 1987b. Of Yuppies and housing: Gentrification, social restructuring and the urban dream. *Environment and Planning D: Society and Space* 5(2):151–72.

———, and Williams, P., eds. 1986. *Gentrification of the city*. London: Allen & Unwin.

Society and Space. 1986. Special issue on Los Angeles. *Environment and Planning D: Society and Space* 4(3):249–390.

Society and Space. 1987. Reconsidering social theory: A debate. *Environment and Planning D: Society and Space* 5(4):367–434.

Society and Space. 1988. Special issue on cultural geography. *Environment and Planning D: Society and Space* 6(2):115–228.

Soja, E. 1980. The socio-spatial dialectic. *Annals of the Association of American Geographers* 70(2):207–25.

———. 1986. Taking Los Angeles apart: Some fragments of a critical human geography. *Environment and Planning D: Society and Space* 4(3):255–76.

———. 1989. *Post-modern geographies*. London: Verso.

———; Morales, R.; and Wolff, G. 1983. Urban restructuring: An analysis of social and spatial change in Los Angeles. *Economic Geography* 59(2):195–230.

Stimpson, C. et al, eds. 1981. *Women and the American city*. Chicago: University of Chicago Press.

Stone, M. 1975. The housing crisis, mortgage lending and class struggle. *Antipode* 7(2):22–37.

Storper, M. 1982. The spatial division of labor: Technology, the labor process and the location of industries. Unpublished Ph.D. diss., University of California, Berkeley.

———. 1984. Who benefits from industrial decentralization? Social power in the labor market, income distribution, and spatial policy in Brazil. *Regional Studies* 18(2):143–64.

———. 1985a. Oligopoly and the product cycle: Essentialism in economic geography. *Economic Geography* 61(3):260–82.

———. 1985b. The spatial and temporal constitution of social action: A critical reading of Giddens. *Environment and Planning D: Society and Space* 3(3):407–24.

———, and Walker, R. 1983. The theory of labor and the theory of location. *International Journal of Urban and Regional Research* 7(1):1–41.

———, and ———. 1984. The price of water. Monograph, Institute of Governmental Studies, University of California, Berkeley.

———, and Scott, A. 1986. Contemporary realities and theoretical tasks. In *Work, production, territory,* eds. A. J. Scott and M. Storper, pp. 3–15. Boston: Allen & Unwin.

———, and Christopherson, S. 1987. Flexible specialization and regional industrial agglomerations: The case of the U.S. motion picture industry. *Annals of the Association of American Geographers* 77(1):104–17.

———, and Scott, A. 1988. The geographical foundations and social regulation of flexible production complexes. In *Territory and social reproduction,* eds. J. Wolch and M. Dear. Boston: Allen & Unwin. In press.

———, and Walker, R. 1989. *The capitalist imperative: Territory, technology and industrial growth.* New York: Blackwell. In press.

Taylor, M., and Thrift, N., eds. 1982. *The geography of multinationals.* New York: St. Martin's Press.

Taylor, P. 1982. A materialist framework for political geography. *Transactions of the Institute of British Geographers* 7:15–34.

———. 1985. *Political geography: World economy, nation state and locality.* New York: Longman.

———. 1987. The paradox of geographic scale in Marx's politics. *Antipode* 19(3):287–306.

Thrift, N. 1981. Owners' time and own time: The making of a capitalist time consciousness, 1300–1880. In *Space and time in geography: Essays dedicated to Torsten Hagerstrand,* ed. A. Pred, pp. 56–84. Lund, Sweden: Glerup.

———. 1983. On the determination of social action in space and time. *Environment and Planning D: Society and Space* 1(1):23–57.

Urry, J. 1986a. Capitalist production, scientific management and the service class. In *Production, work, territory,* eds. A. Scott and M. Storper, pp. 41–66. Boston: Allen & Unwin.

———. 1986b. Locality research: The case of Lancaster. *Regional Studies* 20:233–42.

———. 1987. Some social and spatial aspects of services. *Environment and Planning D: Society and Space* 5(1):1–118.

Vail, D. 1982. Family farms in the web of the community: Exploring the rural political economy of the United States. *Antipode* 14:26–38.

Vogeler, I. 1981. *The myth of the family farm.* Boulder, CO: Westview Press.

Walker, R. 1974. Urban ground rent: Building a new conceptual framework. *Antipode* 6(1):51–58.

———. 1975. Contentious issues in Marxian value and rent theory: A second and longer look. *Antipode* 7(1):31–54.

———. 1978. The transformation of urban structure in the 19th century United States and the beginnings of suburbanization. In *Urbanization and conflict in market societies,* ed. K. Cox, pp. 165–213. Chicago: Maaroufa Press.

———. 1981. A theory of suburbanization: Capitalism and the construction of urban space in the United States. In *Urbanization and urban planning in capitalist societies,* eds. M. Dear and A. Scott, pp. 383–430. New York: Methuen.

———. 1985a. Is there a service economy? The changing capitalist division of labor. *Science and Society* 49:42–83.

———. 1985b. Technological determination and determinism: Industrial growth and location. In *High technology, space and society,* ed. M. Castells, pp. 226–64. Beverly Hills: Sage Publications.

———. 1988a. The dynamics of value, price and profit. *Capital and Class* 35:147–81.

———. 1988b. The geographical organization of production systems. *Environment and Planning D: Society and Space.* In press.

———. 1988c. Machinery, labour and location. In *The transformation of work?,* ed. S. Wood. London: Hutchinson. In press.

———, and Storper, M. 1978. Erosion of the Clean Air Act of 1970: A study in the failure of government regulation and planning. *Boston College Environmental Affairs Law Review* 7(2):189–258.

———; ———; and Widess, E. 1979. The limits of environmental control: The saga of Dow in the Delta. *Antipode* 11(2):1–16.

———, and Heiman, M. 1981. Quiet revolution for whom? *Annals of the Association of American Geographers* 71:67–83.

———, and Storper, M. 1981. Capital and industrial location. *Progress in Human Geography* 5(4):473–509.

———, and ———. 1982. A guide for the Ley reader of Marxist criticism. *Antipode* 14(1):38–43.

———, and Greenberg, D. 1982. Post-industrialism and political reform in the city: A critique. *Antipode* 14(1):17–32.

———, and Williams, M. 1982. Water from power: Water supply and regional growth in the Santa Clara Valley. *Economic Geography* 58(2):95–119.

Wallerstein, I. 1974. *The modern world-system*. New York: Academic Press.

Warde, A. 1988. Industrial restructuring, local politics and the reproduction of labour power: Some theoretical considerations. *Environment and Planning D: Society and Space* 6(1):75–96.

Warf, B. 1988. Regional transformation, everyday life, and Pacific Northwest lumber production. *Annals of the Association of American Geographers* 78(2):326–46.

Warren, B. 1980. *Imperialism: Pioneer of capitalism*. London: Verso.

Watts, H. 1981. *The branch plant economy*. London: Longman.

Watts, M. 1983. *Silent violence: Food, famine and peasantry in northern Nigeria*. Berkeley: University of California Press.

———. 1984. State, oil and accumulation. *Environment and Planning D: Society and Space* 2(4):403–28.

———. 1986. Drought, environment and food security. In *Drought and hunger in Africa,* ed. M. Glantz, pp. 171–212. Cambridge: Cambridge University Press.

———. 1987. Powers of production—geographers among the peasants. *Environment and Planning D: Society and Space* 5(2):215–31.

———. 1988. Deconstructing determinism: Marxism, development theory and a comradely critique of *Capitalist World Development*. *Antipode* 20(2):142–68.

———, and Bassett, T. 1985. Crisis and change in African agriculture: A comparative study of the Ivory Coast and Nigeria. *African Studies Review* 28(4):3–27.

Webber, M. 1987a. Profits, crises and industrial change 1: Theoretical considerations. *Antipode* 19:307–28.

———. 1987b. Quantitative measurement of some Marxist categories. *Environment and Planning A* 19:1303–21.

———. 1987c. Rates of profit and interregional flows of capital. *Annals of the Association of American Geographers* 77:63–75.

———. 1988. Profits, crises and industrial change 2: Canada 1952-1981. *Antipode* 20:1–32.

———, and Foot, S. 1984. The measurement of unequal exchange. *Environment and Planning A* 16:927–47.

———, and Rigby, D. 1986. The rate of profit in Canadian manufacturing. *Review of Radical Political Economics*. 18:33–35.

Weiner, D., et al. 1985. Land use and agricultural productivity in Zimbabwe. *Journal of Modern African Studies* 23(2):251–85.

Wekerle, G.; Peterson, R.; and Morley, D., eds. 1980. *New space for women*. Boulder, CO: Westview Press.

Westcoat, J. 1984. *Integrated water development*. University of Chicago Department of Geography Research Paper No. 210. Chicago: University of Chicago Press.

Williams, P. 1976. The role of institutions in the inner London housing market: The case of Islington. *Transactions of the Institute of British Geographers* 1:72–82.

———. 1978. Building societies and the inner city. *Transactions of the Institute of British Geographers* 3:72–82.

Williams, R. 1973. *The country and the city*. London: Chatto & Windus.

Williamson, O. 1975. *Markets and hierarchies*. New York: Free Press.

Wisner, B. 1977. Man-made famine in eastern Kenya. In *Land-use and African development,* eds. P. O'Keefe and B. Wisner, pp. 194–215. London: International African Institute, Report No. 5.

———. 1985. Making ends meet: Food, fuel and water need conflicts in rural development perspective. *Rural Systems* 3(2):105–20.

Wolch, J. 1987. Voluntary organizations and the state: Lessons from the Greater London Council. Paper presented to the annual meeting of the Canadian

Association of Geographers, McMaster University, Hamilton, Canada, September 5.

Women and Geography Study Group. 1984. *Geography and gender*. London: Hutchinson.

Worster, D. 1985. *Rivers of empire*. New York: Pantheon.

Zelinsky, W.; Monk, J.; and Hanson, S. 1982. Women and geography: A review and prospectus. *Progress in Human Geography* 6(3):317–66.

Zukin, S. 1982. *Loft living: Culture and capital in urban change*. Baltimore: Johns Hopkins University Press.

The Urban Problematic

Sallie A. Marston | George Towers |
Martin Cadwallader | Andrew Kirby

This chapter has modest goals when compared to the field it aims to document. Over three-quarters of the American population is now urban (employing standard definitions), and almost any empirical analysis, and much theoretical conjecture, must in some way touch upon the realities of urban life. Any attempt to provide a systematic documentation of this work would stretch the chapter to impossible lengths. So, indeed, would an effort to overview the many ways that have been tried to distinguish the urban realm from rural life. Here, the approach is simple, but not necessarily noncontroversial. It assumes that urbanization and social change go hand in hand, and that modern society is essentially urban, at least within the U.S., which is the focus of this chapter.

THE HISTORICAL LEGACY OF URBAN GEOGRAPHY

American urban geography was slow to develop, for two reasons. In the first instance, the dominant discourse that developed in the early decades of this century was essentially both regional and exceptionalist in nature. Although the systematic investigation of urban phenomena was not excluded from geography, there could be no effort made to develop general principles of urban organization (as were developed in Germany by Christaller, for example), or to join the debate on the nature of social organization within urban settings. As Platt observed, the city was simply "another item in the regional pattern" (Platt 1931, 52). A little more privilege was given to urban study by Huntington and Carlson, who noted that "urban geography [is] an important phase of regional geography" (Huntington and Carlson 1933, 401). They continued: "the subject includes such topics as the location of cities, their size, growth and functions; the density of population of cities, and their relation to the hinterland, or surrounding

---

The authors would like to thank those who contributed to the production of this chapter, including Bill Clark, Larry Bourne, and the editors. As always, responsibility for this piece rests with the authors.

country, upon which they largely depend for their food supplies, raw materials for their industries, and customers for their products" (1933, 401).

In short, the study of the city and its functions was placed firmly within the regional inventory. This did not exclude entirely the attributes of cities themselves, but examination of internal organization was to be restricted to physical features: "no phase of urban geography is of more importance than that of urban land utilization" (1933, 401).

In contrast to the regional emphasis, and the second reason for American urban geography's slow development, was the concurrent development of Sauerian cultural geography. Insofar as this looked much more to Boasian anthropology for its inspiration, it was to all intents and purposes *anti*-urban in its orientation (Solot 1986). In consequence, the two most-powerful influences upon American geography either ruled out the study of urban affairs, or laid down extremely restrictive menus for scholarly activity.

The implications of this legacy are clear. The extracts quoted above are exactly contemporary with Christaller's work on central-place theory, and concurrent with the widely cited work on urban ecology undertaken in Chicago by Park, Burgess, and coworkers. It is little surprise that any urban geography undertaken had little impact either within or outside the discipline, and that geographers were to be net borrowers from individuals like Ernest Burgess and Homer Hoyt.

The legacies of the Chicago school of urban sociology in the 1940s have been far-reaching, rather like the legacy of the Chicago school of economics some decades later. Geographers borrowed extensively from the spatial logic of the ecologists, without, it seems, paying attention to the sociological underpinnings of the research (e.g., see Mayer 1979). In some ways, this borrowing was indeed more sterile than the morphological mapping so frequently undertaken by German and some British geographers.

Urban morphology is an explicit description of the physical fabric of the city; nothing more, nothing less. In contrast, the random dismemberment of Park, Burgess, and Wirth's research led inevitably to serious errors. In the first case, the orthodoxy of concentric zones, sectors, and wedges became total (Fellmann 1957). It is still hard to find an urban-geography textbook that does not replicate these ideal types, and generations of students have plodded earnestly around small towns throughout North America (and, it must be said, Europe), trying to find the concentric rings.

Much more serious is the divorce of spatial imagery from the underlying models of social organization that were invoked by the ecologists (Scott 1980). Harvey pointed this out two decades ago, and his cogent analysis still stands (1973, 131). As he noted, the interpretation of Chicago was one in which social organization was seen to flow from spatial competition: that is to say, short-run competition led to long-run ecological sorting. This assumption has been recycled numerous times, and may be seen in many subsequent types of analysis. There is a direct link, for instance, between the ecological model and the bid-rent models developed by a number of researchers in geography and urban economics in the 1950s and beyond. To repeat, such models saw social organization as stemming from spatial competition, rather than vice versa.

Harvey is thus correct when he traces direct links from Burgess through to the urban geography of the 1960s. The latter was powerful in the extreme, in terms of its ability to dissect the spatial form of the city, but it was—with analytical hindsight—limited. As Harvey observed,

> ... the main thrust of the Chicago school was necessarily descriptive. This tradition had an enormously powerful influence over geographic thinking, and although the techniques of description have changed somewhat (factorial ecology replacing human ecology), the essential direction of the work has not changed greatly. (Harvey 1973, 131)

This is to say: the sheer complexity of the city provoked a technical response (empirical analysis of patterns plus model-building), rather than a theoretical thrust (identification of social processes and the nature of the city). This does not, of course, represent a dismissal of this research; indeed, as the following section indicates, a number of interesting insights have emerged from this literature.

ANALYTICAL URBAN GEOGRAPHY

During the last 25 years, urban geography has been characterized by the use of abstract modes of thought, especially those associated with statistics and mathematics (Cadwallader 1985). Particular emphasis has been given to model-building, which represents a pattern-seeking viewpoint that stresses recurrent connections and interrelationships. Such models can serve a variety of functions. First, they are often designed to explicate relationships between exogenous and endogenous variables, thus allowing one to predict future values of the endogenous variables. Second, models provide a framework for defining, collecting, and ordering information. Third, models are used to generate and substantiate empirical hypotheses that are significant for some larger theoretical structure. We can identify a number of contexts within which such work has been undertaken.

SYSTEMS OF CITIES

Generally speaking, urban geographical analysis has emphasized either the system of cities or the internal structure of cities, and this distinction will be maintained in the present overview. Investigation of the system of cities tests them as points in space, at either a national or regional level, while investigation of the internal structure of cities focuses upon the spatial arrangement of places and activities within those cities.

The search for regularities in the organization and behavior of a system of urban settlements has been dominated by questions concerning location, size, growth rates, and interaction. Locational questions were derived initially from the lead of central-place theory (King 1984), but have also been interpreted from a more-statistical perspective. In particular, Dacey (1966) has discarded the economic baggage of central-place theory to treat the location pattern of settlements as the outcome of one or more probabilistic processes. Similarly, Curry (1967) has also introduced the operation of chance into the system.

Questions concerning the distribution of city sizes within an urban system have been of persistent interest. Within this context, the so-called rank-size rule has often been used as a benchmark from which to measure deviations (Bourne, Sinclair, and Dziewonski 1984). The most commonly occurring deviation seems to be a primate distribution, where an abnormally large proportion of a country's urban population resides in the largest city. Despite the prolific documentation of national city-size distributions, however, efforts to explain the occurrence of rank-size—or other distributions—have been largely unconvincing (Carroll 1982). Level of economic development, political structure, and history of urbanization have all received attention as potential determinants of city-size distributions, but the empirical evidence is ambiguous.

Of perhaps greater theoretical significance are the investigations of urban growth and decline, where the focus on process is more explicit. One approach to urban change suggests that growth or decline is exogenously determined: for example, the export-base concept implies that increased demand in the export sector of a city's economy will lead to increased employment and purchasing power. If this increased purchasing power is directed at locally produced goods and services, then multiplier effects are set in motion that will generate further growth (Pred 1977).

An alternative viewpoint argues that urban growth or decline is controlled primarily by a set of endogenous mechanisms, and it has long been suggested that there is a strong relationship between size and growth (Robson 1973). Larger cities were thought to possess a comparative advantage due to their greater industrial mix, political power, fixed capital investment, and economies of scale. But the so-called size-ratchet effect has fallen into some disrepute since the onset of counterurbanization, which was first appreciated widely in the U.S. during the 1970s. Explanations of this phenomenon of metropolitan decline and population deconcentration are as yet rather ad hoc (Berry 1973; Bourne 1980a), but there is obviously a need to reconceptualize the whole process of urban change.

One such reconceptualization has been provided by neo-Marxist theories of uneven development, in which regional growth and decline is considered to be an inevitable manifestation of the inherent crises and instability of capitalism. Indeed, some scholars see uneven development as not only an inevitable consequence of capitalism, but also as a necessary prerequisite for capital accumulation (Bradbury 1985). Capital accumulation within a region depends upon the rate of profit of the firms located there, and such accumulation cannot occur at an even pace in both space and time (Browett 1984). Rather, capital tends to accumulate in some places as opposed to and at the expense of others. Behind this pattern of uneven development and a restructuring urban system lies the logic of what Smith (1984, 148), has called the "seesaw" movement of capital, by which capital tends to move back and forth between developed and underdeveloped areas. Documentation of such linkages requires careful analysis of interaction patterns within the urban system. For example, how are economic impulses transmitted through a system of cities, and how do the impact and timing of such impulses vary from one city to another? (Gaile and Hanink 1985).

An encouraging development in the context of systems of cities has been the recent work of the International Geographical Union Commission on Urban Systems in Transition (Borchert, Bourne, and Sinclair 1986). In particular, the Commission's work has reflected the increasing importance of cross-cultural comparisons of urbanization (Brunn and Williams 1983). More significantly, however, there has been a firm commitment to the development of a broad set of coherent organizational structures and concepts which can be used to explicate the underlying processes of urban change. For example, Simmons (1986) categorizes and evaluates a series of economic, demographic, and institutional processes that can be investigated in a variety of national contexts.

The traditional framework for explaining the distribution of land use and land value in urban areas revolves around the concept of "highest and best use" and the methodology of bid-rent curves (Muth 1985). Other recent research has attempted to adapt this standard monocentric model to the polycentric configuration of many modern cities, for as metropolitan form has undergone significant changes during the past few decades, economic activities have become concentrated increasingly in suburban nucleations (Erickson 1986). Despite the continued interest in patterns of land use and land value, exemplified by Thrall's recent (1987) geometric interpretation using the consumption theory of land rent, this general approach has been criticized for its extremely restrictive assumptions. The roles of externalities, government intervention, and historical inertia in the built environment have not received the attention they deserve (inter alia see Marston and Kirby 1988).

A closely related line of investigation has involved the construction of models designed to describe the distribution of population density within cities. Although a variety of density functions has since been tested, Clark (1951) was the first to provide convincing empirical evidence that population density tends to decline in an exponential fashion with increasing distance from the central business district (CBD). Newling (1969) later generalized much of the earlier work within a quadratic exponential model that reflected the density crater immediately surrounding the CBD. He also suggested more dynamic evolutionary schemes which received some empirical validation from the study of Toronto by Latham and Yeates (1970).

Other analyses have attempted to capture the polycentric structure of cities by using trend-surface models (Haggett and Bassett 1970). These studies of urban structure have, however, done scant justice to the variety of underlying phenomena. Such processes have been given more explicit consideration in the study of individual land-use categories like retailing, industry, and residential location. The retail structure of urban areas was first extensively described by Berry (1963) and his associates working in Chicago. They postulated three main elements: specialized areas, ribbons, and retail nucleations. The retail nucleations were subjected to more detailed examination, using the principles of central-place theory to generate various hierarchical and spacing models (Warnes and Daniels 1979).

The idea of an intrametropolitan central-place hierarchy is still visible in the most recent literature on retail structure (Morrill 1987), but attention has shifted toward the construction of models of consumer spatial behavior. Such models are aimed at explicating the underlying decision-making process, and often involve the use of learning frameworks (Golledge and Stimson 1987).

Traditional treatments of industrial location within cities tended to classify industries, and then comment on their locational tendencies (Pred 1964). Recent work has been more theoretically oriented, emphasizing the essentially interdependent nature of locational decision and choice of production technique. For example, Scott (1982) argues convincingly that under capitalism, the locational patterns of urban industry can be divided into two main categories: labor-intensive firms tend to locate toward the center of the urban land market, while capital-intensive firms are attracted by the relatively cheap land inputs at the peripheral locations. The increasing substitution of capital for labor has thus generated a more decentralized locational pattern.

RESIDENTIAL STRUCTURE AND MIGRATION BEHAVIOR

Residential structure has been of continuing concern to urban geographers, with the factorial ecological studies of the 1960s suggesting that there are underlying dimensions of residential differentiation that have particular spatial manifestations (for reviews, see Goddard and Kirby 1976; Johnston 1971). Criticisms of the factorial ecological approach have revolved around such technical issues as the use of different types of rotations (Davies 1984), and a number of issues of interpretation, including the arbitrary choice of cities for analysis and/or comparison, and the fact that the resulting social areas do not necessarily constitute cohesive communities. (That is, such areas are relatively homogeneous with respect to certain specified variables, such as income, education, and family size, but they might not be characterized by a high degree of internal interaction.) Attempts to calibrate the amount of internal interaction have focused on activity patterns associated with the workplace, friends, and clubs (Everitt 1976).

This latter genre of work, which considers patterns of social behavior explicitly, has been enriched considerably by the behavioral approach to urban geography; it emphasizes the role of cognitive mapping and individual decision-making processes. Lynch's work (1960) on urban images had an important influence on the field of cognitive mapping, but the recent contributions by geographers have been methodologically more sophisticated. Techniques such as multidimensional scaling have been used to explore systematically the differences between cognitive and physical representations of urban environments (Golledge and Stimson 1987). Indeed, because of its preoccupation with measurement and highly formalized methodology, the behavioral approach has been viewed as merely an appendage to the spatial-positivist tradition (Duncan 1987). In fairness, however, behavioral geographers have argued convincingly for a more process-oriented approach, incorporating such concepts as cognition, learning, information processing, and attitude formation (Golledge and Rushton 1984). Perhaps a more important criticism focuses on the implied assumption of subject-object separation, whereby the world is separated into an objective world of things and a subjective world of the mind, implying that the observer is somehow separated from the observed.

Work on spatial-choice behavior in urban environments has been equally sophisticated in the methodological sense (Preston 1987). Within the context of consumer behavior, the use of discrete-choice models, which are intrinsically related to a family of statistical models involving the analysis of categorical data, represents a substantial contribution (Wrigley and Longley 1984).

However, greater attention has been paid to migration within cities, where the decision-making process has been conceptualized traditionally in terms of three stages: the decision to move, the search for available alternatives, and the evaluation of those alternatives. The initial decision to move has been related to residential stress and the duration-of-residence effect, by postulating a trade-off between dissatisfaction and inertia (Huff and Clark 1978). By way of distinction, models of residential search have focused upon information sources, the length of search activity, and the spatial pattern of search activity (Clark 1982). Attempts to model the actual choice process involved in residential mobility have included using the conjoint-measurement technique, information-integration theory, log-linear models, and the elimination-by-aspects model (Cadwallader 1986).

More temporary movement patterns within cities—i.e., daily—have been considered within the context of traffic forecasting models, where the issues of trip generation and attraction, trip distribution, modal choice, and trip assignment have received particular attention (Daniels and Warnes 1980). Geographers have been involved especially with trip-distribution models, and Wilson's family of spatial interaction models (1971), based on concepts derived from statistical mechanics, has been particularly influential.

The comparative statics approach, which has been dominant in urban geography, does not address the process of change itself. Dynamic models, on the other hand, describe how new kinds of structures emerge, and identify the critical thresholds, or bifurcation points, at which the system switches to a new trajectory. The focus is on multiple equilibria and various space-time trajectories, rather than on the transformation between times t and t+1. Such models are right at the cutting edge of contemporary model-building in human geography.

EXPLANATION AND UNDERSTANDING IN URBAN GEOGRAPHY

Despite the advances that have been achieved in construction, calibration, and testing of urban models, it has also been recognized that many such models fail to do justice to the different institutional structures that have molded that development, and the historical settings of urban development. The first of these criticisms was developed explicitly by British geographers a decade ago. This move led first to a reassessment of simple models of human agency (such as blaming the victim: Gray 1975), and rapidly linked with a general move away from urban pattern analysis—first toward managerialism, and then on toward more complex and socially grounded analyses (Johnston 1980).

This is not the place to trace this history (see Badcock 1985). Rather, we might simply contrast the situation within French or British urban geography and that obtaining in the U.S. It is clear that, while the subdiscipline here is in relatively good shape—as measured by, for instance, its own eponymous journal[1]—the challenge offered by a decade of sociologically informed work has only just begun to be felt. The most obvious indication of this is in the field of housing research. A good deal of study has been undertaken upon what used to be termed "nonfarm" housing, although much of this has remained in the rather antiseptic area of preference functions, rent gradients, and other extensions of the new urban economics (Bourne 1980b; Knox 1987b; Palm 1981). This is changing belatedly, as a new generation of researchers is reexamining the operation of the housing market.

One example of this is the work on gentrification. Although this phenomenon may not be of great quantitative import in terms of population movements, its significance lies in the way that its study has provided a springboard to the highly complex residential and economic restructuring that is taking place in cities like New York. Research by Smith (1984), Chouinard (1988), Fincher (1984), Lauria and Knopp (1985), Pratt (1986), Rose (1984), and others has in consequence begun to sketch some parts of a new urban geography.

---

[1] The journal *Urban Geography* is edited by Brian Berry, James Wheeler, and Robert Lake, and published by V. Winston.

The same progress can be seen in the field of service provision. A large component of this work has emphasized the issues of accessibility, and the attendant questions of efficiency, equality, and equity. It also has emphasized the empirical organization of spatial forms at the expense of social processes, the description of lattices at the expense of the study of implications. This field is slowly reorienting itself to take account of changes in the broad sociopolitical climate, and to comprehend the role of services—in the broadest sense—in the operation of the urban economy, both past (Knox, Bohland, and Shumsky 1984) and present (Scarpaci, Infanto, and Gaete 1988). Wolpert has examined the role of voluntarism under the New Federalism, while Geiger and Wolch have examined the functioning of the voluntary sector as a "shadow state" in the Los Angeles economy. Dear and Wolch have examined how the changed political-economic climate has had an impact upon the urban homeless (Dear and Wolch 1987; Geiger and Wolch 1986; Wolpert and Reiner 1984).

Inevitably, changes in the research agenda may be at the expense of some long-established concerns. Since the early 1980s, and until rather recently, urban geographic work paid far less attention to issues of race, ethnicity, and segregation than it did in the 1970s and 1960s. Much of the research conducted has tended to extend the predisposition toward description and measurement of racial segregation that was established during the previous two decades (e.g., Darden 1983; Jakubs 1986). Still, there are numerous examples of attempts to go beyond description and to attempt compelling explanations. A provocative example is Western's work on Capetown (1981), which examines both the structural forces that perpetuate the apartheid city and the reality of black existence in the restricted townships.

Consistent with this approach to racial segregation, founded upon institutional explanation, is some exciting work that has emerged in recent years. For example, Palm (1985) has examined the role that black real-estate agents play in sorting American urban residential space, while Clark (1984) has looked at the role of judges in attempting to understand "who is to blame for racial segregation." Shifting focus to Latin America, Scarpaci, Infanto, and Gaete (1988) have shown how the urban-planning process has systematically reinforced racial segregation in Santiago, Chile.

While not focusing exclusively on racial segregation, Agnew, Mercer, and Sopher's (1984) edited collection, *The City in Cultural Context,* provides insights into the state of current research on the structuring influence of ethnic and cultural forces on urban space. More recently, Marston (1988) has attempted to link the forces of everyday life with the constraints and opportunities of an evolving political and economic context, in a study of the emergence of political consciousness among a nineteenth-century ethnic group. Yet, despite these (and other) exceptions, a perusal of the geographic journals substantiates assertions that attention to race and ethnicity declined during the early 1980s. It is likely that, as the Reagan administration placed less emphasis on urban social issues and more on urban economic development, the research agenda in geography shifted likewise (Clarke, Redburn, and Buss 1985).

However, it should be noted that, for the nineties, there appears to be renewed and spreading interest in race, ethnicity, and the city, particularly with respect to the debate surrounding the emergence of an urban underclass (Clark 1987; Hughes 1987), and comparative ethnicity (Johnson and Oliver 1988). The 1988 annual meeting of the AAG devoted several sessions to urban ethnic and racial issues. There is also abundant evidence of increased attention to issues of urban space and gender (for a bibliography, see Holcomb 1986) and sexual preference (Lauria and Knopp 1985).

The second criticism—that much of this research lacks a historical understanding—can again be traced to the orthodoxies of the 1930s. It is clear that to study phenomena in a historical setting was anathema, and the specter of an urban historical geography was explicitly condemned by William Morris Davis:

> A precaution must be suggested in order that the study of cities . . . shall be given a truly geographical character and thus avoid becoming too historical . . . geographical study presents their areal development only in so far as its effects are visible in their present form. (Davis 1932, 225)

Naturally, an emphasis upon visible phenomena ruled out a historical dimension to urban research, and the emergence of a recognizable urban historical geography has labored under this burden, as we shall see below.

URBAN HISTORICAL GEOGRAPHY

In this section, we turn our attention to recent work in urban historical geography. Despite a tightly specified topic, we are in reality addressing a loosely knit literature, for this is a borrower field, using a historical perspective primarily to explore a wide range of concepts developed elsewhere within urban geography. For the sake of order, we shall divide our discussion along the lines already established above, making a distinction between research focusing on interurban issues, and that focused on intraurban themes.

Interurban Research

Two broad topical subcategories can be identified, of which the first examines the genesis of urban systems, and the second the relative location of cities within these urban systems. Conceptually, a similar bifurcation can be made, for many if not most urban historical researchers are interested primarily in explaining only their own subject matter. They seek to understand a particular case, and not to test sweeping generalizations that account for the urban system, much less derive these sorts of models themselves. Conversely, a second but smaller group of workers is intrigued by a broader view of interurban themes and uses historical settings to evaluate the validity of generalized models from a wider urban theory—specifically, central-place theory and Vance's mercantilist hypothesis (Vance 1970).

As an example of our first theme (i.e., the development of urban systems), Burghardt (1979) and Winters (1981) have applied models from urban geography to historical settings. Burghardt's work on the relationship between the road system of Roman Pannonia and the incipient network of cities sheds light on the power of central-place theory (CPT) to explain the development of urban systems. In Pannonia, cities were originally established to serve military strategy, and in consequence, site factors determined urban location. Only later, when a linked system of strategically located cities was in place, did situational factors (principally distance) become important. This study suggests that, as CPT implies, later rounds of urban development may be due to situational and economic factors, while site characteristics and noneconomic rationales may be of greater significance in explaining the placement of the original urban nodes.

Going one stage further, Winters examines the urban system of medieval Mali, exploring both CPT and Vance's model of trade relationships (1981). He demonstrates that trade was more important in shaping the early system of Malian cities, but also finds that noneconomic factors—notably religion and politics—had more to do with the patterning of cities than either set of economic predictors. However, it should be noted that other workers interested in the genesis of urban systems have not made such explicit reference to theory. For instance, Graham's report on postconquest Ireland, and Slater's on medieval Staffordshire, focus exclusively on understanding these particular cases (Graham 1979; Slater 1985).

As with the researchers focusing on the creation of urban systems, studies dealing with the trade relationships of cities with other cities and their hinterlands can be divided between theory-testing and more idiographic concerns. Perhaps the best example of the former is the examination of colonial Guadalajara's credit hinterland (Greenow 1984). In testing dependency theory's claim that the domination of rural regions by colonial cities was achieved via one-sided credit-lending relationships, Greenow found that this was not quite the case in this region. Although borrowers were concentrated in outlying areas and creditors in the city, their relationship was much more equitable than had been presumed—the city was just as dependent upon the payment of rural interest, as the rural areas were upon receiving credit. In addition, many loans were never repaid, resulting in a sort of "urban subsidy" for rural areas: city and countryside were in a relationship of mutual interdependence, not one of unilateral dependence.

An example of research concerned less explicitly with assessing existing models, but which might nevertheless be useful in understanding the general process by which cities come to dominate their hinterlands, is Marks' study of Baltimore and nearby St. Mary's County, Maryland. She describes the changes in life in the county as a result of Baltimore's growth, identifying technological change (specifically the introduction of the steamboat) as the catalyst behind St. Mary's loss of economic self-sufficiency (Marks 1982).

Urban historical geographers' research into interurban issues thus has had its successes, but there exist lacunae. As mentioned, most of the recent research in this area has been limited to interpretations of particular and isolated cases. Therefore, this literature does little to increase our understanding of wider processes (in either space or time), and is of interest almost exclusively to specialists. While other urban historical geographers who are interested in interurban issues have given much attention to general issues, their work has not expanded to fill the voids in the field. Efforts such as those of Burghardt and Winters can inform existing theory and lead to important modifications, but they do little to develop the subfield as a whole. Instead, they refine *urban* geography, and relegate *urban historical* geography to its role as a testing ground of past events. Until the subfield is able to generate intrinsic forms of explanation, its integrity will remain clouded, a theme continued in our discussion of work done on intraurban issues.

Intraurban Research

Historical investigation of intraurban issues has produced significantly more research than that into interurban matters, and consequently the subject matter is significantly more diverse. However, for the sake of parsimony, only three broad subareas of study will be discussed—urban morphology, urban landscape symbolism, and intraurban interaction.

As with interurban research, there are several prominent conceptual schools informing large portions of this work. Unsurprisingly, much of the work done on urban morphology draws its conceptualization from the persistent models developed by human ecologists. Marxism has had a smaller yet significant impact here, but a leading role with respect to work on "symbolic landscapes." Intraurban interaction is the least-avowedly theoretical of the three areas. For example, much of its work has been concerned with the spread of epidemics (Shannon and Cromley 1982), although other researchers have employed the precepts of the Lund group's time-geography to investigate questions relating to movement in city space.

Urban morphological studies are concerned essentially with the description and explanation of patterns of urban land use. The bulk of historical work in this area continues to be carried out via the constructs developed in human ecology by Park, his colleagues, and disciples such as Hawley and Duncan. Naturally, ecological models frame the occupance of urban areas in ecological terms, although two strands of argument are in evidence. First, there are models like Burgess' concentric-zone model that try to account for the spatial distribution of urban land uses. Second, there are "succession" models that seek to identify the stages of land use through which urban areas pass. These models are applicable to a wide variety of morphological topics, from suburbanization to ghettoization. Deskins, for one, provides a useful example of their application to the latter. Examining the evolution of Detroit's black ghetto, he formulates a model for ghetto emergence that is based on a selective synthesis of available ecological frameworks (Deskins 1981).

The materialist examination of urban morphology has been inspired by the pioneering efforts of David Harvey and his students, and has focused on the way in which unequal class relationships and patterns of capital accumulation intersect with the spatial form of the city. For example, Harvey (1985a) has demonstrated how the spatial restructuring of Paris during the 1850s and 1860s was congruent with the intentions of the Second Empire to facilitate capital accumulation. The freeing of Paris' clogged street network and the reorganization of its housing market was at once a response to, and a catalyst for, a more rapid circulation of capital and investment of surplus capital. Similarly, Walker's study of the genesis of American suburbanization demonstrates the relationship between suburban space and the desire for both separate and unequal class reproduction, and the profitable reinvestment of surplus capital (Walker 1981).

Historical studies of urban morphology outside of these two research traditions have also provided useful insights. For instance, Ward's study of segregation in Leeds sheds some light on the timing of class-based segregation, relative to the onset of industrial capitalism (1980). Ward finds that the upper class had isolated itself by this period; conversely, the spatial differentiation of the working class did not occur until finance capitalism took hold. These findings, in contrast to prior interpretations of the causes of segregation, have illuminated not only the relationship between perception and urban developments, but also that between capitalism and the *timing* of segregation. In short, work on urban morphology can be valuable in evaluating and enlightening existing theories and for offering new points of view on social and economic processes—although it has, once again, not caused urban historical geography to move far beyond being a backcloth for external theories.

A second area of topical concern within this intraurban set of issues has been the study of urban symbolism. Recently, geographers have devoted considerable energy to assessing information carried by the urban landscape, and to interpreting concurrent meanings. For the most part, research into this subject has also been informed by Marxism. Harvey, again focusing on Paris, has added to this literature. In "Monument and Myth" (1979), he showed how the Basilica du Sacre Coeur, associated with a reactionary faction of Catholicism, came to symbolize the interests of the far right. After the defeat of the Commune in 1871, the rightist government, despite protests from the left, was adamant that the Basilica be constructed atop the commanding hill of Montmartre, from which it dominated the Parisian landscape. Harvey shows clearly how pieces of the urban landscape can take on class-based meanings, and come in consequence to symbolize class struggle.

It should be pointed out that Marxists are not the only historically inclined geographers studying the symbolism of urban landscapes.[2] Rowntree and Conkey (1980), for instance, have applied a biologically and ecologically derived model of stress to their work on the historic-preservation movement in Salzburg, Austria. In it, they maintain that the movement to preserve the old city can be seen as an attempt to lessen the stress of increasing industrialization and urban development.

Another branch of research within intraurban historical geography's third topical category has investigated the flow of people through the urban environment. Time-geography has been employed to focus upon institutionally defined "projects," societal goals that require the participation of a number of individuals, and personal trajectories through time and space. Pred, a leading proponent of this approach, has looked at the ways in which individual and household paths have changed in American cities as a result of industrialization (1981, 1984). The projects of early industrial capitalism, which demanded long hours from their participants and rewarded them with low wages, introduced many serious constraints into workers' lives, making it difficult for them to coordinate household tasks and spend their little leisure time in pleasurable ways.

Although intraurban research encompasses a larger and more diverse body of literature than does interurban research, it suffers from the same difficulties. It has, for the most part, limited itself to the evaluation and elaboration of external models, and as in the case of interurban research, it has reduced historical settings to little more than a passive medium. However, some workers have taken the explanatory power of time into account, and have hinted at the potential for urban historical geography to develop an indigenous set of conceptual constructs.

An example of such innovative work is provided by Olson (1979). Building on Harvey's earlier explorations into the geographic expression of capital accumulation and investment, Olson focuses upon cycles of investment in mid-nineteenth century Baltimore, measuring these 15-year cycles of investment by their social and spatial effects. Olson is careful to point out that existing social and spatial conditions guide

---

[2]British geographer Cosgrove has also been involved in connecting the symbols associated with landscape to the social order (1984), and while occupying a Marxist territory, he takes a different tack from Harvey. He selects the landscape as a whole, not just its symbolically charged pieces, as his subject matter. In addition, he looks first to explain the landscape's relation to class-based *ideologies,* while Harvey places an overriding emphasis on class *struggle.* Cosgrove links the rise of capitalism to the replacement of use values by exchange values in the determination of economic wealth of the landscape, while concomitantly, the transition from feudalism to capitalism led to the alienation of the individual from the landscape.

the course of investment. For example, investment in the spatial structure of Baltimore was likely to be concentrated in the more profitable suburbs and central business district (CBD). By adopting this cyclical view, essential to the understanding of the investment process, Olson connects the past with the present. The latter becomes the latest manifestation of processes, the understanding of which is found via historical analysis. Time becomes both an essential object of explanation and a necessary ingredient in the explanatory process. The recognition that time, as well as space, is the subject matter of historical geography points the way toward an independent urban historical geography that is not only willing to make sense of the past, but contributes to the analysis of the present as well.

This lack of a concept of history has been noted as a hallmark of historical geography since its founding (Guelke 1982). Tracing the roots of the subdiscipline back to Hartshorne, Sauer, Darby, and Clark, Guelke charges all of them with ignoring the meaning of history while studying geographies of the past. By refusing to direct their attention to the "meaning of human actions of geographical interest," and instead, viewing history as simply synonymous with the past, scholars were encouraged "to see human activity as spectacle rather than as an historical creation" (1982, 12). In his plea for geographers to reject the practice of reconstructing period geographies, Guelke asserts the need to relate geographical activity to the social and institutional complexes within which it is embedded. While Guelke's emphasis on the importance of ideas to understanding the nature of human geographical change has been disparaged by some (e.g., Kearns 1984), his book does provide an important criticism of contemporary research in historical geography.

Within the last decade, there have been some successful attempts to recognize the critical importance of taking history into account when trying to understand past urban geographies. We turn next to examine some of the research in historical geography which has given philosophical weight to both history and geography.

In his much-cited work on suburbanization in the U.S., Walker (1978, 1981) regards the transformation of the urban landscape as a product of the structural conditions of capitalist production, particularly the two processes of accumulation and class struggle. His theory of suburbanization is an attempt to show "how the logic of capital in the abstract would tend to create a 'suburbanized' form of the built environment" (1981, 384). Walker argues that the city must be constructed and reconstituted continually to facilitate capital accumulation and to maintain capitalist social relations. In short, he goes way beyond simply describing the changes in the nineteenth and early twentieth century urban landscape that occurred through suburbanization. He attempts instead to explain the new urban form by reference to the structure and logic of capitalist industrial production during its early and corporate periods.

The suburbanization process that began in the U.S. in the nineteenth century had three major characteristics, according to Walker: spatial differentiation, decentralization, and waves of successive and distinct urbanization. During the incipient industrialization of the American economy, entrepreneurial capitalist production relations reorganized the spatial relations of the work and domestic spheres, undermining the traditional occupational household that had maintained during the mercantile period of American cities. Later, with the undoing of production and consumption relations, came the separation of production, selling, and warehousing, and the eventual emergence of a CBD. Eventually, and as early as the 1830s in New York and Boston, the

CBD itself began to be differentiated according to function, predicated on the rationalization of economic activity in the production sphere and the social division of labor in the consumption sphere. Following the spatial differentiation of the mercantile city came the decentralization of the industrial city.

Walker's model of decentralization "rests on three pillars: (i) diminishing restraints on location of all kinds (generalization of capital); (ii) push-pull forces between uses at the center and periphery, with capital working at both ends; and (iii) the way the property circuit propels the whole process" (1981, 395). Rejecting the conventional wisdom that cheap land, transportation improvements, and the need for more space-extensive land uses explain American suburbanization, Walker maintains that the rise of the industrial bourgeoisie, a complex and multifaceted process, was a far more important cause. Propelled by the changing dynamics of accumulation evolving in this nineteenth-century context, as well as being repelled by the negative aspects of production activities and associated residential working-class districts, the industrial bourgeoisie sought an exit from the urban industrial landscape to an Arcadian ideal on the fringe.

In his attention to the waves of capital accumulation and the accompanying changes in the form of urbanization, Walker demonstrates how the temporal dynamics of the economy translate into changes in urban spatial organization. His periodization of accumulation cycles treats history as an important dynamic process, while at the same time recognizing the impact of changing historical forces on the urban landscape. By linking spatial differentiation, decentralization, and the waves of accumulation to the changing dynamics and logic of capitalist production relations, Walker demonstrates a sensitivity to history and the complexity of historical time, and a rejection of the understanding of historical geography of the American city as simply the reconstruction of past urban forms.

Ward (1982, 1984) has also acknowledged the importance of history to past urban geographies. In a paper on the American ethnic ghetto (Ward 1982), he argues that changes in the division of labor that occurred in the transition from mercantile to industrial capitalism, and eventually to finance capitalism, altered the impact of foreign migration on urban residential differentiation. In attempting to understand the complexity of the American ghetto as it altered over time, Ward insists that we recognize the importance of the ethnic division of labor—including the differences in the arrival dates of various immigrant groups—to the changing demands of the capitalist labor market. As he states it:

> The transition from mercantile to industrial capitalism during the middle decades of the nineteenth century and the transition from industrial to finance capital towards the turn of the century created distinctive stratified labor markets in which ethnic groups were unevenly distributed. (1982, 268)

Further on, he illustrates this point:

> Certainly, the once deprived status of the Irish-Americans was transformed in the late nineteenth century when changes in the division of labor made necessary by the new scale of both public and private enterprise opened avenues for their social mobility. Irish-Americans became the indispensable supervisors and managers of the new unskilled labor force recruited from southern and eastern Europe. (1982, 270)

These critical changes in the segmentation of the labor market, which usually paralleled changes in the organization of industrial-production activities, had important effects on the composition and function of the ghetto for different immigrant groups. Thus for Ward, the ghetto is not simply a fossilized artifact of nineteenth-century urban landscapes, but a spatial form and process rooted in changing historical circumstances.

In a more recent paper on inner-city slums at the turn of the nineteenth century, Ward (1984) shows how important actors, the Progressives, viewed and attempted to address the social problems associated with the impoverished residential areas of the inner city. As with the previous work discussed, this essay allows the importance of history to geography by showing how historically situated actors understood the spatial arrangements of urban society and acted—albeit in a misguided way—to change them.

In an analogous paper that examines the social force and material effect of the idea of Chinatown in turn-of-the-century Vancouver, Anderson (1987) asserts the importance of social forces in defining place. In a unique case study of a "Western landscape type," she argues for recognition of the multiple reality of place. Anderson's objective in the paper is not to provide yet another description of an ethnic ghetto, but rather to conceptualize Chinatown "as a social construction with a cultural history and a tradition of imagery and institutional practice that has given it a cognitive and material reality in and for the West" (1987, 581).

As yet another example of how a sense of history can provide a compelling understanding of geography, Anderson shows how Western ideas about Chinatown provided the foundation for an institutionalized racism that was sanctioned and enforced by the state. In arguments reminiscent of Guelke (1982), she conceptualizes Chinatown as a historically specific idea, as well as an empirical reality. As she states it: "racial ideology has been materially embedded in space . . . and it is through 'place' that it has been given a local referent, become a social fact, and aided in its own reproduction" (584). Balancing the notion of Chinatown as both an idea and a social fact, Anderson shows how civic authorities in Vancouver manipulated mental categories about race, in order to define people of Chinese origin, as well as their residential space. We quote at some length:

> In the Vancouver case, "Chinatown" accrued a certain field of meaning that became the justification for recurring rounds of government practice in the ongoing construction of both the place and the racial category. Indeed the state has played a particularly pivotal role in the making of a symbolic (and material) order around the idiom of race in Western societies. By sanctioning the arbitrary boundaries of insider and outsider and the idea of mainstream society as "white," the levels of the state have both "enforced" and "propagated" a white European hegemony. (Anderson 1987, 584–5)

In this extract, Anderson contends not only that powerful social institutions have generated, in her words, a western landscape type, but also how the existence of Chinatown reinforced the racial ideology that begat it.

The examples discussed above remind us that there are complex and important historical forces that have generated the urban landscapes of the past (and of course, the present). As a result of the enduring and pervasive ahistorical legacy of "the founding fathers" of Anglophone urban geography, contemporary urban historical geography needs to break out of the limiting conditions imposed by them, to produce truly "historical" geography and not simply "old" geography (see also the chapter on historical geography). By so doing, it will also allow us to provide a greater depth to efforts to understand the social nature of the city, to which we now return.

THE URBAN QUESTION

As we have seen, the urban sociological tradition owes a major debt to the initial formulations produced by the Chicago ecological school back in the 1920s, and has never really overcome the major problems faced by that tradition. These issues—which have all-too-obvious implications for urban geography—are well scrutinized by Scott (1980, 66–73), who notes that for the ecologists, the city was an a priori category with its own dynamics, tensions, and social relations, somehow apart from and indeed reshaping the social fabric as a whole.

Unsurprisingly, there have been a series of reactions against such a position, although they have not always produced compatible results. Castells' early work, written in French and not translated until later, asked "y-a-t-il une sociologie urbaine?", a question which specifically confronted the meanderings of then-current urban sociology. His reformulation of the urban question was in a very specific mold—emphasizing the nature of collective consumption—but it represented a reemphasis upon the existence of the urban category as a distinct (sociospatial) unit with explicit functional characteristics (Castells 1977).

Saunders, also a sociologist, notes that there exists "a choice between sociological non-urban theories and urban non-sociological theories" (1981, 257). By this he means that the phenomena that might be studied "sociologically" are rarely, if ever, restricted to urban areas, or as he puts it, "social processes cannot be confined within particular locations" (1981, 9). Conversely, the things about cities that may be interesting—in which category he includes specifically the circulation of capital—cannot be examined via sociological constructs. We might well quibble that Saunders is prizing apart aspects of knowledge here (political economy and sociology) and loosely interweaving terms (urban, spatial). This notwithstanding, he is at least consistent in his commitment to a nonspatial (although not an aspatial) perspective on urban analysis, i.e., a perspective that sees cities simply as arenas in which more general processes are resolved.

Saunders' monograph is thus but one recent attempt to deal with the fundamental question of whether we study cities as arenas or objects. (Others have followed; e.g., see Gottdiener 1985.) Urban geographers, it might be supposed, would have developed a clear notion of the city as a "real object," to employ Castells' now-rejected Althusserian language, an explicit sociospatial unit. This is, though, not the case. As argued above, the ecological legacy was, to the geographer, a spatial one, and not an explicitly social one. In consequence, the question "is there a (so to speak) geographie urbaine?" has only been asked intermittently, and indeed has not been able to counteract a powerful and widespread push away from the urban question. Saunders, for one, argues that the city has no social significance in societies that postdate the capitalist transformation:

> ... as Weber saw, as Durkheim saw, and as Giddens clearly sees, the smallest discrete spatial unit which can be taken as the basis for sociological analysis in contemporary capitalist societies is that defined by the territorial boundaries of the nation state. (Saunders 1983, 237)

This proposition is based upon the lead of the "founding fathers" like Weber. Yet, it is implausible on a series of levels:

1. It wishes away, rather than confronts, large areas of the literature, such as Castells' assertion that "the city is society".
2. It assumes a total decline in the authoritative role exerted by cities in class societies.
3. The nation state is elevated—as the smallest spatial unit that can be recognized and analyzed—to a theoretical position that cannot be sustained.

Geographical contributions to this basic discourse on the nature of the urban question have been, for the most part, imitative of those already evolving in other disciplines. Indeed, it is indicative that the recent discussion on the future of urban studies has centered so explicitly upon the efficacy of *method* (Marxism, realism, or some alternative), rather than the ability to say anything about the evolving nature of cities in advanced societies (Society and Space 1987; Cadwallader 1988). While it would be naive to assume that these were separate issues, and that by calling for more empirical work on cities we could in some simple way make "progress"—however that might be defined—it is not accidental that so much of the research that takes place below the scale of the nation-state has shifted to examination of the locality rather than the city.

Quite simply, the locality holds out an opportunity to say something substantive about matters that have been thrown up by theoretical analysis, without falling into the endless debates that center around urban definition. In large measure, this constitutes a failing of will that comes close to that displayed by those who dispense with the city in entirety. The study of the locality is, let there be no mistake, quite crucial to a necessary reorientation of social science. But the same is true of the city. To walk away from urban studies is to ignore a socially constructed phenomenon of massive import.

To search for rural-urban contrasts is to miss the point: in the last two centuries we have moved from a particular form of sociospatial organization which dominated the landscape, to another which has almost entirely replaced it. This is enough of a justification for urban study in itself. But as David Harvey reminds us, this metamorphosis has not been a simple project. He notes:

> ... the urbanization of capital is an objectification in the landscape of that intersection between the productive force of capital investment and the social relations required to reproduce an increasingly urbanized capitalism. But this implies that we should look also at the implications for political consciousness of such processes. The "urbanization of consciousness" has, I therefore submit, to be taken as a real social, cultural, and political phenomenon in its own right. (Harvey 1985a, xxviii)

Looking at the city from this vantage point suggests a number of ways in which urban geography can evolve. First, it is clear that a historical consciousness is crucial. No interpretation of an individual city can proceed—as W. M. Davis dictated, with all the insight and finality that a geomorphologist could bring to this task—as an existential exercise. Conversely, and perhaps more crucially, no study of social evolution could neglect the role of urban economies, first as mercantile, and then as industrial centers.

Second, we must view the city as an evolving thing, a sedimented collection of practices that are being recreated constantly. Talk of postindustrial cities is nonsensical; no social, political, and economic relation can move into a new phase without pulling with it a massive baggage from its collective past. This realization has been spelled out with some success by Michael Dear and his colleagues in their analysis of Los Angeles. While they look forward to some postmodern creation, there is no suggestion that such a city form would be divorced from what has gone before (Dear 1986; Knox 1987a; *Society and Space* 1986).

As a third imperative, we would argue that our studies of the component parts of urban life must pay more attention to the social meaning of what we study. In the urban setting, we can identify briefly a number of efforts to do just that. The first example is provided by Allen Scott, whose recent work on industrial change suggests that there is an explicit reorganization taking place at the urban scale. He notes that this has particular implications for urban research: "cities are implicated in the reproduction of the social relations of capitalism in complicated, contradictory and unpredictable ways; and in turn, those ways that are given in history produce widely varying versions of the urban process" (Scott 1986, 34).

A second instance is offered by Lauria and Knopp, in their work on the gay community in Minneapolis. They use a social constructivist lens through which to interpret the way in which this minority community has both consolidated itself spatially within certain neighborhoods, and employed its political acumen to maintain and increase political influence (Lauria and Knopp 1985).

A third example is to be found, despite Harvey's strictures, in the city as a legal unit. Very little work has been done in the past on law as an interpretable text, and Clark has shown with great originality the possibilities—and imperatives—of this field (Clark 1985).

A fourth example is in the context of local economic development. In the past decade, we have seen a necessary move to reincorporate the city into regional, national, and global economies, thereby allowing us to view the former, not as a self-contained entity, but as enmeshed in broader currents of capital movement. Obvious examples are research on regional changes (Sunbelt, Snowbelt), and urban places as global cities. In part, though, the pendulum has swung back to the individual city and to the strategies that it can employ in initiating local economic-development strategies (e.g., see Kirby and Lynch 1987; Soja, Morales, and Wolff 1983).

Much of this research works within the growth-machine framework developed in sociology, and examines the political, economic, and environmental issues that surround growth and decline within specific communities. We do not, of course, claim these examples to be the only representatives of important research (nor even without fault and beyond criticism). They do, however, indicate some of the intriguing possibilities.

SUMMARY AND CONCLUSIONS

In this overview, we have been selective, and have attempted to provide a sketchmap of some of the issues within urban geography, rather than a compendium of everything ever written. We have argued that the field has been burdened by a crippling historical legacy, which it has only begun to cast off in recent years. It is developing a historical consciousness, has sophisticated analytical skills, and is moving into new and challenging areas of substantive research, like law. While it may not yet possess the reputation of urban politics, urban geography has certainly overtaken urban sociology in terms of sophistication, and is now poised to make significant contributions to American urban studies.

REFERENCES

Agnew, J.; Mercer, J.; and Sopher, D. 1984. *The city in cultural context.* London: George Allen & Unwin.

Anderson, K. 1987. The idea of Chinatown: The power of place and institutional practice in the making of a racial category. *Annals of the Association of American Geographers* 77:580–98.

Badcock, B. 1985. *Unfairly structured cities.* Cambridge: Blackwell.

Berry, B. J. L. 1963. *Commercial structure and commercial blight: Retail patterns and processes in the city of Chicago.* University of Chicago Department of Geography Research Paper No. 85. Chicago: University of Chicago.

———. 1973. *The human consequences of urbanization.* London: Macmillan.

Borchert, J.; Bourne, L.; and Sinclair, L., eds. 1986. *Urban systems in transition.* Netherlands Geographical Studies No. 16, Utrecht.

Bourne, L. 1980a. Alternative perspectives on suburban decline and population deconcentration. *Urban Geography* 1:39–52.

———. 1980b. *The geography of housing.* London: Arnold.

———; Sinclair, R.; and Dziewonski, K., eds. 1984. *Urbanization and settlement systems: International perspectives.* New York: Oxford University Press.

Bradbury, J. 1985. Regional and industrial restructuring processes in the new international division of labor. *Progress in Human Geography* 9:38–63.

Browett, J. 1984. On the necessity and inevitability of uneven development under capitalism. *International Journal of Urban and Regional Research* 8:155–77.

Brunn, S., and Williams, J. 1983. *Cities of the world.* New York: Harper & Row.

Burghardt, A. F. 1979. The origin of the road and city network of Roman Pannonia. *Journal of Historical Geography* 5:1–20.

Cadwallader, M. 1985. *Analytical urban geography.* Englewood Cliffs, NJ: Prentice-Hall.

———. 1986. Migration and intra-urban mobility. In *Population geography: Progress and prospect,* ed. M. Pacione, pp. 257–83. London: Croom Helm.

———. 1988. Urban geography and social theory. *Urban Geography* 9(3):227–51.

Carroll, G. 1982. National city-size distributions: What do we know after 67 years of research? *Progress in Human Geography* 6:1–43.

Castells, M. 1977. *The urban question.* London: Arnold.

Chouinard, V. 1988. Explaining local experiences of state formation: The case of cooperative housing in Toronto. *Environment and Planning D: Society and Space.* In press.

Clark, C. 1951. Urban population densities. *Journal of the Royal Statistical Society* A114:490–96.

Clark, G. L. 1984. Who's to blame for racial segregation? *Urban Geography* 5:193–209.

———. 1985. *The Cities and the judges.* Chicago: University of Chicago Press.

Clark, W. A. V., ed. 1982. *Modelling housing market search.* London: Croom Helm.

———. 1987 School desegregation and white flight. *Social Science Research* 16(3):211–28.

Clarke, S. E.; Redburn, F. S.; and Buss, T. 1985. The local impacts of economic change. *Urban Affairs Quarterly* 21(2):139–41.

669

Cosgrove, D. 1984. *Social formation and symbolic landscape*. New York: Barnes & Noble Books.

Curry, L. 1967. Central places in the random spatial economy. *Journal of Regional Science* 7:217–38.

Dacey, M. 1966. A probability model for central place locations. *Annals of the Association of American Geographers* 56:549–68.

Daniels, P., and Warnes, A. 1980. *Movement in cities: Spatial perspectives on urban transport and travel*. New York: Methuen.

Darden, J. T. 1983. Sharing residential space in the 1920s. *Ethnic and Racial Studies* 6:237–45.

Davies, W. K. D. 1984. *Factorial ecology*. Aldershot: Gower.

Davis, W. M. 1932. A retrospect of geography. *Annals of the Association of American Geographers* 22(4):211–30.

Dear, M. J. 1986. Postmodernism and planning. *Environment and Planning D: Society and Space* 4(3):367–84.

———, and Wolch, J. 1987. *Landscapes of despair*. Cambridge: Polity Press.

Deskins, D. R. 1981. Morphogenesis of a black ghetto. *Urban Geography* 2:95–114.

Duncan, J. 1987. Review of urban imagery: Cognitive mapping. *Urban Geography*. 8:264–72.

Erickson, R. 1986. Multinucleation in metropolitan economies. *Annals of the Association of American Geographers* 76:331–46.

Everitt, J. 1976. Community and propinquity in a city. *Annals of the Association of American Geographers* 66:104–16.

Fellmann, J. D. 1957. Pre-building growth patterns in Chicago. *Annals of the Association of American Geographers* 47(1):59–82.

Fincher, R. 1984. Identifying class struggle outside commodity production. *Environment and Planning D: Society and Space* 2(3):209–28.

Gaile, G. L., and Hanink, D. M. 1985. Relative stability in American metropolitan growth. *Geographical Analysis* 17(4):341–48.

Geiger, R. K., and Wolch, J. R. 1986. A shadow state? Voluntarism in metropolitan Los Angeles. *Environment and Planning D: Society and Space* 4(3):351–66.

Goddard, J. B., and Kirby, A. M. 1976. *An introduction to factor analysis*. Norwich: Geo Books.

Golledge, R., and Rushton, G. 1984. A review of analytic behavioural research in geography. In *Geography and the urban environment: Progress in research and applications,* eds. D. Herbert and R. Johnston, Volume VI, pp. 1–43. Wiley: New York.

———, and Stimson, R. 1987. *Analytical behavioural geography*. London: Croom Helm.

Gottdiener, M. 1985. *The social production of urban space*. Austin: University of Texas Press.

Graham, B. 1979. The evolution of urbanism in Ireland. *Journal of Historical Geography* 5:111–25.

Gray, F. 1975. Non-explanation in urban geography. *Area* 7 228–35.

Greenow, L. 1984. City and region in the credit market of late colonial Guadalajara. *Journal of Historical Geography* 10:263–78.

Guelke, L. 1982. *Historical understanding in geography: An idealist approach*. Cambridge: Cambridge University Press.

Haggett, P., and Bassett, K. 1970. The use of trend-surface parameters in inter-urban comparisons. *Environment and Planning A* 2:225–37.

Harvey, D. W. 1973. *Social justice and the city*. London: Arnold.

———. 1979. Monument and myth. *Annals of the Association of American Geographers* 69:362–81.

———. 1985a. *Consciousness and the urban experience*. Baltimore: Johns Hopkins University Press.

———. 1985b. *The urbanization of capital*. Baltimore: Johns Hopkins University Press.

Holcomb, B. 1986. Women in the city. *Urban Geography* 5(3):247–54.

Huff, J., and Clark, W. A. V. 1978. Cumulative stresses and cumulative inertia: A behavioral model of the decision to move. *Environment and Planning A* 10:1101–19.

Hughes, M. A. 1987. Moving up and moving out: Confusing ends and means about ghetto dispersal. *Urban Studies* 24:503–17.

Huntington, C. C., and Carlson, F. A. 1933. *The geographical basis of society*. New York: Prentice-Hall.

Jakubs, J. F. 1986. Recent racial segregation in the US SMSAs. *Urban Geography* 7:146–63.

Johnson, J. H., Jr., and Oliver, M. L. 1988. *Ethnic dilemmas in comparative perspective*. University of California. Mimeo.

Johnston, R. J. 1971. *Urban residential patterns: An introductory review*. New York: Praeger.

———. 1980. On the nature of explanation in human geography. *Transactions of the Institute of British Geographers* 5(4):402–12.

Kearns, R. 1984. Review of Guelke 1982, Historical understanding in geography: An idealist approach. *Environment and Planning D: Society and Space* D2(1):116–17.

King, L. 1984. *Central place theory*. Beverly Hills: Sage Publications.

Kirby, A. M., and Lynch, K. A. 1987. A ghost in the growth machine. *Urban Studies* 24:587–96.

Knox, P. L. 1987a. The social production of the built environment: Architects, architecture and the postmodern city. *Progress in Human Geography* 11(3):354–77.

———. 1987b. *Urban social geography*. Harlow: Longman.

———; Bohland, J.; and Shumsky, N. L. 1984. Urban development and the provision of personal services. In *Public service provision and urban development,* eds. A. Kirby, P. L. Knox, and S. Pinch, pp.152–75. New York: St. Martins.

Latham, R., and Yeates, M. 1970. Population density growth in metropolitan Toronto. *Geographical Analysis* 2:177–85.

Lauria, M., and Knopp, L. 1985. The role of gay communities in the urban renaissance. *Urban Geography* 6:152–69.

Lynch, K. 1960. *The image of the city*. Cambridge: MIT Press.

Marks, B. E. 1982. Rural response to urban penetration. *Journal of Historical Geography* 8:113–27.

Marston, S. A. 1988. Neighborhood and politics: Irish ethnicity in 19th century Lowell, Massachusetts. *Annals of the Association of American Geographers* 78(3):414–32.

———, and Kirby, A. M. 1988. Urbanization, industrialization and the social creation of a space economy. *Urban Geography* 9(4):358–75.

Mayer, H. M. 1979. Urban geography and Chicago in retrospect. *Annals of the Association of American Geographers* 69(1):114–18.

Morrill, R. 1987. The structure of shopping in a metropolis. *Urban Geography* 8:97–128.

Muth, R. 1985. Models of land-use, housing, and rent: An evaluation. *Journal of Regional Science* 25:593–606.

Newling, B. 1969. The spatial variation of urban population densities. *Geographical Review* 59:242–52.

Olson, S. H. 1979. Baltimore imitates the spider. *Annals of the Association of American Geographers* 69:557–74.

Palm, R. I. 1981. *The geography of American cities*. New York: Oxford University Press.

———. 1985. Ethnic segregation of real estate agent practice in the urban housing market. *Annals of the Association of American Geographers* 75:58–68.

Platt, R. S. 1931. An urban field study. *Annals of the Association of American Geographers* 21(1):52–74.

Pratt, G. 1986. Housing tenure and social cleavages in urban Canada. *Annals of the Association of American Geographers* 76(3):366–80.

Pred, A. 1964. The intrametropolitan location of American manufacturing. *Annals of the Association of American Geographers*. 54:165–80.

———. 1977. *City-systems in advanced economies: Past growth, present processes, and future development options*. New York: Wiley.

———. 1981. Production, family, and free-time projects. *Journal of Historical Geography* 7:3–36.

———. 1984. Structuration, biography formation, and knowledge: Observations on port growth during the late mercantile period. *Environment and Planning D: Society and Space* 2(3):251–76.

Preston, V. 1987. Spatial choice behavior in urban environments. *Urban Geography* 8:374–79.

Robson, B. 1973. *Urban growth: An approach*. London: Methuen.

Rose, D. 1984. Rethinking gentrification. *Environment and Planning D: Society and Space* 2(1):47–74.

Rowntree, L. B., and Conkey, M. W. 1980. Symbolism and the cultural landscape. *Annals of the Association of American Geographers* 70:459–74.

Saunders, P. 1981. *Social theory and the urban question*. London: Hutchinson.

———. 1983. Social theory and the urban question: A response to Paris and Kirby. *Environment and Planning D: Society and Space* 1(2):234–40.

Scarpaci, J.; Infanto, R. P.; and Gaete, A. 1988. Planning residential segregation: The case of Santiago, Chile. *Urban Geography* 9:19–36.

Scott, A. J. 1980. *The urban land nexus and the state*. London: Pion.

———. 1982. Locational patterns and dynamics of industrial activity in the modern metropolis. *Urban Studies* 19:111–41.

———. 1986. Industrialization and urbanization: A geographical agenda. *Annals of the Association of American Geographers* 76(1):25–37.

Shannon, G. W., and Cromley, R. G. 1982. Philadelphia and the yellow fever epidemic of 1798. *Urban Geography* 3:335–70.

Simmons, J. 1986. The urban system: Concepts and hypotheses. In *Urban systems in transition,* eds. J. Borchert, L. Bourne, and R. Sinclair, pp. 23–31. Netherlands Geographical Studies No. 16, Utrecht.

Slater, T. R. 1985. The urban hierarchy in medieval Staffordshire. *Journal of Historical Geography* 11:115–37.

Smith, N. 1984. *Uneven development: Nature, capital, and the production of space.* Oxford: Blackwell.

Society and Space. 1986. Los Angeles: Capital of the late twentieth century. *Environment and Planning D: Society and Space* 4(3):249–390.

———. 1987. Reconsidering social theory: A debate. *Environment and Planning D: Society and Space* 5(4):367–434.

Soja, E. W.; Morales, R.; and Wolff, G. 1983. Urban restructuring: An analysis of social and spatial change in Los Angeles. *Economic Geography* 59:195–230.

Solot, M. 1986. Carl Sauer and cultural evolution. *Annals of the Association of American Geographers* 76(4):508–20.

Thrall, G. 1987. *Land use and urban form: The consumption theory of land rent.* New York: Methuen.

Vance, J. 1970. *The merchant's world.* Englewood Cliffs, NJ: Prentice-Hall.

Walker, R. 1978. The transformation of urban structure in the nineteenth century and the beginnings of suburbanization. In *Urbanization and conflict in market societies,* ed. K. Cox, pp. 165–211. London: Methuen.

———. 1981. A theory of suburbanization: Capitalism and the construction of urban space in the United States. In *Urbanization and urban planning in capitalist society,* eds. M. Dear and A. Scott, pp. 383–429. London: Methuen.

Ward, D. 1980. Environs and neighbors in the "Two Nations": Residential differentiation in nineteenth century Leeds. *Journal of Historical Geography* 6:133–62.

———. 1982. The ethnic ghetto in the United States: Past and present. *Transactions of the Institute of British Geographers,* NS, 7:257–75.

———. 1984. The progressives and the urban questions: British and American responses to the inner city slums 1880–1920. *Transactions of the Institute of British Geographers,* NS, 9:299–314.

Warnes, A., and Daniels, P. 1979. Spatial aspects of an intrametropolitan central place hierarchy. *Progress in Human Geography* 3:384–406.

Western, J. 1981. *Outcast Capetown.* London: Allen & Unwin.

Wilson, A. 1971. A family of spatial interaction models, and associated developments. *Environment and Planning A* 3:1–32.

Winters, C. 1981. The urban system in medieval Mali. *Journal of Historical Geography* 7:341–55.

Wolpert, J., and Reiner, T. 1984. Service provision by the not-for-profit sector. *Economic Geography* 60:28–37.

Wrigley, N., and Longley, P. 1984. Discrete choice modelling in urban analysis. In *Geography and the urban environment: Progress in research and applications,* Volume VI, eds. D. Herbert and R. Johnston, pp. 45–94. New York: Wiley.

Geographic Perspectives on Women

Eve Gruntfest

Open nearly any human geography text published before 1975, and you will find that men are representative of the species. Generalizations made about the human condition are drawn largely from men's experiences. Vast descriptions of economic development, urban spatial analysis, early settlement history, and sense-of-place research assume that the masculine view speaks for the human view.

Within the last 15 years, more articles and books have recognized the role that gender plays in human spatial behavior. A new feminist geography has emerged in many nations, including the United Kingdom (Women and Geography Study Group of the Institute of Behavioral Geography 1984; Momsen and Townsend 1987; Bowlby et al. 1982); the Netherlands; Spain; Italy; France; Australia; Taiwan; India (Bagchi 1981, 1987); Brazil; and elsewhere (Ardener 1981). The potential now exists for addressing questions that have eluded human geographers because of their previously narrow focus. The greatest amount of feminist geography has been published in the U.S. and Canada. This research takes three major directions, discussed below.

First there is a focus upon expanding our understanding of women in the world. For many years, women were officially invisible. Great strides have been made in our understanding of the roles women play, the availability of data on women, and in making these data available for research and teaching to a degree similar to the availability of information on men.

The second strand of research has two components: (1) integrating information on women into the larger body of geographic literature, thus making women an integral part of human geography, and (2) describing women as a disadvantaged group. Descriptive studies tended to come first. These studies documented inequalities be-

---

The author appreciates the effort of many active members of the Geographic Perspectives on Women Specialty Group who helped in the preparation of this chapter. Janice Monk, Jacquelyn Beyer, Susan Hanson, David Lee, Suzanne MacKenzie, and Marie Truelove offered rich comments on earlier drafts. The author is also grateful for the editorial assistance of Carole Huber, Veronica Pillsbury, and Renate Speer at the University of Colorado at Colorado Springs.

tween women and men, and indicated that women often face spatial constraints that place them at a disadvantage relative to men.

The integration idea is more associated with the goals of feminist criticism (e.g., Monk and Hanson 1982), or with revision of teaching (e.g., Monk 1984b). Integration calls for studies to focus not just on the "household head," or to assume that both sexes are represented if only one sex has been interviewed. Gender becomes one of the variables of interest when researching any issue in human geography. The title of Monk and Hanson's (1982) article, "On Not Excluding Half of the Human in Human Geography," captures the essence well.

The third trend, and perhaps the area having the greatest potential to influence theories in geography, is the emerging feminist theory. The early phases of feminist geography in North America critiqued human geography for its failure to examine women's lives, and included descriptive studies of differences in women's and men's spatial behavior and responses to the environment. This established gender as an important aspect of human experience which should be incorporated into geographic research.

Feminist geography argues against gender-blindness. It argues for the study of gender relations, as well as the study of women. Theory developed to show the significance of gender relations, especially in urban geography, emphasizes the constraints exposed by capitalism and patriarchy. Current North American emphasis has responded to limitations of the structural approach, in which women are cast as passive victims. Focus is now upon women as active agents in society, with the objective of learning how they view the world, what decisions they make, and how they deal with the constraints imposed by larger social forces (Monk 1988):

> Since its appearance in the 1970s as an academic expression of the women's movement, feminist geography has become more complex in the questions it asks and the approaches it employs, reflecting an increasingly sophisticated scholarship. For each of us, however, our approach to feminist research is shaped by our personal world views and situations and by the problems and nature of our own societies and their cultural and intellectual traditions. (Monk 1988, 1)

This chapter takes a representative sample of recent contributions in feminist geography. It is not all-inclusive. The topics considered here—gender in a global perspective, gender and work, gender and the landscape, methodological applications, gender-balancing in the curriculum, and new directions—illustrate the main areas of research, and can be drawn directly and explicitly into ongoing research and the classroom.

Most importantly, this chapter shows how our discipline suffers intellectually if we fail to appreciate the role that gender plays in human spatial interaction. We call for the incorporation of the concepts and data on gender, and for new work to enrich our discipline creatively by more accurately reflecting human behavior at an individual and societal level.

GENDER IN A GLOBAL PERSPECTIVE

A recently published collection of papers (Momsen and Townsend 1987) and three innovative atlases of women (Seager and Olson 1986; Shortridge 1987; Gibson and

Fast 1986) provide excellent illustrations of spatial variation in the lives of women, both globally and within a North American context. Seager and Olson (1986) not only show what is happening between women and men, and among women themselves, but where:

> By mapping the world of women, patterns are revealed that are usually obscured in statistical tables or in narratives. The similarities and differences, the continuities and contrasts among women around the world are best shown by—literally—mapping out their lives . . . what we do see, however, is that everywhere women are worse off than men: women have less power, less autonomy, more work, less money, and more responsibility. Women everywhere have a smaller share of the pie [;] if the pie is very small (as in poor countries) women's share is smaller still. Women in rich countries have [a] higher standard of living than do women in poor countries but nowhere are women equal to men. (Seager and Olson 1986, 7)

Great differences exist among women, country by country and region by region (Andrews 1982; Lee, Stewart, and Winter 1979). Many geographers have written about women and development issues. For example, Hayes (1986) focused on women and wood fuel in Kenya, and Carney (in press) examined women and rice production in Gambia. From the Seager and Olson atlas (1986, 7–8) we learn that:

- In Afghanistan, 4% of eligible girls are enrolled in secondary school; in Australia 88% are.
- In Angola, fewer than 1% of adult women have access to contraceptives; in Belgium 78% do.
- Women in Ghana bear an average of over six children; women in West Germany fewer than two.
- In Jamaica, the maternal mortality rate stands at 106 mothers' deaths for every 100,000 births; in Norway there are fewer than 8 deaths per 100,000 births. (Incidentally, maternal mortality statistics are much more difficult to find than infant mortality figures.)

There are gender-related differences within countries. Since women live longer than men in North America, one of the public-policy issues for the future concerns old, old women, those 85 years and older (Morrow-Jones 1986). Women live longest in the Plains states and in the West, with the notable exceptions of Alaska and Nevada. Another small pocket of longevity occurs in five New England States. In contrast, women's average lifetime is shorter in the South and selected portions of the industrial metropolitan northeast. The shortest lifetime is in South Carolina (72.3 years), closely followed by Mississippi and Louisiana. A woman's chance of living a longer life is increased substantially by living in North Dakota, where the average lifetime is 77 years. What environmental factors or lifestyle choices account for these differences (Shortridge 1987, 107–197)?

Simple conclusions cannot be drawn about the relationship between "development" and women's situation. Among countries with the highest ratios of women to men enrolled in universities are the German Democratic Republic (141 women:100 men), Poland (123:100), Kuwait (122:100), the Philippines (120:100), Lesotho (153:100), and Brazil (116:100). Substantially lower ratios are recorded in most areas of western Europe (Monk 1988, 2). These data raise questions about the ways in which cultural values influence gender roles and women's opportunities.

EMERGING PERSPECTIVES ON GEOGRAPHIC INQUIRY

Equality and provisions for women's needs vary widely from one country to another. Data deficiencies and incongruities exist. Some information can be found on almost all topics, but with the exception of a few standard indicators, information on women is conspicuously absent from conventional sources. A large share of the data for the atlas *Women in the World* was gleaned from feminist writing in small presses, newsletters from women's organizations the world over, and "alternative" journals of various descriptions.

The official invisibility of women perpetuates the myth that what women do is less important, less noteworthy, less significant. Women are made invisible by policies and priorities that discount the importance of collecting information about them. Although the United Nations Decade for Women (1975–85) resulted in a considerable increase in international information on women, women are still not generally included in the information mainstream (Seager and Olson 1986, 9).

GENDER AND WORK

To a greater extent than for men, women's lives have centered on the personal or on the private sphere of home and family. Conversely, in industrial and postindustrial societies, men's lives have been more closely associated with the "public" sphere of making money and politics. Geographers have excluded activities that take place in the home from their studies. Urban geography has begun to address the implications of the separation of men's and women's work. (A special issue of *Geoforum* has been devoted to gender relations in urban space—Christopherson 1989; England 1989; Jones and Kodras 1989; Marston and Saint-Germaine 1989; Pratt and Hanson 1989; Regulska 1989; Rose and Chicoine 1989; Smith 1989.) Over 70% of women are in the paid labor force in the USSR, eastern Europe, and southeast Asia, and over 50% in the U.S. and China, but in Middle Eastern countries the rates fall to about 15% (Dixon 1981).

Early research on the roles that women play analyzed and planned for women's growing labor-force participation. It concentrated especially on studying the spatial constraints produced by married women's dual roles as housewives and wage earners. Wives had more restricted activity patterns than their husbands, and wage-earning wives had more restricted activity patterns than fulltime housewives. The empirical conclusions documented the legitimacy and importance of treating women as a geographic subgroup in their own right, and laid the basis for questioning the assumptions of many geographic models. But the concentration on women's spatial constraints per se also limited this research to documenting rather than explaining these constraints.

Then, writers took women's restricted activity spaces as a starting point, and attempted to analyze them as an expression and reinforcement of women's restricted social position. Much of this literature saw the gender division of labor as a space-structuring force, and saw spatial form as reinforcing women's restricted social position. There is now increasing empirical evidence for the importance of understanding women's environmental constraints as one set of forces contributing to women's restricted social position. By and large, women were presented as victims of gender roles and expectations, and not as actual or potential creators of environments.

A 1988 special issue of *Urban Geography* builds on a series of papers cosponsored by the Specialty Group at the Association of American Geographers annual meeting in Portland. The emphasis is on the spatial dimensions of sex segregation in employment. "Each paper documents striking differences between women and men . . . Should the goal be to eliminate sex segregation in employment?" asks guest editor Susan Hanson.

> Despite differences in theoretical groundings and methodological approaches, a unifying theme emerges: the interdependence of social, economic and locational factors which explain women's shorter journey to work. (Giuliano 1988, 203)

Four particular areas for further research are recommended:

1. The absence of a consistent explanation for women's shorter work trips
2. The relationship between gender-based spatial segregation and equal opportunity in the workplace
3. The relative impact of spatial segregation
4. The evaluation of research results in the context of further changes in mobility

A solid basis is now available for the development of a more complete understanding of the characteristics, extent, and implications of gender-based spatial labor-market segmentation (Hanson and Johnston 1985; Johnston-Anumonwo 1988; Hanson and Pratt 1988; Rutherford and Wekerle 1988; Villeneuve and Rose 1988).

Within the urban context, geographers are also beginning to examine the suburbanization of clerical work in American metropolitan areas, the growing importance of work in electronic cottages, demographic shifts resulting from reductions in household size, and the complex transportation issues (Holcomb 1984a, 1984b; Fox 1983, 1985; Gober 1980; Fincher 1987; Brooker-Gross and Maraffa 1985; Mackenzie 1985, 1986; Marston 1986; Mackenzie and Rose 1982; Miller 1983).

GENDER AND LANDSCAPE

In creating landscapes, women can express their social and personal identities. Traditional landscape research, however, emphasized men's interaction with landscape, and ignored women's representation of and interaction with the landscape (Monk 1984a). Research on women and landscape provides an opportunity to test theories about the relationships between gender roles and place identity, and to extend the bounds of traditional landscape research in the discipline from exterior landscapes, long associated with men, to interior landscapes (especially of the home), the domain women have had more opportunity to shape (Loyd 1975; Hayden 1981, 1984).

Because traditional research has investigated primarily mens' response to the landscape, a great deal of women's writings, including diaries, are ignored. Recent interdisciplinary studies, as well as geographic works, reveal that women's visions of the American West historically varied greatly from men's, e.g., Schlissel (1982), Wilkinson (1979), Cragg (1980), and Kolodny (1984). Whereas men often wanted to conquer, possess, and subdue the land, nineteenth-century women were more inclined to regard the landscape as a sanctuary to be made into a new home, providing new security in often unknown surroundings.

These visions are not uniform, however, as revealed in the recent book edited by Norwood and Monk (1987), *The Desert Is No Lady: Southwestern Landscapes in Women's Writing and Art*. They show how ethnicity, historic period, and "insider-outsider" status are associated with differences among women, even though women's responses still differ fundamentally from men's. Their work also extends the definition of "art" to include crafts, such as weaving and pottery, traditionally done by women and ignored by the "fine" art community. This illustrates again how different sources need to be studied to understand women's expressions.

Twentieth-century women writers born in the arid Southwest—Anglo-Americans, Mexican-Americans, and American Indians—have placed value on adapting to an environment of scarcity. When they write metaphorically of the desert landscape as a woman, they admire her strength and resistance to human transformation, or they see her as an old and wise woman who teaches us how to live. Rather than valuing empty and heroic landscapes, these writers find beauty and meaning in the ordinary and everyday: the flower that survives in the asphalt parking lot, as a reminder of the people who once lived on the site; the simple plastered adobe homes or the peeling painted messages left by children on the walls of public-housing projects (Norwood and Monk 1987).

Monk (1984a) calls for two directions for future research on women and landscape: (1) more studies of landscape responses that recognize the diversity of women, expanding on the early emphasis on middle class, Anglo-American women; and (2) studies of women in different contexts. Work to date has focused on women in the American West, especially in the nineteenth century, and twentieth-century North American cities and suburbs.

METHODOLOGICAL APPLICATIONS

We have noted the difficulties of obtaining adequate statistics on women's lives, and the efforts in landscape studies to turn to such sources as diaries and folk arts to discover women's own visions and interpretations of experience. Several examples are provided by Monk (1988). She notes that geographers dealing with contemporary urban issues are also turning to new methods, particularly to open-ended qualitative interviews to learn how women frame their own realities and develop strategies for daily living.

For example, England (1987) conducted open-ended interviews with women clerical workers in suburban areas near Columbus, Ohio. Using statistical analysis, she established the local growth of women's clerical employment. She then conducted in-depth interviews to explore the strategies used to handle the dual roles of employee and housewife, and whether the length of the journey to work was critical in their decision-making. These interviews reveal complex interactions among individual, marital, family, and work histories, including such themes as the need for income in the past and projected for the future, assessments of career prospects, evaluations of child-care options, and housing preferences. England provides extended quotations in which women describe different strategies.

Hanson and Pratt (1988) have collected the same sort of qualitative data on a larger sample that is representative of the population—including all occupation types, and men.

Turning to studies of how women feel about and shape environments, we again find scholars using in-depth interviews to discover women's perspectives. Geraldine Pratt's conversations with upper-middle-class women in Vancouver reveal how their sense of self relates to their environments. She contrasts two groups—those from old, established families, and those who are socially mobile and cosmopolitan in their origins.

To discover whether and how the gender of a respondent is critical in survey research, Hansen (1988) conducted open-ended interviews with husbands and wives in the same households in Hermosillo, Mexico. Analyzing their stories about migration decisions, she found differences in every household in male and female interpretations of their family's move. Careful assessment of language usage often revealed that men expressed a greater sense of autonomy and choice, and women expressed a greater sense of obligation to the family.

Similarly, Christopherson (in progress) is studying contributions of both husbands and wives to household income formation in nonmetropolitan communities. Hanson and Pratt have compared income and employment of women and men in the same households, to question theories of the homogeneous class composition of urban neighborhoods (1988).

GENDER-BALANCING THE CURRICULUM

Geographers are actively providing methods for gender-balancing our teaching. Coming of age for this endeavor is indicated by the 1982 Zelinsky, Monk, and Hanson article, the 1982 Monk and Hanson contribution, and the 1983 Mazey and Lee resource paper (Association of American Geographers). Before choosing a text, geographers should carefully review the language, graphics, content, and references cited. Many ideas are now available for integrating gender content into courses. Among those who have participated in faculty development projects designed to integrate the new research into their teaching are Brooker-Gross and Eborall (1988), Monk (1984b, 1988), Larimore (1978), McDowell and Bowlby (1983), Mackenzie (1986), and Rengert and Monk (1982).

Courses aimed at gender-balancing are offered at many universities (Drake 1983); only a few are mentioned here:

- At Southern Illinois University, Bagchi's course "Women and the Global Environment" examines the sociocultural framework of the sexual division of labor in various geographic regions, under different economic and social systems. The course emphasizes female roles in societies that are adjusting to development and culture change.
- At Cornell University, Christopherson teaches "Women and Urban Living—Implications for Planning."
- At the University of Colorado at Colorado Springs, Beyer initiated a class entitled "Women's Space, Women's Place, Women's Role in Changing the Face of the Earth."
- Brooker-Gross co-teaches an urban geography class that combines fiction and nonfiction sources (Brooker-Gross and Eborall 1988).

Geographers are urged to consult Lee's comprehensive bibliography of all works in known geography publications that make reference to women or gender issues (1988).

Teaching of traditional topics must be transformed by incorporating gender perspectives, and by adding new topics that illuminate women's lives. At the simplest level, exercises can be introduced that ask students to count the representation of women and men in pictures in their textbooks, to see who is presented as central or peripheral, and to look at whether they have been presented in active or passive roles. Language usage can be analyzed in a similar way. Are workers always described as "he"? Is the "family farm" presented only as a unit where the farmer and his sons do the work, and hand the property down through the generations?

> No geography that hopes to motivate students to create a "better world" and prepare them to bring that world into being can presume it will achieve its goals if it leaves out half of humanity, fails to challenge gender inequities and does not show women and girls as valid and valuable sources of information and important contributors to society. . . . A curriculum that presents the world through the eyes of both genders, like a stereoscopic perspective does not come into focus for all of us instantly. To "engender" the new geography calls for a transformation of our vision so that we see the world in stereoscopic perspective and create a curriculum that serves the interests of women as well as men. (Monk and Williamson-Fien 1986, 97)

NEW DIRECTIONS

Over the last 15 years, research in feminist geography has clearly documented that gender influences human experience of space and place around the world. Much of this research has shown that women globally suffer from many inequalities—of income, education, power, and property rights, despite their heavy responsibilities for and contributions to economic and social life.

As the body of research grows, however, we are learning not only about the difficulties that women face, but about how they cope with constraints and shape their own lives and those of others. Research is revealing how their strategies and experiences vary among places. But it is important that we develop understanding of both the similarities and the differences among women (Monk 1988). We need to see how variables including class, ethnicity, life stage, sexual preference, and the specifics of place interact with gender to create diversity among women (Monk 1988, 1).

Two early phases of feminist geography in North America involved (1) criticism of human geography for its failure to examine women's lives and (2) descriptive studies of differences in women's and men's spatial behavior and responses to the environment. These phases were concerned with establishing gender as an important aspect of human experience that should be incorporated into geographic research. Additionally, the descriptive studies documented inequalities between women and men in many realms of life. A third phase sought a theoretical framework for analyzing these inequalities. Current work is emphasizing women as social agents who are dealing with constraints and expressing their own values.

Monk and Hanson's (1982) evaluation of the content of human geography identified five types of problems, when the literature is viewed from the perspective of women's lives:

1. Many research questions are inadequately or incompletely framed. Scholars write about areas of life in which women and men both participate, but men's experiences are presumed to represent both.
2. Such omissions lead to the development of "gender-blind" theories.
3. When women are discussed, authors often assume stereotypical and fixed gender roles. They additionally may bring a Western bias to discussions of women's lives in other cultures.
4. Research themes that directly address women's lives are neglected, such as child care, women's legal rights, or unpaid work.
5. Brief recognition may be given to gender differences, but their significance is dismissed in making generalizations.

In addition, Monk has noted that women outside of urban settings in capitalist developed nations have been slighted in theory-building efforts, and that we have devoted most of our effects to traditional academic purposes—explaining how things are or came to be, not how they might be changed. These are important issues for North American geographers to address (Monk 1988).

Gender studies provide the opportunity to extend geography's theoretical base and its applied value. Ten years ago, because the materials were not available, the excuse for leaving women out of the human-geography equation had some merit. This excuse no longer holds.

REFERENCES

Andrews, A. C. 1982. Towards a status of women index. *Professional Geographer* 34:24–31.

Ardener, S. ed. 1981. *Women and space ground rules and social maps.* New York: St. Martin's Press.

Bagchi, D. 1981. Women in agrarian transition in India: Impact of development. *Geografiska Annaler* 63B:95–107.

———. 1987. Rural energy and the role of women. In *Geography of gender in the Third World,* eds. J. Momsen and J. Townsend. Albany: State University of New York Press.

Bowlby, S.; Foord, J.; McDowell, L.; and Momsen, J. 1982. Environment, planning and feminist theory: A British perspective. *Environment and Planning A* 14:711–16.

Brooker-Gross, S., and Eborall, D. 1988. A fully human urban social geography: Combining fact and fiction. Paper presented at the Association of American Geographers, Phoenix, Arizona.

———, and Maraffa, T. 1985. Commuting distance and gender among nonmetropolitan university employees. *Professional Geographer* 37:303–10.

Carney, J. n.d. Struggles over crop rights and labor within contract farming households in a Gambian irrigated rice project. *Journal of Peasant Studies* 15:3. In press.

Christopherson, S. n.d. Household income formation in non-metropolitan Arizona. Supported by the Aspen Institute. In press.

———. 1989. Urban space and the feminization of work. *Geoforum.* In press.

Cragg, B. 1980. Mary Hallock Foote's images of the Old West. *Landscape* 24:42–47.

Dixon, R. 1981. Jobs for women in rural industry and services. In *Invisible farmers: Women and the crisis in agriculture,* ed. D. C. Lewis. Washington: U.S. Agency for International Development.

Drake, C. 1983. Teaching about third world women. *Journal of Geography* 82:163–69.

England, K. V. L. 1987. The employment of women: Clerical work in Ohio. Paper presented at the Association of American Geographers, Portland, Oregon.

———. 1989. Production and reproduction of labor. *Geoforum.* In press.

Fincher, R. 1987. Social theory and the future of urban geography. *Professional Geographer* 39:9–12.

Fox, M. 1983. Working women and travel: The access of women to work and community facilities. *Journal of the American Planning Association* 49:156–70.

———. 1985. Access to workplaces for women: New community designs in the USA. *Ekistics* 310:69–76.

Gibson, A., and Fast, T. 1986. *Women's atlas of the United States*. New York: Facts on File Publications.

Giuliano, G. 1988. Commentary: Women and employment. *Urban Geography* 6:203–8.

Gober, P. 1980. Shrinking household size and its effect on urban population density patterns: Case study of Phoenix, Arizona. *Professional Geographer* 32:55–62.

Hansen, E. R. 1988. Mexican women and the decision to migrate: Multiple respondents in household studies. Master's thesis, Department of Geography, University of Arizona.

Hanson, S., and Johnston, I. 1985. Gender differences in workshop length: Explanations and implications. *Urban Geography* 6:193–219.

———, and Pratt, G. 1988. Spatial dimensions of the gender division of labor in a local labor market. *Urban Geography* 9:180–202.

Hayden, D. 1981. *The grand domestic revolution: A history of feminist designs for American homes, neighborhoods, and cities*. Cambridge: MIT Press.

———. 1984. *Redesigning the American dream*. New York: W. W. Norton.

Hayes, J. J. 1986. Not enough wood for women: How modernization limits access to resources in the domestic economy of rural Kenya. Ph.D. diss., Department of Geography, Clark University.

Holcomb, B. 1984a. Women in the city. *Urban Geography* 5:247–54.

———. 1984b. Women in the rebuilt urban environment: The United States experience. *Built Environment* 10:18–24.

Johnston-Anumonwo, I. 1988. The journey to work and occupational segregation. *Urban Geography* 9:138–54.

Jones, J. P., and Kodras, J. E. 1989. Spatial and gender division of labor in metropolitan areas. *Geoforum*. In press.

Kolodny, A. 1984. *The land before her*. Chapel Hill: University of North Carolina Press.

Larimore, A. 1978. Humanizing the writing of cultural geography textbooks. *Journal of Geography* 77:183–85.

Lee, D. 1988. Women in geography: A comprehensive bibliography. Boca Raton, FL: Atlantic University.

———; Stewart, R.; and Winter, D. G. 1979. The geography of female status: A world-wide view. *Transition* 9:2–7.

Loyd, B. 1975. Women's place, man's place. *Landscape* 20:10–13.

McDowell, L. 1989. Diversity and change, feminist theory on geography. *Geoforum*. In press.

———, and Bowlby, S. 1983. Teaching feminist geography. *Journal of Geography in Higher Education* 7:97–107.

Mackenzie, S. 1985. No one seems to go to work anymore: Women redesignating and redesigning the city. *Canadian Woman Studies* 6:5–8.

———. 1986. Feminist geography in focus. *Canadian Geographer* 30:268–70.

———, and Rose, D. 1982. On the necessity for feminist scholarship in human geography. *Professional Geographer* 34:220–23.

Marston, S. 1986. Putting women on the agenda: A new perspective for geographers. Special Issue of *Urban Resources* 3:60–61.

———, and Saint-Germain, M. 1989. Changing urban form and women's role in neighborhood politics. *Geoforum*. In press.

Mazey, M. E., and Lee, D. 1983. *Her space, her place: A geography of women*. Washington: Association of American Geographers.

Miller, R. 1983. The Hoover in the garden: The middle class women and suburbanization 1850–1920. *Environment and Planning D: Society and Space* 1:73–87.

Momsen, J. H., and Townsend, J. 1987. *Geography of gender in third world*. Albany: State University of New York Press.

Monk, J. 1984a. Approaches to the study of women and landscape. *Environmental Review* 13:1.

———. 1984b. Human diversity and perceptions of place. In *Perceptions of people and places through media. Papers of the Symposium of the Commission of Geographical Education,* ed. H. Haubrich, pp. 45–67. Freiburg, Federal Republic of Germany: International Geographical Congress.

———. 1988. On not excluding half of the human world. In *Proceedings of the Conference on Feminist Geography,* ed. Lia Karsten, pp. 23–33. Amsterdam: University of Amsterdam.

———, and Hanson, S. 1982. On not excluding half of the human in human geography. *Professional Geographer* 34:11–23.

———, and Williamson-Fien, J. 1986. Stereoscopic visions: Perspectives on gender-challenges for the geography classroom. In *Teaching geography for a better world,* eds. J. Fien and R. Gerber, pp. 186–220. Brisbane: Australian Geography Teachers Association and Jacaranda Press.

Morrow-Jones, H. A. 1986. The geography of housing: Elderly and female households. *Urban Geography* 7:263–69.

Norwood, V., and Monk, J., eds. 1987. *The desert is no lady: Southwestern landscapes in women's writing and art.* New Haven: Yale University Press.

Pratt, G., and Hanson, S. 1989. Occupational segregation and the life cycle: a geographic perspective. *Geoforum.* In press.

Regulska, J. 1989. Women's participation in local government: Grassroots initiatives and failures. *Geoforum.* In press.

Rengert, A. C., and Monk, J. eds. 1982. *Women and spatial change: Learning resources for social science courses.* Dubuque, IA: Kendall/Hunt.

Rose, D., and Chicoine, N. 1989. The daycare center: A neighborhood resource for whom? The case of Montreal's old and new "inner city." *Geoforum.* In press.

Rutherford, E. M., and Wekerle, G. 1988. Captive rider, captive labor: Spatial constraints and women's employment. *Urban Geography* 9:116–37.

Schlissel, L. 1982. *Women's diaries of the westward journey.* New York: Schocken Books.

Seager, J., and Olson, A. 1986. *Women in the world: An international atlas.* New York: Simon & Schuster.

Shortridge, B. G. 1987. *Atlas on American women.* New York: Macmillan.

Smith, S. J. 1989. Gender relation of housing consumption. *Geoforum.* In press.

Villeneuve, P., and Rose, D. 1988. Gender and the separation of employment from home in metropolitan Montreal, 1971–1981. *Urban Geography* 9:155–79.

Wilkinson, N. 1979. Women on the Oregon Trail. *Landscape* 23:42–47.

Women and Geography Study Group of the Institute of Behavioral Geography. 1984. *Geography and gender, an introduction to feminist geography.* London: Huchinson.

Zelinsky, W.; Monk, J.; and Hanson, S. 1982. Women and geography. *Progress in Human Geography* 6:3.

ANALYSIS AND DISPLAY OF
GEOGRAPHIC PHENOMENA

Cartography

A. Jon Kimerling

Cartography has undergone tremendous changes over the last two decades. The forces of a technological revolution in society of unprecedented proportions has swept mapping into the modern information age. Demand has grown dramatically for spatial information presented graphically in ways that take advantage of the speed, power, and display flexibility of modern electronics. This has led to the use of automated methods which have shattered the age-old image of the cartographer hunched over a drafting table, carefully crafting maps by hand. The revolution in cartography has been launched on one hand by large outlays of public funds directed at modernizing military and civilian mapping agencies, and on the other by the commercialization of the data, software, and hardware needed for electronic mapping.

Cartographers have been transformed into geographic-information scientists who build and maintain digital cartographic databases, design maps on interactive workstations, and oversee the construction of maps by computer-controlled printers and plotters. Dramatic improvements in the quantity and quality of environmental data collected in computer-usable form have enlarged the realm of mappable phenomena. Cartographers can now map the Earth's surface repeatedly in intricate detail from Earth-observation satellite data; plot the location of a vehicle on a display map with the aid of inertial- or global-positioning navigation systems; or portray simultaneously alternative versions of the myriad cultural characteristics of our nation and world, as reflected in the growing volume of statistical data (Monmonier 1985).

Technological revolution has forced cartographic researchers to broaden the definitions of maps and cartography, to treat the cartographic process with greater mathematical rigor, and to evaluate the effectiveness of maps as instruments of graphic communication. Definitions that restrict maps to paper or similar products, three-dimensional models, and globes have been expanded to include the electronic maps of the computer age, and the mental maps that guide our everyday activities. The

---

Thanks to Harold Moellering for his role in structuring the analytical facet of this chapter. Phillip C. Muehrcke's suggestions improved the draft manuscript considerably, as did the editorial corrections of Mark S. Monmonier and the editors.

fifteen-year-old definition of cartography as "the art, science and technology of making maps, together with their study as scientific documents and works of art" (Multilingual Dictionary of Technical Terms in Cartography 1973, 1) is directed toward what might be called the *map production* and *historical* facets of cartography.

Broader definitions, such as "the science and technology of communicating spatial information by means of maps," reflect the *communication* facet of cartography. In contrast, computer-oriented definitions, such as "an information transfer process that is centered about a data base which can be considered, in itself, a multifaceted model of geographic reality," given by Starr and Guptill (pers. com. 1984), reflect the *analytical* facet of the field. Research in these four facets of cartography is varied in scope, rigor, design, and application. But in all its variations, this research has greatly enriched a profession that is steeped in tradition and convention.

An academic *discipline* of cartography has emerged, complete with recently introduced journals devoted entirely to cartographic research, new textbooks focusing on individual facets of the field, and a proliferation of conferences and symposia on specialized research topics. Vigorous research efforts, coupled with broader academic and professional employment opportunities, have fostered the rapid expansion of academic and professional education in cartography.

Cartography in the U.S. is still taught mostly in geography departments, and this long association with geography has done much to shape the character of cartographic education (Dahlberg and Jensen 1986). Half of the over 500 colleges and universities offering cartography teach only an introductory course, whereas fewer than 10 schools have full-fledged programs at the undergraduate or graduate level. This educational setting has steered cartographic research toward small-scale thematic mapping.

The early graduate centers of cartographic instruction and research that emerged at the University of Kansas, University of Washington, and University of Wisconsin–Madison have in the last decade been joined by cartography programs at such institutions as the Ohio State University, Syracuse University, University of South Carolina, University of Maryland, Michigan State University, University of California–Santa Barbara, and several others. At these schools, rigorous programs of study and research in one or more facets of cartography are found, along with graduate-level coursework in such closely allied mapping sciences as surveying, remote sensing, and geographic information systems (GIS). Regardless of program size, older production-oriented cartography courses are being restructured or replaced by new courses which emphasize the communication and analytical facets of the field and their relationship to high-technology map production.

COMMUNICATION AND DESIGN FACET

Judging by the refereed journal articles devoted to cartography, *cartographic communication* is undoubtedly the dominant research thrust of the last two decades, with improved map design its *raison d'être*. Olson (1983) notes that a dictionary definition of cartographic communication would be the use of maps as a medium for imparting knowledge between beings capable of knowledge, and able to transmit and receive information. Cartography is thus seen as a communication science, and the cartographer as an expert in displaying spatial information.

Arthur H. Robinson is the undisputed father of communication and design research. Petchenik clearly recognizes the revolutionary nature of Robinson's first book:

> It is possible that not all cartographers would consider the publication of *The Look of Maps* (1952) as the decisive event separating the modern era of research cartography from the thousands of years of mapping that preceded it. Yet if we examine the characteristics of recent decades of research in cartography we can find stated explicitly in this book all of the fundamental assumptions that shaped that research as well as the major goals the research has been organized to achieve (Petchenik 1983, 38).

Perhaps the book's most prophetic phrases were that:

> The development of design principles based on objective visual tests, experience, and logic; the pursuit of research in the physiological and psychological effects of color; and investigations in perceptibility and readability in typography are being carried on in other fields . . . such a movement in cartography cannot fail to materialize. (Robinson 1952, 13–14)

The Look of Maps was the outline for the academic rebirth of cartography, and Robinson's many graduate students, beginning in the mid-1950s, were in large part responsible for the shift in research emphasis from the production of maps to communication and design theory. The picture is incomplete, however, without recognition of the pioneering research efforts of George F. Jenks, and the contributions of his master's and Ph.D. students who figure so prominently in current perceptual research and the other facets of modern cartography.

Cartographic Communication Models

The late 1960s to mid-1970s marked an era of communication model building. Formulation of cartographic communication models is a logical consequence of viewing cartography as a communication science, and the map as a communication system (Robinson and Petchenik 1975). Communication models were not meant to copy or imitate the communication process exactly, but rather to contain the major components of the communication system and to illustrate the connections among components. Such models still provide an organizational structure for cartographic research and education, as well as the map-production process.

Muehrcke's (1972) diagram of the cartographic processing system (Figure 1) is a good example of early, simplified models based upon the source-channel-recipient view of electronic information communication. He characterized the cartographic process as an "active feedback system" involving a series of information transformations from the real world to raw data (T_1), to the map made from raw data (T_2), and finally to the user's mental image of the map, gained through map reading (T_3). With this model, Muehrcke organized clearly the important activities and research areas in cartography at the most general level. The map-reading transformation was seen as an integral part of cartography, and the cartographer's task was to design maps so that the user's mental image of the map would faithfully resemble the raw data: $T_3 = (T_2)^{-1}$.

Robinson and Petchenik's (1975) Venn diagram summarizing the cognitive elements in cartographic communication (Figure 2) epitomizes a number of more complex models designed to capture the essence of cartographic communication. The

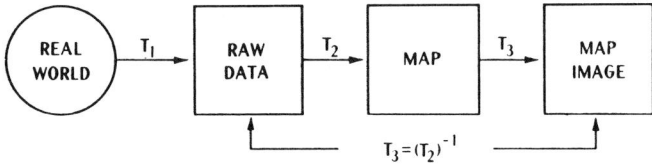

Figure 1
Muehrcke's diagram of the cartographic processing system (Robinson, A. H., and Petchenik, B. B., 1975. The map as a communication system. *The Cartographic Journal* 12:10)

overlapping rectangles illustrate the true complexity of cartographic information assimilation, where incorrect conceptions of the milieu could be mapped by the cartographer or held by the map user (percipient); where a significant fraction of mapped information could (and in many cases, should already) be known by the percipient; where unintended or unrelated information could be inferred; and where information might never be comprehended due to poor map design, mediocre map-use skills, or an inadequate understanding of the environment being mapped. The implication is that communication of environmental information through map interpretation varies widely from person to person, and is as much a matter of motivation, past map-use experience, and a good understanding of the milieu, as it is of good map design. These human characteristics interact to continually challenge and frustrate researchers who conduct map-perception experiments.

These and similar models still have considerable pedagogical value, but concurrent attempts to interrelate the components of cartographic communication through use of more complex Venn diagrams and set-theory terminology (Morrison 1976) have proven less successful. Model building has not completely died, but more recent efforts have been directed at particular aspects of the overall process. Gilmartin's (1985) diagram of map design (Figure 3) best illustrates the trend toward capturing

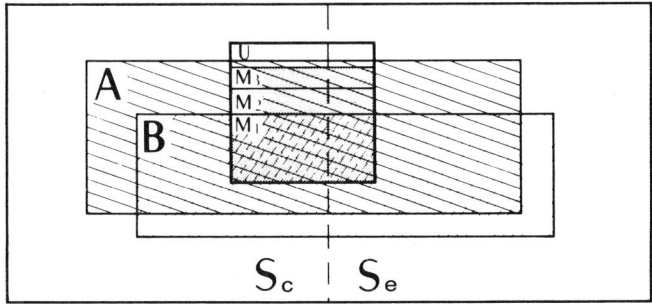

Figure 2
Robinson's Venn diagram summarizing the cognitive elements in cartographic communication. Key: S_c—correct conception of the milieu; S_e—erroneous conception of the milieu; A—conception of the milieu held by cartographer; B—conception of the milieu held by percipient; M—map prepared by cartographer and viewed by percipient; M_1—fraction of M previously conceived by percipient; M_2—fraction of M concerning S newly comprehended by percipient, a direct increment: M_3—fraction of M not comprehended by percipient; U—increase in conception of S by percipient not directly portrayed by M, but which occurs as a consequence of M, an unplanned increment (Robinson, A. H., and Petchenik, B. B., 1975. The map as a communication system. *The Cartographic Journal* 12:11)

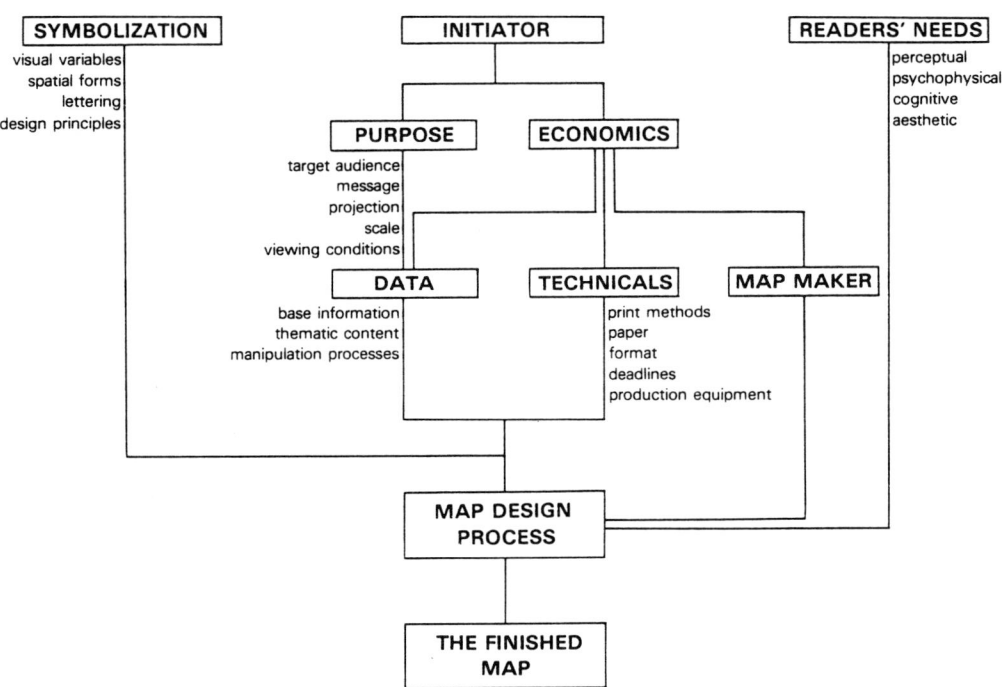

Figure 3
Gilmartin's general model of map design influences (Gilmartin, P. P., 1985. The design of journalistic maps: Purposes, parameters, and prospects. *Cartographica* 22:5)

in detail the interplay of user limitations, map purpose, economic constraints, data considerations, technical factors, and symbolization possibilities that more fully characterizes the cartographic process.

Map Symbolization Theory

Symbolization is central to cartographic communication, linking communication models to map design. M-L. Hsu succinctly summarizes modern views on two theoretical aspects of symbolization—symbol selection and symbol design:

> Selection involves the process of determining, on the basis of the cartographer's conception of the phenomenon to be mapped, the best symbol, or symbols, to signify the phenomenon. . . . Symbol design is a delicate exercise in which the cartographer consciously discerns, among other things, the visual variables of the map symbol—shape, size, color [hue, chroma], value, pattern, and orientation—and the attributes of the phenomenon assigned to these variables, taking into account the quality and scale of measurement of the data. (1979, 123)

Implicit in these definitions is the notion that cartographers design maps to symbolize phenomena, not the data sets describing different phenomena. The spatial nature of phenomena depends above all on map purpose, which among many things may be: "(1) to provide information concerning what, and often how much, exists at different locations and (2) to map the characteristics of a phenomenon to reveal spatial order and organization, such as regionalization, connectivity, movements, or correlation" (M-L. Hsu 1979, 118–19).

Symbol selection and design rules based on tradition and personal preference have been supplanted by guidelines which are based on the idea that geographical phenomena can be conceived as point distributions, linear networks, homogeneous areas, or continuous surfaces viewed as having volume. Point, line, and area symbol design now involves selecting appropriate visual variables for the levels of measurement—nominal, ordinal, interval, or ratio—at which the phenomena are to be depicted (Chang 1978). Levels of measurement also need to be considered from the map reader's point of view (Muehrcke 1976), particularly since ordinal, interval, and ratio-level symbolization is identical, and is only distinguishable through reference to labels or to the map legend.

This organizational framework for map symbolism is directly tied to the concepts of semiology, the science of signs that stems from the field of structural linguistics. The application of these linguistic concepts to cartography has been championed by the French cartographer Bertin (1978). But, other than the visual-variables concept, his works have received little attention in the U.S., with the exception of Wood and Fels (1986). The spirit of the semiotic approach is seen in their statement:

> Every map is at once a synthesis of signs and a sign in itself: an instrument of depiction— of objects, events, places—an instrument of persuasion . . . like any other sign, it is the product of codes: conventions that prescribe relations of content and expression in a given semiotic circumstance. The codes that underwrite the map are as numerous as its motives, and as thoroughly naturalized within the culture that generates and exploits them. (Wood and Fels 1986, 54)

Experimental Research

The communication paradigm suggests research foci on accurate information-encoding through symbol design, and on understanding the process of map reading and interpretation. Robinson (1977, 164) has outlined four primary questions amenable to experimental research:

1. How can the tasks one pursues in using a map be facilitated, such as search made quicker, calculations made surer, analysis made easier, etc.?
2. How can geographic material be best presented for effective map comparison, such as for hypothesis generation, the recognition of geographical change through time, etc.?
3. How can diverse geographical information be presented with the least visual confusion?
4. What kinds of graphic magnitude scales should be employed for the portrayal of geographical variations in qualitative and quantitative characteristics?

Corollary questions of equal significance to map design include:

1. What generalizations can be made regarding classes of the map-reading public [differences in perceptual abilities due to education, age, sex, etc.] . . .
2. At what age levels are the various basic concepts of cartography . . . [scale, coordinate systems, symbolism, etc.] . . . capable of being understood?
3. What are the most efficient forms of instruction in "map reading"?

Various research methods have been applied to these problems, the most important being psychophysical experiments dealing with individual map symbols, cognitive investigations treating maps as a whole, and task-oriented studies of map use.

Psychophysical Experiments

The question of graphic magnitude scales dominated research from the mid-1950s until the beginning of this decade, in large part due to the adoption of psychophysical testing methods. Psychophysics is a branch of behavioral psychology "directed toward identifying a 'psychophysical function' which describes the quantitative relationship between the magnitude of a physical stimulus and the magnitude of the corresponding perceptual experience" (Gilmartin 1981a, 10). The stimulus-response power function $R = KS^n$ is widely accepted today.

Psychophysical testing has worked well with certain map symbols, while producing inconclusive results at best with others. Perhaps the most widely used test results come from color-perception experiments, particularly those dealing with the achromatic dimension of color that is termed *value*. The Munsell color system equal-value scale, where equal perceptual increments of grayness were determined over 80 years ago to be a power function of surface reflectance, has been found to best describe the perception of gray-tones produced by screen tints (Kimerling 1975), line printers (Smith 1980), and plotters (Peterson 1979; Slocum and McMaster 1986). However, gray-tone perception is a complex visual process, since both the background gray tone (Cox 1980) and the dot texture of tones (Kimerling 1985; Smith 1987) change the appearance of gray areas, to the extent that white or black backgrounds and coarse or fine dot textures produce significantly different power function equations.

Most psychophysical experiments have been attempts to determine power function or other equations relating the perceived size of point symbols to their measured area, width, or height. Such equations are central to *apparent magnitude symbol scaling,* the designing of graduated symbols that appear to be twice, three times, or any other number of times the size of an anchor symbol.

The long-observed phenomenon of circle-size underestimation was first described by a power function equation in the mid-1950s, and the black graduated circle has since become the "white rat" of perceptual research—poked and prodded from many angles over the past three decades. It was soon found that, as with the equal-value gray scale, different power functions were obtained under different experimental conditions, specifically:

1. differences in testing methods,
2. the specific wording of testing instructions,
3. the size(s) and number of legend circles,
4. the map and legend circle size range,
5. the sequence in which size judgments are made (Chang 1980b),
6. the distance from legend circles to map symbols,
7. the legend design (Fraczek 1984),
8. the density and size of adjacent circles on the map (Gilmartin 1981a), and
9. the method of portraying overlapping circles (Groop and Cole 1978).

Chang (1980a, 161) summarizes the body of research on this subject by stating: "The stimulus-response relationship for circles is fairly complex, and any correction in map design based on one psychophysical study alone is of limited value, especially given the incomparability between the conditions of the experiment and of real map use."

Graduated circles and gray tones have been the mainstay of psychophysical research, but other map symbols have also received at least cursory examination over the last decade. For example, a tendency to underestimate the number of dots on dot maps has been found (Provin 1977; Olson 1977), along with a correction equation. Readers appear to perceive height differences, and not volume differences, when viewing stepped statistical surface maps (Cuff and Bieri 1979). Minimal size differences for easy discrimination of map lettering have also been determined (Shortridge 1979), as have just-noticeable differences (JNDs) for sectored-pie graphs (Slocum 1981).

Limitations of Psychophysical Studies

The more that cartographers attempted to develop map design principles based upon psychophysical experiments, the more apparent became the inherent limitations of this approach. The most rigorous critique (Petchenik 1983) focuses upon:

1. results that bear little relation to real map use, because of the researcher's need to hold everything constant except the symbol being studied;
2. the almost exclusive use of thematic maps in tests, which may limit use of test results to this map form;
3. an inadequate recognition of the effect of motivation on map use, and the nature of spontaneous task performance where summaries of results from users show wide variations in individual and average responses that mask the fact that two or more different underlying processes of judgment were used;
4. the widely known fact that test results vary with the specific question asked, or task specified (Shortridge and Welch 1980);
5. the relatively uniform nature of the test subjects used, and the related lack of attention to test-subject differences;
6. a general lack of attention to the meaning of test maps, and the fact that users who bring previous knowledge to a map-use problem may find almost any graphic symbol transparent to actual meaning.

To this, Gilmartin (1981b, 11) adds that information about how people perceive specific cartographic symbols:

> . . . while certainly useful, has it limitations. Even if we know the answer to every psychophysical question and could, theoretically, design the ultimate map from a psychophysical standpoint, we still would know little about how people read maps in the sense of learning, remembering, and making use of them.

Disillusionment with purely psychophysical studies has lead to an interest in applying cognitive psychology to cartography. Olson (1983, 278) notes that:

> The very trend away from the psychophysical and towards cognitive study with maps is tied to the rethinking of map use tasks, and while psychophysical studies can still be useful for deriving certain guidelines for map design, it is the more cognitive study that is attracting interest in cartography . . .

Cognitive cartographic research can also be viewed as a natural outgrowth of the need to explain the wide variation in individual responses to psychophysical test questions.

There appears to be no "typical" map user upon which stimulus-response equations can be based.

Cognitive Research

Cognitive psychology is broader in scope and experimental method than psychophysics, encompassing such topics as sensation, perception, mental imagery, short- and long-term memory, recognition and recall, problem solving, and learning (Olson 1979). These are components of what cognitive psychologists term the "information processing" approach to cognition. Visual information processing-system models that show the linkages among these aspects of cognition have appeared in recent cartographic literature (Dobson 1979a; Eastman 1985; Peterson 1987). Underlying these models is the firm belief that the map reader is an active player in the map-reading process. Hence, the cartographer is more a facilitator than a communicator, supplying the spatial context of features and the relationships among them.

There is an inherent difference between the purpose of cognitive research in cartography, and its purpose in psychology. In cartography, cognitive processes are important to our understanding of map perception and comprehension, but the map and its design are the central concern. "To the cognitive psychologist, on the other hand, it is the *mental process* that is the center of attention; a map may be a tool by which to unravel the mysteries of such processes, but the focus is a mental rather than physical phenomenon" (Olson 1979, 40).

Cartographic research of a purely cognitive nature has dealt primarily with the "visual mental image" formed when viewing a map. Early studies (Cole 1981) introduced two broad categories for memory—recognition and recall—and associated testing methods. Recognition involves selection from a set of mapped distributions the closest match to a memorized distribution. Recall involves the physical reconstruction of the distribution from memory. As one would expect, errors in matching are much greater with recall experiments.

A few subsequent recall experiments have been slanted toward map design (Kulhavy, Schwartz, and Shaha 1982; Patton and Slocum 1985) and toward gender-based differences in ability to recall geographic information from mental images formed by reading text and/or maps (Gilmartin 1986). While intellectually interesting, studies of this nature appear to be of limited utility in map design, since cartographers generally assume that the map is available for use and does not have to be remembered.

Reaction time has also been used in cognitive experiments. Reaction time plays a major role in the current debate over whether information about objects is coded and remembered as a series of verbal propositions, or as images, or both. Steinke and Lloyd (1983) conducted a shape-recognition experiment which supported the argument that geographic information is remembered as visual images. A later and more-rigorous study with graduated circles (Lloyd and Steinke 1985a) suggests that most subjects appeared to memorize symbol sizes as an ordered list, rather than as a mental map image. A third study with qualitative point symbols (Lloyd and Steinke 1985b) supports a "dual-coding theory" in which geographic information is stored in memory both as map images and verbal propositions.

One of the few reaction-time studies aimed specifically at map design concerns the effect of lettering visual-cue redundancy on the discrimination of town sizes on maps (Shortridge and Welch 1982). Assuming that a shorter reaction time implies easier discrimination of different town sizes, variations in two visual cues (e.g., size and boldness) were found to be far better than variations in just one.

The trend away from psychophysical toward cognitive research is clear, but at least one researcher (Gilmartin 1981b, 9) has argued for an integrated research approach, since in reality:

> Perceptual experiments are seldom *purely* psychophysical or purely cognitive. They may be primarily one or the other, depending upon the questions to be answered by the research.

Several recent studies have employed both psychophysical and cognitive testing methods to gain a more complete understanding of map reading than either method alone can provide. A good example is MacEachren's (1982) use of magnitude estimation and recall tasks to study how map complexity influences the relative effectiveness of choropleth and isopleth maps. The only significant difference in effectiveness between the methods was improved memory of general patterns with isopleth maps.

Other Types of Communication Research

Map Comparison. Recent investigations of human map-comparison abilities can be traced to an interest, beginning in the 1950s and 1960s, in relating the visual and statistical properties of maps. This led to visual map-comparison studies of the similarity of choropleth maps. Steinke and Lloyd (1981, 13) summarize several studies by stating:

> Three key variables—the similarity of the overall map patterns (correlation), the similarity of the distribution of the spatial units within the gray-scale map classes (blackness), and the class similarity of the neighboring spatial units (complexity)—have generally been identified as the most important attributes used by map readers to determine the similarity of choropleth maps.

Cartographers have also experimented with alternative mapping techniques to improve map comparison. Monmonier (1976) notes that continuously shaded (unclassed) choropleth maps may have the greatest potential for map comparison, because the "fuzzy" image of correlation inherent in classed maps is eliminated. He later suggests that the visual comparison of two statistical maps may be "ill-advised" (Monmonier 1979), and that bivariate maps, where two distributions are portrayed simultaneously, may depict geographic correlation more clearly.

Carstensen (1984, 1986) has compared the computer-produced unclassed bivariate choropleth map with two classed choropleth maps depicting the same information. He concluded that bivariate maps naturally depict more geographic detail, but that the differences in cross-hatch shape used to depict differences in correlation on his maps were not understood as well as traditional choropleth map symbolism.

Eye-Movement Studies

Psychologists use the experiment methods of verbal protocols, brain-activity study, and eye movements to infer relationships between cognition and neuromotor activity. In the early 1970s, George F. Jenks and his students conducted the seminal eye-movement detection and recording studies in cartography. The rationale for such studies was that "Eye movements are an outward manifestation of visual/cognitive processing and by examining 'how' the eye moves, we can gain important information about the workings of this crucial process by which the reader confronts and understands the map" (Castner and Eastman 1985).

Steinke (1987, 68) summarizes the path taken over the last decade when he states:

> Significant changes have occurred over the years in the kinds of questions cartographers have tried to answer using eye movement recordings. Initially they were concerned with simply trying to document the map reading activity. It was believed (or hoped) that knowing where a person looked on a map, how long was spent looking at various parts of the map, and determining the sequence in which parts of the map were viewed would in and of itself, somehow, lead to solutions to cartographic problems [e.g., Dobson 1979b; Antes, Chang, and Mullis 1985]. When it became obvious that the understanding of the map reading process would require more than an analysis of eye movements, i.e., that map reading involved cognitive processes as well as perceptual and that the deciphering of the perceptual activity was not necessarily a surrogate measure of the cognitive activity, cartographers became a bit disillusioned with the technique. . . . Several of the more recent eye movement studies have been successful because they employed tasks that were carefully selected to match the research objectives. . . . A second major reason that some of the recent research has been more successful is the fact that it has a much stronger theoretical base than the earlier work [e.g., Castner and Eastman 1984; Dobson 1980].

Task-Oriented Research

Task-oriented research, the outgrowth of applying human-factors engineering concepts to cartography, focuses on measuring the efficiency of map reading and analysis. Robinson (1977) notes the two popular ways of conducting this type of research:

> . . . (1) by testing subjects with various sorts of exercises, recording and analyzing the results, and (2) by "interviewing" subjects while they are performing the tasks.

He lists the major types of tasks:

> Task-oriented research is most appropriately aimed at problems of cartographic design having to do with such operations as search (for names, symbols, etc.) or the application of skills (reading scales, estimating quantities, etc.). (Robinson 1977, 167)

Measuring the relative efficiency of map reading and analysis is a popular research approach. Smith (1977), for example, showed that a standard topographic map and an orthophotomap of the same area differed little in terms of symbol readability and performance of tasks involving locating, counting, verifying, and comparing. Along these same lines, but more closely tied to real map use, is DeLucia's (1979) comparison of conventional line maps and high-oblique photomaps for public-planning information display. Here, visualization tasks such as locate-identify, delimit area, and verify were completed in the same time on both map forms, but with higher accuracy on the photomap.

The effects of education, culture, and sex upon map-use abilities has been examined. The first cartographic study of sex difference in map-use skills (Gilmartin and Patton 1984) showed males to be significantly better than females in learning and recalling geographic information from a text, with or without the aid of reference maps. However, no differences in road-map reading skills were found. A more comprehensive examination (Chang and Antes 1987) showed (1) males outperforming females in topographic map reading, (2) Taiwanese males and females outperforming North Dakota males in topographic map reading, and (3) better male performance in both cultures. These studies underline the importance of environmental factors, as well as the apparent linkage between gender and the visualization of three-dimensional images from two-dimensional topographic and reference maps.

Personal Experience

Research based on personal experience goes hand in hand with perceptual experiments, as many aspects of communication and design are not well suited to scientific investigation. Insights gained from years of challenging and often frustrating work on major mapping projects are valuable, even though they may be hard to apply to other mapping projects. The few studies along this line detail the problems involved in compiling and designing a specific type of map.

Petchenik (1977), for example, points out the very real problems of data availability, quality, and consistency encountered when compiling the *Atlas of Early American History*. In a later paper on school-textbook maps, she observes that "Map design for children is largely a matter of values, values that cannot be determined or applied on the basis of the methods and techniques of scientific research" (Petchenik 1985a, 26). Publishing-firm managers, not cartographers, make the basic decisions as to map content and design, with the values held by state textbook adoption committee members fully in mind.

The most penetrating paper in this vein, however, is Gersmehl's (1981) examination of maps he has used in landscape interpretation. Conceptual problems in compiling soil and vegetation maps are brought to light, such as whether vegetation should be conceived as a biotic continuum or as discrete habitats, or the fact that each aspect of a phenomenon has an intrinsic scale of variation which may not be constant throughout a region. A plea is also made for innovative map designs that enable the reader to see the ambiguities and uncertainties in mapped distributions.

ANALYTICAL FACET

Analytical cartography encompasses the mathematical concepts and methods underlying cartography, and their application in map production and the solution of geographic problems. Tobler (1976) first outlined the scope of analytical cartography in his syllabus for a course of that name. His outline contained many of the topics of current analytical cartography, including:

cartographic data models
digital cartographic data-collection methods and standards
coordinate transformations and map projections
geographic data interpolation
analytical generalization
numerical map analysis and interpretation

Herein lies the conceptual and mathematical foundation for the spectacular rise in the quality of computer-assisted mapping and geographic-information systems over the last decade. The integration of cartographic and statistical methods within a GIS holds great potential for geographers (see the chapter on GIS).

Flowing through these topics is a transformational concept of cartography (Tobler 1979b) in which the entire process of making and using a map can be viewed as a sequence of information transformations. Moellering's (1984) concept of real and virtual maps links transformations to cartographic products. Cartographic products are organized into:

1. *Real maps,* which are directly viewable as cartographic images that have a permanent tangible reality.
2. Directly viewable *virtual maps–type 1,* which lack permanent tangibility, such as CRT map images.
3. *Virtual maps–type 2,* which are not viewable as a cartographic image, but have permanent tangible reality, such as gazetteers.
4. *Virtual maps–type 3,* which are neither viewable nor have permanent tangible reality, such as disk files or mental maps.

Cartographic methods and map-use activities are the agents of transformation between real and virtual maps, whether virtual-map types 2 and 3 are viewed as maps or as real and virtual sources of cartographic data.

Cartographic Data Models

Underlying analytical cartography are different ways of modeling reality, which can be viewed as levels of data organization. Peuquet (1984) identifies four levels used in database design and construction, whereas Nyerges (1981, 7–8) notes that computer data processing has evolved into a management-systems approach based upon six levels:

Information Reality—observations that exist as ideas about geographical entities and their relationships.

Information Structure—a formal model that specifies the organization of phenomena in reality. It includes entity sets, plus the types of relationships that exist among those entity sets.

Canonical Structure—a model of data that represents the inherent structure of these data, and hence is independent of individual applications of the data and of the software or hardware mechanisms employed to represent and use the data.

Data Structure—a description elucidating the logical structure of data accessibility in the canonical structure. There are access paths dependent on explicit links, i.e., resolved through pointers, and others that are independent of links, i.e., resolved through other forms of reference.

Storage Structure—an explicit statement of the nature of links, expressed in terms of diagrams that represent cells, linked and contiguous lists, levels of storage medium, etc. The storage structure includes indexing that indicates how stored fields are represented, and in what physical sequence the records are stored.

Machine Encoding—a machine representation of data, including the specification for addressing (absolute, relative, or symbolic), data compression, and machine code.

A systematic approach to the design of computer-mapping systems has emerged from these concepts (Moellering 1984). User goals and needs define system data and data input, processing, analysis, and output requirements. The logical design of the system follows, wherein are specified the command structure, real and virtual map transformations and associated algorithms, and each level of organization in the overall model. A similar "top-down" approach has been advocated for the design of cartographic databases (Calkins and Marble 1987).

Information Structures

American cartographers have made major contributions to the theory of geographical information structures. Standardization of terminology, an ongoing effort, has culminated in agreement of the National Committee on Digital Cartographic Data Standards (NCDCDS 1988). Cartographic *entities* (real-world phenomena), *objects* (digital representations of entities), *attributes* (characteristics of entities) and other fundamental terms are precisely defined. Standard definitions for cartographic objects are particularly important, in light of the confusing range of terms currently used for each object (e.g., point, node, line segment, string, arc, link, chain, ring, area, polygon, pixel, grid cell, and so on).

The first comprehensive comparison of data structures (Peucker and Chrisman 1975) introduced vector, tessellation, and three-dimensional structures. The use of topological encoding (beginning and ending chain nodes; polygons to the left and right of chains) was advocated to eliminate redundant data, as well as to capture the contiguity relationships among polygons in vector structures.

Nontopologic object-by-object data structures, called "spaghetti models," cannot be used for most types of GIS spatial analysis, since inner polygon boundaries are digitized twice, automatic error-checking is impossible, and connectivity and contiguity relationships are absent. Models of this type are limited in use to reproducing some or all of the original objects at various map scales, and with different map symbols—the basic functions of many computer-assisted mapping systems. A deeper understanding of topology (White 1984a, 1984b) sheds light on the solution of many current data structure and analysis problems.

Meanwhile, ever-more sophisticated topologically based data structures continue to be developed and applied by others in a variety of computer environments. Examples are the Census Bureau's application-oriented TIGER (Marx 1986) and the U.S. Geological Survey's DLG data exchange structure (Luman 1987).

Tessellation models are regular or irregular subdivisions of space into "tiles" of a specific shape. Cartographers are well acquainted with regular tessellations where (column, row) adjacency implies spatial contiguity, such as square or rectangular grids (raster structure), quadrilaterals, Landsat pixels, hexagons, and triangles. According to the sampling theorem, regular tessellation tiles should be not more than one half the size of the smallest entity. This often results in unacceptably large data sets, yet larger tiles cannot capture small but important entities. This dilemma has prompted cartographers to experiment with nested tessellation models (Peuquet 1984). The quadtree model is most popular, partly due to past experience with Public Land Survey sections, quarter sections, and further quarterings.

The best-known irregular tessellation is the triangular irregular network (TIN) where (x, y, z) data points form triangle vertices, giving a three-dimensional tessellation which, if designed correctly, will closely match a real surface. Efforts continue to maximize the efficiency of triangular network creation.

Time-consuming conversions between raster and vector data structures will continue to challenge cartographers and GIS users because each structure has unique advantages (see chapter on GIS). Peuquet describes and compares state-of-the art algorithms for conversion from raster to vector (1981a) and vector to raster (1981b). The former is basically a sorting and intricate line-following problem; the latter is intimately tied to raster-line thinning. These techniques for very fast conversion of scan-digitized maps into vector format for data processing, and then back into raster format for plotting or screen display, have been incorporated into major military and civilian computer-mapping systems.

The trend toward structuring cartographic data for analytical data manipulations, as well as for graphic display, has lead to experiments with models normally associated with attributes, such as the hierarchical, relational (Cox, Alfred, and Rhind 1980), and network structures. The recently introduced hypergraph-based data structure (HBDS) is even more general and appears to offer the most potential. However, it is still very much in the research-and-development stage (Rugg 1984).

Cartographic Data Quality

Appropriately structured data are highly valued, but may well be low in quality. Positional accuracy of digitized map information, a key aspect of quality, is linked in the sales literature to digitizing hardware precision and accuracy, with practical guidelines for digitizing found in most system manuals. The implication that better manual-digitizing equipment produces more accurate databases must be weighed against evidence showing that human error in stream-mode line digitizing overshadows machine limitations (Jenks 1981). Errors tend to be systematic in direction and magnitude, but can be reduced through positive feedback training and costly line-editing procedures. This unavoidable loss in quality has led major mapping organizations to purchase costly line-following digitizers and scanning devices. These reduce, but do not eliminate, positional errors (Monmonier 1982b).

Other equally important factors affecting quality include the nature of geodetic control, classification accuracy, and temporal inconsistencies among data (Chrisman 1984). Cartographic-data users presently are lucky to receive any information regarding these and other important aspects of data quality. A Digital Cartographic Data Quality standard has been developed to alleviate this unfortunate situation. The standard is in the form of a "truth in labeling" quality report (NCDCDS 1988). It contains:

1. A *lineage* section that describes the source materials from which the data were derived, and the methods of derivation, including all coordinate transformations involved in producing the final digital files.
2. A *positional accuracy* section that gives a measure of accuracy as obtained from deductive measurements, internal evidence, comparisons to source materials, or comparisons with independent sources of higher accuracy.

3. An *attribute accuracy* section stating that accuracy can be measured by deductive estimates, by tests based on independent samples, or by tests based on polygon overlay.
4. A *logical consistency* section that describes the fidelity of encoded relationships, as determined from tests of valid values, general tests for graphic data, and specific topological tests.
5. Information about selection criteria, definitions used, and other relevant mapping rules.

Cartographic Data Transfer Standards

Collection of data in formats and geocoding conventions unique to particular mapping systems have emerged as a major hindrance to the efficient sharing of information among cartographers. The National Committee on Digital Cartographic Data Standards has just published a spatial-data transfer specification that will "provide a mechanism for the transfer of digital spatial information between noncommunicating parties using dissimilar computer systems, preserving the meaning of the information . . . ." (NCDCDS 1988, 37). The standard covers the transfer of vector, grid, attribute, and other ancillary data; is media-independent and extendable to new spatial information; and is based on other existing standards, where appropriate.

Rectangular-Coordinate and Map-Projection Transformations

Coordinate transformations and map projections, the oldest analytical subjects, are still very much in the research limelight. This is due in large part to the universal use of digital computers for lightning-fast coordinate computations, making possible transformations and projections that would have been computationally unfeasible several decades ago. Least-squares solutions to 3-parameter affine coordinate transformations have been used for several decades to convert digitizer values to UTM or State Plane Coordinates, to register Landsat pixels to UTM grids, and for many similar problems. Sprinsky (1987) introduced 4-, 5- and 6-parameter affine equations for digitizer-to-rectangular-coordinate conversions for source maps on nonstable materials, noting that folded maps should be digitized tile by tile for best results.

Potentially more accurate is piecewise rubber sheeting, where control points are used to divide the sheet to be transformed into triangles. Affine equations are computed for each triangle, and any point falling within a triangle is transformed by the corresponding equation (White and Griffin 1985). "Simplicial" coordinates have been used to simplify the problem of finding which triangle a point falls in, and to simultaneously compute the transformation coefficients (Saalfeld 1985). "Zipper" and other rubber-sheeting methods have been devised to match edges of adjoining digital map sheets (Beard and Chrisman 1988).

New map projections with useful properties continue to be derived, adding to the basic projections developed over the last 2000 years and introducing new classes of projections to our taxonomy. For example, Tobler (1986b) has introduced the polycylindric projection class, completing the poly-family. Fascination with viewing Earth from space has prompted the derivation of an orthographic-like projection that shows essentially the entire Earth (Delucia and Snyder 1986), as well as perspective projections showing the Earth as it would be seen from the side window of a horizontally positioned spacecraft (Snyder 1981). Projections for continent-size areas have also been improved in various ways.

Equations have been derived for complex projections heretofore constructed graphically, such as the Miller oblated stereographic projection used by the American Geographical Society and Defense Mapping Agency for Africa, Europe, Asia, and Australasia (Sprinsky and Snyder 1986). Minimum-error projections for specific areas have been invented, with Snyder's (1984) projection for all 50 states being the best example. The use of equations based on the ellipsoid to lessen scale and shape distortion for national and smaller areas is also advocated (Snyder 1978). Another interesting innovation is projections that resemble magnifying glasses, enlarging an area of interest (Snyder 1987) and providing a way to have a higher symbol density in the area of interest (Monmonier 1977).

The problem of mathematically determining the similarity of map projections to each other and to the sphere has been tackled (Tobler 1986a). A simple map-coordinate-based difference measure has been derived from Tissot's Indicatrix as an alternative to scale and angular deformation values (Brooks and Roberts 1986). Similar methods have been used to compare ancient and modern maps, as well as to compare maps drawn from memory with conventional maps.

Statistical cartograms with data collection-unit areas adjusted to be proportional to distribution values can be viewed as mass distributing (pycnomirastic) map projections. Tobler developed the mathematical theory for such projections in the early 1970s, and continues to search for computationally simple approximations to their complex equations, calling one approach "pseudo-cartograms" (Tobler 1986c). Non-contiguous cartograms are far easier to derive, since countries represented on a Mercator or any other world projection can be enlarged or reduced in area by simple scale changes, and centered in their original location (Tobler 1978).

Geographic Data Interpolation

Cartographers are continually faced with the problem of *spatial prediction*. Surmising what is happening between sample points normally involves data interpolation. Interpolation is a cartographic operation that is central (1) to converting from one set of data-collection units to another (e.g., zip codes to counties), (2) to automated contouring from irregular control points, or (3) to resampling of Landsat pixels for orthophotomap production.

Cartographers have joined scientists from many fields in developing and evaluating interpolation methods. In a comprehensive review of interpolation, Lam (1983) proposed a convenient way of classifying the wide variety of methods, beginning with whether data are *point* or *areal* in nature (the latter refers to values for areas represented by a centroid location). Exact point-interpolation methods (where control-point values are preserved) are well known, involving distance weighting, interpolating polynomials, and the finite-difference method. Research has shifted to bidirectional spline interpolation for finding gridded control points from isoline maps (Legates and Willmott 1986), and to evaluating the distinctly different statistical method known as Kriging, the only method where confidence estimates can be obtained.

Approximate point-interpolation methods, such as the familiar power series trend surfaces, have been joined by Fourier series for regularly undulating surfaces, least-squares fitting with splines, and distance-weighted least-squares. Areal-interpolation methods are identical to point methods, if the surface volume is not being preserved. Volume-preserving methods, such as the grid overlay methods well-known from remote sensing and GIS, are on firmer theoretical ground. Tobler's (1979a) pycnophylactic (volume preserving) method for isopleth mapping is distinctly different, in that it assumes the existence of a smooth density function. These methods are based on different assumptions—really theories—about the spatial nature of reality, and a particular method must be selected with this in mind.

Other factors affecting interpolation accuracy are beginning to be examined. One finding important for small-scale isoline maps is that spherically based interpolation to a regular grid of latitude/longitude points prior to contouring produces more accurate maps than planar interpolation to a grid of (x, y) map coordinates, particularly at map edges (Willmott, Rowe, and Philpot 1985). Control-point density also affects interpolation accuracy. MacEachren and Davidson (1987) have derived from empirical measurements a power-function equation that predicts the mean interpolated-point accuracy for a given sample size, for surfaces of varying complexity.

Analytical Generalization

The process of cartographic generalization is understood to include simplification, classification, and symbolization. Simplification—the determination and possible exaggeration of important features, and elimination of unwanted detail—has received much attention, because it has been historically the most subjective of the three elements.

Simplification. Almost all research has dealt with simplification of lines defined by coordinate strings. In an excellent review of automated line generalization, McMaster (1987) organizes research into studies dealing with elimination of detail by point reduction, smoothing of lines by "planing" away small irregularities, and displacement where features are shifted (so as to prevent line overlap).

McMaster notes three general approaches to point elimination. Initially, independent point algorithms were devised, where 1/nth of all coordinates are systematically or randomly eliminated. The second approach, local processing, uses characteristics of immediate neighboring points. (The first such methods eliminated points that are too closely spaced, where there is minimal angular change between coordinate pairs, or where the perpendicular distance from a second point to a straight line between the first and third points is less than a user-specified threshold—Jenks 1981.) The third approach treats the line as a whole when eliminating detail, with the best example being the well-known Douglas-Peucker (Poiker) perpendicular-distance algorithm.

Recent research has focused on the evaluation of point-elimination algorithms. One characteristic that can be used to rate methods, particularly since people appear to pick the same points as "critical" on a line, is to maintain the "critical" points that either are crucial to preserving a line's unique form, or are seen as positions of importance (Marino 1979). E. R. White (1985) compared Marino's critical points with those selected by various algorithms, finding the Douglas-Peucker to give the best match of points. McMaster (1986) takes another tack by developing 30 mathematical measures of either linear attributes or linear displacement for evaluating algorithm efficiency. Again, the Douglas-Peucker algorithm performed best, making it the recommended method on all grounds except computation time.

The previous methods treat geography as geometry, and further progress hinges on also understanding the intrinsic geometric characteristics of a line that gives it geomorphic character. Buttenfield (1987) has taken the first step, by developing a method for quantifying the information contained in digitized lines. This could provide a method for segmenting lines and adjusting tolerance parameters in simplification algorithms to each segment, based on "structural signatures." Another intriguing possibility is to store a simplified data set, and then regenerate detail in generic terms, possibly through application of fractals (Dutton 1981). Although violating the fundamental principle that lost information cannot be recreated, "generic ungeneralization" may have value for small-scale thematic maps where positional accuracy is not critical.

Classification. Cartographers have long been concerned with methods for classifying quantitative and qualitative data, particularly for choropleth maps. Publication of the now-familiar Jenks' Optimal Method in 1971 essentially ended research into new methods for determining class intervals on single-variable choropleth maps. Interest has shifted toward qualitative classification, and researchers with joint interests in remote sensing and GIS have been quick to embrace image-processing approaches to classification (S-Y. Hsu 1979; Jensen 1981, and many others covered in the chapter on remote sensing). This trend should continue as remote sensor and traditional cartographic data become increasingly merged in GIS, stimulating further research into large-scale thematic mapping of qualitative information.

Symbolization. Links among the communication, map production, and analytical facets of cartography are perhaps strongest in the area of symbolization, as mathematical guidelines for symbol design, taken from communication research, are being incorporated into specific computer-mapping systems, and may be a key component of future "expert" cartographic systems. However, research has been, and continues to be, carried out independent of the communication paradigm. A prime example is the mapping of geographic movement such as migration, using data in "from-to" tables. Tobler (1987) has experimented with new forms of lines and surface-flow symbols to show one-way and two-way migrations.

Symbolizing the probabilistic allocation of demand to facilities from new location-allocation models is an equally challenging problem. The computing and placing of networks of graduated circles and interconnecting lines appears to be best accomplished by viewing the problem both in communication and analytical terms, with solution maps designed on interactive graphic-design systems having analytical capabilities (Allard and Hodgson 1987).

Research into area symbolization, having dealt primarily with algorithms for filling in areas with colors or patterns, is closely tied to map production. Cartographers have joined computer scientists in developing new algorithms for placing line-shading or cross-hatching within vector-defined polygons. The goal is to develop methods that are relatively insensitive to polygon topology and the density of shade lines, while being efficient in the use of storage and CPU time (Cromley 1984).

Nearly all forms of continuous-surface symbolization are inherently mathematical, and have been computerized over the last 20 years. This even includes relief shading, the apparent domain of the artist. The basic mathematical theory of surface illumination was developed by European cartographers in the 1950s, and has been expanded considerably by computer-graphics specialists (Horn 1982). Current research is centered upon finding the best way to combine shading with homogeneous area coloring, layer tinting, or other map symbols (D. White 1985). Fractal simulation of topographic relief is another promising avenue (Clarke 1988). Algorithms for methods that are old in concept but practical only by computer are also being worked out, such as continuous-density shading by randomly or quasirandomly placed dots (Lavin 1986).

Recent advances in computer storage, computation speed, and data-transfer rates to CRT terminals have made it possible to introduce animation as a form of symbolism (Moellering 1980). One possibility is a 3-D perspective map that can be rotated and enlarged at will by the user, allowing the surface to be "explored" fully. The goal, nearly achieved by military contractors, is to allow exploration of natural-appearing surfaces such as those colored by aerial photo information. This approach stands in contrast to efforts in recent years that have been directed at determining the optimal viewing point for static perspective maps (Monmonier 1978).

A second option is a static-background map with certain symbols animated, such as approaching fronts on a weather map. By viewing changing spatial phenomena dynamically, map users may be better able to formulate hypotheses as to the cause of change. The research challenge is both theoretical and technical, for cartographers must have a deep understanding of system hardware as well as a firm theoretical understanding of how animation should be carried out.

Not only has animated cartography come of age; an era of interactive mapping and map use has begun. Map users can now establish a dialogue with geographic databases through a graphic map interface. Different numerical map analyses can be requested, and results displayed instantly. This relieves the map of its traditional information-storage function, which is shifted to the database where it more properly belongs.

Numerical Map Analysis

Map analysis—making measurements concerning entities represented by symbolized objects on maps—is an important activity in many disciplines. The basic map-analysis methods mentioned in the communication facet were developed decades ago, and current research in large part deals with extension, refinement, and evaluation of methods, in light of computer computation power.

Area computation is a good example. The coordinate method of area computation was devised by mathematicians long ago, but the basic idea of adding and subtracting trapezoids has been extended to spheroidal triangles, giving a computation method for areas outlined by geodetic coordinates (Kimerling 1984). Shape measurement, a problem that has challenged cartographers and others for decades, is being refined by harmonic analysis in the complex-number domain. This approach supplements the older single-parameter measures of compactness. Dual-axis Fourier shape analysis (Moellering and Rayner 1982) is the most sophisticated of these, being independent of scale, translation, rotation, and starting point.

The rapid growth of cartographic data in raster form has directed research toward adapting map-analysis methods to raster data. Peuquet (1979) clearly states the key research question for both raster and vector data-processing, including map analysis: "Relative efficiency of various raster and vector algorithms has never been quantified in a cartographic context. The availability of such knowledge would allow the building of efficient raster-oriented systems . . . ." Digital elevation model (DEM) data-analysis algorithms are a good example, since various methods exist for slope degree and aspect determination. Now being studied is the far-more-difficult problem of formulating algorithms for determining drainage networks (Mark 1984) and delineating drainage basins (Mark, Dozier, and Frew 1984).

Numerical analysis of map similarity, usually in concert with communication studies, has interested cartographers for over 25 years. Researchers have realized that high statistical-correlation values do not always translate into reader perception of high similarity between maps. Statistical measures of *agreement,* such as the KHAT from psychology and remote sensing, are now under study (Carstensen 1987) since they appear to be well tuned with our visual impressions of map similarity.

Another serious map-comparison problem is data incompatibility due to collection from different geographic points or areal units. Matson (1985) reviews existing solutions and proposes a gridless surface-generation strategy using the SYMAP distance-weighted interpolation algorithm to give estimated values at common locations from which correlation and regression coefficients can be computed.

A little-researched topic that will be of increasing concern to specialists in cartography, remote sensing, and GIS is measuring the accuracy of map-analysis results. The mathematical basis for evaluating the accuracy of one method—digital overlay analysis (Newcomer and Szajgin 1984)—has been presented in probability-theory terms. Similar formulations for other methods are sorely needed to complement statistical-sampling procedures currently in use.

MAP PRODUCTION FACET

The large and ever-growing body of communication and analytical literature masks the importance of map production, which is *the* function of cartography as seen by the general public. To be sure, many communication and analytical studies have been directed toward map production directly or indirectly, through map-design recommendations. But map production has a narrower meaning here, encompassing manual and computer-assisted map construction and reproduction techniques, mathematical map construction guidelines, and the planning, marketing, and legal aspects of mapping projects. The body of academic literature on map production is surprisingly small, considering the vast public and private expenditures of money and time during the last decade for the modernizing of map production. Unfortunately, many advances often are described only in contract reports and proprietary in-house documents.

Manual Map Construction

Cartographers have recently begun to augment textbook overviews of manual construction techniques with more detailed descriptions and comparisons. For example, Gilman's (1981) description and comparison of manual and photomechanical relief-shading methods used in federal mapping agencies brings to light many unpublished in-house experiments. Cartographic applications of new photomechanical equipment and materials, such as the Diffusion Transfer Process (Graves and DesRivieres 1979) and custom-made symbol-transfer sheets (Turner 1983), have also been documented. Detailed explanations have been published of specific techniques such as component and final color separation (Olson 1985), chokes and spreads to modify line weights (Hodler and Doyon 1984), and improved methods for producing large relief models (Ryerson 1983). Photomechanical mosaicking methods for creating a satellite-image map of Illinois have been described in detail (Dahlberg, Luman, and Stohr 1986), and will undoubtedly be applied in other states and countries.

This emphasis on technological aspects of map construction is understandable in this period of rapid change, but the human side of construction is often forgotten. Equal time should be given to understanding the internal and external circumstances that affect human use of technology. Muehrcke (1982) categorizes both internal and external circumstances. Internal circumstances include value traps (value rigidity, ego, anxiety, boredom, and impatience), truth traps (indeterminate answers) and psycho-motor traps (inadequate tools and materials, bad surroundings, and muscular insensitivity). External circumstances include unrealistic client demands, inadequate finances, incomplete data, and poor equipment. He argues that, in teaching cartography, greater stress should be placed on integrating analytical and intuitive-holistic thinking in map design and production, so that cartographers can function better under these circumstances.

Tactual maps for the blind have been studied from the design and construction viewpoint, beginning in the 1950s with John C. Sherman and his students at the University of Washington. The many manual-construction methods employed over the last 30 years for raised-surface tactual maps have recently been summarized and compared (Turner and Sherman 1986). The trend is toward computer-assisted tactual-map construction, and map reproduction by raised-surface printing plates. For example, the near-future prospect of interactive tactual mapping with braille-dot-matrix-like display devices has been explored (Lai 1985).

Mathematical Map Construction Guidelines

Cartographers have joined graphic-arts specialists in developing mathematical equations for various aspects of lithographic map printing, particularly the printing of graytones and colors using screen tints. The basic problem is to convert colors specified in the perceptually based Munsell or physically based CIE system into process-color screen tints (Kimerling 1980). A combination of equations from probability theory, colorimetry, and graphic arts allows this to be done for ideal printing conditions, although in practice inconsistencies in printing reduce their productive power. The same is true for printed graytones, where printing variations reduce the utility of equal-value grayscale equations (Monmonier 1980).

Interest has now shifted to the mathematics of color specification for high-resolution color monitors. Here the problem is to specify sets of red-green-blue (RGB) or similar color coordinates for the types of color progressions normally used in cartography (Sibert 1980). Work on value, chroma, and tint progressions is already underway (Edwards and Batson 1980).

Computer-Assisted Map Construction

Computer-assisted map construction has evolved to the point that some have predicted the virtual disappearance of manual mapmaking by the end of the century (Monmonier 1985). Researchers recognizing this profound technological transition have focused on the technical aspects of constructing maps on CRT displays, printers, plotters, and other virtual and real display devices. The inexpensive dot-matrix printer, for example, has been studied at depth, and dot arrangements for different graytones and like patterns have been determined (Groop and Smith 1982; Plumb and Slocum 1986). Similar studies with laser, ink jet, and higher-resolution printers will soon follow.

A second approach to understanding new technology involves studying the sequence of procedures used to construct a particular type of map. Summaries of data input, manipulation, and display procedures employed in mapping projects that range from line-plotter city street maps (Carstensen 1985) to shaded-relief maps on color monitors (Judd 1986) have proven useful to those contemplating similar projects. Valuable also are observations as to why certain mapping systems have prospered or wilted with use; the demise of the Domestic Information Display System (DIDS) is a prime example (Cowen 1984).

Map Marketing

Surprisingly little attention has been given to the role of the marketplace in map production, despite the fact that most maps are made to be sold to individuals or organizations. Petchenik explains that marketing is not generally the domain of academic cartographers, and that commercial firms operating in a highly competitive price-sensitive environment cannot afford to share marketing secrets. In a call for marketing research, she notes:

> We cannot continue to focus our research exclusively on "users," when users and buyers are not synonymous. We need to know more, for example, about the complex set of relationships that links an individual map-buying taxpayer with an individual map-using foot soldier, a relationship that includes the Congress, the Army, and hierarchies of government employees whose job it is to spend the limited allocated funds on a limited amount of mapping. (1985b, 16)

Or, as Monmonier notes:

> . . . cartographic education must recognize that, in communicating geographic information, public policy is a more significant factor than the eye-brain system . . . Mapping has advanced to the state where the collection of many types of geographic information is technically possible and the principal constraints are institutional and political, not perceptual and cognitive. (1982a, 99)

Production Planning

Very few cartographers have attempted to link map design and construction with other managerial concerns, even though such integrated discussions are extremely valuable to the practicing cartographer. Loy's insights (1980) gained from personal experience with the *Atlas of Oregon* form the most comprehensive guide to production planning. He notes that atlas planning entails "establishing a rationale for the atlas; obtaining funding; developing a staffing plan; acquiring equipment; developing the table of contents; designing both the book and individual pages; establishing a registry system; and choosing a means of creating linework, area colors, and lettering" (Loy 1980, 105). This ordering of activities will hold true for most manual mapping projects.

Petchenik's (1977) memoir from the *Atlas of Early American History* project touches on these same topics, but stresses design and production problems unique to mapping historical phenomena. Data quality is a prime question here, since many interesting distributions and events may not be "mappable." In other instances, mappability may be as much a matter of the political nature of data access as of data quality; witness the lack of recent detailed maps of Albania, South Africa, and Vietnam.

Similar accounts of projects involving mapping in a computer-assisted production-planning environment are needed, since new problems and procedures are undoubtedly faced. For example, Monmonier (1981) introduces computerized production-planning methods, such as sheet layout codes and computer programs for displaying schematic layouts of map sheets. These methods are likely to gain wide acceptance in the near future.

Legal Aspects

Recent court rulings which have been costly to cartographic firms have stirred interest in the often neglected legal aspects of cartography. First, copyright infringement is a concern of private map producers. The Copyright Law of 1978, as described by Cerny (1978), spells out requirements for copyright registration, the penalties for infringement, and the fact that the effort of compilation, not the originality of expression, is being protected. Maps in digital form have much in common with computer software, and appear to be protected by the Computer Software Act of 1980 (Monmonier 1985).

Liability for inaccuracies is a second legal issue of growing concern to private and public map producers, since producers have recently been held liable for accidents attributed to missing or incorrect map information. Errors in digital cartographic databases will soon be a legal problem as well as a managerial and technical one, forcing the adoption of the previously mentioned digital cartographic data-quality standards.

A third important legal aspect is the use of maps as evidence in court cases, which some term "forensic" cartography. Cartographers must better understand those aspects of map production which are likely to be challenged in court by lawyers who are aware of the selection and generalization inherent in map design and construction.

HISTORICAL FACET

The last decade has witnessed a growing interest in the history of cartography. Although new maps of historical significance have been uncovered, and existing maps have been examined at greater depth, the major advances have been conceptual. A new theoretical framework for the historical study of cartography has emerged, assembled largely by American and British academicians. Woodward's suggested framework (1974) for the study of the history of cartography, and Blakemore and Harley's later monograph (1980) focusing on concepts in the history of cartography, make clear the distinction between *historical cartography* and the *history of cartography*.

Historical cartography is founded on a view of maps as a storehouse of geographic information to be used in the reconstruction of past events or landscape features. Study of the *content* of maps—i.e., the information about the environment communicated by assemblages of map symbols—is at the heart of historical cartography.

The *history of cartography* is distinctly different, in that it involves:

> . . . the study of maps, mapmakers, and mapmaking techniques in their human context through time. The field includes the study of creation of the maps by survey, compilation, engraving, printing, and coloring; the history of their distribution by publishing and selling; and the history of their acquisition, care, cataloging, and use. (Woodward 1974, 114)

Woodward's map production-oriented definition and Blakemore and Harley's broader definition are based upon the map-communication paradigm. In it, the history of cartography "is concerned with the development of maps of all kinds as formal systems for the communication of spatial information, and it focuses on the nature, structures, distribution, and significance of cartographic language within past societies" (Blakemore and Harley 1980, 13). This redefines and redirects a field self-criticized as lacking a logical structure, standardized terminology, and a general definition.

Another conceptual contribution of note is the methodology and organizational framework for the history of thematic cartography advocated by MacEachren (1979). He divides the evolution of thematic cartography according to methods of point, line, and area symbolization. Within each division, historical developments are further classified by whether the data symbolized are qualitative or quantitative, and by whether physical or cultural phenomena are represented. This methodology is appealing, because it corresponds with the modern organizational framework for cartographic symbolization, but its widespread adoption has yet to be seen.

Shifts in research approaches in response to the new conceptual framework can be seen by looking at recent trends in the literature. The *bio-bibliographical* approach, dominant since the nineteenth century, is still used to document the lives and output of maps by individual cartographers (Thrower 1978; Snyder 1979). The most ambitious American study in this vein is Ristow's *American Maps and Mapmakers* (1985). However, Ristow and others are now placing more emphasis on historical links among cartographers and on the development of "schools of mapmaking," such as the Chicago school of commercial cartography (Conzen 1984).

Interest continues in identifying the *earliest surviving maps* of a particular society, geographic area, historical event, or mapping method. Hsu's description (1978) of Han Dynasty maps is a classic example of this approach. Post-Renaissance cartography is currently receiving the greatest attention, exemplified by Robinson's account (1982) of the scientist-cartographers and historical events responsible for the sudden appearance of small-scale thematic maps in the nineteenth century.

Maps are beginning to be viewed as *physical artifacts*. Methods of map production and their impact on the content and appearance of early maps are receiving increased attention as a discrete research area. A classic work along this line is Woodward's examination (1977) of wax engraving and its influence on nineteenth-century American cartography. Pearson (1983) also treats maps as physical artifacts, but concentrates on the series of technical innovations during the nineteenth century that mechanized the creation of patterns, tones, and colors.

A related but distinctly different aspect of the history of cartography is the effect of political and managerial decisions upon the areas mapped in the past, as well as on the content of early maps. The historical development of U.S. national mapping policy (Southard 1983; Edney 1986) has received the greatest scrutiny, particularly the division of civilian and military mapping authority, the consolidation of topographic surveys into the USGS, and legislation expanding or contracting the authority of mapping agencies.

The accuracy of early maps has long interested cartographers. Woodward's examination (1987) of the manuscript, engraved, and typographic traditions of map lettering focuses on *chronometric accuracy,* since he finds that physical differences in lettering can be used to date maps. American research has tended to focus on *geodetic and planimetric accuracy,* following Tobler's pioneering (1966) least-squares regression analysis of geometric distortions on Medieval maps, relative to modern positions. Lloyd and Gilmartin (1987) have expanded the regression approach to include analysis-of-variance of directional-error vectors for the South Carolina coastline, as depicted on historical maps.

The final and most recent research trend is the iconographic. Iconography, the study of metaphorical, poetic, or symbolic meanings of artwork—including maps—is an obvious meeting ground of art and cartography. Gilmartin (1984, 39) notes three levels of subject matter or meaning through which a work of art can be analyzed, according to art historian Erwin Panofsky:

1. primary or natural subject matter—in cartography corresponding to the study of map marks symbolizing spatial phenomena independent of their geographic location; thus viewing map symbols as "artistic motifs";
2. secondary or conventional subject matter—connecting the "cartographic motifs" from the first level with specific themes, concepts or meaning through study of their location on maps and their verbal content (lettering);
3. intrinsic meaning or content—ascertaining those underlying principles, which reveal the basic attitude of a nation, a period, a class, a religious or philosophical persuasion . . . .

Woodward's examination (1985) of Medieval mappaemundi is an excellent example of iconographic analysis at the second and third levels. He explores several themes, including the type of reality represented by the maps, the way the map center changed as the Middle Ages developed, and the relationship between concepts of the Earth's sphericity and the graphic constraints on the mappaemundi.

This section cannot be concluded without mention of Harley and Woodward's monumental *History of Cartography* project, with the first volume now published (1987). This is a grand attempt to outline the history of cartography in both Western and Eastern cultures, based on:

> First, a catholic definition of "map"; second, commitment to a discussion of the manifold technical processes that have contributed to the form and content of individual maps; third, recognition that the primary function of cartography is ultimately related to the historically unique mental ability of map-using peoples to store, articulate, and communicate concepts and facts that have a spatial dimension; and fourth, the belief that since cartography is nothing if not a perspective on the world, a general history of cartography ought to lay the foundations, at the very least, for the world view of its own growth. (1987, xviii)

No better statement can be made of the goals and facets of the new approaches to the history of cartography.

RESEARCH PROSPECTS

The intellectual fervor reflected in the research of the last 10 years will continue to be sustained by technological innovation coupled with new map forms, new subjects to map, and a new relationship between map and map user. Steady improvements in integrated circuitry insure the continual introduction of microcomputers with greater memory capacities, higher computation speeds, and higher-resolution color monitors. Ever-more robust mapping software will be introduced, *allowing program users having little understanding of cartographic principles to instantly become mapmakers*. This will also occur with geographic information systems, where the map will fill two roles, both as database interface and information visualization.

The specter looms of poorly designed maps by the millions. An integrated research program is needed in which psychophysical and other stimulus-response testing directly on display monitors will provide graphic magnitude scales for specification of value and chroma progressions, line widths, lettering and point symbol sizes, and so on. Such stimulus-response equations will provide acceptable symbol-selection defaults, and noncartographers may be restricted to using these.

But stimulus-response experiments are not enough—cognitive aspects of new display technology must be investigated as well. The order in which cartographic objects are placed on the display screen may enhance or retard the memorization, recognition, and recall of geographical information. Flashing or animated map symbols may have similar effects.

Taking a broader view, there may be cognitive differences between printed and electronic maps due to different viewing conditions, the interactive approach to map display, and other ergonomic factors. Marrying ergonomic studies of attention span, eye strain, and other psychological and physiological effects of screen viewing with cognitive and psychophysical experiments appears to be a promising research path (Monmonier 1985). It may well be that the impact of automation on map use will be more significant than its impact on map construction.

This strategy of carrying out communication experiments that provide the analytical basis for producing map symbols illustrates the linkages among the several facets of cartography. These will grow stronger in the coming years, as new display technologies are marketed. Communication aspects of animated video-disk maps, digital holograms, and stereoscopically viewable color monitors must be investigated, in concert with research into the mathematical theory underlying image creation. These new display devices hold great promise for the production of realistic 3-D perspective landscape maps from digital elevation-model data and satellite-image data. Real-time surface exploration from a continually changing viewpoint is the ultimate goal. Cartographers, remote sensing specialists, and computer scientists are beginning to work out the analytical and communication aspects of this problem.

Closely related is the development of automated vehicle-navigation aids. Cartographers are beginning to study the differences among commercially available systems, but the most crucial component, the user interface, is understood the least (McGranaghan, Mark, and Gould 1987). The basic question of whether to use continuously updated maps and written or verbal instructions is not completely resolved, although empirical results from cognitive studies suggest that verbal presentations may be more effective. The role of automated navigation aids in learning the spatial structure of a place—be it an orienteering course, or the New York City street system that so challenges taxi drivers—is only beginning to be explored. (However, a rich flight-simulation literature exists.)

The transition to digital cartography in many governmental agencies was undertaken to increase productivity, to reduce the subjective element in map and chart production, and to increase internal consistency in map series. A parallel transition to more computer-oriented employees has led to serious declines in the number of experienced cartographers who oversee map production. To capture the corporate expertise which is in danger of being lost, experimental "expert" map-design systems are being developed for the selection, design, and placement of map symbols (Bossler et al. 1988). The large knowledge base and many rules needed for the placement of point symbols and lettering (Ahn and Freeman 1983) hints at the complexity of future systems. Research into how to code so much information—past experience, communication-study results, and technical limitations—into expert system knowledge bases and inference engines will challenge analytical cartography for years to come.

Analytical cartography, remote sensing, and geographic information systems have always shared data sources, data-manipulation methods, and output techniques. The common concern with mapping and spatial analysis is bringing about an integration of research and researchers in these three fields, as problems inherent to large-scale thematic mapping are attacked. Interest in *large-scale thematic cartography* will rise dramatically, as geographic-information systems become a standard tool of environmental planning.

Mapping *policy* is a little-studied area of future importance (Monmonier 1985). Topics of mutual concern are copyright restrictions; suppression of data for security, privacy, and other reasons; and coordination of data exchange. Policy issues are as important to professional geography as they are to society at large.

Future integration of cartography into the realm of automated geographic information systems is inevitable, and the impact of this development upon geography will be felt for decades. Radical change will occur in what is considered mappable, how it is mapped, and the way in which maps are used. The effort needed to mold and fine-tune the mapping process to suit geographers' needs in the coming information age will involve resolution of an ever-widening array of mapping problems. It all spells a bright research future for cartography.

REFERENCES

Ahn, J., and Freeman, H. 1983. A program for automatic name placement. *Proceedings of the Auto-Carto Six Conference,* Ottawa, Canada, 16–21.

Allard, L., and Hodgson, M. J. 1987. Interactive graphics for mapping location-allocation solutions. *American Cartographer* 14:49–60.

Antes, J. R.; Chang, K-T; and Mullis, C. 1985. The visual effect of map design: An eye-movement analysis. *American Cartographer* 12:143–55.

Beard, K., and Chrisman, N. R. 1988. Zipper: A localized approach to edgematching. *American Cartographer* 15:163–72.

Bertin, J. 1978. Theory of communication and theory of the graphic. *International Yearbook of Cartography* 28:118–26.

Blakemore, M. J., and Harley, J. B. 1980. Concepts in the history of cartography. *Cartographica Monograph* no. 26., 120 p.

Bossler, J. D.; Pendleton, D. L.; Swetnam, G. F.; Zitalo, R. L.; Schwarz, C. R.; Alper, S.; and Danley, H. P. 1988. Knowledge-based cartography: The NOS experience. *American Cartographer* 15:149–61.

Brooks, W. D., and Roberts, C. E. 1986. A mathematical analysis and comparison of three circular world graticules. *Cartographica* 23(3):28–41.

Buttenfield, B. P. 1987. Automating the identification of cartographic lines. *American Cartographer* 14:7–20.

Calkins, H. W. and Marble, D. F. 1987. The transition to automated production cartography: Design of the master cartographic database. *American Cartographer* 14:105–19.

Carstensen, L. W. 1984. Perceptions of variable similarity on bivariate choropleth maps. *Cartographic Journal* 21(2):23–39.

———. 1985. Computer-assisted compilation and drafting/planning for the Green Bay, Wisconsin area street map. *Cartographica* 22(1):93–105.

———. 1986. Hypothesis testing using univariate and bivariate choropleth maps. *American Cartographer* 13:231–51.

———. 1987. A measure of similarity for cellular maps. *American Cartographer* 14:345–58.

Castner, H. W., and Eastman, J. R. 1984. Eye-movement parameters and perceived map complexity: 1. *American Cartographer* 11:107–17.

———, and ———. 1985. Eye-movement parameters and perceived map complexity: 2. *American Cartographer* 12:29–40.

Cerny, J. W. 1978. Awareness of maps as objects for copyright. *American Cartographer* 5:45–56.

Chang, K-T. 1978. Measurement scales in cartography. *American Cartographer* 5:57–64.

———. 1980a. Circle size judgement and map design. *American Cartographer* 7:155–62.

———. 1980b. Visual estimation of graduated circles. *Canadian Cartographer* 14:130–38.

———, and Antes, J. R. 1987. Sex and cultural differences in map reading. *American Cartographer* 14:29–42.

Chrisman, N. R. 1984. The role of quality information in the long-term functioning of a geographic information system. *Cartographica* 21(243):79–87.

Clarke, K. C. 1988. Scale-based simulation of topographic relief. *American Cartographer* 15:173–81.

Cole, D. G. 1981. Recall vs. recognition and task specificity in cartographic psychophysical testing. *American Cartographer* 8:55–66.

Conzen, M. P. 1984. *Chicago mapmakers*. Chicago: Chicago Historical Society.

Cowen, D. J. 1984. Rethinking DIDS: The next generation of interactive color mapping systems. *Cartographica* 21(2&3):89–92.

Cox, C. W. 1980. The effects of background on the equal value gray scale. *Cartographica* 17:53–71.

Cox, N. J.; Alfred, B. K.; and Rhind, D. W. 1980. A relational data base system and a proposal for a geographical data type. *Geoprocessing* 1:217–29.

Cromley, R. G. 1984. The peak-pit-pass polygon line-shading procedure. *American Cartographer* 11:70–79.

Cuff, D. J., and Bieri, K. R. 1979. Ratios and absolute amounts conveyed by a stepped statistical surface. *American Cartographer* 6:157–68.

Dahlberg, R. E., and Jensen, J. R. 1986. Education for cartography and remote sensing in the service of an information society. *American Cartographer* 13:51–71.

———; Luman, D. E.; and Stohr, C. J. 1986. Creation of a satellite image map of Illinois: A case study of technology transfer and linkage. *Cartographica* 23(4):14–28.

DeLucia, A. A. 1979. An analysis of the communication effectiveness of public planning maps. *Canadian Cartographer* 16:168–80.

———, and Snyder, J. P. 1986. An innovative world map projection. *American Cartographer* 13:165–67.

Dobson, M. W. 1979a. The influence of map information on fixation location. *American Cartographer* 6:51–65.

———. 1979b. Visual information processing during cartographic communication. *Cartographic Journal* 16:14–20.

———. 1980. The influence of the amount of graphic information on visual matching. *Cartographic Journal* 17:26–32.

Dutton, G. 1981. Fractal enhancement of cartographic line detail. *American Cartographer* 8:23–40.

Eastman, J. R. 1985. Cognitive models and cartographic design research. *Cartographic Journal* 22:95–101.

Edney, M. H. 1986. Politics, science, and government mapping policy in the United States, 1800–1925. *American Cartographer* 13:295–306.

Edwards, K., and Batson, R. M. 1980. Preparation and presentation of digital maps in raster format. *American Cartographer* 7:39–49.

Fraczek, I. 1984. Cartographic studies on graduated symbol map perception. *International Yearbook of Cartography* 24:75–84.

Gersmehl, P. J. 1981. Maps in landscape interpretation. *Cartographica* 18(2):79–115.

Gilman, C. R. 1981. The manual/photomechanical and other methods for relief shading. *American Cartographer* 8:41–53.

Gilmartin, P. P. 1981a. Influences of map content on circle perception. *Annals of the Association of American Geographers* 71:253–58.

———. 1981b. The interface of cognitive and psychophysical research in cartography. *Cartographica* 18:9–20.

———. 1984. The Austral continent on 16th century maps: An iconological interpretation. *Cartographica* 21:38–52.

———. 1985. The design of journalistic maps: Purposes, parameters and prospects. *Cartographica* 22(4):1–18.

———. 1986. Maps, mental imagery, and gender in the recall of geographical information. *American Cartographer* 13:335–44.

———, and Patton, J. C. 1984. Comparing the sexes on spatial abilities: Map use skills. *Annals of the Association of American Geographers* 74:605–19.

Graves, F. W., and DesRivieres, D. L. 1979. Cartographic applications of the diffusion transfer process. *American Cartographer* 6:107–15.

Groop, R. E., and Cole, D. 1978. Overlapping graduated circles: Magnitude estimation and method of portrayal. *Canadian Cartographer* 15:114–22.

———, and Smith, R. M. 1982. Matrix line printer maps. *American Cartographer* 9:19–24.

Harley, J. B., and Woodward, D., eds. 1987. *Cartography in prehistoric, ancient, and medieval Europe and the Mediterranean*. Vol. 1 of *The history of cartography*. Chicago: University of Chicago Press.

Hodler, T. W., and Doyon, R. 1984. Spreads and chokes: Cartographic applications and procedures. *American Cartographer* 11:63–69.

Horn, B. K. P. 1982. Hill shading and the reflectance map. *Geoprocessing* 2:65–146.

Hsu, M-L. 1978. The Han maps and early Chinese cartography. *Annals of the Association of American Geographers* 68:45–60.

———. 1979. The cartographer's conceptual process and thematic symbolization. *American Cartographer* 6:117–27.

Hsu, S-Y. 1979. Automation in cartography with remote sensing methodologies and technologies. *Canadian Cartographer* 16(2):183–94.

Jenks, G. F. 1981. Lines, computers, and human frailties. *Annals of the Association of American Geographers* 71:1–10.

Jensen, J. R. 1981. Urban change detection mapping using Landsat digital data. *American Cartographer* 8:127–47.

Judd, D. B. 1986. Processing techniques for the production of an experimental computer-generated shaded-relief map. *American Cartographer* 13:72–79.

Kimerling, A. J. 1975. A cartographic study of equal value gray scales for use with screened gray areas. *American Cartographer* 2:119–27.

———. 1980. Color specification in cartography. *American Cartographer* 7:139–53.

———. 1984. Area computation from geodetic coordinates on the spheroid. *Surveying and Mapping* 44:343–51.

———. 1985. The comparison of equal-value gray scales. *American Cartographer* 12:132–42.

Kulhavy, R. W.; Schwartz, N. H.; and Shaha, S. H. 1982. Interpretative framework and memory for map features. *American Cartographer* 9:141–47.

Lai, P-C. 1985. Moving towards an interactive tactual mapping environment. *Cartographic Journal* 22:102–5.

Lam, N. S-N. 1983. Spatial interpolation methods: A review. *American Cartographer* 10:129–49.

Lavin, S. 1986. Mapping continuous distributions using dot-density shading. *American Cartographer* 13:140–50.

Legates, D. R., and Willmott, C. J. 1986. Interpolation of point values from isoline maps. *American Cartographer* 13:308–23.

Lloyd, R., and Gilmartin, P. 1987. The South Carolina coastline on historical maps: A cartometric analysis. *Cartographic Journal* 24:19–26.

———, and Steinke, T. R. 1985a. Comparison of quantitative point symbols/The cognitive process. *Cartographica* 22(1):59–77.

———, and ———. 1985b. Comparison of qualitative point symbols: The cognitive process. *American Cartographer* 12:156–68.

Loy, W. G. 1980. State atlas creation. *American Cartographer* 7:105–21.

Luman, D. E. 1987. Applying USGS digital line graph data in a microcomputer environment. *American Cartographer* 14:321–44.

MacEachren, A. M. 1979. The evolution of thematic cartography/A research methodology and historical review. *Canadian Cartographer* 16:17–33.

———. 1982. The role of complexity and symbolization method in thematic map effectiveness. *Annals of the Association of American Geographers* 72:495–513.

———, and Davidson, J. V. 1987. Sampling and isometric mapping of continuous geographic surfaces. *American Cartographer* 14:299–320.

McGranaghan, M.; Mark, D. M.; and Gould, M. D. 1987. Automated provision of navigation assistance to drivers. *American Cartographer* 14:121–38.

McMaster, R. B. 1986. A statistical analysis of mathematical measures for linear simplification. *American Cartographer* 13:103–17.

———. 1987. Automated line generalization. *Cartographica* 24(2):74–111.

Marino, J. S. 1979. Identification of characteristic points along naturally occurring lines: An empirical study. *Canadian Cartographer* 16(1):70–80.

Mark, D. M. 1984. Automated detection of drainage networks from digital elevation models. *Cartographica* 21 (2&3):168–78.

———; Dozier, J.; and Frew, J. 1984. Automated basin delineation from digital elevation data. *Geoprocessing* 2:299–311.

Marx, R. W. 1986. The TIGER system: Automating the geographic structure of the United States Census. *Government Publications Review* 13:181–201.

Matson, J. T. 1985. Applying the SYMAP algorithm for surface compatibility and comparative analysis of areal data. *American Cartographer* 12:114–22.

Moellering, H. 1980. The real-time animation of three dimensional maps. *American Cartographer* 7:67–75.

———. 1984. Real maps, virtual maps and interactive cartography. In *Spatial statistics and models,* eds. G. L. Gaile and C. J. Willmott, pp. 109–32. Dordrecht: D. Reidel.

———, and Rayner, J. N. 1982. The dual axis Fourier shape analysis of closed cartographic forms. *Cartographic Journal* 19:53–59.

Monmonier, M. S. 1976. Modifications of the choropleth technique to communicate correlation. *International Yearbook of Cartography* 18:160–64.

———. 1977. Nonlinear reprojection to reduce the congestion of symbols on thematic maps. *Canadian Cartographer* 14(2):35–47.

———. 1978. Viewing azimuth and map clarity. *Annals of the Association of American Geographers* 68:180–95.

———. 1979. An alternative isomorphism for mapping correlation. *International Yearbook of Cartography* 21:79–88.

———. 1980. The hopeless pursuit of purification in cartographic communications: A comparison of graphic arts and perceptual distortions of graytone symbols. *Cartographica* 17(2):24–39.

———. 1981. Automated techniques in support of planning for the National Atlas. *American Cartographer* 8:161–68.

———. 1982a. Cartography, geographic information, and public policy. *Journal of Geography in Higher Education* 6(2):99

———. 1982b. *Computer-assisted cartography: Principles and prospects.* Englewood Cliffs, NJ: Prentice-Hall.

———. 1985. *Technological transition in cartography.* Madison, WI: University of Wisconsin Press.

Morrison, J. L. 1976. The science of cartography and its essential processes. *International Yearbook of Cartography* 16:84–97.

Muehrcke, P. C. 1972. *Thematic cartography.* Washington: Commission on College Geography Resource Paper no. 19, Association of American Geographers.

———. 1973. *Multilingual dictionary of technical terms in cartography*, ed. E. Meynen. Wiesbaden: Franz Steiner Verlag.

———. 1976. Concepts of scaling from the map reader's point of view. *American Cartographer* 3:123–41.

———. 1982. An integrated approach to map design and production. *American Cartographer* 9:109–22.

NCDCDS. 1988. Proposed standard for digital cartographic data. *American Cartographer* 15:1–142.

Newcomer, J. A., and Szajgin, J. 1984. Accumulation of thematic map errors in digital overlay analysis. *American Cartographer* 11:58–62.

Nyerges, T. L. 1981. Cartographic information modeling as a basis for cartographic data base structures.

2nd International Hypergraph Based Data Structures Symposium, Richmond, VA.

Olson, J. M. 1977. Rescaling dot maps for pattern enhancement. *International Yearbook of Cartography* 17:125–37.

———. 1979. Cognitive cartographic experimentation. *Canadian Cartographer* 16(1):34–44.

———. 1983. Future research directions in cartographic communication and design. In *Graphic communication and design in contemporary cartography,* ed. D. R. F. Taylor, pp. 257–84. New York: Wiley.

———. 1985. Component-color and final-color separation in mapping. *Cartographica* 22(3):61–69.

Patton, J. C. and Slocum, T. A. 1985. Spatial pattern recall/an analysis of the aesthetic use of color. *Cartographica* 22(3):70–87.

Pearson, K. S. 1983. Mechanization and the area symbol: Cartographic techniques in 19th century geographical journals. *Cartographica* 20:1–34.

Petchenik, B. B. 1977. Cartography and the making of an historical atlas: A memoir. *American Cartographer* 4:11–28.

———. 1983. A mapmaker's perspective on map design research 1950-1980. In *Graphic communication and design in contemporary cartography,* ed. D. R. F. Taylor, pp. 37–68. New York: Wiley.

———. 1985a. Facts or values: Basic methodological issues in research for educational mapping. *Cartographica* 22(3):20–42.

———. 1985b. Maps, markets and money: A look at the economic underpinnings of cartography. *Cartographica* 22(3):7–19.

Peterson, M. P. 1979. An evaluation of unclassed crossed-line choropleth mapping. *American Cartographer* 6:21–37.

———. 1987. The mental image in cartographic communication. *Cartographic Journal* 24:35–40.

Peucker (Poiker), T. K., and Chrisman, N. R. 1975. Cartographic data structures. *American Cartographer* 2:55–69.

Peuquet, D. J. 1979. Raster processing: An alternative approach to automated cartographic data handling. *American Cartographer* 6:129–39.

———. 1981a. An examination of techniques for reformatting digital cartographic data. Part 1. The raster-to-vector process. *Cartographica* 18(1):34–48.

———. 1981b. An examination of techniques for reformatting digital cartographic data. Part 2. The vector-to-raster process. *Cartographica* 18(3):21–33.

———. 1984. A conceptual framework and comparison of spatial data models. *Cartographica* 21(4):66–113.

Plumb, G. A., and Slocum, T. A. 1986. Alternative designs for dot-matrix printer maps. *American Cartographer* 13:121–33.

Provin, R. W. 1977. The perception of numerousness on dot maps. *American Cartographer* 4:111–25.

Ristow, W. W. 1985. *American maps and mapmakers.* Detroit: Wayne State University Press.

Robinson, A. H. 1952. *The look of maps: An examination of cartographic design.* Madison, WI: University of Wisconsin Press.

———. 1977. Research in cartographic design. *American Cartographer* 4:163–69.

———. 1982. *Early thematic mapping.* Chicago: University of Chicago Press.

———, and Petchenik, B. B. 1975. The map as a communication system. *Cartographic Journal* 12:7–14.

Rugg, R. D. 1984. Building a hypergraph-based data structure. *Cartographica* 21(2&3):179–87.

Ryerson, C. C. 1983. Improved methods of reproducing large relief models. *American Cartographer* 10:151–57.

Saalfeld, A. 1985. A fast rubber-sheeting transformation using simplicial coordinates. *American Cartographer* 12:169–73.

Shortridge, B. G 1979. Map reader discrimination of lettering size. *American Cartographer* 6:13–20.

———, and Welch, R. B. 1980. Are we asking the right questions? *American Cartographer* 7:19–23.

———, and ———. 1982. The effect of stimulus redundancy on the discrimination of town size on maps. *American Cartographer* 9:69–80.

Sibert, J. L. 1980. Continuous-color choropleth maps. *Geoprocessing* 1:207–13.

Slocum, T. A. 1981. Analyzing the communication efficiency of two-sectored pie graphs. *Cartographica* 18(3):53–65.

———, and McMaster, R. B. 1986. Gray tone versus line plotter symbols: A matching experiment. *American Cartographer* 13:151–64.

Smith, C. A. 1977. A test concerning the relative readability of topographic and orthophotomaps. *American Cartographer* 4:133–43.

Smith, R. M. 1980. Improved areal symbols for computer line-printer maps. *American Cartographer* 7:51–57.

———. 1987. Influence of texture on perception of gray tone map symbols. *American Cartographer* 14:43–47.

Snyder, J. P. 1978. Equidistant conic map projections. *Annals of the Association of American Geographers* 68:373–83.

———. 1979. The Erskine-DeWitt maps. *Surveying and Mapping* 39:33–49.

———. 1981. The perspective map projection of the Earth. *American Cartographer* 13:165–67.

———. 1984. A low-error conformal map projection for the 50 states. *American Cartographer* 13:253–61.

———. 1987. "Magnifying-glass" azimuth map projections. *American Cartographer* 14:61–68.

Southard, R. B. 1983. The development of U.S. national mapping policy. *American Cartographer* 10:5–15.

Sprinsky, W. H. 1987. Transformation of positional geographic data from paper-based map products. *American Cartographer* 14:359–66.

———, and Snyder, J. P. 1986. The Miller oblated stereographic projection for Africa, Europe, Asia, and Australasia. *American Cartographer* 13:253–61.

Steinke, T. R. 1987. Eye movement studies in cartography and related fields. *Cartographica* 24:40–73.

———, and Lloyd, R. E. 1981. Cognitive integration of objective choropleth map attribute information. *Cartographica* 18(1):13–23.

———, and ———. 1983. Images of maps: A rotation experiment. *Professional Geographer* 35:455–61.

Thrower, N. J. W. 1978. *The compleat plattmaker: Essays on chart, map and globe making in England in the 17th and 18th centuries*. Los Angeles: University of California Press.

Tobler, W. R. 1966. Medieval distortions: The projections of ancient maps. *Annals of the Association of American Geographers* 56:351–60.

———. 1976. Analytical cartography. *American Cartographer* 3:21–31.

———. 1978. A proposal for an equal area map of the entire world on Mercator's projection. *American Cartographer* 5:149–54.

———. 1979a. Smooth pycnophylactic interpolation for geographic regions. *Journal of the American Statistical Association* 74(367):519–36.

———. 1979b. A transformational view of cartography. *American Cartographer* 6:101–6.

———. 1986a. Measuring the similarity of map projections. *American Cartographer* 13:135–39.

———. 1986b. Polycylindric map projections. *American Cartographer* 13:117–20.

———. 1986c. Pseudo-cartograms. *American Cartographer* 13:43–50.

———. 1987. Experiments in migration mapping by computer. *American Cartographer* 14:155–63.

Turner, E. 1983. New imaging materials assist in map drafting. *American Cartographer* 10:73–76.

———, and Sherman, J. C. 1986. The construction of tactual maps. *American Cartographer* 13:199–218.

White, D. 1985. Relief modulated thematic mapping by computer. *American Cartographer* 12:62–67.

White, E. R. 1985. Assessment of line-generalization algorithms using characteristic points. *American Cartographer* 12:17–28.

White, M. S. 1984a. Automated cartography and how mathematics helps. *Cartographica* 21(2&3):148–59.

———. 1984b. Technical requirements and standards for a multipurpose geographic data system. *American Cartographer* 11:15–26.

———, and Griffin, P. 1985. Piece wise linear rubber-sheet map transformation. *American Cartographer* 12:123–31.

Willmott, C. J.; Rowe, C. M.; and Philpot, W. D. 1985. Small-scale climate maps: A sensitivity analysis of some common assumptions associated with grid-point interpolation and contouring. *American Cartographer* 12:5–16.

Wood, D., and Fels, J. 1986. Designs on signs: Myth and meaning in maps. *Cartographica* 23(3):54–103.

Woodward, D. 1974. The study of the history of cartography: A suggested framework. *American Cartographer* 1:101–15.

———. 1977. *The all-American map*. Chicago: University of Chicago Press.

———. 1985. Reality, symbolism, time, and space in medieval world maps. *Annals of the Association of American Geographers* 75:510–21.

———, ed. 1987. *Art and cartography*. Chicago: University of Chicago Press.

Mathematical and Statistical Analysis in Human Geography

John Odland | Reginald G. Golledge | Peter A. Rogerson

Geographers who have been involved in the development of statistical and mathematical methods have been especially concerned with problems of developing methods for using geographic information and geographic concepts in formal ways. Many of these problems are shared with other disciplines, but the need for specialized approaches in geography arises from the nature of geographic information and geographic concepts.

Geographic concepts derive much of their meaning from reference to location and place, and geographic information is set apart from other forms of information because much of its value in understanding reality is derived from its relation to some system of locational coordinates. This fact also implies that geographic information contains substantial amounts of data about relations among objects, as well as information about the objects themselves. Similarly, geographic concepts gain much of their value from their capacity to support explanations of realities in explicitly locational contexts.

Statistical or mathematical models developed in other disciplines often do not accommodate analyses of geographic data or geographic concepts. The geographer's basic operation of arraying data on a map, or in some map-like form, often reveals locational relationships which enrich the information content of the data, and may also be the focus of geographical investigations. Basic statistical theory may indicate the criteria that must be fulfilled in order to measure these locational relationships numerically and test hypotheses about them in reliable ways, but conventional statistical methods are often insufficient to fulfill these criteria for geographic data.

For example, patterns on a map may suggest that the data for various locations are interdependent, because they are related to data values at other locations. This indicates that the data values may depend on geographic processes which extend over space, and also that statistical investigations that fail to account for these processes are likely to be incomplete and misspecified. One of the major accomplishments of statistical geography has been the development of methods for investigating locational relationships in ways that are consistent with statistical theory, and the extension of these methods to an array of different problems.

Similar situations occur in the application of mathematical (as opposed to statistical) formalisms in geographic research. Mathematical modeling requires the basic components of reasoning to be expressed in symbolic terms, but offers an array of theorems for manipulating those components and provides frameworks for examining the implications and limitations of a line of reasoning in ways that are thorough and unambiguous. Many of the fundamental ideas used in mathematical models require enrichment to function in analyses of geographical problems. For example, the idea of a general equilibrium is widely used to obtain solutions for systems that consist of interacting entities. The interacting entities may be abstract "decision-makers" in mathematical social science, and in the case of an economic equilibrium, a range of possible behaviors for each decision maker is defined a priori, sometimes in the form of a production or consumption function. An equilibrium solution identifies sets of values, such as prices, that emerge from interactions among the decision makers and mediate the behavior of each decision maker.

The idea of a general equilibrium can be applied in explicitly locational contexts in which the ranges of behavior for decision-makers are conditioned by their localization in particular places, and interactions among them are affected by spatial separation. The process of solving for an equilibrium is complicated by these locational effects, but the result reflects the particular effects of localization and spatial interaction.

Statistical and mathematical models in human geography have developed along lines dictated by the central role of location and spatial relations in geographic reasoning, but the need for methods which account for space and location is by no means limited to the research activities of geographers. Professional statisticians have shown a growing interest in the problems of analyzing spatial data (e.g., see Ripley 1981, 1984; Upton and Fingleton 1985). Location and space play a central role in the formal models developed in regional and urban economics (for reviews see Nijkamp 1986; Mills 1987). And spatial statistics have important applications in many fields, including epidemiology (Pocock, Cook, and Shaper 1982), biology (Sokal and Oden 1978), and archaeology (Hodder and Orton 1976). The development of formal methods for analyzing locational patterns and spatial processes is especially crucial to the development of research in geography, however, because the concepts of place and location are crucial to the distinctive ways in which geographers explain human and physical realities (Morrill 1985, 1987). Progress has occurred in a variety of areas that reflect the diversity of research in human geography, and has often resulted from the need to deal with particular research problems, rather than from efforts to develop general frameworks. Nonetheless, the totality of these contributions represents some important progress in an ongoing project of developing a set of theoretical and operational approaches that are suited to the analysis of geographic questions in terms of the formal languages of mathematics and statistics.

STATISTICAL MODELS FOR SPATIAL STRUCTURE AND SPATIAL VARIATION

Geographers who began to apply statistical methods in the 1950s and 1960s soon recognized that special problems were associated with the application of classical statistical methods to the analysis of geographic data (Cliff and Ord 1969; Gould 1970). Many of these problems occurred because the available statistical methods had been developed for investigations in which phenomena were isolated from particular contexts, including their spatial (and temporal) contexts. Geographers were usually concerned with investigating phenomena within their spatial contexts, where some of the assumptions of classical statistical methods, such as independence of events, did not hold.

For example, geographers were often concerned with identifying and explaining spatial patterns and the interdependencies associated with those patterns, but the very existence of a spatial pattern in a set of data indicated that inferences could not be based on methods that required independence among the observations in a sample. The interdependence of observations across locations, which was often the focus of geographic research, seemed to preclude the application of statistical approaches founded on assumptions about independence.

The restrictive assumptions of classical statistics serve to simplify calculations, but more generally, they reflect the usual scientific approach of focusing investigations by constructing models that isolate phenomena from particular contexts. The theoretical emphases of other disciplines may make it reasonable to isolate phenomena from their locational or spatial contexts, but these contexts are usually central in geographic research. Consequently, geographers found it necessary to develop methods for "spatial statistics" by enlarging or modifying statistical methods for investigating phenomena that are explicitly situated in locational contexts (Gaile and Willmott 1984). Most of these developments have occurred within the frameworks of conventional statistical theory, although quantitative geographers have also developed approaches based on other less-restrictive frameworks, and have begun to extend the development of spatial statistics to include methods for analyzing categorical data.

Spatial Autocorrelation and Spatial Autoregression

The problem of spatial interdependence or spatial autocorrelation is central in most applications of statistical methods to geographic data, either as a focus of investigation in itself, or as a factor which complicates investigations of other hypotheses. General discussions of the problem are available at both advanced levels (Anselin 1988b; Bennett and Haining 1985; Cliff and Ord 1981; Griffith 1980, 1987; Haining 1980; Upton and Fingleton 1985), and in elementary presentations (Griffith 1988b; Odland 1988b). Much of the statistical work on the problem can be summarized by referring to the spatial regressive-autoregressive model, written (as is usual in geography) in terms of simultaneous expectations (Anselin 1988b):

$$y = \rho W_1 y + X\beta + \epsilon$$
$$\epsilon = \lambda W_2 \epsilon + \nu.$$

An alternative formulation in terms of conditional expectations has received relatively less development (see Besag 1974; Cliff and Ord 1981). The model consists of a function that describes the variation in y, a vector of values for a variable distributed over a set of locations; and a distribution for the associated errors, ϵ, in the second equation. The matrix X contains values for a set of exogenous variables at the same locations, while β is a set of parameters and ν is a vector of independently and identically distributed random variables. The remaining terms are critical in spatial analysis: W_1

and W_2 are matrices whose entries are measures of proximity for each pair of locations, while ρ and λ are numerical measurements of spatial autocorrelation.

The conventional regression model of classical statistics amounts to the special case where the constraints $\rho = 0$, $\lambda = 0$ hold, so that:

$$y = X\beta + \nu$$

Inferences about spatial data made on the basis of this simplified but conventional version of the model are likely to be mistaken, because of bias in the estimated standard errors for the elements of β (Anselin and Griffith 1988) when values for ρ and λ are incorrectly taken as zero. Application of the model in a spatial framework requires specifications for W_1 and W_2, as well as estimates of ρ and λ, along with estimates of β. Likelihood estimation is necessary, and this complication has probably discouraged widespread application of the model. However, procedures which make it possible to estimate these parameters with commonly available statistical packages have recently been made available (Anselin 1986; Griffith 1988a).

Parameter estimates for this model are specific to a particular division of space into a finite set of locations or subregions, and to the assignments of the values in W_1 and W_2, which measure proximities among the subregions. The sensitivity of parameter estimates to particular aggregations of a space into subregions has been demonstrated by Openshaw (1984), and complete resolution of this problem may require the transformation of data values into the domain of a continuous space (Tobler and Kennedy 1985). The values in W_1, W_2 have generally been assigned on the basis of simple measures of proximity, although it is widely recognized that rules for assigning these values should be based on considerations of underlying spatial processes (Gatrell 1983). Anselin (1984) has devised a method for comparing alternative weighting matrices, by treating them as nonnested hypotheses.

The measurement of autocorrelation in a single variable (as opposed to autoregression) can be carried out with a simpler form of the model, in which exogenous variables are omitted:

$$y = \rho W y + \nu$$

so that the value at any location is decomposed into a component which is attributed to spatial pattern or spatial autocorrelation, $\rho W y$, and a strictly local component, ν. The parameter ρ is then a numerical measurement of spatial autocorrelation in a single variable. Tests for the relatively simple hypothesis $\rho = 0$ under general specifications of W have now been available for more than 15 years (Cliff and Ord 1973, 1981) for both continuous and categorical data, and for regression residuals.

Current investigations center on the more general—and more difficult—problems of obtaining reliable numerical estimates of ρ, making reliable inferences on the basis of those estimates, and diagnosing particular misspecifications in the model. Griffith has summarized these problems (1980, 1987) under a set of requirements for a "theory of spatial statistics" which would enlarge the range of statistical theory to address the problems of estimation and hypothesis testing within spatial or locational contexts. Many of the complications in estimation are related to the limited sample sizes typically available for spatial analysis, and the effects of boundaries (see Griffith and Amrhein 1983; Griffith 1983). Further problems arise from the possibility that data are generated by processes that are heterogeneous over space, a situation that may sometimes be remedied by specifying more complex models in systematic ways, as suggested by the expansion method of Casetti (1972; Casetti and Jones 1987). There are many possible sources of misspecification in models for spatial data, even when the most current methods are applied, so the development of an array of diagnostic tests for both autocorrelation and heterogeneity constitutes an important long-term project in this area (Anselin 1987, 1988a).

Analysis of Point Patterns

Models of spatial autocorrelation or spatial autoregression apply to spatial patterns which are formed by the variation of data over a set of locations. But statistical investigations of the patterns formed by basic geometric entities—points, lines, and areas—also have a long tradition in geography (Getis and Boots 1978; Boots and Getis 1988). The usual approach to analyzing these patterns has been to postulate particular processes that might account for the locations of pattern elements, such as points; derive some implications of the process for features of the pattern, such as the distances between points, in the form of a probability distribution; and, finally, compare the implications of this model with the observed pattern. Relatively complex models may be necessary where patterns have multiple determinants and develop in heterogeneous environments (Odland and Barff 1982). Possibilities for thorough examination of the spacing in point patterns have been enhanced by the recent development of second-order methods, which accommodate tests based on the variation (or second moment) of the distances between pattern elements, rather than mean distances (Ripley 1981; Getis 1984).

Methods for the analysis of patterns formed by lines and areas have not developed as rapidly (see Boots 1984, 1985). But, since points, lines, and areas form a basic framework for organizing data in geographic-information systems, the need for basic approaches to measuring and analyzing the information conveyed by patterns of these geometric elements may become urgent in the near future. Statistics for other elements of spatial pattern, especially directional statistics (Mardia 1975), have also had relatively little development within geography. Most directional statistics have been developed for data that are treated as unit vectors (Gaile and Burt 1980), so that statistical analyses deal only with direction, and not with magnitude.

Inference in Less-Restrictive Frameworks

Many of the restrictive assumptions of classical statistics which become problematic in applications to spatial analysis (such as normality and independence in sampling distributions) make it possible to economize on computation, by allowing the properties of the associated statistics to be derived through general mathematical analyses. However, the computing equipment available in the 1980s makes calculations relatively inexpensive, and a number of geographers have turned to methods that require extensive calculations but do not require restrictive assumptions about sampling distributions (Costanzo 1983).

Knudsen (1987) distinguishes three categories of computation-extensive approaches: distributional methods, randomization methods, and bootstrap methods. Randomization methods have been extensively developed to address hypotheses about spatial pattern (Hubert and Golledge 1981; Hubert, Golledge, and Costanzo

1981; Hubert et al. 1985a, 1985b), and much of this work can be summarized in terms of a generalized cross-product statistic:

$$\Gamma = \Sigma\Sigma \, W_{ij}Y_{ij}$$

where elements of the matrices W_{ij} and Y_{ij} are measures of similarity between i and j. For example, the elements of W_{ij} may be measures of the relative locations of objects or regions indexed by i and j, while Y_{ij} may be some measure of the similarity of data values for i and j. The statistic Γ is then a general measure of the similarity between the two matrices, and where W_{ij} contains measures of relative location, Γ is a measure of covariation between the relative locations or proximities and data values at the locations.

Expectations about the distribution for Γ under particular definitions of W_{ij} and Y_{ij} are needed to perform significance tests, and the analytic derivation of such sampling distributions generally requires restrictive assumptions about the sampling distributions underlying the data values. These assumptions typically restrict the possibilities to a relatively limited set of cases. However, alternative distributions can be obtained through randomization procedures, in which values of the observations are permuted over the locations to yield a distribution for the values of Γ under the assumption of a sampling framework in which the set of data values is fixed but their distribution over locations is treated as a random variable. The approach accommodates virtually unlimited definitions of W_{ij} and Y_{ij}, and is not limited to applications in spatial analysis. But it is particularly useful in that context, where hypothesis testing is often focused upon a pattern of data values over a set of fixed locations.

A further set of computation-intensive methods is based on assumptions about distributions for the data, as opposed to the classical assumption of a sampling distribution for the process that generates the data. Extensive computation can then be used to analyze the behavior of various statistics that might be calculated for the data, under alternative assumptions about the error characteristics of the data. For example, Knudsen and Fotheringham (1986) use this approach to identify the behavior of various statistics for matrix comparison under alternative levels of error.

Methods for Analyzing Categorical Data and Latent Variables

The initial focus of quantitative geography on classical statistical methods influenced not only the kinds of methods that geographers selected to analyze data, but also the ways in which they thought about data and measurement. The statistical methods initially adopted in geography were adapted to analyses of data on ratio or interval scales, and "measurement" usually referred to the collection of such data. Geographers have, over the last 15 or more years, realized that meaningful measurements are often nonmetric or categorical, and methods for analyzing such data have come to be widely applied.

The interest in categorical data may be traced to the research focus on behavioral geography and spatial choice within the discipline (Golledge and Rushton 1984); the development by statisticians of a systematic array of methods for analyzing categorical data (Nelder and Wedderburn 1972; Bishop, Fienberg, and Holland 1975); and the availability of the necessary computational algorithms (O'Brien 1983; O'Brien and Wrigley 1980). The development of these methods in geography has now reached the point where there is a standard textbook on geographic applications (Wrigley 1985).

Logit models and log-linear models have been more widely used in geography than other approaches to the analysis of categorical data. Logit models are related to theories of choice-making at extensive margins, and that makes them especially suited to the analysis of spatial choice—a topic that is discussed in a subsequent section of this chapter. Log-linear models provide a very general framework for analyzing multidimensional contingency tables which has been applied to such obviously geographic problems as housing choice (Deurloo, Dieleman, and Clark 1988) and the temporal stability of migration flows (Aufhauser and Fischer 1985). The array of possible models for noncontinuous data also includes models for latent variables (Folmer and Nijkamp 1984), and an emerging set of models in econometrics for the analysis of models with truncated or censored variables (Maddala 1983).

Categorical data are no less likely than continuous data to exhibit spatial dependence, and the effects of spatial dependence are likely to lead to mistaken inference in these models as well. Little work has been done on the effects of spatial dependence in categorical models, other than Fingleton's modifications of inferential tests (1983a, 1983b, 1986) for the log-linear model in the presence of spatial dependence.

ANALYSIS OF SPATIAL INTERACTION

The relations between spatial structure or spatial patterns and various kinds of spatial interaction are central issues in geographic research (Bennett and Haining 1985; Bennett, Haining, and Wilson 1985), and recent developments in research on flows and interactions over space have taken place in two fairly distinct areas. The first is the modeling of interregional flows, which have often been analyzed in the form of matrices. The second area is defined by a series of models of spatial interaction; these have direct implications for the behavior of spatial structures.

Models of Interregional Flows

The logical bases of spatial-interaction models have been substantially enlarged through the derivation of the gravity model formulation on the basis of stochastic utility maximization (Anas 1983; Fotheringham 1986), as well as information-theoretic formulations (Wilson 1970) and the development of other very general formulations for spatial interaction (Tobler 1983; Dorigo and Tobler 1983). Much of the recent research on these models has centered on statistical problems, especially the effects of spatial structure on the parameter estimates for interaction models, and on problems associated with the dynamics of flow patterns.

Statistical Problems in Spatial Interaction

Specification problems associated with the spatial pattern of origins and destinations have been a major area of development for statistical models of spatial interactions. The evolution of our understanding of this source of misspecification is recorded in the interchange among Cliff, Martin, and Ord (1974, 1975), and Curry (1972), Curry, Griffith, and Sheppard (1975), and Sheppard, Griffith, and Curry (1976). This history is summarized by Sheppard (1984), and some recent developments in specification tests are presented by Baxter (1985, 1987).

Fotheringham (1983) deals with the problem by adding an accessibility term to the interaction model. The parameter associated with this term provides for an estimate of the effects of proximity among alternative destinations, as either a competition effect (which reduces interactions with destinations that are near to alternative destinations), or an agglomeration effect (which increases interactions with destinations that are close to alternatives). The origin-specific production-constrained version of the model, in which all parameters are specific to an origin, is:

$$I_{ij} = [\Sigma m_j(\Sigma m_k d_{jk}^\sigma)^\delta d_{ij}^\beta] O_i m_j (\Sigma m_k d_{jk}^\sigma)^\delta d_{ij}^\beta$$

where I_{ij} is interaction, m_j is the attractiveness of a destination, d_{ij} is distance, and O_i is the number of interactions associated with the origin. The term in parentheses $(\Sigma m_k d_{jk}^\sigma)$, where the summation is over destinations other than j, is a measure of the accessibility of j to alternative destinations. The improved specification in this "competing destinations model" makes it possible to obtain estimates of the distance parameters β, which are independent of particular configurations of origins and destinations.

Dynamics of Flow Patterns

There has also been considerable research in recent years on the dynamics of flow patterns. Much of this work has been motivated by an interest in understanding changes in aggregate flow patterns, but some of the modeling efforts also have clear links with the dynamics of choice processes. Directions for the development of dynamic models are suggested by the structural resemblance between matrices of place-to-place interactions and matrices of transition probabilities, but the current generation of models transcends some of the untenable assumptions of simple Markov processes, especially the assumption of stationarity.

The Markov model may be generalized into a more realistic and flexible form by assuming that a constant causative matrix maps a flow pattern for one time period, expressed as transition probabilities, into a flow pattern for the next time period (Rogerson and Plane 1984; Plane and Rogerson 1986). This model may be expressed as:

$$P_t = P_{t-1} C$$

where P_t is the transition matrix at time t, P_{t-1} is the transition matrix at time $t-1$, and C is the causative matrix, which is assumed constant over time. The elements of C may be interpreted as dynamic measures of competition. In the case of migration, they reflect region j's changing ability to compete with region i for migrants. Plane and Rogerson (1985) also demonstrate how migration probabilities may be updated to reflect changes in economic conditions that may have occurred at destinations. They employ an origin-constrained model of the form:

$$p_{ij}^t = p_{ij}^{t-1} m_i (x_j^t / x_j^{t-1})^\gamma$$

where

$$m_i = \Sigma (x_k^t / x_k^{t-1})^\gamma$$

and where x_j is a measure of the attractiveness of destination j, p_{ij} is the probability of making a transition between i and j, and γ is a parameter that measures the elasticity of the transition probability with respect to changes in the measure of attractiveness.

Both of these methods for modeling change in flow patterns are characterized by parameters that summarize the nature of change. The latter method is also related to individual choice behavior. Rogerson (1984) demonstrates that this formulation is equivalent to an incremental logit model of choices (Koppelman 1983), which captures the effects of changes in destination characteristics on individual choice probabilities. These probabilities take the form:

$$p_j^{t+1} = \frac{p_j^t \exp(\beta\{x_j^t - x_j^{t-1}\})}{\Sigma p_k^t \exp(\beta\{x_k^t - x_k^{t-1}\})}$$

where x_j is an attribute of destination j, and the parameter β measures the sensitivity of choice probabilities to attribute changes. Odland (1988a) has also used incremental logit models to examine the effects of regional demographic and economic change on relocation probabilities.

Spatial Interaction and Spatial Structure

There has also been a substantial amount of research carried out during the last few years on the relationships between spatial processes, manifested as various kinds of spatial interactions, and the consequent effects on spatial structure. Much of this work indicates that processes which seem to operate in relatively straightforward ways when abstracted from space (such as price determination in markets) may have much more complex implications when they are analyzed in explicitly spatial contexts.

Conditions for the uniqueness and stability of equilibria may be especially problematic in the case of price determination for a set of localized but interacting markets. Prices for labor and for many consumer goods are determined in localized markets, which interact with other localized markets for the same good. The behavior of prices in this situation may be summarized, in a simple way, in terms of probabilities for price changes (see Bennett and Haining 1985):

$$\pi(x_{i,t+1} = p_b | x_{i,t} = p_a) = \exp[\alpha + \Sigma \beta_{ij} x_j].$$

The left-hand side is the probability that a price at location i changes from p_a to p_b in some interval of time, and this probability depends on a parameter α, which is independent of location, as well as interactions with the prices at a set of neighboring locations. This type of model corresponds to a Markov random field or an Ising model (Isham 1981; Kinderman and Snell 1980), and it is also a model of the changes that occur in a map of localized prices, as the price at each locality adjusts to prices at other localities.

Simple solutions correspond to situations where this adjustment process converges to a stable and unique map of prices for any set of parameters. However, Haining has shown (1983, 1984, 1985) that processes of this kind may be characterized by nonlinear dynamics in which transitions between alternative equilibria are possible, and that these may be associated with perturbations that originate in limited subregions of the space, but eventually affect conditions over very extensive regions. Haining (1984) also notes that these transitions become more frequent as intersite interactions become relatively more important in determining price changes.

Curry has addressed similar problems in his research on labor markets, and has demonstrated how spatial structure in the form of geographic differentiation in occupations may result from the diffusion of recruitment information (Curry 1982). He has also examined how inefficiencies in spatial pricing may result from the differential availability of information and spatially differentiated interactions among decision-makers in spatial contexts (Curry 1984, 1985).

The consequences of these inefficiencies are especially important in the case of labor markets, and Curry has employed Ising models to examine the results of interactions between workers on a uniform lattice. By making the employment status of an individual a function of labor demand and the employment status of neighbors, he demonstrates how geographic differentiation may result from the interactions among neighbors, with isolated patches of similar economic well-being emerging in ways that depend on the strength of the interactions. In this case, the process of wage determination in spatial frameworks is affected by inefficiencies that are not accounted for in nonspatial models, but which may lead to persistent interregional differences in incomes.

These analyses indicate that the processes that mediate human affairs, such as price-setting in markets, may have substantially different implications when they are situated in explicitly spatial contexts. Many of the results, such as shifts toward alternative equilibria under localized perturbations, are related to general structures in nonlinear dynamics. These are a class of mathematical models that have recently been applied to explain a wide range of physical phenomena, such as the transition from laminar to turbulent flow and the spatial structure of chemical concentrations in reaction-diffusion systems, as well as processes in social systems (dePalma and Lefevere 1983; Weidlich and Haag 1983). Many of these ideas are related to the pioneering research of Priogine on nonlinear dynamics (see Nicolis and Priogine 1977).

The realization that complex dynamics arise directly from the nonlinearities and interdependencies that are usually inherent in spatial frameworks (Wilson 1981) has given rise to an array of investigations on the dynamics of urban systems (Clarke and Wilson 1983), migration patterns (Kanaroglou, Liaw, and Papageorgiou 1986a, 1986b), interregional trade (Curry 1986), and retailing systems (Fotheringham and Knudsen 1986). These developments signal a growing shift toward explanations in geography that are based on models which are not only dynamic, but also evolutionary (see Griffith and Lea 1983, 1–11). That is, these models not only accommodate changes over time, but involve trajectories which lead to irreversible transformations of the system. The emphasis in these approaches is not on the traditional concerns with the dynamics of response to exogenous impacts, but on the internal logic of change within a system, including the possibility of multiple trajectories of system development and critical transitions where the system adopts new trajectories.

Abstract models of dynamics in spatial systems using approaches as dissimilar as difference equations (Rogerson 1985) and cellular automata (Couclelis 1988) indicate that relatively simple underlying rules for system behavior may manifest themselves as patterns of observable behavior which are extremely complicated. Mathematical investigations of these types of systems may portend a need for some new perspectives on empirical and statistical analyses, because observations of such systems, over the limited time periods and restricted arrays of exogenous circumstances that are typically available, are unlikely to be sufficient for thorough empirical analyses using traditional approaches.

MODELS IN ECONOMIC LOCATION THEORY

Mathematical modeling has been the major approach to analyzing the traditional problems of location theory in the last two decades, including problems in the location of production (Beckman and Thisse 1986), investigations of land-use patterns (Jones 1984a, 1984b; Papageorgiou 1976; Thrall 1987), and analyses of central-place systems (Mulligan 1983, 1984). Models have generally been formalized in terms of the microeconomic theory of the household or the firm, and analyzed as mathematical optimization problems whose solutions correspond to partial or general market equilibria.

Research on these problems by geographers is not strongly distinguished from work by regional scientists and urban and regional economists, but the distinctive problems of formalizing locational problems in economic terms stem from the explicit recognition of the role of space and location in economic behavior. When conditions for production and consumption are formalized in spatial frameworks, they are characterized by nonconvexities, and by spatial variations in conditions for production or consumption. Nonconvexities are an essential feature of economic-location theory, because strict convexity would imply that the levels of production and consumption activities for all decision-makers would be undifferentiated in any space where exogenous endowments of natural resources were homogeneous. Although some form of nonconvexity is logically necessary for the localization of activities in an abstract space, this feature makes some aspects of traditional approaches to economic analysis problematic in spatial contexts, including the conditions for equivalence between competitive equilibria and social optima.

Geographers involved in economic-location theory have been especially concerned with problems associated with agglomeration economies (Mullally and Papageorgiou 1978; Papageorgiou 1983), and with the effects of externalities which assume spatial distributions. The analysis of spatial externalities has been especially challenging, because of their acknowledged importance in urban land use and because they seem to require analysis within the framework of general rather than partial equilibrium. Papageorgiou (1978a, 1978b) has applied a general class of mathematical optimization models to examine the relations between competitive spatial equilibria and patterns of land use for systems that are characterized by externalities that diffuse over space.

Economic-location theory within geography has been concerned mainly with private-sector decisions, but some recent developments may portend the development of a location theory for the public space economy, as outlined by Lea (1981). Papageorgiou (1987a, 1987b) has recently enlarged the analysis of spatial externalities to examine spatial public goods and to analyze the types of public policies that are necessary to maintain patterns of land use which are optimal under different theories of distributive justice. However, most modeling of public-sector operations to date has centered on practical models for the location of systems of centralized public facilities. These models are generally formulated as mathematical programming models to determine the locations of facilities such that service levels are maximized under constraints on the costs of facilities; or facility costs are minimized under constraints on minimum service levels (Leonardi 1981; Church and Roberts 1983; Revelle 1987).

Variants of these models have been produced for a wide variety of circumstances, including situations where demands are sensitive to the locations of facilities, and where a series of facility locations must be scheduled over time (Leonardi 1981). The problems of equity that are inherent in the location of centralized facilities have been examined (Bigman and ReVelle 1978; Lea 1979), but a body of general location theory sufficient to analyze the welfare economics of a broad range of public-sector activities remains to be developed.

QUANTITATIVE ANALYSIS OF SPATIAL CHOICES AND SPATIAL BEHAVIOR

Behavioral geography emerged as an important component of human geography in the last quarter century, and the development of research on spatial behavior enlarged the range of quantitative methods applied in geography. The explanation of spatial behavior on the basis of its antecedents, including the cognition of spatial environments and the formation of preferences, became an important focus of research, and one which led to new methods of measurement and analysis, such as multidimensional scaling. Developments related to behavioral geography also included efforts to identify appropriate frameworks for integrating models of decision-making into explanations of spatial processes, and this has led to an array of research efforts centered on models of spatial choice.

Models of Spatial Choice

A decision process can be said to result in "spatial choices" in any situation where the decision-makers and the alternatives are distributed in space. But geographers have largely been concerned with decision-making in situations where the choices are also mutually exclusive or discrete. That is, the choice of a residential location, travel route, or migration destination implies that the decision-maker occupies that alternative at some specific time, to the exclusion of other alternatives in the choice set. The mutually exclusive nature of most spatial choices, which is one case of the nonconvexities mentioned above, means that the conventional marginal analysis of microeconomics is not applicable, and the development of general frameworks for analyzing discrete choice has been a necessary prerequisite for systematic analyses of spatial choice.

Random-Utility Models. A logically coherent set of operational models for discrete choices emerged in the 1970s, based on a logical foundation in random-utility theory (McFadden 1981; Fischer and Nijkamp 1985; Wrigley 1982). Random-utility models have close relations with general statistical models for discrete data, and their development was also stimulated, at least initially, by applications to spatial choice, particularly choices among transportation alternatives (Domencich and McFadden 1975; Hensher and Johnson 1981). Random-utility models have since been widely applied to other types of spatial choice, such as migration and residential location.

Random-utility theory is based on the same fundamental assumption as marginalist-utility theory: that the information processing of decision-makers can be summarized as an algebraic function, which represents an assignment of utilities to the alternatives in a choice set. It departs from marginalist theory because of the further assumption that the utilities assigned to discrete alternatives can be partitioned into a deterministic or "systematic" component and a random variable. The systematic component can consist of a function $v(z_{ij})$, which relates the utility of each alternative to a set of abstract characteristics of the alternatives and the decision-maker, provided that the decision-maker's information processing is not contingent on the particular set of alternatives being considered. The second component is a random variable, e_{ij}. The utility of an alternative j, for decision-maker i, is then the random variable:

$$u_{ij} = v(z_{ij}) + e_{ij}.$$

The random component of utility has usually been rationalized in the discrete-choice literature as representing unavailable information about the determination of utilities but it may also be interpreted as a representation of stochastic instability in the determination of utilities (Wrigley 1982).

The formulation of utilities as random variables makes it possible to analyze the choice of alternative j in terms of the probability of choosing j over other members of a choice set called A. If $v(z_{ij})$ is a linear function with parameters β, and if θ is a set of parameters that describes the distribution of the error term, then the probability that individual i will select alternative j is:

$$P(j|z_{ij},\beta,\theta) = P(u_{ij} > u_{ik})$$

for all k included in A, $j \neq k$.

The key to specifying a statistical model for these choices is the selection of a distribution for the e_{ij}. The multivariate normal distribution is an obvious candidate, because that form allows the variances and covariances of the errors to be unrestricted. A multivariate normal distribution for the e_{ij} leads to a multivariate probit model for the choice probabilities (Daganzo 1979). The probit model provides a very general form for discrete choice models, but computational problems have limited the use of the probit model in empirical research, although there have been a few applications to geographic problems (Van Lierop, 1986).

Models that are more tractable in terms of computations can be derived on the basis of more restrictive specifications of the error distribution. McFadden (1978, 1981) has shown how a series of discrete-choice models can be derived from members of a family of generalized extreme-value distributions. The simplest case occurs when the errors are assumed to be independently and identically distributed with the double-exponential distribution. The choice probabilities then assume the form of the multinomial logit distribution:

$$P(j|z_{ij}) = \frac{\exp(\beta z_{ij})}{\Sigma \exp(\beta z_{ij})}.$$

This model has been widely used in categorical data analysis. It can be estimated using conventional likelihood methods (Wrigley 1985). The model is readily interpreted, because it makes the probability of choosing alternative j a ratio of the observable utility of j to a sum of the utilities of all of the alternatives in the choice set, and estimates of β can be used to test hypotheses about the effect of various characteristics of the alternatives and the decision-maker on the choice probabilities.

Luce (1959) derived a choice model with the same functional form, without reference to utility maximizing. That model depends on Luce's axiom of independence from irrelevant alternatives, which is equivalent to the assumption of independent and identically distributed errors. This assumption is essential for the relatively simple, and computationally tractable, form of the multinomial logit model. The assumption is also extremely restrictive, however, because it implies equal rates of substitution between all pairs of alternatives (Fischer and Nijkamp 1985; Wrigley 1982; Hensher and Johnson 1981). This condition is unlikely to hold for the complex decisions, or series of decisions, that are characteristic of spatial-choice problems—such as the choice of a residential location (see Clark and Van Lierop 1986) or the decision to leave an origin and migrate to one of several possible destinations (Kanaroglou, Liaw, and Papageorgiou 1986a). Consequently, geographers investigating these kinds of choice problems have sought alternative models which retain a logical basis in utility theory, but accommodate complex decision processes.

Models for Choice in Complex Spatial Frameworks. Alternative functional forms that are computationally tractable can be obtained on the basis of less-restrictive assumptions about the distributions of the unobserved error in utilities. The form which has been most widely applied in analyses of spatial choice is the nested multinomial logit model. This model allows a more complex error structure to be introduced, through a partitioning of the choice set into a nested hierarchy of alternatives.

Consider the analysis of residential choice as an example (Onaka and Clark 1983). The discrete alternatives are dwellings which are located in neighborhoods, and their utilities may depend on characteristics of the neighborhoods, z_k, as well as characteristics of the individual dwellings, z_{jk}, so that the utility of dwelling j in neighborhood k can be written as

$$u(j,k) = v(z_{jk}) + v(z_k) + e_{jk}$$

where the index for the individual decision-maker is omitted for simplicity.

The multinomial logit model could be applied if the e_{jk} was independently distributed, but that assumption will not hold if dwellings within a neighborhood are closer substitutes than dwellings in different neighborhoods, even when characteristics of neighborhoods are included in the model. The hypothesis that some sets of alternatives are closer substitutes than others amounts to a hypothesis about the structure of the covariances of the e_{jk}, and in this case, the hypothesized structure of the covariances corresponds to a regionalization of the alternatives. (That is, the covariances may be larger for pairs of alternatives in the same neighborhood than for pairs of alternatives in different neighborhoods.) This hypothesis about the covariance structure can be incorporated into a nested multinomial logit model, and tested along with hypotheses about the influence of dwelling characteristics and neighborhood characteristics on choice probabilities.

McFadden (1981) has shown how the nested model can be derived as one of a family of models based on a generalized extreme-value distribution for the unobserved components of utility. Extensions to hierarchies with multiple levels are straightforward. The approach provides a compromise between the generally intractable problem of estimating the matrix of error covariances for all the alternatives, and the overly simple assumption of independence in the errors. The general approach of partitioning a choice set into a hierarchy corresponds to a regionalization of alternatives in many geographical applications. It also suggests hypotheses about the structure of decision-making in complex spatial environments.

The nested logit model has been applied to situations that involve decisions to leave an established location and select a destination from an array of alternatives, including interregional migration (Liaw and Ledent 1987; Odland and Ellis 1987) and residential mobility (Clark and Onaka 1985). Problems of misspecification in the implied structure of error covariances in these models are a serious concern, but a series of specification tests for the nesting structure has recently been developed by Horowitz (1983, 1987).

Multiattribute Preference Models

Research on multiattribute preference models addresses the preference or utility orderings which are antecedents to the overt spatial choices. These preference orderings are usually investigated on the basis of laboratory experiments, in contrast to the analyses of revealed behavior that are usual in investigations of spatial choice. An individual's preferences over the members of some choice set are usually represented as a utility function, which is defined over a set of attributes associated with each alternative:

$$U_i = U(a_{1i}, a_{2i}, \cdots a_{ni})$$

where the a's represent quantities of each attribute associated with alternative i, and the functional form may be either additive or multiplicative. Total utility is interpreted as a sum or product of independent "part-worths" (or "part-utilities") associated with each attribute and the utility function is treated as ordinal, so that it need only specify the orderings of the alternatives.

Multiattribute preference models have their roots in several different theoretical frameworks. Initial work by Menchick (1972), Louviere (1974), and Knight and Menchick (1976) drew heavily on Anderson's information-integration theory (1974). Anderson argued that preferences were established during the process of integrating information, and this integration was undertaken via simple algebraic rules such as addition or multiplication, and considerable attention has been paid to the problems of establishing appropriate combination rules (Louviere 1976). Other theoretical influences have derived from random-utility theory, and both the theory and techniques of conjoint measurement and functional measurement have been influential (Louviere 1982).

The distinction between preference and choice remains a difficult problem. Louviere and Woodworth (1983) suggested a way to combine preference and choice into a single model, by using a fractional factorial experimental design to combine multiattribute choice with multiattribute preference. Such a combination can occur if individuals choose among a constant number of alternatives on the basis of a uniformly perceived quantity of attributes, decomposed into the same number of levels. Given these assumptions, a multinomial logit model can be used to predict the composition of choice sets. Thus, choice sets containing different attributes and levels can be constructed from complex factorial designs, with individuals being asked to select a single alternative from each choice set. Utilities can then be estimated on the basis of the selection process. Alternative procedures may require subjects to estimate how they would share their patronage among a set of different alternatives, or to allocate a quantity of fixed resources among experimentally defined choice sets.

Longitudinal Analysis of Spatial Choice and Spatial Behavior

The analysis of spatial choice and spatial behavior is likely to include a growing component of longitudinal analysis in the future. Research on the longitudinal analysis of individual behavior has grown rapidly in social science since 1980, with the development of methods (Heckman and Singer 1986; Tuma and Hannen 1984), and the increasing availability of longitudinal data. In the geographical literature, Pickles and Davies (1984) argue that the analysis of individual behavior over time is an essential prerequisite for adequate understanding of aggregate geographic phenomena, because cross-sectional analyses cannot capture the critical effects of heterogeneity and individual histories on decision-making.

The research team of Crouchley, Davies, and Pickles has investigated a series of basic problems in applying longitudinal analysis in geographic contexts (Crouchley, Davies, and Pickles 1982; Davies, Crouchley, and Pickles 1982, 1983; Pickles 1983; Pickles, Davies, and Crouchley 1982). The core of these problems is the interaction between heterogeneity among individuals and event-history effects. Heterogeneity implies that, although a process may be well specified, the parameters may vary across individuals. Event histories often imply violations of the Markov property, so that transition probabilities depend on a series of previous states, because of cumulative inertia where the duration of time spent in a state (or location) affects the probability of moving to other states (Pickles 1983). Because heterogeneity and event-history effects interact, simultaneous estimation of the effects is necessary.

The influences of nonstationarity, heterogeneity, and event histories probably cannot be untangled without appropriate longitudinal models and longitudinal data. Further, the development of longitudinal models may make it possible to enlarge basic research questions in useful ways. For example, cross-sectional analyses of spatial mobility have been organized around questions about why people move (at a particular time), and why they choose certain destinations. Longitudinal frameworks suggest more general and more trenchant versions of these questions, extending to the reasons why people occupy particular locations for various lengths of time, and why they select particular sequences of locations over periods of their personal histories.

The limited availability of longitudinal data is the greatest obstacle to widespread application of longitudinal analyses. The methods require not only data on individuals, but data on the same individuals over extended periods of their personal histories. Consequently, most of the available data has been gathered through long-term institutional efforts, such as the National Longitudinal Surveys and the Panel Study of Income Dynamics. Geographers have rarely organized or participated in such institutionalized programs.

Geographers also face especially demanding sampling problems, because sample sizes must often be very large to represent spatially heterogeneous populations, and sampling must extend for prolonged periods to support analyses of relatively infrequent events, such as migration, that are of particular interest in geography. Even so, the existing record of applications to problems—such as shopping behavior (Halperin 1988; Wrigley and Dunn 1984), industrial location (O'Farrell and Crouchley 1987), and residential mobility (Pickles 1983; Pickles and Davies 1985)—serves to demonstrate the utility of longitudinal approaches.

COMPUTATIONAL-PROCESS MODELS

The structure of computational-process models contrasts with that of the models discussed above, because their logic is expressed in terms of computer programs rather than sets of algebraic equations. A computational-process model may, for example, address the way that an individual interacts with an environment during a complex task such as wayfinding (Couclelis 1986b). Computational-process models give this task an explicit step-by-step description in the form of a computer program. The program may include information about the immediate spatial environment where the task occurs; information in the individual's permanent knowledge structure, which provides a general basis for interpreting objects, actions, and events in the immediate environment; and a set of decision rules that, when implemented, allow wayfinding to occur. The model simulates the information-processing activities of an individual (or a group of interacting individuals), not merely to predict behavior, but to investigate relationships between behavior and a cognitive representation of the world (Clark and Smith 1985; Smith, Clark, and Cotton 1984).

Computational-process models provide concrete representations of knowledge structures, knowledge-accessing processes, and cognitive architectures (Smith, Pelligrino, and Golledge 1982). The key objective is to develop general representations of how permanent knowledge structures guide the decisions and actions of individuals who must interpret objects, actions, and events in complex environments, and respond by making decisions. Although the models may have application to a very wide range of research problems (Smith 1984), much of the current work focuses on issues of determining the type and nature of stored knowledge, the nature of the mechanisms that stimulate the retrieval of stored information, and the processes that operate on a knowledge base to produce new knowledge (Golledge, Hendriks, and Lensink 1984).

Computational-process models are a departure from more-familiar modeling frameworks, and their formulation as computer programs may seem to have no wider theoretical basis than an analogy between human information processing and automated computation. Couclelis (1985, 1986a) has recently clarified the theoretical status of computational-process models in terms of a "discrete-structure hierarchy," which formalizes the relations between set theory and logic (which occupy the top of the hierarchy), through modern algebra and the theories of modeling and computation, to computer algorithms. The connections between the levels of this hierarchy indicate that any computational-process model can, at least in principle, be transformed into a representation in the language of a higher level.

Situating computational-process models within this hierarchy not only establishes their general theoretical status with respect to representations in other languages but also places them in a general framework that establishes relations among mathematical models and axiomatic theories, the theory of computation, and operational computer programs. This introduction of this very general computational perspective on the manipulation of symbols may enlarge the range of concepts that geographers are able to express in terms of formal languages, and clarify the limits of various kinds of modeling approaches.

For example, the idea of "becoming" is beyond the range of traditional mathematical formalisms, because it implies the emergence of something that was not implicit in the premises of a logical calculus. But Couclelis (1986a) shows how it can be expressed in terms of symbolic manipulations, which are equivalent to mappings between integer spaces. These kinds of insights may make it possible to clarify the kinds of concepts that can be expressed and examined in terms of formal models, and to identify the limitations of various modeling approaches. At the least, widespread attention to this kind of general framework will assure the validity of Couclelis's statement that "From now on a much subtler level of argument . . . will be required to help draw the line between what in the human world can and cannot be expressed in a scientific model" (1986a, 8).

GENERAL CONCEPTS OF SPACE

Much of the research in quantitative geography has been concerned with adapting mathematical and statistical approaches to make them more suitable for analyses in which space and location are central to the logic of the investigation. Mathematical languages have been chosen because they offer (or seem to offer) clear definitions of terms and operations, so that when results are obtained the origins and definitional bases of the results are also clear. These languages are also restrictive, and once problems are expressed in terms of a certain kind of mathematics, it is likely that the nature of the mathematical system will restrict and direct the way that problems are defined and analyzed (Gould 1984). These restrictions extend to the treatment of space in mathematical and statistical models. That treatment has usually been a very conventional one, based on notions of a physically objective space based on standard geometries.

Conceptions of space are a fundamental issue in geography (Couclelis and Gale 1986; Sack 1980) and quantitative geographers have recently begun to formalize geographic problems in terms of alternative mathematical frameworks, such as fuzzy sets (Leung 1987) and fractal geometry (Batty and Longley 1986). Recognition of the restrictive nature of many of the basic languages used in quantitative models has also led a number of geographers to develop formal models in terms of languages in which the incorporation of basic concepts such as distance and proximity is not a foregone conclusion. Efforts in this direction include research using very general and nonrestrictive approaches to symbolizing and categorizing information, including Q-analysis (Gould 1981) and variants of Q-analysis, such as Galois-lattices (MacGill 1985; Couclelis et al. 1987). The work on computational-process models also raises issues of appropriate treatment of terms such as "in front of," "next to," or "to the right of," which are not readily expressed in conventional mathematical notation.

Geographers have generally developed methods of analysis by incorporating concepts of space within existing methods. This approach assures that the meaning and interpretation of these basic concepts are limited by the possibilities for definition within the mathematics that lie behind the method. Investigations of alternative ways of formalizing spatial concepts indicate a growing awareness of the range of possibilities for expressing spatial concepts in symbolic ways, and a growing appreciation of how the definitions of basic concepts govern the results of an analysis.

INTEGRATION WITH GEOGRAPHIC INFORMATION SYSTEMS

The development of mathematical and statistical approaches in geography can reach deeply into the theoretical basis of the discipline. But it is also a practical matter, and the needs for systematic analyses of geographic data are growing very rapidly with the development of geographic information systems (GIS). There has been, up to now, little effort to coordinate developments in quantitative methods with GIS, but the need to integrate the two will become urgent in the near future as GIS becomes even more important, not only in support of management but also in support of scientific research.

The development of GIS has to date been concerned primarily with the storage, retrieval, and display of spatially referenced data (see the chapter on geographic information systems). Although many current systems are very sophisticated in performing these tasks, they have virtually no built-in capability for carrying out even basic operations in spatial statistics. More important, the protocols for symbolizing and storing information are often developed without attention to the requirements of spatial statistics and geographic modeling.

Goodchild (1988) notes that the complexity of the data required for spatial analyses inhibits the development of integrated software. For example, the analysis of a spatial-interaction model may require destination data in the form of points, origin data in the form of areas, and trip data in the form of area-point pairs. GIS does offer the potential to permit such complex data representations, but the development of suitable procedures has only begun.

There are several ways in which the integration of quantitative methods with GIS may help to close the gap between theory and practice in spatial analysis. First, software for spatial analysis and spatial statistics could be integrated within GIS. Computer software for spatial analysis has generally been developed for the limited purposes of particular research projects, with virtually no development of comprehensive packages for routine application. GIS will provide very large and comprehensive databases that will stimulate much wider and more routine application of the various methods of spatial analysis, especially if appropriate software can be integrated with the information system.

The provision of ready access to consistent, comprehensive databases of very wide geographic scope will also facilitate empirical testing and simulation, and may lead geographers to carry out more comprehensive empirical analyses on much broader geographic scales than in the past. Consistency in the databases and development of routines for tasks (such as spatial aggregation and the identification of network attributes) will be critical for such large-scale investigations.

Beyond the potential role in facilitating applications of GIS, spatial statistics and spatial analysis may be directly useful in the development of GIS procedures and algorithms. For example, methods of spatial autocorrelation are useful in accounting for errors in digitizing (Griffith and Amrhein 1988), and in coping with missing information (Haining, Griffith, and Bennett 1984). At a more general level, the development of large-scale, general-purpose geographic information systems may eventually have to be guided by some set of principles that could amount to a "theory of spatial information." The existing body of knowledge derived from spatial analysis is the most promising starting point for developing such principles.

There is a tremendous potential inherent in the integration of GIS, spatial statistics, and spatial modeling, especially if this relationship is a symbiotic one in which information systems facilitate the use and improvement of quantitative methods, and spatial modeling leads to improvements in information systems. The rapid development of GIS indicates that many of the results of this relationship will be realized in the near future.

DIRECTIONS FOR THE FUTURE

There has been a reasonably successful record of developing statistical methods and mathematical models that are suited to the distinctive research problems of geography. In looking to the future, it is tempting to try to identify gaps in the existing literature where progress may build on earlier developments in an incremental way. Incremental progress is important, but eventually leads to diminishing returns unless the process is refreshed by modifications of theory or expansion of the areas of potential application. While some promising areas for incremental progress can be identified, the work on mathematical modeling and quantitative analysis is broad enough for the identification of some more general changes.

One of the most obvious directions for future development is in the general area of spatial statistics, which needs further development and advertising before its full significance will be realized within and beyond the disciplinary borders of geography. Computational difficulties still limit the applications of methods such as spatial autoregression, which should be routine in geographic research. Incorporation of spatial statistics into convenient application packages would help to bring the methods into general use.

Theoretical developments are equally important, and the subfield of spatial statistics appears to have matured beyond a stage of adapting generalized statistical methods to the exigencies of geographical research, to a point where development of a coherent body of principles for measurement and inference with spatial data appears to be on the horizon. Presentation of spatial statistics as part of a coherent body of statistical theory, associated with accessible computation routines, is important if the methods of spatial analysis are not to languish in geography, or be reinvented in other disciplines. The recent text by Anselin (1988b) is a substantial step in this direction.

Models of spatial behavior have developed rapidly, but are still largely restricted to analyses of decision-making by individuals in exogenous environments. A more general understanding of the importance of spatial behavior in social processes is likely to emerge if attention shifts toward research on the ways in which interactions among decision-makers transform the environment. The microsimulation models of land markets by Anas (1982), and labor markets by Amrhein and MacKinnon (1988), offer one direction for this kind of research.

Practical applications of the methods of spatial analysis in decision-support systems are likely to become widespread in the near future. But the extent of these applications will depend on the integration of the methods with geographic-information systems. Rushton's recent work on integrating location-allocation models with information systems may be a harbinger of progress in this area (Densham and Rushton 1988). The combination of comprehensive databases, flexible accessing systems, and robust mathematical models should lead to a much wider application of methods that were developed for academic research to planning and decision problems in both the public and private sectors.

Many of the models discussed in this chapter, including longitudinal analysis and nonlinear dynamics, have been important areas of development, not only in human geography but in the social sciences generally. Geographers have been relatively insensitive to theoretical developments in other social sciences, with the possible exception of economics. The development of mathematical modeling is accelerating in other social sciences, in ways that often parallel developments in geography (Bartholomew 1983, 1984), and mathematical models may provide bridges across the chasms that separate conceptual thinking in different disciplines. It is likely that research in human geography will be influenced to an increasing degree by developments in other social sciences, at least in the area of mathematical modeling.

Perhaps the most general results of the research on statistical and mathematical models stem from their relations to geographic theory. The construction of these models, and their application to hypothesis testing, has made it possible for geographers to evaluate and refine at least some of the basic ideas in the discipline, by putting them into the formal terms that are required for mathematical modeling. The experience of constructing these models has started to reveal some of the limitations of particular approaches, as well as the rich variety of formalisms that are possible.

REFERENCES

Amrhein, C. G., and MacKinnon, R. D. 1988. A microsimulation model of a spatial labor market. *Annals of the Association of American Geographers* 78:112–31.

Anas, A. 1982. *Residential location markets and urban transportation*. New York: Academic Press.

——. 1983. Discrete choice theory, information theory and multinomial logit and gravity models. *Transportation Research B* 17:13–23.

Anderson, N. H. 1974. Information integration theory: A brief survey. In *Contemporary developments in mathematical psychology*, eds. D. H. Krantz, R. C. Atkinson, R. C. Luce, and P. Suppes, pp. 236–301. San Francisco: W. H. Freeman.

Anselin, L. 1984. Specification tests on the structure of interaction in spatial econometric models. *Papers of the Regional Science Association* 54:165–82.

——. 1986. Estimation and model validation of spatial econometric models using the GAUSS microcomputer statistical software. Working Paper, Department of Geography, University of California, Santa Barbara.

——. 1987. Model validation in spatial econometrics: A review and evaluation of alternative approaches. *International Regional Science Review*. In press.

——. 1988a. Lagrange multiplier test diagnostics for spatial dependence and spatial heterogeneity. *Geographical Analysis* 20:1–17.

——. 1988b. *Spatial econometrics: Methods and models*. Dordrecht: Martinus Nijhoff.

——, and Griffith, D. A. 1988. Do spatial effects really matter in regression analysis? *Papers of the Regional Science Association*. In press.

Aufhauser, E., and Fischer, M. M. 1985. Log-linear modelling and spatial choice. *Environment and Planning A* 17:931–52.

Bartholomew, D. J. 1983. Some recent developments in social statistics. *International Statistical Review* 41:1–9.

——. 1984. Recent developments in non-linear stochastic modelling of social processes. *Canadian Journal of Statistics* 12:39–52.

Batty, M., and Longley, P. A. 1986. The fractal simulation of urban structure. *Environment and Planning A* 18:1143–79.

Baxter, M. 1985. Misspecification in spatial interaction models: Further results. *Environment and Planning A* 17:673–78.

———. 1987. Tests for misspecification in models of spatial flows. *Environment and Planning A* 19:1153–60.

Beckman, M. J., and Thisse, J. F. 1986. The location of production activities. In *Handbook of regional and urban economics,* Vol. 6, *Regional economics,* ed. P. Nijkamp, pp. 21–96. Amsterdam: North-Holland.

Bennett, R. J., and Haining, R. P. 1985. Spatial structure and spatial interaction: Modelling approaches to the statistical analysis of geographical data. *Journal of the Royal Statistical Society.* A148:1–36.

———; Haining, R. P.; and Wilson, A. G. 1985. Spatial structure, spatial interaction, and their integration. *Environment and Planning A* 17:625–46.

Besag, J. E. 1974. Spatial interaction and the statistical analysis of lattice systems. *Journal of the Royal Statistical Society.* B36:192–236.

Bigman, D., and ReVelle, C. 1978. The theory of welfare considerations in applied public facility location problems. *Geographical Analysis* 10:229–40.

Bishop, Y. M. M.; Fienberg, S. E.; and Holland, P. W. 1975. *Discrete multivariate analysis: Theory and practice.* Cambridge: MIT Press.

Boots, B. 1984. Evaluating principal eigenvectors as measures of network structure. *Geographical Analysis* 16:270–75.

———. 1985. Size effects in the spatial patterning of nonprincipal eigenvectors of planar networks. *Geographical Analysis* 17:74–81.

———, and Getis, A. 1988. *Spatial point pattern analysis.* Beverly Hills: Sage Publications.

Casetti, E. 1972. Generating models by the expansion method: Applications to geographic research. *Geographical Analysis* 4:81–91.

———, and Jones, J. P. 1987. Spatial aspects of the productivity slowdown: An analysis of U.S. manufacturing data. *Annals of the Association of American Geographers* 77:76–88.

Church, R. L., and Roberts, K. L. 1983. Generalized coverage models and public facility location. *Papers of the Regional Science Association.* 53:117–35.

Clark, W. A. V., and Onaka, J. L. 1985. An empirical test of a joint model of residential mobility and housing choice. *Environment and Planning A* 17:915–30.

———, and Smith, T. 1985. Production system models of residential search behavior: A comparison of behavior in computer-simulated and real-world environments. *Environment and Planning A* 17:555–68.

———, and Van Lierop, W. F. J. 1986. Residential mobility and household location modelling. In *Handbook of regional and urban economics,* Vol. 6, *Regional Economics,* ed. P. Nijkamp, pp. 97–132. Amsterdam: North-Holland.

Clarke, M., and Wilson, A. G. 1983. The dynamics of urban spatial structure: Progress and problems. *Journal of Regional Science* 23:1–18.

Cliff, A. D., and Ord, J. K. 1969. The problem of spatial autocorrelation. In *Studies in regional science,* ed. A. J. Scott, pp. 25–55. London: Pion.

———, and ———. 1973. *Spatial autocorrelation.* London: Pion.

———, and ———. 1981. *Spatial processes: Models and applications.* London: Pion.

Cliff, A. D.; Martin, R.; and Ord, J. K. 1974. Evaluating the friction of distance parameter in gravity models. *Regional Studies* 8:281–86.

———; ———; and ———. 1975. Map pattern and friction of distance parameters. *Regional Studies* 9:285–88.

Costanzo, C. M. 1983. Statistical inference in geography: Modern approaches spell better times ahead. *Professional Geographer* 35:158–64.

Couclelis, H. 1985. Cellular worlds: A framework for modeling micro-macro dynamics. *Environment and Planning A* 17:585–96.

———. 1986a. Artificial intelligence in geography: Conjectures on the shape of things to come. *Professional Geographer* 38:1–11.

———. 1986b. A theoretical framework for alternative models of spatial decision and behavior. *Annals of the Association of American Geographers* 76:95–113.

———. 1988. Of mice and men: What rodent populations can teach us about complex spatial dynamics. *Environment and Planning A* 20:99–110.

———, and Gale, N. 1986. Space and spaces. *Geografiska Annaler.* B68:1–12.

———; Golledge, R. G.; Gale, N.; and Tobler, W. 1987. Exploring the anchor-point hypothesis of spatial cognition. *Journal of Environmental Psychology* 7:99–122.

Crouchley, R.; Davies, R. B.; and Pickles, A. R. 1982. Dynamic models of shopping behaviour: Testing the linear learning model and some alternatives. *Geografiska Annaler* B64:27–33.

Curry, L. 1972. A spatial analysis of gravity flows. *Regional Studies* 6:131–47.

———. 1982. Recruitment as diffusion and the spatial structure of occupations. *Journal of Regional Science* 22:479–98.

———. 1984. Inefficiency of spatial prices using the thermodynamic formalism. *Environment and Planning A* 16:5–16.

———. 1985. Inefficiencies in the geographical operation of labour markets. *Regional Studies* 19:203–15.

———. 1986. Trade as spatial interaction and central places. In *Transformations through space and time,* eds. D. A. Griffith and R. P. Haining, pp. 27–58. Dordrecht: Martinus Nijhoff.

———; Griffith, D.; and Sheppard, E. 1975. Those gravity parameters again. *Regional Studies* 9:289–96.

Daganzo, C. F. 1979. *Multinomial probit, the theory and its application to demand forecasting.* New York: Academic Press.

Davies, R. B.; Crouchley, R.; and Pickles, A. R. 1982. A family of hypothesis tests for a collection of short series events with an application to female employment participation. *Environment and Planning A* 14:603–14.

———; ———; and ———. 1983. Some methods for the testing and estimation of dynamic models which use panel data. *Environment and Planning A* 15:1475–88.

Densham, P., and Rushton, G. 1988. Decision support systems for locational planning. In *Behavioural modelling in geography and planning,* eds. R. G. Golledge and H. Timmermans, pp. 56–90. London: Croom Helm.

dePalma, A., and Lefevere, C. 1983. Individual decision-making in dynamic collective systems. *Journal of Mathematical Sociology* 9:103–24.

Deurloo, M. C.; Dieleman, F. M.; and Clark, W. A. V. 1988. Generalized log-linear models of housing choice. *Environment and Planning A* 20:55–70.

Domencich, T. A., and McFadden, D. 1975. *Urban travel demand: A behavioral analysis.* Amsterdam: North-Holland.

Dorigo, G., and Tobler, W. 1983. Push-pull migration laws. *Annals of the Association of American Geographers* 73:1–17.

Fingleton, B. 1983a. Independence, stationarity, categorical spatial data and the chi-squared test. *Environment and Planning A* 15:483–99.

———. 1983b. Log-linear models with dependent spatial data. *Environment and Planning A* 15:801–13.

———. 1986. Analyzing cross-classified data with inherent spatial dependence. *Geographical Analysis* 18:48–61.

Fischer, M. M., and Nijkamp, P. 1985. Developments in explanatory discrete spatial data and choice analysis. *Progress in Human Geography* 9:515–51.

Folmer, H., and Nijkamp, P. 1984. Linear structural equation models with latent variables and spatial correlation. In *New developments in spatial data analysis,* eds. G. Bahrenberg, M. Fischer, and P. Nijkamp, pp. 163–71. Gower: Aldershot.

Fotheringham, S. 1983. A new set of spatial interaction models: The theory of competing destinations. *Environment and Planning A* 15:15–36.

———. 1986. Modelling hierarchical destination choice. *Environment and Planning A* 18:401–18.

———, and Knudsen, D.C. 1986. Modeling discontinuous change in retailing systems: Extensions of the Harris-Wilson framework with results from a simulated urban retailing system. *Geographical Analysis* 18:295–312.

Gaile, G. L., and Burt, J. E. 1980. *Directional statistics.* Norwich: Geo Abstracts.

———, and Willmott, C. J. 1984. *Spatial statistics and models.* Dordrecht: D. Reidel.

Gatrell, A. C. 1983. *Distance and space: A geographical perspective.* Oxford: Clarendon.

Getis, A. 1984. Interaction modelling using second-order analysis. *Environment and Planning A* 16:173–84.

———, and Boots, B. 1978. *Models of spatial processes.* Cambridge: Cambridge University Press.

Golledge, R. G., and Rushton, G. 1984. A review of analytic behavioural research in geography. In *Geography and the urban environment,* eds. D. T. Herbert and R. J. Johnston, pp. 1–44. New York: Wiley.

———; Hendriks, P.; and Lensink, E. 1984. Spatial choice models: The problem of determining the dimensions. Paper presented at the Netherlands Geographic Conference, April, 1984.

Goodchild, M. F. 1988. A spatial analytic perspective on geographic information systems. *International Journal of Geographical Information Systems*. In press.

Gould, P. R. 1970. Is *Statistix Inferens* the geographical name for a wild goose? *Economic Geography* 46:439–48.

———. 1981. A structural language of relations. In *Future trends in geomathematics,* eds. R. G. Craig, and M. L. Labovitz, pp. 281–313. London: Pion.

———. 1984. Statistics and human geography: Historical, philosophical, and algebraic reflections. In *Spatial statistics and models,* eds. G. L. Gaile and C. J. Willmott, pp. 17–32. Dordrecht: D. Reidel.

Griffith, D. A. 1980. Towards a theory of spatial statistics. *Geographical Analysis* 12:325–29.

———. 1983. The boundary value problem in spatial statistical analysis. *Journal of Regional Science* 23:377–87.

———. 1987. Toward a theory of spatial statistics: Another step forward. *Geographical Analysis* 19:69–82.

———. 1988a. Estimating spatial autoregressive model parameters with commercial statistical packages. *Geographical Analysis* 20:176–86.

———. 1988b. *Spatial autocorrelation: A primer*. Washington: Association of American Geographers.

———, and Amrhein, C. 1983. An evaluation of correction techniques for boundary effects in spatial statistical analysis. *Geographical Analysis* 15:352–60.

———, and ———. 1988. GIS and statistical quality control. *Proceedings of the International Geographical Information Systems Symposium*. In press.

———, and Lea, A. C. 1983. *Evolving geographical structures*. The Hague: Martinus Nijhoff.

Haining, R. 1980. Spatial autocorrelation problems. In *Geography and the urban environment, Progress in Research and Applications,* eds. D. T. Herbert and R. J. Johnson, pp. 1–44. New York: Wiley.

———. 1983. Modelling intraurban price competition. *Journal of Regional Science* 23:517–28.

———. 1984. Testing a spatial interacting markets hypothesis. *Review of Economics and Statistics* 66:576–83.

———. 1985. The spatial structure of competition and equilibrium price dispersion. *Geographical Analysis* 17:231–42.

———; Griffith, D.; and Bennett, R. 1984. A statistical approach to the problem of missing data using a first-order Markov model. *Professional Geographer* 35:338–45.

Halperin, W. 1988. Current topics in behavioural modeling of consumer choice. In *Behavioral modelling in geography and planning,* eds. R. G. Golledge and H. Timmermans, pp. 1–26. London: Croom Helm.

Heckman, J. J., and Singer, B. 1986. Econometric analysis of longitudinal data. In *Handbook of econometrics,* Vol. III, eds. Z. Griliches and M. D. Intriligator, pp. 1689–1763. Amsterdam: North-Holland.

Hensher, D., and Johnson, L. 1981. *Applied discrete choice modelling*. London: Croom Helm.

Hodder, I., and Orton, C. 1976. *Spatial analysis in archaeology*. London: Cambridge University Press.

Horowitz, J. 1983. Statistical comparison of non-nested probabalistic discrete choice models. *Transportation Research B* 17:319–50.

———. 1987. Specification tests for nested logit models. *Environment and Planning A* 19:395–402.

Hubert, L., and Golledge R. G. 1981. A heuristic model for the comparison of related structures. *Journal of Mathematical and Statistical Psychology* 29:190–241.

———; ———; and Costanzo, C. M. 1981. Generalized procedures for evaluating spatial autocorrelation. *Geographical Analysis* 13:224–23.

———; ———; ———; and Gale, N. 1985a. Measuring association between spatially defined variables: An alternative procedure. *Geographical Analysis* 17:36–46.

———; ———; ———; and ———. 1985b. Tests of randomness: Unidimensional and multidimensional. *Environment and Planning A* 17:373–86.

Isham, V. 1981. An introduction to spatial point processes and Markov random fields. *International Statistical Review* 49:21–43.

Jones, D. W. 1984a. A land use model with a constant-utility spatially variant wage. *Geographical Analysis* 16:121–33.

———. 1984b. Nonland factor markets in the Thunen model. *Papers of the Regional Science Association*. 54:43–57.

Kanaroglou, P.; Liaw, K. L.; and Papageorgiou, Y. Y. 1986a. An analysis of migratory systems: 1. Theory. *Environment and Planning A* 18:913–48.

———; ———; and ———. 1986b. An analysis of migratory systems: 2. Operational framework. *Environment and Planning A* 18:1039–60.

Kinderman, R., and Snell, J. L. 1980. *Markov random fields and their applications*. Providence, RI: American Mathematical Society.

Knight, R., and Menchick, M. 1976. Conjoint preference estimation for residential policy land use evaluation. In *Spatial choice and spatial behavior,* eds. R. G. Golledge and G. Rushton, pp. 135–56. Columbus, OH: Ohio State University Press.

Knudsen, D. C. 1987. Computer-intensive significance testing procedures. *Professional Geographer* 39:208–15.

———, and Fotheringham, A. S. 1986. Matrix comparison, goodness-of-fit, and spatial interaction modelling. *International Regional Science Review* 10:127–47.

Koppelman, F. 1983. Predicting transit ridership in response to transit service changes. *Journal of Transport Economics and Policy* 109:548–64.

Lea, A. C. 1979. Welfare theory, public goods, and public facility location. *Geographical Analysis* 11:218–39.

———. 1981. Public facility location models and the theory of impure public goods. *Sistemi Urbani* 3:345–90.

Leonardi, G. 1981. A unifying framework for public facility location problems—Part I: A critical overview and some unsolved problems. *Environment and Planning A* 13:1001–28.

Leung, Y. 1987. On the imprecision of boundaries. *Geographical Analysis* 19:125–51.

Liaw, K. L., and Ledent, J. 1987. Nested logit model and maximum quasi-likelihood method: A flexible methodology for analyzing interregional migration patterns. *Regional Science and Urban Economics* 17:67–88.

Louviere, J. 1974. Predicting the response to real stimulus objects from abstract evaluation of the attributes: The case of trout streams. *Journal of Applied Psychology* 59:572–77.

———. 1976. Information processing theory and functional measurement in spatial behavior. In *Spatial choice and spatial behavior,* eds. R. G. Golledge and G. Rushton, pp. 211–48. Columbus, OH: Ohio State University Press.

———. 1982. Applications of functional measurement to problems in spatial decision making. In *Proximity and preference: Problems in the multidimensional analysis of large data sets,* eds. R. Golledge and J. Rayner, pp. 191–214, Minneapolis: University of Minnesota Press.

———, and Woodworth, G. 1983. Design and analysis of simulated consumer choice and allocation experiments: An approach based on aggregate data. *Journal of Marketing Research* 20:350–67.

Luce, R. D. 1959. *Individual choice behavior*. New York: Wiley.

McFadden, D. 1978. Modelling the choice of residential location. In *Spatial interaction theory and planning methods,* eds. A. Karlqvist, L. Lundqvist, F. Snickars, and J. Weibull, pp. 76–96. Amsterdam: North-Holland.

———. 1981. Econometric models of probabalistic choice. In *Structural analysis of discrete data with econometric applications,* eds. C. F. Manski and D. McFadden, pp. 198–272, Cambridge: MIT Press.

MacGill, S. M. 1985. Structural analysis of social data: A guide to Ho's Galois lattice approach and a partial respecification of Q-Analysis. *Environment and Planning A* 17:1089–1109.

Maddala, G. S. 1983. *Limited-dependent and qualitative variables in econometrics*. Cambridge: Cambridge University Press.

Mardia, K. V. 1975. Statistics of directional data. *Journal of the Royal Statistical Society* B37:349–53.

Menchick, M. 1972. Residential environmental preferences and choice: empirically validating preference measures. *Environment and Planning A* 4:455–58.

Mills, E. S. 1987. *Urban economics*. Vol. 2 of *Handbook of regional and urban economics*. Amsterdam: North-Holland.

Morrill, R. L. 1985. Some important geographical questions. *Professional Geographer* 37:263–70.

———. 1987. A theoretical imperative. *Annals of the Association of American Geographers* 77:535–41.

Mullally, H., and Papageorgiou, Y. Y. 1978. Spatial non-convexities. *Environment and Planning A* 10:37–42.

Mulligan, G. F. 1983. Central place populations: A microeconomic consideration. *Journal of Regional Science* 23:83–92.

———. 1984. Agglomeration in central place theory: A review of literature. *International Regional Science Review* 9:1–42.

Nelder, J. A., and Wedderburn, R. W. W. 1972. Generalized linear models. *Journal of the Royal Statistical Society* A135:370–84.

Nicolis, G., and Priogine, I. 1977. *Self-organization in non-equilibrium systems*. New York: Wiley.

Nijkamp, P. 1986. *Regional economics*. Vol. 1 of *Handbook of regional and urban economics*. Amsterdam: North-Holland.

O'Brien, L. G. 1983. Generalized linear modelling using the GLIM system. *Area* 15:327–36.

———, and Wrigley, N. 1980. Computer programs for the analysis of categorical data. *Area* 12:263–68.

Odland, J. 1988a. Sources of change in the process of population redistribution in the United States, 1955-1980. *Environment and Planning A* 20:789–809.

———. 1988b. *Spatial autocorrelation*. Newbury Park, CA: Sage Publications.

———, and Barff, R. 1982. A statistical model for the development of spatial patterns: Applications to the spread of housing deterioration. *Geographical Analysis* 14:326–39.

———, and Ellis, M. 1987. Disaggregate migration behavior and the volume of interregional migration. *Geographical Analysis* 19:111–23.

O'Farrell, P. N., and Crouchley, R. 1987. Manufacturing plant closures: A dynamic survival model. *Environment and Planning A* 19:313–29.

Onaka, J., and Clark, W. A. V. 1983. A disaggregate model of residential mobility and housing choice. *Geographical Analysis* 15:287–304.

Openshaw, S. 1984. Ecological fallacies and the analysis of areal census data. *Environment and Planning A* 16:17–32.

Papageorgiou, Y. Y. 1976. *Mathematical land use theory*. Lexington, MA: Heath.

———. 1978a. Spatial externalities: 1. Theory. *Annals of the Association of American Geographers* 68:465–76.

———. 1978b. Spatial externalities: 2. Applications. *Annals of the Association of American Geographers* 68:477–92.

———. 1983. Models of agglomeration. *Sistemi Urbani* 3:391–410.

———. 1987a. Spatial public goods: 1. Theory. *Environment and Planning A* 19:331–52.

———. 1987b. Spatial public goods: 2. Applications. *Environment and Planning A* 19:471–92.

Pickles, A. R. 1983. The analysis of residence histories and other longitudinal data: A continuous time Markov model incorporating exogenous variables. *Regional Science and Urban Economics* 13:271–85.

———, and Davies, R. B. 1984. Recent developments in the analysis of movement and recurrent choice. In *Spatial statistics and models,* eds. G. L. Gaile and C. J. Willmott, pp. 321–43. Dordrecht: D. Reidel.

———, and ———. 1985. The longitudinal analysis of housing careers. *Journal of Regional Science* 25:85–101.

———; ———; and Crouchley, R. 1982. Heterogeneity, non-stationarity, and duration-of-stay effects in migration. *Environment and Planning A* 14:615–22.

Plane, D. A, and Rogerson, P. A. 1985. Economic-demographic models for forecasting interregional migration. *Environment and Planning A* 17:185–98.

———, and ———. 1986. Dynamic flow modeling with interregional dependency effects: An application to structural change in the U.S. migration system. *Demography* 23:91–104.

Pocock, S. J.; Cook, D. G.; and Shaper, A. G. 1982. Analyzing geographic variation in cardiovascular mortality: Methods and results. *Journal of the Royal Statistical Society* A145:313–41.

Revelle, C. 1987. Urban public facility location. In *Urban economics,* Vol. 2 of *Handbook of urban and regional economics,* ed. E. S. Mills, pp. 1053–96. Amsterdam: North-Holland.

Ripley, B. D. 1981. *Spatial statistics*. New York: Wiley.

———. 1984. Spatial statistics: Developments 1980-3. *International Statistical Review* 52:141–50.

Rogerson, P. A. 1984. New directions in the modelling of interregional migration. *Economic Geography* 60:111–21.

———. 1985. Disequilibrium adjustment processes and chaotic dynamics. *Geographical Analysis* 17:185–98.

———, and Plane, D. A. 1984. Modeling temporal change in flow matrices. *Papers of the Regional Science Association* 54:147–64.

Sack, R. D. 1980. *Conceptions of space in social thought*. Minneapolis: University of Minnesota Press.

Sheppard, E. 1984. The distance-decay gravity model debate. In *Spatial statistics and models,* eds. G. L. Gaile and C. J. Willmott, pp. 367–88. Dordrecht: D. Reidel.

———; Griffith, D. A.; and Curry, L. 1976. A final comment on mis-specification and autocorrelation in those gravity parameters. *Regional Studies* 10:337–39.

Smith, T. R. 1984. Artificial intelligence and its applicability to geographical problem solving. *Professional Geographer* 36:147–58.

———; Clark, W. A. V.; and Cotton, J. W. 1984. Deriving and testing production system models of sequential decision-making behavior. *Geographical Analysis* 16:191–222.

———; Pellegrino, J.; and Golledge, R. G. 1982. Computational process modelling of spatial cognition and behavior. *Geographical Analysis* 14:305–25.

Sokal, R. R., and Oden, N. L. 1978. Spatial autocorrelation in biology 2: Some biological applications of evolutionary and ecological interest. *Biological Journal of the Linnean Society* 10:229–49.

Thrall, G. I. 1987. *Land use and urban form*. New York: Methuen.

Tobler, W. 1983. An alternative formulation for spatial-interaction modelling. *Environment and Planning A* 15:693–703.

———, and Kennedy, S. 1985. Smooth multidimensional interpolation. *Geographical Analysis* 17:251–57.

Tuma, N. B., and Hannen, M. T. 1984. *Social dynamics: Models and methods*. New York: Academic Press.

Upton, G., and Fingleton, B. 1985. *Point pattern and qualitative data*. Vol. 1 of *Spatial data analysis by example*. New York: Wiley.

Van Lierop, W. F. J. 1986. *Spatial interaction modelling and residential choice analysis*. Aldershot: Gower.

Weidlich, W., and Haag, G. 1983. *Concepts and models of a quantitative sociology: The dynamics of interacting populations*. Berlin: Springer-Verlag.

Wilson, A. G. 1970. *Entropy in urban and regional modelling*. London: Pion.

———. 1981. *Catastrophe theory and bifurcation: Applications to urban and regional systems*. London: Croom Helm.

Wrigley, N. 1982. Quantitative methods: Developments in discrete choice modelling. *Progress in Human Geography* 6:547–62.

———. 1985. *Categorical data analysis for geographers and environmental scientists*. New York: Longman.

———, and Dunn, R. 1984. Stochastic panel-data models in urban shopping behaviour: 1. Purchasing at individual stores in a single city. *Environment and Planning A* 16:629–50.

Remote Sensing

John Jensen | James Campbell | Jeff Dozier | Jack Estes |
Michael Hodgson | C. P. Lo | Kamlesh Lulla | James Merchant |
Ray Smith | Doug Stow | Alan Strahler | Roy Welch

Remote sensing is a dynamic, multidisciplinary technology used by geographers to obtain detailed spatial information about the Earth and human activities. Remote sensing is unique in that it can be used to collect fundamental biophysical data, unlike other techniques such as cartography, geographic-information systems, and statistics, which rely on data that are already available. Remote sensing-derived biophysical data can be transformed into information by using analog or digital image-processing techniques. Remote sensing-derived information is critical to the successful modeling of numerous natural processes (e.g., watershed runoff) and cultural processes (e.g., land-use conversion at the urban fringe). In fact, many models that rely on spatially distributed information cannot function without remote-sensing data.

Geographers should understand the fundamentals of remote sensing in order to use the technology properly in their scientific investigations. Many geographers are experts in remote sensing, and have been intimately involved in the development of remote-sensing theory. In fact, most of the remote-sensing education in the world today is conducted within geography departments. This chapter summarizes the contributions made by geographers to the advancement of remote sensing in America.

A DEFINITION OF REMOTE SENSING

The term *remote sensing* was coined in the early 1960s by geographers at the Office of Naval Research to describe the process of obtaining data by use of both photographic and nonphotographic instruments (Simonett et al. 1983). Today, remote sensing is formally defined as "the measurement of some property of an object of interest by a sensor that is not in direct physical contact with the object" (adapted from Colwell 1983, Fussell, Rundquist, and Harrington 1986, and Curran 1987). Most remote-sensing systems function by measuring electromagnetic energy that is propagated by electromagnetic radiation (EMR) at a velocity of approximately 3×10^8 m/sec (the speed of light) from a source (usually the sun) through space, and then reflected or reradiated from the object of interest back to the remote sensor. Therefore, EMR represents a high-speed communications link between the sensor and remotely located objects. Changes in the amount and properties of the EMR become, upon detection by the remote sensor, a valuable source of data about the object (Jensen 1983a).

THE REMOTE-SENSING PROCESS

What do geographers do with the remotely sensed data? This is best explained by reviewing the remote-sensing process (Figure 1). Geographers first select a scientific methodology based on (1) inductive logic (observation, classification, generalization, and theory formulation), (2) deductive logic (based on a statement of the problem, theory, hypothesis, observation, and verification-falsification), or (3) technological logic (identify human need, evaluate theories, design an appropriate plan, and apply it). When using induction, remotely sensed images and ancillary data are interpreted to yield theory based on generalizations. When using deduction, remotely sensed images and data are used to verify or falsify hypotheses as a means of improving theory (Curran 1987). A browse through any remote-sensing journal reveals that many researchers are not in pursuit of knowledge through either inductive or deductive logic. Rather, remotely sensed data are often used to solve applied problems by means of a technological approach. There is debate as to how each methodology used in remote sensing produces scientific knowledge (Fussell et al. 1986; Curran 1987).

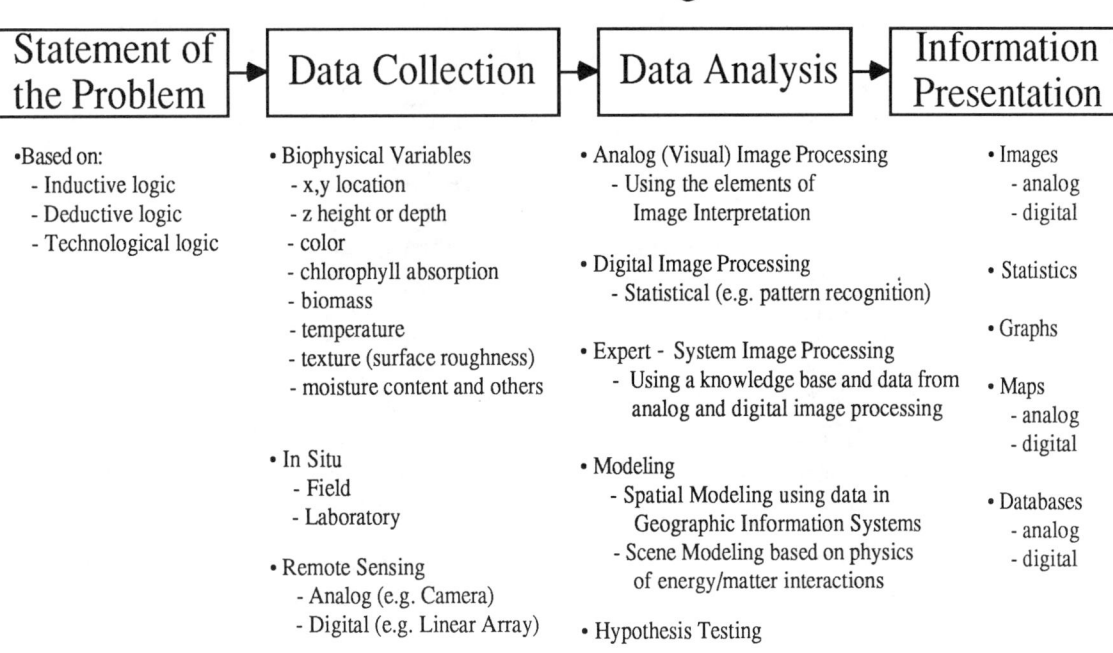

Figure 1
The remote-sensing process

Data Collection

Once the problem is stated and the appropriate logic specified, the data to be collected are identified. Usually, only some of the data necessary to address a particular problem can be derived from remote sensing. Jensen (1983a) identified biophysical variables that can be remotely sensed, including an object's x, y location, elevation (height), color, chlorophyll-absorption characteristics, biomass, temperature, surface roughness (texture), and moisture content. The biophysical characteristics of an object can then be evaluated in conjunction with other data to yield hybrid information, e.g., vegetation stress, nominal scale land use/cover information, or socioeconomic characteristics.

Remote measurement of biophysical variables such as moisture content and surface temperature will become as important to basic and applied research as is the current preoccupation with remote sensing as predominantly a land-use/cover mapping tool. The ordinal and interval scaled biophysical data are more easily incorporated into models that accept spatially distributed data. Studies cited in this chapter identify advances made by geographers in the collection of biophysical and hybrid data, using remote sensing.

It is usually necessary to collect in situ data to calibrate the remotely sensed data. Geographers pioneered the development of in situ data-collection techniques as described in the chapter, "Ground Investigations in Support of Remote Sensing," in the *Manual of Remote Sensing* (Dozier and Strahler 1983). Geographers interested in the application of remote sensing should have a good background in a systematic area of geography, e.g., biogeography, soils, etc., wherein they have learned appropriate field techniques.

Both analog and digital remote-sensor data can be collected. Current and proposed remote-sensing systems and their spatial, spectral, and temporal characteristics are summarized in Table 1. Color and color infrared aerial photography continue to be important analog remote-sensing systems for many geographic applications. In addition, high-resolution digital remote-sensing systems, such as the Daedalus DSS-1268 multispectral scanner (MSS) and NASA's Thermal Infrared Multispectral Scanner (TIMS) mounted onboard aircraft, continue to provide high spatial and spectral resolution multispectral digital remotely sensed data. Unfortunately, the data are very expensive per km^2.

Satellite remote-sensor systems, however, provide a wealth of remotely sensed data that is relatively inexpensive per km^2 (Lulla 1983). The sensors have progressed from multispectral scanning systems (Landsat MSS), to more advanced scanning systems (Landsat Thematic Mapper), to the use of linear-array technology (SPOT HRV sensors). New area-array technology will be used in the HIRIS (High Resolution Imaging Spectrometer) and MODIS (Moderate Resolution Imaging Spectrometer) on board the Earth Observing System (EOS) space station. Note (Table 1) that instead of the usual 4 to 7 bands found in the traditional MSS and linear-array sensor systems, there will be 192 bands in the spectral region from 0.4 to 2.5 μm for the HIRIS and 95 bands from 0.4 to 12.0 μm for the MODIS sensor systems. The HIRIS sensor will collect data at a spatial resolution of 30 × 30 m, while the MODIS will collect data at 1 and 0.5 km spatial resolutions using its two sensor systems.

Significant improvements will also take place in microwave remote sensing. The EOS HMMR (High Resolution Multifrequency Microwave Radiometer) will consist of a number of sensors and will be capable of providing detailed biophysical information on atmospheric temperature, water, snow cover, sea-ice age and extent, precipitation amount and distribution, and surface soil moisture at spatial resolutions from 10 to 50 km. Active microwave remote sensing will be improved with the launch of the EOS Synthetic Aperture Radar (SAR), operating at three wavelengths (L, C, and X) and at multiple polarizations. For many Earth-science applications, it has a spatial resolution of approximately 40 × 40 m. The new sensor systems are optimized to gather biophysical information, although they can be used to obtain land use/cover data. Data from these remote-sensor systems are critical to the successful modeling of Earth ecosystems (Estes 1987).

Data Analysis

Once the in situ data, remotely sensed data, and all ancillary information are collected, analysis is necessary to judge their significance, and to accept or reject hypotheses. This analysis is performed by:

1. Analog (visual) image processing of remotely sensed data
2. Digital image processing
3. Expert-system approach
4. GIS approach, and/or
5. True scene modeling

Analog Image Processing. Geographers extract information from remotely sensed images using visual-interpretation techniques based on the fundamental elements of image interpretation including size, shape, shadow, color (tone), texture, pattern, site, and association. The definitive work on this topic was the *Manual of Photographic Interpretation,* published by the American Society of Photogrammetry in 1960 (Colwell 1960). Subsequently, several works by geographers have included chapters on how to perform visual-image interpretation (Richason 1978; Townshend 1981; Estes, Hajic, and Tinney 1983; Holz 1985; Avery and Berlin 1985; Campbell 1983, 1987; Curran 1985; Lo 1986a). They have shown how to identify an object, measure it, and judge its significance, without investigating the psychological aspects of image interpretation. But knowledge about "image understanding" may eventually be required if expert systems are to mimic the actions of the photointerpreter. There will be a resurgence in the art and science of visual photointerpretation as the new digital remote-sensor systems provide images of higher spatial resolution. These images will lend themselves to visual photointerpretation techniques.

Digital Image Processing. Significant advances in digital image-processing techniques have taken place since the launch of the Landsat MSS in 1972. Methods of image restoration, enhancement, and classification are summarized in the *Manual of Remote Sensing* (Colwell 1983) and numerous articles. Geographers have made significant contributions to the development of digital image-processing techniques, which are summarized in the chapters by Townshend (1981), Estes, Hajic, and Tinney (1983), Campbell (1987), and Jensen (1986). Unfortunately, some of the algorithms are at best simplistic. For example, classification algorithms have been primarily per-pixel in nature, not taking into account the statistical variability of the terrain immediately sur-

Table 1
Current and proposed remote-sensing systems and their major characteristics

Remote-Sensing System	RESOLUTIONS								Spatial in Meters	Temporal in Days
	Spectral									
	B	G	R	NIR	MIR	TIR	M			
Aircraft										
Panchromatic film	0.4-------0.7								variable	variable
Color film	0.4-------0.7								variable	variable
Color infrared film		0.5-----------------0.9							variable	variable
Daedalus DSS 1260	3	2	2	2	—	2	—		variable	variable
Daedalus DSS 1268	1	1	2	2	2	2	—		variable	variable
Thermal IR Multispectral Scanner [TIMS]	—	—	—	—	—	6	—		variable	variable
Satellite										
NOAA-9 AVHRR LAC	—	—	1	1	1	2	—		1100	14.5/day
NOAA-K, L, M (proposed)	—	—	1	1	2	2	—		1100	14.5/day
Landsat Multispectral Scanner (MSS)	—	—	1	1	2	—	1		79	16-18
Landsat Thematic Mapper (TM)	1	1	1	1	2	1	—		30[a]	16
Landsat Enhanced Thematic Mapper (ETM):	1	1	1	1	2	1	—		30[a]	16
Panchromatic	—	0.5-----1-----0.9			—	—	—		13 x 15	16
EOSAT STAR (proposed)	—	1	1	1	—	—	—		5	pointable
EOSAT Sea Wide Field Sensor (SeaWiFS) (proposed) ocean color	1	2	1	2	—	2	—		1.13[b]	1 day

Instrument						
Shuttle Imaging Radar (SIR-C) (proposed)	—	—	—	—	3	40	unknown
SPOT HRV Multispectral:	—	—	1	1	—	20	pointable
Panchromatic	—	0.51 — 0.73	—	—	—	10	pointable
SMS/GOES Series (East and West)	—	0.55 — 0.72	—	1	—	700	0.5/hr
EOS High Resolution Imaging Spectrometer (HIRIS)—minerals, soils, vegetation, water	0.4 ————— 192 bands ————— 2.5				—	30	pointable
EOS Moderate Resolution Imaging Spectrometer:							
(MODIS-T)—ocean color, chlorophyll	0.4 ——— 60 bands ——— 1.0				—	1000	pointable
(MODIS-N)—ecosystems, climate, ocean	0.4 ——— 35 bands ——— 12.0				—	500	—
EOS High Resolution Multifrequency Microwave Radiometer [HMMR]:							
Advanced Microwave Sounding Unit							
(AMSU-A)—temperature	—	—	—	—	15	50,000	1–3
(AMSU-B)—water vapor	—	—	—	—	5	15,000	1–3
Advanced Mechanically Scanned Radiometer (AMSR)—snow cover, sea ice, rain	—	—	—	—	6	10,000	1–3
Electronically Scanner Thinned Array Radiometer (ESTAR)—surface soil moisture	—	—	—	—	1	10,000	1–3
EOS Synthetic Aperture Radar (SAR)[c]	—	—	—	—	3	40, 10–120	—

Note: [a]The TM thermal channel has 120 × 120 m spatial resolution

[b]A 4.5-kilometer resolution will be synthesized on board to provide Global Area Coverage (GAC)

[c]L-band (1.25), C-band (5.3), X-band (9.6) Ghz; multiple polarization HH,VV,VH, HV for L and C bands

rounding the pixel in question. Although there are contextual classifiers, most are inefficient and elementary in nature. More robust algorithms are required which take into account the context of an individual pixel and all pixels within a certain threshold distance from it.

Research is also needed on how to analyze imaging spectrometry data. The spectral reflectance/emittance curves associated with each pixel can be compared with a data bank of curves. This will require new methods of spectral analysis that compare curve shapes. Thus, to improve the classification of remotely sensed data, it is necessary to improve per-pixel classification algorithms, contextual algorithms, and curve-matching techniques.

Computers and algorithms that process data in parallel will be responsible for additional breakthroughs in digital image-processing research, especially for imaging-spectrometry data. In a typical per-pixel classification algorithm, each pixel of each line of each band is passed to the CPU serially. Depending upon the number of rows, the number of columns, and the number of bands, this can become computationally intensive. Conversely, each pixel in a line (or each band) can be sent to an individual CPU, if parallel processing is possible. Parallel processing becomes very important when one realizes that the new generation of imaging spectrometers such as HIRIS will have as many as 192 individual spectral bands, high spatial resolution, and higher radiometric accuracy (12 bits per pixel). We are at the forefront of a significant increase in data dimensionality which can only be analyzed efficiently through parallel processing.

Expert Systems. Current digital image-processing techniques rely primarily on image tone (color) and occasionally texture for the interpretation of remotely sensed data. They rarely incorporate any of the heuristic rules and knowledge that an expert uses when interpreting an image. Artificial intelligence (AI) is the study of how to make computers do things that, at the moment, people do better (Rich 1983). The area of AI that has the greatest potential in remote sensing is the use of expert systems to perform tasks requiring a great deal of specialized knowledge that most people would not possess (Estes, Sailer, and Tinney 1986).

Expert remote-sensing systems can be used to interpret an image, and/or to place all the information contained within an image in its proper context with other ancillary data in order to extract more valuable information. In the first case, collateral data and rules used by an expert might be used by novices to more accurately interpret a remotely sensed image. In the second case, an expert system might be used to conduct crop-yield forecasting or to predict population growth. Geographers with good systematic training and an understanding of remote sensing may make significant contributions in this area.

Geographic Information Systems. An oversight of those who promote the integration of remote-sensing information and GIS is their frequent assumption that the flow of data should be unidirectional, i.e., from the remote-sensing system to the GIS. But the "backward" flow of ancillary data from the GIS to the remote-sensing system can be very valuable. For example, land-cover mapping using remotely sensed data has been significantly improved by incorporating topographic information from digital-terrain models (Strahler 1981) and by incorporating other GIS data layers (Hutchinson 1982; Hodgson 1987).

The interface between GIS and remote-sensing systems is functional but weak (Jensen and Christensen 1986). Each technology suffers from a lack of critical support that could be provided by the other. GISs need timely, accurate updating of the spatially distributed variables in the database, which remote sensing can often provide. Remote sensing can benefit from access to accurate ancillary information to improve classification accuracy and other types of modeling. Such synergy is critical if successful expert remote-sensing systems are to be developed. Indications suggest this relation is improving.

Scene Modeling. While most geographers directly use the raw digital brightness values found in remotely sensed data, others attempt to model these values until they understand the relation between the energy incident to and exiting the target. The latter approach requires scene modeling, where the atmospheric path radiance, target conditions, and remote-sensing detector geometry and characteristics are all known. Successful modeling will predict how much radiant flux in certain wavelength intervals should exit a particular object, even without actually sensing the object. When the model's prediction is the same as the sensor's measurement, then we have modeled the relation correctly. When this takes place, the scientist will have a greater appreciation for energy-matter interactions in the scene and may be able to extend the technology to other regions with confidence.

REMOTE-SENSING APPLICATIONS

The remote-sensing process (Figure 1) can be applied to various types of terrain. The following sections summarize how geographers remotely sense soil and rocks, vegetation, water resources (terrestrial, coastal, and marine), and urban land use/cover data.

Remote Sensing of Soil and Rocks

The use of remote sensing to derive information about the Earth's surface soils and rocks requires diverse knowledge, ranging from the expected (e.g., topography, soil color) to the unexpected (e.g., plant physiology, atmospheric scattering). Some surface materials may be directly recognized by their distinctive spectral properties. In other instances, it is necessary to indirectly derive their identities and properties (Campbell 1987). For example, soils are not defined solely by their surface characteristics, but also by the soil profile, formed by a distinctive series of subsurface horizons. The soil profile is not subject to direct remote sensing because its defining characteristics are not visible to the sensor. Such features can be mapped only by indirect means, using those characteristics that are visible to infer those that are not. This dichotomy between direct and indirect derivation of information from remotely sensed data is fundamental to understanding the application of remote sensing to the study of soil and rocks.

Indirect Remote Sensing of Soil and Rocks. The indirect approach yields knowledge of classes of soil, rock, or terrain, rather than knowledge of the specific properties of these classes. It was developed in the context of manual image interpretation, where analysts integrate topography, slope, drainage, vegetation, and land-use data to derive information on patterns of soil and rock (Campbell and Edmonds 1984). Interpretation techniques include overlay (the use of transparencies of drainage, topography, and vegetation), photomorphic mapping (recognition of regions of distinctive photographic appearance), and geobotany (interpreting plant type and condition to derive geologic information).

In the digital domain, the indirect approach lends itself to use of coarse-resolution remote-sensor data, in which reflectances of soil, plant cover, and topographic shadowing may be mixed into a complex spectral composite (Weismiller, Persinger, and Montgomery 1977). Individual components of the composite cannot be resolved, but may be distinct enough to permit distinguishing among broad classes of soil and rock. This approach leads to maps of soil associations and/or terrain complexes with broadly defined landscape units. Detailed results have been produced by combining spectral data with collateral information, e.g., digital terrain data. Results are not comparable in detail or accuracy to a conventional soil survey map, but some specific units may be delineated with good accuracy, and the representation of broad-scale units may be satisfactory.

Direct Remote Sensing of Soil and Rocks. The direct approach usually attempts to quantitatively estimate properties such as mineral content, percentage of organic matter, iron oxide, or soluble salts. Direct remote sensing depends upon pure spectral-response patterns which are uncontaminated by the effects of atmosphere, shadowing, vegetation cover, or mixtures with different materials. Optimum results are obtained using high-resolution spectral, spatial, and radiometric remote-sensor data.

The direct approach requires knowledge of the spectral properties of the target, so research has also been devoted to obtaining spectroradiometer measurements of Earth-surface materials either in situ or in the laboratory (Condit 1970). Baumgardner et al. (1985) summarized relations between soil spectra and soil properties including moisture, particle size, iron oxides, soluble salts, organic matter, mineral composition, and parent material. Some of the direct approaches to remote sensing of the Earth's soils and rocks are summarized below.

Hydrothermally altered rocks may be identified directly, based on the amount of reflectance in the near- and middle-infrared portions of the spectrum (Buckingham and Sommer 1983). For example, Davis, Berlin, and Chavez (1987) found that Thematic Mapper bands 1, 4, and 5 could be used to distinguish red cone basalts from gray flow basalts and sedimentary country rocks for three volcanic fields in the southwestern U.S.

The *soil brightness line* is a concept used to identify contributions of soil and vegetation to the brightness of mixed pixels. Given spectral measurements from bare soil in the red and near-infrared regions, a line is defined by a diagonal continuum in multidimensional data space. Wet soils lie at the end of the continuum nearest the origin, while the upper end of the soil line is formed by reflectance from drier soils. Vegetated pixels tend to be dark in the red region because of the absorption of red light for photosynthesis, but bright in the infrared region due to strong reflection of infrared radiation by mesophyllic tissue. Therefore, vegetated pixels occupy a distinctive region of the data space, separated from the pixels along the soil brightness line. Partially vegetated pixels occupy intermediate positions with respect to the soil brightness line. Richardson and Wiegand (1977) measured the distance to the line and called it the Perpendicular Vegetation Index (PVI), a measure of the degree to which a pixel's brightness is determined by reflectance from living vegetation.

Geobotany is the science of deriving lithological information by direct measurement of spectral properties of the plant canopy (Lulla 1985). Plant growth and vigor are sensitive not only to the availability of the major plant nutrients, but also to micronutrients and other elements that may be present only in trace amounts. Some elements present in the geologic substratum are incorporated into plant tissue and affect the spectral properties of the canopy, or change the timing of foliation. One example is the blue shift, i.e., a change in position of the edge of the chlorophyll-absorption region toward shorter blue wavelengths (Chang and Collins 1983). This geobotanical indicator is very subtle, requiring knowledge of the local geologic and botanical conditions. It has been recorded using high-resolution data and may not be detectable using coarse remote-sensor data.

Future Developments. Research in remote sensing of soil and rocks will focus upon uses of high-resolution data (HIRIS and SAR in Table 1) that permit application of the direct approach in a rigorous manner, using knowledge of spectra to identify specific classes of soil and rocks, and perhaps to measure some properties. Application of the indirect approach will also benefit from the availability of imagery of finer detail, although it should be noted that many earlier applications of the indirect approach have taken advantage of the composite signatures generated by use of coarse-resolution imagery.

Remote Sensing of Vegetation Resources

Geographers have contributed greatly to our understanding of how to remotely sense the type, extent, and condition of natural and cultural vegetation. During the last decade, geographic research on remote sensing of vegetation has emphasized the regional, continental, and global perspectives afforded by Landsat and the NOAA satellites. However, much important work having site-specific focus continues to be carried out, as new technologies such as videography and the airborne imaging spectrometer (AIS) are investigated.

Natural Vegetation. The launch of Landsat-1 in 1972 made it possible to examine vegetation on a regional, continental, or even global scale, and to monitor natural and human-induced vegetation change. The capabilities of the Landsat sensors for mapping forest vegetation have been particularly well studied by geographers. Walsh (1980) used MSS data to classify coniferous vegetation in Crater Lake National Park, Oregon with an accuracy of over 88%. Slope angle, slope aspect, and cover-type variation were important variables influencing spectral reflectance (Walsh 1987). Strahler (1981) developed methods for improving MSS-based classification of forest vegetation in northern California by stratification of the data, using image texture and digital terrain data.

Hutchinson (1982) identified techniques for using ancillary data (e.g., elevation, slope, aspect) to enhance digital classification of natural vegetation. Thematic Mapper simulator data were used by Franklin (1986) and Peterson et al. (1986) to estimate conifer basal area and biomass. Franklin et al. (1986) also reported methods for using

MSS data to produce softwood timber volume estimates having a standard error comparable to in situ measurement techniques. Capabilities of Landsat for forest-condition assessment (e.g., insect damage) and monitoring of deforestation were documented by Williams and Miller (1979) and Williams and Nelson (1986).

Attention has also been directed toward rangeland and semiarid vegetation. Warren and Hutchinson (1984) found that satellite data could be used to assess total vegetation cover and the ratio of shrub to grass cover, both important indicators of change in range condition. Frank (1984, 1985) used textural, albedo, and vegetation indexes based on red light/near-infrared reflectance relations to classify semiarid terrain in Utah. He concluded that the indices appear to be particularly useful for identifying integrated terrain units, combinations of vegetation, and lithologic units.

Working in more moist environments, Walker et al. (1982) used Landsat data to prepare a 1:250,000-scale map of arctic tundra vegetation cover for the Arctic National Wildlife Refuge, Alaska. Jensen et al. (1986) and Hodgson et al. (1988) used Landsat data and aircraft-borne MSS data to identify and monitor wetland communities in South Carolina. Thermal infrared data obtained simultaneously were shown to be useful for relating vegetation type to water temperature. Place (1985) found Seasat radar imagery, because of its sensitivity to moisture, to be a useful complement to conventional sources for mapping Atlantic coastal forested wetlands.

Aerial photography has been used for more than 50 years in site-specific studies of vegetation. Geographers continue to make great use of photography, but also are exploring newer remote-sensing techniques. For example, Lulla et al. (1987) and Lusch and Sapio (1987) reported success in using airborne videography to compute vegetation indices and to monitor gypsy moth defoliation. Rundquist (1985) used 128-channel AIS data to examine relations between spectral reflectance and topographic/moisture/vegetation gradients in the Nebraska Sand Hills. Working both in the laboratory and in the field, Ripple (1986) documented variations in plant spectral reflectance attributable to leaf cover, moisture content, and soil background reflectance.

A number of geographers are currently involved in an extraordinary exercise being carried out on the Konza Prairie Natural Area, Kansas, under the auspices of the International Satellite Land Surface Climatology Project (ISLSCP). The First ISLSCP field experiment is an intensive assessment of the degree to which remote sensing can be used to yield quantitative information concerning land-surface climatological conditions (Schmugge and Sellers 1986).

There is currently great activity in global inventory and monitoring of vegetation resources (Estes and Starr 1986). Goward (1987) and Goward et al. (1987) found that vegetation index measurements derived from the NOAA Advanced Very High Resolution Radiometer (AVHRR) were well correlated with primary productivity and seasonality patterns of major biomes. Thermal data from the AVHRR have been employed by Ambrosia and Brass (1988) to examine wildfires and to assess their effects on global biogeochemical cycling.

Cultural Vegetation. Geographic research on remote sensing of cultural vegetation has been directed largely toward agricultural environments. Loveland and Johnson (1983) used Landsat data to map irrigated crops in north-central Oregon. The resulting digital maps were then placed in a GIS, along with terrain, soils, and hydrographic data, to predict suitability for irrigation development and to estimate water and power demand. Mohler et al. (1986) developed techniques for using multitemporal Landsat MSS data to estimate spring small-grains acreage for areas in the northern Great Plains.

On a continental scale, Johnson, van Dijk, Sakamoto (1987) used vegetation indices computed from AVHRR data to estimate sorghum and millet yields in Africa. They suggest that such information could be a great aid to early warning of potential food shortages.

Yool et al. (1985) used a GIS to model the brushfire hazard in southern California. Landsat MSS data were used to provide information on brush fuel type and density. Vegetation data were integrated in the GIS with data describing fire history, rainfall, and topography, and the database was used to model brushfire hazard on a site-specific basis.

Future Prospects and Research Needs. NASA's EOS initiative will provide a broad array of new sensors with which to observe global vegetation (EOS Science Steering Committee 1987). For example, the SAR will provide routine multifrequency, multipolarization, multiple-incidence-angle radar imagery of the entire globe. Remote sensing in the microwave region has great potential for providing unique information on plant moisture and canopy morphology (Richards, Sun, and Simonett 1987). Perhaps most importantly, it will be possible to monitor the vegetation in the tropics through cloud cover.

Remote Sensing of Water Resources

Terrestrial Water Resources. Geographers have been active in the remote sensing of terrestrial water supply, water demand, watershed-runoff characteristics, and water quality, all of which are critically important to numerous water-resource models. For example, snow mapping is one of the major inputs to the modeling of water supply and climate. Dozier (1984) found that snow and clouds could be discriminated using remotely sensed imagery in the 1.55 to 1.75 µm region. Also, near-infrared Landsat and NOAA meteorological satellite data can be used to distinguish new snow from that which has been melted (Dozier, Schneider, and McGinnis 1981). Dozier and Marks (1987) corrected snow-reflectance measurements for atmospheric effects and estimated grain size. Unfortunately, there are two major shortcomings in the present use of satellite visible and infrared data to monitor the Earth's snow cover: (1) it is difficult to measure snow-water equivalence, and (2) snowmelt often occurs in cloudy weather. Foster et al. (1984) describes how passive microwave remote sensing may solve these problems.

Geographers have pioneered techniques to assess agricultural and urban water demand. Agricultural water-demand prediction requires modeling of (1) individual crop types, (2) evapotranspiration characteristics of the crops, and (3) antecedent soil-moisture conditions (Estes, Jensen, and Tinney 1978; Jensen and Chery 1980; Loveland and Johnson 1983). Although initial water-demand prediction results are encouraging, additional research is required. Urban water-demand estimation requires knowledge of the individual building units within an urban area and the average consumptive use per day (Jensen et al. 1977).

Watershed characteristics such as soil moisture, runoff, and the sediment characteristics of the water are important hydrologic measurements which can be determined by using remote sensing. For example, Blanchard and O'Neill (1983) used a time series of microwave data immediately after a rainstorm to estimate saturated hydraulic conductivity in the soil. Schmugge, O'Neill, and Wang (1986) found passive microwave emission in the 1.4 GHz region to be highly correlated with soil moisture in the top 5 cm of the soil. Goward and Taranik (1986) summarized the use of multispectral thermal remote sensing for estimating soil moisture and evapotranspiration from bare soil and vegetated surfaces.

Slack and Welch (1980) developed remote-sensing techniques to derive USDA Soil Conservation Service watershed runoff-curve numbers from Landsat data. Additional research revealed that the amount of soil erosion per watershed could be estimated by using large-scale stereoscopic photogrammetric techniques (Welch and Jordan 1983). Digital terrain models of the watershed on multiple dates were differenced to compute the volume of sediment lost to sheet and rill erosion across the entire field or from concentrated flow erosion (Thomas, Welch, and Jordan 1986).

Thermal characteristics of lakes and streams have been studied, using thermal-infrared sensors, by Jensen et al. (1986). Lake eutrophication characteristics, including suspended sediment and chlorophyll concentrations, have been studied using (1) Landsat satellite data and chromaticity techniques (Campbell 1987), and (2) high-resolution MSS data and scene-modeling techniques which take into account atmospheric-path radiance (Ramsey and Jensen 1988). Rundquist et al. (1987) used thermal-infrared data to document the "flow-through" model for lakes in Nebraska's Sand Hills and Landsat MSS data to compute a relation between rainfall and seasonal lake-surface area.

Coastal and Marine Water Resources. Coastal studies encompass the nearshore hinterland including plains, wetlands, estuaries, sea cliffs, coastal dunes, and the shore zone (Stow and Estes 1983). Marine studies investigate the open water and ice found seaward of the coastal zone. Space and time scales of coastal/marine processes vary such that remotely sensed data are required for a variety of spatial and temporal resolutions and areal extents (Smith et al. 1987). The interdisciplinary nature of these studies and techniques requires geographers to interact with oceanographers, coastal/ocean engineers, marine biologists, aquatic ecologists, geologists, and physicists. Geographers have the ability to integrate complex phenomena into a spatial framework using remotely sensed data, and to provide new perspectives in which to understand coastal/marine processes (see the chapter on coastal and marine studies).

Remote sensing in the coastal zone is used to inventory and model geomorphology, shore-protection conditions, and estuaries. Major areas of technical development have been in shoreline change, bathymetric mapping, and analyses of radiation transfer in estuarine and nearshore waters. Coastal geomorphology and shore-protection studies require data that have high spatial and temporal resolution. Historical and short-time lapse aerial photography are ideal (Jackson and Rosenfeld 1987). Accurate spatial-registration methods must be used to assess long-term erosion rates and short-term morphodynamics (Leatherman 1983).

Stereoscopic photointerpretation techniques provide height information, and thus better morphological interpretations. Ground-penetrating radars show promise for assessing depth information in terms of stratigraphic relationships of deposited coarse sediments (Leatherman 1987). Water-depth information is becoming more measurable through the use of active and passive microwave techniques (Schmugge, O'Neill, and Wang 1986). The spatial resolution of Landsat and SPOT sensor systems make satellite data viable for (but still limited to) regional-scale studies of coastline morphology.

Remote sensing provides spatial information on the biophysical properties of water and on the circulation and mixing of surface estuarine waters. Biophysical properties such as temperature, chlorophyll pigments, suspended sediments, and salinity exhibit large ranges and complex spatial structures (Lavoie et al. 1985); these can be inventoried using remote-sensing techniques (Miles, Stow, and Jones 1988). These same properties serve as natural tracers for studying the surface circulation of estuarine and shelf waters (Stow 1987b). Circulation may be inferred by examining tracer patterns on a single image (Klemas et al. 1977), by tracking tracer features on time-sequential images that are spatially registered, or by combining tracer gradient information from time-sequential images with numerical flow models (Stow 1987a).

Geographers utilize remote-sensing techniques for the study of physical and biological oceanography, for the mapping of water quality and pollution, and for determining the distribution and dynamics of sea ice. Physical oceanographic studies by geographers are primarily associated with the use of AVHRR infrared imagery. For example, Zheng, Klemas, and Huang (1984) used AVHRR data to study the dynamics of the slope water off New England and its influence on the Gulf Stream, and Gagliardini et al. (1984) used Landsat MSS, AVHRR, and Nimbus Coastal Zone Color Scanner (CZCS) to study the La Plata River in South America.

Studies of living marine resources in the sea have also made use of the CZCS and AVHRR imagery. Smith and Baker (1982) used CZCS data to show that concurrent ship and satellite data can be used for the quantitative definition of an oceanic habitat descriptor (chlorophyll concentration) for modeling of the marine environment. This work was extended to convert satellite chlorophyll data to primary production (Smith, Eppley, and Baker 1982). CZCS imagery has also been used to investigate the spatial variability (Smith, Zhang, and Michaelsen 1988) and temporal variability (Michaelsen, Zhang, and Smith 1988) of pigment biomass in the California Current system.

Aircraft, Shuttle Imaging Radar (SIR), and Landsat MSS data have been used for the detection and mapping of surface oil slicks (Estes, Crippen, and Starr 1985). Klemas (1980) showed how estuarine and coastal fronts influence pollutant dispersion by capturing oil slicks and other pollutants that are concentrated in surface films, and then used Landsat imagery to map the location, type, and extent of estuarine fronts over all portions of the tidal cycle.

Passive microwave radiometers and imaging radars are used routinely by scientists for sea-ice observations and studies. The value of microwave observations for these studies stems from their all-weather capability and the large contrast between the microwave-brightness temperature or radar-backscatter properties of sea ice and open water. Barry (1983) extensively reviewed the use of passive microwave sensors for remote sensing of snow and ice distribution, sea-ice dynamics, and the detection of icebergs.

Future Developments. Additional research is required to understand the interaction of electromagnetic energy and the phenomena found in hydrologic environments.

ANALYSIS AND DISPLAY OF GEOGRAPHIC PHENOMENA

The new visible and near-infrared imaging spectrometry and passive microwave remote sensors on board the EOS platform will provide valuable biophysical information for water-resource modeling.

Remote Sensing of Urban Land Use/Cover

Land use describes society's activities on the land. Land cover describes the biophysical materials covering the landscape. Unlike land cover, land use must be inferred from indirect evidence. In urban areas, land use/cover is heterogeneous because of the high intensity of land utilization caused by the juxtaposition of different activities within a small geographic area. Also, human activities occur in both horizontal and vertical dimensions (e.g., multistory buildings), which complicates image analysis. Changes in the city occur so rapidly that updating of land use/cover information is required at frequent intervals.

Classification Schemes. The most widely adopted land use/cover classification scheme developed for use with remotely sensed data was created by geographers and others at the U.S. Geological Survey (Anderson et al. 1976). Remotely sensed data having a specific spatial resolution can be used to provide land use/cover data at a certain level of detail. For example, to extract Level 3 or 4 urban land use/cover information from remotely sensed data, it is necessary to have imagery with a spatial resolution of ≤ 3 meters (Welch 1982). Generally, when conducting urban land use/cover mapping, spatial resolution is more important than spectral resolution—e.g., the number and size of spectral bands recorded by the remote sensor (Jensen 1983b).

Urban Land Use/Cover Mapping Using a Variety of Remote-Sensor Systems. High-spatial-resolution aerial photography is preferred for the extraction of land use/cover information in the city. Geographers and others at the U.S. Geological Survey have successfully used black and white aerial photography at 1:80,000 scale and color-infrared photography at 1:60,000 with photointerpretation keys to produce urban land use/cover maps for the entire U.S. (Jensen 1983b). Aerial photography has also been used for detecting land-use changes within cities and for generating change statistics (Lindgren 1985). Stereoscopic aerial photography permits a three-dimensional model of the city to be obtained, which can be used to develop an urban information system (Lo 1981).

The extraction of urban land use/cover information from aircraft platforms using thermal-infrared multispectral scanners has been successful but limited, primarily due to the cost of data acquisition. However, general thermal characteristics of individual houses (especially roof-top characteristics) can be routinely monitored using thermal-infrared scanning systems if the environmental and sensor-system parameters are used to calibrate the remotely sensed data (Jensen 1983b).

Geographers have been involved in the application of active microwave remote sensing of urban land cover. Henderson and Anuta (1980) used X-band and K-band real-aperture radar images from different areas of the U.S. and found that settlement detectability was significantly influenced by population size, radar look direction, and depression angle. Henderson and Wharton (1980) produced a Level I urban land use/cover map of Denver, Colorado, by interpreting Seasat SAR images of 1:500,000 scale. Digital analysis of the data could not consistently be used to identify Level I land-use categories. Shuttle Imaging Radar (SIR) data were found to be effective in detecting settlement patterns in the Chinese environment, but less effective for similar environments in the U.S. (Lo 1986b). At the present time, no single configuration of microwave sensor systems is ideal for urban land cover classification (Haack 1984), because the relation between radar backscatter signal and the various land use/cover types is still not fully understood (Bryan 1983).

Because of the relatively poor spatial resolution of Landsat MSS data (79 by 79 m), collateral information is usually required to assist in detecting urban land use/cover. Also, urban land use/cover change detection using Landsat digital data cannot provide results as accurate as those produced by the visual interpretation of large-scale aerial photographs. Nevertheless, this remains an important area of remote sensing, primarily because digital remote-sensing systems are beginning to acquire data with spatial resolutions approaching that of medium-scale aerial photography. Studies by Jensen (1981) and Fung and Le Drew (1987) document the theory and application of change-detection algorithms applied to digital data.

The availability of Landsat TM and SPOT data have produced a surge of research to identify the utility of these data for urban land use/cover mapping. Colwell and Poulton (1985) found that SPOT data provided significantly more urban land use/cover information compared with Landsat MSS and TM data. However, recent studies have encountered problems extracting urban information from the higher-spatial-resolution imagery (Haack, Bryant, and Adams 1987). Digital classification of the urban environment must incorporate more of the cues associated with manual image analysis such as texture, pattern, shape, site, and association.

Future Developments. Urban land use/cover mapping is dependent on high-spatial-resolution data such as aerial photography, combined with a visual-interpretation approach. Interest in nonphotographic imagery such as radar data has been strong, despite less-than-satisfactory results. The digital approach in urban land use/cover mapping using Landsat MSS, TM, and SPOT data has received significant attention by geographers. However, more elements of the visual-interpretation process must be incorporated into the analysis if significant improvement in digital land use/cover mapping is to become a reality.

REMOTE SENSING'S RELATIONS WITH CARTOGRAPHY AND GIS

Satellite remote-sensing systems, digital image processing, and digital database technologies are converging, resulting in significant improvements in the creation of topographic, thematic, and image map products (Welch 1987). For example, most topographic and image maps of 1:100,000 scale or larger are produced today using photogrammetric techniques applied to stereoscopic aerial photographs at scales of 1:20,000 to 1:60,000. The photo resolution at these scales is approximately 20 to 30 line pairs/mm for low-contrast objects, which is equivalent to ground resolutions of 3 to 0.6 m, and permits better than 95% of the required map detail to be compiled directly from the photographs. Thus, the base maps upon which all thematic mapping takes place continue to be derived from remotely sensed data.

In the U.S., cartographers frequently use the root-mean-square-error (RMSE) to measure map accuracy in x, y, and z. The RMSE values can be used to determine whether National Map Accuracy Standards (NMAS) for map products have been met. If meeting national standards is important, NMAS standards greatly influence decisions on appropriate scales for cartographic products that can be produced from remotely sensed image data. For example, many geographic applications require topographic data and the ability to generate 3-D terrain models (Dubayah and Dozier 1986). If x, y, and z terrain coordinates can be extracted from the pixels of an image data set, it may be possible to develop a cartographic database from which planimetric, topographic, thematic, and image maps can be produced by computer techniques. However, the recovery of z-coordinate values from satellite data to accuracies compatible with NMAS is a challenging task, and requires the use of stereoscopic image data of high resolution, recorded at base-to-height ratios ≥ 0.5 (Petrie 1970). Satellite data are best suited to meeting the requirements for topographic maps of 1:50,000 scale or smaller.

In addition to NMAS standards, there is the concept of completeness. The completeness of information on a map prepared from digital-image data is primarily dependent on spatial resolution, as defined by the size of the instantaneous-field-of-view (IFOV) or pixel. For example, representative IFOVs for the Landsat and SPOT sensors are compared to average urban plot sizes for various countries of the world in Figure 2. Urban terrain contains high-frequency detail on satellite image data, and several IFOVs or pixels per land parcel are required before a feature can be reliably identified (Welch 1982). Interpretability improves as the spatial resolution increases, and further gains may be made by using multispectral images and viewing the data in stereo.

Image maps offer geographers an alternative to line maps for many cartographic applications. Experimental image maps and mosaics ranging from 1:250,000 to

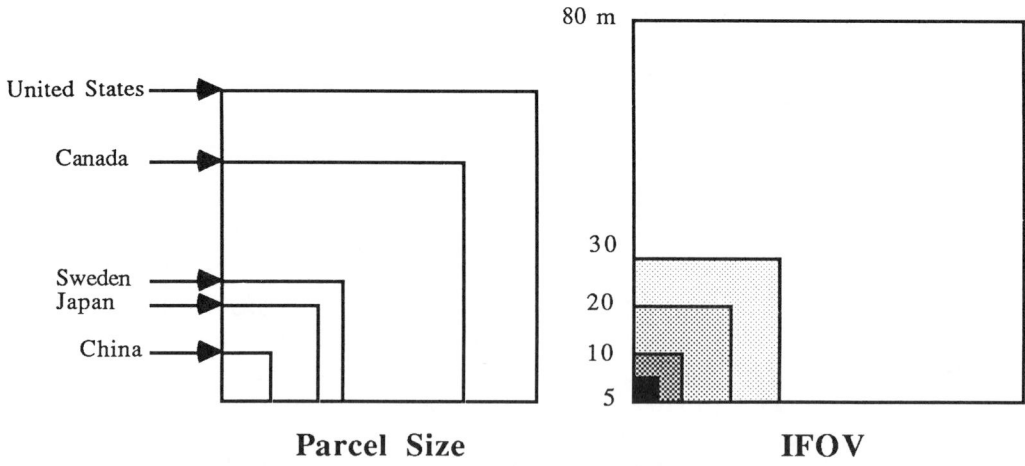

Figure 2
Comparison of urban-land parcel size with the instantaneous-field-of-view (IFOV) of the Landsat MSS (80 m), Landsat TM (30 m), SPOT HRV MSS (20 m), and SPOT HRV Panchromatic (10 m) sensor systems (after Welch 1982)

1:1,000,000 scale were initially produced by USGS personnel from Landsat-1 MSS film images and later from digital image data in computer-compatible tape (CCT) formats. More than 40 satellite image maps have now been produced, including both Landsat MSS (1:250,000 and 1:500,000 scale) and TM (1:100,000) products (Jannace and Ogrosky 1987). The technology has also been adapted by others to produce image maps for specific geographic regions (Dahlberg, Luman, and Stohr 1986). Image maps at scales of 1:25,000 are possible with the improved resolution of SPOT data (Welch 1985). Because image map products can be produced for a fraction of the cost of conventional line maps, they provide the basis for national map series oriented toward the exploration and economic development of the less developed areas of the world, most of which have not been mapped at scales of 1:100,000 or larger.

Geocoded images are becoming increasingly popular, and represent a refinement of the basic image-map process. A geocoded image is obtained by rectifying a digital image to create a data set in which each pixel has an integer dimension (usually in multiples of 5 or 10 m), and is aligned with a standard map-coordinate system (Welch and Usery 1984). Geocoded images are likely to become a major source of information for revising conventional line maps in developed countries. Also, the use of geocoded image data as the primary reference layer in a GIS database is very appealing to geographers, especially when base maps are out-of-date (Welch, Jordan, and Ehlers 1985). A modification of this approach is to merge image data recorded by different sensors with a topographic map, to facilitate revision of the map or extraction of thematic information. The merging of images and maps to form a cartographic database requires that all source material be in digital format, rectified to a standard coordinate reference system, and resampled to a common pixel dimension (Welch 1984). For example, Hallada (1986) created a merged image of the Chernobyl reactor site in the Soviet Union, and Welch and Ehlers (1987) employed similar techniques to merge images of Atlanta, Georgia.

Future Developments

Geographers will increasingly use remotely sensed data in combination with GIS technologies to derive topographic, thematic, and image map products. Factors that will accelerate development include the availability of high-resolution satellite-image data in stereoscopic formats, the anticipated increase in the application of GIS technology, and the growth of educational programs in cartography and remote sensing.

ARTIFICIAL INTELLIGENCE IN REMOTE SENSING

Remote sensing specialists seeking to improve the classification of digital images have begun to examine techniques broadly classified as artificial intelligence (AI), defined as "the art and science of making machines perform tasks which would require intelligence if done by humans" (Estes, Sailer, and Tinney 1986). Most AI research in remote sensing has focused upon expert (or knowledge-based) systems. Expert systems are computer programs that address problems generally considered to be solvable only by human experts who have extensive knowledge about the specific problem being examined. Successful use of expert systems requires that human expert knowledge (facts and rules) about a specific domain of problems be specified and encoded. Human experts must also specify the logic used to draw conclusions. Once the knowledge-base and reasoning strategies have been developed, the system may draw upon an array of numeric, symbolic, and logical "tools" to solve a problem.

Geographic Research: Applying Expert Systems in Remote Sensing

The efforts geographers have made to characterize and classify image-interpretation procedures are of fundamental importance in developing expert systems for application in remote sensing. Campbell (1978) developed a classification of image-interpretation methods. Similarly, Estes, Hajic, and Tinney (1983) identified processes and procedures used in both visual and digital image analysis. Building on these initial efforts, a number of geographers have examined numerical techniques for spatial analysis of images that more closely emulate spatial cues used in visual-image interpretation. For example, Hsu and Burright (1980) and Hodgson and Lloyd (1986) evaluated texture measures, and Gurney and Townshend (1983) provided an overview of the use of context in the classification of digital images and defined the first topology of context measures.

Geographers have also tried to improve digital-image classification using logical strategies to incorporate ancillary data into the classification procedure. For example, Jensen (1978) described a layered approach to land use/land cover classification. This multistep procedure began with segmentation of an image into solid, liquid, and mixture (solid/liquid) strata. Then, in subsequent steps, pixels falling within each stratum were classified via logical-decision "filters." Merchant (1984, 1985) developed a strategy for image classification that involved both spectral and spatial analyses of digital-image data in a fashion designed to emulate certain aspects of visual-image interpretation. He stratified Landsat TM data into water, vegetation, and nonvegetated pixels. Patches of similar land-cover composition were then identified via a "region-growing" algorithm. Classification concluded by analyzing patch size and neighborhood characteristics (e.g., regional diversity and interspersion). Previously inseparable land-use classes were discriminated using this spectral/spatial logic.

Wang et al. (1983) employed "spatial reasoning" to extract ridges, valleys, and stream drainage from Landsat data. Spectral information, an illumination model, and expert logic were used. Hodgson (1987) and Hodgson et al. (1988) developed a logical strategy for improving classification of wetland and old fields using shadow removal and rule-based reclassification procedures. Significant improvements in vegetation classification over results obtained using conventional per-pixel spectral classification methods were obtained.

In other studies, Hsu (1985) evaluated AI techniques for analysis of synthetic-aperture radar data. Enslin, Ton, and Jain (1987) demonstrated how a GIS can be used to help guide image classification. Working with Landsat TM data, they used spatial-image-segmentation procedures, a GIS, and logical rules to identify changes in forest lands and detect new oil and gas wells in Michigan.

Future Developments

To implement expert systems to classify remotely sensed data, greater emphasis must be placed upon knowledge engineering, the process of eliciting knowledge from experts. Experts must specify the attributes, the ways in which they are used, and the underlying logic that defines the order in which they are applied and inferences that might be drawn. Because image interpretation is founded upon application of geographic expert knowledge, geographers (especially those who specialize in remote sensing) should contribute substantially to research in knowledge engineering.

Parallel processing and design of "neural networks" will make feasible many types of spatial-image analyses (e.g., contextual and neighborhood analyses) that are impractical to implement in conventional serial modes of data processing. Similarly, as user-friendly symbolic processing software (i.e., nonnumeric) becomes more widely available, image analysts will be provided new opportunities to explore methods for data analysis that operate on objects or entities defined in images.

Remote-sensing research will be further enhanced by increasingly strong linkages between image processing and GIS. Geographers should examine not only ways in which information derived via remote sensing can be incorporated and used in GIS, but also how such systems can be used to improve image classification through application of the logic and techniques of artificial intelligence (Enslin, Ton, and Jain 1987). Although geographers have made contributions in remote sensing of benefit to AI research, most AI research in image analysis is being conducted by others (e.g., see May 1987 issue of *IEEE Transactions on Geoscience and Remote Sensing*). There is a need for greater involvement by geographers.

MODELING IN REMOTE SENSING

Models used in remote sensing range from conceptual and schematic to physical and deterministic. The focus of this section, however, is upon physical, deterministic models used by geographers in remote sensing. Some of these are spatial in nature, while others are independent of neighborhood and location. Much modeling work in remote sensing has been done by scientists other than geographers, e.g., agronomists, atmospheric physicists, statisticians, foresters, and optical scientists. Most of the truly spatial work, however, has been done by or with geographers.

Model Framework

Strahler, Woodcock, and Smith (1986) describe a framework for modeling in remote sensing. Basically, a remote-sensing model has three components:

1. A *scene model,* which specifies the form and nature of the energy and matter within the scene and their spatial and temporal order
2. An *atmospheric model,* which describes the interaction between the atmosphere and the energy entering and being emitted from the scene
3. A *sensor model,* which describes the behavior of the sensor in responding to the energy fluxes incident upon it and in producing the measurements that constitute the image

Strahler and his colleagues characterize the remote sensing problem as ". . . inferring the order in the properties and distributions of matter and energy in the scene from the set of measurements comprising the image." They suggest that ". . . whether explicit or not, scene inference implies the application of a remote sensing model, in that assumptions must always be made concerning the ground scene, atmosphere, and sensor. The problem of scene inference, then, becomes a problem of model inversion in which the order in the scene is reconstructed from the image and the remote sensing model" (p. 123).

Two types of scene models are recognized: discrete and continuous. In the discrete model, the matter in the scene is taken to consist of discrete objects. The discrete model thus assumes that there are boundaries in the scene where the properties of matter change abruptly over space or through time. Conversely, in the continuous model, the changes in matter are taken to be everywhere continuous in space or time. These two basic types of models condition the approach to information extraction.

For the discrete case, two further types of models are recognized: H- and L-resolution. An H-resolution model is defined as one in which the objects in the scene are larger than the resolution cells; the L-resolution model presents the opposite case. In the H-resolution model, the spatial arrangement of the objects in the scene can be detected directly because individual objects can be identified. In the L-resolution case, the objects are smaller than the resolution cells and are not individually detectable. Thus, the spatial arrangement of the objects can only be detected indirectly.

Remote-sensing models can be contrasted another way, as being either deterministic or empirical. A deterministic remote-sensing model utilizes the physical laws of electromagnetic radiation and describes real processes of energy and matter interaction. An empirical remote-sensing model associates observed sensor measurements with objects or properties in the scene, typically in a statistical fashion. In reality, the terms deterministic and empirical describe endpoints of a continuum of models. Models that are basically deterministic are often formulated with empirical components, and vice versa. The advantage of deterministic models is that they lead to a more direct understanding of the nature of the scene, sensor, and atmosphere, and their joint interactions.

Remote-sensing models may also be either invertible or noninvertible. An invertible model is one in which some of the properties or parameters that characterize the elements are unknown, but can be inferred from the remotely sensed measurements either singly or taken as an image. In a noninvertible model, such estimation of the parameters is not possible or desirable. An invertible model must be a complete remote-sensing model in that scene, atmosphere, and sensor models are required even if they are not explicitly specified. This arises because scene inference begins with the measurements and ends with the scene. The distinction between invertible and noninvertible models may be somewhat arbitrary. This arises because modern analytical techniques allow inversion of rather complex models which are driven by many parameters, given appropriate side constraints, such as least squares or maximum likelihood. Thus, models that are not explicitly formulated to be invertible can often be inverted under proper conditions.

Examples of Modeling in Remote Sensing

Illustrative of the type of deterministic, physically based models in remote sensing that have been recently developed by geographers are the atmospheric models of Dozier and Frew (1981) and Li, Wan, and Dozier (1987). In the former paper, the authors developed wavelength-dependent atmospheric-correction methods for digital imagery taken from mountainous terrain. The correction methods required precise registration of the digital image to an accurate digital-terrain model, and utilized estimates of atmospheric aerosol and water-vapor content in a physical model to yield surface-exitance data. The latter paper relied on a three-component model for upwelling radiation as seen by a satellite sensor. Using radiative-transfer theory and computer codes, the authors derived an atmospheric point-spread function and used a deconvolution procedure to retrieve the radiance of the surface.

In another example, Stow (1987b) investigated the use of fluorescent dye, chlorophyll pigments, suspended sediments, surface temperature, and salinity as tracers to reveal surface-flow patterns in nearshore marine waters. This appraisal required accurate modeling of atmospheric effects, flow patterns, and surface-interaction mechanisms.

Another area of physical modeling in remote sensing that has concerned geographers is the remote sensing of snow. Dozier and Warren (1982) used radiative-transfer theory to establish the importance of view angle on thermal emissivity of snow. In a later study, Davis, Dozier, and Chang (1987) used snow-microstructure and snow-wetness parameters in a microwave-emission model that combined radiative-transfer methods with Mie theory. In studies more directly related to applications, Dozier and Marks (1987) and Dozier (1987) combined atmospheric and snow-reflectance models with digital terrain data to develop typical spectral signatures for several classes of snow covers as they might appear in Thematic Mapper imagery.

Geographers have also been active in modeling remotely sensed images of plant canopies. Li and Strahler (1986) devised a series of geometric-optical models of conifer forests as three-dimensional cones on a contrasting background. Some of their models have been oriented toward inversion of Landsat, TM, or SPOT data to yield height and spacing of trees from interpixel variance of brightness values within a timber stand. Franklin and Strahler (1989) recently extended the invertible work to ellipsoidal canopies in African woodlands. In the radar domain, Richards, Sun, and Simonett (1987) developed a three-component model that accounted for volume scattering of leaves and branches, surface scattering from the ground, and corner reflection from trunks for forest canopies. On quite a different temporal and spatial scale, Goward (1987) developed models for predicting and mapping mean monthly air temperature, precipitation, and the amount of photosynthetically active radiation absorbed by vegetation cover for large areas (e.g., North America) from AVHRR Vegetation Index data.

Another area of modeling in remote sensing that is being developed by geographers is that of spatial modeling and spatial statistics of images. Recent work by Jupp, Strahler, and Woodcock (1988) provides a theoretical background for the study of spatial structure in digital images. Their report shows how the spatial autocorrelation functions of digital images depend on the underlying spatial autocorrelation pattern of the scene and the averaging effect of sensing within a finite field of view. In related work, Woodcock and Strahler (1987) investigated how local variance (variance of brightness values within a small neighborhood) is related to the resolution cell size of an image and the size of objects in the scene. More recently, Woodcock, Strahler, and Jupp (1988) explored the use of variograms in remote sensing as tools to characterize such ground-scene parameters as the size distribution and density of objects.

Future Developments

The empirical relations that have sustained many applications of remote sensing in the past are based on the spatial structure and characteristics of the matter contained within the remotely sensed scene, how that matter interacts with electromagnetic energy, and how that energy interacts with a sensor to form a digital image. Models enhance our understanding of that interaction, and the more deterministic the model, the more intensely it focuses attention upon what is really happening in the remote-sensing process. Thus models—especially deterministic models—play a key role, not only in the development of the field, but in how researchers think about the remote-sensing problem. Geographers are perhaps most aware of the spatial aspects of models, and it is in this area that geographers can be most helpful in advancing the field.

THE FUTURE OF REMOTE SENSING IN GEOGRAPHY

Four polar-orbiting platforms are projected to be launched as part of the EOS space station in the 1990s. These platforms will produce more than one terabyte of information per day. Geographers will have high-resolution spatial, spectral, and temporal satellite data available for essentially all areas of the globe. The systems used to process and analyze remotely sensed data will become faster and more powerful. In addition, geographers will expand their use of ancillary data, and will integrate elements of modeling, artificial intelligence, and the use of GIS into the processing of remotely sensed data. An integrating concept will be telescience, which allows a scientist to be networked to remote-sensing space systems, data archives, and other researchers. Telescience will allow geographers to accomplish the type of interdisciplinary, multi-institutional, multinational research required to truly achieve an improved understanding of environmental processes at scales from regional to global.

Remote-Sensing Science and Applications

Research continues to yield basic new knowledge about the Earth and to expand our understanding of the applications of remote sensing. Major new programs aimed at the study of global change are being initiated by organizations such as the International Council of Scientific Unions. One program focuses upon Earth Systems Science, which attempts to achieve a scientific understanding of the entire Earth system at a global scale. Remote sensing combined with surface observations will be employed within these studies in an information-system context (NASA 1988). Federal agencies that fund remote sensing, i.e., NASA, NOAA, USGS, and NSF, are supporting these efforts. Applications of remote-sensing technology will also remain strong at local and regional levels, funded by local, state, and federal agencies. The NASA remote-sensing commercialization program, which calls for joint NASA and user funding of commercially oriented research, will also provide expanded opportunities for research and employment of geographers.

REMOTE SENSING AND THE DISCIPLINE OF GEOGRAPHY

Geographers have made significant contributions to the field of remote sensing and will continue to do so. What is less clear is the impact that remote sensing has had upon the discipline of geography. That remote sensing can be employed for measurement, mapping, monitoring, and modeling of environmental phenomena and processes is undeniable. However, the majority of geographers do not use remote sensing in their research. They fail to realize that remote-sensing technology has the potential to be a powerful tool in accomplishing the very mission of geography itself, i.e., it provides the means to achieve an improved understanding of humanity's relationship to planet Earth.

Other disciplines have seen this, and the Earth System Science initiative is the result. In this initiative, remote sensing is destined to play a significant role. Yet few geographers were involved in the development of this initiative, which should be a core focus of our discipline.

Perhaps this lack of initiative on the part of the leadership of our discipline can be correlated with the fact that most of the geographers who are conducting remote-sensing research do not publish their results in geography journals. Rather, they publish them in remote-sensing journals, which most geographers do not read. Therefore, the majority of geographers are unaware of the tremendous strides that remote-sensing technology has made in the last two decades, and how it can be applied to their research.

Remote sensing can facilitate improved measurement of a significant range of biophysical, geochemical, and socioeconomic phenomena of use to both physical and cultural geographers. Remote sensing allows us to monitor change and can provide a level of input to models that is far more accurate and based less on assumed variables than was previously possible. The future of remote sensing within the discipline depends on how its specialists educate geographers to understand remote sensing's potential and how to apply it to geographic problems. To conclude, we can do no better than to quote the final paragraph from Estes, Jensen, and Simonett's article, "Impacts of Remote Sensing on U.S. Geography," published in 1980:

> The exploitation of the improved or unique information available to the geographer via the application of remote sensing techniques has barely begun. Yet, when thoughtfully analyzed it can be seen to provide the geographer with significant improvements in the quantity, quality, and timeliness of data required. As more geographers become aware of the significant implications of remote sensing for providing such data, the true impact of this technique in the discipline will be felt. Remote sensing, like cartography, is approaching such a state of technology and a body of coherent knowledge and theory that it can almost be viewed as a discipline in and of itself. What is required to increase the impact of remote sensing in geography is a concerted effort on the part of geographers and others who specialize in remote sensing to conduct their research thoughtfully so as to more effectively impact their disciplines.

This statement is still true today.

REFERENCES

Ambrosia, V. G., and Brass, J. A. 1988. Thermal analysis of wildfires and effects on global ecosystem cycling. *Geocarto International–A Multidisciplinary Journal of Remote Sensing* 3:29–40.

Anderson, J. R.; Hardy, E. E.; Roach, J. T.; and Witmer, R. E. 1976. A land use and land cover classification system for use with remote sensor data. *U. S. Geological Survey Professional Paper* No. 964.

Avery, T. E., and Berlin, G. L. 1985. *Interpretation of aerial photographs*. Minneapolis, MN: Burgess Publishing.

Barry, R. G. 1983. Research on snow and ice. *Review of Geophysics and Space Physics* 21:765–76.

Baumgardner, M. F.; Silva, L. F.; Biehl, L. L.; and Stoner, E. R. 1985. Reflectance properties of soil. *Advances in Agronomy* 38:1–44.

Blanchard, B. J., and O'Neill, P. F. 1983. Estimation of the hydraulic character of soils with passive microwave systems. *Proceedings, Conference on Advances in Infiltration, American Society of Civil Engineers,* p. 215.

Bryan, M. L. 1983. Urban land use classification using synthetic aperture radar. *International Journal of Remote Sensing* 4:215–33.

Buckingham, W. F., and Sommer, S. E. 1983. Mineralogical characterization of rock surfaces formed by hydrothermal alteration and weathering–Application to remote sensing. *Economic Geology* 78:664–74.

Campbell, J. B. 1978. A geographical analysis of image interpretation methods. *Professional Geographer* 30:264–69.

———. 1983. *Mapping the land: Aerial imagery for land use information*. Washington: Association of American Geographers.

———. 1987. *Introduction to remote sensing*. New York: Guilford Press.

———, and Edmonds, W. T. 1984. The missing geographic dimension to soil taxonomy. *Annals of the Association of American Geographers* 74:83–97.

Chang, S., and Collins, W. 1983. Confirmation of the airborne bio-geophysical mineral exploration technique using laboratory methods. *Economic Geology* 78:723–36.

Colwell, R. N., ed. 1960. *Manual of photographic interpretation*. Falls Church, VA: American Society of Photogrammetry.

———, ed. 1983. *Manual of remote sensing*. 2d ed. Falls Church, VA: American Society for Photogrammetry and Remote Sensing.

———, and Poulton, C. 1985. SPOT simulation imagery for urban monitoring: A comparison with Landsat TM and MSS imagery and with high altitude color infrared photography. *Photogrammetric Engineering and Remote Sensing* 51:1093–1101.

Condit, H. R. 1970. The spectral reflectance of American soils. *Photogrammetric Engineering and Remote Sensing* 36:955–66.

Curran, P. J. 1985. *Principles of remote sensing*. New York: Longman.

———. 1987. Remote sensing methodologies and geography. *International Journal of Remote Sensing* 8:1255–75.

Dahlberg, R. E.; Luman, D. E.; and Stohr, C. J. 1986. Creation of a satellite image map of Illinois: A case study of technology transfer and linkage. *Cartographica* 23:14–28.

Davis, P. A.; Berlin, G. L.; and Chavez, P. S. 1987. Discrimination of altered basaltic rocks in the southwestern United States by analysis of Landsat Thematic Mapper data. *Photogrammetric Engineering and Remote Sensing* 53:45–55.

Davis, R. E.; Dozier, J.; and Chang, A. T. C. 1987. Snow property measurements correlative to microwave emission at 35 GHz. *IEEE Transactions on Geoscience and Remote Sensing* GE-25:751–57.

Dozier, J. 1984. Snow reflectance from Landsat-4 Thematic Mapper. *IEEE Transactions on Geoscience and Remote Sensing* GE-22:323–30.

———. 1987. Remote sensing of snow characteristics in the southern Sierra Nevada. Large scale effects of seasonal snow cover. *Proceedings, Vancouver Symposium, IAHS Pub. No. 166,* pp. 305–14.

———, and Frew, J. 1981. Atmospheric corrections to satellite radiometric data over rugged terrain. *Remote Sensing of Environment* 11:191–205.

———, and Marks, D. 1987. Snow mapping and classification from Landsat Thematic Mapper data. *Annals of Glaciology* 9:97–103.

———, and Strahler, A. H. 1983. Ground investigations in support of remote sensing. In *Manual of remote sensing,* ed. R. N. Colwell, pp. 959–86. Falls Church, VA: American Society for Photogrammetry and Remote Sensing.

———, and Warren, S. G. 1982. Effect of viewing angle on the infrared brightness temperature of snow. *Water Resources Research* 18:1424–34.

———; Schneider, S. R.; and McGinnis, D. F. 1981. Effect of grain size and snow pack water equivalence on visible and near-infrared satellite observations of snow. *Water Resources Research* 17:1213.

Dubayah, R. O., and Dozier, J. 1986. Orthographic terrain views using data derived from digital elevation models. *Photogrammetric Engineering and Remote Sensing* 52:509–18.

EOS Science Steering Committee. 1987. From pattern to process: The strategy of the Earth Observing System. Washington: NASA.

Enslin, W. R.; Ton, J.; and Jain, A. 1987. Land cover change detection using a GIS-guided, feature-based classification of Landsat Thematic Mapper Data. *Proceedings, American Society for Photogrammetry and Remote Sensing* 6:108–20.

Estes, J. E. 1987. Remote sensing and biomes special issue. *Geocarto International–A Multidisciplinary Journal of Remote Sensing* 2:3–4.

———; Jensen, J. R.; and Tinney, L. R. 1978. Remote sensing of agricultural water demand information: A California study. *Water Resources Research* 14:170–76.

———; ———; and Simonett, D. S. 1980. Impacts of remote sensing on U.S. geography. *Remote Sensing of Environment* 10:43–80.

———; Hajic, E. J.; and Tinney, L. R. 1983. Fundamentals of image analysis: Visible and thermal infrared data. Chapter 24 in *Manual of remote sensing*, ed. R. N. Colwell, pp. 987–1125. Falls Church, VA: American Society for Photogrammetry and Remote Sensing.

———; Crippen, R. E.; and Starr, J. L. 1985. Natural oil seep detection in the Santa Barbara Channel, California with shuttle imaging radar. *Geology* 13:282–84.

———; Sailer, C.; and Tinney, L. R. 1986. Applications of artificial intelligence techniques to remote sensing. *Professional Geographer* 38:133–41.

———, and Starr, J. L. 1986. Support for global science: Remote sensing challenge. *Geocarto International–A Multidisciplinary Journal of Remote Sensing* 1:3–14.

Foster, J. L.; Hall, D. K.; Chang, A. C.; and Rango, A. 1984. An overview of passive microwave snow research and results. *Reviews of Geophysics and Space Physics* 22:195.

Frank, T. D. 1984. The effect of change in vegetation cover and erosion patterns on albedo and texture of Landsat images in a semiarid environment. *Annals of the Association of American Geographers* 74:393–407.

———. 1985. Differentiating semiarid environments using Landsat reflectance indexes. *Professional Geographer* 37:1;36–46.

Franklin, J. 1986. Thematic mapper analysis of coniferous forest structure and composition. *International Journal of Remote Sensing* 7:1287–1301.

———; Logan, T. L.; Woodcock, C. E.; and Strahler, A. H. 1986. Coniferous forest classification and inventory using landsat digital terrain data. *IEEE Transactions on Geoscience and Remote Sensing* GE-24:139–49.

———, and Strahler, A. H. 1989. Invertible canopy reflectance modeling of vegetation structure in semi-arid woodland. *IEEE Transactions on Geoscience and Remote Sensing* 26:809–25.

Fung, T., and Le Drew, E. 1987. Application of principal components analysis to change detection. *Photogrammetric Engineering and Remote Sensing* 53:1649–58.

Fussell, J.; Rundquist, D.; and Harrington, J. A. 1986. On defining remote sensing. *Photogrammetric Engineering and Remote Sensing* 52:1507–11.

Gagliardini, D. A.; Karszenbaum, H.; Legeckis, R.; and Klemas, V. 1984. Application of Landsat MSS, NOAA/TIROS AVHRR, and Nimbus CZCS to study the La Plata River and its interaction with the ocean. *Remote Sensing of Environment* 15:21–36.

Goward, S. N. 1987. Evaluating North American net primary productivity with satellite observation. *Advances in Space Research* 7:165–74.

———; Dye, D.; Kerber, A.; and Kalb, V. 1987. Comparisons of North and South American Biomes from AVHRR observations. *Geocarto International–A Multidisciplinary Journal of Remote Sensing* 2:27–39.

———, and Taranik, J. V. 1986. *Commercial applications and scientific research requirements for thermal infrared observations of terrestrial surfaces*. Washington: NASA.

Gurney, C., and Townshend, J. R. G. 1983. The use of contextual information in the classification of remotely sensed data. *Photogrammetric Engineering and Remote Sensing* 49:55–64.

Haack, B. N. 1984. L- and X-band and cross-polarized synthetic aperture radar for investigating urban environments. *Photogrammetric Engineering and Remote Sensing* 50:331–40.

———; Bryant, N.; and Adams, S. 1987. An assessment of Landsat MSS and TM data for urban and near-urban land-cover digital classification. *Remote Sensing of Environment* 21:201–13.

Hallada, W. 1986. Cover photo. *Photogrammetric Engineering and Remote Sensing* 52.

Henderson, F. M., and Anuta, M. 1980. Effects of radar system parameters, population, and environmental modulation settlement visibility. *International Journal of Remote Sensing* 1:137–51.

———, and Wharton, S. W. 1980. Seasat SAR identification of dry climate urban land cover. *International Journal of Remote Sensing* 1:293–304.

Hodgson, M. E. 1987. The use of knowledge-based reclassifiers in mapping land cover classes. Ph.D. diss., University of South Carolina, Columbia.

———, and Lloyd, R. E. 1986. A cognitive measure of texture in imagery. *Proceedings, American Society for Photogrammetry and Remote Sensing*, pp. 406–16.

———; Jensen, J. R.; Pinder, J.; Collins, B. S.; and Mackey, H. E. 1988. Vegetation mapping using high resolution MSS Data: Canopy reflectance and shadow problems encountered and possible solutions. *Proceedings, American Society for Photogrammetry and Remote Sensing* 4:32–39.

Holz, R. K., ed. 1985. *The surveillant science: Remote sensing of the environment*. New York: Wiley.

Hsu, S. 1985. Applications of stable structure theory and artificial intelligence approaches to feature extraction and terrain analysis with synthetic aperture radar image data. *Proceedings, American Society for Photogrammetry and Remote Sensing*, pp. 499–508.

———, and Burright, R. G. 1980. Texture perception and the RADC/Hsu texture features. *Photogrammetric Engineering and Remote Sensing* 46:1051–58.

Hutchinson, C. F. 1982. Techniques for combining Landsat ancillary data for digital classification improvement. *Photogrammetric Engineering and Remote Sensing* 48:123–30.

Jackson, P. L., and Rosenfeld, C. L. 1987. Erosional changes at Alsea Spit, Waldport, Oregon. *Oregon Geology* 49:55–59.

Jannace, R., and Ogrosky, C. 1987. Cartographic programs and products of the U.S. Geological Survey. *American Cartographer* 14:197–202.

Jensen, J. R. 1978. Digital land cover mapping using layered clarification logic and physical composition attributes. *American Cartographer* 5:121–32.

———. 1981. Urban change detection mapping using Landsat digital data. *American Cartographer* 8:127–47.

———. 1983a. Biophysical remote sensing. *Annals of the Association of American Geographers* 73:111–32.

———, 1983b. Urban/suburban land use analysis. In *Manual of remote sensing*, 2d ed., ed. R. N. Colwell, pp. 1571–1666. Falls Church, VA: American Society for Photogrammetry and Remote Sensing.

———. 1986. *Introductory digital image processing: A remote sensing perspective*. Englewood Cliffs, NJ: Prentice-Hall.

———; Estes, J. E.; Bowden, L. W.; and Tinney, L. R. 1977. Remote sensing of water demand information. *Geographical Review* 67:322–34.

———, and Chery, D. L. 1980. Landsat crop identification for watershed water balance determinations. *International Journal of Remote Sensing* 1:345–59.

———; Hodgson, M. E.; Christensen, E. J.; Mackey, H. E.; and Sharitz, R. 1986. Remote sensing inland wetlands: a multispectral approach. *Photogrammetric Engineering and Remote Sensing* 52:87–100.

———, and Christensen, E. J. 1986. Solid and hazardous waste disposal site selection using digital geographic information system techniques. *Science of the Total Environment* 56:265–76.

Johnson, G. E.; van Dijk, A.; and Sakamoto, C. M. 1987. The use of AVHRR data in operational agricultural assessment in Africa. *Geocarto International–A Multidisciplinary Journal of Remote Sensing* 2:41–60.

Jupp, D. L. B.; Strahler, A. H.; and Woodcock, C. E. 1988. Autocorrelation and regularization in digital images: I. Basic theory, and II. Simple image models. *IEEE Transactions on Geoscience and Remote Sensing*. In press.

Klemas, V. 1980. Remote sensing of coastal fronts and their effects on oil dispersion. *International Journal of Remote Sensing* 1:11–18.

———; Davis, G.; Lackie, J.; Whelan, W.; and Tornatore, G. 1977. Satellite, aircraft and drogue studies of coastal currents and pollutants. *IEEE Transactions on Geoscience Electronics* GE-15:97–108.

Lavoie, A.; Bonn, F.; Dubois, J. M.; and El-Sabh, M. I. 1985. Structure thermique et variablite du courant de surface de l'estuaire maritime du Saint-Laurent a l'aide d'images du satellite HCMM. *Canadian Journal of Remote Sensing* 11:70–84.

Leatherman, S. P. 1983. Shoreline mapping: A comparison of techniques. *Shore and Beach* 51:28–33.

———. 1987. Coastal geomorphological applications of ground penetrating radar. *Journal of Coastal Research* 3:397–99.

Li, S.; Wan, Z. M.; and Dozier, J. 1987. A component decomposition model for evaluating atmospheric effects in remote sensing. *Journal of Electromagnetic Waves and Applications* 4:323–47.

———, and Strahler, A. H. 1986. Geometric–optical modeling of a conifer forest canopy. *IEEE Transactions on Geoscience and Remote Sensing* GE-23:705–21.

Lindgren, D. T. 1985. *Land use planning and remote sensing.* Boston: Martinus Nijhoff.

Lo, C. P. 1981. A photogrammetric urban information system using the Cp-1 plotter. *Photogrammetric Record* 57:311–29.

———. 1986a. *Applied remote sensing.* New York: Longman Scientific & Technical.

———. 1986b. Human settlement analysis using Shuttle Imaging Radar-A data: An evaluation. In *Remote sensing for resource development and environmental management,* ed. M. J. Damen, pp. 841–45. Rotterdam: A. A. Balkema.

Loveland, T. R., and Johnson, G. E. 1983. The role of remotely sensed and other spatial data for predictive modeling; The Umatilla, Oregon example. *Photogrammetric Engineering and Remote Sensing* 49:1183–92.

Lulla, K. 1983. The Landsat satellites and selected aspects of physical geography. *Progress in Physical Geography* 7:1–45.

———. 1985 Some observations on geobotanical remote sensing and mineral prospecting. *Canadian Journal of Remote Sensing* 11:17–38.

———; Kramber, W.; Skelton, D.; and Mausel, P. 1987. Video-based vegetation indices. *Proceedings, Workshop on Aerial Photography and Videography.* Falls Church, VA: American Society for Photogrammetry and Remote Sensing, pp. 270–9.

Lusch, D., and Sappio, F. J. 1987. Mapping gypsy moth defoliation in Michigan. *Proceedings, Workshop on Aerial Photography and Videography.* Falls Church, VA: American Society for Photogrammetry and Remote Sensing, pp. 261–69.

Merchant, J. W. 1984. Using spatial logic in classification of Landsat TM data. *Proceedings, Pecora 9 Symposium.* Falls Church, VA: American Society for Photogrammetry and Remote Sensing, pp. 378–85.

———. 1985. Employing geographic reasoning in landscape mapping. *Proceedings, Pecora 10 Symposium.* Falls Church, VA: American Society for Photogrammetry and Remote Sensing, pp. 557–66.

Michaelsen, J.; Zhang, X.; and Smith, R. C. 1988. Variability of pigments biomass in the California Current system as determined by satellite imagery. Part 2. Temporal variability. *Journal of Geophysical Research.* In press.

Miles, M.; Stow, D. A.; and Jones, J. P. 1988. Incorporating spatially-varying parameter methods into remote sensing-based water quality modeling. In *Expansion methods in geography,* eds. E. Casseti and J. P. Jones. In press.

Mohler, R. J.; Palmer, W. F.; Baker, T. C.; and Bizzell, R. M. 1986. An automatic procedure for estimating spring small grains acreage from Landsat data. *Professional Geographer* 38:375–82.

NASA. 1988. *Earth system science: A closer view.* Washington: NASA.

Peterson, D. L.; Westman, W. E.; Stephenson, N. L.; Ambrosia, J. A.; Brass, J. A.; and Spanner, M. A. 1986. Analysis of forest structure using Landsat Thematic Mapper data. *IEEE Transactions on Geoscience Electronics* GE-24:113–21.

Petrie, G. 1970. Some considerations regarding mapping from earth satellites. *Photogrammetric Record* 6:590–624.

Place, J. L. 1985. Mapping of forested wetland: Use of Seasat radar images to complement conventional sources. *Professional Geographer* 37:463–69.

Ramsey, E., and Jensen, J. R. 1988. A comparison of original aircraft MSS and generated surface water reflectance images as predictors of lake water quality indicators. *Proceedings, American Society for Photogrammetry and Remote Sensing,* pp. 39–50.

Rich, E. 1983. *Artificial intelligence.* New York: McGraw-Hill.

Richards, J. A.; Sun, G. Q.; and Simonett, D. S. 1987. L-band radar backscatter modeling of forest stands. *IEEE Transactions on Geoscience and Remote Sensing* GE-25:487–98.

Richardson, A. J., and Weigand, C. L. 1977. Distinguishing vegetation from soil background. *Photogrammetric Engineering and Remote Sensing* 43:1541–52.

Richason, B. F., ed. 1978. *Introduction to remote sensing of the environment.* Dubuque, IA: Kendall/Hunt.

Ripple, W. J. 1986. Spectral reflectance relationships to leaf water stress. *Photogrammetric Engineering and Remote Sensing* 52:1669–75.

Rundquist, D. R. 1985. Preliminary evaluation of AIS spectra. *Proceedings, Airborne Imaging Spectrometer Data Analysis Workshop.* Pasadena, CA: Jet Propulsion Lab, pp. 74–78.

———; Lawson, M. P.; Queen, L.; and Cerveny, R. S. 1987. The relationship between summer-season rainfall events and lake-surface area. *Water Resources Bulletin* 23:493–508.

Schmugge, T. J., and Sellers, P. 1986. *Experimental plan for the first ISLSCP field experiment.* Greenbelt, MD: Laboratory for Terrestrial Physics, NASA Goddard Space Flight Center.

———; O'Neill, P. E.; and Wang, J. 1986. Passive microwave soil moisture research. *IEEE Transactions on Geoscience and Remote Sensing* GE-24:12–20.

Simonett, D. S.; Reeves, R. G.; Estes, J. E.; Bertke, S. E.; and Sailer, C. 1983. Development and principles of remote sensing. In *Manual of remote sensing,* 2d ed., ed. R. N. Colwell, pp. 1–35. Falls Church, VA: American Society for Photogrammetry and Remote Sensing.

Slack, R., and Welch, R. 1980. Soil conservation service runoff curve number estimates from Landsat data. *Water Resources Bulletin* 16:887–93.

Smith, R. C., and Baker, K. S. 1982. Oceanic chlorophyll concentrations as determined by satellite (Nimbus-7 Coastal Zone Color Scanner). *Journal of Marine Biology* 66:269–79.

———; Eppley, R. W.; and Baker, K. S. 1982. Correlation of primary production as measured aboard ship in Southern California coastal waters and as estimated from satellite chlorophyll images. *Journal of the Marine Biological Association* 66:282–88.

———; Brown, O. B.: Hoge, F. E.; Baker, K. S.; Evans, R. H.; Swift, R. N.; and Esaias, W. E. 1987. Multiplatform sampling (ship, aircraft and satellite) of a Gulf Stream warm core ring. *Applied Optics* 26:2068–81.

———, Zhang, X., and Michaelsen, J. 1988. Variability of pigment biomass in the California current system as determined by satellite imagery. Part 1. Spatial variability. *Journal of Geophysical Research.* In press.

Stow, D. A. 1987a. Numerical derivation of a hydrodynamic surface flow field from time sequential remotely sensed data. *Remote Sensing of Environment* 23:1–22.

———. 1987b. Remotely-sensed tracers for hydrodynamic surface flow estimation. *International Journal of Remote Sensing* 8:261–78.

———, and Estes, J. E. 1983. Status of remote sensing for coastal zone management. *Proceedings, Coastal Zone 1983,* American Society of Civil Engineers, San Diego, CA: pp. 565–74.

Strahler, A. H. 1981. Stratification of natural vegetation for forest and rangeland inventory using Landsat digital imagery and collateral data. *International Journal of Remote Sensing* 2:15–41.

———; Woodcock, C. E.; and Smith, J. A. 1986. On the nature of models in remote sensing. *Remote Sensing of Environment* 20:121–39.

Thomas, A. W.; Welch, R.; and Jordan, T. R. 1986. Quantifying concentrated-flow erosion on cropland with aerial photogrammetry. *Journal of Soil and Water Conservation* 41:249–52.

Townshend, J. R. G., ed. 1981. *Terrain analysis and remote sensing.* London: Allen & Unwin.

Walker, D. A.; Acevedo, W.; Everett, K. R.; Gaydos, L.; Brown, J.; Webber, P. J. 1982. Landsat-assisted environmental mapping in the Arctic National Wildlife Refuge, Alaska. *CRREL* Report No. 82-37. Hanover, NH: U.S. Army Cold Regions Research and Engineering Laboratory, 52 pp.

Walsh, S. J. 1980. Coniferous tree species mapping using Landsat data. *Remote Sensing of Environment* 9:11–26.

———. 1987. Variability of Landsat MSS spectral responses of forests in relationship to stand and site characteristics. *International Journal of Remote Sensing* 8:1289–99.

Wang, S.; Elliott, D. B.; Campbell, J. B.; Erich, R. W.; and Haralick, R. M. 1983. Spatial reasoning in remotely sensed data. *IEEE Transactions on Geoscience and Remote Sensing* GE-21:94–101.

Warren, P. L., and Hutchinson, C. F. 1984. Indicators of rangeland change and their potential for remote sensing. *Journal of Arid Environments* 7:107–26.

Welch, R. 1982. Spatial resolution requirements for urban studies. *International Journal of Remote Sensing* 3:139–46.

———. 1984. Merging Landsat 4 and SIR-A image data in digital formats. *Photomethods* (July):11–12.

———. 1985. Cartographic potential of SPOT image data. *Photogrammetric Engineering and Remote Sensing* 51:1085–91.

———. 1987. Integration of photogrammetric, remote sensing and database technologies for mapping applications. *Photogrammetric Record* 12:409–28.

———, and Jordan, T. R. 1983. Analytical non-metric close-range photogrammetry for monitoring stream channel erosion. *Photogrammetric Engineering and Remote Sensing* 49:367–74.

———, and Usery, E. L. 1984. Cartographic accuracy of Landsat-4 MSS and TM image data. *IEEE Transactions on Geoscience and Remote Sensing* GE-22:281–88.

———; Jordan, T. R.; and Ehlers, M. 1985. Comparative evaluations of geodetic accuracy and cartographic potential of Landsat-4 and Landsat-5 thematic mapper image data. *Photogrammetric Engineering and Remote Sensing* 51:1249–62.

———, and Ehlers, M. 1987. Merging multiresolution SPOT HRV and Landsat TM data. *Photogrammetric Engineering and Remote Sensing* 53:301–3.

Weismiller, R. A.; Persinger, I. D.; and Montgomery, O. L. 1977. Soil inventory from digital analysis of satellite and topographic data. *Journal of the Soil Science Society of America* 41:1166–70.

Williams, D. L., and Miller, L. D. 1979. *Monitoring forest canopy alteration around the world with digital analysis of Landsat imagery*. Greenbelt, MD: NASA Goddard Space Flight Center.

———, and Nelson, R. F. 1986. Use of remotely sensed data for assessing forest stand conditions in the Eastern United States. *IEEE Transactions on Geoscience and Remote Sensing* GE-24:130–38.

Woodcock, C. E., and Strahler, A. H. 1987. The factor of scale in remote sensing. *Remote Sensing of Environment* 21:311–32.

———; Strahler, A. H.; and Jupp, D. L. B. 1988. The use of variograms in remote sensing. I and II-scene models, simulated images, and real digital images. *Remote Sensing of Environment* 25:323–80.

Yool, S. R.; Eckhardt, D. W.; Estes, J. E.; and Cosentino, M. J. 1985. Describing the brushfire hazard in southern California. *Annals of the Association of American Geographers* 75:417–30.

Zheng, Q.; Klemas, V.; and Huang, N. E. 1984. Dynamics of the slope water off New England and its influence on the Gulf Stream as inferred from satellite IR data. *Remote Sensing of Environment* 15:135–53.

Geographic Information Systems

Nicholas R. Chrisman | David J. Cowen |
Peter F. Fisher | Michael F. Goodchild | David M. Mark

A Geographic Information System is an integrated package for the input, storage, analysis, and output of spatial information. Of these four components, the geographer would probably see analysis as the most significant, implying as it does the ability to use a digital computer to explore, manipulate, and examine spatial data in ways that go far beyond conventional capabilities using paper maps. To Abler (1988, 137), "GIS are simultaneously the telescope, the microscope, the computer, and the xerox machine of regional analysis and synthesis." Thus, from a geographical perspective, we might define a GIS as a system for support of spatial analysis, in analogy to the role of such well-known packages as SAS and SPSS in supporting statistical analysis.

The term GIS was first used in the mid-1960s, but it was not until the 1980s that a combination of reduced hardware costs, accumulated software development, and rapidly developing applications led to the explosion of interest that we see today. In a market economy, one measure of the importance of a sector is gross revenue. Ten years ago no such measure was available for the GIS industry, but it would be surprising if the sector accounted for more than a few million dollars. Recently, a Dataquest survey placed 1986 revenue at $115 million and estimated 1991 revenue at $464 million. While the connections between gross revenue in industry and academic activity are far from clear, such rates of increase inevitably place pressures on the research and training sectors generally. The new industry is directly related to the power of GIS software to integrate and synthesize all forms of spatial data, extending the computer well beyond the roles of electronic filing cabinet and automated drafting system. Because of its concentration on the integration and presentation of diverse forms of spatial information, GIS can be viewed as a set of processes and procedures that approach the heart of modern geography (Abler 1988).

As a collection of tools, geographic information systems can serve a number of applications. While geographers have been prominent in developing these tools, their efforts have been combined with those of others having backgrounds in a broad range of disciplines including computer science, surveying engineering, landscape architecture, remote sensing, soil science, forestry, and environmental studies. Such multidisciplinary interest creates a vigorous conceptual background for GIS research.

This chapter is structured as follows:

- ☐ The first section reviews the origins of GIS, from intellectual roots which stretch far back into geography and cartography and theories of spatial information.
- ☐ The second section reviews definitions of GIS, and attempts to place conceptual bounds on what is clearly a loosely defined assortment of technologies and applications.
- ☐ The third section examines the forces that drive current interest in GIS, from industry, applications, and academic disciplines.
- ☐ A fourth section reviews the current status of the field through its applications, using a series of case studies.
- ☐ The final section discusses current and future trends.

We hope these several cross-cutting and overlapping themes will clarify the significance of this burgeoning, multidisciplinary field.

ORIGINS OF GIS

Though geographers study diverse phenomena, the common characteristics of spatial information create some generality in the geographer's methods. Consider Nystuen's (1963) fundamental spatial concepts of distance, direction, orientation, and connectivity in the modern era of computers and GIS. Berry's 1964 article "Approaches to Regional Analysis: A Synthesis" reduced geographical analysis to 10 major operations that can be performed on a geographical matrix of places and variables. These can be further reduced to studies of the nature of a single spatial distribution, the covariance of different distributions, the distribution of the same phenomena through time, locational inventories, areal differentiation, and changing spatial distributions. Both Nystuen and Berry, along with others working during the same period, created many of the conceptual foundations for GIS systems as we now know them. However, the direction of their work led toward a different set of quantitative methods and the development of spatial analysis.

Despite roots in geographic theory, early GIS development actually grew out of the demands of practical projects in land-use planning, forestry, and other applied disciplines. Systems were developed to automate a number of conceptually simple but labor-intensive tasks, such as map overlay and area measurement, and to take advantage of the relative ease with which an automated system could edit and redraft a simple map. There is usually a high initial cost associated with building a digital map database, but once the database is created an automated system offers enormous potential for map analysis and display.

The range of possible forms of spatial analysis is vast, and consequently there has been a relatively rapid increase in the analysis and modeling capabilities offered by GIS vendors. This is especially clear in cases where software vendors have tried to target more than one area of application: one system (ESRI's ARC/INFO) now includes more than 800 different commands to create, store, analyze, and display spatial information. Spatial analysis is potentially far more complex than more conventional statistical or financial analysis, and in this sense GIS is only now beginning to implement the vast range of techniques that were developed during the heyday of spatial analysis in the 1960s and 1970s.

Because the potential range of GIS operations is so large, there have been several efforts to develop taxonomies, and to establish meaningful bounds. One example (Berry 1987) is based on a widely used grid cell-based program, the Map Analysis Package (MAP), which has appeared in numerous versions. MAP provides the following fundamental operations: reclassification of map categories, overlaying maps, measuring distance, establishing connectivity, and characterizing cartographic neighborhoods (Tomlin 1983; Berry 1987). Within the limitations of a grid-cell structure, it has been demonstrated that all of Nystuen's original concepts can be implemented with MAP. Furthermore, by combining functions, it has been possible to simulate the classic location-theory problems of Weber, von Thünen and Christaller.

In the more elaborate vector-based or polygon-based systems, the same operations are carried out on digital records which represent the points, lines, or areas of the original source maps (Goodchild 1987b). Buffer operations can be used to extend specified features by given distances, e.g., to represent zones of impact. Concepts such as connectivity are implemented on networks of connected linear features.

The tremendous commercial growth of the field can be directly related to the evolution of successful solutions to a wide range of geographical data-handling problems. These solutions have stemmed from diverse fields including mathematics, geometry, photogrammetry, operations research, civil engineering, image processing, surveying, and computer science. With current computer-based methods, maps, images, and surveys can be transformed into spatially registered layers of information. In other words, the concept of a geographic matrix envisioned by Berry (1964) a quarter-century ago, in which an infinite range of spatial information (rows) is stored for every place (columns), is now a reality. Even though he suggested that such a matrix might be more of a nightmare than a dream, Berry clearly saw the potential of such large-scale organized data-collection and storage efforts. The operations required to construct the matrix, which were once considered difficult or even unsolvable, are now routinely handled by inexpensive desktop computers.

In addition to the creation, storage, retrieval, and display of a variety of spatial data, GIS now include powerful procedures to analyze the union and intersection of diverse layers of information, to create new information and produce answers to queries needed for planning and decision-making. Integration of diverse data is the key GIS capability that was not clearly foreseen in the early spatial-analysis literature.

Within many organizations, GIS have become integral parts of everyday management practice and a multimillion-dollar yearly expense. According to a recent (1985) policy study in the U.S. Department of the Interior:

> The number of different cartographic and earth-science data types that could be analyzed was [previously] limited by the user's ability to mentally integrate several types and produce new maps from selected combinations of data sets. GIS technology permits cost-effective analysis of the cartographic and earth-science data types using economic and scientific models that consider the many factors needed to make policy decisions, develop resource management plans, or make resource assessments and appraisals.

Over the last decade, the demand for these systems has expanded beyond the federal natural-resource agencies into the state-and-local-government arena. Applications now cover a full range of inventory, monitoring, planning, and regulatory functions (Tomlinson 1987; Department of the Environment 1987). For example, timber companies are creating and maintaining GIS files to manage massive tracts of timber. Utility companies are planning and monitoring extensive networks of electrical, water, and sewerage services. Tax assessors are maintaining accurate cadastral systems and automatically determining property values. State agencies are regulating growth, planning for infrastructure improvements, monitoring environmental impacts, and drawing new political districts. Developers are estimating demand for new projects and planning actual site improvements with GIS technology. Of course, the military is using GIS to analyze terrain and optimize strategies for movements of troops and equipment. Federal agencies are managing timber resources, national parks, oil leases, game reserves, wetlands, and hazardous-waste dumps.

Development of a holistic spatial data-handling system has required analysis of the nature of geographical data and the operations that geographers perform. This process includes reappraisal of the nature of maps and the geometric view of geography as proposed by Bunge (1961). Maps are the geographer's most useful model of the Earth (Board 1967) and form the primary source for spatial databases. Since maps are two-dimensional objects, they can be reproduced by the simple elements of plane geometry. The problems associated with building spatial databases involve the conversion of these analog models into their digital counterparts, but most of these problems have been solved for several years (Peucker and Chrisman 1975). Although digitizing remains the most time-consuming and expensive part of the process, the difficult problems are often more a matter of efficiency and accuracy of the source materials than of technical know-how.

Creation of a detailed, accurate, and current geographical matrix of the type envisioned by Berry is only one of the elements contributing to a full understanding of the origins of GIS. In the remainder of this section, some others are discussed. They include the early work of the Bureau of the Census in developing suitable data structures; the impetus provided by rapid development of remote-sensing technology and the deluge of digital information that it produced; the trend toward automation in cartography and the consequent availability of digital cartographic databases; and the development of spatial applications of operations research and modern location theory. These are only some of the forces that have helped to pull GIS together, and to create a new feeling of common interest among the disciplines that use spatial information.

DIME Files

Starting with the 1970 census, it became apparent that the Bureau of the Census needed the ability to geocode automatically, in order to reference mailing addresses to specific areas on the Earth's surface. The Geographical Base File/Dual Independent Map Encoding (GBF/DIME) system, or DIME, was a rudimentary digital street map of the major metropolitan areas of the country. It provided the means to automatically assign a street address to a street segment and to interpolate a specific pair of geographical coordinates for each address. This form of urban geocoding provided the basis for converting any set of addresses into a truly spatial data set or input into a GIS.

Perhaps more significantly, DIME also introduced some basic topological constructs (Corbett 1978) that have proven valuable well beyond census applications. A simple cartographic object such as a line or a bounded area can be coded as a string of coordinate pairs, and thus represented digitally in a spatial database in what is termed a *cartographic data structure*. However, in this form, the object is relatively unsuited to complex spatial analyses, since all of the relations between it and other objects must be determined by computation, such as connectedness, proximity, adjacency, and intersection. A topological data structure is defined as one that encodes such relations directly. DIME was the first widely recognized topological data structure, although precursors can be found in several systems: it allowed the user to determine directly the identity of streets meeting at an intersection without searching for matching coordinates or calculating distances.

Remote Sensing

Advances in satellite-based remote-sensing technology have also had major impact on the development of GIS. A key to the successful integration of several layers of geographic information is the positional and attribute accuracy of the data. The satellite-based Global Positioning System now provides an affordable means to enhance the geodetic control network, and thus to improve the registration of maps and photographs to specific coordinate systems. Multispectral scanning systems with resolutions of 20 meters now provide excellent and timely sources of digital data that can be rectified and registered to the Earth's surface. Remote-sensing software technology concentrates on the geometric rectification and on the conversion of multispectral data into classified information that can be merged with other layers in a GIS.

Improvements in sensors and in image-processing software have moved the field far beyond the stage of generating interesting pictures. As remote sensing becomes involved in analytical applications, there is an increasing awareness of the artificiality of the boundary between it and GIS. For example, Fussell, Rundquist, and Harrington (1986) raised the following questions:

> What will be the role of remote sensing in the developing trend toward Geographic Information Systems (GIS) technology? Is our future role to be reduced to providing input to GIS activities?

Development of GIS should not reduce the importance of remote sensing, but offer broader horizons for application, and new incentives for improvements in sensors and interpretation techniques.

Automation in Cartography

While GIS developed in parallel to remote sensing, an important component of GIS developed directly from computer-assisted cartography; i.e., many GIS programs developed from software for thematic mapping. At the same time, many traditional cartographers realized that GIS provides a basis for greater utility of maps and spatial information. Costs of creating a spatial database can be offset by the benefits that result from subjecting the database to new kinds of analysis and applications.

Muller, a cartographer by training, views GIS as a unifying force for geography, by serving as a way "to bridge the gap between research and application within the discipline of geography." He is optimistic about the ability of GIS to improve the overall image of the discipline:

> The multidisciplinary nature of geography has more often been considered as a sign of weakness rather than strength, particularly from the point of view of other professionals who keep questioning the fuzzy boundaries which delimit the areas of competency of geographers. The application of GIS, if successful, will upgrade the image of geography by demonstrating both the advantages of the multidisciplinary, holistic approach and the irrelevance of clear delimitations between geography and other connected disciplines. (Muller 1985, 43)

Operations Research and Modern Location Theory

GIS offers the ability to link the data-collection process firmly to the analytical process, and nowhere is this clearer than in the merging of GIS technology with location theory. Geography has a long tradition of involvement in the development and testing of theories of location. Processes leading to spatial patterns of agricultural land use, human settlement, and industry form the basis for the three classical theories of location of, respectively, von Thünen, Christaller, and Weber. Much effort has been expended in the past three decades searching for patterns that support these theoretical constructs, particularly in human settlement, and in adapting them to the constraints and distortions of real spatial systems.

Most work in location theory has been in the formulation of propositions about decision-making, and the development of logical consequences of these propositions in specific spatial contexts. The role of empirical tests has been in indirect evaluation of the truth of the initial propositions and the correctness of the logical development. Emphasis has thus been on the explanation of observed pattern. In recent years, however, a separate tradition has developed, concerned more with the use of location theory in the design of spatial systems. The term *location analysis* has often been used to distinguish this more pragmatic tradition from location theory.

Suppose, for example, that we wish to select a number of suitable sites in a city for the establishment of fire stations. We might decide that the best set of locations would be the one that provides a fire service that could reach every building in the city from the nearest fire station in the least possible time. The term *location-allocation* is used to describe problems where locations must be sought for a number of facilities from which service is provided to a dispersed pattern of demand, using some defined objective. The set of possible applications is large; for a recent selection see Ghosh and Rushton (1987). Location-allocation provides an excellent example of the power of GIS in spatial analysis. It deals with multiple types of spatial objects, using points to represent selected sites and areas to represent demand zones, and must deal also with attributes of the relations between objects, including distances, travel times, and allocations. Because the set of criteria that affects a given locational decision can never be fully enumerated, it is desirable to carry out site selections interactively.

Optimum locations can be modified by the constraints of zoning, site availability, local traffic, etc., and the effects of modifications can be evaluated.

Several authors have suggested embedding location-allocation within a GIS as a spatial decision support system (SDSS), which would use graphics, text display, and interaction to provide a complete environment for location analysis. An SDSS might be carried into the field to allow on-site access to demographic databases, or it might be used to allow a decision-maker to explore a remote site without incurring the cost of travel. Goodchild (1984) described the use of the ILACS package for interactive, graphic location-allocation in retail site selection. A system for location-allocation on microcomputers was developed by Goodchild and Noronha (1983), and a spatial decision support system for location-allocation has been designed and built by Densham and Armstrong (1987).

As a technique of mathematical optimization, location-allocation falls within the broad field of operations research (OR). In addition to central facilities location, a number of other OR methods have direct application to spatial problems and relevance to GIS. Without the ability of GIS to handle large volumes of data, it would be (and was) necessary to simplify a problem by making a number of assumptions, or by aggregating information in a relatively small number of places or greatly generalized network.

A comprehensive digital representation of a transportation network can be used to solve a number of optimization problems, including selection of the shortest route between a given origin and a destination, design of optimum routing for delivery vehicles and buses, load balancing in electrical distribution networks, and development of transportation systems. Programs have been developed in all of these areas; for a recent review of microcomputer vehicle scheduling and delivery software, see Golden and Bodin (1986). Recently these procedures have begun to be incorporated into the standard set of functions included in a GIS (Lupien, Moreland, and Dangermond 1987). Such systems are now being incorporated into vehicle navigation systems that manage fleets from the office or can actually be placed in automobiles and trucks. The ability to process street addresses is being utilized extensively by the very competitive overnight delivery services which must optimize their routes between locations in complex urban settings.

In all of these areas, the same arguments for integration with GIS arise: GIS can provide the spatial data input and output functions necessary for housekeeping, and the ability to process multiple types of spatial objects and attributes of the relations between them. Any optimization problem can be regarded as interfacing with a GIS database in the same fundamental way; objects and associated attributes are selected from the database, used to support some analytic function, and the results are returned to the database as new attributes of existing objects, or in some cases as new objects. For example, the input to a shortest-path problem might be a specification of origin and destination as attributes of node objects, together with impedances as attributes of link objects. The path might be returned as a new attribute of either links or nodes, or both.

Despite this, integration of optimization procedures and GIS databases remains incomplete. In part, this is because the problems of analysis have been regarded as more challenging than those of database creation and maintenance, and in part, it is because relatively little work in optimization reaches the stage of application, and relatively little encounters the problems of real implementation. By offering to make implementation easier, GIS may remove many of the impediments to the use of OR techniques in spatial analysis.

DEFINING GIS

In many ways, the emergence of geographic information systems is the most recent trend in a steady evolution of geographic techniques since World War II. It also reflects the general improvement in computing power over the same period. In the broadest sense, a GIS implements the processes required to convert geographic data into more useful information. Geographic data must be systematically gathered, spatially registered, stored, analyzed, generalized, retrieved, and displayed. Geographic information systems represent the integration of all of these separate operations into a single automated environment. Since even early computer-mapping systems such as SYMAP provided many of these capabilities, it is necessary to distinguish modern full-fledged GIS from earlier forms of automated geography.

From the viewpoint of the consumer of GIS technology in the late 1980s, the most important distinction lies in the operations that are unique to GIS. Any type of computer-oriented geographic analysis must begin with database development. Cartography and, recently, remote sensing provide the basic raw material for the generation of the digital spatial database. The process of converting analog maps into their digital equivalents is considered the digitization stage. In many cases, cartographers or draftspersons simply wish to create the digital product for selective retrieval and display at different scales and with different symbology.

In other words, a surprising amount of digital cartography is merely electronic drafting. Such operations are increasingly handled with computer-aided design (CAD) systems in much the same manner as any other drafting operation. The net result of this process is an efficient way to create and maintain graphic drawing files. Although CAD systems have greatly increased the productivity of base-mapping operations, they do not fully meet the needs of computer mapping or GIS functions (Dueker 1987; Cowen 1987).

The key to more sophisticated geographical data processing lies in the attributes that describe and differentiate the geographic entities on the map. As Goodchild (1987a, 327) suggests:

> It is often observed that the element which distinguishes geographical information systems from other forms of spatial data handling activity, such as automated cartography and remote sensing, is an emphasis on analysis. Although there is clearly a very great variation in capability from one system to another, the ability to manipulate spatial data into different forms and to extract additional meaning from them is at the root of current interest in GIS technology.

Goodchild characterizes much of the research of the past three decades as an attempt to find an optimal model of spatial information. If the data are organized in a rectangular matrix or flat file structure in which each row of a data matrix corresponds to a different geographical entity, then it is easy to examine relations between the columns or variables of the matrix using standard statistical procedures. The adaptation of a full range of multivariate statistical procedures provides an appropriate set of tools for handling most of these operations, i.e., if the data are organized by the same discrete set of regions, such as census tracts, cities, counties, states, or nations. Standard statistical tools, such as correlation, regression, and factor analysis, can treat the data as a flat file, the equivalent of Berry's geographic matrix. Modern statistical packages now even include computer-mapping capabilities, but the combination does not constitute a GIS.

At a minimum, an automated choropleth mapping system provides a linkage between such a database and a graphical-display system. Even the earliest computer-mapping systems, such as SYMAP, allowed the user automatically to assign symbology to geographical entities on the basis of attributes or variables in a database. Over the last couple of decades, these mapping systems have evolved into highly sophisticated programs which even allow flexible relational linkages between the attribute database and the file containing the coordinates of the boundaries of the geographical entities. By taking advantage of significant improvement in computer graphics devices, these systems provide professional-quality statistical maps. With the recent release of personal-computer versions of the popular mainframe systems for both statistical analysis and choropleth mapping, this level of geographic analysis is readily available for even modest-size organizations.

Although linking a database to the graphical representation of geographic entities enables the researcher to address geographic queries relating to the spatial distribution of a single variable or the covariation of several variables, such systems are limited to the predefined set of spatial entities in the database. In other words—regardless of the sophistication of the database, the mapping procedures, and the output device—the user is only retrieving and displaying information that has been previously stored. The dependency on existing attribute data provides the basis for distinguishing a GIS from a computer mapping system.

For example, tax-assessment files contain a variety of information about individual land parcels. The local government can choose the level of sophistication as it encodes the data. At the simplest level, the computer is used to automate the accounting and the mailing list. The technology required is no different from any commercial billing application. But parcels have a spatial component, so there may be pressure to automate the functions performed in drafting tax maps. A CAD system can provide the capability to automate the drafting tasks. By adding sophistication to the parcel descriptions, the assessor could generate statistical maps relating to average market value, size of lot, year of construction of buildings, or any other parcel attribute. This step would require integrating the billing database with the geometric description. The assessor may be satisfied at this point.

However, developers and planners may raise many other questions, such as which parcels are located in the 100-year flood plain, what is the distance to the nearest fire hydrant, or in which school district is a particular parcel located? These and hundreds of other ad hoc queries involve integration of data from different sources. Unless each parcel has been coded with these relations, a simple mapping system is inadequate. According to one source, planners in New York City estimate that 85% of all information they handle is associated with some geographic entity (Stiefel 1987). The power of a GIS resides in its ability to integrate all of these diverse sources, thus synthesizing new information. This concept of map overlay was popularized by Ian McHarg and has been widely used by planners to create suitability and capability models, but it was not explicitly recognized in the early work in quantitative geography as the crucial key that it has become.

Polygon overlay provides a basis for automatically updating and expanding the original parcel database as needed to respond to ad hoc queries. Instead of attribute records for a base layer (such as parcels), the GIS maintains distinct map layers that can be spatially registered with each other. Most importantly, the GIS can address every possible combination of spatial entities. Geographers and other spatial decision-makers should not be forced to work with arbitrary statistical units such as counties or census tracts, and artificially aggregate information into such units.

GIS provides a tool that geographers have always needed, the power to synthesize spatial information. It is important to realize that the recent success of the field is related to this important function. It is also important for geographers to recognize how this function separates true GIS systems from computer-aided design, remote sensing, statistical analysis, and computer mapping.

DRIVING FORCES

The rise of geographic information systems, while impressive, should be placed in context. Growth of GIS has much to do with general driving forces that are reshaping our global society and its economy. This section reviews the wider social context of the current interest in GIS.

Hardware Advances

The development of general-purpose computing hardware has proceeded at a rate unprecedented in almost any other technology. There is no need to chronicle the increased speed and capacity of modern computers, compared to those of 30, or 20, or even 10 years ago. Prices for computing power have provided a major driving force for all forms of automation throughout society. GIS are thus a part of this broader phenomenon, but the special characteristics of geographic information have made its development seem even more rapid.

For a long period in scientific research, computers meant numerical computation. The display devices available for computer results were quite crude, consisting of line printers and teletypes. While these crude devices were used for map production in early systems, the results were clearly inferior to manual cartography. Before digital databases could become economical, computer graphics peripherals had to be created. Some devices, like the digitizing table, were developed for cartographic use, and then slowly found a more general market. Some other developments, like the laser line follower, did not gain a sufficiently broad market to become widespread.

Considering the relatively tiny number of cartographic production facilities, the invention of new hardware was most frequently generated by other market demands. The pen plotter and television CRT have become the most common devices for graphic output. Fortunately for the interests of GIS, the general marketplace has changed dramatically in its demand for graphics. Now a sophisticated microcomputer is controlled mostly by graphic metaphors rather than typed commands. While a graphic interface may not offer new theoretical insight, it increases consumer demand for high-performance graphics, widening the demand base and lowering the price for those, like geographers, to whom graphics are indispensible.

Software Development

While hardware advances seem inexorable, raw processing speed and graphic resolution do not directly solve real problems. Hardware is controlled by software, which is a peculiar invention of the past 20 years. Writing programs for some problems simply involves the translation of known mathematical algorithms into computer instructions. Designing and writing GIS software, however, involves considerable abstract thinking and endless attention to detail.

In the past, university geography departments did not adequately support such efforts. Corporations attempted to hire programmers for their mathematical skills and then make them work with the mapping specialists. This model, tried many times, has not been as fruitful as the geographer who learns programming and can integrate the two modes of thought. The history of two centers of early GIS software development has been recently documented for the Harvard Laboratory for Computer Graphics and Spatial Analysis (Chrisman 1988) and the Experimental Cartography Unit (Rhind 1988).

Software development in the early period occurred in many organizations, and with varying levels of quality control. Although many of the early developments became production systems, they also embodied numerous fundamental concepts. Recently, the GIS consumer has become more sophisticated, demanding well-engineered solutions which can be installed and operated with minimal risk of failure. Two models currently dominate the industry. One is a raster solution in which all information is reduced to a common grid, and held as independent layers. The other is a vector solution in which point, line, and area objects are topologically structured, and their attributes are stored in a relational database-management system. While vector-based systems dominate in some sectors, raster-based software dominates in others, so there will always be a need to translate between the two fundamental approaches.

Digital Data Sources

The creation of digital data is a major expense and barrier to establishing a GIS. As more and more GIS become operational, the availability of digital data affects adoption by others. Digital mapping is becoming common enough within federal agencies and other traditional data providers for publicly available data to become an expectation, rather than an unusual event. The legal treatment of information in federal and local agencies often makes it difficult to recover full development costs, but open availability does foster multiple use. Other countries, for example the United Kingdom, place official information under copyright and other restrictions. This creates potential barriers to GIS development, but allows more efficient cost recovery.

One possibility for the development of digital data is through the private sector. Some large companies, such as Phillips Petroleum, have marketed databases developed internally, and some small companies have been founded specifically to create digital databases. Geographic Data Technology, founded by an originator of DIME files, has a goal, shared by some competitors, of providing current address-matching functions nationwide. Nevertheless, this company must depend on Census Bureau contracts to develop the public databases. If recently developed navigational equipment becomes successful in the automobile or package-delivery markets, a consumer demand for digital data could expand these enterprises dramatically.

Support Groups

Any new development, such as GIS, benefits from the existence of a support network. Professional organizations play a role in publicizing a new technology and sharing new developments. GIS have links to a number of organizations due to the multidisciplinary origins of the participants. Many important individuals in the GIS community are geographers and members of the AAG. However, the nature of the annual meeting and the orientation of geographic journals have not provided the best forum for this specialized and somewhat tool-oriented topic. Recently, members of the AAG have created the AAG GIS Specialty Group, which has sponsored a number of conferences directly oriented toward GIS research. One group with a long-term record of supporting GIS development is the IGU Commission on Geographical Data Sensing and Processing. They have organized a number of technical symposia aimed at the GIS research community.

Another technical conference with a broader focus has been the string of AUTO-CARTO events. Originally held separately, these have become a biennial feature of the joint meeting of the American Congress on Surveying and Mapping (ACSM) and American Society of Photogrammetry and Remote Sensing (ASPRS). These two organizations also provide outlets for GIS publications through their journals, *The American Cartographer* and *Photogrammetric Engineering and Remote Sensing*. Another organization, the Urban and Regional Information Systems Association (URISA), has a long history of annual meetings with a varied technical content, including many GIS issues.

While the AAG is primarily composed of academic geographers, ACSM includes many government cartographers and private surveyors, ASPRS includes governmental and private photogrammetrists along with remote-sensing specialists of all descriptions, and URISA includes many data-processing and mapping staff in local government. Many GIS experts are members of two, three, or all four organizations. Fortunately, these four organizations have recognized that each one plays a part in a larger process, and they have agreed to cooperate on joint conferences and other ventures.

CURRENT STATUS AND SELECTED CASE STUDIES

Public Sector

As with most cartographic developments, GIS have been heavily dependent on the public sector. One of the earliest uses of the term was for the Canadian Geographic Information System in the mid-1960s (Tomlinson, Calkins, and Marble 1976). This pioneer effort has been emulated by many other agencies at different levels of government.

Federal Government. A recent interagency report (Federal Interagency Coordinating Committee on Digital Cartography 1987) found that 28 organizations within the federal government either use or intend to use GIS, and that 11 of these are currently using them extensively. Most of the national investment in GIS-related activities during the next decade will be in development of the digital databases needed to use the GIS technology. Early adoption of data exchange, content, and accuracy standards will assist in the efficient transfer and use of data throughout the federal community and beyond. Of particular importance is the recognition by agencies that instituting standards may initially increase the effort and expense of creating digital databases. Hence, the benefits must be considered over a period of years, and government wide.

One example of federal cooperation is the development of cartographic databases to respond to the needs of the 1990 census. The Bureau of the Census joined with the U.S. Geological Survey to create a National Digital Cartographic Database of 1:100,000-scale Digital Line Graph (DLG) files that represent complete nationwide digital coverage of transportation and hydrography. The Bureau will utilize these digital maps to conduct the 1990 census, and will attach street names and address ranges to the road segments. Most significantly, the project represents the merging of both cartographic and GIS needs. Digital files can be converted easily into analog models (maps) while also serving as the basis for numerous transportation and hydrography layers for GIS applications. A result will be the ability to locate any street address in the nation with accuracy.

State Government. Federal structure in the U.S. decentralizes authority. The 50 state governments have many different needs for geographic information within their many agencies. Some years ago, there was a tendency to create centralized state geographic information systems, often constructed to respond to contentious issues such as environmental protection. At least two-thirds of the states created such an agency in the 1970s, but very few have survived intact. Although the early GIS software promised more than it could deliver, the fault was more in the administration of these systems.

While many states have made solid beginnings in the adoption of GIS, perhaps the most long-term success is the Land Management Information System housed in the Minnesota State Planning Office (Hsu et al. 1975). This agency developed from a university research laboratory 20 years ago into a service center for a variety of individual state agencies. While the group has kept its GIS technology current, the stability of the operation probably depends on human resources, such as dedicated staff and flexible management.

Natural-resource management in a state government, or in any organization, requires massive spatial inventories which have been developed for many years on maps and air photographs. The analytical functions of a GIS, particularly polygon overlay, parallel the procedures performed manually. Under these conditions, the conversion to an automated GIS is simply a natural evolution toward higher labor productivity.

Recent developments in GIS have had substantial impact on state agencies. One common theme is a concentration of Departments of Natural Resources (or equivalent) which may include environmental regulation, but often include direct sources of income such as forest management, coal mining, or other primary industries. The use of GIS for forestry management has been a particular growth area (Tomlinson 1987).

Local Government. Most land-related decisions are not made at the federal or state level, since municipalities and counties have the most-direct involvement in these decisions. Unfortunately, these units of government are dispersed and often underfunded. However, the large number of local government agencies, coupled with their demands for detailed geographic databases, will offer one of the largest markets for GIS development.

One pilot project for a county GIS, the Dane County Land Records Project, occurred recently in Wisconsin (Niemann et al. 1987). This project focused upon a natural-resource problem, soil erosion, but in developing the layers required to solve that problem, many related topics had to be addressed. Custodians of the layers included a number of federal agencies, the state Department of Natural Resources, and separate units inside the county. GIS technology was demonstrated by integrating information from the land-records system, natural-resources maps, and interpretations of satellite images. Integration is heavily dependent on a framework of geodetic coordinates, an area where surveying technology has changed rapidly. While a primary task involved these technology demonstrations, the project also developed cooperative efforts to coordinate the different agencies in a noncentralized network.

Local governments around the country are anxious to improve the efficiency of their tax-mapping and assessment procedures. Some have invested heavily in automated systems of varying complexity, from tabular databases through computer-aided mapping to complex systems having analytical functions. Some systems involve extensive ground control and accurate aerial-photograph bases to register property boundaries, utility lines, and other features. When directly linked to assessor's records, these systems provide a precise way to analyze individual parcels for a wide variety of assessment and planning operations. Many of the systems installed in local government use GIS technology, but they are developed under labels such as Land Information Systems (LIS) or multipurpose cadastres. Modernization of local land records has captured attention at the National Research Council (1983) and in a number of state studies (for example, Wisconsin Land Records Committee 1987).

For public or private utility companies (see below), accurate digital map bases provide the means for maintaining and planning facilities, or AM/FM (Automated Mapping/Facilites Management). From the geographer's perspective, LIS and AM/FM are specialized subcategories of a larger topic because GIS are independent of both scale and subject matter. Precise large-scale mapping and land-information systems set high standards for GIS technology. At the local level of government, the path from the technicians to the decision-makers is relatively short, and there is particular attention to cost-effectiveness and legal defensibility.

Private Sector

There is little functional difference between a public-sector or private-sector GIS. Some applications such as forestry and utility management remain essentially identical, whether performed for government or industry. There are a few differences in the motivation for the systems, however. While the private sector tends to make decisions based on profit margins, public agencies have to justify a GIS on grounds that include noneconomic issues of equity.

Utilities (public or private) use GIS capabilities to manage their extensive distribution networks and facilities; the term AM/FM is commonly used to describe GIS applications in this area. A single utility may manage a network of hundreds of thousands of kilometers, with tens of millions of consumers. Perhaps the most significant application is in retrieving records on facilities that may be impacted by construction. Several thousand activities a day may potentially affect the distribution network, and in each case it is necessary to query the system to identify all facilities in the immediate area of the activity. In addition, it may be necessary to identify all customers whose supply may be interrupted by the activity, and to notify them.

As a relatively large-scale example, British Gas serves roughly 25 million customers in the U.K., and has recently initiated a GIS acquisition in the $100 million range, to provide capabilities in each of its divisions and regions. Digitization of the distribution network will require input of information which is currently stored on roughly 250,000 1:1250-scale map sheets. It is hoped that the base-map information will be available to the utility in digital form, allowing gas facilities to be captured against a digital "backcloth." Once operational, the system should be capable of graphic display of any part of the distribution network. In addition, users should be able to query facilities within a given distance of any indicated point. More sophisticated features include the automatic generation of work orders and customer notifications of service interruptions, optimal routing and scheduling of maintenance crews, and predictions of future demands based on socioeconomic data.

GIS offers numerous potential applications in retailing (an almost exclusively private function) because of the importance of location and spatial relationships in the profitability of any retail establishment. Retail analysis relies heavily on map information to summarize demographic distributions from sources such as the census, along with retail sites and customer locations. Many retailers make use of simple microcomputer mapping packages (for an evaluation, see Day 1986).

By way of example, Compusearch Marketing and Social Research Ltd. is a Canadian company specializing in locational analysis for retailers. Much of its business consists of simple evaluations of potential sites against census data. Rings are constructed at fixed distances from the site, and superimposed on the boundaries of census reporting zones to allow tabulations of particular market segments.

Other GIS applications include the frequent need to overlay reporting zones from different census years or from other agencies in order to make data spatially compatible. Customer addresses must be matched through postal codes to geographic coordinates, and then summarized by census zones or by distance from retail site. Location-allocation techniques, discussed above, are increasingly important in guiding site-selection decisions. Finally, more sophisticated spatial analyses include the calibration of spatial-interaction models to predict customer flows from origin neighborhoods to retail sites, and the effects of changes in the retail environment.

FUTURE TRENDS AND ISSUES

NSF National Center for Geographic Information and Analysis

A recent development with a major impact on the GIS field and geography at large was a solicitation, issued by the National Science Foundation (NSF) in June of 1987, calling for proposals to establish a "National Center for Geographic Information and Analysis" (Abler 1987a). Issuance of the solicitation has increased the visibility of GIS within geography, and of geography within academia; it also has caused an unusually high rate of new academic positions to be created and equipment to be granted in geography and related departments.

The NCGIA solicitation states (National Science Foundation 1987, 2):

The goals of the NCGIA are to:

- advance the theory, methods, and techniques of geographic analysis based on geographic information systems (GIS) in the many disciplines involved in GIS research;
- augment the nation's supply of experts in GIS and geographic analysis in participating disciplines;
- promote the diffusion of analysis based on GIS throughout the scientific community; and
- provide a central clearinghouse and conduit for disseminating information regarding research, teaching, and applications.

The goals thus span research, education, technology transfer, and dissemination of information. After a discussion of these goals, the same page of the solicitation identifies five general problem areas that should form focal areas for research at the NCGIA:

Whereas each applicant institution should propose a plan suitable for its individual strengths and missions, research programs should address several of the following general problems:

- Improved methods of spatial analysis and advances in spatial statistics;
- A general theory of spatial relationships and database structures;
- Artificial intelligence and expert systems relevant to the development of geographic information systems;
- Visualization research pertaining to the display and use of spatial data; and
- Social, economic, and institutional issues arising from the use of GIS technology.

From this list of research problems, it seems clear that the NCGIA is intended to have a broad mission within geography and related disciplines, and to go well beyond common GIS concerns. Within geography, the topic of the first problem clearly falls within the area normally associated with mathematical models and quantitative methods. The use of "Geographic Information and Analysis" for the title of the Center signals substantial concern with these topics.

The second and third research problems address topics that fall solidly within GIS as narrowly defined, although this research involves substantial interaction with computer science. The fourth problem reflects a general initiative within NSF and American science toward "Visualization in Scientific Computing" (ViSC; see McCormick 1987). Within geography, visualization would most clearly fall within cartography. The fifth research problem has received less attention from GIS technologists within geography, though institutional concerns (see below) are growing.

In August 1988, after a multistage review process, NSF awarded NCGIA to a consortium including the State University of New York at Buffalo and the University of Maine, with primary headquarters at the University of California at Santa Barbara. Funding was set at $1.1 million for each of the first five years. This venture represents an opportunity for the whole geographic discipline to increase its national research visibility. The NCGIA will require cooperation between the GIS community and other, longer-established research themes within geography, and related science and engineering disciplines.

International Geosphere Biosphere Program

The International Geosphere Biosphere Program (IGBP) is a major scientific initiative under the auspices of the International Council of Scientific Unions. Its aim is to integrate research on many of the major systems affecting the Earth, including atmospheric and ocean circulation, with studies of the interaction of these systems with the biosphere. The potential use of GIS technology in supporting IGBP and global science has been described in a series of papers edited by Mounsey (1988). Of particular concern are the problems of integrating data from different sources, at various scales and levels of resolution; the need for improved methods of representing global data in digital databases; and problems associated with error and uncertainty in large global models. Databases that will support global science in the coming decades will likely be of unprecedented size, and will require orders-of-magnitude improvement in our current abilities to handle very large spatial data sets.

Artificial Intelligence

Research related to the implementation of Artificial Intelligence (AI) in GIS is at present patchy, with many disparate pieces of work being done in a variety of locations. Researchers working in AI aim to program computers to execute tasks so that the input is similar to that given to humans and the final product is indistinguishable from what a human would produce. A number of areas of AI research are of potential use to the geographical community in general and GIS researchers in particular (Robinson and Frank 1987).

Vision research involves the segmentation and understanding of scenes via feature extraction, an area that remote-sensing scientists are beginning to explore and that is likely to become important to GIS workers for automated data checking, feature extraction, and presentation (de Simone 1986). The use of natural language to describe geographic phenomena, and particularly to pose queries to GIS, is being studied with the intention of improving the human interface (Robinson and Wong 1987). Knowledge-based techniques are also being applied to the area of database management (Smith et al. 1987), and to the use of analytical functions available within GIS (White and Morse 1987). Finally, researchers are working on improving the graphic output from GIS by developing cartographic expert systems (Nickerson and Freeman 1986).

Spatial Statistics

Because a computer is a precise processor of information, it is to be expected that problems will arise in processing and interpreting imprecise spatial data (Chrisman 1987; Blakemore 1984). For example, the spurious-polygon problem (Goodchild 1978) is an artifact of inaccurate data. Analysis of spatial information with GIS techniques has prompted renewed interest in the modeling of error in spatial data, particularly for complex line and area objects. The eventual objectives of this research are the development of measures of error, techniques for tracking error through GIS operations, and measures of uncertainty which can be output in association with each GIS product.

Spatial Languages

Vector and raster systems represent space respectively in the form of objects and images. Yet our ways of understanding and working with spatial information do not necessarily match either method. Geographical directions are often given using language that has no direct application within conventional GIS. In the past few years we have seen renewed interest in abstract concepts of spatial language, with the ultimate aim of improving the ability of GIS to communicate with users. This research lies at the interface between computer science, with its concerns for structuring digital information, and cognitive psychology, with its theories of spatial learning and processing (Head 1984; Mark, Svorou, and Zubin 1987).

Institutional Issues and Societal Outcomes

A GIS, as a system, involves more than software and hardware. Successful application depends on the ability of people in organizations to process information and to make decisions. The structure of institutions has not been a major topic of discussion during past phases of GIS development. However, in the current phase of implementation and operational use, institutional issues have gained visibility. For instance, the early design for many systems imposed a central database manager, often due to the high capital costs of early equipment. Resistance to a centralized solution might be seen as resistance to any form of progress, but there are legitimate concerns hidden in the bureaucratic warfare. Many current systems are designed to fit into existing institutional frameworks, and to foster cooperation between equal organizations. This more complex design shows a sensitivity to the human aspects of the technology. Future efforts should also study the interaction of GIS technology with social goals. Equal access to information and equitable treatment of individuals or groups should be considered along with benefit/cost economics. GIS offer the chance to reorganize the handling of spatial information, so perhaps they can help reach more-general goals.

GIS and Geography

Although many of the key individuals involved in GIS may come from other disciplines or perspectives, there is little doubt that GIS will affect the discipline of geography in many ways. Jerome Dobson (1983) foresees a form of "automated geography" with GIS capabilities handling the details. Ron Abler, past Director of the Geography and Regional Science Program of the National Science Foundation, sees an important role for GIS in the future:

> The analysis and processing capabilities inherent in GIS could help resolve some long-standing dilemmas in geographical analysis. Heretofore one way to estimate the operations of complex, multivariate systems over a large area was to scale-up manageable studies of interactions among the variables in a microregion. Another way was to integrate separate topical studies of the individual variables over regional, continental and even global scales.

They could therefore be the catalyst needed to dissolve the regional-systematic and human-physical dichotomies that have long plagued geography. (Abler 1987a, 323)

In his presidential address to the Association of American Geographers, Abler discussed the role of maps and their relation to GIS:

> Geographers think in and through maps, and changes in mapping capabilities will affect the ways we think about the world. Many geographers will speak to the basic research needed to develop geographic information systems. Many more of us will soon find GIS indispensable to our thinking, teaching, and practice. (Abler 1987b, 515)

At the same time, it is evident that the relation between GIS as a technology and the discipline of geography is far from clear. GIS do not yet appear in any of the standard treatments of the philosophy of the discipline, and many geographers clearly view them as another in a series of technical tools (Jordan 1988). We lack textbooks (but see Burrough 1986; Marble, Peuquet, and Calkins 1985) and an adequate supply of faculty with sufficient training to teach GIS.

The appeal of GIS as a technology appears to lie in five major areas:

- The ability to display spatial information at a wide range of scales, and without regard to map-sheet boundaries
- The ability to overlay maps of different themes, and from different sources, and to analyze their relations
- the ability to compute spatial relations, such as "overlaps" or "is next to," from locational information
- The ability to display aspects of spatial information that are difficult to show on conventional maps, such as uncertainty, time dependence, flows and interactions, and three dimensions
- The ability to carry out complex modeling in an integrated environment which provides access to spatial objects and their attributes

In essence, a GIS database is a formal model of spatial information, and as such plays a fundamental role in geographical analysis. Whether this role will be recognized, and GIS regarded as fundamental to the discipline of geography, or whether GIS will be relegated to the level of another technical fad, remains to be seen.

REFERENCES

Abler, R. F. 1987a. The National Science Foundation National Center for Geographic Information and Analysis. *International Journal of Geographical Information Systems* 1:303–26.

———. 1987b. What shall we say? To whom shall we speak? *Annals of the Association of American Geographers* 77:511–24.

———. 1988. Awards, rewards and excellence: Keeping geography alive and well. *Professional Geographer* 40:135–40.

Berry, B. J. L. 1964. Approaches to regional analysis: A synthesis. *Annals of the Association of American Geographers* 54:2–11.

Berry, J. K. 1987. Fundamental operations in computer-assisted map analysis. *International Journal of Geographical Information Systems* 1:119–36.

Blakemore, M. 1984. Generalization and error in spatial databases. *Cartographica* 21:131–39.

Board, C. 1967. Maps as models. In *Models in geography,* eds. R. J. Chorley and P. Haggett, pp. 671–725. London: Methuen.

Bunge, W. 1961. *Theoretical geography*. Lund Studies in Geography. Lund, Sweden: Gleerup.

Burrough, P. 1986. *Principles of geographical information systems for land resource assessment*. Oxford: Clarendon Press.

Chrisman, N. R. 1987. The accuracy of map overlays: A reassessment. *Landscape and Urban Planning* 14:427–39.

———. 1988. The risks of software innovation: A case study of the Harvard Lab. *American Cartographer* 15:291–300.

Corbett, J. 1978. *Topological principles in cartography*. Technical Report No. 48. Washington: US Bureau of the Census.

Cowen, D. J. 1987. GIS vs CAD vs DBMS: What are the differences? In *GIS-87: Proceedings, Second Annual International Conference on Geographic Information Systems,* pp. 46–56. Falls Church, VA: ASPRS/ACSM.

Day, C. O. 1986. Putting your business on the map. *PC Magazine* 5:219–34.

Densham, P. J., and Armstrong, M. P. 1987. A spatial decision support system for locational planning: Design, implementation and operation. *Proceedings, AutoCarto 8,* pp. 112–19. Falls Church, VA: ASPRS/ACSM.

Department of the Environment. 1987. *Handling geographic information: Report of the Committee of Enquiry chaired by Lord Chorley*. London: HMSO.

de Simone, M. 1986. Automatic structuring and feature recognition for large scale digital mapping. *Proceedings, AutoCarto London* 1:86–105.

Dobson, J. 1983. Automated geography. *Professional Geographer* 35:135–43.

Dueker, K. J. 1987. Geographic information systems and computer-aided mapping. *Journal of the American Planning Association* 53:383–90.

Federal Interagency Coordinating Committee on Digital Cartography. 1987. *Coordination of digital cartographic activities in the federal government*. Fifth annual report of FICCDC. Washington: Office of Management and Budget.

Fussell, J.; Rundquist, D.; and Harrington, J. A. 1986. On defining remote sensing. *Photogrammetric Engineering and Remote Sensing* 52:1507–11.

Ghosh, A., and Rushton, G., eds. 1987. *Spatial analysis and location-allocation models*. New York: Van Nostrand Reinhold.

Golden, B. L., and Bodin, L. 1986. Microcomputer-based vehicle routing and scheduling software. *Computers and Operations Research* 13:277–85.

Goodchild, M. F. 1978. Statistical aspects of the polygon overlay problem. *Harvard Papers on Geographic Information Systems* 6. Reading, MA: Addison-Wesley.

———. 1984. ILACS: A location-allocation model for retail site selection. *Journal of Retailing* 60:84–100.

———. 1987a. A spatial analytic perspective on geographical information systems. *International Journal of Geographical Information Systems* 1:327–34.

———. 1987b. Towards an enumeration and classification of GIS functions. *Proceedings, IGIS '87*. Falls Church, VA: ASPRS/ACSM.

———, and Noronha, V. T. 1983. *Location-allocation for small computers*. Monograph No. 8. Iowa City, IA: Department of Geography, University of Iowa.

Head, C. G. 1984. The map as natural language: A paradigm for understanding. *Cartographica* 21:1–32.

Hsu, M. L.; Kozar, K.; Orning, G. W.; and Streed, P. G. 1975. Computer applications in land-use mapping and the Minnesota Land Management Information System. In *Display and analysis of spatial data,* eds. J. C. Davis and M. J. McCullagh, pp. 298–310. London: Wiley.

Jordan, T. G. 1988. President's column: The intellectual core. *Newsletter of the Association of American Geographers* 23(5):1.

Lupien, A. D.; Moreland, W. H.; and Dangermond, J. 1987. Network analysis in geographic information systems. *Photogrammetric Engineering and Remote Sensing* 53:1417–22.

McCormick, B., ed. 1987. *Visualization in scientific computing*. Report of the NSF Panel on Graphics, Image Processing and Workstations. Washington: National Science Foundation.

Marble, D. F.; Peuquet, D.; and Calkins, H. W., eds. 1985. *Basic readings in GIS*. Amherst, NY: Spad Systems.

Mark, D. M.; Svorou, S.; and Zubin, D. 1987. Spatial terms and spatial concepts: Geographic, cognitive and linguistic perspectives. *Proceedings, IGIS '87*. Falls Church, VA: ASPRS/ACSM.

Mounsey, H., ed. 1988. *Building databases for global science*. London: Taylor & Francis.

Muller, J. C. 1985. Geographic information systems: A unifying force for geography. *Operational Geographer* 8:41–43.

National Research Council. 1983. *Procedures and standards for a multipurpose cadastre*. Washington: National Academy Press.

National Science Foundation. 1987. *National Center for Geographic Information and Analysis*. Directorate for Biological, Behavioral, and Social Sciences. Washington: National Science Foundation.

Nickerson, B. G., and Freeman, H. 1986. Development of a rule-based system for automatic map generalization. *Proceedings, Second International Symposium on Spatial Data Handling,* Seattle, pp. 50–64. Buffalo, NY: International Geographical Union Commission on Geographical Data Sensing and Processing.

Niemann, B. J., Jr.; Sullivan, J. G.; Ventura, S. J.; Chrisman, N. R.; Vonderohe, A. P.; Mezera, D. F.; and Moyer, D. D. 1987. Results of the Dane County Land Records Project. *Photogrammetric Engineering and Remote Sensing* 53:1371–78.

Nystuen, J. D. 1963. Identification of some fundamental spatial concepts. In *Spatial analysis: A reader in statistical geography,* eds. B. J. L. Berry and D. F. Marble, pp. 35–41. Englewood Cliffs, NJ: Prentice-Hall.

Peucker [Poiker], T. K., and Chrisman, N. R. 1975. Cartographic data structures. *American Cartographer* 1:55–69.

Rhind, D. 1988. Personality as a factor in the development of a discipline: The example of computer-assisted cartography. *American Cartographer* 15:277–89.

Robinson, V. B., and Frank, A. U. 1987. Expert systems for geographic information systems. *Photogrammetric Engineering and Remote Sensing* 53:1435–41.

———, and Wong, R. N. 1987. Acquiring approximate representations of some spatial relations. *Proceedings, Auto-Carto 8,* pp. 604–22. Falls Church, VA: ASPRS/ACSM.

Smith, T. R.; Peuquet, D.; Menon, S.; and Agarwal, P. 1987. KBGIS-II: A knowledge-based geographical information system. *International Journal of Geographical Information Systems* 1:149–72.

Stiefel, M. 1987. Mapping out the difference: Amory Geographic Information System. *Klein Computer Graphics Review* (Fall):73–87.

Tomlin, C. D. 1983. A map algebra. *Proceedings, Harvard Graphics Week.*

Tomlinson, R. F. 1987. Current and potential uses of geographical information systems. *International Journal of Geographical Information Systems* 1:203–18.

———; Calkins, H. W.; and Marble, D. F. 1976. *Computer handling of geographical data*. Paris: UNESCO Press.

U.S. Department of the Interior. 1985. *GIS implementation planning report*. Washington: GIS Technology Implementation Planning Committee.

White, W. B., and Morse, B. W. 1987. ASPENEX: An expert system interface to a geographic information system for aspen management. *AI Applications in Natural Resource Management* 1:49–53.

Wisconsin Land Records Committee. 1987. *Final report*. Madison, WI: Institute of Environmental Studies, University of Wisconsin-Madison.

Name Index

Abler, R. F., 10, 14–17, 359, 776, 791, 793–94
Abrahams, A. D., 80–82, 86–88, 114,
Abudaud, A. S., 532
Ackerman, E., 357, 359
Adam, D., 36
Adams, C. C., 31
Adams, J. S., 359
Adams, R. A., 399
Adams, R. B., 550
Adams, S., 761
Adams, T., 353
Aglietta, M., 591
Agnew, J. A., 162, 229, 588, 593, 596, 598, 601, 604–6, 658
Aguado, E. D., 113
Ahn, J., 713
Aiken, C. S., 341, 369
Aiken, R. S., 518, 520
Aitken, S. C., 210, 221–23
Ajzen, I., 225
Akbar, M. J., 374
Akhtar, R., 427
Al-Khameri, S., 506
Albers, P. C., 178, 243
Albokhair, Y., 533
Alexander, C. S., 143, 144, 494, 497
Alexander, D., 414
Alexander, J. W., 305
Alexander, L. M., 141, 142, 150
Alford, J. J., 337
Alfred, B. K., 700
Alijani, B., 52, 533
Allan, N. J. R., 345, 529
Allard, L., 704
Allen, J., 584, 633
Allen, J. C., 103, 416

Allen, J. P., 241
Allen, J. R., 76, 88, 141, 145, 146
Allton, D., 391
Alonso, W., 269, 359, 366, 372
Alt, B. T., 52
Altman, I., 219, 452
Alvic, D. R., 102
Alvine, M. J., 296
Ambrose, P., 621
Ambrosia, V. G., 756
Amedeo, D., 225
Amin, R., 526
Amin, S., 627, 628
Amrhein, C. G., 722, 737, 738
Anas, A., 725, 738
Anderson, B., 374, 603
Anderson, D. A., 119, 127
Anderson, J., 606–8, 760
Anderson, K. J., 212, 214, 569, 634, 664
Anderson, N. H., 733
Anderson, R., 370, 430, 434
Anderson, W., 97
Andressen, B., 574
Andrew, C., 631
Andrews, A. C., 675
Andrews, J. T., 55, 556
Angel, D. P., 623, 625
Angulo, J. J., 489
Ankerle, G., 213
Annenkov, V. V., 553–55
Annis, S., 434, 435
Anselin, L., 308, 721–23, 738
Antes, J. R., 696, 697
Anthes, R. A., 49
Antonini, G. A., 494
Anuta, M., 760
Arbona, S., 427, 433

Archer, C., 567
Archer, J. C., 583, 594, 595, 597, 598
Ardener, S., 673
Arey, D. J., 122
Armento, B. J., 14
Armstrong, M. J., 430
Armstrong, M. P., 782
Armstrong, R. W., 430, 433, 434
Armstrong, W., 479
Arnborg, L., 142
Arnell, N. W., 415
Arnfield, A. J., 49, 50
Arnold, F., 273
Arreola, D. D., 227, 493
Artibise, A. F. J., 564
Asaolu, A., 427
Aschmann, H., 241, 242
Ashley, G. M., 115
Astin, A. W., 14
Atkinson, B. W., 49
Aufhauser, E., 725
Augelli, J. P., 488, 495–97, 603
Ausubel, J. H., 58, 413, 415, 418
Avedon, E. M., 390
Avery, T. B., 749

Babcock, M., 125
Babitsky, T. T., 326
Bach, R. L., 277
Bach, W., 50, 59
Badcock, B., 657
Bagchi, D., 673, 679
Bahre, C. J., 32
Bailey, R. G., 35
Bailey, W., 427, 429
Bailie, J. G., 394

797

Baird, J., 221
Baker, A., 633
Baker, D. G., 55, 58
Baker, K. S., 759
Baker, R., 410
Baker, V. R., 83
Baker, W. L., 34, 35
Bale, J., 398, 399
Ball, J. M., 17
Ballard, K. P., 270
Ballard, P. L., 496
Ballas, D. J., 241, 243
Balling, R. C., 49, 55, 59
Baltensperger, B. H., 167, 169, 336
Band, L. E., 81
Bantock, P. R., 434, 435
Baougher, A. H., 105
Bar-El, R., 373
Barai, D., 527
Baran, P., 627
Barber, G. M., 316, 318, 322, 324, 325
Barbour, M. G., 29
Bare, J. K., 15
Barff, R., 723
Barker, M. L., 389, 495
Barnbrock, J., 621, 626
Barnes, D. F., 103, 416
Barnes, K., 101, 224
Barnes, T., 626
Barnum, H. G., 371
Barr, B. M., 548, 550
Barrett, E., 492
Barrett, W., 492
Barrows, H. H., 193, 194
Barrows, T. S., 15, 18
Barry, R. G., 50–53, 56, 57, 113, 759
Barsch, D., 83
Barth, F., 197
Bartholomew, D. J., 739
Bartlein, P. J., 36, 55, 56, 115
Barton, B., 164
Bashenina, N. V., 71
Bashshur, R. W., 434
Bassett, K., 633, 655
Bassett, T. J., 201, 469, 472, 473, 477, 628, 629
Bassin, M., 548
Bater, J., 548, 549
Batey, P. W. J., 308
Batson, R. M., 707
Batty, M., 736
Bauer, B. O., 145, 146
Baum, W. A., 48
Baumann, D. D., 117, 120, 121, 224, 415
Baumgardner, M. F., 754
Baxter, M., 725
Bayliss-Smith, T., 521
Beale, C. L., 267, 268, 352
Bean, F. D., 274

Beard, K., 701
Beatty, S. W., 34, 35, 37
Beauchamp, Y., 430, 431
Beaumont, J. R., 105
Beauregard, R., 634, 638
Becker, B. K., 605
Becker, J. M., 7
Becker, R. J., 244
Beckman, M. J., 729
Beeley, B. W., 532
Beesley, K. B., 566
Begovich, C., 323
Behr, M., 264
Beiswenger, R. E., 32, 36, 122
Belec, J., 621
Belisle, F. J., 494
Bendavid-Val, A., 373
Benhart, J. E., 370
Bennett, C. F., Jr., 30, 31
Bennett, J. W., 198
Bennett, R. J., 374, 721, 725, 727, 737
Bensel, R. F., 597
Berberian, J. H., 58
Berberian, M., 413, 415, 418
Berechmann, J., 326
Berentsen, W. H., 371, 551, 552
Berger, J., 630
Bergman, E. F., 369
Bergman, R. W., 200, 201, 491
Berlin, G. L., 749, 754
Berman, M., 634
Bernard, F. E., 375, 471, 477
Berry, B. J. L., 13–14, 266, 269, 311, 317, 352, 356, 357, 359–63, 368, 370, 373, 376, 410, 527, 654, 655, 657, 777, 778, 784
Berry, D., 334
Berry, J. K., 778
Berry, L. G., 104, 475, 482
Bertin, J., 691
Besag, J. E., 721
Best, A., 483
Bethell, L., 173, 174
Beyer, J., 673, 679
Beyers, W. B., 296
Beyers, W. H., 369
Bhardwaj, S. M., 427, 429, 434, 525
Bhaskar, R., 588
Bianchi, G., 434
Bieri, K. R., 693
Bigman, D., 730
Billick, I. H., 430
Birdsall, S. S., 453, 461, 573
Birkbeck, C., 494
Birkin, M., 343
Birks, J. S., 273
Bishop, B. C., 529
Bishop, Y. M. M., 724
Biswas, A. K., 59
Bjorklund, E. M., 210, 223

Black, W. R., 318, 321–23, 325
Blackbourn, A., 568
Blackwell, S., 453
Bladen, W. A., 416, 517, 528–31, 533
Blaikie, P., 202, 375, 413, 471, 472, 483, 629
Blake, G. H., 531, 584, 610
Blakemore, H., 497
Blakemore, M., 710, 792
Blanchard, B. J., 758
Blasing, T. J., 54
Blaut, J. M., 218, 228, 606, 608, 619, 629, 634
Blazar, W., 295
Blenman, M., 231
Blick, J. D., 151
Bloch, M., 161
Blouet, B. W., 497, 609
Blouet, E. M., 497
Blowers, A., 100
Bluestone, B., 363
Blynn, G., 521
Board, C., 779
Boddy, M., 621, 633, 636
Bodin, L., 782
Boeckh, J. L., 434, 439
Boehm, R. G., 9, 15, 151
Bogue, D. J., 352
Bohland, J. R., 245, 433–35, 437, 453–56, 458, 461, 658
Bohn, D., 242
Bohning, W. R., 272
Boland, J. J., 120, 121
Bolton, R., 106
Bond, A. R., 18, 548, 549, 551, 556
Bondy, P., 231
Bonine, M. E., 531–33
Boots, B., 723
Borchert, J. R., 13, 78, 295, 296, 340, 341, 368, 369, 410, 556, 654
Borts, G., 269
Boserup, E., 198, 470–71
Boss, T. R., 32
Bossler, J. D., 713
Botts, H., 177
Boulding, E., 418
Boulding, K. E., 16
Bourne, L., 651, 653, 654, 657
Bourne, R., 374
Bowden, M., 60, 169, 181
Bowlby, S., 320, 631, 673, 679
Bowler, I. R., 343, 344
Bowman, I., 31, 172, 353
Bowonder, B., 224
Box, E. O., 35, 36
Boyce, D., 317
Boyd-Bowman, P., 173
Boyer, M. C., 354
Boyle, M. J., 101
Bradbury, D. E., 32

Bradbury, J., 623, 630, 654
Braden, K. E., 548, 550
Bradley, R., 311
Bradley, R. S., 51–53, 57
Bragaw, D. H., 7, 14
Brana-Shute, C., 274
Brana-Shute, R., 274
Branch, M. C., 352, 367
Branscomb, L. M., 16
Brass, J. A., 756
Braudel, F., 161, 173, 633
Brazel, A. J., 52
Brea, J. A., 280
Breitbart, M., 631, 632
Brenner, M. E., 326
Brenner, R., 627
Briggs, J., 338, 344
Briggs, R., 311
Brinkmann, W. A. R., 54–56, 113
Britton, R. A., 388, 392
Brklacich, M., 344
Broadbent, G., 213
Brodrick, J. R., 105
Brodsky, H., 434
Broek, J. O. M., 507
Bromley, R., 371, 373, 493–95, 497, 628
Bromley, R. D. F., 494, 497
Brook, G. A., 114
Brooker-Gross, S., 229, 677, 679
Brookes, I. A., 115
Brookfield, H. C., 194, 195, 198, 199, 202, 375, 413, 472, 483, 629
Brooks, E., 126, 127, 129, 131
Brooks, W. D., 702
Browett, J., 628, 654
Brown, A. V., 78
Brown, B., 395
Brown, B. J., 34, 52, 420
Brown, C. L., 296
Brown, D. A., 35, 337
Brown, D. E., 33
Brown, J. A., 49
Brown, L. A., 260, 271, 273, 277–81, 494
Brown, M. A., 103, 104, 225
Brown, M. D., 51
Brown, P., 198, 199
Brown, R. E., 158, 180
Brownill, S., 621, 631
Brubacher, J. S., 2, 9, 10
Brubaker, D. L., 2, 7
Brummell, A. C., 260
Brunn, S. D., 101, 271, 415, 583, 602, 609, 610, 654
Brush, G. S., 35, 37
Brush, J. E., 507, 526, 527
Brush, S., 201, 205
Brussock, P. P., 78
Bryan, M. L., 761
Bryant, C. R., 334

Bryant, N., 761
Bryant, P. T., 353
Bryson, R. A., 56, 530
Buckingham, W. F., 754
Buckley, W., 198
Buge, C., 245
Bunge, W., 360, 619, 779
Bunker, S. G., 494
Bunting, B. T., 567
Bunting, T. E., 230
Burbank, B. K. R., 415, 418
Burchell, R. W., 363
Bureau, L., 570
Burgess, E., 652, 661
Burgess, R., 628
Burgess, W. E., 194
Burghardt, A. F., 573, 602, 604, 659
Burk, J. H., 29
Burnett, A., 583, 585, 604
Burnett, K. P., 222, 320, 322
Burns, K., 303
Burns, L. S., 357
Burright, R. G., 764
Burrough, P., 794
Burt, J. E., 50, 59, 222, 261, 723
Burton, I., 194, 410, 412, 418–420
Buss, T., 658
Bustamante, J., 273
Busteed, M. A., 585
Butler, D. R., 33, 34, 58, 80, 81, 598
Butler, L. M., 129, 130, 339
Butler, R. W., 390, 393, 395
Butlin, R. A., 178
Butt, P. L., 431, 434
Buttel, F., 630
Buttenfield, B. P., 704
Buttimer, A., 227
Butzer, K. W., 14, 55, 59, 84, 88, 193–95, 198, 204, 211
Byerts, T. O., 452
Byrne, R., 36, 37, 54

Cadwallader, M. T., 218, 221, 653, 656, 667
Caine, N., 76, 80, 83
Calkins, H. W., 699, 787, 794
Calzonetti, F. J., 97, 99, 100, 102
Cameron, B., 82
Cameron, D. M., 101, 471, 472
Campbell, J. B., 749, 753, 754, 758, 764
Campbell, W. H., 530
Cañadas, L., 201
Cannon, J., 568
Cantor, T. M., 338
Capel, H., 604
Carey, G. W., 430, 431
Carleton, A. M., 51–53, 60

Carlson, A., 241, 242
Carlson, F. A., 651
Carlyle, W. J., 339, 567
Carneiro, R., 197
Carney, J., 473, 476, 629, 675
Caro, R. A., 353
Carr, C. J., 195, 200
Carroll, G., 654
Carroll, T., 495
Carruthers, G., 358
Carson, R. T., 123
Carstensen, L. W., 695, 706, 708
Carter, J. E., 260
Carter, T. R., 58
Cartin, K. F., 399
Casetti, E., 294, 369, 723
Castells, M., 593, 594, 632, 637, 666
Castner, H. W., 696
Catalano, A., 621
Catau, J., 262
Catchpole, A. J. W., 55
Caviedes, C. N., 53, 115, 496, 497
Cecil, R. G., 575
Cerny, J. W., 709
Cerveny, R. S., 49, 59
Chadwick, R. A., 388
Chakrapani, K., 427
Chakravarti, A. K., 521, 522, 524–26, 528
Chalmers, J., 100
Chan, K. W., 512
Chang, A. T. C., 767
Chang, C., 513
Chang, H. H., 146
Chang, K-T., 692, 696, 697
Chang, S., 755
Chang, S-D., 510, 511, 514
Changnon, S. A., 54, 58, 59, 113
Chao, T. K., 513
Chapdelaine, M., 570
Chapman, B. R., 335
Chapman, M., 280, 519
Chardon, R., 492
Charnes, A., 370
Chavez, P. S., 754
Cheema, G. S., 372
Chen, C-K., 98, 323
Chen, R. S., 418
Cherry, D. L., 757
Cheung, M., 303
Chicoine, N., 676
Child, J., 496
Chisholm, G. G., 301
Chisholm, M., 371
Chorley, R. J., 195, 317
Chouinard, V., 619, 636, 637, 657
Choy, D., 392, 393
Chrisman, N. R., 699–701, 779, 786, 792
Christaller, W., 352, 393, 395, 651

799

Christensen, D. E., 355
Christensen, E. J., 753
Christopher, A. J., 172
Christopherson, S., 306, 621, 623, 628, 631, 633, 676, 679
Chubb, H. R., 391
Chubb, M., 391
Church, M., 115, 556
Church, R. L., 102, 324, 434, 729
Churchill, R. A., 229
Churchill, R. R., 571
Churchill, W., 245
Cirrincione, J. M., 8
Clark, A. H., 30, 158, 160, 165, 168, 177, 180
Clark, C., 655, 663
Clark, G. L., 270, 292, 293, 306, 361, 363, 364, 368, 369, 434, 585–87, 592, 600, 620, 622–24, 626, 633, 636, 658, 668
Clark, J. L. D., 15, 18
Clark, J. R., 323, 533
Clark, M. J., 415
Clark, S., 434
Clark, T. J., 634
Clark, W. A. V., 222, 260–62, 651, 656, 725, 732, 733, 735
Clark, W. C., 414, 419
Clarke, A., 392
Clarke, C., 245
Clarke, D. L., 198
Clarke, D. R., 434
Clarke, D. W., 339
Clarke, G. C., 494
Clarke, G. M., 115
Clarke, J., 572
Clarke, K. C., 705
Clarke, M., 728
Clarke, S. E., 658
Clarke, W. C., 199
Clarkson, J. D., 194, 198
Clawson, D. L., 490, 494, 495, 520
Clawson, M., 370, 373, 388, 401
Cleaveland, M. K., 55
Cleaves, E. T., 76
Clem, R., 549
Clements, F. E., 31
Cleveland, C. J., 95, 99
Cliff, A. D., 352, 721, 722, 725
Clifton, J. A., 178
Cline-Cole, R. A., 229
Coase, R., 623
Coates, B. E., 371
Coates, D. R., 76, 86, 87
Cobban, J. L., 518, 519
Cobbe, J., 274
Cockburn, C., 599, 600, 636
Cocklin, C. R., 103, 335, 339, 341, 343
Coffey, B., 570
Coffey, W. J., 296

Cohen, S. B., 18, 142, 507, 531, 532, 583, 603, 609, 610
Cohen, S. J., 52, 58, 571
Cohen, Y. S., 229
Cohon, J. L., 100
Cole, D., 692, 694
Cole, D. N., 32
Coleman, J. M., 35
Colenutt, R., 621
Colinvaux, P., 29, 37
Collins, C., 394
Collins, R., 161
Collins, W., 755
Colten, C. E., 181
Colwell, R. N., 746, 749
Concord, C. M. S., 455
Condit, H. R., 754
Conkey, L. E., 33, 36, 55
Conkey, M. W., 212, 662
Conkling, E. C., 162, 361
Connell, J., 279
Conway, D., 273, 274, 522, 527
Conzen, K. N., 167, 181
Conzen, M. P., 166–68, 176, 181, 710
Cook, D. G., 720
Cook, E., 97, 100
Cook, N. D., 492
Cooke, P., 619, 631, 633, 638
Cooke, R. U., 71, 73, 76, 180, 414
Cooper, S. H., 355
Coppock, J. T., 387
Corbett, J., 780
Corbridge, S., 162, 375, 628, 629, 638
Cordes, L. S., 397
Corey, K. E., 517
Corrigan, P., 52, 60
Cosgrove, D. E., 159, 209, 212–15, 228, 630, 634, 662
Costa, F. J., 371, 509, 533
Costa, J. E., 72, 76, 77, 85, 88
Costanza, R., 99
Costanzo, C. M., 723–24
Costello, M. A., 519
Costello, M. P., 519
Cotton, J. W., 735
Couclelis, H., 219, 220, 222, 223, 728, 735–36
Cousins, L. R., 230
Covello, V. T., 224
Cowen, D. J., 708, 783
Cowles, H. C., 31
Cox, C. W., 692
Cox, K. R., 219, 223, 229, 585, 588, 592, 598, 599, 619–21, 634–37
Cox, N. J., 700
Cragg, B., 677
Craig, A. K., 493
Crane, R. G., 52, 53
Crary, D., 507

Crawford, N. C., 114
Cressey, G. B., 506, 507, 509, 511, 531, 550
Cribier, F., 452, 455
Crippen, R. E., 759
Cromartie, J. B., 271, 430, 434, 437
Cromley, E. K., 230, 434, 437, 456, 458, 461
Cromley, R. G., 292, 303, 661, 704
Cronon, W. J., 204
Crooker, R. A., 520
Crossley, J. C., 343
Crouchley, R., 734
Crowley, J. M., 32
Crush, J., 473, 479–80, 483, 628, 629
Csikszentmihalyi, M., 213
Cuff, D. J., 97, 693
Cui, G., 511
Cullen, B. T., 104, 117
Cumberland, J. H., 352, 358
Cummings, H., 245
Cunningham, O. R., 434
Cunningham, S. M., 371
Curran, P. J., 746, 747, 749
Currey, B., 418, 427, 429, 430
Currey, D. R., 55
Curry, L., 326, 653, 725, 728
Curry-Roper, J. M., 243
Curtice, J., 597
Curtis, F. A., 104
Curtis, J. T., 38
Cutter, S. L., 101, 222, 224, 225, 418, 430, 432, 596
Cybriwsky, R., 516
Czerniak, R. J., 244

Dacey, M. F., 358, 360, 653
Daganzo, C. F., 731
Dagodag, W. T., 273
Dahlberg, R. E., 687, 707, 763
Dahmann, D. C., 269
Daiches, S., 454
Damron, J. E., 119
Dando, W. A., 430, 547
Dangermond, J., 782
Daniels, P. W., 295–97, 655, 657
Daniels, S., 634
Danielson, M. N., 368
Danta, D. R., 551, 552
Darby, H. C., 158, 663
Darden, J. T., 658
Darmody, R. G., 81
Daroozi, E., 552
Datel, R. E., 176, 227, 229
DaVanzo, J., 271
Davgun, S., 524, 525
David, T. G., 230
Davidson, J. V., 703
Davidson, W. V., 489, 497
Davies, J. A., 49

Davies, M., 584, 606
Davies, R. B., 230, 734
Davies, W. K. D., 656
Davis, A. M., 55
Davis, D., 141, 146
Davis, D. H., 506, 509, 510
Davis, F. W., 37
Davis, J. K., 59
Davis, M., 622, 634
Davis, P. A., 754
Davis, R. E., 767
Davis, W. M., 2, 3, 10, 31, 353, 659, 667
Dawsey, C. B., 494
Day, C. O., 790
Day, M., 497
Day, M. J., 73, 80
de Blij, H. J., 337, 482, 483, 583
de Kleine, R., 496
de la Blache, P. V., 161, 361
de Laubenfels, D. J., 29
de Lisle, D., de G., 568
de Simone, M., 792
de Souza, A., 15, 16, 483, 573
de Vorsey, L., 181, 242
de Witte, M., 468
Dear, M. J., 214, 245, 363, 364, 434, 438, 439, 585–88, 592, 600, 620, 622, 635–38, 658, 668
DeAre, D., 267, 268
DeFanti, T. A., 51
DeJanvry, A., 628
DeJong, G. G., 271
Delson, R. M., 492
DeLucia, A. A., 696, 701
Demars, S. E., 146, 394
Demko, G. J., 17, 372, 547–49, 552–55
Denevan, W. M., 32, 173, 192, 194, 197, 198, 201, 203, 204, 245, 490, 491, 495, 497
Dennis, R., 171
Densham, P. J., 738, 782
dePalma, A., 728
der Heijden, V., 221
Derr, P. G., 101
Desbarats, J., 222, 229, 271, 274, 275, 325, 520
Deskins, D. R., 661
Despain, A. M., 50
DesRivieres, D. L., 707
Detro, R. A., 146
Deurloo, M. C., 261, 725
Dewey, K. F., 52
Dey, B., 122, 530
Dhussa, R. C., 524, 527
Diaz, H. F., 53, 54
Dickason, C., 341
Dickason, D. G., 506, 518, 521, 522, 527
Dickenson, J. P., 492, 496

Dickinson, J. C., 493, 494
Dieleman, F. M., 261, 725
Dienes, L., 97, 98, 549–52, 555, 556
Dikshit, R. D., 606
Dilley, R. S., 389
Dillon, C. L., 296
Dilsaver, L. M., 181
Dingemans, D. J., 176, 227
Dixon, J. C., 78, 81
Dixon, R., 676
Dobson, J. E., 100, 793
Dobson, M. W., 694, 696
Dodge, S. L., 35
Doeppers, D. F., 518, 519
Dogan, M., 373
Doherty, J., 483, 619
Doig, J. W., 368
Dolan, L. S., 80, 81
Dolan, R., 145
Domencich, T. A., 325, 730
Domosh, M., 212, 215
Donley, M. W., 241
Doolittle, W. E., 192, 201, 204, 205, 491, 495
Dorigo, G., 725
Dorn, R. I., 55, 80
Dort, W., 81
Doucet, M. J., 171, 564
Doughty, R. W., 32, 227
Douglas, J. N. H., 610
Downie, M. C., 453
Downing, T., 472
Downs, R. M., 218, 221, 230
Downton, M. W., 59
Doyon, R., 707
Dozier, C. L., 174
Dozier, J., 49, 706, 748, 757, 762, 766, 767
Drake, C., 519, 679
Drake, J. J., 114–16
Driever, S. L., 491
Driver, H. E., 178
Drucker, P. F., 338
Drysdale, A., 531, 532, 603
Dubayah, R. O., 762
Dubin, R., 323
Dudycha, D., 396
Dueker, K. J., 783
Dufournaud, C. M., 103
Dunbar, G. S., 16, 493
Duncan, J. S., 212–14, 218, 228, 242, 638, 656
Duncan, N. G., 212–14
Duncan, S. S., 600, 601
Dunford, D. J., 18
Dunlap, R. E., 225
Dunleavy, P., 597, 598
Dunn, R., 734
Dunnell, R., 211
Durkheim, E., 667

Durrenberger, R., 369
Dury, G. H., 76
Dutt, A. K., 371, 426–28, 517, 518, 523–26, 531
Dutta, H. M., 426, 427, 531
Dutton, G., 704
Dworkin, J. M., 121, 131
Dyck, H., 576
Dynneson, T. L., 15
Dzerdzeevski, B., 53
Dziegielewski, B., 122
Dziewonski, K., 653

Eagle, T. C., 222
Eagleson, P. S., 116
Earickson, R. J., 430
Earle, C. V., 163, 164, 167–69, 172, 177
Easley, J. A., Jr., 3, 7
Easterling, W. E., 59, 113
Easterly, E., 150
Eastman, J. R., 694, 696
Easton, D., 595
Eborall, D., 679
Ede, K., 478
Edmonds, W. T., 754
Edmondson, B., 400
Edncy, M. H., 711
Edungbola, L., 427
Edwards, C., 492
Edwards, K., 707
Ehlers, M., 763
Ehrenberg, R. E., 175
Ehrlichman, J. D., 366
Eichen, M. A., 102
Eidt, R. C., 491
Eighmy, T. H., 373
Eilers, E. H., 35, 36
Eitzen, D. S., 399
Elbow, G. S., 490, 493
Eliot Hurst, M. E., 316
Ellen, R., 196
Ellet, C., 319
Elliott, L. H., 19
Elliott-Fisk (Elliott), D. L., 34, 53, 55, 82, 566
Ellis, M., 733
Elo, I. T., 268
Elton, W., 30
Emel, J. L., 126–29, 131
Enders, W. T., 371, 434
Eney, A. B., 117
Engerman, S., 172
England, J., 51, 52
England, K. V. L., 631, 676, 678
English, P. W., 531
English, R., 7
Ennis, P., 598
Enslin, W. R., 764, 765
Entrikin, J. N., 211
Eppley, R. W., 759

801

Erb, R., 575
Erickson, R. A., 293, 297, 547, 655
Ernst, J. A., 164
Escobar, B., 36
Escobar, M., 492
Espenshade, E., 554
Estes, J. E., 749, 752, 756–59, 763, 764, 769
Evans, G. W., 230
Everitt, J., 493, 656
Ewald, U., 492
Ewel, J. J., 34
Ewers, H-J., 373
Ewing, G. D., 389
Ewing, G. O., 322
Eyre, J. D., 506, 515, 516
Eyton, J., R., 399

Fabbri, P., 147
Fainstein, S., 622
Fairbridge, R. W., 57
Fairweather, M., 574
Faludi, A., 361
Farhi, A., 626
Farmer, R., 8
Farrell, B. H., 397
Farrell, R. T., 8
Farrington, I. S., 203
Fast, T., 674–75
Faust, K., 326
Fawcett, J. T., 271
Feagin, J., 621
Fedor, T. S., 548, 551, 552
Feldman, M., 621, 636
Fellmann, J. D., 652
Fels, J., 691
Fergany, N., 274
Fernie, J., 97, 101
Fesenmaier, D. R., 390, 391, 398
Fielding, A., 267
Fielding, G. J., 322, 326
Fienberg, S. E., 724
Fik, T. J., 222
Fincher, R., 593, 619, 620, 635, 637, 657, 677
Finger, F. G., 51
Fingleton, B., 720, 721, 725
Fink, J. L., III, 456, 458
Fink, L. D., 17
Fischer, C. S., 262
Fischer, M. M., 725, 730, 732
Fischhoff, B., 223, 224
Fishbein, M., 225
Fitzsimmons, M., 343, 619, 629–31
Flad, H., 245
Fleisher, P. J., 77
Fleri, E., 55
Flohn, H., 56
Florida, R., 621, 625, 636
Florin, J. W., 425, 573
Flowerdew, R., 260
Flynn, C. B., 452

Foggin, P., 429, 430
Foglesong, R. E., 354
Folmer, H., 725
Foord, J., 631, 632
Foot, S., 626
Foote, K. E., 182, 212–14, 227, 231
Foote, S., 325
Foraie, J., 245
Forbes, D. K., 279, 280, 375, 468, 469, 483
Ford, D. C., 114
Ford, L. R., 175, 181, 219, 227, 389, 493
Forde, C. D., 197
Foresta, R. A., 338
Forward, C. N., 566
Fosberg, F. R., 30
Foster, H. D., 125, 415, 430
Foster, J. L., 757
Foster-Carter, A., 627
Fotheringham, A. S., 269, 270, 319, 322, 724–26, 728
Fournier, J. P., 147
Fowler, G. L., 100, 105
Fox, H., 115
Fox, M., 677
Fraczek, I., 692
Francaviglia, R. V., 176
Frank, A. G., 623
Frank, A. U., 792
Frank, T. D., 756
Franke, R. W., 477, 478
Franklin, J., 35, 37, 755, 767
Frech, P., 434, 456, 461
Frederic, P., 350, 572
Fredrich, B., 32
Freeman, D. B., 179, 241, 243, 479, 481
Freeman, D. H., 434
Freeman, H., 713, 792
Freeman, T. W., 354
French, R. A., 548
Frenkel, R. E., 32, 34
Frenkel, S., 480
Frew, J., 706, 766
Frey, A., 352
Frey, W. H., 268
Frey-McClung, V., 453, 461
Friedland, W., 630
Friedlander, S. C., 102
Friedmann, H., 628, 630
Friedmann, J., 277, 351, 356–60, 372, 376
Fryer, D. W., 517, 518
Fuchs, R. J., 372, 547, 549, 551, 552, 554, 555
Fudge, C., 636
Fuggle, R. F., 35
Fuguitt, G. V., 267, 268
Fuller, D. L., 242
Fuller, G., 488
Fung, T., 761

Furuseth, O. J., 334–36, 340, 341
Fussell, J., 746, 747, 780

Gade, D. W., 32, 229, 489, 492, 496, 570
Gaete, A., 658
Gage, G., 434
Gagliardini, D. A., 759
Gaile, G. L., 95, 277, 320, 324, 373, 479, 494, 513, 654, 721, 723
Gale, N., 37, 221, 736
Galloway, G. E., 124, 131
Galloway, J. H., 492
Garcia, A., 489
Gardiner, V., 88
Gardner, D. P., 2
Gardner, J., 36, 58
Garrett, M. J., 16, 17
Garrison, W. L., 317, 360, 362
Gatrell, A. C., 399, 722
Geddes, P., 353, 361, 376
Geertz, C., 197
Geiger, R. K., 637, 658
Geisler, C. C., 246
Georgianna, T. D., 99, 338, 340, 341
Gerhard, D., 172
Gerhard, P., 173, 492
Gerlach, J., 147
Gerlach, R., 169
Gersmehl, P. J., 30, 35, 337, 340, 341, 344, 697
Gerson, R., 115
Gertler, M. S., 292, 294, 306, 368, 620, 623–26, 633
Gesler, W. M., 425, 427, 429, 430, 434, 435, 437
Getis, A., 37, 723
Ghosh, A., 326, 781
Giardino, J. R., 75, 80, 81
Gibbins, R., 369
Gibbs, J., 266
Gibson, A., 674–75
Gibson, J., 548
Gibson, K., 626
Giddens, A., 161, 212, 588, 633, 667
Gier, J., 632
Gilbert, A. G., 229, 372, 483, 494, 634
Gilbert, W. H., 244
Gill, A., 221, 222
Gillard, Q., 370
Gillespie, W. M., 319
Gillis, R. P., 569
Gilman, C. R., 706
Gilmartin, P. P., 689, 692–96, 711
Ginsburg, N., 507, 513, 518
Giuliano, G., 322, 326, 677
Glacken, C. J., 412
Glantz, M., 60, 470
Glaser, B. G., 374

802

Glasmeier, A., 299, 369, 623
Glassie, H., 175, 176
Glassner, M. I., 150, 496, 532, 583
Glick, B., 430
Glickman, N., 308
Gober, P. (Gober-Meyers), 262–65, 277, 278, 434, 437, 453, 455, 677
Goddard, J. B., 373, 656
Godfrey, B. J., 494, 634
Godfrey, M. A., 334
Godon, D., 429, 430
Goetz, A. R., 280
Goheen, P. G., 171, 572
Gohstand, R., 548, 554
Golant, S. M., 231, 262, 452–58, 461
Gold, J. R., 218, 219, 221
Goldberg, M. A., 564, 565
Golden, B. L., 782
Goldstein, H. A., 311
Goldstein, S., 279
Golledge, R. G., 218–23, 230, 258, 655, 656, 723–24, 735
Gonzalez, A., 495
Gonzalez Casanova, P., 373
Good, C. M., 373, 427, 434, 437
Goodchild, M. C., 324
Goodchild, M. F., 82, 391, 737, 778, 782–84, 792
Goode, J. P., 192
Goodey, B., 221, 245
Goodlett, J. C., 35
Goodman, B. M., 56
Goodman, J. J., 179
Goodman, J. M., 9, 179, 240, 241, 243–46
Goodman, R., 375
Goodwin, G. C., 241, 243
Goodwin, M., 600, 601
Goodwin, R. F., 151
Gorbachev, M., 550
Gordon, B. L., 32
Gordon, J. J., 229
Gordon, P., 267
Gordon, R. J., 434, 437
Gordon, S. I., 100
Gore, C., 373–75
Gosling, L. A. P., 519
Gottdiener, M., 213, 368, 666
Gottlieb, R., 631
Gottmann, J., 266, 295, 583, 604
Goudie, A., 179
Gould, M. D., 712
Gould, P. R., 172, 218, 319, 359, 371, 399, 584, 721, 736
Goward, S. N., 758, 767
Gradus, Y., 371
Graebner, N. A., 374
Graeme, H., 418
Graf, W. L., 71, 72, 76, 77, 80–83, 85, 88, 114–17, 180, 244

Graff, T. O., 453
Graham, B., 660
Graham, J., 619, 633, 638
Gramsci, A., 601, 607
Granger, O. E., 54, 55, 116, 337, 497
Graumlich, L. J., 36, 37, 55
Graves, F. W., 707
Gray, E., 335
Gray, F., 657
Green, G., 341
Green, J. L., 368
Green, M., 98, 295, 296, 550
Greenberg, D., 622
Greenberg, M. R., 370, 415, 430–33
Greene, D. L., 98, 99, 105, 322, 323
Greene, G. M., 49
Greenland, D., 60, 430, 432
Greenow, L. L., 492, 660
Greenwood, B., 88, 145, 146
Greenwood, M. J., 278
Greer, C., 510, 513
Greer, D. C., 241
Gregor, H., 335, 336
Gregory, D., 620, 632, 633
Gregory, K. J., 71
Gregson, N., 632, 633
Greis, N. P., 115
Gribb, W. J., 244
Griffin, D. W., 355
Griffin, E., 389, 493
Griffin, P., 701
Griffith, D. A., 368, 721, 722, 725, 728, 737
Griffith, P., 430
Grim, R. E., 175
Grima, A. P. L., 119, 121
Gritzner, C. F., 9, 14–17, 176
Groop, R. E., 692, 708
Gros, S., 126, 129, 339
Grossman, J. B., 274
Grossman, L., 192, 200, 629
Grosvenor, G. M., 19
Groves, P. A., 159, 167, 170, 178, 241
Gruntfest, E. C., 102, 124, 126, 418
Guelke, L., 159, 172, 663, 664
Guest, A., 271
Guitterrez, P. R., 273
Gully, J. L. M., 172
Gupta, A., 115, 116
Gurnell, A. M., 415
Gurney, C., 764
Gustafson, N. C., 369
Gustke, L., 396
Guyette, S., 246
Guyot, A. H., 3

Haag, G., 728
Haack, B. N., 761
Haas, J. D., 6

Haas, J. E., 122, 412
Haberkorn, G., 271
Hack, J. T., 35
Haddock, K., 427
Hadjimichalis, C., 588
Hadley, K. S., 29
Hadley, R. F., 77
Hafner, J. A., 506, 519, 520
Hagerstrand, T., 161, 162, 276, 359, 369, 623
Haggett, P., 317, 352, 655
Haglund, D. K., 567, 574
Haigh, M. J., 16
Haila, A., 621
Haining, R., 721, 725, 727, 737
Hajic, E. J., 749, 764
Hakkert, A. S., 434
Hale, G. A., 203
Hall, C., 99, 371, 493
Hall, G. B., 434, 435, 438, 439, 455, 565
Hall, P., 162, 299, 300, 342, 352, 354, 355, 362, 364, 369, 372–73, 583, 623
Hall, R. B., 506, 509
Hall, R. B., Jr., 515
Hallala, W., 763
Hallman, S. P., 453, 461
Halperin, W., 734
Halpin, J., 55
Hamilton, F. E. I., 290, 291, 549
Hamley, W., 103
Hamnett, C., 565, 621
Hancock, J. A., 452
Hand, I., 367
Handcock, W. G., 166
Handmer, J., 414
Hanham, R. Q., 431, 434, 438, 453, 454
Hanink, D. M., 292, 302, 303, 324, 494, 654
Hanna, D., 171
Hannen, M. T., 734
Hannon, B., 105
Hansen, E. R., 679
Hansen-Bristow, K. J., 34, 35
Hanson, N. R., 210
Hanson, P., 322, 456
Hanson, R., 361, 368
Hanson, S., 222, 318, 322, 323, 329, 622, 631, 673, 674, 676–81
Hanten, E. W., 512
Hanushek, R., 260
Happel, S. K., 460
Harding, D. M., 414
Hare, F. K., 48, 58, 59
Hargrove, E. C., 367
Harker, P. T., 322
Harley, J. B., 709–11
Harman, J. R., 35, 52, 54, 533
Harmon, J., 399, 574
Harmon, R. S., 55

803

Harnapp, V. R., 495, 498
Harper, R. A., 15–17
Harper, S., 343
Harries, K. D., 59
Harrington, J. A., 52, 119, 746, 780
Harrington, J. W., Jr., 294, 303
Harris, B., 355, 361, 363
Harris, C. D., 14, 335, 515, 516, 547, 549, 554, 556
Harris, D. R., 198
Harris, R., 171, 565, 621
Harris, R. C., 162, 163, 166, 167, 172, 498, 572, 573
Harris, R. J., 261, 273
Harris, W. W., 610
Harrison, B., 355, 363
Harrison, M., 245
Harrison, P. D., 203, 470, 474, 491
Hart, J. F., 334–37, 339, 341, 369
Hart, M. G., 71
Hart, R., 230
Hartnett, S., 177
Hartshorne, R., 156–58, 362, 582–83, 602, 663
Hartsock, N., 632
Harvey, D. W., 228, 305, 362–65, 584, 588–95, 599, 601, 608, 610, 619–22, 624, 625, 629, 632–34, 652, 653, 661, 662, 667
Harvey, L. E., 37
Harwell, M. A., 412, 418
Hauptman, L. M., 172
Hausladen, G., 489, 549
Hawke, S. D., 7, 8
Hawley, A. H., 194
Hawley, F., 399
Hay, A. M., 316, 373
Hay, D., 373
Hay, J. E., 49, 53
Hayden, D., 631, 677
Hayes, J. J., 675
Hayes, J. T., 50
Haynes, K. E., 99, 269, 319, 338, 340, 341
Hays, F. B., 353
Hayuth, Y., 151
Head, C. G., 793
Healy, R. G., 337, 370
Heatwole, C. A., 147, 149, 391
Hecht, M. E., 241
Hecht, S. B., 32, 201, 494, 628–31
Hechter, M., 605
Heckman, J. J., 734
Heckscher, E., 293
Hecock, R. D., 16, 17, 148
Heffernan, W., 341
Heidenrich, C. E., 178, 241
Heim, R., Jr., 52
Heiman, M. K., 353, 370, 619, 621, 630, 631
Heimlich, R. E., 341
Heinitz, E. F., 34

Hekstra, G. P., 58
Helburn, N., 6, 17, 20
Helburn, S. W., 6
Hemming, J., 173, 375
Henderson, F. M., 760
Henderson, M. L., 243
Henderson-Sellers, A., 48–50
Hendriks, P., 735
Henkel, R., 244
Henretta, J., 163
Henry, N. F., 434
Hensher, D., 222, 730, 732
Hepple, L., 610
Hepworth, M., 295, 296
Herbert, D. T., 458
Herkstra, G. P., 418
Herman, G. P., 49
Hertzog, S., 171
Herwitz, R., 35, 36
Heskin, A., 262
Heth, C., 246
Hewes, L., 168, 243
Hewings, G. J. D., 291, 306, 309
Hewitt, K., 410, 418, 629
Hickcox, D. H., 99, 245
Hicks, D., 311
Hidore, J. H., 533
Hiebert, P., 33, 35
Higgins, B., 178, 245
Higman, B. W., 173
Hill, A., D., 17, 18
Hill, M., 227, 231
Hilliard, S. B., 167, 169, 243
Hills, T. L., 33
Hillsman, E. L., 95, 102
Hiltner, J., 453, 456, 461
Hiraoka, M., 495
Hirschboeck, K. K., 56, 114, 115, 124
Hirschman, A. O., 359
Hirshberg, T. W., 177
Hite, J. C., 334, 338, 345
Hoare, A., 97
Hobbs, E. R., 32
Hobgood, J. S., 49
Hoch, C., 593, 621
Hodder, I., 210, 211, 720
Hodge, D. C., 224
Hodge, G., 355
Hodge, J. S. C., 371
Hodgson, G., 626
Hodgson, M. E., 752, 756, 764
Hodgson, M. J., 704
Hodler, T. W., 707
Hoetink, H., 498
Hoffman, G. W., 97, 372, 549, 551, 552, 604
Hoffman, J., 325
Hoffman, R., 163, 164
Hoffpauir, R., 526
Hogan, T. D., 460
Hogg, H. C., 35

Hohenemser, C., 412
Hohenemser, H., 224
Holcomb, B., 229, 631, 658, 677
Holdgate, M. W., 224
Holdsworth, D., 570, 573
Holland, P. G., 35, 36
Holland, P. W., 724
Holland, S., 375
Holliday, V. T., 56, 84
Holloway, J., 586
Holly, B. P., 296
Holmes, J. H., 172, 621, 623
Holton, G., 15
Holtzclaw, G. D., 318, 325
Holz, R. K., 749
Hooson, D., 547, 550, 551
Hoover, E. M., 263
Horn, B. K. P., 705
Horn, J., 36
Hornbeck, D., 167, 177, 242
Hornsby, S., 569
Horowitz, J. L., 320–23, 733
Horsfield, K., 82
Horsman, R., 178
Horst, O. H., 490
Horton, F., 361
Hosier, R. H., 103, 476, 481
Houghton, D. D., 57
Hourihan, K., 229
House, J. W., 310, 584
Hovinen, G. R., 393
Howard, A. D., 76
Howard, P., 229
Howarth, D. A., 48, 53
Howe, A. L., 453, 461
Howe, H. V., 142
Howes, D. R., 455, 460
Hoy, D. R., 491, 494, 495
Hoyt, H., 652
Hsieh, C-M., 515
Hsu, M-L., 510, 511, 515, 690, 691, 710, 788
Hsu, S-Y., 704, 764
Hua, C., 372
Huang, N. E., 759
Hubbard, R., 14
Hubert, L., 723–24
Hudman, L. E., 275, 397
Hudson, B., 604
Hudson, J. C., 1, 16, 166, 168, 170, 172
Hudson, R., 369, 585, 593, 633
Huff, J. O., 222, 260, 261, 322, 656
Hughes, J., 295
Hughes, M. A., 658
Hugill, P., 396
Hugo, G. J., 280, 427, 429, 430, 453, 455
Huke, R. E., 507, 508, 519–20, 530
Hummon, D. M., 226

Hunter, J. M., 426, 427, 430, 432–34, 439, 496
Huntington, C. C., 651
Huntington, E., 31, 160, 506, 507, 509
Hupp, C. R., 33, 35
Huriot, J-M., 626
Husband, E., 219
Husbands, W. C., 394, 395
Hutchinson, C. F., 752, 755
Hutchinson, D. M., 229
Hutchinson, T. C., 412, 418
Hutton, T., 296
Hyma, B., 393, 427, 434, 437
Hymer, S., 623

Ickes, H., 356
Iijima, S., 529, 531
Infanto, R. P., 658
Ingalls, G. L., 609
Ingram, D. R., 434
Innes, F., 430
Inskeep, E. L., 355
Ironside, R. G., 568
Irvine, K. N., 115, 116
Isard, W., 13, 308, 310, 352, 356–59, 361, 376, 410
Isham, V., 727
Issel, H. L., 243
Isserman, A. M., 99, 269, 553
Ives, J. D., 34, 83, 514, 529
Ives, P., 83

Jackson, A. L., 52, 60
Jackson, D. W., 80
Jackson, E. L., 102, 104, 229, 389
Jackson, J. B., 212, 215
Jackson, J. C., 518
Jackson, J. K., 368
Jackson, J. N., 355
Jackson, P., 159, 209, 215, 334, 634, 758
Jackson, R. H., 275, 370
Jackson, W. A. D., 515, 547, 548, 550, 551, 583
Jacobs, W. C., 57
Jacobs, W. R., 178
Jain, A., 764, 765
Jakle, J. A., 169, 396
Jakubs, J. F., 658
James, P. E., 3, 4, 14, 15, 31, 489, 497
Janelle, D. G., 322, 324, 585
Janiskee, R. L., 181, 387
Janke, R. A., 241
Jannace, R., 763
Jansen, S. D., 105
Janson, B. N., 104
Jaworski, E., 151
Jefferson, M., 316
Jenkins, A., 610
Jenkins, J. R. G., 606

Jenks, G. F., 688, 695, 700, 703, 704
Jensen, J. R., 687, 704, 747–49, 753, 756–58, 760, 761, 764, 769
Jensen, M., 352
Jensen, R. C., 306, 308, 309
Jensen, R. G., 547, 549–51
Jett, S. C., 178, 241–43
Jocher, K., 362
Joerg, W. L. G., 172
Johannessen, C. L., 489, 491, 496, 497
Johannessen, C. W., 32
Johnson, D. L., 37, 55, 114, 116, 200
Johnson, D. M., 54
Johnson, D. V., 32
Johnson, E. A. J., 372, 373
Johnson, G., 181
Johnson, G. E., 756, 757
Johnson, H. B., 168
Johnson, I., 222
Johnson, J., 496
Johnson, J. D., 243
Johnson, J. H., 101, 104, 224, 225, 261, 271, 415, 658
Johnson, K., 629
Johnson, L., 632, 730, 732
Johnson, P. G., 567
Johnson, W. C., 33, 37, 84
Johnston, I., 677
Johnston, K., 623, 629
Johnston, R. J., 71, 210, 211, 264, 301, 302, 361, 368, 371, 373, 468, 483, 584, 585, 588, 594, 595, 597–601, 605, 607, 609, 636, 656, 657
Johnston, T., 341
Johnston-Anumonwo, I., 677
Jojola, T. S., 245
Jones, B. G., 373, 478, 479
Jones, D. W., 95, 99, 103, 342, 370, 729
Jones, E. L., 161
Jones, H. R., 258
Jones, J. P., 676, 723, 759
Jones, J. R., 55, 82
Jones, P. D., 57
Jones, R. C., 272–75, 279, 494
Jones, S. B., 583
Jones, W. D., 506
Jordan, T. G., 16, 17, 166, 169, 176, 211, 212, 794
Jordan, T. R., 758, 763
Jorgensen, J. G., 245
Jose, C., 244
Joseph, A. E., 434, 435, 439, 455, 565
Joshi, T. R., 522
Jourabchi, M., 97
Judd, D. B., 708
Judge, M. A., 14

Jumper, S. R., 9, 17
Jupp, D. L., 767
Juvik, J. O., 32

Kahn, S., 17
Kahneman, D., 224
Kalkstein, L. S., 52, 59, 60
Kalnicky, R. A., 53, 54
Kanaroglou, P., 455, 728, 732
Kansky, K. J., 317
Kaplan, L., 32
Karan, P. P., 15, 17, 229, 416, 506–33
Karaska, G. J., 372, 373
Kardos, A., 176
Karmarkar, P. R., 529
Karn, V., 455
Karunaratne, N. D., 309
Kasarda, J. D., 373
Kasperson, J. X., 224, 412, 415
Kasperson, R. E., 100, 101, 224, 410, 413, 415, 583
Kassas, M., 224
Kates, R. W., 13, 16, 58, 59, 101, 122, 123, 125, 194, 214, 224, 410, 412–16, 418–20, 472
Katz, R. W., 55, 60
Katzman, M. T., 374
Kay, J., 32, 35, 179, 241, 243
Kay, P. A., 55, 116, 566
Keables, M. J., 114
Kearney, M. S., 37, 55
Kearns, R. A., 434, 439, 663
Keat, R., 588
Keddie, P. D., 337
Keeler, T. E., 327
Keely, C. B., 272
Keen, R. A., 52
Keller, C., P., 393
Kellerman, A., 342
Kellman, M. C., 30, 33–35
Kellogg, W. W., 50
Kelly, K., 530
Kemp, W., 97
Kennedy, S., 430, 434, 722
Kenney, M., 625
Kent, R. B., 494, 495
Kenzer, M., 210
Keogh, B., 397
Kessler, S., 479
Key, J., 52
Keyes, D. L., 104
Keys, P., 105
Keyser, G. L., 150
Khan, C. C., 524
Kiladis, G. N., 51–53
Kimber, C. T., 32, 35
Kimerling, A. J., 692, 705, 707
Kincheloe, J. L., 14, 15
Kinderman, R., 727
King, D. C., 7
King, L., 653

805

Kingkade, W., 271
Kingsbury, R., 550
Kipnis, B. A., 229
Kirby, A. M., 414, 600, 601, 622, 633, 636, 655, 656, 668
Kirchheer, E. C., 483
Kirchhofer, W., 52
Kirchner, J. A., 494
Kirkby, A. V. T., 195, 201
Kirkby, D., 172
Kirkby, M. J., 82, 83
Kirman, J. M., 3, 4
Kirn, J. R., 131
Kirn, T. J., 295, 296, 369
Kish, G., 17, 550, 552
Kitching, G., 469
Klein, S. F., 15, 18
Kleinpenning, J. M. G., 498
Klemas, V., 759
Klinger, L. F., 36
Klink, K., 35
Kliot, N., 583, 603, 604, 610
Klodawsky, F., 631
Kloos, H., 426, 427, 429, 430, 432
Knapp, G. W., 192, 201, 203, 491
Knapp, J., 551
Knapp, L., 231
Knapp, R. G., 172, 510, 515
Knetsch, J. L., 388
Kniffen, F., 163, 175
Knight, C. G., 200, 468, 469, 475
Knight, D. B., 244, 369, 573, 583, 584, 602, 603, 605–7
Knight, R., 733
Knopp, L., 637, 657, 658, 668
Knox, J. C., 56, 84, 114, 116
Knox, J. L., 53
Knox, P. L., 371, 373, 427, 429, 433, 434, 437, 622, 634, 636, 657, 658, 668
Knudsen, D. C., 322, 323, 723, 724, 728
Kobrin, F., 271
Kodras, J. E., 277, 676
Kofman, E., 209, 584, 605, 607
Kohler, M. A., 113
Kohn, C. F., 15, 17, 18
Kolars, J., 533
Kolodny, A., 677
Komar, P. D., 77
Konrad, V. A., 179, 181, 241, 570, 572, 575
Kontuly, T., 258, 267
Koo, H. P., 265
Kopec, R. J., 18, 453, 461
Koppelman, F., 727
Kornhauser, D. H., 510, 515, 516
Kosambi, M., 527
Kotlyakov, V. M., 415, 419, 420, 555
Kouba, L. J., 337
Kracht, J. B., 9

Krampen, M., 213
Krausse, G., 141, 518
Krenz, M., 60
Kritz, M. M., 272, 273
Kroeber, C. B., 172
Kromm, D. E., 116, 128, 131
Kruks, S., 477
Krumme, G., 292
Krummel, J. R., 103
Küchler, A. W., 30, 35
Kulhavy, R. W., 694
Kuivinen, K. C., 55
Kuklinski, A., 359
Kulikoff, A., 163
Kulka, T., 389
Kulkarni, G. S., 529
Kumar, B., 530
Kureth, E., 399
Kutzbach, J. E., 56
Kvale, K. M., 117, 426, 427, 490

La Duke, W., 245
Laborde, J. M., 430
Lagopoulos, A., 213
Lai, P-C., 707
Laity, A. L., 369
Laity, J. E., 81
Lake, R. W., 260, 657
Lakshmanan, T. R., 97, 105, 106, 372
Lall, A., 527
Lam, N. S-N., 37, 430, 702
Lamb, P. J., 58
Land, S. W., 340
Landale, N. S., 271
Landrum, J. W., 434
Landsberg, H., 57, 58
Lanegran, D. A., 9
Langlois, A., 565
Lansing, J. B., 271
Lapping, M. B., 334
Lappo, G. M., 554, 555
Larimore, A., 679
Larsen, R. D., 355
Lasserre, J-C., 573
Latham, R., 655
Lattimore, O., 506, 509, 515
Laudan, L., 317
Lauria, M., 231, 600, 601, 621, 636, 637, 657, 658, 668
Lavin, S., 705
Lavoie, A., 759
Law, C. M., 453–55
Lawrence, H. W., 335, 342, 394
Lawrence, P., 468
Lawrie, D. H., 50
Lawson, M. P., 55
Lawson, V. A., 277, 278, 281, 494, 495
Lawton, M. P., 452, 456
Layton, R. I., 334, 336

Lazewski, T., 245, 246
Lea, A. C., 728–30
Leatherman, S. P., 78, 758, 759
LeBlanc, R. G., 569
Ledent, J., 266, 307, 733
LeDrew, E. F., 52, 761
Lee, D., 673, 675, 679, 680
Lee, M. A. B., 34
Lee, Y., 513
Lefebvre, H., 632
Lefevere, C., 728
Legates, D. R., 702
Legreid, A. M., 167
Leighly, J., 194, 210
Leinbach, R., 292
Leinbach, T. R., 322, 326, 518, 519
Leitner, H., 636
Lemmon, J. J., 130
Lemon, J. T., 162–64, 167
Lengel, J. G., 3, 7, 8
Lenk, C., 35
Lensink, E., 735
Lentnek, B., 490
Lenz, R., 427, 433
Leonardi, G., 729
Leone, M., 211
Lerner, G., 632
Leung, C. K., 513
Leung, Y., 736
Leveen, P., 630
Levi, M., 605
Levine, G., 171
Leving, G. L., 151
Lew, A. A., 518
Lewin, K., 325
Lewis, G. K., 151, 369
Lewis, G. M., 179, 241
Lewis, J. E., 391
Lewis, K. E., 181
Lewis, L. A., 475, 482
Lewis, M. E., 341
Lewis, N. D., 426, 427, 432
Lewis, P. F., 167, 175, 176, 211, 227, 335, 369
Lewis, P. G., 533
Lewis, R. A., 547, 549
Lewis, R. D., 171
Lewis, T. R., 337
Lewthwaite, G. R., 337, 338
Ley, D., 214, 219, 228, 229, 245, 296, 564, 565, 593, 638
Li, S., 766, 767
Liaw, K-L., 455, 728, 732, 733
Libbee, M. J., 5, 14, 15, 17, 525
Licate, J. A., 492
Lichtenstein, S., 224
Lichty, R. W., 38
Lieber, S. R., 271, 390, 391, 398
Liebowitz, R., 551
Ligocki, C., 14, 15
Lim, G. C., 355, 367
Lim, L. Y. C., 519

806

Lime, D. W., 389, 398
Lindgren, D. T., 760
Lins, H. F., 98, 338
Linsley, R. K., 113
Lintz, C., 84
Linz, J., 603
Liossatos, P., 626
Lipietz, A., 591, 624, 628
Lipka, J., 246
Lipton, M., 277, 373
Little, J., 631
Liu, C., 514
Liu, J. C., 396
Liu, J. T., 105
Liu, K. B., 37
Liverman, D. M., 58, 60, 224, 415, 418
Lizarraga-Arciniega, J. R., 77
Lloyd, P. E., 292
Lloyd, R., 221, 694, 695, 711, 764
Lloyd, W. J., 77, 130
Lo, C. P., 510, 511, 513–15, 517, 518, 749, 760, 761
Lo, F., 360
Lockhart, J., 174
Lockwood, J. G., 49
Lodrick, D. O., 526
Lofgren, G. R., 54
Logan, B., 478
Logan, J. R., 593, 621, 636
Lombardi, C. L., 498
Lombardi, J. V., 498
Lonergan, S. C., 95, 100, 102, 103, 339, 343
Long, L., 262, 267, 268
Longley, P., 222, 656, 736
Lonsdale, R. E., 172, 488, 549
Lord, J., 262
Lorenz, D. C., 33, 34
Lorenz, E. W., 53
Los, M., 361
Losch, A., 319, 327
Louder, D. R., 241, 569
Loukissas, P. J., 392
Louviere, J. J., 320, 322, 733
Loveland, T. R., 756, 757
Lovell, G. W., 173
Lovell, W. G., 492, 493
Lovering, J., 633
Lovingood, P. E., Jr., 391
Lowe, J. C., 316
Lowell, B. L., 274
Lowenthal, D., 218, 227
Lowry, I. A., 270, 276, 496
Loy, J., 399
Loy, W. G., 708, 709
Loyd, B., 215, 677
Lucas, R. C., 389
Luce, R. D., 731
Luckman, B. H., 37, 55
Ludlow, L., 344
Luft, E. R., 114

Luk, S-H., 80, 114
Lulla, K., 29, 35, 37, 748, 755
Luman, D. E., 699, 707, 763
Lund, I. A., 52
Lund, S. W., 76
Lundgren, J. O. J., 397
Lupien, A. D., 782
Lura, R., 474
Lusch, D., 756
Luten, D., 97
Luther, E., 36
Luther, J., 344
Lydolph, P., 547, 549, 550, 555
Lynch, K., 218, 656, 668

Ma, L. J. C., 506, 509–12, 514, 518
Mabin, A., 628
Mabogunje, A. L., 279, 371, 375
McAllister, I., 597–98
McAllister, P. E., 241
MacArthur, R. H., 28, 29
McCann, L.D., 171, 573
McCarthy, J., 366
McCarthy, K. F., 267
McCartney, D. M., 370
McCarty, H., 554
McCarty, J., 619, 637
McCaslin, R., 456, 461
McCloy, J., 147
McConnell, H., 399
McConnell, J. E., 294, 302
McCormick, B. H., 51, 791
McCoy, W. D., 55
McCune, S. B., 507, 509, 510, 517
McCusker, J. J., 163
McDonald, D. B., 398
MacDonald, G. M., 37
McDonald, J. N., 29, 36, 37, 242
McDowell, B. D., 367
McDowell, L., 631, 679
McDowell, P. F., 84
MacEachren, A. M. 695, 703, 710
McElroy, C., 274
Macey, S. M., 103, 225, 458, 461
McFadden, D., 325, 730–32
McFarlane, A., 59
McGee, T. G., 479, 519
MacGill, S. M., 736
McGinnis, D. F., 757
McGovern, C., 492
McGranaghan, M., 712
McGreevy, P., 570
McGregor, K. M., 59
McHugh, K. E., 271
McIlwraith, T. F., 573
Macinko, G., 333, 344, 370
McIntire, E. G., 242, 245, 246
McIntire, W. G., 144, 151
McIntosh, C. B., 168
McIntosh, R. P., 31
Mackay, D. B. 221

MacKaye, B., 353, 359, 361, 376
McKee, J. O., 179, 241, 243, 244, 246
Mackenzie, F., 473, 476
McKenzie, R. D., 194
Mackenzie, S., 619, 622, 631, 632, 673, 677, 679
Mackett, R. L., 222
Mackinder, H., 353, 609
MacKinnon, R. D., 270, 316, 738
McKnight, T. L., 32, 336
McLafferty, S., 326, 434, 435, 438
MacLaughlin, J. G., 604, 607
Mcluskey, J. M., 76, 80
McManis, D., 169
McMaster, R. B., 692, 703
McMurry, K. C., 387
McNee, R., 305
McNeill, W. H., 161
McNulty, M., 479
Macpherson, A. G., 569
McQuillan, D. A., 166, 167, 181
McTeer, J. H., 14
Maddala, G. S., 725
Madden, M., 308
Maddock, T., 127, 128
Maddox, R. A., 49
Madsen, D. B., 55
Magilligan, F. J., 114
Mahacek-King, V. L., 114
Maier, E., 150
Mair, A., 619, 634–37
Majeed, A., 374
Malanson, G. P., 29, 33, 35
Malecki, E. J., 298, 299, 369
Mallory, W. E., 570
Malone, T. F., 420
Malthus, R., 473
Manabe, S., 49
Mandel, R. D., 80
Mann, P. C., 102
Manners, G. M., 97–99
Manners, I. R., 98, 179, 337
Mannion, J. J., 167, 176
Manogaran, C., 525
Mansfeld, Y., 229
Manson, G., 7, 8, 17, 18
Manzo, J. T., 243, 399
Maraffa, T. A., 229, 677
Marble, D. F., 317, 320, 360, 361, 699, 787, 794
Marcus, M. G., 77, 80, 84, 180
Marcus, W. A., 80, 117
Mardia, K. V., 723
Marin, L. M., 356
Marino, J. S., 703
Mark, D. M., 81, 82, 115, 706, 712, 793
Marks, B. E., 660
Marks, D., 757, 767
Markusen, A., 293, 299, 300, 352, 363, 366, 369, 623

Marran, J. F., 9
Marrero y Artiles, L., 174
Marsden, T., 630
Marsh, B., 99, 170, 171, 369
Marsh, G. P., 176, 193, 353, 376
Marshall, J. U., 355
Marston, R. A., 77, 78, 80–82
Marston, S. A., 413, 414, 622, 629, 637, 655, 658, 676, 677
Martell, G., 547
Marti, B., 141, 151
Martin, C. J., 31, 33
Martin, C. W., 37, 114
Martin, E. B., 483
Martin, G. J., 3, 193
Martin, N., 471
Martin, R., 371, 725
Martinson, T., 490
Marts, M., 131
Marx, K., 161, 162, 585, 586, 588, 589, 599, 607
Marx, R. W., 699
Mason, C. M., 292
Mason, R. J., 341
Mass, C. F., 51
Massam, B. H., 352
Massey, D. S., 273, 344, 345, 363, 584, 588, 620–24, 631–33, 635
Massie, R. K., 374
Mastin, J. F., 226
Masyk, J., 397
Mather, C., 15, 17, 416, 523–25, 529, 573
Mather, E. C., 369
Mather, J. R., 35, 58, 113, 122, 420
Mathewson, K., 203, 491, 493
Mathieson, A., 397
Matley, I. M., 149, 387, 547, 552
Matson, J. T., 706
Matthews, E., 35
Matthews, G. J., 166, 498
Matthews, O. P., 131
Matzerath, H., 373
Matzke, G., 427, 474
Mayer, H. M., 316, 317, 323, 357, 652
Mayer, J. D., 426, 427, 430, 432–35, 437, 438
Mayfield, R. C., 507
Mayo, W. L., 3, 8
Mazey, M. E., 679
Meade, M. S., 425–27, 429, 430, 432, 434, 531
Mealor, W. T., Jr., 337
Mearns, L. O., 55
Meddeb, N., 105
Meegan, R., 363, 623
Meehl, G. A., 53
Meentemeyer, V., 30, 36, 567
Mehedinti, S., 552
Meierding, T. C., 82

Meinig, D. W., 157, 159–64, 166, 169–72, 176, 178, 180, 211, 212, 227, 243, 498, 571, 603
Melamid, A., 149, 532, 533
Menard, R. R., 163
Menchick, M., 733
Mensch, G., 162
Mera, K., 372
Mercer, D., 606
Mercer, J., 229, 564, 565, 593, 658
Merchant, J. W., 764
Merlin, M. D., 32
Merrens, H. R., 163, 164
Merrett, C., 572
Merriam, C. H., 31
Merrifield, J. D., 99, 370
Mescon, T., 397
Messerli, B., 514
Meyer, D., 169
Meyer, J. W., 454–56, 458, 461
Meyer, R., 261
Meyer-Arendt, K., 146, 147, 393
Meyerson, A., 621
Michaelsen, J., 759
Michel, A., 523
Michelson, W., 262
Michener, J. A., 18
Micklin, P. P., 548, 554
Mielke, M. W., 36
Miezkowski, S. T., 388
Mikesell, M. W., 17, 172, 179, 198, 199, 606
Miles, E. J., 571, 572, 575
Miles, M., 759
Mileti, D. S., 415
Miller, B. A., 102
Miller, D. H., 58, 113, 172
Miller, G. H., 434, 439
Miller, G. J., 368
Miller, J. P., 33, 112, 115
Miller, L. D., 756
Miller, M. M., 84
Miller, R., 171, 677
Mills, E. S., 720
Mills, H. H., 81
Millward, H., 569, 585
Mines, R., 273
Minghi, J. V., 532, 583
Minghst, K. A., 610
Minkel, C. W., 497
Minnich, R. A., 33–35, 37
Mintz, Y., 113, 116
Misra, R. P., 371, 373
Mitchell, B., 125, 362
Mitchell, G., 6
Mitchell, J. F. B., 50
Mitchell, J. K., 223, 224, 410, 412–15, 417, 418
Mitchell, K., 570
Mitchell, L., 141

Mitchell, L. S., 387, 391, 400–402
Mitchell, P., 575
Mitchell, R. B., 322
Mitchell, R. C., 123
Mitchell, R. D., 159, 162–64, 167, 168, 170, 172, 178, 241
Miyanishi, K., 33, 35
Miyazawa, K., 308
Modelski, G., 609
Moellering, H., 698, 699, 705
Mohler, R. J., 568, 757
Moley, R., 356
Mollenkopf, J., 364
Molotch, H., 591, 621, 636
Momsen, J. H., 496, 629, 673, 674
Monk, J., 17, 20, 229, 230, 631, 673–75, 677–81
Monmonier, M. S., 686, 695, 700, 702, 705, 707–9, 712–13
Monroe, C. B., 527
Monteverdi, J., 54
Montgomery, O. L., 754
Montz, B. E., 224, 418
Moodie, D. W., 243
Mookherjee, D., 371, 524, 527
Moon, H. E., Jr., 335, 340
Mooney, H. A., 38
Moore, E. G., 260, 261, 265, 271
Moore, H. E., 362
Moore, J., 374
Moore, R. C., 230
Moore, T. G., 179, 241, 242
Moore-Milroy, B., 631
Moos, A. I., 588
Morales, R., 622, 633, 668
Moreland, W. H., 782
Morgan, J. R., 520
Morgan, K., 623
Morin, L., 575
Moritz, R. E., 51, 52
Morley, D., 631
Morrill, R. L., 16, 17, 172, 269, 319, 359, 360, 368, 371, 489, 594, 655, 720
Morrill, R. W., 230
Morris, A., 497
Morris, C., 52
Morris, M. A., 150, 496
Morris, S. E., 75
Morrison, J. L., 689
Morrison, P. A., 267, 271
Morrissett, I., 4, 6, 8, 14, 16
Morrow-Jones, H. A., 9, 264, 675
Morse, B. W., 792
Morse, J., 2
Moryadas, S., 316
Moses, R., 353
Moses, T., 53
Mosher, A. T., 373
Mosley, W. H., 439
Mossa, J., 77

Most, B. A., 610
Mote, V. L., 547, 548, 550, 551, 554
Moulaert, F., 623
Mounsey, H., 792
Moustapha, A. F., 533
Mower, R. D., 491
Moyle, P. B., 243
Muckleston, K. W., 125
Muehrcke, P. C., 707
Mueller, E., 271
Mueller, J. E., 76
Mullally, H., 729
Muller, E. K., 167, 168
Muller, J. C., 781
Muller, P., 266
Muller, R. A., 52, 58, 60, 113
Mulligan, G. F., 322, 326, 729
Mullis, C., 696
Mumford, L., 353, 359, 361, 376
Munn, R. E., 414, 419
Munro, D. S., 49
Murauskas, G. T., 597
Murck, B. W., 103
Murdie, R. A., 434
Murgatroyd, L., 631, 633
Murie, A., 261
Murphey, R., 507, 512
Murphy, A. B., 606
Murphy, M. E., 492
Murphy, P. E., 391–93, 395, 397, 574
Murray, S., 244
Murray, W. B., 273
Murton, B. J., 530
Muschett, F., 432
Musk, L. F., 57
Muth, R., 655
Myers, D., 263
Myrdal, G., 359

Nairn, T., 605
Napton, D., 340, 344
Narayan, R. K., 525
Narcho, R. J., 246
Naro, N. P. S., 174
Natoli, S. J., 17, 18, 20
Natraj, V. K., 371
Nechamen, W., 124
Neils, E., 245
Neilson, B., 399
Nelder, J. A., 724
Nellis, D. M., 336–39, 344
Nellis, M. D., 119
Nelson, F. E., 81, 566
Nelson, G. B., 439
Nelson, H., 305
Nelson, J. G., 397
Nelson, K., 306, 622, 623, 631
Nelson, R. F., 756
Newby, H., 630

Newcomer, J. A., 706
Newcomer, R., 452
Newell, D., 572
Newling, B., 655
Newman, J. L., 200, 427, 429, 430, 474
Newman, M., 367
Newman, R., 274
Newson, L. A., 173, 492
Newton, K., 593
Newton, M. B., 163, 176
Newton, P., 264, 453, 461
Newton, T. G., 125
Nicholson, S., 482
Nicholson, S. E., 53, 54, 56
Nickerson, B. G., 792
Nickling, W. G., 52
Nicolis, G., 728
Niemann, B. J., Jr., 789
Nietschmann, B. Q., 195, 201, 202, 491, 494, 496
Nijkamp, P., 106, 720, 725, 730, 732
Nkemdirim, L. C., 55, 567
Noble, A. G., 506, 510, 511, 518, 523–25, 527, 531, 533, 534, 570
Noble, W. A., 506, 524, 525
Noma, E., 221
Norcliffe, G. B., 481
Nordstrom, K. F., 76, 80, 141, 146, 149
Noronha, V. T., 782
North, R. N., 550, 551
Norton, R. D., 292
Norton, W., 157, 172, 178, 210, 568
Norwood, V., 230, 678
Nostrand, R. L., 178
Noyelle, T. J., 295
Nunez, M., 49
Nyerges, T. L., 998
Nystuen, J. D., 222, 320, 360, 777

Oakey, R. P., 300
Oberlander, T. M., 80
O'Brien, D. W., 215
O'Brien, L. G., 724
O'Brien, M. J., 181
Obudho, R. A., 373
Occhietti, S., 56
O'Connor, J., 635
Odell, P. R., 97, 98
Oden, N. L., 720
Odland, J., 721, 723, 727, 733
O'Donnell, T., 182
Odum, E. G., 31, 38
Odum, H. W., 352, 359
O'Farrell, P. N., 734
Ogrosky, C., 763
Ohlin, B., 293, 301
Ohta, R. J., 456

O'hUallachain, B., 269, 294, 302
Oi, W. Y., 323
Ojala, C. F., 399
Oke, T. R., 47, 49
O'Keefe, P., 413, 629, 633
O'Kelly, M., 320, 324, 326
Oldakowski, R. K., 271, 455
O'Leary, J. F., 33, 37
Oliver, J. E., 57, 60, 418
Oliver, M. L., 261, 658
Olmstead, C. W., 15
O'Loughlin, J., 583, 584, 594, 610
Olson, A., 674–75, 676
Olson, J. M., 687, 693, 694, 707
Olson, S., 171, 413, 662, 663
Olsson, G., 584
Olswig-Whittaker, L. N. D., 35
Olwig, K., 630, 631
Olyphant, G. A., 75
O'Mara, J., 164
Omernik, J. M., 37, 117
Omner, R. E., 166, 167
Onaka, J. L., 222, 261, 732, 733
O'Neill, A. V. O., 38
O'Neill, P. E., 758, 759
Oosterhaven, J., 308
Openshaw, S., 100, 101, 722
Orchard, J. E., 506, 509
Ord, J. K., 721, 722, 725
Oriard, M. V., 399, 400
O'Riordan, T., 100, 117, 410
Orme, A. R., 70, 71, 88, 144
Ormrod, R. K., 454–55, 492
O'Rourke, P. A., 49, 50
Orridge, A. W., 605
Ortiz, A., 494
Ortner, S. B., 211
Orton, C., 720
Osborne, B. S., 171, 172
Osei-Kwame, P., 595, 596
Oshiro, K. K., 336, 506, 515–17
Osleeb, J. P., 98, 99, 151, 549
Ostergren, R. C., 166, 167, 169
Osterkamp, W. R., 33
O'Suliivan, P., 318, 322, 325
O'Tuathail, G., 610
Outcalt, S. I., 49
Oviatt, C. G., 55
Owens, S., 104

Pabst, D. L., 8, 9
Pacione, M., 373
Paddison, R., 604, 606
Padilla, S. M., 357
Padoch, C., 201
Page, G., W., 370
Pallone, R., 267, 269
Palm, R. I., 16, 125, 222, 224, 261, 264, 419, 564, 657, 658
Palma, G., 627

809

Palmer, G. B., 172
Palmieri, R., 524, 526
Panciera, S. E., 114
Panditharatna, B. L., 527
Pannell, C. W., 510, 511, 513–15
Panofsky, E., 711
Panter-Brick, K., 374
Papageorgiou, Y. Y., 728, 729, 732
Papathanassopoulos, E., 104
Park, R. E., 194
Park, S., 373
Parker, A. J., 30, 33–37
Parker, D. J., 123, 414, 418
Parker, K. C., 34, 36
Parker, L. M., 244
Parker, W. H., 609
Parkinson, C. L., 48–50
Parnell, H., 319
Parr, J. B., 361
Parry, M., 201
Parry, M. L., 58, 418
Parson, C. G., 75
Parsons, A. J., 80, 86, 114
Parsons, J. J., 31, 32, 210, 215, 489, 491, 492, 495, 497
Pasqualetti, M. J., 95, 100–103, 412
Patel, D. I., 368
Patil, B. R., 373
Patrick, J. J., 7, 8
Patrick, R., 126
Patterson, K. D., 427
Pattison, W. D., 14, 17, 194, 412
Patton, D. J., 17
Patton, J. C., 694, 696
Pauer, G., 517, 518, 531
Paul, A. H., 574
Paul, B. K., 427, 429, 434
Paulhus, L. J., 113
Peace, S., 456
Peach, C., 245
Peake, L., 631
Pearce, D. G., 392, 394
Pearson, K. S., 710
Pease, J. R., 334, 335, 343, 344
Pedersen, P. O., 359, 371
Peet, R., 293, 305, 589, 619, 623, 628, 629, 633
Peippo, J., 142
Pelczarski, S. G., 415, 418
Pellegrino, J. W., 222, 735
Pelzer, K. J., 507
Pendleton, R. L., 507
Penning-Rowsell, E. C., 123, 414
Pennington, C. W., 32, 489, 496
Pepper, D., 100, 610
Perez, F. L., 35
Perloff, H. S., 354, 356–58, 360, 376
Perroux, F., 360
Perry, A. H., 51, 52
Persinger, I. D., 754

Petchenik, B. B., 687, 688, 693, 708, 709
Peters, G. L., 337
Peterson, B. E., 105
Peterson, D. L., 755
Peterson, M. P., 692, 694
Peterson, R., 631
Petnick, J., 144
Petrella, R., 359
Petrie, G., 762
Petzold, D. E., 117
Peuker (Poiker), T. K., 699–703, 779
Peuquet, D. J., 698, 699, 706, 794
Phillips, A. G., 260
Phillips, D. R., 434, 435, 453
Phillips, J. D., 114, 115
Philpot, W. D., 703
Picciotto, S., 586
Pickett, S. T. A., 33
Pickles, A. R., 230, 734
Pickles, J., 620, 628
Pico, R., 357
Pielke, R. A., 48, 49
Pierce, J. T., 334–36, 341
Pigram, J. J., 338, 339
Pijawka, K. D., 100, 325, 410, 412
Pillsbury, R. R., 163, 176, 241, 243, 399, 400
Pinch, S., 636
Piore, M., 272
Pipkin, J. S., 320, 322
Pitt, D. G., 230
Pitts, W. D., 29
Pitty, A. F., 71
Place, J. L., 756
Place, S. E., 201
Plane, D. A., 267, 269, 270, 323, 369, 726
Platt, R. H., 124–26, 130, 131, 333, 334, 370, 411, 415, 418
Platt, R. S., 651
Plaut, T., 334
Plotkin, S., 621
Plumb, G. A., 708
Pocock, D., 227
Pocock, S. J., 720
Pohl, T. W., 369
Poiker (Peuker), T. K., 699–703, 779
Polenske, K. R., 306
Polese, M., 296
Ponczyniski, J. J., 81
Popper, F. J., 430, 431
Porteous, J. D., 226, 496, 576
Porter, C. F., 370
Porter, P. W., 192, 193, 195, 198, 200, 468, 473, 475, 481, 482
Porter, R. N., 246
Portugali, J., 603
Posey, C., 58

Posner, J. L., 495
Potter, K. W., 115, 116
Poulton, C., 761
Pounds, N. J. G., 583, 602
Powell, J. M., 54
Powers, C. F., 117
Pozorski, T., 491
Pratt, G., 565, 621, 622, 657, 676–79
Pratt, S. G., 453, 461
Pred, A. R., 159, 171, 222, 295, 305, 368, 369, 588, 620, 633, 654, 655, 662
Prentice, I. C., 36
Preobrazhenskiy, V. S., 554
Prescott, J. R. V., 610
Preston, D. A., 497
Preston, V. A., 221, 222, 224, 230, 656
Preuss, P. W., 430
Preziosi, D., 213
Price, E. T., 176
Price, J. M., 164
Price, K. A., 369
Price, L. W., 35, 36, 83
Price, R. L., 567
Priddle, G. B., 391
Prigogine, I., 728
Prothero, R. M., 280
Provin, R. W., 693
Pruitt, E. L., 142
Prunty, M. C., 337, 369
Pryde, P. R., 100, 102, 103, 334, 548, 555
Psuty, N. P., 88, 141, 144
Pudup, M. B., 483, 498, 619, 630, 634
Pulliam-DiNapoli, L., 176
Pulsipher, L. M., 173, 492, 493
Purdue, R., 396
Putnam, R. G., 568
Pye, V., 126
Pyle, G. F., 425, 427, 433, 434
Pyle, L. A., 335, 337, 343
Pyrdol, J. J., 99

Quarles, J., 126
Quayson, T., 397
Quigley, J., 260

Radford, J. P., 167
Radwan, A. E., 325
Rahman, A., 509, 522, 523, 531
Raitz, K., 400
Rajogopal, R., 117, 118, 129
Ralston, B. A., 322, 324
Ramesh, A., 427, 434, 437
Ramsey, E., 758
Randolph, J. C., 100
Rannie, W. F., 55
Raphael, C., 49

Raphael, N., 151
Rapkin, C., 322–23, 362
Rapoport, A., 213
Rappaport, R. A., 195, 198
Ratick, S. J., 99, 106, 151
Ratzel, F., 194
Raup, H. A., 149
Raup, H. M., 30
Raveche, H. J., 50
Ravenstein, E. G., 276
Ray, A. J., 165, 178, 179, 241, 243
Ray, D. M., 162
Ray, M., 361
Rayner, J. N., 48, 49, 219, 221, 705
Reagan, R., 310, 311, 355
Recker, W. W., 221
Reclus, E., 361
Redburn, F. S., 658
Redclift, M., 631
Reed, R. R., 518, 520
Reed, W., 151
Rees, J., 291, 292, 297, 299, 300, 310, 311, 369
Rees, J. D., 32
Rees, P. W., 17
Reeves, R. W., 76, 241
Regulska, J., 552, 676
Reichert, J. S., 273
Reider, R. G., 55, 84
Reiner, T., 658
Reitan, C. H., 54
Relph, E., 226
Remillard, M., 119
Rengert, A. C., 679
Rengert, G., 224, 229
Renner, G. T., 361, 362
Renwick, W. H., 115
ReVelle, C., 729, 730
Rey, L., 426, 427, 432
Reynolds, D. R., 585, 601
Rhind, D. W., 700, 786
Rhoads, B. L., 76, 82, 114, 115
Ricardo, D., 342
Rice, J. G., 166–69
Rich, E., 752
Richards, J. A., 757, 767
Richards, P., 471, 629
Richardson, A. J., 754
Richardson, B. C., 174, 494
Richardson, D. B., 244
Richardson, H. W., 307, 372
Richardson, P., 631
Richason, B. F., 749
Richmond, A. H., 607
Rickert, J. E., 393
Ricketts, P. J., 412
Riddell, J. B., 468, 469, 471, 478, 479, 483
Riebsame, W. E., 59, 60
Rieck, R. L., 82
Rigby, D., 626

Rimmer, P. J., 316, 317, 375, 468, 469
Ring, M. L., 455
Ring, N., 14
Ripley, B. D., 720, 723
Ripple, W. J., 756
Ristow, W. W., 710
Ritchie, J., 119
Ritter, D. F., 71, 88
Roach, T. R., 569
Robbins, J. C., 323, 325
Robert, A. S., 519
Roberts, C. E., 702
Roberts, C. F., 507
Roberts, K. L., 729
Roberts, R. S., 126, 129, 130, 339, 340, 341
Robertson, J., 31
Robinson, A. H., 688, 691, 696, 710
Robinson, D. J., 173, 492–94, 497
Robinson, J. L., 354, 572, 575, 576
Robinson, M. E., 101
Robinson, P., 3, 4
Robinson, V. B., 792
Robson, B., 654
Robson, P., 456
Rochberg-Halton, E., 213
Rodenburg, E. E., 392
Roderick, H., 571
Rodgers, A. L., 550, 551
Rodwin, L., 358
Roemer, J., 626
Rogers, A., 269, 453–54
Rogers, G., 28, 31, 33, 34, 37, 575
Rogers, J. C., 52–54
Rogerson, C. M., 172, 483
Rogerson, P. A., 269, 270, 323, 368, 726–28
Rogge, J. R., 473, 474, 573, 606
Rokkan, S., 374, 585
Rollison, P. A., 104
Romanoff, E., 307
Romey, W. D., 570
Romsa, G., 231
Rondinelli, D. A., 266, 372–74, 494
Rooney, H. F., Jr., 569
Rooney, J. F., 398, 399, 431, 434
Rooney, J. G., 241
Roosevelt, F. D., 356, 357
Rose, D., 621, 631, 657, 676, 677
Rose, H., 555
Rose, M. H., 366
Rose, R., 597–98
Roseman, C. C., 258, 262, 271, 273, 455
Rosen, P. S., 76, 80
Rosen, S., 3, 9
Rosenberg, M. W., 434
Rosenblood, L., 395
Rosenfeld, C. L., 73, 81, 758
Rosenthal, L. D., 583

Ross, T. E., 179, 241, 242, 245
Rossi, P., 260, 271
Roundy, R., 147, 427, 474
Rowe, C. M., 49, 113, 116, 703
Roweis, S. T., 364, 365, 584, 621
Rowland, R., 549
Rowles, G. D., 227, 231, 452–61
Rowntree, L. B., 55, 210, 212, 662
Rowntree, R. A., 32
Roy, A. G., 82
Roy, P., 373
Rubin, B., 176
Rudd, D. M., 453
Ruddle, K., 32, 372, 373
Rudzitis, G., 452–54
Rugg, R. D., 551, 700
Rumble, H. E., 3
Rumney, T., 337, 574
Rundquist, D., 746, 756, 758, 780
Rushton, G., 218–20, 434, 435, 493, 656, 724, 738, 781
Russel, C. S., 122
Russell, J. A., 181
Russell, R. J., 142, 144
Russwurm, L. H., 334, 556
Rutherford, E. M., 677
Rutherford, T., 621
Rutter, R. A., 8
Ryan, B., 390
Ryerson, C. C., 707

Saalfeld, A., 701
Saarinen, T. F., 218, 219, 225, 369
Sack, R. D., 241, 588, 604, 736
Sadalla, E. K., 221, 245
Sadler, D., 593
Saegert, S., 215
Saffron, S., 246
Sage, G., 399
Sagers, M. J., 98, 549–52, 556
Sailer, C., 752, 763
Saint-Germain, M., 676
Salih, K., 360
Salisbury, N. E., 76
Salt, J., 273
Salter, C. L., 9, 510, 515
Saltzman, B., 48, 50
Salvatore, D., 278
Samatar, A., 477, 629
Sambrook, S. L., 434, 439
Sami, A., 527
Samuels, M. S., 219, 610
Sanders, R. A., 569
Sanders, R. L., 273, 278, 279
Sanderson, M., 49, 59, 571
Sandoval, A., 494
Sangwan, C., 524
Sappio, F. J., 756
Sardido, S., 530
Sartre, J. P., 95, 277, 320
Sassen, S., 634

811

Sauer, C. O., 30, 32, 33, 157, 158, 160, 161, 175, 178–80, 194, 197, 210–12, 214, 218, 489, 498, 582, 663
Sauer, J. D., 29, 32, 35
Saunders, P., 599, 600, 638, 666, 667
Sauressig-Schreuder, Y., 166, 167
Savage, M., 598, 599
Sawyer, S. W., 102, 103, 122
Sayer, A., 302, 587, 588, 599, 600, 620, 623–26, 631, 634, 635
Sayre, G. S., 102
Scarpaci, J. L., 434, 494, 658
Schaffer, R., 621
Scharfer, G., 49
Schattschneider, E. E., 596
Scheibe, F., 119
Scheidt, R. J., 452
Schelenken, J. A., 241, 243
Schenk, R., 55
Schertzer, W., 49
Schinkel, D. R., 389
Schlegel, F. M., 35, 36
Schlenker, J. A., 179
Schlereth, T. L., 175
Schlesinger, M. E., 50
Schlichting, J., 547
Schlissel, L., 677
Schmandt, J., 571
Schmid, J. A., 32
Schmidt, C. G., 513
Schmudde, T., H., 18
Schmugge, T. L., 756, 758, 759
Schneider, D., 434
Schneider, S. H., 55, 418
Schneider, S. R., 757
Schoeller, P., 229
Schoenberger, E., 294, 623–25
Schönfeld-Leber, B., 243
Schraml, L. A., 277
Schugart, H. H., Jr., 38
Schuler, H. J., 221
Schumacher, J. A., 59
Schumm, S. A., 38, 75, 88
Schutz, E., 294
Schwab, M., 323
Schwartz, A. R., 151
Schwartz, N. H., 694
Schwartz, S. B., 174
Schwartzberg, J., 520, 524, 530
Schwendeman, J. R., 10
Scivener, D., 567
Scobie, J., 174
Scott, A. J., 293, 300, 303, 306, 363–65, 368, 584, 619–26, 633–35, 652, 655, 666, 668
Scott, D., 426, 427, 432
Scott, E., 475
Scully, M., 18
Sdasyuk, G. V., 420
Sea, T., 516

Seager, J., 215, 674–76
Seamon, D., 210, 211, 219, 227, 229
Searle, M. S., 229
Seccombe, I. J., 273
Sechrist, R. P., 274
Segoe, L., 124
Seley, J. E., 370
Sell, J. L., 219, 221, 225, 230
Sell, R. R., 271
Sellers, P. J., 49, 756
Sellers, W. D., 49
Selya, R. M., 434, 513, 514
Semple, E. C., 160, 506, 509
Semple, R. K., 295, 296
Sender, J., 468
Serow, W. J., 452
Sewell, W. R. D., 131
Shabad, T., 97, 547–51, 554–56
Shaha, S. H., 694
Shaikh, A., 626
Shair, I., 533
Shamekh, A., 532
Shamim, A. S., 524
Shankmann, D., 34
Shannon, G. W., 430, 433, 434, 437, 439, 456, 458, 460–61, 661
Shantz, H. L., 31
Shaper, A. G., 720
Sharma, K. K., 524
Sharwood, P., 453, 461
Shaw, R. P., 565
Shear, J. A., 122
Shelley, F. M., 101, 399, 583, 594, 596–98, 601
Shelton, M. L., 58, 59, 113, 122
Sheppard, E., 322, 323, 326, 619, 620, 626, 627, 725
Sherman, D. J., 76, 88, 145, 146
Sherman, J. C., 707
Sheskin, I. M., 98, 322, 323, 461
Shinar, A., 229
Shine, K. P., 48, 49, 50
Short, J. L., 337
Short, J. R., 585, 600
Shortridge, B. G., 674, 675, 693, 695
Shortridge, J. R., 369
Shrestha, M. N., 528, 529
Shrestha, N. E., 506, 508, 509, 522, 527, 531
Shrestha, N. R., 274
Shroder, J. F., 75, 80
Shryock, H. S., 276
Shuldiner, P., 323
Shumsky, N. L., 433, 434, 437, 658
Sibert, J. L., 707
Siddiqi, A. H., 528, 530
Siegel, J. S., 276
Siemens, A. H., 491
Sien, C. L., 518
Silberfein, M., 477, 479

Simard, A., 430, 431
Simmonett, D. S., 746, 757, 767, 769
Simmons, J., 260, 654
Simmons, T., 242
Simon, D., 480
Simoons, F. J., 30, 243, 524, 526
Simpson-Housley, P., 570
Sims, J. H., 120, 224, 415
Sinclair, C. A., 273
Sinclair, L., 654
Sinclair, R., 566, 653
Singer, B., 734
Singer, M., 213
Singh, S., 524, 529–31
Singh, T., 54
Singleton, F., 548
Sjaastad, L., 269
Skaggs, R. H., 55, 58, 122
Skeldon, R., 279
Skindlov, J. A., 52
Skinner, G. W., 512
Skocpol, T., 586
Slack, R., 119, 758
Slater, D., 498, 628, 637
Slater, T. R., 660
Slaymaker, O., 566
Slocum, T. A., 692–94, 708
Sloggett, G., 341
Slovic, P., 223, 224
Sly, D. F., 98, 368
Smagorinsky, J., 48, 49
Smale, B. J. A., 395
Smale, S., 328
Smil, V., 97, 102, 103, 513, 514
Smit, B. R., 103, 335, 341, 343, 344
Smith, A. D., 603
Smith, B., 274
Smith, B. W., 453, 456, 461
Smith, C., 36, 37, 358
Smith, C. A., 352, 373, 696
Smith, C. J., 431, 434, 438, 439
Smith, C. T., 497
Smith, D. A., 151
Smith, D. C., 390
Smith, D. M., 290, 368
Smith, E. G., Jr., 336
Smith, G. C., 221, 222, 230, 456, 458
Smith, G. E., 602, 606
Smith, J., 35
Smith, J. A., 765
Smith, J. H., 494
Smith, J. M. B., 36
Smith, J. R., 301
Smith, K., 57
Smith, L., 149
Smith, M., 6, 622
Smith, N., 16, 32, 103, 589, 619–22, 625, 631, 634, 637, 638, 654, 657
Smith, R. C., 758, 759

Smith, R. H. T., 373
Smith, R. L., 229
Smith, R. M., 692, 708
Smith, R. V., 387, 389, 400
Smith, S. J., 676
Smith, S. L. J., 391, 395, 468
Smith, T. R., 222, 260, 261, 735, 792
Smith, W. R., 335
Smith-Fowler, H., 439
Snead, R. E., 530, 531
Snell, J. L., 727
Snow, R. W., 431, 434
Snyder, J. P., 701, 702, 710
Snyder, P. Z., 245
So, F. S., 367
Soderstrom, E. J., 224
Soesilo, A., 325
Soffer, A., 532
Soile, D. G., 245
Soja, E. W., 359, 371, 588, 622, 632–34, 668
Sokal, R. R., 720
Solomon, B. D., 97, 99, 101, 103, 413
Solomon, S., 244
Solot, M., 210, 652
Sommer, S. E., 754
Sommers, L. M., 98, 117
Sonka, S. T., 58
Sonnenfeld, J., 218, 222, 246
Sopher, D. E., 229, 520, 523, 524, 658
Sorenson, C. J., 55, 80, 81
Sorenson, J. H., 95, 100, 101, 224, 225
Sorenson, J. R., 415
Sorrentino, A., 102
South, S., 262
Southard, R. B., 711
Southworth, F., 95, 104, 105, 320, 322, 326
Spate, O. H. K., 498
Spayne, R. W., 55
Speare, A., 271, 455
Spector, A., 631
Spelt, J., 573
Spencer, A. H., 221
Spencer, J. E., 203, 506, 507, 509, 518, 520
Spencer, V. E., 178, 241–43, 245
Speth, W., 210
Spetz, D., L., 8, 9
Spodek, H., 527
Spooner, D., 99, 102
Sprinsky, W. H., 701, 702
Spurlock, C. W., 434
Spyrou, M., 146
Stack, C. B., 271
Stackhouse, L. L., 59
Stadler, S., 59
Stafford, H. A., 291, 294

Stahl, C. W., 273
Stahle, D. W., 55
Stake, R. E., 3, 7
Stalley, M., 353
Stanback, T. M., Jr., 295
Stanfield, C. A., 393
Stanislawski, D., 489, 492
Stankey, G. H., 398
Stanley, W. K., 151
Stanley, W. R., 532
Stansfield, C., 147
Stanton, T. H., 245
Stapleton, C. M., 264, 265
Staplin, L. J., 221
Stark, D., 244
Starr, H., 610
Starr, J. L., 756, 759
Stea, D., 218, 245
Stebelsky, I., 547, 548, 551
Steed, M., 597
Steffen, J. O., 172
Steila, D., 113, 122
Stein, J., 269
Steiner, R., 244, 336
Steinitz, M., 212, 213
Steinke, T. R., 694–96
Stelter, G. A., 564
Stephenson, L. K., 130
Stephenson, R. A., 144
Stephenson, R. J., 434
Stern, E., 322, 324
Sternberg, H. O'R., 32, 77, 103, 415, 489, 491
Sternberg, R., 494, 496
Sternlieb, G., 295, 363
Steward, J. H., 197
Stewart, G. H., 32, 34
Stewart, R., 675
Stewart, T. R., 59
Steyn, D. G., 35, 49
Stiefel, M., 784
Stillwaggon, E. M., 245
Stimpson, C., 621, 631
Stimson, R. J., 219, 220, 655, 656
Stock, R., 426, 427, 434, 435, 439, 474, 475
Stocking, M., 470
Stockton, C. W., 55
Stoddard, R. H., 14, 522, 525
Stoddart, D. R., 195, 375
Stohr, C. J., 707, 763
Stohr, W. B., 372
Stokes, D., 598
Stokols, D., 219
Stoltman, J. P., 3–6, 9
Stone, J. R. N., 308
Stone, M., 621
Stopher, P., 323
Storper, M., 293, 306, 363, 620, 622–26, 628, 630, 631, 633–35
Stough, R. R., 325, 393
Stow, D. A., 146, 758, 759, 767

Strahler, A. H., 35, 37, 748, 752, 765, 767
Strahler, A. N., 75
Strand, P. J., 274
Strauss, A. L., 374
Streiner, D. L., 434, 439
Stroud, H., 397
Stuart, L. C., 30
Sturm, R., 7
Stutz, F. P., 9
Suckling, P. W., 49, 55, 59
Sugden, D. E., 87, 574
Sukhwal, B. L., 509, 520, 528
Sullivan, D., 460
Sullivan, J. A., 456, 461
Sun, C., 114
Sun, G. Q., 757, 767
Sundaram, K. V., 373
Superka, D. P., 3, 7, 8
Susman, P., 294, 413
Sussman, C., 353
Sutcliffe, B., 468
Sutherland, C. L., 513
Sutton, I., 241, 243, 244
Suwarno, B., 519
Svenson, O., 223
Svorou, S., 793
Swain, G. W., Jr., 8, 14
Swann, M. M., 492
Swanson, F. J., 78
Swanson, J. C., 274
Swarts, S. W., 246
Swartz, R. D., 461
Swearingen, W. D., 488
Sweet, J. K., 9
Switzer, T. J., 6–8, 15
Symanski, R., 274, 275, 373
Szajgin, J., 706

Taaffe, E. J., 172, 319, 550, 554
Tan, G., 52
Tan, K. C., 511, 512
Taranik, J. V., 758
Tarleton, L. F., 53
Tasker, G. D., 115
Tata, R. J., 494, 496
Taylor, A. H., 34, 36
Taylor, A. K., 222
Taylor, D. R. F., 372, 476
Taylor, J. E., 278
Taylor, J. G., 59, 221, 225
Taylor, J. L., 244
Taylor, L., 274
Taylor, P. J., 162, 468, 583–87, 592, 594–98, 604, 605, 609, 623, 628, 637
Taylor, S. M., 221, 224, 427, 429, 434, 438, 439
Tchakerain, V. P., 144
Teetzen, M. L., 430
Teleki, P., 552
Templer, O. W., 127, 128, 130, 131

Tenner, E., 16
Terich, T. A., 77
Terjung, W. H., 49, 50
Teye, V. B., 393
Thatcher, M., 371
Thisse, J. F., 729
Thom, B. G., 35
Thom, D., 471
Thomas, A. W., 758
Thomas, F. H., 119
Thomas, M., 426, 427
Thomas, M. D., 298, 301
Thomas, M. K., 57, 60
Thomas, R. M., 2, 7
Thomas, R. N., 494
Thomas, W. L., Jr., 179, 353, 510, 520
Thompson, C. A., Jr., 6
Thompson, E. P., 633
Thompson, G. L., 244, 326
Thompson, L. G., 55
Thompson, R., 36
Thompson, S. I., 172
Thorn, C. E., 81, 82, 86, 87
Thornthwaite, C. W., 35, 48, 113, 122
Thouez, J. P., 430–32
Thrall, G. I., 655, 729
Thrift, N., 588, 597, 620, 623, 628, 632, 633, 635
Thrower, N. J. W., 710
Tietze, W., 527
Timberlake, L., 468, 470
Timmermans, H., 219, 221, 222
Tinney, L. R., 749, 752, 757, 763, 764
Tirtha, R., 528
Tobin, G. A., 119, 125, 129, 415
Tobler, W. R., 13, 321, 322, 360, 410, 697, 698, 701–4, 711, 722, 725
Tomlin, C. D., 778
Tomlinson, R. F., 779, 787, 788
Ton, J., 764, 765
Took, L., 455, 460
Townroe, P. M., 292
Townsend, J., 629, 673, 674
Townshend, J. R. G., 749, 764
Toy, T. J., 77, 80
Trabaud, L., 33
Treps, L., 454–55
Tretyakova, A., 98
Trewartha, G. T., 258, 259, 506, 507, 509, 530
Tricart, J. L. F., 71
Trimble, S. W., 76, 115, 169, 181, 339, 340
Troll, C., 38
Troughton, M. J., 568
Truelove, M., 673
Tuan, Y-F., 221, 369, 413, 509
Tucker, J. L., 6

Tugwell, R., 356–57, 359, 366, 376
Tuma, N. B., 734
Turner, B. L., II, 192, 195, 201, 203, 205, 212, 375, 412, 416, 491
Turner, E., 241, 707
Turner, F. J., 162, 172
Tversky, A., 224

Ulack, R., 517–19
Ullman, E. L., 316, 317, 320, 323, 360
Ulrich, R. S., 231
Unger, M., 532
Upton, D., 213
Upton, G., 720, 721
Urban, D. L., 38
Urry, J., 588, 593, 622, 633
Urwin, D. W., 374
Urzua, R., 279
Usery, E. L., 763

Vail, D., 630
Vakamudi, R., 527
Valdez, A., 273
Vale, T. R., 30, 33–36
Valencia, M. J., 520
Valimont, K. M., 59
Vanderhill, B. G., 172, 568
van der Wusten, H., 602, 610
van Dijk, J., 308
VanDoren, C. S., 391, 396, 401
van Liere, K. D., 225
van Lierop, W. F. J., 260, 261, 731, 732
van Loon, H., 53, 54
Van Otten, G. A., 244, 246
Vance, J. E., Jr., 164, 305, 319, 322, 326, 369, 659, 660
Var, T., 396, 397
Vaucher, M. E., 567
Veblen, A. T., 35
Veblen, T. T., 32–36, 492, 497
Venkatesan, D., 55
Venugopal, G., 524, 527, 531
Vermeer, D. E., 201, 469, 475, 481, 482
Vernon, R., 262
Vick, S. G., 75
Villeneuve, P., 677
Vincent, J. A., 453
Vining, D. R., 267, 269
Virden, M., 454
Visser, S., 342, 343
Vitek, J. D., 71, 75, 76, 81, 86-88, 114
Vogeler, I., 242, 343, 344, 630
Vogelsang, R., 267
Vogt, B. M., 415
von Thunen, J. H., 342, 353
Vonnegut, N., 31

Vozikis, V., 397
Vuicich, G., 3, 6, 17

Wachs, M., 329
Wacker, P. O., 164, 175
Waddell, E., 195, 200
Wagner, M., 221
Wagner, P. L., 199, 242
Wahlquist, W. L., 9
Walker, D., 367, 568
Walker, E., 6
Walker, G., 566
Walker, H. J., 71, 77, 88, 141, 142, 144
Walker, R., 116, 228, 293, 295, 306, 620–26, 630, 631, 633–36, 661, 663, 664
Wall, G., 391, 393, 394, 396, 397
Wallace, I., 572
Wallach, B., 337, 340, 529, 520
Waller, P. W., 373
Waller, R. E., 59
Wallerstein, I., 157, 160–62, 171–73, 586–87, 592, 595, 609, 627, 637
Walling, D. E., 115
Walmsley, D. J., 219, 229
Walsh, S. J., 114, 755
Walter, H., 31, 36
Walton, J., 632
Wan, Z. M., 766
Wang, J., 758, 759
Wang, S., 764
Wanmali, S., 521, 529
Ward, D., 59, 158, 166, 167, 171, 178, 661, 664, 665
Ward, P., 494
Ward, R. M., 17, 334, 340
Warde, A., 633, 634
Warf, B., 214, 367, 630, 633
Warnes, A. M., 231, 452–55, 460–62, 655, 657
Warntz, W., 319
Warren, B., 627, 628
Warren, P. L., 756
Warren, S. G., 767
Warrick, R. A., 58, 60
Washington, W. M., 48–50
Wasilchick, J., 224, 229
Waterman, S., 604., 606
Waterstone, M., 117, 129
Watkins, J., 454
Watts, D., 173, 493
Watts, H., 623
Watts, M., 192, 195, 202, 469–72, 474, 619, 628–30, 633, 638
Watts, S., 474, 480, 481
Watts, S. J., 427
Waylen, P. R., 114, 115, 124
Weaver, C., 359, 360
Weaver, J. R., 171, 324
Weaver, R., 52

814

Webb, J. L., 427
Webb, T., III., 36, 55, 56
Webb, W. P., 172
Webber, M., 619, 620, 626, 627
Weber, M., 161, 667
Wedderburn, R. W. W., 724
Weerasinghe, G. M. S., 522
Weidlich, W., 728
Weigand, C. L., 754
Weigend, G., 142, 483
Weightman, B. A., 240, 245
Weightman, D., 394
Weil, C., 117, 427, 430, 437, 494
Weil, R., 243, 244
Weinbert, D., 260
Weiner, D., 473, 629
Weinstein, C. S., 230
Weirich, F. H., 76, 80, 114, 115, 339
Weismiller, R. A., 754
Weiss, E. T., Jr., 7
Weiss, I. R., 6, 8
Wekerle, G. R., 215, 229, 631, 677
Welch, R., 119, 758, 760–63
Welch, R. A., 511, 514, 515
Welch, R. B., 693, 695
Wellington, A. M., 319
Wells-Parker, E., 434
Wepfer, A., 399
Werner, C., 317, 319, 324
Wernsted, F. L., 518
West, N. C., 147, 149, 391, 392
West, R. C., 144, 489, 497
Westbrook, M., 104
Westcoat, J. L., Jr., 126, 131, 630
Wester, L., 32
Western, J., 480, 658
Westerveld, H., 221
Westman, W. E., 29, 31, 33, 35–37
Westover, T. N., 224
Wharton, S. W., 760
Wheatley, P., 515, 516
Whebell, C. F. J., 359, 606
Wheeler, J. O., 296, 316, 657
White, A. L., 104
White, C. L., 362
White, D., 705
White, D. A., 52
White, E. R., 703
White, G. F., 6, 13, 14, 16, 17, 112, 117, 119, 122–24, 128, 193, 194, 218, 224, 410–12, 416, 418–20, 555
White, M. S., 699, 701
White, P. J., 33
White, R., 244
White, R. B., 178, 264, 369
White, R. R., 480–81
White, S. E., 258, 271
White, W. B., 792
Whiteman, J., 292, 368, 620, 623, 624, 626, 633

Whitmore, T. M., 204
Whitney, H. A., 573
Whitney, J. B. R., 103
Whittaker, R. J., 38
Whyte, A. V., 117, 410
Widess, E., 630, 631
Wiens, H. J., 507, 509, 515
Wilbanks, T. J., 17, 95, 97, 100, 106, 107
Wilds, S., 336
Wiley, K. B., 6, 8
Wilhite, D. A., 59
Wilken, G. C., 173, 201, 491
Wilkie, J. R., 494
Wilkie, R., 494, 495
Wilkinson, N., 677
Willekens, F. J., 269
Williams, A., 394
Williams, C. H., 603, 605–7
Williams, D. L., 756
Williams, J., 50, 54, 510, 513, 515, 654
Williams, J. L., 241
Williams, L. S., 266, 497
Williams, M., 214, 630, 636
Williams, M. J., 116
Williams, P., 621, 622, 638
Williams, R., 630
Williamson, J. G., 359
Williamson, O., 637
Williamson-Fien, J., 680
Willis, R., 2, 9, 10
Willmott, C. J., 35, 47–49, 52, 95, 113, 116, 702, 703, 721
Wilson, A. G., 269, 325, 328, 343, 657, 725, 728
Wilson, E. O., 28, 29, 34
Wilson, J. A., 416
Wilson, J. P., 224
Wilson, J. R., 530
Wilson, M. F., 49
Winchell, D. G., 244–46
Windley, P., 452
Winkler, J. A., 113
Winston, B. J., 6–8, 14
Winter, D. G., 675
Winters, C., 172, 480, 659, 660
Winters, H. A., 82
Wirth, L., 652
Wirtshafter, R. M., 102
Wiseman, R., 262, 452–55
Wishart, D. J., 167, 179, 241, 246
Wisner, B., 245, 413, 415, 419, 474, 476, 477, 628, 629
Withington, W. A., 518–20
Witt, T. S., 102
Wittick, R. I., 494
Wixman, R., 549
Wohlenberg, E. H., 368
Wohlwill, J. F., 452
Wolch, J. R., 434, 438, 439, 637, 658

Woldenberg, M. J., 82, 86, 87
Wolf, E. R., 161, 172
Wolfe, R. I., 387
Wolff, G., 622, 633, 668
Wolforth, J., 575
Wolman, A., 117, 118
Wolman, M. G., 13, 14, 33, 112, 115–18, 410
Wolpert, J., 13, 260, 271, 410, 555, 585, 658
Wonders, W. C., 573
Wong, R. N., 792
Woo, M. K., 114, 115, 124, 566
Wood, C. H., 277
Wood, D., 230, 691
Wood, E. F., 115
Wood, J., 164
Wood, R. C., 368
Wood, W., 519
Wood, W. B., 427, 428
Woodcock, C. E., 765, 767
Woodring, P., 8, 18
Woodward, D., 709–11
Woodward, J., 454
Woodworth, G., 733
Works, M. A., 490, 495
Worster, D., 366, 631
Wright, A., 550
Wright, G. S., 5
Wright, L. D., 35
Wrigley, N., 361, 656, 724, 730–32, 734
Wronski, S. P., 7, 14
Wulff, R., 277
Wyckoff, W., 170, 489
Wyman, W. D., 172
Wynn, G., 572

Yamaguichi, D. K., 73
Yamamoto, S., 495
Yanarella, E., 583
Yapa, L. S., 371, 522
Yarnal, B., 51–54, 567
Yeates, M., 573, 655
Yetman, N. R., 399
Yeung, Y., 518, 519
Yevyevich, V. M., 116
Yitayew, M., 127
Yool, S. R., 757
York, R. A., 225
Yorty, R., 430, 432
Yoshioka, G. A., 35
Young, B., 574
Young, C. W., 177, 389
Young, G. J., 49
Young, K. R., 34
Young, W. J., 97

Zaniewski, K., 552
Zavattero, D., 104
Zdorkowski, T., 59
Zedong, M., 376, 514

Zeigler, D. J., 101, 224, 225, 415
Zelinsky, W., 98, 163, 166, 215, 241, 267, 268, 272, 273, 279, 280, 368, 394, 565, 569, 570, 603, 631, 679
Zhang, X., 759
Zheng, Q., 759
Ziegenfus, R., 432
Zimmerman, R. C., 35
Zinnes, J. L., 221
Zisheng, Q., 34, 36
Zonn, L. E., 227, 453, 455
Zube, E. H., 221, 225, 226, 230, 453
Zubin, D., 793
Zukin, S., 634
ZumBrunnen, C., 548–51, 555
Zurick, D. N., 529

Subject Index

Aboriginal
  agriculture, 497
  relics, 491–92
Abortion, 434
Abstract flow analysis, 321–22
Accessibility, 261, 270, 279, 319, 324, 326, 342, 345, 515, 518, 519, 525, 658
Acid rain, 59, 102, 117, 344, 571
Activity
  patterns, 373
  spaces, 320, 322–23
Afghanistan, 272, 531
Africa, 172, 173, 195, 200–201, 468–87, 508–9, 609, 629
  East, 470, 472, 476, 481
  West, 470, 472, 481
Age, 229, 230, 276, 323, 511
Aged, 451–62
  activity
    daily, 460
    patterns, 452, 455–56
    spaces, 456
  behavior
    mental functioning and relationships, 456, 457–58
  environment, 451–52
    assessments, 457
    barriers, 456
    experiences, 455
    individually defined, 452
    neighborhood, 453, 456–57
    residential, 455–59
    rural, 453
    stresses, 451, 455
    suburban, 453, 455
    urban, 453, 459
  expectations and aspirations, 458
  health
    Alzheimer's disease, 460
    psychological well-being, 458
  housing options, 454
    age-segregated retirement housing, 453
    home ownership, 454–55
  migration
    aging-in-place, 454
    history, 459
    mobility patterns, 453–54
    model schedules, 454
    net processes and rates, 454
    residential mobility, 452–55, 458–59
  perceptions, 452, 456–58
  population concentrations, 451
    old-old, 451
    South, 453
    Sun Belt, 459
  residences
    concentrations, 453
    locations, 452–53
    metropolitan, 454
    patterns, 452–53
    settings, 455–56
    spatial segregation, 461
  services
    church, 458
    community facilities, 455
    delivery, 461
    public transportation, 456
Aggregation, 221, 270, 320
Agriculture, 58, 351
  abandoned landforms, 203–4
  aboriginal, 497
  agrarian question, 628
  change, 173, 472–73, 517
  commercial, 58, 172
  contemporary, 333–50
  Corn Belt, 340
  devolution, 169
  evolution, 471

Agriculture, *continued*
  frontier, 495
  gender, 680
  indigenous, 173, 200, 201, 469, 475
  intensification, 198, 200, 201, 203, 204
  involution, 197
  issues, 495
  land preservation, 370
  modernization, 515
  origins, 160
  prehistoric, 203
  reform, 491, 495, 507
  shifting, 197, 201, 520
  Soviet, 547
  structure, 336–37
  subsistence, 172, 278, 490
  sustainable, 472
  systems, 491, 630
  traditional, 491, 495
  urban, 342
Agroclimatology, 482
Agropolitan development, 360
Air
  flow trajectories, 53
  pollution, 60
  shed, 60
Alabama, 274
Albania, 553
Algeria, 478
Amazon, 201, 203, 374, 489, 491, 494, 497, 628, 630, 631
American education, 1–20
Americans. *See also* Native Americans
  Asian, 178
  Black, 178, 271
  Hispanic, 178
  Indians, 178
  restless, 262
Anchorpoint theory, 222
Andes, 203
Angola, 272
Antarctica, 556
Apartheid, 480, 658
Appalachia, 99, 270, 271, 310, 367, 634
Arab
  cities, 533
  states, 532, 604
Architecture, 165, 213
  postmodern, 634
Archival documents, 158, 174–75
Arctic, 556, 557
Areal differentiation, 156, 157, 166, 258, 583
Arizona, 274
Artificial intelligence, 791–92
Asia, 172, 203, 273, 360, 506–45
  Central, 507, 550
  East, 304, 508–17
  South, 507, 508, 520–31
  Southeast, 303, 507, 508, 517–20, 627, 628
  Southwest, 507, 508, 531–33
  studies, 506–9
Assimilation, 166, 169
  ethnocultural, 167

Association of American Geographers, 157, 172, 179, 192, 258, 291, 321, 333, 375, 490, 546, 548, 552–54, 556, 584, 619, 658, 787
Atlas Graphics, 177
Atlas planning, 708–9
Atlases, 498, 550, 556
Atmosphere, 118
  carbon dioxide, 50, 56
  circulation
    Hadley regime, 54
    intertropical convergence zone, 54
  circulation shift, 53–54, 60
  ozone depletion, 59, 102
Australian school, 194
Austria, 371, 662
Autocad, 177
Autocorrelation
  serial, 308
  spatial, 308
Awards
  Bryan, K., 80, 86
  Gilbert, G. K., 80–81, 85
  Nystrom, W. J., 85

Backcountry, 163, 164
Baltimore, 660, 663
Bangladesh, 521, 523
Basic needs approach, 475, 476, 529
Behavior
  aged, 456–58
  commuter, 229
  consumer, 655
  corporate, 291, 375
  demographic, 511
  evacuation, 225
  human, 219–23, 266, 268, 474, 597
  international, 609–10
  investment, 313
  irrational, 471
  marital, 511
  migration, 656–57
  political, 598
  search, 223, 270
  shopping, 222, 326
  social, 656
  spatial, 325
    choice, 656
  travel, 325
Belgium, 602, 606
Belize, 274, 494
Berkeley school, 160, 193, 194, 199, 210, 629
Bhutan, 523, 529, 531
Biogeography, 28–38, 338
  animal, 32, 36
  classical, 28–29
  development of, 30–31
  methodological research, 37
  physiographic (plant), 35
  urban, 32
  zoogeography, 30, 36–37
Biosphere, 30
Biotechnology, 334, 345
Blacks, 341

Bloody Taylorism, 628
Boston, 215, 299, 311, 663
Bottom up approach, 475, 476
Boundaries, 165, 532, 667
    historical, 175, 181
    territorial, 586
Bowman's pioneer fringe, 172
Braudelian world view, 173
Brazil, 201, 359, 371, 374, 491, 492, 494, 496, 553, 605, 628
    colonial, 173
Brunei, 518
Buenos Aires, 174
Bulgaria, 553
Burma, 518, 520

California, 274, 299, 300, 310, 334, 338, 369, 630
Cambodians, 274
Cameroon, 471
Canada, 100, 164–66, 171, 296, 305, 321, 334–35, 551, 606, 636–37
    bibliographies, 574
    borders, 572
    cities, 564
    cultural geography, 569–70
    economy, 566
        agriculture, 568–69, 572
        fish, 571
        free trade, 572, 576
        industry, 568–69, 573
        mining, 572
        oil and gas, 574
        shipping, 573
        tourism, 570, 574
    education, 574–75
    environment, 566, 568, 571
        arctic, 566–67
        biogeography, 566–68, 575
        flood control, 567
        forests, 569
        geomorphology, 567
        glaciers, 567
        hydrology, 566
        landscape, 570
        perception, 570–71
        periglacial, 566
        permafrost, 566
        prairies, 572–73
        precipitation, 567
        prehistory, 572
        rock glaciers, 567
        soil, 567
        water supplies, 571
        weather, 571
    ethnic groups, 569, 571
    geography departments, 574–76
    historical geography, 571–72
    housing, 565, 572
    minorities
        French-Canadians, 569
        Japanese-Canadians, 570
        Mennonite farms, 568
        racism, 569
        Ukrainians, 570
    regional geography, 572–74
    social geography, 564–66, 569–70
        class-consciousness, 565
        gentrification, 564–65
        minorities, 568–70
    transportation, 574
    urban geography, 564–66
        architecture, 570
        inner-city, 565
        landscape, 573
        migration, 565
        residential environs, 565, 573
        urban systems, 574
        urban-rural fringe, 566
Capital, 293, 294, 298, 301, 305, 364–65
    accumulation, 586, 591, 595, 596, 599–601, 610, 625, 628, 654, 661, 662, 664
        flexible, 628
        uneven, 507
    flight, 499, 591
    flows, 621, 626
    internationalization of, 605, 628
    labor relationships, 305
    Marxist theory of, 622
    movements, 294
    role of, 336
Capitalism, 159, 161, 162, 167, 173, 266, 268, 292, 306, 343, 359–60, 363, 474, 572, 586, 591, 592, 610, 621–22, 624–38, 654, 655, 663, 666, 668
    agrarian, 473
    globalization of, 375
    legitimation function, 586
    necessary relations, 587, 599
    penetration, 495
    transnational, 494
Carbohydrate frontier, 160, 161
Careers
    historical geography, 180–82
Caribbean, 160, 161, 173, 174, 273, 274, 489, 491, 493, 494, 497
    climates, 497
Carnegie-Mellon University, 360
Carrying capacity, 197, 199, 528
    human, 471
Cartography, 6, 492, 515, 686–713, 761–63, 783
    analytical, 697, 706, 713
        affine equations, 701
        coordinate transformation, 697, 701–2
        generalization, 697, 703–4
        gray-scale equations, 707
        mathematical construction, 707
        mathematics of color separation, 707
        rubber-sheeting, 701
        shape analysis, 705
        tessellation models, 699–700
    animation, 705
        map symbols, 712
    atlas planning, 708–9
    cartograms, 702
    cognitive research, 693–95
        experiments, 712
        investigation, 691

Cartography, *continued*
  psychology, 693–95
  communication studies, 691, 704, 706, 710
  computer-assisted, 248, 344, 355, 708, 780–81
    area computation, 705
    interactive mapping, 705, 708
    interpolation, 702–3
    line generalization, 703
    numerical map analysis, 697, 705–6
    point-elimination algorithm, 703
    software, 177, 783–84
    storage structure, 698
    SYMAP, 783–84
    vector-to-raster conversion, 700
    vehicle navigation, 712
  data (information), 705
    control point density, 703
    geodetic accuracy, 711
    interpolation, 702–3
    model, 697–99
    positional accuracy, 700
    quality, 700–701
    reality, 698
    structure, 698–700, 780
    topological encoding, 699
  digital
    elevation, 706, 712
    hologram, 712
    three-dimensional maps, 705, 712
    virtual map, 698
  early maps, 710
    accuracy, 711
    Chinese, 515
  education, 687
  experimental
    circle-size estimation, 692–93
    color perception, 692
    dot underestimation, 693
    eye-movement studies, 695–96
    gray-tone perception, 692–93
    stimulus response, 692, 712
  expert systems, 704, 713
  geographic information systems (GIS), 712–13
  historical, 709–11
  legal aspects, 709
  map (mapping), 710
    accuracy standards, 762
    bathymetric, 758
    bivariate, 695
    comparison, 695
    construction, 706–8
    design, 689
    image, 762
    lettering discrimination, 693
    line, 762
    marketing, 708
    overlay, 777, 784–85
    photomorphic, 754
    production, 706–9
    projection, 697, 701–2
    reading, 696–97
    similarity, 706
    symbolization, 690–91
    tactual, 707
    video-disk, 712
    virtual, 698
  photomechanical
    equipment, 707
    materials, 707
    mosaicking, 707
  psychophysical studies, 692–93, 712
  relief models and shading, 705–8
  semiology, 691
  sex differences in map use, 696
  statistical, 702, 706
    fractals, 704–5
  symbolization, 704–5
    apparent magnitude scaling, 692–93
    area, 704
    continuous surface, 705
    design, 690–91
    selection, 690–91
  visual aspects, 694, 696, 791
Caste system, 524
Central place
  hierarchy, 512
  systems, 164, 361, 373–74, 512, 529
  theory, 295, 305, 481, 652, 653, 659
Chernobyl, 96, 101, 550, 556
Chesapeake tobacco colonies, 164
Chicago, 655
  school, 193–95, 652, 653, 666
  University of, 356–58, 360
  urban sociology, 194, 652, 653
Chickens, black-boned, 490
Children, 228, 230, 274
  environments, 230
  utility of, 474
Children's Environmental Response Inventory, 230
Chile, 358–59, 497, 658
China, 371, 372, 496, 507–15, 517, 534
  ideology, 515
Choice
  housing market, 261
  neighborhood, 261
  sequential models, 261
Cities. *See also* Urban, 170–71
  building, 171
  capital, new, 494
  colonial, 173, 480, 492, 660
    capital, 518
    port, 507, 526
  inner, 267, 361, 552, 621, 665
  legal units, 668
  primate, 266, 371, 518, 519, 653
  secondary, 372, 519
  size distribution, 361, 371–73
  socialist, 552
  systems, 305, 519, 653–54, 659, 660
Civil society, 586, 587, 607
Clark University, 619
Climate
  anomalies, 53, 54
  change, 47, 54–58, 60, 107, 114, 116, 122, 133, 335, 344
  control, 48

Climate, *continued*
  data sources, 51–52, 54–56
  dynamics, 47, 51, 53–54
  extremes, 59, 113, 122–23
    nuclear winter, 59
  fluctuations, 55
  future, 59
  greenhouse, 56–57
  impact
    assessment, 57
    studies, 58, 60
  instruments
    radiometer, 51
  legislation
    National Climate Program Act, 57
  modeling, 47–51
    boundary conditions, 49, 53
    elements, 48–49
    energy-balance, 48–50
    general circulation (GCM), 48–50
    hydroclimatic, 58, 114
    intransitivity, 53
    parameterization, 48, 50
    radiative-convective, 50
  perception, 59
  prediction, 48
  proxy indicators, 55–56
  reconstruction, 55–56
  regimes, 53
  regions
    Pakistan, 530
    Soviet Union, 547
  seasonality, 56–57
  teleconnections, 53
  variability, 34, 55
Climatology, 36, 38, 47–61, 481–82, 552
  applied, 47, 57–60
  classification procedures, 52–53
  diagnostic analysis, 52
  forensic, 60
  future research, 61
  heat budgets, 49–50
  hydroclimatology, 113
  methods, 51–52
    map-correlation, 52
  synoptic, 47, 51–53
  technoclimatology, 58–60
Coastal environs
  accessibility, 147
  beaches
    barrier, 142, 145
    condominiums, 149
    residential development, 142
    sediment transport, 145
  boating, 147, 149
    ramps, 149
  boundaries, 150
    disputes, 150
    marine, 142, 149–51
  dunes, 149
  fish, 150
    fisheries, 142
    fishing, 147
    shellfishing, 149
  geomorphology, 142–46
  legislation
    Coastal Barriers Resources Act, 144
    Coastal Zone Management Act, 144
    freedom of navigation, 149–50
    National Environmental Policy Act, 144
    territorial sea, 150
  management
    Coast Guard, 148
    dredging, 148
    National Park Service, 144
    Office of Ocean/Coastal Resources Management, 144
    port development, 142, 151
    World Court, 150
  recreational areas
    Gateway National Recreation Area, 145
    marinas, 149
  regions
    Cape Cod, 144–45
    Cape Hatteras, 144
    Louisiana's coastline, 144
    Mississippi River Delta, 145
    Sandy Hook, 144–45
  resources
    extraction, 150
    gas, 150
    management, 151
    oil, 150
  tourism, 146, 149
    communities, 146
    developments, 149
  urban, 146, 151
  wetlands, 149, 151
Cognition, 192, 196, 212, 218, 220, 228
  dissonance, 224, 225
  images, 221
  mapping, 656, 693–95
  spatial, 219–23
  urban, 229
Colombia, 276, 489
Colonialism, 162–64, 477, 479, 480
  administrative policies, 172
  British, 178
  cities, 493, 660
  French, 178
  Iberians, 173
  internal, 605
  port cities, 507, 526
  sources, 492
Colorado, 369
Communication, 266, 276
  cartographic, 691, 704, 706, 710
  nonverbal, 213
  transportation trade-offs, 324
Community, 351
  attitudes, 225
  development, 360
  minority, 668
  participation, 230
Commuting, 334–35, 361, 517
  patterns, 326

Comparative advantage, 268, 293, 337, 338, 626, 654
Computers. *See also* Cartography, computer-assisted; GIS
   in historical geography 176–77
   information processing, 735
Conflict, 633
   boundary, 532, 610
   class, 591, 595, 626
   ethnic, 525
   labor, 595
   locational, 592, 593
   neighborhood, 637
   peasant-herder, 472
   residential, 262
   resolution, 358
   urban social, 585, 593
Connecticut, 310
Conservation, 32
   energy, 103–5
Constraints, 222
   institutional, 261
Contextual analysis, 221, 222, 225, 271, 483, 597
Contras, 489
Core-periphery, 161, 269, 277, 359–60, 477, 595, 605, 620, 623, 664
   semi-periphery, 161, 605
Corn Belt, 340
Cornell University, 355
Corporations, 363
   multinational, 294, 371
   transnational, 303, 304, 603, 628
Costa Rica, 201, 276, 278, 281, 371, 493, 495
Counterurbanization, 259, 266–68, 361, 654
Crime, 224, 524
Crisis, 468
   food, 469
   woodfuel, 477
Critical literacy theory, 227
Cross-class alliance, 590–91, 595, 599, 609
Cuba, 174, 489
Culture, 227, 342, 489
   adaptation, 197, 198
   areas, 195, 197, 493
   behavior, 193, 202
   change, 175
   conflict, 229
   contact, 160
   core, 197
   ecology, 172, 178, 192–205, 209, 211, 212
      diachronic, 199, 203
      synchronic, 199, 202, 205
   evolution, 197, 198
   hearth thesis, 163
      diffusion, 163
   material, 167, 176, 526
   preservation, 529
   processes, 195
   relativism, 499
   role of, 196
   selection, 198
   traits, 225
   values, 226, 227, 229
   world, 7
Curriculum
   elementary, 2
   materials, 7
   national, 2
   secondary school, 3, 7
   university, 14
Cybernetics, 195, 196

Dacca, 372
Dallas, 297
Debt, 488, 499, 629
Decentralization, 267, 372, 491, 526, 624, 663, 664
Decision theory, 224
Decision-making, 219, 223, 260, 270, 358, 369, 371, 665
   government, 352
   individual, 278
   locational, 295
   public policy, 224, 225
   sequential, 261
   socialist, 372
   spatial, 222
Degrees conferred, 10–11
Delphi method, 343
Demography, 195, 204, 258
   equation, 276
   housing, 263
   multiregional, 269
   transition theory, 275, 516, 528
Dependency theory, 274, 277, 627, 628
Des Moines, 264
Desegregation, 262
Desert, 345
   desertification, 59, 514
   encroachment, 470
Destinations
   characteristics, 269
   competing, 270
   rural, 278
Detroit, 661
Development, 627–29
   agropolitan, 360
   cattle, 522
   community, 360
   definition, 351
   economic, 163, 199, 201, 259, 275, 281, 297, 303, 305, 310, 511, 515, 528, 529, 551, 556
   family planning, 529
   grass roots, 372
   hydroelectric, 496
   planning, 475–78, 494
   political dimensions, 495–96
   regional, 305, 351–86, 474, 494, 550–52, 628, 633
      general theory, 359
   rural, 373, 476, 479, 521, 522
      appropriate, 475–76
      integrated, 477–78
   uneven, 158, 159, 167, 343, 371, 474, 475, 477, 605, 610, 626, 652
   urban, 621, 628, 658
   from within, 476
Diffusion, 163, 278, 279, 359, 478–79
   cattle-ranching, 490
   computers, 177

Diffusion, *continued*
  dairy cow, 517
  fuel efficient stoves, 476
  historical, 175
  modernization, 359, 371
  network, 319
  routes, 175
  Sikhism, 524
Direction
  bias, 221, 263, 527
  trade, 302
Disabled, 228, 231
Disaster, 410–12
  avalanche, 33
  fire, 33–34, 37
  flood, 33, 113–15, 125
  nuclear winter, 59
Disease. *See also* Health, 204, 473–75
  chronic, 429–33
    mental disability, 438, 441
  contagious, 426–28
    AIDS, 426, 428, 439
    influenza, 433
    scrub typhus, 432
    tuberculosis, 427
  control, 474
  diffusion, 433
  ecology, 425–33
  endemic areas, 427
  environmental factors
    urban, 429–31, 437, 439
    urbanization, 428, 432–33
  malnutrition, 428–29
  mental illness, 438–39, 441
  morbidity, 429–30
  noncontagious
    cancer, 430–31
    cardiovascular, 429–30
    hypertension, 429
    multiple sclerosis, 432
  parasites
    guinea-worm, 427
    schistosomiasis, 426
  race, 430
  toxin-induced
    pesticides, 432
    ciguatera, 432
  vector-borne, 426–28
    malaria, 426, 432
Distance, 269, 279, 334, 342, 358, 513, 527, 530, 584, 624, 632, 655, 659
  bias, 221
  decay, 513
  long average haul, 550
  marriage, 525
Divergent tendencies, 157, 158
Domestication, 32
Double duping, 596
Drought, 58, 169, 472, 481–82, 629

Earth Systems Science, 768
Ecology, 29–31, 38, 195, 196, 241
  animal, 36
  Boserupian, 470, 471
  change, 168, 497
  considerations, 223–25
  cultural, 32, 172, 178, 192–205, 211, 212, 243, 629
  disease, 425–33
  factorial, 653
  geographical, 28–29
  human, 194, 362, 411, 471, 474, 520, 653, 661
  landscape, 75, 77–78, 89
  mountain, 529
  political, 471
  social theory of dryland, 471
  sustainability, 195
  urban, 652
Economics, 8, 14, 270, 272, 274, 290, 352, 354, 357, 358, 363, 608
  agglomerative, 729
  neoclassical, 276, 277, 290, 584, 585, 622, 627
  price determination, 727
  urban, new, 657
  urban-transit, 326
  welfare, 730
Economy
  of affection, 473
  agglomeration, 266, 268, 292, 295, 306, 624
  areal patterns, 305
  external, 293, 623–24
  global, 338
  growth, 303
  information, 309
  of scale, 266, 268, 336, 372, 654
  sectoral analysis, 306
  Soviet, 551
  suprastate, 605
  transformation, 174
  urbanization, 295
  world, 588, 591, 595
Ecosystem, 30, 36, 77–78
  structure, 35–36
  function, 35–36
Ecuador, 201, 203, 281, 371
Education, 323, 351
  American, 1–20
  cartographic, 687
  citizenship, 6
  development, 7
  general, 16
  geomorphological, 88–89
  higher, 1, 9–14
  international, 7
  liberal, 16
  multi-cultural, 7
Efficiency, 20
  economic, 328, 372, 658
Egypt, 204, 276, 532
El Niño/southern oscillation, 53, 60
El Salvador, 272, 274, 275
Elderly, 230, 262, 272
Empirical research, 221–22
  empiricism, 499
Energetics school, 198, 201

Energy, 95–107, 323, 338–40, 548–49, 552
  agencies
    research
      National Research Council, 101
      Oak Ridge National Laboratory, 106
    regulatory
      Nuclear Regulatory Commission, 101
      U.S. Department of Energy, 98
      U.S. Environmental Protection Agency, 106
  analysis
    econometric, 98
    integrated, 105–6
    net energy, 99, 103
    Strategic Environmental
      Assessment System (SEAS), 106
  cogeneration, 105
  conservation, 103–5, 225
    high-occupancy vehicle, 104
    home, 103
    industrial, 105
    transportation, 104–5
  education
    Boston University, 106
  electricity
    hydroelectric, 103
    trade, 102
  fossil fuels
    British coal industry, 99
    coal, 99–100
    crude oil, 95, 98
    gasoline demand, 98
    natural gas, 98
    petroleum, 95, 98–99, 106–7
  geothermal, 103
  nuclear, 96, 100–101, 107
    Chernobyl, 96, 101
    decommissioning, 101
    emergency planning, 101
    opposition, 100
    radioactive-waste disposal, 101, 107
    Three-mile Island (TMI), 96–97, 100–101
  planning, 476–77
    emergency, 101
    land-use, 104
    power-plant siting, 100–102
    transportation, 105
  policy, 96
    oil embargo, 95, 107
  regions
    Appalachian Region, 99
    Canada, 100
    developing countries, 106
    Kenai Peninsula, Alaska, 98
    Middle East, 98
    North Sea, 98
    Soviet Union, 98
  renewable sources, 59, 102–3
    biomass, 103
    solar, 102
    synthetic fuels, 100
England, 300, 593
Entrepreneurship, 297, 300, 301

Environment, 351, 353, 358, 370, 629–31
  biophysical, 58–59
  change, 176, 375
    human-induced, 179
  conservation, 360, 361, 482, 491, 548
  crusade, 179
  degradation, 169, 470–72, 494, 514, 528, 529, 629
  determinism, 160
  impact, 370
  influence, 176
  learning, 221
  management, 225, 482, 548
  marine, 520
  modification, 555
  movement, 630–31
    Green, 631
  past, 179
  perception, 168, 169, 193, 199, 209, 212, 215, 218–32, 513, 530
    social aspects, 229
  preserves, 548
  quality, 224
  relationship, 345
  risk, 224
  rural, 338
  science, 352
  socioeconomic, 59
  study of (using GIS), 777
  symbolism, 213, 227
  threats, 224
  transport, aspects, 323
  urban, 513
Epistemologies, 210–12, 317, 584, 585, 587, 588, 600
Equity, 20, 305, 372, 476, 658
  spatial, 317, 319
Ethnicity, 166, 167, 525, 599, 607, 658, 664, 680
  conflict, 525
  demographic changes, 549
  frontiers, 493
  groups, 272, 275
  identities, 368
  minority areas, 549
  values, 229
  variation, 175
Ethnomethodology, 227
Ethnoscience, 200, 475
  biology, 32
  ecology, 32
Eurocentrism, 177, 179
Europe, 300, 368, 507, 609, 633–35, 664
  East, 372, 546–62
  economic history, 275
  expansion, 160, 197
  Western, 625
European community, 593, 609
Evapotranspiration, 49
Exceptionalism, 162, 651
Exchanges
  Eastern Europe-U.S., 553
  Soviet-U.S., 553–57
Expert systems, 704, 713, 752, 764, 791

824

Famine, 202, 472, 473, 629
Farmers
   marginal, 338
   part-time, 336
Farming
   collective, 495
   crisis, 334
   dryland, 339
   efficiency, 523
   ethnocultural practice, 168
   family, 169, 343, 680
   fragmented, 336
   hobby, 334
   no-till, 339, 340
   size, 336
   traditional, 199
   wives, 517
Farmland
   conversion, 334–36, 340
   loss of, 335, 375
   marginal, 339
   ownership, 337
   protection, 334–36
Feminism. *See also* Gender, 20, 229, 306, 620–21, 638, 673–74, 680
Feminist methodology, 678–79
Fertility, 258, 275, 511, 528
   control, 475
   decline, 528
Field studies, 174, 176, 193, 195
Fields
   raised, 491
   sunken, 491
Fire, 33–34, 37
Flood, 33, 115
   analysis, 114
   coastal damage, 144
   control structures, 124
   flash, 124
   hazard, 123–26, 132
   human adjustment, 124–25
   human impact, 125
   insurance, 125
Flows, 362
   abstract, analysis, 321–22
   analysis, general, 323
   capital, 491, 621, 626
   channelization, 273
   commodity, 317, 320, 327, 373, 550
   energy, 323
   modeling, 321
   trade, 316
Food
   consumption, 547
   crisis, 469
   export crop, 477
   habits, 495, 526
   internationalization, 334, 628
   preferences, 525
   production, 193, 195, 202, 521–23
   relief, 472
   self-sufficiency, 517

Fordism, 625, 628
   post-Fordism, 625
Forests, 338, 339, 353, 514, 548, 630, 631
   Boreal, 56
   deforestation, 489, 514, 529
   GIS, uses of, 777, 779
   structures, 497
Fossil fuels, 549
France, 625
Frontier
   American, 166–72
   evolution, 168
   farmers, 201
   resource, 605
   thesis, 162, 172
   urban systems, 172
Functionalism, 158, 197, 198, 583, 586–88, 595
   dualism, 475

Gambia, The, 473
Gemeinschaft-gesellschaft, 163
Gender, 215, 229, 599, 621, 629, 631, 632, 673–74, 680–81
   access, 473
   balance methodology, 679
   balanced curriculum, 674, 679–80
   class, 680
   culture, 679
   division of labor, 306
   environment, 680
      American West, 677–78
      cities, 677–78
      desert, 678
      North American cities and suburbs, 678
      Southwest U.S., 678
   ethnicity, 680
   global perspective, 674–76
   landscape, 674, 677–78
      interior, 677
      exterior, 677
   life stage, 680
   methodology, 674, 678–79
   migration, 679
   residence
      family farm, 680
      households, 676
   spatial behavior, 674
   work, 674, 676–77
      labor force, 306, 676
      segregation, 677
Gentrification, 229, 621, 622, 637, 657
Geographers
   British, 261, 494, 635, 652, 657
   Conference of Latin American, 490, 498
   European, 273
   French regional, 161, 361, 362, 610, 633, 657
   German, 652
   university, 15
Geographical Education National Implementation Project (GENIP), 18–19

Geographic information systems (GIS), 116, 119, 248, 712–13, 723, 737–38, 752–53, 761, 765, 768, 776–94
  agencies
    Canadian GIS, 787
    Census Bureau, 787
    Dane County, Wisconsin, 789
    Land Management Information System, 788
    U.S. Geologic Survey, 788
  artificial intelligence, 791–92
  computer
    aided design, 783
    cartography, 780–81
    graphics, 785
    hardware, 785
    science, 776
  data
    Digital Line Graph, 788
    DIME, 779–80, 786
    Geographic Data Technology, 786
    Global Positioning System, 780
    National Digital Cartographic Database, 788
    Phillips Petroleum, 786
    topologic structure, 780
  journals (professional)
    American Cartographer, 787
    Photogrammetric Engineering and Remote Sensing, 787
  operations research, 781–82
  private companies
    Compusearch, 790
    Geographic Data Technology, 786
    Phillips Petroleum, 786
  remote sensing, 752–53, 761–63
  research organizations
    National Center for Geographic Information and Analysis, 790–92
  societies and conferences (professional)
    American Congress on Surveying and Mapping, 787
    American Society of Photogrammetry and Remote Sensing, 787
    Association of American Geographers, 787
    AUTO-CARTO, 787
    IGU Commission on Geographical Data Sensing and Processing, 787
  software, 786
    ARC/INFO, 177, 777
    Canadian GIS, 787
    Experimental Cartography Unit, 786
    expert systems, 791
    Harvard Laboratory for Computer Graphics, 786
    Land Management Information System, 788
    Map Analysis Package, 778
    SYMAP, 783–84
    vehicle navigation system, 782
  theory
    geographical matrix, 777–78
    spatial concepts, 752–53, 761, 765, 768, 777
Geography
  Africa, 468–87
  air age, 6
  alliances, 19
  applied, 178, 193, 344
    historical, 180–82
  Asian, 506–45
  atheoretical, 209, 210, 627
  behavioral, 117, 156, 171, 210, 212, 218–32, 260, 269–71, 290, 317, 318, 656, 724, 730
    analytical, 219, 221, 223
    measures, 121
    research, 218
    theory construction, 223
  of capitalism, 588, 589, 592
  Chinese, 510–15
  commercial, 3, 301
  cultural, 209–15, 227, 232, 242–43, 496, 515, 523, 534, 631, 634
    atheoretical heritage, 210
    diachronic, 214
    humanistic, 210
    "new", 209
  East European, 546–62
  economic, 97, 159, 160, 173, 290–315, 351–86, 496, 515, 528, 557, 622–26
  education, 1–20
    renaissance, 18–19
  electoral, 594–99
  feminist, 673–74, 680
  future, 376
  global, 6
  historical, 156–82, 211, 227, 232, 241, 259, 490, 491, 515, 530, 548–49, 663
    of capitalism, 588, 592
    macro-scale, 160–62
    physical geography, 176, 179–80
    urban, 305, 651–53, 659–66
  illiteracy, 18
  in an Urban Age, 6
  industrial, 290–315, 549–50, 590, 622–26
    new, 622–26, 632
  information systems (GIS), 177, 343, 354, 498
  medical, 117, 474, 475, 530
  overlooked departments, 16
  physical, 469, 481, 483, 497, 530
  political, 333, 369, 531, 582–618, 635
    definitions, 582–84
    international, 609–10
    new, 582, 584–89, 592, 594, 600, 601, 610
    place-based, 589
  population, 158, 258–89, 510–11, 528, 548–49
  quantitative, 358–61
  radical, 162, 584, 588, 589, 610–11, 619–50
  reformist development, 468–69
  regional, 6, 483, 517, 518, 632–35, 651
    "new", 483, 498, 625, 632, 634
  of representation, 594
  requirements, 9
  rural, 333–50
  school, 1
  social, 171, 173, 480, 552
  socialist, 305, 362, 363, 619–50

826

Geography, *continued*
  soils, 334
  Soviet, 546–62
  space-time, 221, 222
  status of, 1, 19
  teaching, 15
    effective, 17
    teacher training, 20
  thought, Soviet, 547
  time, 222, 369, 633, 661, 662
  transportation, 316–32, 550
    definition, 316
    holistic, 318–21, 329
  of underdevelopment, 468–69
  undergraduate, 17
  urban, 305, 306, 326, 361, 515, 527, 548, 585, 592, 610, 621–22, 651–72
    analytical, 653–57
    gender, 676–77
    historical, 651–53, 659–66
    "new", 657
  of voting, 594, 599
    influences, 594
  of women, 631–32
  world, 6
Geomorphology, 38, 73, 70–89
  aeolian, 77, 80–81, 87
    wind tunnel, 80
  arid environs, 76, 80, 87
    deserts, 80
  biogeochemical, 76
  coastal, 77, 80, 142–46, 151
    barrier islands, 73, 82
    beaches, 73, 76–77
    deltas, 73
    estuaries, 73
  dating methods
    dendrogeomorphology, 80
    isotopes, 80
    radiocarbon, 81, 84
    rock varnish, 80
    stratigraphic, 81
  debris flows, 73, 80–81
    landslides, 73
  fluvial, 76–77, 114
    alluvial fans, 80
    channel, 77, 81–82
    deltas, 73
    regions, 72–73, 76, 78, 82, 87
    turbidity, 80
  geobotany, 754–55
  glacial, 73, 75–76, 80–82, 84, 87
    till, 81
  historical, 71
  methods
    field research, 71, 78–80, 88
    laboratory, 78–81
    mapping, 73, 81
    modeling, 70–71, 82
    numerical, 78–79, 81–82
    photogrammetry, 77
    remote sensing, 81
    simulation, 82
    statistical fractals, 82–83
    stochastic, 82
    terrain analysis, 77
    trend surface, 82
  mountain
    alpine, 75–76
    regions, 72, 77, 83, 86
  periglacial environs, 73, 81, 86
    solifluction, 81
  sediment, 115
    cascade, 76
    study of, 73, 75, 77, 81, 84
  societies (professional)
    International Geographical Union, 86
  soil, 70, 73, 81, 83
  tectonic, 73, 77, 81, 87
  theory
    castastrophism, 73, 76, 81, 83, 87
    deterministic, 82
    paradigms, 88–89
    uniformity, 75–76
  thresholds, 75–77, 82
Geophagy, 496
Geopolitics, 490, 496, 517, 609–10
Georgia, 274
Germany, 651
  East (GDR), 371, 551–53
Gerontologists, 451
Ghana, 276
Ghettos, 661, 664, 665
  ethnic, 664, 665
    Chinatown, 665
GIS. *See* Geographic information systems
Glaciers
  mass balance, 52
Glaciology, 73, 75–76, 80–82, 84, 87
Glasnost, 546
Global
  change, 555
  climatic change, 107, 224
  environmental problems, 224
  food model, 60
  geography, 6
  market, 342, 345
  studies, 7
  trade, 303
  warming, 343, 344
Goldenweiser's principle, 161
Great Britain, 338, 360, 363, 371, 498, 525, 598, 620, 632, 633, 636, 637
Great Plains, 336, 369
Greece, 532
Green revolution, 371, 495, 520–23
Greenhouse climate, 56–57, 102
Groundwater, 339
Group Areas Act, 480
Growth center/pole theory, 277, 305, 359, 372–73, 512, 625
Growth machine, 591, 668
Guatemala, 173, 274, 493, 494, 496
Guidelines for Geographic Education, 18

827

Hacienda, 173, 492
Hadley circulation, 482
Halifax, 171
Harvard University, 354
Hazards. See also Disaster, 181, 194, 223, 225, 333, 410–21, 497, 630
  acts of God, 415
  disasters, 410–12
  exposure, 415–16
  global, 419
  institutions
    international, 417
    research units, 414
  management, 412, 420
    anticipatory strategies, 417
    assessment of alternatives, 411
    intervention, 411
    mitigation, 416–17
    policy, 411
    post audits, 418
    preparedness, 416
    prevention, 416–17
    program evaluation, 418
    research, 415
    response, 415–16
    structural engineering, 411
    sustainable development, 420
    use of floodplains, 411
    use of information, 417–18
  natural, 412, 472, 630
    climate, 419, 477
    drought, 113, 122–23
    flood, 114–15, 123–26, 132, 411
  perception, 169, 225
  research
    assessment, 415
    communication, 415
    decision-making, 414
    disaster, 412
    initiatives, 417
    integrated theory, 419
    interdisciplinary, 412, 418–20
    theories, 412–13
  risk, 412, 415–16, 432
    floods, 411
    uncertainty, 417
    vulnerability, 415–16
  social, 224, 420
  technological, 224, 412, 530, 629
    toxic waste, 630–31
    transporting
    materials, 325
Health. See also Disease, 351
  accidents, 431–32
  air pollutants, 430, 432
  cultural influences, 425–26
    behavior, 426
    diet, 431
    lifestyles, 429
  environmental factors, 425
    air pollutants, 430, 432
    groundwater, 432
    toxins, 430, 432

Malaysians, 430
  mortality, 430, 440
    infant, 428–29
  nutrition, 440
    famine, 428–29
  occupation, 429, 431
    female unemployment, 429
  Oklahoma, 431
  poverty, 426
  public-health model, 431
  Quebec, 431
  research gaps, 439–42
  services, 435–38
    chiropractor, 437
    for-profit hospitals, 438
    Health Maintenance Organization, 437
    hospital closure, 438
    housing program, 439
    location of, 433–39
    mental-health care, 438–39
    research on, 433–39
    secondary medical care, 435–38
  third world, 426, 429, 435, 440–41, 475
    Bangladesh, 429
    Guatemala, 435
    Jamaica, 429
    modernization, 426
    Nigeria, 435
  traditional medicine, 436
  tropics, 426
  substance abuse, 431, 433
    alcohol, 431, 433
    cigarettes, 433
  violence (crime), 431–32, 440–41
Hegemony, 160, 162, 215, 493, 592, 593, 600, 620
  ideological, 601
High school
  Geography Project, 6
  teachers, 15
  vocational, 3
Hill stations, 520
Historical materialism, 584, 588, 599, 607, 631–33, 638
  geographic, 631–32
Historiography
  African, 483
  controversies, 158
History, 5, 14, 345, 488
  perspective, 203–4, 277, 278, 662
  preservation, 175, 176, 181, 227, 229
  transportation, 326
  urban, 480
Holland Land Company, 170
Homeless, 658
Honduras, 494
Hong Kong, 513, 534
Honors, 12–14
  American Academy of Arts & Science, 14
  Guggenheim Memorial Foundation, 14
  National Academy of Sciences, 12–14
Households, 263–65, 472, 676
  change, 264
  composition, 263–65
  intra-urban, 263

Household, *continued*
  segregated, 453
  size, 263–64
  structure, 263, 492
Housing, 171, 215, 636, 637, 657, 661
  choice, 221
  rural, 515
  sites and services, 480
  urban, 259–65
HUD, 262
Hudson's Bay Company, 179
Human/biota interaction, 32
Humanism, 159, 194, 210–11, 215, 226, 227, 343, 473
Hungary, 551–53
Hydrology, 113, 124
  global, 116
  modeling, 114–15
  regionalization, 115–16
  snow, 113

Identity, layers of, 602–3
Illinois, University of, 357
Impacts
  analysis, 324
  climate, 507
    assessment, 472
  environmental, 370, 548
  reciprocal, 320–21
  tourism, 529
  weather, 547
Imperialism, 161, 173
  Leninist, 161
  Meinig's, 161, 172
Inca, 203
India, 276, 361, 371, 373, 374, 507, 513, 517, 521–30, 534
Indians. *See* Native Americans
Indonesia, 197, 279, 361, 518–20
Industrialization, 171, 300, 516, 528, 549
  city-based, 512
  deindustrialization, 625, 637
  small town, 549
  territorial, 624, 633
Industry, 351
  change, 290
  copper, 492
  decentralization, 292, 293
  innovation, 297–300
  iron and steel, 549
  location, 269, 292, 305, 481
  organization, 306
  petrochemical, 550
  production, 303
  regional analysis, 306–10
  regional change, 292
  regional policy, 310
  rural, 335
  state policy, 312
  structure, 268
  synthetic rubber, 549
Inequalities
  income, 522
  interregional, 359

  regional, 359, 368–72, 478, 528
  social, 585
Informal sector, 280, 480, 481, 628
Information, 301
  economy, 309
  flows, 171, 192, 193, 196, 270, 276
  networks, 299
  technology, 295, 296
  theory, 195, 270
Innovation, 198, 200, 359
  industrial, 297–300
Institutional
  constraints, 261
  contexts, 261
  perspectives, 156
Intertropical convergence zone, 54, 483
Interdisciplinary
  connections, 197–99
  exploration, 221
International Geographical Union, 291, 553, 554, 584, 654
International Geosphere Biosphere Program (IGBP), 78, 792
Interpretation
  conjunctive, 169
  disjunctive, 169
Interstate Highway System, 362, 366, 374
Introduced animals, 32
Investment
  decisions, 294
  federal, 368
  flows, 303
  foreign, 301–4, 309, 311, 368
  regional strategy, 551
Iran, 531, 533
Ireland, 637, 660, 664
  Northern, 374
Irrigation, 204, 336, 339, 521, 530
  systems, 491
Isolation, 172, 199, 340, 584, 602
Israel, 371, 507, 531–32, 534
Italy, 625
Ivory Coast, 472, 473, 477
Ivy League, 16

Jamaica, 173
Japan, 160, 172, 311, 507–9, 515–17, 534
Jordan, 532

Kampuchea, 518
Kansas, 336
Kentucky, 271
Kenya, 276, 373, 471–74, 477, 479, 482
Kondratieff waves, 162, 168
Korea, 359, 507–9, 515–17, 533, 534
  War, 508

Labor, 266, 269, 293, 303–6, 364–65, 590, 593
  capital relationships, 305
  control, 473
  demand, 306
  division of, 363, 597, 623

Labor, *continued*
   ethnic, 167, 664
   gender, 306
   international, 303, 620
   social, 623
   spatial, 293, 622–24, 633
   territorial, 588
  flexibility, 306
  force, 304
  location, 303
  markets, 267, 277, 291, 303–6, 590, 623, 625, 631, 664
  migrant, 473
  mobility, 304
  pools, 299
  relations, 306, 620
  theory of production, 293
Land
  alienation, 473
  degradation, 59, 202, 471, 629
  holding patterns, 167
  information systems, 789
  management, 471, 491
  market, 621
  ownership, 336
  preservation, 370
  redistribution, 473
  reform, 278
  rent, 513, 621, 626, 655
  virgin, 547
Land use, 129, 351, 362, 511, 515
  change, 585
  competition, 341
  conflict, 334
  information, 344
  patterns, 370
  planning with GIS, 777
  regulation, 621
  rent, 342–43
  rural, 333–50
  urban, 652, 661
Landscape, 31, 38, 176, 177, 211–13, 215, 225, 227, 282, 493, 496, 498
  appreciation, 496
  architecture, 354, 776
  Asian, 507
  capitalist production, 590
  changes, 32, 34
  cultural, 175, 176, 194, 195, 213–15, 226, 515
  cultural transformations, 158
  ecology, 38
  evaluation, 221
  evolution, 343
  interpretation, 634
  natural, 158, 180, 226, 630
    ideological, 631
  perception, 219, 225–28
  recreational, 490
  regional, 634
  rural, 226
  scale, 35
  symbolism, 212, 213, 661

  urban, 175, 265, 498, 547, 634, 662, 663, 665
    industrial, 664
Laos, 274, 518
Latin America, 172–74, 201, 203, 488–505, 508–9, 627, 658
Law
  international, 609
  land, 353
  of refraction, 319
  water, 339
Lebanon, 374
Lesotho, 477
Life-cycle, 261, 263, 265
  analysis, 231
  participation, 230
Linguistic analogies, 212, 213
Linkages, 302, 304, 309
  hazards-social processes, 472
  interfirm, 300
  interurban, 305
  nation-nation, 309
  region-nation, 309
  state government, 367
Local
  autonomy, 599–602, 636
  boosterism, 591
  class struggle, 592, 601
  development, economic, 668
  growth coalitions, 636
  state, 599–602
Location
  analysis, 324, 357, 361, 781, 790
  branch plant, 293
  change, 291
  cities, relative, 659
  conflict, 592
  corporate headquarters, 296
  decision-making, 295
  industrial, 292, 305, 310
  investment, 303
  labor, 303
  manufacturing employment, 293
  theory, 219, 301, 302, 306, 316, 356, 358, 433, 481, 622, 624, 729–30, 778, 781–82
    agricultural, 342, 343
London, 296, 300, 636
  University College, 360
Long cycle theory, 609
Longitudinal analysis, 734
Los Angeles, 268, 300, 368, 520, 658, 668
Love-hotel zone, 516

Malaysia, 518–20
Mali, 660
Management, 293, 294
  metropolitan, 353
  resource, 193, 202, 226
  science, 352
  soil, 337, 344
  water, 344
Mapping. *See* Cartography

Marginalization, 202, 475
Marine environs, 150
  boundaries, 142, 149–51
  navigation, 149–51
  pollution, 148
  recreation, 142, 146–49
  regions, 149–50
  research, 150
    Office of Naval Research (ONR), 141–42
  resources, 141
  transportation, 142, 151–52
Maritimes, 171
Market, 163, 165, 172, 364, 373–74
  area, 296, 325
  export, 296
  gardening, Chinese, 520
  imperfections, 303
  labor, 267, 277, 291, 303–6, 623–25, 631
    urban, 590
  land, 621
  localized, 727
  periodic, 373, 481
  relationships, 296
  shifts, 337
  structure, 296
  world, 337–38, 345
Marxism, 156, 159, 228, 277, 343, 362–65, 469, 547,
    585, 586, 588, 589, 599, 619–38, 661, 662, 667
  capital, theory of, 622
  French Regulationist, 625, 628
  historicism, 161
  neoMarxism, 214, 654
  post-Marxism, 637, 638
  structural, 212, 607, 627
Maryland, 660
Mass movements, 33
Massachusetts, 213
Mathematics, 719–39
  fractal geometry, 736
  information theory, 725
  nonlinearities, 728
  programming model, 729
  Q-analysis, 736
Mauritania, 482
Maya, 203
Mental mapping, 221, 513
Mercantilism, 164, 659
Meteorology, 6
Metropolis
  giant, 372, 512
Metropolitan dominance, 171
Metropolitanization, 266
  anti-, 549
Mexico, 173, 204, 273–76, 278, 489, 492, 493, 495, 628,
    629, 660
Mexico City, 372
Microcomputers, 498
Middle East, 273, 531–33
Migration, 163–67, 172, 175, 258–89, 371, 479, 480,
    490, 492, 549, 629, 656–57
  aged, 453–54, 459
  chain effect, 276, 280

Chinese, 519
circulation, 273, 275, 280, 519
core-periphery, 269, 277
cyclical underpinnings, 281
decision-making, 270
development interrelation, 276
economic determinants, 270
emigration, 204, 273, 275, 489
  British, 163
  Portuguese, 173
  Spanish, 173
fields, 166
forecasting, 269
immigration, 166
  labor force, 171
impacts, 274–75
internal, 259, 269–72
international, 272–75, 664
interregional, 269
interstate, 269
push-pull factors, 273, 278–79
refugee, 273, 274
return, 271
rural-urban, 512, 519
seasonal labor, 516–17
selective, 277, 597
stem-family, 271
stepwise, 527
streams, 264
temporal change, 270
undocumented, 273
wage-labor, 273, 473
Milwaukee, 555
Minneapolis, 264, 668
Minnesota, University of, 619
Minorities, 177–79, 261, 262, 272
  geographic perspectives, 20
Mobility
  constraints, 261
  decisions, 230
  directionality, 263
  elderly, 262
  impairment, 231
  intra-urban, 264–65
  labor, 305
  residential, 259–64
  transition, 273, 279
  upward, 262
Models, 306–10
  agropolitan, 360
  analytical, 307
  bid-rent, 652, 655
  cartographic, 688–91, 698–706
  categorical, 261, 656
  causative matrix, 270
  climatic, 47–51
  competing destinations, 726
  computational process, 735–36
  concentric-zone, 661
  dependency, 620
  discrete choice, 221, 222, 318, 320, 325, 656
  ecological, 652

831

Models, *continued*
  econometric, 308
    spatial, 308
  economic base, 307
  elination-by-aspects, 656
  entropy-maximization, 320
  equilibrium, 309, 327
  export base, 654
  flow, 321
  forecasting, traffic, 657
  geomorphic, 70–71, 82
  goal-programming, 343
  gravity, 276, 316, 320, 325
  hierarchical locational decision-making, 295
  housing-disequilibrium, 260
  housing-search, 260
  input-output, 307–9, 343, 595
  Ising, 727
  lag structures, 270
  land rent, 342–43
  life-span, 231
  log linear, 656, 725
  logit, 725, 727, 731–33
  longitudinal, 734
  Maoist urban, 512
  market imperfection, 303
  market-oriented locational decision-making, 295
  Markov, 269, 270
  mathematical, 626–27
  multiattribute preference, 733
  nonequilibrium, 309
  nonlinear dynamic systems, 327
  nonstationary forecasting, 270
  non-Weberian cost, 290
  plant productivity, 36
  population density, 655
  preference, 221
  probit, 731
  Program Planning, 517
  public policy, 290
  push-pull, 479
  random utility, 730–32
  regional analytical, 307, 309
  regression-autoregression, 721–22
  remote sensing, 765–68
  robust, 261
  sequential-choice, 261
  shift-share analysis, 270, 308
  social-cleavage, 596–98
  spatial interaction, 269, 270, 320, 321, 325–26, 657
  specification of, 308
  state-derivation, 586, 587
  stress-inertia, 260
  succession, 661
  Thunen land use, 103
  time-series, 308
  trend-surface, 655
  trip distribution, 657
  trip generation, 322–23
  trip-stop chaining, 320
  urban structure, 493
  utility-trade-off, 260
  verification, 270
Modernization, 163, 200, 202, 371, 478, 496, 511, 515, 518, 525, 529, 605, 622
  agriculture, 515
  neomodernization, 469, 479
  postmodernization, 468, 622, 634, 638
  school, 476
Mongolia, 515
Mongols, 507
Monsoon (Asian), 53
Montreal, 171
Montserrat, 493
Mortality, 258, 275
Moscow, 549, 554, 555
Mountain, 345
  deforestation, 514
  ecology, 529
Mozambique, 477

National disintegration, 602–9
National Geographic Society, 19
  Education Foundation, 19
Nationalism, 602–8, 610, 629
  Basque, 603, 606
  bourgeois, 607–8
  necessary relations, 602, 606, 607
Nations
  agenda, 20
  building, 374, 602–9
  unity, 374
Native American Specialty Group, 246
Native Americans, 165, 173, 177–79, 246
  claims cases, 178
  culture, 242–43
    change, 243
    ecology, 243
    ethnicity, 245, 247
    history, 239, 246
    landscape, 242
    material, 242–43
  demography, 245, 248
    migration, 245, 248
  energy resources, 244
  environment
    data, 244, 248
    impact assessment, 244
    land use, 240, 243–44
  government
    tribal, 240
    tribal sovereignty 240, 247–48
  management
    land tenure, 240, 243–44
    land use, 240, 243–44
    policy, 244
    strategies, 247
    tribal planning, 244
  research, 239, 247–48
    behavioral, 246
    historical geography, 241
    humanistic description, 245
  research methods
    computer cartography, 248

Native Americans, *continued*
    geographical information systems, 248
    geotechniques, 248
    standard of living
        housing, 244–45
        poverty, 245
    urban Indians, 245
Nature-society interrelationships, 193, 195, 199, 202, 223–25, 266, 472, 475, 483, 529, 583
Neighborhoods, 262, 265, 353, 373, 592, 593, 598, 631, 634, 637, 668
Neocolonialism, 324
Neoconservatism, 214
Nepal, 274, 522, 526–31
Netherlands, The, 371
    banking houses, 170
Network, 319, 515
    analysis, 317
    connectivity, 494, 515
    growth and change, 319, 324, 327
    optimal location, 324, 327
    structure, 324, 518
New Deal, 355, 356, 362, 366
New England, 163, 164, 204
New Federalism, 311, 658
New Guinea, 195, 198–200
New Jersey, 366
New York, 170, 215, 268, 274, 296, 303, 311, 353, 356, 366, 368, 369, 555, 657, 663
New Zealand, 338
Newfoundland, 165
Newly industrialized countries, 266, 605, 627, 628
Nicaragua, 173, 174, 201
Niger, 478
Nigeria, 202, 371, 374, 472, 474, 476, 481, 606
Nile Valley, 204
"Noble Savage," 178
Nonmetropolitan
    land, 333
    turnaround, 267
North America, 162–72, 211, 275, 333–45, 369, 620, 625, 635, 636
    closing of mind, 489
North Carolina, 274
    University of, 362
Northern Heartland, 369
Nuclear power, 224
    plant accidents, 224
Nutrient cycles, 35–36

Ogallala aquifer, 338, 341
Ohio, 510
Older people. *See* Aged
Operation Bootstrap, 356
Optimization
    methods, 324, 343
    systems, 319
Orange County, 300, 625
Ordination, 35
Organizational structure
    firm, 296

Pacific rim, 304
Pakistan, 197, 507, 521–23, 528–30
Paleoenvironment, 36
    climate, 497
    climatology, 36
    ecology, 36
    interpretations, 55
    modeling, 48
    palynology, 36–37
Palestinians, 532
Palynology, 36–37, 55–56, 81–82
Paris, 592, 622, 633, 634, 636, 661, 662
Pastoralism, 200, 201, 472
PC Arc/Info, 177
Peasants, 628, 629
    dwellings, 515
    economies, 165, 278
Pennsylvania, 163, 169, 170
    University of, 357, 359, 361
Perestroika, 550, 551, 557
Periodicals
    Annals, AAG, 72–73
    Geographical Abstracts, 73–74
    Landscape Magazine, 215
    Professional Geographer, 72
Persian Gulf states, 532
Personal Travel Survey, 328
Peru, 203, 276, 279, 489, 491, 492
Phenomenology, 156, 210, 211, 227
Philadelphia, 170, 554
    Social History Project, 177
Philippines, 518–20
Phoenix, 263–65
Picdmont, 369
Place, 632
    based analysis, 588, 599, 610
    based politics, 590
    character of, 369
    sense of, 228, 277, 589, 591, 601, 602, 610
    ties, 271
Planning
    definition, 351
    development, 475–78, 494
    energy, 576–77
    environmental, 369
    family, 528
    five year, 550
    grass roots, 359
    land use, 372
    national, 370–72
    natural resource, 343
    Program Model, 517
    regional, 351–86, 518
    residential, 229
    retail location, 326
    rural, 344
        development, 469, 519
    socialist, 557
    spatial, 469
    theory
        Marxist, 365
        new, 363

Planning, *continued*
   transport, 317, 325
      urban, 326
   urban, 354, 518, 549, 658
      transactional, 517
Pluralism, 71, 89, 209, 232, 355, 364, 583, 585
Poland, 551, 552
Polar regions, 345
Political economy, 202–3, 316, 343, 369, 469, 475, 483, 495, 522, 629, 630, 636, 666
   cultural ecology links, 629
Political science, 5–6, 14, 272, 352, 355, 356, 363, 369, 496, 597
Politicians
   pork-barrel, 366
Politics, 660
   local, 594, 596–99, 637
   parties, 594–99
   of place, 624
   of power, 595
   of support, 595
   territorial, 636
   urban, 591, 594, 595, 600, 601, 621
Pollution, 375, 513, 530, 548
   air, 60
   energy-related, 102
   noise, 224
   non-point, 340
   water, 117, 548
Population, 258–89, 368, 510, 470–75
   change, 492
   characteristics, 264
   concentration, 266
   density, 259
   dispersion, 266
   distribution, 259
   exurban, 337
   growth, 471, 473, 507
   Indian, 173
   mobility, 479, 519
   morbidity rate, 472
   mortality rate, 472
   natural increase, 165, 276
   pressure, 471, 523, 528
   pyramid, 276
Populism, 169, 469, 475, 606
   neopopulism, 468, 469
Ports
   colonial cities, 507
   treaty, 507
Positivism, logical, 156, 159, 161, 198, 210–12, 220, 226, 328, 362, 488, 584
   postpositivism, 211, 582
Post-Sputnik era, 6
Postmodernism, 214, 215
Poverty, 341, 368, 477, 499
Precipitation, 481–82
   acid, 59, 102, 117, 344, 571
Preferences
   modeling, 221
   personal, 229
Prince Edward Island, 158, 159

Problematic inquiry, 214–15
Producer services, 294–97
Product cycle, 292, 298, 299, 624
Production, 305
   agricultural, 343
   articulation of modes of, 469, 474–75, 478
   capitalist, 588
   commodity, petty, 630
   farm, 630
   flexibility of, 623
   food, 521–23
   information-laden, 306
   internationalization of, 303
   macrogeography, 625–26
   mesogeography, 623–25
   microgeography, 622–23
   modes of, 627, 628, 632
   organization, 306
   peasant, 469
   of places, 589–92, 610, 625
   relations, 475
   restructuring, 469, 473
   rice, 508, 517
   sugar, 492
   transactions-intensive, 306
   wheat, 517
Productivity (plant)
   crop, 50
   growing season, 55
   modeling, 36
   primary, 36
Profit
   cycle, 292–93
   maximization, 337
Public policy, 259, 260, 262, 268, 281, 301, 353, 356, 364, 370
   analysis, 310–12
   formation, 344
   housing, 267, 636, 637
   immigration, 275
   industrial, de facto, 310
   local, 299, 300
   macroeconomic, 311
   monetary, 311
   national, debate, 551
   one child, 511
   regional, 299, 310
      indirect, 310
      industrial, 310
      transport, 318, 325
   research, 310
   rural, 340–42
   state, 300, 312
      industrial, 312
   tax, 311
   urban, 310, 522
      national, 357, 366
      Soviet, 549
      transport, 326
Puerto Rico, 356, 357
Puget Sound, 296

Quantitative
  economic analysis, 356
  methods, 626–27
  revolution, 158, 175, 211, 583, 584
Quaternary, 36, 71, 73–75, 83–85, 89
Quebec City, 170
Quotidian environments, 215

Race, 323, 456, 599, 620, 634, 665
  mixing, 492
Railroads, 170
  Baikal-Amur (BAM), 550, 554
  Soviet, 550
Rainforests, 197
Real estate
  agents, 261
  analysts, 265
Realism, 587, 588, 620, 624, 635, 667
Reasoned action, theory of, 225
Refugees, 474
Regional perspective, 158, 173
Regional science, 307, 357, 358, 360–61, 624
Regionalism. *See* Geography, regional, 361, 369, 493, 602
  "born again," 489
  "new," 483, 498
Regionalization, 352
Regions, 351
  American agricultural, 168
  analysis, 493–94
  class alliances, 589–92, 626
  climate-based, 507
  concept, 352
  cultural, 163, 634
  definition of, 352
  development, 305, 351–86, 494, 550–52, 628, 633
  differences, 305
  economic impacts, 299
  economy, 309
  entrepreneurial, 301
  growth, 301
  hierarchy of, 352
  industrial analysis, 306–10
  inequalities, 359, 478
  investment strategy, 551
  landscapes, 634
  linkages, 309
  organic natural, 362
  planning, 351–86, 518
  public policy, 299, 310
  rural resource, 340
  savanna, 473
  size of, 352
  systems, 309
  transformation, 633–34
Religion, 169, 494, 525, 526, 660
  Buddhist, 275, 525
  Calvinist, 534
  Hindu, 524–26
  Moslem, 533
Remote sensing, 36–37, 81, 116, 119, 142, 151, 248, 338, 344, 355, 568, 746–69, 777, 780, 783

applications, 753–61
  agriculture, 756–57
  arctic tundra, 756
  brush fuel type, 757
  chlorophyll, 759
  coastal and marine water, 758–59
  cultural vegetation 756–57
  estuarine water, 758–59
  forests, 755, 757
  global inventory, 756
  runoff, 757–58
  sediment, 757–59
  snow cover, 757
  soil and rock, 753–55
  soil moisture, 757–58
  temperature, 758–59
  urban land cover, 760–61
  vegetation, 755–757
  water resources, 757–59
artificial intelligence, 752, 763–65, 768
biophysical, 748, 769
classification
  algorithms, 752
  schemes, 760
commercialization, 768
data
  analysis, 749, 752–53
  ancillary, 764, 768
  collection, 747–49
  geocoding, 763
  satellite, 768
  SPOT, 763
digital
  image maps, 762
  image processing, 749, 752
  parallel processing 752, 765
expert systems, 752, 764
geographical information systems (GIS), 752–53, 761, 765, 768
image
  analysis, 765
  interpretation, 764
  maps, 762
  processing, 749, 752
  spectrometry, 752
modeling, 765–68
  atmosphere, 765–67
  deterministic, 766
  discrete scene, 766
  empirical, 766
  example of, 766–68
  framework, 765–66
  invertible, 766
  marine environs, 759
  noninvertible, 766
  plant canopy, 767
  scene, 753, 765–66
  sensor, 765
  snow reflectance, 767
  spatial, 767
  surface-flow pattern, 767
National Map Accuracy Standards, 762

Remote sensing, *continued*
  photogrammetric techniques, 761
  platforms
    EOS space station, 768
    radiation, 746
      in estuarine water, 758
      in nearshore water, 758
    resolution, 748, 750–51, 768
Research and development, 298, 475
  corporate, 299
Residential
  choice, 222
  differentiation, 167
  structure, 171, 656–57
Resource, 470–75
  access to productive, 473
  assessment, 335
  assignment of, 352
  base, 549, 550
  biotic, 548
  depletion, 375
  distribution of, 476
  energy, 513, 514
  labor, 549
  management, 193, 202, 226, 357, 469, 471, 482, 548
    household, 629
  natural, 333, 338–39, 344
    conservation, 361, 362
    interpretations, 344
  water, 333, 338–39, 513
Restructuring, 368, 623, 624
  debate, 306
  economic, 369, 593
  energy planning, 476
  firm, 268
  household, 265
  industrial, 293, 294, 301, 303, 304, 307, 631, 633, 634
  linguistic factors, 524
  production, 469, 473
  regional, 269
  social relations, 601
  Soviet (perestroika), 550–51, 557
  spatial, 661
  transportation, 335
  urban, 335, 622
Rio de Janeiro, 174
Risk, 474, 514
  analysis, 292, 342
  management, 494
Ritual, 198, 526
River basins, 366
Rostow's "take-off," 168
Routing, 317, 325, 327
Ruling coalition, 591–93, 595
Rural. *See also* Planning
  freehold empire, 167–70
  nonfarm sector, 481
Rural-urban, 480, 511, 527, 667
  change, 259, 518
  colonial urban dominance, 660
  distinctions, 276
  migration, 266, 273, 278, 280, 512, 519
  relations, 478–79
  urban dominance, 531

Sacred cows, 524, 526
Sahel, 470, 472, 478, 482
Saint Lawrence, 165
San Diego, 555
San Francisco, 263, 300
Sao Paulo, 372
Satellites, 491, 511
  imagery, 515
Saudi Arabia, 532–34
Scale, 341, 342, 375, 637
  returns to, 266, 268, 372, 654
Scotland, 300, 593
Seattle, 297, 555
Sectionalism, 596, 597
Sedimentation, 340
Segregation, 480
  class-based, 661
  racial, 658
  residential, 494, 621
  spatial, 628
Self-determination, 602, 603, 606–8
Semiotics, 212–14, 227
Senegal, 480, 481
Service
  centers, 295
  courses, 17
  provision, 658
  sector growth, 295
  structures, 295
Settlement, 163, 164, 170, 204, 259, 266, 268, 532, 653
  abandoned, 491
  ethnic, 166
  immigrant, 166
  patterns, 515
  rural, 166, 167, 341
    non-farm, 338
  squatter (spontaneous), 480, 519, 527
  systems, 490, 549, 555
Shanghai, 372
Sierra Leone, 478–80
Silicon Valley, 299, 300, 625
Singapore, 534
Sinkiang, 515
Slavery, 164, 169, 174, 341
Slums and ghettos, 167
Small towns, 268, 341, 344, 479, 512, 652
  industrialization, 549
Snowbelt, 369, 668
Social
  areas, 656
  behavior, 656
  control, 588, 592
  groups, 218, 229
  issues, 317
  networks, 229
  organization, 229
  relevance, 281
  studies, 2, 3

Social, *continued*
  K-12, 4
  new, 6
  survey tools, 343
  theory, 209, 478, 483, 489, 582, 584–89, 593, 594, 632, 639
  "new," 210
Sociology, 6, 8, 14, 252, 260, 272, 274, 352, 355, 357, 363, 588, 592, 630, 633, 667
  political, 597
  urban, 652, 666
Soil, 70, 73–74, 81, 83, 338, 339
  conservation, 340–41
  deterioration, 164, 169, 180, 181
  erosion, 340, 341, 344, 375, 514, 547, 629
  management, 482
  science using GIS, 777
Somalia, 477
South Africa, Republic of, 324, 473, 479, 480, 482, 628, 658
South Carolina, 163
Sovereignty, 587, 602, 610
Space
  action, 222
  activity, 222
Space-time, 599
  autonomy, 324
Spain, 204
Spatial
  analysis, 219, 225, 273, 291, 476, 633, 777, 791
  behavior, 730, 734, 738
  choice, 724, 730–34
  competitive equilibria, 729
  decision support systems, 782
  dependence, 725
  externalities, 729
  flow dynamics and patterns, 726–28
  geography, 198
  hierarchies, 623
  interaction, 258, 269, 270, 316, 319, 320, 325–26, 328, 657, 725–28
  laws, 158
  languages, 793
  organization, 476
  pattern, 29, 35, 305, 721
  planning, 469
  processes, 720
  public goods, 729
  scale, 29, 34–35
  science, 220, 632
  statistics, 720–23, 738, 792
  structures, 498
  utility maximization, 222
Species turnover, 29
Spread-backwash effects, 277
Sri Lanka, 374, 521–22, 525, 527, 531
Staples
  thesis, 163
  trades, 165
State, 629, 635–37
  apparatus, 635
  authoritarian, 496
  capitalist, 362, 586–88, 592, 595, 635
  colonial, 477
  control, 2
  Department of, 553
  derivation model, 586, 587
  discourse on the, 636
  intervention, 472, 475
  landlocked, 496
  local, 599–602, 635
  nation, 498, 667
  policy, 300, 312
  predatory, 479
  rational bureaucratic, 161
  role of, 310, 363, 478
  socialist, 365, 477
  structure, 477
  theories of, 583, 585–88, 594, 599
  welfare, 637
Statistics, 220, 720–25
  cartographic, 702, 704–6
  categorical, 724–25
  computationally intensive, 724
    bootstrap methods, 723
  directional, 723
  distributional methods, 723–24
  fractal geometry, 736
  fuzzy sets, 736
  Galois-lattices, 736
  inferential, 259
  Markov processes, 726–27
  measurement, 724
  randomization methods, 723–24
  regression-autoregression, 721–22
  spatial, 720–21, 738, 792
    autocorrelation, 721–23
    autoregression, 723
    data, 720
    theory of, 722
  stochastic utility maximization, 725
  transfer functions, 55
Structural change, 292–93
Structural-functionalism, 212
Structuralism, 159, 211, 212, 227, 277–79, 633
  Post-structuralism, 211, 212, 215, 628, 636, 638
Structuration, 156, 159, 202, 212, 489, 588, 600, 620, 633
Structure/agency debate, 584, 588–89
Structured coherence, 590–91, 593
Subsistence economies, 197, 200, 203
Suburbanization, 262, 264, 266, 267, 621, 661, 663
Suburbs, 292, 297, 355, 361, 362, 366, 368, 549, 636
Sunbelt, 366, 368, 369, 459, 668
Swaziland, 473, 479–80
Switzerland, 606
Symbolic interactionism, 227
Symbolism, 228, 229, 515, 518, 526, 603, 630, 661, 662
Syracuse, University of, 490
Syria, 532, 533
Systems
  integration, 202
  theory, 195, 196, 583, 609

Taiwan, 276, 509–15, 534
Teachers
  certification, 8
  preparedness, 6
  training, 8
Teaching
  philosophy and practice, 17
  quality, 19
Technology, 293–95, 336, 591, 623, 624, 627
  appropriate, 373
  capital-saving, 342
  change, 294
  computer, 309
  entrepreneurship, 297
  European, 507
  flexible, 625
  high, 297–301, 369
  information, 295, 296
  rice, 520
  systems, 224
Telecommunication, 291, 295
Tennessee Valley Authority, 358, 359, 367
Territory, 588, 602, 603–9, 635
  boundaries, 586, 667
  claims, 610
  remolding, 374
Texas, 275, 310
  University of, 360
Textbooks, 7, 258, 352, 497, 551, 583, 652
  adoption, 7
Thailand, 518–20
Theory
  location, 729–30
  spatial statistics, 722
Third World, 193, 199, 202, 205, 266, 277, 281, 360, 361, 368, 373, 473, 488–505, 506–45, 596, 610, 627–30
  transportation, 326
Tibet (Xizang), 372, 515, 526
Time-space, 161–62
  distanciation, 161
Tokyo, 372
Toronto, 171, 655
Tourism, 229, 270
Trade, 309, 660
  commodities, 495
  directional, 302
  flows, 316
  foreign, 548
  fur, 165, 179
  global, 303
  interfirm, 304
  international, 301–4, 327, 551
  interregional, 373
  terms of, 471
  theory, 326
Transactionalism, 223
Transportation, 266–67, 302, 305, 316–32, 372, 373, 518
  Census of, 328
  deregulation, 324
  environmental aspects, 323
  freight rates, 550
  history, 326
  hubs, 324
  infrastructure, 353
  modal-choice, 318, 320
  networks, 319, 515
    connectivity, 515
    growth, 319
    structure, 518
  nodes, 516
  regulation, 328
  routes, 319, 325
  urban, 318
Travel
  analysis, 222
  disaggregate, studies, 323
  patterns, 319–20, 322–23
Tree
  invasion, 34
  line, 34
  rings, 55–56
Trinidad, 494
Turkey, 531–33

U.C.L.A., 354, 357
Uganda, 478
United Kingdom, 305, 370, 597, 605, 630
Urban
  agglomerations, giant, 372–73
  analysis, 493–94
  bias, 373, 479
  built environment, 226, 585, 592, 593, 621, 636, 655
  definition, 511, 667
  development, 305, 621, 628
  ecology, 652
  environment, 513
  evolution, 164
  functions, 373, 494
  geocoding, 780
  growth, 344, 489, 491
  hierarchies, 372
  historical, 480–81
  homeless, 658
  inertia, 171
  institutions, 170
  intraurban patterns, 305
  land use, 652, 661
  landscape, 175, 634, 662, 663, 665
    industrial, 664
  morphology, 652, 661
  networks, 171
  planning, 354, 518, 549, 658
  policy, 305
    Soviet, 549
  politics, 591, 594, 600, 601, 621
  primacy, 372
  "question," 666–68
  radical critique, 305
  regional change, 300
  renaissance, 637
  renewal, 490

Urban, *continued*
  restructuring, 622
  size, optimal, 372, 373
  social areas, 527
  social conflicts, 585
  social movements, 592, 593, 637
    gay, 637, 668
  studies, 229
  symbolism, 662
  systems, 305, 359, 368, 511, 552, 653, 654, 659, 660
  transportation, 318, 326–27
Urbanization, 170, 266, 300, 344, 534, 555, 621, 622, 654 664
  African, 480–81
  Asian, 509
    South, 526
    Southeast, 518
  capitalist, 589
  Chinese, 511, 512
  Hungarian, 552
  Japanese, 516
  U.S.S.R., 373, 374, 546–62

Value
  exchange, 170, 626
  labor, 626
  surplus, 589
  theory, 626
  use, 170
Vancouver, 214, 296, 665
Vegetation
  disturbance, 37
    regimes, 33
  dynamics, 33
  environment relations, 34–35
  geobotany, 754–55
  growing season, 55
  mapping, 30, 35
  productivity, 36
  succession, 33–34
Venezuela, 276, 359, 494
Vietnam, 275, 518, 520
Virginia, 213

Washington, University of, 360, 362
Washington, D.C., 357, 553, 554
Water, 338–40, 630
  balance (budget), 58, 77, 113, 482
  estuarine, 118, 142
  groundwater, 114, 126–30, 133
  hydrologic cycle, 71
  hydrology, 73
    global, 116
  irrigators, 128–29
  lake levels, 113
  law, 339
  marine, 118
  pollution, 117, 548
  problems, 554
  rivers, 118, 145
  transfer, 339, 514, 548
  urban, 119–23
Water resources, 58, 513, 519
  conservation, 119–22
  critical issues
    flood hazard, 125–26
    future directions, 132–34
    groundwater quality, 129–30
    hydrology, 115
    international focus, 117–18
  data sources, 119
    remote sensing, 116, 119
  demand
    forecasting, 120–22
    management, 119–21
  education, 120
  extremes
    drought, 113, 122–23
    flood, 114–15, 123–26, 132
  institutions, 127
  legal aspects, 130–32
    impact analysis, 130–31
    individual rights, 130
    methodology, 131
    property rights, 127, 130
    theories, 131
  management, 116, 119–23, 132–33, 366, 514, 523
    demand, 119–21
    drought, 122–23
    groundwater, 127–29
    regulation, 118, 120
    strategies, 127
  modeling, 114, 116, 132–33
    explanatory, 132
    groundwater, 114
    hydroclimatic, 114
    hydrologic, 114–15
    predictive, 132
    surface processes, 114
  policy, 126–27, 133–34
    groundwater, 128–29
  quality, 116–19
    groundwater, 129–30
    monitoring, 117
    planning, 117–18
    standards, 123
  reliability, 123
  theory, 125–26, 132
    development, 132
    relevance, 132
Weather
  modification, 48
  singularities, 53
Wilderness, 370, 373, 375
  areas, 226
Wisconsin school, 160
Women. *See also* Gender, 165, 178, 215, 228, 262, 272, 473, 477, 620, 629, 631, 673, 679
  aged, 454
  economic development, 675

Women, *continued*
  market, 481
  oppression of, 620, 629, 631
  role of, 629
  status of, 528
  traders, 481
Woodfuel, 476, 477
World Bank, 469, 477, 480, 481, 602

World system analysis, 160–62, 172, 173, 498, 586, 587, 595, 605, 608, 609, 627, 637
World War II, 6, 354, 371, 507, 517

Yokohama, 372

Zimbabwe, 473, 476
Zoogeography, 30, 36–37

DATE DUE

JAN 0 6 1993			

Demco, Inc. 38-293